ELECTRIC CIRCUITS

FOR ENGINEERING TECHNOLOGY

R. E. RIDSDALE

British Columbia Institute of Technology

McGraw-Hill Book Company

GREGG DIVISION

New York • St. Louis • Dallas • San Francisco • Auckland • Düsseldorf
Johannesburg • Kuala Lumpur • London • Mexico • Montreal • New Delhi
Panama • Paris • São Paulo • Singapore • Sydney • Tokyo • Toronto

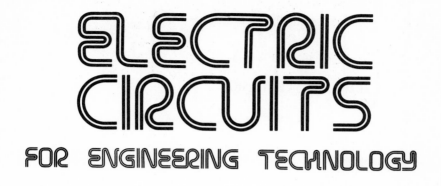

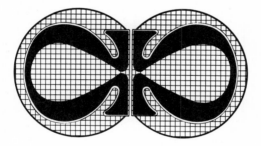

Library of Congress Cataloging in Publication Data

Ridsdale, R   E
  Electric circuits for engineering technology.

  Bibliography: p.
  Includes index.
  1. Electric circuits. 2. Electric networks.
I. Title.
TK454.R52   621.319'2   75-16230
ISBN 0-07-052937-X

ELECTRIC CIRCUITS FOR ENGINEERING TECHNOLOGY

234567890 KPKP 7832109876

The editors for this book were Gordon Rockmaker and Zivile K. Khoury, the designer was Tracy A. Glasner, and the production supervisor was Robert Smith.
It was set in Times Roman by Progressive Typographers.
Printed and bound by Kingsport Press, Inc.

Contents

Preface

This book is designed as a basic text in electric circuits for engineering technicians and technologists.

In the electric circuits context, the successful achievement of learning can be measured in terms of circuit consciousness or feeling. The needs of the learner will determine the level to which this consciousness is to be developed. For electrical craft workers a basic knowledge of a limited range of circuit configurations and quantities will usually suffice, leaving them free to concentrate on the acquisition of the many practical skills that are their specialty. Design engineers must be thoroughly conversant with advanced circuit theory; they will usually have engineering technicians and craft workers available to develop, construct, and test the designs. Engineering technicians, however, must be fed a balanced diet of the practical and the theoretical in their training, since they operate in the realm between the craft workers and the engineers. They must have a good practical sense to command the respect of craft workers whose work they may supervise and even participate in. Their theoretical knowledge must be adequate to interpret the written, verbal, mathematical, and symbolic language of the engineer, to perform some design functions, and to analyze problems.

The balance between theory and practice in this text lies in:

1. *The text material.* The development of principles by logical discussion and derivation forms the basis on which the subject matter is then related to practical issues.
2. *The examples.* In each section of the book theoretical examples demonstrate, as simply as possible, principles and techniques. Further examples refer to situations met in practical situations, some simple and some complex.
3. *The problems.* Drill exercises at the end of each chapter test and reinforce knowledge of the text material. Practical circuit problems add realism.

The practical view is enhanced by the use of commercial component values. The internationally standardized approved values are used in most of the examples and problems. Real circuit values rarely result in neat numbers in calculations. While simple, tidy problem solutions could be obtained by selecting given data so that it works out nicely, continual avoidance of genuine but awkward situations gives a false impression of the ease of obtaining solutions and leaves the learner ill-prepared to deal with the problems that will occur on the job.

This book supplies the theoretical needs of the engineering technician by its approach through principles, definitions, derivations, and relationships.

For the most part, engineering terminology and symbols approved by international authority are used throughout this volume. The units are SI (MKSA); quantity and unit symbols conform to "Letter Symbols for Quantities Used in Electrical Science and Electrical Engineering" (USAS Y10.5-1968); graphic symbols conform to IEEE specification no. 315, March 1971, "IEEE Standard and American National Standard Graphic Symbols for Electrical and Electronics Diagrams" (ANSI Y32.2-1971).

Although they are considered American standards, these conventions have truly worldwide recognition, national differences occurring only in specification numbers.

Unit symbols are used for all quantities in calculations, even though such a routine is too laborious for use in practical electrical work. This total adoption of unit symbols results in a good learning situation, because it provides a rationale for the units in which the answers are given.

From a realistic viewpoint, accuracy in calculating needs to be no greater than that of the available data. Tolerances in electric components and measurements are rarely better than 1 percent, so three significant figures are considered adequate for most electrical situations. But for learning it is often desirable to solve a given problem several times using different techniques. Identical answers encourage confidence in the validity of the techniques. The use of three significant figures is sometimes inadequate to ensure that answers are identical, whereas four significant figures are appreciably better in this regard. Coupled with the fact that four-figure mathematical tables are readily available, the learning advantage is accepted as the justification for the adoption of four significant figures (including angles to 0.01°) throughout this text.

The volume is organized into four parts, each concerned with a fairly specific area of material. Part I is introductory and may not be required by persons with a background in modern high school science. The principles of dc circuits are used in dealing with ac circuits, so the contents of Parts II and III are prerequisites for Part IV.

No attempt will be made to suggest to teachers how they should schedule the text material. Many factors will determine the sections to be selected; the order of presentation and the timing will be chosen to suit the purposes of the complete learning program, even perhaps ignoring the remarks above concerning prerequisite subject matter.

It is recommended that students embarking on a program of self-study progress through the text in chapter sequence.

This is a book about electric circuits. It is not intended to teach mathematics, physics, or circuit devices. These subjects are certainly closely allied to the study of circuits, but they are adequately covered in a number of books, some of which are listed in the bibliography. To have included a detailed treatment in this text would have meant diluting the circuit content. However, brief surveys of some significant topics are given in the appendix.

No textbook is ever the exclusive effort of its author, and this book is certainly no exception. Many people have contributed ideas, comments, and criticisms, all of inestimable value. I must select a few for special mention: Ed Gaspard, collaborator in authorship and colleague; Earl McConechy, colleague and critic; Dave Tyrrell, consulting editor until his wife's health forced his resignation; Herb Adams, mentor; and Pegi Ridsdale, proofreader, grammar corrector, and long-suffering, wonderful wife.

R. E. RIDSDALE

Terminology

In all walks of life specialized terminology is used. However, it is not always graced by the title terminology. Sometimes it is called jargon, and sometimes slang. It always refers to the use of language in a manner which suits particular conditions or special activities. Often the origins of specialized terms are steeped in history, so that the people who use them do not know why they do so. Often such terms are common words used in an uncommon manner.

This use of specialized language for particular purposes applies in the arts, in the professions, and in the technologies. To an artist a medium is a liquid with which he or she mixes pigments. To a person interested in the occult a medium refers to one who apparently can communicate with the dead. The most usual meaning of "medium" is an intermediate value or position, as in high, medium, and low.

Each technology uses words in an exclusive manner. You may be familiar with some examples of this. In mechanical technology a tap is a tool for cutting screw threads. In hydraulics a tap is a device for controlling fluid flow. In electrical technology a tap is a connection point in a circuit.

There is little doubt that the initial difficulty in learning any technology is the apparently overwhelming profusion of words, symbols, and formulas with which one is assailed. Electrical and electronic technology is certainly no exception.

Unfortunately, little comfort can be given. There is no way of avoiding the issue, because terminology is the language of communication within the industry. Learning a new language is never easy; repeated exposure and use produce the necessary familiarity. The only consolation that can be offered to you is the fact that many people before you have faced this task with dismay and have then gone on to master electrical and electronic language, usually without too much difficulty.

The history and background of electrical technology produce some conflicting views on what is the correct terminology. Electric circuits have direct links to physical science. Therefore many people believe that only terminology which is scientifically accurate should be used. But those working in the electrical industry frequently do not maintain this stern allegiance. They use terms which are, strictly speaking, incorrect or even nonsensical. A striking example of this is "ac voltage," an abbreviation for "alternating current voltage," a meaningless term embracing two different quantities. But everyone who works with electricity knows what is implied by "ac voltage," so almost everyone uses this strictly ridiculous term without trouble.

Different people in the industry sometimes use different terminology. Construction electricians frequently use the term "amperage." Some engineers would be contemptuous of this looseness of language and correct it to "current." Others object to "current," and prefer "rate of flow of charge."

Some terminology problems have historical or geographical origins. What we now call "radio" used to be known as "wireless." What is called a "tube" in North America is called a "valve" in England. A "transistor" was, and still is, a semiconductor equivalent of a "tube." It is also a personal radio receiver.

There are even problems related to grammar. "Stereo" is an adjective denoting a two-channel system of sound reproduction. It is also used as a noun, "a stereo" being a type of electronic sound equipment.

In writing an electrical textbook, what can be done about all this? This author adopts the attitude that most people who study electric circuits do so for very practical reasons. Their need to become familiar with electrical and electronic terminology has the objective of communicating with those who work in the industry. Therefore the terminology used in this book is that which is most common. Thus some incorrect, but universal, terms such as "ac voltage" are used, as are terms that are strictly accurate. When a loose but conventional term is first introduced, it is accompanied by the correct term.

ONE
Some Fundamentals

INTRODUCTION

Electric circuits—what are they? They are systems of conductors and elements connected together so that electric current can exist.

What is a system? What is a conductor? What is an element? What is current? What is needed for the existence of current?

The original question produced five further questions. This text will help you to answer these questions and others that have not yet arisen.

Electrical technology is very closely associated with physical science and mathematics. The nature of electricity has its roots in physics, and the laws governing circuit behavior are physical laws. These laws are usually stated in the language of mathematics. The development of electrical concepts from physical laws depends on manipulation of the mathematical relationships between quantities. While it is certainly not the aim of this book to teach physics and mathematics, it is necessary that some physical and mathematical ideas be dealt with as an introduction to the behavior of electric circuits. Thus Part I introduces some of the terms, symbols, equations, and methods you will meet frequently in succeeding parts. For whatever time and effort you spend in assimilating this material, you will be amply repaid by the ease with which you will be able to handle circuit principles and applications.

Because the terms "parameter" and "variable" are related to the basic elements of which electric circuits are fashioned, it is logical that we start our investigation of circuits by looking at these quantities. This investigation is followed by an introduction to some familiar basic elements and the relationships between them.

Many quantities, units, and relationships are introduced in Part I, and they are summarized in Tables 2-1, 2-2, and 2-3.

1

Variables, Parameters, and Units

Principles and techniques used extensively in electrical work were originated by Maxwell, Laplace, and other mathematicians. Mathematical methods are used to determine the quantities involved in electric circuits. It is not surprising, therefore, to find that mathematical terms such as "equation" and "square root," and symbols such as = and $\sqrt{\ }$, are common in electrical language.

Many of the relationships between electrical quantities can be expressed in the form of simple equations of the type $y = kx$ (one quantity equals the product of two others). Applications of the equation simply use different symbols for different quantities. The most common electrical version of this equation is $V = IR$, relating voltage, current, and resistance (Ohm's law).

Usually, in the Ohm's law equation two of the quantities (voltage and current) are variables; that is, their values can vary over a wide range. The other quantity (resistance) is a parameter of nominally fixed value.

Variables are always specified in terms of units. Parameters too are usually associated with units, but sometimes they are dimensionless numbers (numbers without units).

In your study of electric circuits, you will become familiar with many variables, parameters, and units. An understanding of electric circuits depends on this familiarity.

1-1 VARIABLES

A *variable* is a quantity which can readily be made to change value.

Continuing to use Ohm's law as a convenient, frequently used equation, without concern for its electrical meaning, let us look at voltage and current.

In any circuit, it is possible to have a wide range of voltage and current values. Thus, in the Ohm's law equation, V (voltage) and I (current) are variables.

It is common practice to use a graphical representation of a range of voltage and current values, as in Fig. 1-1.

In Fig. 1-1a, the values of current are plotted horizontally (on the X axis). By convention this implies that current is the independent variable, its values having been deliberately selected. The values of voltage are plotted vertically (on the Y axis), implying that this quantity is the depen-

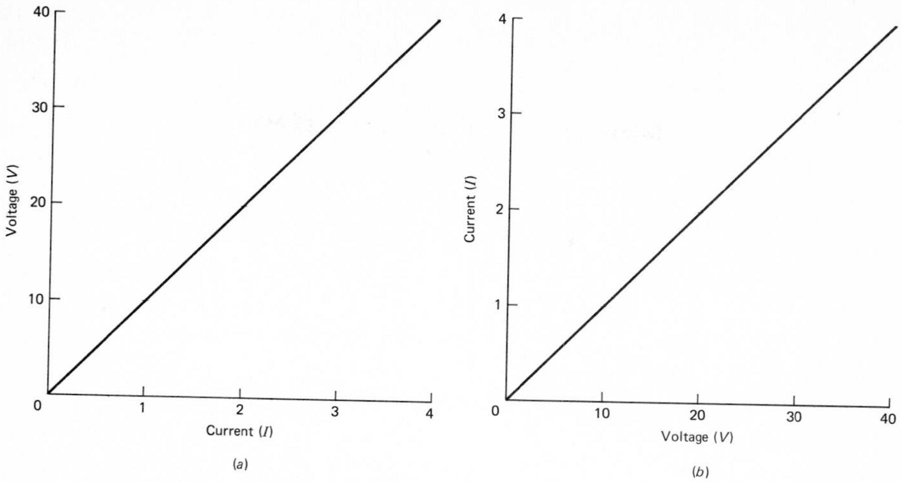

Fig. 1-1 Linear graphs of voltage versus current.

dent variable, its values being dependent on those selected for current.

The graph in Fig. 1-1*b* shows voltage as the independent variable and current as the dependent variable.

Both graphs in Fig. 1-1 depict the same electrical situation. The choice of presentation will be determined by the purpose of the graph.

Voltage and current are the most common electrical variables. Some others are power, energy, charge, and time.

1-2 PARAMETERS

A *parameter* is a constant, the value of which characterizes the relationship between two variables.

In electrical terms, a parameter is usually regarded as a circuit characteristic which defines the circuit's behavior.

Perhaps the word constant should have been placed in quotes. Constant means unvarying, but some circuit parameters can change value under operating conditions. Figure 1-1 relates to truly constant parameters in linear circuit operation (the "curves" of the graphs are straight lines). In any such situation the value of the parameter is given by the slope of the curve (the angle of the graph line from the horizontal). The numerical value of the parameter is determined by the ratio of any chosen value of the dependent variable to that of the related independent variable.

In Fig. 1-1*a*, the parameter has a value of 10, any value of voltage V being 10 times the corresponding value of current I. The equation for this graph is $V = IR$ where R is the parameter resistance, a circuit characteristic.

The equation for Fig. 1-1*b* is $I = GV$, where G is conductance, another circuit parameter. The value of G in this case is $\frac{1}{10}$ or 0.1.

Resistance (voltage divided by current) and conductance (current divided by voltage) are different parameters even when, as in Fig. 1-1, the voltage and current values are the same. Some other common circuit parameters are capacitance, impedance, and phase angle.

1-3 UNITS

Units reveal the magnitudes, or sizes, of quantities. Most numbers associated with electric circuits have no meaning unless accompanied by units.

A *unit* is a specific value of a quantity with which other quantities of the same type can be compared.

Is 1000 larger than 10? Obviously it is. But is 1000 miles larger than 10 hours? Clearly it is not. The mile, a unit of distance, cannot be directly compared with the hour, a unit of time; they represent different types of quantities. However, there can be a relationship between miles and hours. For instance, a car may travel 120 miles in 2 hours; but this merely introduces miles per hour, the unit for another quantity, velocity or speed.

In technical work, it is important to distinguish clearly between a quantity and the unit in which it is specified. The word "mileage" is used by most of us instead of the correct term "distance," when referring to how far a car travels. Such terminology is harmless enough in social conversation among people who all speak English. But for technological uses, which may involve international information exchange, it is unacceptable to add "age" to the end of a unit name. A great effort has been made by international standardizing authorities to eliminate the confusion which can arise as a result of basing the name of a quantity on the term used for its unit.

Of course, there seems to be an exception to every rule. The term "voltage" is so firmly entrenched in electrical language that it has won official approval. But "amperage" for "current" and "wattage" for "power" are not acceptable, even though they are used in the industry.

Many units bear the names of internationally famous scientists. Thus you will encounter such unit names as ampere, hertz, ohm, joule, and newton. Volt is an abbreviation of Volta, another scientist's name. The mho, the unit for conductance for many years, has been officially changed to the siemens (named after the scientist von Siemens). Therefore, in this book we will refer to the siemens as the unit of conductance.

1-4 SYSTEMS OF UNITS

There are several different unit systems in existence. A quantity can be represented in more than one unit system and can have several different units. To illustrate this, let us consider some units of length or distance:

Statute mile	1609.35 meters
Nautical mile	1852 meters
Foot	0.3048 meter
Rod	5.029 meters

There are other distance units in use throughout the world for various purposes. To reduce the confusion in modern communication caused by such a proliferation of different units for the same quantity, the meter has been agreed upon as the international standard, although we know that it has not yet displaced other units.

The reasons for the existence of various unit systems are historical or geographical, going back to the early need to measure. The meter was originally based on a calculation of one-ten-millionth of the surface distance from pole to equator. The nautical mile originated as 1 minute of longitude measured at the equator. A foot was the length of a man's foot, a dubious standard.

Despite attempts to encourage the exclusive use of MKS units, there are three unit systems in use in electrical work. Each of them is named in terms of its units for length, mass, and time.

MKS	meter-kilogram-second
FPS (British)	foot-pound-second
CGS	centimeter-gram-second

The FPS and CGS systems have not been approved for electrical purposes. But many charts and tables have been compiled over the years with these units, and people have become so familiar with them that there is a natural reluctance to change. The end of the use of these two systems is not in sight, and so it is necessary that we recognize their existence.

1-5 MKSA UNITS

The MKS unit system is extended, for electrical purposes, to include the ampere. It is then known as the MKSA system. With the units for distance (or length), mass, time, and current defined, it is possible to define all other electrical units. Because of its widespread international use and its decimal base, the MKSA unit system was agreed on by the Conference Générale des Poids et Mesures (General Conference on Weights and Measures, CGPM) in 1960. It was designated SI, Système International d'Unités (International System of Units). Since that time most, if not all, scientific and engineering organizations throughout the world have endorsed the adoption of SI units. The Institute of Electrical and Electronics Engineers has given its recommendation for the universal adoption of SI units in electrical technology.

All the variables and parameters in this text will be associated with the relevant SI units.

If you already have some electrical knowledge, you need not fear that you will have difficulty learning an unfamiliar unit system. Most of the traditional electrical units, including volt, ampere, ohm, farad, and henry, are MKSA units and are SI units.

1-6 DEFINITIONS OF MKSA UNITS

Recent advances in technology are shown by the changes which have taken place in the definitions of basic units. As recently as 1960, the meter was defined in terms of the length of a platinum-iridium bar kept at the International Bureau of Weights and Measures. Measuring between two marks on a piece of metal, however stable, is not good enough for many present-day purposes.

The modern definition of the meter refers length to atomic radiation and results in a defined meter almost exactly the same as that defined by the platinum-iridium bar. Any organization which can justify the elaborate equipment needed is able to set up the modern standard independently of the International Bureau of Weights and Measures and be sure that it complies with the definition of the meter.

Mass is not so easy to define by direct measurement. Advantage is taken of the fact that, at any place on earth, two bodies which weigh the same on a balance have the same mass. A standard kilogram weight is maintained at the International Bureau of Weights and Measures. Other weights can be compared with it.

The precise definition of the second, the SI basic unit of time, has still not been agreed on. In the meantime, most authorities accept the use of instruments, such as a cesium clock, for time standardization. These instruments utilize molecular resonance, a physical phenomenon of accurately known time characteristics. The ultimate definition of the second will probably be expressed in terms of molecular resonance and, as such, it will be a refined method of specifying the second we all now use.

Two current-carrying conductors exert a force on each other. The magnitude of this force depends on the magnitudes of the currents. This principle forms the basis of the modern electromagnetic definition of the SI basic unit of current, the ampere.

Definitions of SI basic units are given in the appendix.

Two additional SI units are of interest in electric circuits, the kelvin and the radian.

Temperature is a variable which can affect the behavior of electrical materials and components. The SI basic unit of temperature is the kelvin. The degree Celsius, which has the same unit value as the kelvin, is commonly used. Kelvin temperatures can be converted to degrees Celsius[1] by subtracting 273.15. A temperature of 0 degrees Celsius is the same as 273.15 kelvins.

Example 1-1 The relationship between degrees Fahrenheit and degrees Celsius is:

$$\text{Temperature in degrees Celsius} = (\text{temperature in degrees Fahrenheit} - 32) \times \tfrac{5}{9}$$

Express normal room temperature, 68 degrees Fahrenheit, in degrees Celsius and in kelvins.

Solution

$$\text{Temperature in degrees Celsius} = (68 - 32) \times \tfrac{5}{9} = 36 \times \tfrac{5}{9} = \mathbf{20\ degrees\ Celsius}$$

$$\text{Temperature in kelvins} = \text{temperature in degrees Celsius} + 273.15$$

$$= 20 + 273.15 = \mathbf{293.15\ kelvins}$$

Therefore 68 degrees Fahrenheit is the same as 20 degrees Celsius or 293.15 kelvins.

Plane angles (angles between lines) are often important in electric circuits. For instance, phase relationships in ac circuits are usually specified in terms of angles. The radian is the SI unit for plane angle. There are 2π radians in one complete revolution.

The degree is frequently used in preference to the radian. There are 360 degrees in one revolution. Thus, 2π radians = 360 degrees.

Example 1-2 How many degrees equal 1 radian?

Solution (See Fig. 1-2.)

$$2\pi \text{ radians} = 360 \text{ degrees}$$

$$1 \text{ radian} = \frac{360}{2\pi} \text{ degrees}$$

$$1 \text{ radian} = \mathbf{57.3\ degrees}$$

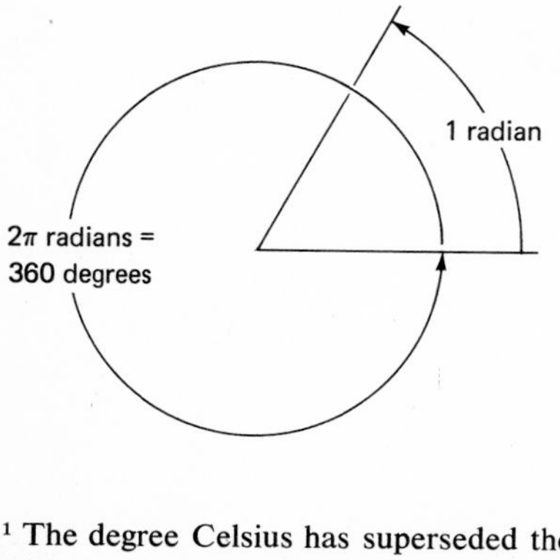

Fig. 1-2 Example 1-2: the radian.

[1] The degree Celsius has superseded the degree centigrade. They are identical in value, the name change being for purposes of international standardization.

Example 1-3 How many radians are there in a right angle (90 degrees)?
Solution (See Fig. 1-3.)

$$360 \text{ degrees} = 2\pi \text{ radians}$$

$$90 \text{ degrees} = \frac{360}{4} \text{ degrees} = \frac{2\pi}{4} \text{ radians}$$

$$90 \text{ degrees} = \frac{\pi}{2} \textbf{ radians}$$

Usually the answer is left in the form $\pi/2$ radians, although it can be written as 1.571 radians.

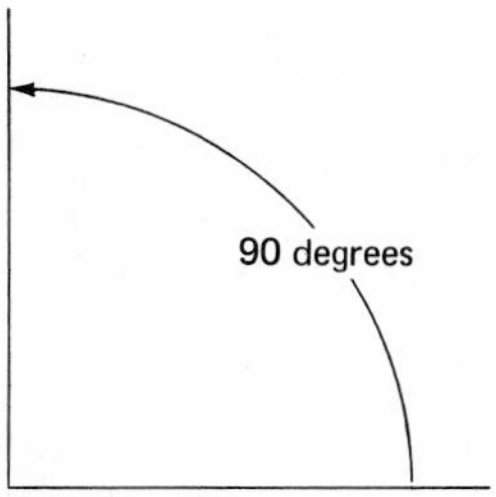

Fig. 1-3 Example 1-3: the right angle.

1-7 SYMBOLS

It is laborious to write words repeatedly, particularly long ones. Language difficulties make words undesirable for information exchange. Symbols are short and international.

To illustrate the effectiveness of symbols, Examples 1-1 and 1-2 are repeated in the more usual symbolic form.

Example 1-1 (Repeated using abbreviations) The relationship between degrees Fahrenheit (θ°F) and degrees Celsius (t°C) is

$$t°\text{C} = (\theta°\text{F} - 32) \times \tfrac{5}{9} \tag{1-1}$$

Express normal room temperature, 68°F, in degrees Celsius and in kelvins.
Solution

$$t = (68°\text{F} - 32) \times \tfrac{5}{9} = 36 \times \tfrac{5}{9} = \mathbf{20°C}$$

The temperature in degrees Celsius (t°C) and in kelvins (T K) is given by

$$T\text{ K} = t°\text{C} + 273.15 \tag{1-2}$$

$$T = 20°\text{C} + 273.15 = \mathbf{293.15\ K}$$

Example 1-2 (Repeated using abbreviations) How many degrees (°) equal 1 radian (1 rad)?

Solution

$$2\pi \text{ rad} = 360° \tag{1-3}$$

$$1 \text{ rad} = \frac{360°}{2\pi}$$

$$1 \text{ rad} = \mathbf{57.3°}$$

A quantity symbol is a single letter, in some cases Greek, with a subscript added when required. Some examples are

I = current

v = instantaneous voltage

I_o = output current

ω = angular velocity or angular frequency

P_{av} = average power

ϕ_1 = phase angle number one

A unit symbol is a letter, a group of letters, or a special sign. A few typical examples are

K = kelvin

cm = centimeter

P = pascal

Ω · m = ohm meter

°C = degree Celsius

There is a total of 100 capital and lowercase Roman (ordinary) and Greek letters. There are many more scientific and engineering quantities and units that require symbols. To minimize confusion, *italic* (sloping) letters are used for quantities, and roman (upright) letters for units. It is therefore important to distinguish between, say, m for mass and m for meter.

The symbols for the SI quantities and units referred to in this chapter are listed in Table 1-1.

Other units which have appeared in this chapter but which are not SI units are listed in Table 1-2.

The Greek letters used in these tables are

ρ = lowercase rho

ϕ = lowercase phi

θ = lowercase theta

ω = lowercase omega

Ω = capital omega

Table 1-1 Symbols for SI quantities and units

Quantity	*Symbol*	*Unit*	*Symbol*
angular velocity or angular frequency	ω	radian per second	rad/s
conductance	G	siemens	S
current*	I	ampere	A
distance†	s	meter	m
mass	m	kilogram	kg
plane angle‡	ϕ	radian	rad
power*	P	watt	W
pressure	p	pascal	P
resistance†	R	ohm	Ω
resistivity	ρ	ohm meter	Ω · m
temperature	T	kelvin	K
time	t	second	s
voltage*	V	volt	V

* Lowercase i, v, or p may be used as the symbols for current, voltage, or power, respectively, when it is desired to indicate instantaneous values, or magnitudes at specified instants of time. Subscripts may be used to identify specific quantities (for example, P_o may mean power at the output terminals).

† Multiples and fractions of SI units are frequently used. Centimeter (cm) and kilohm (kΩ) are examples.

‡ Other Greek letters may be used as symbols for plane angle. ϕ is shown because it is commonly used for phase angle.

Table 1-2 Symbols in Chap. 1 which are not SI

Quantity	*Symbol*	*Unit*	*Symbol*
plane angle*	ϕ	degree	°
temperature	t	degree Celsius	°C
temperature	θ	degree Fahrenheit	°F

* Other Greek letters may be used as symbols for plane angle. ϕ is shown because it is commonly used for phase angle.

1-8 SUPER- AND SUBUNITS

Primary SI or MKSA units are not always of a convenient size. For example, the meter is a satisfactory unit in which to express the length of a table, but for specifying distances between cities the kilometer (1000 m) is more useful. Units larger or smaller than the primary unit may be convenient for use in electric circuit work.

A superunit is greater than the primary unit, the kilometer being an example. A subunit is smaller than the primary unit, for instance, the millisecond, $\frac{1}{1000}$ s. It has been found convenient to place super- and subunits on a scale such that adjacent values are spaced by a ratio of 1000. The list so derived, with the standard prefixes, is given in Table 1-3.

Table 1-3 Prefixes for use with SI units

Prefix	*Symbol*	*Relationship to primary unit**
tera	T	unit $\times\ 10^{12}$ or $\times$ 1 000 000 000 000
giga	G	unit $\times\ 10^{9}$ or $\times$ 1 000 000 000
mega	M	unit $\times\ 10^{6}$ or $\times$ 1 000 000
kilo	k	unit $\times\ 10^{3}$ or $\times$ 1000
unit		unit $\times\ 10^{0}$ or $\times$ 1
milli	m	unit $\times\ 10^{-3}$ or $\times$ 0.001
micro	μ	unit $\times\ 10^{-6}$ or $\times$ 0.000 001
nano	n	unit $\times\ 10^{-9}$ or $\times$ 0.000 000 001
pico	p	unit $\times\ 10^{-12}$ or $\times$ 0.000 000 000 001
femto	f	unit $\times\ 10^{-15}$ or $\times$ 0.000 000 000 000 001
atto	a	unit $\times\ 10^{-18}$ or $\times$ 0.000 000 000 000 000 001

* For numbers with more than four digits on either side of the decimal point (greater than 9999 or smaller than 0.0001) it is recommended by international standard that spaces be used to separate *triads*, groups of three digits, commencing at the decimal point. Examples: 12 345 or 0.123 45. This use of spaces supersedes the use of commas where this is the practice.

Typical examples of the use of Table 1-3 are

$$10 \text{ microseconds} = 10\ \mu\text{s} = 10 \times 10^{-6}\ \text{s} = 10^{-5}\ \text{s}$$

$$50 \text{ kilovolts} = 50\ \text{kV} = 50 \times 10^{3}\ \text{V} = 5 \times 10^{4}\ \text{V}$$

$$6.8 \text{ megohms} = 6.8\ \text{M}\Omega = 6.8 \times 10^{6}\ \Omega$$

$$200 \text{ picofarads} = 200\ \text{pF} = 200 \times 10^{-12}\ \text{F} = 2 \times 10^{-10}\ \text{F}$$

$$4 \text{ gigahertz} = 4\ \text{GHz} = 4 \times 10^{9}\ \text{Hz}$$

The centimeter, cm, is a direct decimal submultiple of the meter ($\frac{1}{100}$ m). However, it is preferable to call 1 cm either 10^{-2} m or 0.01 m.

ARITHMETIC USING SUPER- AND SUBUNITS

All calculations involving units can be carried out in scientific notation or standard form. This is the expression of any number as a number between 1 and 10 multiplied by a power of 10 (e.g., $123.4 = 1.234 \times 10^2$). This notation has been widely used, as it simplifies writing and calculation methods and helps to reduce errors.[1] A further simplification can be achieved by using super- and subunits and applying arithmetic methods directly to them. The following examples demonstrate the technique.

1. mega × milli = kilo

$$4\ \text{M}\Omega \times 5\ \text{mA} = 20\ \text{kV} \qquad (4 \times 10^6\ \Omega \times 5 \times 10^{-3}\ \text{A} = 20 \times 10^3\ \text{V})$$

2. unit/micro = mega

$$\frac{10\ \text{V}}{5\ \mu\text{A}} = 2\ \text{M}\Omega \qquad (10\ \text{V} \div 5 \times 10^{-6}\ \text{A} = 2 \times 10^6\ \Omega)$$

3. milli squared/kilo = nano

$$\frac{(5\ \text{mV})^2}{5\ \text{k}\Omega} = 5\ \text{nW} \qquad [(5 \times 10^{-3}\ \text{V})^2 \div 5 \times 10^3\ \Omega = 5 \times 10^{-9}\ \text{W}]$$

4. square root of mega = kilo

$$\sqrt{1\ \text{W} \times 1\ \text{M}\Omega} = 1\ \text{kV} \qquad (\sqrt{10^0\ \text{W} \times 10^6\ \Omega} = 10^3\ \text{V})$$

5. reciprocal of mega = micro

$$\frac{1}{1\ \text{MHz}} = 1\ \mu\text{s} \qquad (1 \div 10^6\ \text{Hz} = 10^{-6}\ \text{s})$$

Note that the result should be 1 μs/cycle, but the form given (1 μs) is that most commonly used.

1-9 SUMMARY OF EQUATIONS

$t°\text{C} = (\theta°\text{F} - 32) \times \frac{5}{9}$ (1-1)

$T\ \text{K} = t°\text{C} + 273.15$ (1-2)

$2\pi\ \text{rad} = 360°$ (1-3)

[1] For a review of arithmetic using powers of 10 there are many good mathematics books (for example, N. M. Cooke and H. F. R. Adams, "Basic Mathematics for Electronics," 3d ed., McGraw-Hill Book Company, New York, 1970).

EXERCISE PROBLEMS

Section 1-5

1-1 Express the following temperatures in kelvins.

a. 0°C (*ans.* 273.15 K)
b. 0°F (*ans.* 255.4 K)
c. 100°C (*ans.* 373.15 K)
d. 32°F (*ans.* 273.15 K)
e. 212°F (*ans.* 373.15 K)
f. 150°F (*ans.* 338.7 K)
g. −40°C (*ans.* 233.15 K)
h. −40°F (*ans.* 233.15 K)

1-2 Express the following temperatures in degrees Celsius.

a. −60°F
b. 120°F
c. 283 K
d. 390 K

1-3 Express the following temperatures in degrees Fahrenheit.

a. 0 K (*ans.* −459.7°F)
b. 300 K (*ans.* 80.33°F)
c. 25°C (*ans.* 77°F)
d. −25°C (*ans.* −13°F)

1-4 Express the following angles in degrees.

a. 0.5 rad
b. 2.1 rad
c. 0.785 rad
d. 4.71 rad

1-5 Express the following angles in radians.

a. 150° (*ans.* 2.618 rad)
b. 30° (*ans.* 0.5235 rad)
c. 990° (*ans.* 17.28 rad)
d. 1° (*ans.* 0.01745 rad)

Section 1-8

1-6 Express the following in kilounits and megaunits.

a. 150 000 V
b. 5 000 000 W
c. 680 000 Ω
d. 400 000 000 Hz

1-7 Express the following in milliunits and microunits.

a. 0.0001 V (*ans.* 0.1 mV, 100 μV)
b. 0.025 A (*ans.* 25 mA, 25 000 μA)
c. 0.001 25 S (*ans.* 1.25 mS, 1250 μS)
d. 0.000 01 W (*ans.* 0.01 mW, 10 μW)

1-8 Express the following as numbers between 1 and 1000 followed by super- or subunits.

a. 2 800 000 000 Hz
b. 0.000 005 5 V
c. 0.000 000 000 01 F
d. 0.000 000 1 s
e. 0.000 02 H
f. 300 000 000 Ω

1-9 Express the values given in Probs. 1-6, 1-7, and 1-8 in scientific notation. (*ans.* Problem 1-6: *a.* 1.5×10^5 V; *b.* 5×10^6 W; *c.* 6.8×10^5 Ω; *d.* 4×10^8 Hz. Problem 1-7: *a.* 1×10^{-4} V; *b.* 2.5×10^{-2} A; *c.* 1.25×10^{-3} S; *d.* 1×10^{-5} W. Problem 1-8: *a.* 2.8×10^9 Hz; *b.* 5.5×10^{-6} V; *c.* 1×10^{-11} F; *d.* 1×10^{-7} s; *e.* 2×10^{-5} H; *f.* 3×10^8 Ω)

Miscellaneous problems on units

1-10 The temperature rise of a semiconductor device is from 25 to 150°C. Express this temperature rise in:

a. Kelvins
b. Degrees Fahrenheit

1-11 A wheel is turning at a rate of 600 revolutions per minute; express this in terms of radians per second. (*ans.* 62.83 rad/s)

1-12 Electrical energy supplied to your home has a frequency which can be expressed as 60 cycles per second. Express this frequency in terms of radians per second if 1 cycle is represented by one complete revolution.

2

Basic Electrical Quantities and Relationships

A *system* is an orderly arrangement of elements into a useful whole.

All electrical systems have certain features in common:

1. There must be an interchange of energy in order for the system to be of practical use.
2. There must be a part of the system where energy can be put to use.
3. There must be a part of the system where energy can be introduced, because it cannot be taken out for use unless it is first put in.

For a complete electric circuit three subsystems are necessary: the *generator,* where energy enters the system; the *load,* where energy is usefully employed; and the *network,* which conveys the energy from the generator to the load.

It must be possible to specify, predict, and measure energy levels within the system. Many quantities are involved in this process. This chapter is concerned with some of these quantities and the relationships between them, as a foundation for the investigation of electric circuits.

2-1 ENERGY AND WORK

Electric circuits exist for the purpose of transferring energy and doing work. The electrical utility which supplies your home sells you electrical energy. It is able to do so because it converts other forms of energy to electrical energy. When you use electricity, you put electrical energy to work, converting it to some other form of energy.

Figure 2-1 shows a hydroelectric system.

Fig. 2-1 Hydroelectric system.

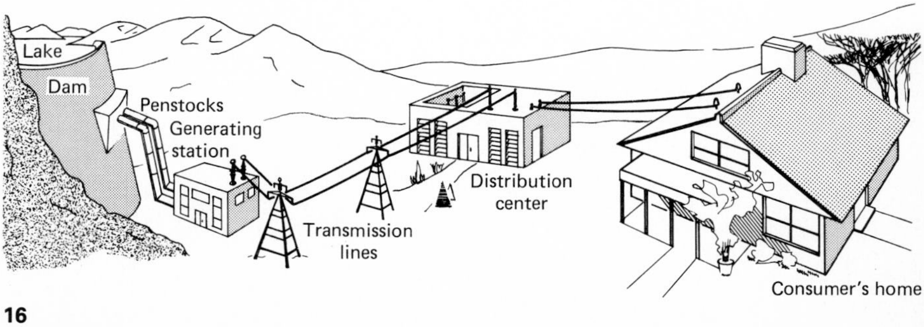

A lake holds a large volume of water, maintained high above sea level by a dam. The water thus stored possesses *potential energy,* the energy of position. The water is directed down into the generating station through pipes called penstocks. The falling water has *kinetic energy,* the energy of motion. The generators produce electrical energy when the kinetic energy of the water causes the shafts of the generators to turn. Energy is converted from potential energy to kinetic energy to electrical energy. None of these energy forms is visible or otherwise capable of direct perception by our senses. But all are evident through what they can do. Perhaps this best explains what is meant by energy: It causes things to happen; we are aware of the existence of energy through what it does.

Energy is sometimes defined as the ability to do work. This is just another way of saying that energy causes things to happen, because work in the physical sense is a quantity by which action is measured. The electrical energy from the generating station is conducted by transmission lines to centers which then distribute it to houses, factories, hospitals, farms, etc. Here the electrical energy is used—put to work. A heater converts electrical energy to heat energy. A light bulb converts electrical energy to light energy. An electric motor converts electrical energy to mechanical energy. Energy is converted; work is done.

Figure 2-2 shows this energy and work system in block diagram form.[1]

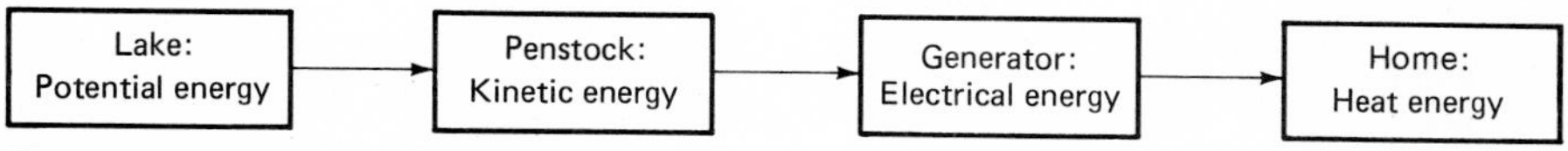

Fig. 2-2 Energy, from lake to home.

Energy and work are closely allied. Energy is the ability to do work. Work is the result of converting energy. The amount of energy expended is the amount of work done.

Both energy and work have the same symbol, W, and the same SI unit, the joule. The symbol for joule is J.

When a force is applied to an object causing the object to move, work is done. The amount of work done is the force multiplied by the distance moved:

$$W = Fs \tag{2-1}$$

where W = energy, J*
F = force, N
s = distance, m

[1] A block diagram is a drawing showing the major components of a system as blocks of various shapes; it depicts the overall action in a simple manner.
* W is energy expressed in joules, J. Statements of this type will be abbreviated as shown.

Writing Eq. (2-1) in unit form gives

$$\mathrm{J} = \mathrm{N} \cdot \mathrm{m} = \frac{\mathrm{kg} \cdot \mathrm{m}}{\mathrm{s}^2} \times \mathrm{m} = \frac{\mathrm{kg} \cdot \mathrm{m}^2}{\mathrm{s}^2} \qquad (2\text{-}2)$$

In Eq. (2-2) all the terms are units: J is joule, N is newton, m is meter, kg is kilogram, and s is second.

It should be noted that *s* (italic) in Eq. (2-1) represents the quantity distance, while s (roman) in Eq. (2-2) represents the unit second.

The term $(\mathrm{kg} \cdot \mathrm{m})/\mathrm{s}^2$ in Eq. (2-2) is the MKSA dimension of the newton, the SI unit of force.

The MKSA dimension of the joule, the SI unit of energy or work, is, as shown in Eq. (2-2), $(\mathrm{kg} \cdot \mathrm{m}^2)/\mathrm{s}^2$.

ENERGY LOSS

Whenever energy is converted, all of it is converted to other forms, but not all to the intended form. An electric motor converts electrical energy to mechanical energy; this is the intent. But the motor becomes warm; some heat energy is produced. A stove element converts electrical energy to heat energy, but you can see the element glow; some light energy is produced. Energy converted to an unintended form represents an energy loss. The amount of energy loss W_L is the difference between the total amount of energy W_T and the amount of energy converted as intended, the useful energy W_U:

$$W_L = W_T - W_U \qquad (2\text{-}3)$$

The total energy is sometimes called the input energy W_i, the useful energy being the output energy W_o:

$$W_L = W_i - W_o \qquad (2\text{-}4)$$

EFFICIENCY

The ratio of the output energy to the input energy is frequently of interest. This ratio is called the *efficiency* and is given by the equation

$$\eta = \frac{W_o}{W_i} \qquad (2\text{-}5)$$

In Eqs. (2-3) to (2-5), W_L is energy loss, W_T and W_i are total and input energies, and W_U and W_o are useful and output energies. All the energy terms are in joules. The symbol η for efficiency is the lowercase Greek letter eta. As efficiency is the ratio of two similar quantities, η has no unit; it is a dimensionless number.

2-2 GENERATOR AND LOAD

Electrical energy enters the circuit because of the conversion of energy from some other form to the electrical form. The potential energy of stored water, the caloric energy of oil or gas, the nuclear energy of atoms, and the chemical energy of acids are all used for conversion to electrical energy. A *generator* is an item of equipment or a system in which this conversion of energy to the electrical form is carried out. It may be large and elaborate, as in a nuclear power station. It may be as small and simple as a flashlight battery. Energy source is another name for a generator.

The light energy of a television tube, the sound energy of a radio, the chemical energy of a plating bath, and the mechanical energy of an electric hammer are the results of conversion from the electrical form. This conversion of electrical energy to some other form takes place in a *load.* A load is sometimes called a termination. (See Fig. 2-3.)

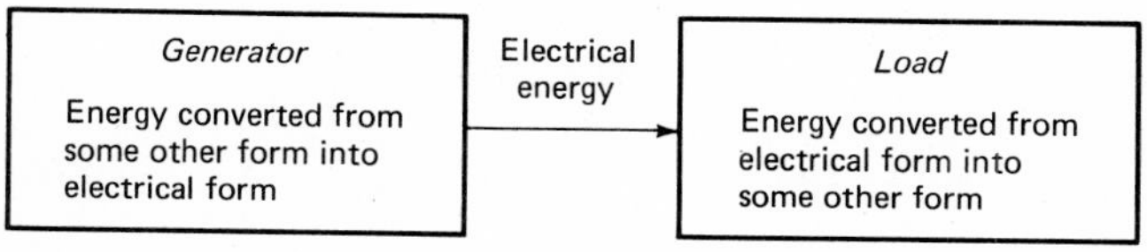

Fig. 2-3 Energy conversions in generator and load.

Application of the term "load" can be confusing. Everything connected to a generator constitutes a load on the generator because all the electrical energy generated is converted to other forms of energy. The total circuit consists basically of three sections: generator, network, and load. (See Fig. 2-4.)

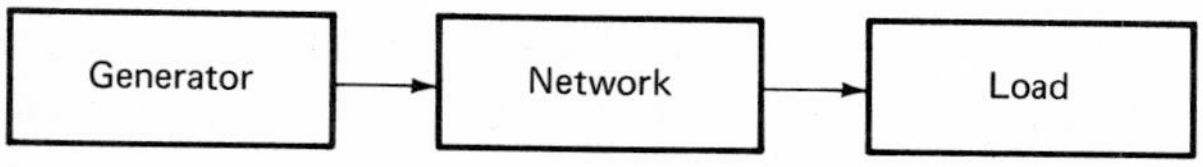

Fig. 2-4 Basic electric system.

In this case the network, which connects the generator to the load, may convert some of the electrical energy to heat, leaving the load to convert less energy than was produced by the generator. It is perhaps best to view this situation in two ways.

In Fig. 2-5*a* the network and the load are together considered to be the load on the generator, taking all the energy output. This situation might be described by saying that, from the generator's viewpoint, the load includes the network.

Fig. 2-5 (*a*) Network + load = total generator load; (*b*) network + generator = total generator.

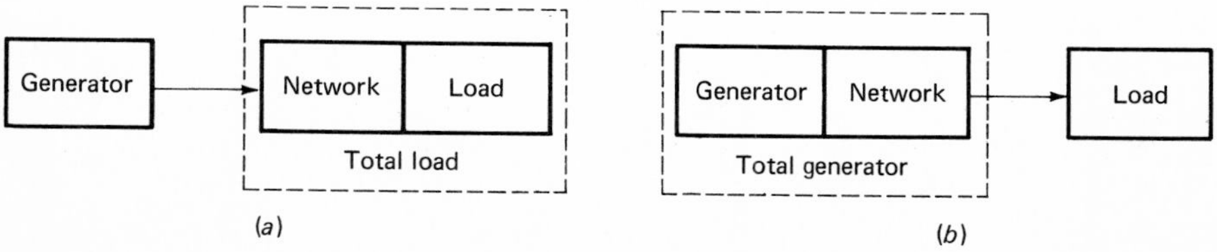

In Fig. 2-5*b* the generator and the network are together considered as the total generator, the load taking all the energy output. Another way of describing this concept is to say that the load sees the generator and the network as if the combination were a single energy source.

2-3 CHARGE

Charge is the medium through which electrical energy is conducted by the atoms of a substance. The amount of charge held by an electron is usually considered the fundamental quantity of charge. The charge of an electron has been measured and found to be extremely small—too small for practical use as an electrical unit. A much larger unit, the coulomb, symbol C, is the SI unit of charge.[1] The charge of a single electron is $1.602\ 03 \times 10^{-19}$ C, so that 1 C represents the total charge of approximately 6.242×10^{18} electrons. To enable you to more clearly visualize what a large number this is, think of it as about $6\frac{1}{4}$ million million million electron charges.

How did these numbers arise? Why is the coulomb not specified as, say, 10^{18} electron charges? Like some other SI units, the coulomb was a unit of known magnitude before the charge of the electron was accepted as a basis of electricity. The number of electron charges in a coulomb was determined fairly recently by using modern measuring techniques. The other SI units to which the coulomb is related would have to be changed if the value of the coulomb were to be changed.

Charge is sometimes called quantity of electricity, giving rise to its standard symbol Q.

As an illustration of the availability of electrical charge, consider the metal copper, a good material for electrical conduction. Copper contains approximately 8.54×10^{28} free electrons per cubic meter. This means that there is about 1.37×10^{10} (13.7 thousand million) C of charge available in a cubic meter of copper. If this number surprises you, remember that a cubic meter is a large volume; a cubic meter of copper weighs 8890 kg (about 10 tons).

Example 2-1 How many free electrons exist in a copper rod $\frac{1}{4}$ in. in diameter and 6 in. long? What charge does this number of free electrons represent?

Solution (See Fig. 2-6.)

Fig. 2-6 Example 2-1: dimensions of a copper rod.

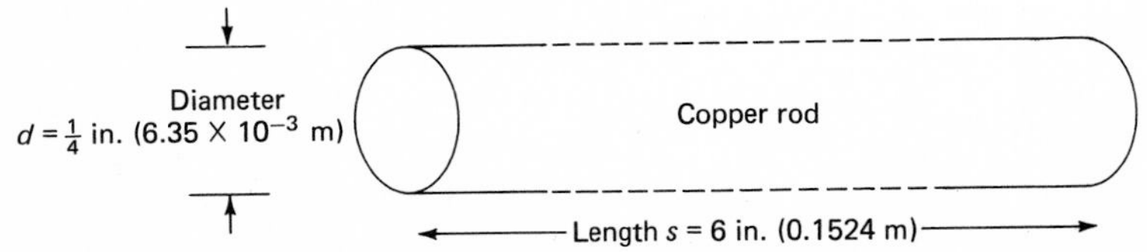

[1] The symbol C (roman) for the unit coulomb should not be confused with the symbol C (italic) for the quantity capacitance.

The cross-sectional area of the rod is

$$A = \pi r^2 = \pi(3.175 \times 10^{-3}\ \text{m})^2 = 3.167 \times 10^{-5}\ \text{m}^2$$

The volume of copper in the rod is

$$V = As = 3.167 \times 10^{-5}\ \text{m}^2 \times 0.1524\ \text{m} = 4.827 \times 10^{-6}\ \text{m}^3$$

The number of free electrons equals the number of free electrons per unit volume times the volume. There are 8.54×10^{28} free electrons per cubic meter of copper, and so the number of free electrons in the copper rod is

$$8.54 \times 10^{28} \times V = 8.54 \times 10^{28}\ \frac{\text{electrons}}{\text{m}^3} \times 4.827 \times 10^{-6}\ \text{m}^3$$

The number of free electrons in the rod = $\mathbf{4.122 \times 10^{23}}$.

The charge in coulombs equals the number of free electrons divided by the number of electrons per coulomb:

$$Q = \frac{4.122 \times 10^{23}\ \text{electrons}}{6.242 \times 10^{18}\ \text{electrons/C}}$$

$$Q = \mathbf{6.604 \times 10^{4}\ C}$$

The copper rod in Example 2-1, about the size of an ordinary pencil, contains close to 1 million million million million free electrons, the equivalent of approximately 66 000 C of charge. The very large number of electrons involved in familiar situations is illustrated by considering that, when an ordinary 100-watt light bulb is lit steadily for 1 day (24 hours), the number of electrons which pass through it is about the same as the number of electrons in the copper rod.

2-4 VOLTAGE

Whenever energy is converted to or from the electrical state, movement of charge is involved. Voltage is the factor relating the electrical energy and the charge.

Figure 2-7 shows a basic electric circuit consisting of a generator connected to a load by two conductors. Energy enters the circuit at the generator, is transferred to the load by the conductors, and leaves the circuit at the load.

Fig. 2-7 Electrical energy associated with charge movement.

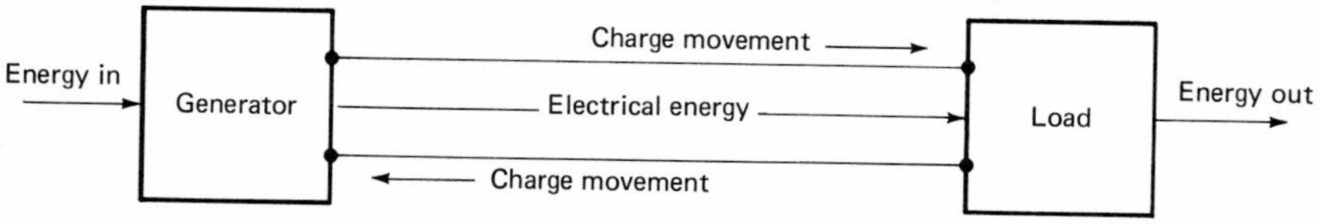

ELECTROMOTIVE FORCE

At the generator external energy is converted to electrical energy. This conversion is associated with a voltage which causes charge movement to take place. The voltage of the generator is sometimes called *electromotive force,* abbreviated emf. "Electromotive force" is a useful term because it conveys the idea of a force which causes electrons to move. It is important to realize, though, that voltage is not a true force, since its unit does not have the same MKSA dimensions as the newton, the SI unit of force.

POTENTIAL DIFFERENCE

At the load the charge movement causes electrical energy to be converted to other forms (heat, light, mechanical, etc.). This energy conversion at a load occasionally results in energy storage, as when the load is a battery under charge. Usually, however, the energy is dissipated or released from the circuit at the load, mainly in the form of heat. A voltage across a load is sometimes referred to as *voltage drop* or *potential difference,* pd.[1] (See Fig. 2-8.)

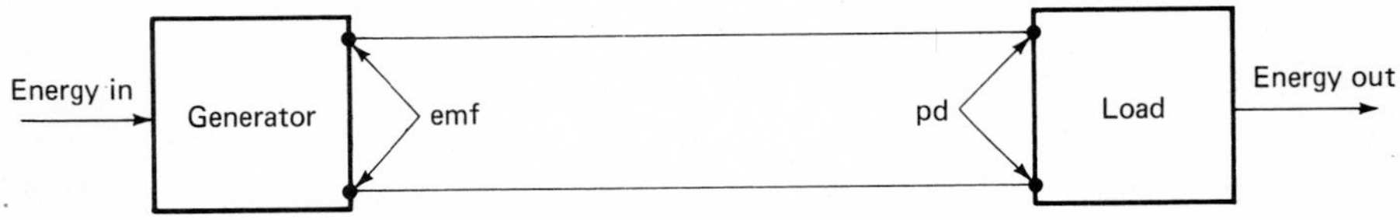

Fig. 2-8 Electrical energy associated with voltage.

The two interpretations of voltage can be summarized as follows.

Electromotive force is a voltage associated with *generated energy.* *Potential difference* is voltage associated with *dissipated energy.*

The general symbol for voltage is V. The symbol V may be used whenever it is unnecessary to distinguish between the voltages of generation and dissipation.

The specific symbol for emf is E. The symbol E only refers to voltage of generation.

The specific symbol for pd is V. This symbol is used to refer to voltage of dissipation where E is used for voltage of generation.

It should be clearly understood that, while V can be used for all voltages, E is to be used *only* for generated voltages associated with electrical energy sources.

The unit for either type of voltage, emf or pd, is the volt, symbol V. The volt is defined by the statement:

[1] Potential differences will be referred to as pds.

If between two points in a circuit 1 J of energy conversion is associated with the movement of 1 C of charge, 1 V exists between those two points.

This definition gives rise to the equation

$$W = VQ \qquad \text{or} \qquad W = EQ \tag{2-6}$$

where W = energy or work, J
V = generalized voltage (emf or pd), V
E = emf, V
Q = charge, C

Inserting unit symbols into Eq. (2-6) gives

$$\mathrm{J} = \mathrm{V} \cdot \mathrm{C} \tag{2-7}$$

where J = joules
V = volts
C = coulombs

It can be seen from Eq. (2-7) that 1 V is 1 J/C.

Example 2-2 A typical 12-V car battery can store a charge of 2×10^5 C. What amount of energy does this charge represent?

Solution

$$W = EQ \tag{2-6}$$

$$W = 12 \text{ V} \times 2 \times 10^5 \text{ C} = \mathbf{2.4 \times 10^6 \text{ J}}$$

Thus the automobile battery of Example 2-2 can store about $2\frac{1}{2}$ million J of energy for the operation of the car's electrical equipment.

Example 2-3 What would be the emf of a battery which is capable of storing 2.4×10^6 J of energy (as in Example 2-2) but does so with a stored charge of 4×10^5 C, double that of the charge in Example 2-2?

Solution

$$W = EQ \tag{2-6}$$

$$E = \frac{W}{Q} = \frac{2.4 \times 10^6 \text{ J}}{4 \times 10^5 \text{ C}} = \mathbf{6 \text{ V}}$$

The result of Example 2-3 implies that, to store the same amount of energy as a 12-V battery, a 6-V battery must store twice as much charge.

CLOSED AND OPEN CIRCUITS

Energy can only be converted to and from the electrical state if charge actually moves. To accomplish this charge movement, a complete or *closed circuit* is necessary as in Fig. 2-9.

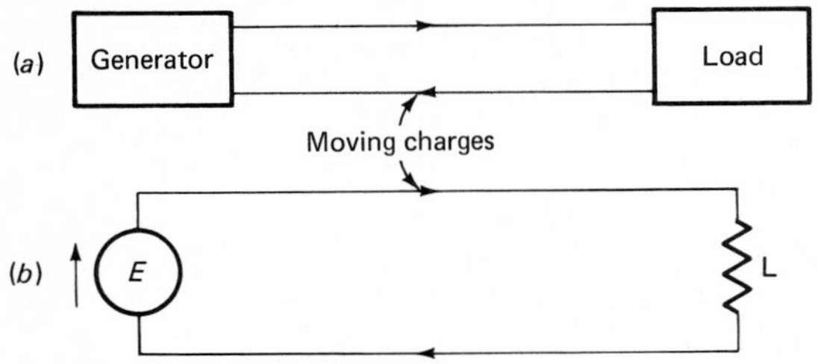

Fig. 2-9 Circuit complete for charge movement. (*a*) Block diagram; (*b*) schematic diagram.

Figure 2-9*a* is a block diagram of a generator connected to a load. Figure 2-9*b* is a schematic diagram[1] showing the generator as a voltage source *E* connected to the load L.

In Fig. 2-10*a* an open switch results in an *open circuit*. Charge movement cannot take place through an open switch, hence there cannot be any energy conversion. The generator emf *E* is present, ready to cause charge movement, but there is no load pd.

In Fig. 2-10*b* the switch is closed, producing a closed circuit in which charge can move. The generator emf *E* is still present, and now the pd *V* also exists.

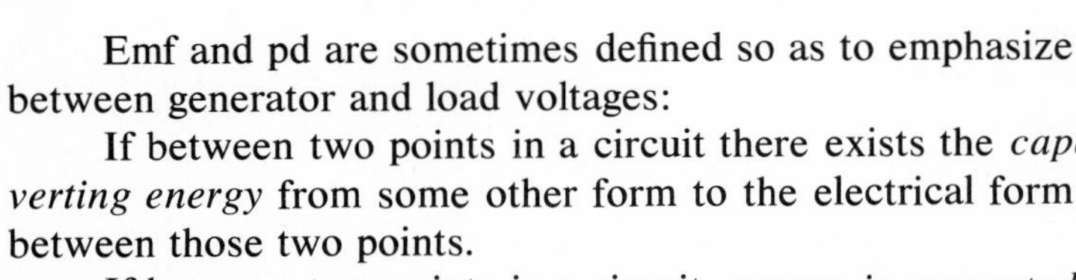

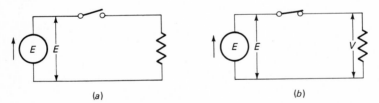

Fig. 2-10 (*a*) Open circuit; (*b*) closed circuit.

Emf and pd are sometimes defined so as to emphasize this difference between generator and load voltages:

If between two points in a circuit there exists the *capability for converting energy* from some other form to the electrical form, an *emf* exists between those two points.

If between two points in a circuit, *energy is converted* from the electrical form to some other form, a *pd* exists between those two points.

The fact that a generator emf can exist when the circuit it feeds is open leads to the frequently used terms "open-circuit generator voltage" and "open-circuit emf."

[1] A schematic diagram is a drawing showing the circuit elements and connections in a standard universal form, each element having a distinct symbol.

POLARITY

The terminals of a generator may be assigned the polarities positive and negative (+ and −). Negative electron charges move from the negative terminal around the circuit to the positive terminal, as depicted in Fig. 2-11.

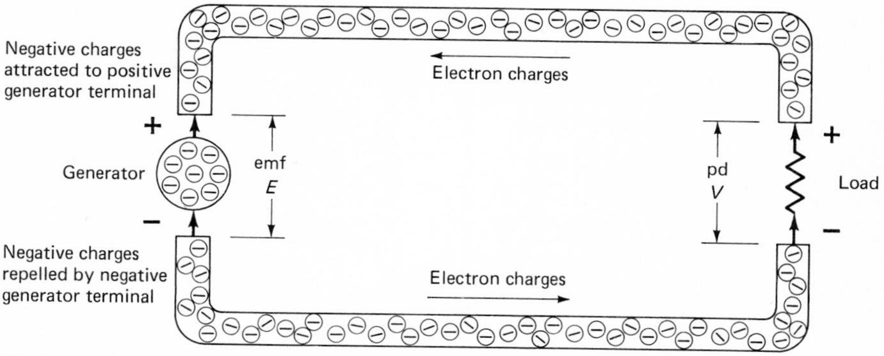

Fig. 2-11 Charges move within a circuit.

Figure 2-11 illustrates the principle of electrostatics whereby particles with polarities of opposite sign (+ and −) are attracted to each other, while particles with polarities of the same sign (+ and +, or − and −) are repelled by each other. In Fig. 2-11 the negative electron charges are attracted to the positive terminal of the generator and are repelled by the negative generator terminal. In this case the combined action results in the charges moving around the circuit in a counterclockwise direction.

When the generator polarity is reversed, as in Fig. 2-12, the attracting and repelling forces cause the charge movement to become clockwise around the circuit. This means that a reversal of generator polarity results in a reversal of the direction of charge movement.

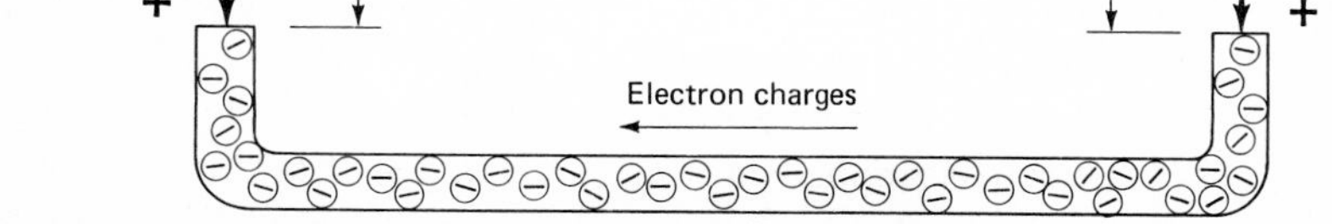

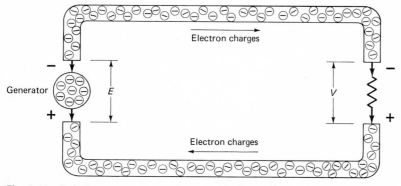

Fig. 2-12 Polarity reversal causes current direction reversal.

Arrows (↑) next to generator symbols can be used to indicate polarity, the arrowhead pointing in the direction of the positive terminal. It is sometimes said that the arrows point in the direction of potential rise.

In semiconductor technology (the basis of transistor operation) it is said that conduction may take place as a result of the movement of negative-charge carriers (electrons) or positive-charge carriers (holes). If this concept is accepted, no problem is created since all circuit relationships are the same for either type of carrier. The only distinction is that negative and positive charges will move in opposite directions, although this movement direction is rarely of interest.

In Figs. 2-11 and 2-12 the pd V is shown as having the same polarity as the emf E.

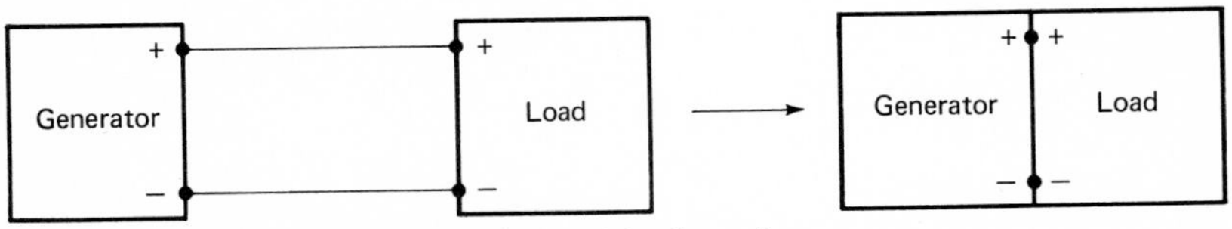

Fig. 2-13 The polarities of generator and load are the same.

If the distance between the generator and the load is gradually reduced, they will eventually be directly joined, as in Fig. 2-13. Here it can clearly be seen that the touching terminals of the generator and the load must have the same polarity.

The subject of polarity is dealt with more fully in Chap. 3, Sec. 3-4.

2-5 FORCES AND CHARGE MOTION

Charges available for movement are called *mobile charges* or *charge carriers*. Their presence in a material depends on its nature and its environment. Metals, which are normally good conductors, have a significant quantity of charge carriers available under ordinary environmental conditions. Semiconductors, which also are metals, need special conditions (such as the presence of strong light, heat, or nuclear radiation) or special chemical treatment before they can readily have charge carriers available for movement. Materials such as plastics or ceramics are insulators which ordinarily do not possess many mobile charges.

Mobile charges can be either negative or positive. They can be caused to move by drift, diffusion, or recombination mechanisms. Except in the internal operation of semiconductor devices, such as transistors, only the movement of negative electron charges by the drift process is significant. This text is not concerned with semiconductor operation, so the main discussion will be restricted to the drift of negative-charge carriers. Some information on diffusion and recombination is given in the appendix.

Whenever an emf E is applied in a circuit, such as between the two metal plates in Fig. 2-14, an electric field is developed. An electric field is sometimes called a force field because it is a region in which any charge is subjected to a force tending to make it move. The lines shown between the plates represent the electric field; their positive-to-negative direction is the standard convention; their parallel, equidistant positions imply a uniform strength of field.

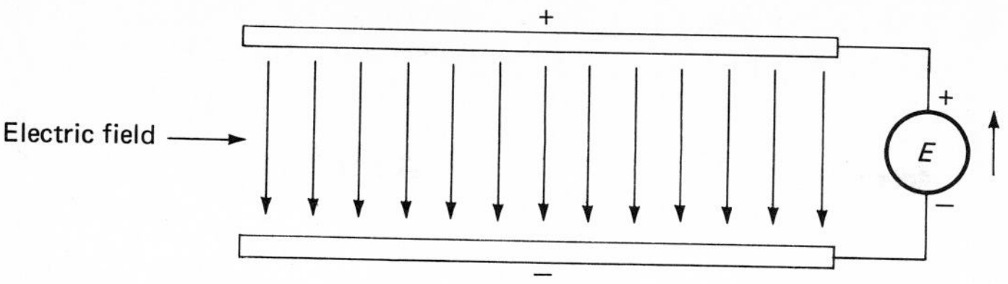

Fig. 2-14 Voltage produces an electric field.

When a negative charge enters the electric field, it is attracted toward the positive plate and repelled away from the negative plate. If there is nothing to obstruct it, the charge will move toward the positive plate, as in Fig. 2-15.

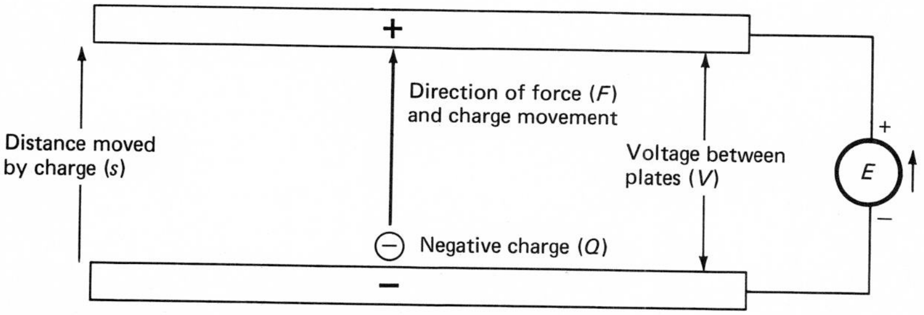

Fig. 2-15 Charge moves toward positive polarity.

If the force causes the charge to move across the field, work will be done. The amount of work is given by Eq. (2-1):

$$W = Fs \tag{2-1}$$

Combining this with Eq. (2-6):

$$W = VQ \tag{2-6}$$

$$Fs = VQ$$

$$F = \frac{VQ}{s} \tag{2-8}$$

where F = force, N
V = voltage, V
Q = charge, C
s = distance, m
W = work, J

Thus charges will tend to move within an electric field under the influence of a real force. This force, for a given charge, is determined by the applied voltage and the distance through which the charges are required to move.

Example 2-4 In the television picture tube shown in Fig. 2-16, an electron beam projected from an electron gun causes the picture to appear on the screen. The force imparted to the electron charges by a high voltage is one factor which determines how bright the television picture will be. If 20 kV, applied across a 1-cm gap in the electron gun, produces a uniform field, what force is experienced by a charge of 1.6×10^{-10} C (the equivalent of 1000 million electrons)?

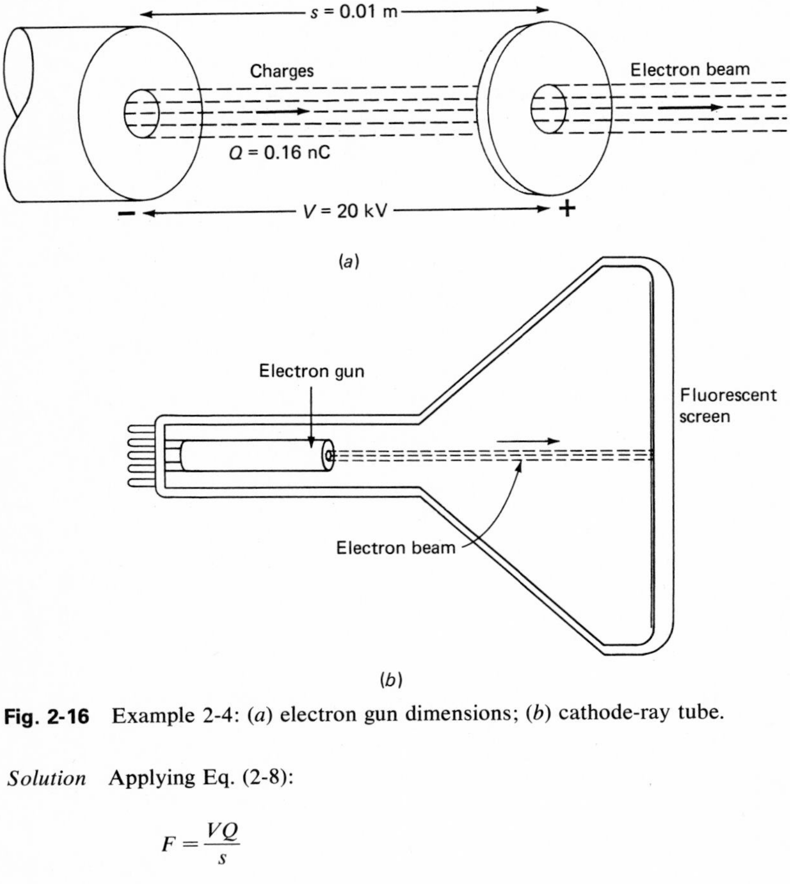

Fig. 2-16 Example 2-4: (*a*) electron gun dimensions; (*b*) cathode-ray tube.

Solution Applying Eq. (2-8):

$$F = \frac{VQ}{s} \tag{2-8}$$

$$= \frac{20 \text{ kV} \times 0.16 \text{ nC}}{0.01 \text{ m}} \quad \text{or} \quad \frac{2 \times 10^4 \text{ V} \times 1.6 \times 10^{-10} \text{ C}}{10^{-2} \text{ m}}$$

$$F = \mathbf{320\ \mu N} \quad \text{or} \quad \mathbf{3.2 \times 10^{-4}\ N}$$

One newton is equivalent to 0.225 lb. Therefore 3.2×10^{-4} N is equivalent to about $\frac{1}{1000}$ oz. This is indeed a very small force, but is quite enough to move 1000 million electron charges.

Electric field strength is defined as the force acting on a charge of one coulomb.

The symbol for electric field strength is E.*

* The symbol E is used for the quantities emf and electric field strength. Although these quantities may be related, they are not the same.

With 1 C of charge, Eq. (2-8) becomes

$$F = \frac{V}{s}$$

F is the force acting on a charge of 1 C and is therefore the electric field strength. The equation can be written as

$$E = \frac{V}{s} \tag{2-9}$$

where V = voltage, V
s = distance, m

The unit for electric field strength E is volt per meter (V/m). It can also be expressed as newton per coulomb (N/C).

Example 2-5 The porcelain insulators used on high-voltage power transmission towers are intended to prevent charge movement (leakage) from the transmission lines to the metal tower structure. On a certain 500-kV transmission line, an insulator is 2.5 m long (about 8 ft). If the electric field between the ends of the insulator is uniform, what is the strength of the electric field tending to cause leakage? (See Fig. 2-17.)

Fig. 2-17 Example 2-5: (*a*) outline sketch of a transmission tower; (*b*) insulator.

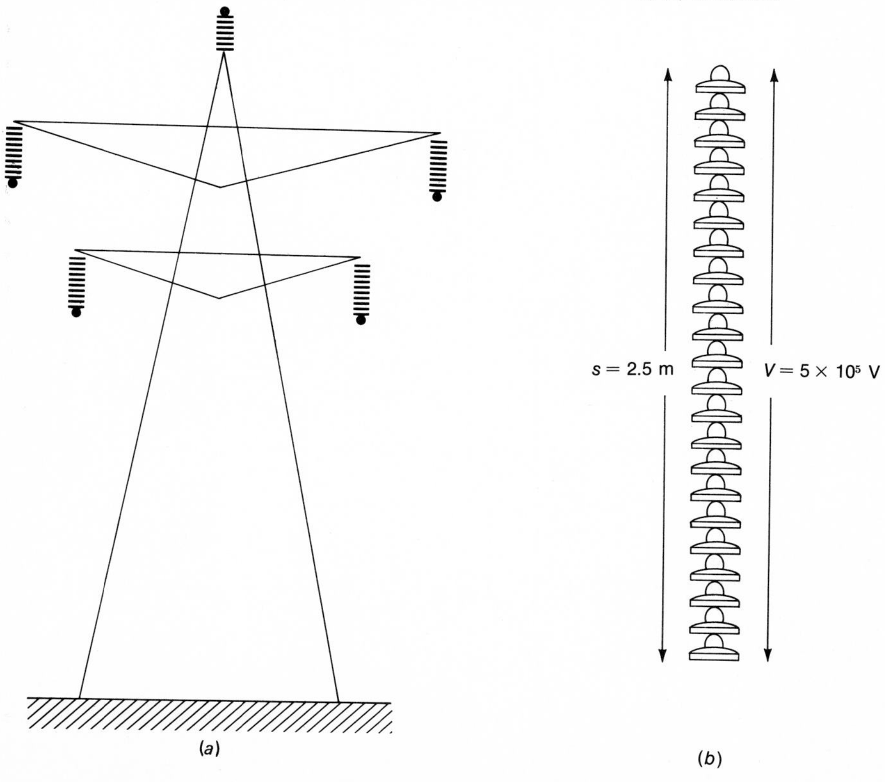

Solution Applying Eq. (2-9):

$$E = \frac{V}{s} \tag{2-9}$$

$$= \frac{500 \text{ kV}}{2.5 \text{ m}}$$

$$E = \mathbf{200\ kV/m}$$

This seems to be a high electric field strength, and it is high enough to cause appreciable leakage in some circumstances. But if you compare 200 kV/m with the field strength in Example 2-4, you will find that it is only one-tenth as great. In Example 2-4 the voltage is definitely lower than in Example 2-5, but it is the much smaller distance through which the voltage acts which results in the higher electric field strength in Example 2-4.

2-6 CURRENT

A conductor contains electron charges free to move. In the absence of an applied voltage and its resulting electric field, these free charges move within the conductor as a result of the forces existing between atoms.

As depicted in Fig. 2-18, the residual charge movement has no net direction; it is a random movement.

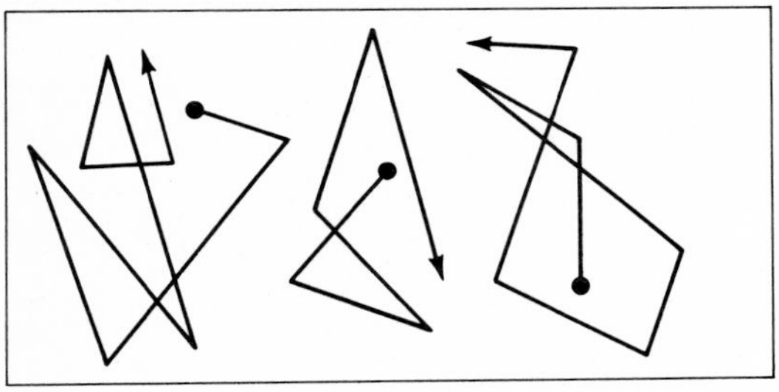

Fig. 2-18 Random movement of charges results in no current.

When a voltage is applied across the conductor, the attraction and repulsion of the negative charges results in a net direction of charge movement; a tendency exists for the charges to move in one general direction. Figure 2-19 shows this as a net movement from left to right.

Fig. 2-19 Net drift of charges in one direction results in a current.

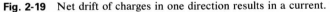

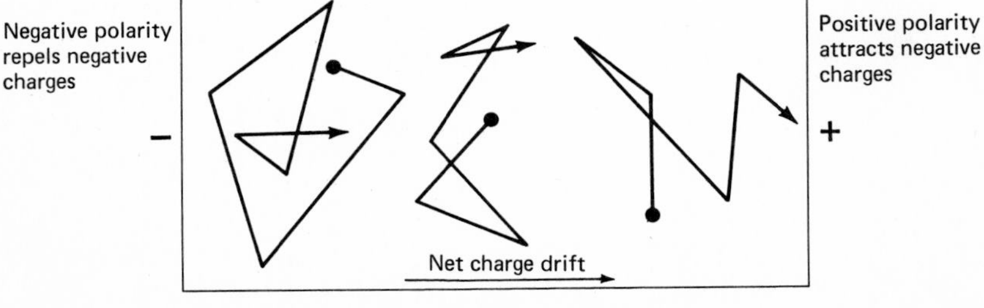

This drift of charges in one net direction is called an electric current.

It is necessary to distinguish between "a current" and "current."

A current is a phenomenon. It is the existence of a net drift of charges through a circuit.[1]

Current is a quantity. It is the magnitude of the net drift of charges through a circuit.

The symbol for current is I, the initial letter of the word "intensity"—intensity of charge drift.

Current is the rate of charge drift—the average amount of charge moving past a given point in a circuit each second. From this description we obtain the equation

$$I = \frac{Q}{t} \tag{2-10}$$

where I = current, A
Q = charge, C
t = time, s

Equation (2-10) is frequently rearranged into product form for ease of memorization

$$Q = It \tag{2-11}$$

THE AMPERE

The SI unit of current is the ampere, symbol A.*

Substituting unit symbols into Eq. (2-10):

$$\mathrm{A} = \frac{\mathrm{C}}{\mathrm{s}} \tag{2-12}$$

One ampere is one coulomb per second. Similarly, it can be seen that the coulomb, the SI unit of charge, has the MKSA dimension ampere · second.

Now that the MKSA dimensions of the joule and the coulomb have been obtained, the dimensions of the volt can be derived. From Eqs. (2-7) and (2-12),

$$\mathrm{J} = \mathrm{V} \cdot \mathrm{C} \tag{2-7}$$

and

$$\mathrm{A} = \frac{\mathrm{C}}{\mathrm{s}} \tag{2-12}$$

[1] This duality is not uncommon. For example, "a voltage" is a phenomenon, while "voltage" is a quantity. But, for some obscure reason, it is only "current" that seems to cause a communication problem.

* The ampere, one of the four basic MKSA units, is referred to in Sec. 1-6 and defined in the appendix.

is obtained:

$$V = \frac{J}{A \cdot s}$$

Combining this with Eq. (2-2):

$$J = \frac{kg \cdot m^2}{s^2} \qquad (2\text{-}2)$$

$$V = \frac{kg \cdot m^2}{s^3 \cdot A} \qquad (2\text{-}13)$$

CURRENT DIRECTION

The direction of a current is defined as the direction in which *positive-*charge carriers drift through a circuit.

Before the electron theory of current conduction became popular, it was believed that current was always due to the motion of positively charged particles of matter. With the advent of the belief that current is due to the movement of negative electron charges, a determined effort was made to change the definition of current direction. The controversy was so great that there was a period during which the electrical industry was split into two factions. One of them preferred the conventional definition of current direction (positive-charge carriers), while the other group

Fig. 2-20 Conventional current direction diagrams. (*a*) Pictorial; (*b*) schematic.

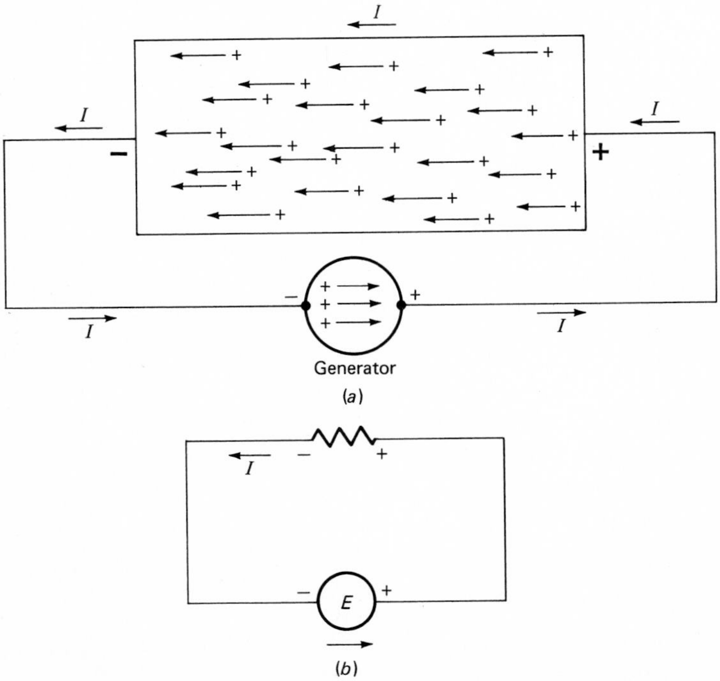

adopted a definition involving electron-flow direction (negative-charge carriers). To prevent the difficulties presented by having two definitions of current direction, the conventional definition was made the standard; it is used throughout this text.

Figure 2-20 shows what is meant by the conventional definition of current direction.

Positive charges are attracted toward the negative side of the applied emf and are repelled away from the positive side. The current direction is from the positive generator terminal, through the circuit, to the negative generator terminal. The current completes its path around the circuit inside the generator, from its negative terminal to its positive terminal.

Figure 2-20*b* is the schematic equivalent of Fig. 2-20*a*. The arrow below the generator symbol indicates the direction of the resulting current.

CURRENT VELOCITY

Just as it was necessary to distinguish between "a current" and "current," so it is necessary to distinguish between "velocity of propagation of electrical energy" and "velocity of an electric current."

When a circuit is open, as in Fig. 2-21*a*, there is no current. After the circuit is closed, as in Fig. 2-21*b*, there will be current I throughout the circuit.

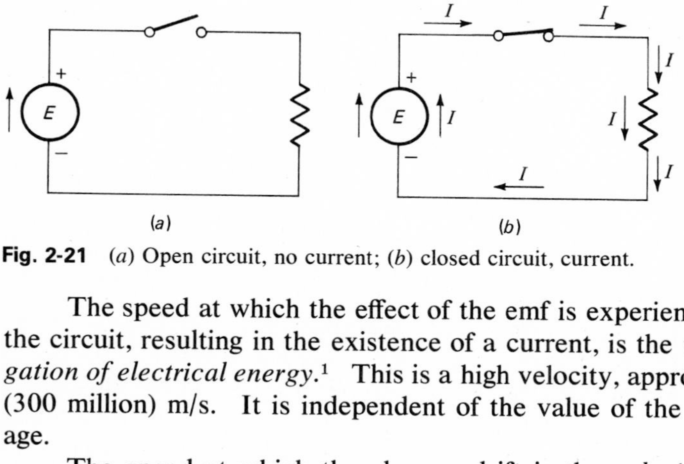

Fig. 2-21 (*a*) Open circuit, no current; (*b*) closed circuit, current.

The speed at which the effect of the emf is experienced at all parts of the circuit, resulting in the existence of a current, is the *velocity of propagation of electrical energy.*[1] This is a high velocity, approximately 3×10^8 (300 million) m/s. It is independent of the value of the current and voltage.

The speed at which the charges drift is the *velocity of the current.* This is quite low, typically fractions of a meter per second. This velocity is influenced by the value of the current and other characteristics of the circuit.

Example 2-6 If 1 ampere of current exists in a copper conductor 1 cm^2 in cross-sectional area and 10 km long, what is the velocity of the current? (See Fig. 2-22.)

[1] The velocity of propagation of electrical energy, or electromagnetic waves, is the "speed of light." This velocity has the symbol c and is 2.99793×10^8 m/s.

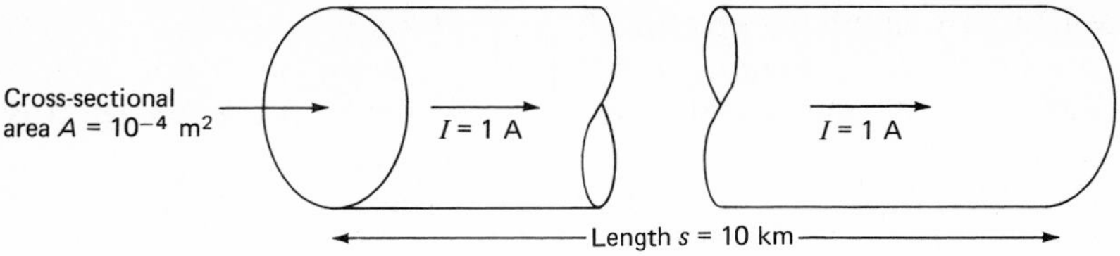

Fig. 2-22 Example 2-6: dimensions of conductor.

Solution The volume of the copper conductor is 1 m^3. This volume of copper contains approximately 8.54×10^{28} free electrons, which is equivalent to 1.37×10^{10} C of available charge carriers.

A current of 1 A is 1 C of charge per second. At this rate, 1.37×10^{10} C represents a time of 1.37×10^{10} s. In that time, all the available charge will have drifted through the conductor. As the charge drifts through, it is replaced by charge moving out of the energy source. Therefore the total charge of 1.37×10^{10} C will, in effect, travel the full 10-km length of the conductor in 1.37×10^{10} s. Velocity is distance per unit time. The velocity of drift can therefore be calculated:

$$v = \frac{s}{t}$$

$$= \frac{10^4 \text{ m}}{1.37 \times 10^{10} \text{ s}}$$

$$\mathbf{v = 7.3 \times 10^{-7} \text{ m/s} = 0.73\ \mu\text{m/s}}$$

The velocity 0.73 μm/s is very low, roughly 30 millionths of an inch per second. It only takes 33 millionths of a second for the current to start to move at all points along the entire length of the conductor in Example 2-6, but it would take 13.7 thousand million seconds, about 400 years, for charge to travel from one end of the conductor to the other, a distance of about 6 miles.

A conductor of 1 cm^2 cross-sectional area would normally be used to carry currents closer to 200 A than to 1 A. With 200 A the velocity would be increased 200-fold to 146 μm/s, still quite low.

2-7 POWER

Power is the rate of converting energy or the rate of doing work. The symbol for power is P. The SI unit for power is the watt, symbol W.*

From the definition of power the following equation can be derived:

$$P = \frac{W}{t} \qquad (2\text{-}14)$$

where P = power, W
W = energy or work, J
t = time, s

* It is easy to confuse W, the symbol for the unit watt, with W, the symbol for the quantity energy. To avoid this confusion do not use the word "wattage" when you mean "power." Use P for power, not W for wattage.

Equation (2-14) is often expressed in product form:

$$W = Pt \tag{2-15}$$

From Eq. (2-14) it can be seen that 1 watt is 1 J/s. Applying this to Eq. (2-2), we obtain a unit equation for the watt:

$$\mathrm{J} = \frac{\mathrm{kg \cdot m^2}}{\mathrm{s^2}} \tag{2-2}$$

$$\mathrm{W} = \frac{\mathrm{kg \cdot m^2}}{\mathrm{s^3}} \tag{2-16}$$

Equation (2-16) gives the MKSA dimension of the watt.

Example 2-7 An electric heater is rated at 1.5 kW. How much heat energy does it give out each hour (3600 s)?

Solution

$$W = Pt \tag{2-15}$$

$$= 1.5\ \mathrm{kW} \times 3.6\ \mathrm{ks}$$

$$W = \mathbf{5.4\ MJ}$$

Example 2-8 In a hydroelectric generating station, water falls 100 m through the penstocks to the generators. If the output power of the generators is 100 MW, how much water enters the penstocks each second? Assume that the water falls freely and that the generating station is 100 percent efficient.

Solution

$$W = Pt \tag{2-15}$$

$$= 100\ \mathrm{MW} \times 1\ \mathrm{s}$$

$$= 100\ \mathrm{MJ} \qquad (10^8\ \mathrm{J})$$

Applying Eq. (2-1):

$$W = Fs \tag{2-1}$$

$$F = \frac{W}{s}$$

$$= \frac{100\ \mathrm{MJ}}{100\ \mathrm{m}}$$

$$F = 1\ \mathrm{MN} \qquad (10^6\ \mathrm{N})$$

When a mass falls freely, the force[1] involved is given by

$$F = mg_n \tag{2-17}$$

where F = force, N

m = mass, kg

g_n = standardized free-fall acceleration, 9.807 m/s^2

[1] The force due to gravitational acceleration acting on a mass of 1 kg is called a "kilogram force."

The mass of water entering the penstocks each second will be

$$m = \frac{F}{g_n} = \frac{1\ \text{MN}}{9.807\ \text{m/s}^2}$$

$$= 10^6\ \frac{\text{kg}\cdot\text{m}}{\text{s}^2} \times \frac{1}{9.807}\ \frac{\text{s}^2}{\text{m}}$$

$$m = 1.019 \times 10^5\ \text{kg}$$

Water weighs 997.9 kg/m³. So the volume of water entering the penstocks each second is

$$V = \frac{1.019 \times 10^5\ \text{kg}}{9.979 \times 10^2\ \text{kg/m}^3}$$

$$V = \mathbf{102.1\ m^3}$$

Example 2-8, in which several different equations are applied, shows that, if a dam releases water through penstocks so that it falls 100 m (a little over 300 ft), approximately 100 m³ (about 100 000 kg) of water per second is required to generate 100 million watts of electric power if the system is 100 percent efficient. Of course, no system is perfect; more energy has to be put in than can be taken out. For a hydroelectric station 40 percent is a high efficiency. At this figure, the equivalent of 250 MW of power must be put into the system for an output of 100 MW, which means 250 cm³ of water enter the penstocks each second. If this does not seem to be much water to generate 100 MW of power, enough to supply a medium-sized city, think of it as approximately 5000 million gallons of water per day.

ENERGY, POWER, AND TIME

It is sometimes convenient to express energy in terms of current and time. Combining Eqs. (2-6) and (2-11):

$$W = VQ \qquad (2\text{-}6)$$

$$Q = It \qquad (2\text{-}11)$$

Therefore,

$$W = VIt \qquad (2\text{-}18)$$

where W = energy, J
V = voltage, V
I = current, A
t = time, s

By combining Eqs. (2-14) and (2-18),

$$P = \frac{W}{t} \qquad (2\text{-}14)$$

$$W = VIt \qquad (2\text{-}18)$$

we can obtain one of the most well-known formulas in electrical technology:

$$P = VI \tag{2-19}$$

Almost everyone who has studied or worked with electric circuits knows that watts are volts times amps. This is a statement of Eq. (2-19) in unit form, amps being the commonly used abbreviation for amperes.

There are two energy units, based on the watt, which are extensions of the joule. They are the watthour (Wh) and the kilowatthour (kWh). A joule is a wattsecond, so

$$1\ \text{Wh} = 1\ \text{W} \times 3600\ \text{s}$$

$$1\ \text{Wh} = 3600\ \text{J} = 3.6\ \text{kJ} \tag{2-20}$$

$$1\ \text{kWh} = 1000\ \text{Wh}$$

$$1\ \text{kWh} = 3.6 \times 10^6\ \text{J} = 3.6\ \text{MJ} \tag{2-21}$$

Example 2-9 How much current is drawn by a 120-V 100-W light bulb when it is lit?
Solution

$$P = VI \tag{2-19}$$

$$I = \frac{P}{V}$$

$$= \frac{100\ \text{W}}{120\ \text{V}}$$

$$I = \mathbf{0.8333\ A}$$

Example 2-10 A house is heated by an oil furnace with a heat energy output of 30 000 J/s. The furnace is to be replaced by electric heaters. What current must the electrical service to the house be capable of supplying for these heaters alone if the supply voltage is 240 V?
Solution

$$W = VIt \tag{2-18}$$

$$I = \frac{W}{Vt}$$

$$= \frac{3 \times 10^4\ \text{J}}{240\ \text{V} \times 1\ \text{s}}$$

$$I = \mathbf{125\ A}$$

BRITISH THERMAL UNIT AND HORSEPOWER

The British thermal unit (Btu) is an FPS unit of heat energy. As it is not an MKSA unit, it is currently not recommended for use. But it is still used extensively for many applications, including the rating of oil furnaces. One Btu is equivalent to 1055 J, so that the furnace in Example 2-10 has a heat energy output of approximately 100 000 Btu/h, a typical value for a three-bedroom house. Thus, to replace a 100 000-Btu furnace with electric heat operating from an ordinary domestic 240-V supply, requires 125 A of current.

The horsepower (hp) is another FPS unit which, although not approved for modern use, is still widely used. It will probably be many years before it becomes common to speak of an automobile of "150 kilowatts" instead of one of "200 horsepower."

$$1 \text{ hp} = 746 \text{ W} \tag{2-22}$$

Example 2-11 A 2-hp electric motor has an efficiency of 80 percent. What is (*a*) its electric power input in watts, (*b*) its energy consumption per day in watthours and in kilowatthours, and (*c*) the current-carrying capacity of its connecting wire, if it is to operate from a supply of 120 V?

Solution

a.

$$1 \text{ hp} = 746 \text{ W} \tag{2-22}$$

$$2 \text{ hp} = 1492 \text{ W}$$

The rated power of an electric motor is the mechanical power its rotating shaft is capable of producing. From Eqs. (2-5) and (2-14),

$$\eta = \frac{W_o}{W_i} = \frac{P_o}{P_i}$$

where η = efficiency (0.8 in this example)
P_o = output power (2 hp or 1492 W)
P_i = electric input power, W

$$P_i = \frac{P_o}{\eta} = \frac{1492 \text{ W}}{0.8}$$

$$P_i = \mathbf{1865 \text{ W}}$$

b. Applying Eq. (2-15):

$$W = Pt \tag{2-15}$$

$$W_i = 1865 \text{ W} \times 24 \text{ h} = \mathbf{44\,760 \text{ Wh}}$$

$$W_i = 1.865 \text{ kW} \times 24 \text{ h} = \mathbf{44.76 \text{ kWh}}$$

c. Applying Eq. (2-19):

$$P = VI \tag{2-19}$$

$$I = \frac{P_i}{V}$$

$$= \frac{1865 \text{ W}}{120 \text{ V}}$$

$$I = \mathbf{15.54 \text{ A}}$$

Thus this 2-hp motor will require an input power of 1865 W, will utilize 44 760 Wh or 44.76 kWh of electrical energy every day (if it is operated continuously), and will need to be connected to the electric system by wire capable of continuously carrying 15.54 A.

Example 2-12 A portable transistor radio has a 9-V battery with a usable capacity of 120 mA · h. The average current required by the radio is 20 mA. (*a*) How long can the radio be operated? (*b*) How much power does the radio require? (*c*) How much energy is the battery capable of supplying? (*d*) What charge is initially available from the battery?

Solution

a. The capacity of a battery is a product of current and time. Thus capacity is the same as quantity (Q).

$$Q = It \qquad (2\text{-}11)$$

$$t = \frac{Q}{I} = \frac{120\ \text{mA} \cdot \text{h}}{20\ \text{mA}} = 6\ \text{h}$$

$$t = 6\ \text{h} = \mathbf{2.16 \times 10^4\ s} \quad \text{or} \quad \mathbf{21.6\ ks}$$

b.

$$P = VI \qquad (2\text{-}19)$$

$$P = 9\ \text{V} \times 20\ \text{mA} = \mathbf{180\ mW} \quad \text{or} \quad \mathbf{0.18\ W}$$

c.

$$W = Pt \qquad (2\text{-}15)$$

$$W = 180\ \text{mW} \times 21.6\ \text{ks} = \mathbf{3888\ J}$$

d. Capacity is the same as quantity or charge:

$$Q = 120\ \text{mA} \cdot \text{h} = 120\ \text{mA} \times 3.6\ \text{ks} = \mathbf{432\ C}$$

This miniature battery will operate the radio for 6 h at an average current of 20 mA. In doing so, it will supply the radio with 3888 J of energy (1.08 Wh) while discharging 432 C of charge. While the radio is being operated, 0.18 W of power will be drawn from the battery. A small portion will be given off as audio (sound) power, but most of it will be given off in the form of heat.

2-8 CONDUCTANCE AND RESISTANCE

Whenever an emf is applied to a closed circuit, a current exists. But the same emf does not always produce the same current.

In Fig. 2-23*a* the 100-V emf causes a current of 10 A, while in Fig. 2-23*b* the same 100-V emf produces a current of only 1 A. The circuit in Fig. 2-23*a* has the greater ability to conduct current. To describe this ability, we say that this circuit has a higher conductance.

Conductance is the ability of a circuit to permit a current to result from an applied emf. The symbol for conductance is G. The SI unit of conductance is the siemens, symbol S.

The circuit in Fig. 2-23*b* has the greater tendency to resist current. It is said to have higher resistance.

Resistance is the ability of a circuit to restrict current. The symbol for resistance is R. The SI unit of resistance is the ohm, symbol Ω (capital Greek letter omega).

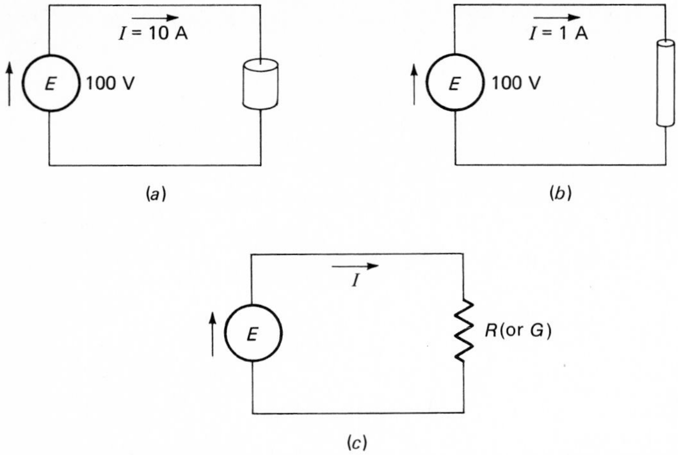

Fig. 2-23 (*a*) High conductance, low resistance; (*b*) low conductance, high resistance; (*c*) schematic diagram of a resistive circuit.

The circuits in Fig. 2-23*a* and *b* can both be represented by the schematic diagram in Fig. 2-23*c*. The schematic symbol for a circuit element gives no indication of the value of the element. The schematic symbol for resistance (or conductance, which has the same symbol) does not suggest its resistance value, just as the generator symbol gives no information about the emf.

If a circuit has high conductance, it must have low resistance, and vice versa; that is,

$$G = \frac{1}{R} \quad \text{and} \quad R = \frac{1}{G} \tag{2-23}$$

Example 2-13 An electric heater has a conductance of 50 mS when hot. What is its resistance?
Solution

$$R = \frac{1}{G} \tag{2-23}$$

$$= \frac{1}{50\ \text{mS}} = \frac{1}{5 \times 10^{-2}\ \text{S}}$$

$$R = \mathbf{20\ \Omega}$$

Example 2-14 The cable connecting a television set to its antenna has a resistance of 10 Ω/100 m length. If the cable is 20 m long, what is its conductance?
Solution The resistance of the cable equals the resistance per unit length multiplied by the length:

$$R = \frac{10\ \Omega}{100\ \text{m}} \times 20\ \text{m}$$

$$R = 2\ \Omega$$

$$G = \frac{1}{R} \tag{2-23}$$

$$= \frac{1}{2\ \Omega}$$

$$G = \mathbf{0.5\ S}$$

OHM'S LAW

Perhaps the most important principle in electrical technology is Ohm's law, which for steady direct current is usually stated: In any electric circuit, the current is directly proportional to the applied emf and inversely proportional to the total circuit resistance.

This statement can be written in the form of an equation usually known as Ohm's law:

$$I = \frac{V}{R} \quad \text{or} \quad V = IR \quad \text{or} \quad R = \frac{V}{I} \tag{2-24}$$

The Ohm's law equation can also be expressed in terms of conductance as

$$I = VG \quad \text{or} \quad V = \frac{I}{G} \quad \text{or} \quad G = \frac{I}{V} \tag{2-25}$$

In Eqs. (2-24) and (2-25), V = voltage (emf or pd), V
I = current, A
R = resistance, Ω
G = conductance, S

The statements at the beginning of this section represent a physical viewpoint of conductance and resistance—admitting and restricting current. We must look to Ohm's law to obtain a circuit view of these quantities and their units:

$$R = \frac{V}{I} \tag{2-24}$$

Resistance is the ratio of voltage to current; 1 ohm is 1 V/A.

$$G = \frac{I}{V} \tag{2-25}$$

Conductance is the ratio of current to voltage; 1 S is 1 A/V.

In unit form Eq. (2-24) is

$$\Omega = \frac{\text{V}}{\text{A}} \tag{2-26}$$

Combining Eqs. (2-13) and (2-26),

$$\Omega = \frac{\text{kg} \cdot \text{m}^2}{\text{s}^3 \cdot \text{A}} \times \frac{1}{\text{A}}$$

$$\Omega = \frac{\text{kg} \cdot \text{m}^2}{\text{s}^3 \cdot \text{A}^2} \tag{2-27}$$

which is the MKSA dimensions of the ohm.

By writing Eq. (2-25) in unit form, we derive the MKSA dimensions of the siemens

$$\text{S} = \frac{\text{A}}{\text{V}} \tag{2-28}$$

$$\text{S} = \frac{\text{s}^3 \cdot \text{A}^2}{\text{kg} \cdot \text{m}^2} \tag{2-29}$$

Example 2-15 An emf of 100 V is applied across a resistance of 20 Ω. What is the current?

Solution (See Fig. 2-24.)

Fig. 2-24 Example 2-15: circuit.

$$I = \frac{E}{R} \tag{2-24*}$$

$$I = \frac{100 \text{ V}}{20\ \Omega} = \mathbf{5\ A}$$

* In circuit equations E and V are usually interchangeable.

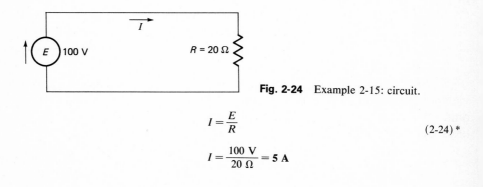

Example 2-16 A current of 100 mA exists in a 5-kΩ resistance. What is the pd across the resistance?
Solution (See Fig. 2-25.)

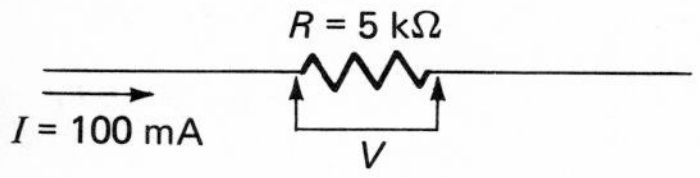

Fig. 2-25 Example 2-16: conditions associated with the resistance.

$$V = IR \qquad (2\text{-}24)$$

$$V = 100 \text{ mA} \times 5 \text{ k}\Omega = \mathbf{500\ V}$$

Example 2-17 If 3 mA is drawn from a 9-V battery, what is the circuit resistance?
Solution (See Fig. 2-26.)

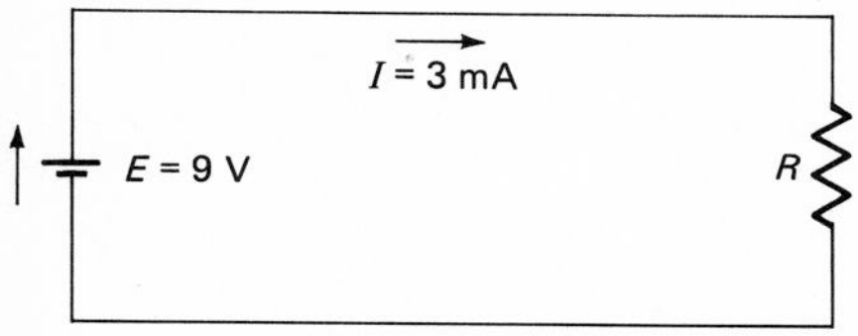

Fig. 2-26 Example 2-17: circuit.

$$R = \frac{E}{I} \qquad (2\text{-}24)$$

$$R = \frac{9 \text{ V}}{3 \text{ mA}} = \mathbf{3\ k\Omega}$$

Example 2-18 An electric motor armature has a conductance of 5 mS. What current will be produced in it by a 125-V emf?
Solution (See Fig. 2-27.)

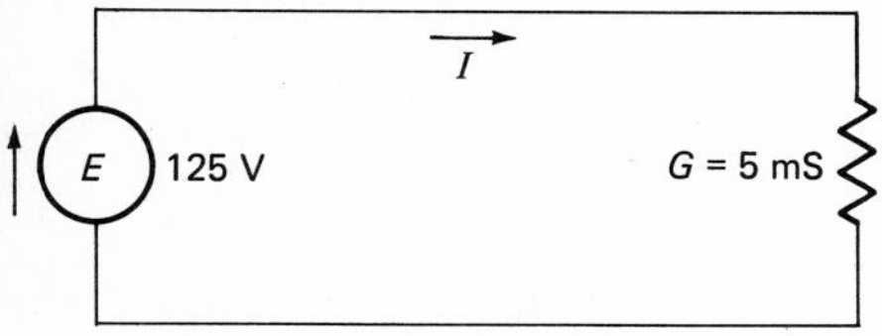

Fig. 2-27 Example 2-18: circuit.

$$I = EG \qquad (2\text{-}25)$$

$$I = 125 \text{ V} \times 5 \text{ mS} = \mathbf{625\ mA} \quad \text{or} \quad \mathbf{0.625\ A}$$

Example 2-19 What is the conductance of a 100-W 120-V light bulb when it is lit?
Solution

$$P = VI \tag{2-19}$$

$$I = \frac{P}{V} = \frac{100\ \text{W}}{120\ \text{V}} = 0.8333\ \text{A}$$

$$G = \frac{I}{V} \tag{2-25}$$

$$= \frac{0.8333\ \text{A}}{120\ \text{V}} = 6.944\ \text{mS}$$

POWER IN TERMS OF RESISTANCE

Combining Ohm's law [Eq. (2-24)] and the formula for power calculation [Eq. (2-19)] leads to two other very useful formulas.

Substituting for V from Eq. (2-24) into Eq. (2-19),

$$V = IR \tag{2-24}$$

$$P = VI \tag{2-19}$$

$$= IR \times I$$

$$P = I^2R \tag{2-30}$$

Substituting for I from Eq. (2-24) into Eq. (2-19),

$$I = \frac{V}{R} \tag{2-24}$$

$$P = VI \tag{2-19}$$

$$= V \times \frac{V}{R}$$

$$P = \frac{V^2}{R} \tag{2-31}$$

Example 2-20 What is the conductance of a 100-W 120-V light bulb? (This is the same question as that in Example 2-19.)
Solution

$$P = \frac{V^2}{R} \tag{2-31}$$

$$R = \frac{V^2}{P}$$

Therefore, as $G = 1/R$,

$$G = \frac{P}{V^2}$$

$$= \frac{100 \text{ W}}{(120 \text{ V})^2}$$

$$G = \mathbf{6.944 \text{ mS}}$$

Example 2-21 What minimum power rating is required for a resistor which is to operate in the collector circuit of a transistor, if its resistance is to be 1 kΩ and the collector current is 50 mA?

Solution

$$P = I^2R \tag{2-30}$$

$$P = (50 \text{ mA})^2 \times 1 \text{ k}\Omega = \mathbf{2.5 \text{ W}}$$

The resistor must be capable of dissipating a power of at least 2.5 W.

GRAPHICAL VIEW OF OHM'S LAW

Plotting values of voltage against current gives a graph of conductance and resistance.

If voltage is the independent variable and current the dependent variable, the slope of the curve will represent the conductance.

In Fig. 2-28, at all points on curve *a* the ratio I/V is 1:10 (for example, when $V = 20$ V, $I = 2$ A). Therefore curve *a* represents a conductance of 0.1 S. Similarly, curve *b* represents a conductance of 0.2 S (when $V = 20$ V, $I = 4$ A). The slope of the curve (its steepness) indicates the magnitude of the conductance.

If the current is the independent variable and the voltage the dependent variable, the slope of the curve will represent resistance.

Fig. 2-28 Graphs showing conductances. (*a*) 0.1 S; (*b*) 0.2 S.

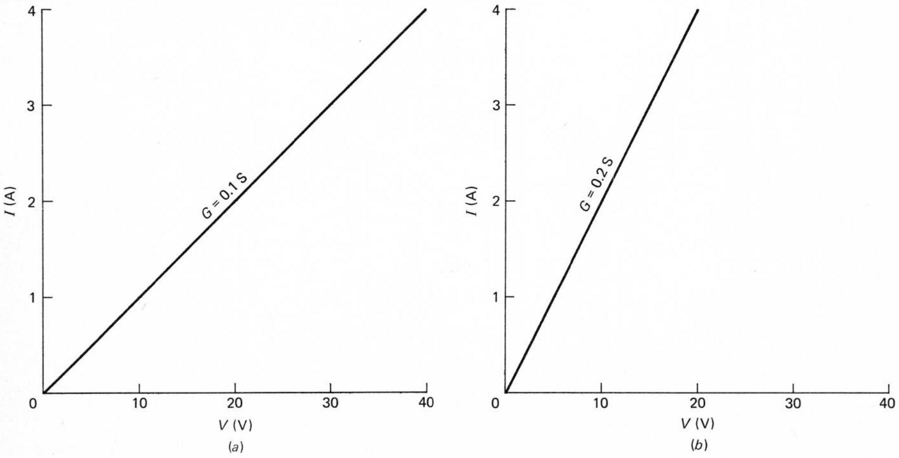

In Fig. 2-29, at all points on curve *a* the ratio V/I is 10, indicating a resistance of 10 Ω. Curve *b* represents a resistance of 20 Ω. The resistance is proportional to the slope of the curve.

The curves in Figs. 2-28*a* and 2-29*a* both represent the same circuit element since a conductance of 0.1 S is the same as a resistance of 10 Ω. The curves in Figs. 2-28*b* and 2-29*b* both have slopes greater than those of their *a* counterparts. They do *not* represent the same circuit elements, because higher conductance means lower resistance (Fig. 2-28*b*) and higher resistance means lower conductance (Fig. 2-29*b*).

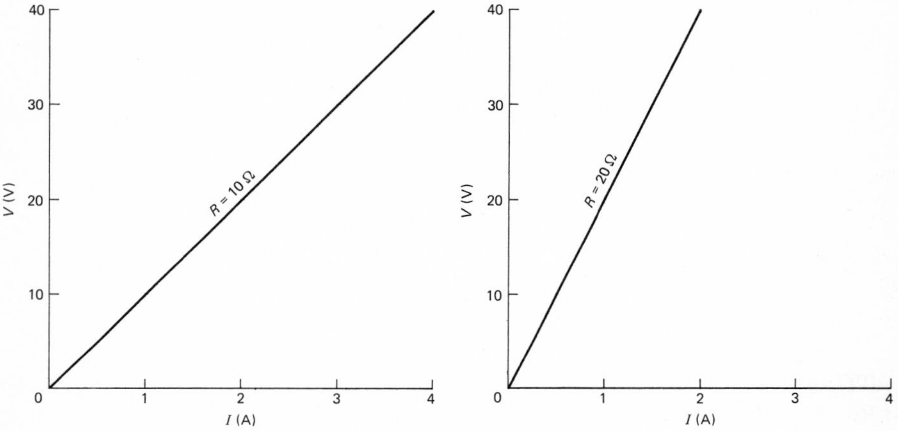

Fig. 2-29 Graphs showing resistances. (*a*) 10 Ω; (*b*) 20 Ω.

The curves in Figs. 2-28 and 2-29 represent linear conductances and resistances. In each case the current and voltage vary in direct proportion to each other. In this text we are only concerned with such linear elements. However, it should be realized that nonlinear elements are quite common. In fact, resistors, which are devices designed to provide the quantity resistance, are slightly nonlinear.[1] When a resistor passes current, it heats up. The rise in temperature causes the resistance to change. Well-designed resistors, correctly chosen for their application, do not show this effect to a significant degree. We are quite justified in considering that, in practical circuits, resistors can be dealt with as constant-value resistance elements and thus as linear devices.

[1] The appendix gives some information on nonlinear resistance.

2-9 SUMMARY OF QUANTITIES, UNITS, AND RELATIONSHIPS

Table 2-1 Quantities and units

Quantity	*Symbol*	*Unit*	*Symbol*	*MKSA equivalent*
energy or work	W	joule	J	$\frac{kg \cdot m^2}{s^2}$
emf pd or voltage drop	E V	volt	V	$\frac{kg \cdot m^2}{s^3 \cdot A}$
electric field strength	E	volt per meter	V/m	$\frac{kg \cdot m}{s^3 \cdot A}$
charge or quantity	Q	coulomb	C	$A \cdot s$
current	I	ampere	A	A
force	F	newton	N	$\frac{kg \cdot m}{s^2}$
resistance	R	ohm	Ω	$\frac{kg \cdot m^2}{s^3 \cdot A^2}$
conductance	G	siemens	S	$\frac{s^3 \cdot A^2}{kg \cdot m^2}$
power	P	watt	W	$\frac{kg \cdot m^2}{s^3}$
distance or length	s	meter	m	m
mass	m	kilogram	kg	kg
time	t	second	s	s
velocity	v	meters per second	m/s	$\frac{m}{s}$
acceleration	a	meters per second squared	m/s^2	$\frac{m}{s^2}$
temperature	T t	kelvin degree Celsius	K °C	
plane angle	$\phi, \theta, \ldots$	radian degree	rad °	
efficiency	η			
capacitance*	C	farad	F	$\frac{A^2 \cdot s^4}{kg \cdot m^2}$
inductance*	L	henry	H	$\frac{kg \cdot m^2}{s^2 \cdot A^2}$

* Capacitance and inductance, together with their units, farad and henry, are included in this table for the sake of completeness, although these topics are not covered in the text until Chaps. 8 and 9.

Table 2-2 Relationships between quantities

Relationship	*Equation*	*Relationship*	*Equation*
$t°\text{C} = (\theta°\text{F} - 32)\frac{5}{9}$	(1-1)	$F = mg_n$	(2-17)
$T\ \text{K} = t°\text{C} + 273.15$	(1-2)	$W = VIt$*	(2-18)
$2\pi\ \text{rad} = 360°$	(1-3)	$P = VI$*	(2-19)
$W = Fs$	(2-1)	$G = 1/R$*	(2-23)
$W_L = W_T - W_U$	(2-3)	$R = 1/G$*	(2-23)
$W_L = W_i - W_o$	(2-4)	$I = V/R$*	(2-24)
$\eta = W_o/W_i$	(2-5)	$V = IR$*	(2-24)
$W = VQ$ or EQ*	(2-6)	$R = V/I$*	(2-24)
$F = VQ/s$	(2-8)	$I = VG$*	(2-25)
$E = V/s$†	(2-9)	$V = I/G$*	(2-25)
$I = Q/t$	(2-10)	$G = I/V$*	(2-25)
$Q = It$*	(2-11)	$P = I^2R$*	(2-30)
$P = W/t$	(2-14)	$P = V^2/R$*	(2-31)
$W = Pt$*	(2-15)		

* These relationships are of great importance and should be memorized.
† E is electric field strength, not emf.

Table 2-3 Relationships between units

Relationship	*Equation*	*Relationship*	*Equation*
$\text{J} = \frac{\text{kg} \cdot \text{m}^2}{\text{s}^2}$	(2-2)	$\Omega = \frac{\text{kg} \cdot \text{m}^2}{\text{s}^3 \cdot \text{A}^2}$	(2-27)
$\text{J} = \text{V} \cdot \text{C}$	(2-7)	$\text{S} = \frac{\text{A}}{\text{V}}$	(2-28)
$\text{A} = \frac{\text{C}}{\text{s}}$	(2-12)	$\text{S} = \frac{\text{s}^3 \cdot \text{A}^2}{\text{kg} \cdot \text{m}^2}$	(2-29)
$\text{V} = \frac{\text{kg} \cdot \text{m}^2}{\text{s}^3 \cdot \text{A}}$	(2-13)	$\text{N} = \frac{\text{kg} \cdot \text{m}}{\text{s}^2}$	
$\text{W} = \frac{\text{kg} \cdot \text{m}^2}{\text{s}^3}$	(2-16)	$\text{F} = \frac{\text{A}^2 \cdot \text{s}^4}{\text{kg} \cdot \text{m}^2}$*	(8-11)
$1\ \text{Wh} = 3600\ \text{J}$	(2-20)		
$1\ \text{kWh} = 3.6 \times 10^6\ \text{J}$	(2-21)	$\text{H} = \frac{\text{kg} \cdot \text{m}^2}{\text{s}^2 \cdot \text{A}^2}$*	(9-14)
$1\ \text{hp} = 746\ \text{W}$	(2-22)		
$\Omega = \frac{\text{V}}{\text{A}}$	(2-26)		

* Capacitance and inductance, together with their units, farad and henry, are included in this table for the sake of completeness, although these topics are not covered in the text until Chaps. 8 and 9.

2-10 SUMMARY OF EQUATIONS

$W = Fs$	(2-1)	$F = mg_n \quad (g_n = 9.807 \text{ m/s}^2)$	(2-17)
$\text{J} = \text{N} \cdot \text{m} = \text{kg} \cdot \text{m}^2/\text{s}^2$	(2-2)	$W = VIt$	(2-18)
$W_L = W_T - W_U$	(2-3)	$P = VI$	(2-19)
$W_L = W_i - W_o$	(2-4)	$1 \text{ Wh} = 3.6 \text{ kJ}$	(2-20)
$\eta = W_o/W_i$	(2-5)	$1 \text{ kWh} = 3.6 \text{ MJ}$	(2-21)
$W = VQ$ or $EQ \quad (E = \text{emf})$	(2-6)	$1 \text{ hp} = 746 \text{ W}$	(2-22)
$\text{J} = \text{V} \cdot \text{C}$	(2-7)	$G = 1/R$ or $R = 1/G$	(2-23)
$F = VQ/s$	(2-8)	$I = V/R$ or $V = IR$ or $R = V/I$	(2-24)
$E = V/s \quad (E = \text{field strength})$	(2-9)	$I = VG$ or $V = I/G$ or $G = I/V$	(2-25)
$I = Q/t$	(2-10)	$\Omega = \text{V}/\text{A}$	(2-26)
$Q = It$	(2-11)	$\Omega = \text{kg} \cdot \text{m}^2/(\text{s}^3 \cdot \text{A}^2)$	(2-27)
$\text{A} = \text{C}/\text{s}$	(2-12)	$\text{S} = \text{A}/\text{V}$	(2-28)
$\text{V} = \text{kg} \cdot \text{m}^2/(\text{s}^3 \cdot \text{A})$	(2-13)	$\text{S} = \text{s}^3 \cdot \text{A}^2/(\text{kg} \cdot \text{m}^2)$	(2-29)
$P = W/t$	(2-14)	$P = I^2R$	(2-30)
$W = Pt$	(2-15)	$P = V^2/R$	(2-31)
$\text{W} = \text{kg} \cdot \text{m}^2/\text{s}^3$	(2-16)		

EXERCISE PROBLEMS

Section 2-4

2-1 How many coulombs of charge must pass through a load while it is converting 17 280 J of electrical energy to heat energy if the supply emf is 24 V? (*ans.* 720 C)

2-2 How many joules of energy are dissipated by a 6-V lamp bulb while passing 1500 C of charge?

2-3 What is the voltage developed across a load if 52 J of energy are dissipated while a movement of 130 C of charge occurs? (*ans.* 0.4 V)

2-4 If it requires 90 kJ of electrical energy to operate a 120-V toaster for $1\frac{1}{2}$ min, how many coulombs of charge must pass through the toaster during this period?

2-5 Find the amount of energy converted to heat by a stove element during a period in which 3×10^4 C of charge pass through the element when the supply voltage is 240 V. (*ans.* 7.2 MJ)

2-6 If the dissipation of energy in Prob. 2-3 occurred in 1 h, what amount of energy was dissipated per second?

2-7 With a line voltage of 120 V a television set dissipates, mainly as heat, 630 kJ of energy each hour.

a. How many coulombs of charge must pass through the set during this period? (*ans.* 5250 C)

b. How many coulombs of charge must pass through the set per second? (*ans.* 1.458 C)

c. How many joules of energy are dissipated per second? (*ans.* 175 J)

Section 2-5

2-8 A charge of 0.34 μC is attracted by a potential of 230 V at a distance of 0.24 cm. What is the attractive force?

2-9 A charge of 5 mC is repelled by a force of 2 kN. If the distance is 1 mm, what voltage is acting on the charge? (*ans.* 400 V)

2-10 Two plates of an electronic device are separated by a distance of 0.1 mm, contain a charge of 150 μC, and have an applied voltage of 150 V.
a. Find the electric field strength.
b. Find the force experienced by the charge.

2-11 If the plates in Prob. 2-10 are moved to one-tenth their previous separation and all else remains the same, what will be the new value of the electric field strength? (*ans.* 15 MV/m)

2-12 The electric field strength between two points 1 in. (2.54 cm) apart is found to be 11.8 kV/m. What is the pd between these two points?

2-13 A force of 0.64 N acts over a distance of 0.25 m on a charge of 8×10^{-6} C. Find the electric field strength? (*ans.* 80 kV/m)

Section 2-6

2-14 What current is present when 5 kC of charge pass a point in a circuit in 10 s?

2-15 What length of time is required to accumulate a charge of 0.8 mC if the average current is 25 mA? (*ans.* 32 ms)

2-16 What quantity of charge passes a point in a circuit in 1 min if the average current is 75 μA?

2-17 In a 120-V light bulb, a charge of 0.5 C passes a given point each second. What current does the bulb draw? (*ans.* 0.5 A)

2-18 A 120-V heating element has a current of 8.33 A.
a. How many coulombs of charge pass a given point in the heater each second?
b. How many coulombs pass a given point each hour?

2-19 A 12-V car battery storing a charge of 200 kC was completely discharged in 24 h. If the rate of discharge was constant, what current was drawn from the battery? (*ans.* 2.315 A)

2-20 A 6-V battery storing 2.4 MJ of energy was completely discharged in 12 h. If the rate of discharge was constant, what current was drawn from the battery?

2-21 A 120-V heating element converts energy at the rate of 1500 J/s. What is the current in the element? (*ans.* 12.5 A)

Section 2-7

2-22 How much time is required for a device with a rating of 600 W to do 2.7×10^5 J of work?

2-23 How much time is required for a transistor dissipating 50 mW to convert 1000 J of energy? (*ans.* 2×10^4 s or 5.556 h)

2-24 What is the power of a motor which does 1.44 MJ of work per hour?

2-25 What is the output power rating of a radio transmitter which radiates 108 kJ of energy per hour? (*ans.* 30 W)

2-26 A 120-V light bulb is rated at 200 W. How much energy does it convert in 24 h? State your answer in:
a. Joules
b. Watthours
c. Kilowatthours

2-27 A 12-V battery has 7 A drawn from it for 30 min. How much energy, in watthours, has been delivered by the battery? (*ans.* 42 Wh)

2-28 The heating element of an electric dryer is rated at 3 kW and is supplied from a 240-V source.
a. What current is drawn by the dryer?
b. How many joules of energy are converted to heat each hour?

2-29 How long will it take a 120-V electric motor drawing 1.5 A to consume 100 kWh of electrical energy? (*ans.* 555.6 h)

2-30 A 120-V motor, operating at 70 percent efficiency has a rated output of 1 hp. How long will it take this motor to do 10^6 J of work?

Section 2-8

2-31 What is the conductance of the following preferred-value resistors?

a. 18 Ω (*ans.* 55.56 mS)
b. 470 Ω (*ans.* 2.128 mS)
c. 1200 Ω (*ans.* 0.8333 mS)
d. 56 kΩ (*ans.* 17.86 μS)
e. 2.2 MΩ (*ans.* 0.4545 μS)
f. 3 mΩ (*ans.* 333.3 S)

2-32 What is the resistance of devices with the following conductances?

a. 1.2 mS *d.* 470 S
b. 2.5 S *e.* 8000 μS
c. 33.3 μS *f.* 0.06 S

2-33 An electronic instrument has a conductance of 25 μS, what is its resistance? (*ans.* 40 kΩ)

2-34 No. 2 copper wire has a resistance of 0.513 Ω/km. What is the conductance of 12 km of this wire?

2-35 Find the conductance of a resistor, in a transistor radio, which has a current of 220 μA and a pd of 5.2 V. (*ans.* 42.3 μS)

2-36 The input resistance of an oscilloscope is 1 MΩ. What is its input conductance?

2-37 Find the resistance of a device which has a pd of 300 V and a conductance of 4.55 mS. (*ans.* 219.8 Ω)

2-38 Find the voltage drop across an 18-kΩ resistor which has a current of 0.5 mA.

2-39 Determine the value of resistance in the following examples:

a. $E = 300$ V; $I = 2$ A (*ans.* 150 Ω)
b. $E = 24$ V; $I = 0.3$ A (*ans.* 80 Ω)
c. $V = 10$ V; $I = 260$ A (*ans.* 0.0385 Ω)
d. $V = 3.4$ V; $I = 1.9$ mA (*ans.* 1789 Ω)
e. $E = 12$ V; $I = 30.8$ μA (*ans.* 389.6 kΩ)

2-40 Determine the values of conductance for Prob. 2-39*a* to *e*.

2-41 Determine the voltage in the following examples:

a. $I = 3$ A; $R = 27$ Ω (*ans.* 81 V)
b. $I = 15$ mA; $R = 27$ kΩ (*ans.* 405 V)
c. $I = 30$ mA; $G = 350$ μS (*ans.* 85.71 V)
d. $R = 82$ Ω; $I = 8$ mA (*ans.* 0.656 V)
e. $G = 45$ μS; $I = 75$ μA (*ans.* 1.667 V)

2-42 Determine the current in the following examples:

a. $G = 25$ mS; $E = 80$ V *d.* $V = 240$ V; $R = 1500$ Ω
b. $G = 10$ S; $V = 6$ V *e.* $E = 9$ V; $G = 0.05$ S
c. $R = 180$ Ω; $V = 1.57$ V

2-43 Find the current in a circuit consisting of a 12-V battery and a 100-Ω load. (*ans.* 120 mA)

2-44 The input resistance of an amplifier is 50 Ω. If the input voltage is 75 μV, what current is developed?

2-45 A current of 20 mA is required in a circuit which has a supply voltage of 15 V. What value of resistance must be used? (*ans.* 750 Ω)

2-46 How much greater than 100 Ω must a resistor be to result in a current of 0.75 A in a circuit being supplied from a 120-V source?

Miscellaneous problems on electrical quantities

2-47 A 12-V battery charger is set for charging a battery at a rate of 6 A. How many coulombs of charge does the battery accumulate in the first 5 min? (*ans.* 1800 C)

2-48 A 120-V motor draws 4 A of current. If the motor is 80 percent efficient,

a. How many joules of mechanical energy are developed by the motor each second?

b. How many watts of mechanical power are developed?

c. How many watts of heating power are developed?

2-49 A $\frac{1}{2}$-hp motor operating from a 120-V supply is 70 percent efficient. How long will it take this motor to produce 10 kWh of work? (*ans.* 26.81 h)

2-50 A $\frac{1}{2}$-hp motor operating from a 120-V supply is 65 percent efficient. What current does it draw from the supply?

2-51 A 120-V soldering iron is rated at 30 W. What is its operating resistance? (*ans.* 480 Ω)

2-52 Find the conductance of a resistor which dissipates 1.8 W of power and has a current of 6 A.

2-53 Determine the voltage drop across a device which has a conductance of 0.166 S and dissipates 29 W of power. (*ans.* 13.22 V)

2-54 An electric radio supplied from a 120-V source consumes 75 W of power. What is its equivalent resistance?

2-55 What is the power rating of a machine which can do 1.73×10^8 J of work per day? Express your answer in watts and in horsepower. (*ans.* 2002 W, 2.684 hp)

2-56 If electrical energy costs 1.25 cents/kWh, how long would it take a 40-W electric light bulb to consume 25 cents worth of energy?

2-57 A voltmeter has a resistance of 5 MΩ on the 250-V range. When the meter reads 150 V, what is

a. The current through the meter? (*ans.* 30 μA)

b. The power dissipated by the meter? (*ans.* 4.5 mW)

c. The conductance of the meter? (*ans.* 0.2 μS)

2-58 An ammeter on the 100-μA range has a resistance of 2500 Ω. When the meter reads 65 μA, what is

a. The voltage drop across the meter?

b. The power dissipated by the meter?

c. The conductance of the meter?

TWO

DC Resistive Circuits

INTRODUCTION

Most of the quantities involved in electric circuits, together with their units and symbols, were dealt with in Part I. The succeeding chapters will utilize this information and apply it repeatedly to enable you to develop the required degree of familiarity with electrical fundamentals.

In Part I we dealt with generators with fixed polarities which resulted in fixed current directions—currents of apparently constant value. Such is not always the case in electric circuits. Currents which always have a single direction are called *direct currents* (*dc*). We also deal with circuits in which direct currents change value at times (*varying dc*). Very often we have to deal with a generator which regularly changes its polarity, resulting in regular reversals of current direction. The term *alternating current* (*ac*) is associated with these reversals of direction.

Circuits in which the currents are direct and of constant value (*steady dc*) seem to be easiest to understand. They minimize the number of different concepts which have to be visualized. Fortunately all the circuit principles which apply to steady dc also apply to varying dc and to ac. Calculations are more complicated with varying and alternating currents, but the fact remains that much can be learned about all circuits by investigating the effects of steady dc (usually, simply called "dc").

An understanding of circuit behavior is acquired largely by problem solving: by analysis and synthesis. *Analysis* implies calculating the values of quantities (currents, voltages, powers, resistances, etc.) in a specified circuit. *Synthesis* means designing a circuit which will result in specified quantities.

To solve a problem effectively, certain steps are necessary.

1. Understand what is known: the given data.
2. Understand the problem: the required data.
3. Determine which of the given data are relevant to the problem, and eliminate those data which are not relevant.
4. Draw a schematic diagram when applicable to the situation.
5. Write down all the relevant data, given and required, in a form which is helpful for problem solving. Preferably the data should be presented in either one or both of the following ways:
   a. Listed.
   b. Shown on the schematic diagram.

6. Write down the solution, proceeding logically from the known data to the required data.

Here is some general advice on problem solving.

1. Use correct symbols and accepted terms.
2. Use SI units and identify them with correct symbols.[1]

Check that each stage of your solution is reasonable, using the rule: If it is reasonable, it may not be right, but if it is not reasonable, it cannot be right.

[1] As previously mentioned, some quantities may be specified in units which are not SI. You will rarely meet such cases in this text.

3

Series Circuits

Circuits are formed by interconnecting elements into various patterns or configurations. The most common of these configurations are series and parallel circuits. Series circuits are, for most people, the easiest to understand.

To many persons the characteristics of series circuits seem obvious—so obvious that there is little point in spending time dealing with them. It is necessary to resist the urge to dismiss series circuits lightly. They are so important that their thorough mastery is essential. This mastery must be developed until the fine points of their functioning become intuitive knowledge. Do not let the simplicity of the series circuit connection deter you from checking every example and solving every problem.

Your understanding of series circuits must become such that you can with absolute certainty identify a series configuration within a complex network in a diagram or in actual manufactured equipment, although this is not always easy. You must be able to estimate the current, the voltage, and the power in series networks, and to know how these quantities vary.

Use your powers of observation and imagination to see and to visualize the many practical applications of series circuits.

3-1 THE MEANING OF SERIES CIRCUIT

Elements are said to be connected in *series* when they are joined together end to end, as in a chain. Figure 3-1 shows a series circuit in block diagram form.[1]

Fig. 3-1 Block diagram of series circuit.

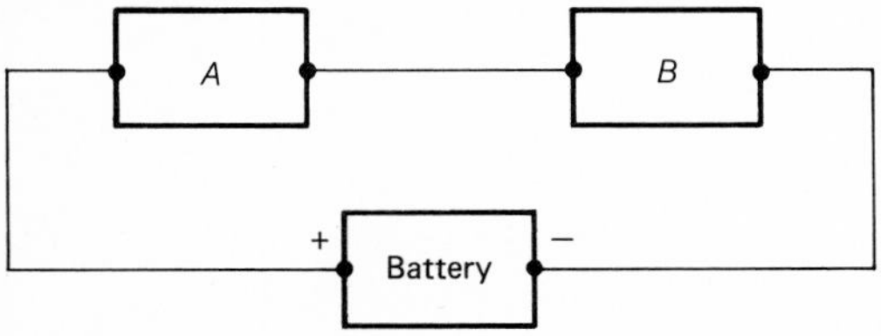

[1] In Figs. 3-1 and 3-2 the generator is shown as a battery. If we choose to ignore the fact that practical batteries are imperfect, a battery can be considered an energy source (generator) with an emf of constant polarity and value, which gives rise to steady direct currents.

In a series circuit any one end, or *terminal,* of an element can only be connected to one other element.

The term "cascade" is an alternative for "series."

3-2 CHARACTERISTICS OF SERIES CIRCUITS

The major characteristics of a series circuit are:

1. There is one current at all parts of the circuit.
2. Elements have individual voltages.
3. Voltages are additive.
4. Emf equals the sum of the element voltages (pds).
5. Resistances are additive.
6. Powers and energies are additive.

Each of these characteristics will now be examined in detail. The principles involved will be discussed, and examples given to demonstrate their application.

One current In a series circuit there is only one possible current path.

In Fig. 3-1 you will see that the current path can only be from the battery's positive terminal, through element *A*, and then through element *B* back to the negative battery terminal.

A schematic diagram, Fig. 3-2, the equivalent of Fig. 3-1, shows elements *A* and *B* as resistances R_A and R_B presented in conventional symbolic form.

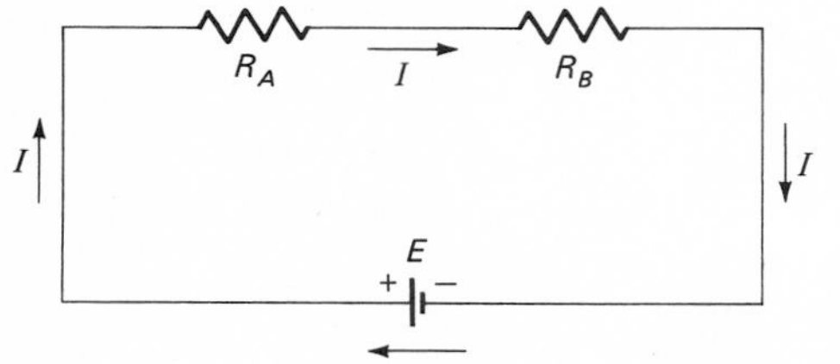

Fig. 3-2 One current in a series circuit.

In Fig. 3-2 the only possible path for the current I is from the battery E, through R_A and R_B in turn, and then back to the battery.

The current at any point in the circuit equals the amount of charge which moves past that point in 1 s. The charge movement is a continuous sequence whereby electron charges move progressively from atom to atom. The number of charges moving is constant throughout the circuit.

An obvious corollary of this constant-current characteristic is that in a series circuit it is only possible to measure one current value. If the battery delivers 1 A of current to R_A, then the same 1 A will exist between R_A and R_B and from R_B back to the battery. What might not be so obvious is that the same 1 A of current will exist within resistances R_A and R_B and within the battery. It is easy to measure the 1 A of current between the elements, but it is not easy to measure it within them.

Voltages individual The current I in Fig. 3-2 develops individual pds across each of the resistance elements. These are shown in Fig. 3-3 as the voltages V_A and V_B.

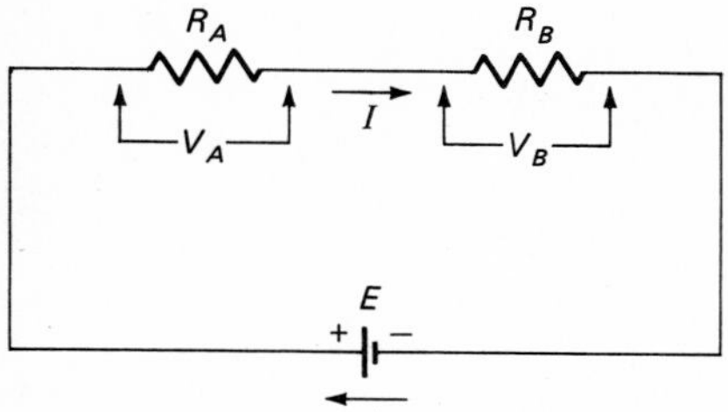

Fig. 3-3 Individual element pds in a series circuit.

By applying Ohm's law [Eq. (2-24)], the values of these pds can be determined:

$$V = IR$$
$$V_A = IR_A$$
$$V_B = IR_B$$

Example 3-1 A series circuit consisting of three resistances, 2, 8, and 20 Ω, connected to a battery has a current of 2 A. What pd exists across each resistance?
Solution (Fig. 3-4)

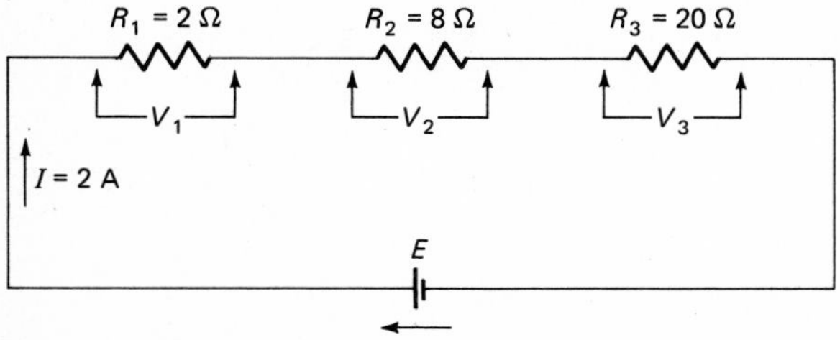

Fig. 3-4 Example 3-1: circuit.

Applying Eq. (2-24),

$$V = IR$$
$$V_1 = 2\text{ A} \times 2\ \Omega = \mathbf{4\ V}$$
$$V_2 = 2\text{ A} \times 8\ \Omega = \mathbf{16\ V}$$
$$V_3 = 2\text{ A} \times 20\ \Omega = \mathbf{40\ V}$$

It should be noted from Example 3-1 that the voltages in a series circuit are proportional to the resistance. $R_2 = 4 \times R_1$; so $V_2 = 4 \times V_1$. $R_3 = 10 \times R_1$; so $V_3 = 10 \times V_1$.

Voltages additive In Fig. 3-5 the pds V_A and V_B are shown with their polarities. With reference to the negative end of the circuit, point x is V_B volts positive. With reference to point x, the positive end of the circuit is V_A volts positive. Therefore the positive end is $V_B + V_A$ volts more positive than the negative end. These statements are summed up by the equation

$$V_T = V_A + V_B$$

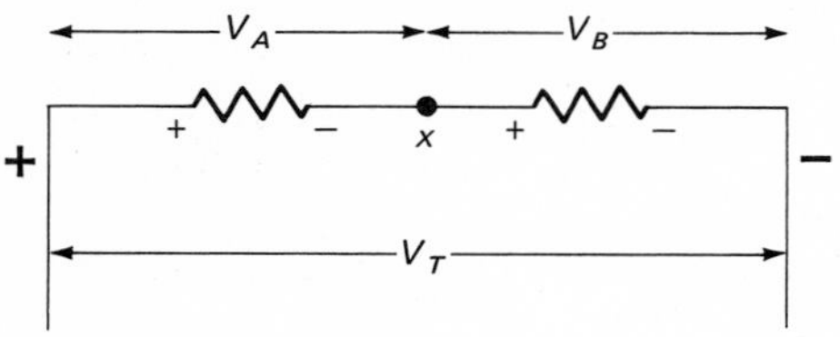

Fig. 3-5 Series circuit voltages are additive.

This relationship is not limited to circuits with just two elements. The total voltage across any number of resistances in series is the sum of the voltages across each of the resistances. This can be written as

$$V_T = V_1 + V_2 + V_3 + \cdots + V_n \qquad (3\text{-}1)$$

Example 3-2 A series network has a current of 5 A. The network consists of three elements, 10, 20, and 50 Ω. What is the total voltage across the circuit?
Solution (Fig. 3-6)

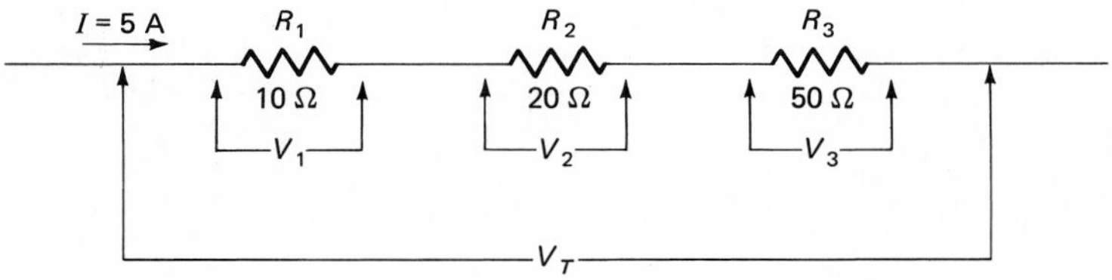

Fig. 3-6 Example 3-2: data.

Applying Eq. (2-24),

$$V = IR$$

$$V_1 = IR_1 = 5\ \text{A} \times 10\ \Omega = 50\ \text{V}$$

$$V_2 = IR_2 = 5\ \text{A} \times 20\ \Omega = 100\ \text{V}$$

$$V_3 = IR_3 = 5\ \text{A} \times 50\ \Omega = 250\ \text{V}$$

Applying Eq. (3-1),

$$V_T = V_1 + V_2 + V_3 \qquad (3\text{-}1)$$

$$V_T = 50\ \text{V} + 100\ \text{V} + 250\ \text{V} = \mathbf{400\ V}$$

Emf equals the sum of the pds (Kirchhoff's voltage law) When the series group of elements[1] comprises the total network across the generator, Eq. (3-1) can be extended to include the emf.

Referring to Fig. 3-7, we can see that the battery emf and the total series pd must be equal because they are connected across the same points.

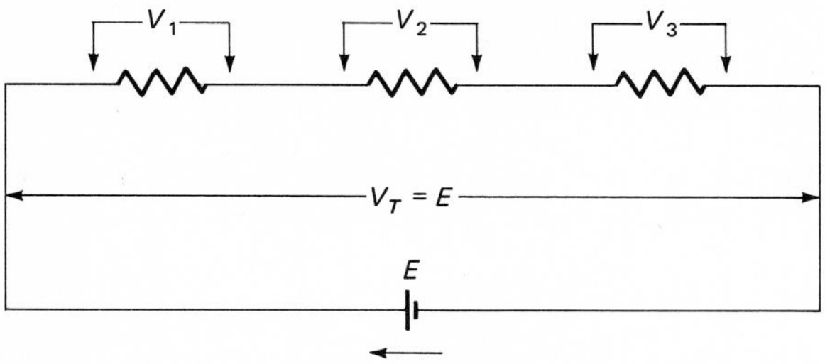

Fig. 3-7 Series circuit emf equals the sum of the element pds.

Therefore,

$$E = V_T$$
$$E = V_1 + V_2 + V_3 + \cdots + V_n \qquad (3\text{-}2)$$

Kirchhoff's voltage law is usually stated: In any closed loop the algebraic sum of the emfs equals the algebraic sum of the pds.

Any continuous conducting path is a *closed loop*. The circuit in Fig. 3-7 is a closed loop, and Eq. (3-2) is the related Kirchhoff's law expression.

The *algebraic sum* is the sum taking into account the polarities.

Example 3-3 Four resistances are connected in series. Their resistance values are 1 kΩ, 5 kΩ, 2.5 kΩ, and 500 Ω. The pd across the 5-kΩ resistance is 10 V. If the battery supplying energy to the network is connected directly across the resistance chain, what is its emf?

Solution

Fig. 3-8 Example 3-3: circuit.

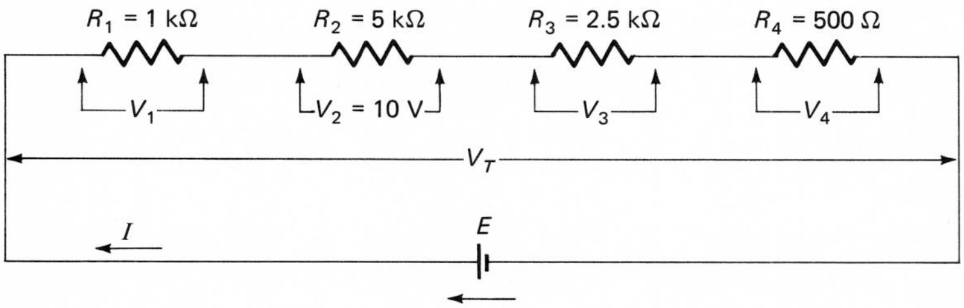

[1] It is usual to consider only those elements in a schematic diagram which are separately shown. When the characteristics of some part of a circuit are significant for circuit operation, these characteristics must be shown as distinct circuit elements. For example, if the conductors connecting elements have resistances which significantly influence the circuit currents, then resistance elements must be shown to account for these conductors.

As with most problems, there are several ways of obtaining a solution. Two methods will be shown.

METHOD 1 Applying Ohm's law, first to obtain the value of the circuit current, and then to obtain the values of the element pds,

$$I = \frac{V_2}{R_2} = \frac{10 \text{ V}}{5 \text{ k}\Omega} = 2 \text{ mA}$$

$$V_1 = IR_1 = 2 \text{ mA} \times 1 \text{ k}\Omega = 2 \text{ V}$$

$$V_2 = 10 \text{ V} \qquad \text{(given data)}$$

$$V_3 = IR_3 = 2 \text{ mA} \times 2.5 \text{ k}\Omega = 5 \text{ V}$$

$$V_4 = IR_4 = 2 \text{ mA} \times 500 \ \Omega = 1 \text{ V}$$

$$V_T = V_1 + V_2 + V_3 + V_4 \tag{3-1}$$

$$= 2 \text{ V} + 10 \text{ V} + 5 \text{ V} + 1 \text{ V} = 18 \text{ V}$$

$$E = V_T \tag{3-2}$$

$$= \mathbf{18\ V}$$

METHOD 2 Resistance ratios can be used to obtain pd values. As an example, consider V_1 and V_2 in Fig. 3-8:

$$\frac{V_1}{V_2} = \frac{IR_1}{IR_2} = \frac{R_1}{R_2}$$

Therefore $$V_1 = V_2 \frac{R_1}{R_2} \qquad \text{or} \qquad V_2 = V_1 \frac{R_2}{R_1}$$

Applying this principle leads to the solution of Example 3-3 without calculating the value of the common current. The following treatment is perhaps no more simple than Method 1, but greater accuracy is usually assured if the number of evaluations is minimized.

$$V_1 = V_2 \frac{R_1}{R_2} = 10 \text{ V} \frac{1 \text{ k}\Omega}{5 \text{ k}\Omega} = 2 \text{ V}$$

$$V_2 = 10 \text{ V} \qquad \text{(given data)}$$

$$V_3 = V_2 \frac{R_3}{R_2} = 10 \text{ V} \frac{2.5 \text{ k}\Omega}{5 \text{ k}\Omega} = 5 \text{ V}$$

$$V_4 = V_2 \frac{R_4}{R_2} = 10 \text{ V} \frac{0.5 \text{ k}\Omega}{5 \text{ k}\Omega} = 1 \text{ V}$$

$$E = V_T = V_1 + V_2 + V_3 + V_4 = 2 \text{ V} + 10 \text{ V} + 5 \text{ V} + 1 \text{ V} = \mathbf{18\ V}$$

Resistances additive If, from an electrical measurement viewpoint, two networks are the same, they are said to be *equivalent*.

If the ammeters and voltmeters in Fig. 3-9*a* and *b* indicate the same currents and voltages,[1] then network *A* is equivalent to network *B*.

[1] "Ammeter" is an abbreviation of "ampere meter"; an *ammeter* is an instrument for measuring the magnitudes of currents. Similarly, a *voltmeter* is an instrument for measuring voltage magnitudes.

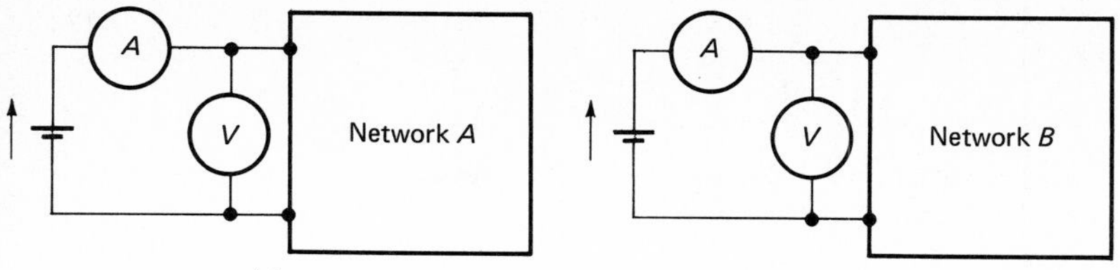

(a) (b)

Fig. 3-9 Equivalent networks.

Figure 3-10 shows network A as consisting of three resistances in series, while network B is just one resistance. If the currents and voltages measured at the terminals are the same in each case, the single resistance R_T is equivalent to the total of the three resistances R_1, R_2, and R_3 in series.

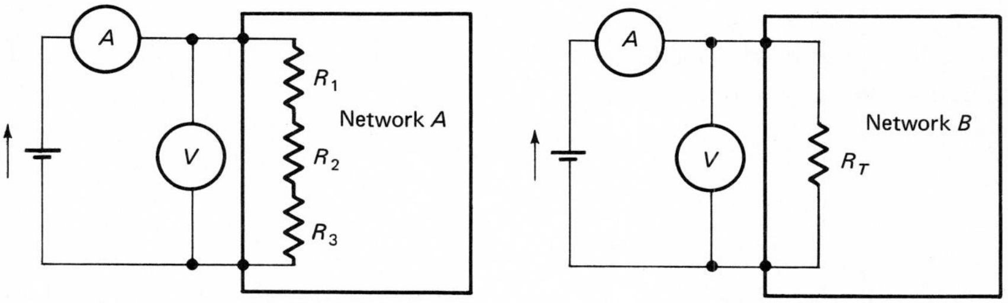

(a) (b)

Fig. 3-10 One resistance equivalent to three in series.

The equivalent resistance of a network is often called the *total resistance,* hence the frequently used symbol R_T.

"Equivalent" does not literally mean "the same as." If we take a number of resistors which are physically the same size and use them to construct two networks, the resulting network A, with three resistors, will be capable of dissipating more power than network B, with only one resistor. Thus the two networks may be equivalent from the total resistance viewpoint, but not from that of power-handling capability.

In any specific case, it is essential that the conditions of equivalence be clearly understood. The terms of the equivalence must be stated or implied by the situation.

When we say that R_T in Fig. 3-10b is the equivalent of R_1, R_2, and R_3 in Fig. 3-10a, we only mean that the total resistance is the same in each case. Any relationship which naturally follows from this (for example, the same emf results in the same current) will also be true.

Figure 3-11a and b represents equivalent networks. With the same current I the total voltage V_T will be the same.

Applying Ohm's law to Fig. 3-11a,

$$V_T = IR_T$$

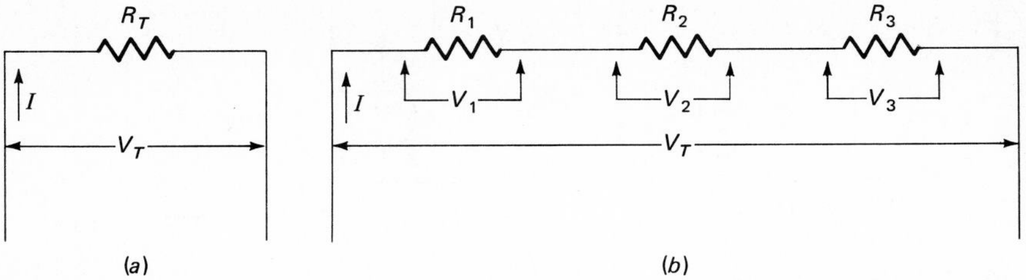

Fig. 3-11 Single-element network (*a*) to be equivalent to three-element network (*b*).

Applying Eq. (3-1) and Ohm's law to Fig. 3-11*b*,

$$V_T = V_1 + V_2 + V_3 \qquad (3\text{-}1)$$

$$V_T = IR_1 + IR_2 + IR_3 = I(R_1 + R_2 + R_3)$$

We have said that the total voltage V_T is the same for each circuit if the currents are the same, so that

$$IR_T = I(R_1 + R_2 + R_3)$$

$$R_T = R_1 + R_2 + R_3$$

This relationship applies to any number of elements connected in series:

$$R_T = R_1 + R_2 + R_3 + \cdot \cdot \cdot + R_n \qquad (3\text{-}3)$$

The total resistance of a number of resistances connected in series is the sum of the individual resistances.

Example 3-4 What is the total resistance of four resistances connected in series if their individual resistance values are 1 MΩ, 1.5 MΩ, 150 kΩ, and 50 000 Ω?
Solution (Fig. 3-12)

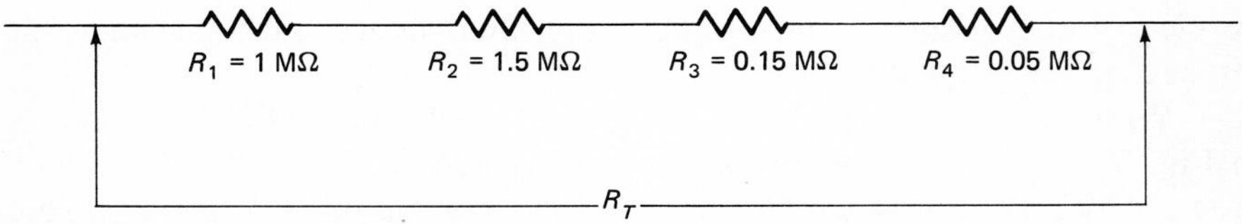

Fig. 3-12 Example 3-4: data.

$$R_T = R_1 + R_2 + R_3 + R_4 \qquad (3\text{-}3)$$

$$R_T = 1\ \text{M}\Omega + 1.5\ \text{M}\Omega + 0.15\ \text{M}\Omega + 0.05\ \text{M}\Omega = \mathbf{2.7\ M\Omega}$$

Example 3-5 What resistance, connected in series with a 1.8-kΩ resistance will result in a total resistance of 2 kΩ?
Solution (Fig. 3-13)

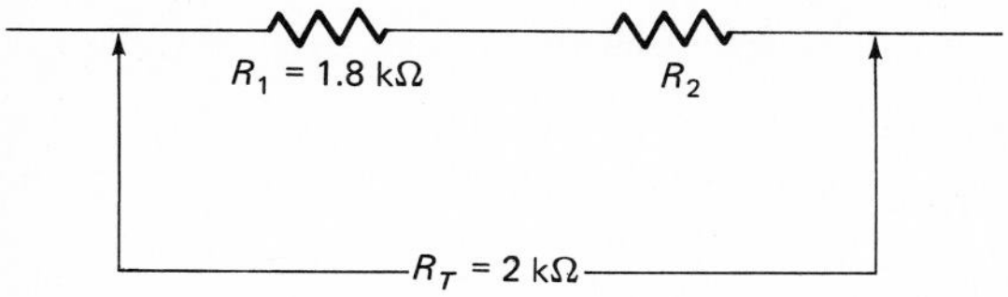

Fig. 3-13 Example 3-5: data.

$$R_T = R_1 + R_2 \tag{3-3}$$

$$R_2 = R_T - R_1 = 2\ \text{k}\Omega - 1.8\ \text{k}\Omega$$

$$R_2 = \mathbf{0.2\ k\Omega} \quad \text{or} \quad \mathbf{200\ \Omega}$$

Powers and energies additive Referring back to Eq. (2-30) ($P = I^2R$), we see that, when the current in a number of elements is the same, the power dissipated by each element is directly proportional to its resistance value. Furthermore, combining Eq. (2-30) with Eq. (2-15) ($W = Pt$) gives $W = I^2Rt$, showing that, over a stated time period, the energy converted by each of the elements will be directly proportional to its resistance value. Hence, since the total equivalent resistance of a number of resistances in series is equal to the sum of the individual resistances, the total power dissipated and the total energy converted will be equal to the sums of the individual powers and energies.

$$P_T = P_1 + P_2 + P_3 + \cdots + P_n \tag{3-4}$$

$$W_T = W_1 + W_2 + W_3 + \cdots + W_n \tag{3-5}$$

Example 3-6 Resistors of 10, 20, and 100 Ω are connected in series across an emf of 26 V. What power is dissipated by each resistor, and what total power does the battery supply to the network?
Solution (Fig. 3-14)

Fig. 3-14 Example 3-6: circuit.

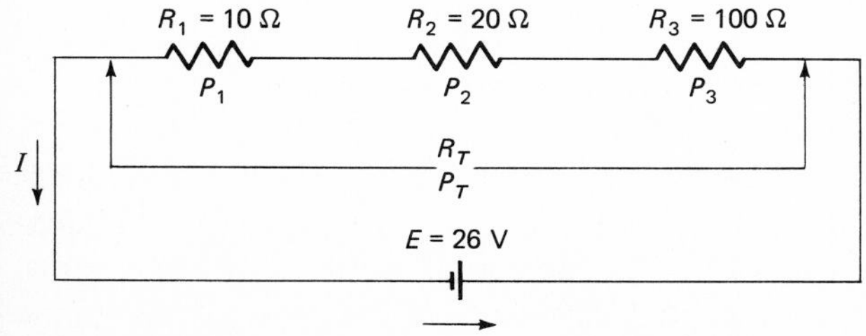

$$R_T = R_1 + R_2 + R_3 \tag{3-3}$$

$$R_T = 10\ \Omega + 20\ \Omega + 100\ \Omega = 130\ \Omega$$

$$I = \frac{E}{R_T} = \frac{26\ \text{V}}{130\ \Omega} = 0.2\ \text{A} \quad \text{or} \quad 200\ \text{mA}$$

$$P_1 = I^2R_1 = (0.2\ \text{A})^2 \times 10\ \Omega = \mathbf{0.4\ W}$$

$$P_2 = I^2R_2 = (0.2\ \text{A})^2 \times 20\ \Omega = \mathbf{0.8\ W}$$

$$P_3 = I^2R_3 = (0.2\ \text{A})^2 \times 100\ \Omega = \mathbf{4\ W}$$

$$P_T = P_1 + P_2 + P_3 \tag{3-4}$$

$$P_T = 0.4\ \text{W} + 0.8\ \text{W} + 4\ \text{W} = \mathbf{5.2\ W}$$

The total power could have been calculated using the emf and the current, as in Eq. (2-19):

$$P_T = EI$$

$$P_T = 26\ \text{V} \times 0.2\ \text{A} = \mathbf{5.2\ W}$$

Example 3-7 Two 120-V 60-W light bulbs are connected across a 240-V supply. What total energy is converted per hour?
Solution (Fig. 3-15)

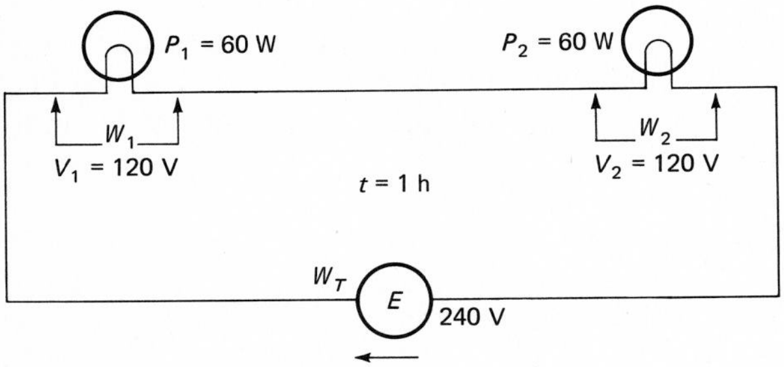

Fig. 3-15 Example 3-7: circuit.

For each bulb Eq. (2-15) can be applied:

$$W_1 = W_2 = P_1t = P_2t$$

$$W_1 = W_2 = 60\ \text{W} \times 1\ \text{h} = 60\ \text{Wh} \quad \text{or} \quad 216\ \text{kJ}$$

$$W_T = W_1 + W_2 \tag{3-5}$$

$$W_T = 2 \times 60\ \text{Wh} = \mathbf{120\ Wh} \quad \text{or} \quad \mathbf{432\ kJ}$$

The solution given for Example 3-7 is only valid if the bulbs are identical (an unlikely event in practice). If the bulbs are assumed to be identical, then the 240-V applied emf will divide equally between the two bulbs, giving each a pd of 120 V, the rated voltage for 60 W of power. If the bulbs are not identical, the problem becomes extremely difficult to solve. Light bulbs are nonlinear, being of much greater resistance when lit than when switched off. A detailed mathematical or graphical knowledge of the characteristics of the bulbs is necessary for the solution of a simple series circuit with two different light bulbs.

3-3 CALCULATIONS FOR SERIES CIRCUITS

The calculations involved in any type of circuit format are concerned with the application of electrical relationships to the variables and parameters of that circuit.

The quantities typical of series circuits are shown in Fig. 3-16, and are then listed with some of the common techniques used for calculating them. Examples are given to demonstrate how the various methods can be used, the method selected depending on the information available. Note that in the examples more data are given than would be normal in a problem, as several methods for calculating the value of each quantity are shown.

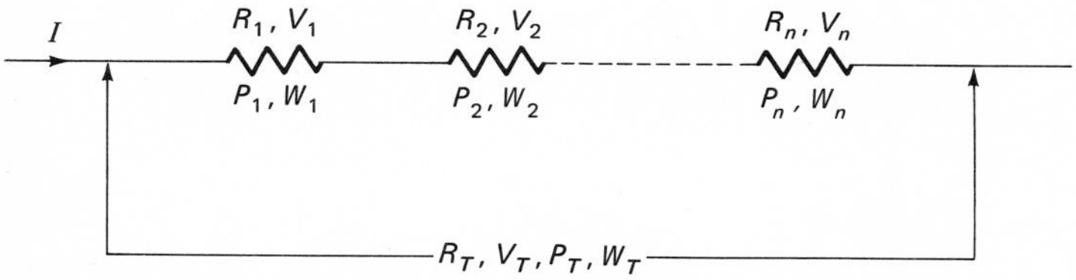

Fig. 3-16 Generalized symbols for a series circuit.

Current *I* The circuit current can be calculated by applying Ohm's law to the complete network or to any part of it:

$$I = \frac{V_T}{R_T} \quad \text{or} \quad \frac{V_1}{R_1} \quad \text{or} \quad \frac{V_2}{R_2} \quad \text{or} \quad \cdots \quad \text{or} \quad \frac{V_n}{R_n}$$

Example 3-8 Find the current in the circuit in Fig. 3-17.

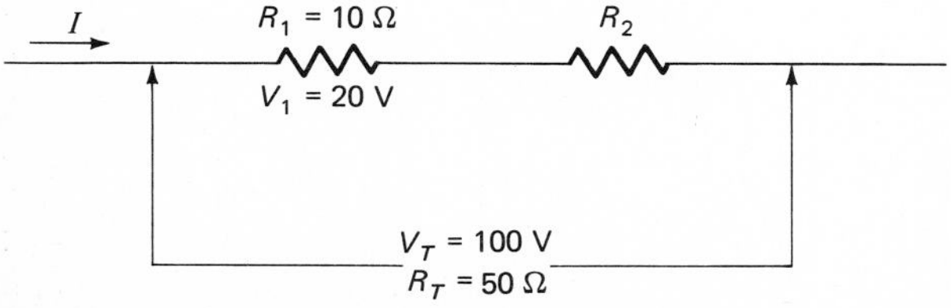

Fig. 3-17 Example 3-8: data.

Solution

METHOD 1 Applying Ohm's law to the total circuit,

$$I = \frac{V_T}{R_T} = \frac{100\ \text{V}}{50\ \Omega} = \mathbf{2\ A}$$

METHOD 2 Applying Ohm's law to one element (any element may be selected),

$$I = \frac{V_1}{R_1} = \frac{20\ \text{V}}{10\ \Omega} = \mathbf{2\ A}$$

Total voltage V_T The total voltage across the network can be calculated by adding the individual element voltages or by the application of Ohm's law:

$$V_T = V_1 + V_2 + \cdots + V_n \qquad (3\text{-}1)$$

or

$$V_T = IR_T$$

The Ohm's law method can be extended to a resistance ratio form:

$$V_T = IR_T = \frac{V_1}{R_1} R_T = V_1 \frac{R_T}{R_1}$$

The resistance ratio equation leads to a general equation:

$$V_T = V_1 \frac{R_T}{R_1} \quad \text{or} \quad V_2 \frac{R_T}{R_2} \quad \text{or} \quad \cdots \quad \text{or} \quad V_n \frac{R_T}{R_n} \qquad (3\text{-}6)$$

Example 3-9 Find the total voltage in the circuit in Fig. 3-18.

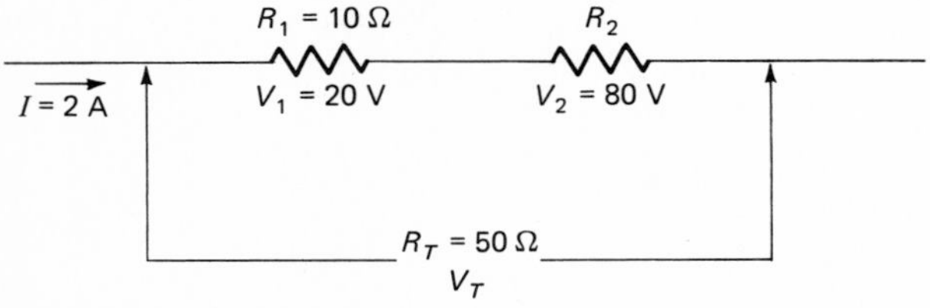

Fig. 3-18 Example 3-9: data.

Solution

METHOD 1 Sum of the pds

$$V_T = V_1 + V_2 \qquad (3\text{-}1)$$

$$V_T = 20\ \text{V} + 80\ \text{V} = \mathbf{100\ V}$$

METHOD 2 Applying Ohm's law to the complete circuit,

$$V_T = IR_T = 2\ \text{A} \times 50\ \Omega = \mathbf{100\ V}$$

METHOD 3 Resistance ratio

$$V_T = V_1 \frac{R_T}{R_1} \tag{3-6}$$

$$V_T = 20\ \text{V}\ \frac{50\ \Omega}{10\ \Omega} = \mathbf{100\ V}$$

Individual voltages $V_1, V_2, \ldots, V_n$ The voltage across an individual element can be calculated by the direct application of Ohm's law:

$$V_1 = IR_1$$

or by extending this equation to a resistance ratio form by substituting a suitable ratio V/R for I.

$$V_1 = \frac{V_T}{R_T} R_1$$

or $\quad V_1 = V_T \frac{R_1}{R_T} \quad$ or $\quad V_2 \frac{R_1}{R_2} \quad$ or $\quad \cdots \quad$ or $\quad V_n \frac{R_1}{R_n} \quad$ (3-7)

The individual voltages V_2 to V_n, if unknown, can be calculated using similar methods.

Example 3-10 What is the value of voltage V_1 in the circuit in Fig. 3-19?

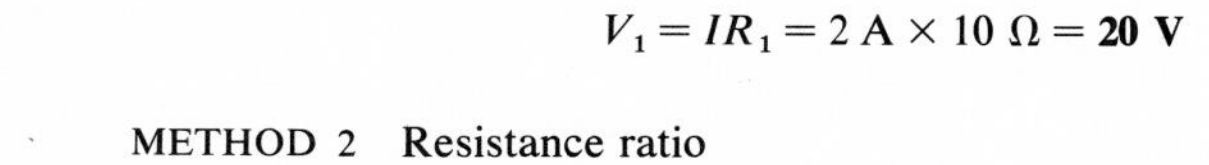

Fig. 3-19 Example 3-10: data.

Solution

METHOD 1 Ohm's law

$$V_1 = IR_1 = 2\ \text{A} \times 10\ \Omega = \mathbf{20\ V}$$

METHOD 2 Resistance ratio

$$V_1 = V_2 \frac{R_1}{R_2} \tag{3-7}$$

$$V_1 = 80\ \text{V}\ \frac{10\ \Omega}{40\ \Omega} = \mathbf{20\ V}$$

Total resistance R_T The total network resistance can be calculated by adding the individual resistances or by the application of Ohm's law:

$$R_T = R_1 + R_2 + \cdots + R_n \tag{3-3}$$

or

$$R_T = \frac{V_T}{I}$$

Expressions for R_T in terms of voltage ratios can be obtained by substituting appropriate ratios R/V for $1/I$ in the Ohm's law equation:

$$R_T = V_T \frac{1}{I} = V_T \frac{R_1}{V_1}$$

$$R_T = R_1 \frac{V_T}{V_1} \quad \text{or} \quad R_2 \frac{V_T}{V_2} \quad \text{or} \quad \cdots \quad \text{or} \quad R_n \frac{V_T}{V_n} \qquad (3\text{-}8)$$

Example 3-11 What is the total resistance of the circuit in Fig. 3-20?

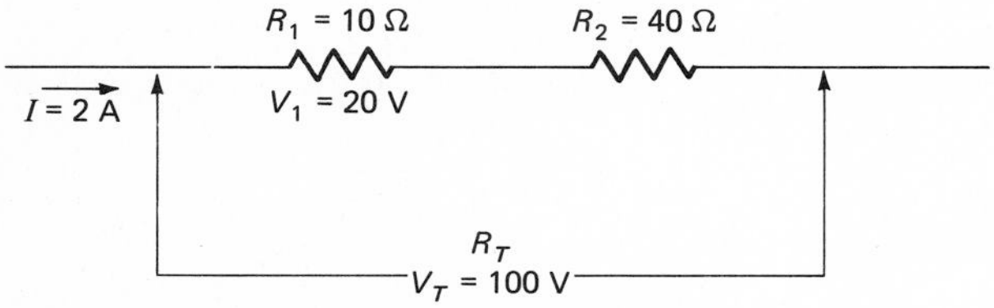

Fig. 3-20 Example 3-11: data.

Solution

METHOD 1 Sum of the resistances

$$R_T = R_1 + R_2 \qquad (3\text{-}3)$$

$$R_T = 10\ \Omega + 40\ \Omega = \mathbf{50\ \Omega}$$

METHOD 2 Ohm's law

$$R_T = \frac{V_T}{I} = \frac{100\ \text{V}}{2\ \text{A}} = \mathbf{50\ \Omega}$$

METHOD 3 Voltage ratio

$$R_T = R_1 \frac{V_T}{V_1} \qquad (3\text{-}8)$$

$$R_T = 10\ \Omega\ \frac{100\ \text{V}}{20\ \text{V}} = \mathbf{50\ \Omega}$$

Individual resistances R_1, R_2, . . . , R_n The individual resistances of the elements can be calculated by Ohm's law:

$$R_1 = \frac{V_1}{I}$$

or by extending this equation to a voltage ratio form by substituting a convenient ratio R/V for $1/I$:

$$R_1 = R_T \frac{V_1}{V_T} \quad \text{or} \quad R_2 \frac{V_1}{V_2} \quad \text{or} \quad \cdots \quad \text{or} \quad R_n \frac{V_1}{V_n} \qquad (3\text{-}9)$$

Similar methods can be used for calculating the individual resistances R_2 to R_n.

Example 3-12 What is the value of resistance R_1 in the circuit in Fig. 3-21?

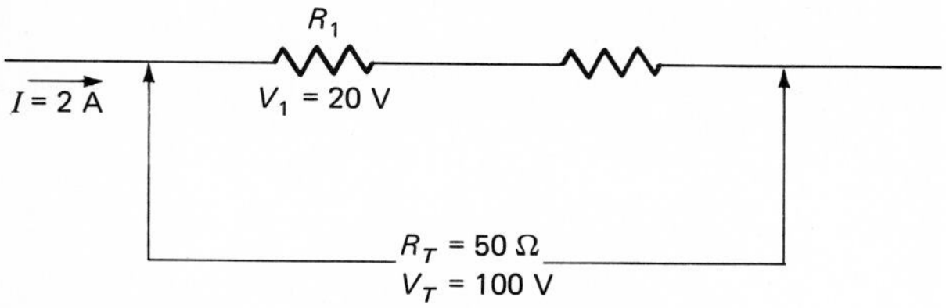

Fig. 3-21 Example 3-12: data.

Solution

METHOD 1 Ohm's law

$$R_1 = \frac{V_1}{I} = \frac{20\ \text{V}}{2\ \text{A}} = \mathbf{10\ \Omega}$$

METHOD 2 Voltage ratio

$$R_1 = R_T \frac{V_1}{V_T} \tag{3-9}$$

$$R_1 = 50\ \Omega\ \frac{20\ \text{V}}{100\ \text{V}} = \mathbf{10\ \Omega}$$

Total power P_T The total power can be calculated by summing the individual powers or by applying Ohm's law to the complete network:

$$P_T = P_1 + P_2 + \cdots + P_n \tag{3-4}$$

or, from Eq. (2-19),

$$P_T = V_T I$$

Substituting IR_T for V_T, and V_T/R_T for I, gives two other equations for P_T [refer to Eqs. (2-30) and (2-31)].

From Eq. (2-30),

$$P_T = I^2 R_T$$

or, from Eq. (2-31),

$$P_T = \frac{V_T^2}{R_T}$$

The total power can be expressed in terms of resistance ratios by noting that power is directly proportional to resistance when the circuit current is constant, as in a series circuit.

$$\frac{P_T}{P_1} = \frac{I^2R_T}{I^2R_1} = \frac{R_T}{R_1}$$

$$P_T = P_1\frac{R_T}{R_1} \quad \text{or} \quad P_2\frac{R_T}{R_2} \quad \text{or} \quad \cdots \quad \text{or} \quad P_n\frac{R_T}{R_n} \tag{3-10}$$

Example 3-13 What is the total power dissipated in the series network in Fig. 3-22?

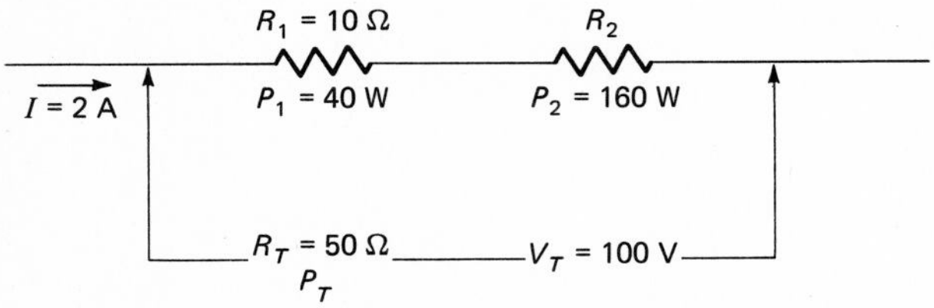

Fig. 3-22 Example 3-13: data.

Solution

METHOD 1 Sum of the powers

$$P_T = P_1 + P_2 \tag{3-4}$$

$$P_T = 40\text{ W} + 160\text{ W} = \mathbf{200\text{ W}}$$

METHOD 2 Product of voltage and current

$$P_T = V_TI = 100\text{ V} \times 2\text{ A} = \mathbf{200\text{ W}}$$

METHOD 3 Current and resistance
From Eq. (2-30),

$$P_T = I^2R_T$$

$$P_T = (2\text{ A})^2 \times 50\ \Omega = \mathbf{200\text{ W}}$$

METHOD 4 Voltage and resistance
From Eq. (2-31),

$$P_T = \frac{V_T^2}{R_T}$$

$$P_T = \frac{(100\text{ V})^2}{50\ \Omega} = \mathbf{200\text{ W}}$$

METHOD 5 Resistance ratio

$$P_T = P_1\frac{R_T}{R_1} \tag{3-10}$$

$$P_T = 40\text{ W}\,\frac{50\ \Omega}{10\ \Omega} = \mathbf{200\text{ W}}$$

Individual powers $P_1, P_2, \ldots, P_n$ The power dissipated by an individual element can be calculated by applying techniques similar to those used for the total power in the last section. Methods for calculating the power in R_1 are given, but they can be applied to any of the other elements R_2 to R_n.

$$P_1 = V_1 I \quad \text{or} \quad I^2 R_1 \quad \text{or} \quad \frac{V_1^2}{R_1}$$

$$P_1 = P_T \frac{R_1}{R_T} \quad \text{or} \quad P_2 \frac{R_1}{R_2} \quad \text{or} \quad \cdots \quad \text{or} \quad P_n \frac{R_1}{R_n} \tag{3-11}$$

Example 3-14 What is the power dissipated by resistance R_1 in the circuit in Fig. 3-23?

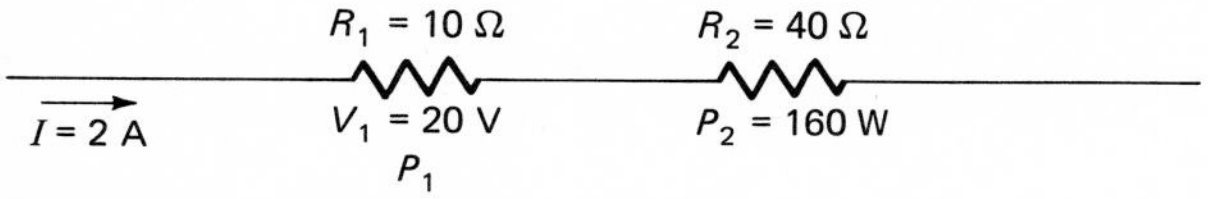

Fig. 3-23 Example 3-14: data.

Solution

METHOD 1 Product of voltage and current

$$P_1 = V_1 I = 20 \text{ V} \times 2 \text{ A} = \mathbf{40\ W}$$

METHOD 2 Current and resistance

$$P_1 = I^2 R_1 = (2 \text{ A})^2 \times 10\ \Omega = \mathbf{40\ W}$$

METHOD 3 Voltage and resistance

$$P_1 = \frac{V_1^2}{R_1} = \frac{(20 \text{ V})^2}{10\ \Omega} = \mathbf{40\ W}$$

METHOD 4 Resistance ratio

$$P_1 = P_2 \frac{R_1}{R_2} \tag{3-11}$$

$$P_1 = 160 \text{ W} \frac{10\ \Omega}{40\ \Omega} = \mathbf{40\ W}$$

Energies The methods for calculating power, given in the last two sections and demonstrated in Examples 3-13 and 3-14, can all be applied to energy. It is only necessary to multiply any power by the appropriate time period to obtain the relevant energy.

Series circuit ratios Ratios are used frequently in series circuit calculations, because the common value of current in all parts of the circuit makes resistance, voltage, power, and energy ratios the same for any two parts of the circuit.

Comparing the quantities relevant to the whole circuit and to R_1 gives relationships which can be applied generally:

$$\frac{V_T}{V_1} = \frac{IR_T}{IR_1} = \frac{R_T}{R_1}$$

$$\frac{P_T}{P_1} = \frac{I^2R_T}{I^2R_1} = \frac{R_T}{R_1}$$

$$\frac{W_T}{W_1} = \frac{I^2R_Tt}{I^2R_1t} = \frac{R_T}{R_1}$$

Therefore

$$\frac{R_T}{R_1} = \frac{V_T}{V_1} = \frac{P_T}{P_1} = \frac{W_T}{W_1} \tag{3-12}$$

This ratio concept can be applied to pairs of elements, for example, $R_2/R_1 = P_2/P_1$.

Example 3-15 Using the ratio of R_T to R_1 in Fig. 3-24, determine (*a*) the total voltage, (*b*) the total power, and (*c*) the total energy used in 2 hours.

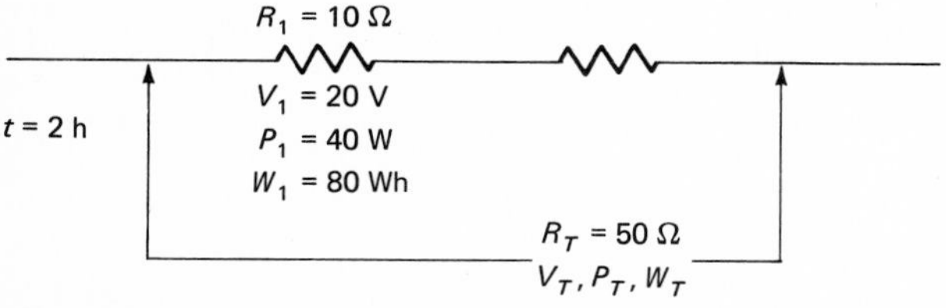

Fig. 3-24 Example 3-15: data.

Solution The ratio of R_T to R_1 is

$$\frac{R_T}{R_1} = \frac{50\ \Omega}{10\ \Omega} = 5$$

This ratio is shown in parentheses in the following solutions.

a. $V_T = (5)V_1 = 5 \times 20\text{ V} = \mathbf{100\ V}$

b. $P_T = (5)P_1 = 5 \times 40\text{ W} = \mathbf{200\ W}$

c. $W_T = (5)W_1 = 5 \times 80\text{ Wh} = \mathbf{400\ Wh}$

3-4 RELATIVE POTENTIAL

The discussion on series circuits thus far has not dwelt on the importance of voltage polarities. They are of particular significance in determining relative potentials.

A *relative potential* is the voltage existing at one point in a circuit with respect to that at another point (usually called the reference).

In Fig. 3-25 is shown a two-element series network with the most negative end, point A, shown as the reference. With respect to point A, both points B and C are positive.

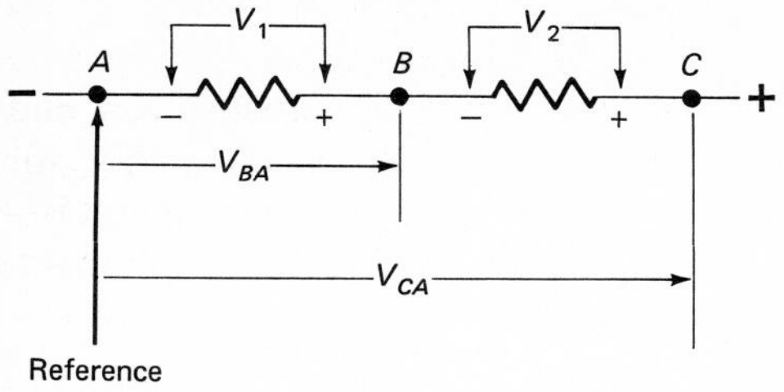

Fig. 3-25 Relative potential, reference at point A.

$$V_{BA}^{*} = +V_1$$

$$V_{CA} = +(V_1 + V_2)$$

Figure 3-26 shows the same network as that of Fig. 3-25, but the reference chosen is at point C.

With respect to point C, both points A and B are negative.

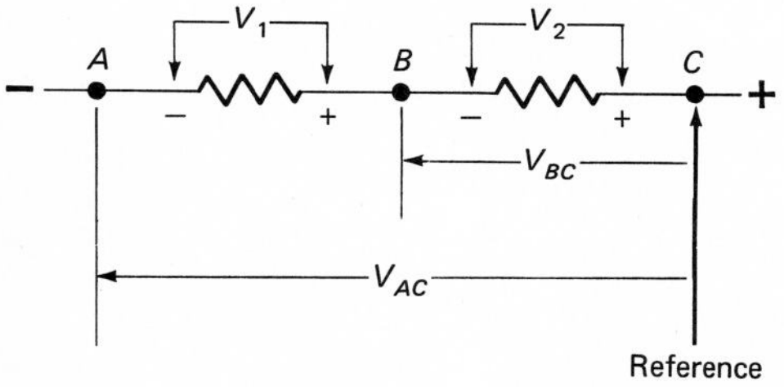

Fig. 3-26 Relative potential, reference at point C.

$$V_{BC} = -V_2$$

$$V_{AC} = -(V_1 + V_2)$$

The reference point is shown at point B in Fig. 3-27. Point C is positive with respect to point B, but point A is negative with respect to point B.

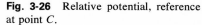

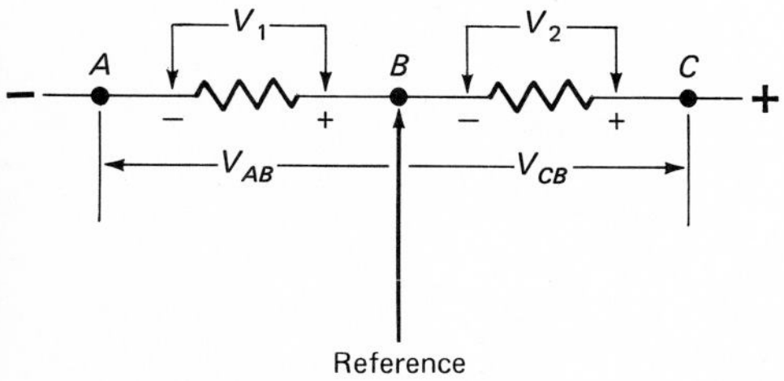

Fig. 3-27 Relative potential, reference at point B.

$$V_{CB} = +V_2$$

$$V_{AB} = -V_1$$

* It will be seen that V_{BA} refers to the voltage at B with respect to that at A. This is the convention used in electronics and which is adopted in this book.

The reference can be chosen at will to be at any point in a circuit. It is selected to suit the situation. Frequently the reference point is called *ground*. Ground is a rather difficult notion to comprehend (despite the fact that many persons use the term quite freely). Originally, it meant a point in a circuit literally connected to the ground (the earth), either for safety in power systems, or for efficient communication in radio transmission and reception. This meaning still exists, but it is quite usual today for "ground" to mean a circuit point connected to a relatively large metal object, such as the chassis, frame, or cabinet supporting the equipment. "Common" is another term used as an alternative to "ground."

Where the reference point is considered to be grounded, or commoned, it is usual to state relative potentials as voltages which are simply positive or negative. It is implicitly understood that they are with respect to ground, or common.

Figure 3-28 is the same as Fig. 3-27, except for the ground symbol at the reference connection, point *B*. The relative potentials at points *A* and *C* can be written as

$$V_A = -V_1$$

$$V_C = +V_2$$

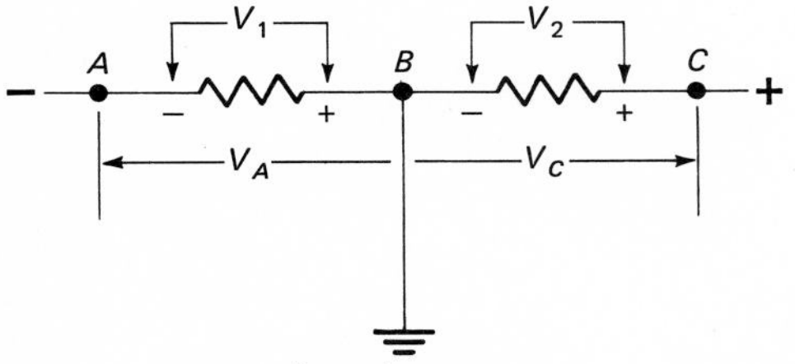

Fig. 3-28 Relative potential with respect to ground.

Example 3-16 What are the relative potentials at points *B*, *C*, and *D* with respect to point *A* in Fig. 3-29?

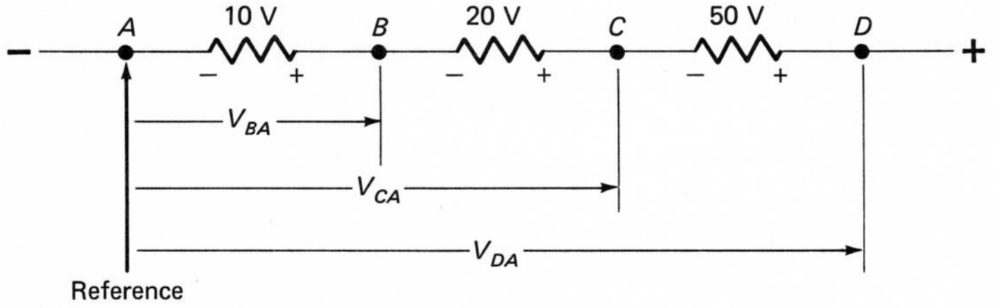

Fig. 3-29 Example 3-16: data.

Solution

$$V_{BA} = \mathbf{+10\ V}$$

$$V_{CA} = +(20\text{ V} + 10\text{ V}) = \mathbf{+30\ V}$$

$$V_{DA} = +(50\text{ V} + 20\text{ V} + 10\text{ V}) = \mathbf{+80\ V}$$

Example 3-17 What are the relative potentials at points A, B, and C with respect to point D in Fig. 3-30?

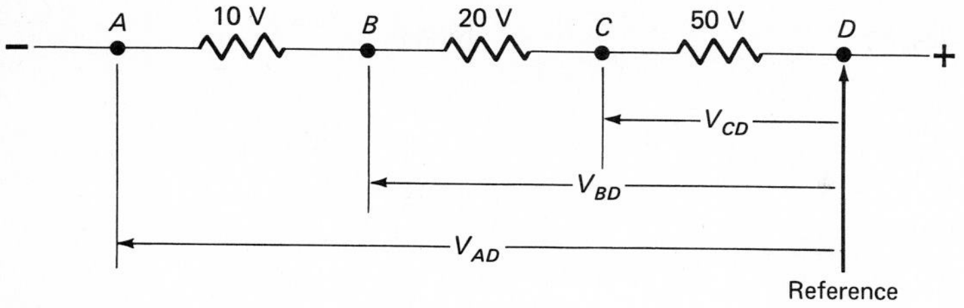

Fig. 3-30 Example 3-17: data.

Solution

$$V_{AD} = -(50\text{ V} + 20\text{ V} + 10\text{ V}) = \mathbf{-80\ V}$$

$$V_{BD} = -(50\text{ V} + 20\text{ V}) = \mathbf{-70\ V}$$

$$V_{CD} = \mathbf{-50\ V}$$

Example 3-18 What are the relative potentials at points A, B, and D with respect to point C in Fig. 3-31?

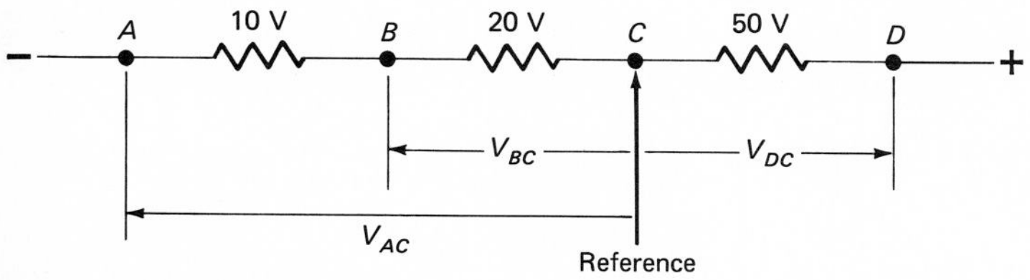

Fig. 3-31 Example 3-18: data.

Solution

$$V_{AC} = -(20\text{ V} + 10\text{ V}) = \mathbf{-30\ V}$$

$$V_{BC} = \mathbf{-20\ V}$$

$$V_{DC} = \mathbf{+50\ V}$$

Example 3-19 What are the potentials at points A, C, and D in Fig. 3-32?

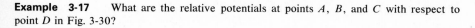

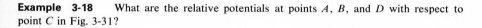

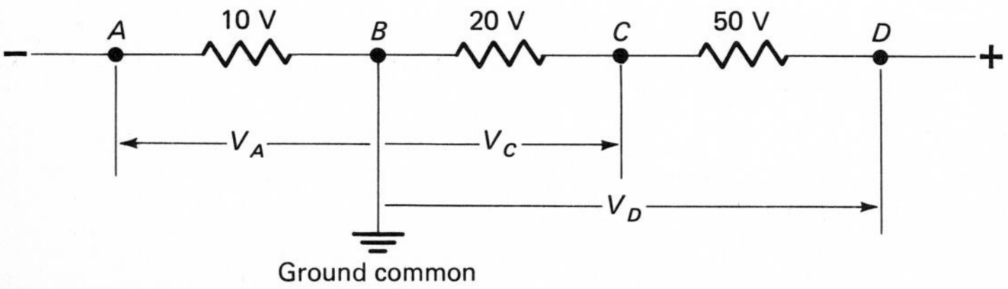

Fig. 3-32 Example 3-19: data.

Solution

$$V_A = \mathbf{-10\ V}$$
$$V_C = \mathbf{+20\ V}$$
$$V_D = \mathbf{+70\ V}$$

Example 3-20 What are the relative potentials in the circuit in Fig. 3-33?

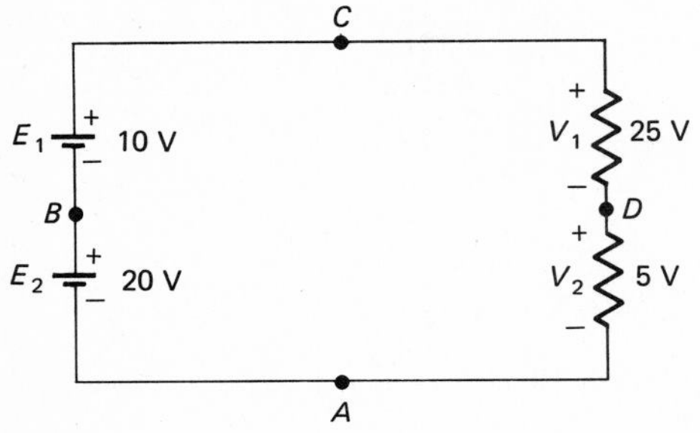

Fig. 3-33 Example 3-20: circuit.

Solution
Reference at point A

$$V_{BA} = +E_2 = \mathbf{+20\ V}$$
$$V_{CA} = +(E_1 + E_2) = \mathbf{+30\ V}$$
$$V_{DA} = +V_2 = \mathbf{+5\ V}$$

Reference at point C

$$V_{AC} = -(E_1 + E_2) = \mathbf{-30\ V}$$
$$V_{BC} = -E_1 = \mathbf{-10\ V}$$
$$V_{DC} = -V_1 = \mathbf{-25\ V}$$

Reference at point B

$$V_{AB} = -E_2 = \mathbf{-20\ V}$$
$$V_{CB} = +E_1 = \mathbf{+10\ V}$$
$$V_{DB} = V_{AB} + V_{DA} = -E_2 + V_2 = -20\ \text{V} + 5\ \text{V} = \mathbf{-15\ V}$$

or
$$V_{DB} = V_{CB} + V_{DC} = +E_1 - V_1 = +10\ \text{V} - 25\ \text{V} = \mathbf{-15\ V}$$

Reference at point D

$$V_{AD} = -V_2 = \mathbf{-5\ V}$$
$$V_{BD} = V_{CD} + V_{BC} = +V_1 - E_1 = +25\ \text{V} - 10\ \text{V} = \mathbf{+15\ V}$$

or
$$V_{BD} = V_{AD} + V_{BA} = -V_2 + E_2 = -5\ \text{V} + 20\ \text{V} = \mathbf{+15\ V}$$
$$V_{CD} = V_1 = \mathbf{+25\ V}$$

Note that a relative potential can be determined by commencing at the reference and moving around the circuit in one direction, either clockwise or counterclockwise, until the point of interest is reached. The values and polarities of each of the voltages are noted in turn and then added algebraically to obtain the result.

Example 3-21 Figure 3-34 shows the conditions which determine the base bias of a transistor in an amplifier circuit. What are the potentials at points B, C, and E with respect to ground? What is the value of the base-to-emitter bias V_{BE}?

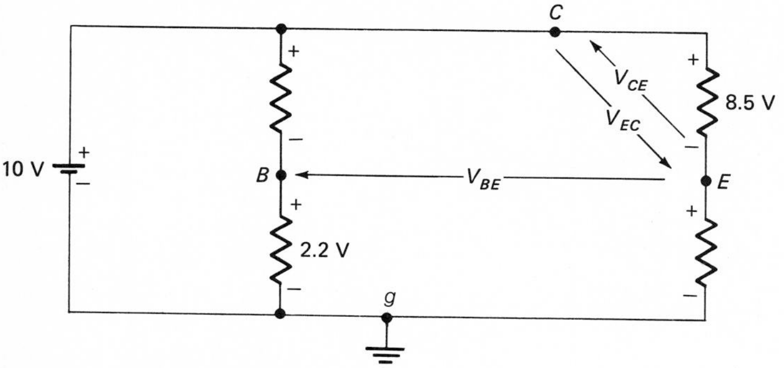

Fig. 3-34 Example 3-21: circuit.

Solution
The potential at point B is

$$V_B = \mathbf{+2.2\ V}$$

The potential at point C is

$$V_C = \mathbf{+10\ V}$$

The potential at point E is

$$V_E = V_C + V_{EC} \quad \text{or} \quad V_C - V_{CE}$$

$$V_E = +10\ \text{V} - 8.5\ \text{V} = \mathbf{+1.5\ V}$$

The base-to-emitter voltage V_{BE} is

$$V_{BE} = V_{Bg} + V_{gE} = 2.2\ \text{V} - 1.5\ \text{V} = \mathbf{+0.7\ V}$$

3-5 APPLICATIONS OF SERIES CIRCUITS

The following examples show a few of the vast number of series circuit applications which exist in practice.

SERIES DROPPING RESISTOR

It is often necessary to operate a device, or a part of a circuit, at a voltage below that of the supply emf. This lower voltage can be obtained by the use of a series dropping resistor, so that the pd across this resistor accounts for the unwanted voltage.

The characteristics of the series resistor can be determined by using the various methods developed for dealing with series circuits. The method used for any particular situation depends on the known data. Some examples will best illustrate typical techniques.

Example 3-22 A light bulb dissipates 100 W when operated at 100 V. What are the required characteristics for a series resistor to operate this light bulb at 100 V if the supply is 125 V?

Solution (Fig. 3-35)

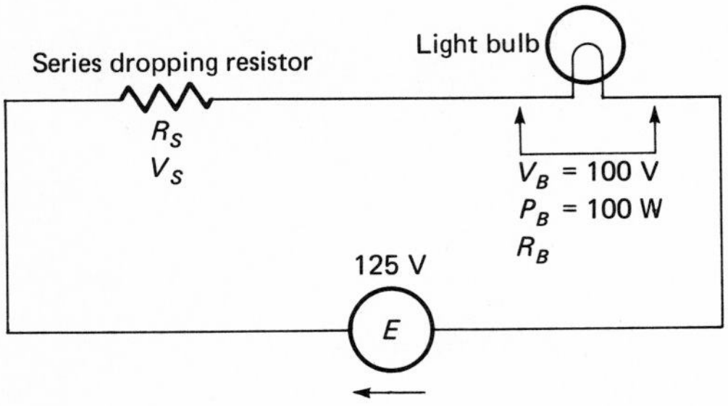

Fig. 3-35 Example 3-22: circuit.

$$P_B = \frac{V_B^2}{R_B}$$

$$R_B = \frac{V_B^2}{P_B} = \frac{(100\ \text{V})^2}{100\ \text{W}} = 100\ \Omega$$

From Eq. (3-2),

$$V_S = E - V_B$$

$$V_S = 125\ \text{V} - 100\ \text{V} = 25\ \text{V}$$

From Eq. (3-9),

$$R_S = R_B \frac{V_S}{V_B}$$

$$R_S = 100\ \Omega\ \frac{25\ \text{V}}{100\ \text{V}} = \mathbf{25\ \Omega}$$

From Eq. (3-12),

$$\frac{P_S}{P_B} = \frac{V_S}{V_B}$$

$$P_S = P_B \frac{V_S}{V_B} = 100\ \text{W}\ \frac{25\ \text{V}}{100\ \text{V}} = \mathbf{25\ W}$$

The series dropping resistor should be of 25-Ω resistance and be capable of dissipating 25 W of power.

Example 3-23 A telephone relay, designed to operate from a 24-V dc supply, normally draws a current of 10 mA. What series resistor is necessary to operate this relay, with its correct current, from a supply voltage which may range between 24 and 48 V?
Solution (Fig. 3-36)

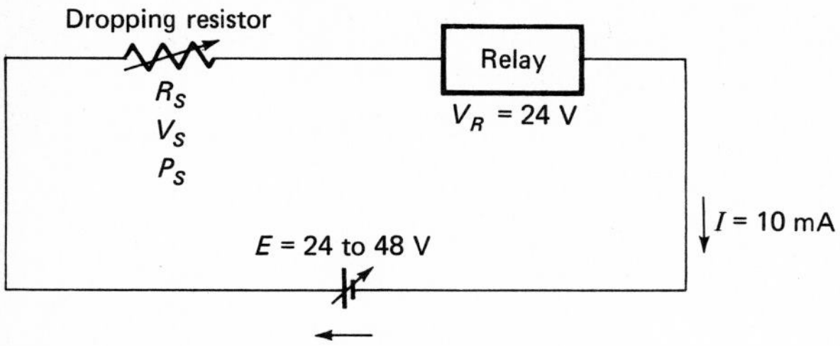

Fig. 3-36 Example 3-23: circuit.

From Eq. (3-2),

$$V_S = E - V_R$$

When $E = 24$ V, $V_S = 24\text{ V} - 24\text{ V} = 0\text{ V}$

$$R_S = \frac{V_S}{I} = \frac{0\text{ V}}{10\text{ mA}} = \mathbf{0\ \Omega}$$

$$P_S = V_S I = 0\text{ V} \times 10\text{ mA} = \mathbf{0\ W}$$

When $E = 48$ V, $V_S = 48\text{ V} - 24\text{ V} = 24\text{ V}$

$$R_S = \frac{V_S}{I} = \frac{24\text{ V}}{10\text{ mA}} = \mathbf{2.4\ k\Omega}$$

$$P_S = V_S I = 24\text{ V} \times 10\text{ mA} = \mathbf{0.24\ W}$$

In this situation, where the emf can have any value between 24 and 48 V, a variable resistor, or rheostat, will be needed for the series dropping resistor. The rheostat will have a resistance range of 0 to 2.4 kΩ, and a power-handling capability of at least $\frac{1}{4}$ W.

Example 3-24 The speed of an electric train can be varied by changing the voltage applied to its motor. A 12-V supply is available, and it is desired to vary the train voltage between 4 and 10 V. When running with 10 V, the motor draws 0.5 A. The resistance of the motor remains essentially constant over a wide current range. What kind of series resistor is necessary to control the train speed?
Solution (Fig. 3-37)

Fig. 3-37 Example 3-24: data.

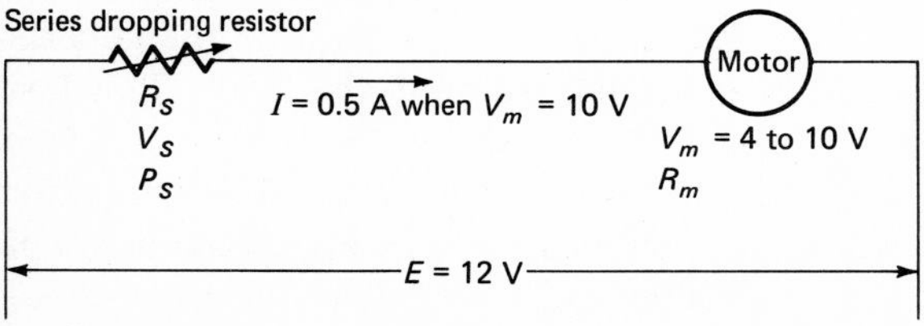

When the motor voltage V_m is 10 V,

$$R_m = \frac{V_m}{I} = \frac{10\text{ V}}{0.5\text{ A}} = 20\ \Omega$$

From Eq. (3-2),

$$V_S = E - V_m$$

$$V_S = 12\text{ V} - 10\text{ V} = 2\text{ V}$$

From Eq. (3-9),

$$R_S = R_m \frac{V_S}{V_m}$$

$$R_S = 20\ \Omega \frac{2\text{ V}}{10\text{ V}} = \mathbf{4\ \Omega}$$

$$P_S = V_S I = 2\text{ V} \times 0.5\text{ A} = \mathbf{1\ W}$$

When the motor voltage is 4 V, from Eq. (3-2),

$$V_S = E - V_m$$

$$V_S = 12\text{ V} - 4\text{ V} = 8\text{ V}$$

From Eq. (3-9),

$$R_S = R_m \frac{V_S}{V_m}$$

$$R_S = 20\ \Omega \frac{8\text{ V}}{4\text{ V}} = \mathbf{40\ \Omega}$$

$$P_S = \frac{V_S^2}{R_S} = \frac{(8\text{ V})^2}{40\ \Omega} = \mathbf{1.6\ W}$$

To control the speed of the train in Example 3-24, a rheostat must be used. Its resistance range will need to be from 4 to 40 Ω. It will dissipate 1 W of power when it is used at its minimum resistance of 4 Ω. It must be rated at 1.6 W or greater, so that it can handle the power when it is used at its maximum value of 40 Ω. When the rheostat has the same resistance as the motor, 20 Ω, it actually dissipates 1.8 W, the maximum possible value for this circuit.

GENERATOR INTERNAL RESISTANCE

When current is supplied to a load, energy is dissipated within the generator, heating it up, and so this energy is not passed on to the load. This lost energy is accounted for in terms of a resistance within the generator. It is easy to visualize the resistance of the wire within an automobile generator or the resistance of the electrolyte in a battery. There are components of generator resistance, however, which are not obvious, the friction of bearings being an example. The *internal resistance* of a generator refers to the total energy loss expressed as an equivalent resistance. Internal resistance can be considered either a series element or a parallel element. Both are valid, and either can be adopted in a particular situation. In this text you will meet both viewpoints. In this chapter we will deal with internal resistance as a series element. In Chap. 4 the parallel view of internal resistance will be investigated.

According to the series internal resistance viewpoint, the generator is made up of two components. One of the components is an emf source (it is not unusual for the emf source itself to be called ideal voltage source, ideal voltage generator, or just generator, and labeled accordingly in diagrams). The other is the internal resistance, which is regarded as a completely separate element connected in series with the emf source. Figure 3-38 shows such a generator connected to a load.

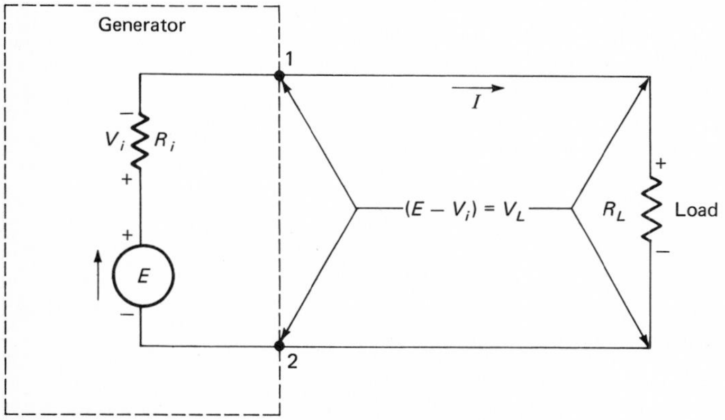

Fig. 3-38 Generator internal resistance.

The voltage across the load V_L must be the same as the voltage appearing across the generator terminals 1 and 2.

When a load is present, as shown in Fig. 3-38, the current I will cause a potential difference V_i to be developed across the generator internal resistance R_i.

$$V_i = IR_i$$

The current direction through R_i is the same as that through the load R_L; in this case, clockwise. Therefore the pds V_i and V_L will have ordinary series circuit polarities, as in Fig. 3-38. Thus Eq. (3-2) applies, as in any series circuit:

$$E = V_i + V_L \tag{3-13}$$

or

$$V_L = E - V_i$$

It will be seen that $E - V_i$ is the voltage at the generator terminals 1 and 2. This voltage is called the terminal pd or the terminal voltage of the generator. The *terminal voltage* of a generator is the voltage across its terminals when it is supplying current to a load.

If the load is disconnected, there will be no current, so the pd across the internal resistance $V_i = IR_i$ will be zero. If this is so, then the voltage at the generator terminals will equal the emf E. The voltage at the terminals of a generator when it is supplying no current to a load is called the *open-circuit emf.*

Equation (3-13) can be written as

$$\text{Terminal voltage} = \text{open-circuit emf} - \text{internal resistance pd}$$

Example 3-25 The open-circuit emf of an automobile battery is 13.2 V. If its internal resistance is 0.01 Ω, what will be (*a*) the starter motor voltage if its current is 300 A, (*b*) the voltage across the headlights if the total electric load at night is 120 W at a nominal[1] 12 V?
Solution
a. Starter motor voltage (Fig. 3-39)

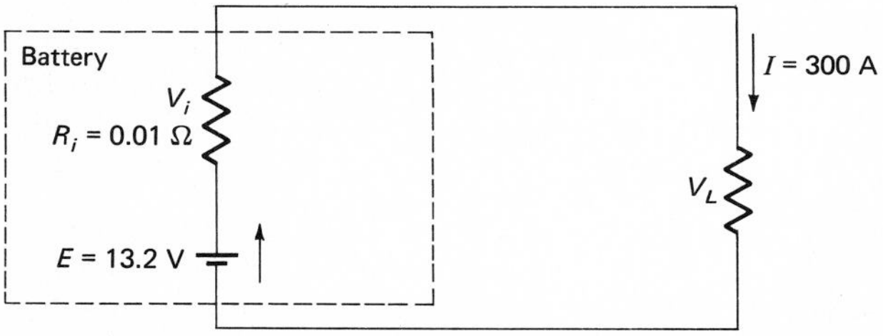

Fig. 3-39 Example 3-25: starter circuit conditions.

$$V_i = IR_i = 300 \text{ A} \times 0.01\ \Omega = 3 \text{ V}$$

$$V_L = E - V_i \qquad (3\text{-}13)$$

$$V_L = 13.2 \text{ V} - 3 \text{ V} = \mathbf{10.2\ V}$$

b. Voltage across headlights (Fig. 3-40)

Fig. 3-40 Example 3-25: lighting circuit conditions.

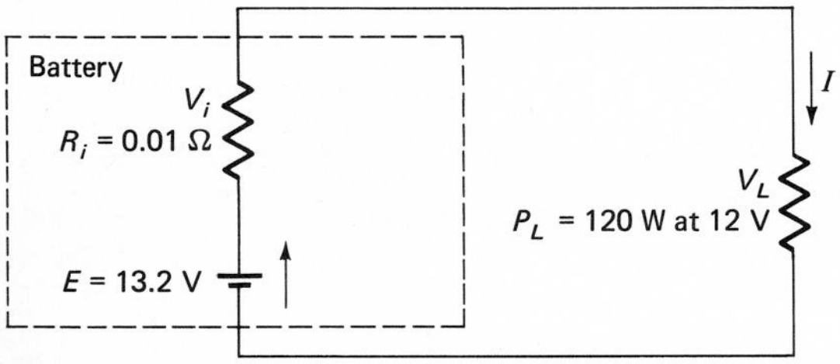

[1] The word "nominal" is used frequently in electrical work. It means "in name only." A *nominal* value is an assigned value, the actual existing value perhaps being slightly different. The amount of tolerable variation from the nominal value is the *tolerance*.

$$P_L = V_L I$$

If V_L is nominally 12 V,

$$I = \frac{P_L}{V_L} = \frac{120\ \text{W}}{12\ \text{V}} = 10\ \text{A}$$

$$V_i = IR_i = 10\ \text{A} \times 0.01\ \Omega = 0.1\ \text{V}$$

$$V_L = E - V_i \tag{3-13}$$

$$V_L = 13.2\ \text{V} - 0.1\ \text{V} = \mathbf{13.1\ V}$$

The values in Example 3-25 are typical of an automobile electric system. The automobile battery, nominally 12 V, is made up of six cells, each nominally of 2 V emf. A fully charged 2-V cell in good condition has an actual emf of about 2.2 V, so that six cells connected in series result in a total battery emf of about 13.2 V. An internal resistance of 0.01 Ω is of the correct order of magnitude for moderate temperatures. It is higher for winter temperatures. The load imposed by the starter motor is severe, about 300 A being common. Thus if the battery resistance is high, as a result of its being in poor shape or because of low winter temperatures, the voltage across the starter motor might be low (as low as 6 V), making it difficult to start the car. The ordinary operating load, even at night with the headlights on high beam, should not reduce the load voltage below 12 V if the battery and the electric system in general are in good condition.

You will see that, although the headlight load was specified as 120 W at a nominal 12 V, it actually receives 13.1 V. This 13.1 V means that the power taken by the lighting system is higher than 120 W. Such deviations of voltage from the nominal are common in automotive systems; therefore they must be designed to operate satisfactorily over a wide voltage range.

CONDUCTOR RESISTANCE

Up to this point it has been assumed that the various circuit elements are connected together by perfect conductors, conductors which have no resistance and therefore do not influence the values of current and voltage in the circuit.

For many applications good physical circuit design aims at ensuring that conductor resistances are so low, compared to the resistances of the elements they interconnect, that they can be ignored. The amount of resistance considered significant will depend on the circuit application, but a fairly general rule is: Conductor resistance is significant if it exceeds 1 percent of the lowest of the element resistances.

When the resistance of a conductor is significant to the circuit operation, it must be treated as a separate element so that its characteristics can be taken into account.

In Fig. 3-41 is a generator, and its internal resistance, connected to a load by two conductors which have significant resistance.

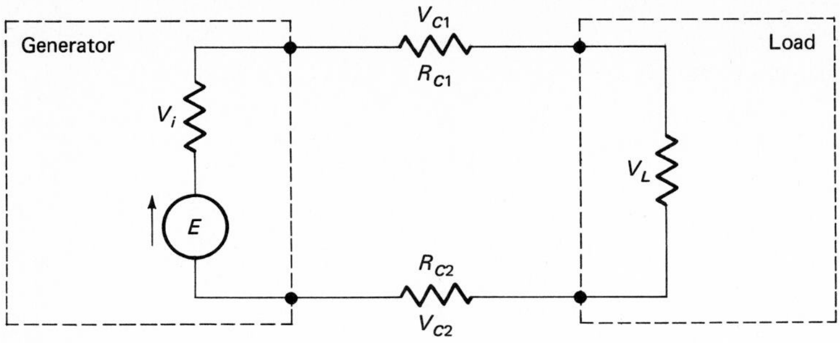

Fig. 3-41 Resistance of connecting leads.

If the conductor resistances R_{C1} and R_{C2} were not present, from Eq. (3-13),

$$V_L = E - V_i$$

With the conductor resistances,

$$V_L = E - V_i - V_{C1} - V_{C2}$$

or

$$V_L = E - (V_i + V_{C1} + V_{C2}) \tag{3-14}$$

Since the objective of the circuit is to provide load current and voltage, V_i, V_{C1}, and V_{C2} are wasted voltages; they contribute nothing useful to the circuit operation. Therefore the conductor resistances (and the generator internal resistance) are often called *circuit losses.*

It is common practice, in setting up a circuit, to aim at producing a load voltage of maximum and constant value. The presence of significant conductor resistance can prevent the realization of this objective.

The examples which follow will show some of the effects of conductor resistance.

Example 3-26 A battery of 10 V emf and 1 Ω internal resistance is connected to a 100-Ω load. What is the load voltage (*a*) if the conductor resistance can be ignored, and (*b*) if each conductor connecting generator and load has a resistance of 9.5 Ω?

Solution

a. (Fig. 3-42)

Fig. 3-42 Example 3-26: circuit without conductor resistance.

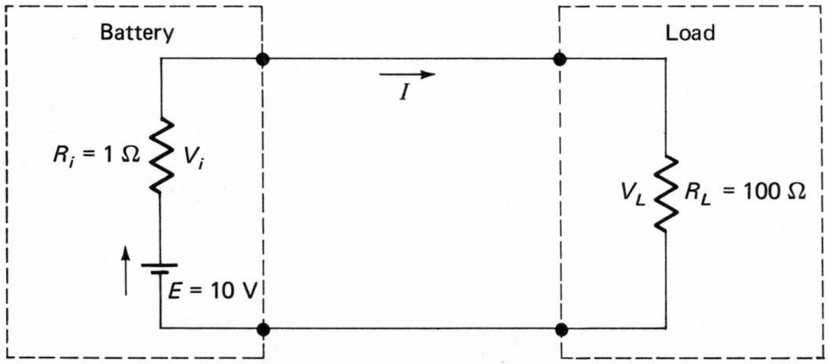

From Eq. (3-3),

$$R_T = R_i + R_L$$

$$I = \frac{E}{R_T} = \frac{E}{R_i + R_L} = \frac{10\ \text{V}}{1\ \Omega + 100\ \Omega} \simeq 0.1\ \text{A}^*$$

$$V_i = IR_i \simeq 0.1\ \text{A} \times 1\ \Omega \simeq 0.1\ \text{V}$$

$$V_L = E - V_i \simeq 10\ \text{V} - 0.1\ \text{V} \simeq \mathbf{9.9\ V}$$

b. (Fig. 3-43)

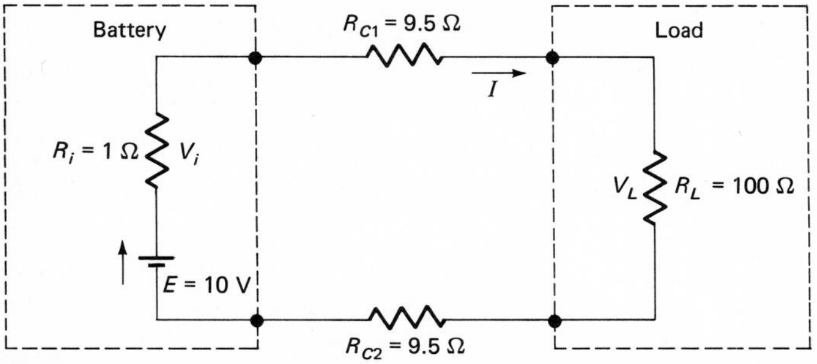

Fig. 3-43 Example 3-26: circuit with conductor resistance.

From Eq. (3-3),

$$R_T = R_i + R_L + R_{C1} + R_{C2}$$

$$I = \frac{E}{R_T} = \frac{E}{R_i + R_L + R_{C1} + R_{C2}}$$

$$I = \frac{10\ \text{V}}{1\ \Omega + 100\ \Omega + 9.5\ \Omega + 9.5\ \Omega} = \frac{10\ \text{V}}{120} = 0.08333\ \text{A}$$

$$V_L = E - (V_i + V_{C1} + V_{C2}) \tag{3-14}$$

$$V_L = E - I(R_i + R_{C1} + R_{C2}) = 10\ \text{V} - 0.08333\ \text{A}\ (1\ \Omega + 9.5\ \Omega + 9.5\ \Omega)$$

$$V_L = 10\ \text{V} - 1.666\ \text{V} = \mathbf{8.334\ V}$$

In Example 3-26, if V_L had been calculated using Ohm's law ($V_L = IR_L$), the result would have been 8.333 V. The difference between 8.333 and 8.334 V is due to the number of significant figures used in the computation. The current I was calculated to four significant figures (0.08333 A), and the correct value of V_L to four significant figures is 8.333 V.

Example 3-27 What is the percentage voltage reduction due to the conductor resistances in Example 3-26?

* The symbol ≃ means "approximately equal to." In electric circuit work, if two quantities are within about 1 percent, they are usually considered to be approximately equal to each other.

Solution

The voltage reduction due to the conductor resistances is

$$V_C = V_{C1} + V_{C2} = I(R_{C1} + R_{C2})$$

$$V_C = 0.08333\ (9.5\ \Omega + 9.5\ \Omega) = 1.583\ \text{V}$$

The voltage with zero conductor resistance (from Example 3-26*a*) is

$$V_L = 9.9\ \text{V}$$

The percentage voltage reduction due to the conductors is

$$\frac{\text{Voltage reduction}}{\text{Voltage with zero conductor resistance}} \times 100 = \frac{V_C}{V_L} \times 100 = \frac{1.583\ \text{V}}{9.9\ \text{V}} \times 100$$

Voltage reduction $\simeq$ **16 percent**

It is evident from Example 3-27 that conductor resistances can make appreciable percentage differences in available load voltage. It is the relationship between the load resistance and the total conductor resistance which determines the severity of this effect. Had the load resistance been 1000 Ω instead of 100 Ω in Example 3-27, the percentage voltage reduction due to the 9.5-Ω conductors would have been only 1.86 percent.

Example 3-28 What would be the effect on the starter motor voltage in Example 3-25 if the total conductor resistance in the leads connecting the motor to the battery were 0.1 Ω?
Solution In Example 3-25, the current I was 300 A, while the motor voltage V_L was 10.2 V. (See Fig. 3-44)

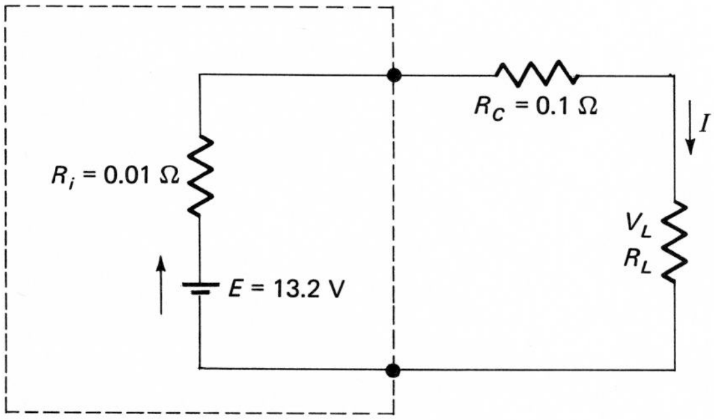

Fig. 3-44 Example 3-28: circuit.

The starter motor resistance R_L will be

$$R_L = \frac{V_L}{I} = \frac{10.2\ \text{V}}{300\ \text{A}} = 0.034\ \Omega$$

From Eq. (3-3),

$$R_T = R_i + R_c + R_L$$

$$R_T = 0.01\ \Omega + 0.1\ \Omega + 0.034\ \Omega = 0.144\ \Omega$$

$$I = \frac{E}{R_T} = \frac{13.2\ \text{V}}{0.144\ \Omega} = 91.67\ \text{A}$$

$$V_L = IR_L = 91.67\ \text{A} \times 0.034\ \Omega = \mathbf{3.117\ V}$$

Again it is demonstrated in Example 3-28 that even an apparently small amount of conductor resistance, $\frac{1}{10}$ Ω, can seriously affect circuit operation. Although 0.1 Ω does not seem to be a high resistance, it is high when compared with the load resistance of 0.034 Ω. It is this comparative resistance which matters, rather than the absolute number of ohms.

Example 3-29 A 100-V generator of negligible internal resistance is connected to a load which varies in resistance from 10 to 100 Ω. What is the load voltage if the total conductor resistance is (*a*) insignificant, and (*b*) 10 Ω?

Solution

a. Figure 3-45*a* and *b* shows the extremes of the load resistance with zero conductor resistance. Clearly, in both cases, as there are no losses, the load voltage will equal the generator emf.

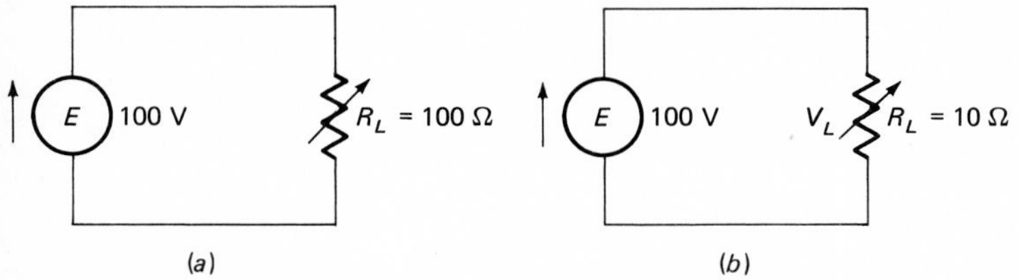

Fig. 3-45 Example 3-29: circuit without conductor resistance.

$$E = V_L$$

$$V_L = \mathbf{100\ V}$$

b. In Fig. 3-46*a*, the load resistance is at its maximum value of 100 Ω.

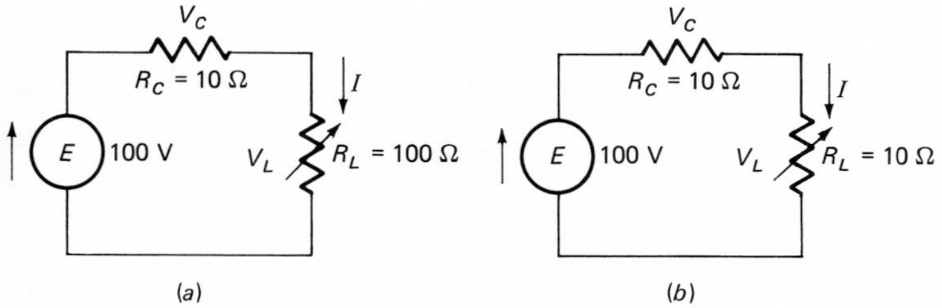

Fig. 3-46 Example 3-29: circuit with conductor resistance.

$$I = \frac{E}{R_c + R_L} = \frac{100\ \text{V}}{10\ \Omega + 100\ \Omega} = 0.9091\ \text{A}$$

$$V_L = IR_L = 0.9091\ \text{A} \times 100\ \Omega = \mathbf{90.91\ V}$$

In Fig. 3-46*b*, the load resistance is at its minimum value of 10 Ω.

$$I = \frac{E}{R_c + R_L} = \frac{100\ \text{V}}{10\ \Omega + 10\ \Omega} = 5\ \text{A}$$

$$V_L = IR_L = 5\ \text{A} \times 10\ \Omega = \mathbf{50\ V}$$

The significance of Example 3-29 is that, if the conductor resistance is negligible, a load fed from a constant-voltage source will have a constant voltage across it, even though the load resistance varies. If, on the other hand, the conductor resistance is significant, the load voltage will vary as its resistance varies. The circuit with the significant conductor resistance is said to have poor voltage regulation because the load voltage, which is meant to be constant, varies considerably.

VOLTMETER MULTIPLIERS

Usually a voltmeter consists of a basic meter movement with a resistor connected in series with it.

The meter movement measures the current flowing through it, and is thus an ammeter. The position of its pointer on the calibrated scale is directly proportional to the current. A common type of movement used for voltmeters has a full-scale deflection of 50 μA, meaning that its pointer will be deflected to the end of the calibrated scale with a current of 50 μA. As an electric circuit element, the movement has a resistance R_m, a full-scale current I_m, and a full-scale voltage V_m. These components are shown in Fig. 3-47.

Figure 3-47*a* shows a convenient, and frequently used, symbol for a meter movement; it is the equivalent of, and will be used as an alternative to, the resistance element viewpoint of a movement shown in Fig. 3-47*b*.

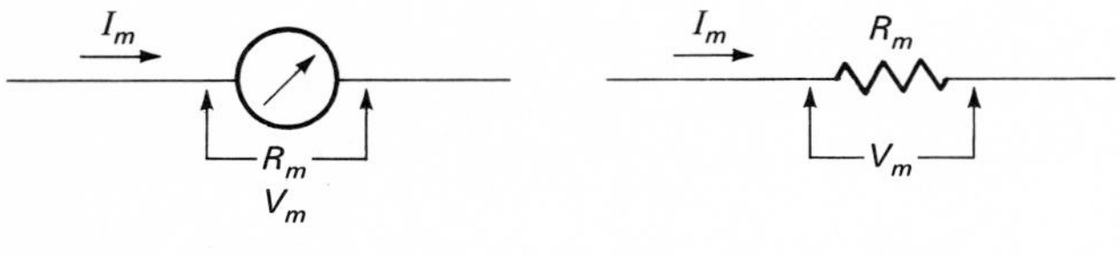

Fig. 3-47 Meter movement as a circuit element.

Example 3-30 A meter has a full-scale current of 50 μA and a resistance of 2 kΩ. What voltage will produce a full-scale deflection?

Solution (Fig. 3-48)

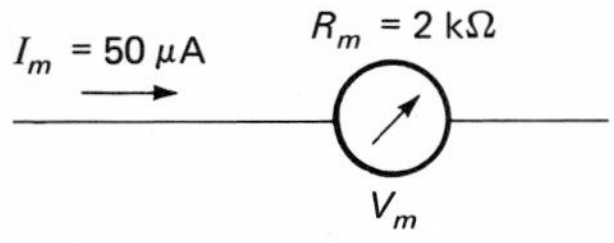

Fig. 3-48 Example 3-30: data.

$$V_m = I_m R_m = 50\ \mu\text{A} \times 2\ \text{k}\Omega = \mathbf{0.1\ V}$$

The pointer of this 50-μA movement will therefore be deflected to the end of the calibrated scale if $\frac{1}{10}$ V is applied across it.

Because the full-scale condition represents the safe limit to which the movement can be operated, and because the range of a meter is specified in terms of its full-scale deflection, only full-scale values are considered in calculations. However, it should be noted that, since voltage is directly proportional to current, deflections other than full-scale will be produced by voltages proportionally less than the full-scale voltage. In the case of the basic voltmeter in Example 3-30, 0.05 V will produce a half-scale reading.

As you can see, the basic movement is of limited value as a voltmeter. If it is limited to a maximum voltage of $\frac{1}{10}$ V, as in Example 3-30, it cannot be of much use. A resistor connected in series with the movement will result in a useful voltmeter, and this arrangement is typical of most commercial voltmeters. In this application the series resistor is called a *multiplier*.

Figure 3-49 shows the basic equivalent network of a voltmeter.

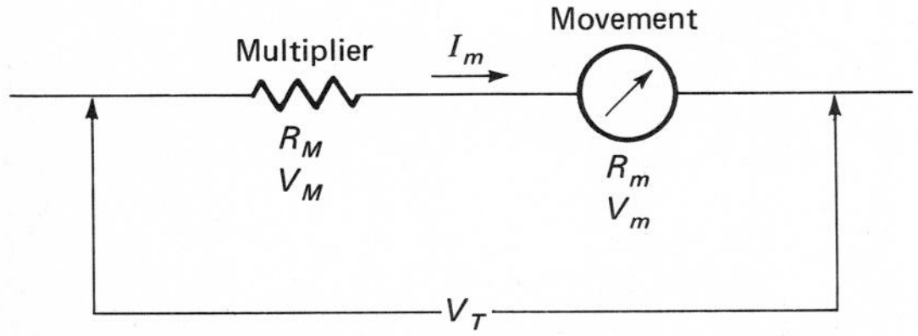

Fig. 3-49 Basic voltmeter.

$$V_T = V_M + V_m \tag{3-15}$$

By applying Ohm's law to this relationship, meters can be designed for measuring virtually any range of voltage.

Example 3-31 It is desired to make a voltmeter with a range of 0 to 1 V (1 V full-scale), using a 50-μA 2-kΩ, movement. What value of multiplier resistor will be needed?
Solution (Fig. 3-50)

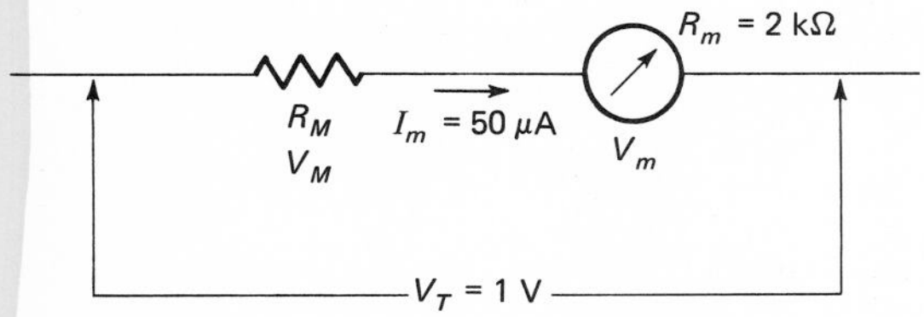

Fig. 3-50 Example 3-31: data.

Only the full-scale conditions need to be considered, as all other conditions are directly proportional to them. From Example 3-30,

$$V_m = 0.1 \text{ V}$$

From Eq. (3-15),

$$V_M = V_T - V_m$$

$$V_M = 1 \text{ V} - 0.1 \text{ V} = 0.9 \text{ V}$$

From Eq. (3-9),

$$R_M = R_m \frac{V_M}{V_m}$$

$$R_M = 2\ \text{k}\Omega \times \frac{0.9\ \text{V}}{0.1\ \text{V}} = \mathbf{18\ k\Omega}$$

Example 3-32 Using the same 50-μA 2-kΩ movement as in Examples 3-30 and 3-31, design a voltmeter with three ranges, 1, 10, and 100 V, full-scale.

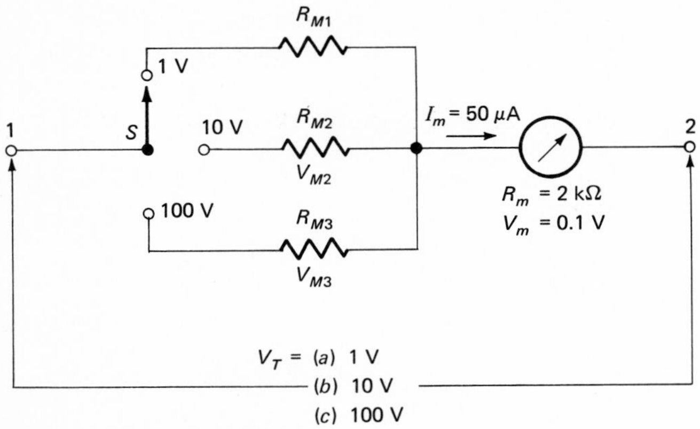

Fig. 3-51 Example 3-32: data.

Solution Figure 3-51 shows the movement with three multipliers capable of being connected, one at a time, in series with the movement by means of the switch S. For any of the three switch positions, one series circuit will exist between the voltmeter terminals 1 and 2.

a. For 1 V full-scale, the multiplier R_{M1} will be in series with the movement. This range is the same as that in Example 3-31.

$$R_{M1} = \mathbf{18\ k\Omega}$$

b. For 10 V full-scale, from Eq. (3-15),

$$V_{M2} = V_T - V_m$$

$$V_{M2} = 10\ \text{V} - 0.1\ \text{V} = 9.9\ \text{V}$$

From Eq. (3-9),

$$R_{M2} = R_m \frac{V_{M2}}{V_m}$$

$$R_{M2} = 2\ \text{k}\Omega \times \frac{9.9\ \text{V}}{0.1\ \text{V}} = \mathbf{198\ k\Omega}$$

c. For 100 V full-scale, from Eq. (3-15),

$$V_{M3} = V_T - V_m$$

$$V_{M3} = 100\ \text{V} - 0.1\ \text{V} = 99.9\ \text{V}$$

From Eq. (3-9),

$$R_{M3} = R_m \frac{V_{M3}}{V_m}$$

$$R_{M3} = 2\ \text{k}\Omega \times \frac{99.9\ \text{V}}{0.1\ \text{V}} = \mathbf{1998\ k\Omega} \text{ or } \mathbf{1.998\ M\Omega}$$

VOLTMETER SENSITIVITY

In Example 3-32, the total resistance $R_M + R_m$ divided by the full-scale voltage V_T is the same for each range. This relationship is not coincidental, but is bound to be true because in any series circuit the current is the same throughout. Therefore the ratio voltage/resistance (or resistance/voltage) will be the same for the entire voltmeter as for the movement alone. The quantity ohms per volt, in voltmeter specifications, is known as the *sensitivity* of the instrument. A 50-μA movement will always result in a voltmeter sensitivity of 20 kΩ/V, as in our examples. This is a higher sensitivity than the 1 kΩ/V which would be obtained by using a movement with a full-scale current of 1 mA (1000 μA).

AMMETER LOADING

An ammeter always has to be connected as a series circuit element so that it will be directly in the path of the current being measured.

The network in Fig. 3-52*a* draws a current I_1 from the battery. To measure this current, an ammeter must be connected in series with the network, as shown in Fig. 3-52*b*. Of course, the ammeter can be connected on either side of the network, as the current will be the same. The schematic diagram in Fig. 3-53 is the equivalent of Fig. 3-52*b*.

Obviously, the intention in using the ammeter is to measure the current I_1 in the network, as in Fig. 3-52*a*.

$$I_1 = \frac{E}{R_N}$$

Fig. 3-52 (*a*) Network connected to a battery; (*b*) ammeter connected to read the network current.

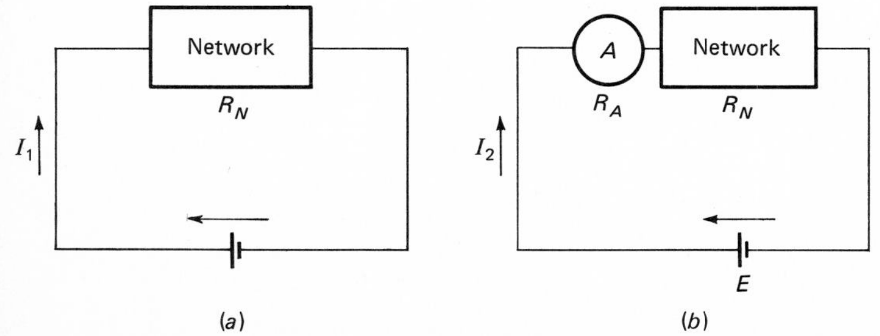

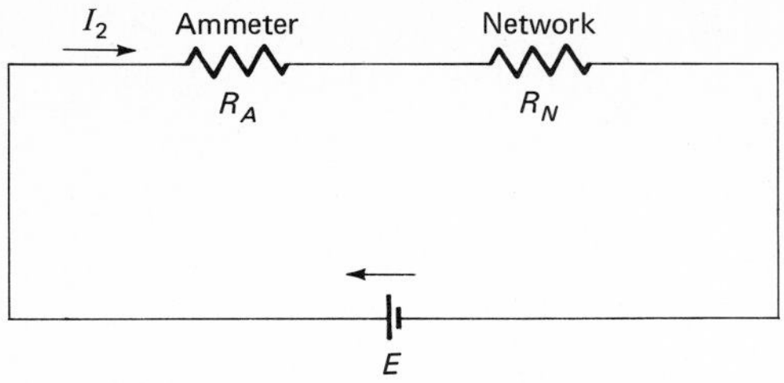

Fig. 3-53 Schematic of ammeter connection.

But the existence of the ammeter modifies this current, reducing it to the value I_2, as in Fig. 3-52*b* or 3-53

$$I_2 = \frac{E}{R_A + R_N}$$

This interference with the quantity being measured is a characteristic of all measuring instruments and is unavoidable. What has to be done in practice is to reduce the disturbance, which is usually called *loading,* by the choice of an instrument of acceptable characteristics. An ammeter should be chosen so that its resistance introduces only a tolerable amount of measurement error. If the ammeter resistance R_A is 1 percent of the network resistance R_N, it will introduce an error of about 1 percent.

Example 3-33 The resistance of a network connected across a 1-V supply is 40 kΩ. The 50-μA 2-kΩ meter movement in Examples 3-30 to 3-32 is to be used as an ammeter to measure the current in the network. What percentage of error is caused by the presence of the ammeter?
Solution

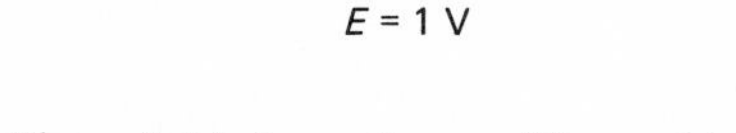

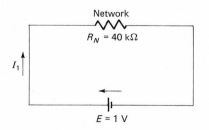

Fig. 3-54 Example 3-33: circuit without ammeter.

Figure 3-54 shows the conditions which exist without the meter. The current to be measured will be

$$I_1 = \frac{E}{R_N} = \frac{1\ \text{V}}{40\ \text{k}\Omega} = 2.5 \times 10^{-5}\ \text{A or } 25\ \mu\text{A}$$

This current of 25 μA should result in half-scale deflection of the meter pointer.

In Fig. 3-55, the meter movement is shown as an ammeter, introducing 2 kΩ of series resistance into the circuit.

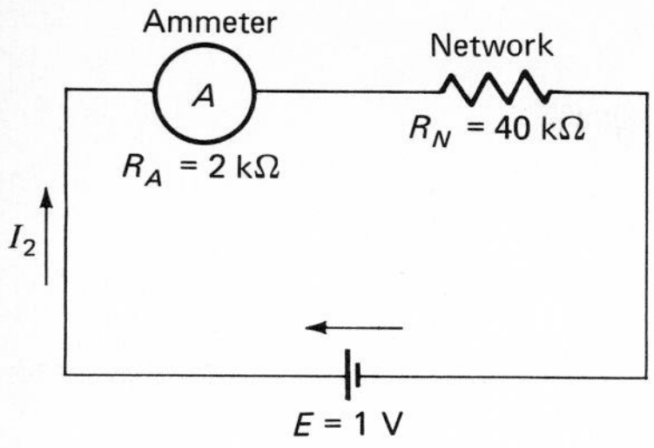

Fig. 3-55 Example 3-33: circuit with ammeter.

$$I_2 = \frac{E}{R_A + R_N} = \frac{1\ \text{V}}{2\ \text{k}\Omega + 40\ \text{k}\Omega} = 2.38 \times 10^{-5}\ \text{A or } 23.8\ \mu\text{A}$$

The percentage error due to the movement resistance will be

$$\text{Percentage error} = \frac{\text{current error}}{\text{correct current}} \times 100 = \frac{I_1 - I_2}{I_1} \times 100$$

$$= \frac{25\ \mu\text{A} - 23.8\ \mu\text{A}}{25\ \mu\text{A}} \times 100 = \frac{1.2\ \mu\text{A}}{25\ \mu\text{A}} \times 100$$

Percentage error = **4.8 percent**

If this basic error of 4.8 percent is intolerable, then an ammeter of lower resistance will have to be used.

EFFECT OF THE SERIES CONNECTION OF A VOLTMETER

A voltmeter is a high-resistance device connected in parallel with a network in which the voltage is to be measured. If the voltmeter were connected in series instead of in parallel, a gross measurement error would result, as illustrated by the following example.

Example 3-34 A series circuit consists of a 100-V emf source and two resistances, 1 and 4 kΩ. What is (*a*) the voltage across the 1-kΩ resistance, (*b*) the voltage across this resistance if a 20-kΩ/V, 100-V, full-scale deflection voltmeter is connected to the 1-kΩ resistance in series instead of in parallel, and (*c*) the voltage reading in situation (*b*)?

Solution

Fig. 3-56 Example 3-34: circuit. (*a*) Without voltmeter; (*b*) with voltmeter.

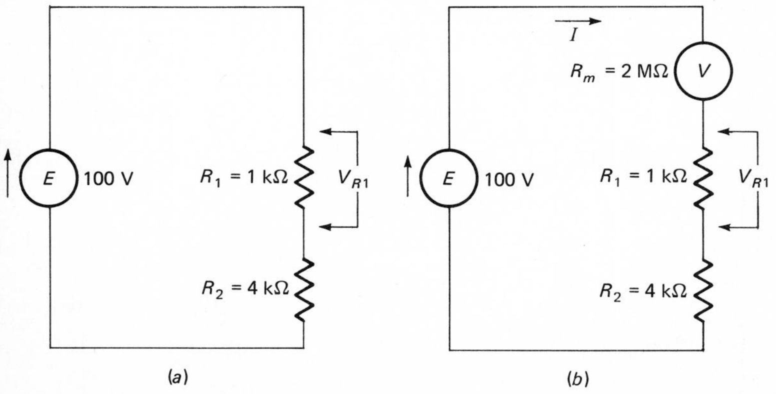

a. Refer to Fig. 3-56*a*. From Eq. (3-7),

$$V_{R1} = E\frac{R_1}{R_T} = E\frac{R_1}{R_1 + R_2}$$

$$V_{R1} = 100\text{ V}\frac{1\text{ k}\Omega}{1\text{ k}\Omega + 4\text{ k}\Omega} = \mathbf{20\ V}$$

b. Refer to Fig. 3-56*b*. The voltmeter resistance R_m equals its sensitivity in ohms per volt multiplied by the range setting in use.

$$R_m = 20\text{ k}\Omega/\text{V} \times 100\text{ V} = 2\text{ M}\Omega$$

From Eq. (3-7),

$$V_{R1} = E\frac{R_1}{R_T} = E\frac{R_1}{R_1 + R_2 + R_m}$$

$$V_{R1} = 100\text{ V}\frac{1\text{ k}\Omega}{1\text{ k}\Omega + 4\text{ k}\Omega + 2\text{ M}\Omega} \simeq \mathbf{0.05\ V}$$

c. Refer to Fig. 3-56*b*.

$$I = \frac{E}{R_T} = \frac{100\text{ V}}{2.005\text{ M}\Omega} \simeq 50\ \mu\text{A}$$

As 50 μA is the full-scale deflection current of the movement in a 20-kΩ/V voltmeter, the voltmeter will read approximately **100 V.**

In Example 3-34 the objective is measurement of the 20-V pd of the resistance. However, the presence of the incorrectly connected voltmeter reduces this voltage to 50 mV. Furthermore the voltmeter indicates 100 V. Obviously the situation is meaningless for the measurement objective.

UNLOADED VOLTAGE DIVIDER

The *unloaded* voltage divider is a series network which can be used to feed to other networks a number of different voltages, all derived from a single voltage source. There is a constraint on the use of the unloaded voltage divider which, if rigidly imposed, would render it useless for practical application; namely, that the loads (networks) fed by the voltage divider can draw no current from it. Fortunately, the principle of the unloaded voltage divider holds within useful limits if the loads fed by it draw very small currents.

First, the principle will be developed assuming no current loading.

In Fig. 3-57 is a simple voltage divider with two output voltages V_1 and V_2, both derived from the emf E. No loads are connected to the output terminals, the output voltages V_1 and V_2 being directly derived from the resistance network R_1 and R_2.

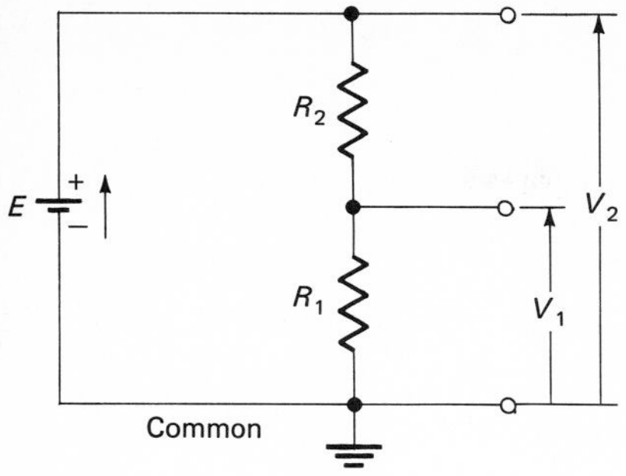

Fig. 3-57 Unloaded voltage divider.

$$V_2 = E$$

If no current is drawn by the V_1 output lead, from Eq. (3-7),

$$V_1 = E\frac{R_1}{R_T} = E\frac{R_1}{R_1 + R_2}$$

This concept can be applied to voltage dividers with any number of outputs and is called the "voltage divider principle."

$$V_n = E\frac{R_n}{R_T} \tag{3-16}$$

where V_n = voltage between the output terminal and ground
E = network supply voltage
R_n = resistance between the output terminal and ground
R_T = total resistance of the voltage divider

Example 3-35 What are the output voltages of the unloaded voltage divider in Fig. 3-58?

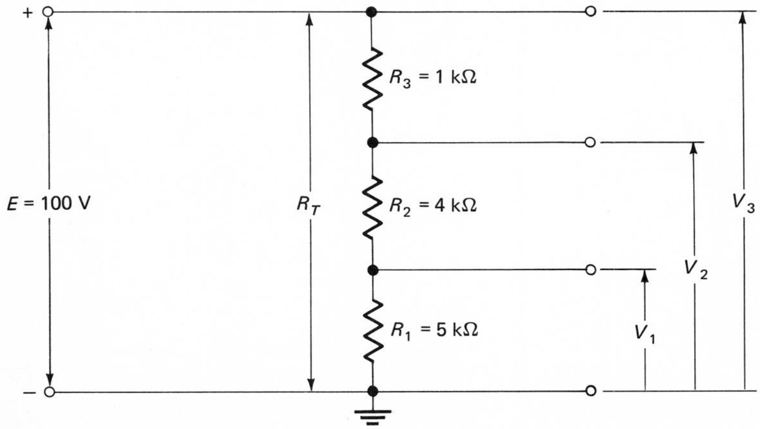

Fig. 3-58 Example 3-35: network.

Solution

$$R_T = R_1 + R_2 + R_3 \tag{3-3}$$

$$R_T = 5\ \text{k}\Omega + 4\ \text{k}\Omega + 1\ \text{k}\Omega = 10\ \text{k}\Omega$$

From Eq. (3-16),

$$V_1 = E\frac{R_1}{R_T}$$

$$V_1 = 100\ \text{V} \times \frac{5\ \text{k}\Omega}{10\ \text{k}\Omega} = \mathbf{50\ V}$$

From Eq. (3-16),

$$V_2 = E\frac{R_1 + R_2}{R_T}$$

$$V_2 = 100\ \text{V} \times \frac{9\ \text{k}\Omega}{10\ \text{k}\Omega} = \mathbf{90\ V}$$

$$V_3 = E = \mathbf{100\ V}$$

Example 3-36 What are the resistances of the elements of the unloaded voltage divider in Fig. 3-59 if the divider chain (the resistances of the divider) draws a current of 0.1 A from the supply?

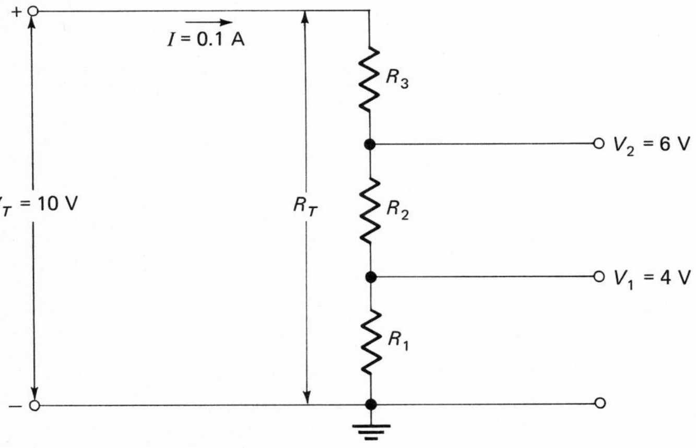

Fig. 3-59 Example 3-36: network.

Solution

$$R_T = \frac{V_T}{I} = \frac{10\ \text{V}}{0.1\ \text{A}} = 100\ \Omega$$

From Eq. (3-9),

$$R_1 = R_T\frac{V_1}{V_T} = 100\ \Omega \times \frac{4\ \text{V}}{10\ \text{V}} = \mathbf{40\ \Omega}$$

From Eq. (3-9),

$$R_1 + R_2 = R_T \frac{V_2}{V_T} = 100\ \Omega \times \frac{6\ \text{V}}{10\ \text{V}} = 60\ \Omega$$

$$R_2 = (R_1 + R_2) - R_1 = 60\ \Omega - 40\ \Omega = \mathbf{20\ \Omega}$$

$$R_3 = R_T - (R_1 + R_2) = 100\ \Omega - 60\ \Omega = \mathbf{40\ \Omega}$$

Example 3-37 What are the output voltages of the unloaded voltage divider in Fig. 3-60?

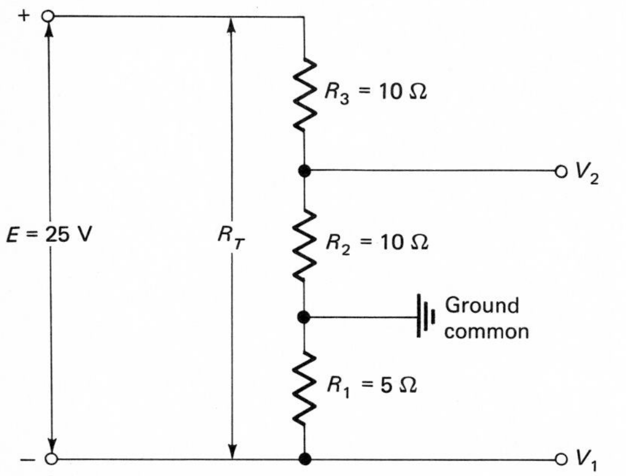

Fig. 3-60 Example 3-37: network.

Solution V_1 will be negative with respect to the ground reference, but V_2 will be positive.

$$R_T = R_1 + R_2 + R_3 \quad (3\text{-}3)$$

$$R_T = 5\ \Omega + 10\ \Omega + 10\ \Omega = 25\ \Omega$$

$$V_1 = -E \frac{R_1}{R_T} \quad (3\text{-}16)$$

$$V_1 = -25\ \text{V} \times \frac{5\ \Omega}{25\ \Omega} = \mathbf{-5\ V}$$

$$V_2 = +E \frac{R_2}{R_T} \quad (3\text{-}16)$$

$$V_2 = +25\ \text{V} \times \frac{10\ \Omega}{25\ \Omega} = \mathbf{+10\ V}$$

In designing an unloaded voltage divider to provide specified output voltages, it is necessary to decide how much current the divider itself will be permitted to draw. If the loads are literally currentless, as in Examples 3-35 and 3-37, the divider current can be kept down to a low value, hence avoiding high power dissipation. But, if the loads do draw small currents from the outputs, then the divider current (called the *bleeder current*) must be large with respect to the load currents.

Example 3-38 Design a voltage divider to provide outputs of 6 V at 10 mA and 10 V at 5 mA from a power supply with an output of 12 V.

Solution By making the bleeder current I_b at least 100 times the largest output current, the circuit can be virtually considered to be an unloaded voltage divider, the load currents having negligible effect on the output voltages.

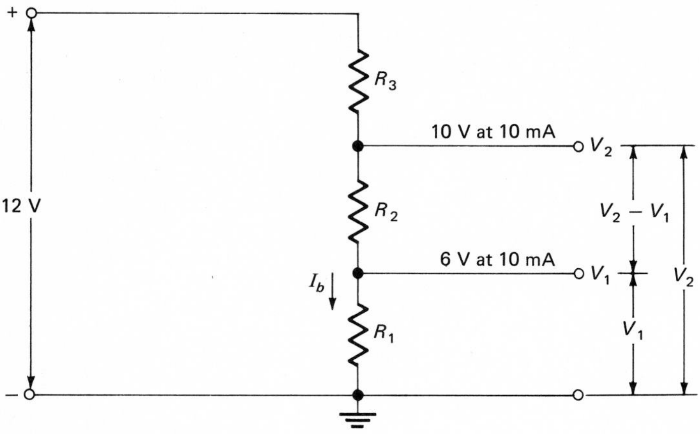

Fig. 3-61 Example 3-38: network.

Let I_b be 100 times the largest output current ($100 \times 10\text{ mA} = 1\text{ A}$).

$$R_1 = \frac{V_1}{I_b} = \frac{6\text{ V}}{1\text{ A}} = \mathbf{6\ \Omega}$$

From Eq. (3-9),

$$R_2 = R_1 \frac{V_2 - V_1}{V_1} = 6\ \Omega \times \frac{4\text{ V}}{6\text{ V}} = \mathbf{4\ \Omega}$$

From Eq. (3-9),

$$R_3 = R_1 \frac{E - V_2}{V_1} = 6\ \Omega \times \frac{2\text{ V}}{6\text{ V}} = \mathbf{2\ \Omega}$$

It is important for you to note the voltages in Example 3-38 (Fig. 3-61). The voltages V_1 and V_2 are output voltages with respect to ground. R_2 is connected between the V_1 and V_2 outputs. So the voltage across R_2 is $V_2 - V_1$. It is quite wrong to think it should be V_2, a typical error when first introduced to circuits of this type.

VOLTAGE REGULATION

It is common for the voltage at a load to be different from the value desired. It is also common for a load voltage to vary when it is desired that it be constant. Both of these aberrations can be attributed to three basic causes:

1. The presence of resistances other than the load resistance. These resistances include the internal resistance of the energy source and the resistance of the connecting conductors (often called *line resistance*).
2. The source emf being different from its nominal value.
3. The load resistance being different from its nominal value, or the load demand being different from its nominal value. The load *demand* is the current or power taken by the load, both of which are dependent on the load resistance and the load voltage.

Varying load voltage will be due to one or more of the factors just mentioned, the respective quantities being subject to variation.

The degree to which the actual load voltage differs from the desired value is called *voltage regulation.* A low value of regulation, or good regulation, indicates a small difference.

In general, voltage regulation is given by the ratio

$$\frac{\text{Voltage deviation from that desired}}{\text{Voltage desired}}$$

However, voltage regulation can be expressed in different ways, according to the circumstances. Two examples are

1. Load or output voltage steady; desired load voltage to be the source emf. Figure 3-62 shows the conditions.

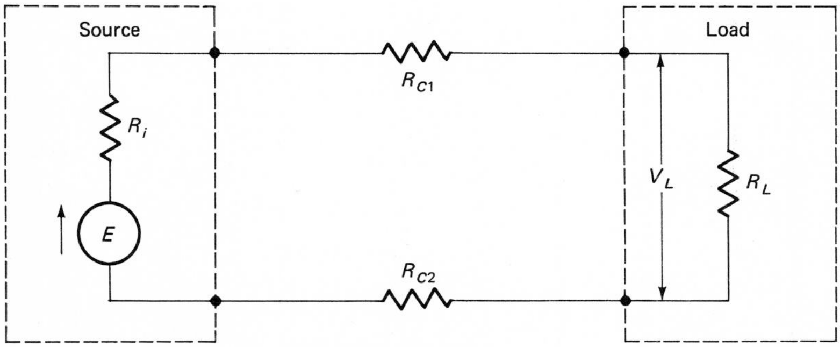

Fig. 3-62 Voltage regulation, load voltage steady.

$$\text{Voltage regulation (percent)} = \frac{E - V_L}{E} \times 100$$

E is the source emf or what the load voltage would be if R_i, R_{c1}, and R_{c2} were not present. V_L is the actual load voltage.

Example 3-27 illustrates this case. The answer to Example 3-27 might have been "voltage regulation = 16 percent."

2. Load voltage varies.

In Fig. 3-63 variation of emf and load resistance is indicated, these being the most likely variables. This situation might be described by a statement such as "the output voltage will not change by more than x volts for an output current variation between y and z amperes." In this case,

$$\text{Voltage regulation (percent)} = \frac{\text{maximum variation in } V_L}{\text{nominal } V_L} \times 100$$

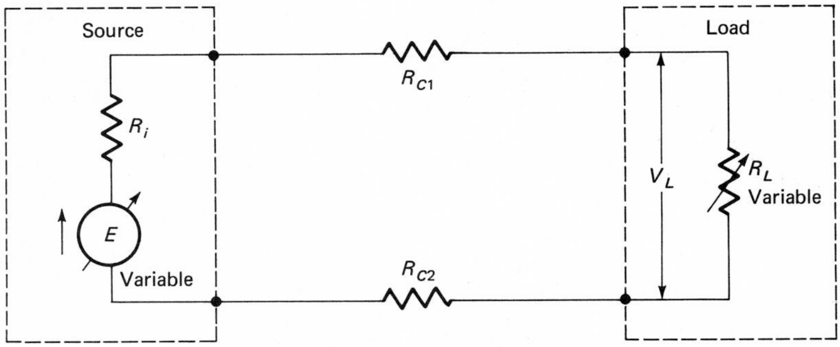

Fig. 3-63 Voltage regulation, load voltage varying.

Example 3-29 illustrates load voltage variation, the regulation being 50 percent with a 10-Ω line resistance and the nominal load voltage equal to the 100-V emf.

Example 3-39 The maximum output rating of an electronic power supply is 12 V at 5 A. The specification gives the voltage regulation as 1 percent for load currents between 1 and 5 A. What is the emf and the effective internal resistance of the power supply?

Solution

$$\text{Voltage regulation} = \frac{\text{maximum variation in } V_L}{\text{nominal } V_L}$$

$$\text{Maximum variation in } V_L = \text{voltage regulation} \times \text{nominal } V_L$$

$$= 1 \text{ percent of } 12 \text{ V} = 0.12 \text{ V}$$

It is specified that $V_L = 12$ V when the output current is 5 A. It is reasonable to assume that V_L will rise to 12 V + 0.12 V = 12.12 V as the current drops to 1 A.

Fig. 3-64 Example 3-39: circuit. (*a*) When the current is 5 A; (*b*) when the current is 1 A.

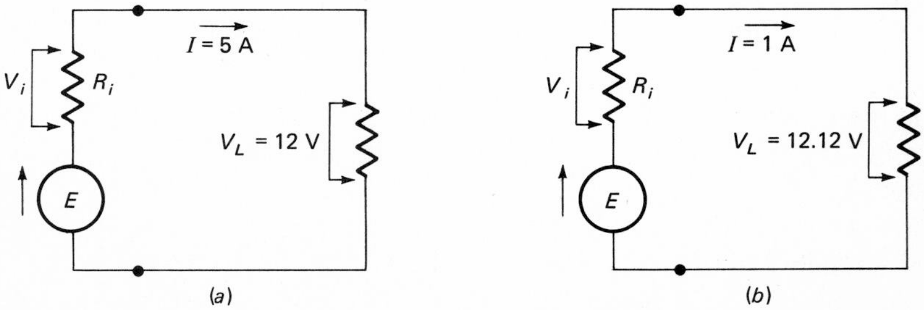

Applying Eq. (3-13) ($E = V_i + V_L$),
When $I = 5$ A (Fig. 3-64*a*),

$$V_i = IR_i = 5R_i$$

$$E = 5R_i + 12 \tag{A}$$

When $I = 1$ A (Fig. 3-64*b*),

$$V_i = IR_i = R_i \qquad (1 \text{ A} \times R_i)$$

$$E = R_i + 12.12 \tag{B}$$

The emf E is a constant for any energy source, so the right sides of Eqs. (A) and (B) are equal.

$$5R_i + 12 = R_i + 12.12$$

$$R_i = \mathbf{0.03\ \Omega}$$

Substituting for R_i in Eq. (A),

$$E = 5R_i + 12 = 5 \times 0.03 + 12 = \mathbf{12.15\ V}$$

VOLTAGE SOURCES IN SERIES

When voltage sources (generators) are connected in series, their emfs are added algebraically (taking their polarities into account) and their internal resistances are added.

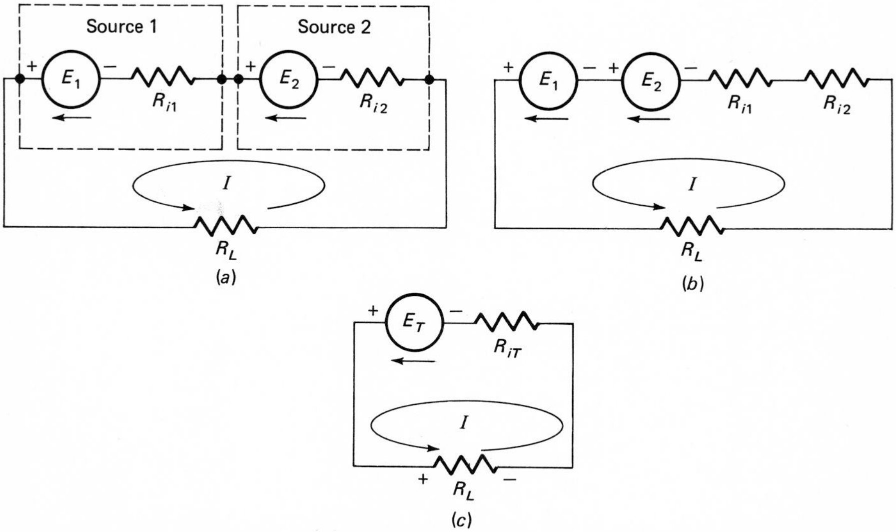

Fig. 3-65 Voltage sources in series aiding. (*a*) Individual sources; (*b*) emfs grouped and internal resistances grouped; (*c*) equivalent circuit.

Figure 3-65*a* shows two voltage generators connected in *series aiding,* their polarities being such that their emfs can be added together. In Fig. 3-65*b* the two sources are shown with their emfs grouped together and their internal resistances similarly grouped. A single source which is the equivalent of the original two generators is shown in Fig. 3-65*c*.

$$R_{iT} = R_{i1} + R_{i2}$$
$$E_T = E_1 + E_2$$

When the polarities of generators are such that each tends to produce opposite direction currents, the connection is called *series opposing*. This type of connection is shown in Fig. 3-66, the diagrams in the sequence being the series opposing equivalents of those in Fig. 3-65.

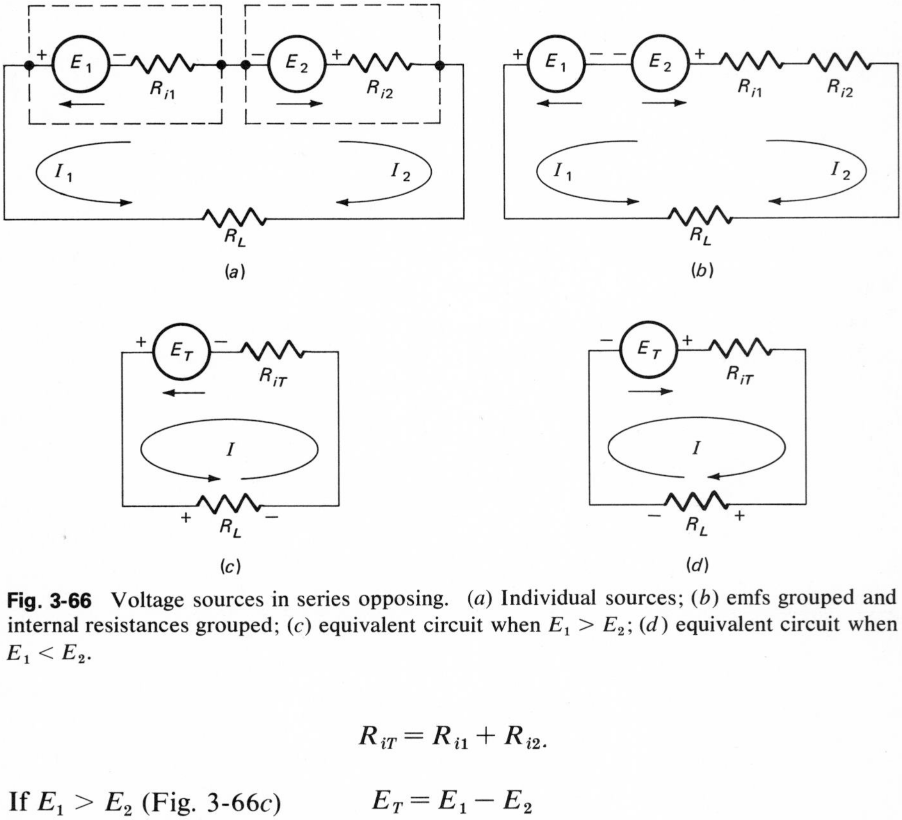

Fig. 3-66 Voltage sources in series opposing. (*a*) Individual sources; (*b*) emfs grouped and internal resistances grouped; (*c*) equivalent circuit when $E_1 > E_2$; (*d*) equivalent circuit when $E_1 < E_2$.

$$R_{iT} = R_{i1} + R_{i2}.$$

If $E_1 > E_2$ (Fig. 3-66*c*) $\qquad E_T = E_1 - E_2$

If $E_1 < E_2$ (Fig. 3-66*d*) $\qquad E_T = E_2 - E_1$

Example 3-40 A telephone exchange is operated by a battery consisting of 24 cells connected in series aiding, each cell being of 2 V nominal emf. When the cells are fully charged, their emfs are 2.2 V and their internal resistances are 0.1 mΩ. The equipment requires 48 V at 1000 A. Counter emf (cemf) cells are available to be connected in series opposing with the main battery to reduce its effect when fully charged. They can be switched out of action when the main battery loses some of its charge during operation. When the main battery is fully charged, how many cemf cells are needed to provide, as near as possible, 48 V at the load without exceeding this voltage? Assume that the cemf cells are the same as the main battery cells and that conductor resistance can be ignored.

Solution (Fig. 3-67)

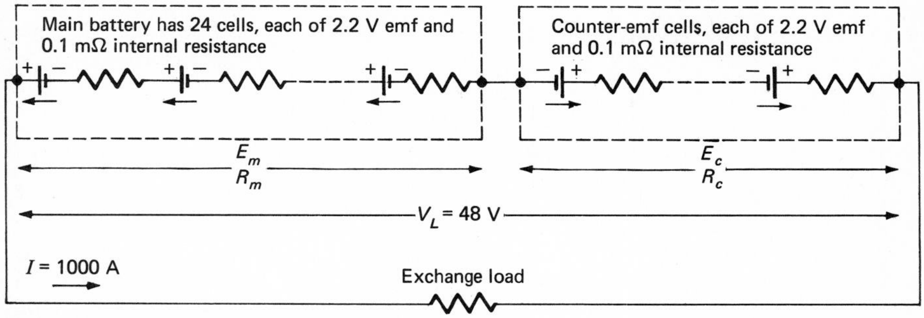

Fig. 3-67 Example 3-40: circuit.

MAIN BATTERY

$$E_m = 24 \times 2.2 \text{ V} = 52.8 \text{ V}$$

$$R_m = 24 \times 0.1 \text{ m}\Omega = 2.4 \text{ m}\Omega$$

CEMF BATTERY

$$E_c = C \times 2.2 \text{ V}$$

$$R_c = C \times 0.1 \text{ m}\Omega$$

The net emf will equal the difference between the main and cemf battery emfs:

$$E = E_m - E_C = (52.8 - 2.2 \text{ C}) \text{ V}$$

The total internal resistance will equal the sum of the main and cemf battery resistances:

$$R_i = R_m + R_c = (2.4 + 0.1 \text{ C}) \text{ m}\Omega$$

The pd across the net internal resistance will be

$$V_i = IR_i = 1000 \text{ A } (2.4 + 0.1 \text{ C}) \text{ m}\Omega = (2.4 + 0.1 \text{ C}) \text{ V}$$

The load voltage will be given by

$$V_L = E - V_i \qquad (3\text{-}13)$$

$$48 \text{ V} = (52.8 - 2.2 \text{ C}) \text{ V} - (2.4 + 0.1 \text{ C}) \text{ V}$$

$$48 = (52.8 - 2.4) - (2.2 + 0.1) \text{ C} = 50.4 - 2.3 \text{ C}$$

$$C = 1.044$$

If one cemf cell were used, its effect would not be great enough to reduce the load voltage to 48 V. Therefore **two** cemf cells would have to be used.

3-6 SUMMARY OF EQUATIONS

$$V_T = V_1 + V_2 + V_3 + \cdots + V_n \tag{3-1}$$

$$E = V_1 + V_2 + V_3 + \cdots + V_n \tag{3-2}$$

$$R_T = R_1 + R_2 + R_3 + \cdots + R_n \tag{3-3}$$

$$P_T = P_1 + P_2 + P_3 + \cdots + P_n \tag{3-4}$$

$$W_T = W_1 + W_2 + W_3 + \cdots + W_n \tag{3-5}$$

$$V_T = V_1R_T/R_1 \quad \text{or} \quad V_2R_T/R_2 \quad \text{or} \quad \ldots \quad \text{or} \quad V_nR_T/R_n \tag{3-6}$$

$$V_1 = V_TR_1/R_T \quad \text{or} \quad V_2R_1/R_2 \quad \text{or} \quad \ldots \quad \text{or} \quad V_nR_1/R_n \tag{3-7}$$

$$R_T = R_1V_T/V_1 \quad \text{or} \quad R_2V_T/V_2 \quad \text{or} \quad \ldots \quad \text{or} \quad R_nV_T/V_n \tag{3-8}$$

$$R_1 = R_TV_1/V_T \quad \text{or} \quad R_2V_1/V_2 \quad \text{or} \quad \ldots \quad \text{or} \quad R_nV_1/V_n \tag{3-9}$$

$$P_T = P_1R_T/R_1 \quad \text{or} \quad P_2R_T/R_2 \quad \text{or} \quad \ldots \quad \text{or} \quad P_nR_T/R_n \tag{3-10}$$

$$P_1 = P_TR_1/R_T \quad \text{or} \quad P_2R_1/R_2 \quad \text{or} \quad \ldots \quad \text{or} \quad P_nR_1/R_n \tag{3-11}$$

$$R_T/R_1 = V_T/V_1 = P_T/P_1 = W_T/W_1 \tag{3-12}$$

$$E = V_i + V_L \text{ or } V_L = E - V_i \tag{3-13}$$

$$V_L = E - (V_i + V_{C1} + V_{C2}) \tag{3-14}$$

$$V_T = V_M + V_m \tag{3-15}$$

$$V_n = ER_n/R_T \tag{3-16}$$

EXERCISE PROBLEMS

Section 3-3

Resistors R_1 and R_2 are part of a transistor biasing circuit. The current to the transistor base is very small and will be ignored.

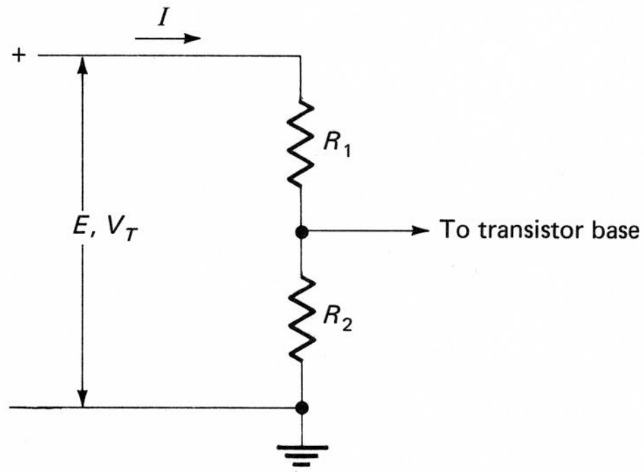

Fig. 3-1P Problems 3-1 to 3-6: network.

3-1 Find the voltage across each resistor when

a. $R_1 = 12$ kΩ, $R_2 = 5.6$ kΩ, $I = 680$ μA (*ans.* 8.16 V, 3.808 V)
b. $R_1 = 8.2$ kΩ, $R_2 = 4.7$ kΩ, $I = 775$ μA (*ans.* 6.355 V, 3.643 V)
c. $R_1 = 56$ kΩ, $R_2 = 18$ kΩ, $I = 0.19$ mA (*ans.* 10.64 V, 3.42 V)
d. $R_1 = 5.6$ kΩ, $R_2 = 1.8$ kΩ, $I = 2.7$ mA (*ans.* 15.12 V, 4.86 V)
e. $R_1 = 1.8$ kΩ, $R_2 = 1.8$ kΩ, $I = 2.7$ mA (*ans.* 4.86 V, 4.86 V)
f. $R_1 = 27$ kΩ, $R_2 = 4.7$ kΩ, $I = 0.57$ mA (*ans.* 15.39 V, 2.679 V)

3-2 In Prob. 3-1 find the total pd and the applied emf for each case.

3-3 In Prob. 3-1 find the total resistance for each case.
(*ans*. *a*. 17.6 kΩ; *b*. 12.9 kΩ; *c*. 74 kΩ; *d*. 7.4 kΩ; *e*. 3.6 kΩ; *f*. 31.7 kΩ)

3-4 In Prob. 3-1 find the power dissipated by each resistor and the total power for each case.

3-5 When $E = 15$ V, $R_1 = 3.9$ kΩ, $R_2 = 1.2$ kΩ, find I. (*ans*. 2.941 mA)

3-6 When $I = 1$ mA, $V_1 = 2.2$ V, $V_2 = 6.8$ V, find R_T.

3-7 A voltmeter consists of a 99-kΩ resistor in series with a 1000-Ω meter movement. If the current in the resistor is 99 μA, what is the current in the meter movement? (*ans*. 99 μA)

3-8 A 2000-Ω resistor and a 3000-Ω resistor form a series circuit with a 50-V supply. If the current in the 2000-Ω resistor is 10 mA, what is the current in the 3000-Ω resistor?

3-9 A 1000-Ω resistor and a 2000-Ω resistor are in series. If the pd across the 1000-Ω resistor is 50 V, what is the current in the 2000-Ω resistor? (*ans*. 50 mA)

3-10 The 0.625-Ω internal resistance of a generator is in series with a 100-Ω load. If the voltage drop caused by the internal resistance is 7.5 V, what is the voltage across the load?

3-11 An amplifier producing a signal voltage of 25 V and having an output resistance of 8 Ω is connected to an 8-Ω speaker. Find

a. The current in the amplifier output resistance. (*ans*. 1.563 A)
b. The current in the speaker. (*ans*. 1.563 A)
c. The power dissipated by the output resistance. (*ans*. 19.55 W)
d. The total power dissipated by the circuit. (*ans*. 39.08 W)

3-12 Eight Christmas tree lamps, each with a resistance of 37.5 Ω (when lit), are connected in series across a 120-V supply. Find

a. The total resistance.
b. The current through the circuit.
c. The pd across each lamp.
d. The total power dissipated.
e. The power dissipated by each lamp.
f. The energy used by each lamp per hour.
g. The total energy used per hour.

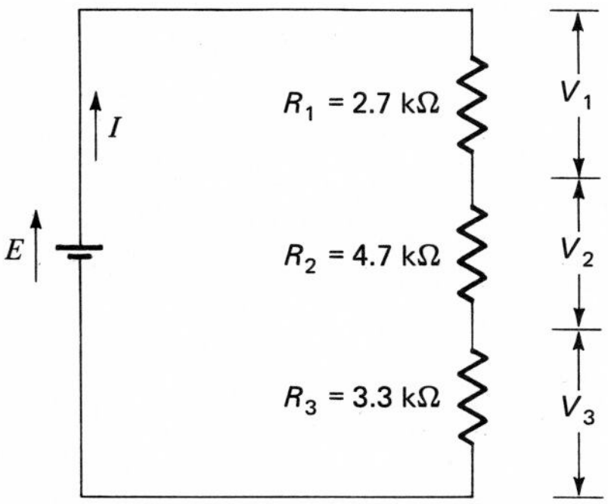

Fig. 3-2P Problem 3-13: circuit.

3-13 Find the current in the circuit in Fig. 3-2P for each of the following cases.

a. $V_1 = 11$ V (*ans*. 4.074 mA)
b. $V_2 = 78$ V (*ans*. 16.6 mA)
c. $V_3 = 66$ V (*ans*. 20 mA)
d. $E = 220$ V (*ans*. 20.56 mA)

3-14 The open-circuit voltage of a generator is 120 V, and its internal resistance is 0.3 Ω. It is connected to a 14.4-Ω heating element through an extension cord. Find the current in the circuit if the terminal voltage is 117.6 V when the heater is connected.

3-15 For the conditions stated in Prob. 3-14, find

a. The voltage across the heating element. (*ans.* 115.2 V)

b. The resistance of the extension cord. (*ans.* 0.3 Ω)

c. The power dissipated by the heating element. (*ans.* 921.6 W)

d. The power lost by the internal resistance and the extension cord. (*ans.* 38.4 W)

3-16 The terminal voltage of a generator is 117 V when a load is connected. The total resistance of the connecting wires is 0.506 Ω, and the voltage across the load is 113.4 V. What is

a. The circuit current?

b. The resistance of the load?

3-17 A 100-μA movement with an internal resistance of 2500 Ω is connected in series with a 10 000-Ω load. When the meter reads 60 μA, what is

a. The applied voltage? (*ans.* 0.75 V)

b. The voltage drop across the meter? (*ans.* 0.15 V)

c. The voltage drop across the load? (*ans.* 0.6 V)

3-18 In Prob. 3-17 the meter is removed and the circuit reconnected with the same applied voltage. Find

a. The voltage across the load.

b. The current through the load.

3-19 A 100-μA meter movement with an internal resistance of 2500 Ω is connected in series with a 270-kΩ load. The voltage drop across the load is found to be 24 V. What is

a. The voltage across the meter? (*ans.* 0.2222 V)

b. The percentage error introduced by the meter? (*ans.* 0.9174 percent)

3-20 Using the circuit for Prob. 3-13, find the voltage across R_1 and R_3 when the voltage across R_2 is 22 V.

3-21 Using the circuit for Prob. 3-13, find the individual voltages, using ratios, when the emf is 24 V. (*ans.* 6.056 V, 10.54 V, 7.402 V)

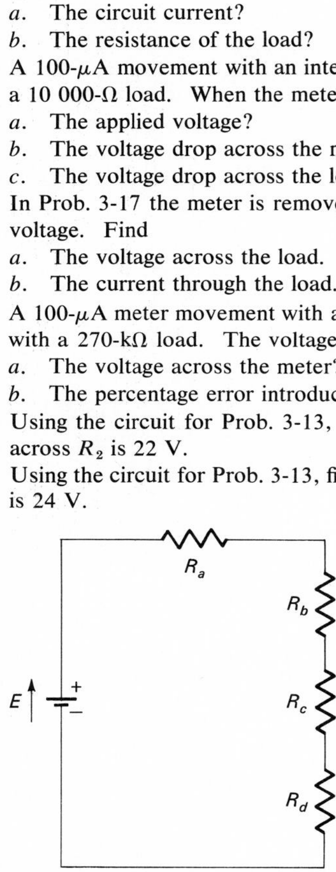

Fig. 3-3P Problems 3-22 to 3-24: circuit.

3-22 When $R_a = 680\ \Omega$, $V_a = 3.6$ V, $E = 18$ V, find R_T.

3-23 When $R_c = 4.7$ kΩ, $V_c = 4.2$ V, $V_b = 7.5$ V, $E = 18$ V, find R_T. (*ans.* 20.14 kΩ)

3-24 From the data given in Prob. 3-23, find R_b.

3-25 A meter movement with a full-scale deflection of 1 mA and an internal resistance of 115 Ω is used in series with a resistor to form a voltmeter. What must the total series resistance be for the meter to read full-scale when the voltmeter is placed across 5 V? (*ans.* 5 kΩ)

3-26 A circuit with an applied emf of 17 V has two series resistors R_1 and R_2 with a total resistance of 680 Ω. If $V_1 = 14$ V, what is the resistance of

a. R_1? *b.* R_2?

3-27 An electric saw has an effective resistance, under full load, of 11 Ω. It is fed from a 120-V supply through a 30-m extension cord. If the terminal voltage of the supply is 120 V and the voltage at the saw is 111.2 V, what is the resistance of the extension cord? (*ans.* 0.87 Ω)

3-28 A 6-V radio is operated from a 12-V supply through a series dropping resistor. If the radio requires 30 W of power for normal operation, what is

a. The current in the circuit?
b. The total power drawn from the supply?
c. The amount of power dissipated by the dropping resistor?

3-29 A 120-V generator with an internal resistance of 0.2 Ω feeds a 1-kW heater through a 15-m extension cord which has a total resistance of 0.4 Ω. Find

a. The power dissipated by the generator. (*ans.* 13.89 W)
b. The power dissipated by the extension cord. (*ans.* 27.78 W)

3-30 For the circuit in Prob. 3-29, if a short circuit occurs at the end of the extension cord, find

a. The short-circuit current.
b. The power dissipated in the generator.
c. The power dissipated in the extension cord.

Section 3-4

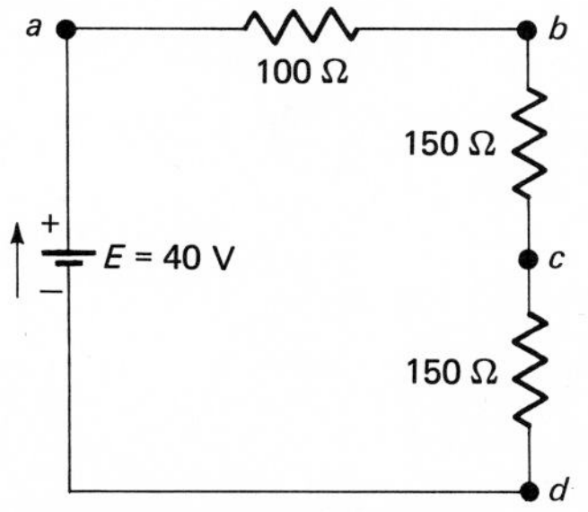

Fig. 3-4P Problem 3-31: circuit.

3-31 Refer to Fig. 3-4P and find

a.	V_{ab}	(*ans.* +10 V)	*e.*	V_{da}	(*ans.* −40 V)
b.	V_{ac}	(*ans.* +25 V)	*f.*	V_{bd}	(*ans.* +30 V)
c.	V_{cb}	(*ans.* −15 V)	*g.*	V_{bc}	(*ans.* +15 V)
d.	V_{ca}	(*ans.* −25 V)	*h.*	V_{dc}	(*ans.* −15 V)

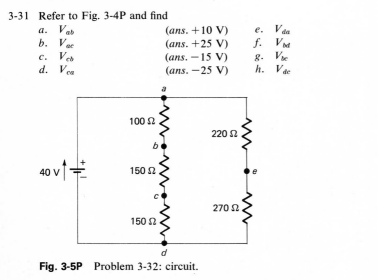

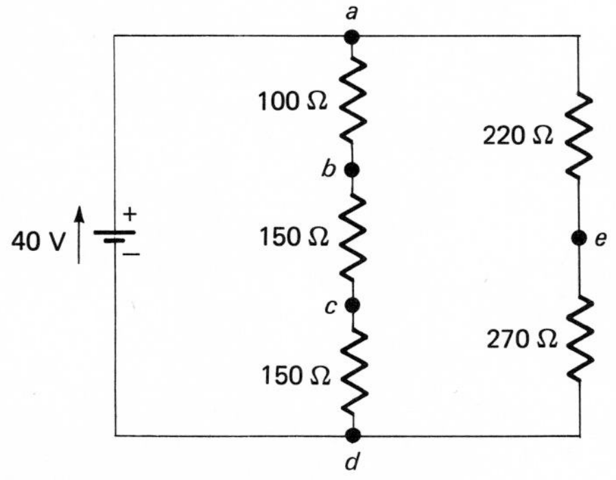

Fig. 3-5P Problem 3-32: circuit.

3-32 Refer to Fig. 3-5P and find

a. V_{be} *b.* V_{ec}

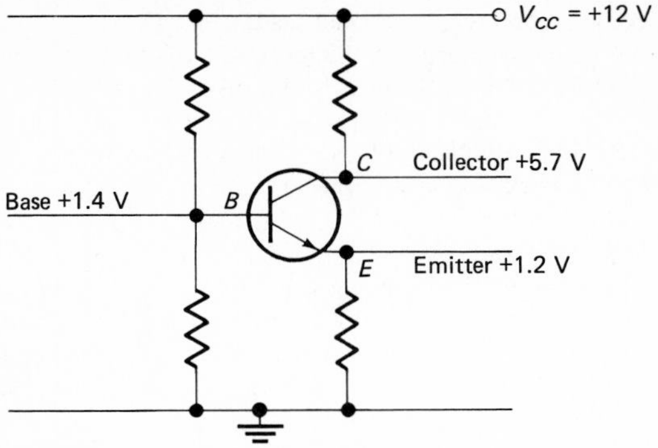

Fig. 3-6P Problem 3-33: circuit.

3-33 From the transistor amplifier circuit in Fig. 3-6P determine

a. The voltage at the collector with respect to the emitter. (*ans.* +4.5 V)

b. The voltage at the base with respect to the emitter. (*ans.* +0.2 V)

c. The voltage at the base with respect to the collector. (*ans.* −4.3 V)

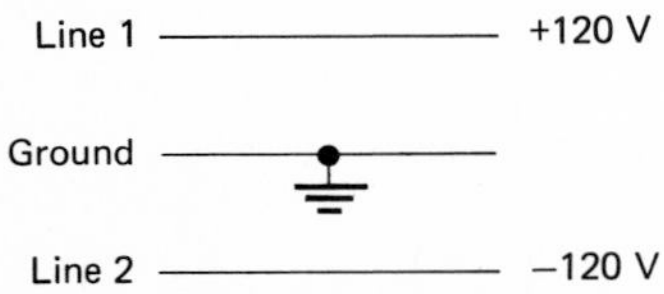

Fig. 3-7P Problem 3-34: three-wire distribution voltages.

3-34 Refer to Fig. 3-7P and determine

a. The voltage at ground with respect to line 1.

b. The voltage at ground with respect to line 2.

c. The voltage at line 2 with respect to line 1.

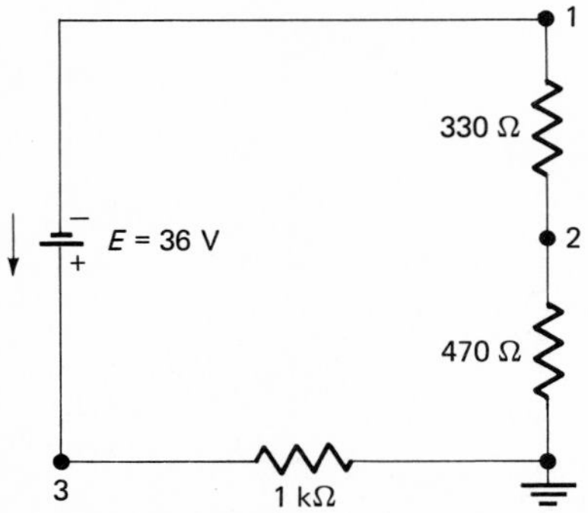

Fig. 3-8P Problem 3-35: circuit.

3-35 Give the magnitude and polarity of the following voltages in the circuit in Fig. 3-8P.

a. V_1 (*ans.* −16 V) *d.* V_{32} (*ans.* +29.4 V)

b. V_2 (*ans.* −9.4 V) *e.* V_{12} (*ans.* −6.6 V)

c. V_3 (*ans.* +20 V) *f.* V_{13} (*ans.* −36 V)

Section 3-5

3-36 A 50-μA meter movement with an internal resistance of 5 kΩ is to be used to make a multirange voltmeter. Find the required multiplier resistances for ranges of

a. 5 V
b. 15 V
c. 50 V

3-37 A 100-μA meter movement with an internal resistance of 850 Ω is equalized to 1000 Ω by adding 150 Ω in series. What values of resistance are required to make this equalized meter movement into a multirange voltmeter with ranges of

a. 1.5 V (*ans.* 14 kΩ)
b. 5 V (*ans.* 49 kΩ)
c. 15 V (*ans.* 149 kΩ)

3-38 When the following meter movements are adapted for use as voltmeters, what is their sensitivity (ohms per volt)?

a. 50 μA with internal resistance of 2 kΩ.
b. 50 μA with internal resistance of 5 kΩ.
c. 100 μA with internal resistance of 850 Ω.
d. 1 mA with internal resistance of 250 Ω.
e. 10 μA with internal resistance of 20 kΩ.

3-39 One volt is applied to a 1000-Ω resistance. If the current is to be measured by a 0 to 1 mA meter movement with an internal resistance of 250 Ω, what current will be indicated by the meter? What percent error does this current reading represent?

(*ans.* 0.8 mA, 20 percent)

3-40 A 100-μA meter movement with an internal resistance of 850 Ω is used to plot the low-current characteristics of a forward-biased semiconductor diode. If, when the applied voltage is 0.18 V, the microammeter reads 100 μA, what is

a. The resistance of the diode?
b. The pd across the diode?
c. The actual current through the diode?

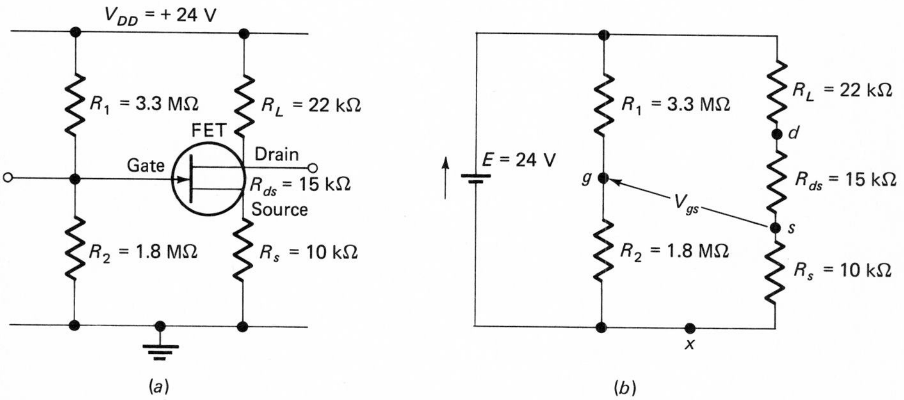

Fig. 3-9P Problem 3-41: (*a*) original circuit; (*b*) equivalent circuit.

3-41 The current drawn by the gate of a field-effect transistor (FET) is essentially zero. What is the relative potential, gate to source, V_{gs}, for the FET in Fig. 3-9P*a*? Figure 3-9P*b* is an equivalent circuit.

(*ans.* 3.36 V)

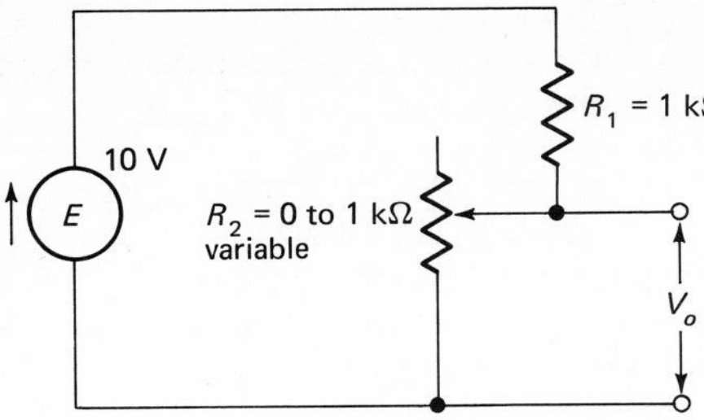

Fig. 3-10P Problem 3-42: circuit.

3-42 Figure 3-10P is the circuit of an unloaded voltage divider using a fixed resistance and a variable resistance. Plot a graph of V_o as R_2 is varied in 100-Ω steps from 0 to 1000 Ω.

4

Parallel Circuits

The parallel circuit is the dual of the series circuit. Each of the characteristics of the series circuit has its parallel complement.

In this chapter the conductance approach is stressed, because the need for it exists in developing parallel circuit principles. It is often more convenient to use resistances in dealing with practical resistive circuit problems, because:

1. Practical resistive circuit elements are resistors. Resistors are specified, made, and bought in resistance values. It would just complicate matters to specify their conductances also.
2. Conductances, usually fractional quantities, cannot be used for calculations as conveniently as resistances.

For many people, the series circuit and its characteristics seem to be easier to visualize than the parallel equivalents. There is no logical reason for this preference; if you will free your mind to accept the fact that parallel circuits are as easy to master as series circuits, you will quickly learn to understand them thoroughly. When you have become familiar with both parallel and series circuits, you will have the essential foundation upon which you can build an understanding of the more complex circuits.

Later chapters will reinforce your familarity with both series and parallel circuits, particularly with regard to the identification of a configuration when it appears in an unfamiliar form within a complex network.

4-1 THE MEANING OF PARALLEL CIRCUIT

When all the elements of a circuit are connected between two common points, they are said to be connected in *parallel*.

In Fig. 4-1 the elements A, B, and C are connected in parallel

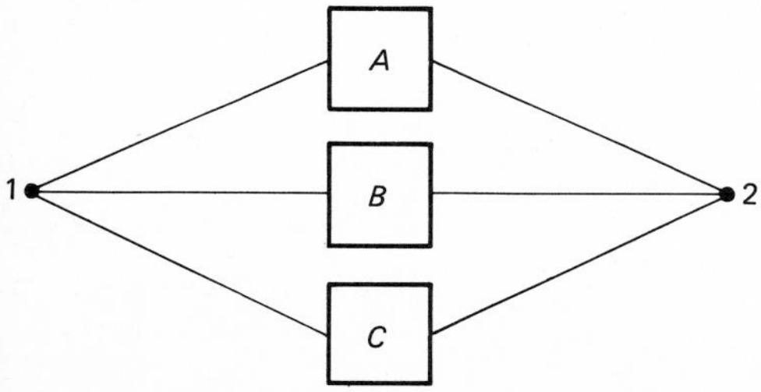

Fig. 4-1 Elements between common points.

between points 1 and 2. Figure 4-2 shows this parallel circuit as it is more likely to appear in electrical or electronic diagrams. The Fig. 4-2 presentation illustrates the meaning of the word "parallel," as all the elements are depicted as being geometrically parallel, as are the conductors between which they are connected.

An alternative term for "parallel" is "shunt."

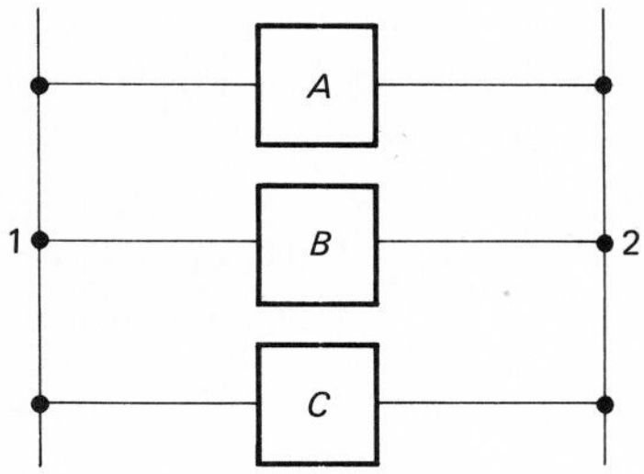

Fig. 4-2 Elements between common conductors.

4-2 CHARACTERISTICS OF PARALLEL CIRCUITS

The most important characteristics of parallel circuits are:

1. There is only one voltage across all parts of the circuit.
2. Elements have individual currents.
3. Currents are additive.
4. Conductances are additive.
5. Resistance reciprocals are additive.
6. Powers and energies are additive.

Each of these characteristics is now dealt with in detail.

One voltage In a parallel circuit, the voltages across each of the elements are the same.

Figure 4-3 is a schematic diagram equivalent of Figs. 4-1 and 4-2. Here the elements are a battery and two resistors. The voltage V between the conductors is the only possible circuit voltage. Therefore the battery emf E_A and the pds across the resistances V_B and V_C are all equal to each other and equal to V. Any voltage measurement in this circuit must be made between the conductors; it does not matter where this measurement

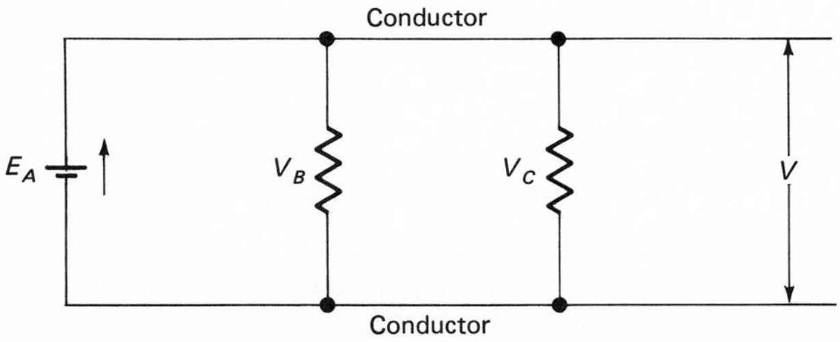

Fig. 4-3 Common voltage across a parallel circuit.

is made along the length of the conductors, since the result will always be the same.

Currents individual Individual currents exist in each of the elements of a parallel circuit. These are the currents I_A, I_B, and I_C in Fig. 4-4.

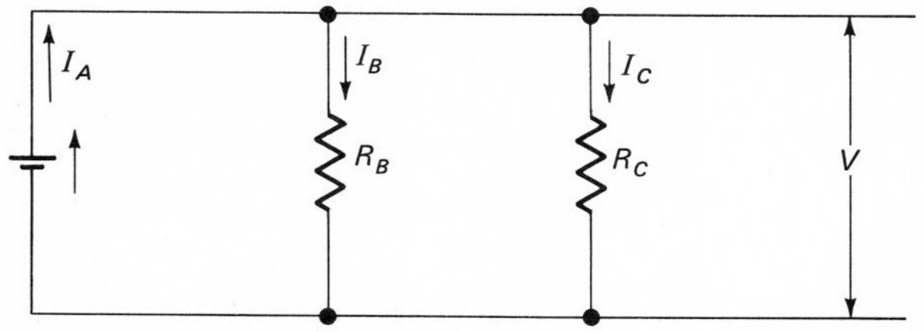

Fig. 4-4 Individual currents in a parallel circuit.

Applying Ohm's law to each of the resistance elements,

$$I_B = \frac{V}{R_B} \quad \text{and} \quad I_C = \frac{V}{R_C}$$

Example 4-1 A battery with a 12-V emf and negligible internal resistance forms a parallel circuit with a network of three resistors, 2, 4, and 6 Ω. What is the current in each resistor?

Solution (Fig. 4-5)

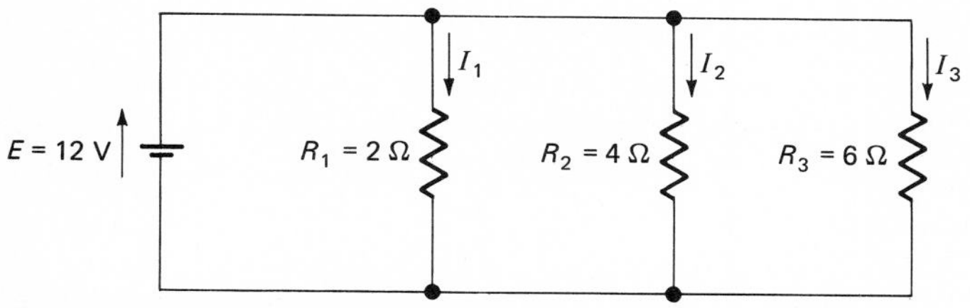

Fig. 4-5 Example 4-1: circuit.

The 12-V emf E is connected to the three resistances R_1, R_2, and R_3. The current in each resistance is related to that voltage by Ohm's law:

$$I_1 = \frac{E}{R_1} = \frac{12\text{ V}}{2\ \Omega} = \mathbf{6\ A}$$

$$I_2 = \frac{E}{R_2} = \frac{12\text{ V}}{4\ \Omega} = \mathbf{3\ A}$$

$$I_3 = \frac{E}{R_3} = \frac{12\text{ V}}{6\ \Omega} = \mathbf{2\ A}$$

Currents additive (Kirchhoff's current law) The current leaving the network at B in Fig. 4-6*a* must be the same as that entering it at A. There is no way in which the total current can change; it cannot collect additional current and so leave greater than it entered; no current can be lost, causing the amount leaving to be less than the amount entering.

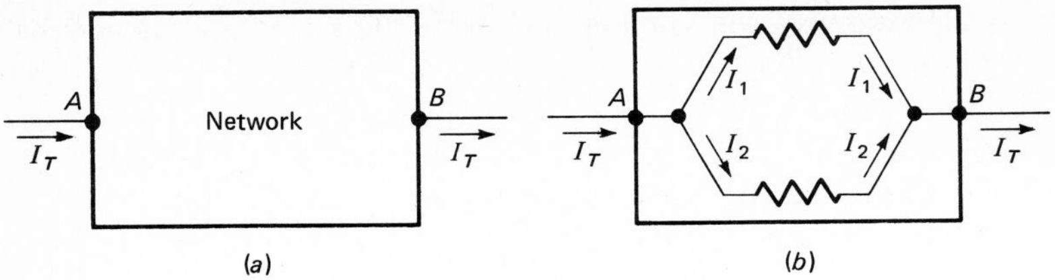

Fig. 4-6 (*a*) Current enters and leaves a network; (*b*) current divides at a branching point, or node, and then recombines.

When the network consists of elements in parallel, as in Fig. 4-6*b*, the current which enters splits up among the elements, in this case into two components I_1 and I_2. The two components recombine and leave as the original current. At both points A and B,

$$I_T = I_1 + I_2$$

This addition of current is a general characteristic of parallel resistive circuits, no matter how many elements are involved. In general form it can be written as

$$I_T = I_1 + I_2 + I_3 + \cdots + I_n \tag{4-1}$$

Kirchhoff's current law is usually stated: The algebraic sum of the currents entering a node equals the algebraic sum of the currents leaving it.

Equation (4-1) is an expression of Kirchhoff's current law. It applies to points A and B in Fig. 4-6, since these points are nodes or conductor junctions.

Example 4-2 What is the total current entering and leaving the three-element resistive network in Example 4-1?
Solution (Fig. 4-7)

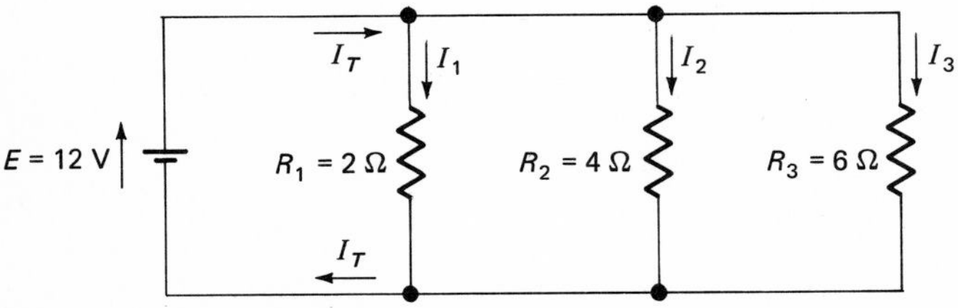

Fig. 4-7 Example 4-2: circuit.

From Example 4-1,

$$I_1 = 6\text{ A} \qquad I_2 = 3\text{ A} \qquad I_3 = 2\text{ A}$$

$$I_T = I_1 + I_2 + I_3 \tag{4-1}$$

$$I_T = 6\text{ A} + 3\text{ A} + 2\text{ A} = \mathbf{11\ A}$$

When, as in Examples 4-1 and 4-2, the total circuit consists of a generator and a parallel network, the total current entering and leaving the network will be the total current drawn from the generator.

Example 4-3 A generator is applied to a parallel network consisting of four resistances, 10, 20, 25, and 50 Ω. The current in the 25-Ω resistance is 4 A. What are the other circuit currents?

Solution (Fig. 4-8)

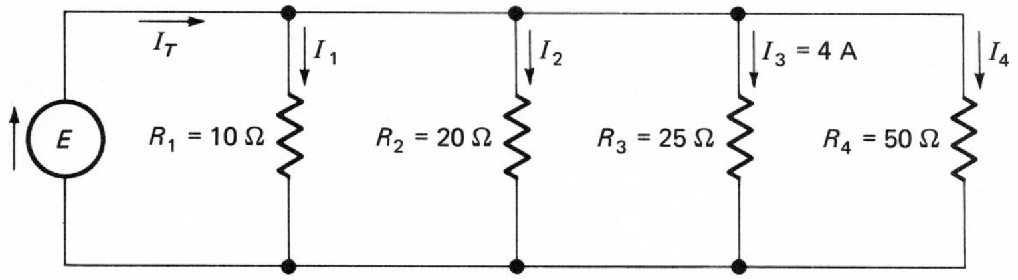

Fig. 4-8 Example 4-3: circuit.

Since this is a parallel network, there is only one circuit voltage; so the pd across R_3 must be the same as the generator emf E.

$$E = I_3R_3 = 4\text{ A} \times 25\ \Omega = 100\text{ V}$$

This emf of 100 V will also be the pd across the other three resistances R_1, R_2, and R_4.

$$I_1 = \frac{E}{R_1} = \frac{100\text{ V}}{10\ \Omega} = \mathbf{10\ A}$$

$$I_2 = \frac{E}{R_2} = \frac{100\text{ V}}{20\ \Omega} = \mathbf{5\ A}$$

$$I_4 = \frac{E}{R_4} = \frac{100\text{ V}}{50\ \Omega} = \mathbf{2\ A}$$

$$I_T = I_1 + I_2 + I_3 + I_4 \qquad (4\text{-}1)$$

$$I_T = 10\text{ A} + 5\text{ A} + 4\text{ A} + 2\text{ A} = \mathbf{21\ A}$$

Conductances additive Figure 4-9*a* shows a network with three parallel elements of conductance G_1, G_2, and G_3. Figure 4-9*b* is its single-element equivalent. The voltages V and the currents I_T are the same in each network, so the conductance G_T is the total equivalent conductance of G_1, G_2, and G_3.

Fig. 4-9 (*a*) Three-element parallel network; (*b*) single-element equivalent.

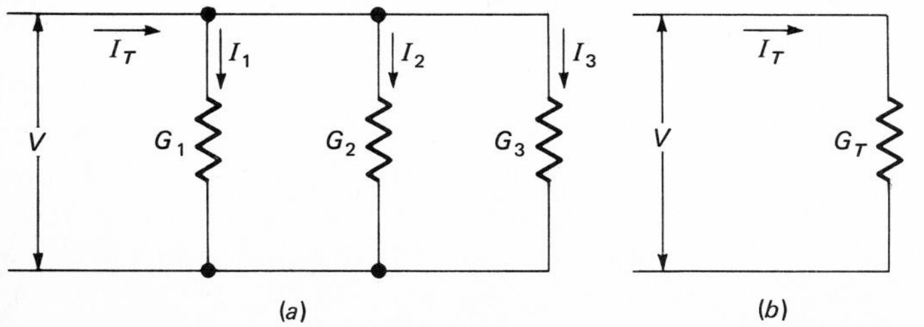

Conductance is the reciprocal of resistance, or $G = 1/R$. Therefore Ohm's law can be written in the form $I = VG$ instead of $I = E/R$. Applying this to Fig. 4-9,

$$I_T = VG_T$$

$$I_1 = VG_1 \qquad I_2 = VG_2 \qquad I_3 = VG_3$$

$$I_T = I_1 + I_2 + I_3 \tag{4-1}$$

$$VG_T = VG_1 + VG_2 + VG_3 = V(G_1 + G_2 + G_3)$$

$$G_T = G_1 + G_2 + G_3$$

This formula applies to any number of parallel elements. Thus,

$$G_T = G_1 + G_2 + G_3 + \cdots + G_n \tag{4-2}$$

The *total equivalent conductance* of a number of elements connected in parallel is the sum of their individual conductances.

Resistance reciprocals additive Because it is usual to express element values in resistance rather than conductance, the resistance equivalent of Eq. (4-2) can be useful.

$$G_T = G_1 + G_2 + G_3 + \cdots + G_n \tag{4-2}$$

$$\frac{1}{R_T} = \frac{1}{R_1} + \frac{1}{R_2} + \frac{1}{R_3} + \cdots + \frac{1}{R_n} \tag{4-3}$$

The *reciprocal of the equivalent resistance* of a number of elements in parallel is the sum of the reciprocals of the resistances of the individual elements.

This statement seems to be quite complicated, but it is actually the same statement as that associated with Eq. (4-2); it is the reciprocal aspect that leads to the complication. This difficulty is one reason for recommending that you think in terms of conductance when dealing with parallel circuits.

You will see that Eq. (4-3) does not directly give the resistance of a parallel circuit; to obtain this, Eq. (4-3) must be inverted.

$$R_T = \frac{1}{1/R_1 + 1/R_2 + 1/R_3 + \cdots + 1/R_n}$$

This equation is only given to illustrate how awkward matters can become if the resistance viewpoint is used throughout in dealing with parallel circuits. The statement which relates to this equation makes this graphically clear: The total equivalent resistance of a number of elements connected in parallel is the reciprocal of the sum of the reciprocals of the resistances of the individual elements.

There is no need to memorize the last equation and its related statement.

Example 4-4 A parallel network consists of three elements with conductances 0.5, 0.25, and 0.2 S. What is the total equivalent (*a*) conductance and (*b*) resistance?

Solution

a.

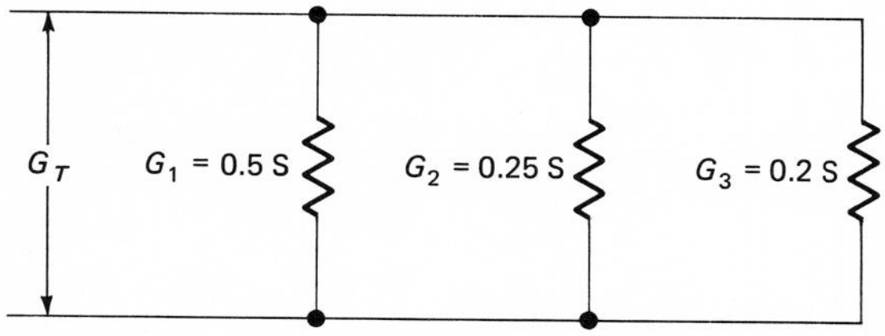

Fig. 4-10 Example 4-4: data in conductance form.

$$G_T = G_1 + G_2 + G_3 \tag{4-2}$$

$$G_T = 0.5\ \text{S} + 0.25\ \text{S} + 0.2\ \text{S} = \mathbf{0.95\ S}$$

b.

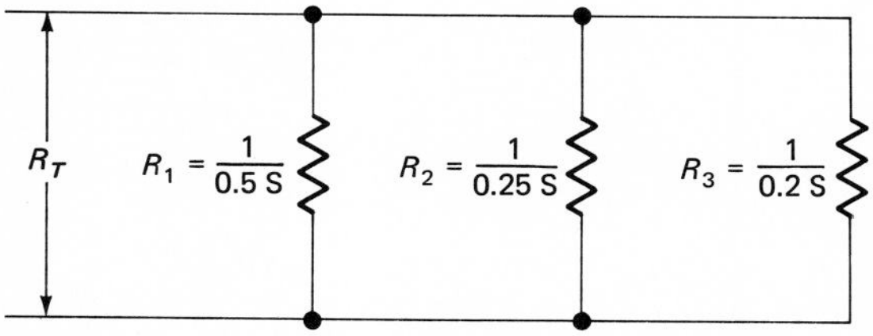

Fig. 4-11 Example 4-4: data in resistance form.

$$R_1 = \frac{1}{G_1} = \frac{1}{0.5\ \text{S}} = 2\ \Omega$$

$$R_2 = \frac{1}{G_2} = \frac{1}{0.25\ \text{S}} = 4\ \Omega$$

$$R_3 = \frac{1}{G_3} = \frac{1}{0.2\ \text{S}} = 5\ \Omega$$

$$\frac{1}{R_T} = \frac{1}{R_1} + \frac{1}{R_2} + \frac{1}{R_3} \tag{4-3}$$

$$\frac{1}{R_T} = \frac{1}{2\ \Omega} + \frac{1}{4\ \Omega} + \frac{1}{5\ \Omega} = \frac{10 + 5 + 4}{20\ \Omega} = \frac{19}{20\ \Omega}$$

$$R_T = \frac{20}{19}\ \Omega = \mathbf{1.053\ \Omega}$$

Example 4-5 A parallel network has elements with resistances of 30, 40, 60, and 120 Ω. What is the network's total equivalent (*a*) resistance and (*b*) conductance?

Solution (Fig. 4-12)

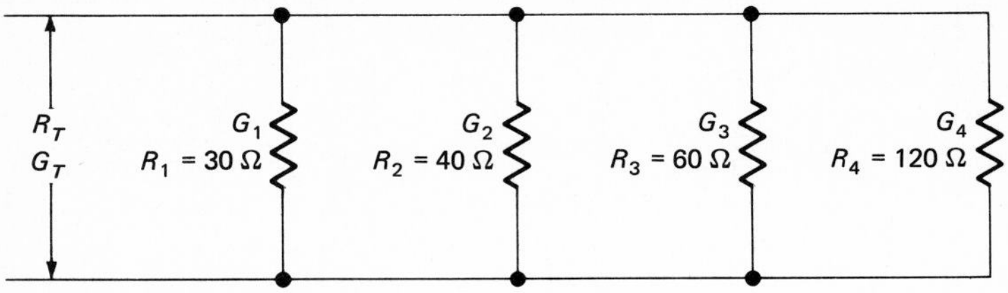

Fig. 4-12 Example 4-5: data.

a.

$$\frac{1}{R_T} = \frac{1}{R_1} + \frac{1}{R_2} + \frac{1}{R_3} + \frac{1}{R_4} \tag{4-3}$$

$$= \frac{1}{30\ \Omega} + \frac{1}{40\ \Omega} + \frac{1}{60\ \Omega} + \frac{1}{120\ \Omega}$$

$$= \frac{4+3+2+1}{120\ \Omega} = \frac{10}{120\ \Omega} = \frac{1}{12\ \Omega}$$

$$R_T = \mathbf{12\ \Omega}$$

b.

$$G_T = G_1 + G_2 + G_3 + G_4 \tag{4-2}$$

$$= \frac{1}{30}\ \text{S} + \frac{1}{40}\ \text{S} + \frac{1}{60}\ \text{S} + \frac{1}{120}\ \text{S}$$

$$G_T = \frac{4+3+2+1}{120}\ \text{S}$$

$$G_T = \mathbf{\frac{1}{12}\ S}$$

The solutions to Examples 4-4*b* and 4-5*b* would have been simpler if reciprocals had been used instead of the demonstration methods shown:

Example 4-4*b*: $$R_T = \frac{1}{G_T} = \frac{1}{0.95\ \text{S}} = 1.053\ \Omega$$

Example 4-5*b*: $$G_T = \frac{1}{R_T} = \frac{1}{12\ \Omega} = \frac{1}{12}\ \text{S}$$

Example 4-6 Four 100-kΩ resistors are connected in parallel. What is the total equivalent (*a*) conductance and (*b*) resistance?
Solution (Fig. 4-13)

Fig. 4-13 Example 4-6: data.

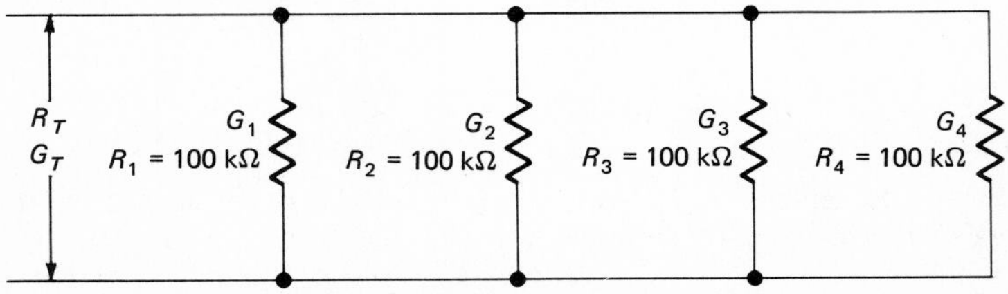

a. Each resistor has a conductance of

$$G = \frac{1}{R} = \frac{1}{100\ \text{k}\Omega} = 10\ \mu\text{S}$$

$$G_T = G_1 + G_2 + G_3 + G_4 \qquad (4\text{-}2)$$

$$= 10\ \mu\text{S} + 10\ \mu\text{S} + 10\ \mu\text{S} + 10\ \mu\text{S}$$

$$G_T = 4 \times 10\ \mu\text{S} = \mathbf{40\ \mu S}$$

b.

$$R_T = \frac{1}{G_T} = \frac{1}{40\ \mu\text{S}} = 2.5 \times 10^4\ \Omega \text{ or } \mathbf{25\ k\Omega}$$

or

$$R_T = \frac{R}{4} = \frac{100\ \text{k}\Omega}{4} = \mathbf{25\ k\Omega}$$

You will see from Example 4-6 that the *total conductance* of a number of equal elements in parallel is the conductance of one element multiplied by the number of elements. Also, the *total resistance* is the resistance of one element divided by the number of elements.

$$G_T = nG \qquad (4\text{-}4)$$

$$R_T = \frac{R}{n} \qquad (4\text{-}5)$$

where G_T = total equivalent conductance
n = number of elements connected in parallel
G = conductance of each equal individual element
R_T = total equivalent resistance
R = resistance of each equal individual element

Powers and energies additive The power dissipated in a resistance is

$$P = \frac{V^2}{R} \qquad (2\text{-}31)$$

Equation (2-31) can be written in conductance form as

$$P = V^2 G$$

From this equation it can be seen that, when elements have the same voltage, the power dissipated by each element will be directly proportional to its conductance value. Therefore, since the total conductance of a parallel network is equal to the sum of the individual conductances, the total power dissipated will be equal to the sum of the powers dissipated by each element. Furthermore, this summation applies to energy, because energy and power are related by time, which will be a constant for any particular situation.

$$P_T = P_1 + P_2 + P_3 + \cdots + P_n \tag{4-6}$$

$$W_T = W_1 + W_2 + W_3 + \cdots + W_n \tag{4-7}$$

It should be noted that Eqs. (4-6) and (4-7), which apply to parallel circuits, are the same as Eqs. (3-4) and (3-5) which give the power and energy relationships for series circuits. The concept can be carried further. In any resistive network, no matter what its configuration, the total power or energy is the sum of the individual powers or energies.

Example 4-7 Resistors of 12, 3, and 24 kΩ are connected in parallel across a 400-V power supply. What power is supplied to the parallel network?
Solution (Fig. 4-14)

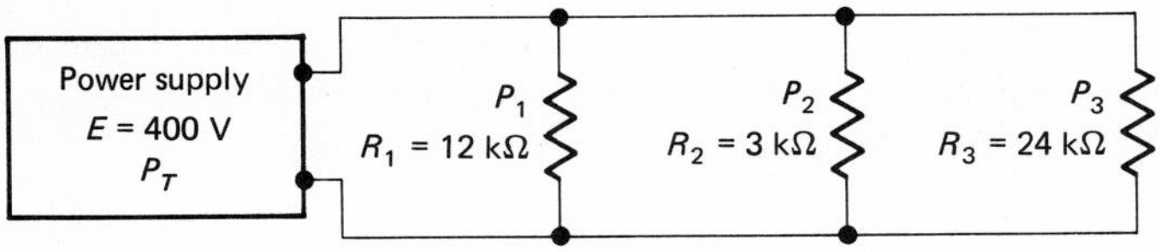

Fig. 4-14 Example 4-7: circuit.

METHOD 1 Sum the powers directly:

$$P_T = P_1 + P_2 + P_3 \tag{4-6}$$

$$= E^2G_1 + E^2G_2 + E^2G_3 = E^2(G_1 + G_2 + G_3)$$

$$= (400 \text{ V})^2 \left(\frac{1}{12 \text{ k}\Omega} + \frac{1}{3 \text{ k}\Omega} + \frac{1}{24 \text{ k}\Omega}\right)$$

$$P_T = 1.6 \times 10^5 \left(\frac{2 + 8 + 1}{2.4 \times 10^4}\right) = \frac{1.6 \times 10^5 \times 11}{2.4 \times 10^4} = \mathbf{73.33 \text{ W}}$$

METHOD 2 Calculate the individual powers and then sum them:

$$P_1 = \frac{E^2}{R_1} = \frac{1.6 \times 10^5 \text{ V}^2}{1.2 \times 10^4 \ \Omega} = 13.33 \text{ W}$$

$$P_2 = \frac{E^2}{R_2} = \frac{1.6 \times 10^5 \text{ V}^2}{3 \times 10^3 \ \Omega} = 53.33 \text{ W}$$

$$P_3 = \frac{E^2}{R_3} = \frac{1.6 \times 10^5 \text{ V}^2}{2.4 \times 10^4 \ \Omega} = 6.667 \text{ W}$$

$$P_T = P_1 + P_2 + P_3 \tag{4-6}$$

$$P_T = 13.33 \text{ W} + 53.33 \text{ W} + 6.667 \text{ W} = \mathbf{73.33 \text{ W}}$$

Example 4-8 Two 120-V 60-W light bulbs are connected to a 120-V supply. What total energy is converted per hour? If the bulbs are to operate normally on the 120-V supply, they will have to be connected in parallel across it.

Solution (Fig. 4-15)

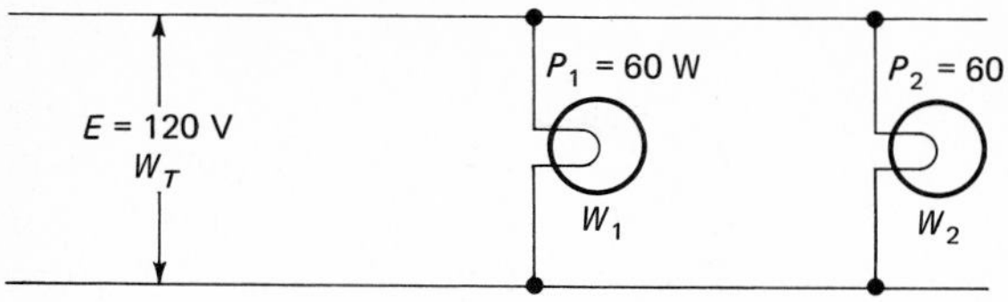

Fig. 4-15 Example 4-8: data.

For each bulb,

$$W_1 = W_2 = P_1 t = P_2 t = 60 \text{ W} \times 1 \text{ h}$$

$$W_1 = W_2 = 60 \text{ Wh} \qquad \text{or} \qquad 216 \text{ kJ}$$

$$W_T = W_1 + W_2 \tag{4-7}$$

$$W_T = 2 \times 60 \text{ Wh} = \mathbf{120\ Wh} \qquad \text{or} \qquad \mathbf{432\ kJ}$$

This example is the parallel equivalent of Example 3-7. The same total energy is converted because, although the supply voltage is different in each case, the bulbs are the same and each is operating normally, dissipating the rated 60 W. The total power is 120 W in each case. (In this example it is 120 V × 1 A; in Example 3-7 it was 240 V × 0.5 A.)

Example 4-9 Two electric baseboard heaters are used to heat a living room. Each operates at 240 V. One is rated at 1.25 kW, and the other at 0.5 kW. What is the cost of operating the heaters for 1 month (30 days) if both are on for an average of 10 h/day and the electricity rate is 2 cents/kWh?

Solution (Fig. 4-16)

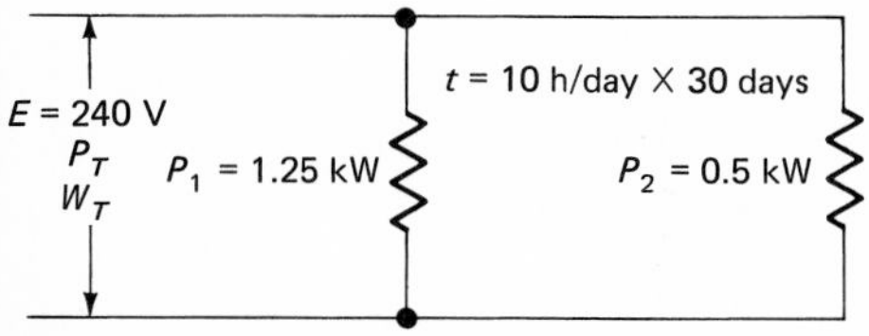

Fig. 4-16 Example 4-9: data.

$$P_T = P_1 + P_2 \tag{4-6}$$

$$P_T = 1.25 \text{ kW} + 0.5 \text{ kW} = 1.75 \text{ kW}$$

For 30 days at 10 h/day,

$$t = 10 \frac{\text{h}}{\text{day}} \times 30 \text{ days} = 300 \text{ h}$$

$$W_T = P_T t = 1.75 \text{ kW} \times 300 \text{ h} = 525 \text{ kWh}$$

$$\text{Cost of electricity} = \text{energy} \times \text{rate} = W_T \times 2 \text{ cents/kWh}$$

$$= 525 \text{ kWh} \times \frac{2 \text{ cents}}{\text{kWh}} = 1050 \text{ cents} \qquad \text{or} \qquad \mathbf{\$10.50}$$

Electrical utilities sell energy, even though they are often called power companies. The cost of the electricity is, as implied in the solution to Example 4-9, obtained by taking the basic rate for an energy unit and multiplying it by the sum of the energies taken by the individual elements (heaters, lamps, etc.).

It has been an industrial practice for many years to calculate electricity supply conditions for domestic loads in kilowatthours, rather than in joules, the SI unit of energy. The reason why the use of the kilowatthour is common is made evident by comparing the energy in Example 4-9 given in kilowatthours with the same amount of energy expressed in joules. 525 kWh is clearly easier to write or say than 1.89×10^9 J.

The figures given in Example 4-9 are typical of what may be involved in heating a living room of about 250 ft² area in winter.

4-3 CALCULATIONS FOR PARALLEL CIRCUITS

Calculations for parallel circuits involve the relationships applicable to single elements, which were dealt with in Chap. 2, and their association with the characteristics inherent in the parallel configuration.

A generalized parallel circuit is shown in Fig. 4-17, together with the quantities which can be calculated. There are many different ways in which calculations can be performed in such a circuit, depending on what data are known and what information is required. The techniques described, and demonstrated by examples, are typical of those frequently used in practice; by observing these methods, you can learn to devise others to fit any parallel circuit needs.

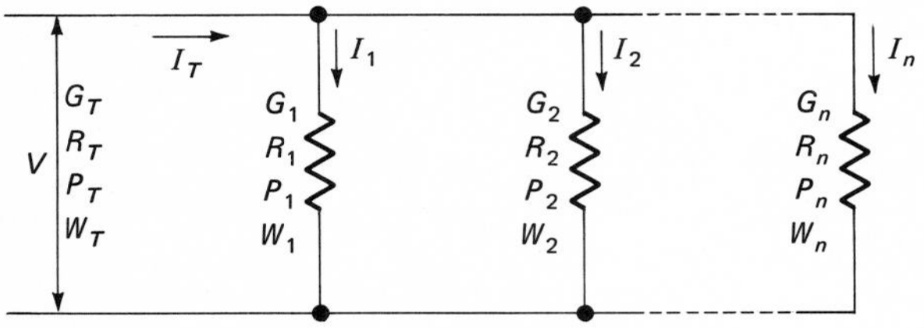

Fig. 4-17 Generalized parallel circuit.

The examples intentionally demonstrate several different ways of achieving the same result, each using different given data. There is no need for you to memorize each of these methods, but you should check them all thoroughly so that you can see how they are derived from basic circuit principles. In this way you will get practice in the use of fast, direct methods of calculating.

Voltage *V* The only voltage in the circuit is V, which appears across each element. It can be calculated by applying Ohm's law to the complete circuit, or to any part of it:

$$V = I_T R_T \quad \text{or} \quad I_1 R_1 \quad \text{or} \quad I_2 R_2 \quad \text{or} \quad \cdots \quad \text{or} \quad I_n R_n$$

Example 4-10 What is the common voltage across the circuit in Fig. 4-18?

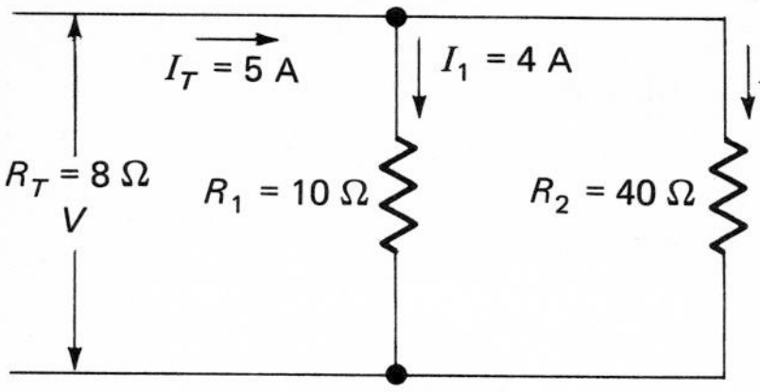

Fig. 4-18 Example 4-10: data.

Solution

METHOD 1 Applying Ohm's law to the entire circuit:

$$V = I_T R_T = 5\ \text{A} \times 8\ \Omega = \mathbf{40\ V}$$

METHOD 2 Applying Ohm's law to one element (any element can be selected):

$$V = I_1 R_1 = 4\ \text{A} \times 10\ \Omega = \mathbf{40\ V}$$

Total current I_T The total current in a parallel circuit can be obtained by adding the individual element currents:

$$I_T = I_1 + I_2 + \cdots + I_n \tag{4-1}$$

or by applying Ohm's law to the complete circuit:

$$I_T = VG_T$$

The latter equation can be converted into conductance ratio or resistance ratio form:

$$I_T = VG_T = \frac{I_1}{G_1} G_T = I_1 \frac{G_T}{G_1}$$

This equation can be written as one of a set of similar equations:

$$I_T = I_1 \frac{G_T}{G_1} \quad \text{or} \quad I_2 \frac{G_T}{G_2} \quad \text{or} \quad \cdots \quad \text{or} \quad I_n \frac{G_T}{G_n} \tag{4-8}$$

Because resistance and conductance are reciprocal quantities ($R = 1/G$), the Ohm's law equations for the total current can be written as

$$I_T = \frac{V}{R_T}$$

or $$I_T = I_1 \frac{R_1}{R_T} \quad \text{or} \quad I_2 \frac{R_2}{R_T} \quad \text{or} \quad \cdots \quad \text{or} \quad I_n \frac{R_n}{R_T} \tag{4-9}$$

Example 4-11 Determine the total current in the circuit in Fig. 4-19.

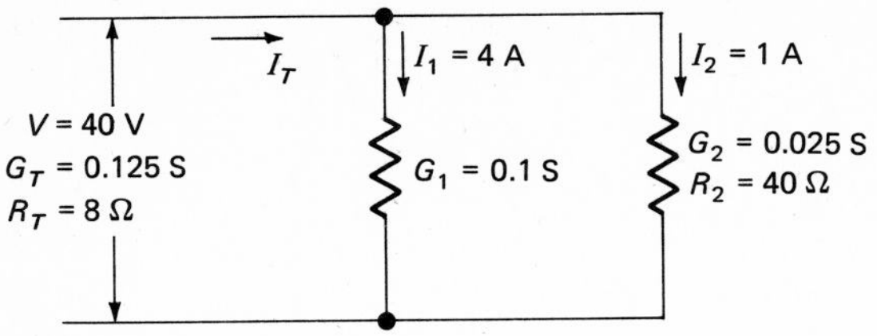

Fig. 4-19 Example 4-11: data.

Solution

METHOD 1 Kirchhoff's current law

$$I_T = I_1 + I_2 \tag{4-1}$$

$$I_T = 4\text{ A} + 1\text{ A} = \mathbf{5\ A}$$

METHOD 2 Ohm's law (conductance form)

$$I_T = VG_T = 40\text{ V} \times 0.125\text{ S} = \mathbf{5\ A}$$

METHOD 3 Conductance ratio

$$I_T = I_2 \frac{G_T}{G_2} \tag{4-8}$$

$$I_T = 1\text{ A}\,\frac{0.125\text{ S}}{0.025\text{ S}} = \mathbf{5\ A}$$

METHOD 4 Ohm's law (resistance form)

$$I_T = \frac{V}{R_T} = \frac{40\text{ V}}{8\ \Omega} = \mathbf{5\ A}$$

METHOD 5 Resistance ratio

$$I_T = I_2 \frac{R_2}{R_T} \tag{4-9}$$

$$I_T = 1\text{ A}\,\frac{40\ \Omega}{8\ \Omega} = \mathbf{5\ A}$$

Individual currents $I_1, I_2, \ldots, I_n$ The currents in each of the elements can be calculated by the application of Ohm's law:

$$I_1 = VG_1 \qquad \text{or} \qquad I_1 = \frac{V}{R_1}$$

Substituting I/G for V will give a conductance ratio form

$$I_1 = I_T \frac{G_1}{G_T} \qquad \text{or} \qquad I_2 \frac{G_1}{G_2} \qquad \text{or} \qquad \cdots \qquad \text{or} \qquad I_n \frac{G_1}{G_n} \tag{4-10}$$

Alternatively, resistance ratio forms can be obtained:

$$I_1 = I_T \frac{R_T}{R_1} \quad \text{or} \quad I_2 \frac{R_2}{R_1} \quad \text{or} \quad \cdots \quad \text{or} \quad I_n \frac{R_n}{R_1} \tag{4-11}$$

The other individual currents I_2 to I_n can be calculated by using similar techniques.

Example 4-12 What is the current I_1 in the circuit in Fig. 4-20?

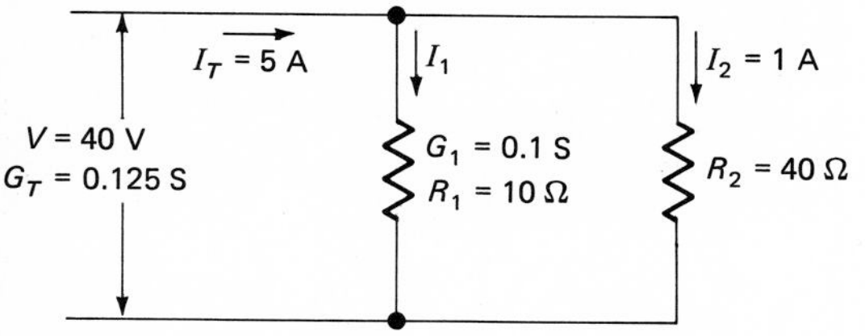

Fig. 4-20 Example 4-12: data.

Solution

METHOD 1 Ohm's law (conductance form)

$$I_1 = VG_1 = 40 \text{ V} \times 0.1 \text{ S} = \mathbf{4\ A}$$

METHOD 2 Conductance ratio

$$I_1 = I_T \frac{G_1}{G_T} \tag{4-10}$$

$$I_1 = 5 \text{ A} \frac{0.1 \text{ S}}{0.125 \text{ S}} = \mathbf{4\ A}$$

METHOD 3 Ohm's law (resistance form)

$$I_1 = \frac{V}{R_1} = \frac{40 \text{ V}}{10\ \Omega} = \mathbf{4\ A}$$

METHOD 4 Resistance ratio

$$I_1 = I_2 \frac{R_2}{R_1} \tag{4-11}$$

$$I_1 = 1 \text{ A} \frac{40\ \Omega}{10\ \Omega} = \mathbf{4\ A}$$

Total conductance G_T The equivalent conductance of the complete network can be obtained by summing the conductances of the individual elements

$$G_T = G_1 + G_2 + \cdots + G_n \tag{4-2}$$

or by applying Ohm's law to the entire network

$$G_T = \frac{I_T}{V}$$

G_T can be expressed in current ratio form by substituting appropriate ratios G/I for $1/V$ in the Ohm's law equations:

$$G_T = \frac{I_T}{V} = I_T \frac{1}{V} = I_T \frac{G_1}{I_1} = G_1 \frac{I_T}{I_1}$$

$$G_T = G_1 \frac{I_T}{I_1} \quad \text{or} \quad G_2 \frac{I_T}{I_2} \quad \text{or} \quad \cdots \quad \text{or} \quad G_n \frac{I_T}{I_n} \tag{4-12}$$

Example 4-13 Find the total conductance of the circuit shown in Fig. 4-21.

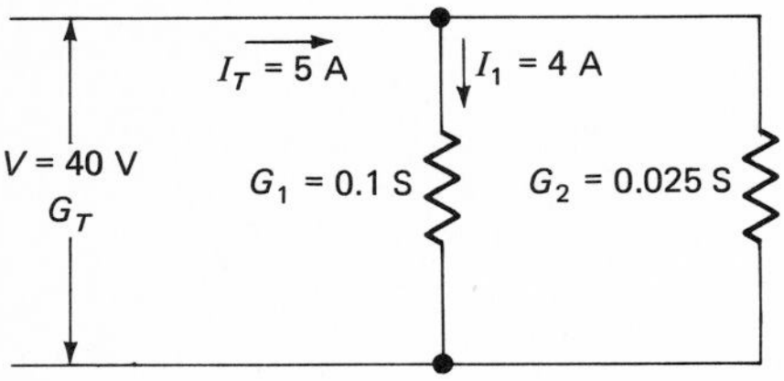

Fig. 4-21 Example 4-13: data.

Solution

METHOD 1 Sum of the conductances

$$G_T = G_1 + G_2 \tag{4-2}$$

$$G_T = 0.1 \text{ S} + 0.025 \text{ S} = \mathbf{0.125\ S}$$

METHOD 2 Ohm's law

$$G_T = \frac{I_T}{V} = \frac{5 \text{ A}}{40 \text{ V}} = \mathbf{0.125\ S}$$

METHOD 3 Current ratio

$$G_T = G_1 \frac{I_T}{I_1} \tag{4-12}$$

$$G_T = 0.1 \text{ S} \frac{5 \text{ A}}{4 \text{ A}} = \mathbf{0.125\ S}$$

Total resistance R_T The total equivalent resistance of a parallel circuit is best calculated by first computing the total conductance and then taking its reciprocal.

Thus the total resistance of the two-element, parallel network in Examples 4-10 to 4-13 will be

$$R_T = \frac{1}{G_T} = \frac{1}{0.125 \text{ S}} = 8\ \Omega$$

Individual conductances G_1, G_2, . . . , G_n The conductance of an individual element can be calculated by taking the reciprocal of its resistance

$$G_1 = \frac{1}{R_1}$$

or by the application of Ohm's law to the element

$$G_1 = \frac{I_1}{V}$$

or by extending the Ohm's law equation into a series of current ratio equations by substituting a suitable ratio G/I for $1/V$

$$G_1 = G_T \frac{I_1}{I_T} \quad \text{or} \quad G_2 \frac{I_1}{I_2} \quad \text{or} \quad \cdots \quad \text{or} \quad G_n \frac{I_1}{I_n} \tag{4-13}$$

Similar methods may be used to calculate the conductances G_2 to G_n.

Example 4-14 What is the value of conductance G_1 in the circuit in Fig. 4-22?

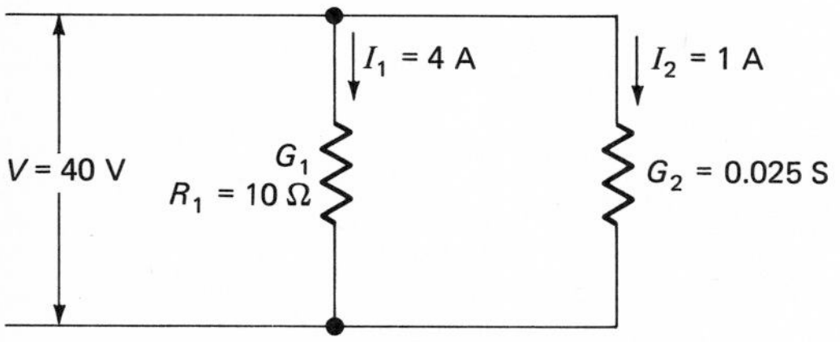

Fig. 4-22 Example 4-14: data.

Solution

METHOD 1 Resistance reciprocal

$$G_1 = \frac{1}{R_1} = \frac{1}{10\ \Omega} = \mathbf{0.1\ S}$$

METHOD 2 Ohm's law

$$G_1 = \frac{I_1}{V} = \frac{4\ \text{A}}{40\ \text{V}} = \mathbf{0.1\ S}$$

METHOD 3 Current ratio

$$G_1 = G_2 \frac{I_1}{I_2} \tag{4-13}$$

$$G_1 = 0.025\ \text{S} \frac{4\ \text{A}}{1\ \text{A}} = \mathbf{0.1\ S}$$

Individual resistances $R_1, R_2, \ldots, R_n$ The resistance of an individual element can be calculated by taking the reciprocal of its conductance

$$R_1 = \frac{1}{G_1}$$

or by the direct use of Ohm's law

$$R_1 = \frac{V}{I_1}$$

or by the indirect use of Ohm's law in the form of current ratios

$$R_1 = R_T \frac{I_T}{I_1} \quad \text{or} \quad R_2 \frac{I_2}{I_1} \quad \text{or} \quad \cdots \quad \text{or} \quad R_n \frac{I_n}{I_1} \tag{4-14}$$

The individual resistances R_2 to R_n can be similarly treated.

Example 4-15 What is the value of resistance R_1 in the circuit in Fig. 4-23?

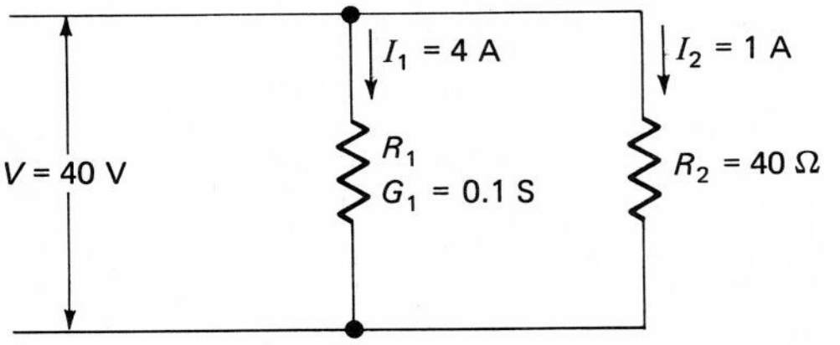

Fig. 4-23 Example 4-15: data.

Solution

METHOD 1 Conductance reciprocal

$$R_1 = \frac{1}{G_1} = \frac{1}{0.1\ \text{S}} = \mathbf{10\ \Omega}$$

METHOD 2 Ohm's law

$$R_1 = \frac{V}{I_1} = \frac{40\ \text{V}}{4\ \text{A}} = \mathbf{10\ \Omega}$$

METHOD 3 Current ratio

$$R_1 = R_2 \frac{I_2}{I_1} \tag{4-14}$$

$$R_1 = 40\ \Omega\ \frac{1\ \text{A}}{4\ \text{A}} = \mathbf{10\ \Omega}$$

Total power P_T The value of the total power in the circuit can be obtained by adding the individual powers

$$P_T = P_1 + P_2 + \cdots + P_n \tag{4-6}$$

or by applying the basic power relationship

$$P_T = VI_T$$

or by combining the basic equation with Ohm's law to give

$$P_T = I_T^2 R_T \qquad \text{or} \qquad \frac{V^2}{R_T}$$

The voltage in a parallel circuit is common, therefore, by deriving an expression for the voltage in any convenient manner and applying this expression to the last equation, power can be expressed in resistance ratio form.

$$P_1 = \frac{V^2}{R_1} \qquad \text{or} \qquad V^2 = P_1 R_1$$

Substituting $V^2 = P_1 R_1$ in the equation $P_T = V^2/R_T$,

$$P_T = \frac{P_1 R_1}{R_T} = P_1 \frac{R_1}{R_T}$$

The same procedure can be applied to any of the other elements, so that

$$P_T = P_1 \frac{R_1}{R_T} \qquad \text{or} \qquad P_2 \frac{R_2}{R_T} \qquad \text{or} \qquad \cdots \qquad \text{or} \qquad P_n \frac{R_n}{R_T} \tag{4-15}$$

Example 4-16 What is the total power dissipated by the network in Fig. 4-24?

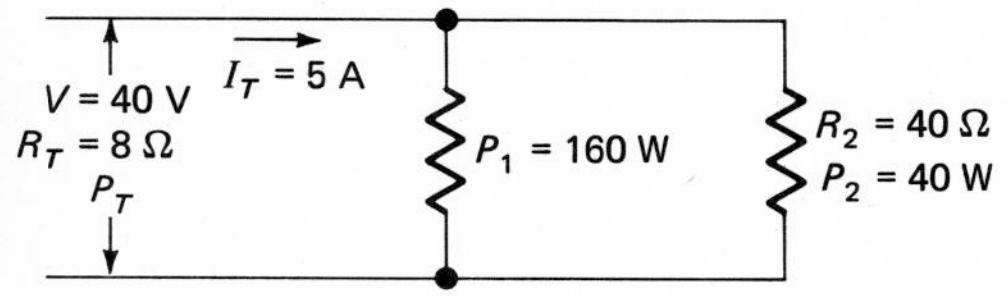

Fig. 4-24 Example 4-16: data.

Solution

METHOD 1 Sum of the powers

$$P_T = P_1 + P_2 \tag{4-6}$$

$$P_T = 160 \text{ W} + 40 \text{ W} = \mathbf{200 \text{ W}}$$

METHOD 2 Current and voltage

$$P_T = VI_T = 40 \text{ V} \times 5 \text{ A} = \mathbf{200 \text{ W}}$$

METHOD 3 Current and resistance

$$P_T = I_T^2 R_T = (5\ \text{A})^2 \times 8\ \Omega = \mathbf{200\ W}$$

METHOD 4 Voltage and resistance

$$P_T = \frac{V^2}{R_T} = \frac{(40\ \text{V})^2}{8\ \Omega} = \mathbf{200\ W}$$

METHOD 5 Resistance ratio

$$P_T = P_2 \frac{R_2}{R_T} \tag{4-15}$$

$$P_T = 40\ \text{W}\ \frac{40\ \Omega}{8\ \Omega} = \mathbf{200\ W}$$

Individual powers P_1, P_2, . . . , P_n The methods demonstrated in Example 4-16 for the total power can be used to obtain the power dissipated by individual elements. The following relationships are for P_1, the power in element R_1, but apply equally well to the other elements R_2 to R_n.

$$P_1 = VI_1 \quad \text{or} \quad I_1^2 R_1 \quad \text{or} \quad \frac{V^2}{R_1}$$

$$\text{or} \quad P_1 = P_T \frac{R_T}{R_1} \quad \text{or} \quad P_2 \frac{R_2}{R_1} \quad \text{or} \quad \cdots \quad \text{or} \quad P_n \frac{R_n}{R_1} \tag{4-16}$$

Example 4-17 What power is dissipated in element R_1 in Fig. 4-25?

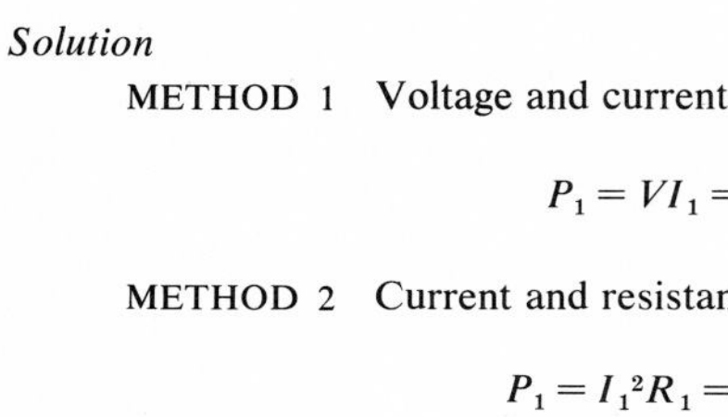

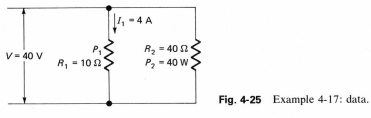

Fig. 4-25 Example 4-17: data.

Solution

METHOD 1 Voltage and current

$$P_1 = VI_1 = 40\ \text{V} \times 4\ \text{A} = \mathbf{160\ W}$$

METHOD 2 Current and resistance

$$P_1 = I_1^2 R_1 = (4\ \text{A})^2 \times 10\ \Omega = \mathbf{160\ W}$$

METHOD 3 Voltage and resistance

$$P_1 = \frac{V^2}{R_1} = \frac{(40\ \text{V})^2}{10\ \Omega} = \mathbf{160\ W}$$

METHOD 4 Resistance ratio

$$P_1 = P_2 \frac{R_2}{R_1} \qquad (4\text{-}16)$$

$$P_1 = 40 \text{ W} \frac{40\ \Omega}{10\ \Omega} = \mathbf{160\ W}$$

Energy Energy values can be calculated by using any of the methods in Examples 4-16 and 4-17 to obtain the appropriate power, and then multiplying by the time, which for any situation will be common to the whole circuit.

RATIOS IN PARALLEL CIRCUITS

When you have acquired a thorough understanding of the relevance of equating the ratios of various values in parallel circuits, much calculating time and effort can be avoided, and your familiarity with circuit behavior will be enhanced.

The ratio concept is developed as follows:

$$\frac{I_T}{I_1} = \frac{VG_T}{VG_1} = \frac{G_T}{G_1}$$

$$\frac{P_T}{P_1} = \frac{VI_T}{VI_1} = \frac{I_T}{I_1}$$

$$\frac{W_T}{W_1} = \frac{P_T t}{P_1 t} = \frac{P_T}{P_1}$$

These ratios can be coupled with the inverse ratio

$$\frac{G_T}{G_1} = \frac{R_1}{R_T}$$

to give the result

$$\frac{G_T}{G_1} = \frac{I_T}{I_1} = \frac{P_T}{P_1} = \frac{W_T}{W_1} = \frac{R_1}{R_T} \qquad (4\text{-}17)$$

Instead of G_1, G_T, etc., the values for any of the elements or the whole network can be used with equal validity.

Example 4-18 Using the ratio G_T/G_1, determine (*a*) the total current, (*b*) the total power, (*c*) the total energy, and (*d*) the total resistance associated with the circuit in Fig. 4-26.

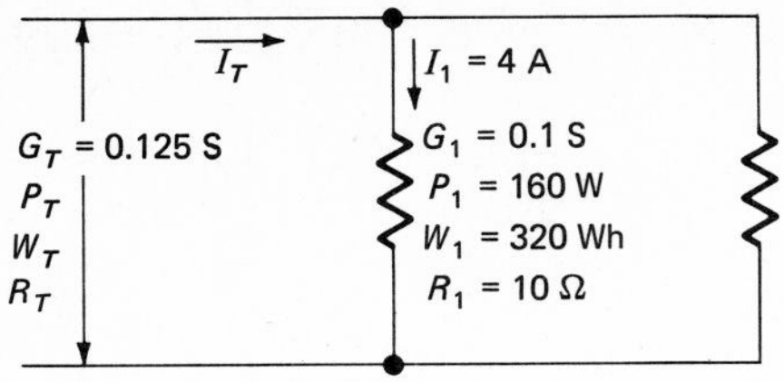

Fig. 4-26 Example 4-18: data.

Solution The ratio G_T/G_1 is given by

$$\frac{G_T}{G_1} = \frac{0.125\ \text{S}}{0.1\ \text{S}} = 1.25$$

This value is shown in parentheses in the following solutions.

a. $$I_T = (1.25)I_1 = (1.25)4\ \text{A} = \mathbf{5\ A}$$

b. $$P_T = (1.25)P_1 = (1.25)160\ \text{W} = \mathbf{200\ W}$$

c. $$W_T = (1.25)W_1 = (1.25)320\ \text{Wh} = \mathbf{400\ Wh}$$

d. $$R_T = \frac{1}{(1.25)}R_1 = \frac{1}{(1.25)}10\ \Omega = \mathbf{8\ \Omega}$$

Two elements in parallel It has been explained why resistance values are used in practice more frequently than conductance values. It has also been made clear that the calculation of resistance values in parallel circuits is a cumbersome procedure. These situations can be made compatible by the use of a simple formula for calculating the total equivalent resistance of a two-element parallel network. (See Fig. 4-27.)

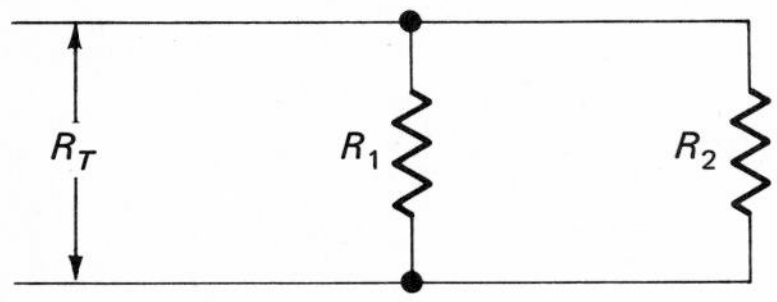

Fig. 4-27 Two parallel elements.

$$\frac{1}{R_T} = \frac{1}{R_1} + \frac{1}{R_2} \tag{4-3}$$

$$\frac{1}{R_T} = \frac{R_2 + R_1}{R_1 \times R_2}$$

$$R_T = \frac{R_1 \times R_2}{R_1 + R_2} \tag{4-18}$$

From Eq. (4-18), it can be seen that the total equivalent resistance of two resistances connected in parallel is their product divided by their sum.

In practical circuits it is more usual to find just two elements in parallel than a larger number. So, for this reason alone, Eq. (4-18) can be useful. But it is possible to use this equation for parallel networks with more than two elements by successively treating them as pairs. The simplicity of the equation, requiring two slide-rule operations (one multiplication and one division) and one addition, makes it quite fast for repetitive use.

Examples 4-19 to 4-21 show how Eq. (4-18) can be used. The resistances selected for these examples are preferred values. This means that they are resistors in the series of values actually manufactured for industrial use, and therefore are representative of a truly practical situation, even though the numbers involved may appear peculiar.

Example 4-19 Two resistors, 22 and 47 Ω, are connected in parallel. What is the total equivalent resistance of the network?
Solution (Fig. 4-28)

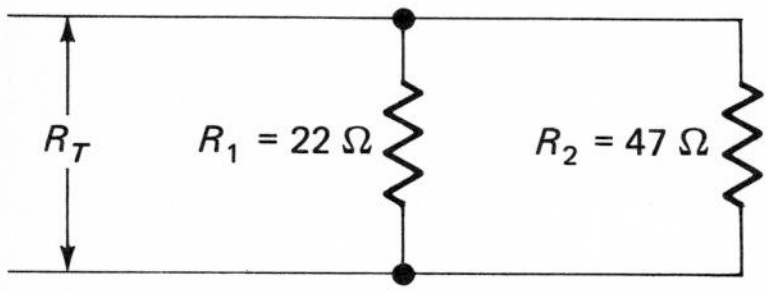

Fig. 4-28 Example 4-19: data.

$$R_T = \frac{R_1 R_2}{R_1 + R_2} \tag{4-18}$$

$$R_T = \frac{22\ \Omega \times 47\ \Omega}{22\ \Omega + 47\ \Omega} = \frac{1034\ \Omega^2}{69\ \Omega} \simeq \mathbf{15\ \Omega}$$

Example 4-20 Three resistors, 56, 100, and 27 kΩ, are connected in parallel. What is the equivalent resistance of the network?
Solution (Fig. 4-29)

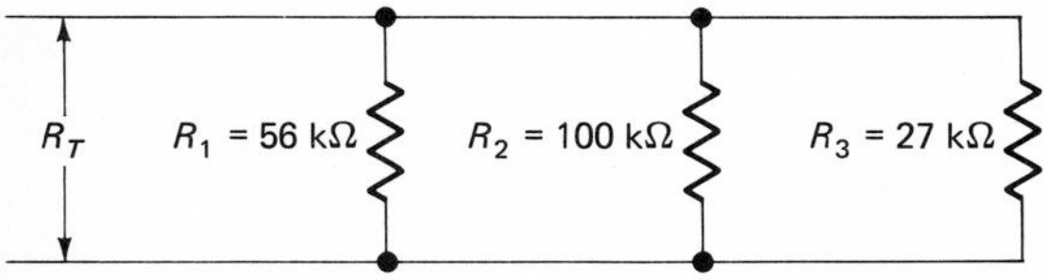

Fig. 4-29 Example 4-20: original data.

Taking R_1 and R_2 as a pair,

$$R_{TA} = \frac{R_1 R_2}{R_1 + R_2} \tag{4-18}$$

$$R_{TA} = \frac{56\ \text{k}\Omega \times 100\ \text{k}\Omega}{56\ \text{k}\Omega + 100\ \text{k}\Omega} = \frac{5600\ (\text{k}\Omega)^2}{156\ \text{k}\Omega} = 35.9\ \text{k}\Omega$$

The network is now reduced to that shown in Fig. 4-30

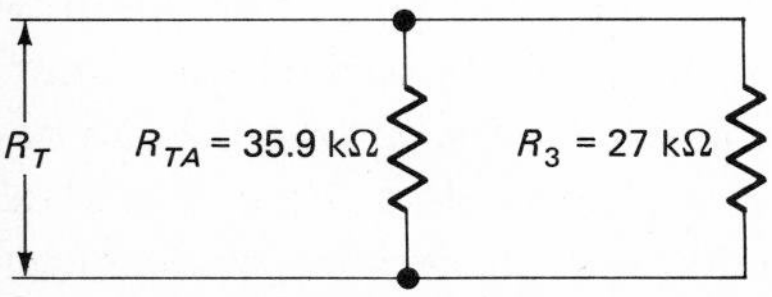

Fig. 4-30 Example 4-20: R_1 and R_2 combined as R_{TA}.

$$R_T = \frac{R_{TA} R_3}{R_{TA} + R_3} \tag{4-18}$$

$$R_T = \frac{35.9 \text{ k}\Omega \times 27 \text{ k}\Omega}{35.9 \text{ k}\Omega + 27 \text{ k}\Omega} = \frac{970 \text{ (k}\Omega)^2}{62.9 \text{ k}\Omega} = \mathbf{15.41 \text{ k}\Omega}$$

Example 4-21 What is the equivalent resistance of a network consisting of four resistors, 12, 15, 18, and 33 Ω, connected in parallel?
Solution (Fig. 4-31)

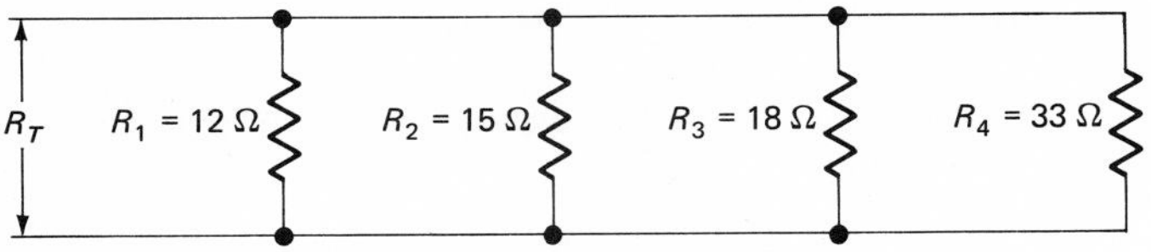

Fig. 4-31 Example 4-21: original data.

Taking R_1 and R_2 as a pair,

$$R_{TA} = \frac{R_1 R_2}{R_1 + R_2} \tag{4-18}$$

$$R_{TA} = \frac{12\ \Omega \times 15\ \Omega}{12\ \Omega + 15\ \Omega} = \frac{180\ \Omega^2}{27\ \Omega} = 6.667\ \Omega$$

Taking R_3 and R_4 as a pair,

$$R_{TB} = \frac{R_3 R_4}{R_3 + R_4} \tag{4-18}$$

$$R_{TB} = \frac{18\ \Omega \times 33\ \Omega}{18\ \Omega + 33\ \Omega} = \frac{594\ \Omega^2}{51\ \Omega} = 11.65\ \Omega$$

The network reduces to that shown in Fig. 4-32

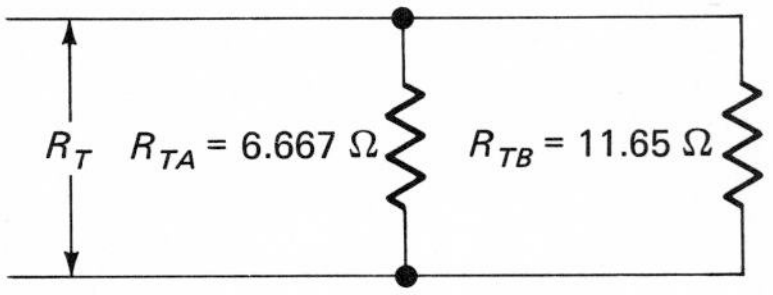

Fig. 4-32 Example 4-21: R_1 and R_2 combined as R_{TA}; R_3 and R_4 combined as R_{TB}.

$$R_T = \frac{R_{TA} R_{TB}}{R_{TA} + R_{TB}} \tag{4-18}$$

$$R_T = \frac{6.667\ \Omega \times 11.65\ \Omega}{6.667\ \Omega + 11.65\ \Omega} = \frac{77.67\ \Omega^2}{18.32\ \Omega} = \mathbf{4.24\ \Omega}$$

The formula for two parallel elements can be arranged in such a manner that it directly gives the value of one of the resistances if the other resistance and the total resistance are both known. This knowledge can be useful in practical situations when it is necessary to find the value of a resistance which will in effect reduce the value of an existing resistance.

Referring to Fig. 4-33,

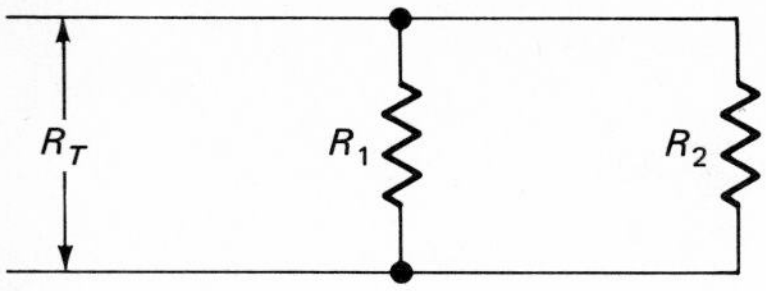

Fig. 4-33 Finding the value of one of a parallel resistance pair.

$$\frac{1}{R_T} = \frac{1}{R_1} + \frac{1}{R_2} \tag{4-3}$$

$$\frac{1}{R_1} = \frac{1}{R_T} - \frac{1}{R_2} = \frac{R_2 - R_T}{R_2 \times R_T}$$

$$R_1 = \frac{R_2 \times R_T}{R_2 - R_T} \tag{4-19}$$

The value of one resistance of a parallel pair is the product of the other resistance and the total resistance divided by the difference between them.

It should be noted that the total resistance of a parallel network is always less than any of its individual resistances (because the conductances are additive). Therefore the total resistance is subtracted from the individual resistance to obtain the difference in Eq. (4-19).

Example 4-22 What resistance should be connected in parallel with a 500-Ω resistor to reduce its effective value to 100 Ω.

Solution (Fig. 4-34)

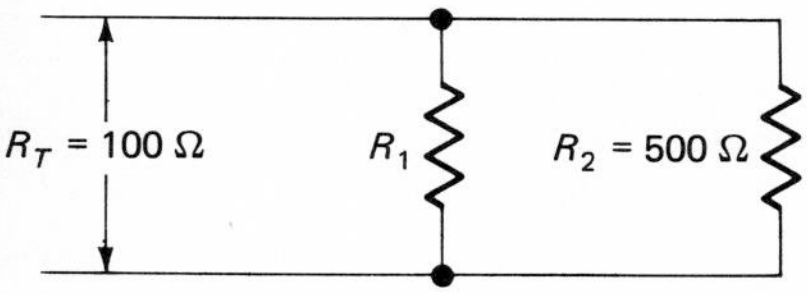

Fig. 4-34 Example 4-22: data.

$$R_1 = \frac{R_2 R_T}{R_2 - R_T} \tag{4-19}$$

$$R_1 = \frac{500\ \Omega \times 100\ \Omega}{500\ \Omega - 100\ \Omega} = \frac{5 \times 10^4\ \Omega^2}{400\ \Omega} = \mathbf{125\ \Omega}$$

Use of reciprocals The reciprocal of a quantity can be determined by the use of the CI or CIF scales on a slide rule or by a table of reciprocals, such as is found in most books of mathematical tables. Thus, if the resistances of the elements of a parallel network are known, the total resistance can be calculated with comparative ease. This resistance reciprocal technique can be applied even if the resistance values do not lend themselves to easy addition of their reciprocals expressed as fractions.

Example 4-23 Calculate the total equivalent resistance of the parallel network in Example 4-21, using Eq. (4-3).

Solution (Fig. 4-35)

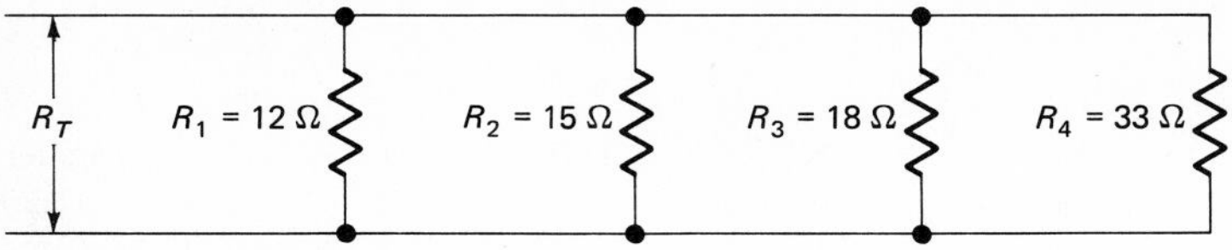

Fig. 4-35 Example 4-23: data.

Using the slide rule to calculate reciprocals,

$$\text{Reciprocal of } R_1 = \frac{1}{R_1} = 0.0833$$

$$\text{Reciprocal of } R_2 = \frac{1}{R_2} = 0.0667$$

$$\text{Reciprocal of } R_3 = \frac{1}{R_3} = 0.0556$$

$$\text{Reciprocal of } R_4 = \frac{1}{R_4} = 0.0303$$

$$\frac{1}{R_T} = \frac{1}{R_1} + \frac{1}{R_2} + \frac{1}{R_3} + \frac{1}{R_4} \tag{4-3}$$

$$\frac{1}{R_T} = 0.0833 + 0.0667 + 0.0556 + 0.0303 = 0.2359$$

$$R_T = \text{reciprocal of } 0.2359$$

$$R_T = \mathbf{4.24\ \Omega}$$

This answer checks with that for Example 4-21. It is left to your discretion which method you prefer when you need to make similar calculations.

4-4 APPLICATIONS OF PARALLEL CIRCUITS

CURRENT DIVIDER

The current entering a parallel circuit divides so that part of the current flows through each of the elements (Fig. 4-36). High-conductance elements will have a greater share of the total current than will those of low conductance.

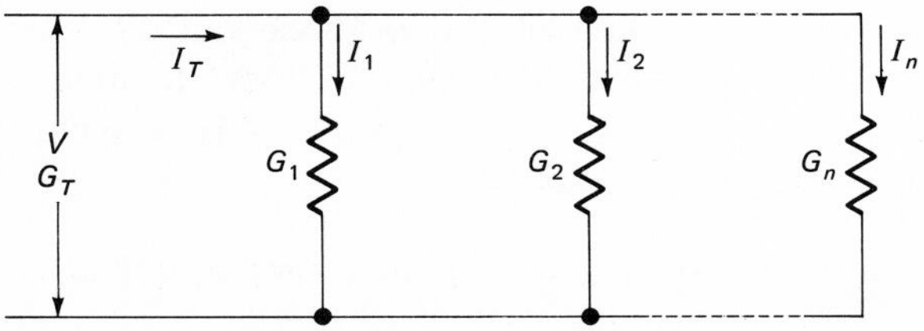

Fig. 4-36 Current division in a parallel network.

$$V = \frac{I_T}{G_T}$$

$$I_1 = VG_1 = \frac{I_T}{G_T} G_1$$

$$I_1 = I_T \frac{G_1}{G_T} \quad \text{or} \quad I_T \frac{G_1}{G_1 + G_2 + \cdots + G_n} \tag{4-20}$$

The result will be similar for any other current I_2 to I_n.

Example 4-24 Three resistors, 10, 25, and 50 Ω, are connected in parallel. If the total network current is 4 A, what is the current in each element?
Solution (Fig. 4-37)

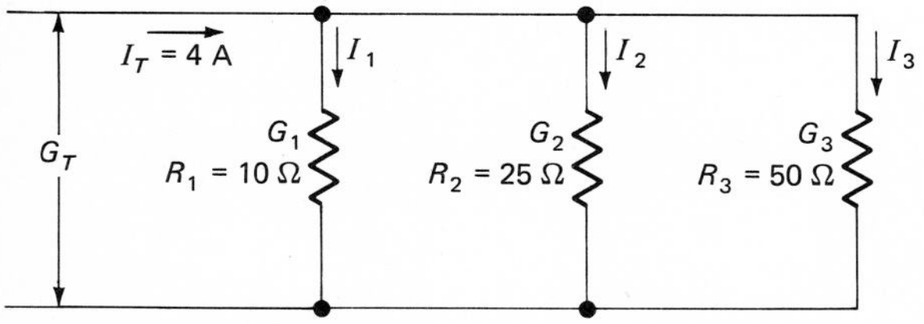

Fig. 4-37 Example 4-24: data.

$$G_1 = \frac{1}{R_1} = \frac{1}{10\ \Omega} = 0.1\ \text{S} \qquad G_2 = \frac{1}{R_2} = \frac{1}{25\ \Omega} = 0.04\ \text{S} \qquad G_3 = \frac{1}{R_3} = \frac{1}{50\ \Omega} = 0.02\ \text{S}$$

$$G_T = G_1 + G_2 + G_3 \tag{4-2}$$

$$G_T = 0.1\ \text{S} + 0.04\ \text{S} + 0.02\ \text{S} = 0.16\ \text{S}$$

$$I_1 = I_T \frac{G_1}{G_T} \tag{4-20}$$

$$I_1 = 4\ \text{A} \frac{0.1\ \text{S}}{0.16\ \text{S}} = \mathbf{2.5\ A}$$

$$I_2 = I_T \frac{G_2}{G_T} = 4\ \text{A} \frac{0.04\ \text{S}}{0.16\ \text{S}} = \mathbf{1\ A}$$

$$I_3 = I_T \frac{G_3}{G_T} = 4\ \text{A} \frac{0.02\ \text{S}}{0.16\ \text{S}} = \mathbf{0.5\ A}$$

Two-element parallel networks, with values given as resistances, are commonly met in practical circuits. A simple formula can be derived for calculating the division of current between the elements. (See Fig. 4-38.)

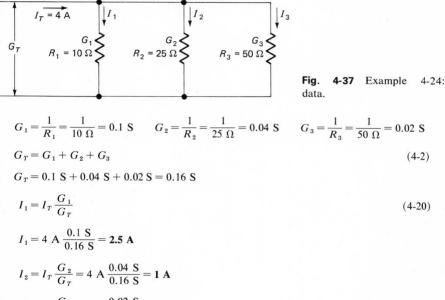

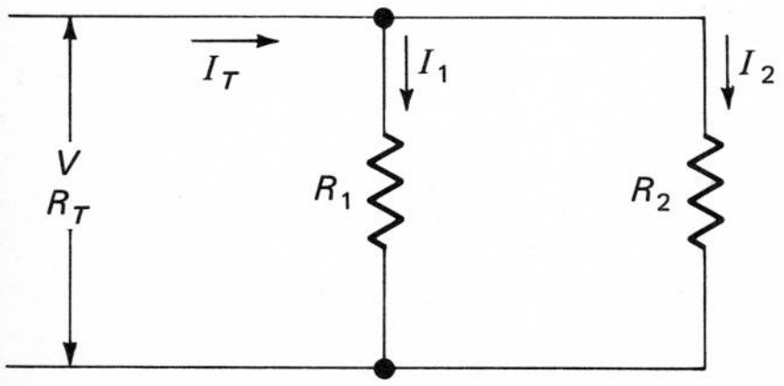

Fig. 4-38 Current division in a two-element parallel network.

From Eq. (4-18),

$$V = I_T R_T = I_T \frac{R_1 R_2}{R_1 + R_2}$$

$$I_1 = \frac{V}{R_1} = I_T \frac{R_1 R_2}{R_1 + R_2} \times \frac{1}{R_1}$$

$$I_1 = I_T \frac{R_2}{R_1 + R_2} \tag{4-21}$$

The formula for I_2 will be similar $[I_T \times R_1/(R_1 + R_2)]$.

In a two-element parallel network, the current in one element is the total current multiplied by the resistance of the other element divided by the sum of the two resistances. With more than two elements this rule does not apply.

Example 4-25 A two-element parallel network has element values of 15 and 47 Ω. If the total network current is 5 A, what is the current in each of the elements?
Solution (Fig. 4-39)

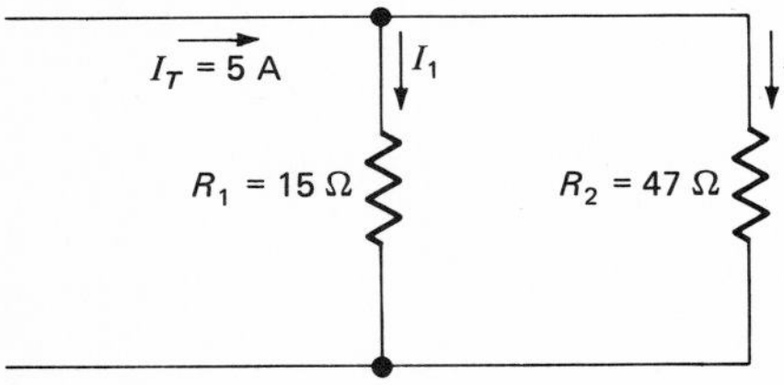

Fig. 4-39 Example 4-25: data.

$$I_1 = I_T \frac{R_2}{R_1 + R_2} \tag{4-21}$$

$$I_1 = 5\text{ A}\,\frac{47\ \Omega}{15\ \Omega + 47\ \Omega} = 5\text{ A}\,\frac{47\ \Omega}{62\ \Omega} = \mathbf{3.79\ A}$$

$$I_2 = I_T \frac{R_1}{R_1 + R_2} \tag{4-21}$$

$$I_2 = 5\text{ A}\,\frac{15\ \Omega}{15\ \Omega + 47\ \Omega} = 5\text{ A}\,\frac{15\ \Omega}{62\ \Omega} = \mathbf{1.21\ A}$$

Example 4-26 A small electric motor has a variable shunt of 5 to 100 Ω connected across it. If the motor has a constant resistance of 10 Ω, and the network is fed from a 1-A constant-current source, what is the range of current in the motor?
Solution (Fig. 4-40)

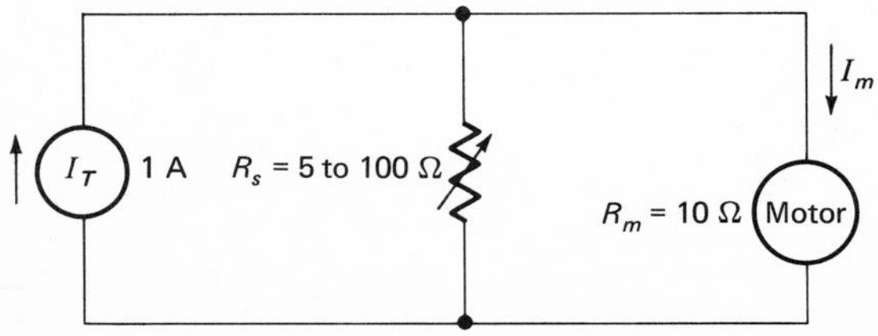

Fig. 4-40 Example 4-26: circuit.

When R_s is at its minimum value of 5 Ω,

$$I_m = I_T \frac{R_s}{R_s + R_m} \tag{4-21}$$

$$I_m = 1\ \text{A} \frac{5\ \Omega}{5\ \Omega + 10\ \Omega} = 1\ \text{A} \frac{5\ \Omega}{15\ \Omega} = \mathbf{0.3333\ A}$$

When R_s is at its maximum value of 100 Ω,

$$I_m = I_T \frac{R_s}{R_s + R_m} \tag{4-21}$$

$$I_m = 1\ \text{A} \frac{100\ \Omega}{100\ \Omega + 10\ \Omega} = 1\ \text{A} \frac{100\ \Omega}{110\ \Omega} = \mathbf{0.9091\ A}$$

The motor current range in Example 4-26 is 0.3333 to 0.9091 A. This means that, as the shunt resistance is varied from $\frac{1}{2}$ the motor resistance to 10 times its resistance, the motor current varies from $\frac{1}{3}$ of the supply current to almost the whole of it. Therefore, when a motor (or any other device) is fed from a constant-current source, its current can be varied by shunting it with a variable resistor, or rheostat. When the lowest resistance value of the rheostat is very much lower than the device resistance, the device current will be low. Conversely, when the highest value of the rheostat greatly exceeds that of the device, the device current will be almost equal to that of the supply.

Example 4-26 gives, in simplified form, a type of circuit often used in industrial control. The constant-current source is typically obtained from a transistor circuit. The shunt resistance is also likely to be a transistor, the resistance of which can be varied by a control voltage.

INTERNAL CONDUCTANCE OF A CURRENT GENERATOR

Current generators are frequently used in electronic circuits. In its ideal state a voltage generator is typified as supplying a constant voltage regardless of the nature of the load. An ideal current generator always supplies a constant current to a load. We saw in Sec. 3-5 that a practical voltage generator does not in fact produce a constant voltage, because its internal resistance reduces its output voltage (terminal pd) by an amount which depends on the load current. Similarly, because of its internal conductance, a practical current generator does not actually supply a constant current to its load.

In Fig. 4-41 is shown a current generator connected to its load. The ideal current source supplies a constant current I to a two-element parallel network consisting of the generator internal conductance G_i and the load G_L. The source current divides between G_i and G_L. The load current can be determined by using the current divider principle:

$$I_L = I \frac{G_L}{G_L + G_i} \tag{4-20}$$

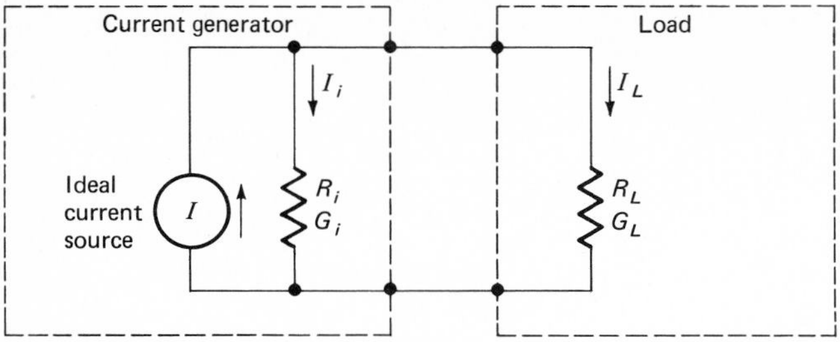

Fig. 4-41 Current generator and load.

or, in resistance form,

$$I_L = I \frac{R_i}{R_L + R_i} \tag{4-21}$$

Kirchhoff's current law can be applied to give another statement of the conditions in a loaded practical current generator:

$$I = I_i + I_L \qquad \text{or} \qquad I_L = I - I_i \tag{4-22}$$

You will see that the current supplied by a practical current generator is a function of its internal conductance (or resistance).

Example 4-27 A current generator has an ideal output of 5 A. If its internal conductance is 0.1 S, what current will it supply to a load of 0.15 S?
Solution (Fig. 4-42)

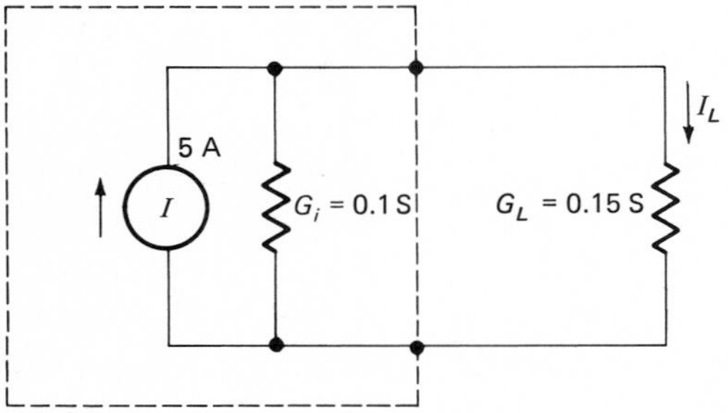

Fig. 4-42 Example 4-27: circuit.

$$I_L = I \frac{G_L}{G_L + G_i} \tag{4-20}$$

$$I_L = 5\text{ A} \frac{0.15\text{ S}}{0.15\text{ S} + 0.1\text{ S}} = 5\text{ A} \frac{0.15\text{ S}}{0.25\text{ S}} = \mathbf{3\text{ A}}$$

Example 4-28 Referring to Example 4-27, what would be the current in the load if its conductance were 10 S?

Solution (Fig. 4-43)

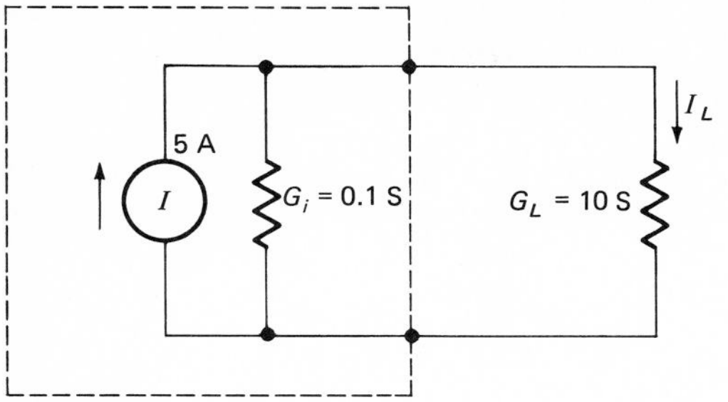

Fig. 4-43 Example 4-28: circuit.

$$I_L = I \frac{G_L}{G_L + G_i} \tag{4-20}$$

$$I_L = 5\text{ A} \frac{10\text{ S}}{10\text{ S} + 0.1\text{ S}} = 5\text{ A} \frac{10\text{ S}}{10.1\text{ S}} = \mathbf{4.95\text{ A}}$$

From Example 4-28, it can be seen that, if the load conductance is high compared to the internal conductance of the generator, the load current will almost equal the current supplied by an ideal current generator. If the load conductance were infinitely high, the load current would literally equal the source current.

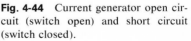

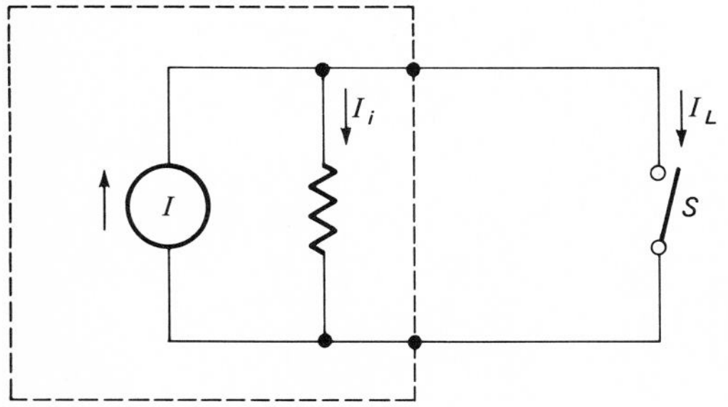

Fig. 4-44 Current generator open circuit (switch open) and short circuit (switch closed).

In Fig. 4-44 the switch S, when open, removes all loading from the current generator. The open-circuit load condition now occurs, with all the source current existing in the internal conductance:

$$I_L = 0$$

$$I_i = I$$

When the switch is closed, a solid connection is formed between the generator terminals. No load element is shown, so it is implied that the load resistance is zero, the load conductance being infinite. This zero load resistance is the short-circuit[1] load condition, all the source current now being in the load.

$$I_L = I$$
$$I_i = 0$$

Example 4-29 A current generator has a variable resistor connected across its terminals, its resistance value ranging from 0 to 10 Ω. If the generator source current is 10 A and its internal resistance is 10 mΩ, what is the load current when the load resistance is (*a*) 0 Ω, (*b*) 1 Ω, and (*c*) 10 Ω?
Solution (Fig. 4-45)

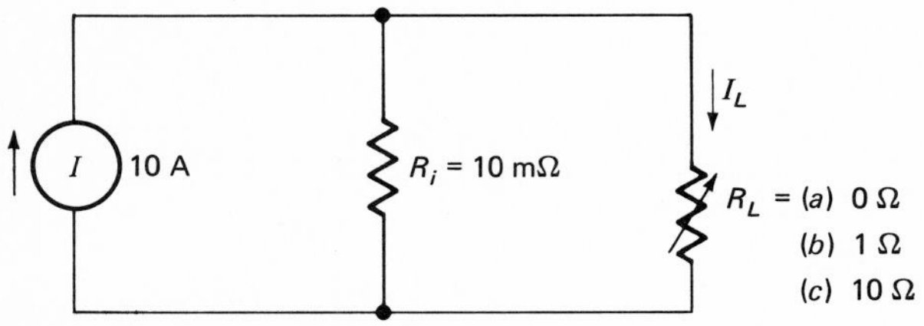

Fig. 4-45 Example 4-29: circuit.

a. When $R_L = 0\ \Omega$,

$$I_L = I \frac{R_i}{R_i + R_L} \tag{4-21}$$

$$I_L = 10\text{ A} \frac{10\text{ m}\Omega}{10\text{ m}\Omega + 0\ \Omega} = \mathbf{10\ A}$$

b. When $R_L = 1\ \Omega$,

$$I_L = I \frac{R_i}{R_i + R_L} \tag{4-21}$$

$$I_L = 10\text{ A} \frac{10\text{ m}\Omega}{10\text{ m}\Omega + 1\ \Omega} = 10\text{ A} \frac{10\text{ m}\Omega}{1.01\ \Omega} = \mathbf{0.099\ A}$$

c. When $R_L = 10\ \Omega$,

$$I_L = I \frac{R_i}{R_i + R_L} \tag{4-21}$$

$$I_L = 10\text{ A} \frac{10\text{ m}\Omega}{10\text{ m}\Omega + 10\ \Omega} = 10\text{ A} \frac{10\text{ m}\Omega}{10.01\ \Omega} = \mathbf{0.01\ A}$$

[1] The term "short circuit" is often misused to imply a faulty condition. Actually it is quite specific and does not mean that there is something wrong. When a short circuit exists between two points in a circuit, the resistance between them is 0 Ω. A short circuit occurs in practice because of a direct connection by a conductor, such as a wire or a metal plate.

As the load resistance varies from a value of 0 Ω until it becomes high relative to the generator internal resistance (in Example 4-29 it becomes 10 Ω, which is 1000 times the internal resistance), the load current decreases from an initial value equal to the basic source current down to a value which is by comparison very low.

Example 4-30 A certain transistor behaves as a current generator with a source current of 10 mA and an internal resistance of 1 kΩ. What load current will exist if the load resistance is (*a*) 100 Ω, (*b*) 1 kΩ, and (*c*) 4 kΩ?
Solution (Fig. 4-46)

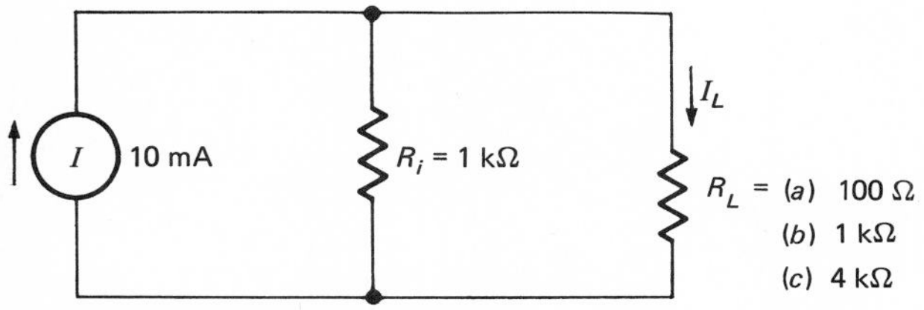

Fig. 4-46 Example 4-30: circuit.

a. When $R_L = 100\ \Omega\ (0.1\ \text{k}\Omega)$,

$$I_L = I \frac{R_i}{R_i + R_L} \qquad (4\text{-}21)$$

$$I_L = 10\ \text{mA} \frac{1\ \text{k}\Omega}{1\ \text{k}\Omega + 0.1\ \text{k}\Omega} = 10\ \text{mA} \frac{1\ \text{k}\Omega}{1.1\ \text{k}\Omega} = \mathbf{9.091\ mA}$$

b. When $R_L = 1\ \text{k}\Omega$,

$$I_L = I \frac{R_i}{R_i + R_L} \qquad (4\text{-}21)$$

$$I_L = 10\ \text{mA} \frac{1\ \text{k}\Omega}{1\ \text{k}\Omega + 1\ \text{k}\Omega} = 10\ \text{mA} \frac{1\ \text{k}\Omega}{2\ \text{k}\Omega} = \mathbf{5\ mA}$$

c. When $R_L = 4\ \text{k}\Omega$,

$$I_L = I \frac{R_i}{R_i + R_L} \qquad (4\text{-}21)$$

$$I_L = 10\ \text{mA} \frac{1\ \text{k}\Omega}{1\ \text{k}\Omega + 4\ \text{k}\Omega} = 10\ \text{mA} \frac{1\ \text{k}\Omega}{5\ \text{k}\Omega} = \mathbf{2\ mA}$$

COMPARISON OF VOLTAGE AND CURRENT GENERATORS

Much has been said about voltage generators and current generators, but what are they in practice? Is a battery a voltage generator or a current generator? What about an automobile generator; what type is it, voltage or current? All real generators are both voltage and current generators. The distinction is a matter of viewpoint; either view can be chosen to suit specific needs.

A discussion of the interchangeability of voltage and current generators was postponed until now, because the principles that have been developed for these generators are needed to show how to perform the conversion from one to the other. The need to do this will arise when you work with different circuit types and analysis methods.

Schematic diagrams for a voltage and a current generator are shown in Fig. 4-47*a* and *b*, respectively. Resistances are shown in both circuits to simplify comparison between them.

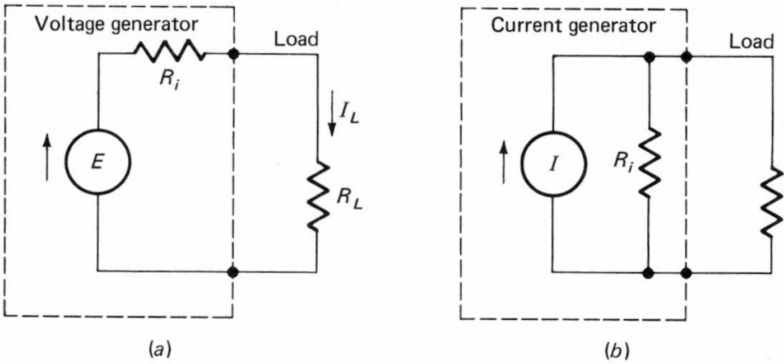

Fig. 4-47 (*a*) Voltage generator; (*b*) current generator.

If the internal resistance is made the same for each generator:
Voltage generator (Fig. 4-47*a*):

$$I_L = \frac{E}{R_i + R_L} \tag{4-23}$$

Current generator (Fig. 4-47*b*):

$$I_L = I \frac{R_i}{R_i + R_L} \qquad \text{(4-24)*}$$

For the generators to be interchangeable, they must be equivalent from the load current viewpoint (if their internal resistances are the same and they give the same current to the same load resistance, they are equivalent).

The expressions for the load currents from Eqs. (4-23) and (4-24) are equated

$$\frac{E}{R_i + R_L} = I \frac{R_i}{R_i + R_L}$$

$$E = IR_i \qquad \text{or} \qquad I = \frac{E}{R_i}$$

* Equation (4-24) is specific to a current-generator-driven load. The same equation appears previously as an application of the two-element current divider [Eq. (4-21)].

Thus, if the emf of the voltage generator equals the product of the source current and internal resistance of the current generator, and their internal resistances are equal, the two generators will be equivalent.

Alternatively, if the source current of the current generator equals the quotient of the emf and internal resistance of the voltage generator, and their internal resistances are equal, the two generators will be equivalent.

GENERATOR LOADING

The loading on a generator is said to be increased when its load current is increased. Therefore a high-conductance load results in a high, or heavy, loading.

Put in resistance terms, a low-resistance load means a heavy loading. This last viewpoint is important, since there seems to be a tendency to regard an increase in load resistance as an increase in loading, the opposite of what is actually true.

When loading is increased, the load voltage goes down for either voltage or current generators.

Voltage generator (Fig. 4-48*a*):

$$V_L = E\frac{R_L}{R_i + R_L} = \frac{E}{R_i/R_L + 1}$$

$$V_L = \frac{E}{R_iG_L + 1} \qquad (4\text{-}25)$$

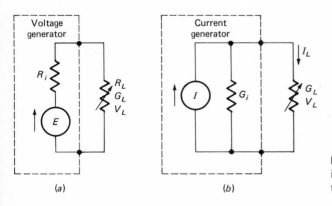

Fig. 4-48 Generator loading. (*a*) Voltage generator; (*b*) current generator.

Current generator (Fig. 4-48*b*):

$$V_L = \frac{I_L}{G_L} = I\frac{G_L}{G_i + G_L} \times \frac{1}{G_L}$$

$$V_L = \frac{I}{G_i + G_L} \qquad (4\text{-}26)$$

From Eqs. (4-25) and (4-26) it can be seen that, if the load conductance G_L is increased, the load voltage will decrease.

AMMETER SHUNT

A moving coil meter movement, such as was referred to in Sec. 3-5 when voltmeter multipliers were discussed, is a current-operated device. As such, it is an ammeter. A typical basic movement, however, is of limited use as an ammeter because the range of current it can handle is too small. The movement which forms the basis of many commercial meters is driven to full-scale deflection with a current of 50 μA. To be able to measure a current in excess of this small value, the movement needs to be combined with a parallel resistor, called a *shunt*.

Figure 4-49 is a schematic diagram of a basic meter movement. The symbol in Fig. 4-49*a* is a representation of a meter face, showing the pointer; this symbol will be used in the ammeter diagrams. The electric circuit equivalent of the movement is the resistance in Fig. 4-49*b*. It is common to specify the full-scale deflection current I_m and the resistance of the movement R_m; its full-scale voltage V_m can be calculated by using Ohm's law.

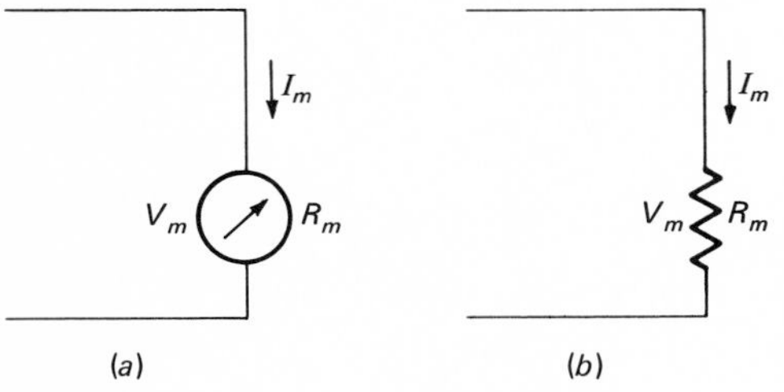

Fig. 4-49 Meter movement as circuit element. (*a*) General meter symbol; (*b*) equivalent resistance.

Figure 4-50 is a schematic diagram of an ammeter. It consists of a meter movement with a shunt R_s connected in parallel with it.

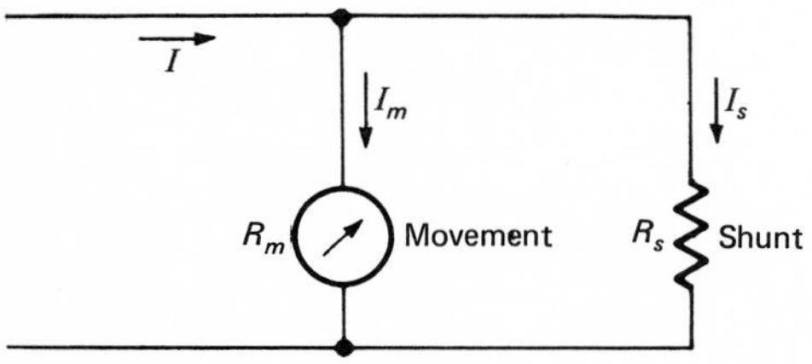

Fig. 4-50 Basic ammeter circuit.

There are several different methods of computing the resistance value of the shunt, some involving formulas to be memorized. As it is unlikely that you will remember special formulas when you need to use them, it is better to use circuit principles. However, the following will show you how one ammeter formula can be developed.

The current to be measured I is divided into two components, the movement current I_m and the shunt current I_s.[1] From Eq. (4-1),

$$I_s = I - I_m$$

From Eq. (4-14),

$$R_s = R_m \frac{I_m}{I_s}$$

Substituting $I - I_m$ for I_s gives a formula for the resistance of the shunt:

$$R_s = R_m \frac{I_m}{I - I_m} \qquad (4\text{-}27)$$

Example 4-31 An ammeter with a full-scale deflection of 1 mA (range of measurement, 0 to 1 mA) is required. What value shunt resistance should be used with a 50-μA 2-kΩ movement?

Solution (Fig. 4-51)

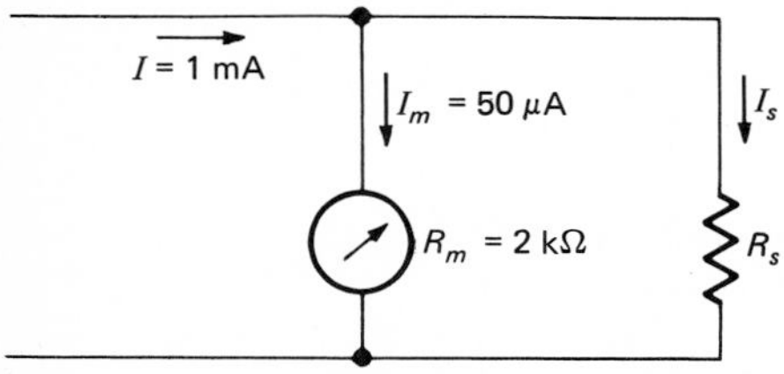

Fig. 4-51 Example 4-31: data.

Applying circuit principles:

$$I_s = I - I_m \qquad (4\text{-}1)$$

$$I_s = 1000\ \mu\text{A} - 50\ \mu\text{A} = 950\ \mu\text{A}$$

$$R_s = R_m \frac{I_m}{I_s} \qquad (4\text{-}14)$$

$$R_s = 2\ \text{k}\Omega\ \frac{50\ \mu\text{A}}{950\ \mu\text{A}} = \mathbf{105.3\ \Omega}$$

Using the formula:

$$R_s = R_m \frac{I_m}{I - I_m} \qquad (4\text{-}27)$$

$$R_s = 2\ \text{k}\Omega\ \frac{50\ \mu\text{A}}{1000\ \mu\text{A} - 50\ \mu\text{A}} = \mathbf{105.3\ \Omega}$$

[1] Full-scale values of current and voltage are normally used in meter computations unless otherwise stated. Full-scale values deflect the meter pointer to the end of the calibrated scale. A meter range, as specified or as indicated by a range selector switch, is also a full-scale value.

As you can see in Example 4-31, if you remember the ammeter formula, its use gives a rapid solution because it saves some writing. The denominator (1000 μA − 50 μA) has to be calculated, but not necessarily written separately, as when applying the principles from the start.

Example 4-32 Using the same movement as in Example 4-31, design an ammeter with three ranges, (*a*) 1, (*b*) 10, and (*c*) 100 mA.

Solution Figure 4-52 shows the movement with three shunts which can be individually connected in parallel with the movement by the switch *S*. For any of the three switch positions, a two-element parallel circuit will exist between the ammeter terminals 1 and 2.

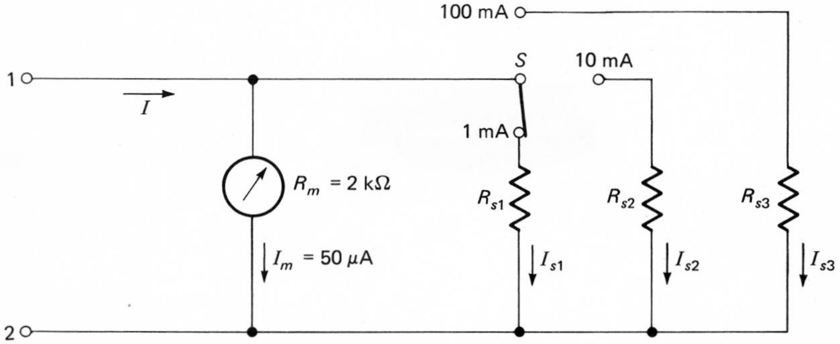

Fig. 4-52 Example 4-32: network.

a. For 1 mA full scale, the shunt R_{s1} will be in parallel with the movement. This range is the same as that in Example 4-31:

$$R_{s1} = \mathbf{105.3\ \Omega}$$

b. For 10 mA (10^4 μA):

$$I_{s2} = I - I_m \qquad (4\text{-}1)$$

$$I_{s2} = 10^4\ \mu\text{A} - 50\ \mu\text{A} = 9950\ \mu\text{A}$$

$$R_{s2} = R_m \frac{I_m}{I_{s2}} \qquad (4\text{-}14)$$

$$R_{s2} = 2\ \text{k}\Omega\ \frac{50\ \mu\text{A}}{9950\ \mu\text{A}} = \mathbf{10.05\ \Omega}$$

c. For 100 mA (10^5 μA):

$$I_{s3} = I - I_m \qquad (4\text{-}1)$$

$$I_{s3} = 10^5\ \mu\text{A} - 50\ \mu\text{A} = 99\ 950\ \mu\text{A}$$

$$R_{s3} = R_m \frac{I_m}{I_{s3}} \qquad (4\text{-}14)$$

$$R_{s3} = 2\ \text{k}\Omega\ \frac{50\ \mu\text{A}}{99\ 950\ \mu\text{A}} \simeq \mathbf{1\ \Omega}$$

COMPARISON OF VOLTMETER AND AMMETER

By referring to Sec. 3-5, it can be seen that a voltmeter is a typical series circuit, conforming to series circuit principles. As the range increases (higher full-scale voltage), the multiplier resistance increases. The overall circuit resistance is directly proportional to the range value.

In this section, it is established that an ammeter is an application of parallel circuit principles. A range increase is associated with a decrease in shunt resistance. The total circuit conductance is directly proportional to the range value.

VOLTMETER LOADING

A voltage is developed across a circuit. To measure this voltage, a voltmeter is connected across the circuit.

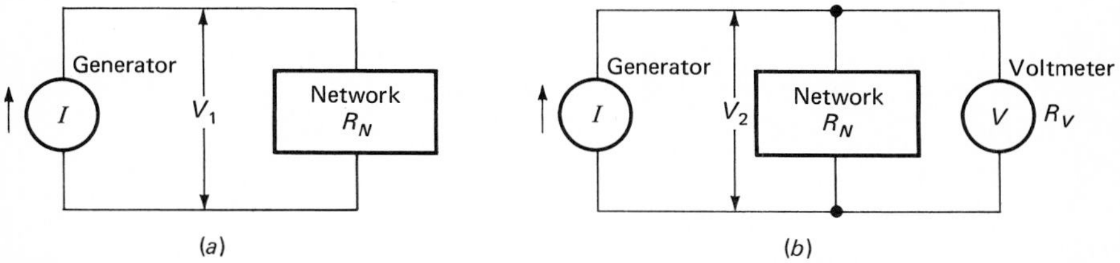

Fig. 4-53 (*a*) Network connected to a current source; (*b*) voltmeter connected to measure the network voltage.

The network in Fig. 4-53*a* will have a pd of V_1 developed across it due to the current I from the constant-current generator. To measure this voltage, a voltmeter must be connected in parallel with the network, as shown in Fig. 4-53*b*. With the voltmeter present, the pd across the network is V_2.

The schematic diagram in Fig. 4-54 is the electrical equivalent of Fig. 4-53*b*.

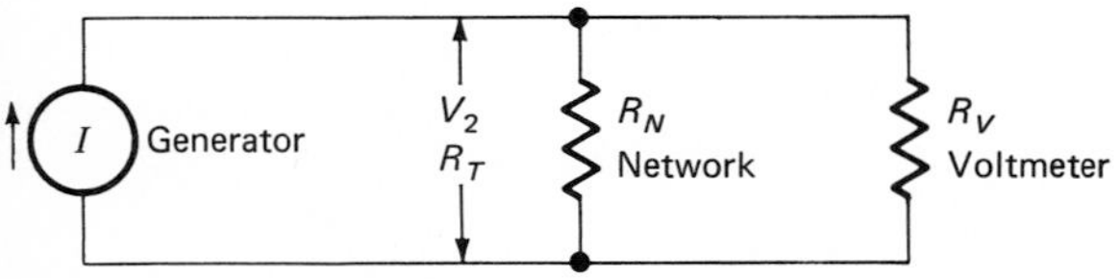

Fig. 4-54 Voltmeter loading.

The voltmeter is intended to measure the voltage across the network, as in Fig. 4-53*a*.

$$V_1 = IR_N$$

The resistance of the voltmeter modifies this voltage, reducing it to

$$V_2 = IR_T = I\,\frac{R_N R_V}{R_N + R_V}$$

Since R_T is less than R_N, V_2 is less than V_N.

This reduction in the true voltage being measured is called *voltmeter loading.* It cannot be avoided, but it can be minimized by selecting a voltmeter of resistance much greater than that of the network. If the voltmeter resistance R_V is 100 times the network resistance R_N, it will introduce an error of about 1 percent.

Example 4-33 A constant-current generator supplies 100 μA to a network of 50-kΩ resistance. A 10-V voltmeter with a resistance of 200 kΩ (20 kΩ/V) is connected across the network to measure its voltage. What percentage error is caused by voltmeter loading?
Solution In Fig. 4-55 is shown the network alone connected to the current generator.

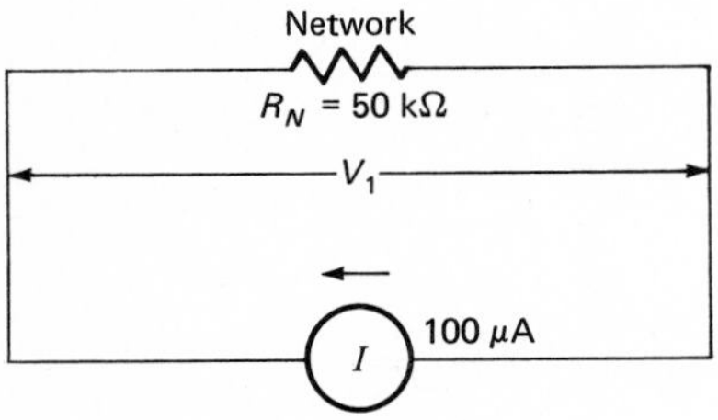

Fig. 4-55 Example 4-33: network connected to the current source.

The voltage to be measured will be

$$V_1 = IR_N = 100\ \mu\text{A} \times 50\ \text{k}\Omega = 5\ \text{V}$$

This 5 V should result in half-scale deflection of the voltmeter pointer.

Figure 4-56 shows the voltmeter resistance R_V connected across the network.

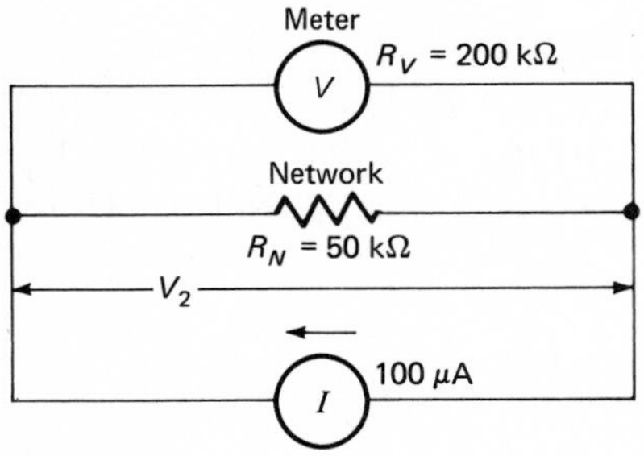

Fig. 4-56 Example 4-33: network voltage measured by the voltmeter.

$$V_2 = I\frac{R_N R_V}{R_N + R_V} = 100\ \mu\text{A}\,\frac{50\ \text{k}\Omega \times 200\ \text{k}\Omega}{50\ \text{k}\Omega + 200\ \text{k}\Omega} = 4\ \text{V}$$

The voltmeter pointer will thus be deflected 40 percent full scale instead of the desired 50 percent.

The percentage error due to the voltmeter resistance will be:

$$\text{Percentage error} = \frac{\text{voltage error}}{\text{correct voltage}} \times 100$$

$$= \frac{V_1 - V_2}{V_1} \times 100 = \frac{5\ \text{V} - 4\ \text{V}}{5\ \text{V}} \times 100$$

Percentage error = **20 percent**

If this 20 percent error is not acceptable, a voltmeter of higher resistance must be used.

EFFECT OF PARALLEL CONNECTION OF AMMETER

An ammeter is a low-resistance device which should be series-connected in the circuit where the current is to be measured. If an ammeter were connected so that it formed a parallel network within the circuit, the measurement would be ineffective. In some cases serious damage to equipment or the meter would result. Example 4-34 demonstrates a situation in which damage would probably occur.

Example 4-34 A 10-V power supply of negligible internal resistance furnishes energy to some electrical equipment. It is known that the equipment should draw 5 A at 10 V, but it is desired to determine how close to 5 A the current really is. A 10-A full-scale ammeter is to be used for the measurement. Its resistance is 0.1 Ω (it has a 1-mA full-scale 1-kΩ movement). This meter is *incorrectly* connected in parallel with the circuit. What is (*a*) the current drawn from the supply, (*b*) the equipment current, and (*c*) the meter current?

Solution A schematic diagram of the circuit with the meter connected in parallel with the equipment load is shown in Fig. 4-57.

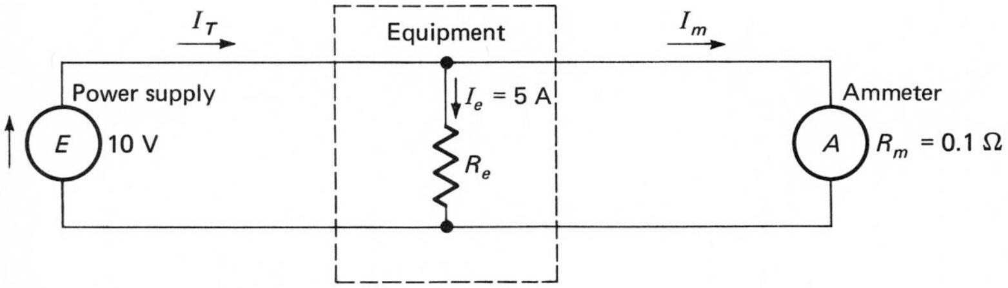

Fig. 4-57 Example 4-34: circuit.

The equipment loading is given as 5 A at 10 V:

$$R_e = \frac{E}{I_e} = \frac{10\ \text{V}}{5\ \text{A}} = 2\ \Omega$$

The total load on the power supply is R_e in parallel with R_m:

$$R_T = \frac{R_e R_m}{R_e + R_m} = \frac{2\ \Omega \times 0.1\ \Omega}{2\ \Omega + 0.1\ \Omega} = 0.0952\ \Omega$$

a. $$I_T = \frac{E}{R_T} = \frac{10\ \text{V}}{0.0952\ \Omega} = \mathbf{105\ A}$$

b. $$I_e = \mathbf{5\ A}$$

c. $$I_m = I_T - I_e = 105\ \text{A} - 5\ \text{A} = \mathbf{100\ A}$$

The objective in Example 4-34 was to measure accurately the 5-A equipment current. This measurement cannot be achieved with the parallel ammeter connection because the ammeter current is 100 A.

It is likely that there will be problems more serious than incorrect meter readings. These may include:

1. The power supply will not be capable of supplying 105 A at 10 V for any extended period of time. If it has fuses or other protective devices,

they may be activated, cutting off the output power. The power supply may be damaged because of excessive current.
2. The meter will be damaged. A 10-A meter will almost certainly be damaged beyond repair by a current of 100 A.

CURRENT SOURCES IN PARALLEL

When current sources are connected in parallel, the total source current is the algebraic sum of the individual currents, and the total equivalent internal conductance is the sum of the individual internal conductances.

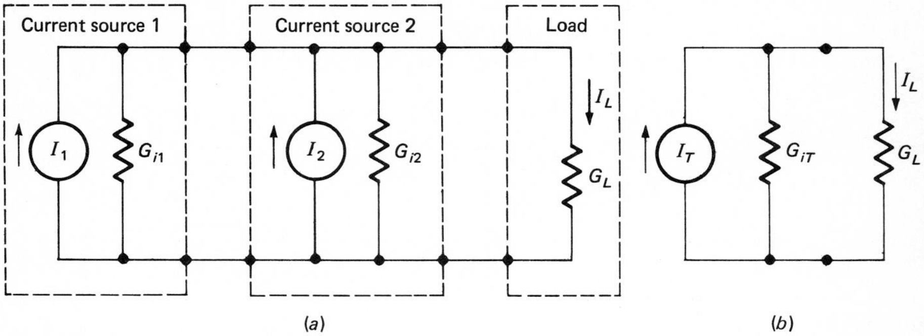

Fig. 4-58 Current sources in parallel aiding.

Figure 4-58*a* shows two current generators, each with its internal conductance, connected in parallel and supplying a load. Each of the sources tends to produce the same direction of current in the load. This connection is called *parallel aiding*. In Fig. 4-58*b* the equivalent single-source circuit is shown.

$$G_{iT} = G_{i1} + G_{i2}$$

$$I_T = I_1 + I_2$$

A *parallel opposing* connection is shown in Fig. 4-59*a*, each current source tending to produce load currents of opposite directions.

$$G_{iT} = G_{i1} + G_{i2}$$

If $I_1 > I_2$ (as implied by the arrows in Fig. 4-59*b*),

$$I_T = I_1 - I_2$$

If $I_1 < I_2$ (the arrows in Fig. 4-59*b* will have directions opposite those shown)

$$I_T = I_2 - I_1$$

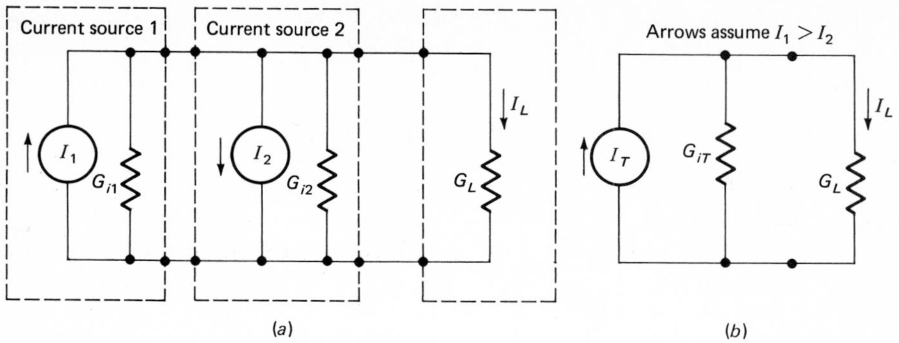

Fig. 4-59 Current sources in parallel opposing.

Example 4-35 An electronic power supply will deliver 10 A if its output terminals are short-circuited. When two of these power supplies are connected in parallel, together they cause a current of 15 A in a 2-Ω load. What is the internal resistance of each power supply?
Solution As each power supply delivers 10 A to a short circuit, it can be regarded as a 10-A current generator. Figure 4-60*a* shows the circuit consisting of two of these generators connected in parallel aiding, the equivalent single-source circuit being given in Fig. 4-60*b*.

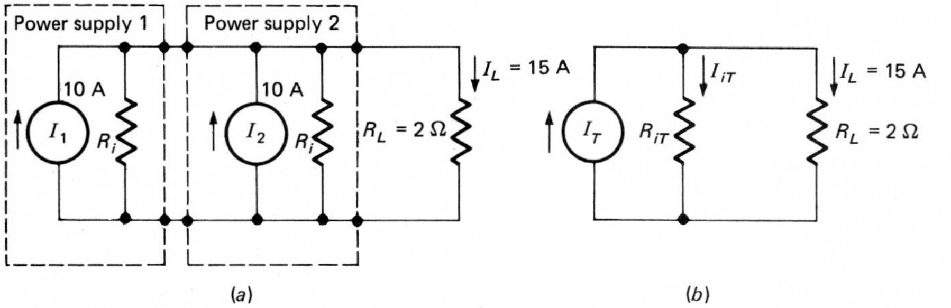

Fig. 4-60 Example 4-35: (*a*) original circuit; (*b*) equivalent circuit.

$$I_T = I_1 + I_2 = 20 \text{ A}$$

$$I_{iT} = I_T - I_L = 20 \text{ A} - 15 \text{ A} = 5 \text{ A}$$

Since $I_{iT} = 5 \text{ A} = \frac{I_L}{3}$, then $R_{iT} = 3R_L$

$$R_{iT} = 3 \times 2\ \Omega = 6\ \Omega$$

R_{iT} is the two individual internal resistances in parallel, so

$$R_i = 2 \times R_{iT} = 2 \times 6\ \Omega = \mathbf{12\ \Omega}$$

Example 4-36 A car will not start because its battery is discharged. The emf is 9.5 V, and the internal resistance is 0.2 Ω. A fully charged battery is jumped across the discharged battery to start the car. The charged battery has an emf of 13.2 V, and its internal resistance is 20 mΩ. If the starter motor draws 300 A at 12 V, (*a*) what starter current is supplied by both batteries together, and (*b*) what is the starter voltage?

Solution The discharged battery with the charged battery boosting it is shown in Fig. 4-61*a*. The current generator equivalent circuit is given in Fig. 4-61*b*.

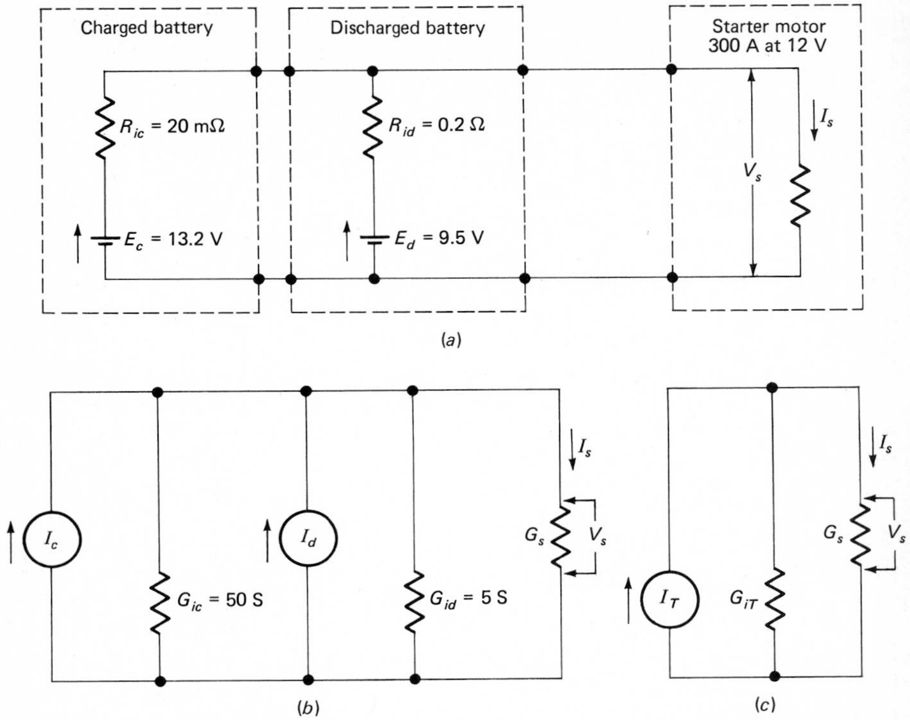

Fig. 4-61 Example 4-36: (*a*) original circuit; (*b*) circuit with voltage sources converted to current sources, (*c*) equivalent single-source circuit.

$$I_c = \frac{E_c}{R_{ic}} = \frac{13.2\text{ V}}{20\text{ m}\Omega} = 660\text{ A}$$

$$I_d = \frac{E_d}{R_{id}} = \frac{9.5\text{ V}}{0.2\ \Omega} = 47.5\text{ A}$$

$$G_s = \frac{300\text{ A}}{12\text{ V}} = 25\text{ S}$$

Figure 4-61*c* shows a single-source equivalent circuit of Fig. 4-61*b*.

$$I_T = I_c + I_d = 660\text{ A} + 47.5\text{ A} = 707.5\text{ A}$$

$$G_{iT} = G_{ic} + G_{id} = 50\text{ S} + 5\text{ S} = 55\text{ S}$$

a.

$$I_s = I_T \frac{G_s}{G_{iT} + G_s} \tag{4-20}$$

$$I_s = 707.5\text{ A}\,\frac{25\text{ S}}{55\text{ S} + 25\text{ S}} = \mathbf{221.1\text{ A}}$$

b.

$$V_s = \frac{I_s}{G_s} = \frac{221.1\text{ A}}{25\text{ S}} = \mathbf{8.844\text{ V}}$$

4-5 SUMMARY OF EQUATIONS

$$I_T = I_1 + I_2 + I_3 + \cdots + I_n \quad (4\text{-}1)$$

$$G_T = G_1 + G_2 + G_3 + \cdots + G_n \quad (4\text{-}2)$$

$$1/R_T = 1/R_1 + 1/R_2 + 1/R_3 + \cdots + 1/R_n \quad (4\text{-}3)$$

$$G_T = nG \quad (4\text{-}4)$$

$$R_T = R/n \quad (4\text{-}5)$$

$$P_T = P_1 + P_2 + P_3 + \cdots + P_n \quad (4\text{-}6)$$

$$W_T = W_1 + W_2 + W_3 + \cdots + W_n \quad (4\text{-}7)$$

$$I_T = I_1G_T/G_1 \text{ or } I_2G_T/G_2 \text{ or } I_nG_T/G_n \quad (4\text{-}8)$$

$$I_T = I_1R_1/R_T \text{ or } I_2R_2/R_T \text{ or } I_nR_n/R_T \quad (4\text{-}9)$$

$$I_1 = I_TG_1/G_T \text{ or } I_2G_1/G_2 \text{ or } I_nG_1/G_n \quad (4\text{-}10)$$

$$I_1 = I_TR_T/R_1 \text{ or } I_2R_2/R_1 \text{ or } I_nR_n/R_1 \quad (4\text{-}11)$$

$$G_T = G_1I_T/I_1 \text{ or } G_2I_T/I_2 \text{ or } G_nI_T/I_n \quad (4\text{-}12)$$

$$G_1 = G_TI_1/I_T \text{ or } G_2I_1/I_2 \text{ or } G_nI_1/I_n \quad (4\text{-}13)$$

$$R_1 = R_TI_T/I_1 \text{ or } R_2I_2/I_1 \text{ or } R_nI_n/I_1 \quad (4\text{-}14)$$

$$P_T = P_1R_1/R_T \text{ or } P_2R_2/R_T \text{ or } P_nR_n/R_T \quad (4\text{-}15)$$

$$P_1 = P_TR_T/R_1 \text{ or } P_2R_2/R_1 \text{ or } P_nR_n/R_1 \quad (4\text{-}16)$$

$$G_T/G_1 = I_T/I_1 = P_T/P_1 = W_T/W_1 = R_1/R_T \quad (4\text{-}17)$$

$$R_T = (R_1 \times R_2)/(R_1 + R_2) \quad (4\text{-}18)$$

$$R_1 = (R_2 \times R_T)/(R_2 - R_T) \quad (4\text{-}19)$$

$$I_1 = I_TG_1/G_T \text{ or } I_TG_1/(G_1 + G_2 + \cdots + G_n) \quad (4\text{-}20)$$

$$I_1 = I_TR_2/(R_1 + R_2) \quad (4\text{-}21)$$

$$I = I_i + I_L \text{ or } I_L = I - I_i \quad (4\text{-}22)$$

$$I_L = E/(R_i + R_L) \quad (4\text{-}23)$$

$$I_L = IR_i/(R_i + R_L) \quad (4\text{-}24)$$

$$V_L = E/(R_iG_L + 1) \quad (4\text{-}25)$$

$$V_L = I/(G_i + G_L) \quad (4\text{-}26)$$

$$R_s = R_mI_m/(I - I_m) \quad (4\text{-}27)$$

EXERCISE PROBLEMS

Section 4-3

Fig. 4-1P Problems 4-1 to 4-3: network.

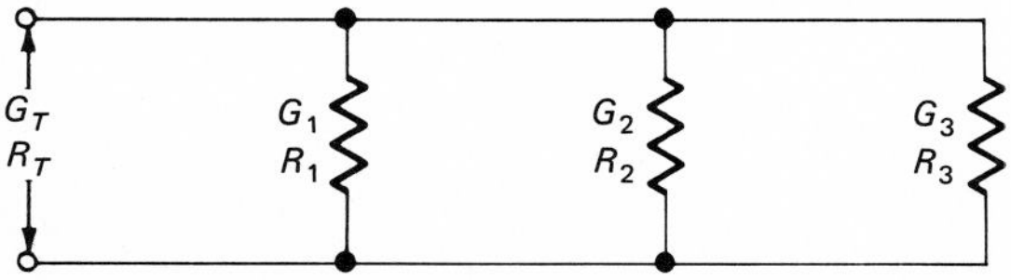

4-1 For the circuit in Fig. 4-1P find the total conductance when

a. $G_1 = 0.1$ S, $G_2 = 0.5$ S, $G_3 = 2$ S (*ans.* 2.6 S)

b. $G_1 = 0.01$ S, $G_2 = 0.015$ S, $G_3 = 0.05$ S (*ans.* 75 mS)

c. $G_1 = 0.001$ S, $G_2 = 5$ mS, $G_3 = 2.5$ mS (*ans.* 8.5 mS)

d. $G_1 = 2$ mS, $G_2 = 6$ mS, $G_3 = 500$ μS (*ans.* 8.5 mS)

e. $G_1 = 2500$ μS, $G_2 = 1000$ μS, $G_3 = 0.25$ mS (*ans.* 3.75 mS)

f. $G_1 = G_2 = G_3 = 0.005\ 56$ S (*ans.* 16.68 mS)

4-2 For the circuit in Fig. 4-1P find the total resistance when

a. $G_1 = 0.1$ S, $G_2 = 0.5$ S, $G_3 = 2$ S

b. $G_1 = 0.001$ S, $G_2 = 0.001\ 22$ S, $G_3 = 0.004\ 55$ S

c. $G_1 = 0.0263$ S, $G_2 = 0.0179$ S, $G_3 = 0.0303$ S

d. $G_1 = 0.833$ mS, $G_2 = 0.37$ mS, $G_3 = 0.556$ mS

e. $G_1 = 0.883$ mS, $G_2 = 370$ μS, $G_3 = 666$ μS

f. $G_1 = G_2 = G_3 = 30.3$ mS

4-3 For the circuit in Fig. 4-1P find the total resistance when

a. $R_1 = 100\ \Omega$, $R_2 = 150\ \Omega$, $R_3 = 180\ \Omega$ (*ans.* 45 Ω)

b. $R_1 = 1000\ \Omega$, $R_2 = 820\ \Omega$, $R_3 = 220\ \Omega$ (*ans.* 147.8 Ω)

c. $R_1 = 47\ \Omega$, $R_2 = 56\ \Omega$, $R_3 = 33\ \Omega$ (*ans.* 14.4 Ω)

d. $R_1 = 1200\ \Omega$, $R_2 = 2700\ \Omega$, $R_3 = 1800\ \Omega$ (*ans.* 568.4 Ω)

e. $R_1 = 560$ kΩ, $R_2 = 1.5$ MΩ, $R_3 = 120$ kΩ (*ans.* 92.72 kΩ)

f. $R_1 = R_2 = R_3 = 18$ kΩ (*ans.* 6 kΩ)

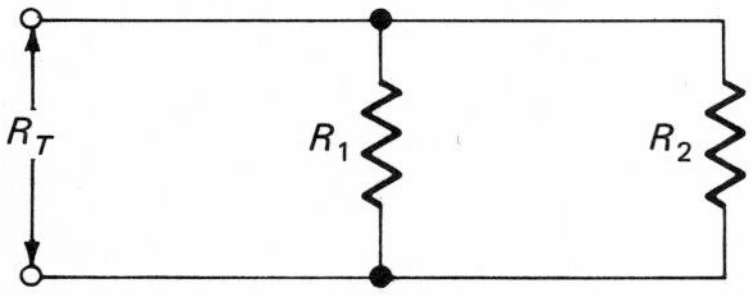

Fig. 4-2P Problems 4-4 and 4-5: network.

4-4 For the circuit in Fig. 4-2P find the total resistance, using the conductance method, when

a. $R_1 = 10\ \Omega$, $R_2 = 22\ \Omega$

b. $R_1 = 2.2$ MΩ, $R_2 = 680$ kΩ

c. $R_1 = 82$ kΩ, $R_2 = 39$ kΩ

d. $R_1 = 1800\ \Omega$, $R_2 = 2.7$ kΩ

e. $R_1 = 680\ \Omega$, $R_2 = 680\ \Omega$

f. $R_1 = 100\ \Omega$, $R_2 = 1200\ \Omega$

4-5 For the circuit in Fig. 4-2P find the total resistance, using the product-over-sum method, when

a. $R_1 = 33\ \Omega$, $R_2 = 82\ \Omega$ (*ans.* 23.53 MΩ)

b. $R_1 = 150$ kΩ, $R_2 = 68$ kΩ (*ans.* 46.79 kΩ)

c. $R_1 = 1200\ \Omega$, $R_2 = 1.2$ kΩ (*ans.* 600 Ω)

d. $R_1 = 15$ kΩ, $R_2 = 150$ kΩ (*ans.* 13.64 kΩ)

e. $R_1 = 390\ \Omega$, $R_2 = 470\ \Omega$ (*ans.* 213.1 Ω)

f. $R_1 = 1.8$ MΩ, $R_2 = 2.7$ MΩ (*ans.* 1.08 MΩ)

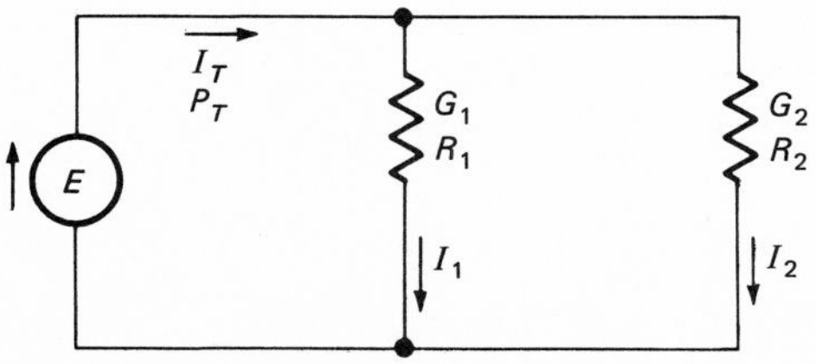

Fig. 4-3P Problems 4-6, 4-7, and 4-29: circuit.

4-6 For the circuit in Fig. 4-3P find the total current and the total power drawn from the generator when

a. $E = 30$ V, $I_1 = 3$ A, $I_2 = 6$ A
b. $E = 60$ V, $I_1 = 6$ A, $I_2 = 5$ A
c. $E = 5$ V, $G_1 = 0.1$ S, $I_2 = 0.1$ A
d. $E = 15$ V, $G_1 = 0.01$ S, $I_2 = 0.1$ A
e. $I_1 = 8$ mA, $G_T = 0.004$ S, $G_1 = 0.001$ S
f. $I_1 = 30$ mA, $G_T = 0.006$ S, $R_1 = 500\ \Omega$
g. $V_1 = 25$ V, $R_2 = 1000\ \Omega$, $I_1 = 50$ mA
h. $E = 12$ V, $R_1 = 22$ kΩ, $R_2 = 15$ kΩ
i. $E = 117$ V, $R_1 = 140\ \Omega$, $R_2 = 28\ \Omega$
j. $V_2 = 6.3$ V, $I_1 = 0.15$ A, $I_2 = 200$ mA
k. $V_1 = 10$ V, $G_T = 0.0129$ S, $R_2 = 220\ \Omega$
l. $V_2 = 15$ V, $G_1 = 0.001\ 22$ S, $R_T = 530\ \Omega$

4-7 For the circuit in Fig. 4-3P find I_1 when

a. $V_1 = 12$ V, $G_1 = 0.005\ 56$ S, $G_2 = 0.066$ S (*ans.* 66.72 mA)
b. $V_2 = 18$ V, $G_1 = 0.005\ 56$ S, $G_2 = 0.0066$ S (*ans.* 100 mA)
c. $E = 48$ V, $I_T = 48$ mA, $I_2 = 4$ mA (*ans.* 44 mA)
d. $E = 120$ V, $R_T = 35\ \Omega$, $R_1 = 70\ \Omega$ (*ans.* 1.714 A)
e. $I_T = 18$ mA, $R_T = 11$ kΩ, $R_1 = 33$ kΩ (*ans.* 6 mA)
f. $I_T = 150$ mA, $R_2 = 15$ kΩ, $I_2 = 45$ mA (*ans.* 105 mA)
g. $I_2 = 45$ mA, $R_1 = 6.8$ kΩ, $R_2 = 3.3$ kΩ (*ans.* 21.84 mA)
h. $V_1 = 100$ V, $R_2 = 1$ kΩ, $I_T = 100$ mA (*ans.* 0 A)

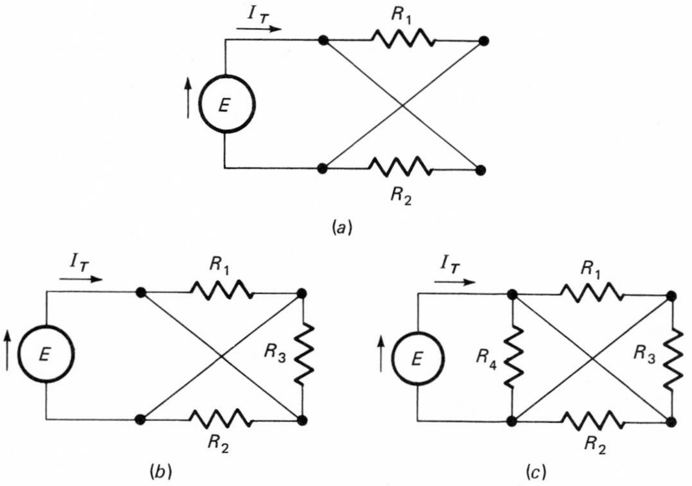

Fig. 4-4P Problems 4-8 to 4-10: circuits.

4-8 Find I_T in Fig. 4-4P*a* when $E = 20$ V, $R_1 = 680\ \Omega$, and $R_2 = 560\ \Omega$.

4-9 Find I_T in Fig. 4-4P*b* when $E = 20$ V, $R_1 = 680\ \Omega$, $R_2 = 560\ \Omega$, and $R_3 = 390\ \Omega$. (*ans.* 116.4 mA)

4-10 Find I_T in Fig. 4-4P*c* when $E = 20$ V, $R_1 = 680\ \Omega$, $R_2 = 560\ \Omega$, $R_3 = 390\ \Omega$, and $R_4 = 820\ \Omega$.

4-11 Four 6-V power sources, each capable of delivering 150 mA, are connected in parallel.

a. What is the total voltage? (*ans.* 6 V)
b. What is the maximum current available? (*ans.* 600 mA)

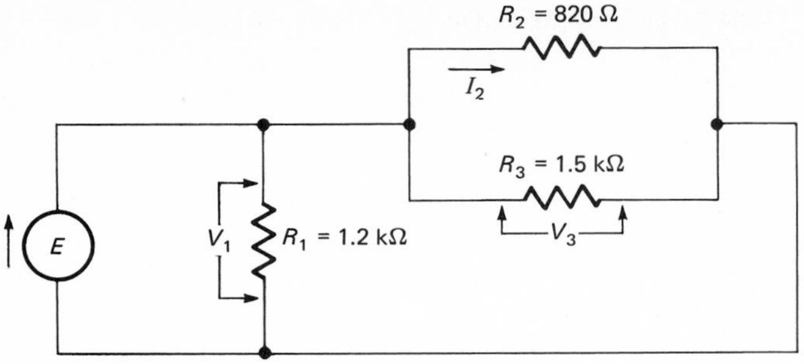

Fig. 4-5P Problems 4-12 to 4-14: circuit.

4-12 Find I in Fig. 4-5P when $V_3 = 12$ V.

4-13 Find I in Fig. 4-5P when $V_1 = 10$ V. (*ans.* 18.86 mA)

4-14 Find I in Fig. 4-5P when $I_2 = 10$ mA.

4-15 Three constant-current sources of 10, 15, and 20 mA are in parallel across a 1000-Ω load. Find

a. The total current supplied to the load. (*ans.* 45 mA)

b. The pd across the load. (*ans.* 45 V)

Section 4-4

4-16 Two current generators are connected in parallel. $I_{g1} = 50$ mA and $I_{g2} = 25$ mA. Each generator has an internal conductance of 100 μS in parallel with its current source. Find the open-circuit voltage.

4-17 Find the current and power that would be supplied to the following loads by the parallel-connected generators in Prob. 4-16.

a. 1000 Ω (*ans.* 62.5 mA, 3.906 W)

b. 5000 Ω (*ans.* 37.5 mA, 7.031 W)

c. 10 kΩ (*ans.* 25 mA, 6.25 W)

4-18 The basic movement of a microammeter has a full-scale deflection of 100 μA and a resistance of 850 Ω. Find the value for a shunt resistance which will give the instrument a full-scale deflection of

a. 10 mA

b. 50 mA

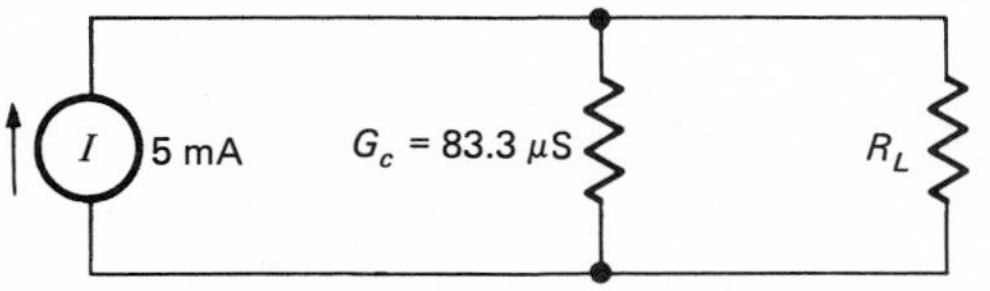

Fig. 4-6P Problems 4-19 to 4-21: circuit.

4-19 The output equivalent circuit of an amplifier employing a type 2N3905 transistor is as shown in Fig. 4-6P. Find the voltage across R_L when R_L is

a. 600 Ω (*ans.* 2.857 V)

b. 1500 Ω (*ans.* 6.667 V)

c. 2200 Ω (*ans.* 9.296 V)

4-20 From the equivalent circuit in Fig. 4-6P, what must the value of R_L be to obtain an output voltage across R_L of 15 V?

MISCELLANEOUS PARALLEL CIRCUIT PROBLEMS

4-21 What is the theoretically highest voltage obtainable across R_L in Fig. 4-6P, and what value must R_L be to obtain this voltage? (*ans.* 60 V, infinite resistance)

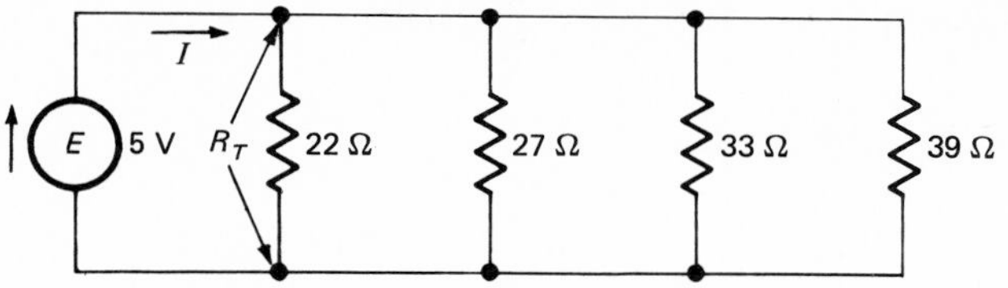

Fig. 4-7P Problems 4-22 and 4-23: circuit.

4-22 Find the total resistance of the circuit shown in Fig. 4-7P.

4-23 A fault has occurred in the circuit in Fig. 4-7P. One resistor is known to be burned out and open circuit. The current I is 564 mA. Which resistor is burned out? (*ans.* 39 Ω)

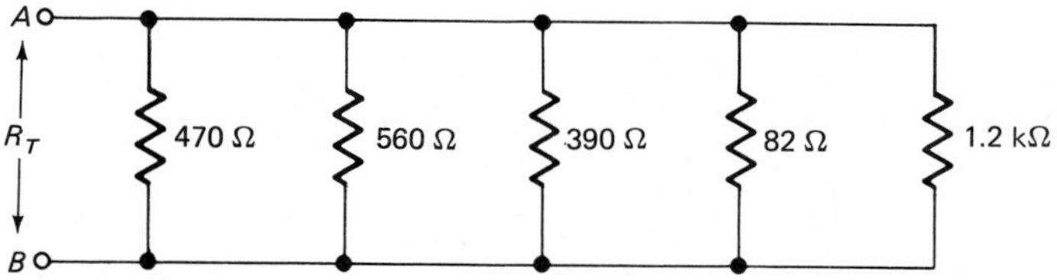

Fig. 4-8P Problems 4-24 and 4-25: network.

4-24 Find the total resistance of the five parallel resistors shown in Fig. 4-8P.

4-25 In the circuit in Fig. 4-8P, if the resistance between A and B measures 59 Ω, which resistor is open circuit? (*ans.* 390 Ω)

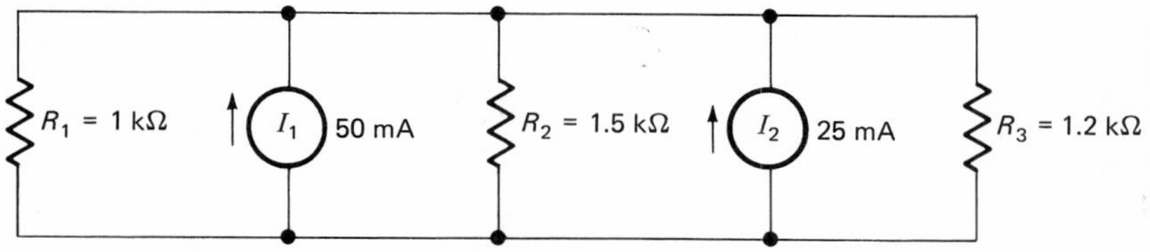

Fig. 4-9P Problems 4-26 to 4-28 and 4-30: circuit.

4-26 For the circuit shown in Fig. 4-9P, find
a. The pd across R_1.
b. The current through R_2.
c. The power dissipated by R_3.

4-27 For the circuit in Fig. 4-9P, what is the fault if the pd across R_1 is 50 V? (*ans.* 1 kΩ, open)

4-28 For the circuit in Fig. 4-9P, what is the fault if the current through R_2 is 13.3 mA?

4-29 In Fig. 4-3P, $R_1 = 270\ \Omega$ and $R_2 = 470\ \Omega$. Both are $\frac{1}{2}$-W resistors. If E is increased from zero, which resistor will reach its maximum power rating first and at what voltage will this power rating be reached? (*ans.* R_1, 11.6 V)

4-30 In Fig. 4-9P each resistor has a rating of 1 W. If I_1 is held constant and I_2 is increased, which resistor will be the first to dissipate 1 W and at what value of I_2 will this dissipation occur?

5

Series-Parallel Circuits

Simple series and parallel networks do not usually constitute the whole of a practical circuit. They are often connected together to form *series-parallel* or *combination* circuits.

There are no characteristics of series-parallel circuits which differ basically from those of the series and parallel circuits described in Chaps. 3 and 4. This chapter is therefore primarily concerned with some practical applications.

5-1 BASIC SERIES-PARALLEL CONFIGURATIONS

There are two basic methods by which series and parallel networks can be combined.

Parallel element groups connected in series In Fig. 5-1, two separate parallel circuits are connected in series with each other. Applying the principles developed in Chaps. 3 and 4,

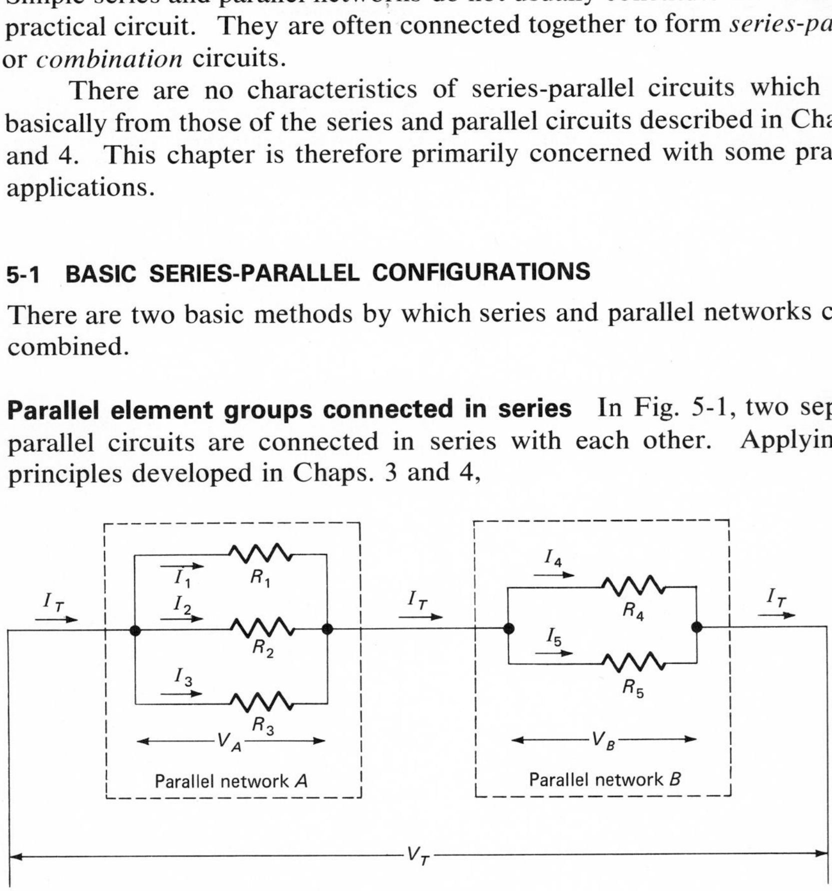

Fig. 5-1 Two parallel circuits connected in series.

1. The total current I_T in each of the parallel networks is the same ($I_T = I_1 + I_2 + I_3$ and $I_T = I_4 + I_5$).
2. The voltage across each element in either one of the parallel groups is the same (V_A is the voltage across R_1, R_2, and R_3; V_B is the voltage across R_4 and R_5).
3. The total voltage across the series-parallel network is the sum of the voltages across the individual parallel groups ($V_T = V_A + V_B$).

The methods for calculating the quantities involved in this type of circuit depend on the known and the unknown data. They will be essentially those that apply to simple series and parallel circuits.

Example 5-1 What is the current in R_3 in Fig. 5-2?

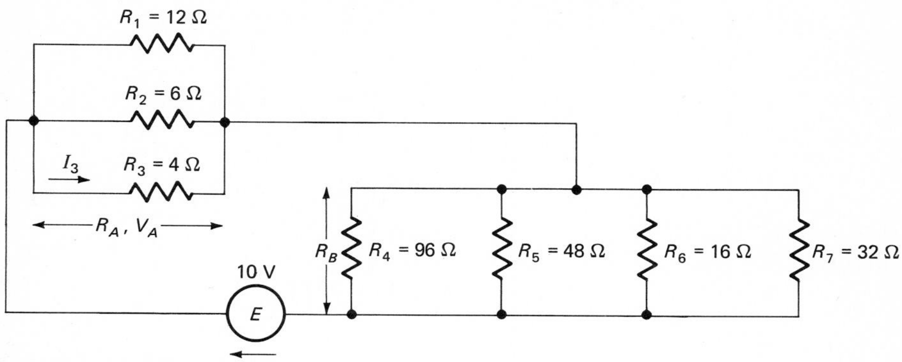

Fig. 5-2 Example 5-1: circuit.

Solution

$$\frac{1}{R_A}=\frac{1}{R_1}+\frac{1}{R_2}+\frac{1}{R_3}=\frac{1}{12\ \Omega}+\frac{1}{6\ \Omega}+\frac{1}{4\ \Omega}=\frac{1}{2\ \Omega}$$

$$R_A = 2\ \Omega$$

$$\frac{1}{R_B}=\frac{1}{R_4}+\frac{1}{R_5}+\frac{1}{R_6}+\frac{1}{R_7}=\frac{1}{96\ \Omega}+\frac{1}{48\ \Omega}+\frac{1}{16\ \Omega}+\frac{1}{32\ \Omega}=\frac{1}{8\ \Omega}$$

$$R_B = 8\ \Omega$$

$$V_A = E\frac{R_A}{R_A+R_B} = 10\ \text{V}\frac{2\ \Omega}{2\ \Omega + 8\ \Omega} = 2\ \text{V}$$

$$I_3 = \frac{V_A}{R_3} = \frac{2\ \text{V}}{4\ \Omega} = \mathbf{0.5\ A}$$

Example 5-2 What is the potential at point A in the circuit in Fig. 5-3?

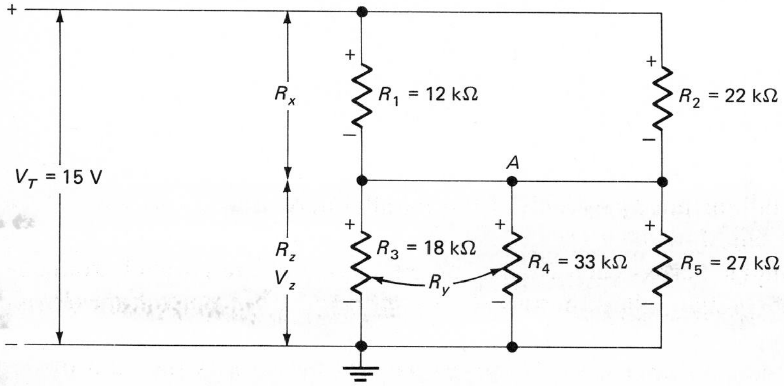

Fig. 5-3 Example 5-2: network.

Solution R_x is the parallel combination R_1 and R_2, R_y is the parallel combination R_3 and R_4, and R_z is the parallel combination R_3, R_4, and R_5.

$$R_x = \frac{R_1 R_2}{R_1 + R_2} = \frac{12\ \text{k}\Omega \times 22\ \text{k}\Omega}{12\ \text{k}\Omega + 22\ \text{k}\Omega} = 7.765\ \text{k}\Omega$$

$$R_y = \frac{R_3 R_4}{R_3 + R_4} = \frac{18\ \text{k}\Omega \times 33\ \text{k}\Omega}{18\ \text{k}\Omega + 33\ \text{k}\Omega} = 11.65\ \text{k}\Omega$$

$$R_z = \frac{R_y R_5}{R_y + R_5} = \frac{11.65\ \text{k}\Omega \times 27\ \text{k}\Omega}{11.65\ \text{k}\Omega + 27\ \text{k}\Omega} = 8.138\ \text{k}\Omega$$

Applying the voltage divider principle,

$$V_z = V_T \frac{R_z}{R_x + R_z} = 15\ \text{V} \frac{8.138\ \text{k}\Omega}{7.765\ \text{k}\Omega + 8.138\ \text{k}\Omega} = 7.676\ \text{V}$$

The potential at point A (with respect to ground) = **+7.676 V**

Series element groups connected in parallel Figure 5-4 shows two separate series networks (R_1, R_2, and R_3, R_4, R_5) connected in parallel. The main characteristics of this series-parallel circuit are:

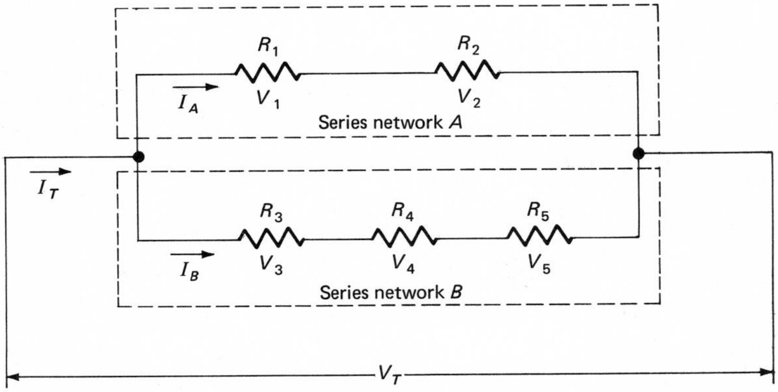

Fig. 5-4 Two series circuits connected in parallel.

1. The total voltage V_T across each of the series networks is the same ($V_T = V_1 + V_2$ and $V_T = V_3 + V_4 + V_5$).
2. The current in each of the elements in either one of the series branches is the same (I_A is the current in both R_1 and R_2; I_B is the current in R_3, R_4, and R_5).
3. The total current I_T is the sum of the branch currents I_A and I_B.

Example 5-3 What is the current I_A in the circuit in Fig. 5-5?

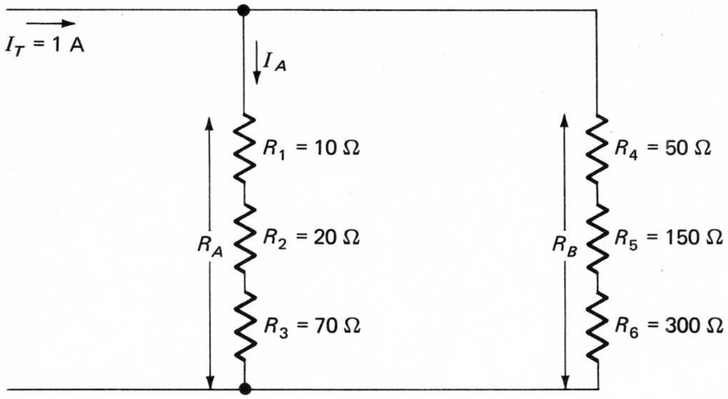

Fig. 5-5 Example 5-3: network.

Solution

$$R_A = R_1 + R_2 + R_3 = 10\ \Omega + 20\ \Omega + 70\ \Omega = 100\ \Omega$$

$$R_B = R_4 + R_5 + R_6 = 50\ \Omega + 150\ \Omega + 300\ \Omega = 500\ \Omega$$

Applying the current divider principle,

$$I_A = I_T \frac{R_B}{R_A + R_B} = 1\ \text{A} \frac{500\ \Omega}{100\ \Omega + 500\ \Omega} = \mathbf{0.833\ A}$$

Example 5-4 What is the emf of the battery in the circuit in Fig. 5-6? The pd across the 6.8-MΩ resistance is 5 V.

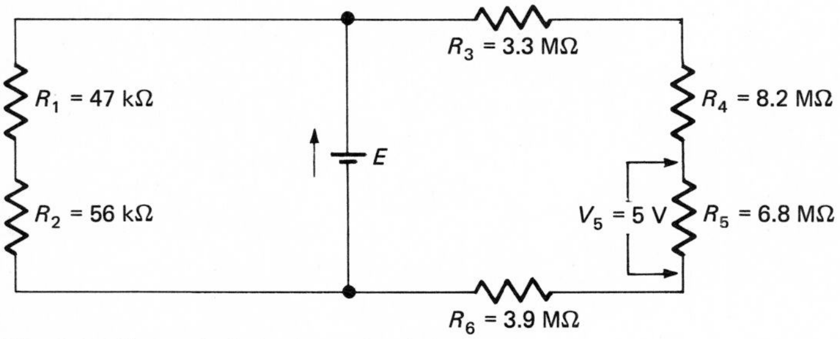

Fig. 5-6 Example 5-4: given circuit.

Solution The emf of the battery E is the voltage across each of the two branch networks. Thus, as there are sufficient data in the branch containing resistances R_3 to R_6, only this branch need be considered, as shown in Fig. 5-7.

$$R_T = R_3 + R_4 + R_5 + R_6 = 3.3\ \text{M}\Omega + 8.2\ \text{M}\Omega + 6.8\ \text{M}\Omega + 3.9\ \text{M}\Omega$$

$$R_T = 22.2\ \text{M}\Omega$$

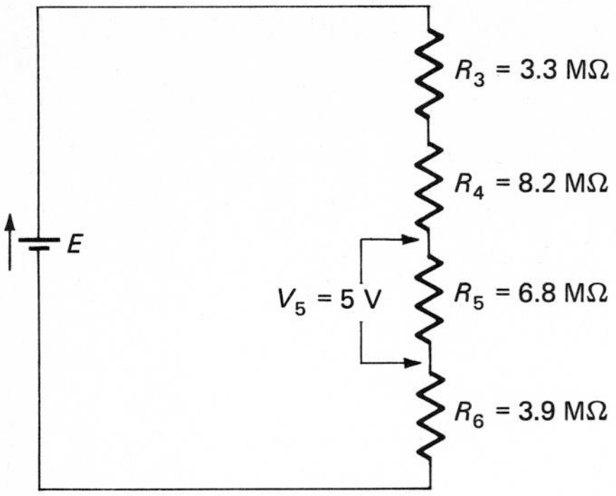

Fig. 5-7 Example 5-4: significant circuit.

Applying ratios,

$$\frac{E}{V_5} = \frac{R_T}{R_5}$$

$$E = V_5 \frac{R_T}{R_5} = 5 \text{ V} \frac{22.2 \text{ M}\Omega}{6.8 \text{ M}\Omega} = \mathbf{16.32 \text{ V}}$$

5-2 APPLICATIONS OF SERIES-PARALLEL CIRCUITS

In practice there is no limit to the variety of combinations of series-parallel networks. The following applications illustrate a few of them.

LOADED VOLTAGE DIVIDER

A voltage divider can be used to apply, to a number of loads, voltages which are known fractions of the output voltage of an energy source. The series circuit voltage divider, described in Chap. 3, is limited in its usefulness because the currents the loads can draw must be very small. Where substantial load currents are required, the voltage divider must be viewed as a series-parallel circuit application.

A loaded voltage divider with one load is shown in Fig. 5-8.

Fig. 5-8 Loaded voltage divider.

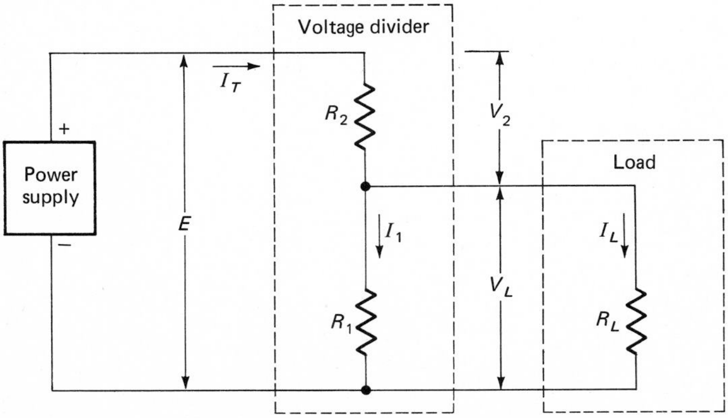

The *power supply* in Fig. 5-8 is the source of energy for supplying the load via the voltage divider chain (R_1 and R_2). The aim in designing the voltage divider is to reduce the supply voltage E to the value required by the load V_L, while maintaining an acceptable level of voltage regulation.

The voltage divider originated as a bleeder resistor, connected across a capacitor to dissipate its energy and thus make it safe to handle when not in use. It was said to bleed current out of the capacitor. The bleeder resistor was often tapped to obtain reduced voltages for various circuit functions. The term *bleeder current* now refers to the lowest current in any of the voltage divider elements. The choice of bleeder current is a compromise; if it is too low, small variations in load current will seriously affect load voltage (poor regulation); if it is too high, much power will be dissipated by the divider (an expense and heating problem). The bleeder current is often chosen arbitrarily as 10 to 20 percent of the total load current. After the bleeder current I_1 has been determined, the design can proceed.

$$I_T = I_1 + I_L$$

The voltage across R_1 is the load voltage V_L. So the voltage across R_2 will be the difference between the power supply voltage and the load voltage:

$$V_2 = E - V_L$$

$$R_1 = \frac{V_L}{I_1}$$

$$R_2 = \frac{V_2}{I_T}$$

If the power ratings for R_1 and R_2 are required, they can be calculated by any of the usual methods (the product of voltage and current, for example).

For voltage dividers supplying more than one load, the technique of design is essentially the same.

Example 5-5 A 12-V battery is to be used to supply three loads with 6 V at 0.75 A, 9 V at 0.5 A, and 12 V at 1 A. Design a suitable voltage divider.

Solution In Fig. 5-9 load R_{L3} is connected directly across the power supply, so its current I_{L3} is not included in the bleeder current calculation.

The total load current supplied by the divider is

$$I_{DL} = I_{L1} + I_{L2} = 0.75\ \text{A} + 0.5\ \text{A} = 1.25\ \text{A}$$

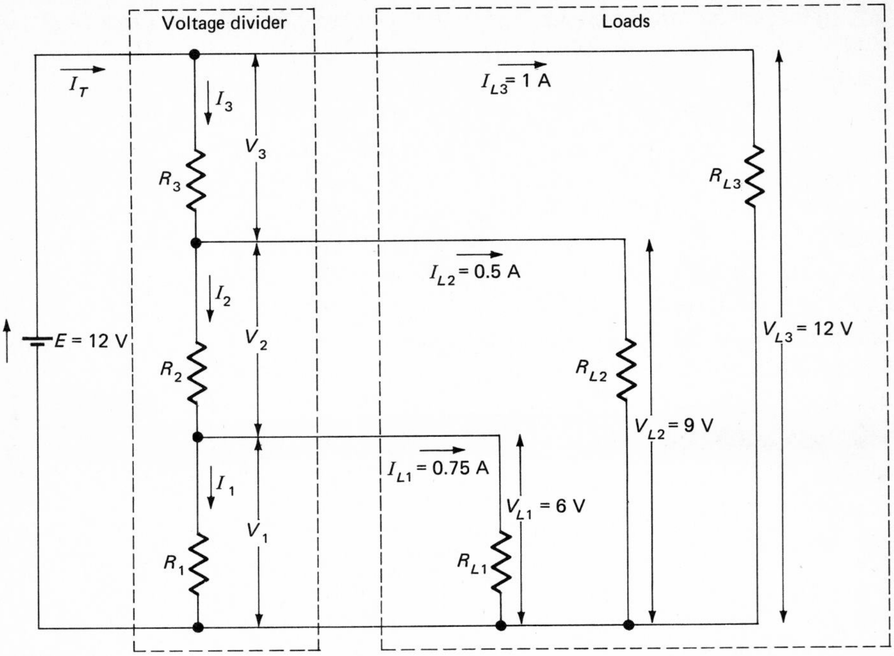

Fig. 5-9 Example 5-5: circuit.

Selecting a bleeder current which is 20 percent of the total divider load current,

$$I_1 = 20\%\ I_{DL} = 0.2 \times 1.25\ \text{A} = 0.25\ \text{A}$$

$$I_2 = I_1 + I_{L1} = 0.25\ \text{A} + 0.75\ \text{A} = 1\ \text{A}$$

$$I_3 = I_2 + I_{L2} = 1\ \text{A} + 0.5\ \text{A} = 1.5\ \text{A}$$

Note particularly in Fig. 5-9 that the voltage across R_1 is the load voltage V_{L1}, while the voltages across R_2 and R_3 are *not* the load voltages V_{L2} and V_{L3}. R_2 and R_3 are not connected directly across the loads; each one is between two load connection points.

$$V_1 = V_{L1} = 6\ \text{V}$$

$$V_2 = V_{L2} - V_{L1} = 9\ \text{V} - 6\ \text{V} = 3\ \text{V}$$

$$V_3 = V_{L3} - V_{L2}\ (\text{or}\ E - V_{L2}) = 12\ \text{V} - 9\ \text{V} = 3\ \text{V}$$

$$R_1 = \frac{V_1}{I_1} = \frac{6\ \text{V}}{0.25\ \text{A}} = \mathbf{24\ \Omega}$$

$$R_2 = \frac{V_2}{I_2} = \frac{3\ \text{V}}{1\ \text{A}} = \mathbf{3\ \Omega}$$

$$R_3 = \frac{V_3}{I_3} = \frac{3\ \text{V}}{1.5\ \text{A}} = \mathbf{2\ \Omega}$$

The power dissipations for R_1, R_2, and R_3 are

$$P_1 = V_1 I_1 = 6\text{ V} \times 0.25\text{ A} = \mathbf{1.5\ W}$$

$$P_2 = V_2 I_2 = 3\text{ V} \times 1\text{ A} = \mathbf{3\ W}$$

$$P_3 = V_3 I_3 = 3\text{ V} \times 1.5\text{ A} = \mathbf{4.5\ W}$$

Example 5-6 A solid-state amplifier requires for its operation +25 V at 1.25 A, −25 V at 1.75 A, and −7.5 V at 0.25 A. What resistance values are needed for a voltage divider to obtain these outputs from a 60-V power supply?

Solution When both positive and negative load voltages are to be supplied from a voltage divider, it is important that the principles of relative potential be borne in mind. These principles were discussed in Chap. 3, Fig. 3-28 and Example 3-19 being relevant. Note that in the present example there is a total of 50 V between the +25-V and the −25-V outputs, and that the voltage across R_4 in Fig. 5-10 is the difference between this total and the power supply emf.

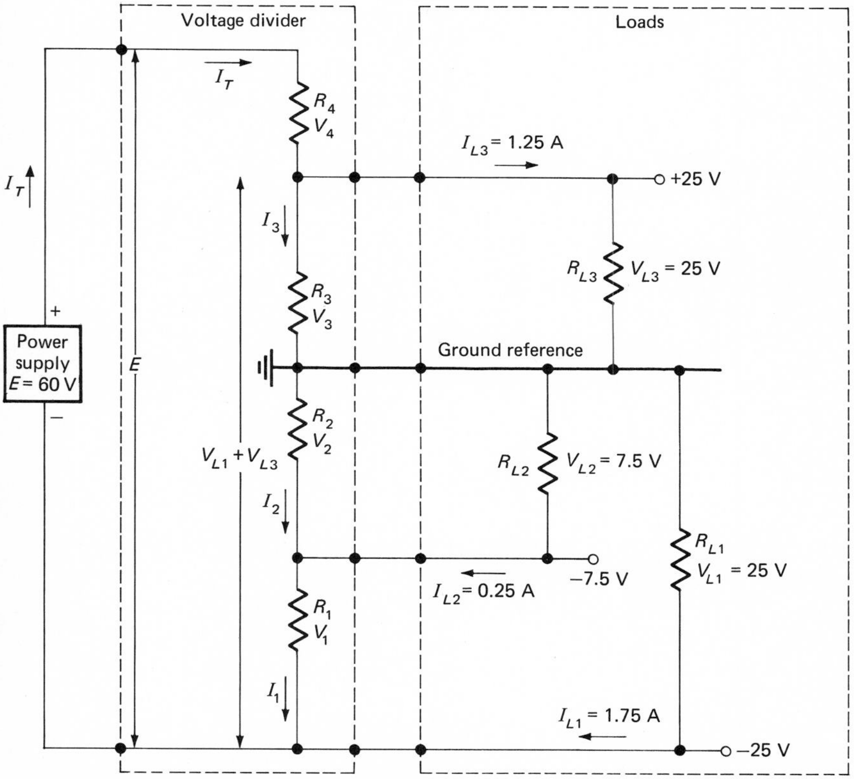

Fig. 5-10 Example 5-6: circuit.

The current directions should also be noted. For instance, the direction of I_{L3}, on the positive side, is out of the voltage divider chain, while that of I_{L1} and I_{L2}, on the negative side, is into the chain.

Again, because of the existence of positive and negative load voltages in this example, it is necessary to determine on which side of the ground reference, positive or negative, the greater total load current occurs. This value is then used to calculate the bleeder current. The *bleeder resistance* is the one connected to the reference on the side of maximum load current.

The total load current on the negative side is

$$I_{t-} = I_{L1} + I_{L2} = 1.75\text{ A} + 0.25\text{ A} = 2\text{ A}$$

The total load current on the positive side is

$$I_{t+} = I_{L3} = 1.25\text{ A}$$

The total negative load current is greater. R_2 and I_2 will be the bleeder resistance and current.

Let the bleeder current be 10 percent of the total negative load current I_{t-},

$$I_2 = 10\% I_{t-} = 0.1 \times 2\text{ A} = 0.2\text{ A}$$

$$I_1 = I_2 + I_{L2} = 0.2\text{ A} + 0.25\text{ A} = 0.45\text{ A}$$

$$I_T = I_1 + I_{L1} = 0.45\text{ A} + 1.75\text{ A} = 2.2\text{ A}$$

$$I_3 = I_T - I_{L3} = 2.2\text{ A} - 1.25\text{ A} = 0.95\text{ A}$$ Note that the current in R_4 is I_T.

$$V_1 = V_{L1} - V_{L2} = 25\text{ V} - 7.5\text{ V} = 17.75\text{ V}$$

$$V_2 = V_{L2} = 7.5\text{ V}$$

$$V_3 = V_{L3} = 25\text{ V}$$

$$V_4 = E - (V_{L1} + V_{L3}) = 60\text{ V} - (25\text{ V} + 25\text{ V}) = 10\text{ V}$$

$$R_1 = \frac{V_1}{I_1} = \frac{17.75\text{ V}}{0.45\text{ A}} = \mathbf{39.44\ \Omega}$$

$$R_2 = \frac{V_2}{I_2} = \frac{7.5\text{ V}}{0.2\text{ A}} = \mathbf{37.5\ \Omega}$$

$$R_3 = \frac{V_3}{I_3} = \frac{25\text{ V}}{0.95\text{ A}} = \mathbf{26.32\ \Omega}$$

$$R_4 = \frac{V_4}{I_T} = \frac{10\text{ V}}{2.2\text{ A}} = \mathbf{4.545\ \Omega}$$

Example 5-7 A 300-V-output electronic power supply, with a maximum current capability of 50 mA, is to be used to operate a vacuum tube amplifier. The amplifier requires +250 V at 40 mA, +100 V at 5 mA, and −10 V at zero current. Determine the resistance values of a suitable voltage divider.

Solution Referring to Fig. 5-11 as the negative load current is zero, the positive load current is greater.

$$I_{t+} = I_{L2} + I_{L3} = 5\text{ mA} + 40\text{ mA} = 45\text{ mA}$$

At the ground common node Kirchhoff's current law must apply. Therefore the sum of the currents entering this node from the positive side must equal the sum of the currents leaving the node from the negative side. Furthermore, each of these sums must equal the current drawn from the supply. Applying this law to the positive side,

$$I_{L2} + I_{L3} + I_2 = I_T$$

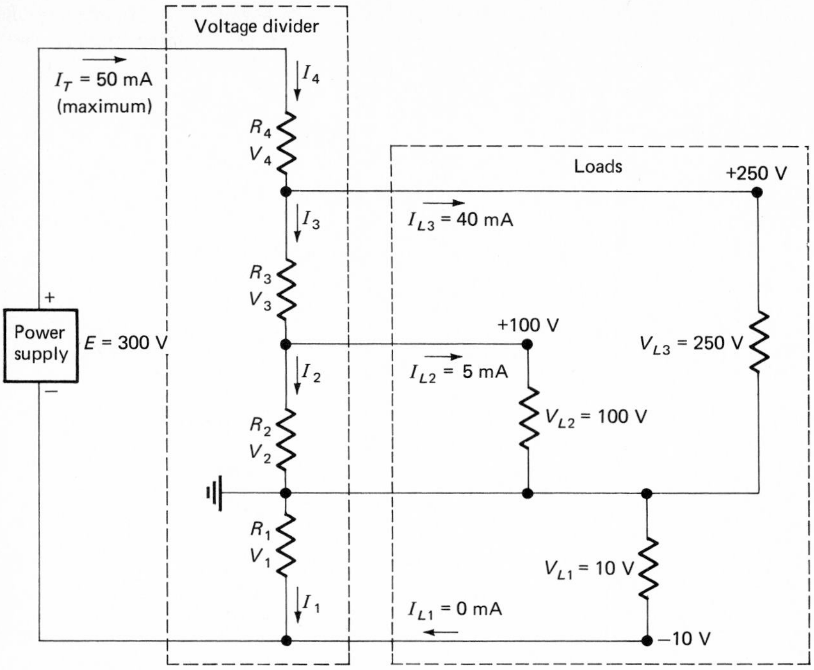

Fig. 5-11 Example 5-7: circuit.

If we use the total power supply current of 50 mA,

$$I_2 = I_T - (I_{L2} + I_{L3}) = 50\text{ mA} - 45\text{ mA} = 5\text{ mA}$$

The current I_2 is the bleeder current in this voltage divider. It is 10 percent of the total supply current.

$$I_1 = I_T - I_{L1} = 50\text{ mA} - 0\text{ mA} = 50\text{ mA}$$

$$I_3 = I_2 + I_{L2} = 5\text{ mA} + 5\text{ mA} = 10\text{ mA}$$

$$I_4 = I_T = 50\text{ mA} \qquad R_4 \text{ is in series with the power supply.}$$

$$V_1 = V_{L1} = 10\text{ V}$$

$$V_2 = V_{L2} = 100\text{ V}$$

$$V_3 = V_{L3} - V_{L2} = 250\text{ V} - 100\text{ V} = 150\text{ V}$$

$$V_4 = E - (V_{L1} + V_{L3}) = 300\text{ V} - (10\text{ V} + 250\text{ V}) = 40\text{ V}$$

$$R_1 = \frac{V_1}{I_1} = \frac{10\text{ V}}{50\text{ mA}} = \mathbf{200\ \Omega}$$

$$R_2 = \frac{V_2}{I_2} = \frac{100\text{ V}}{5\text{ mA}} = \mathbf{20\ k\Omega}$$

$$R_3 = \frac{V_3}{I_3} = \frac{150\text{ V}}{10\text{ mA}} = \mathbf{15\ k\Omega}$$

$$R_4 = \frac{V_4}{I_T} = \frac{40\text{ V}}{50\text{ mA}} = \mathbf{800\ \Omega}$$

Example 5-8 In the circuit in Fig. 5-12, the load represents a transistor switching circuit which draws 1 A at 9 V when it is on and zero current when it is off. What is the load voltage regulation?

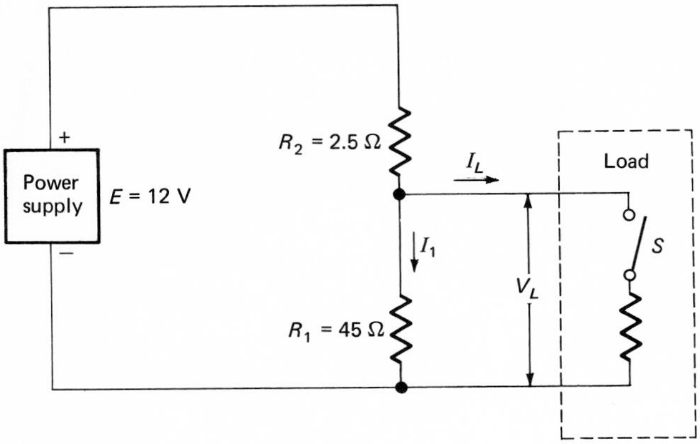

Fig. 5-12 Example 5-8: circuit.

Solution In Fig. 5-12, the switch S simulates the ON-OFF conditions of the load. When the switch is closed (load on),

$$I_L = 1\ \text{A}$$

$$V_L = 9\ \text{V}$$

$$I_1 = \frac{V_L}{R_1} = \frac{9\ \text{V}}{45\ \Omega} = 0.2\ \text{A}$$

I_1 is thus 20 percent of the ON load current.

When the switch is open (load off), I_L is zero and the network across the power supply in effect consists only of R_1 and R_2 in series.

$$V_L = E\frac{R_1}{R_1 + R_2} = 12\ \text{V}\frac{45\ \Omega}{45\ \Omega + 2.5\ \Omega} = 11.37\ \text{V}$$

The ON load voltage is 9 V. It rises to 11.37 V when the load is off. In this application, it is probable that the ON voltage would be regarded as the desired voltage (required to be constant). A significant voltage rise during the OFF situation might be a cause for concern; it could damage the transistors.

The voltage regulation is

$$\text{Percent regulation} = \frac{\text{OFF voltage} - \text{ON voltage}}{\text{ON voltage}} \times 100$$

$$= \frac{11.37\ \text{V} - 9\ \text{V}}{9\ \text{V}} \times 100$$

$$\text{Voltage regulation} = \mathbf{26.33\ percent}$$

Example 5-9 In Example 5-8, what would be the effect on the voltage regulation if the ON bleeder current were increased to 0.5 A (50 percent of the ON load current)?

Solution When the load is on,

$$I_L = 1\ \text{A}$$

$$V_L = 9\ \text{V}$$

$$I_1 = 0.5\ \text{A}$$

$$R_1 = \frac{V_L}{I_1} = \frac{9\ \text{V}}{0.5\ \text{A}} = 18\ \Omega$$

$$I_T = I_1 + I_L = 0.5\ \text{A} + 1\ \text{A} = 1.5\ \text{A}$$

$$V_2 = E - V_L = 12\ \text{V} - 9\ \text{V} = 3\ \text{V}$$

$$R_2 = \frac{V_2}{I_T} = \frac{3\ \text{V}}{1.5\ \text{A}} = 2\ \Omega$$

When the load is off,

$$V_L = E\frac{R_1}{R_1 + R_2} = 12\ \text{V}\ \frac{18\ \Omega}{18\ \Omega + 2\ \Omega} = 10.8\ \text{V}$$

$$\text{Percent regulation} = \frac{\text{OFF voltage} - \text{ON voltage}}{\text{ON voltage}} \times 100$$

$$= \frac{10.8\ \text{V} - 9\ \text{V}}{9\ \text{V}} \times 100$$

Voltage regulation = **20 percent**

In this case, doubling the ON bleeder current improves the voltage regulation by 6.33 percent.

In Examples 5-8 and 5-9 there is another factor which might be of interest. If the transistor switching circuit is operated directly from the 12-V supply, without a voltage divider, the transistor will receive 12 V on and off. There will be zero voltage regulation, but the voltage will always be $33\frac{1}{3}$ percent too high. As the required voltage is lowered below the supply voltage, the regulation problem due to the voltage divider becomes more severe.

DOMESTIC THREE-WIRE POWER DISTRIBUTION

The electrical wiring in most homes is carried out in a three-wire configuration. While it is true that these domestic installations are alternating current, the basic principle underlying three-wire distribution can be seen by regarding it as a series-parallel dc circuit.

Two, virtually separate, 120-V supplies are available to operate the lights and most appliances. Although it depends on the home equipment, typically the water heater, furnace, clothes dryer, and some of the stove elements operate at 240 V, this voltage is derived from the two 120-V supplies in series.

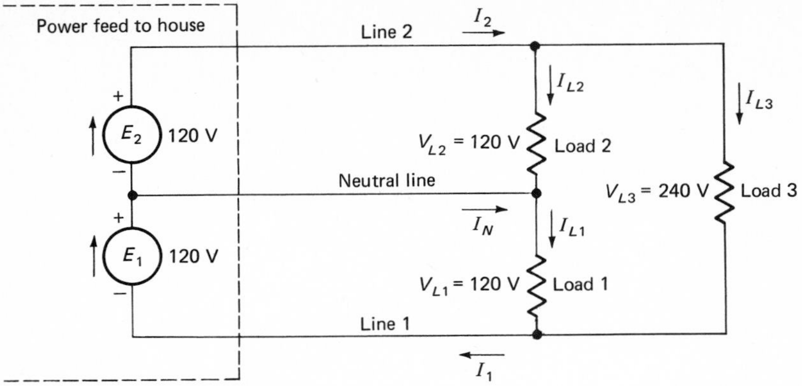

Fig. 5-13 Three-wire power distribution system.

Figure 5-13 shows the equivalent circuit of a three-wire system. The two 120-V generators, connected in series, simulate the power feed to the home. The three wires, or *lines,* provide two 120-V and one 240-V supplies. One of the wires is usually called the *neutral* line because it is connected to both the plus and minus terminals of the supplies and in most systems is grounded. If the two 120-V loads (loads 1 and 2) are the same, the current I_N will be zero (hence *neutral*). The main relationships are

$$V_{L1} = V_{L2} = 120 \text{ V}$$

$$V_{L3} = 240 \text{ V}$$

$$I_1 = I_{L1} + I_{L3}$$

$$I_2 = I_{L2} + I_{L3}$$

If $I_{L1} > I_{L2}$, $I_N = I_{L1} - I_{L2}$ with the direction shown in Fig. 5-13.

If $I_{L2} > I_{L1}$, $I_N = I_{L2} - I_{L1}$ with its direction opposite that shown.

If $I_{L1} = I_{L2}$, I_N will be zero. This is called a *balanced* condition.

Example 5-10 At a particular time, electrical equipment puts the following load on a three-wire supply:

Line 1 (120 V) Six light bulbs, one refrigerator, one radio, and one stove element; total power $P_{L1} = 2.99$ kW.

Line 2 (120 V) Four light bulbs, one washing machine, one television set, and one stove element; total power $P_{L2} = 2.473$ kW.

Lines 1 and 2 (240 V) One clothes dryer and one stove oven; total power $P_{L3} = 5$ kW.

What is the current in each of the three lines?

Solution

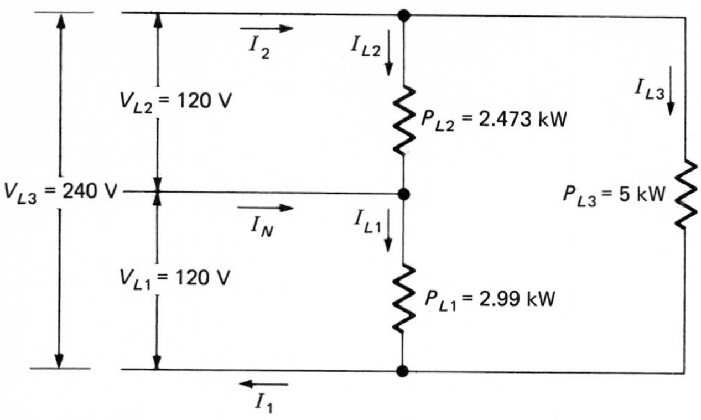

Fig. 5-14 Example 5-10: network.

$$I_{L1} = \frac{P_{L1}}{V_{L1}} = \frac{2990\ \text{W}}{120\ \text{V}} = 24.92\ \text{A}$$

$$I_{L2} = \frac{P_{L2}}{V_{L2}} = \frac{2473\ \text{W}}{120\ \text{V}} = 20.61\ \text{A}$$

$$I_{L3} = \frac{P_{L3}}{V_{L3}} = \frac{5000\ \text{W}}{240\ \text{V}} = 20.83\ \text{A}$$

$$I_1 = I_{L1} + I_{L3} = 24.92\ \text{A} + 20.83\ \text{A} = \mathbf{45.75\ A}$$

$$I_2 = I_{L2} + I_{L3} = 20.61\ \text{A} + 20.83\ \text{A} = \mathbf{41.44\ A}$$

$$I_N = I_{L1} - I_{L2} = 24.92\ \text{A} - 20.61\ \text{A} = \mathbf{4.31\ A}$$

As I_{L1} is greater than I_{L2}, the direction of the neutral line current I_N will be from left to right, as indicated in Fig. 5-14.

VOLTMETER AND AMMETER CONNECTIONS

When it is desired to measure simultaneously both current and voltage, the resulting series-parallel effects may produce significant errors. Even if the instruments were perfectly accurate and could be interpreted correctly, errors would still be introduced.

The measurement of the voltage and current of a single element will illustrate the basic problem.

In Fig. 5-15, the current I and the voltage V are to be measured at the same time. To do this, two types of ammeter and voltmeter connections can be used.

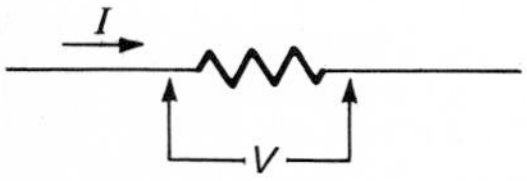

Fig. 5-15 Single element, voltage and current.

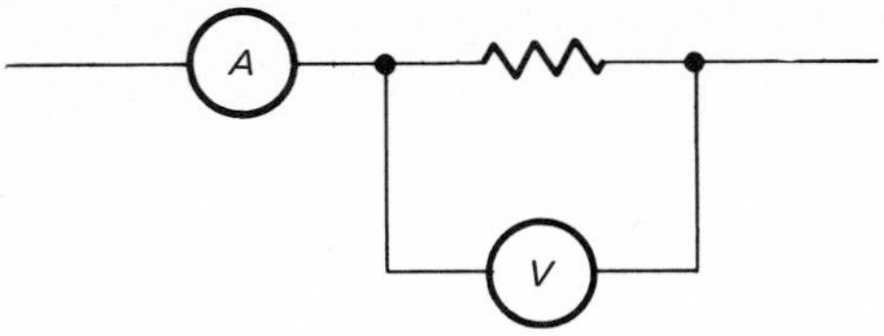

Fig. 5-16 Ammeter measures currents of element and voltmeter.

When the meter connections in Fig. 5-16 are used, the fundamental sources of error are:

1. The resistances of both meters may modify the original current and voltage so that correct measurement of either is not possible.
2. The current in the ammeter includes the current taken by the voltmeter.

If the meters are connected as in Fig. 5-17, there will also be two error sources inherent in the circuit.

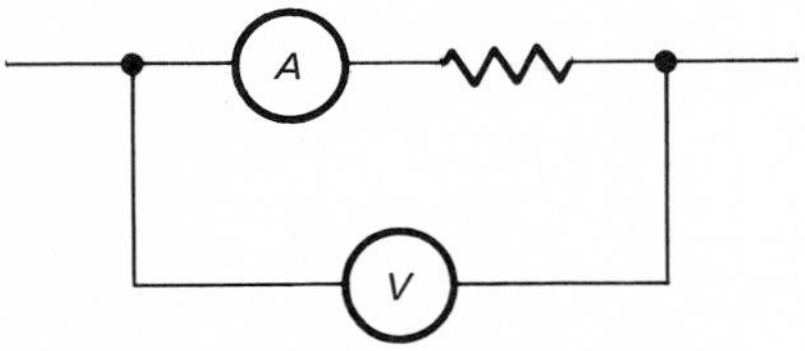

Fig. 5-17 Voltmeter measures pds of element and ammeter.

1. The meter resistances may modify the original current and voltage.
2. The voltage measured by the voltmeter includes the voltage across the ammeter.

The extent and significance of these error sources depends on several factors.

1. The objectives of the measurement, particularly the accuracy sought.
2. The voltmeter resistance; this should be as high as possible.
3. The ammeter resistance; it should be as low as possible.
4. The voltage and current values in the circuit being measured.

Example 5-11 A 10-V battery applied directly across a 1-MΩ resistor produces 10 μA of current. What resistor current and voltage will actually be measured by a 2-kΩ ammeter and a 200-kΩ voltmeter if they are used together?

Solution

Meters connected as in Fig. 5-18*a*:

$$R_p = \frac{RR_V}{R + R_V} = \frac{1\ \text{M}\Omega \times 0.2\ \text{M}\Omega}{1\ \text{M}\Omega + 0.2\ \text{M}\Omega} = 0.1667\ \text{M}\Omega$$

$$R_T = R_A + R_p = 2\ \text{k}\Omega + 0.1667\ \text{M}\Omega = 0.1687\ \text{M}\Omega$$

$$I_1 = \frac{E}{R_T} = \frac{10\ \text{V}}{0.1687\ \text{M}\Omega} = \mathbf{59.28\ \mu A}$$

$$V_1 = E\frac{R_p}{R_T} = 10\ \text{V}\,\frac{0.1667\ \text{M}\Omega}{0.1687\ \text{M}\Omega} = \mathbf{9.881\ V}$$

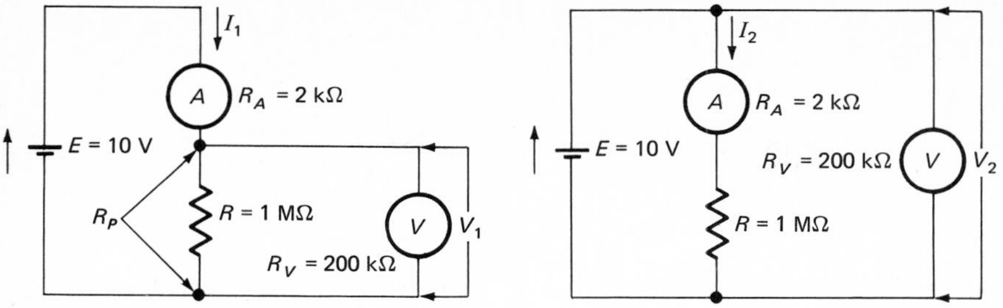

R_P = R and R_V in parallel
R_T = R_A and R_P in series

(a) (b)

Fig. 5-18 Example 5-11: circuits. (*a*) Voltmeter across the resistor; (*b*) ammeter in series with the resistor.

Meters connected as in Fig. 5-18*b*:

$$I_2 = \frac{E}{R + R_A} = \frac{10\text{ V}}{2\text{ k}\Omega + 1\text{ M}\Omega} \simeq \mathbf{10\ \mu A}$$

$$V_2 = E = \mathbf{10\ V}$$

Example 5-12 A 10-Ω resistance is fed with 5 mA from a constant-current source, developing a pd of 50 mV. What would be the current and voltage measured by a 10-Ω ammeter and a 2-kΩ voltmeter both used at the same time?

Solution

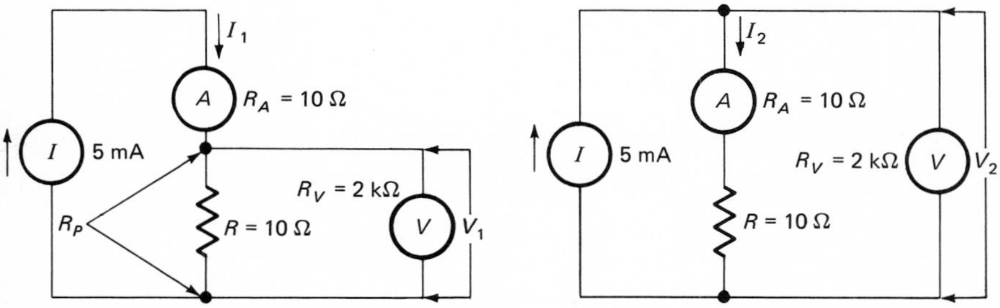

R_P = R and R_V in parallel
R_T = R_A and R_P in series

R_S = R_A and R in series

(a) (b)

Fig. 5-19 Example 5-12: circuits. (*a*) Voltmeter across the resistor; (*b*) ammeter in series with the resistor.

Meters connected as in Fig. 5-19*a*:

$$R_p = \frac{RR_V}{R + R_V} = \frac{10\ \Omega \times 2\text{ k}\Omega}{10\ \Omega + 2\text{ k}\Omega} \simeq 10\ \Omega$$

$$I_1 = I = \mathbf{5\ mA}$$

$$V_1 = IR_p \simeq 5\text{ mA} \times 10\ \Omega \simeq \mathbf{50\ mV}$$

Meters connected as in Fig. 5-19*b*:

$$R_s = R_A + R = 10\ \Omega + 10\ \Omega = 20\ \Omega$$

$$I_2 = I \frac{R_V}{R_V + R_s} = 5\text{ mA}\frac{2\text{ k}\Omega}{2\text{ k}\Omega + 20\ \Omega} \simeq \mathbf{5\ mA}$$

$$V_2 = I \frac{R_V R_s}{R_V + R_s} = 5\text{ mA}\frac{2\text{ k}\Omega \times 20\ \Omega}{2\text{ k}\Omega + 20\ \Omega} \simeq \mathbf{100\ mV}$$

The meter resistances in Examples 5-11 and 5-12 are suitable for voltage and current ranges based on the 50-μA 2-kΩ movement described in Secs. 3-5 and 4-4. In Example 5-11, the meter connections in Fig. 5-18*b* give results which are better than those for Fig. 5-18*a*. The opposite is true of Example 5-12, in which the Fig. 5-19*a* connections give the most accurate basis for measurement.

It is difficult to generalize about the import of these examples, however, you should be careful about the series-parallel connotations of meter connections. Make certain that the current drawn by the voltmeter does not significantly interfere with the ammeter readings, and that the ammeter pd does not seriously upset the voltage measurements.

T AND π NETWORKS

Among the myriad possible circuit configurations, there are a number that occur frequently in practice. Some of these can be viewed as series-parallel circuits.

T and π networks are typically used as *attenuators,* networks for reducing the level of a signal (for example, the audio volume level in broadcasting). They are also often met as sections of more sophisticated circuits.

T and π networks are so called because they are similar in shape to the letter T and the Greek letter π.

T NETWORK

The total resistance across the generator in Fig. 5-20 is

$$R_T = R_1 + \frac{R_3(R_2 + R_L)}{R_2 + R_3 + R_L}$$

Fig. 5-20 T network connected between a source and a load.

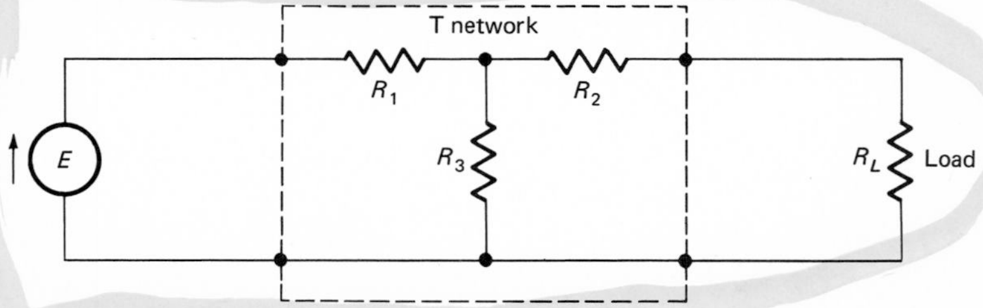

With R_T computed, the total current can be obtained, and any of the other currents and voltages.

Example 5-13 What is the load current and voltage I_L and V_L in the circuit in Fig. 5-21?

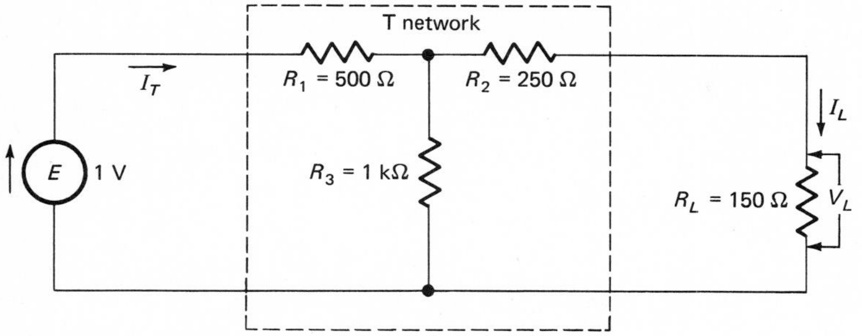

Fig. 5-21 Example 5-13: circuit.

Solution

$$R_T = R_1 + \frac{R_3(R_2 + R_L)}{R_2 + R_3 + R_L} = 500\ \Omega + \frac{1\ \text{k}\Omega(250\ \Omega + 150\ \Omega)}{250\ \Omega + 1\ \text{k}\Omega + 150\ \Omega} = 785.7\ \Omega$$

$$I_T = \frac{E}{R_T} = \frac{1\ \text{V}}{785.7\ \Omega} = 1.273\ \text{mA}$$

$$I_L = I_T \frac{R_3}{R_2 + R_3 + R_L} = 1.273\ \text{mA}\ \frac{1\ \text{k}\Omega}{250\ \Omega + 1\ \text{k}\Omega + 150\ \Omega} = \mathbf{0.9093\ mA}$$

$$V_L = I_L R_L = 0.9093\ \text{mA} \times 150\ \Omega = \mathbf{136.4\ mV}$$

π NETWORK

Since R_1 is connected in parallel with the generator in Fig. 5-22, its conditions can be determined directly; its pd $= E$, and its current $= E/R_1$. For the other elements, the effective total resistance can be computed ignoring R_1:

$$R_x = R_3 + \frac{R_2 R_L}{R_2 + R_L}$$

Fig. 5-22 π network connected between a source and a load.

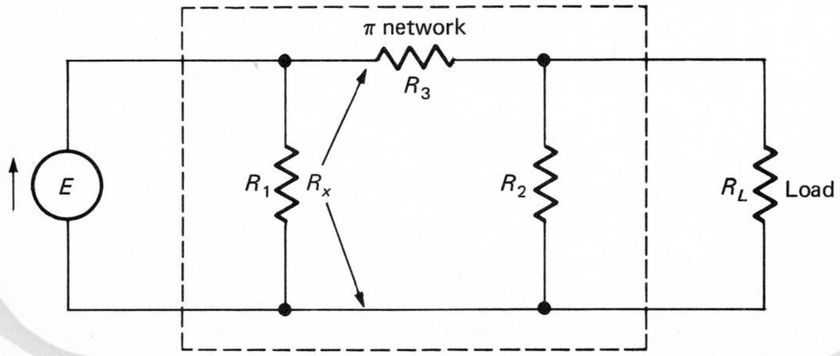

When R_x is known, the voltages and currents in R_2, R_3, and R_L can be calculated.

Example 5-14 What is the load current and voltage in the circuit in Fig. 5-23?

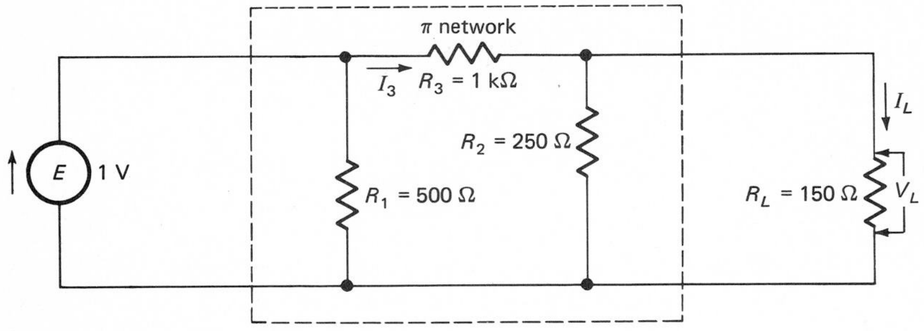

Fig. 5-23 Example 5-14: circuit.

Solution With R_1 omitted, the total resistance across the generator is

$$R_x = R_3 + \frac{R_2 R_L}{R_2 + R_L} = 1\text{ k}\Omega + \frac{250\ \Omega \times 150\ \Omega}{250\ \Omega + 150\ \Omega} = 1.094\text{ k}\Omega$$

$$I_3 = \frac{E}{R_x} = \frac{1\text{ V}}{1.094\text{ k}\Omega} = 0.9141\text{ mA}$$

$$I_L = I_3 \frac{R_2}{R_2 + R_L} = 0.9141\text{ mA}\frac{250\ \Omega}{250\ \Omega + 150\ \Omega} = \mathbf{0.5713\ mA}$$

$$V_L = I_L R_L = 0.5713\text{ mA} \times 150\ \Omega = \mathbf{85.7\ mV}$$

Example 5-15 A digital-to-analog (D-A) converter accepts a number in binary form and gives an output voltage proportional to the number. Figure 5-24 shows a 4-bit D-A converter. The four switches, which in practice would be transistors, store the binary number; +15 V represents 1 and ground is 0. The switches are shown in the position corresponding to the number 5 (in the binary system this is 0101). What is the output voltage V_o?

Fig. 5-24 Example 5-15: given circuit.

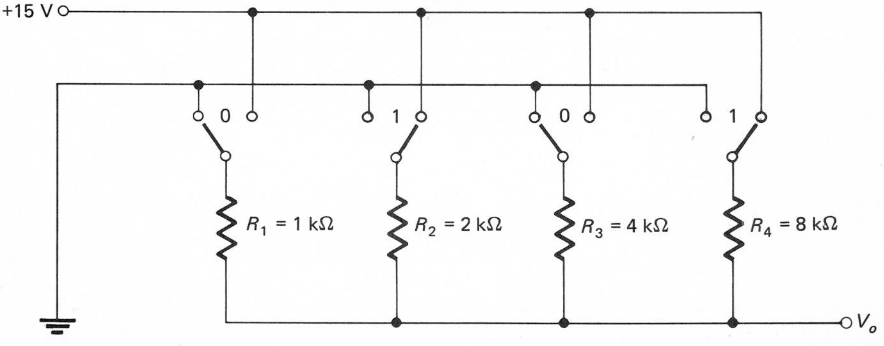

Solution Figure 5-24 can be redrawn as the equivalent circuit in Fig. 5-25.

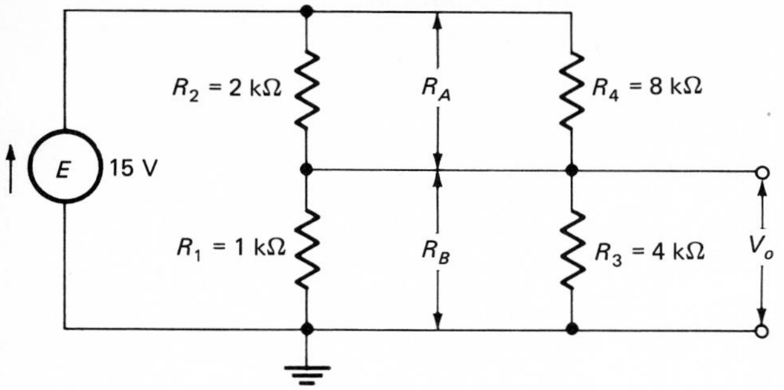

Fig. 5-25 Example 5-15: equivalent circuit.

R_A is R_2 and R_4 in parallel. Similarly, R_B is R_1 and R_3 in parallel.

$$R_A = \frac{R_2 R_4}{R_2 + R_4} = \frac{2\text{ k}\Omega \times 8\text{ k}\Omega}{2\text{ k}\Omega + 8\text{ k}\Omega} = 1.6\text{ k}\Omega$$

$$R_B = \frac{R_1 R_3}{R_1 + R_3} = \frac{1\text{ k}\Omega \times 4\text{ k}\Omega}{1\text{ k}\Omega + 4\text{ k}\Omega} = 0.8\text{ k}\Omega$$

$$V_o = E\frac{R_B}{R_A + R_B} = 15\text{ V}\frac{0.8\text{ k}\Omega}{1.6\text{ k}\Omega + 0.8\text{ k}\Omega} = \mathbf{5\ V}$$

It is not a coincidence that a 5-V output represents the number 5. The supply was selected as 15 V, because the maximum number that can be stored in four binary digits is 15. No matter what number, 0 through 15, is chosen to be stored by the switches in binary form, the output will be that number of volts.

EXERCISE PROBLEMS

Section 5-1

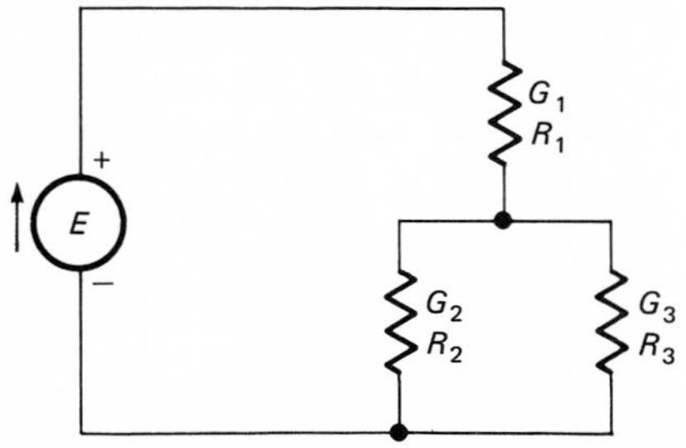

Fig. 5-1P Problems 5-1 and 5-2: circuit.

5-1 For the circuit in Fig. 5-1P, find the total resistance when

a. $R_1 = 180$ kΩ, $R_2 = 1.2$ MΩ, $R_3 = 330$ kΩ (*ans.* 438.8 kΩ)
b. $R_1 = 120$ Ω, $R_2 = 390$ Ω, $R_3 = 680$ Ω (*ans.* 367.9 Ω)
c. $R_1 = 4.7$ kΩ, $G_2 = 178$ μS, $G_3 = 370$ μS (*ans.* 6.525 kΩ)
d. $R_1 = 1500$ Ω, $G_2 = 0.001\ 22$ S, $R_3 = 820$ Ω (*ans.* 1910 Ω)
e. $G_1 = 2$ mS, $G_2 = 0.002\ 13$ S, $G_3 = 1.47$ mS (*ans.* 777.8 Ω)
f. $G_1 = 2.56$ μS, $G_2 = 4.55$ μS, $R_3 = 390$ kΩ (*ans.* 531.2 kΩ)
g. $G_1 = 0.122$ mS, $R_2 = 22$ kΩ, $R_3 = 33$ kΩ (*ans.* 21.4 kΩ)
h. $G_1 = 0.122$ S, $G_2 = 12.2$ mS, $R_3 = 820$ Ω (*ans.* 82.71 Ω)

5-2 For the circuit in Fig. 5-1P, let $R_1 = 47\ \Omega$, $R_2 = 68\ \Omega$, and $R_3 = 27\ \Omega$.
a. Find the current in R_1 when $E = 6$ V.
b. Find the current in R_3 when $E = 10$ V.
c. Find the pd across R_1 when the current in $R_3 = 0.7$ A.
d. Find E when the voltage across $R_2 = 21$ V.
e. Find the current in R_2 when the current in $R_1 = 30$ mA.
f. Find the current in R_1 when the current in $R_2 = 30$ mA.

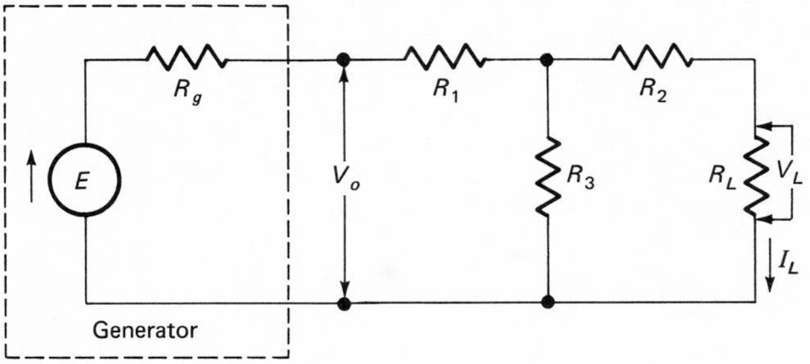

Fig. 5-2P Problem 5-3: circuit.

5-3 For the circuit in Fig. 5-2P, find I_L and V_L when
a. $E = 10$ V, $R_g = 600\ \Omega$, $R_1 = 400\ \Omega$, $R_2 = 400\ \Omega$, $R_3 = 250\ \Omega$, $R_L = 600\ \Omega$
(*ans.* 1.667 mA, 1 V)
b. $E = 5$ V, $R_g = 600\ \Omega$, $R_1 = 0$, $R_2 = 50\ \Omega$, $R_3 = 200\ \Omega$, $R_L = 50\ \Omega$
(*ans.* 5 mA, 0.25 V)
c. $V_o = 3$ V, $R_g = 50\ \Omega$, $R_1 = 0$, $R_2 = 100\ \Omega$, $R_3 = 30\ \Omega$, $R_L = 50\ \Omega$
(*ans.* 20 mA, 1 V)
d. $V_o = 6$ V, $R_g = 50\ \Omega$, $R_1 = 30\ \Omega$, $R_2 = 30\ \Omega$, $R_3 = 100\ \Omega$, $R_L = 50\ \Omega$
(*ans.* 44.78 mA, 2.238 V)
e. $E = 16$ V, $R_g = 600\ \Omega$, $R_1 = 200\ \Omega$, $R_2 = 400\ \Omega$, $R_3 = 800\ \Omega$, $R_L = 1200\ \Omega$
(*ans.* 4 mA, 4.8 V)

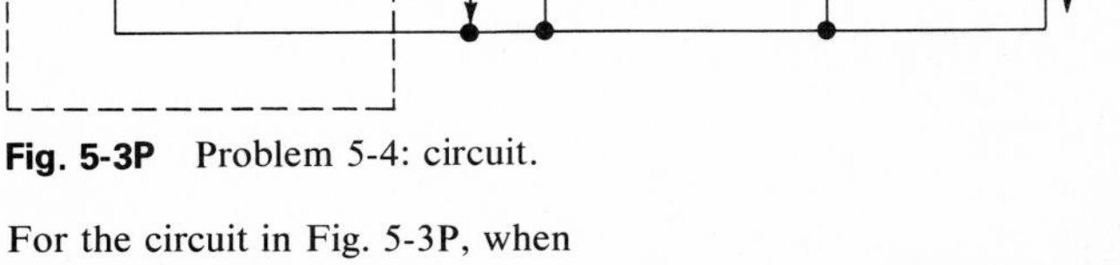

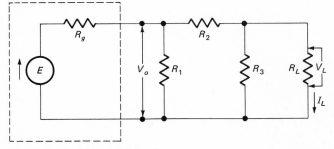

Fig. 5-3P Problem 5-4: circuit.

5-4 For the circuit in Fig. 5-3P, when
a. $E = 10$ V, $R_g = 600\ \Omega$, $R_1 = 900\ \Omega$, $R_2 = 1440\ \Omega$, $R_3 = 900\ \Omega$, $R_L = 600\ \Omega$, find I_L and V_L.
b. $V_o = 6$ V, $R_g = 50\ \Omega$, $R_1 = 230\ \Omega$, $R_2 = 69\ \Omega$, $R_3 = 230\ \Omega$, $R_L = 50\ \Omega$, find I_L and V_L.
c. $E = 16$ V, $R_g = 600\ \Omega$, $R_1 = 1400\ \Omega$, $R_2 = 700\ \Omega$, $R_3 = 2800\ \Omega$, $R_L = 1200\ \Omega$, find I_L and V_L.
d. $E = 20$ V, $R_g = 600\ \Omega$, $R_1 = 360\ \Omega$, $R_2 = 360\ \Omega$, $R_3 = 320\ \Omega$, $R_L = 600\ \Omega$, find V_o.

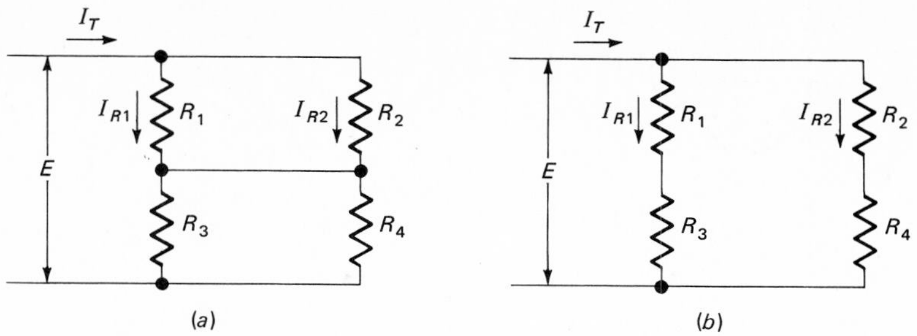

Fig. 5-4P (*a*) Problems 5-5 and 5-7; (*b*) Probs. 5-6 and 5-8.

5-5 For the circuit in Fig. 5-4P*a*, when $R_1 = 150\ \Omega$, $R_2 = 220\ \Omega$, $R_3 = 120\ \Omega$, $R_4 = 270\ \Omega$, and $V_{R4} = 16.9$ V, find
a. I_{R1} (*ans.* 121 mA)
b. I_T (*ans.* 203.4 mA)
c. E (*ans.* 35.04 V)

5-6 Repeat Prob. 5-5 using Fig. 5-4P*b*.

5-7 For the circuit in Fig. 5-4P*a*, when $R_1 = 220\ \Omega$, $R_2 = 470\ \Omega$, $R_3 = 560\ \Omega$, $E = 50$ V, and $V_{R4} = 29$ V, find
a. I_{R1} (*ans.* 95.46 mA)
b. I_{R2} (*ans.* 44.69 mA)
c. I_T (*ans.* 140.1 mA)
d. R_4 (*ans.* 328 Ω)

5-8 Repeat Prob. 5-7 using Fig. 5-4P*b*.

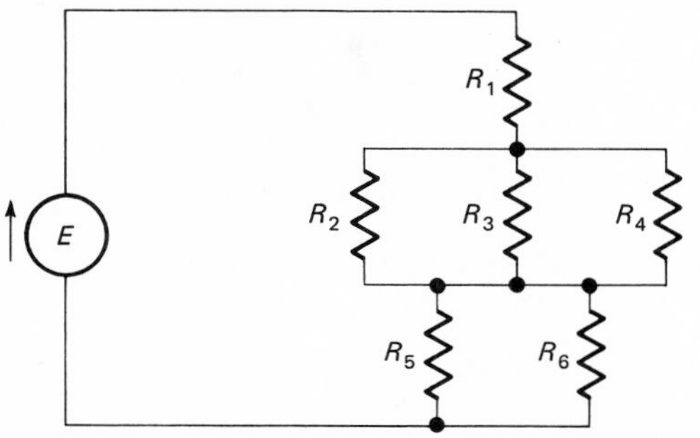

Fig. 5-5P Problems 5-9 to 5-12.

5-9 For the circuit in Fig. 5-5P, find R_T when
a. $R_1 = 12\ \Omega$, $R_2 = 18\ \Omega$, $R_3 = 27\ \Omega$, $R_4 = 18\ \Omega$, $R_5 = 22\ \Omega$, $R_6 = 15\ \Omega$ (*ans.* 27.67 Ω)
b. $R_1 = 680\ \Omega$, $R_2 = 120\ \Omega$, $R_3 = 560\ \Omega$, $R_4 = 470\ \Omega$, $R_5 = 820\ \Omega$, $R_6 = 220\ \Omega$ (*ans.* 935.1 Ω)
c. $R_1 = 120\ k\Omega$, $R_2 = 330\ k\Omega$, $R_3 = 82\ k\Omega$, $R_4 = 470\ k\Omega$, $R_5 = 56\ k\Omega$, $R_6 = 270\ k\Omega$ (*ans.* 224 kΩ)

5-10 Find the current in R_3 in Fig. 5-5P when
a. $E = 25$ V, $R_1 = 82\ \Omega$, $R_2 = 180\ \Omega$, $R_3 = 220\ \Omega$, $R_4 = 68\ \Omega$, $R_5 = 82\ \Omega$, $R_6 = 68\ \Omega$
b. $E = 50$ V and all resistance values remain as in *a*

5-11 Find the voltage across R_5 in Fig. 5-5P when $E = 220$ V, $R_1 = 120\ \Omega$, $R_2 = 150\ \Omega$, $R_3 = 330\ \Omega$, $R_4 = 180\ \Omega$, $R_5 = 82\ \Omega$, $R_6 = 120\ \Omega$. (*ans.* 45.74 V)

5-12 In Prob. 5-11, one resistor is open circuit. If the voltage measured across R_5 is 86.5 V, which resistor is open?

Section 5-2

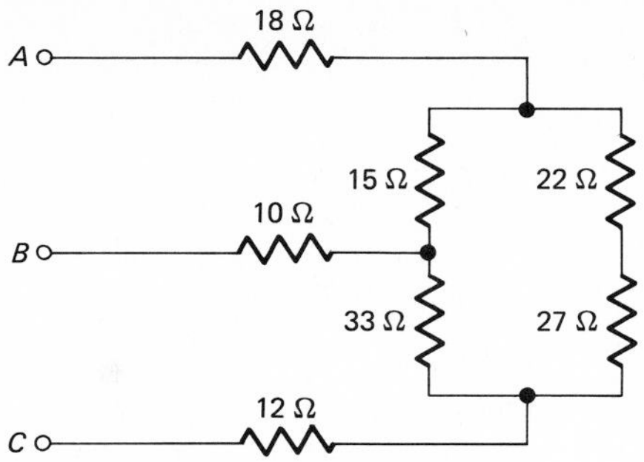

Fig. 5-6P Problem 5-13.

5-13 For the circuit in Fig. 5-6P,

a. With line B open circuit, find the resistance between lines A and C. (*ans*. 54.25 Ω)

b. With line C open circuit, find the resistance between lines A and B. (*ans*. 40.68 Ω)

c. With line A open circuit, find the resistance between lines B and C. (*ans*. 43.77 Ω)

d. With 20 V applied between lines A and C, find the voltage between lines A and B. (*ans*. 9.432 V)

e. With 20 V applied between lines A and B, find the voltage between lines B and C. (*ans*. 7.425 V)

f. With 20 V applied between lines B and C, find the voltage between lines A and C. (*ans*. 13.1 V)

g. With 20 V applied between lines A and C, find the power dissipated by each resistor and the total power dissipated by the circuit.
(*ans*. 2.447 W, 0 W, 1.631 W, 0.7319 W, 0.8983 W, 0.5206 W, 1.145 W, 7.374 W)

h. With 20 V applied between lines A and B, find the power dissipated by each resistor and the total power dissipated by the circuit.
(*ans*. 4.35 W, 2.417 W, 0 W, 0.127 W, 0.156 W, 2.59 W, 0.1907 W, 9.832 W)

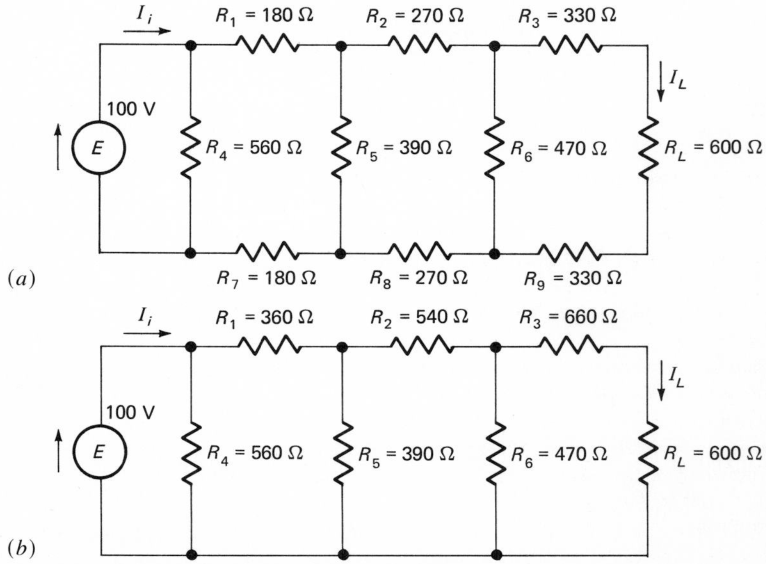

Fig. 5-7P (*a*) Problem 5-14; (*b*) Prob. 5-15.

5-14 For the circuit in Fig. 5-7P*a*, find

a. The equivalent resistance.
b. The input current I_i.
c. The output current I_L.
d. The voltage across R_L.
e. The power dissipated by R_5.

5-15 Repeat Prob. 5-14 using Fig. 5-7P*b*.
(*ans. a.* 296.5 Ω; *b.* 337.2 mA; *c.* 13.21 mA; *d.* 7.926 V; *e.* 4.719 W)

5-16 A 24-V battery is to be used to supply 15 V at 1.6 A to an amplifier system. Design a suitable voltage divider, with resistance values, using a 20 percent bleeder current.

a. Find the power dissipated by each resistor of the voltage divider.
b. Find the voltage supplied to the amplifier from this circuit if the amplifier current dropped to 1.2 A.
c. Calculate the value of a simple series dropping resistor which would supply 15 V at 1.6 A to the amplifier, and then find the voltage to the amplifier when the current drops to 1.2 A.
d. Repeat *b* and *c* for the condition in which the amplifier current drops to 0.8 A.

5-17 A 30-V 1-A power source is to supply three loads: +12 V at 300 mA, −12 V at 300 mA, and −8 V at 200 mA. Design a suitable voltage divider using a bleeder current of 20 percent of the greater of the positive and negative total load currents.
(*ans.* 10 Ω, 40 Ω, 80 Ω, 13.33 Ω)

5-18 Find the three output voltages of the voltage divider designed in Prob. 5-17 when

a. The −8-V load resistance rises to 80 Ω.
b. The −8-V load resistance drops to 25 Ω.

5-19 Power is fed to a house, from a 120-V 125-A power source, by a pair of #4 copper wires each 400 ft long. (#4 copper has a resistance of 0.2485 Ω/1000 ft.) What voltage would be measured at the house if a 1500-W water heater and two 1000-W 120-V stove elements were on. Assume there is no change in the resistance of the heating elements due to minor voltage changes. (*ans.* 114.5 V)

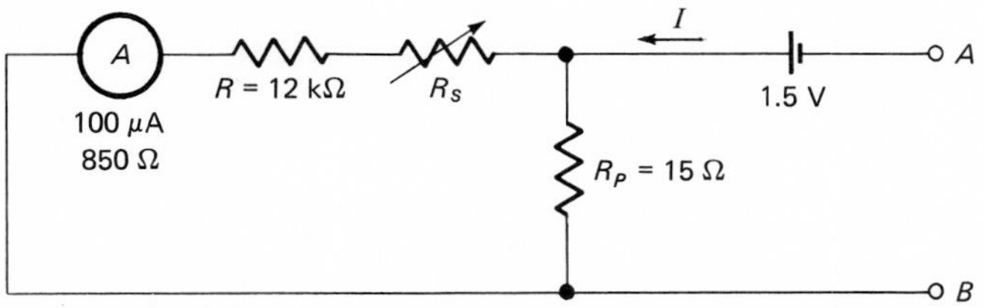

Fig. 5-8P Problem 5-20.

5-20 Figure 5-8P is the circuit for an ohmmeter. To zero the instrument, terminals *A* and *B* are connected together and R_s is adjusted for full-scale deflection (100 μA).

a. What value of R_s is required to obtain full-scale deflection?
b. What is the current *I* (short-circuit current) when *A* and *B* are connected together?
c. What is the current in the microammeter when *A* and *B* are connected together?
d. What current is read on the microammeter when the following resistances are connected between *A* and *B*? (*i*) 5 Ω, (*ii*) 10 Ω, (*iii*) 15 Ω, (*iv*) 20 Ω, (*v*) 25 Ω.

5-21 When measuring the characteristics of a semiconductor device, it is required that the voltage across the device and the current in the device be measured as accurately as possible. For the circuit in Fig. 5-18*a* of the text, $E = 40$ V, ammeter *A* is a 50-μA movement with a resistance of 5 kΩ, voltmeter *V* is an electronic voltmeter with a resistance of 11 MΩ, and resistance *R* represents a reverse-biased semiconductor diode and equals 5 MΩ. Calculate

a. The actual voltage across the diode. (*ans.* 39.94 V)
b. The voltage measured by the voltmeter. (*ans.* 39.94 V)
c. The actual current in the diode. (*ans.* 7.988 μA)
d. The current measured by the ammeter. (*ans.* 11.62 μA)

5-22 Repeat Prob. 5-21 using the circuit in Fig. 5-18*b*.

5-23 For the circuit in Fig. 5-18*a* of the text, $E = 1$ V, ammeter A is a 10-mA movement with a resistance of 25 Ω, voltmeter V is an electronic voltmeter with a resistance of 11 MΩ, and resistance R represents a forward-biased semiconductor diode and equals 100 Ω. Calculate

a. The actual voltage across the diode. (*ans.* 0.8 V)
b. The voltage measured by the voltmeter. (*ans.* 0.8 V)
c. The actual current in the diode. (*ans.* 8 mA)
d. The current measured by the ammeter. (*ans.* 8 mA)

5-24 Repeat Prob. 5-23 using the circuit in Fig. 5-18*b*.

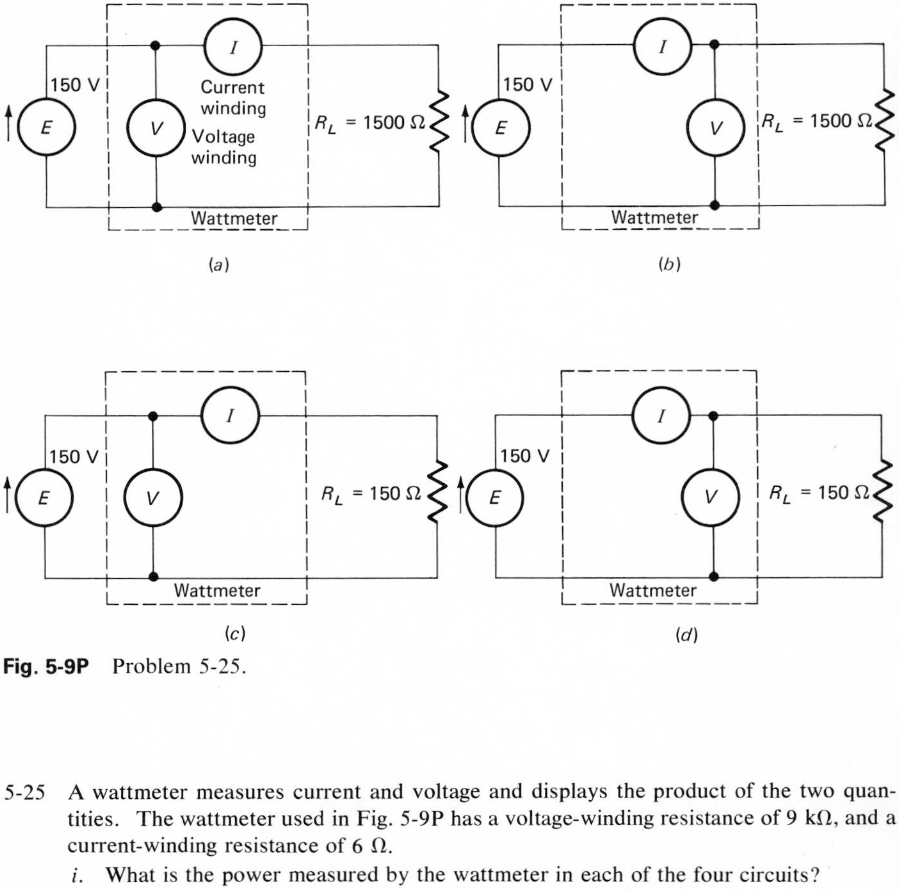

Fig. 5-9P Problem 5-25.

5-25 A wattmeter measures current and voltage and displays the product of the two quantities. The wattmeter used in Fig. 5-9P has a voltage-winding resistance of 9 kΩ, and a current-winding resistance of 6 Ω.

i. What is the power measured by the wattmeter in each of the four circuits?
(*ans. a.* 14.94 W; *b.* 17.33 W; *c.* 144.2 W; *d.* 140.8 W)

ii. What is the actual power dissipated by the load resistance in each of the four circuits? (*ans. a.* 14.88 W; *b.* 14.84 W; *c.* 138.6 W; *d.* 138.5 W)

5-26 The output resistance of a transistor amplifier is 8 Ω. It is required that four equal speakers be connected in series-parallel across the output. What must the resistance of each speaker be so that the series-parallel combination will have the same resistance as the amplifier?

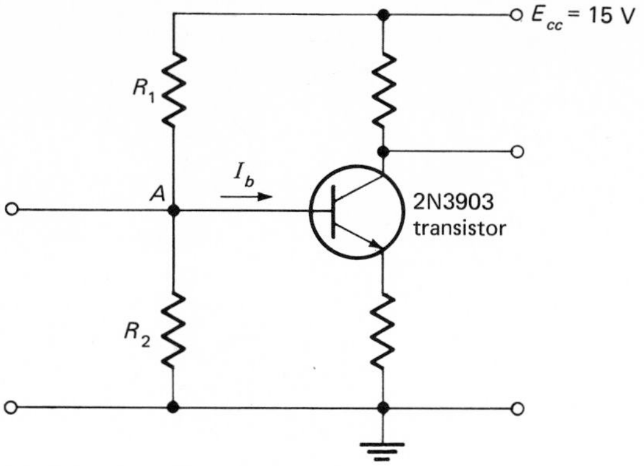

Fig. 5-10P Problem 5-27.

5-27 Figure 5-10P is the circuit of a transistor amplifier. I_b is to be 0.1 mA, and the voltage at point A is to be 6 V with respect to ground. If the bleeder current is to be five times I_b, find R_1 and R_2 (R_1 and R_2 form a voltage divider). (*ans.* 15 kΩ, 12 kΩ)

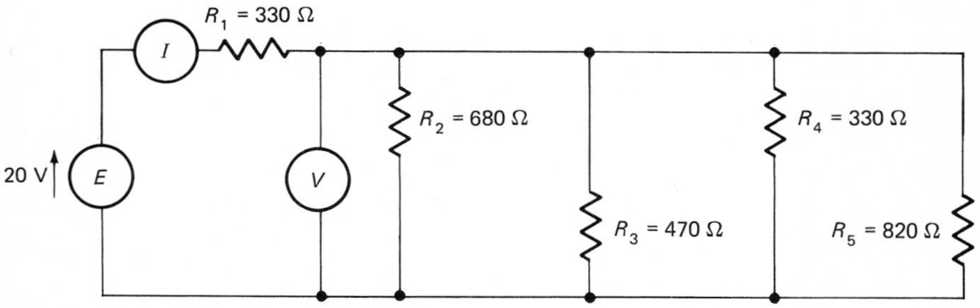

Fig. 5-11P Problem 5-28.

5-28 In the circuit in Fig. 5-11P which resistor is open circuit or disconnected when
a. $V = 7.72$ V?
b. $V = 6.45$ V?
c. $I = 39.6$ mA?
d. $I = 41.1$ mA?

5-29 One resistor has been removed from the circuit in Fig. 5-2 in the text. If I_3 measures 0.394 A, which resistor is missing? (*ans.* R_7)

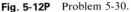

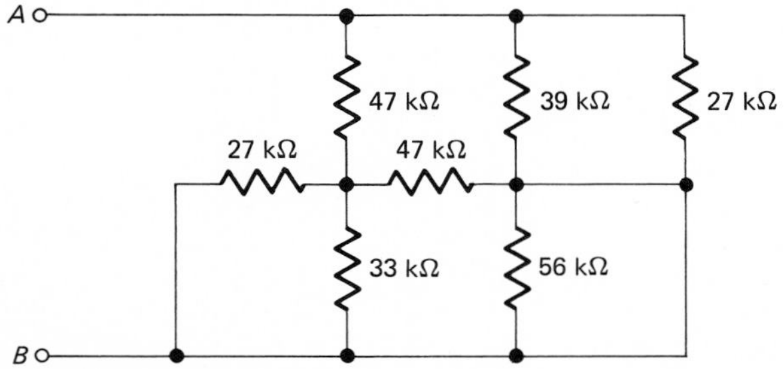

Fig. 5-12P Problem 5-30.

5-30 Find the total resistance between points A and B in Fig. 5-12P.

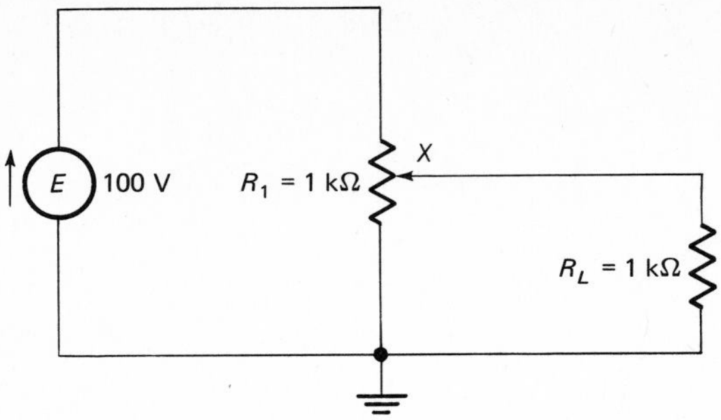

Fig. 5-13P Problem 5-31.

5-31 For the circuit in Fig. 5-13P, find the resistance of R_1 between point X and ground so that the voltage across R_L is 50 V. (*ans.* 618 Ω)

6

Network Analysis by Direct Methods

Network analysis is concerned with methods for determining the quantity values in a network or circuit. There are two general approaches:

1. Direct methods. The network is left in its original form, and its values are calculated. For efficient analysis, these methods are usually restricted to fairly simple circuits. The methods used in Chaps. 3 to 5 were direct.
2. Network reduction. The original network is converted into a more simple equivalent circuit for the rapid calculation of a quantity value, or series of values. Reduction techniques can be applied to simple or complex circuits.

This chapter presents several different direct methods of network analysis. There are always alternate ways of evaluating a circuit quantity; usually one will be quicker or easier than the others for a particular need.

6-1 CURRENT ASSUMPTION

Some networks consist of a generator feeding a load via a long string of elements. The ladder network of Fig. 6-1 is a typical example. It is used extensively in communications as an *attenuator,* a circuit for reducing and controlling signal strength.

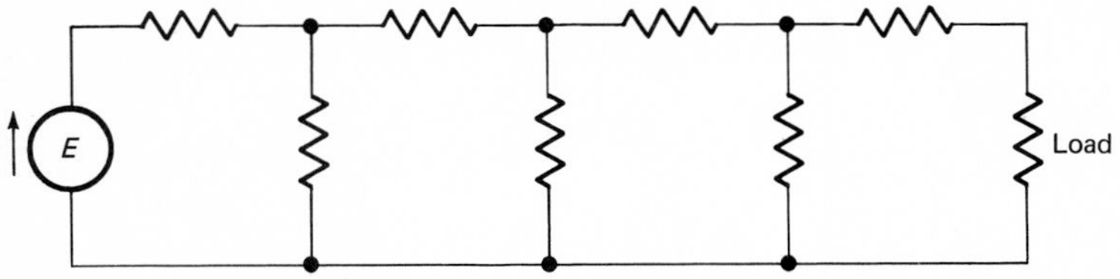

Fig. 6-1 Ladder network.

The currents or voltages in this circuit can be calculated in many ways. In these long-string circuits, with only one generator, the current assumption method is fast and involves only simple arithmetic.

A current in a linear circuit may be assumed to have any desired value if the following conditions are observed:

1. Other currents or voltages derived by using the assumed current value are accepted as valid only in their relationship to that current.
2. Derived currents and voltages are converted to correct values by multiplying by a common ratio.
3. The common ratio is obtained from some known value of current or voltage.

The use of this principle for circuit analysis can best be seen by working examples.

Example 6-1 What are the element currents in the circuit in Fig. 6-2?

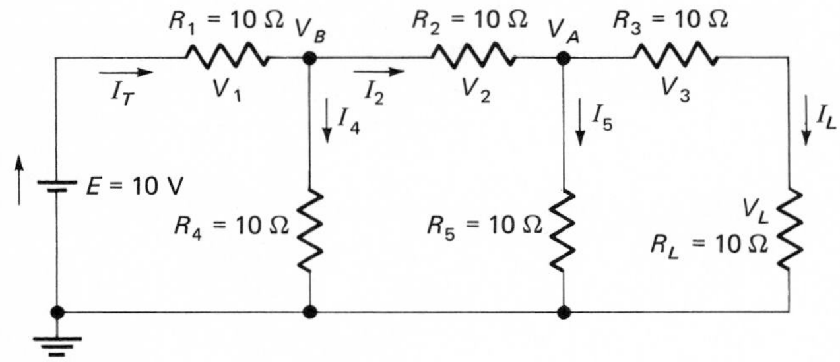

Fig. 6-2 Example 6-1: circuit.

Solution Simplicity is the aim, so an assumed current is chosen which is easy to use. Assume I_L to be 1 A; then,

$$V_L = I_L R_L = 1\text{ A} \times 10\ \Omega = 10\text{ V}$$

We already know that I_L is *not* 1 A, because V_L cannot be 10 V, the same as the generator emf. This does not matter for now, and so we continue:

$$V_3 = I_L R_3 = 1\text{ A} \times 10\ \Omega = 10\text{ V}$$

$$V_A = V_L + V_3 = 10\text{ V} + 10\text{ V} = 20\text{ V}$$

$$I_5 = \frac{V_A}{R_5} = \frac{20\text{ V}}{10\ \Omega} = 2\text{ A}$$

$$I_2 = I_5 + I_L = 2\text{ A} + 1\text{ A} = 3\text{ A}$$

$$V_2 = I_2 R_2 = 3\text{ A} \times 10\ \Omega = 30\text{ V}$$

$$V_B = V_2 + V_A = 30\text{ V} + 20\text{ V} = 50\text{ V}$$

$$I_4 = \frac{V_B}{R_4} = \frac{50\text{ V}}{10\ \Omega} = 5\text{ A}$$

$$I_T = I_4 + I_2 = 5\text{ A} + 3\text{ A} = 8\text{ A}$$

$$V_1 = I_T R_1 = 8\text{ A} \times 10\ \Omega = 80\text{ V}$$

$$E = V_1 + V_B = 80\text{ V} + 50\text{ V} = 130\text{ V}$$

If the load current I_L were 1 A, the battery emf would have to be 130 V. But the emf is only 10 V. In a linear network, all the currents and voltages due to a 130-V battery will be 13 times higher than for a 10-V battery. So, in this example, all the currents are actually $\frac{1}{13}$ of the values obtained from the 1-A assumption for I_L.

$$I_T = \frac{8\text{ A}}{13} = \mathbf{0.6154\ A}$$

$$I_4 = \frac{5\text{ A}}{13} = \mathbf{0.3846\ A}$$

$$I_2 = \frac{3\text{ A}}{13} = \mathbf{0.2308\ A}$$

$$I_5 = \frac{2\text{ A}}{13} = \mathbf{0.1538\ A}$$

$$I_L = \frac{1\text{ A}}{13} = \mathbf{0.0769\ A}$$

A lot of writing is shown for these assumed-current solutions. When these methods are actually used, only the assumed and real values of current and voltage need be noted, not the details of how they are calculated.

Example 6-1 intentionally had simple numbers to demonstrate the technique. All assumed values could have been worked out mentally. Even when the numbers are not as simple, only addition and a few basic slide-rule operations are involved. The following examples have commercial, preferred-resistance values which are arithmetically awkward.

Example 6-2 In the circuit in Fig. 6-3, how much of the generator voltage E is delivered to the load?

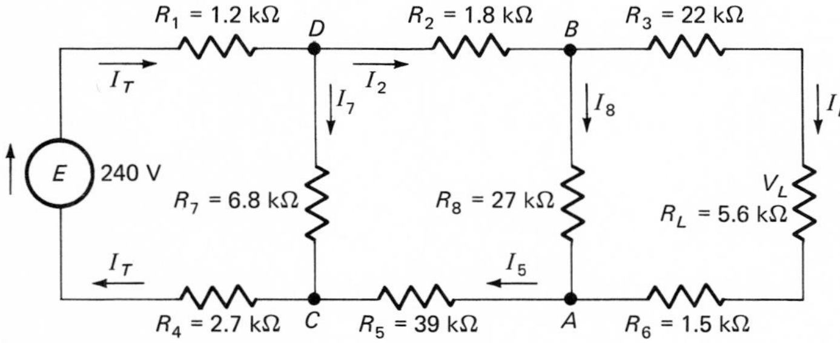

Fig. 6-3 Example 6-2: circuit.

Solution Assuming I_L to be 1 mA, then

$$V_{BA} = I_L(R_3 + R_L + R_6) = 1\text{ mA}(22\text{ k}\Omega + 5.6\text{ k}\Omega + 1.5\text{ k}\Omega) = 29.1\text{ V}$$

$$I_8 = \frac{V_{BA}}{R_8} = \frac{29.1\text{ V}}{27\text{ k}\Omega} = 1.078\text{ mA}$$

$$I_2 = I_5 = I_8 + I_L = 1.078\text{ mA} + 1\text{ mA} = 2.078\text{ mA*}$$

* I_2 and I_5 are exactly the same currents $(I_8 + I_L)$. When elements carry the same current (not just currents of the same value), they may be said to be *virtually in series,* even though they are not connected together.

$$V_{DC} = V_{BA} + I_2(R_2 + R_5) = 29.1 \text{ V} + 2.078 \text{ mA}(1.8 \text{ k}\Omega + 39 \text{ k}\Omega) = 113.9 \text{ V}$$

$$I_7 = \frac{V_{DC}}{R_7} = \frac{113.9 \text{ V}}{6.8 \text{ k}\Omega} = 16.75 \text{ mA}$$

$$I_T = I_7 + I_2 = 16.75 \text{ mA} + 2.078 \text{ mA} = 18.83 \text{ mA}$$

$$E = V_{DC} + I_T(R_1 + R_4) = 113.9 \text{ V} + 18.83 \text{ mA}(1.2 \text{ k}\Omega + 2.7 \text{ k}\Omega) = 187.3 \text{ V}$$

The emf, calculated on an assumed load current of 1 mA, is 187.3 V. But the real emf is 240 V, so the calculated load current (and all the other currents and voltages) are too low by the factor 240 V/187.3 V = 1.281. The real value of the load current is

$$I_L = 1 \text{ mA} \times 1.281 = 1.281 \text{ mA}$$

$$V_L = I_L R_L = 1.281 \text{ mA} \times 5.6 \text{ k}\Omega = \mathbf{7.174 \text{ V}}$$

or 3 percent of the generator voltage E is delivered to the load.

Example 6-3 Figure 6-4 is the schematic diagram of an attenuator in an item of electronic test equipment. Resistor R_4 has become faulty. What power rating is required for the replacement resistor?

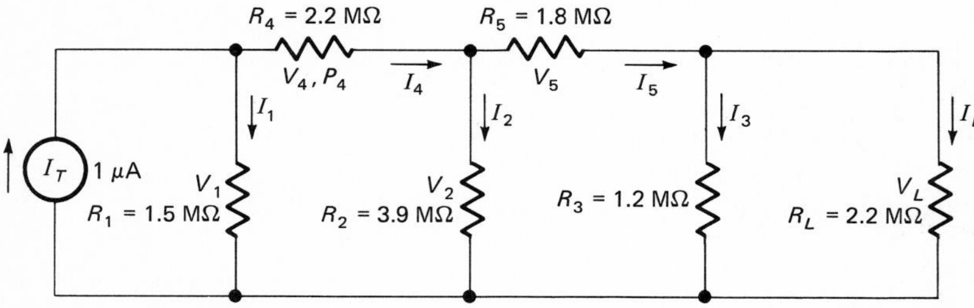

Fig. 6-4 Example 6-3: circuit.

Solution Assume that $I_L = 1\ \mu$A,

$$V_L = I_L R_L = 1\ \mu\text{A} \times 2.2 \text{ M}\Omega = 2.2 \text{ V}$$

$$I_3 = \frac{V_L}{R_3} = \frac{2.2 \text{ V}}{1.2 \text{ M}\Omega} = 1.833\ \mu\text{A}$$

$$I_5 = I_3 + I_L = 1.833 + 1\ \mu\text{A} = 2.833\ \mu\text{A}$$

$$V_5 = I_5 R_5 = 2.833\ \mu\text{A} \times 1.8 \text{ M}\Omega = 5.099 \text{ V}$$

$$V_2 = V_5 + V_L = 5.099 \text{ V} + 2.2 \text{ V} = 7.299 \text{ V}$$

$$I_2 = \frac{V_2}{R_2} = \frac{7.299 \text{ V}}{3.9 \text{ M}\Omega} = 1.872\ \mu\text{A}$$

$$I_4 = I_2 + I_5 = 1.872\ \mu\text{A} + 2.833\ \mu\text{A} = 4.705\ \mu\text{A}$$

$$V_4 = I_4 R_4 = 4.705\ \mu\text{A} \times 2.2 \text{ M}\Omega = 10.35 \text{ V}$$

$$V_1 = V_4 + V_2 = 10.35 \text{ V} + 7.299 \text{ V} = 17.65 \text{ V}$$

$$I_1 = \frac{V_1}{R_1} = \frac{17.65 \text{ V}}{1.5 \text{ M}\Omega} = 11.77\ \mu\text{A}$$

$$I_T = I_1 + I_4 = 11.77\ \mu\text{A} + 4.705\ \mu\text{A} = 16.47\ \mu\text{A}$$

But I_T is actually 1 μA, so all the calculated currents and voltages are too high by the factor 16.47. The real values of V_4 and I_4 are

$$V_4 = \frac{10.35 \text{ V}}{16.47} = 0.6284 \text{ V}$$

$$I_4 = \frac{4.705 \ \mu\text{A}}{16.47} = 0.2857 \ \mu\text{A}$$

The power dissipated by R_4 is

$$P_4 = V_4 I_4 = 0.6284 \text{ V} \times 0.2857 \ \mu\text{A} = \mathbf{0.1795 \ \mu W}$$

This is a very small power. Any commercial resistor will be able to dissipate it without difficulty.

6-2 NODAL ANALYSIS

Nodal analysis and loop analysis (Sec. 6-3) can be regarded as complementary. Nodal analysis relates to current sources and conductances, while loop analysis is concerned with voltage sources and resistances. This duality leads to some general guidelines on when to use nodal analysis.

Note that these are only guidelines, not rigid rules. Such guidelines will aid in choosing the most suitable of several different methods of analyzing the same circuit. A good choice of method can reduce the amount of work and improve the reliability of the results.

Nodal analysis should be considered for a network if

1. The energy sources are predominantly current generators.
2. Voltage values are to be found.
3. There are a number of different quantities to be calculated.
4. There are two or more energy sources.
5. The number of independent nodes is less than the number of loops.

The first step in nodal analysis is to identify the nodes. A *node* is a junction, or connection point, of three or more conductors.

Figure 6-5 is a π network connected between two current generators.

Fig. 6-5 Identifying nodes.

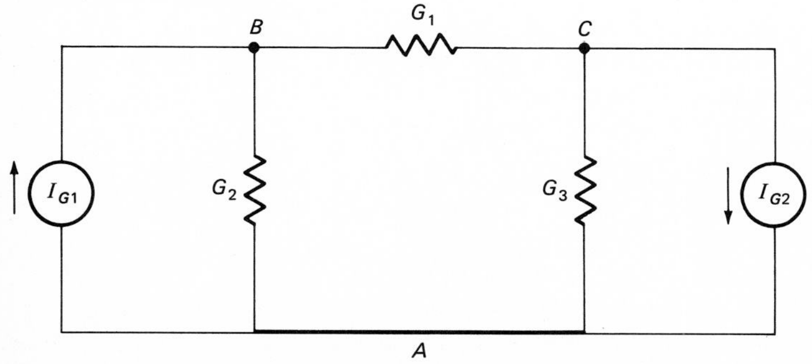

In the circuit in Fig. 6-5, there are three separate nodes. Node A is the junction of four conductors, the bottom connections to the two generators and elements G_2 and G_3. Node B is the junction of three conductors connecting generator I_{G1} to elements G_1 and G_2. Node C is similarly the junction of three conductors.

A reference node is selected. The choice is free; any node can be the reference. It may be convenient to choose for the reference node a circuit common or ground connection. All nodes other than the reference node are called *independent* nodes. If the common line A in Fig. 6-5 is chosen as the reference node, points B and C will be the independent nodes. The voltages at the independent nodes, with respect to the reference node, are labeled V_B and V_C in Fig. 6-6.

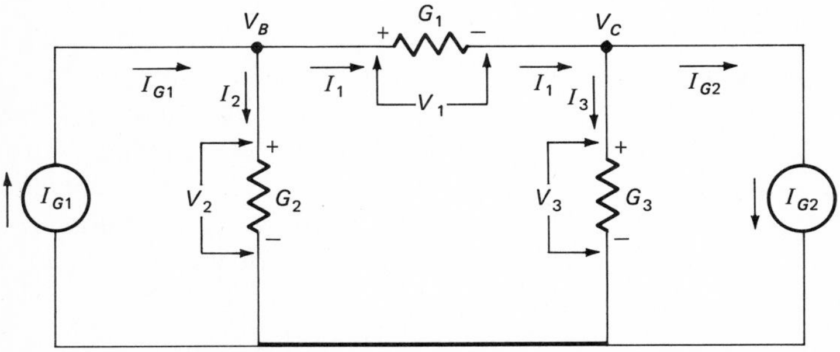

Fig. 6-6 Circuit in Fig. 6-5 with currents and voltages identified.

To aid in the initial understanding of nodal analysis, assumed current directions and element voltage polarities can be indicated. These are shown in Fig. 6-6. When assumptions are made, specific current directions must be consistent with the related voltage polarities ($+ \rightarrow -$ or $- \leftarrow +$). Whether or not assumptions are correct will be established by the analysis results. A positive result confirms an assumption; a negative result indicates that the direction or polarity is opposite that assumed.

Figure 6-6 is the same as Fig. 6-5, except for the addition of current and pd symbols, directions, and polarities.

The conditions existing at the independent nodes are written as equations. The development of the equations for nodes B and C in Fig. 6-6 is as follows:

AT NODE B Generator current I_{G1} is entering the node. Element currents I_1 and I_2 are leaving the node.

The Kirchhoff's current law equation for node B is

$$I_1 + I_2 = I_{G1}$$

All currents, except generator currents, are now expressed in Ohm's law terms, $I = VG$. The pd across G_1 will be $V_B - V_C$, because the assumed direction of current I_1, left to right, implies that V_B is greater than V_C.

$$I_1 = V_1 G_1 = (V_B - V_C)G_1$$

$$I_2 = V_2 G_2 = V_B G_2$$

Therefore the equation for node B is

$$(V_B - V_C)G_1 + V_B G_2 = I_{G1}$$

or

$$V_B(G_1 + G_2) - V_C G_1 = I_{G1} \qquad (6\text{-}1)$$

AT NODE C Element current I_1 is entering the node. Generator current I_{G2} and element current I_3 are leaving the node.

$$I_3 = V_3 G_3 = V_C G_3$$

The equation for node C is

$$I_{G2} + I_3 = I_1$$

$$I_{G2} + V_C G_3 = (V_B - V_C)G_1$$

or

$$V_C(G_1 + G_3) - V_B G_1 = -I_{G2} \qquad (6\text{-}2)$$

Solving the pair of simultaneous equations, Eqs. (6-1) and (6-2), will give the values of V_B and V_C. All other circuit quantities can then be evaluated.

It was assumed that nodes B and C in Fig. 6-6 are both positive with respect to reference node A. The node equations for any circuit can be developed quickly by assuming that all independent nodes are positive with respect to the chosen reference node, and noting what is implied by Eqs. (6-1) and (6-2). For this rapid method there is no need to assign current directions or voltage polarities. The node voltage polarities will be established by the equation solutions. If a solution is positive, the relevant node voltage is positive, and vice-versa. Element polarities and current directions can be determined by applying basic principles and conventions.

WRITING THE EQUATION FOR A NODE

The recommended procedure for writing the equation for a node is as follows. The term "local node" is used to apply to the node for which the equation is being written. The term "remote node" implies a node connected to the local node by a conductance.

Element currents
1. Place on the left side of the equation currents in conductances.
2. Multiply the local node voltage (with respect to the reference) by the sum of the conductances connected to the node. Show the product as a

positive quantity. Ignore elements in series with current sources, as they cannot affect circuit currents.
3. Multiply each remote node voltage by the conductance of the element connecting it to the local node. Show the product as a negative quantity.

Source currents
1. Place source currents on the right side of the equation.
2. Show currents entering the local node as positive quantities.
3. Show currents leaving the local node as negative quantities.
4. If there are no source currents at the local node, show as zero.

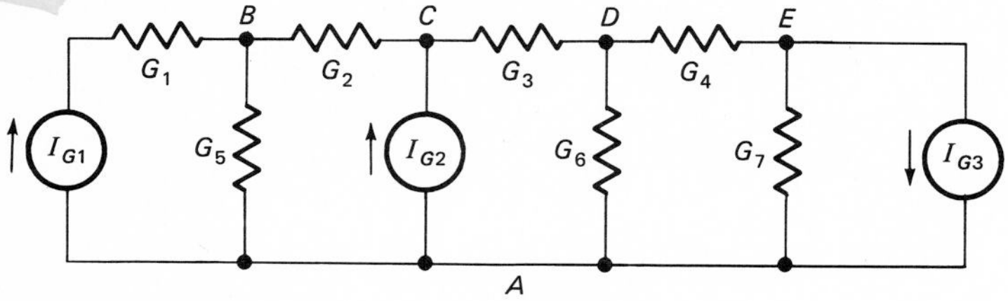

Fig. 6-7 Multisource network with nodes identified.

The circuit in Fig. 6-7 is fairly complex, but its nodal equations can be written rapidly. If point (line) A is the reference node, points B, C, D, and E are independent nodes. Using the steps just described, the node equations are

Node B: $$V_B(G_2 + G_5) - V_C G_2 = I_{G1}$$

Node C: $$V_C(G_2 + G_3) - V_B G_2 - V_D G_3 = I_{G2}$$

Node D: $$V_D(G_3 + G_4 + G_6) - V_C G_3 - V_E G_4 = 0$$

Node E: $$V_E(G_4 + G_7) - V_D G_4 = -I_{G3}$$

Example 6-4 What are the values of each voltage and current in the schematic diagram in Fig. 6-8?

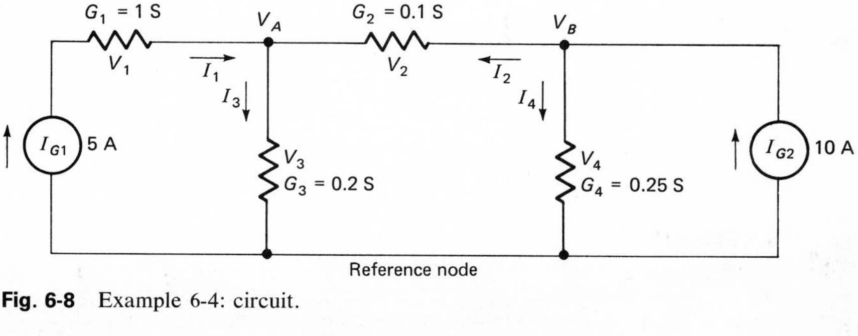

Fig. 6-8 Example 6-4: circuit.

Solution

At node A:

$$V_A(G_2 + G_3) - V_B G_2 = I_{G1}$$

$$V_A(0.1\text{ S} + 0.2\text{ S}) - V_B\, 0.1\text{ S} = 5\text{ A}$$

At node B:

$$V_B(G_2 + G_4) - V_A G_2 = I_{G2}$$

$$V_B(0.1\text{ S} + 0.25\text{ S}) - V_A\, 0.1\text{ S} = 10\text{ A}$$

If units are omitted, the equations reduce to

$$0.3V_A - 0.1V_B = 5$$

$$-0.1V_A + 0.35V_B = 10$$

Solving the pair of simultaneous equations by any convenient method gives the values of V_A and V_B

$$V_A = \mathbf{28.95\ V} = V_3$$

$$V_B = \mathbf{36.84\ V} = V_4$$

$$V_2 = V_B - V_A = 36.84\text{ V} - 28.95\text{ V} = \mathbf{7.89\ V}$$

$$I_2 = V_2 G_2 = 7.89\text{ V} \times 0.1\text{ S} = \mathbf{0.789\ A}$$

$$I_3 = V_A G_3 = 28.95\text{ V} \times 0.2\text{ S} = \mathbf{5.79\ A}$$

$$I_4 = V_B G_4 = 36.84\text{ V} \times 0.25\text{ S} = \mathbf{9.21\ A}$$

$$I_1 = I_{G1} = \mathbf{5\ A}$$

$$V_1 = \frac{I_{G1}}{G_1} = \frac{5\text{ A}}{1\text{ S}} = \mathbf{5\ V}$$

The results of Example 6-4 satisfy the Kirchhoff's law requirements at all three nodes (currents entering equal currents leaving).

Node A:

$$I_{G1} + I_2 = I_3$$

$$5\text{ A} + 0.789\text{ A} \simeq 5.79\text{ A}$$

Node B:

$$I_{G2} = I_2 + I_4$$

$$10\text{ A} \simeq 0.789\text{ A} + 9.21\text{ A}$$

Reference node:

$$I_3 + I_4 = I_{G1} + I_{G2}$$

$$5.789\text{ A} + 9.21\text{ A} \simeq 5\text{ A} + 10\text{ A}$$

The slight inconsistencies in the Kirchhoff checks result from the arithmetic, not from the basic accuracy of the method. Of course the quantities in the example could have been chosen to give exact results, but such neat situations rarely occur in practice.

If the circuit includes voltage generators, these should be converted to equivalent current generators, using the principle established in Sec. 4-4.

This principle is summarized as follows:

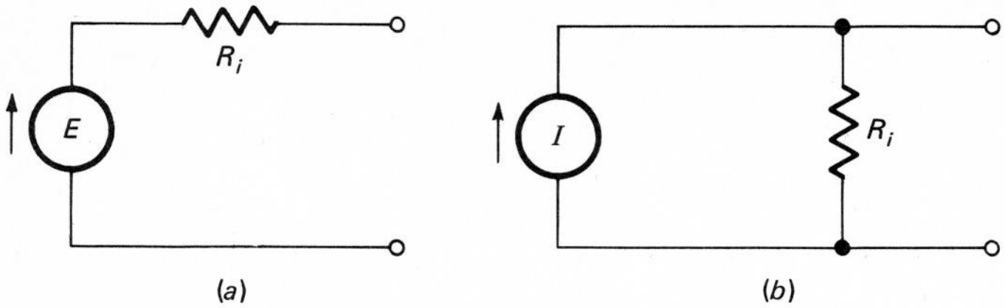

Fig. 6-9 (*a*) Voltage generator; (*b*) current generator.

The current generator in Fig. 6-9*b* will be the equivalent of the voltage generator in Fig. 6-9*a* if

1. The internal resistance R_i of each generator is the same.
2. The source current I of the current generator is related to the voltage generator emf E by the equation

$$I = \frac{E}{R_i}$$

Example 6-5 What is the voltage V_2 in the circuit in Fig. 6-10?

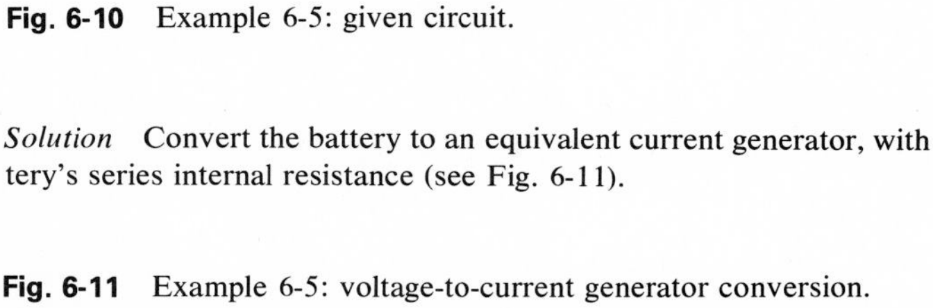

Fig. 6-10 Example 6-5: given circuit.

Solution Convert the battery to an equivalent current generator, with R_1 regarded as the battery's series internal resistance (see Fig. 6-11).

Fig. 6-11 Example 6-5: voltage-to-current generator conversion.

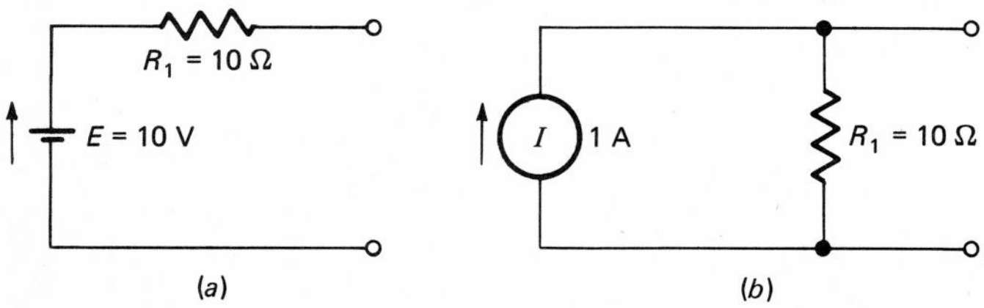

Redraw Fig. 6-10 with all resistance values changed to conductances, omitting R_3, since it is an element in series with a current generator. The results are shown in Fig. 6-12.

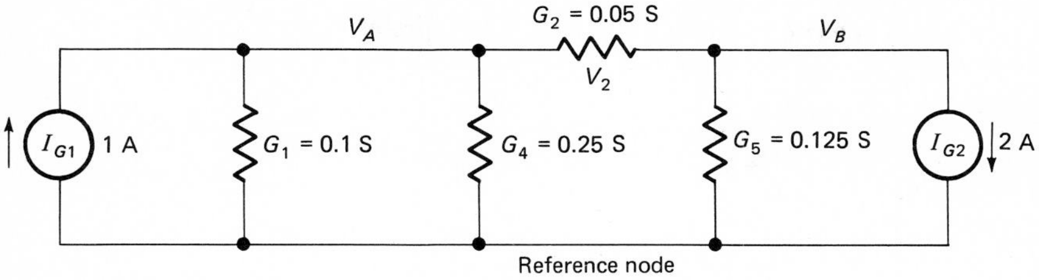

Fig. 6-12 Figure 6-10 with a current generator replacing the battery.

The equations for the two independent nodes are

$$V_A\ (0.1\ \text{S} + 0.25\ \text{S} + 0.05\ \text{S}) - V_B\ (0.05\ \text{S}) = 1\ \text{A}$$

$$V_B\ (0.125\ \text{S} + 0.05\ \text{S}) - V_A\ (0.05\ \text{S}) = -2\ \text{A}$$

Reduce the equations to their simplest forms

$$0.4V_A - 0.05V_B = 1$$

$$-0.05V_A + 0.175V_B = -2$$

The solutions to these equations are

$$V_A = 1.11\ \text{V}$$

$$V_B = -11.1\ \text{V}$$

These solutions mean that the voltages at the ends of element R_2, with respect to the reference node, are as shown in Fig. 6-13.

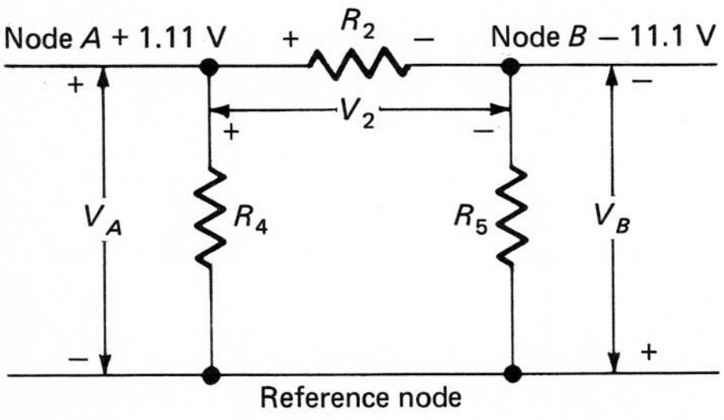

Fig. 6-13 Voltage across R_2 in Fig. 6-10.

$$V_2 = V_A - V_B = 1.11\ \text{V} - (-11.1\ \text{V}) = \mathbf{12.21\ V}$$

The polarity of V_2 will be as shown.

Example 6-6 Figure 6-14 shows the basic elements of an analog adding circuit. Show that the output voltage V_o is approximately proportional to the sum of the input voltages V_1 to V_3.

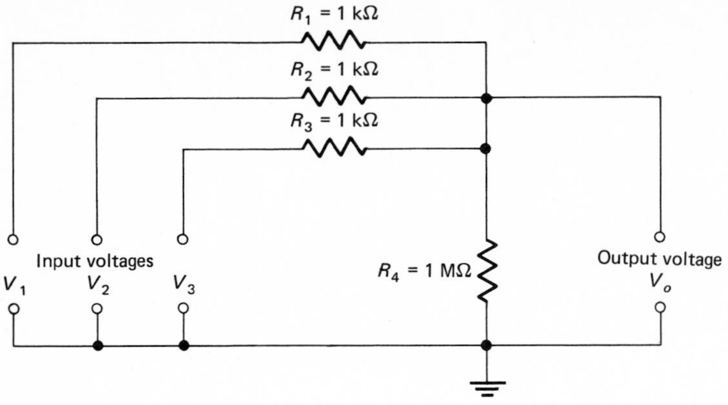

Fig. 6-14 Example 6-6: circuit.

Solution Redrawing the circuit in Fig. 6-14 with the input voltages and resistances converted to current sources and conductances results in the equivalent circuit in Fig. 6-15*a*.

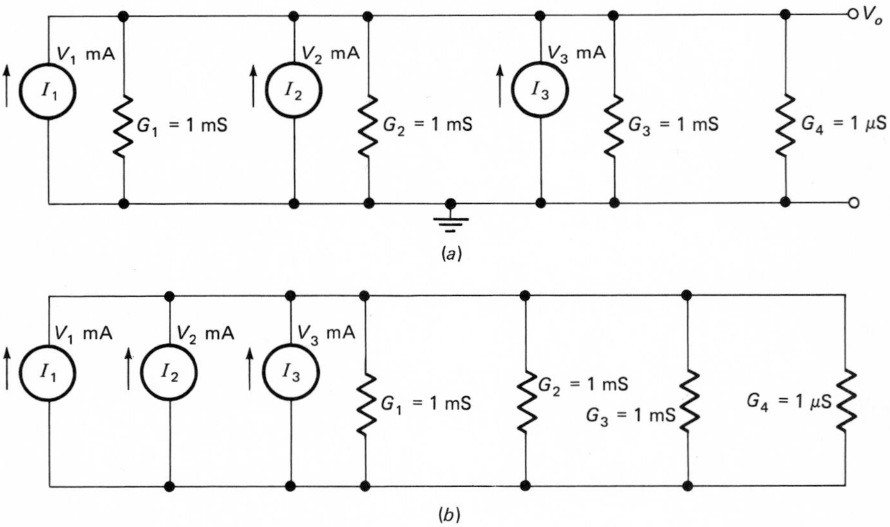

Fig. 6-15 Equivalent circuits of Fig. 6-14.

In Fig. 6-15*a* each of the current sources I_1 to I_3 and associated internal conductances G_1 to G_3 is obtained from a voltage source consisting of the related voltage V_1 to V_3 and resistance R_1 to R_3.

Each of the resistances R_1 to R_3 has the value 1 kΩ, so that each source current has a value in milliamperes numerically equal to its related voltage (e.g., $I_1 = V_1/R_1 = V_1$ volts/1 kΩ $= V_1$ milliamperes).

Each of the conductances G_1 to G_3 is the reciprocal of the related resistance R_1 to R_3, and hence equals 1 mS (e.g., $G_1 = 1/R_1 = 1/1$ kΩ = 1 mS).

The conductance G_4 is the reciprocal of resistance R_4 and has the value 1 μS.

Figure 6-15*b* is the circuit of Fig. 6-15*a* rearranged so that the source currents and source conductances are grouped.

In the equivalent circuits in Fig. 6-15 there is only one independent node, and V_o is its voltage with respect to the reference node. The nodal equation is

$$V_o(G_1 + G_2 + G_3 + G_4) = I_1 + I_2 + I_3$$

$$V_o(1\ \text{mS} + 1\ \text{mS} + 1\ \text{mS} + 1\ \mu\text{S}) = (V_1 + V_2 + V_3) \qquad \text{mA}$$

As 1 μS is very small compared with 1 mS, the quantity in parentheses on the left of the equation is approximately equal to 3 mS.

$$V_o \times 3\ \text{mS} \simeq (V_1 + V_2 + V_3) \qquad \text{mA}$$

$$V_o \simeq \frac{V_1 + V_2 + V_3}{3} \qquad \text{V}$$

In an adding circuit such as that in Fig. 6-14, with any number of inputs, the output voltage will be approximately equal to the sum of the input voltages divided by the number of inputs. Thus the output voltage will be approximately proportional to the sum of the input voltages.

Example 6-7 The circuit in Fig. 6-16 is a transistor switch. It is to be biased by R_1 and R_2 in the ON condition. In this state, the base-to-emitter voltage V_{BE} is to be 0.7 V, and the base current I_B is to be 1 mA. If a value of 1 kΩ is chosen for R_1, what resistance would be suitable for R_2?

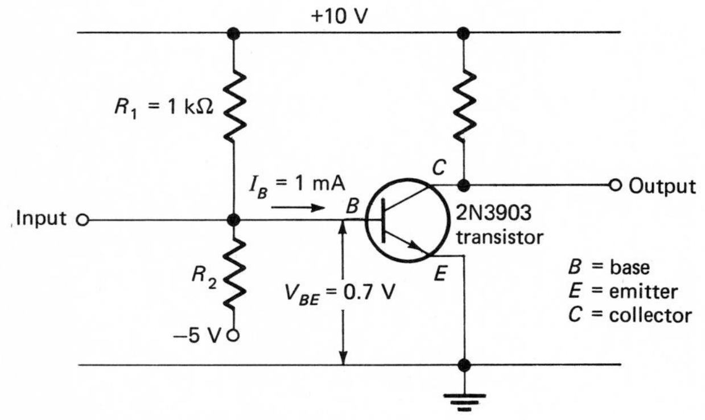

Fig. 6-16 Example 6-7: circuit.

Solution The base-to-emitter resistance is

$$R_{BE} = \frac{V_{BE}}{I_B} = \frac{0.7\ \text{V}}{1\ \text{mA}} = 700\ \Omega$$

The bias circuit can be shown as in Fig. 6-17.

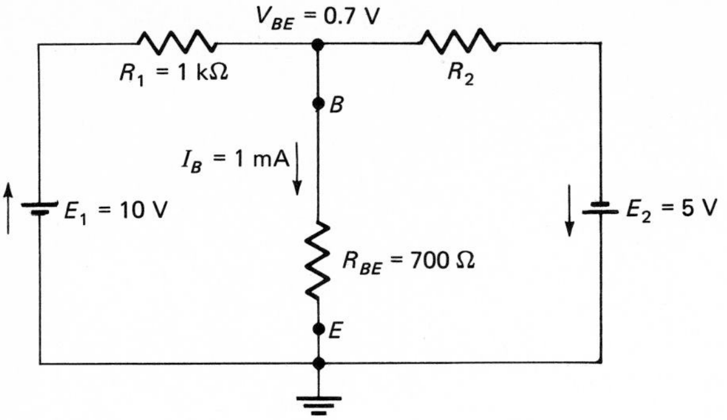

Fig. 6-17 Equivalent circuit of Fig. 6-16.

If E_1 with R_1, and E_2 with R_2 are considered voltage generators converted to current generators, Fig. 6-18 results.

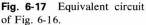

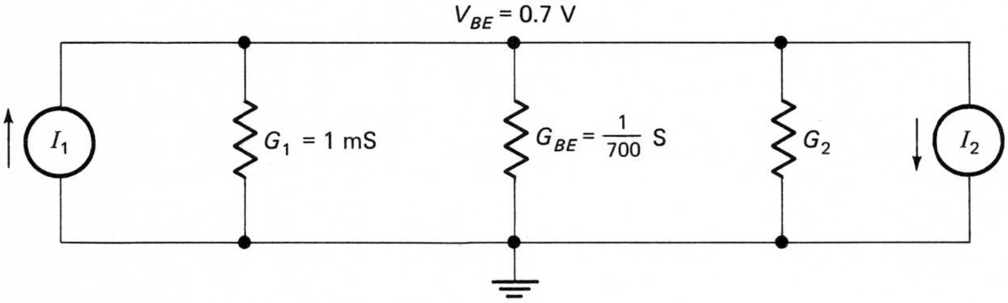

Fig. 6-18 Current-source equivalent of Fig. 6-17.

$$I_1 = E_1G_1 = 10 \text{ V} \times 1 \text{ mS} = 10 \text{ mA}$$

$$I_2 = E_2G_2 = 5 \text{ V} \times G_2 \text{ siemens} = 5G_2 \quad \text{amperes}$$

The equation for the single node is

$$V_{BE}(G_1 + G_{BE} + G_2) = I_1 - I_2$$

$$0.7 \text{ V}(10 \text{ mS} + \tfrac{1}{700} \text{ S} + G_2) = 10 \text{ mA} - 5G_2 \text{ amperes}$$

$$7 \times 10^{-4} + 10^{-3} + 0.7\,G_2 = 10^{-2} - 5G_2$$

$$G_2 = 1.456 \text{ mS}$$

$$R_2 = \mathbf{686.8\ \Omega}$$

The preferred value, 680 Ω, would probably be used.

Example 6-8 The rudder of an airplane is controlled by an electronic servosystem. The rudder's position is relayed to the control circuit as a current magnitude. This current connects to a bridge circuit, as shown in Fig. 6-19. The variable resistor R_5 is controlled by an electric motor so that the input to the amplifier V_{AC} is always close to zero. The control current I_2 has a range of 1 to 6 mA, corresponding to the extreme-left and -right positions of the rudder. The current source resistance is constant at 1 kΩ. What is the required resistance range for R_5 to maintain the bridge in balance with V_{AC} equal to zero?

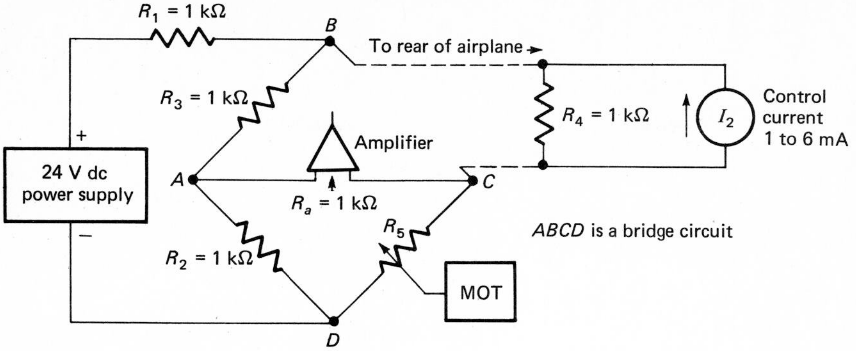

Fig. 6-19 Example 6-8: circuit.

Solution Making D the reference node, converting the power supply (voltage source) and R_1 to a current source, and changing resistances to conductances gives Fig. 6-20, the equivalent circuit.

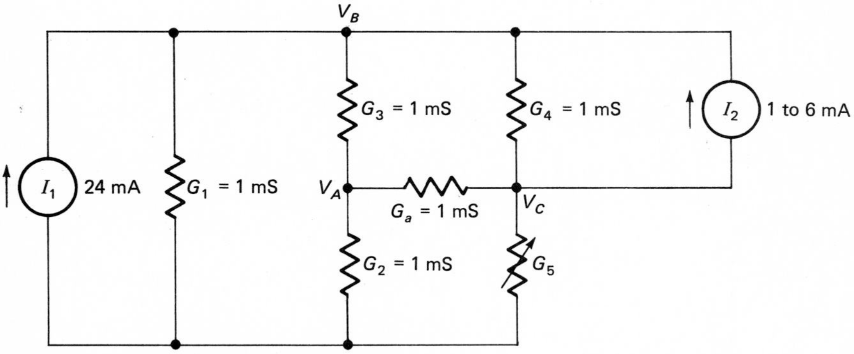

Fig. 6-20 Equivalent circuit of Fig. 6-19.

With the voltages in volts, currents in milliamperes, and conductances in millisiemens, the node equations are

Node A: $$3V_A - V_B - V_C = 0$$

Node B: $$-V_A + 3V_B - V_C = 24 + I_2$$

Node C: $$-V_A - V_B + (2 + G_5)V_C = -I_2$$

These equations apply whether or not the bridge is balanced, and for any value of the control current I_2. In this example, the bridge is balanced, so

$$V_{AC} = 0$$

or $$V_A = V_C$$

Substituting V_A for V_C, the equations become

$$2V_A - V_B = 0 \tag{6-3}$$

$$-2V_A + 3V_B = 24 + I_2 \tag{6-4}$$

$$(1 + G_5)V_A - V_B = -I_2 \tag{6-5}$$

From Eq. (6-3),

$$V_B = 2V_A$$

Substituting $2V_A$ for V_B in Eqs. (6-4) and (6-5),

$$4V_A = 24 + I_2 \tag{6-6}$$

$$(1 - G_5)V_A = I_2 \tag{6-7}$$

Equations (6-6) and (6-7) are valid for any value of I_2, if the bridge is balanced. It remains to insert the values of I_2.

When $I_2 = 1$ mA, we obtain from Eq. (6-6),

$$V_A = \frac{24 + 1}{4} = 6.25 \text{ V}$$

Substituting for V_A and I_2 in Eq. (6-7),

$$6.25(1 - G_5) = 1$$

$$G_5 = 0.84 \text{ mS}$$

$$R_5 = \mathbf{1.19\ k\Omega}$$

When $I_2 = 6$ mA, Eq. (6-6) gives

$$V_A = \frac{24 + 6}{4} = 7.5 \text{ V}$$

Substituting for V_A and I_2 in Eq. (6-7),

$$7.5(1 - G_5) = 6$$

$$G_5 = 0.2 \text{ mS}$$

$$R_5 = \mathbf{5\ k\Omega}$$

Therefore the motor-driven variable resistor R_5 must have a range of 1.19 to 5 kΩ to maintain the bridge in balance while the control current varies between 1 and 6 mA.

Example 6-8 illustrates the use of subunits to simplify equations. When powers of 10 are used, the equations appear to be quite complicated.

6-3 LOOP ANALYSIS

Loop or mesh analysis was used more frequently in the past than nodal analysis. Why this is so is not too clear, as loop analysis often involves more difficult mathematics, and simultaneous equations with more unknowns. Perhaps its popularity has been due to the preference of many people for voltage sources rather than current sources. This attitude probably derives from the fact that, until the advent of semiconductor technology, practical current sources were not common. In any case, loop analysis is a powerful tool that is used extensively. The technique to be described is sometimes called branch circuit analysis.

Some guidelines to help you determine the conditions under which loop analysis is most applicable are:

1. Energy sources predominantly voltage generators
2. Current values to be found
3. A number of quantities to be calculated
4. Two or more energy sources
5. Number of loops less than the number of independent nodes

In implementing loop analysis, the loops must first be identified. A *loop*, or *mesh*, is a closed-circuit path from a point back to that same point.

Figure 6-21 is a T network connected between two voltage sources.

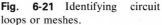

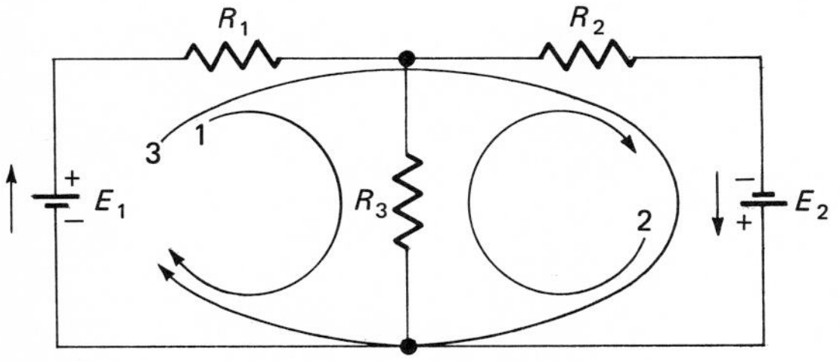

Fig. 6-21 Identifying circuit loops or meshes.

There are three possible loops:

1. From the positive of E_1, through R_1 and R_3, and back to the negative of E_1
2. From the positive of E_2, through R_3 and R_2, and back to the negative of E_2
3. From the positive of E_1, through R_1, R_2, and E_2, and back to the negative of E_1

The number of loops needed for analysis is the minimum which will include each element at least once. Any two of the three loops in Fig. 6-21 embrace all the elements, so only two are needed.

Because of the large variety of circuit configurations, there can be no general rule for deciding which loops to use to give the simplest analysis. Usually, selection of clearly visible loops is the most productive and least likely to result in errors. Therefore loops 1 and 2 are the logical selection for the circuit in Fig. 6-21 (not loops 1 and 3 or 2 and 3).

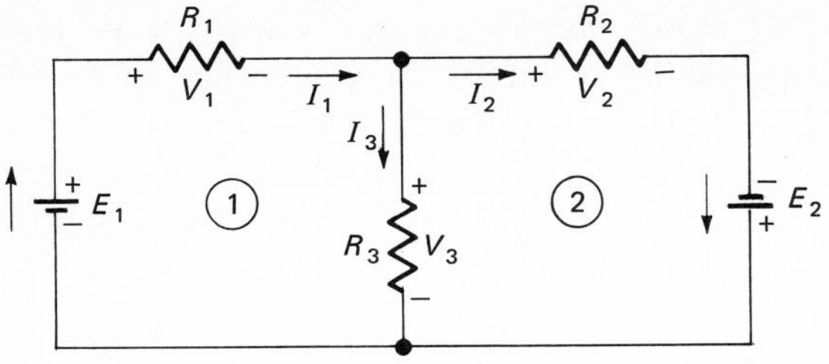

Fig. 6-22 Current directions and voltage polarities within loops.

Figure 6-22 shows the same circuit as in Fig. 6-21, but current directions and voltage polarities are included. For elements R_1 and R_2 these directions and polarities are compatible with the emf polarities, and so are probably correct. The emfs E_1 and E_2 each tend to produce opposite current directions in R_3. Thus the direction and polarity shown for R_3 are assumed. Provided that the current direction and voltage polarity are consistent for each individual element, the analysis results will show whether these assumptions are correct.

Kirchhoff's voltage law is applied to develop an equation for each loop. The algebraic sum of the source emfs equals the algebraic sum of the element pds.

ESTABLISHING THE POLARITY OF A VOLTAGE

The polarity of a voltage for determining an algebraic sum is established as follows.

Element pds

1. A pd is considered to be positive if it is associated with a current of direction compatible with a positive emf.
2. Opposite current directions produce negative pds.

Source emfs

1. If the loop contains one emf, its polarity is considered positive.
2. If there is more than one emf in a loop, one is assumed to be of positive polarity. Any emf of opposing polarity is then negative.
3. If there are no emf sources in a loop, the sum is zero.

In Fig. 6-22 the polarities are

Loop 1	*Loop 2*
E_1 positive	E_2 positive
V_1 positive	V_2 positive
V_3 positive	V_3 negative

In loop 1, I_3 (assumed to be downward) is compatible with the polarity of E_1; V_3 is positive. In loop 2, I_3 is not compatible with the polarity of E_2; V_3 is negative.

The loop equations are

Loop 1: $$E_1 = V_1 + V_3 = I_1R_1 + I_3R_3$$

But $I_3 = I_1 - I_2$ (Kirchhoff's current law),

$$E_1 = I_1R_1 + (I_1 - I_2)R_3$$

or $$(R_1 + R_3)I_1 - R_3I_2 = E_1 \tag{6-8}$$

Loop 2: $$E_2 = V_2 - V_3 = I_2R_2 - I_3R_3$$
$$= I_2R_2 - (I_1 - I_2)R_3$$

or $$-R_3I_1 + (R_2 + R_3)I_2 = E_2 \tag{6-9}$$

The solutions of Eqs. (6-8) and (6-9) will give the values of I_1 and I_2, and hence all the other circuit quantities.

Examination of Eqs. (6-8) and (6-9) leads to a speedy method of writing the loop equations. The technique is known as *Maxwell's cyclic currents,* in honor of the physicist who developed it about 100 years ago. You will see that it bears a similarity to the method of writing nodal equations used in Sec. 6-2. In fact it reinforces the duality of nodal and loop analyses.

The circuit is redrawn in Fig. 6-23 to show I_1 and I_2 as loop circulating currents rather than actual element currents. The directions selected for the currents are not important; the analysis results will show their validity.

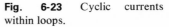

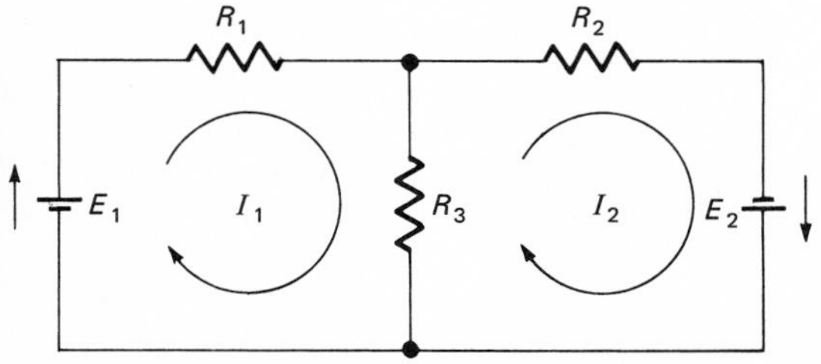

Fig. 6-23 Cyclic currents within loops.

WRITING THE EQUATION FOR A LOOP

In the following procedure for directly writing the equation for a loop, "local" refers to the loop being considered, and "remote" refers to a loop which shares an element with the local loop.

Element voltages

1. Place element voltages on the left side of the equation.
2. Multiply the local loop current by the sum of the resistances in that loop. Show the product as a positive quantity. Ignore any elements in parallel with voltage sources, as they cannot affect the circuit voltages.
3. For an element common to the local loop and some other loop, multiply the remote loop current by the resistance of the common element. Show the product as a positive quantity if the local and remote loop currents tend to be in the same direction in the common element. Show as a negative quantity if the current directions are opposing.

Source emfs

1. Place source emfs on the right side of the equation.
2. Show emfs supporting the direction of the local loop current as positive quantities.
3. Show opposing emfs as negative quantities.
4. If there are no voltage sources in the loop, show as zero.

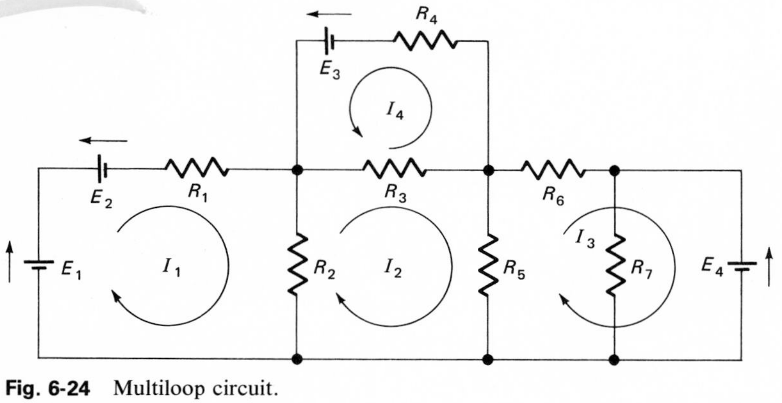

Fig. 6-24 Multiloop circuit.

In Fig. 6-24, element R_7 is ignored because it is in parallel with E_4. The loop equations are

Loop 1: $$(R_1 + R_2)I_1 - R_2I_2 = E_1 - E_2$$

Loop 2: $$(R_2 + R_3 + R_5)I_2 - R_2I_1 + R_3I_4 - R_5I_3 = 0$$

Loop 3: $$(R_5 + R_6)I_3 - R_5I_2 = -E_4$$

Loop 4: $$(R_3 + R_4)I_4 + R_3I_2 = E_3$$

Example 6-9 What are the values of each current and voltage in the circuit in Fig. 6-25?

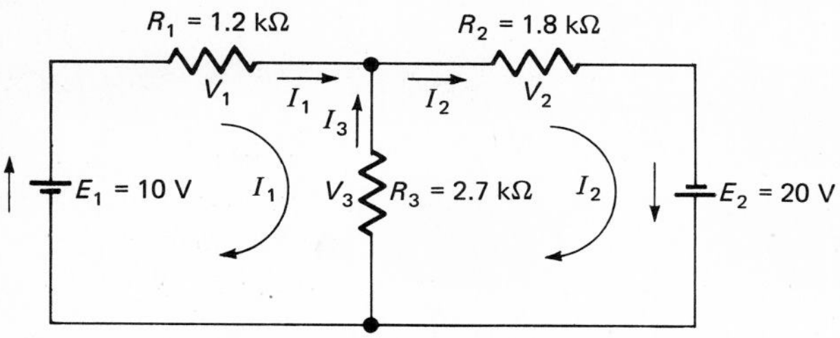

Fig. 6-25 Example 6-9: circuit.

Solution Assigning loop current directions as shown, the equations are

Loop 1:
$$(R_1 + R_3)I_1 - R_3I_2 = E_1$$
$$(1.2\ \text{k}\Omega + 2.7\ \text{k}\Omega)I_1 - 2.7\ \text{k}\Omega I_2 = 10\ \text{V}$$

Loop 2:
$$(R_2 + R_3)I_2 - R_3I_1 = E_2$$
$$(1.8\ \text{k}\Omega + 2.7\ \text{k}\Omega)I_2 - 2.7\ \text{k}\Omega I_1 = 20\ \text{V}$$

Using kilohms, volts, and milliamperes, these equations reduce to

$$3.9I_1 - 2.7I_2 = 10$$
$$-2.7I_1 + 4.5I_2 = 20$$

The solutions of this pair of equations give:

$$I_1 = \mathbf{9.65\ mA}$$
$$I_2 = \mathbf{10.23\ mA}$$
$$I_3 = I_2 - I_1 = \mathbf{0.58\ mA}$$
$$V_1 = I_1R_1 = 9.65\ \text{mA} \times 1.2\ \text{k}\Omega = \mathbf{11.58\ V}$$
$$V_2 = I_2R_2 = 10.23\ \text{mA} \times 1.8\ \text{k}\Omega = \mathbf{18.41\ V}$$
$$V_3 = I_3R_3 = 0.58\ \text{mA} \times 2.7\ \text{k}\Omega = \mathbf{1.566\ V}$$

Points to note in Example 6-9 are:

1. The direction of I_3 is upward, as indicated, because I_2 proves to be greater than I_1.
2. Taking account of the polarity of V_3, Kirchhoff's voltage law is substantiated for each loop within the normal limits of arithmetic accuracy.
3. Omitting the letter k in equations may simplify calculations, but you must be sure that the currents are recognized as milliamperes.
4. I_1 and I_2 are, in addition to being the loop circulating currents, the actual currents in elements R_1 and R_2.

If a circuit includes current generators, they must be changed to equivalent voltage generators for loop analysis. Conductances should be changed to resistances.

The voltage generator in Fig. 6-26*b* will be equivalent to the current generator in Fig. 6-26*a* if

$$E = \frac{I}{G_i} \quad \text{and} \quad R_i = \frac{1}{G_i}$$

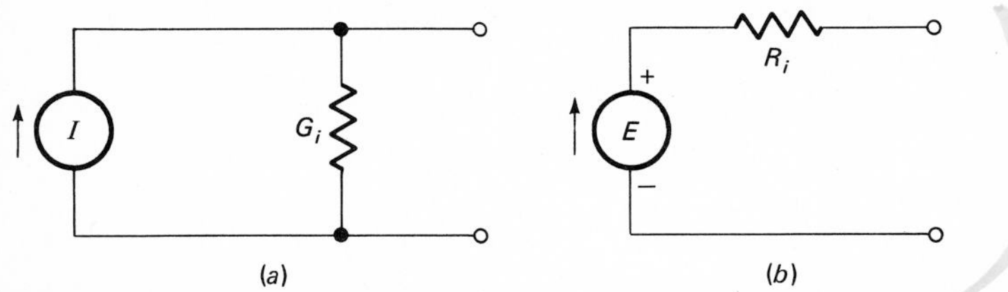

Fig. 6-26 Conversion of a current generator to a voltage generator.

Example 6-10 What is the current supplied by the battery in Fig. 6-27?

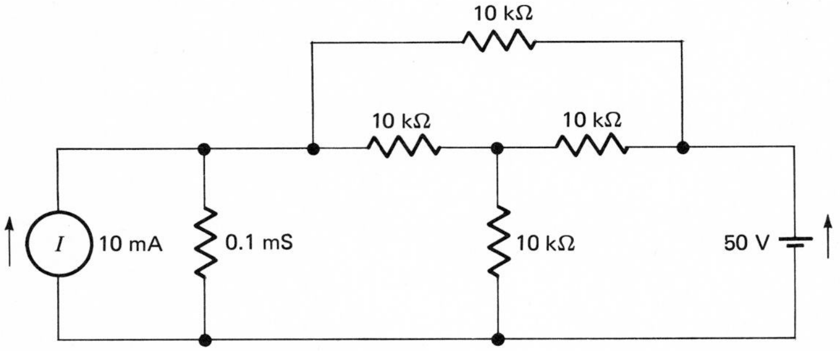

Fig. 6-27 Example 6-10: circuit.

Solution Redraw Fig. 6-27 with the current generator changed to a voltage source and show the loop currents (see Fig. 6-28).

Fig. 6-28 Circuit in Fig. 6-27 with the current generator converted to a voltage generator.

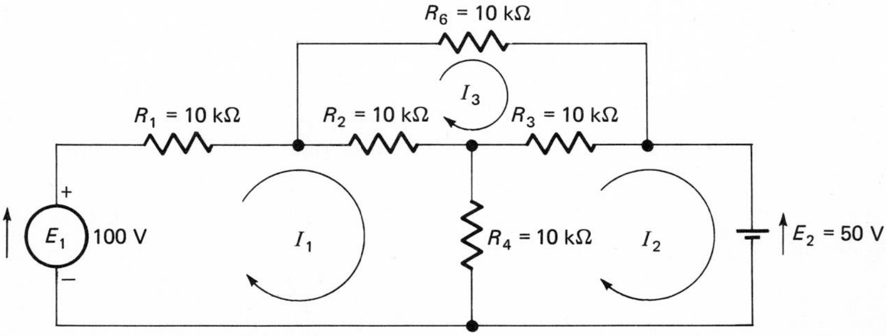

The loop equations are

Loop 1:
$$(R_1 + R_2 + R_4)I_1 - R_4I_2 - R_2I_3 = E_1$$
$$30\ \text{k}\Omega \times I_1 - 10\ \text{k}\Omega \times I_2 - 10\ \text{k}\Omega \times I_3 = 100\ \text{V}$$

Loop 2: The battery emf E_2 has its polarity such that it opposes the indicated direction of the loop current I_2. Thus E_2 is shown as negative in the loop equation

$$(R_3 + R_4)I_2 - R_4I_1 - R_3I_3 = -E_2$$
$$20\ \text{k}\Omega \times I_2 - 10\ \text{k}\Omega \times I_1 - 10\ \text{k}\Omega \times I_3 = -50\ \text{V}$$

Loop 3:
$$(R_2 + R_3 + R_6)I_3 - R_2I_1 - R_3I_2 = 0$$
$$30\ \text{k}\Omega \times I_3 - 10\ \text{k}\Omega \times I_1 - 10\ \text{k}\Omega \times I_2 = 0$$

The solutions of these three simultaneous equations give

$$I_1 = 3.75\ \text{mA} \qquad I_2 = 0 \qquad I_3 = 1.25\ \text{mA}$$

The battery current is the circulating current I_2, which is zero. The 10 mA from the current generator, acting via the resistive bridged-T network, offsets the effect of the 50 V of the battery so that it supplies no current.

This example would have been difficult to solve by nodal analysis. The 50-V battery, without series resistance (nominally perfect), could not be converted into a meaningful equivalent current generator; it would generate infinite current (50 V/0 Ω) which is mathematically difficult to handle.

Example 6-11 What are the currents in the circuit in Fig. 6-29? (The current direction arrows were inserted after the solution was obtained.)

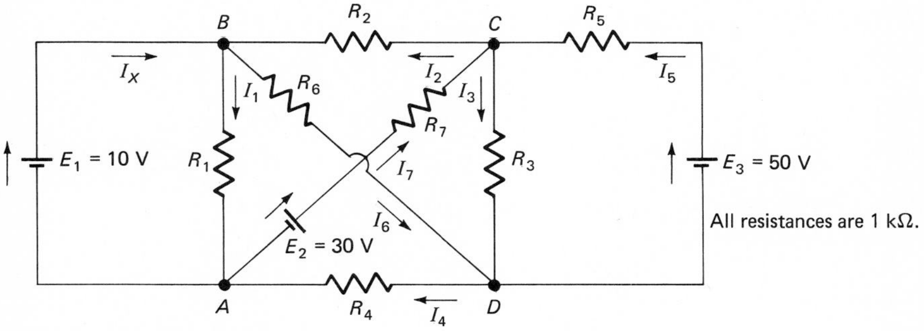

Fig. 6-29 Example 6-11: circuit.

Solution Redraw the circuit with R_1 omitted because it is directly in parallel with a battery (see Fig. 6-30).

The equations are

$$\begin{aligned}
(R_4 + R_6)I_A \quad & -R_6I_B & & = E_1\\
-R_6I_A + (R_2 + R_3 + R_6)I_B \quad & & -R_2I_C \quad & -R_3I_D = 0\\
& -R_2I_B + (R_2 + R_7)I_C & & = E_2 - E_1\\
& -R_3I_B & & + (R_3 + R_5)I_D = -E_3
\end{aligned}$$

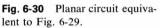

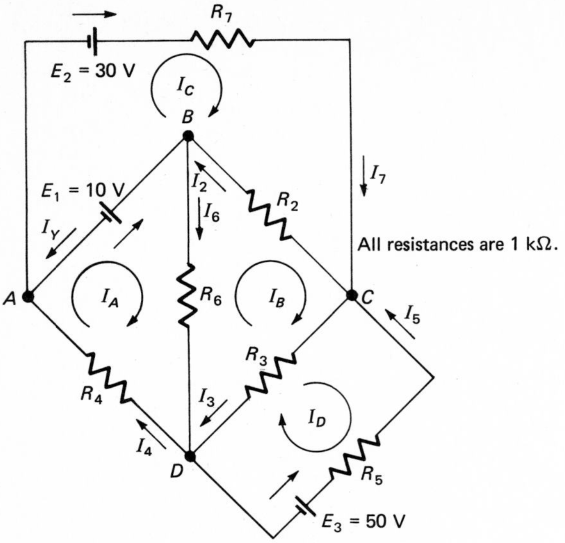

Fig. 6-30 Planar circuit equivalent to Fig. 6-29.

Insert quantities (with units kilohms, volts, and milliamperes):

$$\begin{aligned} 2I_A - I_B \qquad\qquad &= 10 \\ -I_A + 3I_B - I_C - I_D &= 0 \\ -I_B + 2I_C \qquad &= 20 \\ -I_B \qquad + 2I_D &= -50 \end{aligned}$$

The solutions of the equations are

$$I_A = 1\tfrac{2}{3} \text{ mA}$$

$$I_B = -6\tfrac{2}{3} \text{ mA} \qquad (6\tfrac{2}{3} \text{ mA counterclockwise})$$

$$I_C = 6\tfrac{2}{3} \text{ mA}$$

$$I_D = -28\tfrac{1}{3} \text{ mA} \qquad (28\tfrac{1}{3} \text{ mA counterclockwise})$$

The circuit currents are

$$I_1 = \frac{E_1}{R_1} = \frac{10 \text{ V}}{1 \text{ k}\Omega} = \mathbf{10 \text{ mA}} \qquad B \to A$$

$$I_2 = I_C - I_B = 6\tfrac{2}{3} \text{ mA} - (-6\tfrac{2}{3} \text{ mA}) = \mathbf{13\tfrac{1}{3} \text{ mA}} \qquad C \to B$$

$$I_3 = I_B - I_D = -6\tfrac{2}{3} \text{ mA} - (-28\tfrac{1}{3} \text{ mA}) = \mathbf{21\tfrac{2}{3} \text{ mA}} \qquad C \to D$$

$$I_4 = I_A = \mathbf{1\tfrac{2}{3} \text{ mA}} \qquad D \to A$$

$$I_5 = -I_D = -(-28\tfrac{1}{3} \text{ mA}) = \mathbf{28\tfrac{1}{3} \text{ mA}} \qquad D \to C$$

$$I_6 = I_A - I_B = 1\tfrac{2}{3} \text{ mA} - (-6\tfrac{2}{3} \text{ mA}) = \mathbf{8\tfrac{1}{3} \text{ mA}} \qquad B \to D$$

$$I_7 = I_C = \mathbf{6\tfrac{2}{3} \text{ mA}} \qquad A \to C$$

$$I_y = I_C - I_A = 6\tfrac{2}{3} \text{ mA} - 1\tfrac{2}{3} \text{ mA} = \mathbf{5 \text{ mA}} \qquad B \to A$$

$$I_x = I_1 + I_6 - I_2 = 10 \text{ mA} + 8\tfrac{1}{3} \text{ mA} - 13\tfrac{1}{3} \text{ mA} = \mathbf{5 \text{ mA}} \qquad A \to B$$

I_Y, the current in the 10-V battery in Fig. 6-30, has a direction opposing the battery emf E_1. The effect of the 30- and 50-V batteries overcomes that of E_1. But E_1 is *not* the actual 10-V battery current in the original circuit in Fig. 6-29, because element R_1 and current I_1 are omitted from Fig. 6-30. To determine the value of I_X, the real 10-V battery current, it is desirable to look at node B in Fig. 6-29. Node B and its current components are detailed in Fig. 6-31.

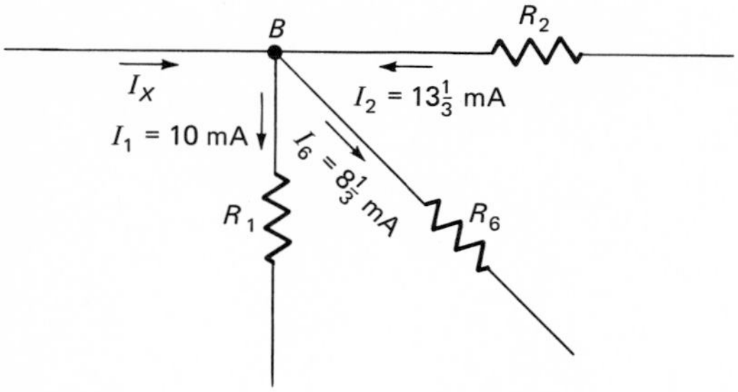

Fig. 6-31 Node B in Fig. 6-29.

The sum of the currents entering node B must equal the sum of the currents leaving it.

$$I_X + I_2 = I_1 + I_6$$

$$I_X = I_1 + I_6 - I_2 = 10\text{ mA} + 8\tfrac{1}{3}\text{ mA} - 13\tfrac{1}{3}\text{ mA}$$

$$I_X = 5\text{ mA}$$

Since I_X is a positive quantity, its assumed direction toward node B is correct.

The result for I_X is valid. Without R_1 the network would supply 5 mA to the battery E_1. R_1 draws 10 mA from the battery. The net result is **5 mA** from the battery.

Significant points which arise from Example 6-11 are

1. It is desirable to redraw a circuit so that it becomes *planar* before attempting an analysis. A planar circuit has no crossing conductors.
2. If the assumed direction of a circulating current is wrong, it will present no problem. The negative values of I_B and I_D expose wrong assumptions about their directions.
3. If an element is omitted because it does not influence the main analysis, great care must be taken to ensure its reinsertion if its presence is important to some of the required data. Element R_1 was not significant in the general analysis, but was important in determining its current I_1 and the current associated with the 10-V battery I_X.
4. It is all right to use sub- and superunits, like milliamperes and kilohms, to simplify the writing and calculations, providing this is remembered when evaluating the circuit quantities.

Example 6-12 The bridge circuit in Fig. 6-32 is the basis of commercial instruments for measuring resistance. The *unknown* (the device or network being measured) is connected between terminals X and Y. The element resistances (R_1, R_2, and R_3) are adjusted to give a balance condition. Balance is identified by a zero-current indication on the meter. The value of the unknown resistance is then read off a scale associated with the variable resistor R_3.

During a particular measurement, the conditions shown in the diagram existed. What is the meter current, and what is the polarity of the voltage across it?

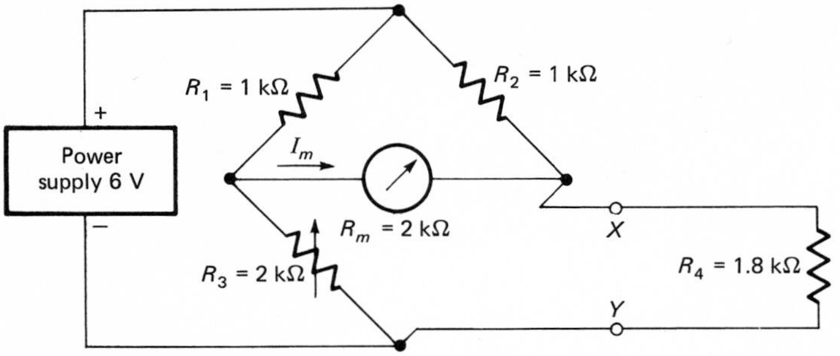

Fig. 6-32 Example 6-12: circuit.

Solution Redraw the circuit as in Fig. 6-33.

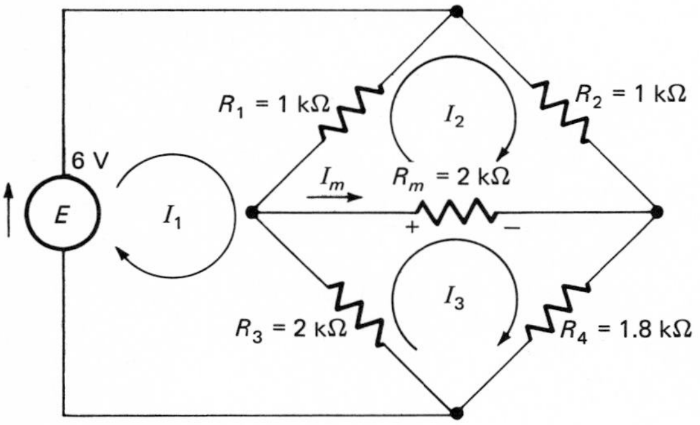

Fig. 6-33 Equivalent circuit of Fig. 6-32.

With the units volts, kilohms, and milliamperes, the loop equations are

$$3I_1 - I_2 - 2I_3 = 6$$

$$-I_1 + 4I_2 - 2I_3 = 0$$

$$-2I_1 - 2I_2 + 5.8I_3 = 0$$

From these equations are obtained:

$$I_1 = 4.144 \text{ mA}$$

$$I_2 = 2.115 \text{ mA}$$

$$I_3 = 2.158 \text{ mA}$$

The meter current I_m is given by

$$I_m = I_3 - I_2 = 2.158 \text{ mA} - 2.115 \text{ mA} = \mathbf{0.043 \text{ mA}} \quad \text{or} \quad \mathbf{43\ \mu A}$$

The direction of the meter current is from left to right, as in the diagrams. Therefore the meter voltage will have the polarity $+ \rightarrow -$, as shown in Fig. 6-33.

Example 6-13 The circuit in Fig. 6-34 is designed to provide base bias for a linear transistor amplifier in a stereo radio. What is the transistor's base-to-ground resistance R_{in} if its base current I_B is 10 μA?

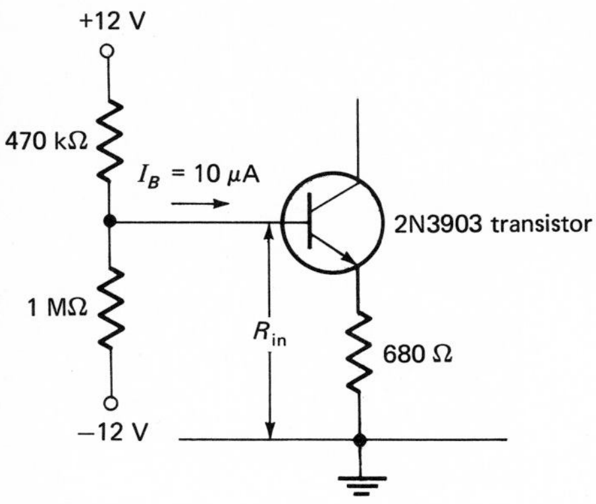

Fig. 6-34 Example 6-13: circuit.

Solution The equivalent of the base-bias circuit is shown in Fig. 6-35.

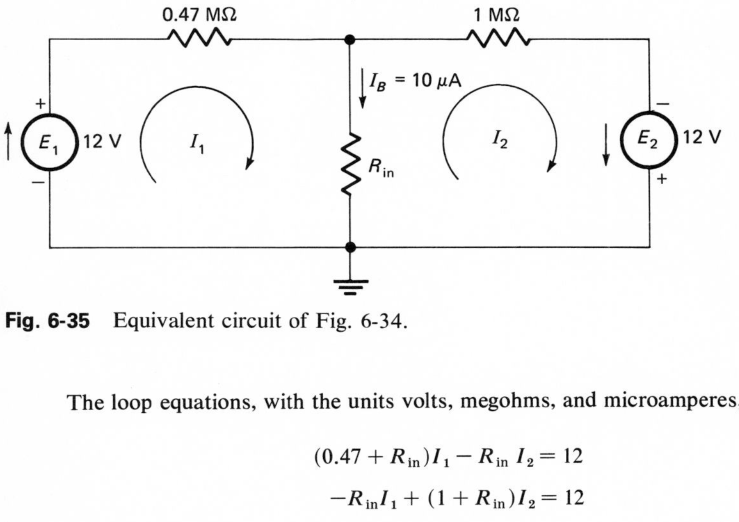

Fig. 6-35 Equivalent circuit of Fig. 6-34.

The loop equations, with the units volts, megohms, and microamperes, are

$$(0.47 + R_{in})I_1 - R_{in}\, I_2 = 12$$
$$-R_{in}I_1 + (1 + R_{in})I_2 = 12$$

As $I_1 = I_2 + I_B$ or $I_2 + 10$ μA, this expression for I_1 can be substituted in the loop equations, giving

$$0.47I_2 + 10R_{in} = 7.3$$
$$I_2 - 10R_{in} = 12$$

Solving these equations for R_{in} results in

$$R_{in} = \mathbf{0.1129\ M\Omega} \qquad \text{or} \qquad \mathbf{112.9\ k\Omega}$$

Example 6-14 A domestic three-wire system with three resistive loads is shown in Fig. 6-36. Each of the lines leading to the house has a resistance of 0.1 Ω. The load powers indicated are based on 120 V and 240 V being present at the loads. What actually is each load power? Assume that the load resistances are not affected by the reduced voltages as a result of line resistance.

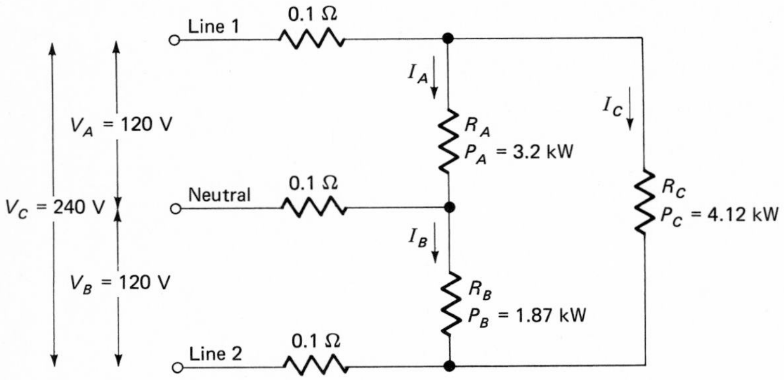

Fig. 6-36 Example 6-14: network.

Solution

$$P = \frac{V^2}{R} \qquad \text{or} \qquad R = \frac{V^2}{P}$$

$$R_A = \frac{V_A^2}{P_A} = \frac{(120\ \text{V})^2}{3.2\ \text{kW}} = 4.5\ \Omega$$

$$R_B = \frac{V_B^2}{P_B} = \frac{(120\ \text{V})^2}{1.87\ \text{kW}} = 7.7\ \Omega$$

$$R_C = \frac{V_C^2}{P_C} = \frac{(240\ \text{V})^2}{4.12\ \text{kW}} = 14\ \Omega$$

The circuit equivalent to Fig. 6-36 is shown in Fig. 6-37.

Fig. 6-37 Equivalent circuit of Fig. 6-36.

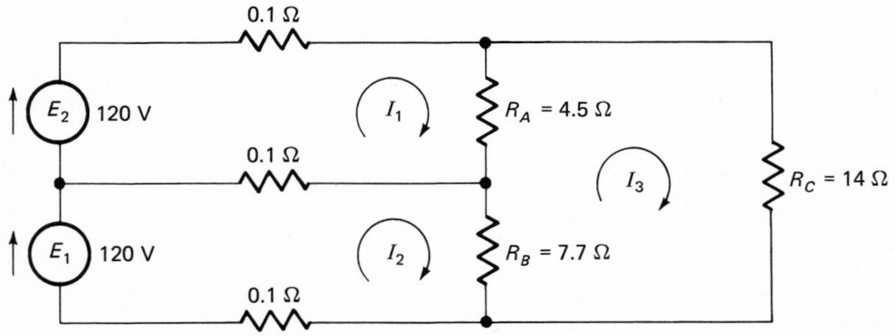

The loop equations are

$$4.7I_1 - 0.1I_2 - 4.5I_3 = 120$$

$$-0.1I_1 + 7.9I_2 - 7.7I_3 = 120$$

$$-4.5I_1 - 7.7I_2 + 26.2I_3 = 0$$

The solutions for these equations are

$$I_1 = 42.2\ \text{A}$$

$$I_2 = 31.81\ \text{A}$$

$$I_3 = 16.57\ \text{A}$$

The load currents are

$$I_A = I_1 - I_3 = 25.63\ \text{A}$$

$$I_B = I_2 - I_3 = 15.24\ \text{A}$$

$$I_C = I_3 = 16.57\ \text{A}$$

The load powers therefore are

$$P_A = I_A^2 R_A = (25.63\ \text{A})^2 \times 4.5\ \Omega = \mathbf{2.956\ kW}$$

$$P_B = I_B^2 R_B = (15.24\ \text{A})^2 \times 7.7\ \Omega = \mathbf{1.788\ kW}$$

$$P_C = I_C^2 R_C = (16.57\ \text{A})^2 \times 14\ \Omega = \mathbf{3.844\ kW}$$

These results imply that the effect of the line resistances, in this case, is to reduce the total power delivered to the loads by about 6 percent.

6-4 SUPERPOSITION THEOREM

The superposition theorem presents an important circuit concept which is of value in understanding dc and ac circuits.

In its effect on circuit quantities, any energy source can be considered separately from other sources. All individual effects can be combined to give the total effect.

This concept can be used to simplify circuit analysis.

Some guidelines on circuits which are most suitable for superposition analysis are

1. More than one energy source
2. Voltage or current sources
3. Simple circuit structure

The need for a simple circuit distinguishes superposition analysis from nodal or loop analysis. While any of the nodal or loop examples could have been solved by superposition, most of them involved circuits too complex for any advantage to have been gained. When any one of the energy sources is considered separately from the others, as described below, the remaining circuit will be series parallel. Where a circuit is suitable for superposition analysis, this method is often preferred because it does not involve simultaneous equations.

There are six basic operations in applying the superposition theorem to circuit analysis.

1. Select one energy source.
2. Consider all other source voltages and currents to be removed.
   a. Source voltages to be short-circuited.
   b. Source currents to be open-circuited.
3. Leave internal resistances of removed sources in the circuit.
4. Determine the current in, or the voltage across, each element. Note the directions and polarities.
5. Repeat 1 to 4 for each source.
6. Algebraically add the individual results.

Numerical examples best illustrate the use of this technique.

Example 6-15 What are the currents in the circuit in Fig. 6-38?

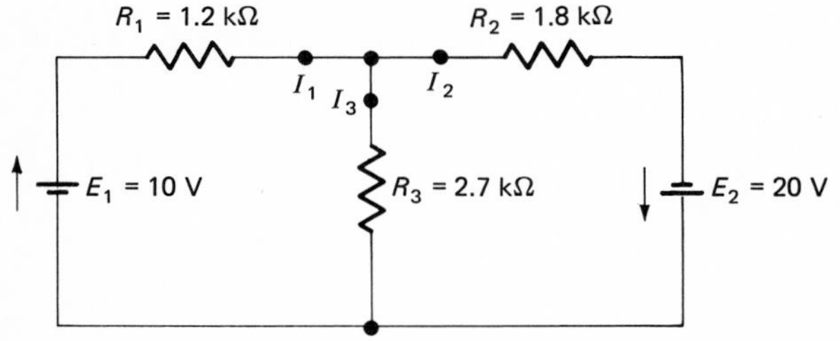

Fig. 6-38 Example 6-15: circuit.

Solution Select source E_1 and redraw the circuit with E_2 removed as in Fig. 6-39.

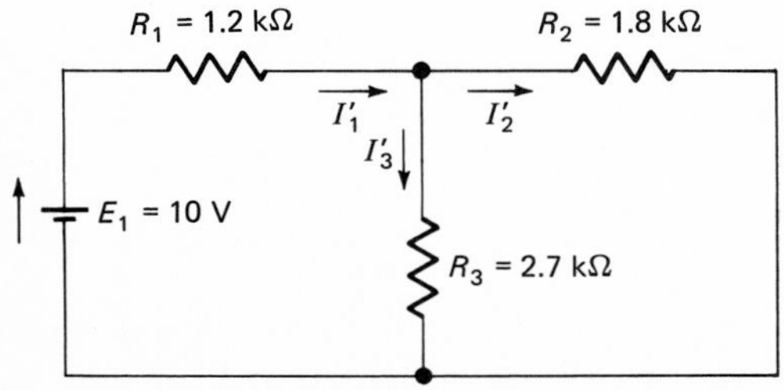

Fig. 6-39 Figure 6-38 with E_2 removed.

Calculate the current components due to E_1.

$$I_1' = \frac{E_1}{R_1 + \dfrac{R_2R_3}{R_2 + R_3}} = \frac{10\text{ V}}{1.2\text{ k}\Omega + \dfrac{1.8\text{ k}\Omega \times 2.7\text{ k}\Omega}{1.8\text{ k}\Omega + 2.7\text{ k}\Omega}} = 4.386\text{ mA}$$

$$I_2' = I_1' \frac{R_3}{R_2 + R_3} = 4.386\text{ mA}\,\frac{2.7\text{ k}\Omega}{1.8\text{ k}\Omega + 2.7\text{ k}\Omega} = 2.632\text{ mA}$$

$$I_3' = I_1' - I_2' = 4.386\text{ mA} - 2.632\text{ mA} = 1.754\text{ mA}$$

Select source E_2 and redraw the circuit with E_1 removed as in Fig. 6-40.

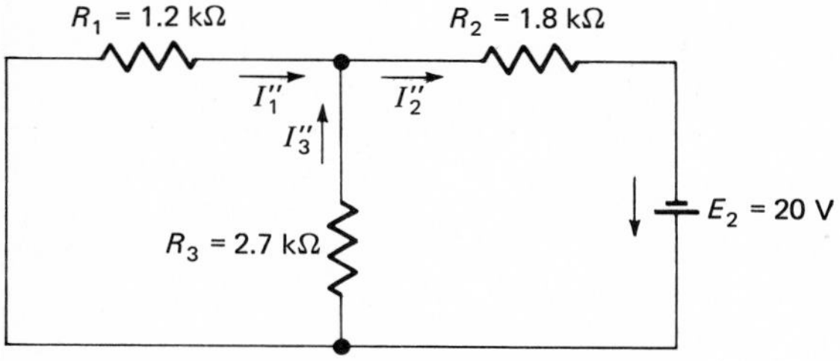

Fig. 6-40 Figure 6-38 with E_1 removed.

Calculate the current components due to E_2.

$$I_2'' = \frac{E_2}{R_2 + \dfrac{R_1R_3}{R_1 + R_3}} = \frac{20\text{ V}}{1.8\text{ k}\Omega + \dfrac{1.2\text{ k}\Omega \times 2.7\text{ k}\Omega}{1.2\text{ k}\Omega + 2.7\text{ k}\Omega}} = 7.602\text{ mA}$$

$$I_1'' = I_2'' \frac{R_3}{R_1 + R_3} = 7.602\text{ mA}\,\frac{2.7\text{ k}\Omega}{1.2\text{ k}\Omega + 2.7\text{ k}\Omega} = 5.263\text{ mA}$$

$$I_3'' = I_2'' - I_1'' = 7.602\text{ mA} - 5.263\text{ mA} = 2.339\text{ mA}$$

The actual current in any element is the sum of its components, taking into account their relative directions.

$$I_1 = I_1' + I_1'' = 4.386\text{ mA} + 5.263\text{ mA} = \mathbf{9.649\text{ mA}} \quad \text{to the right}$$

$$I_2 = I_2' + I_2'' = 2.632\text{ mA} + 7.602\text{ mA} = \mathbf{10.23\text{ mA}} \quad \text{to the right}$$

$$I_3 = I_3'' - I_3' = 2.339\text{ mA} - 1.754\text{ mA} = \mathbf{0.585\text{ mA}} \quad \text{upward}$$

The circuit in Example 6-15 is the same as that in Example 6-9. The results are the same within ordinary arithmetic limits. This shows the validity of the superposition technique.

Example 6-16 What are the currents in the circuit in Fig. 6-41? (The current direction arrows were added after the solution was obtained.)

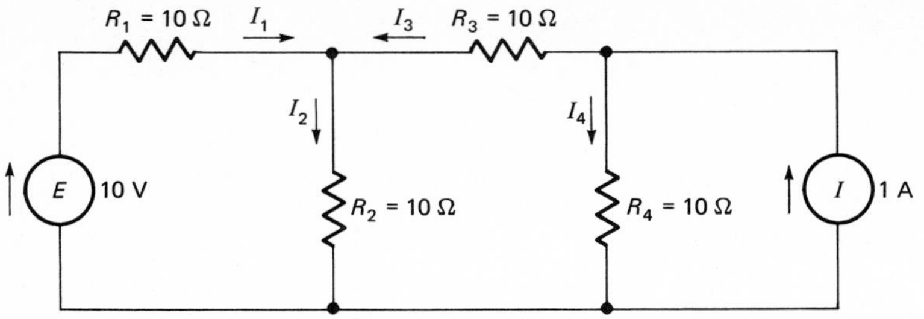

Fig. 6-41 Example 6-16: circuit.

Solution The voltage generator on its own gives Fig. 6-42.

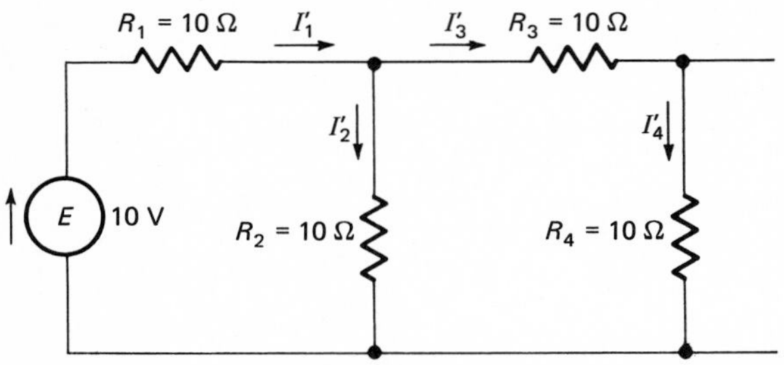

Fig. 6-42 Figure 6-41 with the current generator removed and replaced by its 0-S internal conductance—an open circuit.

$$I'_1 = \frac{E}{R_1 + \dfrac{R_2(R_3+R_4)}{R_2+R_3+R_4}} = \frac{10\text{ V}}{10\ \Omega + \dfrac{10\ \Omega \times 20\ \Omega}{30\ \Omega}} = 0.6\text{ A}$$

$$I'_2 = I'_1\frac{R_3+R_4}{R_2+R_3+R_4} = 0.6\text{ A}\,\frac{20\ \Omega}{30\ \Omega} = 0.4\text{ A}$$

$$I'_3 = I'_4 = I'_1 - I'_2 = 0.6\text{ A} - 0.4\text{ A} = 0.2\text{ A}$$

The current generator on its own gives Fig. 6-43.

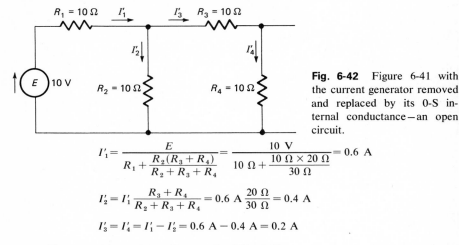

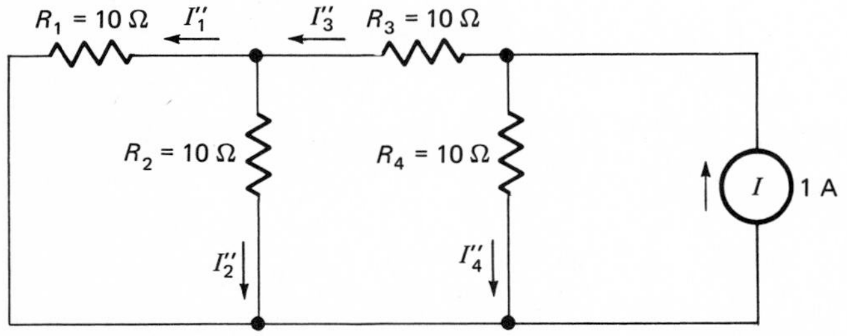

Fig. 6-43 Figure 6-41 with the voltage generator removed and replaced by its 0-Ω internal resistance—a short circuit.

Apply the current divider principle.

$$I''_3 = I\frac{R_4}{R_3 + \dfrac{R_1R_2}{R_1+R_2} + R_4} = 1\text{ A}\,\frac{10\ \Omega}{25\ \Omega} = 0.4\text{ A}$$

$$I''_1 = I''_2 = \frac{I''_3}{2} = 0.2\text{ A}$$

$$I''_4 = I - I''_3 = 1\text{ A} - 0.4\text{ A} = 0.6\text{ A}$$

The total currents are

$$I_1 = I_1' - I_1'' = 0.6\text{ A} - 0.2\text{ A} = \mathbf{0.4\text{ A}}$$

$$I_2 = I_2' + I_2'' = 0.4\text{ A} + 0.2\text{ A} = \mathbf{0.6\text{ A}}$$

$$I_3 = I_3'' - I_3' = 0.4\text{ A} - 0.2\text{ A} = \mathbf{0.2\text{ A}}$$

$$I_4 = I_4' + I_4'' = 0.2\text{ A} + 0.6\text{ A} = \mathbf{0.8\text{ A}}$$

The current directions are as shown in Fig. 6-41.

The superposition treatment can, in some cases, reduce the chance of error.

If Example 6-16 were handled by nodal analysis, the voltage generator would be changed to a current generator. The resulting current generator (Fig. 6-44*b*) would then be incorporated into the original circuit in Fig. 6-41.

Fig. 6-44 Serious errors are possible if a voltage generator is converted to a current generator for nodal analysis and then converted back to a voltage generator with incorrect data interpretation.

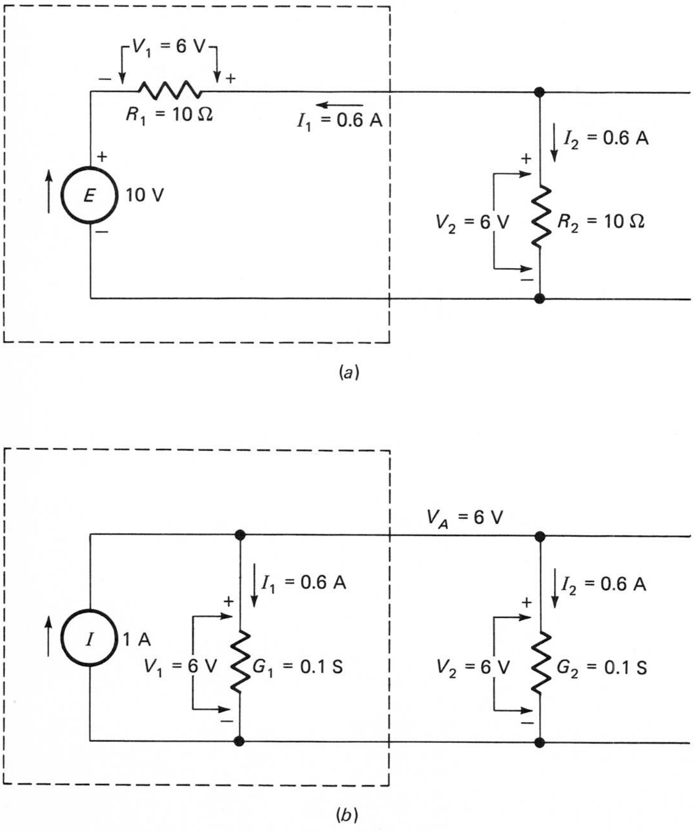

When the nodal equations have been solved, the node voltage V_A in Fig. 6-44*b* is shown to be +6 V (0.6 A/0.1 S). Both I_1 and I_2 in Fig. 6-44*b* are 0.6 A. If these data are now referred back to the voltage generator in Fig. 6-41 without sufficient care, they might appear as shown in Fig. 6-44*a*.

In checking further, it is noted that, if this is so, there are two odd situations:

1. The generator current direction is opposite that of the emf polarity. This result is possible in some circuits, but it does not appear to be correct in this one in which the current source equivalent of the voltage generator has the same source current as the real original current generator.
2. The voltage V_1 across R_1 is 6 V (0.6 A × 10 Ω), with the polarity shown, so that the voltage across R_2 is $E + V_1 = 16$ V. But these data were obtained from a nodal voltage of 6 V across R_2.

Of course the real voltage across R_2 is 6 V (not 16 V), with the polarity shown in Fig. 6-44*a*. This 6 V is the value of the voltage across the generator and R_1 in series, as in Fig. 6-45.

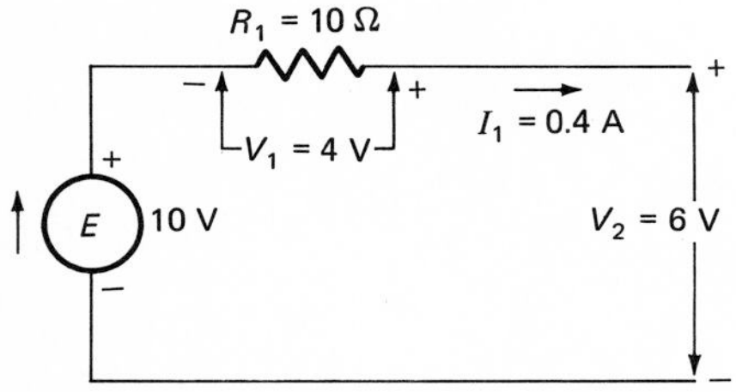

Fig. 6-45 Voltage generator in Fig. 6-41 with correct voltage and current polarities and direction.

The actual voltage across R_1 is $E - V_2 = 10\text{ V} - 6\text{ V} = 4$ V, making its current 0.4 A, with the direction in Fig. 6-45. This result is the same as that given in Example 6-16.

For this circuit nodal analysis is faster than the superposition technique. The superposition method eliminates the error tendency described above. Thus the decision regarding which method you should use could hinge on whether you want speed (nodal analysis) or caution (superposition).

Example 6-17 Figure 6-46 is the circuit of an electronic voltage regulator. Its purpose is to keep the output constant at 20 V, despite changes in the current taken by the load or changes in the input voltage. If the base-to-emitter resistance of transistor Q_1 is 25 kΩ, what is its base current and base-to-emitter voltage?

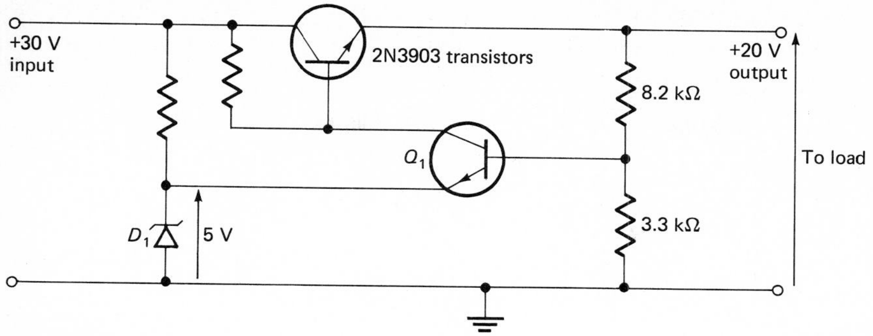

Fig. 6-46 Example 6-17: circuit.

Solution The network which determines the base current I_B and the base-to-emitter voltage V_{BE} of Q_1 is shown in Fig. 6-47.

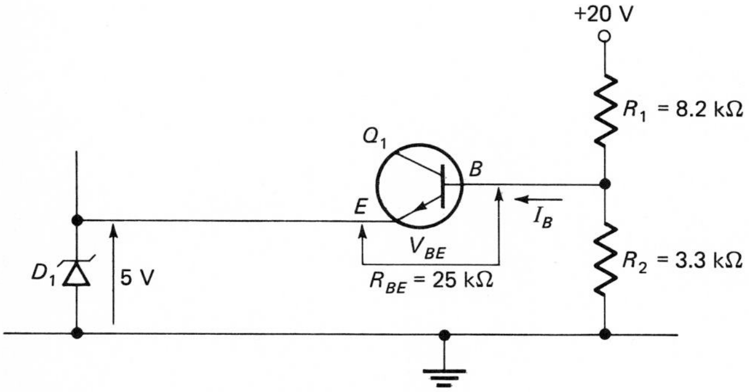

Fig. 6-47 Base-to-emitter circuit of transistor Q_1 in Fig. 6-46.

The zener diode D_1 maintains a steady voltage between the transistor emitter and ground, and is thus a voltage source. Figure 6-47 can be reduced to the equivalent circuit in Fig. 6-48.

Fig. 6-48 Reduced equivalent circuit of Fig. 6-47.

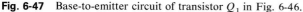

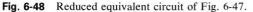

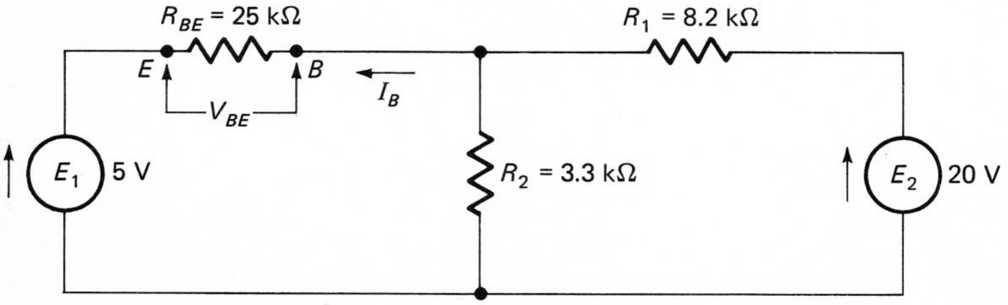

Taking E_1 alone gives Fig. 6-49.

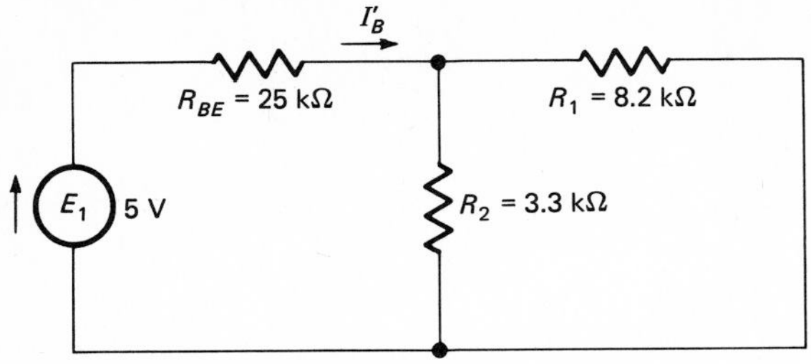

Fig. 6-49 Figure 6-48 with E_1 alone.

$$I'_B = \frac{E_1}{R_{BE} + \dfrac{R_1R_2}{R_1 + R_2}} = \frac{5\ \text{V}}{25\ \text{k}\Omega + \dfrac{8.2\ \text{k}\Omega \times 3.3\ \text{k}\Omega}{8.2\ \text{k}\Omega + 3.3\ \text{k}\Omega}} = 0.1828\ \text{mA}$$

Taking E_2 alone gives Fig. 6-50.

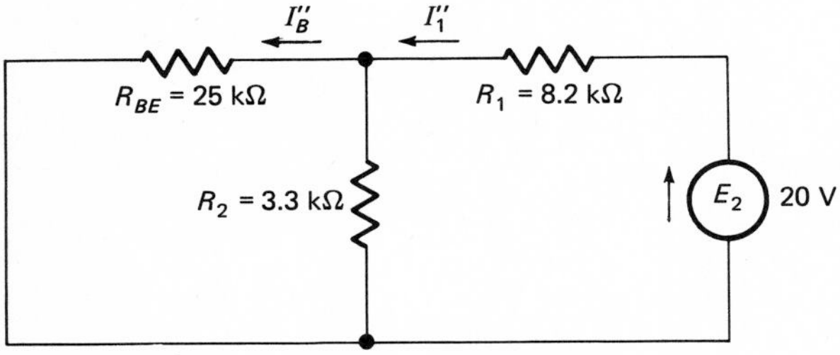

Fig. 6-50 Figure 6-49 with E_2 alone.

$$I''_1 = \frac{E_2}{R_1 + \dfrac{R_2R_{BE}}{R_2 + R_{BE}}} = \frac{20\ \text{V}}{8.2\ \text{k}\Omega + \dfrac{3.3\ \text{k}\Omega \times 25\ \text{k}\Omega}{3.3\ \text{k}\Omega + 25\ \text{k}\Omega}} = 1.799\ \text{mA}$$

$$I''_B = I''_1 \frac{R_2}{R_2 + R_{BE}} = 1.799\ \text{mA}\ \frac{3.3\ \text{k}\Omega}{3.3\ \text{k}\Omega + 25\ \text{k}\Omega} = 0.2098\ \text{mA}$$

$$I_B = I''_B - I'_B = 0.2098\ \text{mA} - 0.1828\ \text{mA} = \mathbf{0.027\ mA}$$

$$V_{BE} = I_BR_{BE} = 0.027\ \text{mA} \times 25\ \text{k}\Omega = \mathbf{0.675\ V}$$

These are typical values for a silicon transistor used in this type of circuit.

The final expression for I_B in Example 6-17 may be used to show how important arithmetic accuracy can be. When it is necessary to subtract two quantities of almost the same value, accuracy can be all-important. For instance, if the values of I'_B and I''_B were calculated to be 0.18 mA and 0.21 mA, the final value of I_B would be 0.03 mA. Thus a maximum calculating error of 1.5 percent would cause a final error of 10 percent.

Example 6-18 A compound-wound electric motor is connected to a 120-V dc supply. The series and shunt field windings have resistances of 10 and 1000 Ω, respectively. The motor's back emf is 100 V, and its armature resistance is 5 Ω. What are the field and armature currents?

Solution The circuit of the electric motor will be as in Fig. 6-51.

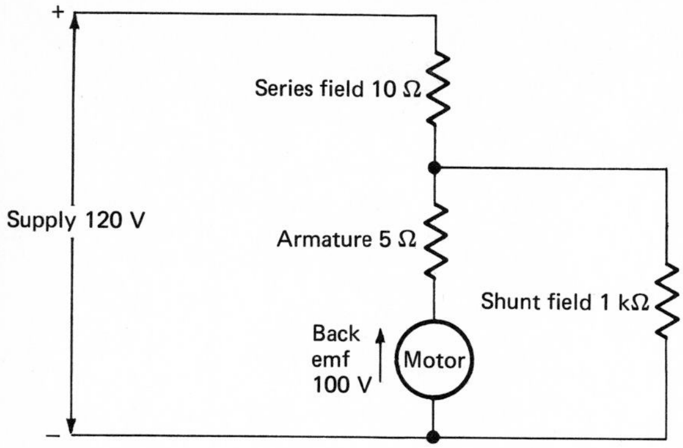

Fig. 6-51 Example 6-18: network.

This diagram can be redrawn as an equivalent circuit (Fig. 6-52).

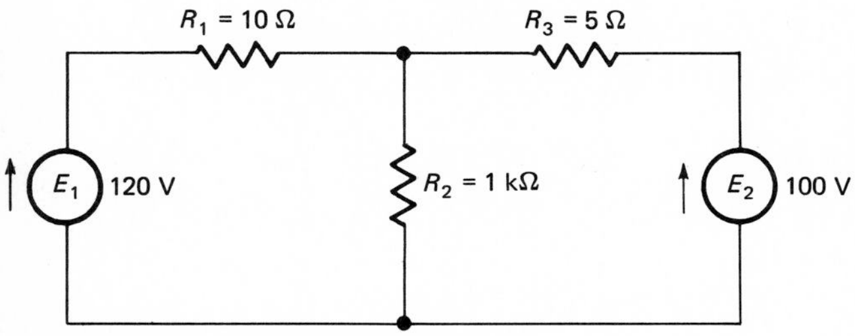

Fig. 6-52 Equivalent circuit of Fig. 6-51.

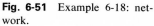

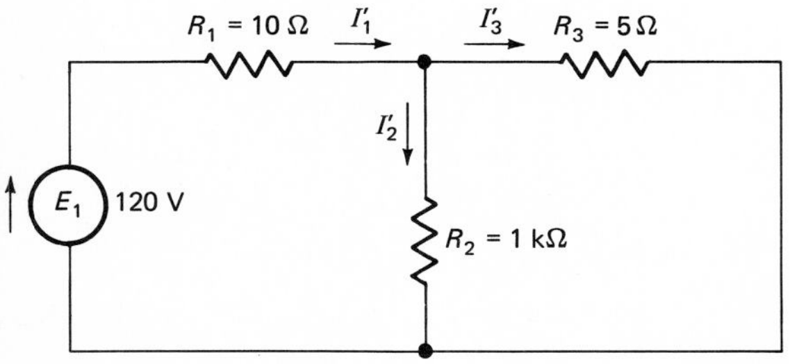

Fig. 6-53 Figure 6-52 with E_2 removed.

With E_1 alone (Fig. 6-53)

$$I_1' = \frac{E_1}{R_1 + \dfrac{R_2 R_3}{R_2 + R_3}} = \frac{120\ \text{V}}{10\ \Omega + \dfrac{5\ \Omega \times 1\ \text{k}\Omega}{5\ \Omega + 1\ \text{k}\Omega}} = 8.013\ \text{A}$$

$$I_2' = I_1' \frac{R_3}{R_2 + R_3} = 8.013\ \text{A}\ \frac{5\ \Omega}{5\ \Omega + 1\ \text{k}\Omega} = 0.0399\ \text{A}$$

$$I_3' = I_1' - I_2' = 8.013\ \text{A} - 0.0399\ \text{A} = 7.973\ \text{A}$$

With E_2 alone (Fig. 6-54)

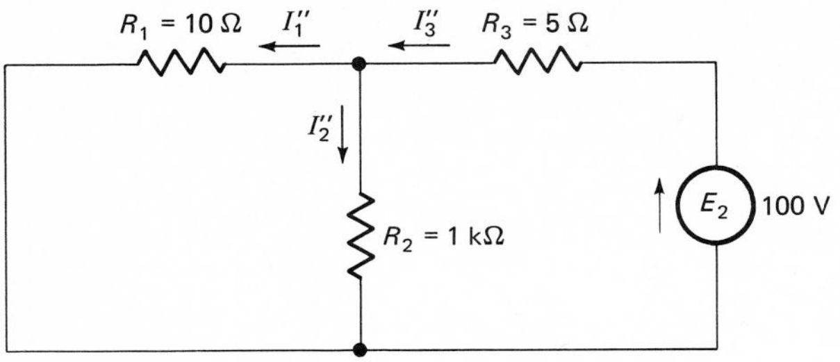

Fig. 6-54 Figure 6-52 with E_1 removed.

$$I_3'' = \frac{E_2}{R_3 + \dfrac{R_1 R_2}{R_1 + R_2}} = \frac{100\ \text{V}}{5\ \Omega + \dfrac{10\ \Omega \times 1\ \text{k}\Omega}{10\ \Omega + 1\ \text{k}\Omega}} = 6.711\ \text{A}$$

$$I_1'' = I_3'' \frac{R_2}{R_1 + R_2} = 6.711\ \text{A} \frac{1\ \text{k}\Omega}{10\ \Omega + 1\ \text{k}\Omega} = 6.644\ \text{A}$$

$$I_2'' = I_1'' - I_3'' = 6.711\ \text{A} - 6.644\ \text{A} = 0.067\ \text{A}$$

$$I_1 = I_1' - I_1'' = 8.013\ \text{A} - 6.644\ \text{A} = 1.369\ \text{A}$$

$$I_2 = I_2' + I_2'' = 0.0399\ \text{A} + 0.067\ \text{A} = 0.1069\ \text{A}$$

$$I_3 = I_3' - I_3'' = 7.973\ \text{A} - 6.711\ \text{A} = 1.262\ \text{A}$$

The motor currents are

Series field (and supply) current $I_1 = \mathbf{1.369\ A}$

Shunt field current $I_2 = \mathbf{0.1069\ A}$

Armature current $I_3 = \mathbf{1.262\ A}$

EXERCISE PROBLEMS

Section 6-1

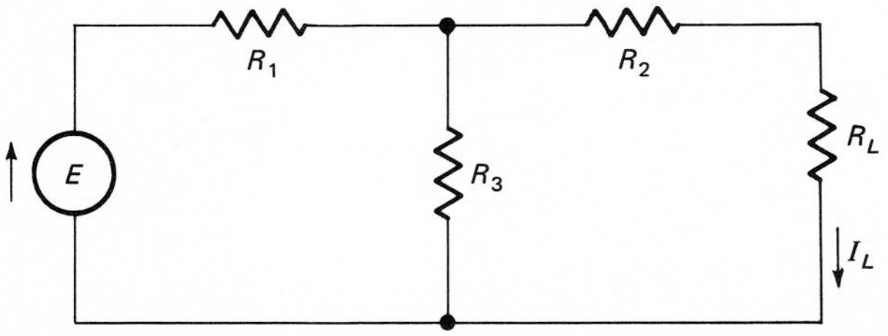

Fig. 6-1P Problem 6-1: circuit.

6-1 For the circuit in Fig. 6-1P, by assuming a current, find I_L when

a. $E = 20$ V, $R_1 = 47\ \Omega$, $R_2 = 68\ \Omega$, $R_3 = 100\ \Omega$, $R_L = 100\ \Omega$ (*ans.* 68.03 mA)

b. $E = 24$ V, $R_1 = 3.3\ \text{k}\Omega$, $R_2 = 3.9\ \text{k}\Omega$, $R_3 = 1.2\ \text{k}\Omega$, $R_L = 1\ \text{k}\Omega$ (*ans.* 1.108 mA)

c. $E = 240$ V, $R_1 = 50\ \Omega$, $R_2 = 25\ \Omega$, $R_3 = 75\ \Omega$, $R_L = 60\ \Omega$ (*ans.* 1.252 A)

d. $E = 6$ V, $R_1 = 120\ \text{k}\Omega$, $R_2 = 270\ \text{k}\Omega$, $R_3 = 82\ \text{k}\Omega$, $R_L = 56\ \text{k}\Omega$ (*ans.* 6.5 μA)

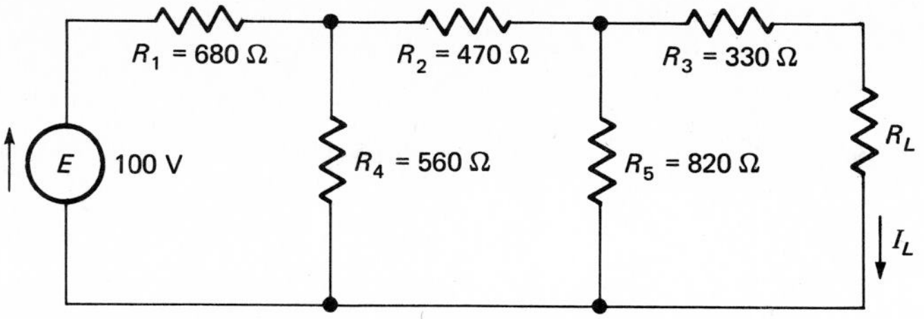

Fig. 6-2P Problems 6-2 and 6-3: circuit.

6-2 For the circuit in Fig. 6-2P, by assuming a current, find I_L when

a. $R_L = 100\ \Omega$

b. $R_L = 1000\ \Omega$

c. $R_L = 1500\ \Omega$

6-3 For the circuit in Fig. 6-2P, by assuming a current, find the current through the 820-Ω resistor when

a. $R_L = 47\ \Omega$ (*ans.* 13.74 mA)

b. $R_L = 150\ \Omega$ (*ans.* 15.43 mA)

c. $R_L = 680\ \Omega$ (*ans.* 20.27 mA)

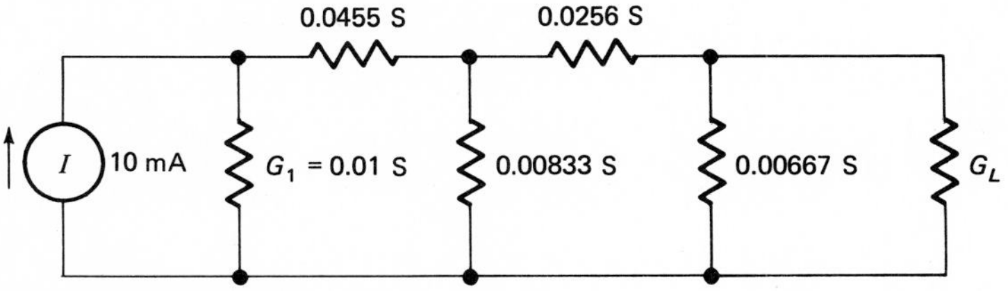

Fig. 6-3P Problem 6-4: circuit.

6-4 For the circuit in Fig. 6-3P, by assuming a current, find the power dissipated by G_1 and G_L when

a. $G_L = 0.01$ S

b. $G_L = 0.0455$ S

c. $G_L = 0.1$ S

Fig. 6-4P Problem 6-5: (*a*) circuit with T network; (*b*) circuit with π network.

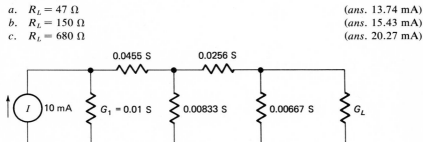

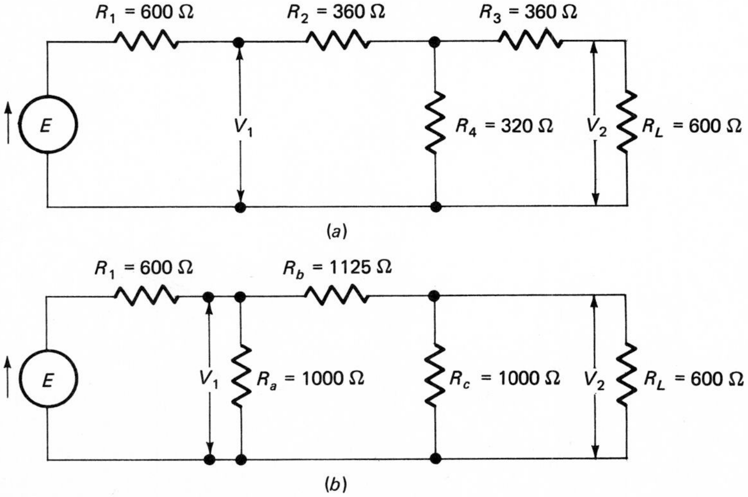

6-5 By assuming a current, find

a. The power dissipation of R_1 and R_L in Fig. 6-4P*a* when $E = 20$ V (*ans.* 166.7 mW, 10.4 mW)

b. The power dissipation of R_1 and R_L in Fig. 6-4P*b* when $E = 20$ V (*ans.* 166.7 mW, 10.4 mW)

c. The voltages V_1 and V_2 in Fig. 6-4P*a* when $E = 100$ V (*ans.* 50 V, 12.5 V)

d. The voltages V_1 and V_2 in Fig. 6-4P*b* when $E = 100$ V (*ans.* 50 V, 12.5 V)

6-6 Solve Prob. 5-2 by the current assumption method.

6-7 By assuming a current in R_2, R_3, or R_4, solve Prob. 5-10. (*ans.* 28.72 mA, 57.45 mA)

6-8 By assuming a current in R_2, R_3, or R_4, solve Prob. 5-11.

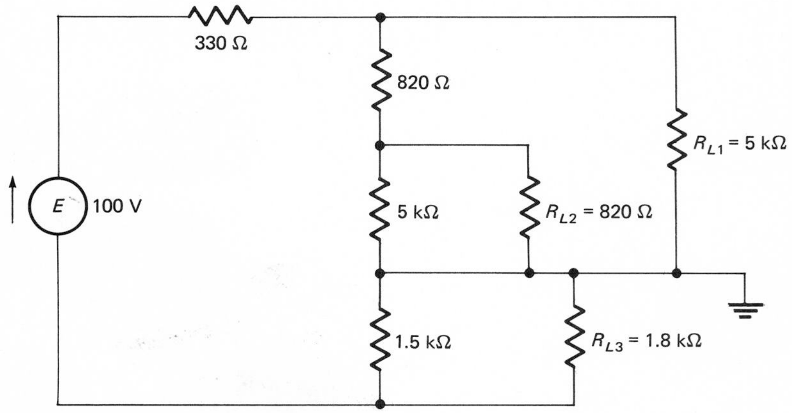

Fig. 6-5P Problem 6-9: circuit.

6-9 By assuming a current, find the three load voltages in the voltage divider circuit in Fig. 6-5P. (*ans.* 50.45 V, 23.3 V, 35.32 V)

Section 6-2

6-10 For Prob. 6-1*a* through *d*, convert the voltage generator *E* and series resistance R_1 to an equivalent current source, convert all resistances to conductances, and solve for I_L using nodal analysis.

6-11 For the circuit in Fig. 6-6 in the text, using nodal analysis, find the current through G_2 when $I_{G1} = 100$ mA, $I_{G2} = 200$ mA, $G_1 = 2.5$ mS, $G_2 = 2$ mS, and $G_3 = 3.33$ mS. (*ans.* 8.3 mA)

6-12 For Prob. 6-2*a* through *c*, convert the voltage generator *E* and the series 680-Ω resistance to an equivalent current source, convert all resistances to conductances, and solve for I_L using nodal analysis.

6-13 Solve Prob. 6-4*a* through *c* using nodal analysis.
(*ans. a.* 1.871 mW, 3.478 mW; *b.* 1.439 mW, 0.2921 mW; *c.* 1.302 mW, 0.182 mW)

6-14 Using nodal analysis and Fig. 6-7 in the text, find the current in each of the conductances when $I_{G1} = 50$ mA, $I_{G2} = 100$ mA, $I_{G3} = 50$ mA, $G_1 = 10$ mS, $G_2 = 5$ mS, $G_3 = 2$ mS, $G_4 = 5$ mS, and $G_5 = G_6 = G_7 = 1$ mS.

6-15 The domestic power system in Example 6-14 and Fig. 6-37 can be redrawn as in Fig. 6-6P. Using nodal analysis, find the powers dissipated by each of the three loads R_A, R_B, and R_C. (*ans.* 2.954 kW, 1.79 kW, 3.844 kW)

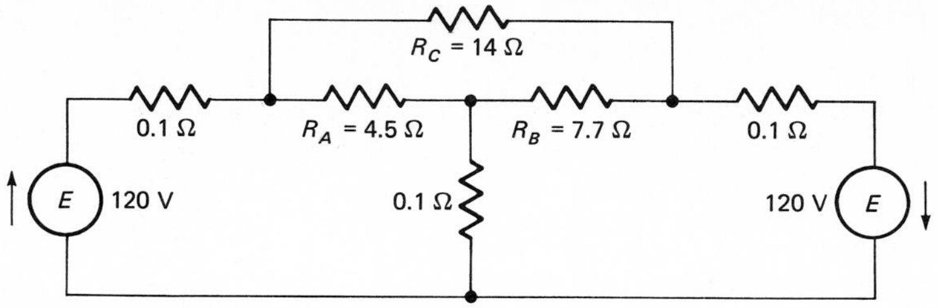

Fig. 6-6P Problem 6-15: circuit.

6-16 Refer to Example 6-7 and Figs. 6-16 and 6-17 in the text. If an input signal applies +2 V to the input terminal (with respect to ground) through a 1-kΩ resistor, what will be the value of I_B? Solve by nodal analysis.

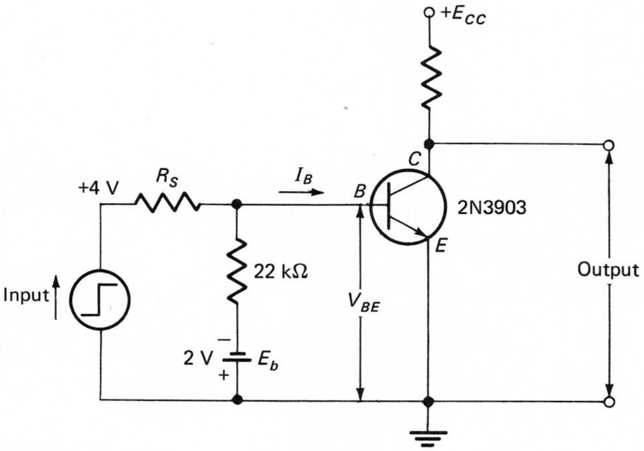

Fig. 6-7P Problems 6-17 to 6-19: circuit.

6-17 The circuit in Fig. 6-7P is a type of transistor switch. When the switch is on, $V_{BE} =$ 0.2 V and $I_B = 150\ \mu$A. Using nodal analysis, find a value for R_S which will allow I_B to be 150 μA when a 4-V positive pulse is applied to the input. To ensure that I_B is at least 150 μA, should the preferred-value resistor selected for R_S be higher than the calculated value? (*ans.* 15.2 kΩ; no, lower)

6-18 If E_b in Prob. 6-17 is increased to 3 V, using nodal analysis find the value to which R_S must be changed for I_B to remain at 150 μA.

6-19 If R_S in Prob. 6-17 is left at the value calculated and E_b is increased to 3 V, to what value must the input pulse be changed for I_B to remain at 150 μA? Solve by nodal analysis. (*ans.* 4.69 V)

Section 6-3

6-20 Solve Prob. 6-1*a* through *d* using loop analysis.

6-21 Solve Prob. 6-3*a* through *c* using loop analysis. (*ans. a.* 13.74 mA; *b.* 15.44 mA; *c.* 20.27 mA)

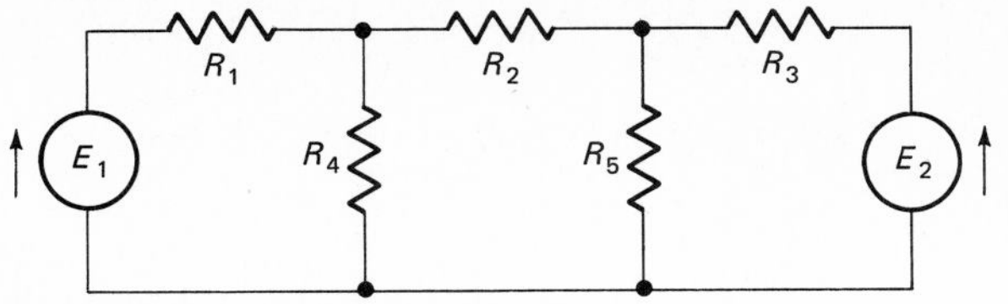

Fig. 6-8P Problems 6-22 and 6-33.

6-22 For the circuit in Fig. 6-8P, using loop analysis, find the direction and magnitude of current through R_2 when

a. $E_1 = 10$ V, $E_2 = 20$ V, $R_1 = 180\ \Omega$, $R_2 = 150\ \Omega$, $R_3 = 120\ \Omega$, $R_4 = 470\ \Omega$, $R_5 = 560\ \Omega$

b. $E_1 = 15$ V, $E_2 = 10$, $R_1 = 4.7$ kΩ, $R_2 = 1.8$ kΩ, $R_3 = 1.8$ kΩ, $R_4 = 2.2$ kΩ, $R_5 = 2.2$ kΩ

c. $E_1 = 40$ V, $E_2 = 50$ V, $R_1 = 12$ kΩ, $R_2 = 27$ kΩ, $R_3 = 22$ kΩ, $R_4 = 18$ kΩ, $R_5 = 33$ kΩ

6-23 Repeat Example 6-12 in the text, finding the meter current by loop analysis when R_1 is 100 Ω and all the other values remain the same. (*ans.* 676 μA, right)

6-24 Repeat Example 6-12 in the text, finding the meter current by loop analysis when R_1 is set to 1000 Ω and R_2 is set to 10 Ω.

6-25 Convert the rudder control current source in Example 6-8 in the text to a voltage source, and using loop analysis find the resistance of R_5 for a balanced condition when the rudder is at mid-position. [Control current $= (6\text{ V} - 1\text{ V})/2\ \Omega = 3.5$ mA.] (*ans.* 2.037 kΩ)

6-26 For Example 6-7 in the text, find R_2 using loop analysis if the −5-V source is changed to −10 V.

6-27 In Prob. 6-26, if an input signal of +2 V is applied through a 1000-Ω resistor, to what value will I_B change, provided R_{BE} remains constant? (*ans.* 1.442 mA)

6-28 In Example 6-14 in the text, if an additional 5-kW load is connected between lines 1 and 2, find the three line currents using loop analysis.

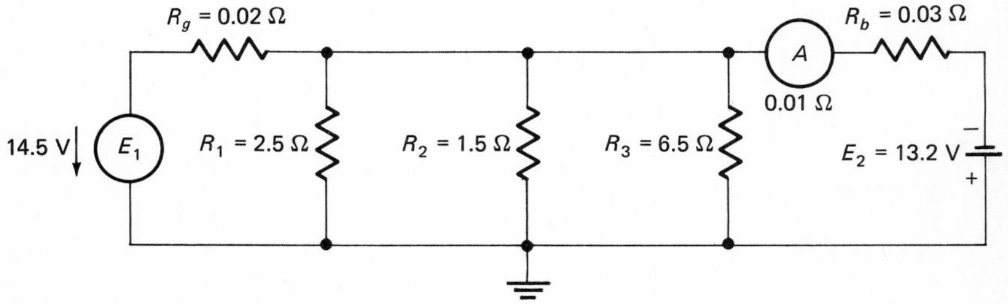

Fig. 6-9P Problem 6-29: circuit.

6-29 The circuit in Fig. 6-9P represents a condition in the electric system of an automobile. E_1 is the generator with an internal resistance $R_g = 0.02\ \Omega$; R_1, R_2, and R_3 represent individual loads being drawn; A is the ammeter with an internal resistance of 0.01 Ω; and E_2 is the battery with an internal resistance R_b of 0.03 Ω. What is the magnitude of the current through the ammeter? Is the battery charging? (*ans.* 16.13 A, yes)

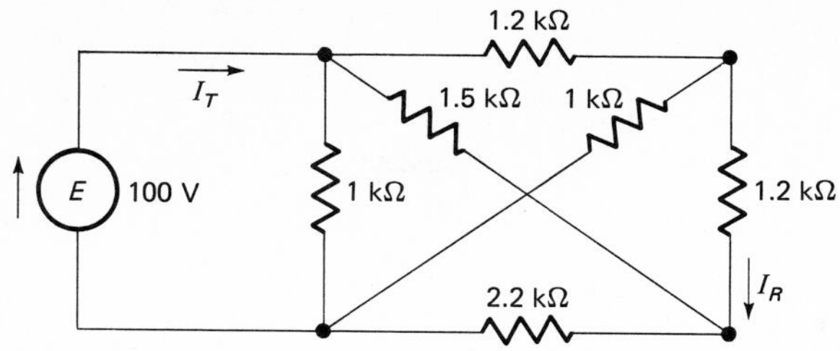

Fig. 6-10P Problem 6-30: circuit.

6-30 Using loop analysis, find I_T and I_R in Fig. 6-10P.

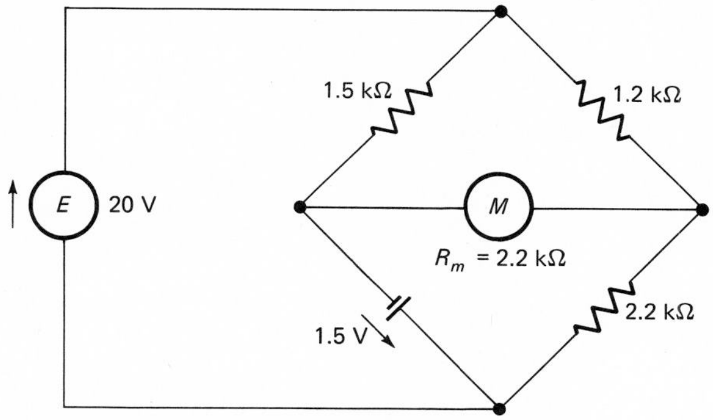

Fig. 6-11P Problem 6-31: circuit.

6-31 Using loop analysis, find the meter current in the bridge circuit in Fig. 6-11P. (*ans*. 4.849 mA)

Section 6-4

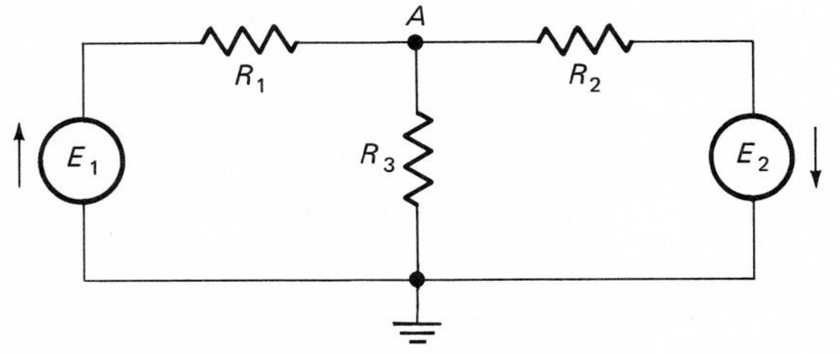

Fig. 6-12P Problem 6-32: circuit.

6-32 Find the magnitude and polarity of the voltage at point A in Fig. 6-12P, using the superposition theorem, when

a. $E_1 = 50$ V, $E_2 = 100$ V, $R_1 = 1$ kΩ, $R_2 = 500$ Ω, $R_3 = 200$ Ω

b. $E_1 = 25$ V, $E_2 = 25$ V, $R_1 = 47$ kΩ, $R_2 = 15$ kΩ, $R_3 = 10$ kΩ

c. $E_1 = 50$ V, $E_2 = 75$ V, $R_1 = 2.2$ kΩ, $R_2 = 3.3$ kΩ, $R_3 = 2.7$ kΩ

6-33 Using the superposition theorem, find the voltage drop across R_5 in Fig. 6-8P when

a. $E_1 = 15$ V, $E_2 = 10$ V, $R_1 = 4.7$ kΩ, $R_2 = 1.8$ kΩ, $R_3 = 2.7$ kΩ, $R_4 = 2.2$ kΩ, $R_5 = 3.3$ kΩ (*ans*. 5.37 V)

b. $E_1 = 100$ V, $E_2 = 200$ V, $R_1 = 1000$ Ω, $R_2 = 2000$ Ω, $R_3 = 1000$ Ω, $R_4 = 1500$ Ω, $R_5 = 1500$ Ω (*ans*. 108.76 V)

6-34 Solve Prob. 6-28 using the superposition theorem.

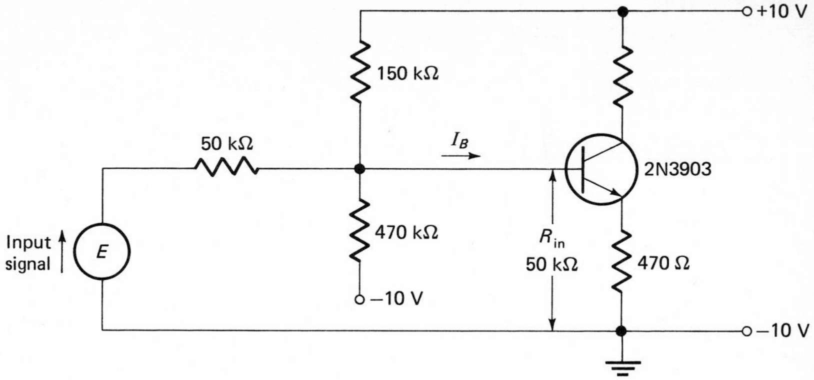

Fig. 6-13P Problem 6-35: circuit.

6-35 In Fig. 6-13P the input signal E varies between $+1$ and $+2$ V. Find the base current I_B, using the superposition theorem, when

a. $E = +1$ V (*ans.* 26.81 μA)

b. $E = +1.5$ V (*ans.* 30.9 μA)

c. $E = +2$ V (*ans.* 35 μA)

Refer to Example 6-13 for hints on drawing an equivalent circuit for Fig. 6-13P.

7

Network Analysis by Equivalent Circuit Methods

You have now been introduced to four methods of analyzing an electric circuit: current assumption, and nodal, loop, and superposition analysis. Which is the best? You have received the message if you realized that there is no *best* method. Each has advantages in certain situations.

It is likely that you will prefer one of the techniques and use it more frequently than the others, even when it is not the most efficient. This is all right, and quite in keeping with the philosophy that it is only through the knowledge of many aspects of a problem that you can make confident, wise decisions. You should be familiar with a number of techniques, each of which can be used for the same purpose. This familiarity gives a degree of circuit consciousness which cannot be obtained through learning how to solve each individual problem with a single preferred method.

The methods of Chap. 6 have one feature in common. They are good for the evaluation of a number of quantities in a circuit, for example, all the currents. The analysis methods to be described now can also be used for this purpose, but excel when the values associated with only one circuit element are called for.

You will see that a number of the examples involve finding the conditions existing in a load. Often the sole purpose of an electric or electronic system is to supply controlled energy to a load. Examples are the radio transmitter supplying energy to an antenna and the hydroelectric station supplying energy to a city.

7-1 T $\leftrightarrow$ π CONVERSIONS

The equivalent circuit methods of network analysis rely on the fact that it is possible to reduce complex networks to simple ones for special purposes. It has already been shown that a two-terminal (single-port)[1] network can be reduced to a single equivalent element. A three-terminal (two-port) network with many elements can be reduced to a three-element network by the use of T $\leftrightarrow$ π conversions. The acceptance of this technique implies the validity of using three-element circuits to represent networks with any number of elements. This permits some general circuit principles to be derived by fairly simple means with the confidence that they even apply to complex circuits.

[1] A *port* is a network location where a connection can be made externally to the network.

T, π, WYE, AND DELTA NETWORKS

Figure 7-1*a* shows a T network and Fig. 7-1*b* a wye network. Both are exactly the same, only the drawing layout is different. Similarly, the π network in Fig. 7-1*c* is exactly the same as the delta network in Fig. 7-1*d*.

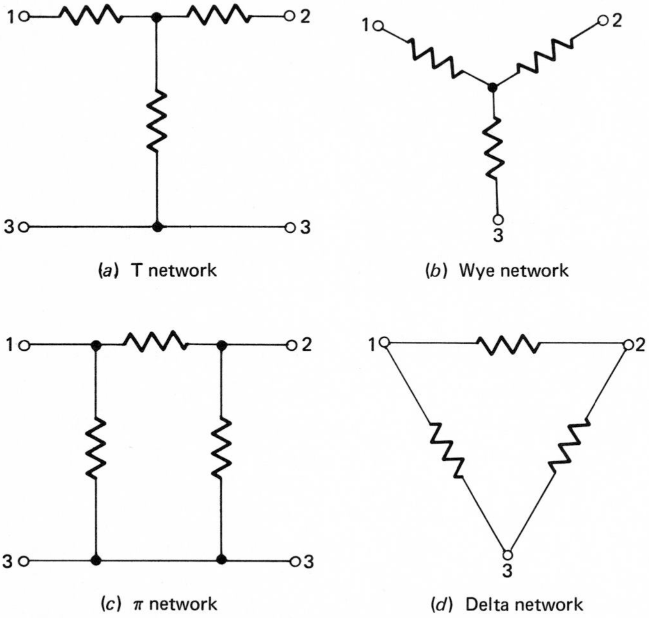

Fig. 7-1 T, wye, π, and delta networks.

The network names are based on the shapes of the English and Greek letters T, Y (wye), π, and Δ (delta). The T or Y format is sometimes called *star* because of its shape.

In electronics the T and π formats are adopted because they conveniently fit the input-output type of external connection so common in that branch of the industry. Typically, terminals 1 and 3 may constitute the input port, while terminals 2 and 3 form the output port. Often terminal 3, the common between the input and output, is grounded.

In electric power systems the wye and delta views are customary because no distinction is made between specific pairs of terminals. None of the terminals is likely to be grounded.

The terms "T" and "π" will be used here because modern industry is predominantly electronic. However, the wye and delta diagrams will be used for some explanations, as they can aid understanding.

EQUIVALENCE

Figure 7-2 shows a T (or wye) and a π (or delta) network which are to be equivalent to each other. For this to be so, it must be impossible to distinguish between them by electrical measurement at the terminals. If the

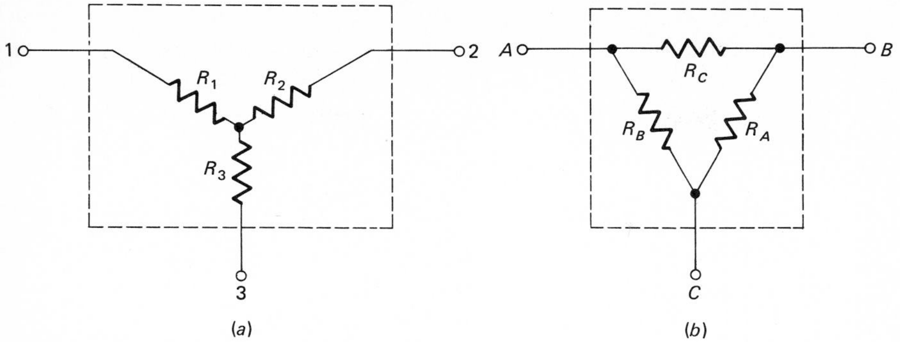

Fig. 7-2 (*a*) T network; (*b*) π network.

resistances between all related pairs of terminals are the same, this equivalence will be established.

The conditions for equivalence will be:

T network, Fig. 7-2a		*π network, Fig. 7-2b*
Resistance 1 to 2	=	Resistance A to B
Resistance 2 to 3	=	Resistance B to C
Resistance 3 to 1	=	Resistance C to A

When one of the terminals of the T network in Fig. 7-2*a* is left disconnected, a series circuit remains between the other two terminals (see Fig. 7-3*a* in which terminal 2 is left disconnected and the circuit between terminals 1 and 3 consists of R_1 and R_3 in series).

If one terminal of the π network in Fig. 7-2*b* is left open, the remaining circuit is series parallel. The circuit between the other two terminals will consist of two resistances in series, this pair being in parallel with the third resistance (see Fig. 7-3*b* in which terminal B is left disconnected and the circuit between terminals A and C is the series combination of R_A and R_C connected in parallel with R_B).

Fig. 7-3 Figure 7-2 redrawn to show equivalent circuits when one terminal is left open.

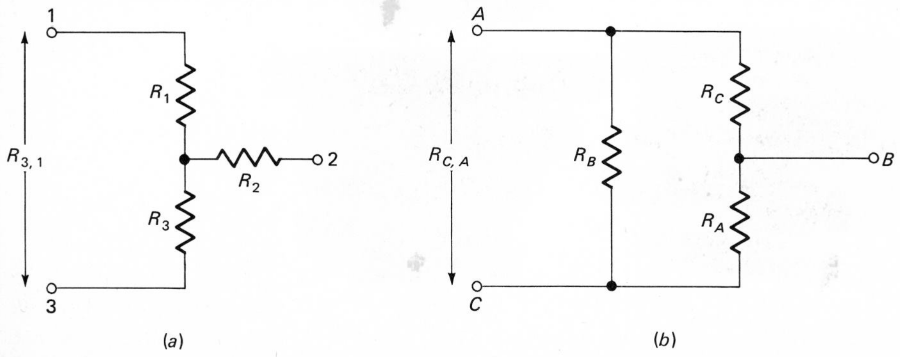

Figure 7-3 compares the resistance between a pair of terminals in each of the networks when the third terminal is left disconnected.

T network: $$R_{3,1} = R_3 + R_1$$

π network: $$R_{C,A} = \frac{R_B(R_C + R_A)}{R_A + R_B + R_C}$$

For the networks to be equivalent, $R_{3,1}$ must equal $R_{C,A}$.

$$R_3 + R_1 = \frac{R_A R_B + R_B R_C}{R_A + R_B + R_C}$$

By repeating this process for the other two pairs of terminals, the three equations for equivalence are obtained.

$$R_1 + R_2 = \frac{R_B R_C + R_C R_A}{\Sigma R_\nabla} \tag{7-1}$$

$$R_2 + R_3 = \frac{R_A R_B + R_C R_A}{\Sigma R_\nabla} \tag{7-2}$$

$$R_3 + R_1 = \frac{R_A R_B + R_B R_C}{\Sigma R_\nabla} \tag{7-3}$$

The symbol ΣR_∇ is used for the sum of the elements in the π network. It is common practice to use Σ (capital Greek letter sigma) to indicate a constantly repeated sum of several terms.

π-TO-T CONVERSION

To convert the π network in Fig. 7-4*a* to its T equivalent, Fig. 7-4*b*, it is necessary to express each of the elements of the T network (R_1, R_2, and R_3) in terms of the elements of the π network (R_A, R_B, and R_C). This is done by manipulating Eqs. (7-1) to (7-3) to give

$$R_1 = \frac{R_B R_C}{\Sigma R_\nabla} \tag{7-4}$$

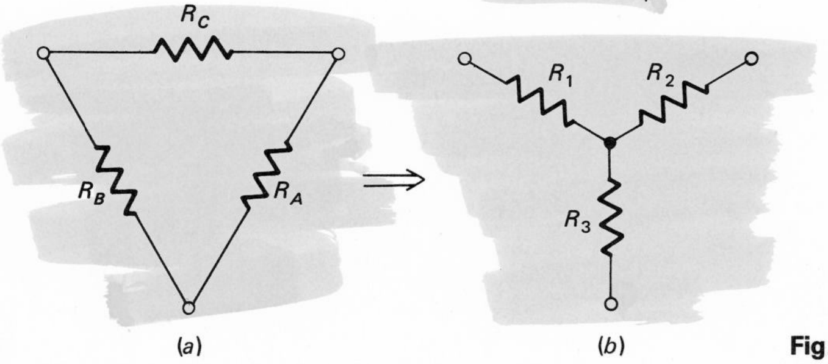

Fig. 7-4 π-to-T conversion.

$$R_2 = \frac{R_C R_A}{\Sigma R_\nabla} \tag{7-5}$$

$$R_3 = \frac{R_A R_B}{\Sigma R_\nabla} \tag{7-6}$$

Note that

1. The denominator ΣR_∇ is the same for all three equations.
2. The numerator for any of the equations is the product of the two π-network elements adjacent to the related terminal. For example, R_3 connects to the bottom terminal of the T network, while R_A and R_B connect to the bottom terminal of the π network.

T-TO-π CONVERSION

To convert the T network in Fig. 7-5*a* to its π equivalent, Fig. 7-5*b*, each of the π-network elements (R_A, R_B, and R_C) must be expressed in terms of the T-network elements (R_1, R_2, and R_3). From Eqs. (7-4) to (7-6) the following equations can be derived:

$$R_A = \frac{R_1R_2 + R_2R_3 + R_3R_1}{R_1} = \frac{\Sigma R_Y}{R_1} \tag{7-7}$$

$$R_B = \frac{\Sigma R_Y}{R_2} \tag{7-8}$$

$$R_C = \frac{\Sigma R_Y}{R_3} \tag{7-9}$$

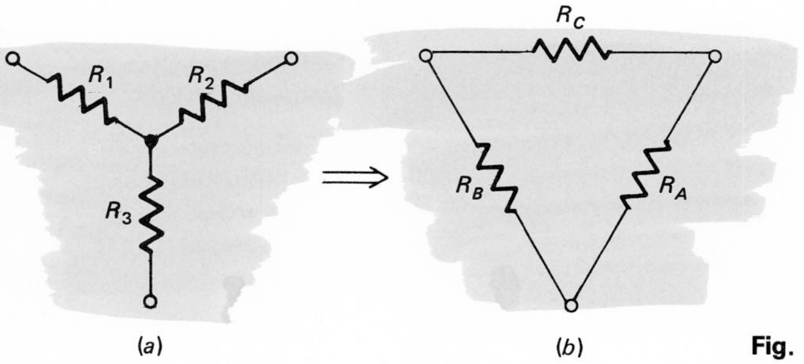

Fig. 7-5 T-to-π conversion.

Note that

1. The numerator is the same for each equation. $\Sigma R_Y = R_1R_2 + R_2R_3 + R_3R_1$ (the sum of the product pairs of the T-network elements).
2. The denominator for any of the equations is the opposite T element. For instance, R_C is at the top of the π network, while R_3 is at the bottom of the T network.

The following examples first show how to perform direct conversions, and then how to reduce complex networks. Some examples of the use of network conversion for circuit analysis are also given.

Example 7-1 Convert the π network in Fig. 7-6*a* to the T format in Fig. 7-6*b*.

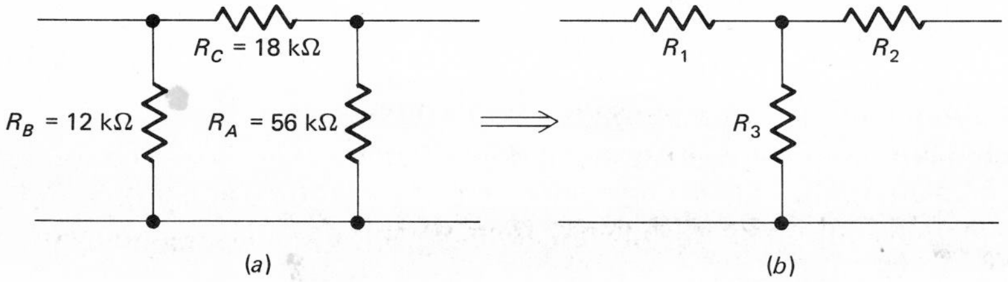

Fig. 7-6 Example 7-1: (*a*) π network; (*b*) equivalent T network.

Solution

$$\Sigma R_\nabla = R_A + R_B + R_C = 56\ \text{k}\Omega + 12\ \text{k}\Omega + 18\ \text{k}\Omega = 86\ \text{k}\Omega$$

$$R_1 = \frac{R_B R_C}{\Sigma R_\nabla} \tag{7-4}$$

$$R_1 = \frac{12\ \text{k}\Omega \times 18\ \text{k}\Omega}{86\ \text{k}\Omega} = \mathbf{2.512\ k\Omega}$$

$$R_2 = \frac{R_C R_A}{\Sigma R_\nabla} \tag{7-5}$$

$$R_2 = \frac{18\ \text{k}\Omega \times 56\ \text{k}\Omega}{86\ \text{k}\Omega} = \mathbf{11.72\ k\Omega}$$

$$R_3 = \frac{R_A R_B}{\Sigma R_\nabla} \tag{7-6}$$

$$R_3 = \frac{56\ \text{k}\Omega \times 12\ \text{k}\Omega}{86\ \text{k}\Omega} = \mathbf{7.814\ k\Omega}$$

Example 7-2 Convert the T network in Fig. 7-7*a* to its π equivalent, Fig. 7-7*b*.

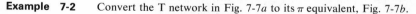

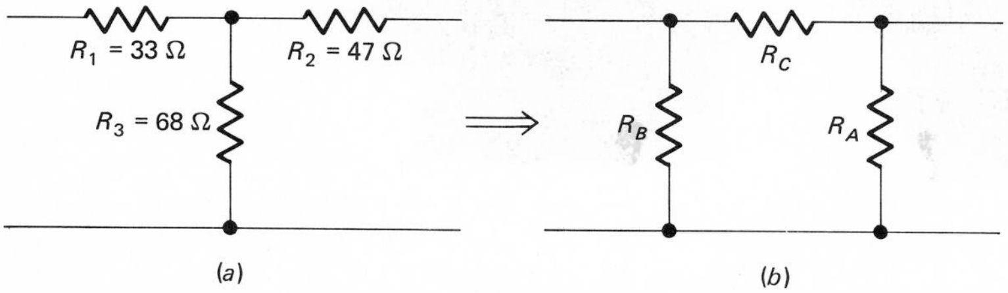

Fig. 7-7 Example 7-2: (*a*) T network; (*b*) equivalent π network.

Solution

$$\Sigma R_Y = R_1R_2 + R_2R_3 + R_3R_1 = 33\ \Omega \times 47\ \Omega + 47\ \Omega \times 68\ \Omega + 68\ \Omega \times 33\ \Omega = 6991\ \Omega^2$$

$$R_A = \frac{\Sigma R_Y}{R_1} \tag{7-7}$$

$$R_A = \frac{6991\ \Omega^2}{33\ \Omega} = \mathbf{211.8\ \Omega}$$

$$R_B = \frac{\Sigma R_Y}{R_2} \tag{7-8}$$

$$R_B = \frac{6991\ \Omega^2}{47\ \Omega} = \mathbf{148.7\ \Omega}$$

$$R_C = \frac{\Sigma R_Y}{R_3} \tag{7-9}$$

$$R_C = \frac{6991\ \Omega^2}{68\ \Omega} = \mathbf{102.8\ \Omega}$$

You will see from the results for Examples 7-1 and 7-2 that the resistances of the T network are lower than those of the equivalent π network. This conclusion is to be expected. Both networks must have the same resistances between related terminal pairs. Each T-network element has another effectively in series with it, giving additional resistance. Each π-network element has elements in parallel with it, making a lower total resistance.

Example 7-3 Reduce the six-element network in Fig. 7-8*a* to an equivalent three-element network, maintaining the original terminals.

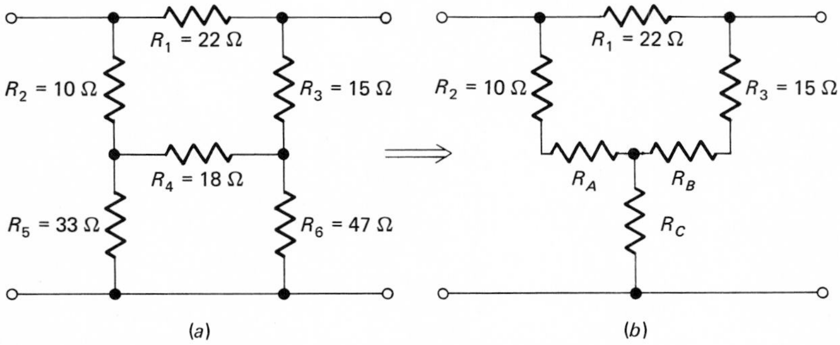

Fig. 7-8 Example 7-3: (*a*) given network; (*b*) first reduction.

Solution

1. Convert the π group R_4, R_5, R_6 into the T network R_A, R_B, R_C in Fig. 7-8*b*.

$$\Sigma R_\nabla = R_4 + R_5 + R_6 = 18\ \Omega + 33\ \Omega + 47\ \Omega = 98\ \Omega$$

$$R_A = \frac{R_4 R_5}{\Sigma R_\nabla} = \frac{18\ \Omega \times 33\ \Omega}{98\ \Omega} = 6.061\ \Omega$$

$$R_B = \frac{R_4 R_6}{\Sigma R_\nabla} = \frac{18\ \Omega \times 47\ \Omega}{98\ \Omega} = 8.633\ \Omega$$

$$R_C = \frac{R_5 R_6}{\Sigma R_\nabla} = \frac{33\ \Omega \times 47\ \Omega}{98\ \Omega} = 15.83\ \Omega$$

2. Two series groups, R_2, R_A and R_3, R_B, are shown as elements R_D and R_E in Fig. 7-9*a*.

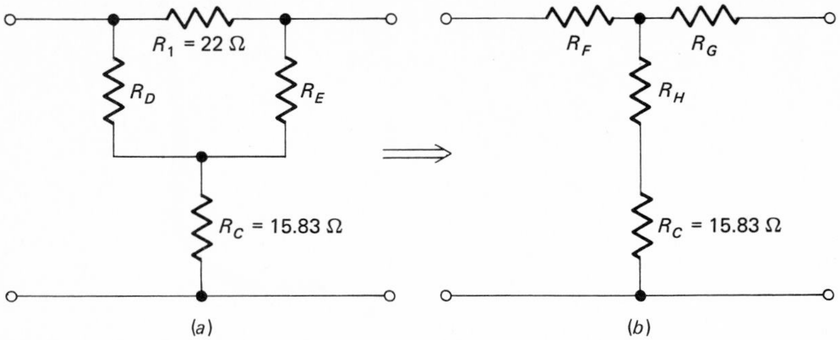

Fig. 7-9 (*a*) Second reduction of Fig. 7-8*a*; (*b*) third reduction of Fig. 7-8*a*.

$$R_D = R_2 + R_A = 10\ \Omega + 6.061\ \Omega = 16.06\ \Omega$$

$$R_E = R_3 + R_B = 15\ \Omega + 8.633\ \Omega = 23.63\ \Omega$$

3. Convert the π group R_1, R_D, R_E in Fig. 7-9*a* into the T network R_F, R_G, R_H in Fig. 7-9*b*.

$$\Sigma R_\nabla = R_1 + R_D + R_E = 22\ \Omega + 16.06\ \Omega + 23.63\ \Omega = 61.69\ \Omega$$

$$R_F = \frac{R_1 R_D}{\Sigma R_\nabla} = \frac{22\ \Omega \times 16.06\ \Omega}{61.69\ \Omega} = \mathbf{5.727\ \Omega}$$

$$R_G = \frac{R_1 R_E}{\Sigma R_\nabla} = \frac{22\ \Omega \times 23.63\ \Omega}{61.69\ \Omega} = \mathbf{8.427\ \Omega}$$

$$R_H = \frac{R_D R_E}{\Sigma R_\nabla} = \frac{16.06\ \Omega \times 23.63\ \Omega}{61.69\ \Omega} = 6.152\ \Omega$$

4. Converting R_C and R_H in series into the single element R_J completes the circuit reduction:

$$R_J = R_C + R_H = 15.83\ \Omega + 6.152\ \Omega = \mathbf{21.98\ \Omega}$$

Figure 7-10 shows the original circuit in Fig. 7-8*a* and the final circuit which is its three-element equivalent. Measurements taken at a pair of terminals of the original network will be the same as those taken at the equivalent pair of terminals of the final network. Also, the relationship between input and output voltages and currents would be the same in each case.

Fig. 7-10 Example 7-3: (*a*) original network; (*b*) final three-element T equivalent network.

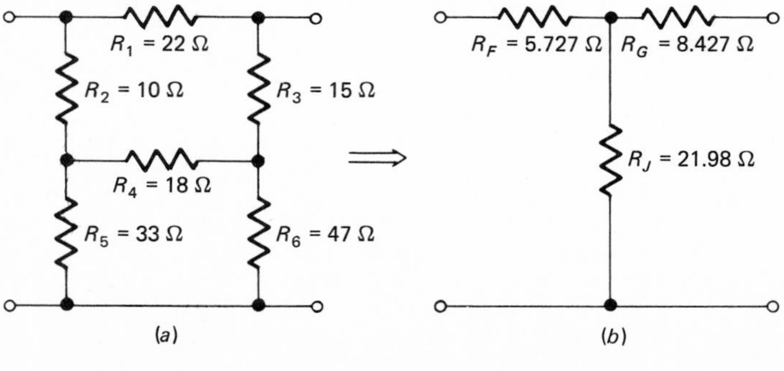

Example 7-4 What is the magnitude and direction of the meter current in the resistance bridge circuit in Fig. 7-11?

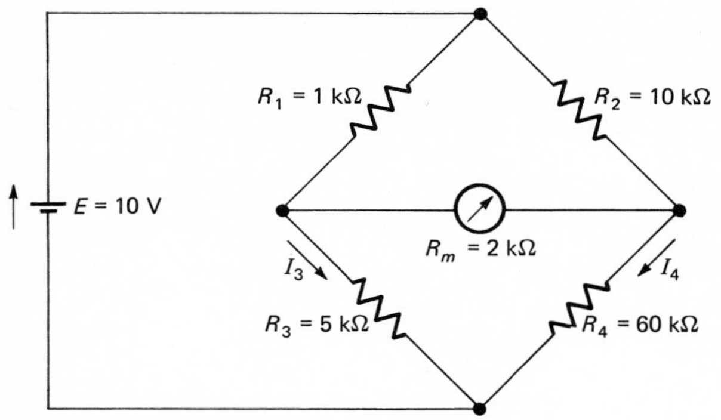

Fig. 7-11 Example 7-4: circuit.

Solution Convert the π network R_1, R_2, R_m into an equivalent T network, as in Fig. 7-12*a*.

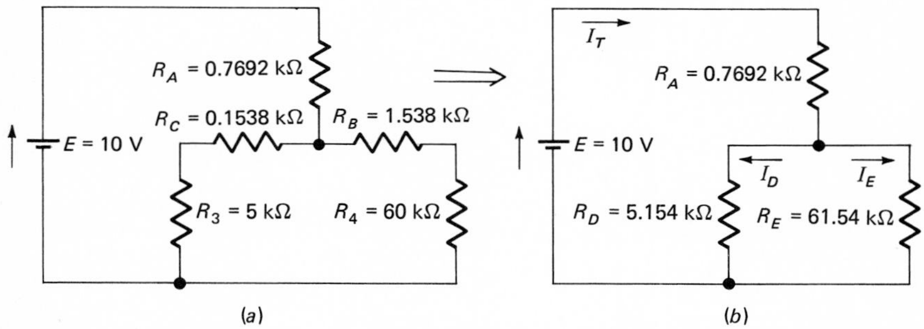

Fig. 7-12 (*a*) First reduction of Fig. 7-11; (*b*) bridge network in Fig. 7-11 reduced to a three-element equivalent network.

$$\Sigma R_\nabla = R_1 + R_2 + R_m = 1\text{ k}\Omega + 10\text{ k}\Omega + 2\text{ k}\Omega = 13\text{ k}\Omega$$

$$R_A = \frac{R_1 R_2}{\Sigma R_\nabla} = \frac{1\text{ k}\Omega \times 10\text{ k}\Omega}{13\text{ k}\Omega} = 0.7692\text{ k}\Omega$$

$$R_B = \frac{R_2 R_m}{\Sigma R_\nabla} = \frac{10\text{ k}\Omega \times 2\text{ k}\Omega}{13\text{ k}\Omega} = 1.538\text{ k}\Omega$$

$$R_C = \frac{R_m R_1}{\Sigma R_\nabla} = \frac{2\text{ k}\Omega \times 1\text{ k}\Omega}{13\text{ k}\Omega} = 0.1538\text{ k}\Omega$$

In Fig. 7-12*b*,

$$R_D = R_3 + R_C = 5\text{ k}\Omega + 0.1538\text{ k}\Omega = 5.154\text{ k}\Omega$$

$$R_E = R_4 + R_B = 60\text{ k}\Omega + 1.538\text{ k}\Omega = 61.54\text{ k}\Omega$$

Figure 7-12*b* shows that the conversion of the π network reduces the original bridge circuit to a series-parallel format. The resistance values of this equivalent series-parallel circuit are now known. The currents in R_D and R_E can be calculated:

$$I_T = \frac{E}{R_A + \dfrac{R_D R_E}{R_D + R_E}} = \frac{10 \text{ V}}{0.7692 \text{ k}\Omega + \dfrac{5.154 \text{ k}\Omega \times 61.54 \text{ k}\Omega}{5.154 \text{ k}\Omega + 61.54 \text{ k}\Omega}} = 1.81 \text{ mA}$$

$$I_D = I_T \frac{R_E}{R_D + R_E} = 1.81 \text{ mA} \frac{61.54 \text{ k}\Omega}{66.69 \text{ k}\Omega} = 1.67 \text{ mA}$$

$$I_E = I_T - I_D = 1.81 \text{ mA} - 1.67 \text{ mA} = 0.14 \text{ mA}$$

The currents I_3 and I_4 in the original bridge circuit (Fig. 7-11) are the same as I_D and I_E, respectively.

$$I_3 = 1.67 \text{ mA} \qquad \text{and} \qquad I_4 = 0.14 \text{ mA}$$

Figure 7-13 shows part of the circuit in Fig. 7-11.

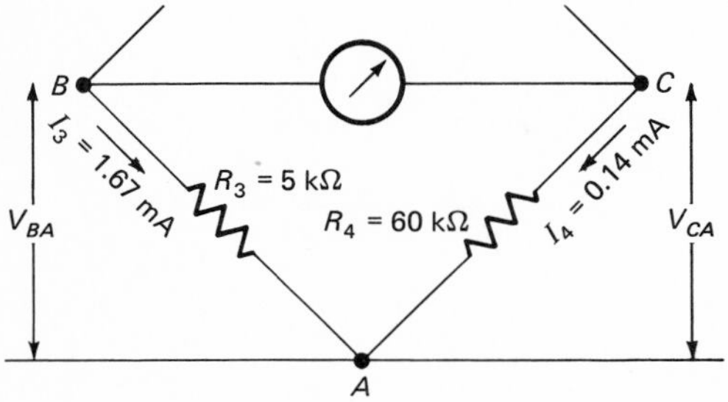

Fig. 7-13 Part of Fig. 7-11 showing the data for resistances R_3 and R_4.

Calculate relative potentials:

$$V_{BA} = I_3 R_3 = 1.67 \text{ mA} \times 5 \text{ k}\Omega = 8.35 \text{ V}$$

$$V_{CA} = I_4 R_4 = 0.14 \text{ mA} \times 60 \text{ k}\Omega = 8.4 \text{ V}$$

Therefore node C is positive with respect to node B, as shown in Fig. 7-14.

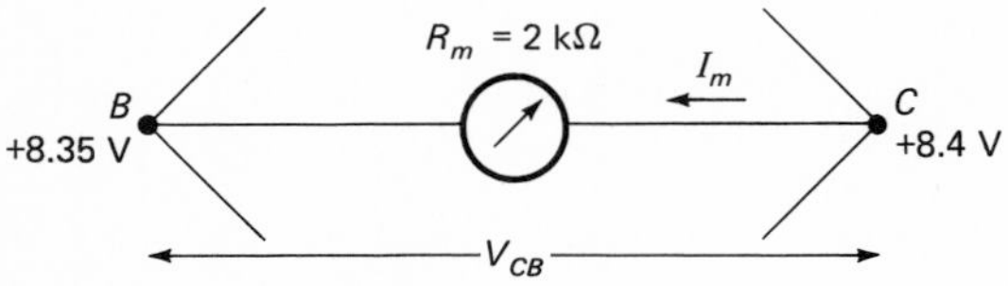

Fig. 7-14 Meter circuit in Fig. 7-11.

The voltage across the meter will be

$$V_{CB} = V_{CA} - V_{BA} = 8.4 \text{ V} - 8.35 \text{ V} = 0.05 \text{ V}$$

The current in the meter will have the direction shown in Fig. 7-14, **right to left.** Its value will be

$$I_m = \frac{V_{CB}}{R_m} = \frac{0.05 \text{ V}}{2 \text{ k}\Omega} = \mathbf{0.025 \text{ mA}} \qquad \text{or} \qquad \mathbf{25\ \mu A}$$

7-2 THEVENIN'S THEOREM

The ultimate in network reduction occurs when a circuit is reduced to just two items: the element about which information is required, and the remainder of the circuit considered as a unit. An example of this type of circuit reduction is depicted in Fig. 7-15.

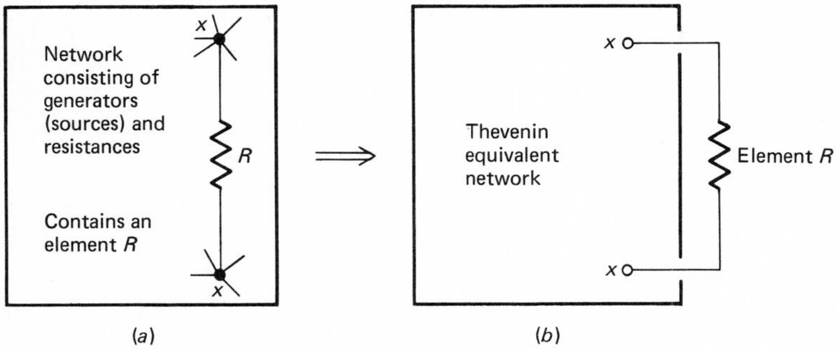

Fig. 7-15 (*a*) Network of unspecified type; (*b*) network with one element isolated.

Thevenin's theorem is concerned with achieving this reduction and then using it to arrive at the current, voltage, or power associated with the element R. Thevenin's theorem states that the current in any resistance R connected to two terminals of a network is the same as if R were connected to a voltage source where

1. The emf is the open-circuit voltage at the terminals of R.
2. The internal resistance is the resistance of the network at the terminals of R, with all other sources replaced by resistances equal to their internal resistances.

THEVENIN EQUIVALENT CIRCUIT

The circuit which results from the Thevenin approach consists of a voltage source with series resistance, as in Fig. 7-16.

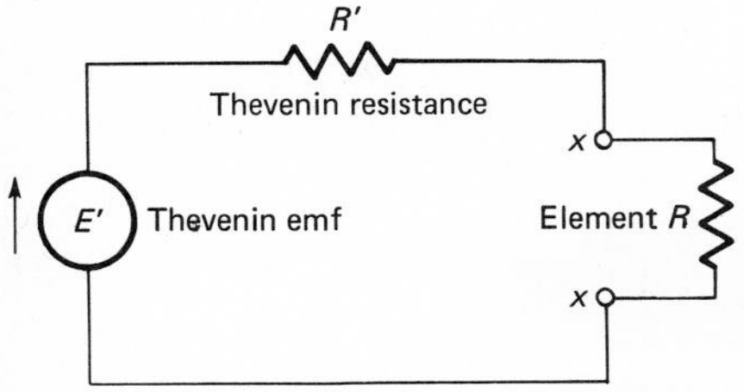

Fig. 7-16 Basic Thevenin circuit.

Referring to Figs. 7-15 and 7-16, the *Thevenin emf* E' is the open-circuit voltage at the x,x terminals (the voltage with R disconnected). The *Thevenin resistance* R' is the resistance at the x,x terminals with these terminals open circuit (R disconnected) and all energy sources in the remaining circuit removed and replaced by their internal resistances.

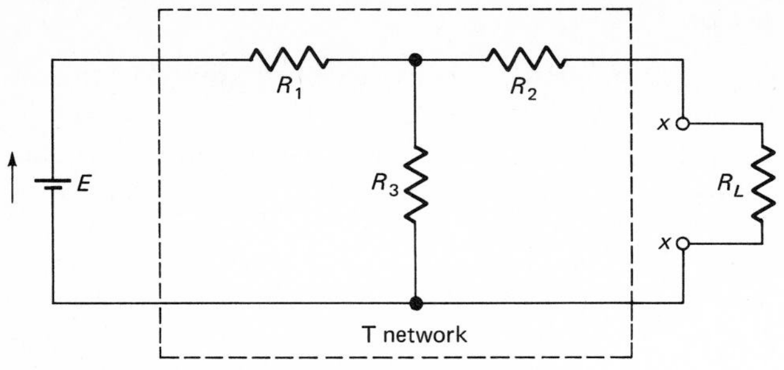

Fig. 7-17 Battery, T network, and load.

To see how the Thevenin voltage and resistance are determined, let us look at a circuit consisting of a battery, a T network, and a load, as in Fig. 7-17. Since a T network represents all networks which can be reduced to a T format, the circuit in Fig. 7-17 is considered typical of all such networks.

To obtain the value of the Thevenin emf E',

1. Disconnect R_L at the x,x terminals.
2. Determine the voltage at the x,x terminals.

When R_L is disconnected (Fig. 7-18), the voltage at the terminals is the same as the voltage across R_3 (since there can be no current in R_2, there is no voltage across it).

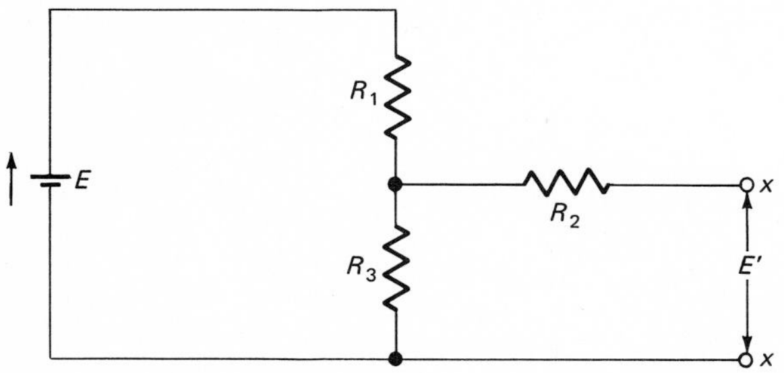

Fig. 7-18 Thevenin voltage in the circuit in Fig. 7-17.

$$E' = E \frac{R_3}{R_1 + R_3} \tag{7-10}$$

To obtain the value of the Thevenin resistance R',

1. Disconnect the load R_L at the x,x terminals.
2. Remove the battery E. As no internal resistance is shown for this battery, it is assumed to be 0 Ω, a short circuit.
3. Determine the resistance between the x,x terminals.

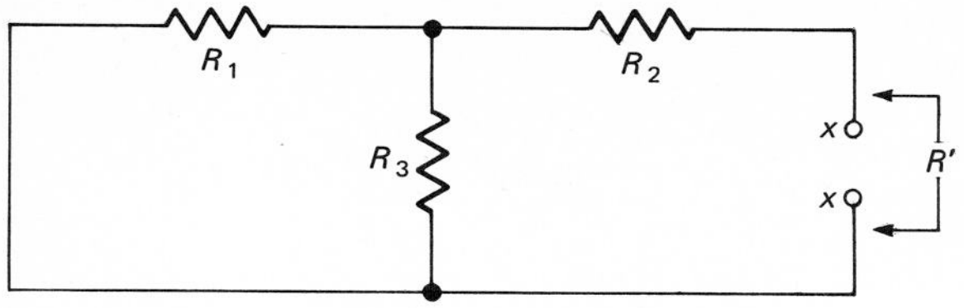

Fig. 7-19 Thevenin resistance in the circuit in Fig. 7-17.

With R_L removed and the battery replaced by a short circuit, the network across the terminals will be series parallel, as shown in Fig. 7-19.

$$R' = R_2 + \frac{R_1 R_3}{R_1 + R_3} \tag{7-11}$$

The Thevenin circuit equivalent to Fig. 7-17 will be as in Fig. 7-20.

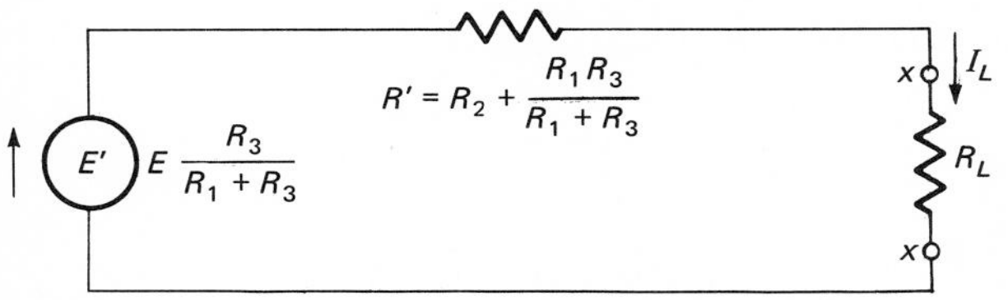

Fig. 7-20 Thevenin equivalent of the circuit in Fig. 7-17.

Although the rather complicated equations which result are not normally used for practical circuit analysis, the following derivation of the load current expression and its check by loop analysis proves the validity of Thevenin's theorem.

$$I_L = \frac{E'}{R' + R_L}$$

Substituting for E' and R' from Eqs. (7-10) and (7-11),

$$I_L = E\frac{R_3}{R_1 + R_3} \div \left[\left(R_2 + \frac{R_1 R_3}{R_1 + R_3}\right) + R_L\right]$$

$$= E\frac{R_3}{R_1 + R_3} \div \frac{R_1 R_2 + R_2 R_3 + R_3 R_1 + R_L(R_1 + R_3)}{R_1 + R_3}$$

$$I_L = \frac{E R_3}{R_1 R_2 + R_2 R_3 + R_3 R_1 + R_L(R_1 + R_3)}$$

The loop analysis check is obtained from Fig. 7-21.

Figure 7-21 is the same circuit as that in Fig. 7-17 with loop circulating currents shown.

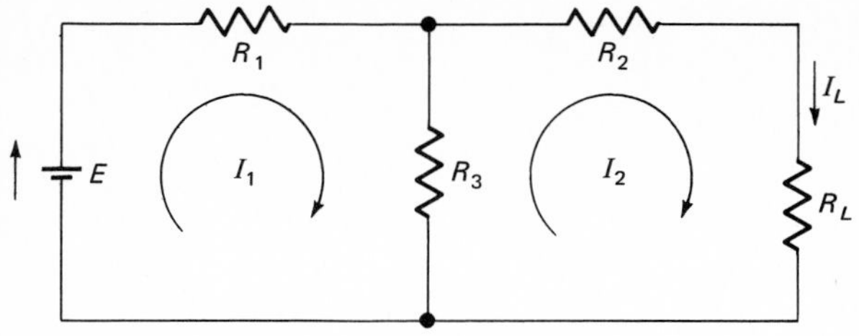

Fig. 7-21 Figure 7-17 with loop currents shown.

The loop equations are

$$(R_1 + R_3)I_1 - R_3I_2 = E \tag{7-12}$$

$$-R_3I_1 + (R_2 + R_3 + R_L)I_2 = 0 \tag{7-13}$$

From Eq. (7-13),

$$I_1 = \frac{(R_2 + R_3 + R_L)I_2}{R_3}$$

Substitute for I_1 in Eq. (7-12),

$$\frac{(R_1+R_3)(R_2+R_3+R_L)I_2}{R_3} - R_3I_2 = E$$

$$\frac{R_1R_2+R_2R_3+R_3R_1+R_L(R_1+R_3)}{R_3} I_2 = E$$

$$I_L = I_2 = \frac{ER_3}{R_1R_2 + R_2R_3 + R_3R_1 + R_L(R_1 + R_3)}$$

The expression for the current I_L is the same using either loop analysis or the Thevenin method. This equality shows that Thevenin's theorem is valid for a T network or any circuit which can be reduced to a T network. It is not feasible to prove the theorem for all possible circuits, but the T network is so general that you can use this method for all linear circuits.

GUIDELINES

Guidelines for judging the suitability of a circuit for Thevenin analysis include:

1. Data for one element are required.
2. That element resistance has various alternate values.
3. There are more voltage sources than current sources.
4. The circuit details are not known.

The last guideline is significant. Provided the Thevenin voltage and resistance can be measured, it is not necessary to know anything about the network. Thus the current, voltage, and power of an element can be determined, even though the equipment with which it is associated is too sophisticated to be analyzed easily.

Numerical examples will demonstrate the application of Thevenin's theorem.

Example 7-5 What is the load current in the circuit in Fig. 7-22?

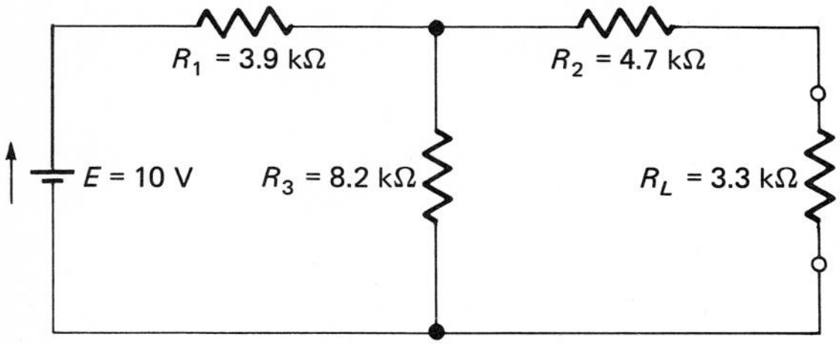

Fig. 7-22 Example 7-5: circuit.

Solution With R_L removed, the voltage across the terminals will be

$$E' = E \frac{R_3}{R_1 + R_3} \tag{7-10}$$

$$E' = 10\ \text{V} \frac{8.2\ \text{k}\Omega}{3.9\ \text{k}\Omega + 8.2\ \text{k}\Omega} = 6.777\ \text{V}$$

With R_L and the battery removed (its internal resistance is zero), the resistance across the terminals will be

$$R' = R_2 + \frac{R_1 R_3}{R_1 + R_3} \tag{7-11}$$

$$R' = 4.7\ \text{k}\Omega + \frac{3.9\ \text{k}\Omega \times 8.2\ \text{k}\Omega}{3.9\ \text{k}\Omega + 8.2\ \text{k}\Omega} = 7.343\ \text{k}\Omega$$

The equivalent Thevenin circuit is as in Fig. 7-23.

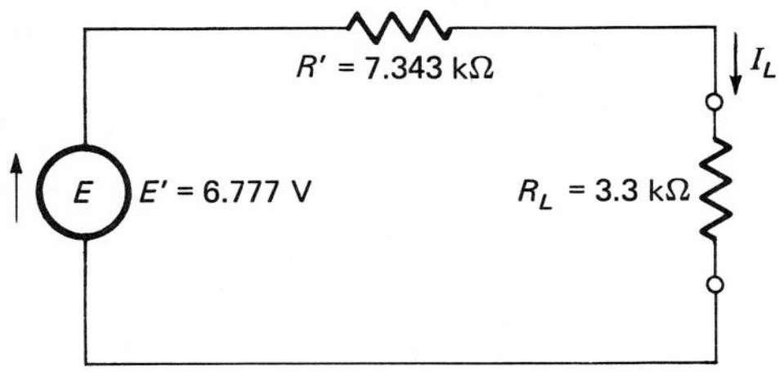

Fig. 7-23 Example 7-5: Thevenin equivalent of Fig. 7-22.

$$I_L = \frac{E'}{R' + R_L} = \frac{6.777\ \text{V}}{7.343\ \text{k}\Omega + 3.3\ \text{k}\Omega} = \mathbf{0.636\ mA}$$

Example 7-6 The bridged-T network in Fig. 7-24*a* can be reduced to the simple T network in Fig. 7-24*b* using preferred-value 10 percent resistors. If the load resistance can be 100 Ω, 1 kΩ, or 10 kΩ, what are the related load currents?

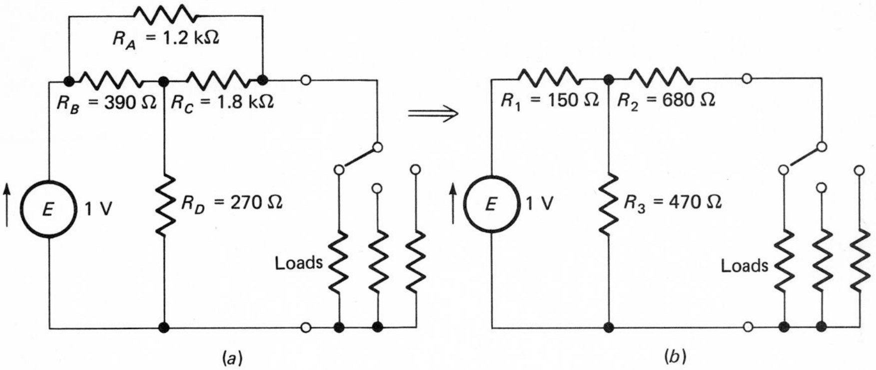

Fig. 7-24 Example 7-6: (*a*) original circuit with a bridged-T network; (*b*) Fig. 7-24*a* with the bridged-T network reduced to a simple T network by π-to-T conversion.

Solution Applying Eq. (7-10) to Fig. 7-24*b*, the Thevenin voltage will be

$$E' = E\frac{R_3}{R_1 + R_3} = 1\text{ V}\frac{470\ \Omega}{150\ \Omega + 470\ \Omega} = 0.7581\text{ V}$$

Applying Eq. (7-11) to Fig. 7-24*b*, the Thevenin resistance will be

$$R' = R_2 + \frac{R_1R_3}{R_1 + R_3} = 680\ \Omega + \frac{150\ \Omega \times 470\ \Omega}{150\ \Omega + 470\ \Omega} = 793.7\ \Omega$$

The equivalent Thevenin circuit is shown in Fig. 7-25.

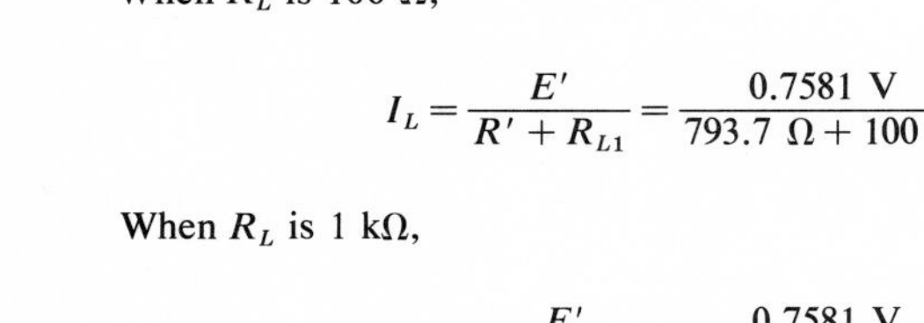

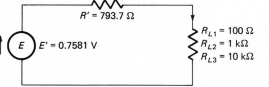

Fig. 7-25 Example 7-6: Thevenin equivalent of Fig. 7-24.

When R_L is 100 Ω,

$$I_L = \frac{E'}{R' + R_{L1}} = \frac{0.7581\text{ V}}{793.7\ \Omega + 100\ \Omega} = \mathbf{0.8483\ mA}$$

When R_L is 1 kΩ,

$$I_L = \frac{E'}{R' + R_{L2}} = \frac{0.7581\text{ V}}{793.7\ \Omega + 1\text{ k}\Omega} = \mathbf{0.4226\ mA}$$

When R_L is 10 kΩ,

$$I_L = \frac{E'}{R' + R_{L3}} = \frac{0.7581\text{ V}}{793.7\ \Omega + 10\text{ k}\Omega} = \mathbf{0.0702\ mA} \quad \text{or} \quad \mathbf{70.2\ \mu A}$$

Example 7-6 illustrates a particular advantage of the Thevenin method of circuit analysis. The Thevenin voltage and resistance have to be calculated just once for any network. When this calculation has been made, a series of values associated with the element of interest can be determined using a simple series circuit. This method is useful in many practical situations.

Example 7-7 An electronic voltage regulator has within its circuit a 1-kΩ 1-W resistor. A circuit designer wishes to reduce the resistor's current by 50 percent by connecting another resistor in series with it. He disconnects the 1-kΩ resistor. With the power on, 125 V is measured across the resistor's circuit terminals. With the power off, 4 kΩ is measured across these terminals. What value series resistor should be used?

Solution Figure 7-26 is a pictorial representation of the conditions measured by the designer. These measurements define the Thevenin voltage and resistance as 125 V and 4 kΩ, respectively. The Thevenin equivalent circuit of the voltage regulator from the viewpoint of the 1-kΩ resistor R_1 is shown in Fig. 7-27*a*. Figure 7-27*b* shows this circuit with a series resistor R_2 added to reduce the current in R_1 by 50 percent.

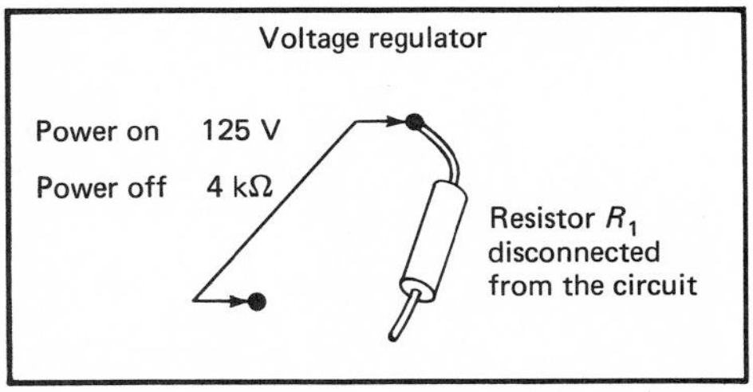

Fig. 7-26 Example 7-7: voltage regulator with resistor disconnected.

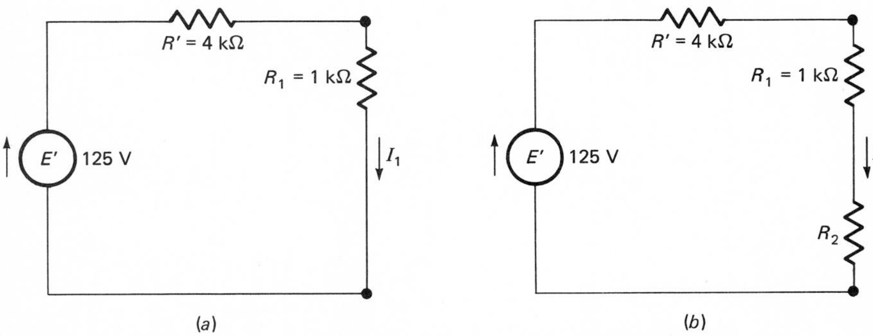

Fig. 7-27 Example 7-7: (*a*) Thevenin circuit of original voltage regulator; (*b*) Thevenin circuit with the current-reducing resistor added.

From the Thevenin equivalent of the original circuit, Fig. 7-27*a*,

$$I_1 = \frac{E'}{R' + R_1} = \frac{125 \text{ V}}{4 \text{ k}\Omega + 1 \text{ k}\Omega} = 25 \text{ mA}$$

The current I_2 in the circuit in Fig. 7-27*b*, which includes the added resistor R_2, is to be 50 percent of the original current I_1, or 12.5 mA.

The total circuit resistance R_T in Fig. 7-27*b* will be

$$R_T = \frac{E'}{I_2} = \frac{125\ \text{V}}{12.5\ \text{mA}} = 10\ \text{k}\Omega$$

$$R_2 = R_T - (R' + R_1) = 10\ \text{k}\Omega - (4\ \text{k}\Omega + 1\ \text{k}\Omega) = \mathbf{5\ k\Omega}$$

In Example 7-7 it was not necessary to know anything about the details of the voltage regulator circuit. By measurement the Thevenin characteristics of the regulator circuit, from the viewpoint of the 1-kΩ resistor, were determined. Once these Thevenin values were known, it was a simple matter to calculate the value of the resistor to be added to meet the designer's requirements.

Example 7-8 What is the meter current in the circuit in Fig. 7-28?

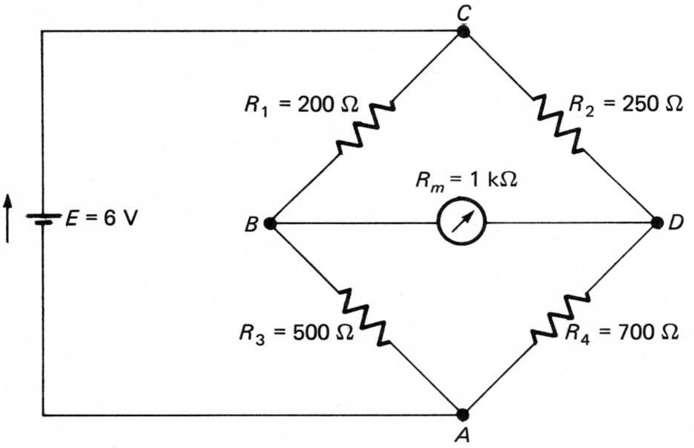

Fig. 7-28 Example 7-8: given circuit.

Solution Redrawing the circuit with the meter removed, to determine the Thevenin emf, gives Fig. 7-29.

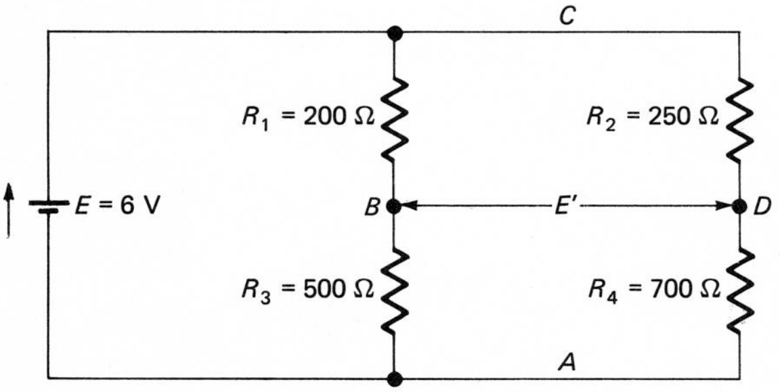

Fig. 7-29 Thevenin voltage in Fig. 7-28.

The Thevenin emf will be the voltage across the open-circuit meter terminals, $V_{DB} = V_{DA} - V_{BA}$

$$V_{DA} = E\frac{R_4}{R_2 + R_4} = 6\ \text{V}\frac{700\ \Omega}{250\ \Omega + 700\ \Omega} = 4.421\ \text{V}$$

$$V_{BA} = E\frac{R_3}{R_1 + R_3} = 6\ \text{V}\ \frac{500\ \Omega}{200\ \Omega + 500\ \Omega} = 4.286\ \text{V}$$

$$E' = V_{DA} - V_{BA} = 4.421\ \text{V} - 4.286\ \text{V} = 0.135\ \text{V}$$

As V_{DA} is greater than V_{BA}, terminal *D* will be positive with respect to terminal *B*.

With the meter terminals open circuit (the meter removed) and the battery removed, the circuit will be as shown in Fig. 7-30*a*. Note the short circuit between points *C* and *A* due to the zero battery resistance.

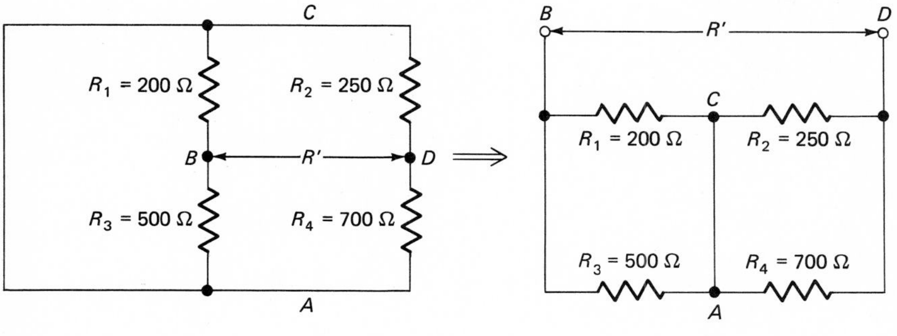

Fig. 7-30 Thevenin resistance in Fig. 7-28. (*a*) Original circuit layout; (*b*) circuit rearranged to show *R'* as the resistance of a series-parallel network.

Figure 7-30*b* shows the circuit with its elements arranged for ease in determining the resistance between terminals *B* and *D*. This is a series-parallel circuit. The resistance *B* to *D* is the Thevenin resistance:

$$R' = \frac{R_1 R_3}{R_1 + R_3} + \frac{R_2 R_4}{R_2 + R_4} = \frac{200\ \Omega \times 500\ \Omega}{200\ \Omega + 500\ \Omega} + \frac{250\ \Omega \times 700\ \Omega}{250\ \Omega + 700\ \Omega} = 327.1\ \Omega$$

The Thevenin equivalent circuit will be as in Fig. 7-31.

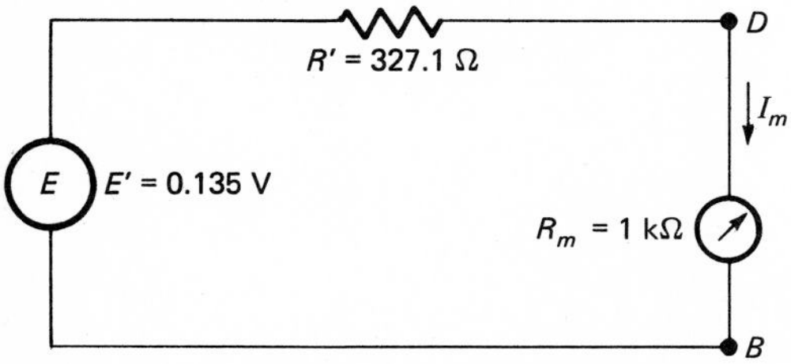

Fig. 7-31 Example 7-8: Thevenin equivalent circuit of Fig. 7-28.

$$I_m = \frac{E'}{R' + R_m} = \frac{0.135\ \text{V}}{327.1\ \Omega + 1\ \text{k}\Omega} = \mathbf{0.1017\ mA} \quad \text{or} \quad \mathbf{101.7\ \mu A}$$

Example 7-9 What power is actually delivered to load R_3 in the three-wire circuit in Fig. 7-32? The nominal load powers are: $P_1 = 2.5$ kW at 120 V, $P_2 = 1.5$ kW at 120 V, and $P_3 = 3, 6, 9,$ or 12 kW at 240 V. Assume that the load resistances do not change appreciably if the load voltages are lower than the rated values.

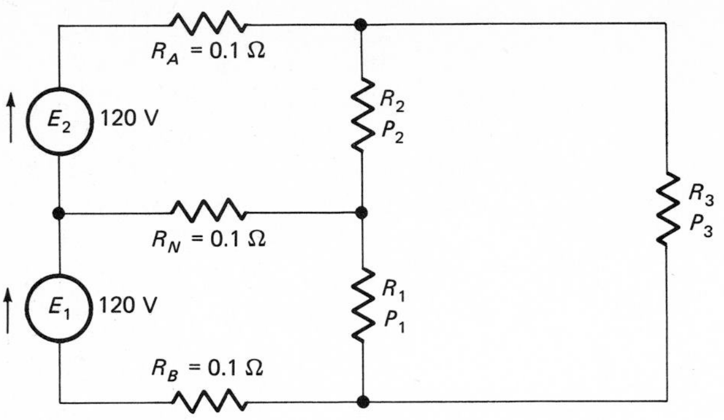

Fig. 7-32 Example 7-9: given circuit.

Solution Calculating the values of the load resistances using the rated load voltages,

$$P = \frac{V^2}{R} \qquad \text{or} \qquad R = \frac{V^2}{P}$$

$$R_1 = \frac{V_1^2}{P_1} = \frac{(120\ \text{V})^2}{2.5\ \text{kW}} = 5.76\ \Omega$$

$$R_2 = \frac{V_2^2}{P_2} = \frac{(120\ \text{V})^2}{1.5\ \text{kW}} = 9.6\ \Omega$$

$$R_3 = \frac{V_3^2}{P_3} = \frac{(240\ \text{V})^2}{3\ \text{kW}} = 19.2\ \Omega$$

$$\text{or} \quad \frac{(240\ \text{V})^2}{6\ \text{kW}} = 9.6\ \Omega \quad \text{or} \quad \frac{(240\ \text{V})^2}{9\ \text{kW}} = 6.4\ \Omega \quad \text{or} \quad \frac{(240\ \text{V})^2}{12\ \text{kW}} = 4.8\ \Omega$$

Removing the load R_3 to obtain the Thevenin voltage gives the circuit in Fig. 7-33.

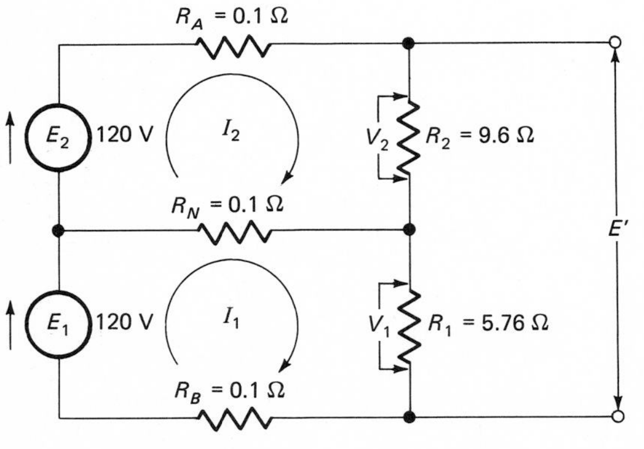

Fig. 7-33 Thevenin voltage in Fig. 7-32.

Loop analysis can be used in determining the Thevenin voltage. The loop equations are

$$5.96I_1 - 0.1I_2 = 120$$

$$-0.1I_1 + 9.8I_2 = 120$$

The solutions to these simultaneous equations are

$$I_1 = 20.34 \text{ A} \qquad \text{and} \qquad I_2 = 12.45 \text{ A}$$

The Thevenin voltage will be

$$E' = V_1 + V_2 = I_1 R_1 + I_2 R_2 = 20.34 \text{ A} \times 5.76\ \Omega + 12.45 \text{ A} \times 9.6\ \Omega = 236.7 \text{ V}$$

Removing the load R_3 and both generators results in the circuit in Fig. 7-34*a*. Converting the π network $R_A R_2 R_N$ into an equivalent T network gives the circuit in Fig. 7-34*b*.

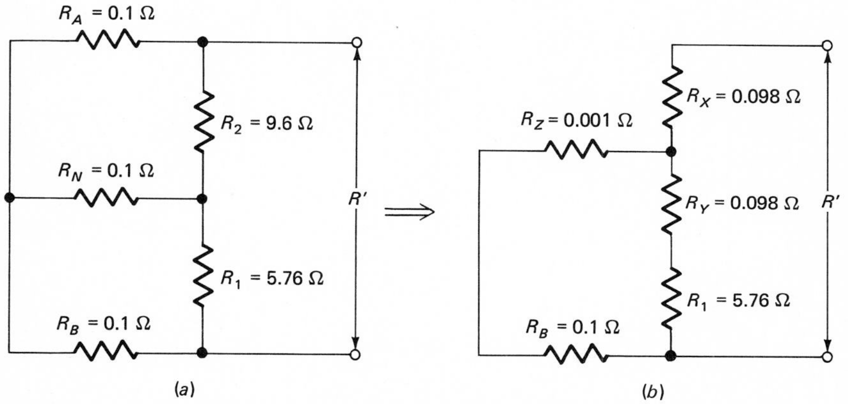

Fig. 7-34 (*a*) Thevenin resistance in Fig. 7-32; (*b*) first reduction of Fig. 7-34*a* by converting the π network R_A, R_2, R_N to the T network R_X, R_Y, R_Z.

$$\Sigma R_\nabla = R_A + R_2 + R_N = 0.1\ \Omega + 9.6\ \Omega + 0.1\ \Omega = 9.8\ \Omega$$

$$R_X = \frac{R_A R_2}{\Sigma R_\nabla} = \frac{0.1\ \Omega \times 9.6\ \Omega}{9.8\ \Omega} = 0.098\ \Omega$$

$$R_Y = \frac{R_2 R_N}{\Sigma R_\nabla} = \frac{9.6\ \Omega \times 0.1\ \Omega}{9.8\ \Omega} = 0.098\ \Omega$$

$$R_Z = \frac{R_A R_N}{\Sigma R_\nabla} = \frac{0.1\ \Omega \times 0.1\ \Omega}{9.8\ \Omega} = 0.001\ \Omega$$

Figure 7-34*b* can now be simplified to the circuit in Fig. 7-35, from which the value of the Thevenin resistance R' can be computed.

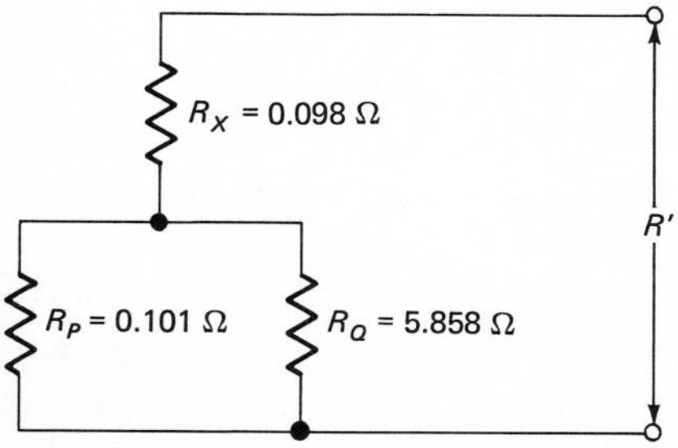

Fig. 7-35 Second reduction of Fig. 7-34*a* by converting the series groups $R_Z R_B$ and $R_Y R_1$ to elements R_P and R_Q.

R_P is R_Z and R_B in series:

$$R_P = R_Z + R_B = 0.001\ \Omega + 0.1\ \Omega = 0.101\ \Omega$$

R_Q is R_Y and R_1 in series:

$$R_Q = R_Y + R_1 = 0.098\ \Omega + 5.76\ \Omega = 5.858\ \Omega$$

$$R' = R_X + \frac{R_P R_Q}{R_P + R_Q} = 0.098\ \Omega + \frac{0.101\ \Omega \times 5.858\ \Omega}{0.101\ \Omega + 5.858\ \Omega} = 0.1973\ \Omega$$

The Thevenin equivalent circuit will be as in Fig. 7-36.

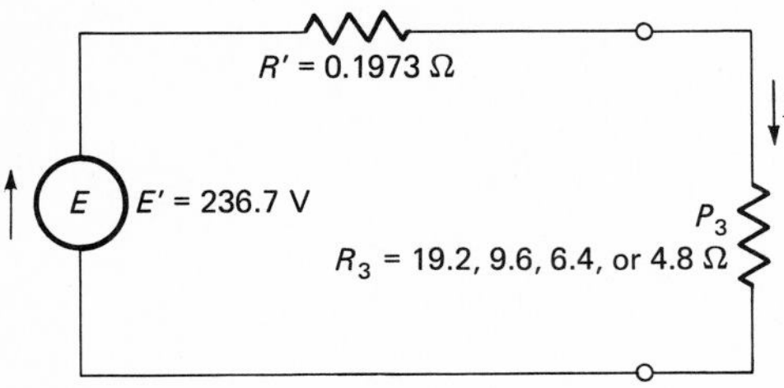

Fig. 7-36 Example 7-9: Thevenin equivalent of Fig. 7-32.

When $R_3 = 19.2\ \Omega$,

$$I_3 = \frac{E'}{R' + R_3} = \frac{236.7\ \text{V}}{0.1973\ \Omega + 19.2\ \Omega} = 12.2\ \text{A}$$

$$P_3 = I_3^2 R_3 = (12.2\ \text{A})^2 \times 19.2\ \Omega = \mathbf{2.858\ kW}$$

When $R_3 = 9.6\ \Omega$,

$$I_3 = \frac{E'}{R' + R_3} = \frac{236.7\ \text{V}}{0.1973\ \Omega + 9.6\ \Omega} = 24.16\ \text{A}$$

$$P_3 = I_3^2 R_3 = (24.16\ \text{A})^2 \times 9.6\ \Omega = \mathbf{5.603\ kW}$$

When $R_3 = 6.4\ \Omega$,

$$I_3 = \frac{E'}{R' + R_3} = \frac{236.7\ \text{V}}{0.1973\ \Omega + 6.4\ \Omega} = 35.88\ \text{A}$$

$$P_3 = I_3^2 R_3 = (35.88\ \text{A})^2 \times 6.4\ \Omega = \mathbf{8.239\ kW}$$

When $R_3 = 4.8\ \Omega$,

$$I_3 = \frac{E'}{R' + R_3} = \frac{236.7\ \text{V}}{0.1973\ \Omega + 4.8\ \Omega} = 47.37\ \text{A}$$

$$P_3 = I_3^2 R_3 = (47.37\ \text{A})^2 \times 4.8\ \Omega = \mathbf{10.77\ kW}$$

In Example 7-9 the work involved in producing the Thevenin equivalent circuit is considerable. If data for a single-value load were needed, this method would not be as good as loop analysis. Where, as in this example, several load values are involved, the initial effort is worthwhile.

You will see that, because of the line resistances, each of the calculated powers is lower than the rated value. You should also note that the percentage power loss becomes greater as the load power increases. When the rated load power is 3 kW, 95.3 percent of this value is developed. When the power is intended to be 12 kW, only 89.8 percent of this rated value appears at the load.

Example 7-10 Figure 7-37 shows part of an electronic circuit. What is the base-to-emitter voltage V_{BE} of the transistor if its base-to-emitter resistance R_{BE} is 800 Ω?

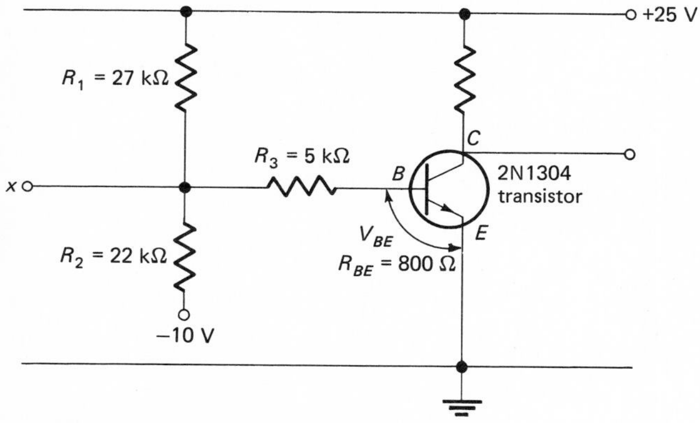

Fig. 7-37 Example 7-10: given circuit.

Solution From the viewpoint of the required information, the circuit can be redrawn as in Fig. 7-38.

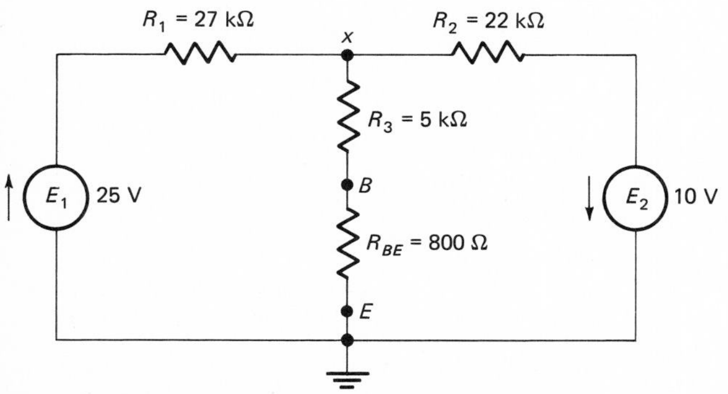

Fig. 7-38 Equivalent base-to-emitter circuit in Fig. 7-37.

Disconnecting R_3 and R_{BE} gives the circuit in Fig. 7-39.

The two generator voltages aid each other, so the current I in Fig. 7-39 will be

$$I = \frac{E_1 + E_2}{R_1 + R_2} = \frac{25\ \text{V} + 10\ \text{V}}{27\ \text{k}\Omega + 22\ \text{k}\Omega} = 0.7143\ \text{mA}$$

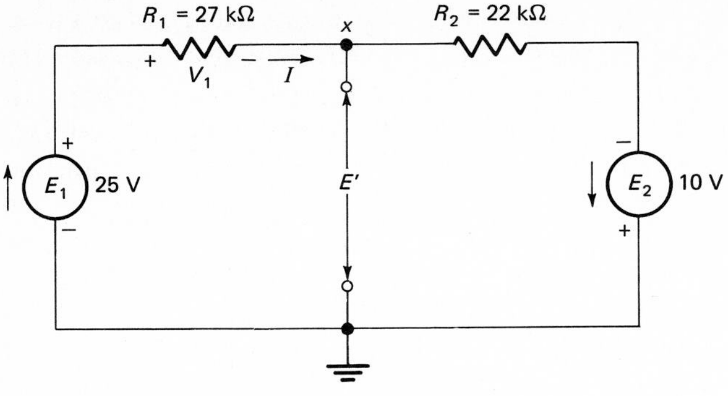

Fig. 7-39 Thevenin voltage in Fig. 7-37.

The pd across R_1 will be

$$V_1 = IR_1 = 0.7143 \text{ mA} \times 27 \text{ k}\Omega = 19.29 \text{ V}$$

The Thevenin voltage is the voltage between point x and ground.

$$E' = E_1 - V_1 = 25 \text{ V} - 19.29 \text{ V} = 5.71 \text{ V}$$

With both generators removed, the Thevenin resistance between point x and ground will be R_1 and R_2 in parallel.

$$R' = \frac{R_1R_2}{R_1 + R_2} = \frac{27 \text{ k}\Omega \times 22 \text{ k}\Omega}{27 \text{ k}\Omega + 22 \text{ k}\Omega} = 12.12 \text{ k}\Omega$$

The Thevenin equivalent circuit will therefore be as in Fig. 7-40.

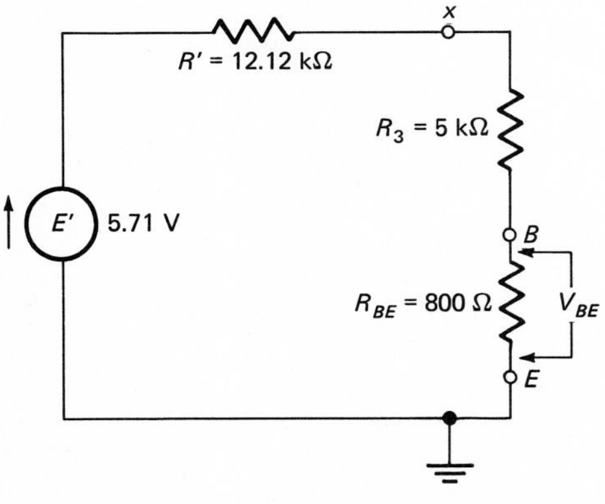

Fig. 7-40 Example 7-10: Thevenin equivalent of Fig. 7-37.

The base-to-emitter voltage will be

$$V_{BE} = \frac{E'R_{BE}}{R' + R_3 + R_{BE}} = \frac{5.71 \text{ V} \times 800 \ \Omega}{12.12 \text{ k}\Omega + 5 \text{ k}\Omega + 800 \ \Omega} = \mathbf{0.2549 \ V}$$

SUCCESSIVE USE OF THEVENIN'S THEOREM

To introduce the Thevenin approach to network analysis, single elements have been isolated for attention. It is not necessary, however, to be so restrictive; complete networks within a circuit can also be isolated. Any two points in a circuit can be chosen, and then all of the circuit on one side of these points can be detached as a unit. The Thevenin equivalent of the circuit on the other side of the points is then determined. By this means, complex circuits can be dealt with in sections to minimize the analysis work.

This successive Thevenin technique is demonstrated by a numerical example.

Example 7-11 What is the value of the current in R_6 in the circuit in Fig. 7-41?

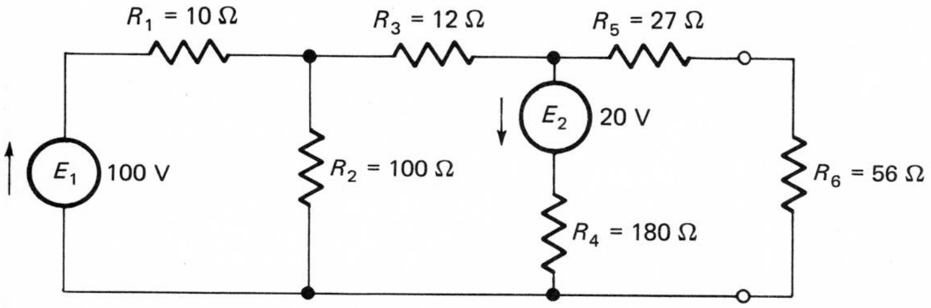

Fig. 7-41 Example 7-11: given circuit.

Solution It is quite feasible to apply Thevenin's theorem directly; resistor R_6 can be removed, and the value of the Thevenin voltage and resistance calculated. The following technique, while requiring a lengthy explanation, is easily applied because the individual calculations are simple. The steps are

1. Disconnect all of the circuit to the right of R_2, leaving a network as in Fig. 7-42*a*.

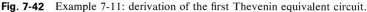

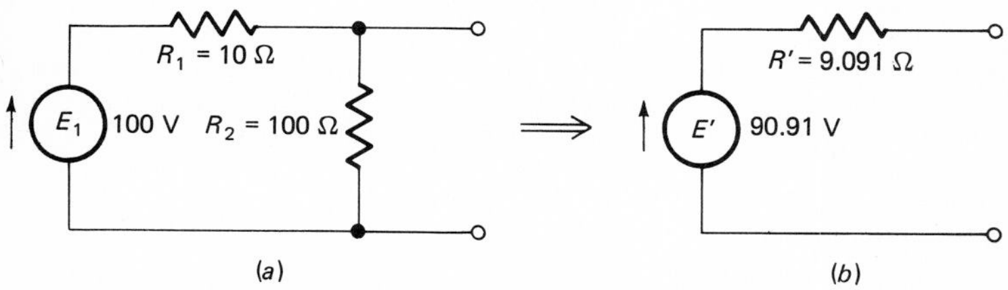

Fig. 7-42 Example 7-11: derivation of the first Thevenin equivalent circuit.

Figure 7-42*b* is the Thevenin equivalent of Fig. 7-42*a*.

$$E' = E\frac{R_2}{R_1 + R_2} = 100\ \text{V}\ \frac{100\ \Omega}{10\ \Omega + 100\ \Omega} = 90.91\ \text{V}$$

$$R' = \frac{R_1R_2}{R_1 + R_2} = \frac{10\ \Omega \times 100\ \Omega}{10\ \Omega + 100\ \Omega} = 9.091\ \Omega$$

2. Add to the first Thevenin circuit elements R_3, R_4, and E_2. Leave the network to the right of R_4 and E_2 disconnected. The circuit in Fig. 7-43*a* results.

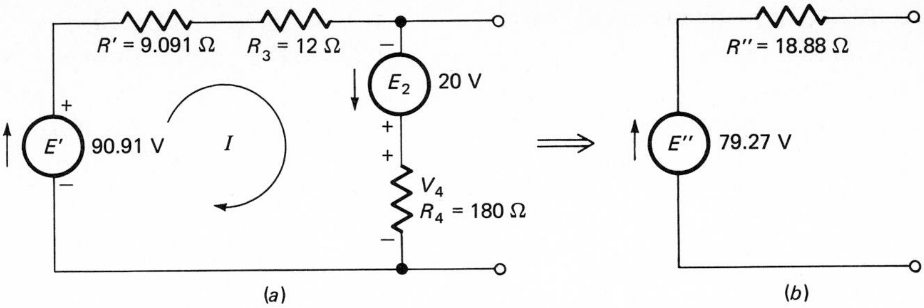

Fig. 7-43 Example 7-11: derivation of the second Thevenin equivalent circuit.

The two generators in Fig. 7-43*a* are aiding each other, so the total emf is $E' + E_2$. The current in the circuit is given by

$$I = \frac{E' + E_2}{R' + R_3 + R_4} = \frac{90.91 \text{ V} + 20 \text{ V}}{9.091\ \Omega + 12\ \Omega + 180\ \Omega} = 0.5515 \text{ A}$$

The pd across R_4 is

$$V_4 = IR_4 = 0.5515 \text{ A} \times 180\ \Omega = 99.27 \text{ V}$$

V_4 and E_2 are of opposing polarity, so that the second Thevenin voltage is the difference between them.

$$E'' = V_4 - E_2 = 99.27 \text{ V} - 20 \text{ V} = 79.27 \text{ V}$$

The second Thevenin resistance is the resistance between the terminals in Fig. 7-43*a*, with both generators removed. It is the series-parallel combination of R_4 in parallel with the series group R', R_3.

$$R'' = \frac{R_4(R' + R_3)}{R_4 + R' + R_3} = \frac{180\ \Omega(9.091\ \Omega + 12\ \Omega)}{180\ \Omega + 9.091\ \Omega + 12\ \Omega} = 18.88\ \Omega$$

3. Connect R_5 to the second Thevenin circuit, Fig. 7-43*b*. This connection gives the circuit in Fig. 7-44*a*. When R_6 is added, as in Fig. 7-44*b*, the Thevenin equivalent of the entire original circuit in Fig. 7-41 results.

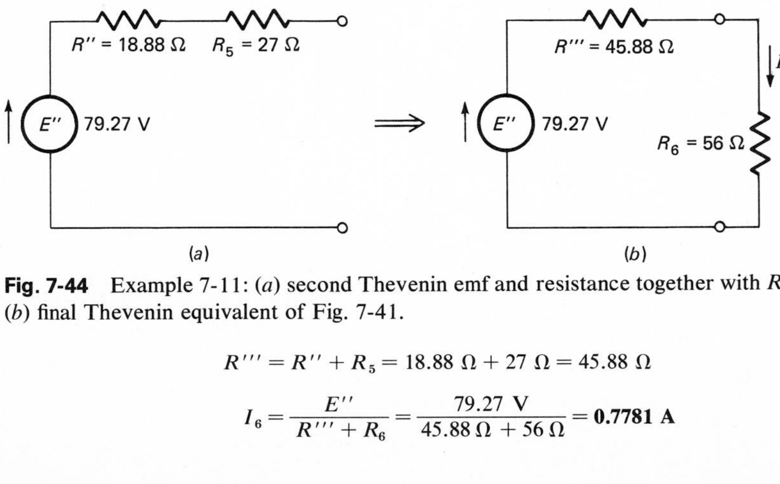

Fig. 7-44 Example 7-11: (*a*) second Thevenin emf and resistance together with R_5; (*b*) final Thevenin equivalent of Fig. 7-41.

$$R''' = R'' + R_5 = 18.88\ \Omega + 27\ \Omega = 45.88\ \Omega$$

$$I_6 = \frac{E''}{R''' + R_6} = \frac{79.27 \text{ V}}{45.88\ \Omega + 56\ \Omega} = \mathbf{0.7781 \text{ A}}$$

7-3 MAXIMUM POWER TRANSFER

In a circuit containing only steady direct currents, maximum power is transferred into an element when its resistance equals the Thevenin equivalent resistance of the remainder of the circuit.

This statement is the dc version of the maximum power transfer theorem or principle.

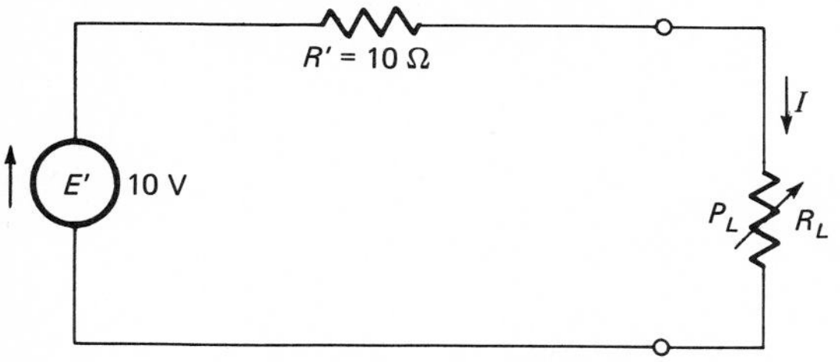

Fig. 7-45 Maximum power transfer. Thevenin equivalent of a circuit with its load resistance variable.

In Fig. 7-45 the Thevenin voltage E' and resistance R' should be considered to have been derived from a circuit which was either simple or complex. The load R_L is the element whose power we wish to maximize:

$$I_L = \frac{E'}{R' + R_L}$$

$$P = I^2 R_L = \frac{(E')^2 R_L}{(R' + R_L)^2} = \frac{(10\text{ V})^2 R_L}{(10\ \Omega + R_L)^2}$$

Computing values of load power related to various values of load resistance R_L gives the results shown in Table 7-1.

Table 7-1

R_L (Ω)	$10^2 R_L$	$10 + R_L$	$(10 + R_L)^2$	$P_L = \frac{10^2 R_L}{(10 + R_L)^2}$ (W)
0	0	10	100	0
5	500	15	225	2.222
7	700	17	289	2.422
8	800	18	324	2.469
9	900	19	361	2.493
9.5	950	19.5	380.25	2.498
10	1000	20	400	2.5
10.5	1050	20.5	420.25	2.499
11	1100	21	441	2.494
12	1200	22	484	2.479
20	2000	30	900	2.222
50	5000	60	3600	1.389

It can be seen from Table 7-1 that the load power P_L is greatest when the load resistance R_L is 10 Ω, the same as the Thevenin resistance R'. Figure 7-46 is a graph using the data in Table 7-1 to show how the power rises to a maximum and then drops off, peaking at the 10-Ω value of R_L.

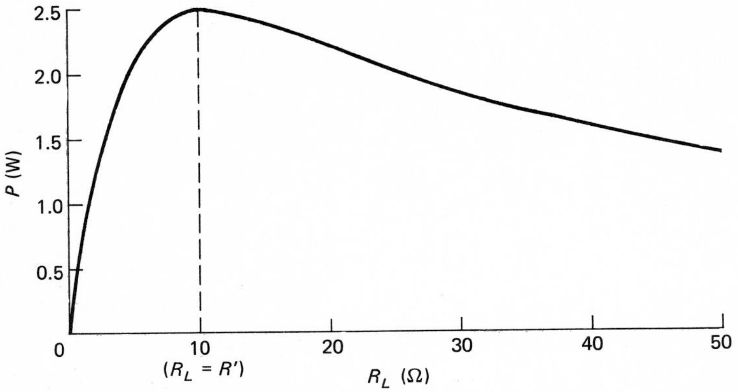

Fig. 7-46 Graph of load power with respect to load resistance in the circuit in Fig. 7-45.

It is of interest to review what influences the load power as the load resistance varies. Referring to Fig. 7-47, which shows the variations in current and load voltage with changing load resistance:

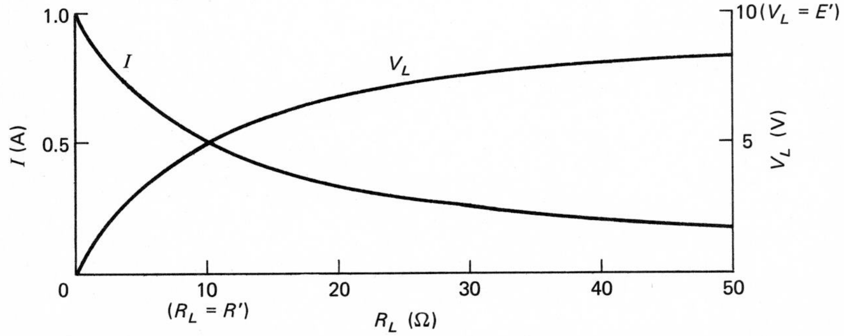

Fig. 7-47 Graphs of current and load voltage with respect to load resistance in the circuit in Fig. 7-45.

1. When R_L is zero, the circuit current has maximum value. However large this current value may be, there can be no voltage across a 0-Ω resistance, so no power can be developed in it.
2. As R_L increases from zero, the current drops but the load voltage rises. The rise of the load voltage from zero causes the power to rise.
3. When R_L becomes equal to R', the current will have dropped to 50 percent of its initial value, and the load voltage will have risen to 50 percent of the emf E'. This equality of generator and load resistances results in maximum load power.

4. As R_L becomes progressively greater than R', the current continues to drop and the load voltage to rise. The rate of change in both current and voltage is then low. The power does not change much, but it does steadily decrease until, if R_L were to become infinity (open circuit), the current and the power would be zero.

The validity of the dc maximum power transfer principle can be demonstrated by a proof which involves some elementary calculus.

In reference to Fig. 7-48, the power in the load P_L is determined as follows:

$$I = \frac{E}{R_1 + R_L}$$

$$P_L = I^2 R_L = \frac{E^2 R_L}{(R_1 + R_L)^2}$$

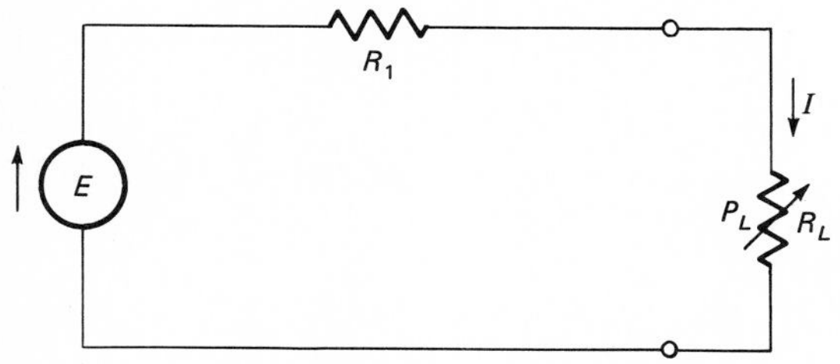

Fig. 7-48 Maximum power transfer. Circuit for a general derivation of conditions.

Differentiating with respect to the variable R_L,

$$\frac{dP_L}{dR_L} = E^2 \frac{(R_1 + R_L)^2 - 2R_L(R_1 + R_L)}{(R_1 + R_L)^4}$$

$$\frac{dP_L}{dR_L} = E^2 \frac{R_1 - R_L}{(R_1 + R_L)^3}$$

The load power P_L will be maximum when this derivative equals zero.

$$E^2 \frac{R_1 - R_L}{(R_1 + R_L)^3} = 0$$

$$R_1 - R_L = \frac{(R_1 + R_L)^3}{E^2} \times 0$$

The right side of the equation is zero, leaving

$$R_1 - R_L = 0$$

or

$$R_1 = R_L$$

Example 7-12 What is the value of load resistance R_L which will result in maximum load power in the circuit in Fig. 7-49?

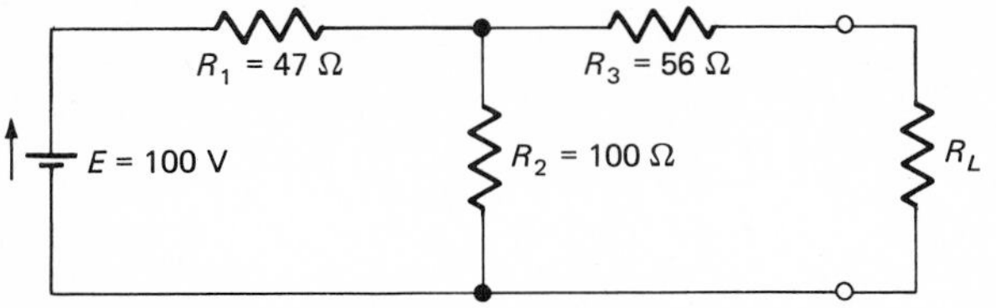

Fig. 7-49 Example 7-12: given circuit.

Solution The Thevenin equivalent circuit is as in Fig. 7-50.

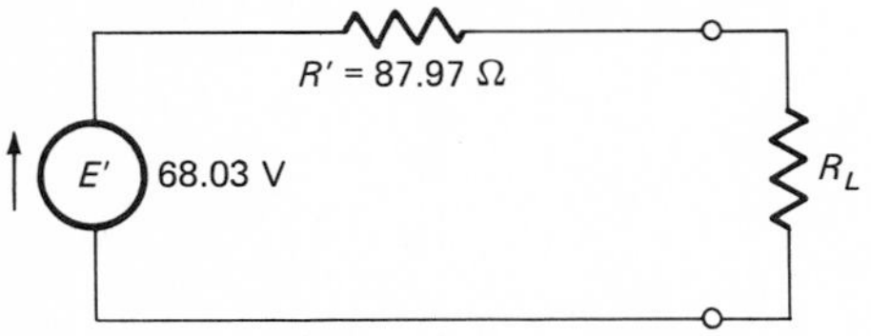

Fig. 7-50 Thevenin equivalent of Fig. 7-49.

For maximum load power,

$$R_L = R' = \mathbf{87.97\ \Omega}$$

Because maximum load power exists when the load resistance equals the Thevenin equivalent resistance of the remainder of the circuit, it naturally follows that in these circumstances,

$$\text{The load voltage } V_L = \frac{E'}{2}$$

$$\text{The load current } I_L = \frac{E'}{2R_L}.$$

$$\text{The load power } P_L = V_L I_L = \frac{(E')^2}{4R_L}$$

Example 7-13 An electronic power supply with an internal resistance of 0.5 Ω delivers 40 V to its output terminals when no load is connected between them. What is (*a*) the value of the load resistance which results in maximum load power, (*b*) the value of the maximum load power, (*c*) the load voltage and current when maximum load power is developed, and (*d*) the total power developed within the power supply?

Solution Whatever its internal circuit is, the power supply can be depicted electrically as the circuit in Fig. 7-51, which is its Thevenin equivalent.

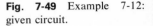

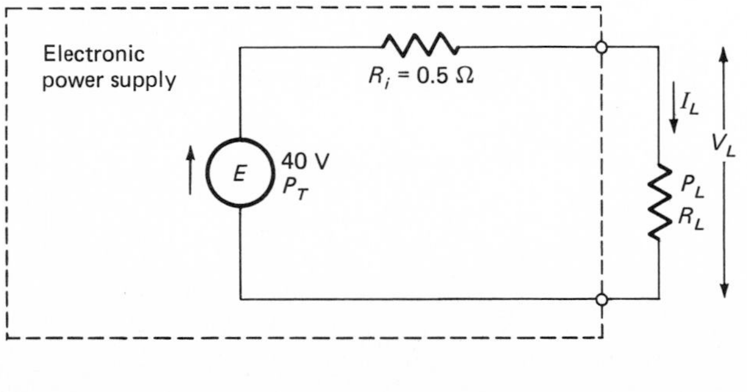

Fig. 7-51 Example 7-13: circuit.

a. For maximum power transfer to the load,

$$R_L = R_i = \mathbf{0.5\ \Omega}$$

b.
$$P_L = \frac{E^2}{4R_L} = \frac{(40\ \text{V})^2}{4 \times 0.5\ \Omega} = \mathbf{800\ W}$$

c.
$$V_L = \frac{E}{2} = \frac{40\ \text{V}}{2} = \mathbf{20\ V}$$

$$I_L = \frac{E}{2R_L} = \frac{40\ \text{V}}{2 \times 0.5\ \Omega} = \mathbf{40\ A}$$

d.
$$P_T = EI = 40\ \text{V} \times 40\ \text{A} = \mathbf{1600\ W}$$

As might be expected, Example 7-13 shows that, when maximum power is transferred to a load, this power is one-half the total power developed within the energy source, the load voltage being one-half the source emf. It is thus evident that criteria other than maximum power transfer must be applied when power efficiency is important.

For purposes such as the distribution of electrical energy, maximum load power is not the condition sought. The distribution system internal resistance is much lower than the load resistance, so as to maintain good voltage regulation (constant output voltage).

The maximum power condition is usually applied in electronic systems in which, to gain certain system performance characteristics, low power efficiency can be tolerated.

7-4 NORTON'S THEOREM

Norton's theorem is the dual of Thevenin's theorem. It too is used to find the values of such quantities as current or voltage associated with one element in a circuit. It too involves the reduction of a network to a two-element equivalent circuit. Norton's theorem is concerned with a current generator equivalent circuit rather than the voltage generator equivalent circuit of Thevenin's theorem.

Norton's theorem states that the current in any conductance G, connected to two terminals of a network, is the same as it would be if G were connected to a current source where

1. The source current is the short-circuit current at the terminals of G.
2. The internal conductance is the conductance at the terminals of G, with all other network sources replaced by conductances equal to their internal conductances.

Norton's theorem is stated in the given form to maintain the practice of regarding current sources and parallel networks from the conductance viewpoint, this view being consistent with the duality of series and parallel circuits. Should you wish to adopt a resistance approach to Norton's theorem, you can of course make the usual substitutions: resistance for conductance, and R for G, in the statements above; $1/R$ for G, and reciprocal values in equations and calculations.

NORTON EQUIVALENT CIRCUIT

In a network, shown in block form in Fig. 7-52*a*, there is an element G about which data are sought. To evaluate these data, the Norton equivalent circuit in Fig. 7-52*b* is developed.

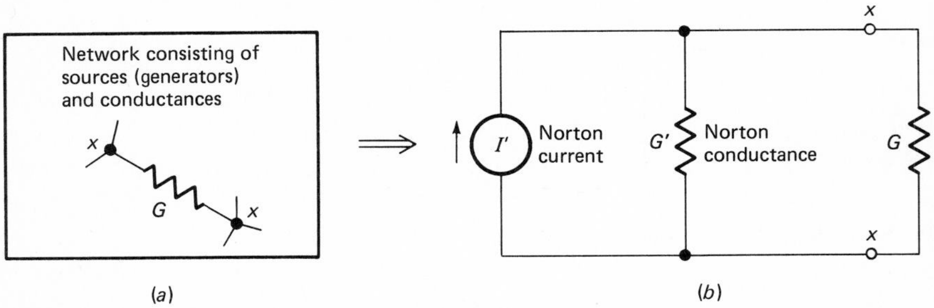

Fig. 7-52 (*a*) Network of unspecified type; (*b*) basic Norton circuit.

The *Norton current* I' is the short-circuit current at the element terminals (the current which would exist between the x,x terminals if they were connected by a good conductor).

The *Norton conductance* G' is the conductance at the x,x terminals with those terminals open-circuited (the element G removed), all sources in the remaining network being removed and replaced by their internal conductances. Note that the Norton conductance is defined in the same manner as the Thevenin resistance. The Norton conductance reciprocal (Norton resistance) is exactly the same as the Thevenin resistance.

Figure 7-53 shows a current source supplying a load through a π network. A π network is representative of all circuits which can be reduced to an equivalent π network, so by applying the Norton approach to Fig. 7-53 all such circuits are covered.

Fig. 7-53 Generator connected to a load through a π network.

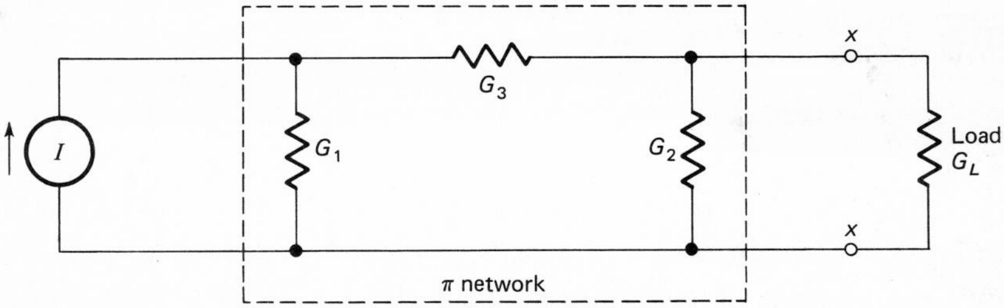

To obtain the value of the Norton current I',

1. Short-circuit G_L at the x,x terminals.
2. Determine the current in the short circuit.

When the load terminals are short-circuited, the result is as shown in Fig. 7-54 (short-circuiting G_L also short-circuits G_2).

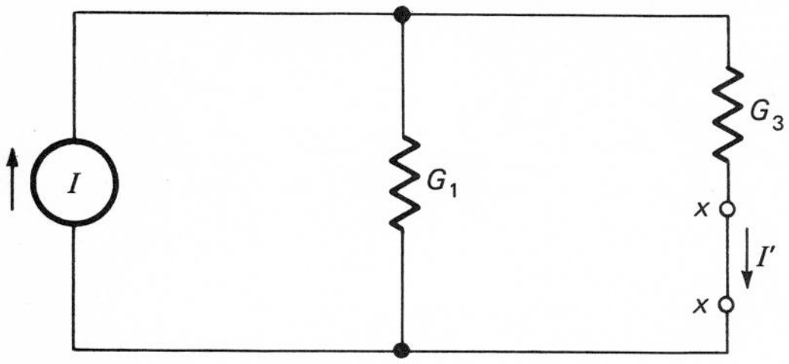

Fig. 7-54 Derivation of the Norton current for the circuit in Fig. 7-53.

The current in G_3 is the current in the short circuit x,x, and is thus the Norton equivalent current.

Applying the current divider principle,

$$I' = I \frac{G_3}{G_1 + G_3} \tag{7-14}$$

To obtain the value of the Norton equivalent conductance G',

1. Disconnect the load G_L at the x,x terminals.
2. Remove the current source I. Since no internal conductance is shown for this source, it is assumed to have zero value, to be an open circuit.
3. Determine the conductance between the terminals.

With G_L and the current source removed, the circuit across the terminals will be series parallel, as in Fig. 7-55.

$$G' = G_2 + \frac{G_1 G_3}{G_1 + G_3} \tag{7-15}$$

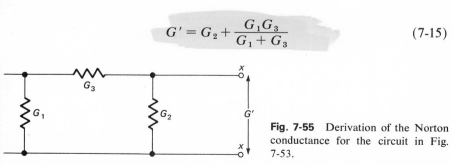

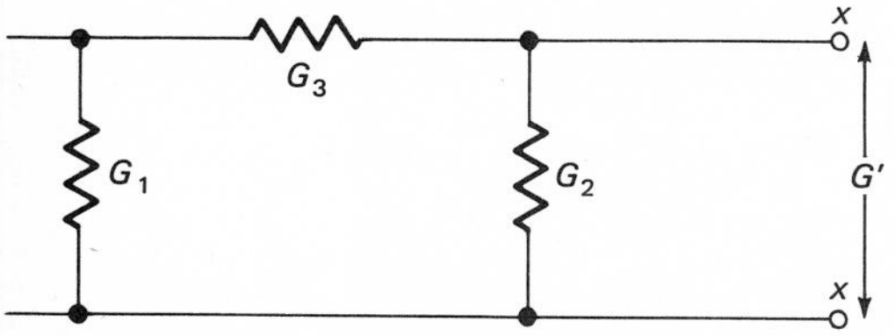

Fig. 7-55 Derivation of the Norton conductance for the circuit in Fig. 7-53.

The Norton equivalent of the circuit in Fig. 7-53 is therefore the one shown in Fig. 7-56.

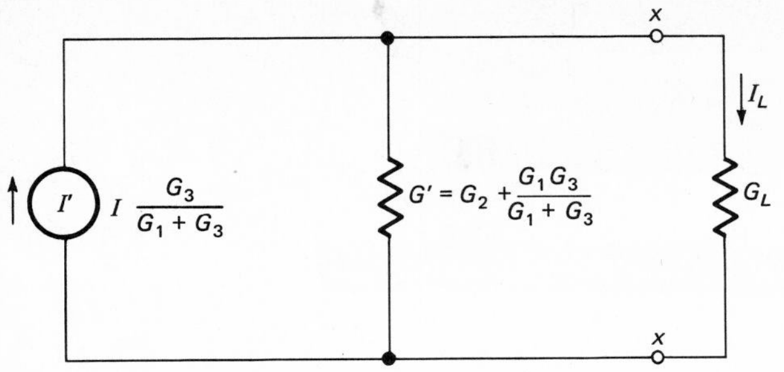

Fig. 7-56 Norton equivalent of the circuit in Fig. 7-53.

G' and G_L form a current divider, so I_L will be

$$I_L = I'\frac{G_L}{G'+G_L} = I\frac{G_3}{G_1+G_3}\,\frac{G_L}{G_2+\dfrac{G_1G_3}{G_1+G_3}+G_L}$$

$$= I\frac{G_3}{G_1+G_3} \times \frac{G_L}{\dfrac{G_1G_2+G_2G_3+G_3G_1}{G_1+G_3}+G_L}$$

$$I_L = \frac{IG_3G_L}{G_1G_2+G_2G_3+G_3G_1+G_L(G_1+G_3)}$$

In any application, after the load current has been calculated, the voltage, power, and energy can be found readily.

Norton's theorem is popular for the analysis of transistor circuits. This is particularly so when the Norton principle is extended from the dc approach established in this section to high-frequency ac techniques. It is thus important that you become convinced that it is reliable. Checking the load current in the circuit in Fig. 7-53 by nodal analysis will help.

Figure 7-53 is redrawn as Fig. 7-57, with the nodal voltages shown. The lower, common conductor is chosen as the reference node.

Fig. 7-57 Figure 7-53 with the nodes identified.

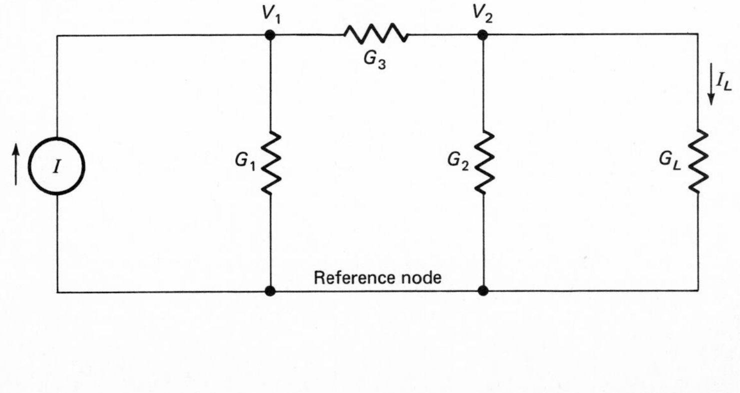

The nodal equations are

$$(G_1 + G_3)V_1 - G_3V_2 = I \tag{7-16}$$

$$-G_3V_1 + (G_2 + G_3 + G_L)V_2 = 0 \tag{7-17}$$

From Eq. (7-17),

$$V_1 = \frac{(G_2 + G_3 + G_L)V_2}{G_3}$$

Substituting for V_1 in Eq. (7-16),

$$\frac{(G_1 + G_3)(G_2 + G_3 + G_L)V_2}{G_3} - G_3V_2 = I$$

$$\frac{[G_1G_2 + G_2G_3 + G_3G_1 + G_L(G_1 + G_3)]\,V_2}{G_3} = I$$

$$V_2 = \frac{IG_3}{G_1G_2 + G_2G_3 + G_3G_1 + G_L(G_1 + G_3)}$$

V_2 is the voltage across the load G_L. The load current can be found by Ohm's law.

$$I_L = V_2G_L = \frac{IG_3G_L}{G_1G_2 + G_2G_3 + G_3G_1 + G_L(G_1 + G_3)}$$

The load current is the same using either Norton or nodal analysis. This exercise validates the correctness of Norton's theorem for a wide variety of circuits, because it was established in Sec. 7-1 that a three-element π network represents all networks which can be reduced to a π equivalent. Norton's theorem is in fact applicable to all linear networks.

GUIDELINES

The suitability of circuits for the application of Norton analysis can be judged by whether they conform to some or all of the following guidelines.

1. Data for one element are required.
2. That element conductance has various alternate values.
3. There are more current sources than voltage sources.
4. The circuit details are not known, but the Norton current and conductance can be measured.

Example 7-14 The circuit in Fig. 7-58 is made up of preferred-value resistors. What is the load current?

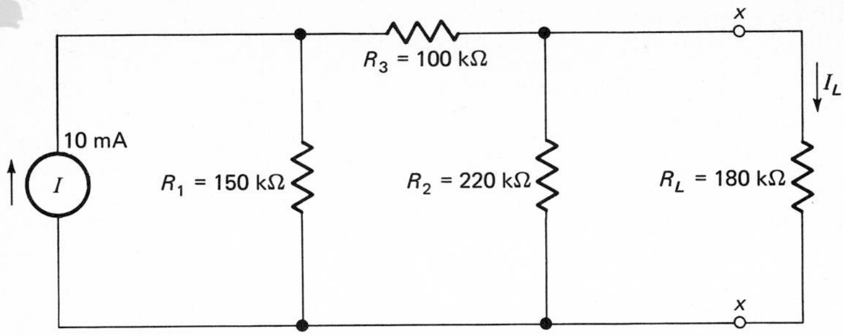

Fig. 7-58 Example 7-14: circuit.

Solution Redrawing the circuit with the resistances converted to conductances gives Fig. 7-59.

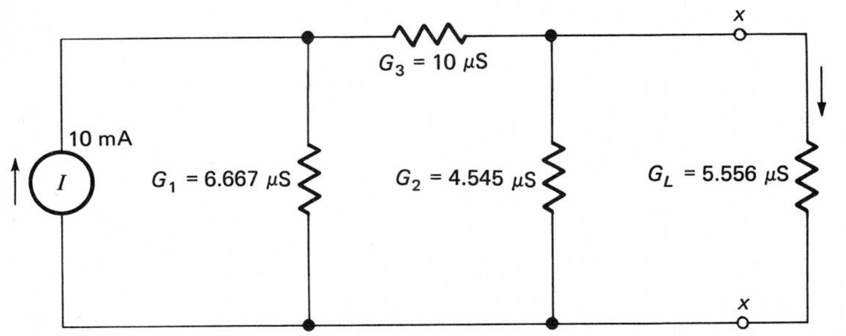

Fig. 7-59 Figure 7-58 with conductance values replacing resistance values.

With the x,x terminals short-circuited, the circuit becomes that in Fig. 7-60.

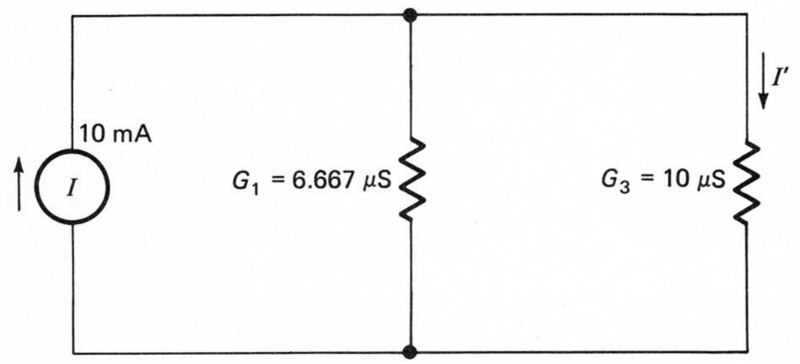

Fig. 7-60 Derivation of the Norton current for the circuit in Example 7-14.

The Norton current will be

$$I' = I\frac{G_3}{G_1 + G_3} = 10\text{ mA}\frac{10\ \mu\text{S}}{6.667\ \mu\text{S} + 10\ \mu\text{S}} = 6\text{ mA}$$

With the load and the generator removed, the circuit will be that in Fig. 7-61.

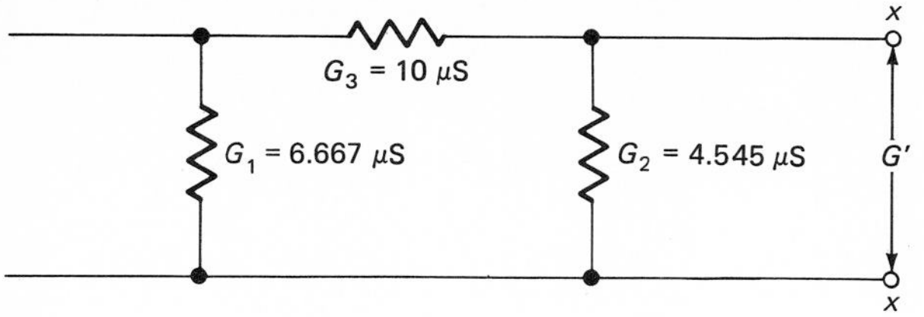

Fig. 7-61 Derivation of the Norton conductance for the circuit in Example 7-14.

The Norton conductance will be

$$G' = G_2 + \frac{G_1 G_3}{G_1 + G_3} = 4.545\ \mu\text{S} + \frac{6.667\ \mu\text{S} \times 10\ \mu\text{S}}{6.667\ \mu\text{S} + 10\ \mu\text{S}}$$

$$G' = 8.545\ \mu\text{S}$$

The Norton circuit equivalent to that in Fig. 7-58 is therefore as shown in Fig. 7-62.

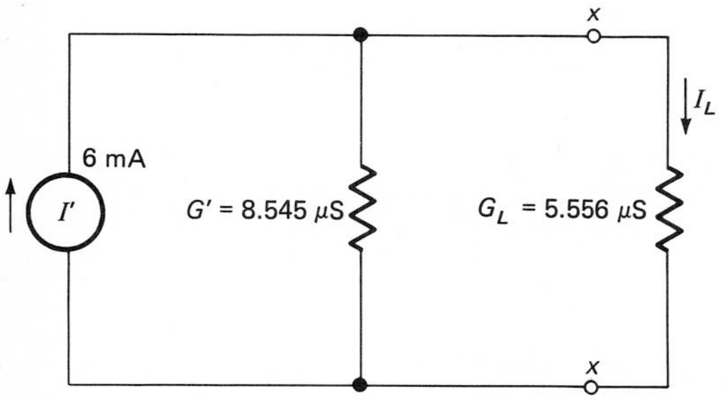

Fig. 7-62 Example 7-14: Norton equivalent circuit.

The load current is

$$I_L = I' \frac{G_L}{G' + G_L} = 6\ \text{mA} \frac{5.556\ \mu\text{S}}{8.545\ \mu\text{S} + 5.556\ \mu\text{S}} = \mathbf{2.364\ mA}$$

Example 7-15 A multielement network reduced to an equivalent π network is shown in Fig. 7-63 connected between a current generator and a set of loads. If the load resistances can be 10 Ω, 100 Ω, or 1 kΩ, what are the related load currents?

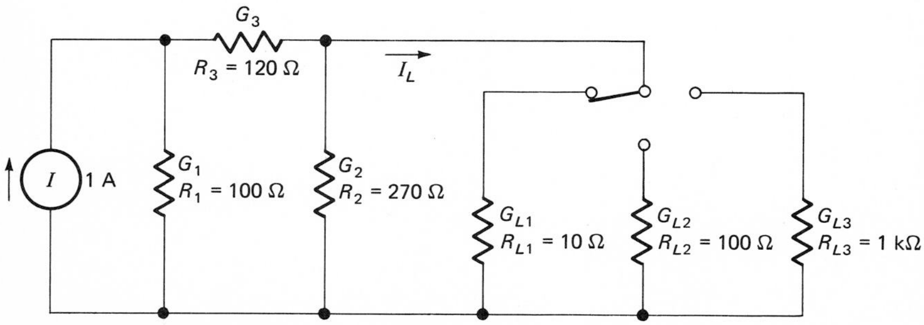

Fig. 7-63 Example 7-15: circuit.

Solution The conductances of the elements are

$$G_1 = \frac{1}{R_1} = \frac{1}{100\ \Omega} = 0.01\ \text{S}$$

$$G_2 = \frac{1}{R_2} = \frac{1}{270\ \Omega} = 0.0037\ \text{S}$$

$$G_3 = \frac{1}{R_3} = \frac{1}{120\ \Omega} = 0.0083\ \text{S}$$

$$G_{L1} = \frac{1}{R_{L1}} = \frac{1}{10\ \Omega} = 0.1\ \text{S}$$

$$G_{L2} = \frac{1}{R_{L2}} = \frac{1}{100\ \Omega} = 0.01\ \text{S}$$

$$G_{L3} = \frac{1}{R_{L3}} = \frac{1}{1\ \text{k}\Omega} = 0.001\ \text{S}$$

The Norton current is

$$I' = I\frac{G_3}{G_1 + G_3} = 1\ \text{A}\ \frac{0.0083\ \text{S}}{0.01\ \text{S} + 0.0083\ \text{S}} = 0.4536\ \text{A}$$

The Norton conductance is

$$G' = G_2 + \frac{G_1G_3}{G_1 + G_3} = 0.0037\ \text{S} + \frac{0.01\ \text{S} \times 0.0083\ \text{S}}{0.01\ \text{S} + 0.0083\ \text{S}} = 0.0082\ \text{S}$$

The Norton equivalent circuit therefore is that in Fig. 7-64.

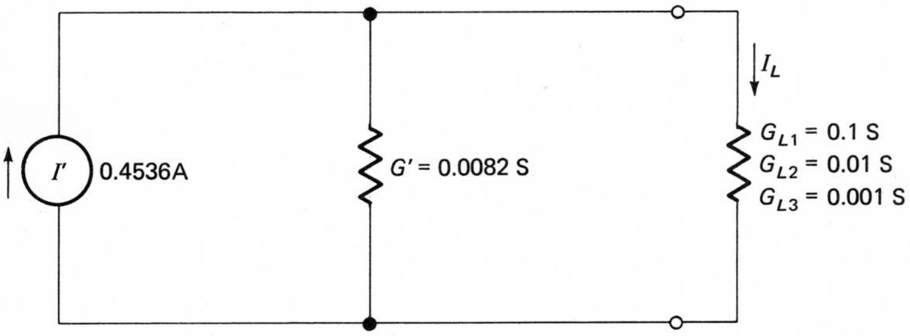

Fig. 7-64 Example 7-15: Norton equivalent circuit.

When G_L is 0.1 S ($R_L = 10\ \Omega$),

$$I_L = I'\frac{G_{L1}}{G_{L1} + G'} = 0.4536\ \text{A}\ \frac{0.1\ \text{S}}{0.1\ \text{S} + 0.0082\ \text{S}} = \mathbf{0.4192\ A}$$

When G_L is 0.01 S ($R_L = 100\ \Omega$),

$$I_L = I'\frac{G_{L2}}{G_{L2} + G'} = 0.4536\ \text{A}\ \frac{0.01\ \text{S}}{0.01\ \text{S} + 0.0082\ \text{S}} = \mathbf{0.2492\ A}$$

When G_L is 0.001 S ($R_L = 1$ kΩ),

$$I_L = I' \frac{G_{L3}}{G_{L3} + G'} = 0.4536 \text{ A} \frac{0.001 \text{ S}}{0.001 \text{ S} + 0.0082 \text{ S}} = \mathbf{0.0493 \text{ A}}$$

You will see from Example 7-15 that, as in the case of the Thevenin approach, Norton's theorem permits the ready calculation of the current in a network element when this element has a range of different resistance values. The Norton current and conductance need only be computed once.

Example 7-16 An electronic power supply has the following measured characteristics at its output terminals: with the terminals connected through a low-resistance ammeter, the meter indicates 0.75 A; with the power switched off, the resistance between the terminals is 100 Ω. What current will this power supply deliver to a 50-Ω load?

Solution Figure 7-65*a* and *b* shows the measured conditions. The current I' and the resistance R' are the values required for a Norton equivalent circuit.

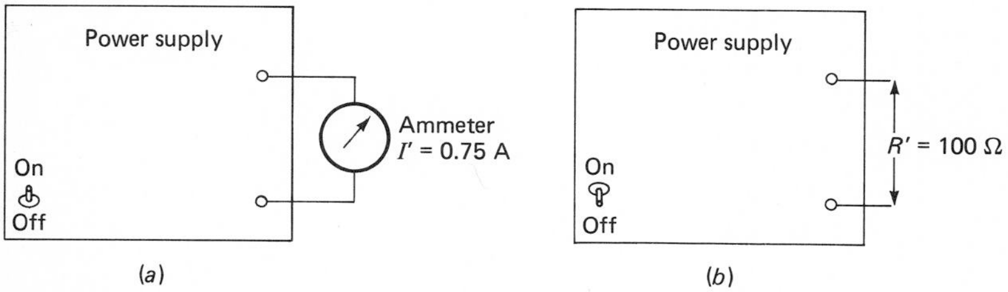

Fig. 7-65 Example 7-16: measuring the Norton characteristics at the power supply terminals. (*a*) Norton current; (*b*) Norton resistance.

This circuit, with the 50-Ω load, is given in Fig. 7-66.

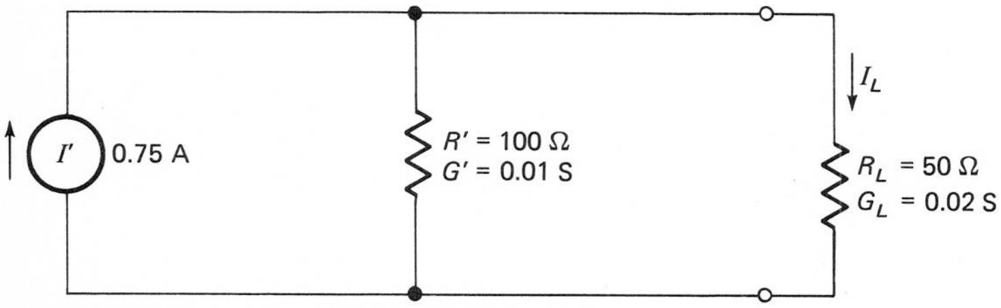

Fig. 7-66 Example 7-16: Norton equivalent circuit.

$$I_L = I' \frac{G_L}{G_L + G'} = 0.75 \text{ A} \frac{0.02 \text{ S}}{0.02 \text{ S} + 0.01 \text{ S}} = \mathbf{0.5 \text{ A}}$$

In Example 7-16, it was unnecessary to know anything about the power supply circuit to determine the load current. If the short-circuit current and the open-circuit resistance at the terminals, as defined by Norton's theorem, can be measured, the current in any element connected between the terminals can be evaluated.

Example 7-17 Two electronic power supplies, each with the same measured values as in Example 7-16, are connected in parallel with each other. The combined output is then applied to a 50-Ω load. What is the load current?

Solution Figure 7-67 shows the initial circuit. Since all elements are connected in parallel, the circuit can be redrawn as in Fig. 7-68.

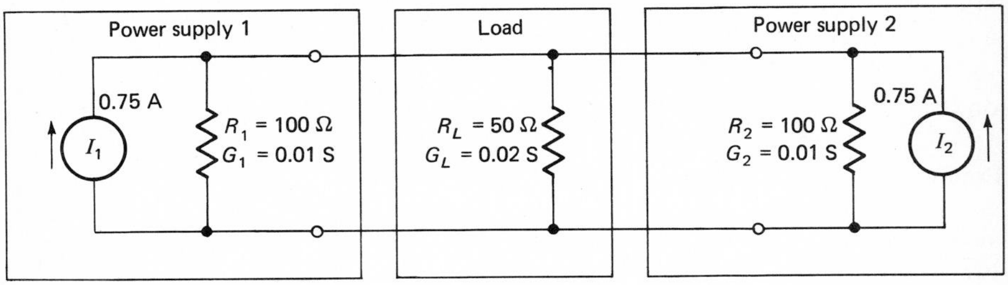

Fig. 7-67 Example 7-17: circuit.

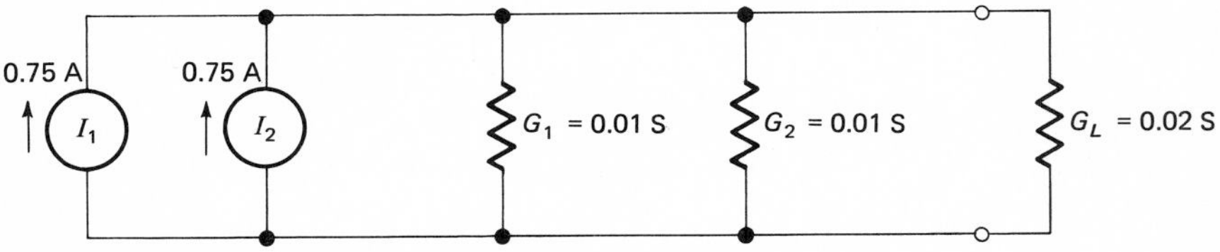

Fig. 7-68 Rearrangement of the components in Fig. 7-67.

The circuit can be further reduced to the Norton equivalent in Fig. 7-69.

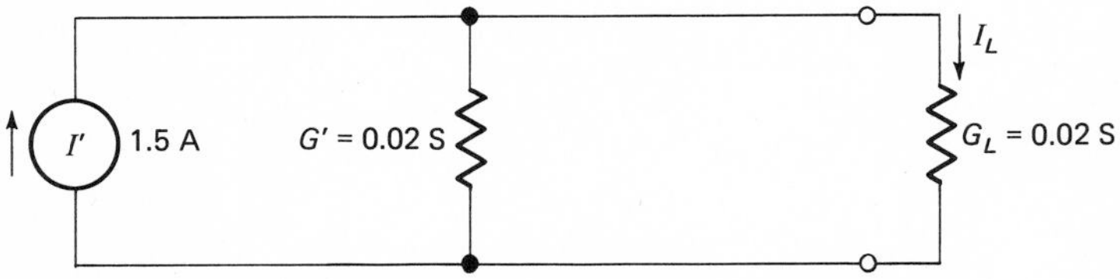

Fig. 7-69 Example 7-17: Norton equivalent circuit.

$$I' = I_1 + I_2 = 1.5 \text{ A}$$

$$G' = G_1 + G_2 = 0.02 \text{ S}$$

$$I_L = I' \frac{G_L}{G_L + G'} = 1.5 \text{ A} \frac{0.02 \text{ S}}{0.02 \text{ S} + 0.02 \text{ S}} = \mathbf{0.75\ A}$$

In Example 7-17 two similar energy sources connected in parallel supply more current to a load than one of the sources alone. This is always true if the sources have the same characteristics.

Example 7-18 What is the base current I_B in the transistor switching circuit in Fig. 7-70? The base-to-emitter resistance R_{BE} can have any value between 125 and 200 Ω.

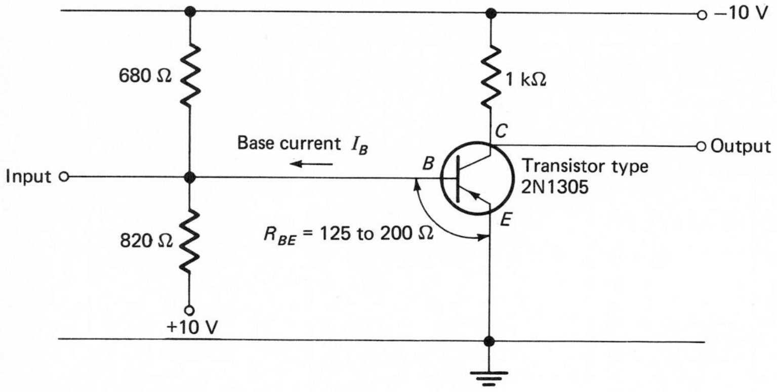

Fig. 7-70 Example 7-18: given circuit.

Solution The equivalent circuit, from the viewpoint of the base-to-emitter resistance, can be redrawn in current generator form as in Fig. 7-71.

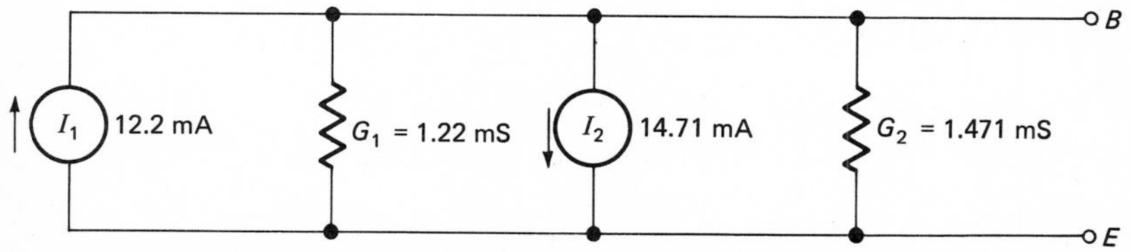

Fig. 7-71 Equivalent base-to-emitter circuit in Fig. 7-70 excluding the transistor.

The positive 10-V supply and the 820-Ω resistor are shown as I_1 and G_1.

$$I_1 = \frac{10 \text{ V}}{820 \ \Omega} = 12.2 \text{ mA}$$

$$G_1 = \frac{1}{820 \ \Omega} = 1.22 \text{ mS}$$

The negative 10-V supply and the 680-Ω resistor are shown as I_2 and G_2.

$$I_2 = \frac{10 \text{ V}}{680 \ \Omega} = 14.71 \text{ mA}$$

$$G_2 = \frac{1}{680 \ \Omega} = 1.471 \text{ mS}$$

The Norton equivalent of Fig. 7-71 is shown in Fig. 7-72.

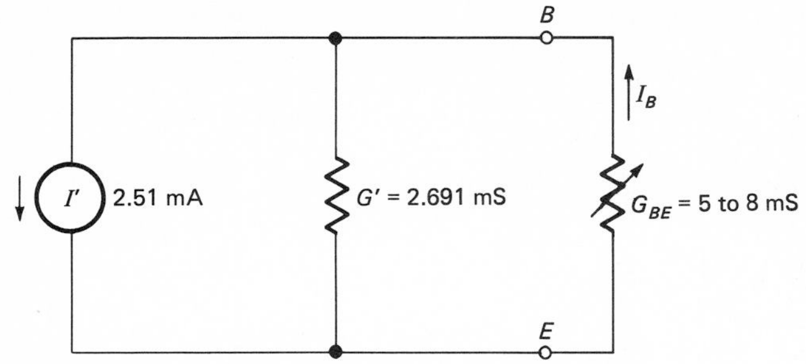

Fig. 7-72 Example 7-18: Norton equivalent circuit.

The currents I_1 and I_2 in Fig. 7-71 are in opposite directions, so that

$$I' = I_2 - I_1 = 14.71\text{ mA} - 12.2\text{ mA} = 2.51\text{ mA}$$

$$G' = G_1 + G_2 = 1.22\text{ mS} + 1.471\text{ mS} = 2.691\text{ mS}$$

When the base-to-emitter resistance is 125 Ω, $G_{BE} = 8$ mS,

$$I_B = I' \frac{G_{BE}}{G_{BE} + G'} = 2.51\text{ mA} \frac{8\text{ mS}}{8\text{ mS} + 2.691\text{ mS}} = \mathbf{1.878\text{ mA}}$$

When the base-to-emitter resistance is 200 Ω, $G_{BE} = 5$ mS,

$$I_B = I' \frac{G_{BE}}{G_{BE} + G'} = 2.51\text{ mA} \frac{5\text{ mS}}{5\text{ mS} + 2.691\text{ mS}} = \mathbf{1.632\text{ mA}}$$

In Example 7-18 the direction of the base current is from emitter to base, making the base negative with respect to the emitter. This polarity is correct for conduction in the PNP transistor shown in Fig. 7-70.

SUCCESSIVE USE OF NORTON'S THEOREM

Where it is considered advantageous as, for instance, in circuits containing a number of current sources separated by series elements, Norton equivalent circuits can be derived in successive steps to obtain the overall Norton current and resistance.

Example 7-19 What is the load current in the circuit in Fig. 7-73?

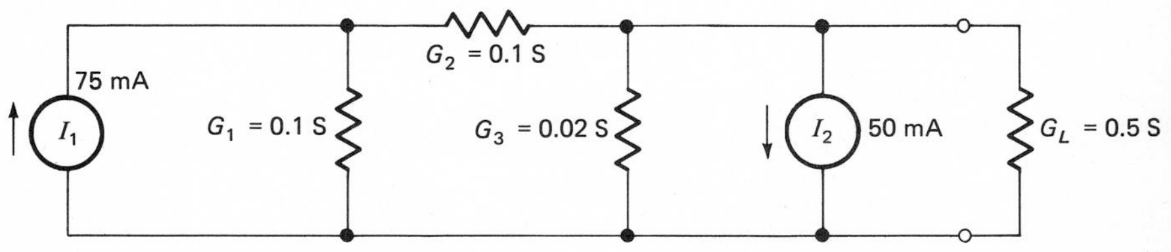

Fig. 7-73 Example 7-19: circuit.

Solution The steps are

1. Remove all of the circuit to the right of G_2, leaving the network in Fig. 7-74*a*.

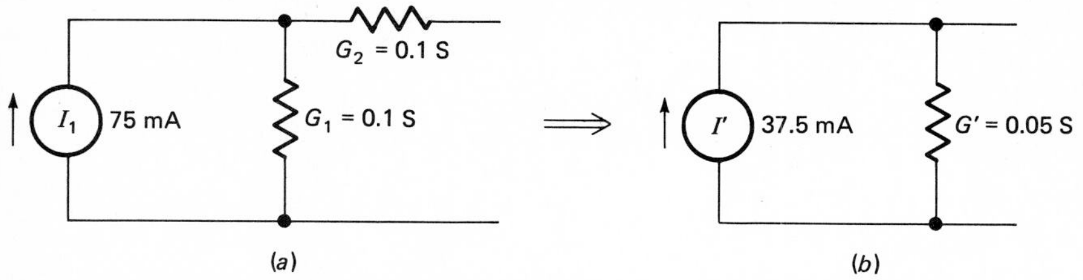

Fig. 7-74 Example 7-19: derivation of the first Norton equivalent circuit.

Figure 7-74*b* is the Norton equivalent of Fig. 7-74*a*. G_1 and G_2, of equal value, will be in parallel when the terminals are short-circuited, so

$$I' = \frac{I}{2} = \frac{75 \text{ mA}}{2} = 37.5 \text{ mA}$$

G_1 and G_2 will be in series when the terminals are open-circuited and the current source removed. Therefore,

$$G' = \frac{G_1}{2} = \frac{G_2}{2} = \frac{0.1 \text{ S}}{2} = 0.05 \text{ S}$$

2. Add to the first Norton circuit the elements G_3, I_2, and G_L. The circuit in Fig. 7-75 results.

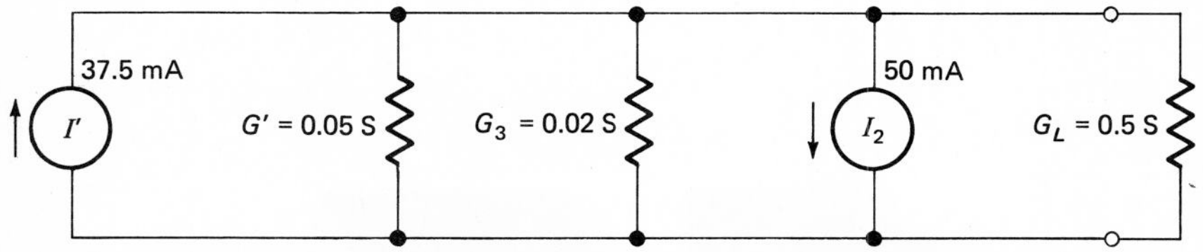

Fig. 7-75 Example 7-19: circuit in Fig. 7-73 with the first Norton circuit replacing I_1, G_1, and G_2.

All the elements in Fig. 7-75 are in parallel with I_1 and I_2. The directions of I_1 and I_2 are opposite. The overall Norton equivalent circuit is as shown in Fig. 7-76.

Fig. 7-76 Example 7-19: the second Norton equivalent circuit gives a final equivalent for the original circuit in Fig. 7-73.

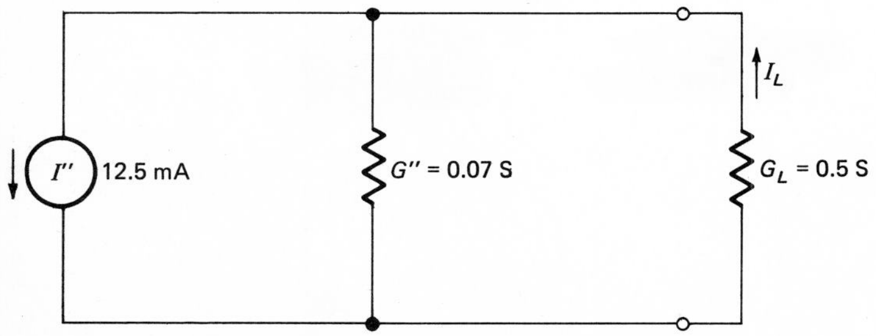

$$I'' = I_2 - I' = 50\ \text{mA} - 37.5\ \text{mA} = 12.5\ \text{mA}$$

$$G'' = G' + G_3 = 0.05\ \text{S} + 0.02\ \text{S} = 0.07\ \text{S}$$

$$I_L = I'' \frac{G_L}{G'' + G_L} = 12.5\ \text{mA} \frac{0.5\ \text{S}}{0.07\ \text{S} + 0.5\ \text{S}} = \mathbf{10.96\ mA}$$

7-5 RELATIONSHIP BETWEEN THEVENIN AND NORTON CIRCUITS

By applying the definitions of the two theorems (which in fact really means applying the basic principles of voltage and current generators), Thevenin and Norton circuits can be interchanged. Figure 7-77*a* shows a Thevenin circuit and Fig. 7-77*b* a Norton circuit. It is implied that both circuits are derived from the same initial circuit and are thus equivalent to each other.

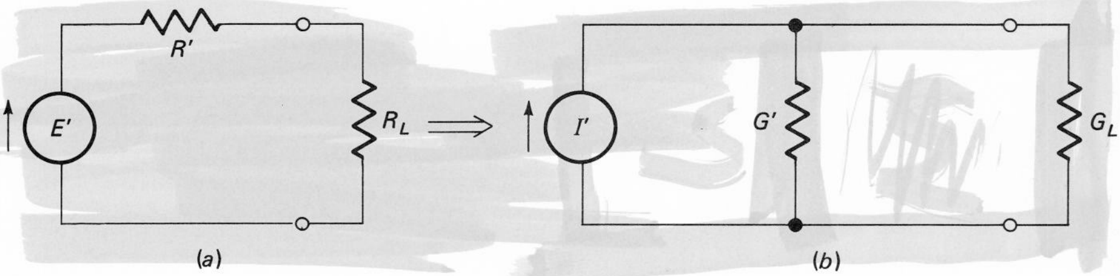

Fig. 7-77 Thevenin and Norton equivalence.

From the definition of the Norton current, it must equal the current in the short-circuited terminals of either circuit in Fig. 7-77.

In Fig. 7-77*a*, the short-circuit current equals E'/R'. The Norton current and conductance in Fig. 7-77*b* can be written using the symbols of the Thevenin circuit in Fig. 7-77*a*:

$$I' = \frac{E'}{R'} \quad \text{and} \quad G' = \frac{1}{R'}$$

From the definition of the Thevenin voltage, it must equal the voltage across the open-circuited terminals of either circuit in Fig. 7-77.

In Fig. 7-77*b*, the open-circuit voltage equals I'/G'. The Thevenin voltage and resistance in Fig. 7-77*a* can be written using the symbols of the Norton circuit in Fig. 7-77*b*.

$$E' = \frac{I'}{G'} \quad \text{and} \quad R' = \frac{1}{G'}$$

These relationships for Thevenin and Norton circuits are the same as those for voltage and current generator equivalence given in Sec. 4-4.

Example 7-20 What are the Norton equivalent values of the Thevenin circuit in Fig. 7-78*a*?

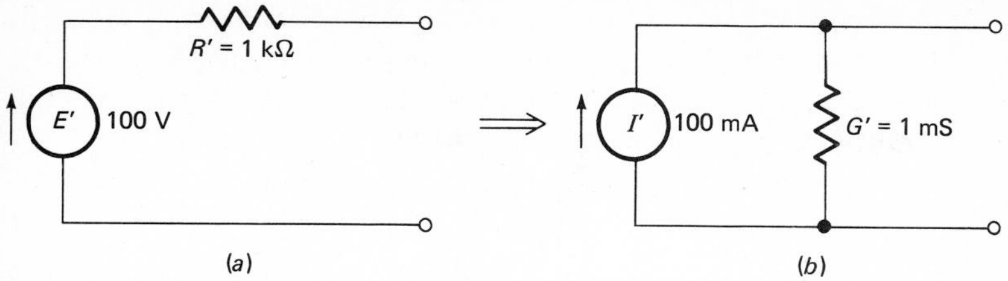

Fig. 7-78 Example 7-20: (*a*) given Thevenin circuit; (*b*) Norton equivalent.

Solution

$$I' = \frac{E'}{R'} = \frac{100 \text{ V}}{1 \text{ k}\Omega} = \mathbf{100 \text{ mA}}$$

$$G' = \frac{1}{R'} = \frac{1}{1 \text{ k}\Omega} = \mathbf{1 \text{ mS}}$$

Example 7-21 What are the Thevenin equivalent values of the Norton circuit in Fig. 7-79*a*?

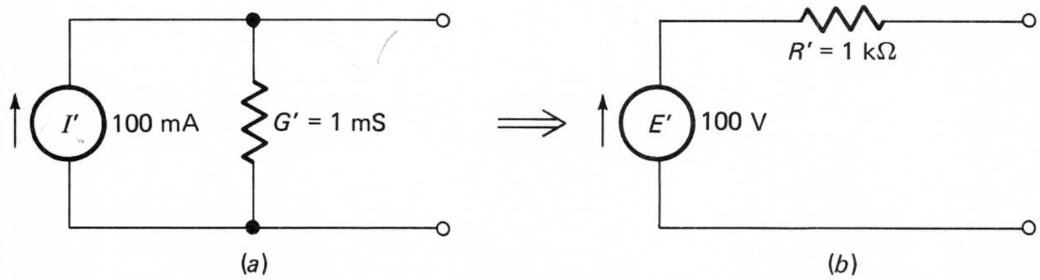

Fig. 7-79 Example 7-21: (*a*) given Norton circuit; (*b*) Thevenin equivalent.

Solution

$$E' = \frac{I'}{G'} = \frac{100 \text{ mA}}{1 \text{ mS}} = \mathbf{100 \text{ V}}$$

$$R' = \frac{1}{G'} = \frac{1}{1 \text{ mS}} = \mathbf{1 \text{ k}\Omega}$$

The results in Example 7-20 are the data given for Example 7-21. The results in Example 7-21 are the same as the data given for Example 7-20. Thus, commencing with the Thevenin voltage source in Fig. 7-78*a*, converting to a Norton current source (Figs. 7-78*b* and 7-79*a*), and then converting again to a Thevenin voltage source (Fig. 7-79*b*) leads back to the original data. This return to the starting conditions after two conversions validates the methods of conversion between the Thevenin and Norton source circuits.

7-6 CONSECUTIVE USE OF THEVENIN AND NORTON CIRCUITS

With some circuits, particularly those which consist of a long chain of elements connecting a load to a generator, it can be helpful to use the successive equivalent circuit technique and apply Thevenin and Norton methods alternately. This can reduce the amount of computing effort, because the calculations can be carried out mentally, a slide rule being used for converting resistance to conductance values.

One numerical example is given showing how to use this analysis method.

Example 7-22 What is the current delivered to the load in the circuit in Fig. 7-80*a* if the load resistance is correct for maximum power transfer?

Fig. 7-80 Example 7-22: (*a*) given circuit; (*b*) to (*e*) steps in deriving the Thevenin equivalent of the circuit in Fig. 7-80*a*.

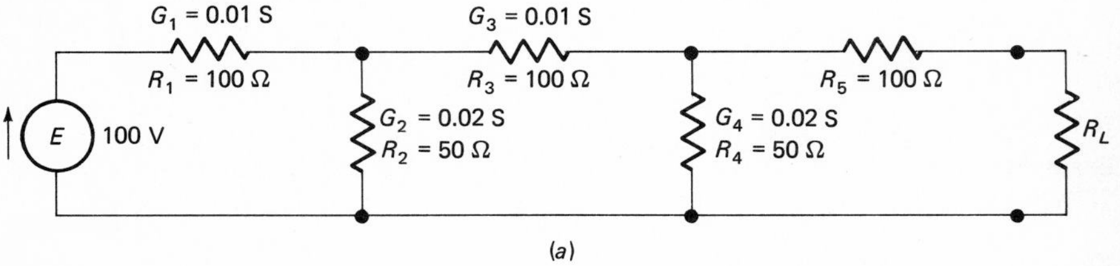

Solution

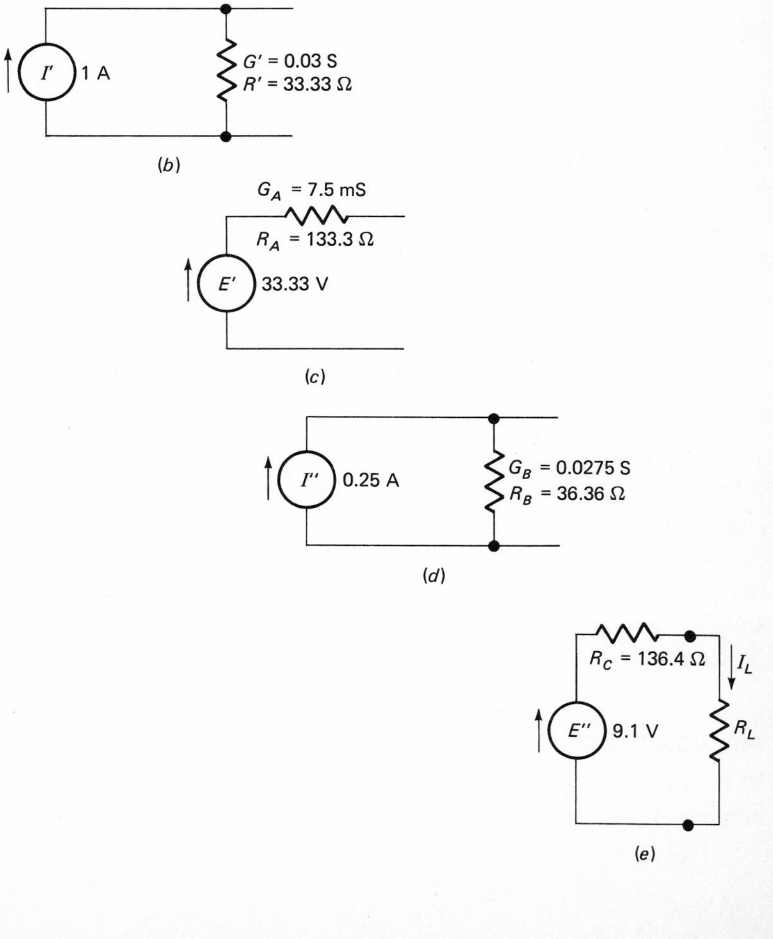

Solution The steps are

1. Change E and R_1 into current source form. Add R_2 written in conductance form. Draw the equivalent Norton circuit (Fig. 7-80*b*).
2. Change the Norton circuit into a Thevenin circuit. Add R_3 to the Thevenin resistance and label the sum R_A (Fig. 7-80*c*).
3. Change the Thevenin circuit into a Norton circuit. Express R_4 and R_A in conductance terms and add G_4 to G_A, calling the sum G_B (Fig. 7-80*d*).
4. Change the Norton circuit into a Thevenin circuit. Add R_5 to the Thevenin resistance, calling the sum R_C (Fig. 7-80*e*). Add the load R_L to complete the overall Thevenin equivalent of the original network in Fig. 7-80*a*. Since $R_C = 136.4\ \Omega$, for maximum power transfer, R_L should also be 136.4 Ω. The load current will be

$$I_L = \frac{E''}{R_C + R_L} = \frac{9.1\ \text{V}}{136.4\ \Omega + 136.4\ \Omega} = \mathbf{0.03336\ mA}$$

7-7 SUMMARY OF EQUATIONS

$\pi \leftrightarrow$ T conversions

$R_1 + R_2 = (R_B R_C + R_C R_A)/\Sigma R_\nabla$ (7-1)

$R_2 + R_3 = (R_A R_B + R_C R_A)/\Sigma R_\nabla$ (7-2)

$R_3 + R_1 = (R_A R_B + R_B R_C)/\Sigma R_\nabla$ (7-3)

$R_1 = R_B R_C/\Sigma R_\nabla$ (7-4)

$R_2 = R_C R_A/\Sigma R_\nabla$ (7-5)

$R_3 = R_A R_B/\Sigma R_\nabla$ (7-6)

$R_A = (R_1 R_2 + R_2 R_3 + R_3 R_1)/R_1 = \Sigma R_Y/R_1$ (7-7)

$R_B = \Sigma R_Y/R_2$ (7-8)

$R_C = \Sigma R_Y/R_3$ (7-9)

Thevenin's theorem

$E' = ER_3/(R_1 + R_3)$ (7-10)

$R' = R_2 + R_1 R_3/(R_1 + R_3)$ (7-11)

$(R_1 + R_3)I_1 - R_3 I_2 = E$ (7-12)

$-R_3 I_1 + (R_2 + R_3 + R_L)I_2 = 0$ (7-13)

Norton's theorem

$I' = IG_3/(G_1 + G_3)$ (7-14)

$G' = G_2 + G_1 G_3/(G_1 + G_3)$ (7-15)

$(G_1 + G_3)V_1 - G_3 V_2 = I$ (7-16)

$-G_3 V_1 + (G_2 + G_3 + G_L)V_2 = 0$ (7-17)

EXERCISE PROBLEMS

Section 7-1

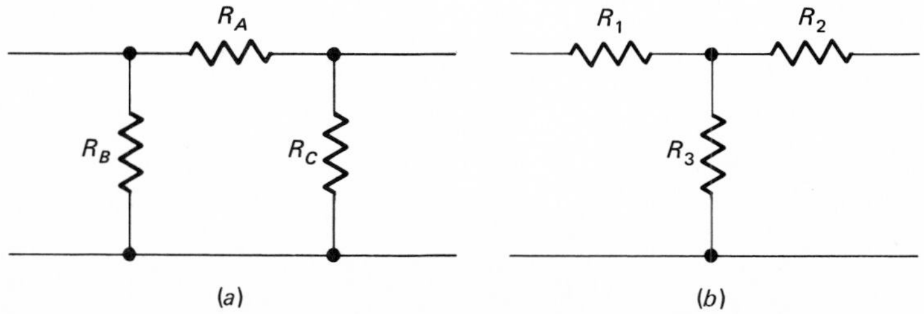

Fig. 7-1P Problems 7-1 and 7-2: (*a*) π network; (*b*) T network.

7-1 Convert the π network in Fig. 7-1P*a* to the T network in Fig. 7-1P*b* when

a. $R_A = 68\ \Omega$, $R_B = 68\ \Omega$, $R_C = 68\ \Omega$ (*ans.* 22.7 Ω, 22.7 Ω, 22.7 Ω)
b. $R_A = 68\ \Omega$, $R_B = 82\ \Omega$, $R_C = 56\ \Omega$ (*ans.* 27.1 Ω, 18.5 Ω, 22.3 Ω)
c. $R_A = 180\ \Omega$, $R_B = 390\ \Omega$, $R_C = 560\ \Omega$ (*ans.* 62.1 Ω, 89.2 Ω, 193.3 Ω)
d. $R_A = 1125\ \Omega$, $R_B = 1000\ \Omega$, $R_C = 1000\ \Omega$ (*ans.* 360 Ω, 360 Ω, 320 Ω)
e. $R_A = 47\ k\Omega$, $R_B = 100\ k\Omega$, $R_C = 47\ k\Omega$ (*ans.* 24.2 kΩ, 11.4 kΩ, 24.2 kΩ)
f. $R_A = 560\ k\Omega$, $R_B = 220\ k\Omega$, $R_C = 330\ k\Omega$ (*ans.* 111 kΩ, 166.5 kΩ, 65.44 kΩ)

7-2 Convert the T network in Fig. 7-1P*b* to the π network in Fig. 7-1P*a* when
a. $R_1 = 68\ \Omega$, $R_2 = 68\ \Omega$, $R_3 = 68\ \Omega$
b. $R_1 = 180\ \Omega$, $R_2 = 390\ \Omega$, $R_3 = 560\ \Omega$
c. $R_1 = 360\ \Omega$, $R_2 = 360\ \Omega$, $R_3 = 320\ \Omega$
d. $R_1 = 1\ M\Omega$, $R_2 = 470\ k\Omega$, $R_3 = 100\ k\Omega$
e. $R_1 = 0.3\ \Omega$, $R_2 = 0.09\ \Omega$, $R_3 = 0.25\ \Omega$
f. $R_1 = 2.51\ k\Omega$, $R_2 = 11.7\ k\Omega$, $R_3 = 7.82\ k\Omega$

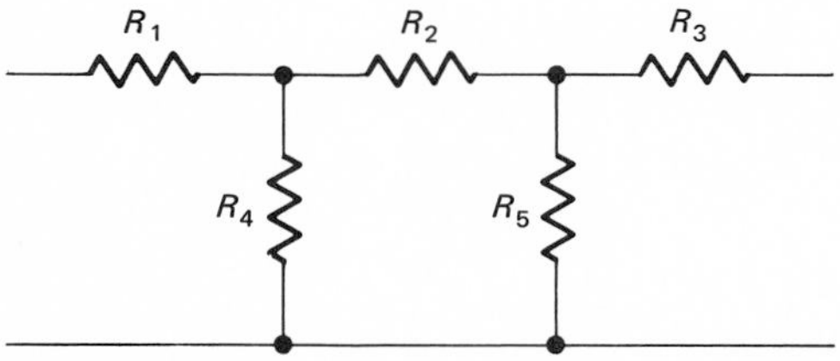

Fig. 7-2P Problems 7-3 and 7-4: network.

7-3 Convert the network in Fig. 7-2P to an equivalent π network when
a. $R_1 = 100\ \Omega$, $R_2 = 220\ \Omega$, $R_3 = 100\ \Omega$, $R_4 = 150\ \Omega$, $R_5 = 150\ \Omega$
(*ans. a.* 945 Ω; 250 Ω; 250 Ω)
b. $R_1 = 560\ k\Omega$, $R_2 = 390\ k\Omega$, $R_3 = 820\ k\Omega$, $R_4 = 560\ k\Omega$, $R_5 = 680\ k\Omega$
(*ans.* 4.596 MΩ, 1.093 MΩ, 1.547 MΩ)

7-4 Convert the network in Fig. 7-2P to an equivalent T network when
a. $R_1 = 47\ \Omega$, $R_2 = 68\ \Omega$, $R_3 = 47\ \Omega$, $R_4 = 33\ \Omega$, $R_5 = 33\ \Omega$
b. $R_1 = 1.8\ k\Omega$, $R_2 = 220\ \Omega$, $R_3 = 1.5\ k\Omega$, $R_4 = 2.7\ k\Omega$, $R_5 = 390\ \Omega$

7-5 Convert the network (R_1 through R_5) in Fig. 6-2P to an equivalent T network and find I_L when
a. $R_L = 120\ \Omega$ (*ans.* 27.3 mA)
b. $R_L = 1200\ \Omega$ (*ans.* 12 mA)
c. $R_L = 1800\ \Omega$ (*ans.* 9.16 mA)

7-6 Convert the T network (R_2, R_3, and R_4) in Fig. 6-4P*a* to an equivalent π network and compare with Fig. 6-4P*b*.

7-7 Convert the π network (R_a, R_b, and R_c) in Fig. 6-4P*b* to an equivalent T network and compare with Fig. 6-4P*a*. (*ans.* same)

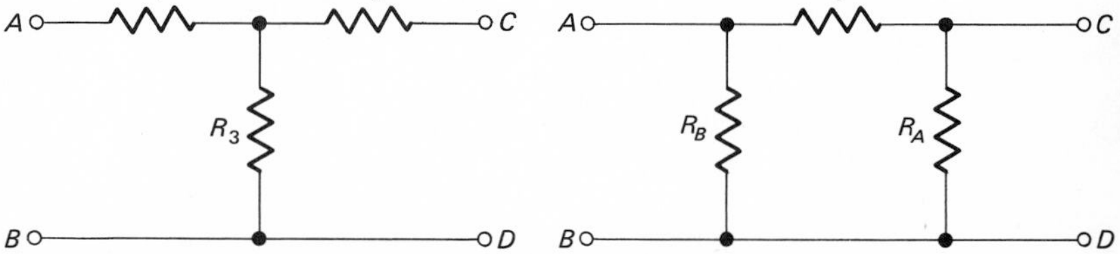

Fig. 7-3P Problem 7-8: (*a*) T network; (*b*) π network.

7-8 Choose any T network previously converted to a π network and use those resistance values for the values in Fig. 7-3P. Under the stated conditions, for both circuits, calculate the resistance between

a. *A* and *B* with *C* and *D* open circuit.
b. *A* and *B* with *C* and *D* short circuit.
c. *C* and *D* with *A* and *B* open circuit.
d. *C* and *D* with *A* and *B* short circuit.
e. *A* and *C* with *A* and *B*, and *C* and *D* open circuit.
f. *A* and *C* with *A* and *B* open circuit, and *C* and *D* short circuit.
g. *A* and *C* with *A* and *B* short circuit, and *C* and *D* open circuit.

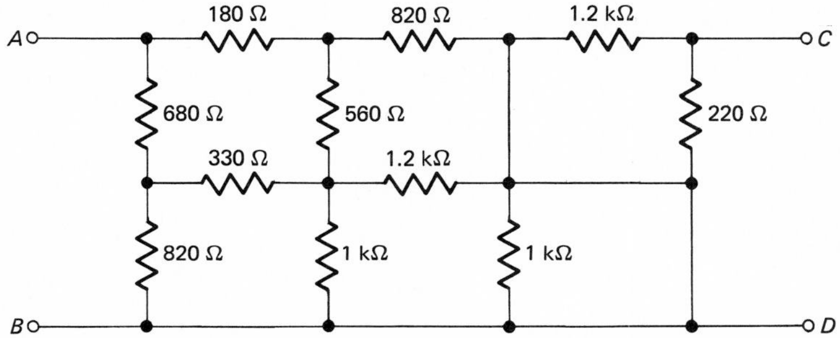

Fig. 7-4P Problem 7-9: network.

7-9 Reduce the circuit in Fig. 7-4P to its minimum. (*ans.* 446.5 Ω)

Section 7-2

7-10 Rework Prob. 6-1*a* through *d* by reducing the circuits to their Thevenin equivalents to find I_L.

7-11 Rework Prob. 6-2*a* through 6-2*c* by reducing the circuits to their Thevenin equivalents to find I_L. (*ans. a.* 27.97 mA; *b.* 13.41 mA; *c.* 10.4 mA)

7-12 Reduce Fig. 6-4P*a* and *b* to equivalent Thevenin circuits and compare.

7-13 Using Thevenin's theorem, find the magnitude and direction of the meter current in Example 7-8 in the text if R_4 is changed to 200 Ω. (*ans.* 1.29 mA, right)

7-14 Reduce the circuit in Fig. 6-10P to its Thevenin equivalent and find the value of I_R.

7-15 Reduce the circuit in Fig. 6-11P to its Thevenin equivalent and find the value of the meter current. (*ans.* 4.85 mA)

7-16 Using Thevenin's theorem, find the power dissipation of R_2 in Example 7-11 in the text.

7-17 Using the ammeter as the element of interest in Fig. 6-9P, find

a. The Thevenin equivalent circuit. (*ans.* 0.9545 V, 0.04952 Ω)
b. The meter current. (*ans.* 16.04 A)

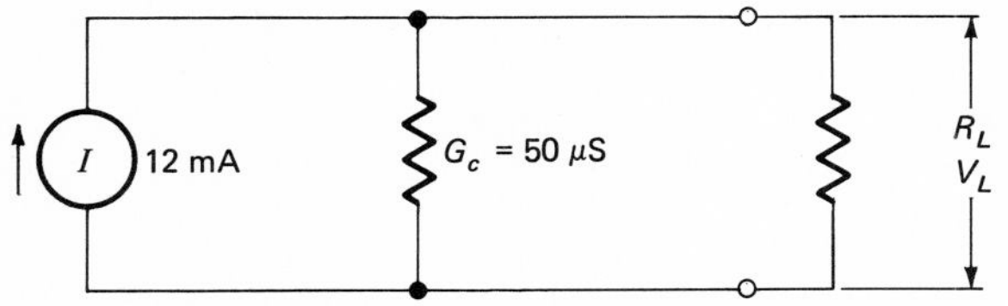

Fig. 7-5P Problem 7-18: circuit.

7-18 Figure 7-5P shows the output equivalent circuit of a transistor amplifier. Convert the circuit to a Thevenin equivalent and find V_L when

a. $R_L = 470$ Ω   *c.* $R_L = 1500$ Ω
b. $R_L = 1000$ Ω   *d.* $R_L = 2700$ Ω

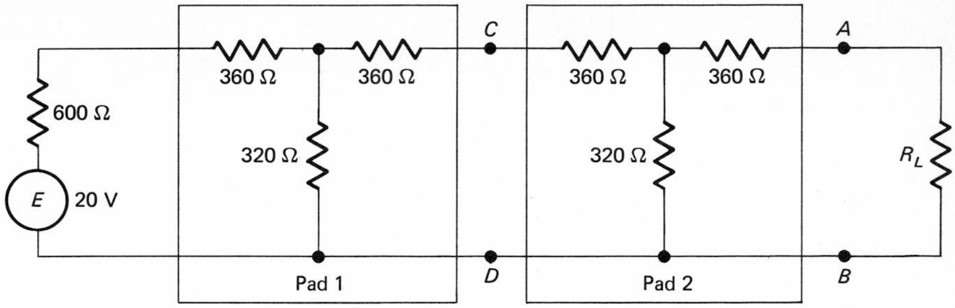

Fig. 7-6P Problem 7-19: circuit.

7-19 The circuit in Fig. 7-6P shows a 600-Ω generator followed by two T pad attenuators.

a. Find the Thevenin equivalent for the circuit at points *C* and *D* with pad 2 disconnected. (*ans.* 5 V, 600 Ω)

b. Find the Thevenin equivalent for the circuit at points *A* and *B* with both pads connected. (*ans.* 1.25 V, 600 Ω)

c. With R_L connected between points *A* and *B*, find the voltage developed across the load when

i. $R_L = 300\ \Omega$ (*ans.* 0.4167 V)

ii. $R_L = 600\ \Omega$ (*ans.* 0.625 V)

iii. $R_L = 1000\ \Omega$ (*ans.* 0.781 V)

d. What is the maximum power obtainable at points *C* and *D*, with pad 2 disconnected? (*ans.* 10.42 mW)

e. What is the maximum power obtainable at points *A* and *B*, with both pads connected? (*ans.* 651 μW)

7-20 The open-circuit voltage output of a generator is 20 V. When a 1000-Ω resistor is connected across the output, the terminal voltage falls to 15.4 V. Find the terminal voltage when a 470-Ω resistor is connected across the output.

7-21 If an input signal of −1 V is applied through a 1000-Ω resistance to the input in Fig. 7-70 of the text, find the base current when the base-to-emitter resistance is 200 Ω. (*ans.* 2.02 mA)

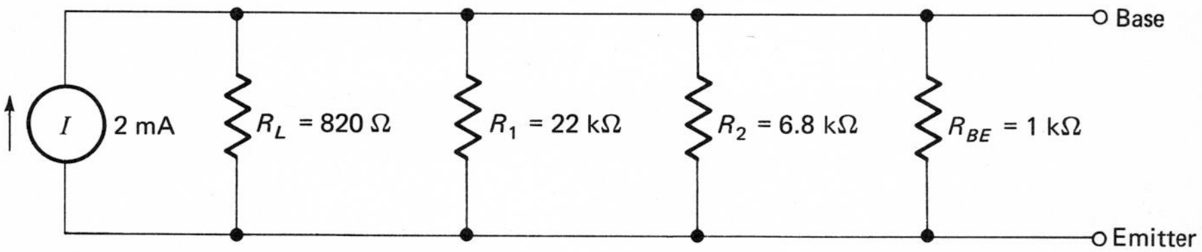

Fig. 7-7P Problem 7-22: circuit.

7-22 The circuit in Fig. 7-7P represents the interstage coupling between two transistor amplifiers. Find the Thevenin equivalent of the circuit with R_{BE} as the element of interest. Find the signal current through R_{BE}.

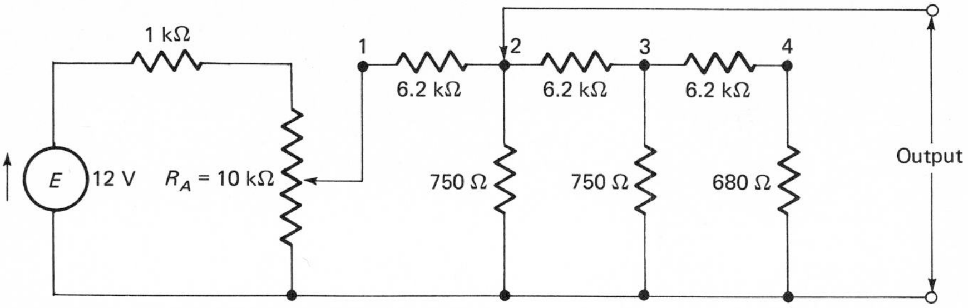

Fig. 7-8P Problems 7-23 and 7-27: circuit.

7-23 Figure 7-8P is the output equivalent circuit of a generator. With R_A set to 5 kΩ, find the equivalent Thevenin circuit when the output is connected to

a. Tap 2. (*ans.* 0.384 V, 628.6 Ω)

b. Tap 3. (*ans.* 0.0378 V, 615.9 Ω)

Section 7-3

7-24 An amplifier has an open-circuit output voltage of 24 V and an output resistance of 8 Ω. How can four 8-Ω speakers be connected to the amplifier so as to obtain maximum power output to the speakers? How much power is delivered to each speaker?

7-25 The open-circuit voltage of a generator is 60 V, and the short-circuit current is 200 mA. What is the maximum power this generator can deliver to a load? (*ans.* 3 W)

7-26 Find the maximum power that the circuit in Fig. 6-3P can deliver to its load G_L. What must be the value of G_L for this maximum power transfer?

7-27 For the circuit in Prob. 7-23, when the variable control R_A is set for maximum voltage, what is the maximum power the generator can deliver to a load from

a. Tap 2? (*ans.* 364 μW)

b. Tap 3? (*ans.* 3.53 μW)

Section 7-4

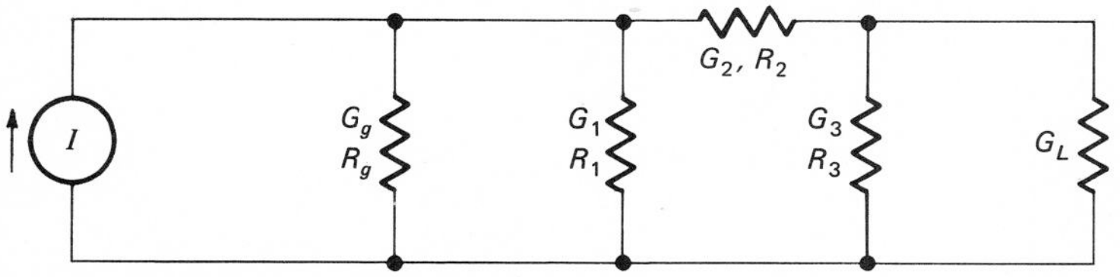

Fig. 7-9P Problem 7-28: circuit.

7-28 Convert the circuit in Fig. 7-9P to its Norton equivalent when

a. $I = 50$ mA, $G_g = 0.05$ S, $G_1 = 0.02$ S, $G_2 = 0.01$ S, $G_3 = 0.02$ S

b. $I = 100$ mA, $G_g = 1.67$ mS, $G_1 = 1$ mS, $G_2 = 890$ μS, $G_3 = 1$ mS

c. $I = 100$ mA, $R_g = 820$ Ω, $R_1 = 390$ Ω, $R_2 = 470$ Ω, $R_3 = 560$ Ω

d. $I = 10$ μA, $G_g = 0$, $G_1 = 1.22$ μS, $G_2 = 0.833$ μS, $G_3 = 0.667$ μS

7-29 Using Fig. 7-53 of the text, find the voltage across G_L, using Norton's theorem when

a. $I = 50$ mA, $G_1 = 0.01$ S, $G_2 = 0.01$ S, $G_3 = 0.01$ S, $G_L = 0.01$ S (*ans.* 1 V)

b. $I = 50$ mA, $G_1 = 1$ mS, $G_2 = 500$ μS, $G_3 = 750$ μS, $G_L = 1$ mS (*ans.* 8 V)

c. $I = 100$ mA, $G_1 = 1.66$ mS, $G_2 = 3.33$ mS, $G_3 = 3.33$ mS, $G_L = 1.66$ mS (*ans.* 10.94 V)

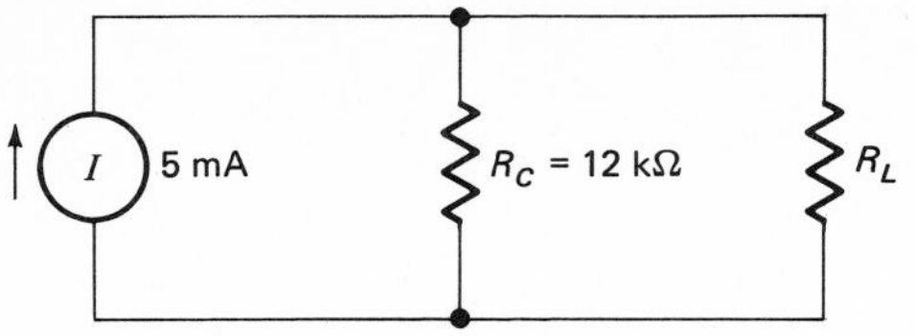

Fig. 7-10P Problem 7-30: circuit.

7-30 For the output equivalent circuit in Fig. 7-10P, find the voltage developed across R_L when
a. $R_L = 820\ \Omega$
b. $R_L = 1800\ \Omega$
c. $R_L = 2700\ \Omega$

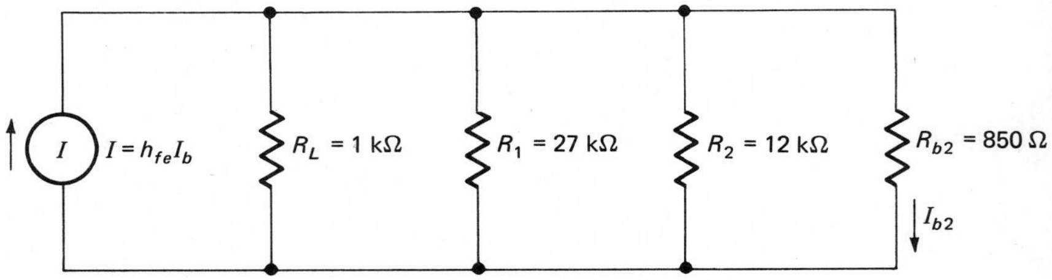

Fig. 7-11P Problem 7-31: circuit.

7-31 Figure 7-11P is the interstage equivalent circuit between two transistor amplifiers. Using Norton's theorem, find I_{b2}, the input current to the second stage, when $h_{fe} = 65$ and
a. $I_b = 6\ \mu A$ (*ans.* 199.8 μA)
b. $I_b = 12\ \mu A$ (*ans.* 399.5 μA)

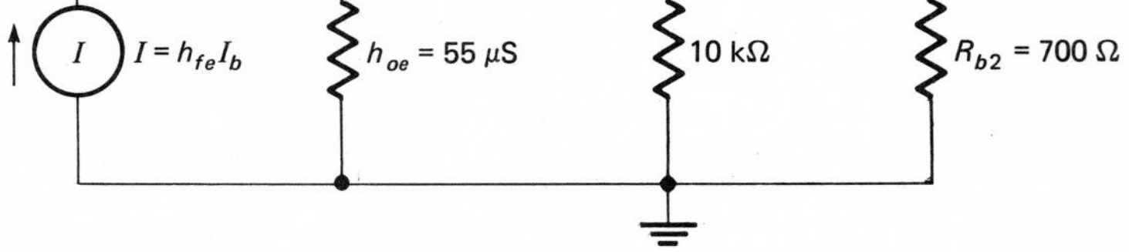

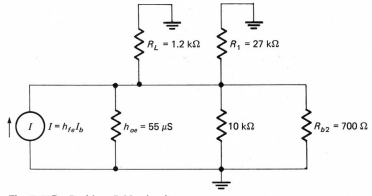

Fig. 7-12P Problem 7-32: circuit.

7-32 Under a given set of conditions, the output circuit of a transistor amplifier can be drawn as in Fig. 7-12P. Find the Norton equivalent circuit and the current through R_{b2}, the input resistance of the next amplifier, when $I_b = 30\ \mu A$ and $h_{fe} = 50$.

7-33 The short-circuit current of a generator is 0.5 A. The open-circuit voltage is 35 V. Find the Norton equivalent circuit. (*ans.* 0.5 A, 14.29 mS)

7-34 The short-circuit current of a generator is 250 mA. If the generator develops 60 V across 720-Ω load, find the Norton equivalent circuit of the generator.

THREE

DC Capacitive and Inductive Circuits

INTRODUCTION

Part II dealt with dc resistive circuits, with the inference that "dc" refers to unvarying values of current and voltage. Initially this dc approach is the most simple method of understanding resistive circuits. It is a completely general approach to resistive circuits, because they behave in exactly the same manner with unvarying direct current, varying direct current, and alternating current.

In capacitive or inductive circuits, it is rare for steady values of current and voltage to be of interest, because the behavior of capacitance and inductance in electric circuits becomes significant only when change is taking place.

Part III involves both varying and nonvarying direct current. Here we are concerned with circuits in which the applied emf always has the same polarity. Within the network connected across the source, currents may change value and direction; pds may change value and polarity.

The force field background for capacitance and inductance is introduced as an aid to the development of capacitive and inductive circuit relationships. Treatment of circuits containing capacitance alone and inductance alone is given in Chaps. 8 and 9. In Chap. 10 the combination of resistance with either capacitance or inductance leads to the important concept of time constant.

8 Capacitive Circuits

A resistance network is completely dissipative. All the energy supplied to it is dissipated into the surrounding environment, mainly in the form of heat. None of this dissipated energy can be recovered by the circuit. The action is unilateral, from energy source to resistance to environment, never vice-versa.

When a circuit contains capacitance or inductance, some of the energy from the source can be stored. The stored energy can be returned to the source. Energy can be stored and then dissipated as heat by resistance elements. Energy can be stored in one element and then transferred to another storage element.

Capacitance and inductance are not important in circuits in which currents and voltages are steady. Thus it was not necessary to consider these quantities in the preceding chapters. From Chap. 10 onward this text is concerned with the investigation of circuits in which currents and voltages are not steady. In such circuits capacitance and inductance play important roles.

This chapter commences with a survey of some principles of electrostatics as an introduction to capacitance as a circuit element.

8-1 CAPACITANCE

When a voltage is applied between points or surfaces, an *electric field* develops. Within this field, free charges experience a force which makes them move. Movement of charge is current, so we can say that the application of voltage causes current. This observation has been made before in this text, so why is it being repeated?

Previously we dealt with voltages applied across resistances which were, to some extent, conductors. Now we intend to investigate some of the effects of voltage applied to insulators. *Insulators* are materials which have few free charges and do not conduct current in the same way metals do.

In Fig. 8-1*a* an open switch prevents any current in the circuit. The switch is shown closed in Fig. 8-1*b*. There would be a steady current in the resistor R, were it not for the insulating material between the two metal plates. Immediately after the switch is closed, there will be an initial circuit current. Its value will be the same as if the resistor were directly connected to the battery.

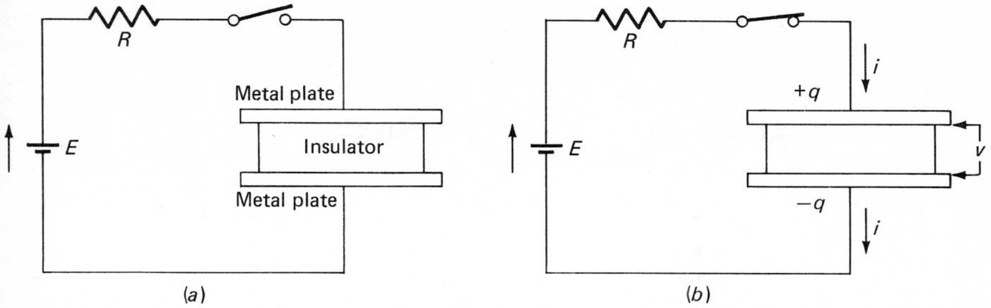

Fig. 8-1 (*a*) An insulator with no applied voltage; (*b*) voltage applied across an insulator develops a charge.

Within the plate-insulator unit there now develops an electric field, a form of energy. Energy has been transferred from the battery to the electric field within the insulator. This action shows up as a charge q at the plates and a voltage v between them. As energy becomes transferred, q and v will increase in value. This process continues, the current decreasing in value and becoming zero when the voltage between the plates equals the battery voltage.

If the switch is now opened, disconnecting the battery from the plate-insulator unit, the electric field energy will remain within the insulator, the voltage between the plates will remain, and so will the charge at each plate.

Capacitance is a measure of the ability to store energy in an electric field. Since it stores energy in an electric field, the metal plate–insulator unit is a *capacitor*. Capacitance used to be called "capacity." Although this term is outmoded, it does convey the idea that a capacitor has the capacity to store energy.

8-2 ELECTRIC FIELD QUANTITIES

The system of two parallel plates with a voltage applied between them was previously shown in Figs. 2-14 and 2-15. It is given again as Fig. 8-2, this time with the plate area suggested. This system is the basic capacitor and is now used to obtain an expression for the capacitance of a capacitor, while introducing some other electric field quantities.

Fig. 8-2 Electric field between two metal plates.

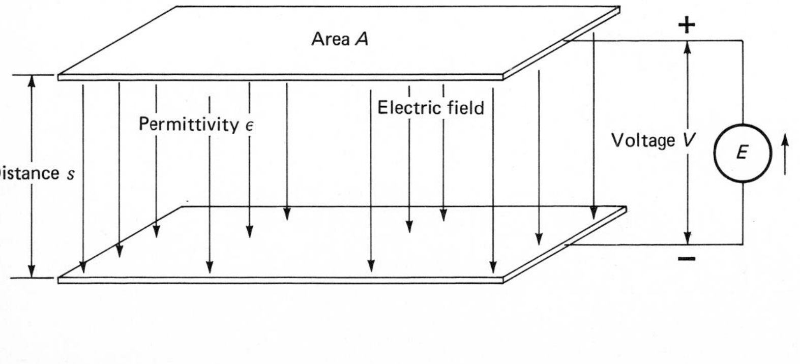

The significant factors influencing the electric field effects in the basic capacitor in Fig. 8-2 are the voltage between the plates V, the plate area A (also the cross-sectional area of the field), the distance between the plates (the width or thickness of the field), and a factor determined by the nature of the insulating material between the plates, the permittivity ϵ.

ELECTRIC FIELD STRENGTH

An electric field is a region in which physical force is experienced by charges. The magnitude of the force acting on a unit charge (1 C) is called the *electric field strength*[1] E. It was referred to in Chap. 2 and given by Eq. (2-9).

$$E = \frac{V}{s} \tag{8-1}$$

where E = electric field strength, V/m or N/C (newtons per coulomb) (these are the same units written in different terms, V/m being the preferred MKS unit)
V = pd between the plates, V
s = distance between the plates, m

ELECTRIC FLUX DENSITY

Charges are induced on conducting materials within an electric field. A conductor with induced charge is subject to forces tending to make it move in the field. The charge-inducing capability of a field is called its *electric flux density* or *electric displacement* D. The magnitude of the flux density of a field is the charge it can induce per unit of field area (or plate area).

$$D = \frac{Q}{A} \tag{8-2}$$

where D = electric flux density, C/m^2
Q = charge, C
A = field cross-sectional area, m^2

PERMITTIVITY

For a given field strength E, which is a function of voltage and field width (or distance between plates), the flux density D will depend on the insulation in which the field exists. This insulation is called the *dielectric*. The factor which determines the relationship between the field strength and the

[1] The international symbol for electric field strength, an italic capital E, is the same as the symbol for emf, a different quantity. If in a particular situation there is likely to be confusion between the two meanings of E, the symbol K is an approved alternative symbol for electric field strength.

flux density is the *permittivity* ϵ of the dielectric material (ϵ is the lowercase Greek letter epsilon).

The relationship between permittivity, field strength, and flux density is

$$\epsilon = \frac{D}{E} \quad \text{or} \quad D = \epsilon E \tag{8-3}$$

In the MKSA (or SI) unit system the magnitude of the permittivity of a material is given as the product

$$\epsilon = \epsilon_v \epsilon_r \tag{8-4}$$

where ϵ = total permittivity, F/m (farads per meter)
ϵ_v = permittivity of a vacuum, a physical constant 8.854×10^{-12} F/m
ϵ_r = relative permittivity of the material

Relative permittivity is a measure of the degree to which a material influences the electric field. It is the ratio of the flux density with that material to the flux density with no material at all, a vacuum. Being the ratio of two similar quantities, ϵ_r has no unit. Relative permittivity was called "dielectric constant" before the adoption of the MKSA system. This term is still in common use.

The unit farad per meter will be referred to a little later.

ELECTRIC FLUX

Electric flux density is the charge which can be induced per unit area of the field. The total induced charge is the *flux*. The flux Ψ is therefore the flux density multiplied by the field area (Ψ is the capital Greek letter psi):

$$\Psi = DA \tag{8-5}$$

Electric flux and charge have the same unit, the coulomb. They are numerically equal:

$$\Psi = Q \tag{8-6}$$

8-3 CAPACITANCE CALCULATION

Using Eqs. (8-1) to (8-6), we can obtain an expression for capacitance in terms of the physical characteristics of a capacitor:

$$Q = \Psi \tag{8-6}$$

Substituting for Ψ from Eq. (8-5),

$$Q = DA$$

Substituting for D from Eq. (8-3),

$$Q = \epsilon EA$$

Substituting for E from Eq. (8-1),

$$Q = \epsilon \frac{V}{s} A$$

Rearranging terms, the basic equation expressing capacitance is obtained:

$$Q = V\left(\frac{\epsilon A}{s}\right) \tag{8-7}$$

Capacitance may be defined as the constant of proportionality between electric charge and voltage. In Eq. (8-7), the expression in parentheses is this constant. Therefore the capacitance of a two-plate system is given by

$$C = \frac{\epsilon A}{s} \tag{8-8}$$

where C = capacitance, F (farads)
ϵ = permittivity $\epsilon_v \epsilon_r$ of the dielectric, F/m
A = field area, m^2
s = distance between plates, m

Equations (8-1) to (8-8), assume that the field and the dielectric between the plates is uniform.

THE FARAD

Equation (8-7) can be written in the form

$$Q = VC \quad \text{or} \quad C = \frac{Q}{V} \tag{8-9}$$

where Q = charge at the plates, C*
V = pd between the plates, V
C* = capacitance, F

* Note the symbol problem. C (italic) is the symbol for the quantity capacitance. C (roman) is the symbol for the unit coulomb. The problem can be resolved by realizing the distinction between quantities and units.

Presented in unit form, Eq. (8-9) leads to the definition of the farad, the MKSA unit of capacitance:

$$\mathrm{F} = \frac{\mathrm{C}}{\mathrm{V}} \tag{8-10}$$

A capacitance of 1 F exists when a charge of 1 C is associated with a pd of 1 V.

THE MKSA DIMENSIONS OF CAPACITANCE

The farad can be expressed in terms of the fundamental MKSA units:

$$C = \frac{Q}{V} \tag{8-9}$$

Substituting for Q from Eq. (2-11),

$$C = \frac{It}{V}$$

Putting this equation in unit form gives

$$\mathrm{F} = \frac{\mathrm{A} \cdot \mathrm{s}}{\mathrm{V}}$$

From (2-13),

$$\mathrm{F} = \mathrm{A} \cdot \mathrm{s} \div \frac{\mathrm{kg} \cdot \mathrm{m}^2}{\mathrm{s}^3 \cdot \mathrm{A}}$$

$$\mathrm{F} = \frac{\mathrm{A}^2 \cdot \mathrm{s}^4}{\mathrm{kg} \cdot \mathrm{m}^2} \tag{8-11}$$

THE FARAD PER METER

The MKSA unit of permittivity is the farad per meter. The validity of this unit can be seen by rearranging Eq. (8-8) so that ϵ is the subject:

$$\epsilon = \frac{Cs}{A}$$

Substituting unit symbols, this equation becomes

$$\frac{\mathrm{F}}{\mathrm{m}} = \frac{\mathrm{F} \cdot \mathrm{m}}{\mathrm{m}^2}$$

From Eq. (8-11) it can be seen that the MKSA dimensions of permittivity are $\mathrm{A}^2 \cdot \mathrm{s}^4/\mathrm{kg} \cdot \mathrm{m}^3$.

Example 8-1 Find the capacitance of a capacitor made from two strips of aluminum foil, each 5 cm wide and 10 m long, separated by a 25-μm thick film of Mylar. The relative permittivity of Mylar is 3.

Solution Commercial capacitors are often made by forming a sandwich of aluminum foil and insulating material and then rolling this sandwich into a cylindrical shape before enclosing it in plastic or a metal can. (Fig. 8-3)

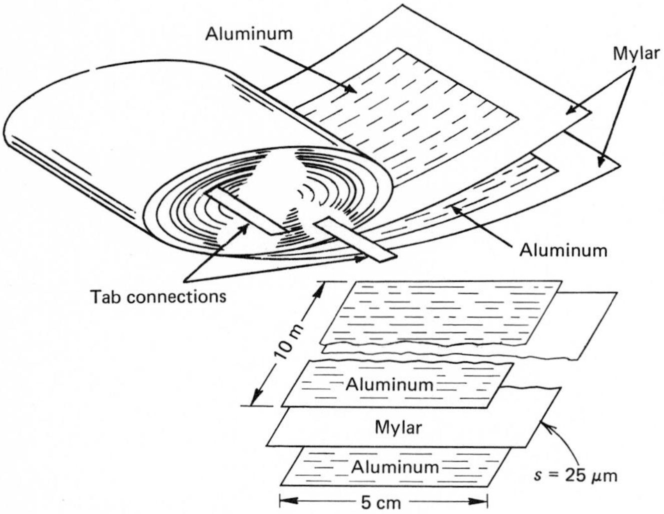

Fig. 8-3 Example 8-1: capacitor dimensions.

Plate area $\qquad A = 5\text{ cm} \times 10\text{ m} = 0.5\text{ m}^2$

From Eqs. (8-8) and (8-4),

$$C = \frac{\epsilon A}{s} = \frac{\epsilon_v \epsilon_r A}{s}$$

$$C = \frac{3 \times 8.854 \times 10^{-12}\text{ F/m} \times 0.5\text{ m}^2}{2.5 \times 10^{-5}\text{ m}} = \mathbf{0.5312\ \mu F}$$

Thus this capacitor made with foil plates about 2 in. wide and 10 yd long, with a dielectric of Mylar film about 1 mil ($\frac{1}{1000}$ in.) thick, has a capacitance of approximately 0.5 μF, a common value.

Example 8-2 The two-gang (two-section) variable capacitor shown in Fig. 8-4 is typical of those used to tune radio receivers. The larger of the two sections consists of 14 moving (rotating) plates and 13 fixed plates. As the moving plates rotate, they mesh with the fixed plates, leaving a 1-mm gap between adjacent plates. The dielectric between the plates is air ($\epsilon_r = 1.0001$). What must be the area of the field between adjacent plates if the maximum capacitance is to be 300 pF?

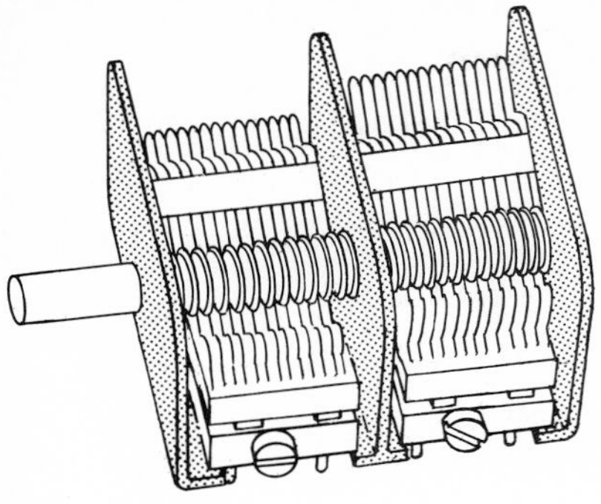

Fig. 8-4 Example 8-2: variable capacitor.

Solution When the moving plates of a variable capacitor are rotated through 180°, the meshing plate area, and hence the electric field area and the capacitance, vary between maximum and minimum values.

Maximum capacitance occurs when the plates are fully meshed, that is, when the field area is greatest. The following calculation is based on the maximum capacitance position of the plates:

$$C = \frac{\epsilon A}{s} \tag{8-8}$$

$$A = \frac{Cs}{\epsilon} = \frac{Cs}{\epsilon_v \epsilon_r} = \frac{300 \text{ pF} \times 1 \text{ mm}}{8.854 \times 10^{-12} \text{ F/m} \times 1.0001}$$

$$A = 0.0339 \text{ m}^2$$

A is the total area between all adjacent plates (assuming that plate and field area are the same). The area between any adjacent pair of plates is the total area *A* divided by the number of adjacent pairs. The number of adjacent pairs of plates of a variable capacitor is the total number of plates minus one. The two outside plates each contribute only one surface to the electric field, whereas all other plates have two effective surfaces.

In this example there are 26 adjacent pairs of plates.

$$A_p = \frac{A}{26} = \frac{0.0339 \text{ m}^2}{26} = \mathbf{0.0013 \text{ m}^2} \qquad \text{or} \qquad \mathbf{13 \text{ cm}^2}$$

Usually the plates of a variable capacitor are a special shape designed to give a desired relationship between the angular position of the moving plates (relative to the fixed plates) and the capacitance. However, the shape is near enough to a semicircle to enable us to visualize the size of the variable capacitor in Example 8-2.

The area of a semicircle is

$$A = \frac{\pi r^2}{2}$$

The radius will be

$$r = \sqrt{\frac{2A}{\pi}} = \sqrt{\frac{2 \times 13 \text{ cm}^2}{\pi}} = 2.877 \text{ cm}$$

Therefore this ordinary radio tuning capacitor will have plates of the order of a 1-in. radius.

8-4 RELATIONSHIPS IN THE CAPACITIVE CIRCUIT

The circuit in Fig. 8-1*b* is shown in schematic form in Fig. 8-5.

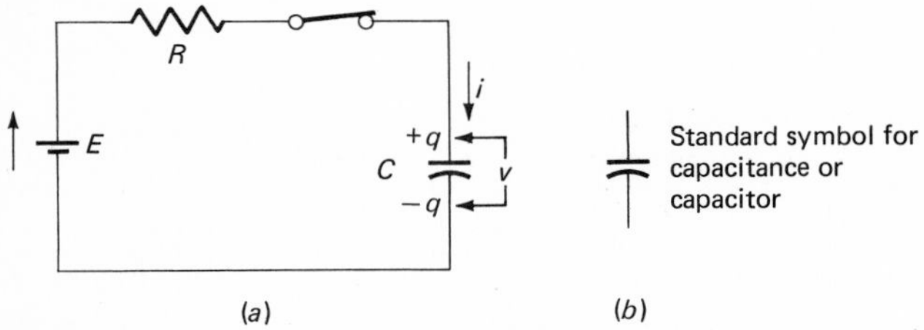

Fig. 8-5 (*a*) Capacitance being charged; (*b*) standard symbol for capacitance or a capacitor.

The charge at the plates of a capacitance at any instant is given by[1]

$$q = Cv \qquad \text{from Eq. (8-9)}$$

where q = instantaneous charge, C
v = instantaneous pd, V
C = capacitance, F

Current is present in a capacitive circuit only if there is charge movement, or if a change in charge magnitude is taking place. A small change in voltage across a capacitance will be accompanied by a small change in its stored charge. Equation (8-9) can be rewritten to show these changes:

$$\Delta q = C \, \Delta v \tag{8-12}$$

where Δq and Δv represent small changes in charge and voltage, respectively. If both sides of Eq. (8-12) are divided by Δt, a small time interval, we have

$$\frac{\Delta q}{\Delta t} = C \frac{\Delta v}{\Delta t} \tag{8-13}$$

[1] The use of lowercase letters for quantity symbols indicates that instantaneous values of varying quantities are being referred to. Equation (8-9) ($Q = VC$), using uppercase symbols for the variables, is the static or fixed-value version of $q = Cv$.

Current is the time rate of change in charge [from Eq. (2-10), $I = Q/t$] or

$$i = \frac{\Delta q}{\Delta t}$$

This expression for current in terms of a small change in charge and a small time interval can be used to write Eq. (8-13) in the form

$$i = C \frac{\Delta v}{\Delta t} \tag{8-14}$$

When this is written in the more common calculus form, it becomes one of the basic electrical equations:

$$i = C \frac{dv}{dt} \tag{8-15}$$

The *instantaneous current* (in amperes) in a capacitive circuit is the capacitance (in farads) times the rate of change in pd (in volts per second).

Example 8-3 The voltage across a 2-μF capacitor rises at a uniform rate from 10 to 60 V in 5 ms. What is (*a*) the change in charge at the capacitor plates and (*b*) the circuit current?
Solution (Fig. 8-6)

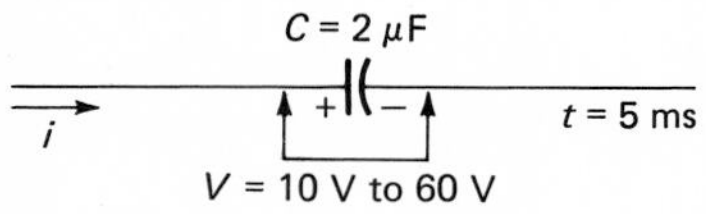

Fig. 8-6 Example 8-3: data.

The change in voltage is

$$\Delta v = 60\ \text{V} - 10\ \text{V} = +50\ \text{V}$$

Δv is positive because the voltage rises from 10 to 60 V, or changes in a positive direction.

a. Applying Eq. (8-12),

$$\Delta q = C\,\Delta v = 2\ \mu\text{F} \times 50\ \text{V} = \mathbf{100\ \mu C}$$

b. Using Eq. (8-15),

$$i = C\frac{dv}{dt}^{*} = 2\ \mu\text{F}\,\frac{50\ \text{V}}{5\ \text{ms}} = 0.02\ \text{A} \qquad \text{or} \qquad \mathbf{20\ mA}$$

* Where the rate of change of voltage is constant (linear), as in Example 8-3, $dv/dt = \Delta v/\Delta t$. It is common practice to assume that Eqs. (8-14) and (8-15) are the same.

The solution could have been written: A 50-V change in 5 ms is a rate of change of 10 000 V/s. Therefore,

$$i = C\frac{dv}{dt} = 2\ \mu\text{F} \times 10\ \text{kV/s} = \mathbf{20\ mA}$$

Example 8-4 If a 20-mA constant-current source is connected in series with an uncharged 2-μF capacitor, what voltage is developed across the capacitor after (*a*) 1 μs, (*b*) 10 ms, and (*c*) 100 ms?

Solution (Fig. 8-7)

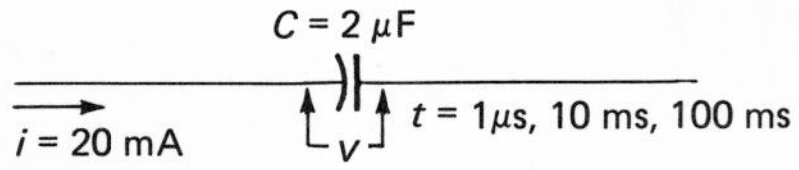

Fig. 8-7 Example 8-4: data.

An uncharged capacitor is one in which no energy is stored ($W = 0$, $Q = 0$, and $V = 0$). Charging is the act of increasing the amount of energy stored, and discharging is the act of decreasing the amount of energy.

Example 8-3 shows that a current of 20 mA is associated with a rate of change in voltage of 10 kV/s across a 2-μF capacitor. In starting from an uncharged state ($v = 0$), the capacitor voltage will become:

a. After time $t = 1\ \mu\text{s}$ ($\Delta t = 1\ \mu\text{s}$),

$$\frac{dv}{dt} = 10\ \text{kV/s}$$

$$\Delta v = 10\ \text{kV/s} \times \Delta t = 10\ \text{kV/s} \times 1\ \mu\text{s} = 0.01\ \text{V}$$

$$v = \mathbf{10\ mV}$$

b. After time $t = 10$ ms ($\Delta t = 10$ ms),

$$\Delta v = 10\ \text{kV/s} \times 10\ \text{ms} = 100\ \text{V}$$

$$v = \mathbf{100\ V}$$

c. After time $t = 100$ ms ($\Delta t = 100$ ms),

$$\Delta v = 10\ \text{kV/s} \times 100\ \text{ms} = 1\ \text{kV}$$

$$v = \mathbf{1\ kV}$$

8-5 CHARGING AND DISCHARGING A CAPACITANCE

CHARGING

It is evident from Eq. (8-15) and Example 8-4 that, when a capacitor becomes charged as the result of a constant current, its rate of change in voltage will also be constant. In other words the capacitor voltage will be directly proportional to time with constant charging current. Figure 8-8 shows this proportionality as a linear relationship between voltage and time.

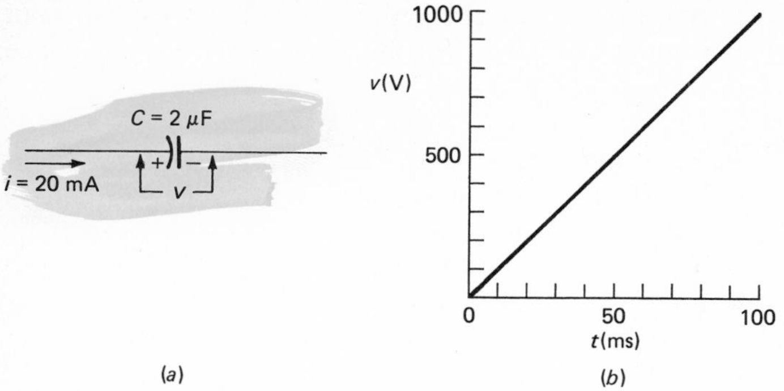

Fig. 8-8 Constant-current charge of a capacitance. (*a*) Circuit conditions; (*b*) graph relating voltage and time.

DISCHARGING

When a capacitor is discharging because it is supplying a constant current to a network, there will also be a linear relationship between the capacitor voltage and time. Now, however, the voltage will decrease because energy is being extracted from the capacitor.

Figure 8-9 shows the linear relationship between decreasing capacitor voltage and time when a 2-μF capacitor is discharged at a constant current of 20 mA, the capacitor having been previously charged to 1 kV.

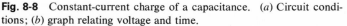

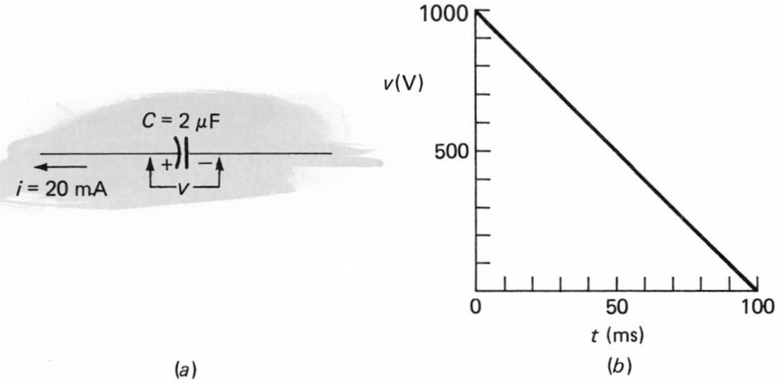

Fig. 8-9 Constant-current discharge of a capacitance. (*a*) Circuit conditions; (*b*) graph relating voltage and time.

Comparison of Figs. 8-8*a* and 8-9*a* shows that the voltage polarities are the same but the current directions are opposite. This observation implies that

1. While charging, a capacitor's voltage polarity will be the same as for a load carrying the same direction of current (current from positive to negative within the load).

2. If the charged capacitor is then discharged, its voltage polarity will not change, but its current direction will; it will be the same as for a voltage source (negative to positive within the source).

The voltage polarities and current directions are consistent with the notion that, when the capacitor is being charged, it is drawing energy from some source, so it is a load; when it is being discharged, it is supplying energy to a network, so it is a source. Remembering this will help you to avoid confusion in dealing with capacitive current directions and voltage polarities. Figure 8-10 shows these charge and discharge conditions in circuit form.

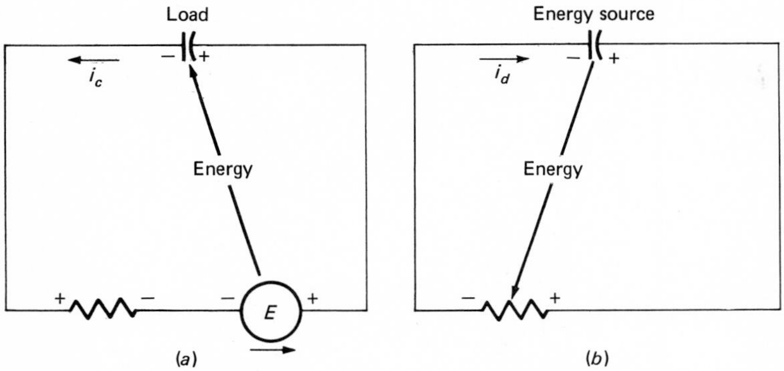

Fig. 8-10 (*a*) Energy direction to the capacitance during charge; the capacitance is a load; (*b*) energy direction from the capacitance during discharge; the capacitance is an energy source.

It is important to note that Fig. 8-10 is only intended to show current directions and voltage polarities. Charging and discharging a capacitance through a resistance will not normally result in a constant current value such as that referred to in Figs. 8-8 and 8-9.

8-6 OPPOSITION TO VOLTAGE CHANGE

Capacitance is sometimes described as the opposition to voltage change. This notion is based on the following reasoning:

$$i = C\frac{dv}{dt} \quad \text{or} \quad \frac{dv}{dt} = \frac{i}{C} \tag{8-15}$$

For a given value of current the rate of change in voltage across a capacitance is inversely proportional to the capacitance value. In more prosaic language, for a stated charge or discharge current, capacitive voltage will change more slowly with a large-value capacitance than with one of small value.

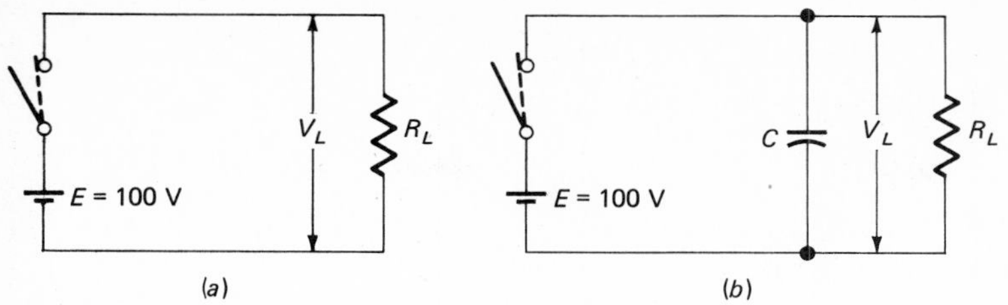

Fig. 8-11 Ripple filter. (*a*) Resistive load; (*b*) capacitor connected across the resistive load.

A practical example of the application of this view of capacitance is the ripple filter.

In Fig. 8-11*a* a 100-V emf is applied to a load through a switch. The switch is closed and opened at regular time intervals. The resulting load voltage is shown in Fig. 8-12*a*. In Fig. 8-11*b* a capacitor of large value is connected across the load. The charge stored in the capacitor maintains load voltage relatively constant, as shown in Fig. 8-12*b*. The details of operation of this circuit are given in Chap. 10.

Fig. 8-12 Load voltage waveforms in Fig. 8-11. (*a*) Without the capacitor; (*b*) with the capacitor.

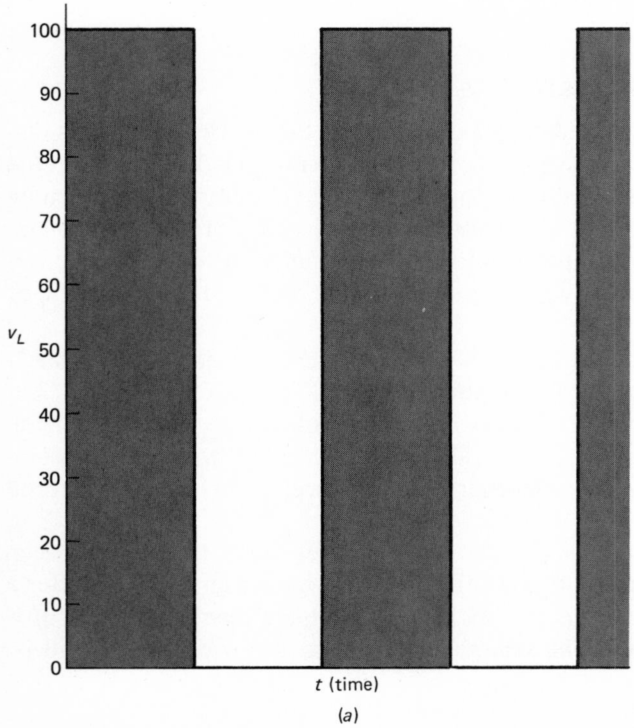

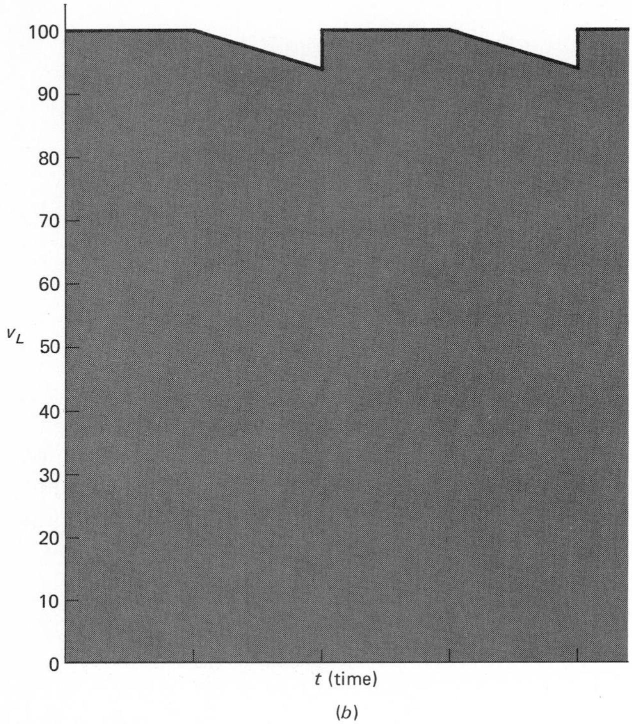

(b)

Fig. 8-12 *(Continued)*

8-7 CAPACITIVE ENERGY STORAGE

Energy is stored in a charged capacitance even if the charging source is disconnected; it is stored in the electric field between the plates (that is, in the dielectric). The energy so stored can be extracted from the capacitance by connecting a circuit to its terminals. If the circuit is resistive, the extracted energy will be dissipated into the environment as heat. If the circuit is capacitive or inductive, the energy will be re-stored in an electric or magnetic field.

When a capacitance is being charged, the energy transfer can be made apparent by current and voltage measurement. The circuit current, existing for a certain time, indicates the amount of charge imparted to the capacitance [$Q = It$, Eq. (2-11)]. This charge, coupled with the pd developed between the plates, is a measure of the energy transferred and stored [$W = VQ$, Eq. (2-6)].

If a capacitance is charged with constant current I, the pd v between its plates will increase at a uniform rate, as shown in Fig. 8-13. If after a time t seconds the current is cut off, the pd will have reached a value of V volts. During this time, the charge imparted to the capacitance will be

$$Q = It \qquad \text{coulombs}$$

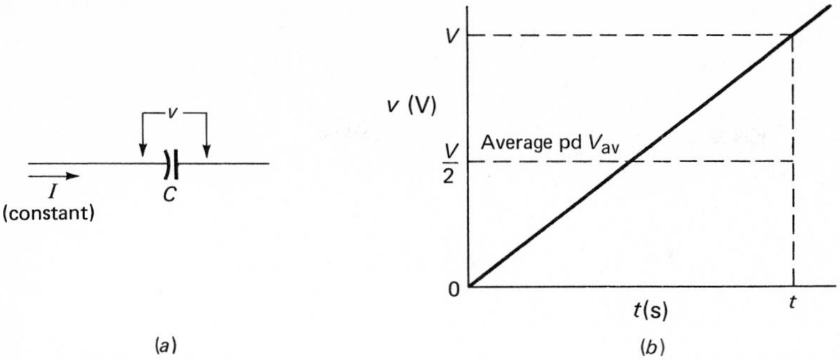

Fig. 8-13 Capacitive energy. (*a*) Circuit conditions; (*b*) graph of voltage with respect to time.

Equation (2-6) ($W = VQ$) gives energy in terms of a constant voltage. When the voltage increases from zero, the energy stored per unit of time also increases from zero. The total energy transferred to the capacitance during a given time will be determined by the average voltage. Since the voltage increases uniformly with constant charge current, the average voltage is one-half the final voltage.

Where the final voltage is V after a time period t seconds, as in Fig. 8-13, the average voltage will be

$$V_{av} = \frac{V}{2}$$

Using Eq. (2-6), the charging energy will be

$$W = V_{av}Q = \frac{V}{2}Q$$

Substituting for Q from Eq. (8-9) gives

$$W = \frac{V}{2}VC$$

$$W = \frac{V^2C}{2} \qquad (8\text{-}16)$$

where W = energy stored in the capacitance, J
V = pd, V
C = capacitance, F

Example 8-5 A 5-μF capacitor is charged until its terminal pd is 100 V. What is the amount of energy stored in it?

Solution (Fig. 8-14)

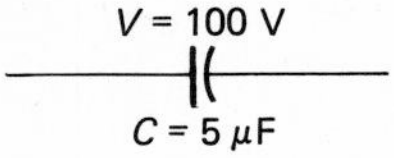

Fig. 8-14 Example 8-5: data.

$$W = \frac{V^2C}{2} \tag{8-16}$$

$$W = \frac{(100 \text{ V})^2 \times 5 \ \mu\text{F}}{2} = \mathbf{0.025 \ J} \quad \text{or} \quad \mathbf{25 \ mJ}$$

Example 8-6 An electronic power supply has a 30 000-μF capacitor in its filter circuit. If it is charged by a constant current of 100 mA until it stores 50 J of energy, (*a*) what is the capacitor pd at the end of the charge and (*b*) how much time does the charging process take?
Solution (Fig. 8-15)

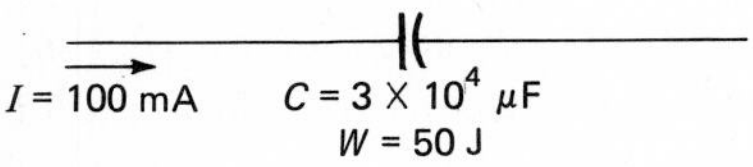

Fig. 8-15 Example 8-6: data.

a.

$$W = \frac{V^2C}{2} \tag{8-16}$$

$$V = \sqrt{\frac{2W}{C}} = \sqrt{\frac{2 \times 50 \text{ J}}{3 \times 10^{-2} \text{ F}}} = \mathbf{57.74 \ V}$$

b.

$$W = VQ$$

$$Q = \frac{W}{V} = \frac{50 \text{ J}}{57.74 \text{ V}} = 0.866 \text{ C}$$

$$Q = It$$

$$t = \frac{Q}{I} = \frac{0.866 \text{ C}}{0.1 \text{ A}} = \mathbf{8.66 \ s}$$

Example 8-7 Figure 8-16 is the circuit of a basic ramp generator (a generator of sawtooth voltages). During the 10-ms periods that the input voltage v_i is negative, the output voltage v_o (which is also the capacitor voltage) rises from approximately zero toward the 25 V of the supply. As v_o rises, the capacitor charges until its voltage at the end of 10 ms is 5 V. If it can be assumed that the increase in v_o is linear from 0 to 5 V, (*a*) what is the charging current, and (*b*) how much energy is stored in the capacitor at the end of the 10-ms charge?
Solution
a. At the end of the 10-ms period, $v_o = 5$ V. Since the voltage increase is assumed to be linear, the rate of change in voltage will be

$$\frac{dv_o}{dt} = \frac{5 \text{ V}}{10 \text{ ms}} = 500 \text{ V/s}$$

For a linear rise in voltage the charging current is constant. From Eq. (8-15),

$$I = C\frac{dv_o}{dt} = 0.1 \ \mu\text{F} \times 500 \text{ V/s} = \mathbf{50 \ \mu A}$$

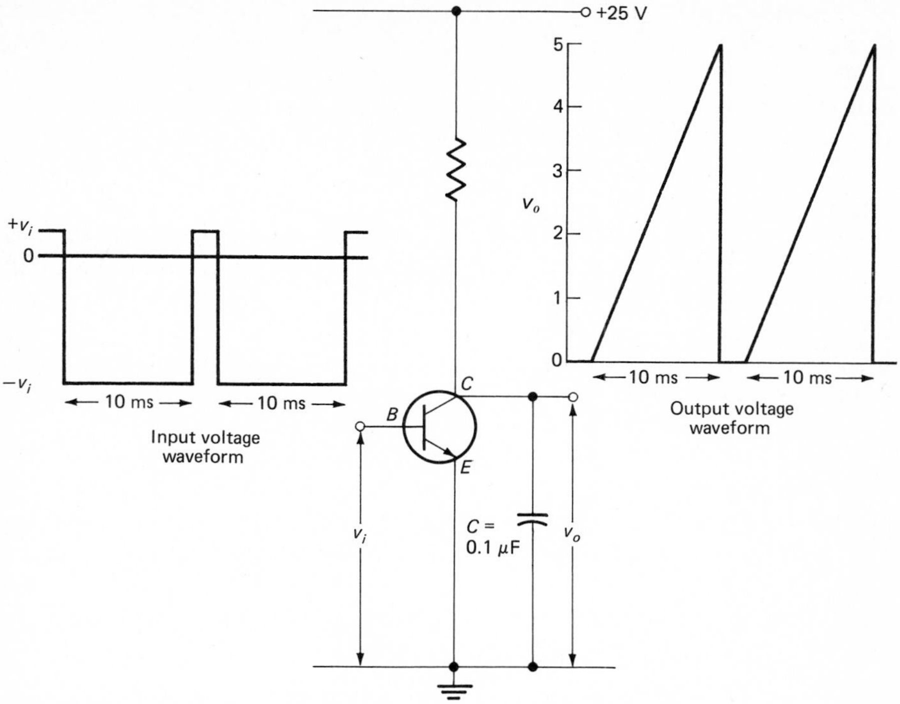

Fig. 8-16 Example 8-7: circuit and input and output voltage waveforms.

b. The energy stored will increase over the 10-ms period, becoming, from Eq. (8-16),

$$W = \frac{V_o^2 C}{2} = \frac{(5\ \text{V})^2 \times 0.1\ \mu\text{F}}{2} = \mathbf{1.25\ \mu J}$$

Ramp generators are widely used to provide constant-velocity electron-beam deflection in cathode-ray tubes. The beam motion causes a spot of light to move steadily across the screen of the tube, usually so fast that it appears to be a line of light. This line of light is called a *trace*.

In an oscilloscope the trace forms the horizontal motion on which is imposed a vertical motion related to the waveform being viewed. The horizontal and vertical motions result in a *display* of the waveform shape.

A radar trace moves from the center of the screen to its edge while being rotated. The screen is totally illuminated by the rotating radial trace. The general illumination is maintained at a low level. A *target* (an object picked up by the radar) causes bright illumination at a point identifying its location.

For television, both horizontal and vertical traces are developed by ramp generators. The combination of these traces results in total screen illumination, the intensity of illumination being controlled to produce the picture.

In these examples of the application of ramp generators, ramp linearity (constant rate of amplitude increase) is desired. The accuracy of waveform presentation, and the resulting measurement accuracy, in an oscilloscope is dependent on the constancy of motion of the horizontal trace. In radar the accuracy of target location can be a function of trace linearity. Television pictures appear to be the wrong shape if the traces are nonlinear.

The charging current will not be constant in a simple circuit such as that in Fig. 8-16, so a cathode-ray tube trace resulting from the ramp output will not have constant velocity. This problem is inherent in charging a capacitor through a resistor and is the subject of Chap. 10. For the practical use of ramp generators, the problem is solved by employing one of two general techniques:

1. If the capacitor charge is limited to a small portion of the source voltage, the ramp can be sufficiently linear for some purposes. For example, it is a matter of personal judgment whether a slightly distorted television picture is tolerable. In Example 8-7 the charge is limited to 20 percent of the available 25 V.
2. Where maximum linearity is required, a constant-current charging source is used. Electronic constant-current sources are commercially designed and constructed.

8-8 CAPACITANCE NETWORKS

All network principles apply to capacitive networks. Obviously these principles must be interpreted in accordance with the capacitance relationships dealt with in Sec. 8-4.

PARALLEL CAPACITANCES

Figure 8-17*a* shows three capacitances in parallel. At a given time instant the capacitances have a common voltage. Over a given time period they experience the same voltage change. Therefore the rate of change of voltage dv/dt is the same for all three capacitances.

The three capacitances in parallel at any instant share a total instantaneous current i_T. The single capacitance C_T in Fig. 8-17*b* is intended to be the equivalent of the parallel combination.

Fig. 8-17 (*a*) Capacitances in parallel; (*b*) single capacitance equivalent.

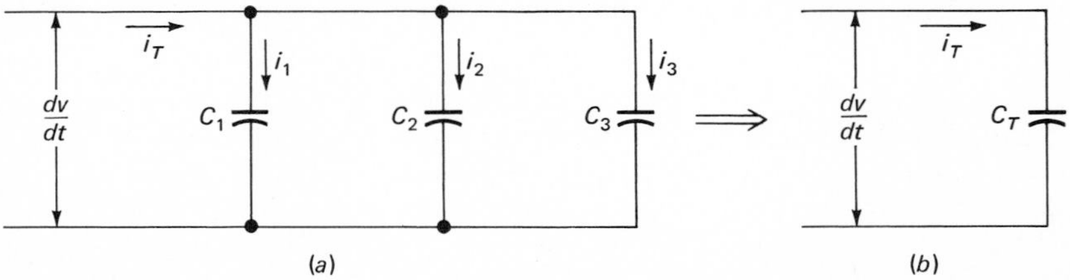

Applying Kirchhoff's current law to Fig. 8-17*a*,

$$i_T = i_1 + i_2 + i_3$$

Substituting the capacitive current expression of Eq. (8-15), we obtain

$$C_T \frac{dv}{dt} = C_1 \frac{dv}{dt} + C_2 \frac{dv}{dt} + C_3 \frac{dv}{dt}$$

Therefore,

$$C_T = C_1 + C_2 + C_3$$

Although this derivation has been applied only to the three capacitances in Fig. 8-17, it will be clear that it can be extended to any number. So, for parallel capacitances

$$C_T = C_1 + C_2 + C_3 + \cdots + C_n \tag{8-17}$$

Example 8-8 Four capacitors, 0.1, 0.22, 0.47, and 0.68 μF, are connected in parallel. What is (*a*) the combined capacitance and (*b*) the total stored energy if the circuit voltage is 500 V?

Solution (Fig. 8-18)

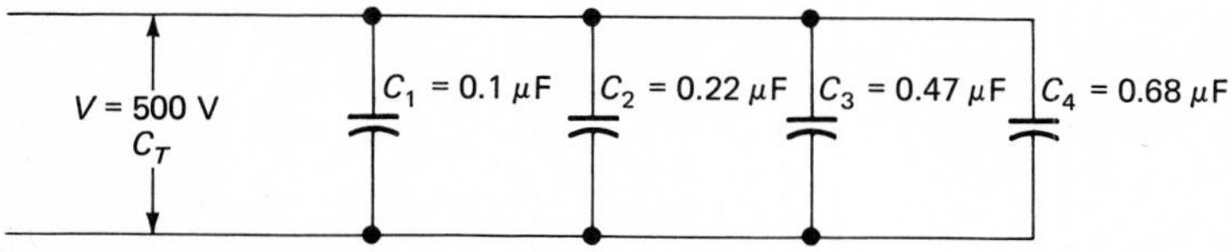

Fig. 8-18 Example 8-8: data.

a. $C_T = C_1 + C_2 + C_3 + C_4$ (8-17)

$C_T = 0.1\ \mu\text{F} + 0.22\ \mu\text{F} + 0.47\ \mu\text{F} + 0.68\ \mu\text{F} = \mathbf{1.47\ \mu F}$

b. $W_T = \dfrac{V^2 C_T}{2}$ (8-16)

$W_T = \dfrac{(500\ \text{V})^2 \times 1.47\ \mu\text{F}}{2} = \mathbf{0.1838\ J}$

In Example 8-8, the energy stored in each capacitor is

$$W_1 = 0.0125\ \text{J}$$

$$W_2 = 0.0275\ \text{J}$$

$$W_3 = 0.0588\ \text{J}$$

$$W_4 = 0.085\ \text{J}$$

Adding these individual energy values gives

$$W_T = 0.1838 \text{ J}$$

This summation gives the same value for W_T as in the example solution, demonstrating the fact that, when capacitances are connected in parallel, the total energy stored is the sum of that stored by each individual capacitance.

SERIES CAPACITANCES

The three series capacitances in Fig. 8-19*a* have the common instantaneous current *i*. At any time instant the total voltage across the series network is the sum of the voltages across each capacitance. Over a given time period the total voltage change will be the sum of the individual capacitance voltage changes. Therefore the rate of change in total voltage will be the sum of the rates of change in the individual capacitance voltages. This is the rate-of-change view of Kirchhoff's voltage law which, in equation form, is

$$\frac{dv_T}{dt} = \frac{dv_1}{dt} + \frac{dv_2}{dt} + \frac{dv_3}{dt}$$

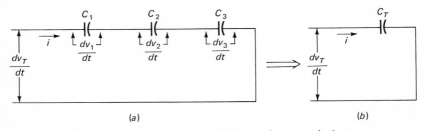

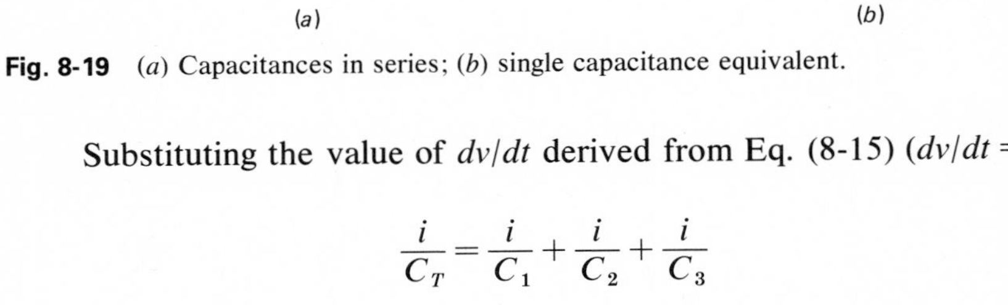

Fig. 8-19 (*a*) Capacitances in series; (*b*) single capacitance equivalent.

Substituting the value of dv/dt derived from Eq. (8-15) ($dv/dt = i/C$),

$$\frac{i}{C_T} = \frac{i}{C_1} + \frac{i}{C_2} + \frac{i}{C_3}$$

or

$$\frac{1}{C_T} = \frac{1}{C_1} + \frac{1}{C_2} + \frac{1}{C_3}$$

This equation can be applied to any number of capacitances connected in series:

$$\frac{1}{C_T} = \frac{1}{C_1} + \frac{1}{C_2} + \frac{1}{C_3} + \cdots + \frac{1}{C_n} \tag{8-18}$$

In applying Eqs. (8-17) and (8-18) to find the total capacitance of a parallel or a series capacitive network, it is only necessary to remember that the rules are opposite those which apply to resistance.

It is interesting to reflect on why it is that capacitances are treated in a manner opposite that in which resistances are treated when totaling their values. Capacitances in parallel are added together; extra parallel capacitances increase the total capacitance, and hence increase the charging current. Conductances in parallel are added together; extra parallel conductances increase the total circuit conductance, and hence increase the current. By relating capacitance and conductance in this manner, it becomes clear that, when total values are considered, the association between capacitance and resistance is one of opposites.

Example 8-9 Repeat Example 8-8 with the capacitances connected in series.
Solution (Fig. 8-20)

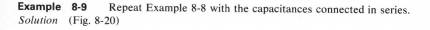

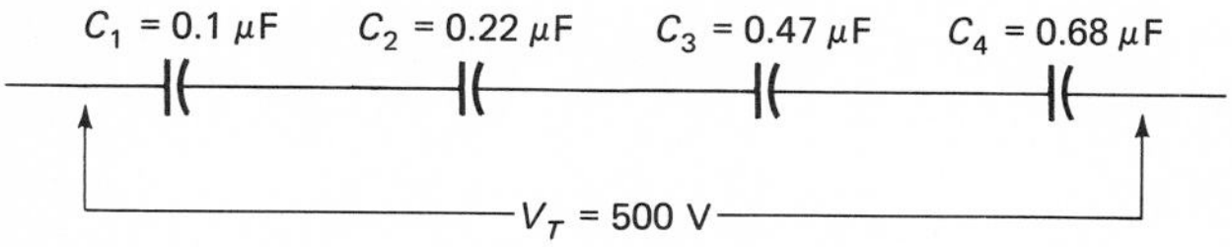

Fig. 8-20 Example 8-9: data.

a.

$$\frac{1}{C_T}=\frac{1}{C_1}+\frac{1}{C_2}+\frac{1}{C_3}+\frac{1}{C_4} \tag{8-18}$$

$$=\frac{1}{0.1\ \mu\text{F}}+\frac{1}{0.22\ \mu\text{F}}+\frac{1}{0.47\ \mu\text{F}}+\frac{1}{0.68\ \mu\text{F}}$$

$$\frac{1}{C_T}{}^{*}=10+4.545+2.128+1.471=18.14$$

$$C_T=\mathbf{0.05513\ \mu F}$$

b. From Eq. (8-16),

$$W_T=\frac{V^2C_T}{2}=\frac{(500\text{ V})^2\times 0.05513\ \mu\text{F}}{2}=\mathbf{6.891\ mJ}$$

Equations (8-9) and (8-16) can be combined to obtain an expression for capacitive energy in terms of charge:

$$W=\frac{V^2C}{2} \tag{8-16}$$

$$W=\frac{V^2C^2}{2C}=\frac{(VC)^2}{2C}$$

* The reciprocal of μF is MF^{-1}, an awkward unit of little practical value. It was omitted to avoid confusion.

Substituting the expression $Q = VC$ from Eq. (8-9),

$$W = \frac{Q^2}{2C} \tag{8-19}$$

Since the current is the same throughout a series circuit, the instantaneous charge ($q = it$) on each capacitance will be the same. Further, since the total equivalent capacitance must have the same current for a given voltage, the total charge must also be the same as the charge held by each individual capacitance.

By taking the total voltage V_T and the total capacitance C_T, the value of the charge on each capacitance in Example 8-9 can be obtained:

$$Q = V_T C_T = 500 \text{ V} \times 0.05513\ \mu\text{F} = 27.57\ \mu\text{C}$$

By using this value of charge in Eq. (8-19), the energy stored in each capacitance of Example 8-9 is found to be

$$W_1 = 3.798 \text{ mJ}$$

$$W_2 = 1.726 \text{ mJ}$$

$$W_3 = 0.8082 \text{ mJ}$$

$$W_4 = 0.5587 \text{ mJ}$$

When these values of stored energy are added, they total 6.891 mJ, the same as the value for the total energy stored in the network. This result leads to the conclusion that, when capacitances are connected in series, the total energy stored is the sum of that stored by each individual capacitance.

It is a general axiom that, for any circuit configuration, the total energy stored in a group of capacitances is the sum of the energies stored in each capacitance.

Example 8-10 The voltage applied to two series capacitors, 0.5 μF and 0.25 μF, changes from 60 to 300 V at a uniform rate in 1 ms. What is the rate of change in voltage across each capacitor?

Solution (Fig. 8-21)

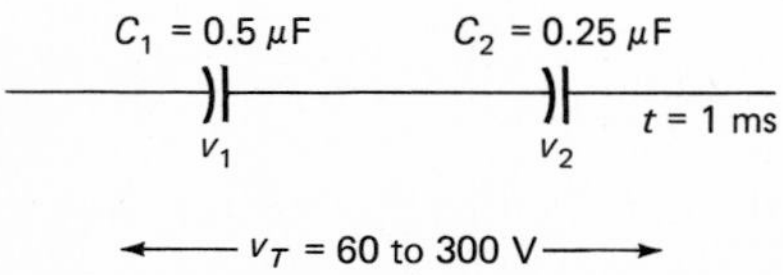

Fig. 8-21 Example 8-10: data.

From Eq. (8-9), the relationship between voltage and capacitance can be seen to be

$$q = Cv$$

$$v = \frac{q}{C} \quad \text{or} \quad v \propto \frac{1}{C} \quad \text{(voltage is inversely proportional to capacitance)}$$

Therefore,

$$\frac{v_1}{v_2} = \frac{C_2}{C_1} = \frac{0.25\ \mu\text{F}}{0.5\ \mu\text{F}} = 0.5$$

or

$$v_2 = 2v_1$$

But

$$v_1 + v_2 = v_T$$

$$v_1 + 2v_1 = v_T$$

$$v_1 = \frac{v_T}{3} \quad \text{and} \quad v_2 = \frac{2v_T}{3}$$

When $v_T = 60$ V,

$$v_1 = \frac{v_T}{3} = \frac{60\ \text{V}}{3} = 20\ \text{V}$$

$$v_2 = \frac{2v_T}{3} = \frac{2 \times 60\ \text{V}}{3} = 40\ \text{V}$$

When $v_T = 300$ V,

$$v_1 = \frac{v_T}{3} = \frac{300\ \text{V}}{3} = 100\ \text{V}$$

$$v_2 = \frac{2v_T}{3} = \frac{2 \times 300\ \text{V}}{3} = 200\ \text{V}$$

As v_T changes at a uniform rate from 60 to 300 V in 1 ms, v_1 changes at a uniform rate from 20 to 100 V in 1 ms, and so the rate of change in v_1 is

$$\frac{dv_1}{dt} = \frac{\Delta v_1}{\Delta t} = \frac{100\ \text{V} - 20\ \text{V}}{1\ \text{ms}} = \mathbf{80\ kV/s}$$

v_2 changes from 40 to 200 V uniformly in 1 ms, so the rate of change in v_2 will be

$$\frac{dv_2}{dt} = \frac{\Delta v_2}{\Delta t} = \frac{200\ \text{V} - 40\ \text{V}}{1\ \text{ms}} = \mathbf{160\ kV/s}$$

In Example 8-10, the sum of the rates of change in voltage across each capacitor ($dv_1/dt + dv_2/dt$) is 240 kV/s. Taking the whole circuit, the rate of change in voltage is

$$\frac{dv_T}{dt} = \frac{300\ \text{V} - 60\ \text{V}}{1\ \text{ms}} = 240\ \text{kV/s}$$

This equality of the rate of change in total voltage and the sum of the individual rates of change in voltage substantiates the previous general statement to this effect.

It should be noted that the rate of change in a quantity can be expressed in terms of the specific change values only if the manner of change is known with certainty. Such expression is most easily accomplished if the rate of change is uniform, the quantities involved (voltage and time, for instance) having a linear (directly proportional) relationship.

Example 8-11 What is the total capacitance of the two-element series network in Fig. 8-22?

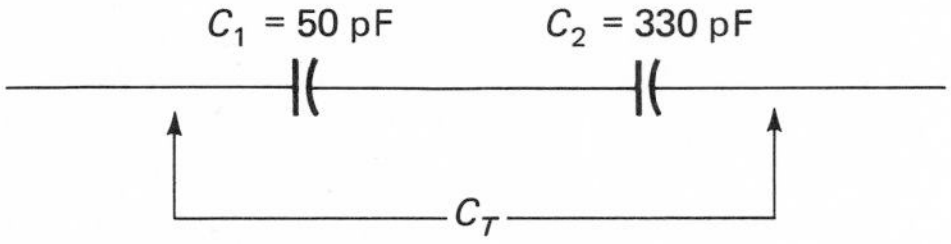

Fig. 8-22 Example 8-11: data.

Solution

$$\frac{1}{C_T}=\frac{1}{C_1}+\frac{1}{C_2} \tag{8-18}$$

$$\frac{1}{C_T}=\frac{C_2+C_1}{C_1C_2}$$

$$C_T=\frac{C_1C_2}{C_1+C_2}=\frac{50\text{ pF}\times 330\text{ pF}}{50\text{ pF}+330\text{ pF}}=\mathbf{43.42\text{ pF}}$$

Example 8-11 shows that the relationship for the total capacitance of two capacitances in series is basically the same as that for the total resistance of two resistances in parallel. They are both given by the product divided by the sum.

Example 8-12 What value of capacitance should be connected in series with a 16-μF filter capacitor to produce a total capacitance of 4 μF?
Solution (Fig. 8-23)

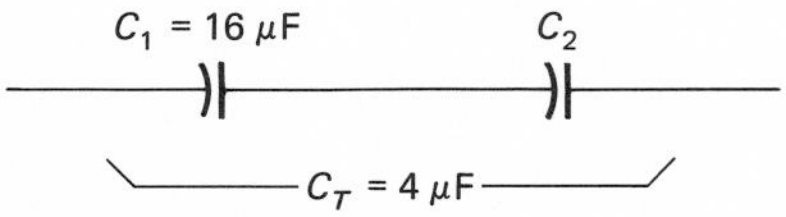

Fig. 8-23 Example 8-12: data.

$$\frac{1}{C_T}=\frac{1}{C_1}+\frac{1}{C_2} \tag{8-18}$$

$$\frac{1}{C_2}=\frac{1}{C_T}-\frac{1}{C_1}=\frac{C_1-C_T}{C_1C_T}$$

$$C_2=\frac{C_1C_T}{C_1-C_T}=\frac{16\ \mu\text{F}\times 4\ \mu\text{F}}{16\ \mu\text{F}-4\ \mu\text{F}}=\mathbf{5.333\ \mu F}$$

Example 8-13 What is the total capacitance of three capacitances, each 0.015 μF, connected (*a*) in parallel and (*b*) in series?
Solution
a. Refer to Fig. 8-24*a*,

$$C_T=C_1+C_2+C_3 \tag{8-17}$$

$$=0.015\ \mu\text{F}+0.015\ \mu\text{F}+0.015\ \mu\text{F}$$

$$C_T=3\times 0.015\ \mu\text{F}=\mathbf{0.045\ \mu F}$$

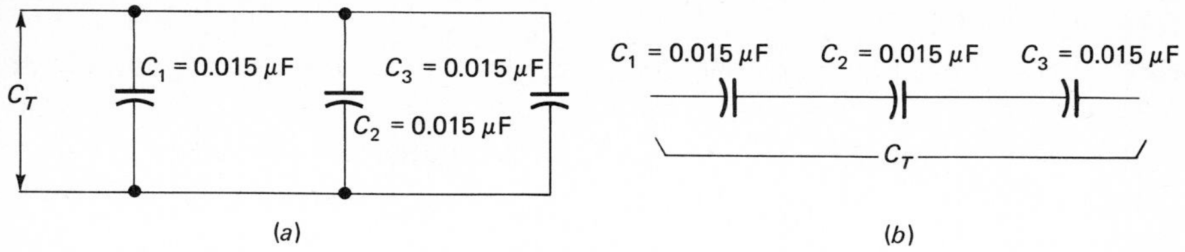

Fig. 8-24 Example 8-13: given data. (*a*) Equal capacitances connected in parallel: (*b*) equal capacitances connected in series.

b. Refer to Fig. 8-24*b*,

$$\frac{1}{C_T} = \frac{1}{C_1} + \frac{1}{C_2} + \frac{1}{C_3} \tag{8-18}$$

$$\frac{1}{C_T} = \frac{1}{0.015\ \mu\text{F}} + \frac{1}{0.015\ \mu\text{F}} + \frac{1}{0.015\ \mu\text{F}} = \frac{3}{0.015\ \mu\text{F}}$$

$$C_T = \frac{0.015\ \mu\text{F}}{3} = \mathbf{0.005\ \mu F}$$

In Example 8-13 it was demonstrated that

1. When several equal-value capacitances are connected in parallel, the total capacitance is the value of any one of them multiplied by the number of capacitances.
2. When several equal-value capacitances are connected in series, the total capacitance is the value of any one of them divided by the number of capacitances.

SERIES-PARALLEL CAPACITANCES

As in the case of resistive circuits, the principles of series and parallel capacitances are applied to series-parallel circuits as the particular situation demands.

8-9 STRAY CAPACITANCE

The impression may have been created that capacitance occurs *only* between parallel plates. As pointed out at the beginning of Sec. 8-1, capacitance occurs whenever there is a voltage between points or surfaces.

Commercial capacitors (devices specifically manufactured to exhibit capacitance) are usually of the parallel-plate form, but capacitance is not restricted to capacitors and always exists in circuits. Unintentional capacitance is usually called *stray capacitance.* Stray capacitances can often be ignored as insignificant in the circuit operation. Equally often such stray capacitances are significant. Whether or not stray capacitance is important depends on the circuit.

STRAY CAPACITANCE IN DEVICES

Resistors, inductors, transformers, vacuum tubes, transistors, and other circuit devices have capacitances associated with them. These capacitances, together with varying voltages, cause undesired currents. Two examples, resistors and transistors, will be briefly described in terms of their stray capacitances.

When there is a current I in the resistor in Fig. 8-25*a*, there is a potential difference V across it. Thus there is an electric field along the resistor's length, and so there is capacitance between its ends. For most commercial resistors the capacitance value is very small, fractions of a picofarad. The effect of so small a capacitance is not great, except at very high ac frequencies.

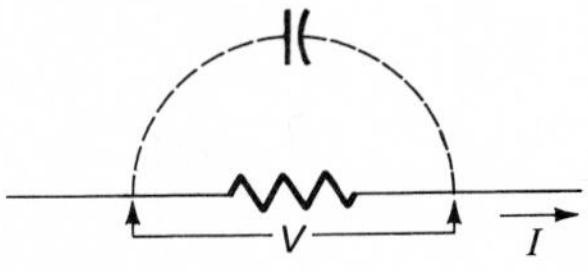

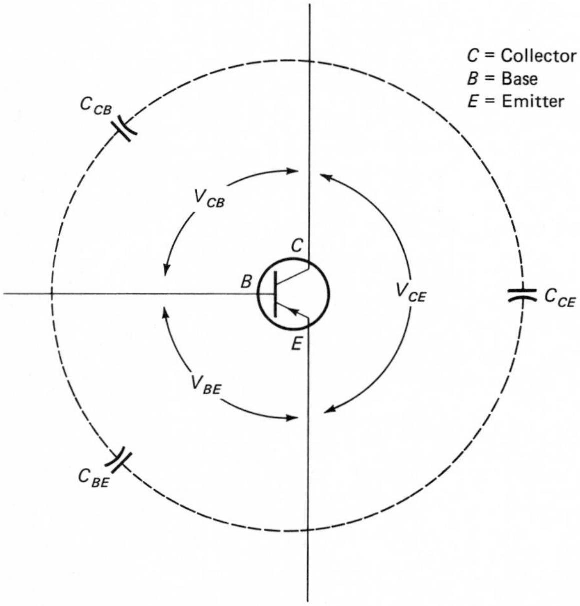

Fig. 8-25 Stray capacitances. (*a*) Resistor; (*b*) transistor.

The operating internal voltages of a transistor are shown in Fig. 8-25*b*. Electric fields exist between the collector, base, and emitter, so capacitances exist there too. They are the collector-to-emitter capacitance C_{CE}, the collector-to-base capacitance C_{CB}, and the base-to-emitter capacitance C_{BE}. Transistor capacitances, although only a few picofarads, can significantly affect circuit performance. These capacitances can be so important that transistor manufacturers specify them in their published data, and circuit designers must determine their effects on equipment performance.

TRANSMISSION LINE AND CABLE CAPACITANCE

Electrical energy is usually transmitted by metallic conductors insulated from each other or from ground. There will always be capacitance in such transmission systems, its effect being small or great depending on the circumstances.

In Fig. 8-26*a* is shown a power line, or a telephone line, with two conductors suspended above the ground. There will be capacitance between the conductors and between each conductor and ground. As it is common for such lines to be very long, there can be considerable capacitance which will influence the line characteristics.

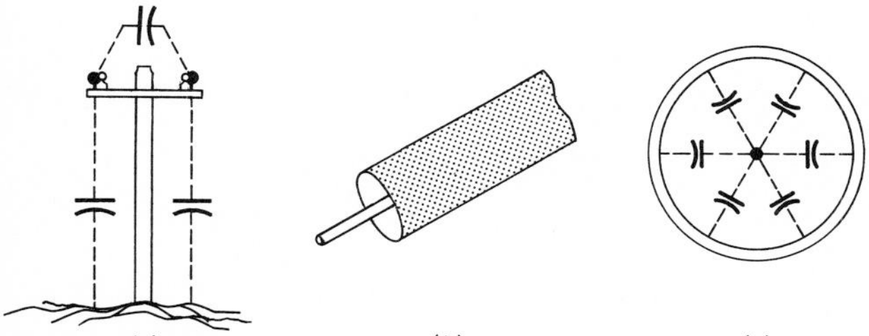

Fig. 8-26 Stray capacitances in lines and cable.

Figure 8-26*b* shows, in pictorial form, a coaxial cable consisting of an inner conductor suspended in the center of an outer tubular conductor. As in Fig. 8-26*c*, there will be capacitance between the two conductors. Coaxial cables are used extensively for high-frequency applications in which capacitance is an important parameter involved in the transmission of the energy.

STRAY CAPACITANCE IN EQUIPMENT

There is always capacitance between conducting surfaces operating at different voltage levels. Therefore, in electrical equipment there is capacitance between all wires, between all conductors on printed circuit boards, between all terminals, between all components, and between all such conducting surfaces and ground. The last example is significant where, as is usual, the equipment is housed in grounded metal enclosures. There are a great many capacitances present in even a simple item of equipment. The circuit designer must be aware of all these capacitances and allow for them in the design.

The significance of stray capacitance will become more apparent during the study of ac circuits in succeeding chapters.

8-10 SUMMARY OF EQUATIONS

Electrostatics

$E = V/s$ (8-1)

$D = Q/A$ (8-2)

$\epsilon = D/E$ or $D = \epsilon E$ (8-3)

$\epsilon = \epsilon_v \epsilon_r$ (8-4)

$\Psi = DA$ (8-5)

$\Psi = Q$ (8-6)

$Q = V(\epsilon A/s)$ (8-7)

$C = \epsilon A/s$ (8-8)

Capacitance

$Q = VC$ (8-9)

$\mathrm{F = C/V}$ (8-10)

$\mathrm{F = A^2 \cdot s^4/kg \cdot m^2}$ (8-11)

$\Delta q = C\ \Delta v$ (8-12)

$\Delta q/\Delta t = C\ \Delta v/\Delta t$ (8-13)

$i = C\ \Delta v/\Delta t$ (8-14)

$i = C\ dv/dt$ (8-15)

$W = V^2C/2$ (8-16)

$C_T = C_1 + C_2 + C_3 + \cdots + C_n$ (8-17)

$1/C_T = 1/C_1 + 1/C_2 + 1/C_3 + \cdots + 1/C_n$ (8-18)

$W = Q^2/2C$ (8-19)

EXERCISE PROBLEMS

Section 8-3

8-1 Find the capacitance of a capacitor made from two strips of foil 2 cm wide and 5 m long using a dielectric of polyethylene 35 μm thick. ($\epsilon_r = 3.3$) (*ans.* 0.0835 μF)

8-2 Find the capacitance of a capacitor made of two disks with diameters of 5.08 cm (2 in.) separated by an air gap of 0.318 cm ($\frac{1}{8}$ in.). ($\epsilon_r = 1.0001$)

8-3 Find the capacitance of a capacitor made of 29 sheets of foil, 1 × 2 cm, separated by 15-μm sheets of mica, with a relative permittivity of 6. Construction is such that both sides of all foil sheets are used except the outside layers. (*ans.* 0.0198 μF)

8-4 A multilayer capacitor to be made from aluminum foil 30 μm thick and sheet Mylar 25 μm thick is to have a capacitance of 0.02 μF ± 5 percent and a dimensional area 2 × 5 cm. Find the minimum number of plates required.

8-5 Two sheets of foil 5 × 5 cm are separated by a Mylar dielectric. If the capacitance is 0.005 μF, what is the thickness of the dielectric? What would be the capacitance if two thickness of the same dielectric were used? (*ans.* 13.28 μm, 0.0025 μF)

Section 8-4

8-6 Find the quantity of charge Q required to produce the given pds across the following capacitors:

a. $C = 1\ \mu F$, $V = 10$ V
b. $C = 0.05\ \mu F$, $V = 50$ V
c. $C = 16\ \mu F$, $V = 300$ V
d. $C = 0.001\ \mu F$, $V = 10$ V
e. $C = 500\ \mu F$, $V = 3$ V
f. $C = 68$ pF, $V = 6$ V
g. $C = 470$ pF, $V = 12$ V
h. $C = 220$ pF, $V = 50\ \mu V$

8-7 What pd would be present across the following capacitors when the charge indicated is present?

a. $C = 0.1\ \mu F$, $Q = 3 \times 10^{-6}$ C (*ans.* 30 V)
b. $C = 220$ pF, $Q = 0.001\ \mu C$ (*ans.* 4.545 V)
c. $C = 15$ pF, $Q = 0.001\ \mu C$ (*ans.* 66.67 V)
d. $C = 2000\ \mu F$, $Q = 0.01$ C (*ans.* 5 V)
e. $C = 10\ 000$ pF, $Q = 1\ \mu C$ (*ans.* 100 V)
f. $C = 0.0001\ \mu F$, $Q = 1$ nC (*ans.* 10 V)

Section 8-5

8-8 A 0.2-μF capacitor being charged from a constant-current source has a pd of 15 V across it after 100 μs. Find the charging current. The capacitor was initially fully discharged.

8-9 Find the value of current from a constant-current source which will charge the 1500-pF capacitor of a linear ramp generator to 0.75 V in 20 μs. (*ans.* 56.25 μA)

8-10 Find the time required to charge a 0.05-μF capacitor from 0 to 20 V from a 15-mA constant-current source.

8-11 Find the time required to increase the voltage from 10 to 12 V across a 20-μF capacitor using a 25-mA constant-current source. (*ans.* 1.6 ms)

8-12 If voltage across a 0.02-μF capacitor is changing at a rate of 50 kV/s, find the instantaneous current.

8-13 If voltage across a 100-pF capacitor is changing at a rate of 10 MV/s, find the instantaneous current. (*ans.* 1 mA)

8-14 If the pd across a 100-pF capacitor is changing at a rate of 10 MV/s, how long will it take for the pd to rise to 20 V from zero?

8-15 A 0.002-μF capacitor discharges at a rate of 100 kV/s for a period of 5 μs. Find the quantity of charge lost by the capacitor during this period. (*ans.* 10^{-9} C)

8-16 A 50-μF capacitor has a pd of 25 V. Find the time required for the capacitor to discharge to 0 V, if the discharge current is constant at 5 mA.

Section 8-7

8-17 In Prob. 8-6, find the energy stored in each capacitor.
(*ans. a.* 50 μJ; *b.* 62.5 μJ; *c.* 0.72 J; *d.* 0.05 μJ; *e.* 2.25 mJ; *f.* 1224 pJ; *g.* 0.0338 μJ; *h.* 275×10^{-21} J)

8-18 In Prob. 8-7, find the energy stored in each capacitor.

8-19 In Prob. 8-8, find the energy stored in the capacitor after

a. 50 μs (*ans.* 5.625 μJ)
b. 100 μs (*ans.* 22.5 μJ)

8-20 In Prob. 8-11, find the energy required to increase the voltage from 10 to 12 V.

8-21 A disk capacitor with a disk separation of 0.318 cm has a capacitance of 5 pF. Find the energy stored, if the capacitor is charged to 2000 V. (*ans.* 10 μJ)

8-22 If after the capacitor in Prob. 8-21 has been charged, the separation of the disks is increased to 0.636 cm without changing the charge Q, find

a. The new value of capacitance.
b. The new voltage across the capacitance.
c. The new quantity of energy stored in the capacitance.

Section 8-8

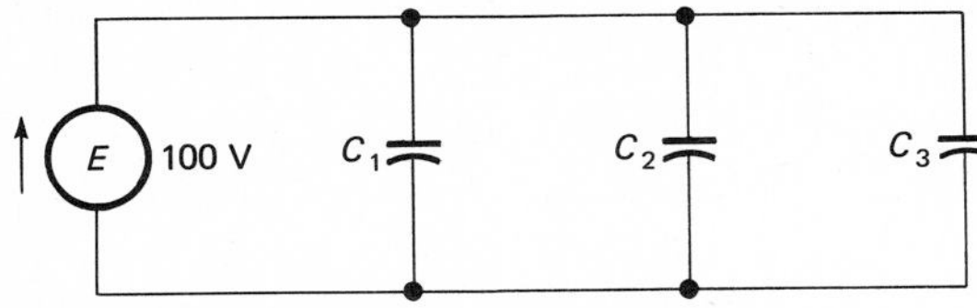

Figure 8-1P Problem 8-23: circuit.

8-23 For the network in Fig. 8-1P find

a. The total capacitance,
b. The total charge stored,
c. The total energy stored,

when

i. $C_1 = 0.5\ \mu F$, $C_2 = 0.5\ \mu F$, $C_3 = 0.5\ \mu F$ (*ans.* 1.5 μF, 150 μC, 7.5 mJ)
ii. $C_1 = 470$ pF, $C_2 = 820$ pF, $C_3 = 220$ pF (*ans.* 1510 pF, 0.151 μC, 7.55 μJ)
iii. $C_1 = 1500$ pF, $C_2 = 0.01\ \mu F$, $C_3 = 0.001\ \mu F$ (*ans.* 0.0125 μF, 1.25 μC, 62.5 μJ)
iv. $C_1 = 15$ pF, $C_2 = 68$ pF, $C_3 = 150$ pF (*ans.* 233 pF, 0.0233 μC, 1.165 μJ)

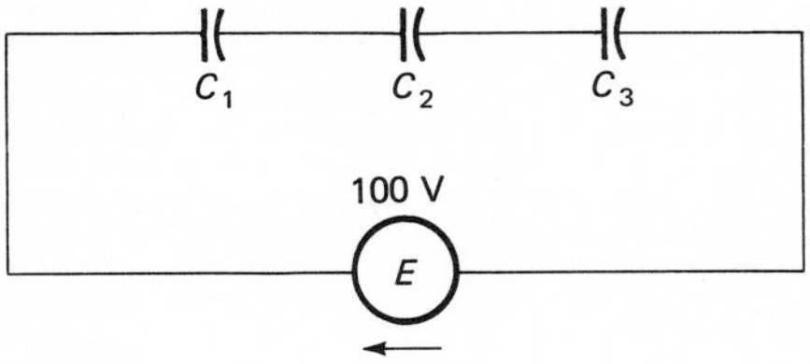

Fig. 8-2P Problem 8-24: circuit.

8-24 Repeat Prob. 8-23 using the network in Fig. 8-2P.

8-25 Using the product-over-sum method, find the total capacitance of two capacitors in series when

a. $C_1 = 0.05\ \mu F$, $C_2 = 0.02\ \mu F$ (*ans.* 0.0143 μF)
b. $C_1 = 680$ pF, $C_2 = 0.001\ \mu F$ (*ans.* 404.8 pF)
c. $C_1 = 220$ pF, $C_2 = 1500$ pF (*ans.* 191.9 pF)
d. $C_1 = 0.0022\ \mu F$, $C_2 = 2200$ pF (*ans.* 1100 pF)
e. $C_1 = 33$ pF, $C_2 = 100$ pF (*ans.* 24.8 pF)

8-26 Two capacitors, $C_1 = 0.05\ \mu F$ and $C_2 = 0.2\ \mu F$, are connected in series across a 50-V source. Find

a. The total capacitance.
b. The total charge stored.
c. The charge stored by C_1.
d. The charge stored by C_2.
e. The pd across C_1.
f. The pd across C_2.

8-27 A capacitive voltage divider is required to sample a 10-kV supply and give an output of 500 V for measurement. If the 500 V is to be read across a 0.1-μF capacitor, what other value of capacitor is required to complete the voltage divider?

(*ans*. 0.00526 pF)

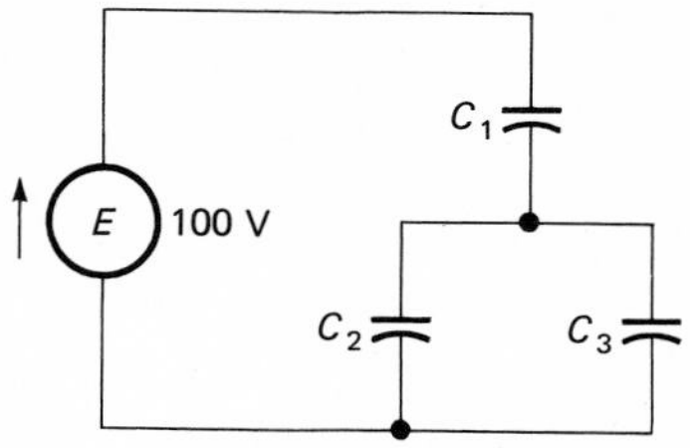

Fig. 8-3P Problem 8-28: circuit.

8-28 Repeat Prob. 8-23 using the network in Fig. 8-3P.

9

Inductive Circuits

Inductance is the dual of capacitance. Like capacitance, inductance can store energy in a force field. It can release this energy for restorage or dissipation. But whereas capacitance stores energy in an electric field associated with voltage, inductance stores energy in a magnetic field dependent on current.

9-1 INDUCTANCE

Whenever there is an electric current, there is a magnetic field. The magnetic field entirely surrounds a current-carrying conductor with a region in which forces may act. These forces can cause magnetized materials, such as iron, to move into alignment with the field direction. They can cause moving charges to change direction of motion, and they can cause charges to move.

The production of a magnetic field around a current-carrying conductor and the effects of this field are called *electromagnetism*. We are concerned with the electric circuit effects of electromagnetism. These effects are of most interest when the current changes value, changing the magnitude of the magnetic field.

Electromagnetic effects are most noticeable when the current-carrying conductor is wound into the form of a coil. In these circumstances, all the magnetic field associated with a long conductor is concentrated into a small space. Such a coil of wire is shown schematically in Fig. 9-1. The switch is open in Fig. 9-1*a*, preventing current in the circuit. When the switch is closed, as in Fig. 9-1*b*, there will be a voltage across the coil and a conducting path through it, but initially there will be no current. The

Fig. 9-1 (*a*) Coil of wire without current; (*b*) coil of wire with current.

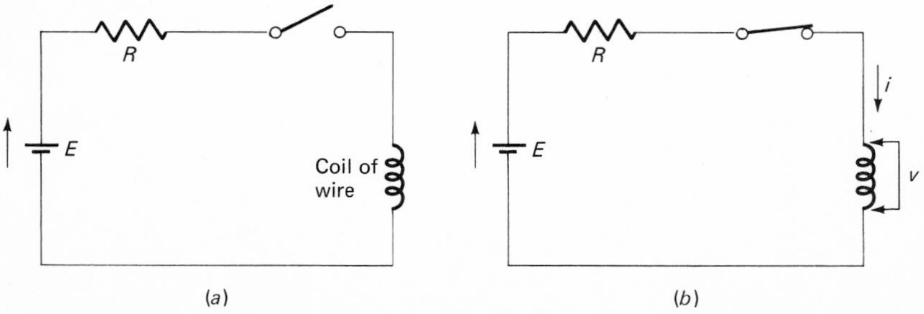

current builds up from zero to a value equal to that which would exist if the resistance R were the only circuit element across the battery. With a current present there is a magnetic field around the coil; it too increases from zero magnitude to a maximum value. The magnetic field thus established is a form of energy which has been transferred from the battery to the coil. This action can be compared with the system in Fig. 8-1, in which energy was transferred from a battery to an electric field between two plates.

	Fig. 8-1	*Fig. 9-1*
Switch open	No current	No current
	No voltage between the plates	No voltage across the coil
	No electric field	No magnetic field
Instant switch closed	Current maximum, equals E/R	No current
	No voltage between the plates	Voltage across the coil maximum, equals E
	No-electric field	No magnetic field
Switch closed but conditions not stabilized	Current decreasing	Current increasing
	Voltage between the plates increasing	Voltage across the coil decreasing
	Electric field increasing	Magnetic field increasing
Switch closed and conditions stabilized	No current	Current maximum, equals E/R
	Voltage between the plates maximum, equals E	Voltage across the coil zero
	Electric field maximum	Magnetic field maximum

There is a duality between what happens in Figs. 8-1 and 9-1. The events in the two circuits are similar but opposite. When conditions have become stabilized, both circuits have maximum energy stored in the form of force fields. The electric field between the capacitor plates is due to maximum voltage. The magnetic field around the coil is due to maximum current.

The ability to store energy in a magnetic field is *inductance*. The coil of wire is an *inductor*. The term "inductance" is based on the ability of a coil of wire to produce an induced emf.

A survey of some electromagnetic quantities including induced emf and inductance leads to a discussion of the behavior of inductive circuits.

9-2 MAGNETIC FIELD QUANTITIES

In developing the background for inductance we are concerned with the production of magnetic fields by electrical means. There is a natural tendency to refer to electromagnetic fields. This urge should be resisted, however, because, in energy propagation, the term "electromagnetic field" has a meaning that is quite different. An electromagnetic field is a force field which has two components simultaneously, one electric and the other magnetic.

The magnetic field of a current-carrying coil of wire is shown in Fig. 9-2*a*. The current produces a magnetic field surrounding the wire along its entire length. Winding the wire into a coil concentrates the magnetic field by confining the total field of a long wire into a small space. Physical forces will be experienced within the field around the coil if magnetic materials or other magnetic fields are introduced or if the intensity of the field is varied.

The significant factors influencing the magnitude of the electromagnetic effects of a coil of wire are indicated in Fig. 9-2*b*. They are the current I, the number of turns of wire N, the length of the coil or the length of the iron core on which it is wound s, the cross-sectional area of the coil or its core A, and a factor determined by the nature of the core material called the permeability μ.

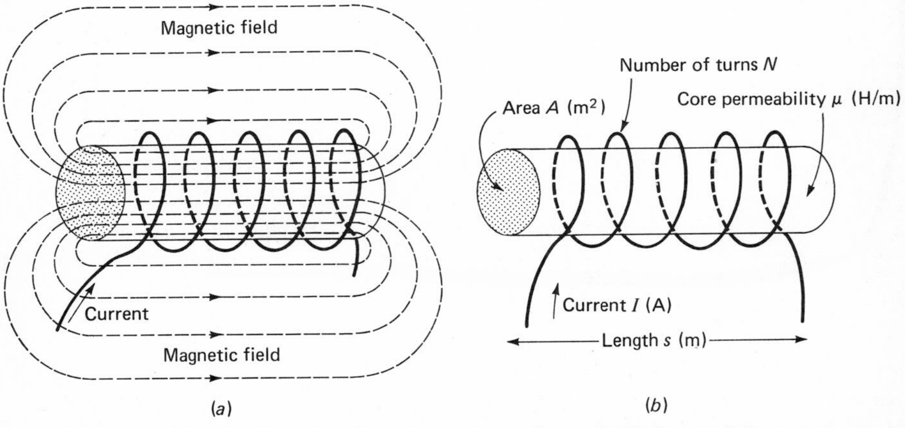

Fig. 9-2 (*a*) Magnetic field surrounding a current-carrying coil; (*b*) factors influencing a magnetic field.

MAGNETOMOTIVE FORCE

Magnetomotive force (mmf) is to a magnetic field what electromotive force (emf) is to an electric field; they are both the basic quantities which determine the field magnitude. A magnetic field is current-dependent and mmf has the ampere as its unit. An electric field is pd-dependent, and emf has the volt as its unit.

Mmf, due to the current in the wire, determines the degree to which electromagnetic effects are produced. Even if the wire is straight (no turns), there is mmf. Forming the wire into a coil increases the mmf, a large number of turns in a given space resulting in a relatively great mmf. This combination of current and turns gives rise to the basic electromagnetic equation

$$F = NI \tag{9-1}$$

where F = mmf, A
I = current, A
N = number of turns of wire

The unit of mmf used to be called the ampere-turn (A · T). While the logic of using this unit is easy to understand, it should be borne in mind that there is no way of expressing the number of turns in the SI-MKSA unit system. The number of turns is a numerical constant with no dimensions (like π). The number of ampere-turns is given in amperes, just as the diameter of a circle πd, is given in meters. References to mmf in ampere-turns are not incorrect, they merely fail to comply with the modern international unit system.

MAGNETIC FIELD STRENGTH

The strength of a magnetic field produced by a current-carrying coil is dependent on the mmf per unit of effective coil length. A short coil has its turns close together, concentrating the field; it has a large mmf per unit length.

$$H = \frac{F}{s} \tag{9-2}$$

where H = field strength A/m
F = mmf, A
s = effective length of the coil, m

This simple relationship assumes that the magnetic field is of constant magnitude throughout the length of the coil. For cylindrical coils of practical length Eq. (9-2) does not quite apply, as the field is somewhat weaker at the ends of the coil. The *effective length* of a coil is the length of a hypothetical coil which is equivalent to the practical coil but which has a constant field magnitude throughout its length. A coil's effective length is less than its actual length.

Combining Eqs. (9-1) and (9-2) gives

$$H = \frac{NI}{s} \tag{9-3}$$

From Eq. (9-3) it can be seen why magnetic field strength is sometimes stated in ampere-turns per meter (A · T/m). As in the case of mmf, it is not incorrect to include turns in the unit. It is just that, by doing so, the international unit system is not maintained.

MAGNETIC FLUX DENSITY

The ability of a magnetic field to produce an effect, such as an induced voltage, is called its *flux density B*. The flux density is influenced by the magnetic field strength H and the nature of the material in the field. The relationship is

$$B = \mu H \tag{9-4}$$

where B = flux density, T (tesla)
H = field strength, A/m
μ = core material permeability, H/m (henry per meter)

PERMEABILITY

The center of a coil is called its *core*. Often the core is made of a magnetic material, an alloy of iron. Sometimes the core is a tube of insulating material on which the wire is wound. Even the tube may be omitted so that the coil becomes entirely self-supporting with nothing but air as the core. The effects produced by the coil's magnetic field are influenced by the nature of the core material. The degree of this influence is called the *permeability* μ of the material (μ is the lowercase Greek letter mu).

In the MKSA system the permeability of a material is given by the product

$$\mu = \mu_v \mu_r \tag{9-5}$$

where μ = permeability, H/m
μ_v = permeability of a vacuum, a physical constant $4\pi \times 10^{-7}$ H/m
μ_r = relative permeability of the material

Relative permeability is the ratio of the flux density with the specified material to the flux density with no material at all, a vacuum. Since μ_r is the ratio of two similar quantities, it has no unit.

Some literature lists the *relative* permeabilities of various materials as their permeabilities. This practice was universal before the adoption of MKSA units. CGS units are still used for magnetic quantities, and in the CGS system μ_v is unity and the total permeability μ is the relative permeability μ_r.

The permeability of most materials, including air, is constant; it does not change with variations in field-strength magnitude. The permeabilities of iron and most of its alloys, however, vary considerably at different field strengths. Also, the relative permeabilities of iron and its alloys are very much higher than those of other materials.

Figure 9-3 shows the shape of a typical *B-H* curve for an alloy of iron.

In Fig. 9-3*a* static values of field strength (H_1, H_2, H_3, and H_4) are projected onto the curve to show the related values of flux density (B_1, B_2, B_3, and B_4). For any pair of values the static permeability is, from Eq. (9-4), given by

$$\mu = \frac{B}{H}$$

The value of μ varies according to the values of *B* and *H* at the various parts of the curve, as indicated in Table 9-1.

Table 9-1 Static permeability (Fig. 9-3*a*)

Part of the B-H curve	B	H	*Static permeability* μ
Lower bend	B_1	H_1	B_1/H_1 (low)
Linear portion	B_2	H_2	B_2/H_2 (high)
Top of linear portion	B_3	H_3	B_3/H_3 (highest)
Horizontal top	B_4	H_4	B_4/H_4 (high)

In Fig. 9-3*b* incremental values of field strength (changes in field strength ΔH_1, ΔH_2, and ΔH_3) are projected onto the *B-H* curve to show the corresponding incremental values of flux density (ΔB_1, ΔB_2, and ΔB_3). For any pair of incremental values, the *incremental permeability* μ_Δ is, from Eq. (9-4), given by

$$\mu_\Delta = \frac{\Delta B}{\Delta H}$$

The values of the incremental permeability also vary for different sections of the *B-H* curve (Table 9-2).

Table 9-2 Incremental permeability (Fig. 9-3*b*)

Part of the B-H curve	ΔB	ΔH	*Incremental permeability* μ_Δ
Lower bend	ΔB_1	ΔH_2	$\Delta B_1/\Delta H_1$ (very low)
Linear portion	ΔB_2	ΔH_2	$\Delta B_2/\Delta H_2$ (high)
Horizontal top	ΔB_3	ΔH_3	$\Delta B_3/\Delta H_3$ (almost zero)

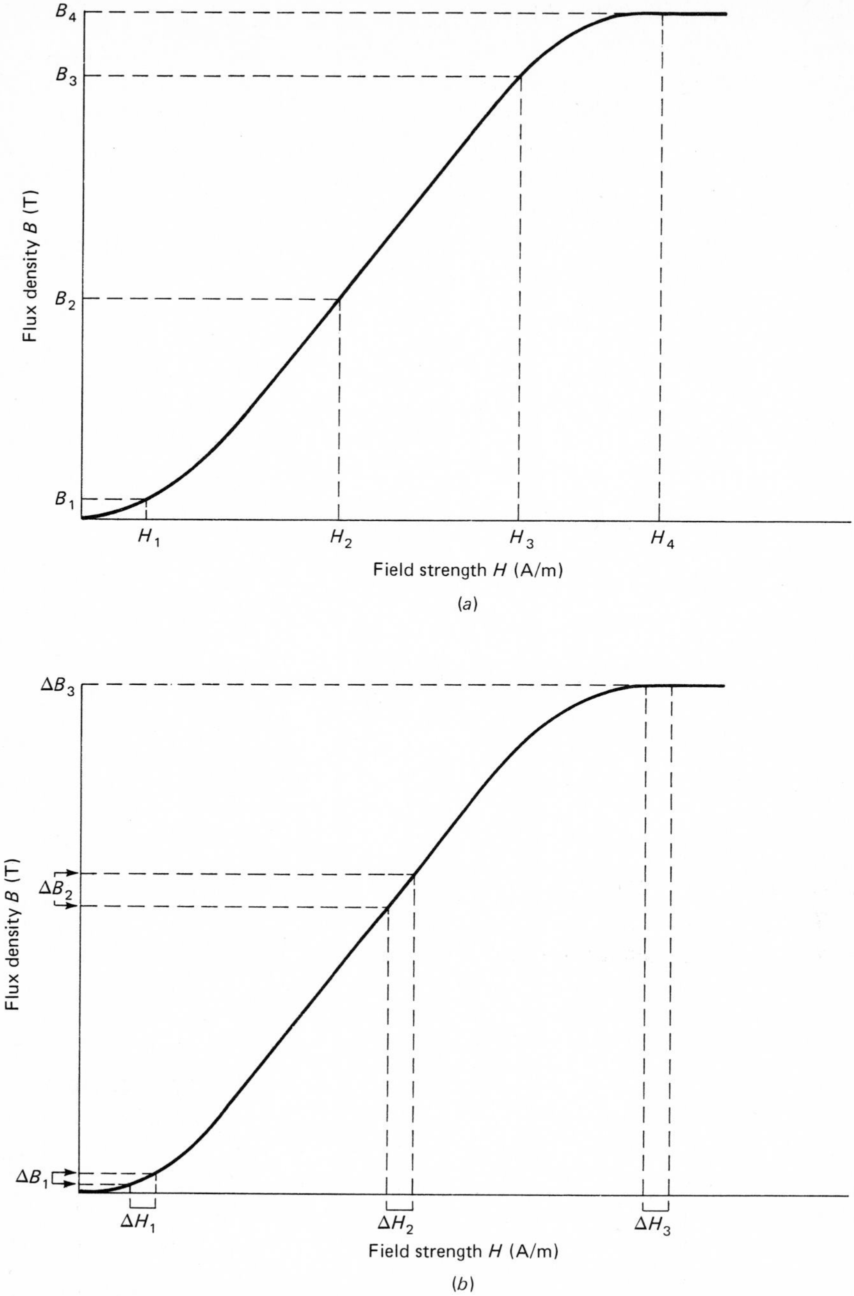

Fig. 9-3 *B-H* curves of iron alloy. (*a*) Showing static values of *B* and *H*; (*b*) showing incremental values of *B* and *H*.

The static and incremental permeabilities are both of practical interest. A device such as a telephone relay (electromagnetic switch) operates through the magnetization of its iron core by direct current. In such a situation the need is for maximum flux density with a steady value of field strength; the static permeability of the core material should be high. The circuit design should ensure that the operating point is high up on the *B-H* curve, near the top of the linear portion, thus giving maximum static permeability. A transformer (used for changing ac voltage) relies for its operation on the variation in its core magnetization produced by an alternating current. The requirement here is for maximum variation in flux density with a varying value of field strength; the incremental permeability of the core material should be high. Design conditions should ensure that operation takes place well above the lower bend of the *B-H* curve. Often it is necessary to operate in the linear portion of the curve to obtain a linear relationship between B and H (constant μ_Δ).

It should be understood that, in applying the magnetic field relationships which follow, the value of μ will depend on the material type, the type of circuit current (dc, ac, ac and dc), and the actual levels of B and H being considered.

MAGNETIC FLUX

The flux density is the concentration of magnetic effect per unit of field area. The total effect over the entire field area is the flux Φ (capital Greek letter phi),

$$\Phi = BA \tag{9-6}$$

where Φ = total flux, Wb (Weber)
B = flux density, T
A = cross-sectional area of the field, m^2

9-3 ELECTROMAGNETIC INDUCTION

Whenever there is relative motion between a magnetic field and a conductor, an emf is induced between the ends of the conductor. The magnitude of the induced emf is proportional to the rate of change in flux linkage.

This statement is Faraday's law of electromagnetic induction. It forms the principle on which inductance and generator action are based.

There are four fundamental methods for achieving relative motion between a magnetic field and a conductor. These are shown pictorially in Fig. 9-4.

In Fig. 9-4*a* a stationary magnetic field is produced by a steady direct current I in a coil which does not move. A conductor is moved across the coil's magnetic field, and an emf e is induced between the ends of the conductor.

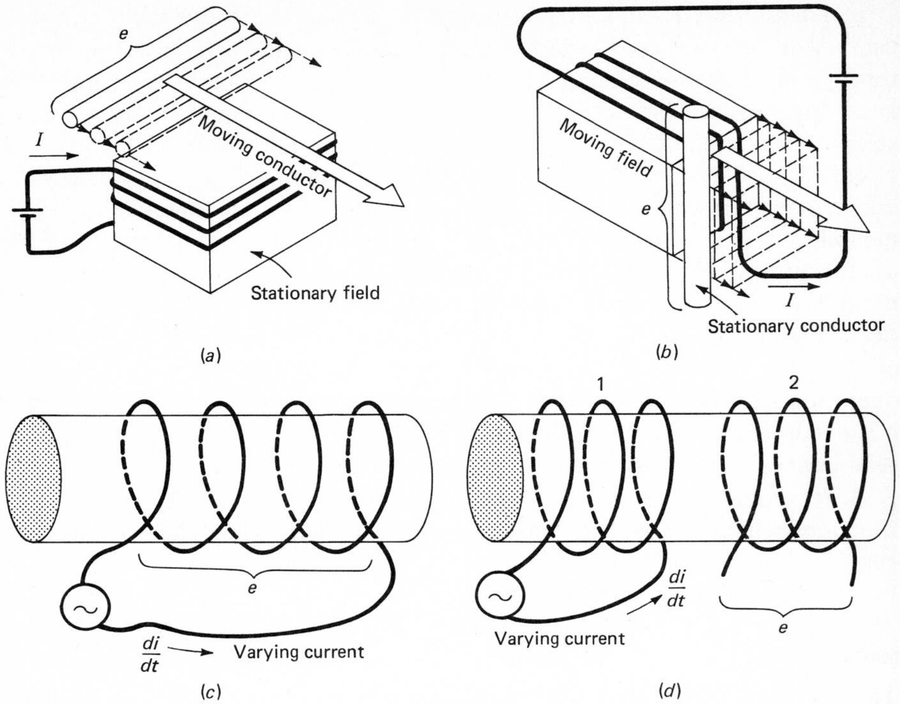

Fig. 9-4 Relative motion between a conductor and a magnetic field.

Figure 9-4*b* shows a situation similar to that in Fig. 9-4*a*, except that here the conductor is held stationary and the coil, with its magnetic field, is moved so that the field cuts across the conductor. Again an emf *e* is induced between the ends of the conductor.

In Fig. 9-4*c* a coil carries a varying current di/dt. This coil itself is a conductor in the varying magnetic field caused by its own current. Relative motion occurs here because the coil is in a field of varying intensity. An emf *e* is induced between its ends.

In Fig. 9-4*d* two coils are wound on a common core. The varying intensity field produced by the varying current di/dt in coil 1 links with the turns of coil 2, so inducing an emf *e* between its ends. Coil 1 also has an induced emf due to its own varying field strength, as in Fig. 9-4*c*.

The methods of producing the induced voltages in Fig. 9-4*a* and *b* form the principle on which many electric generators operate.

In Fig. 9-4*c* and *d* we have the principles of self- and mutual induction. These principles are of great importance in electric circuits. They will be discussed in this chapter and in some succeeding chapters.

The magnitude of the induced emf in each of the four examples of relative motion is given by Faraday's law. Naturally, since Fig. 9-4*a* and *b* deals with actual movement and Fig. 9-4*c* and *d* refers to motion in terms of varying field intensity, the interpretation of Faraday's law is significant.

In the study of electric circuit behavior we are concerned with the self- and mutual induction cases in Fig. 9-4*c* and *d*. Here the interpretation of the induced emf magnitude statement can be put in the form of an equation:

$$|e| = \frac{d\Lambda}{dt} \tag{9-7}^*$$

where $|e|$ = magnitude of the induced emf, V
Λ = flux linkage, Wb
$d\Lambda/dt$ = rate of change in flux linkage, Wb/s

Λ is the capital Greek letter lambda.

A flux linkage is the association of 1 Wb of magnetic flux and one conductor. Therefore, where N conducting turns act together in a magnetic field of flux Φ webers, the total flux linkage is

$$\Lambda = N\Phi \qquad \text{webers} \tag{9-8}$$

The unit for flux linkage is not influenced by the number of turns because, as previously mentioned, this is a dimensionless number for any coil.

9-4 SELF-INDUCTANCE

Equation (9-7) can be rewritten:

$$|e| = \frac{d(N\Phi)}{dt} \tag{9-9}$$

This expression can be written in terms of the self-induced emf in Fig. 9-4*c* by substitution from Eqs. (9-3), (9-4), and (9-6).

Substituting BA for Φ from Eq. (9-6),

$$|e| = \frac{d(NBA)}{dt}$$

Substituting μH for B from Eq. (9-4),

$$|e| = \frac{d(N\mu HA)}{dt}$$

* A quantity symbol enclosed by vertical lines (as in $|e|$) represents the magnitude or scalar value of the quantity. This notation is only used in this text when there is a special need to identify a magnitude. Here we must distinguish between the magnitude of an induced emf (Faraday's law) and its polarity (Lenz' law, Sec. 9.5).

Substituting NI/s for H from Eq. (9-3),

$$|e| = \frac{d(N^2 I \mu A/s)}{dt} \tag{9-10}$$

If the coil's physical characteristics are considered constant, the only variable is current. Equation (9-10) can be written:

$$|e| = \frac{N^2 \mu A}{s} \frac{di}{dt} \tag{9-11}$$

Self-inductance may be defined as the constant of proportionality between the emf induced across a coil and the rate of change in current through it. In Eq. (9-11) the total constant by which di/dt is multiplied is this constant of proportionality. Therefore the inductance of a coil is given by

$$L = \frac{N^2 \mu A}{s} \tag{9-12}$$

Equations (9-1) to (9-12) assume that the magnetic field is concentrated in the core of the coil.

THE HENRY

Equation (9-11) can be written in the form

$$|e| = L \frac{di}{dt} \tag{9-13}$$

In Eqs. (9-10) to (9-13)

$|e|$ = magnitude of the induced emf, V
L = inductance, H
N = number of turns
I = current, A
μ = core permeability, H/m
A = core area, m^2
s = effective coil length, m
di/dt = rate of change in current, A/s

Equation (9-13) is one of the more important electric circuit relationships.

Rearranging the terms of Eq. (9-13) gives the definition of the henry, the MKSA unit of inductance:

$$L = \frac{|e|}{di/dt}$$

An inductance of 1 H exists when an induced emf of 1 V is associated with a rate of change in current of 1 A/s.

THE MKSA EXPRESSION FOR INDUCTANCE

The henry can be expressed in fundamental MKSA units. From Eq. (9-13),

$$L = \frac{|e|}{di/dt} = |e|\,\frac{dt}{di}$$

This can be written in unit form:

$$\mathrm{H} = \mathrm{V} \cdot \frac{\mathrm{s}}{\mathrm{A}}$$

From Eq. (2-13),

$$\mathrm{H} = \frac{\mathrm{kg} \cdot \mathrm{m}^2}{\mathrm{s}^3 \cdot \mathrm{A}} \times \frac{\mathrm{s}}{\mathrm{A}}$$

$$\mathrm{H} = \frac{\mathrm{kg} \cdot \mathrm{m}^2}{\mathrm{s}^2 \cdot \mathrm{A}^2} \tag{9-14}$$

THE HENRY PER METER

From Eq. (9-12) we can obtain

$$\mu = \frac{Ls}{N^2 A}$$

Writing the equation with unit symbols gives

$$\frac{\mathrm{H}}{\mathrm{m}} = \frac{\mathrm{H} \times \mathrm{m}}{\mathrm{m}^2}$$

The number of turns is omitted because it is a numerical constant. Thus the unit henry per meter is appropriate to permeability.

From Eq. (9-14) it can be seen that the MKSA expression for permeability is $\mathrm{kg} \cdot \mathrm{m/s^2} \cdot \mathrm{A}^2$.

9-5 LENZ' LAW

Faraday's law gives the magnitude of an induced emf. Equations (9-7) to (9-13) showed the emf as $|e|$ to emphasize that only the magnitude was referred to. The direction of application, or polarity, of the induced emf is given by Lenz' law.

The direction of a current resulting from an induced emf is such that the change in flux it causes is opposite in direction to the change in flux which induced the emf.

Perhaps the easiest way to interpret Lenz' law for self-inductance is to note that the polarity of the induced emf is opposite that of the driving or source emf.

Figure 9-5 shows the development of this statement from Lenz' law. Figure 9-5*a* contains the driving conditions only. The driving emf e_d is assumed to have the instantaneous polarity shown. The directions of the rates of change in current di_d/dt and flux $d\Phi_d/dt$ due to the source emf e_d are shown by arrows.

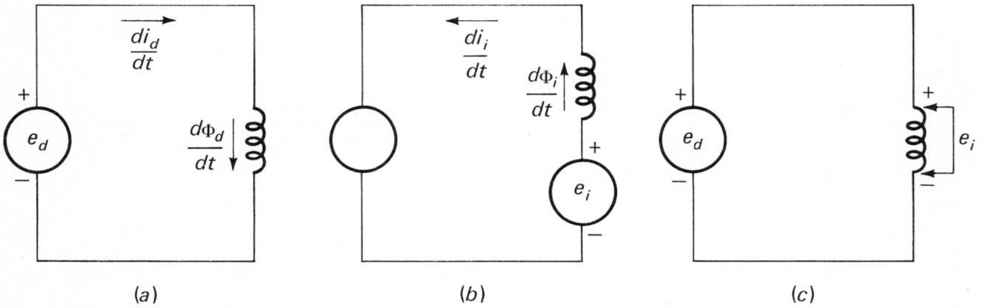

Fig. 9-5 Lenz' law.

In Fig. 9-5*b* the induced conditions are shown alone. The induced emf e_i is shown as a generator voltage with its polarity such as to cause rates of change in current di_i/dt and flux $d\Phi_i/dt$ with directions opposite those of Fig. 9-5*a*.

The driving and induced emf polarities are shown together in Fig. 9-5*c*. You will see that the induced emf polarity is the same as the polarity of any circuit element connected across a generator. Why is Lenz' law needed? In the case of simple self-inductance there is no need for it. The law applies to all electromagnetically induced emfs, and we shall deal with cases, such as the mutually induced emfs in transformers, in which Lenz' law is most helpful in determining the circuit relationships.

Equation (9-13), when written to include the sign (polarity) of e_i becomes

$$e_i = -L \frac{di}{dt} \tag{9-15}$$

This equation form derives from the Kirchhoff's loop equation for Fig. 9-5*c*:

$$E_d = e_i \qquad \text{or} \qquad E_d - e_i = 0 \tag{9-16}$$

Relative to the driving emf E_d, the induced emf e_i is negative.

It is one of the oddities of electrical convention that one often takes care to write the negative sign for an induced emf but rarely ever writes $-RI$ for the pd across a resistor, although it is strictly correct. The negative sign means:

1. A Kirchhoff's loop equation of the form of Eq. (9-16) applies.
2. If the induced emf (or resistor pd) were to cause a circuit current, its direction would be opposite that actually produced by the driving emf.

The term "back emf" is often applied to an induced emf, because it describes this tendency to oppose the driving emf. Although it is not common, back emf is sometimes applied to resistor pd.

Example 9-1 What is the inductance of a 1000-turn coil wound on the iron alloy core shown in Fig. 9-6? The relative incremental permeability of the core material is 500.

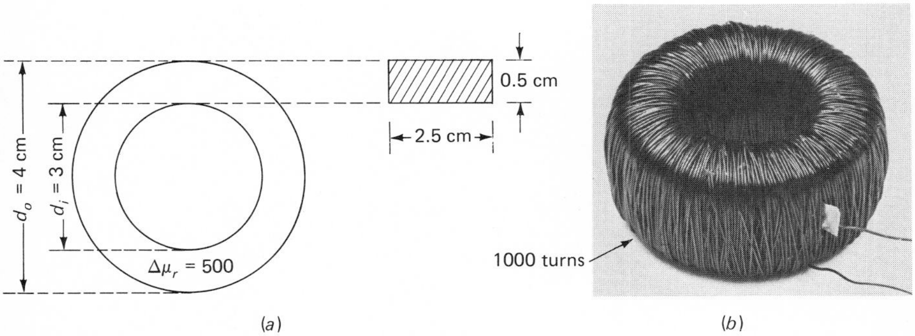

Fig. 9-6 Example 9-1: (*a*) core dimensions; (*b*) toroidal coil.

Solution The core area A will equal the shaded area in Fig. 9-6*a*:

$$A = 2.5\text{ cm} \times 0.5\text{ cm} = 1.25\text{ cm}^2 = 1.25 \times 10^{-4}\text{ m}^2$$

The effective length of the field s will be the average circumference of the core. The average core diameter is

$$d_{\text{av}} = \frac{d_o + d_i}{2} = \frac{4\text{ cm} + 3\text{ cm}}{2} = 3.5\text{ cm}$$

$$s = \pi d_{\text{av}} = \pi \times 3.5\text{ cm} = 0.11\text{ m}$$

Applying Eq. (9-5) to find the value of the core incremental permeability,

$$\Delta\mu = \mu_v \, \Delta\mu_r \tag{9-5}$$

Substituting the value of μ_v ($4\pi \times 10^{-7}$ H/m),

$$\Delta\mu = \mu_v \Delta\mu_r = 4\pi \times 10^{-7} \text{ H/m} \times 500 = 6.283 \times 10^{-4} \text{ H/m}$$

The inductance can be found by substituting values into Eq. (9-12):

$$L = \frac{N^2 \, \Delta\mu A}{s} = \frac{1000^2 \times 6.283 \times 10^{-4} \text{ H/m} \times 1.25 \times 10^{-4} \text{ m}^2}{0.11 \text{ m}} = \mathbf{0.714 \text{ H}}$$

A coil wound on a doughnut-shaped core, as in Example 9-1, is called a *toroid.* A toroidally wound coil has the most efficient magnetic characteristics. Its core, being endless and entirely within the coil winding, concentrates the magnetic field to the maximum degree. A coil of this shape has its measured inductance close to the value calculated using Eq. (9-12). Coils of other shapes may have inductances which differ widely from the calculated values as a result of field inefficiencies.

Example 9-2 A pure 10-mH inductor is supplied with a current which rises at a uniform rate from 20 to 100 mA in 25 μs. What is the magnitude of the emf induced between the ends of the inductor winding?
Solution (Fig. 9-7)

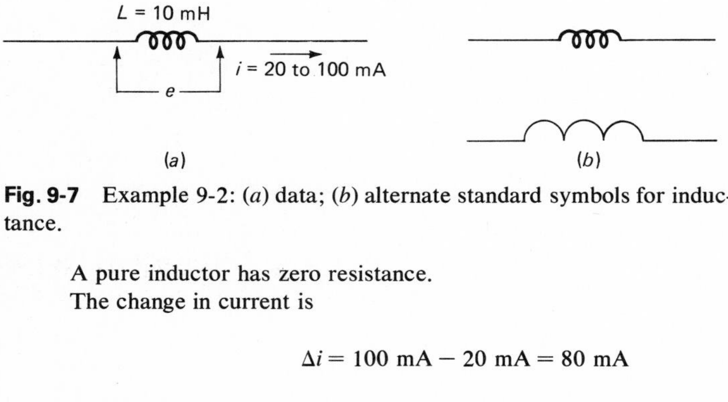

Fig. 9-7 Example 9-2: (*a*) data; (*b*) alternate standard symbols for inductance.

A pure inductor has zero resistance.
The change in current is

$$\Delta i = 100 \text{ mA} - 20 \text{ mA} = 80 \text{ mA}$$

Since the rate of current rise is uniform,

$$\frac{di}{dt} = \frac{\Delta i}{\Delta t} = \frac{80 \text{ mA}}{25 \ \mu\text{s}} = 3.2 \text{ kA/s}$$

$$|e| = L\frac{di}{dt} \tag{9-13}$$

$$|e| = 10 \text{ mH} \times 3.2 \text{ kA/s} = \mathbf{32 \text{ V}}$$

Example 9-3 A constant 32 V is applied across a pure 10-mH inductor which initially has no magnetic field associated with it. What will be its current (*a*) initially, (*b*) after 1 μs, and (*c*) after 10 ms?

Solution (Fig. 9-8)

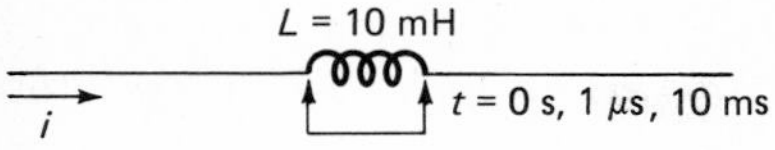

Fig. 9-8 Example 9-3: data.

If an inductor has a constant voltage across it, the rate of change in current through it is also constant ($e \propto di/dt$). With a constant rate of change in current, $di/dt = \Delta i/\Delta t$. Example 9-2 indicates that 32 V is associated with 3.2 kA/s. Thus in the present example the constant 32 V implies a constant 3.2 kA/s. In starting with an uncharged inductor (an inductor with no magnetic field), the current will be

a. Initially, or at $t = 0$ s ($\Delta t = 0$ s),

$$\frac{\Delta i}{\Delta t} = \frac{di}{dt} = 3.2 \text{ kA/s}$$

$$\Delta i = 3.2 \text{ kA/s} \times \Delta t = 3.2 \text{ kA/s} \times 0 \text{ s} = 0 \text{ A}$$

$$i = \mathbf{0\ A}$$

b. After time $t = 1\ \mu$s ($\Delta t = 1\ \mu$s),

$$\Delta i = 3.2 \text{ kA/s} \times 1\ \mu\text{s} = 3.2 \text{ mA}$$

$$i = \mathbf{3.2\ mA}$$

c. After time $t = 10$ ms ($\Delta t = 10$ ms),

$$\Delta i = 3.2 \text{ kA/s} \times 10 \text{ ms} = 32 \text{ A}$$

$$i = \mathbf{32\ A}$$

9-6 CHARGING AND DISCHARGING AN INDUCTANCE

A dictionary definition of the word "charge" is "to load or fill to capacity." Thus a capacitance is charged when it is loaded with stored energy. Similarly an inductance is charged when energy is stored in its magnetic field. Although the term is not universally employed in the inductive context, charge is a useful concept in correlating capacitance and inductance.

The results of charging a capacitance are significantly different from those of charging an inductance. When the energy source is removed, it remains within a capacitance; the capacitance remains charged. Removal of the energy source causes the energy to leave an inductance immediately; the inductance discharges.

CHARGING

An inductor which has no magnetic field associated with it is *uncharged;* it has no stored energy, and it carries no current. *Charging* is the act of increasing the amount of energy stored. Equation (9-13) and Example (9-3) indicate that, when an inductor becomes charged as the result of a constant voltage, its rate of change in current will also be constant. Therefore the inductor's current will be directly proportional to time with constant charging voltage.

Figure 9-9*a* shows the linear increase in current in a 10-mH inductor with a constant 32 V across it. The data were obtained from the solutions to Examples 9-2 and 9-3.

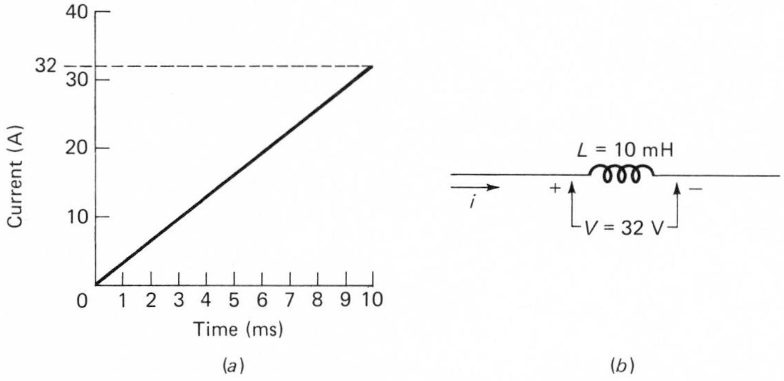

Fig. 9-9 Inductor charging. (*a*) Graph of current with respect to time for a constant voltage; (*b*) schematic diagram.

DISCHARGING

When an inductor is discharging and is supplying a constant voltage to a network, there will also be a linear relationship between the current and time. The current will decrease because energy is being supplied by the inductor's collapsing magnetic field. The conditions are shown graphically in Fig. 9-10*a*.

Fig. 9-10 Inductor discharging. (*a*) Graph of current with respect to time for a constant voltage; (*b*) schematic diagram.

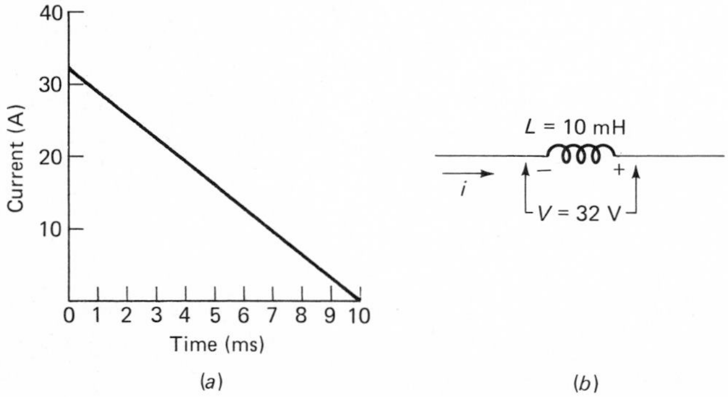

HOLDING INDUCTIVE CHARGE

An inductor can only store energy while it carries current. Thus if the charging source emf is disconnected, the current ceases, the magnetic field collapses, and the energy is released. A comparison of the capacitor and the inductor is given in Fig. 9-11.

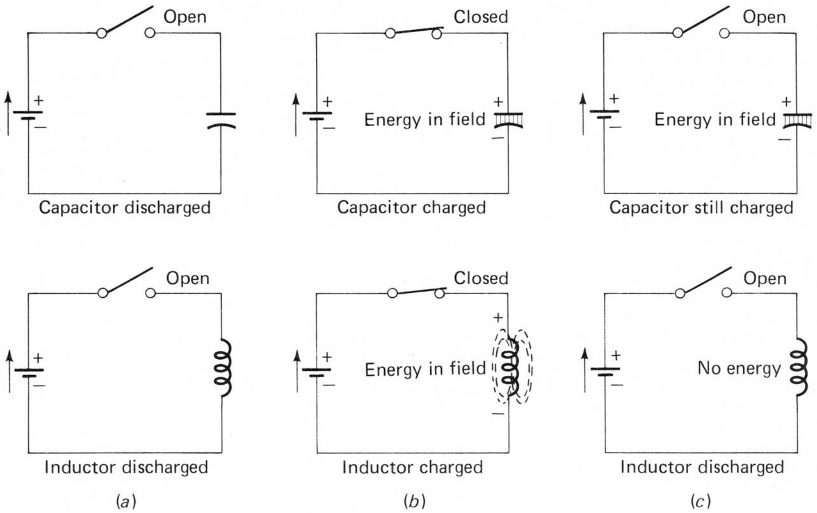

Fig. 9-11 Comparison of a capacitance and an inductance.

Figure 9-11 shows

a. Initially, with the switch open, both the capacitor and the inductor are discharged (no field energy).
b. When the switch is closed, both the capacitor and the inductor become charged (energy is stored in their fields).
c. When the switch is then reopened, the capacitor retains its energy but the inductor does not.

THE COLLAPSING MAGNETIC FIELD

Current is necessary to maintain a magnetic field so, when current in an inductance ceases, the magnetic field collapses.

A magnetic field which increases in intensity causes an induced emf. A magnetic field which decreases in intensity (a collapsing field) also causes an induced emf. However, the rates of change in flux in increasing and decreasing fields are opposite in effective direction and so produce induced emfs of opposite polarity.

Figure 9-12*a* shows a battery connected across an inductance. Current in the direction shown causes a magnetic field to increase in intensity from zero to a maximum value. An induced emf e_a of the indicated polarity exists while the current is increasing.

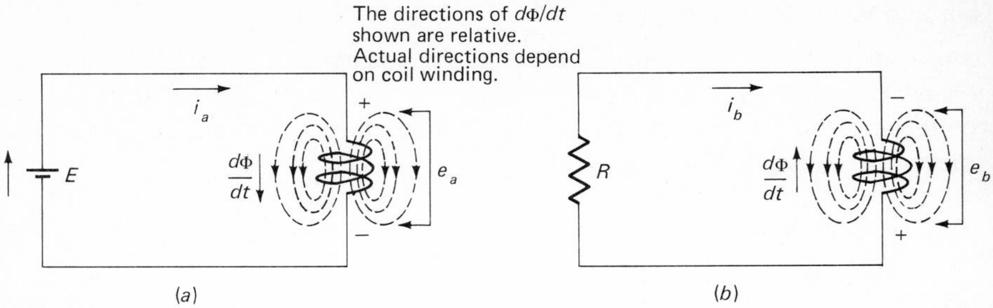

Fig. 9-12 Induced emf polarities. (*a*) Energy from a source to an inductance; (*b*) energy from an inductance to a load.

If, without first becoming open, the circuit changes to that in Fig. 9-12*b*, the removal of the battery stops the original current but a circuit is maintained across the inductance through a resistance. The magnetic field cannot hold up without the battery, so the field collapses. An induced emf e_b, of polarity opposite that of e_a, exists while the field intensity decreases to zero. The induced emf e_b causes a current i_b. It is significant that currents i_a for the increasing field and i_b for the decreasing field have the same direction through the inductance.

This identity of current direction can be seen from the equivalent circuits in Fig. 9-13.

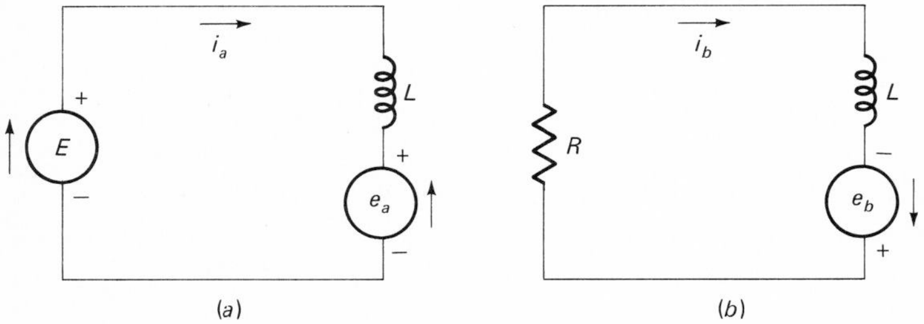

Fig. 9-13 Figure 9-12 shown in schematic diagram form.

Figure 9-13*a* is equivalent to Fig. 9-12*a*. The battery emf E and the induced emf e_a of the increasing field are shown as opposing polarity voltage sources E and e_a.

Figure 9-13*b* is equivalent to Fig. 9-12*b*. Here the induced emf of the collapsing field is shown as the voltage source e_b. The polarities of e_a and e_b are opposite.

The polarity of e_b is similar to that of E; they both produce clockwise currents or currents with a downward direction through the inductor.

The energy supplied by the battery to the increasing intensity field (via the current i_a) will be fully recovered from the collapsing magnetic field (via the current i_b) in the form of heat in the resistance.

To maintain the circuit continuity across the inductance while changing from the battery to the resistance, an arrangement such as that in Fig. 9-14*a* is needed. This diagram shows a make-before-break switch in which the connection between contacts 2 and 3 is completed before that between contacts 1 and 2 breaks. If instead, a transfer or break-before-make switch were used as in Fig. 9-14*b*, so that the battery is disconnected before the resistance is connected, the induced emf polarities and current directions would be basically the same but the values would be unpredictable.

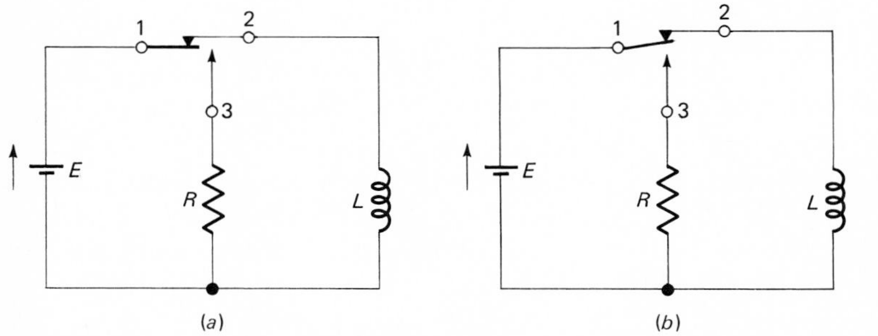

Fig. 9-14 Inductance charge/discharge circuits. (*a*) With make-before-break contacts; (*b*) with break-before-make contacts.

If the circuit across a charged inductance is opened, a theoretically infinite induced emf results because the field collapses in zero time.

$$|e| = \frac{dN\Phi}{dt} \tag{9-9}$$

Assuming a linear relationship, this equation can be written in the form

$$|e| = \frac{\Delta N\Phi}{\Delta t}$$

If $\Delta t = 0$, then

$$|e| = \frac{\Delta N\Phi}{0} = \infty$$

In practice the field does not collapse in zero time, and so the induced emf does not become infinite. The time in which the field collapses does become very short, and the induced emf becomes large enough to develop an arc across the circuit opening. This arc permits current to exist around the circuit, through the open switch contacts.

In Fig. 9-15*a* the battery emf has transferred energy to the inductance via the closed switch, building up a magnetic field.

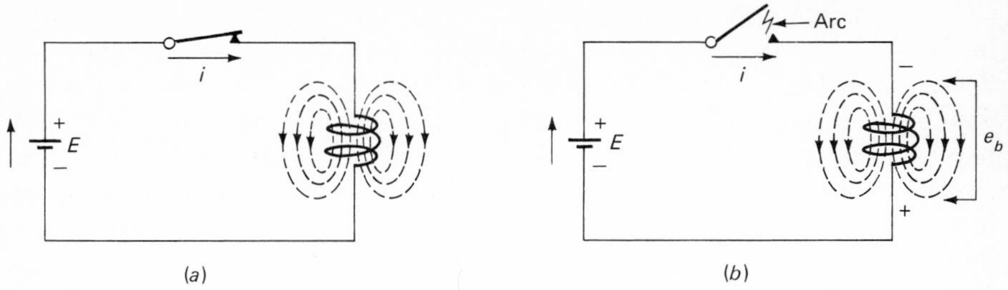

Fig. 9-15 (*a*) Inductance charging through contacts; (*b*) when the contacts open an arc may develop.

Figure 9-15*b* shows the conditions existing at the instant the switch opens. The magnetic field collapses rapidly, inducing a high value of voltage[1] e_b. This high voltage appears across the gap between the switch contacts. The air between the contacts *ionizes* (becomes conductive due to the high voltage breaking down the air's molecular structure to produce mobile charges). Circuit current is maintained through the conducting arc until the energy of the field has dissipated.

The tendency for a collapsing magnetic field to produce a high-voltage arc is, like most phenomena in electric circuits, a blessing or a curse according to the circumstances. In the automobile an induced emf of 20 000 V or more is induced by the collapse of a magnetic field produced by the 12-V electric system. This high induced voltage causes an arc across the gap between the spark plug electrodes. The arc ignites the gas-air mixture to make the motor run. In electrical equipment used for switching inductive circuits, the arc between the opening switch contacts causes the contact surfaces to become pitted and carbonized. This erosion results in a poor electric connection in the switch and eventual destruction of the contact material.

ARC SUPPRESSION

The formation of an arc between opening contacts can be reduced or suppressed by providing a means for the field to collapse more slowly and thus prevent the development of a high value of induced emf. The time taken for a field to collapse becomes greater if the discharge current is higher. Therefore, deliberately providing a path for current across the gap can be used to control the rate of field collapse.

In Fig. 9-16*a* a resistance is connected across the contacts so that, when they open, the resistance will conduct a current as a result of the induced emf of the collapsing field. By suitable choice of resistance, the rate of field collapse, the induced emf, and the pd across the resistance can be limited to a value below that required to produce contact arcing. There is

[1] The polarities of the induced emf e_b and the battery emf E are aiding. Even if they were not, the induced emf magnitude is greater than the battery emf, so the current would flow in the direction indicated.

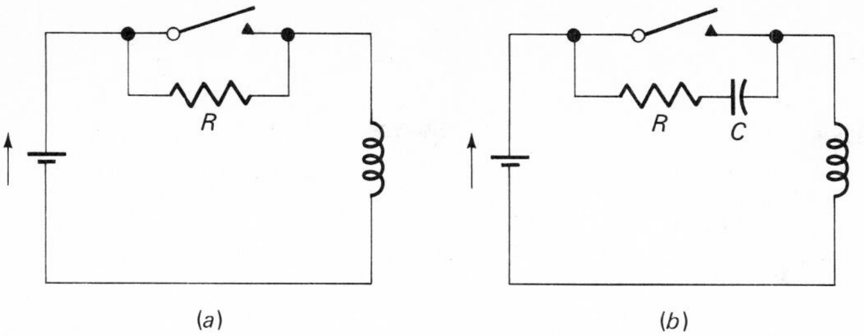

Fig. 9-16 Arc suppression. (*a*) By resistance; (*b*) by resistance and capacitance.

a potential problem here because the circuit is never completely open and the current is never reduced to zero. In some applications this permanent small current is unimportant, provided the resistance value is as high as is possible consistent with arc prevention.

The circuit in Fig. 9-16*b* has a series *RC* circuit across the contacts. This arc-suppression circuit ensures that, when the field energy has completely dissipated in the resistance, the circuit really is open, a capacitance being a dc open circuit. This arrangement has its own problems which may prohibit its use for some applications. These potential problems will become apparent when resonance is dealt with later in the text.

9-7 OPPOSITION TO CURRENT CHANGE

Inductance may be referred to as opposition to current change, based on the following:

$$e = L\frac{di}{dt} \qquad \text{or} \qquad \frac{di}{dt} = \frac{e}{L} \tag{9-13}$$

For a given emf, the magnitude of the rate of change in inductive current is inversely proportional to the inductance value. This proportionality implies that a high value of inductance is related to a low rate of change in current.

This opposition to current change can be viewed in two ways. If the magnitude of the current change is fixed by the circuit conditions, the time required to change is greater with higher inductance. If the time in which the current changes is fixed, the magnitude of the change is smaller with high inductance.

The latter view of inductive opposition to current change is the basis for using an inductor as a *choke*. As an example of this application of inductance, consider the situation shown in Fig. 9-17. A generator producing a steady dc voltage E is connected in series with a generator producing an ac voltage e. The *frequency* of the voltage e (the number of times per second its polarity changes) is fixed. The resulting waveform of current in the load is shown in Fig. 9-17*b*.

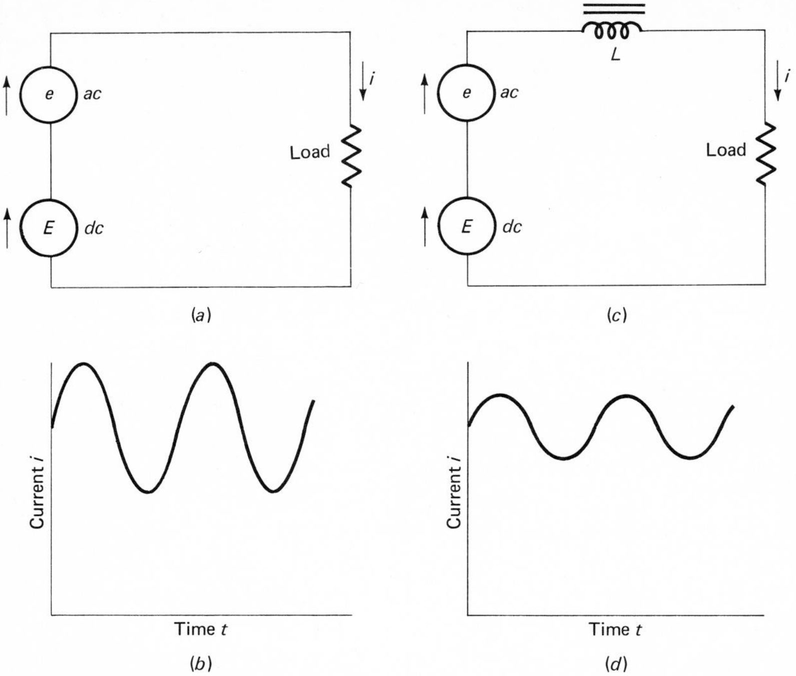

Fig. 9-17 Choke action. (*a*) and (*b*) Large ac component of current without inductance; (*c*) and (*d*) small ac component of current with inductance.

In Fig. 9-17*c* an iron-cored inductor of high inductance L has been added to the circuit in Fig. 9-17*a*. The generator voltages are the same as in Fig. 9-17*a*, so the rapidity of the voltage change, as determined by the voltage e, is the same. The added inductance reduces the rate of change in current. As the rapidity of the change is fixed, the magnitude of the current change is reduced. Figure 9-17*d* shows the effect of the inductor. This choking action is used in electronic power supplies which produce dc outputs from ac power line inputs. The presence of the inductor (choke) helps to ensure a steady dc output relatively free of fluctuations.

9-8 INDUCTIVE ENERGY STORAGE

Energy transferred to the magnetic field of an inductance is stored while the current is maintained. When the current decreases, energy is extracted from the magnetic field. It might be dissipated in a resistor, transferred to a capacitor for storage, or passed back to the supply source.

During the process of charging or discharging an inductance (transferring energy to or from it), a constant induced voltage implies a linear relationship between current and time (refer to Figs. 9-9 and 9-10).

If, over a time period t in Fig. 9-18, the current rises from zero to a value I_m amperes, the energy transferred to the inductance will be related

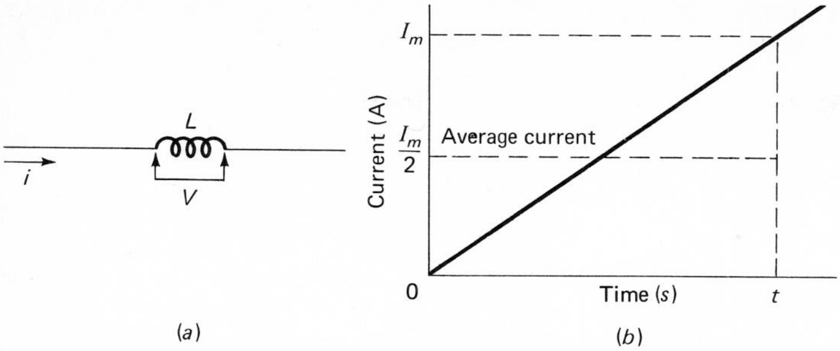

Fig. 9-18 Inductive energy. (*a*) Circuit conditions; (*b*) current-time relationships.

to the average value of that current. Since the current rises in a linear manner, the average value is one-half I_m.

The energy in an electric circuit was given in Eq. (2-18) as

$$W = VIt \tag{2-18}$$

In an inductive circuit with constant voltage, over the time period t seconds, the current will in effect be

$$I = I_{av} = \frac{I_m}{2}$$

Also, from Eq. (9-13),

$$V = |e| = L\frac{di}{dt}$$

Since the current rise is linear, $di/dt = \Delta i/\Delta t$. The change in current Δi is the final current I_m, the change in time Δt being the total time t. Therefore,

$$V = L\frac{I_m}{t}$$

Substituting for I and V in Eq. (2-18) gives

$$W = L\frac{I_m}{t}\frac{I_m}{2}t$$

$$W = \frac{I_m^2 L}{2} \tag{9-17}$$

where W = inductive energy, J
I_m = final current, A
L = inductance, H

Example 9-4 A 10-H, iron-cored inductance is used in the filter circuit of an electronic power supply. What is the amount of energy stored in its magnetic field if the current is approximately constant at 100 mA?
Solution (Fig. 9-19)

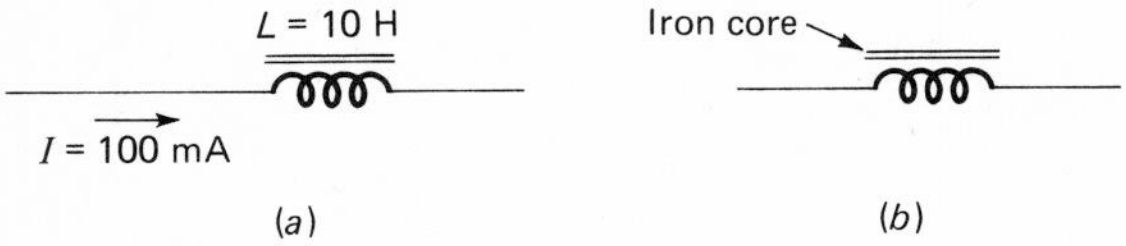

Fig. 9-19 Example 9-4: (*a*) data; (*b*) standard symbol for iron-cored inductance.

If an inductor's current is constant, its value is the equivalent of the value I_m in Fig. 9-18. Thus in this example the current in Eq. (9-17) is 100 mA. From Eq. (9-17),

$$W = \frac{I^2L}{2} = \frac{(100 \text{ mA})^2 \times 10 \text{ H}}{2} = \mathbf{0.05\ J}$$

Example 9-5 When the negative pulse is applied to the base of the transistor in the circuit in Fig. 9-20, the collector current I_c is cut off (reduced to a very low value), the transistor becoming similar to an open switch. If the inductor's field collapses to zero at a uniform rate in 10 μs, (*a*) what amount of energy is dissipated and (*b*) what voltage V_{ce} is developed between the transistor collector and emitter?

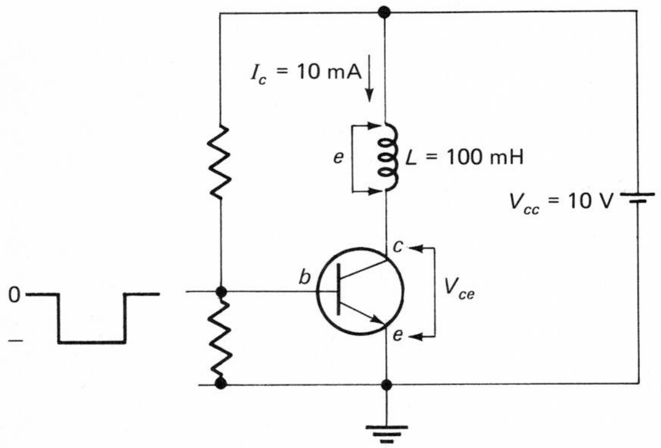

Fig. 9-20 Example 9-5: circuit.

Solution

a. From Eq. (9-17),

$$W = \frac{I_c^2L}{2} = \frac{(10 \text{ mA})^2 \times 100 \text{ mH}}{2} = \mathbf{5\ \mu J}$$

b. The voltage e induced across the inductor by the collapsing field and the collector supply voltage V_{cc} are applied in series across the transistor to form the voltage V_{ce}.

$$e = L\frac{di}{dt} \qquad (9\text{-}13)$$

$$e = 100 \text{ mH} \frac{10 \text{ mA}}{10\ \mu\text{s}} = 100 \text{ V}$$

The polarities of the induced emf e due to the collapsing field and the collector supply voltage V_{cc} will be aiding each other.

$$V_{ce} = e + V_{cc} = 100\text{ V} + 10\text{ V} = \mathbf{110\text{ V}}$$

The transistor selected for use in the circuit in Fig. 9-20 must be able to tolerate 110 V between its collector and its emitter when the inductance discharges. This voltage is quite a severe requirement which many transistors cannot handle even though well able to withstand the 10 mA at 10 V when in the ON condition. The discharge time is important, of course. Design precautions would need to be taken to ensure that the time would not be even less than the 10 μs specified, as this would result in an induced emf greater than 100 V.

9-9 MUTUAL INDUCTANCE

In Fig. 9-21 a varying current in coil 1 produces a magnetic field of varying intensity. This magnetic field links with coil 2, inducing an emf between its ends. This type of action in which a varying quantity in one circuit causes the development of a quantity in a different circuit is called *mutual* or *transfer* action. When, as in Fig. 9-21, a varying current in one circuit causes a voltage to be induced in another circuit, the action is called *mutual induction*. The relationship between the changing current and the induced voltage is given by

$$e_M = -M\frac{di_1}{dt} \qquad (9\text{-}18)$$

where e_M = emf induced in coil 2, V
M = mutual inductance between the windings, H
di_1/dt = rate of change in current in coil 1, A/s

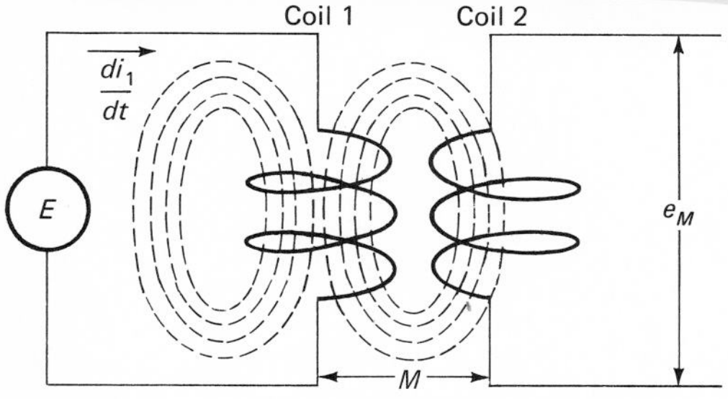

Fig. 9-21 Magnetic field linking two coils.

The negative sign in Eq. (9-18) implies that a mutually induced voltage follows the principle of Lenz' law. The polarity of the induced emf e_M is such that, if the terminals of coil 2 are connected by a conducting network, the current in coil 2 will produce a magnetic field with a flux direction opposing the original flux due to the changing current in coil 1, di_1/dt.

Equations (9-18), $e_M = -M\, di_1/dt$, and (9-15), $e = -L\, di/dt$, should be compared. This comparison will show why M is called mutual inductance. It can be defined as follows:

Mutual inductance is the constant of proportionality between the rate of change in current in one circuit and the resulting emf induced in another circuit.

THE HENRY AS THE MUTUAL INDUCTANCE UNIT

The henry, the unit of self-inductance, is also used for mutual inductance. In the mutual inductance case the henry is defined as follows:

When a rate of change in current of 1 A/s in one circuit induces an emf of 1 V in another circuit, the two circuits have a mutual inductance of 1 H.

The subject of mutual inductance is dealt with in greater detail in later chapters. It is introduced here to help in the understanding of series and parallel inductive circuits in which the magnetic fields of the inductances act on each other.

Example 9-6 Two coils of a current transformer are wound on the same iron core. The mutual inductance between them is 0.5 mH. If the current in the line feeding the input coil changes at the rate of 50 A/s, what emf is measured across the other coil?
Solution (Fig. 9-22)

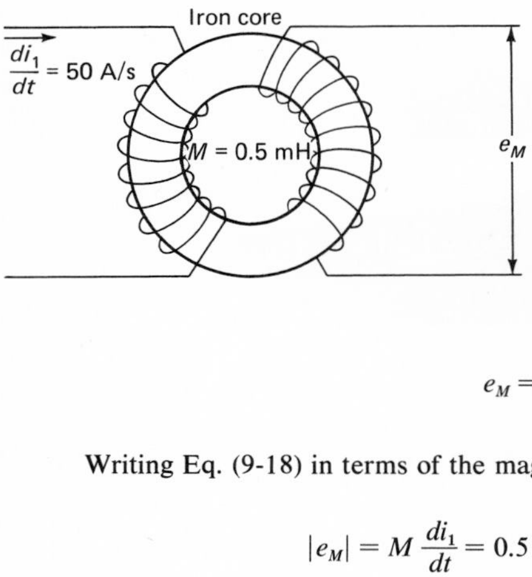

Fig. 9-22 Example 9-6: data.

$$e_M = -M \frac{di_1}{dt} \tag{9-18}$$

Writing Eq. (9-18) in terms of the magnitude of e_M gives

$$|e_M| = M \frac{di_1}{dt} = 0.5 \text{ mH} \times 50 \text{ A/s} = \mathbf{25\ mV}$$

A current transformer is used to measure currents in power circuits where, owing to the presence of high voltages, it is not convenient to measure the currents directly.

Example 9-7 The collector current i_c in the transistor in Fig. 9-23 changes at a uniform rate from 5 to 10 mA in 100 μs. The mutual inductance between the two windings is 100 mH. What is the output voltage v_o?

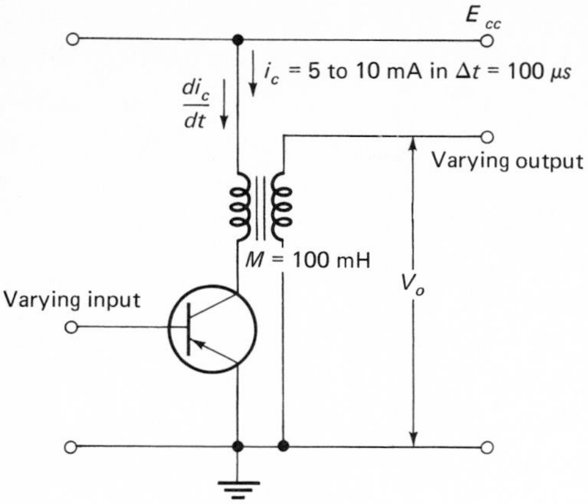

Fig. 9-23 Example 9-7: circuit.

Solution The rate of change in collector current will be

$$\frac{di_c}{dt} = \frac{\Delta i_c}{\Delta t} = \frac{10\text{ mA} - 5\text{ mA}}{100\ \mu\text{s}} = 50\text{ A/s}$$

$$|v_o| = M\frac{di_c}{dt} = 100\text{ mH} \times 50\text{ A/s} = \mathbf{5\ V}$$

9-10 CONTROL OF MUTUAL INDUCTANCE

Different degrees of mutual inductance, from maximum to zero, are needed for various applications. Often, as in the case of transformers, maximum mutual inductance is necessary for the efficient transfer of energy. Coils in the input and output circuits of amplifiers are normally required to have negligible mutual inductance between them for optimum stability of performance. Variable mutual inductance can be used to control energy transfer between a radio transmitter and its antenna.

There are several methods for controlling mutual inductance.

CORE EFFICIENCY

An efficient iron core, such as the core of a toroid-wound (doughnut-shaped) coil, will concentrate the magnetic field into a region within, and close to, the core. Thus any windings on the core will experience high mutual inductance. Nearby coils will not.

In Fig. 9-3*b* it was shown that the incremental permeability of an iron core varies as the level of magnetization changes. Mutual and self-inductances will vary too, so the operating position on the *B-H* curve can be used to control mutual inductance through the control of magnetic flux.

DISTANCE BETWEEN COILS

As the distance between coils increases, the mutual inductance between them decreases.

ANGULAR POSITION OF COILS

In Fig. 9-24*a* two *solenoid-wound* coils (cylindrical) are shown with their axes in line. Their mutual inductance will be relatively high. Figure 9-24*b* shows two similar coils with their axes at right angles. Their mutual inductance will be low (theoretically zero). In any other relative positions, the mutual inductance will be a function of the angles between the coils.

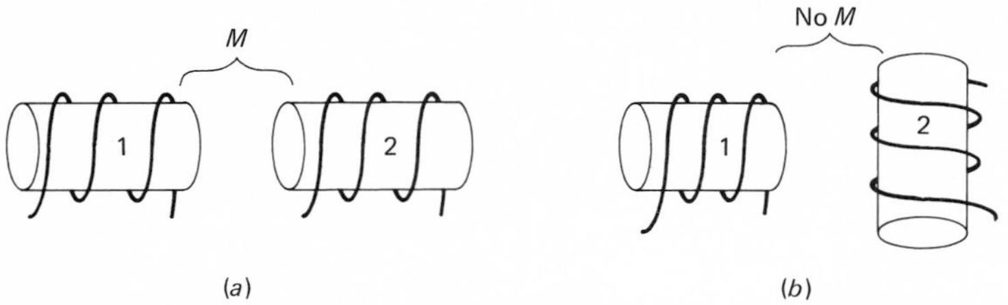

Fig. 9-24 (*a*) Coils in line result in maximum mutual inductance; (*b*) perpendicular coils result in zero mutual inductance.

SHIELDING

A complete enclosure of high-permeability material surrounding a coil will substantially reduce any mutual inductance between it and neighboring circuits.

9-11 INDUCTANCE NETWORKS WITHOUT MUTUAL INDUCTANCE

General network principles apply to inductive circuits. Kirchhoff's laws can be applied directly to series and parallel circuits if there is no mutual inductance. With the exception of transformers, which are dealt with in Chap. 18, practical applications of the quantitative effects of mutual inductance are rare. Thus they are not discussed in this text.

SERIES INDUCTANCES

In Fig. 9-25*a* three inductances are connected in series. Figure 9-25*b* shows the equivalent single inductance. If it is assumed that there is no mutual inductance present, Kirchhoff's voltage law gives

$$e_T = e_1 + e_2 + e_3$$

Fig. 9-25 Series inductances without mutual inductance.

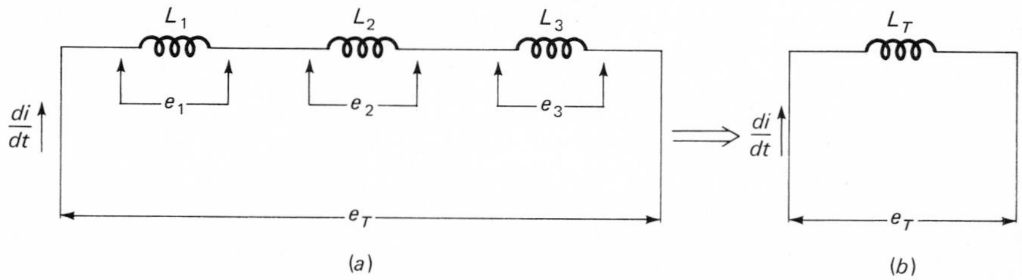

Applying Eq. (9-15),

$$-L_T \frac{di}{dt} = -L_1 \frac{di}{dt} - L_2 \frac{di}{dt} - L_3 \frac{di}{dt}$$

$$L_T = L_1 + L_2 + L_3$$

This relationship can be extended to any number of inductances connected in series without mutual inductance between them.

$$L_T = L_1 + L_2 + L_3 + \cdot \cdot \cdot + L_n \tag{9-19}$$

PARALLEL INDUCTANCES

The three-element parallel inductive circuit in Fig. 9-26*a* has its single-element equivalent as shown in Fig. 9-26*b*. Kirchhoff's current law results in

$$\frac{di_T}{dt} = \frac{di_1}{dt} + \frac{di_2}{dt} + \frac{di_3}{dt}$$

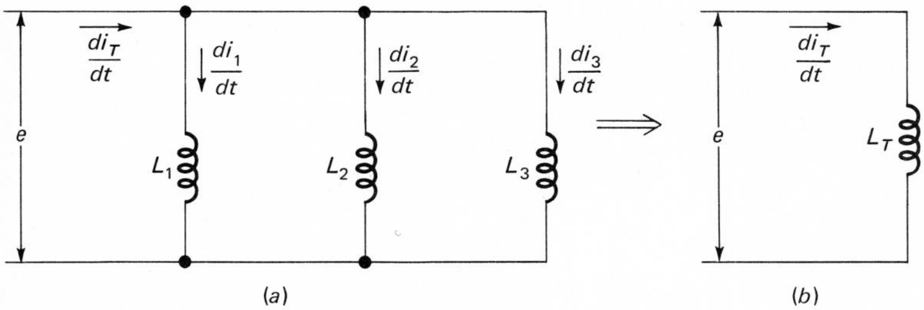

Fig. 9-26 Parallel inductances without mutual inductance.

Rewriting Eq. (9-15) gives an expression for *di*/*dt*:

$$e = -L \frac{di}{dt} \tag{9-15}$$

$$\frac{di}{dt} = -\frac{e}{L}$$

Substituting this expression in each term of the Kirchhoff's current equation,

$$-\frac{e}{L_T} = -\frac{e}{L_1} - \frac{e}{L_2} - \frac{e}{L_3}$$

$$\frac{1}{L_T} = \frac{1}{L_1} + \frac{1}{L_2} + \frac{1}{L_3}$$

This principle applies to any number of elements in parallel:

$$\frac{1}{L_T} = \frac{1}{L_1} + \frac{1}{L_2} + \frac{1}{L_3} + \cdots + \frac{1}{L_n} \tag{9-20}$$

Equations (9-19) and (9-20) show that, if there is no mutual inductance between the elements, the total inductances of series and parallel inductive circuits follow the same rules that apply to total resistance in resistive circuits.

SERIES-PARALLEL INDUCTANCES

Inductances can be connected together in any series-parallel configuration. Provided there is no mutual inductance between the elements, the entire network can be dealt with by considering each series or parallel group separately and then combining the results in a manner similar to that employed for resistive circuits.

Example 9-8 Two inductances, 10 and 20 H, are connected in parallel. If the current in the 10-H inductance changes at a uniform rate from 1 to 5 A in 1 s, what is the rate of change in current in each inductance and in the entire network?

Solution (Fig. 9-27)

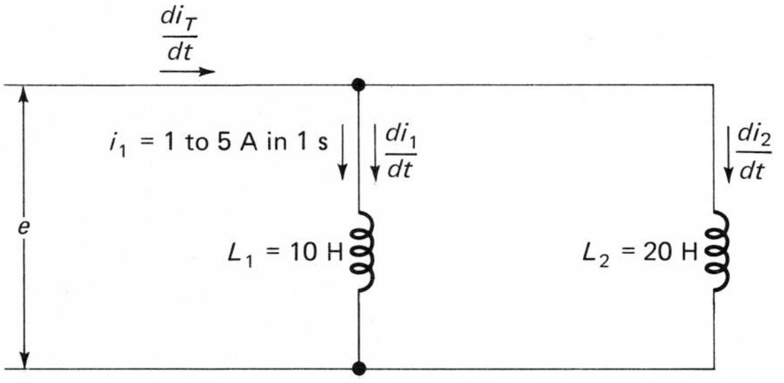

Fig. 9-27 Example 9-8: data.

Since the current changes at a uniform rate in L_1, its voltage must be constant. The same voltage exists across L_2, so its current must change at a uniform rate. This relationship must also apply to the total rate of change in current di_T/dt. Further, because the rates of change in current are constant for each changing current,

$$\frac{di}{dt} = \frac{\Delta i}{\Delta t}$$

For L_1,

$$\frac{di_1}{dt} = \frac{\Delta i_1}{\Delta t} = \frac{5\text{ A} - 1\text{ A}}{1\text{ s}} = \mathbf{4\ A/s}$$

The voltage e is common to both inductances. Therefore,

$$|e| = L_1 \frac{di_1}{dt} \tag{9-13}$$

$$|e| = 10\text{ H} \times 4\text{ A/s} = 40\text{ V}$$

Rewriting Eq. (9-13) for di/dt and substituting the symbols applicable to L_2 gives

$$\frac{di_2}{dt} = \frac{|e|}{L_2} = \frac{40\text{ V}}{20\text{ H}} = \mathbf{2\ A/s}$$

To find the total rate of change in current di_T/dt, the current increments and instantaneous values are first determined.

Equation (9-13) shows that the rate of change in current in an inductive circuit is inversely proportional to the inductance:

$$\frac{di}{dt} \propto \frac{1}{L}$$

For uniform current change rates this means that

$$\frac{\Delta i}{\Delta t} \propto \frac{1}{L}$$

Since, for any circuit situation, the time increment is a constant,

$$\Delta i \propto \frac{1}{L}$$

This expression will apply for all current levels from zero up. Therefore, while the current in L_1 is changing from 1 to 5 A ($\Delta i_1 = 4$ A), the current in L_2 will be changing from 0.5 to 2.5 A ($\Delta i_2 = 2$ A). The latter value agrees with the value of 2 A/s for di_2/dt given above.

At any instant,

$$i_T = i_1 + i_2$$

When $i_1 = 1$ A and $i_2 = 0.5$ A,

$$i_T = 1.5\text{ A}$$

When $i_1 = 5$ A and $i_2 = 2.5$ A,

$$i_T = 7.5\text{ A}$$

Thus

$$\frac{di_T}{dt} = \frac{\Delta i}{\Delta t} = \frac{7.5\text{ A} - 1.5\text{ A}}{1\text{ s}} = \mathbf{6\ A/s}$$

The solution for di_T/dt in Example 9-8 could have been more brief if Eq. (9-20) for the total inductance of a parallel circuit had been used. However, this equation was derived on the assumption that rates of change in current in parallel inductive circuits are additive. The solution given is a demonstration that this assumption is valid ($di_T/dt = di_1/dt + di_2/dt$).

Example 9-9 Four inductances, 10, 25, 50, and 100 mH, are connected in series. What is (*a*) the total inductance and (*b*) the total energy stored when the current is 15 mA?

Solution (Fig. 9-28)

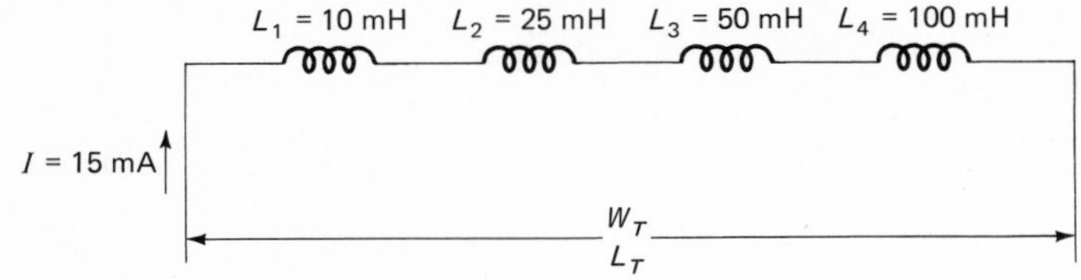

Fig. 9-28 Example 9-9: data.

a.

$$L_T = L_1 + L_2 + L_3 + L_4 \tag{9-19}$$

$$L_T = 10\text{ mH} + 25\text{ mH} + 50\text{ mH} + 100\text{ mH} = \mathbf{185\ mH}$$

b.

$$W_T = \frac{I^2 L_T}{2} \tag{9-17}$$

$$W_T = \frac{(15\text{ mA})^2 \times 185\text{ mH}}{2} = \mathbf{20.81\ \mu J}$$

In Example 9-9, the energy stored in each inductance is

$$W_1 = 1.125\ \mu\text{J}$$
$$W_2 = 2.8125\ \mu\text{J}$$
$$W_3 = 5.625\ \mu\text{J}$$
$$W_4 = 11.25\ \mu\text{J}$$

Adding these energy values gives

$$W_T = W_1 + W_2 + W_3 + W_4 = 20.81\ \mu\text{J}$$

This example illustrates that, when inductances are connected in series without mutual inductance, the total energy stored is the sum of that stored by each individual inductance.

Example 9-10 Repeat Example 9-9 with the inductances connected in parallel. The total network current is to be 15 mA.
Solution (Fig. 9-29)

Fig. 9-29 Example 9-10: data.

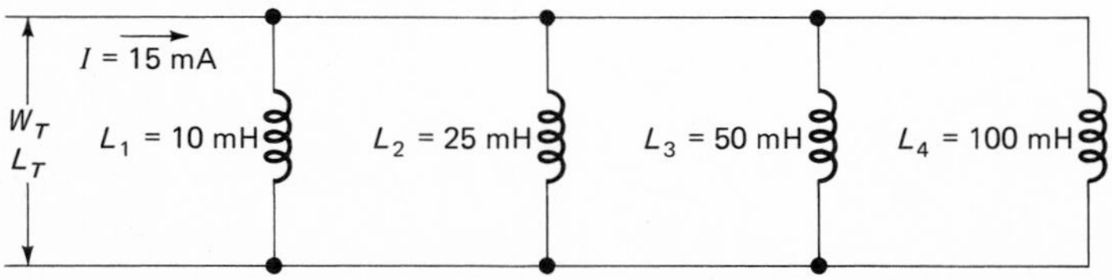

a.
$$\frac{1}{L_T}=\frac{1}{L_1}+\frac{1}{L_2}+\frac{1}{L_3}+\frac{1}{L_4} \tag{9-20}$$

$$\frac{1}{L_T}=\frac{1}{10\text{ mH}}+\frac{1}{25\text{ mH}}+\frac{1}{50\text{ mH}}+\frac{1}{100\text{ mH}}$$

$$=\frac{10+4+2+1}{100\text{ mH}}=\frac{17}{100\text{ mH}}$$

$$L_T=\frac{100\text{ mH}}{17}=\mathbf{5.882\text{ mH}}$$

b. From Eq. (9.17),

$$W_T=\frac{I^2L_T}{2}=\frac{(15\text{ mA})^2\times 5.882\text{ mH}}{2}=\mathbf{0.6617\ \mu J}$$

We have seen from the solution for Example 9-8 that the currents in a parallel inductive circuit are inversely proportional to inductance value, so that

$$\frac{i_1}{i_T}=\frac{L_T}{L_1} \qquad \text{or} \qquad i_1=i_T\frac{L_T}{L_1} \qquad \text{etc.}$$

Using this principle, the energy stored in each inductance in Example 9-10 will be

$$W_1=0.3892\ \mu\text{J}$$
$$W_2=0.1557\ \mu\text{J}$$
$$W_3=0.0779\ \mu\text{J}$$
$$W_4=0.0389\ \mu\text{J}$$

Adding the individual energies gives 0.6617 μJ, the same value as that obtained for the total network energy. The total energy stored in this parallel inductive circuit is the sum of the energies stored in each of the inductances. When this concept is combined with the conclusion drawn from Example 9-9, a general principle is revealed.

For any circuit configuration, the total energy stored in a group of inductances, without mutual inductance, is the sum of the energies stored in each inductance.

Example 9-11 What is the total inductance of a two-element parallel inductive network if the inductances are 5 and 50 μH? There is no mutual inductance present.
Solution (Fig. 9-30)

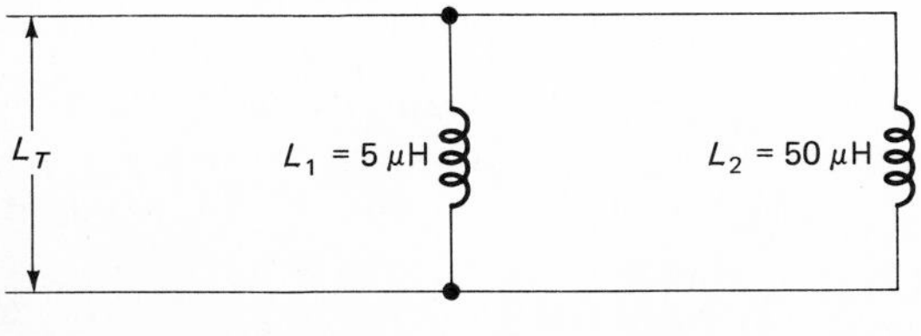

Fig. 9-30 Example 9-11: data.

The total inductance of a number of inductances in parallel follows the same basic equation as for parallel resistances. Therefore the product-divided-by-sum rule can be applied to two inductances in parallel:

$$L_T = \frac{L_1 L_2}{L_1 + L_2} = \frac{5\ \mu\text{H} \times 50\ \mu\text{H}}{5\ \mu\text{H} + 50\ \mu\text{H}} = \mathbf{4.545\ \mu H}$$

Example 9-12 What is the total inductance of three inductances, each 6 H, connected (*a*) in series and (*b*) in parallel? There is no mutual inductance in either case.
Solution

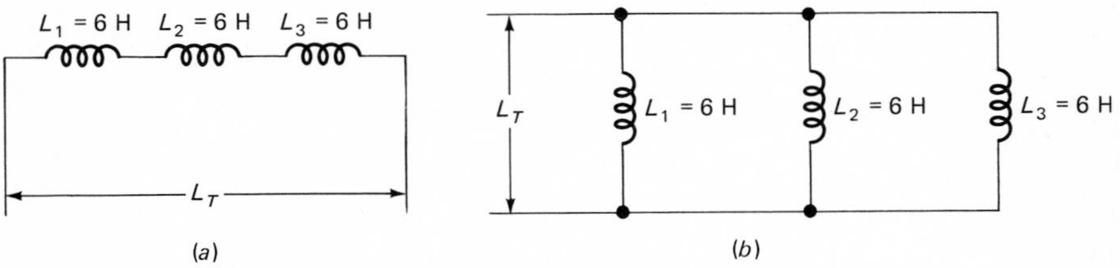

Fig. 9-31 Example 9-12: data. (*a*) Elements in series; (*b*) elements in parallel.

a. Refer to Fig. 9-31*a*,

$$L_T = L_1 + L_2 + L_3 \tag{9-19}$$

$$L_T = 6\ \text{H} + 6\ \text{H} + 6\ \text{H} = 3 \times 6\ \text{H} = \mathbf{18\ H}$$

b. Refer to Fig. 9-31*b*,

$$\frac{1}{L_T} = \frac{1}{L_1} + \frac{1}{L_2} + \frac{1}{L_3} \tag{9-20}$$

$$\frac{1}{L_T} = \frac{1}{6\ \text{H}} + \frac{1}{6\ \text{H}} + \frac{1}{6\ \text{H}} = \frac{3}{6\ \text{H}}$$

$$L_T = \frac{6\ \text{H}}{3} = \mathbf{2\ H}$$

Example 9-12 illustrates the fact that

1. When several equal-value inductances are connected in series, without mutual inductance, the total inductance is the value of any one of them multiplied by the number of inductances.
2. When several equal-value inductances are connected in parallel, without mutual inductance, the total inductance is the inductance of any one of them divided by the number of inductances.

9-12 STRAY INDUCTANCE

Any current produces a magnetic field and, if the current changes value, inductive effects take place. Commercial inductors and transformers consist of conductors wound in coils to enhance inductive effects. But straight conductors, even conductive surfaces, have inductance, and hence have induced voltages which are usually undesirable. Such undesired inductance is called *stray inductance.* Some examples now follow.

EDDY CURRENTS

In Fig. 9-32 is shown an inductor consisting of a coil of wire wound on an iron core. The changing current di/dt produces a changing flux within the core. The core becomes a closed-circuit conductor in a changing magnetic field, so it has an emf induced within it. This induced voltage causes *eddy currents,* which are currents circulating within the core. The large core volume and the relatively low resistivity of iron make the core a good conducting path so, even with a low-value induced voltage, the eddy currents may be substantial. Eddy currents represent power dissipation (I^2R) which contributes to the factors which reduce the inductor's efficiency below the ideal 100 percent. This effect is so significant that elaborate measures are taken to reduce eddy current magnitude. Transformer cores are frequently *laminated,* made of thin sheets of high-resistivity iron insulated from each other. This technique increases the resistance of the conducting paths within the core. The core iron may be pulverized into small particles which are then suspended in an insulating material and molded into the core shape. These *powder cores* further reduce eddy current paths.

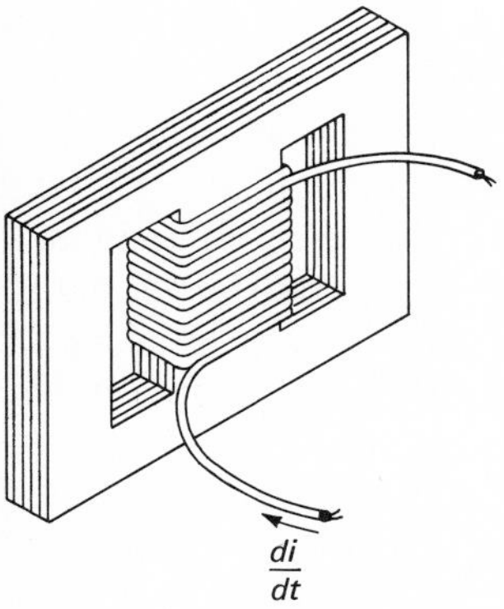

Fig. 9-32 Laminated core inductor.

STRAY INDUCTANCE IN DEVICES

All circuit devices draw current. When the currents change, inductive effects, which can significantly affect performance, become apparent. The stray inductances of resistors and transistors are briefly described. They are typical of unwanted inductance in circuit devices.

A resistor is intended to exhibit only resistance. But a resistor is a conductor so, when its current changes value, it becomes inductive, giving an equivalent circuit similar to that in Fig. 9-33*b*. Except at extremely

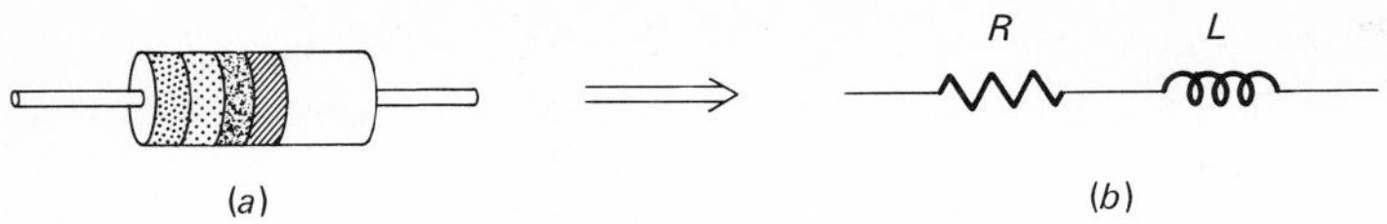

Fig. 9-33 Inductance of a resistor.

high frequencies, resistors correctly selected for their application have insufficient inductance to affect performance seriously.

A transistor has three separate connections, one each to its base, emitter, and collector. Each of the small leads from the connecting pins to the internal elements has inductance, as do the elements themselves. Figure 9-34*b* shows the total effective inductance associated with each connection. These inductances, although small in value, have important effects which can render a particular transistor useless for high-frequency or high-speed work.

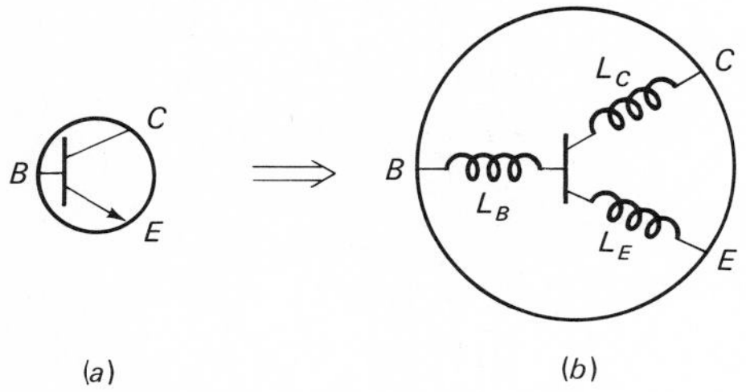

Fig. 9-34 Stray inductances of a transistor.

TRANSMISSION LINE AND CABLE INDUCTANCE

The inductances of the conductors in transmission lines and cables is so important that they must be considered when transmission performance is being determined. Together with its capacitance, the inductance of a line controls the main operating characteristics at high frequencies. The inductance is even involved in the time taken for a signal to be transferred along a line. The latter leads to the deliberate design of delay lines in order to delay the time of arrival at the output of a signal applied at the input. Such delay lines are used in radar and other applications.

If the lines are long enough, it is not necessary for the frequency to be high for the inductive effects to be important. Therefore in long-distance power lines, such as those connecting hydroelectric generating stations with cities, the conductor inductance is great enough to be a factor in the planning of the system.

STRAY INDUCTANCE IN EQUIPMENT

Inductance is inherent in all conductors, including conducting surfaces. All wiring, all components, all devices and interconnections in integrated circuits, and all conductors on printed circuit boards have self-inductance. There is also mutual inductance between all of them wherever they appear in equipment. These inductances are usually of low value, and so their effects are usually small except at high operating frequencies or speeds. As a case in point, modern computers are required to operate at extremely high speeds, often carrying out operations in times of the order of 1 ns. Such speeds necessitate reducing stray inductance to a minimum.

The importance of stray inductance will become more apparent when the relationships between inductance, time, and frequency are dealt with in later chapters.

9-13 SUMMARY OF EQUATIONS

Magnetism

$F = NI$ (9-1)

$H = F/s$ (9-2)

$H = NI/s$ (9-3)

$B = \mu H$ (9-4)

$\mu = \mu_v \mu_r$ (9-5)

$\Phi = BA$ (9-6)

$|e| = d\Lambda/dt$ (9-7)

Electromagnetic induction

$\Lambda = N\Phi$ (9-8)

$|e| = d(N\Phi)/dt$ (9-9)

$|e| = d(N^2 I \mu A/s)/dt$ (9-10)

$|e| = (N^2 \mu A/s)\ di/dt$ (9-11)

$L = N^2 \mu A/s$ (9-12)

Self-inductance

$|e| = L\ di/dt$ (9-13)

$\mathrm{H} = \mathrm{kg} \cdot \mathrm{m}^2/\mathrm{s}^2 \cdot \mathrm{A}^2$ (9-14)

$e_i = -L\ di/dt$ (9-15)

$E_d = e_i$ or $e_d - e_i = 0$ (9-16)

Inductive energy

$W = I_m{}^2 L/2$ (9-17)

Mutual inductance

$e_M = -M\ di_1/dt$ (9-18)

Inductive networks

$L_T = L_1 + L_2 + L_3 + \cdots + L_n$ (9-19)

$1/L_T = 1/L_1 + 1/L_2 + 1/L_3 + \cdots + 1/L_n$ (9-20)

EXERCISE PROBLEMS

Section 9-2

9-1 Find the permeability of an inductor core made from Isoperm 36, which has a relative permeability of 62. (*ans.* 77.91 μH/m)

9-2 Find the permeability of an inductor core made from Sendust, which has a relative permeability of 90 000.

Section 9-4

9-3 Assuming no leakage flux, what is the inductance of an air core coil, 1.5 cm in diameter and 5 cm long, consisting of 100 turns of #32 wire and having a resistance of 538 Ω/km. (*ans.* 44.4 μH)

9-4 Assuming no leakage flux, what is the inductance of a coil of 50 turns wound on a Sendust toroid with a cross-sectional area of 0.5 cm² and a mean coil length of 6 cm (μ_r of Sendust = 90 000).

9-5 Find the voltage applied to an inductance of 100 mH which causes current to rise at the rate of 500 A/s. (*ans.* 50 V)

9-6 Find the voltage across an inductance of 6 H if the current is falling at a rate of 40 A/s.

9-7 Find the rate of change in current in an inductance of 0.2 H if a voltage of 100 V is applied. (*ans.* 500 A/s)

9-8 The primary winding of a blocking oscillator transformer has an inductance of 400 μH. Find the rate of change in current if the full supply voltage of 6 V is across the winding.

9-9 Find the ON time of the blocking oscillator in Prob. 9-8 if the circuit is on during the period while current is rising from 1 to 65 mA. (*ans.* 4.27 μs)

9-10 Current rises from 0 to 150 mA in 25 μs when 15 V is applied across an inductance. Find the value of the inductance.

9-11 Find the time required for current to rise from 100 to 250 mA in an 8-H inductor while an emf of 50 V is applied. (*ans.* 24 mS)

Section 9-8

9-12 In Probs. 9-8 and 9-9, it is desired to double the ON time of the blocking oscillator by doubling the inductance of the transformer windings. If the transformer is wound on an iron alloy toroidal core of 0.5 cm^2 cross-sectional area and a mean length of 6 cm, find the number of turns required (μ_r of the core = 1500).

9-13 Find the energy stored in the magnetic field of the inductor in Prob. 9-10 when the current reaches 150 mA. (*ans.* 28.1 μJ)

9-14 The field windings of a 230-MVA generator have a resistance of 0.15 Ω and an inductance of 1.2 H. When an emf of 145 V produces a field current of 950 A, find the energy stored in the magnetic field of the field windings.

9-15 Find the energy added to the magnetic field of the inductance in Prob. 9-11 when the current rose from 100 to 250 mA under the conditions stated. (*ans.* 0.21 J)

Section 9-9

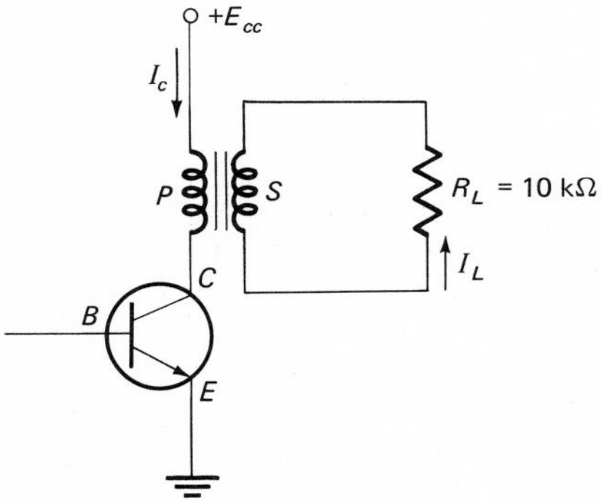

Fig. 9-1P Problem 9-16: circuit.

9-16 For the transformer in the collector circuit in Fig. 9-1P, if the mutual inductance is 2.5 mH and I_c is changing at a rate of 6000 A/s, find I_L and the voltage across R_L.

Sections 9-11 and 9-12

9-17 Two inductors, $L_1 = 250$ mH and $L_2 = 500$ mH, are connected in series with no flux linkage between them and a constant current of 2 A. Find

a. The voltage across each inductor. (*ans.* 0 V, 0 V)
b. The total inductance. (*ans.* 750 mH)
c. The energy stored in each inductor. (*ans.* 0.5 J, 1 J)
d. The total energy stored. (*ans.* 1.5 J)

9-18 Repeat Prob. 9-17 with the inductors in parallel, no flux linkage, and a constant current of 2 A in each inductance.

10

R with *L* or *C* in DC Circuits

The voltage across a capacitance changes at a uniform rate when the current through it is constant. The current in an inductance changes at a uniform rate when the voltage across it is constant. These two statements, which were dealt with in Secs. 8-5 and 9-6, are quite correct but are of limited practical value. It is not usual for capacitive currents or inductive voltages to stay constant; they usually rise or fall because of the presence of resistance in the circuit. The rate of rise and fall is controlled in a precise manner by the values of resistance and capacitance or inductance.

In *RC* or *RL* circuits, the time taken for voltages or currents to reach specific values can be accurately determined. For this reason such circuits are used extensively in practice. Timing circuits for the control of industrial processes, special ac generators called relaxation oscillators, and circuits within a television set for synchronizing the picture are a few of the applications.

While the time effects of *RC* and *RL* circuits are useful, they can also be a source of difficulty. Circuits cannot be manufactured without stray capacitance and inductance. These strays, together with the inevitable circuit resistances, prevent instantaneous response to an applied signal. Such effects as time delay, limitation of maximum operating frequency and speed, and the introduction of waveform distortion can limit circuit performance.

An analysis of the manner in which *RC* and *RL* circuit voltages and currents change leads to equations for the calculation of time relationships. Examples involving the applications mentioned above demonstrate how these equations are used in practice.

10-1 CHARGING CAPACITANCE THROUGH RESISTANCE

Figure 10-1*a* through *d* shows a sequence of events in a series *RC* circuit.

Figure 10-1*a* The capacitance is initially uncharged. With the switch open, the current i, the capacitive voltage v_C, and the resistive voltage v_R are all zero.

Figure 10-1*b* When the switch is first closed, the capacitance is still uncharged, or v_C is still zero, since capacitance opposes an instantaneous

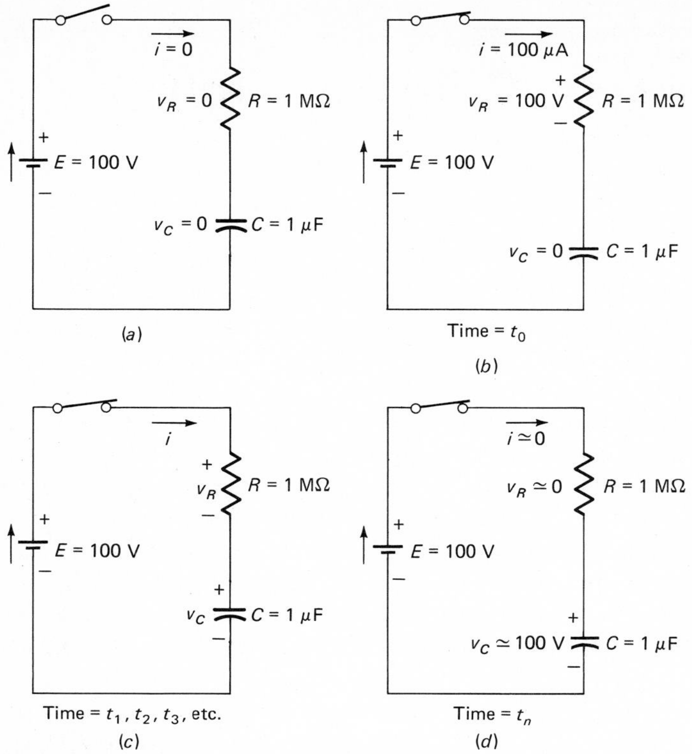

Fig. 10-1 Charging capacitance through resistance.

change in the voltage across its terminals. But Kirchhoff's voltage law must prevail in the closed loop.

$$E = v_R + v_C$$

or
$$v_R = E - v_C = 100\text{ V} - 0\text{ V} = 100\text{ V}$$

Ohm's law must also be satisfied, so there must be current in the resistance.

$$i = \frac{v_R}{R} = \frac{100\text{ V}}{1\text{ M}\Omega} = 100\ \mu\text{A}$$

This is the initial condition on completing the circuit. The events are considered to occur at time t_0.

Figure 10-1c The current causes the capacitance to charge so that, after the initial condition, the voltage v_C will have a value above zero. At this moment we have insufficient information to determine what that value is, except that v_C is greater than zero but must be less than the 100-V emf (the maximum possible voltage in this loop). The situation can be summarized:

$$0 < v_C < 100 \text{ V} \qquad v_C \text{ is greater than zero but less than 100 V}$$

$$v_R = E - v_C = 100 \text{ V} - v_C$$

or $$100 \text{ V} > v_R > 0 \qquad v_R \text{ is less than 100 V but greater than zero}$$

$$i = \frac{v_R}{R} = \frac{100 \text{ V} - v_C}{1 \text{ M}\Omega} = 100\ \mu\text{A} - \frac{v_C}{1 \text{ M}\Omega}$$

or $$100\ \mu\text{A} > i > 0 \qquad i \text{ is less than 100 } \mu\text{A but greater than zero}$$

These conditions are to be considered representative of times t_1, t_2, t_3, etc., between the initial time t_0, in Fig. 10-1*b* and the final time t_n in Fig. 10-1*d*.

Figure 10-1*d* As time goes on, the charge on the capacitance increases, as does its voltage v_C. Ultimately v_C becomes virtually equal to E, which is the maximum possible circuit voltage. This situation can be summarized:

$$v_C \simeq E \simeq 100 \text{ V}$$

$$v_R = E - v_C \simeq 100 \text{ V} - 100 \text{ V} \simeq 0 \text{ V}$$

$$i = \frac{v_R}{R} \simeq \frac{0 \text{ V}}{1 \text{ M}\Omega} \simeq 0 \text{ A}$$

This final event, we say, happens at a time t_n, implying a long time after time t_0.

RC CIRCUIT CHARGE CURVES

Graphical plots of the events in Fig. 10-1 are given in Fig. 10-2.

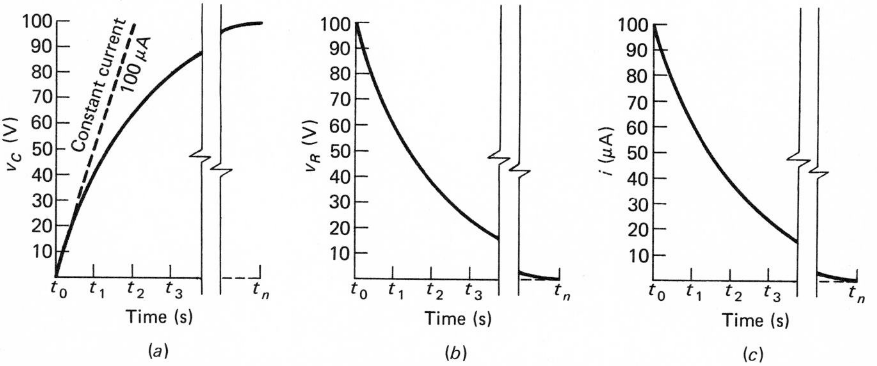

Fig. 10-2 *RC* circuit in Fig. 10-1. Charge curves with respect to time. (*a*) Capacitive voltage; (*b*) resistive voltage; (*c*) current.

WHY ARE THE CURVES CURVED?

With a constant charging current the curve of capacitive voltage with respect to time is a straight line (see Fig. 8-8). In Fig. 10-2*a* the v_C curve really is curved. Obviously then the current is not constant. Figure 10-2*c* shows the current decreasing as time progresses. Why does the current decrease?

To understand why the current decreases, let us take a sequence of events similar to those in Figs. 10-1 and 10-2, but with the assumption that the current stays constant for equal periods of time. In this way the action is frozen, so that calculations can be made without using calculus.

1. At time t_0, the conditions are

$$v_C = 0 \text{ V}$$

$$v_R = 100 \text{ V}$$

$$i = 100 \ \mu\text{A}$$

The current is now assumed to stay at 100 μA for 0.2 s. The charge in the capacitance was zero at time t_0. An amount of charge Δq is now added. From Eq. (2-11),

$$\Delta q = i \ \Delta t$$

$$\Delta q = 100 \ \mu\text{A} \times 0.2 \text{ s} = 20 \ \mu\text{C}$$

The capacitive voltage will rise from zero by an amount Δv_C. From Eq. (8-12),

$$\Delta v_C = \frac{\Delta q}{C}$$

$$\Delta v_C = \frac{20 \ \mu\text{C}}{1 \ \mu\text{F}} = 20 \text{ V}$$

Therefore, at the end of the first 0.2-s period, the capacitive voltage is

$$v_{C1} = 0 + \Delta v_C = 20 \text{ V}$$

2. At time $t_0 + 0.2$ s, the conditions are

$$v_{C1} = 20 \text{ V}$$

$$v_{R1} = E - v_{C1} = 100 \text{ V} - 20 \text{ V} = 80 \text{ V}$$

$$i_1 = \frac{v_{R1}}{R} = \frac{80 \text{ V}}{1 \text{ M}\Omega} = 80 \ \mu\text{A}$$

The conditions at time t_0 are compared with those at the end of the first 0.2-s period in Fig. 10-3. As shown in Fig. 10-3*b*, the polarity of the 20 V

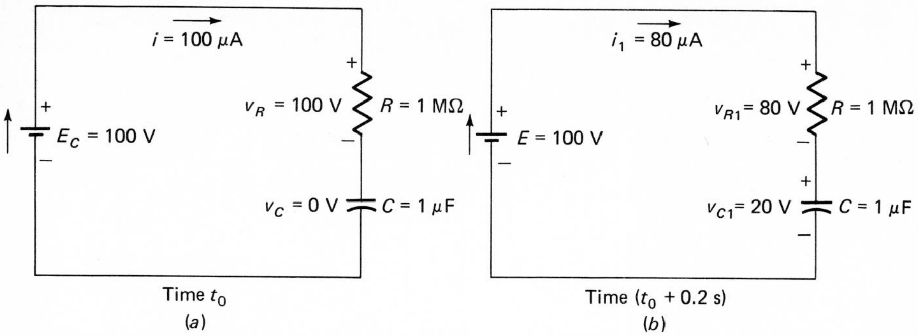

Fig. 10-3 *RC* circuit in Fig. 10-1. (*a*) At time t_0; (*b*) at 0.2 s after time t_0.

of v_{C1} acts in opposition to the 100 V of *E*, causing the net voltage across *R* to become 80 V. The circuit current is now reduced to 80 μA.

Assume that the current now stays at 80 μA for 0.2 s.

$$\Delta q = i_1 \, \Delta t = 80 \ \mu\text{A} \times 0.2 \ \text{s} = 16 \ \mu\text{C}$$

$$\Delta v = \frac{\Delta q}{C} = \frac{16 \ \mu\text{C}}{1 \ \mu\text{F}} = 16 \ \text{V}$$

At the end of the second 0.2-s period, the total capacitive voltage will be

$$v_{C2} = v_{C1} + \Delta v = 20 \ \text{V} + 16 \ \text{V} = 36 \ \text{V}$$

3. At time $t_0 + 0.4$ s, the conditions are

$$v_{C2} = 36 \ \text{V}$$

$$v_{R2} = E - v_{C2} = 100 \ \text{V} - 36 \ \text{V} = 64 \ \text{V}$$

$$i_2 = \frac{v_{R2}}{R} = \frac{64 \ \text{V}}{1 \ \text{M}\Omega} = 64 \ \mu\text{A}$$

Figure 10-4*a* shows the situation at the end of the second 0.2-s period. Assuming that the current stays at 64 μA for 0.2 s,

$$\Delta q = i \, \Delta t = 64 \ \mu\text{A} \times 0.2 \ \text{s} = 12.8 \ \mu\text{C}$$

$$\Delta v = \frac{\Delta q}{C} = \frac{12.8 \ \mu\text{C}}{1 \ \mu\text{F}} = 12.8 \ \text{V}$$

At the end of the third 0.2-s period, the total capacitive voltage will be

$$v_{C3} = v_{C2} + \Delta v = 36 \ \text{V} + 12.8 \ \text{V} = 48.8 \ \text{V}$$

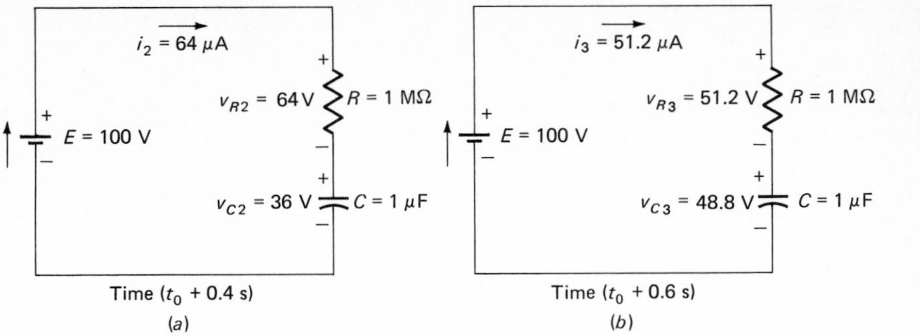

Fig. 10-4 *RC* circuit in Fig. 10-1 at (*a*) 0.4 s after time t_0 and (*b*) 0.6 s after time t_0.

4. At time $t_0 + 0.6$ s, the conditions are

$$v_{C3} = 48.8 \text{ V}$$

$$v_{R3} = E - v_{C3} = 100 \text{ V} - 48.8 \text{ V} = 51.2 \text{ V}$$

$$i_3 = \frac{v_{R3}}{R} = \frac{51.2 \text{ V}}{1 \text{ M}\Omega} = 51.2\ \mu\text{A}$$

Figure 10-4*b* shows the current and voltages at the end of the third 0.2-s period.

The voltages and currents for ten 0.2-s periods after time t_0 are shown in Table 10-1. The values of Table 10-1 are plotted as dots in Fig. 10-5.

Table 10-1 Voltages and current in charging *RC* circuit

t (s)	t_0	0.2	0.4	0.6	0.8	1.0	1.2	1.4	1.6	1.8	2.0
v_C (V)	0	20	36	48.8	59	67.2	73.8	79	83.2	86.6	89.3
v_R (V)	100	80	64	51.2	41	32.8	26.2	21	16.8	13.4	10.7
i (μA)	100	80	64	51.2	41	32.8	26.2	21	16.8	13.4	10.7

Fig. 10-5 Plots of the voltage and current values given in Table 10-1.

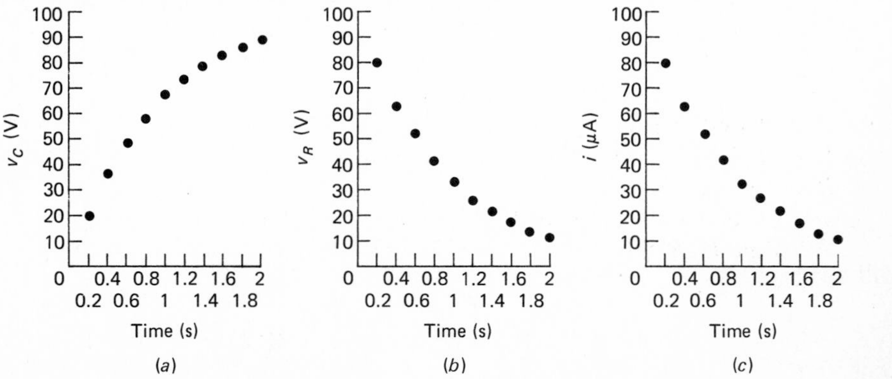

The values in Table 10-1 and Fig. 10-5 are not quite correct. They assume that the current remains constant for 0.2-s periods of time when it actually changes continuously. However, if lines are drawn through the dots, three curves result which are sufficiently similar to those in Fig. 10-2 to indicate that the original curves could represent the true variations in voltage and current with respect to time.

Accurate relationships for the *RC* circuit can only be obtained by applying calculus methods to the loop equations. These methods take account of the continuous current change.

THE EQUATIONS FOR THE CHARGING CIRCUIT

The basic loop equation for the circuit in Figs. 10-1, 10-3, and 10-4 is

$$E = v_C + v_R \tag{10-1}$$

An expression for v_C can be obtained from Eq. (8-15):

$$i = C\frac{dv_C}{dt}$$

$$dv_C = \frac{i}{C}\,dt$$

Integrating, this equation becomes

$$v_C = \frac{1}{C}\int i\,dt$$

Substituting this expression for v_C and iR for v_R in Eq. (10-1) gives

$$E = \frac{1}{C}\int i\,dt + iR$$

This differential equation is solved in the appendix.

The solution is:

$$i = \frac{E}{R}\,e^{-t/RC} \tag{10-2}$$

where i = instantaneous current, A
E = applied emf, V
R = resistance, Ω
e = base of natural logarithms[1] (a numerical constant ≃ 2.718)
t = time from t_0, s
C = capacitance, F

[1] The standard symbol for the base of natural logarithms is *e*. An acceptable alternative is ϵ, the lowercase Greek letter epsilon.

The value of the resistive voltage at any time can be found by substituting the expression for the current in Ohm's law

$$v_R = iR = \frac{E}{R} e^{-t/RC} R$$

$$v_R = Ee^{-t/RC} \tag{10-3}$$

Substituting for v_R in Eq. (10-1) gives an expression for v_C.

$$v_C = E - v_R \tag{10-1}$$

$$= E - Ee^{-t/RC}$$

$$v_C = E(1 - e^{-t/RC}) \tag{10-4}$$

In Eqs. (10-3) and (10-4),

v_R = resistive voltage, V
v_C = capacitive voltage, V
E = applied emf, V
e = base of natural logarithms ≃ 2.718
t = time from t_0, s
R = resistance, Ω
C = capacitance, F

Example 10-1 In the circuit in Figs. 10-1 to 10-4, what are the correct instantaneous values of current and voltages 1 s after time t_0? It has already been noted that the values given in Table 10-1 are not quite correct.
Solution (See Fig. 10-6.)

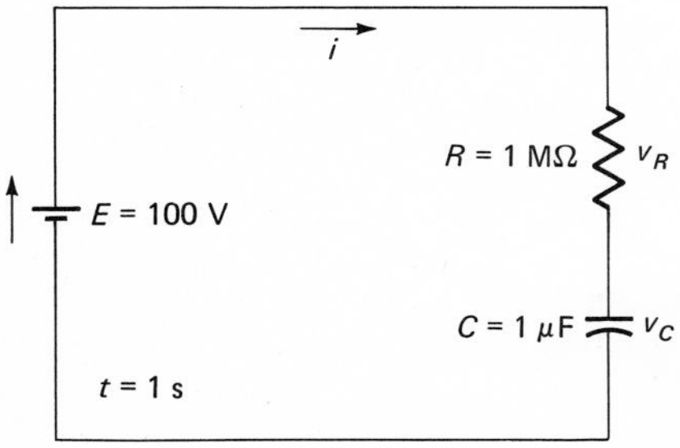

Fig. 10-6 Example 10-1: circuit.

$$i = \frac{E}{R} e^{-t/RC} \tag{10-2}$$

$$i = \frac{100 \text{ V}}{1 \text{ M}\Omega} e^{-1\text{s}/1\text{M}\Omega\times 1\mu\text{F}} = 100 \ \mu\text{A} \ e^{-1} = \frac{100 \ \mu\text{A}}{2.718} = \mathbf{36.79 \ \mu A}$$

$$v_R = iR = 36.79 \ \mu\text{A} \times 1 \text{ M}\Omega = \mathbf{36.79 \ V}$$

$$v_C = E - v_R = 100 \text{ V} - 36.79 \text{ V} = \mathbf{63.21 \ V}$$

10-2 TIME CONSTANT

In Example 10-1, the value of $t = 1$ s was chosen deliberately so that the exponent t/RC would equal 1 and the time would equal the product of resistance and capacitance. This product has a special significance.

RC IN MKSA UNITS

Using the MKSA expressions given previously and summarized in Table 2-3,

$$\Omega = \frac{\text{kg} \cdot \text{m}^2}{\text{s}^3 \cdot \text{A}^2} \tag{2-27}$$

$$\text{F} = \frac{\text{A}^2 \cdot \text{s}^4}{\text{kg} \cdot \text{m}^2} \tag{8-11}$$

The product RC can thus be expressed in fundamental units as

$$\Omega \cdot \text{F} = \frac{\text{kg} \cdot \text{m}^2}{\text{s}^3 \cdot \text{A}^2} \times \frac{\text{A}^2 \cdot \text{s}^4}{\text{kg} \cdot \text{m}^2} = \text{s}$$

Thus the dimension of resistance times capacitance is time. Because of this the *product RC* is called the *time constant* of the circuit.

The time constant is used frequently in electric circuit work and has the symbol τ, the lowercase Greek letter tau.

$$\tau = RC \tag{10-5}$$

where τ = time constant, s
R = resistance, Ω
C = capacitance, F

Resistance values are often stated in multiples of the ohm, and capacitance values are usually given in submultiples of the farad. It is convenient to remember that τ can be expressed in any of the time units listed in Table 10-2.

Table 10-2 Time constants

R	C	τ
Ω	F	s
MΩ	μF	s
kΩ	μF	ms
Ω	μF	μs
MΩ	pF	μs
kΩ	pF	ns

Table 10-1 can be reproduced with correct values and in terms of the circuit time constant ($\tau = RC = 1\ \text{M}\Omega \times 1\ \mu\text{F} = 1$ s) as shown in Table 10-3.

Table 10-3 Universal time constants

t (s)	0	0.2τ	0.4τ	0.6τ	0.8τ	1.0τ	1.2τ	1.4τ	1.6τ	1.8τ	2.0τ
v_C (V)	0	18.2	33	45.1	55.1	63.2	69.9	75.3	79.8	83.5	86.5
v_R (V)	100	81.8	67	54.9	44.9	36.8	30.1	24.7	20.2	16.5	13.5
i (μA)	100	81.8	67	54.9	44.9	36.8	30.1	24.7	20.2	16.5	13.5

When, as in Table 10-3, the values of current and voltage are listed in terms of the time constant τ, rather than in absolute time units, the result is a *universal time-constant* table. Table 10-3 gives the percentage values of v_C, v_R, and i for *any RC* circuit. For example, if the circuit had $R = 47\ \text{k}\Omega$ and $C = 560$ pF, the time constant τ would be 26.32 μs. In this case Table 10-3 can be used where $0.2\tau = 5.264\ \mu$s, $0.4\tau = 10.53\ \mu$s, $0.6\tau = 15.79\ \mu$s, etc.

CHARGE EQUATIONS IN TIME-CONSTANT TERMS

Equations (10-2) to (10-4) can be rewritten with the time constant τ substituted for the product RC.

Equation (10-2) becomes

$$i = \frac{E}{R} e^{-t/\tau} \tag{10-6}$$

Equation (10-3) becomes

$$v_R = Ee^{-t/\tau} \tag{10-7}$$

Equation (10-4) becomes

$$v_C = E(1 - e^{-t/\tau}) \tag{10-8}$$

Equations (10-6) to (10-8) can be reduced to two general equations. As we shall see, these two equations can be applied to the charge and discharge conditions of both RC and RL circuits. The two equations are:

1. For quantities which decrease in value with respect to time,

$$y = Y_{\max} e^{-t/\tau} \tag{10-9}$$

2. For quantities which increase in value with respect to time,

$$y = Y_{\max}(1 - e^{-t/\tau}) \tag{10-10}$$

In using Eqs. (10-9) and (10-10) for any application, the technique is:

1. Decide whether the relevant quantity decreases as time progresses (i or v_R in the case of RC circuit charging) or increases (v_C for RC charging), and select the appropriate equation.
2. Substitute the relevant quantities: $y =$ instantaneous value of the varying quantity, $Y_{max} =$ maximum possible value of the varying quantity, $e =$ base of natural logarithms (2.718), $t =$ time elapsed between t_0 (the time of the fully charged or discharged condition) and the time of interest, and $\tau =$ circuit time constant.

In the case of the RC series circuit when charging, $y = i$ or v_R decreasing [Eq. (10-9)], or $y = v_C$ increasing [Eq. (10-10)], $Y_{max} = E$ for v_R or v_C, or $Y_{max} = E/R$ for i.

UNIVERSAL TIME-CONSTANT CURVES

From the data in Table 10-3 (extended to $t = 5\tau$) the universal time-constant curves in Fig. 10-7 can be drawn. The data for the curves can also be obtained by substituting values into Eqs. (10-9) and (10-10). The curves have been labeled $y = e^{-x}$ and $y = 1 - e^{-x}$, implying that the maximum value Y_{max} is 1 and the exponent x is equal to t/τ.

Fig. 10-7 Universal time-constant curves.

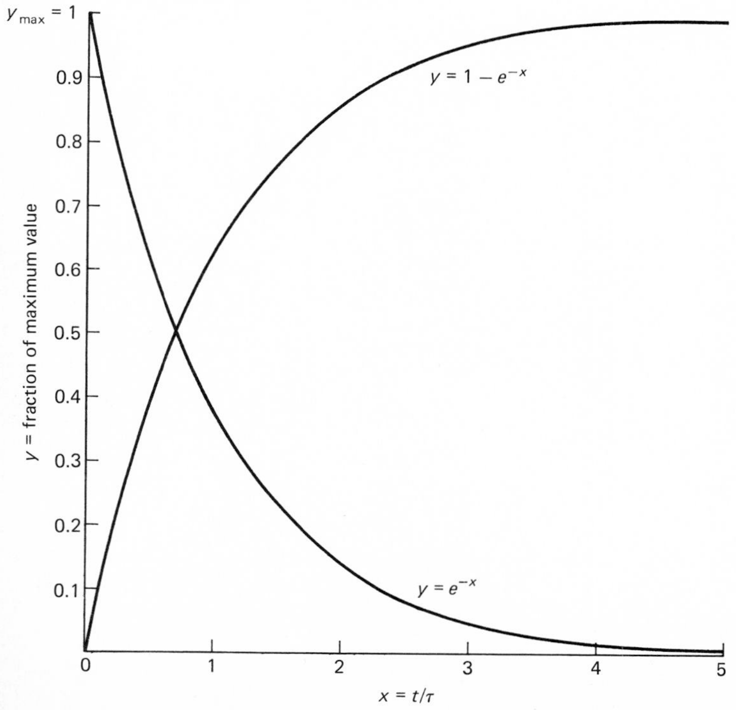

Universal time-constant curves can be used to determine approximate relationships. The horizontal axis x represents time expressed in terms of the time constant. The vertical axis y gives the fraction of the maximum value of current or voltage attained at a specified time. To use the curves, it is first necessary to decide which curve applies to the situation. It is then a simple matter of projection to find the required data. The following examples show how to apply the curves during capacitive charge.

Example 10-2 What is the voltage across the resistor in the circuit in Fig. 10-8 at a time 2.5 ms after the capacitor starts to charge?

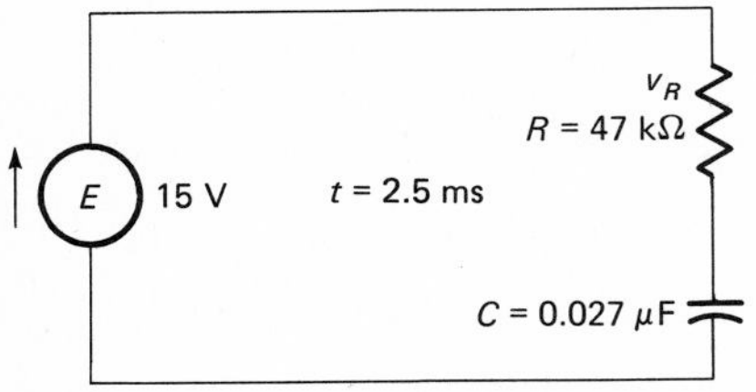

Fig. 10-8 Example 10-2: circuit.

Solution The steps adopted are:

1. *Calculate the time constant:*

$$\tau = RC \qquad (10\text{-}5)$$

$$= 47\ \text{k}\Omega \times 0.027\ \mu\text{F} = 1.27\ \text{k}\Omega \cdot \mu\text{F}$$

From Table 10-2 it can be seen that kΩ · μF = ms, so

$$\tau = 1.27\ \text{ms*}$$

2. *Select the curve:* During charge, the resistive voltage is given by Eq. (10-3), $v_R = Ee^{-t/RC}$, so the decreasing-value curve labeled $y = e^{-x}$ is used.
3. *Project from axis to axis:*

$$x = \frac{t}{\tau} = \frac{2.5\ \text{ms}}{1.27\ \text{ms}} = 1.97$$

From the point on the horizontal axis $x = 1.97$, project onto the curve $y = e^{-x}$ and then onto the vertical axis. The value of y becomes

$$y \simeq 0.14 \text{ of the maximum value of } v_R$$

The maximum value of v_R is $E = 15$ V. Therefore, 2.5 ms after the commencement of charge,

$$v_R \simeq 0.14 \text{ of } 15\ \text{V} \simeq \mathbf{2.1\ V}$$

Example 10-3 Figure 10-9*b* is the equivalent circuit of the input of the transistor pulse amplifier shown in outline in Fig. 10-9*a*. If a 0.5-V pulse is applied to the transistor input, how long does it take for the base-to-emitter voltage to reach 90 percent of the pulse amplitude?

* Three significant figures are adequate when curves such as those in Fig. 10-7 are used.

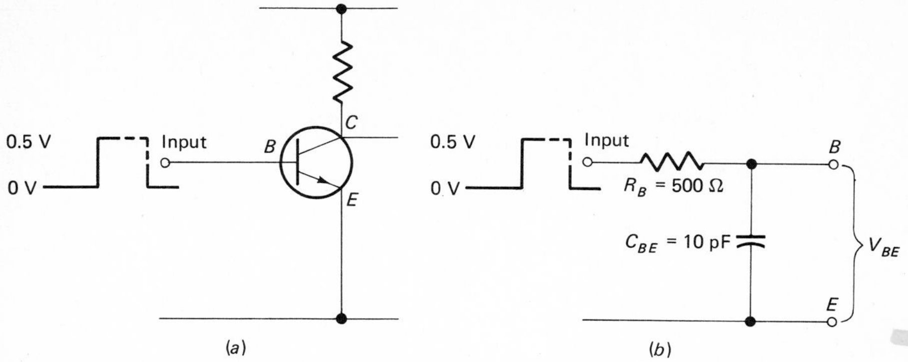

Fig. 10-9 Example 10-3: (*a*) given circuit; (*b*) equivalent base-to-emitter circuit.

Solution The input pulse, rising virtually instantaneously from 0 to 0.5 V, is the equivalent of applying a 0.5-V battery via a switch. Thus the base-to-emitter voltage V_{BE} is the voltage across the base-to-emitter capacitance C_{BE} while it is charging from 0 V toward a maximum of 0.5 V.

1. *Calculate the time constant:*

$$\tau = R_B C_{BE}$$

$$= 0.5\ \text{k}\Omega \times 10\ \text{pF}$$

$$\tau = 5\ \text{ns}$$

2. *Select the curve:* During charge, the capacitive voltage is given by Eq. (10-4), $v_C = E(1 - e^{-t/RC})$, so the increasing-value curve $y = 1 - e^{-x}$ is used.
3. *Project from axis to axis:*

$$y = 0.9 \text{ of maximum}$$

From the 0.9 point on the vertical axis, projecting onto the increasing-value curve and then onto the horizontal axis gives

$$x = \frac{t}{\tau} \simeq 2.3$$

Therefore the time taken for the base-to-emitter voltage v_{BE} to reach 0.9 of the pulse amplitude will be

$$t \simeq 2.3\,\tau \simeq 2.3 \times 5\ \text{ns} \simeq \mathbf{11.5\ ns}$$

In Example 10-3, the 11.5 ns taken for the base-to-emitter voltage to reach 0.9 of the pulse amplitude may seem to be a very small time period. Actually, in many applications of pulse amplifiers, such as high-speed computers, times of this order are much too long. A transistor with the specified input characteristics would be quite unacceptable.

This example can be viewed as an instance of waveform distortion.

In Fig. 10-10*a* is shown the input pulse rise. This pulse is assumed to increase from 0 to 0.5 V in almost zero time. Figure 10-10*b* shows the

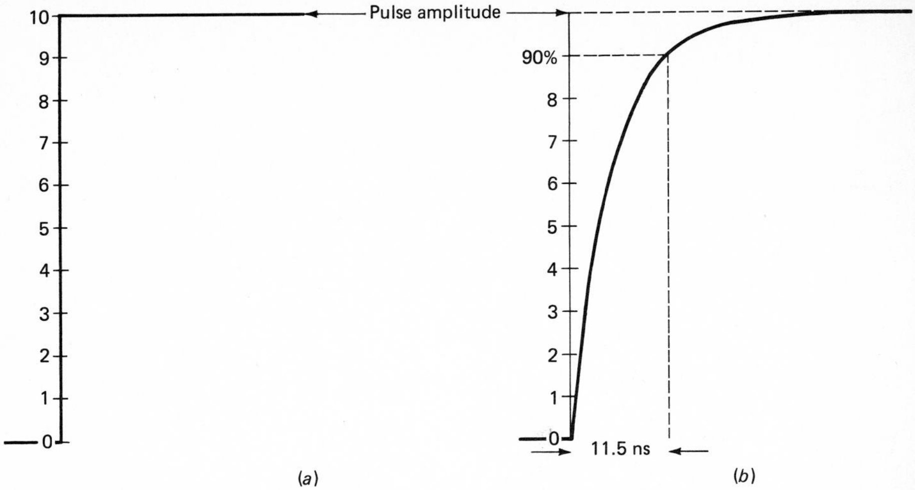

Fig. 10-10 Example 10-3: (*a*) 0.5-V pulse applied to the amplifier input; (*b*) the pulse as it appears between the base and the emitter of the transistor.

shape of the waveform actually appearing between the base and emitter of the transistor.

Example 10-4 The variable resistance and fixed-value capacitance in the circuit in Fig. 10-11 are used to determine the time during which voltage is applied to the electrodes of a welding machine. If this time is that taken for the capacitance to charge from 0 to 10 V, what is the time range covered by the extremes of the resistance values?

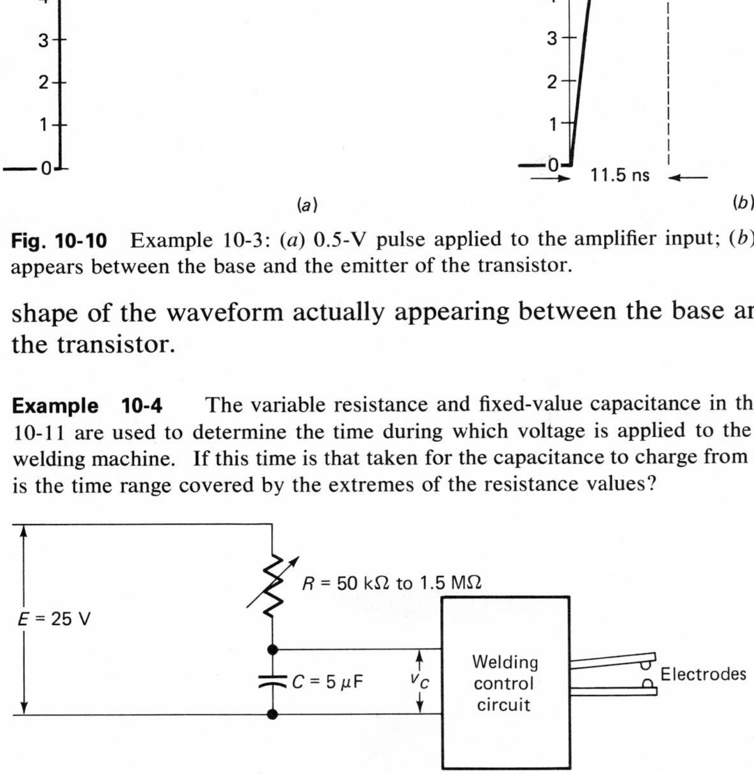

Fig. 10-11 Example 10-4: data.

Solution The voltage required to determine the end of the welding time is

$$v_C = 10 \text{ V}$$

The supply voltage $E = 25$ V is the maximum possible value of v_C, therefore

$$v_C = \frac{10 \text{ V}}{25 \text{ V}} = 0.4 \text{ of maximum}$$

Projecting from $y = 0.4$ onto the $y = 1 - e^{-x}$ curve in Fig. 10-7 and then onto the horizontal axis gives

$$x = \frac{t}{\tau} \simeq 0.5 \qquad \text{or} \qquad t \simeq 0.5\tau$$

When $R = 50\ \text{k}\Omega$,

$$\tau = RC \tag{10-5}$$

$$\tau = 50\ \text{k}\Omega \times 5\ \mu\text{F} = 250\ \text{ms}$$

$$t \simeq 0.5\tau \simeq 125\ \text{ms}$$

When $R = 1.5\ \text{M}\Omega$,

$$\tau = RC = 1.5\ \text{M}\Omega \times 5\ \mu\text{F} = 7.5\ \text{s}$$

$$t \simeq 0.5\tau \simeq 3.75\ \text{s}$$

The welding arc time range can therefore be varied from **0.125** to **3.75 s** by varying the resistance value throughout its range.

10-3 EVALUATION OF TIME

Example 10-4 is typical of situations in which it is desired to evaluate the time period elapsed during capacitance charge. The results given are not very accurate.

Equations (10-9) and (10-10) can be used to refer to some methods commonly adopted to determine time where accuracy greater than that achieved with the curves is essential.

The equations are first rearranged so that the term $e^{-t/\tau}$ becomes the subject:

$$y = Y_{\text{max}} e^{-t/\tau} \tag{10-9}$$

$$e^{-t/\tau} = \frac{y}{Y_{\text{max}}} \tag{10-11}$$

$$y = Y_{\text{max}}(1 - e^{-t/\tau}) \tag{10-10}$$

$$e^{-t/\tau} = \frac{Y_{\text{max}} - y}{Y_{\text{max}}} \tag{10-12}$$

Equations (10-11) and (10-12) indicate that the initial step in evaluating time is to determine a ratio of voltage or current values (subsequently referred to as *the ratio*).

For decreasing-value quantities, determine the ratio as the magnitude of the quantity divided by its maximum possible value [Eq. (10-11)].

For increasing-value quantities, determine the ratio as the difference between the magnitude of the quantity and its maximum value divided by the maximum value [Eq. (10-12)].

The magnitude of the exponent $-t/\tau$ (and hence the time t) can be found by using any of several methods:

Table of exponential functions By locating the value of the ratio [y/Y_{max} or $(Y_{\text{max}} - y)/Y_{\text{max}}$] in an e^{-x} table, the value of t/τ is given by the entry columns.

For Example 10-4 the ratio $(E - V_2)/E$ is 0.6. The closest value to 0.6 in an e^{-x} table is 0.6005. This value relates to 0.51 in the entry columns:

$$e^{-0.51} = 0.6005$$

Therefore $t/\tau \simeq 0.51$. When $\tau = 250$ ms ($R = 50$ kΩ and $C = 5$ μF), $t =$ 127.5 ms.

This method is easy to apply and gives a high degree of accuracy if the value of the ratio can be located in the table.

Table of natural logarithms From Eqs. (10-11) and (10-12),

$$e^{-t/\tau} = \frac{y}{Y_{max}} \quad \text{or} \quad \frac{Y_{max} - y}{Y_{max}}$$

These expressions can be written

$$e^{-t/\tau} = \text{ratio}$$

Therefore,

$$e^{t/\tau} = \frac{1}{\text{ratio}}$$

$$\frac{t}{\tau} = \ln\left(\frac{1}{\text{ratio}}\right)$$

$$t = \tau \ln\left(\frac{1}{\text{ratio}}\right)$$

Since the table of natural logarithms (ln) is a table of logarithms to base e, it gives the value of the exponent t/τ directly. In using the natural logarithm table, the reciprocal of the ratio is found in the entry columns. The exponent t/τ is given in the table.

In Example 10-4 the ratio is 0.6, its reciprocal being 1.667. Find 1.667 in the entry columns of natural logarithm tables. The value of ln 1.667 is 0.511. Therefore $t/\tau = 0.511$. When $\tau = 250$ ms, $t = 127.8$ ms.

This technique is accurate and easy to apply. Because the base of natural logarithms is $e = 2.718$, decimal multipliers expressed in terms of natural logarithms must be used to extend the range of values beyond those given in the tables (usually 1 to 9.999). This might be viewed as a disadvantage.

Table of common logarithms In applying common logarithms (logarithms to the base 10) it is convenient to use the equation form $e^{t/\tau} =$ 1/ratio, thus avoiding negative logarithm characteristics.

$$e^{t/\tau} = \frac{1}{\text{ratio}}$$

$$\frac{t}{\tau}\log e = \log\frac{1}{\text{ratio}}$$

$$\frac{t}{\tau} = \frac{1}{\log e} \times \log\frac{1}{\text{ratio}}$$

Log e (the common logarithm of the base of natural logarithms) is log 2.718 = 0.4343. The reciprocal of 0.4343 is 2.303, so that

$$\frac{t}{\tau} = 2.303 \log\frac{1}{\text{ratio}}$$

Applying this equation to Example 10-4,

$$\frac{t}{\tau} = 2.303 \log 1.667 = 2.303 \times 0.2219 = 0.511$$

When $\tau = 250$ ms, $t = 127.8$ ms.

The common logarithm method is accurate and obviates the problem of decimal multipliers since common logarithms are decimal-based. However, some may object to memorizing the numerical constant 2.303.

Slide rule log-log scales By locating the cursor line over a given number x on the D scale of a slide rule, the value of e^x or e^{-x} can be directly read off the appropriate log-log scale.

If $y = e^x$, then $x = \ln y$. Thus by reversing the above procedure and projecting along the cursor line from a log-log scale point onto the D scale, the exponent of e is directly found. Since in our time-constant relationships the exponent of e is t/τ or $-t/\tau$, log-log scales give the most rapid method of determining time.

Four methods of evaluating time are referred to. Which is the best method? There is no positive answer. You should try each method for every related problem at the end of this chapter. When you are able to apply each method, you will be in a position to decide which to use in any particular situation. A general guide is:

1. If speed is more important than accuracy, use a slide rule.
2. If accuracy is important, use logarithms. Select natural or common logarithms according to your own preference.
3. If speed is important and you have no log-log slide rule available, use the table of exponential functions.
4. For a fast, rough estimate, use the universal time-constant curves.

Finally, if you have access to an electronic calculator with an exponential function capability, forget items 1 to 4 and use the calculator.

There is no special virtue in being able to perform laborious calculations which do not enhance circuit knowledge.

The techniques for evaluating time can be applied to the evaluation of instantaneous voltages and currents (v_R, v_C, and i), the slide rule and calculator methods being particularly useful in this context. These methods were not introduced earlier because, until it becomes necessary to find the value of the exponent of e, the problems involved in the solution of exponential functions do not seem to warrant special attention.

10-4 CHARGING FROM A PARTIALLY CHARGED STATE

A capacitance may charge from an initial condition where it is not fully discharged; where the voltage at time t_0 is not zero. An example will show how this type of situation can be dealt with using the techniques already developed.

Example 10-5 If, in Example 10-4, the welding control circuit does not fully discharge the capacitor after a weld is completed but leaves v_C equal to 2.5 V, what will be the maximum weld time during the next welding cycle?

Solution Figure 10-12*a* shows the conditions existing before the weld cycle starts: $v_C =$ 2.5 V. In Fig. 10-12*b* the welding cycle has been completed, and $v_C = 10$ V. We wish to know how long it takes for v_C to rise from 2.5 to 10 V.

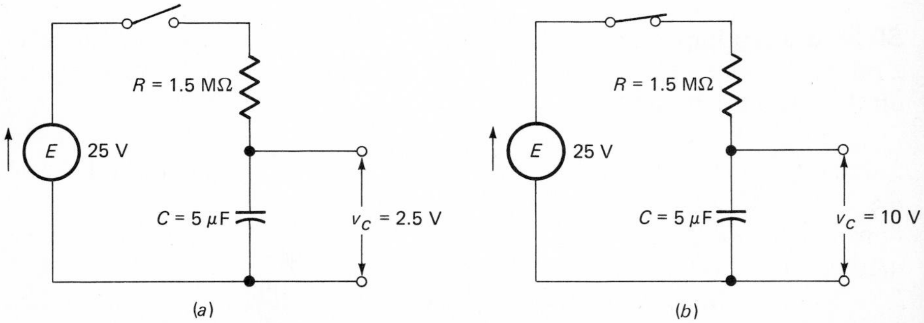

Fig. 10-12 Example 10-5: circuit. (*a*) Before the weld cycle starts; (*b*) at completion of the weld cycle.

The maximum weld time occurs when $R = 1.5$ MΩ or $\tau = 7.5$ s. The time taken for the capacitor to charge from $v_C = 2.5$ V to $v_C = 10$ V will be the difference between the times taken to charge from a fully discharged state to $v_C = 10$ and 2.5 V, respectively. This statement can be written symbolically:

$$\Delta t = t_{10} - t_{2.5}$$

Calculating t_{10}, from Eq. (10-12),

$$e^{-t_{10}/\tau} = \frac{Y_{\max} - y_{10}}{Y_{\max}} = \frac{25\text{ V} - 10\text{ V}}{25\text{ V}} = 0.6$$

$$-\frac{t_{10}}{\tau} = \ln 0.6 \qquad \text{or} \qquad \frac{t_{10}}{\tau} = \ln 1.667$$

But $\tau = 7.5$ s; therefore,

$$t_{10} = 7.5 \text{ s} \times \ln 1.667 = 7.5 \text{ s} \times 0.511 = 3.833 \text{ s}$$

Using a similar technique, $t_{2.5}$ is calculated

$$t_{2.5} = 0.7898 \text{ s}$$

The time taken for the capacitor to charge from 2.5 to 10 V will be

$$\Delta t = 3.833 \text{ s} - 0.7898 \text{ s} = \mathbf{3.043 \text{ s}}$$

10-5 DISCHARGING CAPACITANCE THROUGH RESISTANCE

Figure 10-13*a* through *d* shows a series of events in which a charged capacitance is discharged through a resistance.

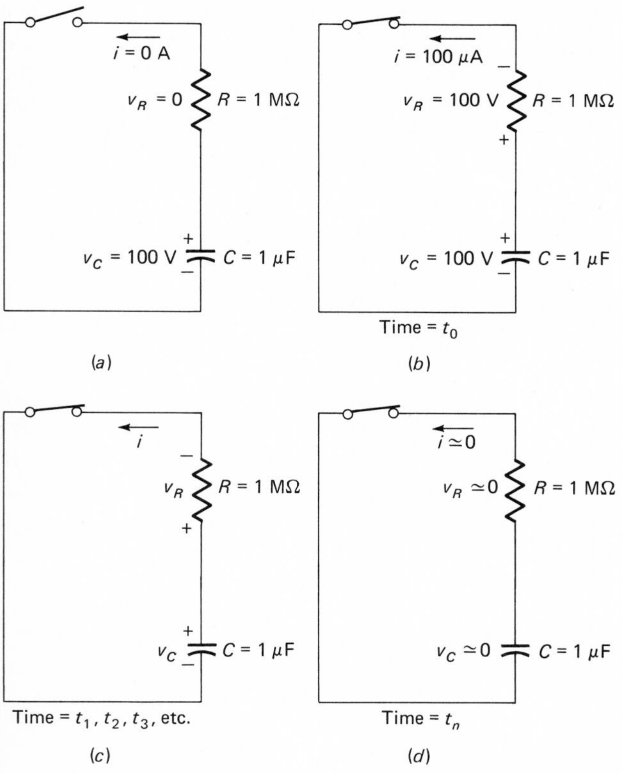

Fig. 10-13 Discharging capacitance through resistance.

During charge (Fig. 10-1) the capacitance is part of the energy source load. During discharge the capacitance *is* the energy source, and the resistance is its load. The discharge current direction follows the normal source-load convention; from capacitance positive through the resistance to capacitance negative.

Figure 10-13*a* The capacitance is assumed to have been charged until $v_C = 100$ V. Since the switch is open, there is no current and therefore no resistive voltage.

Figure 10-13*b* At time t_0, when the switch is first closed, the initial conditions are

$$v_C = 100 \text{ V}$$

$$v_R = v_C = 100 \text{ V}$$

$$i = \frac{v_R}{R} = \frac{100 \text{ V}}{1 \text{ M}\Omega} = 100 \ \mu\text{A}$$

Figure 10-13*c* In supplying current to the circuit, the energy imparted to the capacitance during its charge is extracted. This energy is transferred to the resistance, which dissipates it as heat. As energy leaves the capacitance, its voltage drops toward zero ($W = V^2C/2$ or $V \propto \sqrt{W}$). Since v_R equals v_C and $i = v_R/R$ during discharge, at successive time intervals t_1, t_2, t_3, etc., after the switch is closed (time t_0), both voltages v_C and v_R and the current i decrease toward zero. The voltage and current values can be summarized as follows:

At times t_1, t_2, t_3, etc.,

$$\left.\begin{array}{r} 100 \text{ V} > v_C > 0 \\ 100 \text{ V} > v_R > 0 \\ 100 \ \mu\text{A} > i > 0 \end{array}\right\} \text{Each quantity decreases from its value at time } t_0 \text{ toward zero}$$

Figure 10-13*d* After a period greater than about 5τ, the voltages and current become virtually zero (at 5τ they will each be 0.67 percent of their initial values). Thus at time t_n, a considerable time after t_0, the conditions become

$$v_C \simeq 0 \text{ V}$$

$$v_R \simeq 0 \text{ V}$$

$$i \simeq 0 \text{ A}$$

COMPARISON OF CHARGE AND DISCHARGE

When Fig. 10-13 is compared with Fig. 10-1, the following points can be seen:

1. *a.* v_C increases from zero during charge.
   b. v_C decreases toward zero during discharge.
   c. The polarity of v_C is the same during charge and discharge.

2. *a.* v_R decreases toward zero during charge and discharge.
   b. The polarity of v_R during discharge is opposite that during charge.
3. *a.* *i* decreases toward zero during charge and discharge.
   b. The direction of *i* during discharge is opposite that during charge.

RC CIRCUIT DISCHARGE CURVES

Figure 10-14 shows the discharge curves of v_C, v_R, and *i* with respect to time. The reversals of polarity and direction relative to charge are indicated by the negative values of v_R and *i*.

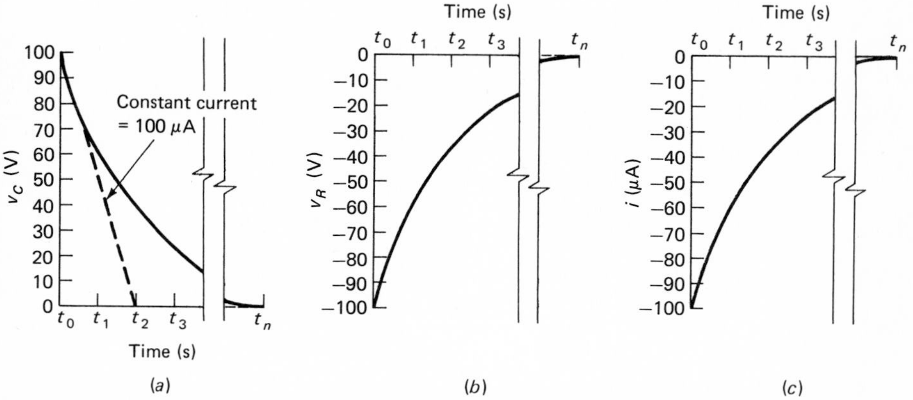

Fig. 10-14 *RC* circuit in Fig. 10-13. Discharge curves. (*a*) Capacitive voltage; (*b*) resistive voltage; (*c*) current.

Curves which follow the same law from a maximum value toward zero are the same shape, whether they represent positive or negative values. The three discharge curves are the same shape. Furthermore, they are the same shape as the decreasing-value charge curves (Fig. 10-2*b* and *c*).

The logic explaining the curvature of the discharge curves is essentially the same as in the case of charging: As the capacitance discharges, its voltage drops. This voltage is the circuit emf during discharge, so the current decreases. The decreasing current slows the rate at which the discharge takes place, thus producing the curvature.

The broken straight line in Fig. 10-14*a* shows the manner in which the capacitive voltage would decrease if the discharge current stayed constant at its initial value of 100 μA. The time it would take for v_C to reach zero at a constant 100-μA discharge current is $t_2 = \tau$ s. A general conclusion follows.

The initial charge held by the capacitance is

$$Q = CV_C \qquad (8\text{-}9)$$

But

$$Q = it \qquad (2\text{-}11)$$

Therefore
$$it = CV_C$$
$$t = C\frac{V_C}{i}$$

V_C is the initial capacitive voltage at the start of discharge; i is the discharge current, here assumed to be constant at a value V_C/R (or $V_C/i = R$). Thus

$$t = CR = \tau$$

This relationship also applies to the straight broken line in Fig. 10-2*a*.

Another way of defining time-constant results is: The *time constant* τ is the time it would take for the capacitive voltage to equal the applied emf during charge, or zero during discharge, if the current were maintained constant at its initial value.

THE EQUATIONS FOR THE DISCHARGE CIRCUIT

The loop equation for the circuit in Fig. 10-13 is

$$v_C = v_R \qquad \text{or} \qquad v_C - v_R = 0 \tag{10-13}$$

Equation (10-13) can be written as:

$$\frac{1}{C}\int i\,dt - iR = 0$$

A calculus solution of this equation, given in the appendix, shows that

$$i = \frac{V_0}{R}\,e^{-t/RC} \tag{10-14}$$

where i = instantaneous circuit current, A
V_0 = initial circuit voltage (the voltage to which the capacitance was charged initially, or v_C at time t_0), V
e = base of natural logarithms $\simeq$ 2.718
R = resistance, Ω
C = capacitance, F
t = time, s

Applying Ohm's law and Eq. (10-14) gives,

$$v_R = iR = \left(\frac{V_0}{R}\,e^{-t/RC}\right)R$$

$$v_R = V_0 e^{-t/RC} \tag{10-15}$$

Substituting for v_R in Eq. (10-13),

$$v_C = v_R = V_0 e^{-t/RC} \qquad (10\text{-}16)$$

By comparing Eqs. (10-14) to (10-16) with Eqs. (10-2) and (10-3), it can be seen that all five are of the same form (they all refer to decreasing values of current or voltage). Therefore the calculation techniques used for i and v_R in the charging circuit can be used for i, v_R, and v_C under discharge conditions. They include the use of

1. The time-constant concept ($\tau = RC$).
2. The universal time-constant table (Table 10-3). When employing this table in the discharge situation, the v_R values are used for v_R and v_C.
3. Figure 10-7. Since only decreasing values of current and voltage exist in the discharge circuit, the curve $y = e^{-x}$ is used for i, v_R, and v_C.
4. The techniques referred to in Sec. 10-3.

Example 10-6 What are the approximate values of current and voltage in the circuit in Fig. 10-15 at $t = 5\ \mu s$ after the switch is closed? The capacitance is initially charged until $v_C = 15$ V.

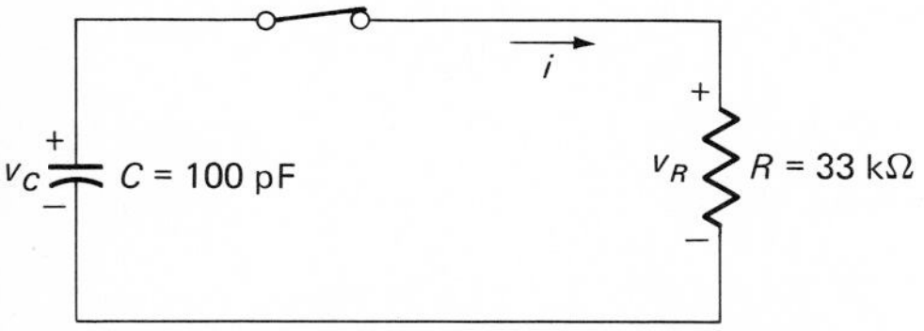

Fig. 10-15 Example 10-6: circuit.

Solution

$$\tau = RC \qquad (10\text{-}5)$$

$$\tau = 33\ k\Omega \times 100\ pF = 3300\ ns = 3.3\ \mu s$$

$$t = 5\ \mu s = \frac{5\ \mu s}{3.3\ \mu s}\tau = 1.52\tau$$

Using the $y = e^{-x}$ curve in Fig. 10-7, projection gives

$$v_C \simeq 22 \text{ percent of } V_0 \simeq 22 \text{ percent of } 15\ V$$

$$v_C \simeq \mathbf{3.3\ V}$$

$$v_R = v_C \simeq \mathbf{3.3\ V}$$

$$i = \frac{V_R}{R} \simeq \frac{3.3\ V}{33\ k\Omega} \simeq \mathbf{100\ \mu A}$$

Example 10-7 A 10-μF capacitor is charged to 100 V. A 100-V 20-kΩ/V voltmeter is connected across the capacitor, and the time of discharge through the voltmeter resistance is measured with a stopwatch. How much time elapses between the voltmeter readings of 90 and 10 V?

Solution (See Fig. 10-16.)

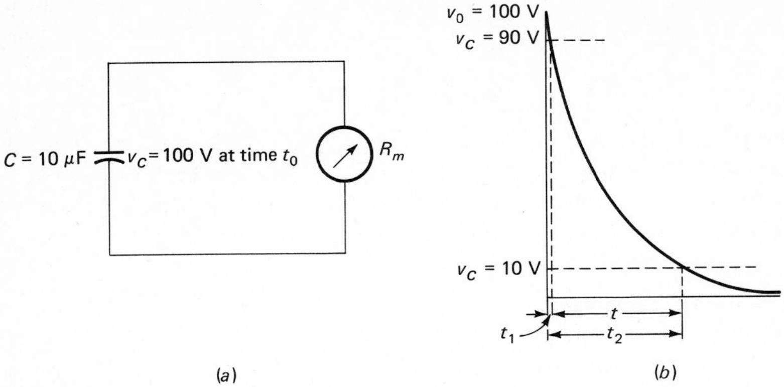

Fig. 10-16 Example 10-7: (*a*) circuit; (*b*) curve of v_C with respect to time.

$$R_m = 100\ \text{V} \times 20\ \text{k}\Omega/\text{V} = 2\ \text{M}\Omega$$

$$\tau = R_m C \qquad (10\text{-}5)$$

$$\tau = 2\ \text{M}\Omega \times 10\ \mu\text{F} = 20\ \text{s}$$

The decreasing-value relationships apply to the current and to both voltages in a discharging RC series circuit. Therefore Eq. (10-11) can be used:

$$e^{-t/\tau} = \frac{y}{Y_{\text{max}}} \qquad (10\text{-}11)$$

$$\frac{t}{\tau} = \ln\left(\frac{Y_{\text{max}}}{y}\right)$$

$$t = \tau \ln\left(\frac{Y_{\text{max}}}{y}\right) = 20\ \text{s} \times \ln\left(\frac{Y_{\text{max}}}{y}\right)$$

The time taken for v_C to decrease from 100 to 90 V is given by

$$t_1 = 20\ \text{s} \times \ln\ (100\ \text{V}/90\ \text{V}) = 20\ \text{s} \times \ln 1.111$$

$$t_1 = 20\ \text{s} \times 0.1053 = 2.106\ \text{s}$$

The time taken for v_C to drop from 100 to 10 V will be

$$t_2 = 20\ \text{s} \times \ln\ (100\ \text{V}/10\ \text{V}) = 20\ \text{s} \times \ln 10$$

$$t_2 = 20\ \text{s} \times 2.303 = 46.06\ \text{s}$$

The time taken for v_C to decrease from 90 to 10 V is the difference between times t_2 and t_1

$$t = t_2 - t_1 = 46.06\ \text{s} - 2.106\ \text{s} = \mathbf{43.95\ s}$$

Example 10-7 could have been solved more rapidly by assuming an initial voltage value $v_C = 90$ V and determining the time taken for v_C to decrease to 10 V.

$$t = 20\ \text{s} \times \ln\ (90\ \text{V}/10\ \text{V}) = 20\ \text{s} \times \ln 9$$

$$t = 20\ \text{s} \times 2.197 = \mathbf{43.94\ s}$$

The difference in the final digits of the two solutions is due to the use of four significant figures in the calculations.

Example 10-8 Figure 10-17*a* is the circuit of a relaxation oscillator used for generating a pulse waveform. As the capacitor C charges through R_1, the voltage v_e rises. When v_e becomes 6 V, the unijunction transistor internal resistance R_e drops suddenly from a very high value to 400 Ω, discharging the capacitor. At this time an output voltage pulse v_o is produced. When v_e drops to 2 V, R_e again becomes very high, allowing the capacitor to recharge. The action automatically repeats itself, so that a series of output pulses occurs at regular time intervals. The resulting waveforms of v_e and v_o are shown in Fig. 10-17*b*. What is the time t between consecutive pulses?

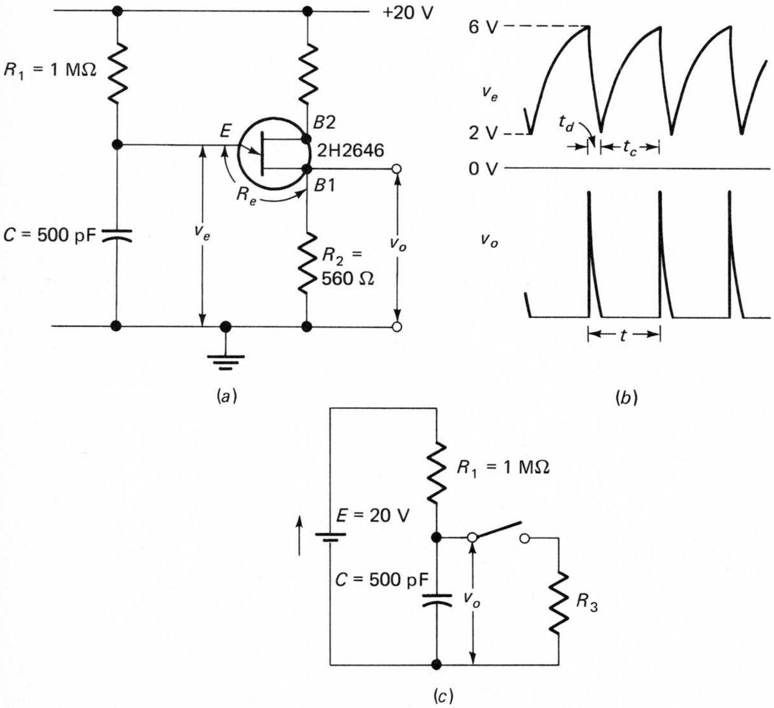

Fig. 10-17 Example 10-8: (*a*) given circuit; (*b*) waveforms of v_e and v_o; (*c*) equivalent circuit.

Solution The complexity of Fig. 10-17*a* can be eliminated for this example by considering the timing circuit equivalent in Fig. 10-17*c*. The switch simulates the transistor action. When the capacitor C is charging to 6 V, the switch is open; for discharging from 6 to 2 V, the switch is closed, connecting resistance R_3 across the capacitor. R_3 is the series equivalent of R_e (400 Ω during capacitor discharge), and $R_2 = 560$ Ω.

While C is charging (switch open in Fig. 10-17*c*), the maximum circuit voltage during charge is the supply voltage:

$$Y_{\text{max}} = 20\ \text{V}$$

The charging resistance is R_1. From Eq. (10-5),

$$\tau_c = R_1 C = 1\ \text{M}\Omega \times 500\ \text{pF} = 500\ \mu\text{s}$$

The time for C to charge from 0 to 6 V is, from Eq. (10-12), given by

$$e^{-t_1/\tau_c} = \frac{Y_{\text{max}} - y}{Y_{\text{max}}}$$

Employing common logarithms,

$$t_1 = 2.303 \times \tau_c \times \log\left(\frac{Y_{\text{max}}}{Y_{\text{max}} - y}\right) = 2.303 \times 500\ \mu\text{s} \times \log\ [20\ \text{V}/(20\ \text{V} - 6\ \text{V})]$$

$$t_1 = 2.303 \times 500\ \mu\text{s} \times 0.155 = 178.5\ \mu\text{s}$$

The time for C to charge from 0 to 2 V is

$$t_2 = 2.303 \times \tau_c \times \log\ [20\ \text{V}/(20\ \text{V} - 2\ \text{V})]$$

$$t_2 = 2.303 \times 500\ \mu\text{s} \times 0.0457 = 52.62\ \mu\text{s}$$

The time for C to charge from 2 to 6 V is

$$t_c = t_1 - t_2 = 178.5\ \mu\text{s} - 52.62\ \mu\text{s} = 125.9\ \mu\text{s}$$

While C is discharging (switch closed in Fig. 10-17*c*),

$$R_3 = R_e + R_2 = 400\ \Omega + 560\ \Omega = 960\ \Omega$$

The discharge circuit is as shown in Fig. 10-18*a*.

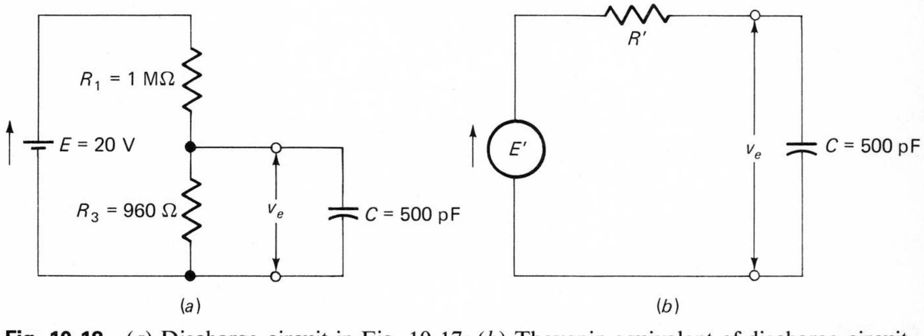

Fig. 10-18 (*a*) Discharge circuit in Fig. 10-17; (*b*) Thevenin equivalent of discharge circuit.

The series-parallel circuit in Fig. 10-18*a* must be converted into a series circuit for direct evaluation of discharge time. The Thevenin principle used with the capacitor as the load can be applied to perform the conversion. Figure 10-18*b* shows the resulting Thevenin equivalent circuit.

$$E' = E\frac{R_3}{R_1 + R_3} = 20\ \text{V}\frac{960\ \Omega}{1\ \text{M}\Omega + 960\ \Omega} = 0.0192\ \text{V}$$

$$R' = \frac{R_1 R_3}{R_1 + R_3} = \frac{1\ \text{M}\Omega \times 960\ \Omega}{1\ \text{M}\Omega + 960\ \Omega} = 959.1\ \Omega$$

Since the Thevenin voltage $E' \simeq 0$ V and the Thevenin resistance $R' \simeq R_3$, for practical purposes it may be considered that C discharges directly into the 960 Ω of R_3. The discharge extends from $v_e = 6$ V to $v_e = 2$ V. From Eq. (10-5),

$$\tau_d \simeq R_3 C \simeq 960\ \Omega \times 500\ \text{pF} \simeq 480\ \text{ns} \simeq 0.48\ \mu\text{s}$$

From Eq. (10-11),

$$e^{-t_d/\tau_d} = \frac{Y_{\max}}{y}$$

$$t_d = 2.303 \times \tau_d \times \log\left(\frac{Y_{\max}}{y}\right) = 2.303 \times 0.48\ \mu\text{s} \times \log\ (6\ \text{V}/2\ \text{V})$$

$$t_d = 2.303 \times 0.48\ \mu\text{s} \times 0.4771 = 0.5274\ \mu\text{s}$$

The time between pulses is the sum of the charge and discharge times:

$$t = t_c + t_d = 125.9\ \mu\text{s} + 0.5274\ \mu\text{s} = \mathbf{126.4\ \mu s}$$

Example 10-9 What will be the current i_C in the circuit in Fig. 10-19*a* at $t = 10$ ms after the switch is closed if the capacitance is initially uncharged?

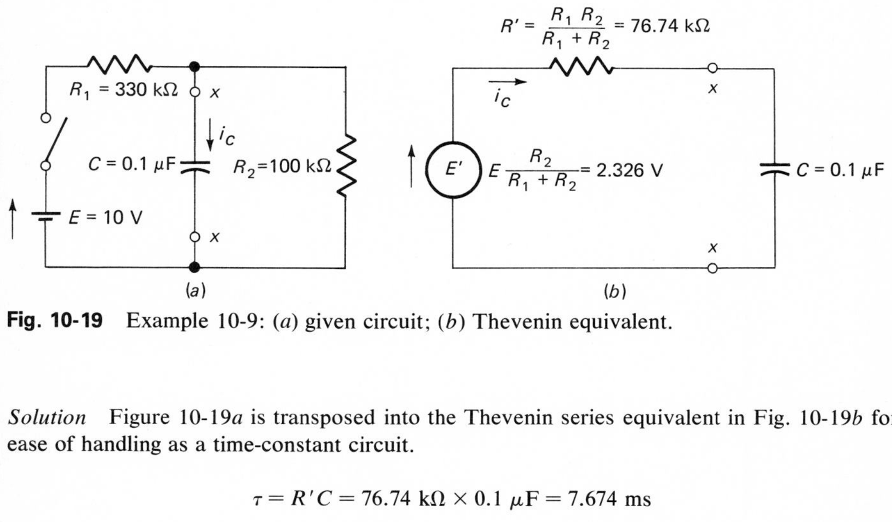

Fig. 10-19 Example 10-9: (*a*) given circuit; (*b*) Thevenin equivalent.

Solution Figure 10-19*a* is transposed into the Thevenin series equivalent in Fig. 10-19*b* for ease of handling as a time-constant circuit.

$$\tau = R'C = 76.74\ \text{k}\Omega \times 0.1\ \mu\text{F} = 7.674\ \text{ms}$$

From Eq. (10-6),

$$i = \frac{E'}{R'}\, e^{-t/\tau} = \frac{2.326\ \text{V}}{76.74\ \text{k}\Omega}\, e^{-10\ \text{ms}/7.674\ \text{ms}} = \mathbf{0.0083\ mA} \quad \text{or} \quad \mathbf{8.3\ \mu A}$$

A review of Example 10-9 from a different approach can aid in understanding these circuits.

1. Maximum capacitive voltage
   The maximum possible voltage which could be developed across the

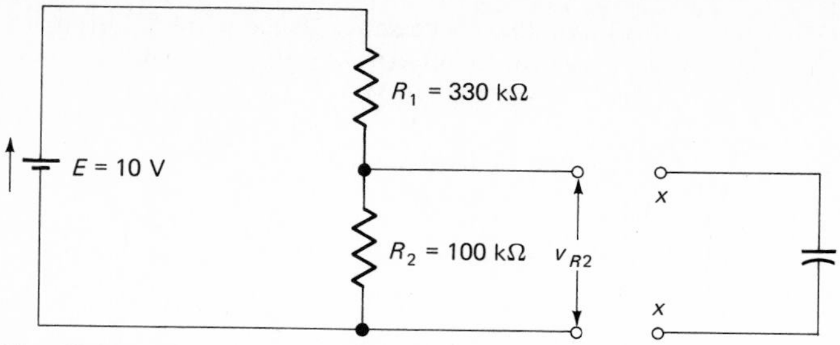

Fig. 10-20 Figure 10-19*a* rearranged.

capacitance is the voltage across R_2 in Fig. 10-20 with the capacitance disconnected:

$$V_{R2} = E\frac{R_2}{R_1 + R_2} = 10\text{ V}\frac{100\text{ k}\Omega}{330\text{ k}\Omega + 100\text{ k}\Omega} = 2.326\text{ V}$$

Thus the voltage of the capacitance during its charge increases from zero toward 2.326 V.

2. Charging resistance. With or without R_2, the available initial charging current will be the same, because $v_C = 0$ V. This is shown in Fig. 10-21.

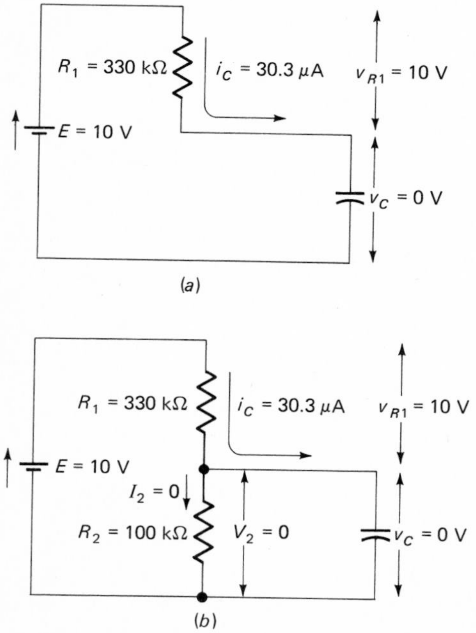

Fig. 10-21 Figure 10-20: initial charging conditions. (*a*) Without R_2; (*b*) with R_2.

If the charge were to be completed ($t \gg 5\tau$), the conditions without and with R_2 would be as shown in Fig. 10-22.

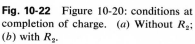

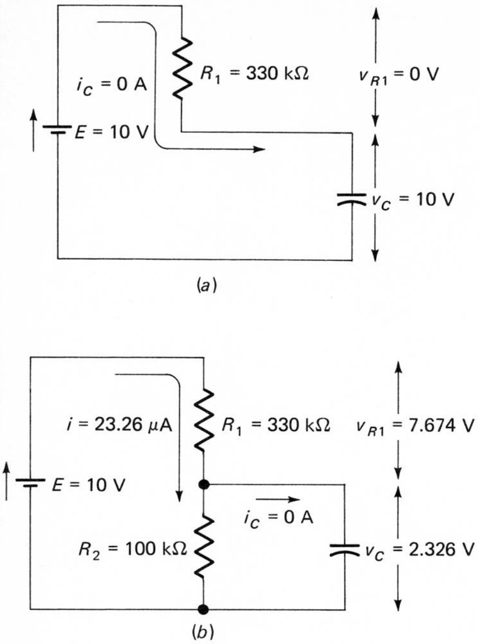

Fig. 10-22 Figure 10-20: conditions at completion of charge. (*a*) Without R_2; (*b*) with R_2.

From the viewpoint of the capacitance terminals, the situation for R_2 will be:

1. Figure 10-21*b*: With the capacitance fully discharged, $v_C = 0$ V and the circuit provides an initial charging current of 0.0303 mA.
2. Figure 10-22*b*: With the capacitance fully charged, $i_C = 0$ A and the capacitive voltage is 2.326 V.

Item 2 confirms that the equivalent charging emf is 2.326 V. Items 1 and 2 together lead to the conclusion that the equivalent charging resistance is equal to 2.326 V/30.3 μA = 76.74 kΩ. This resistance value is that obtained by considering R_1 and R_2 to be effectively in parallel.

10-6 *RC* CIRCUIT WITH APPLIED SQUARE WAVES

Within the broken-line box in Fig. 10-23*a* is a basic square-wave generator. When the switch is at position 1, $v_{in} = 0$ V and the *RC* network is short-circuited so that any charge on the capacitance will be dissipated in the resistance. When the switch is at position 2, $v_{in} = 10$ V is supplied to the *RC* network as a charging voltage. The resulting network supply voltage is shown as the square waveform at the bottom of Fig. 10-23*b*, it being assumed that the switch is automatically operated so that it stays in each position for 1 ms and rapidly switches between the positions.

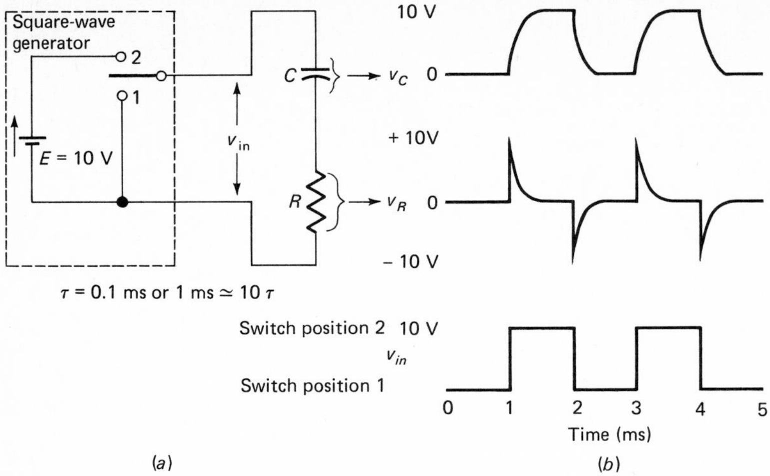

Fig. 10-23 *RC* circuit with applied square waves.

The time constant $\tau = RC$ is chosen to be about 0.1 ms (1 ms $\simeq$ 10τ), so that the resistive and capacitive voltages have time to stabilize between switch operations.

The waveforms of v_R and v_C shown in Fig. 10-23*b* are the result of successive charging and discharging. They are thus composites of Figs. 10-2*a* and 10-14*a* for v_C, and Figs. 10-2*b* and 10-14*b* for v_R.

DIFFERENTIATION

When the applied voltage changes value suddenly, the resistive voltage v_R also changes value suddenly. When the applied voltage suddenly increases from 0 to +10 V (a 10-V positive swing), v_R also swings 10 V in a positive direction. Conversely, when v_{in} drops from +10 to 0 V (a 10-V negative swing) v_R swings 10 V in a negative direction.

If the applied voltage does not change (as, for instance, during the 1-ms periods in Fig. 10-23), the resistive voltage is due only to the capacitive charge conditions stabilizing.

In general, the resistive voltage is determined by the rate and direction of change in the applied voltage. This concept is similar to differentiation in calculus in which rates of change are significant. The resistive voltage v_R is called the *differential voltage.* If the output of an *RC* series circuit is taken from the resistance, the circuit is called a *differentiator,* the time constant usually being not more than 10 percent of the minimum available charge or discharge time.

INTEGRATION

While the applied voltage is at the 10-V level, each increment of capacitive voltage adds to the existing voltage as the capacitance charges. This summation is related to integration in mathematics. The capacitive voltage v_C is called the *integral voltage.* If the output of an *RC* series circuit is taken from the capacitance, the circuit is called an *integrator,* the time constant typically being chosen to be at least 10 times the maximum available charge and/or discharge time.

Example 10-10 The rectangular wave v_{in} in Fig. 10-24*b* is applied to the input of the differentiator (Fig. 10-24*a*). What is the reason for the shape of the output waveform v_{out}?

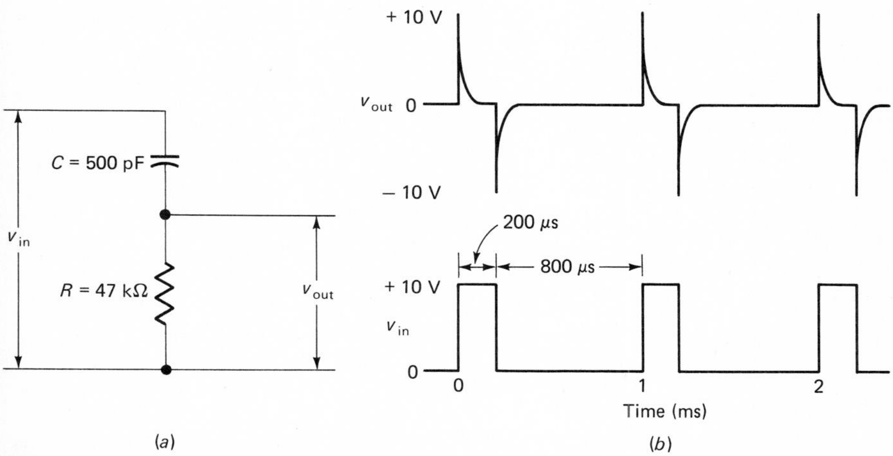

Fig. 10-24 Example 10-10: (*a*) circuit; (*b*) input and output voltage waveforms.

Solution The rectangular wave v_{in} has unequal times for the pulse duration and the period between the pulses. Since the 23.5-μs time constant of the differentiator circuit (47 kΩ × 500 pF) is short compared with the 200-μs pulse duration and with the 800-μs interval between pulses, the capacitor always has time to charge before the input pulse changes level. Thus the output voltage v_{out} is always close to zero prior to an input level change. When the input level changes from 0 to 10 V, a positive swing, the output rises suddenly from 0 to +10 V. When the input level drops from 10 to 0 V, a negative swing, the output *rises* from 0 to −10 V. The output voltage v_{out} consists of spikes which rise almost instantaneously in either a positive or negative direction on an input level change, dropping virtually to zero before another input level change occurs.

Example 10-11 The rectangular wave v_{in} in Fig. 10-25*b* is applied to the input of the integrator (Fig. 10-25*a*). What is the reason for the shape of the output waveform v_{out}?

Solution The 8.2-ms time constant of the integrator is long compared with the 800-μs pulse duration and the 200 μs between the pulses. During an 800-μs period, the capacitor voltage increases by approximately 10 percent of the difference between its voltage at the start of the period and its maximum possible voltage 10 V. During a 200-μs period, the capacitor loses about 2.4 percent of its voltage. Therefore the output voltage v_{out} gradually builds up toward the maximum of 10 V.

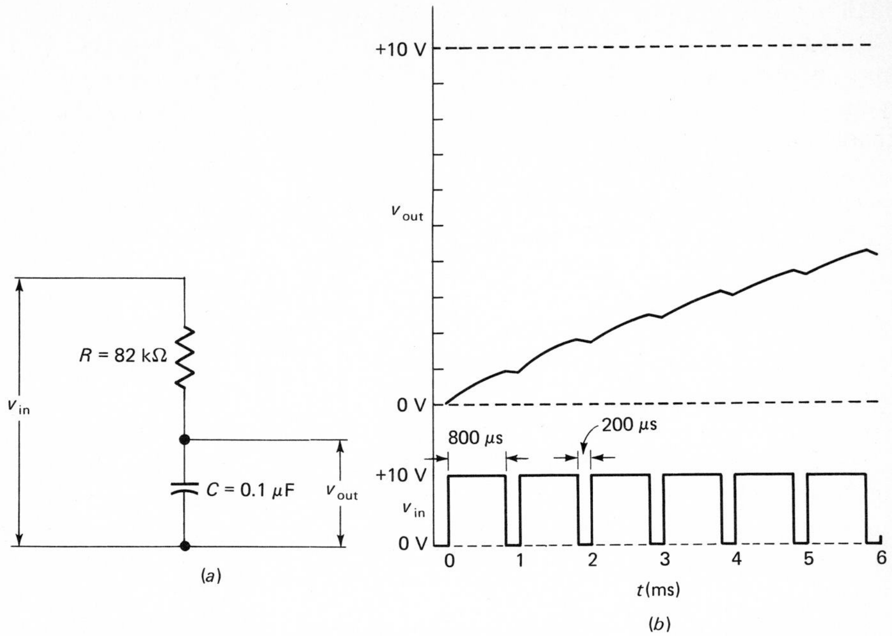

Fig. 10-25 Example 10-11: (*a*) circuit; (*b*) input and output voltage waveforms.

Examples 10-10 and 10-11 typify the operation of circuits in a television set which separate the synchronizing signals to maintain a steady picture.

Short-duration pulses constitute part of the transmitted signal. These enter a short-time-constant differentiator and leave as a series of sharp spikes, as in Example 10-10. These voltage spikes synchronize the horizontal sweep circuit.

Fig. 10-26 Differentiator of Example 10-10. (*a*) Circuit; (*b*) waveforms when the input is the same as in Example 10-11.

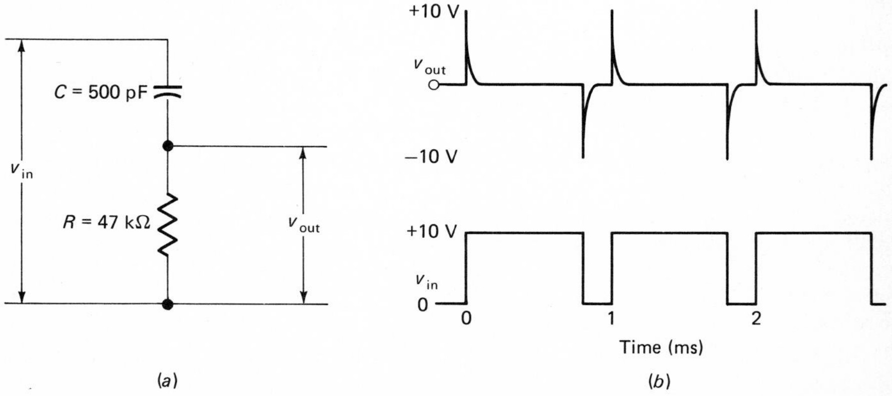

Also included with the transmitted signal are groups of long pulses. Each time one of these groups of pulses enters a long-time-constant integrator, it leaves as one built-up pulse, the action being similar to that in Example 10-11. The output pulses synchronize the vertical sweep circuit. The vertical synchronizing waveform of Fig. 10-25*b* may be simultaneously applied to the differentiator of Fig. 10-24*a*. The resulting series of pulses shown in Fig. 10-26 would be used to maintain horizontal synchronization during the vertical synchronization period.

10-7 CHARGING INDUCTANCE THROUGH RESISTANCE

Figure 10-27 shows a sequence of events in a series *RL* circuit during the charge of the inductance.

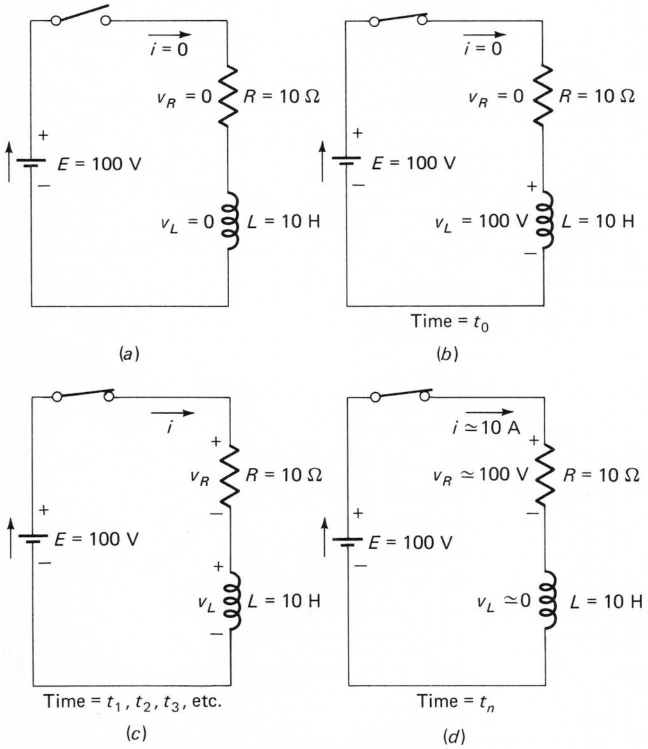

Fig. 10-27 Charging inductance through resistance.

Figure 10-27*a* With the switch open, the current i, the inductive voltage v_L, and the resistive voltage v_R are all zero.

Figure 10-27*b* There is no current at the instant the switch is closed, but there is a maximum rate of change in current. This concept needs some explanation.

Consider a current which, in a period of 1 s, changes from 5 A in one direction to 5 A in the opposite direction. This change is shown graphically in Fig. 10-28.

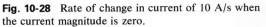

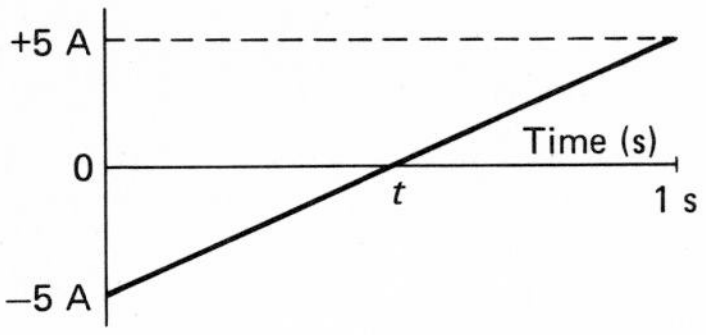

Fig. 10-28 Rate of change in current of 10 A/s when the current magnitude is zero.

The current is changing at the rate of 10 A/s for the whole 1-s period. Therefore it is changing at this rate at time t. At this time, the current value is zero.

In Fig. 10-27*b*, at time t_0, it might be said that the current is *trying* to increase as rapidly as possible from zero value (which certainly existed before the switch was closed) to a maximum value of $E/R = 10$ A. In doing so, it achieves a high rate of increase. This current increase causes an induced emf in the inductance, which opposes the current change (Lenz's law). The initial effect is that the magnitude of the induced emf becomes equal to that of the applied emf and so prevents current. Of course, the zero current value exists only for an immeasurably small time period, since the current must change to produce a voltage.

At time t_0,

$$v_L = E = 100 \text{ V}$$

$$v_R = E - v_L = 100 \text{ V} - 100 \text{ V} = 0 \text{ V}$$

$$i = 0 \text{ A}$$

Figure 10-27c As the current increases, the resistive voltage also increases from zero. Its polarity is such as to reduce the total circuit voltage applied across the inductance. Thus the rate of current rise decreases. As time progresses and v_R becomes higher, the current rise tapers off almost to zero.

The current and voltage values at times t_1, t_2, t_3, etc., are

$$100 \text{ V} > v_L > 0$$

$$0 < v_R < 100 \text{ V}$$

$$0 < i < 10 \text{ A}$$

Figure 10-27*d* Time t_n is to be regarded as sufficiently long after time t_0 for conditions virtually to have become stabilized.

$$v_L \simeq 0 \text{ V}$$

$$v_R = E - v_L \simeq 100 \text{ V} - 0 \text{ V} \simeq 100 \text{ V}$$

$$i = \frac{v_R}{R} \simeq \frac{100 \text{ V}}{10 \ \Omega} \simeq 10 \text{ A}$$

RL CIRCUIT CHARGE CURVES

The curves in Fig. 10-29 summarize the events in Fig. 10-27.

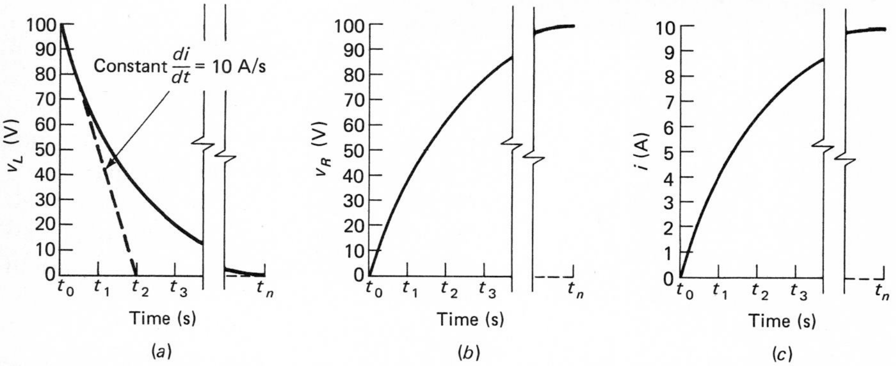

Fig. 10-29 *RL* circuit in Fig. 10-27. Charge curves with respect to time. (*a*) Inductive voltage; (*b*) resistive voltage; (*c*) current.

THE EQUATIONS FOR THE CHARGE CIRCUIT

The loop equation for the circuit in Fig. 10-27 is

$$E = v_L + v_R \tag{10-17}$$

or

$$E = L\frac{di}{dt} + iR$$

From this expression an equation can be obtained for the current in an *RL* circuit during charge. The derivation is given in the appendix.

$$i = \frac{E}{R}(1 - e^{-tR/L}) \tag{10-18}$$

Applying Ohm's law and Eq. (10-18) gives,

$$v_R = iR = E(1 - e^{-tR/L}) \tag{10-19}$$

Substituting for v_R in Eq. (10-17),

$$v_L = E - v_R = E - E(1 - e^{-tR/L})$$

$$v_L = Ee^{-tR/L} \tag{10-20}$$

In Eqs. (10-18) to (10-20),

i = instantaneous current, A
v_R = instantaneous resistive voltage, V
v_L = instantaneous inductive voltage, V
E = supply emf, V
R = resistance, Ω
L = inductance, H
t = time, s
e = base of natural logarithms $\simeq$ 2.718

10-8 DISCHARGING INDUCTANCE THROUGH RESISTANCE

The sequence of events depicted in Fig. 10-30 applies to an inductance when its magnetic field is collapsing and its energy is being dissipated in a resistance. No switch is shown in this circuit because the sequence cannot start with the conditions existing when a charged inductance is connected to an open circuit. Under these conditions the magnetic field would collapse in an indeterminate time.

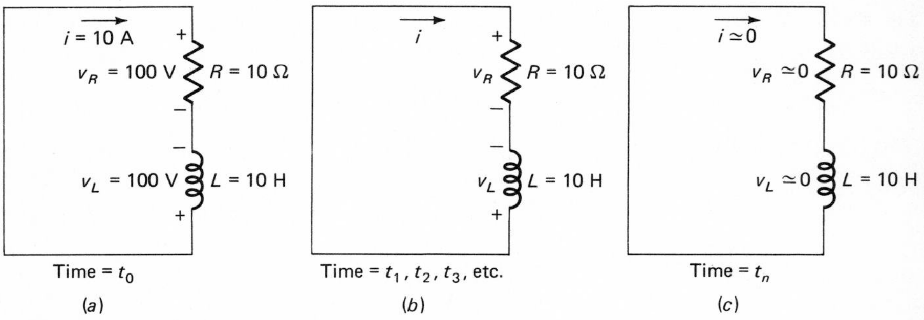

Fig. 10-30 Discharging inductance through resistance.

One way in which the transition from charge to discharge (Fig. 10-27*d* to Fig. 10-30*a*) can take place without the circuit continuity being broken is shown in Fig. 10-31.

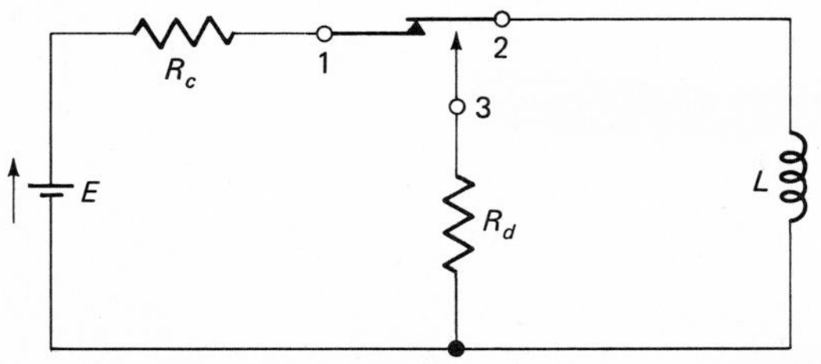

Fig. 10-31 Charge-discharge circuit with make-before-break contacts.

In Fig. 10-31 a make-before-break switch is shown. The circuit by which the inductance L is charged through resistance R_c is completed through contacts 1 and 2. When the switch is operated, contacts 2 and 3 connect together before contacts 1 and 2 open. Thus the discharge circuit through resistance R_d is completed before the charge circuit is broken. The charge and discharge circuits are both connected across the inductance during the time the switch is completing its operation. Therefore the switch operation should be rapid to avoid significant interference with the discharge conditions.

Figure 10-30*a* The inductance is assumed to have its magnetic field developed as in Fig. 10-27*d*. Since the circuit is transformed into that in Fig. 10-30*a* without first becoming open, the collapsing field produces an induced emf of 100 V. The polarity of this emf is opposite that caused by the increasing field during charge. This polarity results in a circuit current in the same direction as that for charge. The current and voltage values at the instant these conditions apply will be, at time t_0,

$$v_L = 100 \text{ V}$$

$$v_R = v_L = 100 \text{ V}$$

$$i = \frac{v_R}{R} = \frac{100 \text{ V}}{10 \ \Omega} = 10 \text{ A}$$

Figure 10-30*b* As the energy of the magnetic field is dissipated in the resistance, the circuit current and voltages decrease toward zero. At times t_1, t_2, t_3, etc.,

$$100 \text{ V} > v_L > 0$$

$$100 \text{ V} > v_R > 0$$

$$10 \text{ A} > i > 0$$

Figure 10-30c When almost all the field energy has been dissipated, the current and voltages will all be approximately zero. At time t_n,

$$v_L \simeq 0 \text{ V}$$

$$v_R \simeq 0 \text{ V}$$

$$i \simeq 0 \text{ A}$$

RL CIRCUIT DISCHARGE CURVES

Figure 10-32 shows how the current and voltages associated with the discharge circuit in Fig. 10-30 vary with respect to time.

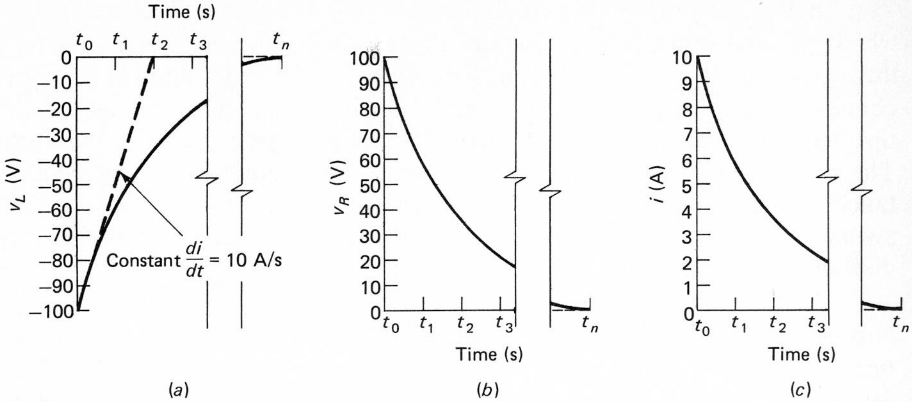

Fig. 10-32 *RL* circuit in Fig. 10-31. Discharge curves with respect to time. (*a*) Inductive voltage; (*b*) resistive voltage; (*c*) current.

THE EQUATIONS FOR THE DISCHARGE CIRCUIT

The loop equation for the *RL* discharge circuit in Fig. 10-30 is

$$v_L = v_R \tag{10-21}$$

or

$$L\frac{di}{dt} - iR = 0$$

An equation for the discharge current can be obtained from this differential equation (see appendix for derivation):

$$i = \frac{V_0}{R}e^{-tR/L} \tag{10-22}$$

Substituting the expression for i in Ohm's law,

$$v_R = iR = V_0e^{-tR/L} \tag{10-23}$$

Substituting for v_R in Eq. (10-18) gives

$$v_L = v_R = V_0e^{-tR/L} \tag{10-24}$$

In Eqs. (10-22) to (10-24),

i = instantaneous current, A
v_R = instantaneous resistive voltage, V
v_L = instantaneous inductive voltage, V
V_0 = initial induced emf produced by the collapsing field, V
L = inductance, H
t = time elapsed, s
e = base of natural logarithms $\simeq$ 2.718

10-9 INDUCTIVE TIME CONSTANT

The meaning of the time constant in *RL* circuits is essentially the same as in *RC* circuits. The evaluation of τ is obtained by equating the exponent tR/L to unity:

$$\frac{tR}{L} = 1$$

$$t = \frac{L}{R}$$

or

$$\tau = \frac{L}{R} \qquad (10\text{-}25)$$

where t = time, s
τ = time constant, s
R = resistance, Ω
L = inductance, H

L/R IN MKSA UNITS

The validity of L/R being considered as time can be seen by substituting the MKSA dimensions from Table 2-3.

$$\text{H} = \frac{\text{kg} \cdot \text{m}^2}{\text{s}^2 \cdot \text{A}^2} \qquad (9\text{-}14)$$

$$\Omega = \frac{\text{kg} \cdot \text{m}^2}{\text{s}^3 \cdot \text{A}^2} \qquad (2\text{-}27)$$

The ratio L/R can be expressed in fundamental units as

$$\frac{\text{H}}{\Omega} = \frac{\text{kg} \cdot \text{m}^2}{\text{s}^2 \cdot \text{A}^2} \div \frac{\text{kg} \cdot \text{m}^2}{\text{s}^3 \cdot \text{A}^2}$$

$$\frac{\text{H}}{\Omega} = \frac{\text{kg} \cdot \text{m}^2}{\text{s}^2 \cdot \text{A}^2} \times \frac{\text{s}^3 \cdot \text{A}^2}{\text{kg} \cdot \text{m}^2} = \text{s}$$

Therefore L/R is correctly regarded as time.

In referring back to Figs. 10-29 and 10-32, it can be seen that straight broken lines are shown representing a constant rate in current change of $E/R = 100$ V/10 Ω = 10 A/s. This rate of change in current is the initial value when charge or discharge commences. The time constant for an *RL* circuit can be expressed in terms of a linear current change, in a manner similar to that adopted for the *RC* circuit.

The time constant τ of an *RL* series circuit is the time it would take for an inductive current to equal the maximum possible value (E/R) during charge, or zero during discharge, if the rate of change in current were maintained constant at its initial value.

10-10 CALCULATIONS FOR *RL* CIRCUITS

Equations (10-18) to (10-20) and Eqs. (10-22) to (10-24) indicate that the shapes of the curves for the *RL* circuit are the same as for the *RC* circuit. Any of the techniques previously applied to *RC* networks can be used for *RL* network problems. The time constant L/R is substituted for *RC*. The methods include:

1. The time-constant concept ($\tau = L/R$).
2. The universal time-constant table (Table 10-3). Apply this table as follows:

a. Inductance charging
For v_L, use v_R row.
For v_R, use v_C row.
For i, use v_C row.

b. Inductance discharging
For v_L, use v_R row.
For v_R and i, use as labeled.

3. The universal time-constant curves (Fig. 10-7)
   a. Inductance charging
   For v_L, use the $y = e^{-x}$ curve.
   For v_R and i, use the $y = (1 - e^{-x})$ curve.
   b. Inductance discharging
   For v_L, v_R, and i, use the $y = e^{-x}$ curve.
4. Natural logarithms.
5. Common logarithms.
6. Table of exponential functions.
7. Slide rule log-log scales.

When calculating circuit values using methods 4 to 7, the general time-constant equations (10-11) and (10-12) are particularly useful. In applying these equations to *RL* circuits the following techniques should be adopted:

Inductance charging:
Use Eq. (10-11), $e^{-t/\tau} = y/Y_{max}$, for v_L.
Use Eq. (10-12), $e^{-t/\tau} = (Y_{max} - y)/Y_{max}$, for i and v_R.
$Y_{max} = E$ for v_L and v_R or E/R for i.

Inductance discharging:
Use Eq. (10-11) for v_L, v_R, and i.
$Y_{max} = Y_0 = I_0 R$ for v_R and v_L.
$Y_{max} = I_0$ for i.

where I_0 is the circuit current when discharge commences.

10-11 COMPARISON OF CHARGE AND DISCHARGE OF *RL* CIRCUITS

By comparing Figs. 10-29 and 10-32, the following can be seen:

1. *a.* v_L decreases toward zero during charge and discharge.
   b. The polarity of v_L during discharge is opposite that during charge.
2. *a.* v_R increases from zero during charge.
   b. v_R decreases toward zero during discharge.
   c. The polarity of v_R is the same during charge and discharge.
3. *a.* i increases from zero during charge.
   b. i decreases toward zero during discharge.
   c. The direction of i is the same during charge and discharge.

Example 10-12 An inductor has a resistance of 5 Ω and an inductance of 1.6 mH. What are the current, resistive voltage, and inductive voltage 1 ms after applying 25 V across the inductor?

The equivalent circuit of the inductor is shown in Fig. 10-33.

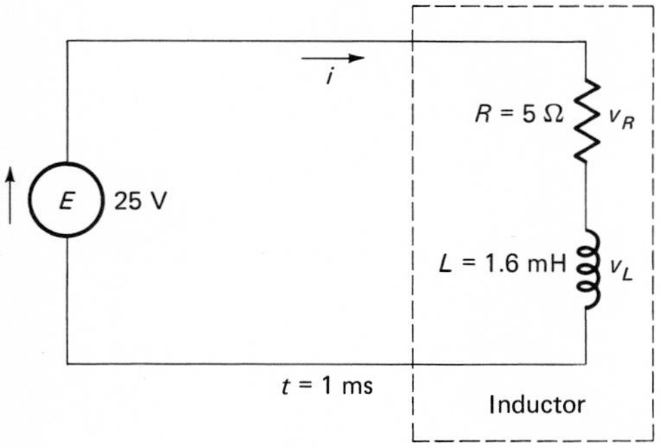

Fig. 10-33 Example 10-12: equivalent circuit of the inductor.

Solution

$$\tau = \frac{L}{R} \tag{10-25}$$

$$\tau = \frac{1.6 \text{ mH}}{5 \ \Omega} = 0.32 \text{ ms}$$

From Eq. (10-18),

$$i = \frac{E}{R}(1 - e^{-t/\tau}) = \frac{25 \text{ V}}{5 \ \Omega}(1 - e^{-1 \text{ ms}/0.32 \text{ ms}}) = \mathbf{4.78 \text{ A}}$$

$$v_R = iR = 4.78 \text{ A} \times 5 \ \Omega = \mathbf{23.9 \text{ V}}$$

$$v_L = E - v_R = 25 \text{ V} - 23.9 \text{ V} = \mathbf{1.1 \text{ V}}$$

Unlike capacitors, which can be made highly efficient, practical inductors usually have substantial energy losses. These losses are commonly represented as equivalent resistances in series with the inductance.

Depending on the application, this resistance can materially affect performance. The time constant of the inductor in Example 10-12 was 0.32 ms. To increase this time constant, with the same inductance, would mean decreasing the resistance. This solution could be an expensive one, requiring a larger wire size and greater bulk.

Example 10-13 A power supply filter inductor has a resistance of 25 Ω and an inductance of 30 H. If the inductor is fully charged and has a current of 0.1 A, what will be the current 1 s after a short circuit is connected across the inductor?

Solution If the fully charged inductor carries a current of 0.1 A, the charged conditions will be

$$i = 0.1 \text{ A}$$

$$v_R = iR = 0.1 \text{ A} \times 25 \ \Omega = 2.5 \text{ V}$$

$$v_L = 0 \text{ V}$$

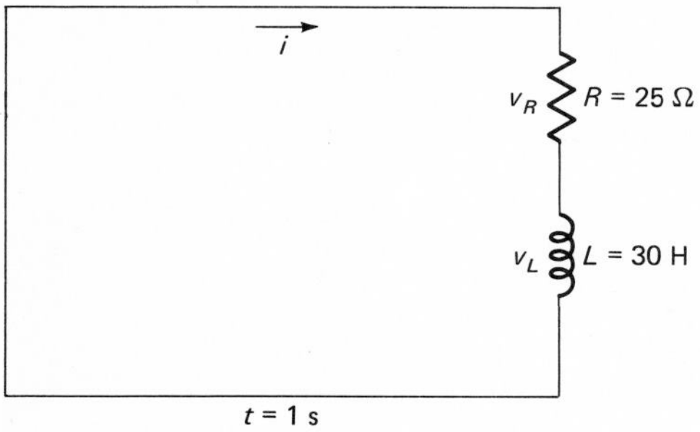

Fig. 10-34 Example 10-13: schematic diagram of the inductor with a short circuit.

If, while these conditions exist, a short circuit is connected across the inductor, as in Fig. 10-34, the initial inductive voltage will equal the resistive voltage in the charged state. Therefore the initial discharge current will equal the final charge current.

$$I_0 = 0.1 \text{ A}$$

$$\tau = \frac{L}{R} \qquad (10\text{-}25)$$

$$\tau = \frac{30 \text{ H}}{25 \ \Omega} = 1.2 \text{ s}$$

$$t = 1 \text{ s} = \frac{1 \text{ s}}{1.2 \text{ s}}\,\tau \simeq 0.83\tau$$

Using the universal time-constant curves in Fig. 10-7, the inductive discharge current follows the law $y = e^{-x}$; so with $t \simeq 0.83\tau$,

$$i \simeq 0.44 I_0 \simeq 0.44 \times 0.1 \text{ A} \simeq \mathbf{44 \ mA}$$

Example 10-14 A telephone relay has a resistance of 100 Ω and an inductance of 0.75 H. It has a 100-Ω resistor connected in parallel with it. How long will it take for the relay current to become 400 mA (*a*) when 48 V is applied across it and (*b*) if the 48 V is disconnected after the inductance has become fully charged?

Solution (See Fig. 10-35.)

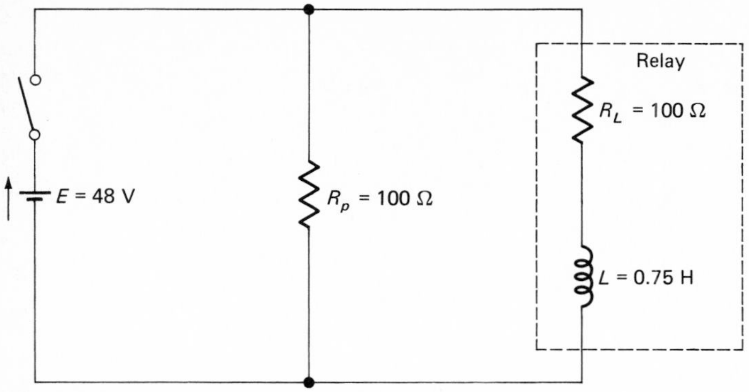

Fig. 10-35 Example 10-14: Circuit.

a. When the switch is closed, 48 V is connected directly across the inductor (R_L and L in series), charging the inductance.

The charge time constant is

$$\tau_c = \frac{L}{R_L} \tag{10-25}$$

$$\tau_c = \frac{0.75\ \text{H}}{100\ \Omega} = 7.5\ \text{ms}$$

The maximum charge current, $I_{\max} = 48\ \text{V}/100\ \Omega = 480\ \text{mA}$. From Eq. (10-12),

$$e^{-t_c/\tau_c} = \frac{I_{\max} - i}{I_{\max}}$$

$$e^{t_c/\tau_c} = \frac{480\ \text{mA}}{480\ \text{mA} - 400\ \text{mA}} = 6$$

$$t_c = \tau_c \ln 6 = 7.5\ \text{ms} \times 1.792 = \mathbf{13.44\ ms}$$

b. Assuming that the charge of the inductance is first completed, when the switch is opened the current will drop from the value

$$I_0 = \text{charging } I_{\max} = 480\ \text{mA}$$

The circuit through which the inductance discharges consists of R_L and R_P in series. The discharge time constant is, from Eq. (10-25),

$$\tau_d = \frac{L}{R_L + R_P} = \frac{0.75\ \text{H}}{200\ \Omega} = 3.75\ \text{ms}$$

From Eq. (10-11),

$$e^{-t_d/\tau_d} = \frac{i}{I_0}$$

$$e^{t_d/\tau_d} = \frac{480\ \text{mA}}{400\ \text{mA}} = 1.2$$

$$t_d = \tau_d \ln 1.2 = 3.75\ \text{ms} \times 0.1823 = \mathbf{0.6836\ ms}$$

10-12 SUMMARY OF EQUATIONS

RC circuits

$E = v_C + v_R$ (10-1)

$i = (E/R)e^{-t/RC}$ (10-2)

$v_R = Ee^{-t/RC}$ (10-3)

$v_C = E(1 - e^{-t/RC})$ (10-4)

$\tau = RC$ (10-5)

$i = (E/R)e^{-t/\tau}$ (10-6)

$v_R = Ee^{-t/\tau}$ (10-7)

$v_C = E(1 - e^{-t/\tau})$ (10-8)

$y = Y_{max}e^{-t/\tau}$ (10-9)

$y = Y_{max}(1 - e^{-t/\tau})$ (10-10)

$e^{-t/\tau} = y/Y_{max}$ (10-11)

$e^{-t/\tau} = (Y_{max} - y)/Y_{max}$ (10-12)

$v_C = v_R$ or $v_C - v_R = 0$ (10-13)

$i = (V_0/R)e^{-t/RC}$ (10-14)

$v_R = V_0e^{-t/RC}$ (10-15)

$v_C = v_R = V_0e^{-t/RC}$ (10-16)

RL circuits

$E = v_L + v_R$ (10-17)

$i = (E/R)(1 - e^{-tR/L})$ (10-18)

$v_R = iR = E(1 - e^{-tR/L})$ (10-19)

$v_L = Ee^{-tR/L}$ (10-20)

$v_L = v_R$ (10-21)

$i = (V_0/R)e^{-tR/L}$ (10-22)

$v_R = iR = V_0e^{-tR/L}$ (10-23)

$v_L = v_R = V_0e^{-tR/L}$ (10-24)

$\tau = L/R$ (10-25)

EXERCISE PROBLEMS

Section 10-1

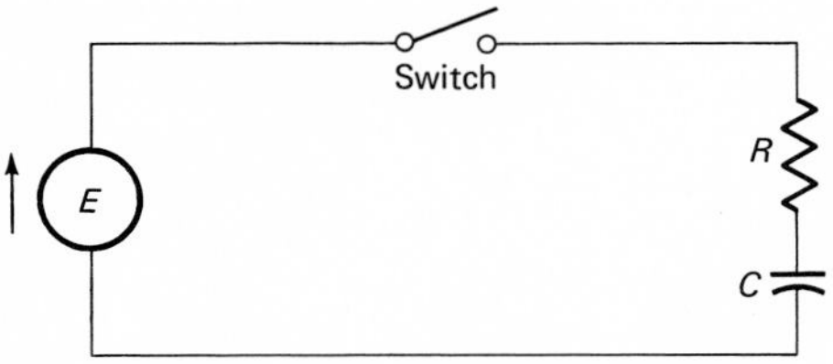

Fig. 10-1P Problems 10-1 to 10-6, 10-10 to 10-12, 10-14, and 10-15: circuit.

10-1 Using Fig. 10-1P with $E = 50$ V, $R = 1$ kΩ, and $C = 10$ μF, find i, v_R, and v_C using Eqs. (10-2) to (10-4) at the following times after the switch is closed.
- *a.* 0 s (*ans.* 50 mA, 50 V, 0 V)
- *b.* 3 ms (*ans.* 37 mA, 37 V, 13 V)
- *c.* 7000 μs (*ans.* 24.8 mA, 24.8 V, 25.2 V)
- *d.* 20 ms (*ans.* 6.77 mA, 6.77 V, 43.23 V)

10-2 Repeat Prob. 10-1 with $E = 12$ V, $R = 100$ kΩ, and $C = 470$ pF, finding i, v_R, and v_C at
- *a.* 5 μs
- *b.* 68 μs
- *c.* 47 μs
- *d.* 0.12 ms

10-3 Repeat Prob. 10-1 with $E = 24$ V, $R = 22$ kΩ, and $C = 1000$ pF, finding i, v_R, and v_C at
- *a.* 110 μs (*ans.* 7.35 μA, 0.161 V, 23.84 V)
- *b.* 0.033 ms (*ans.* 243.4 μA, 5.36 V, 18.64 V)
- *c.* 10 μs (*ans.* 0.692 mA, 15.23 V, 8.77 V)
- *d.* 18 μs (*ans.* 0.481 mA, 10.59 V, 13.41 V)

Section 10-2

10-4 Repeat Prob. 10-1, using the universal time-constant curves, at the following times:
a. 2000 μs
b. 5000 μs
c. 15 ms
d. 8 ms

10-5 Repeat Prob. 10-2, using the universal time-constant curves, at the following times:
a. 190 μs (*ans.* 2.1 μA, 0.21 V, 11.8 V)
b. 15 μs (*ans.* 87 μA, 8.7 V, 3.3 V)
c. 24 μs (*ans.* 72 μA, 7.2 V, 4.8 V)
d. 75 μs (*ans.* 24 μA, 2.4 V, 9.6 V)

10-6 Repeat Prob. 10-3, using the universal time-constant curves, at the following times:
a. 5 μs
b. 20 μs
c. 44 μs
d. 1 μs

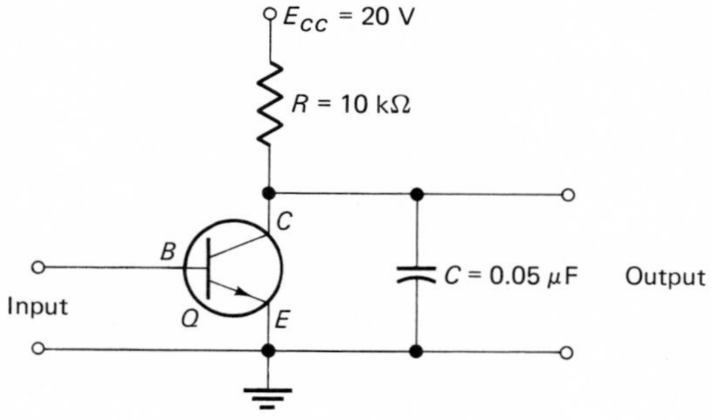

Fig. 10-2P Problem 10-7: circuit.

10-7 Figure 10-2P shows the circuit of a simple ramp generator. *Q* is a transistor switch. When *Q* is on, it is the equivalent of a short circuit across *C*; when *Q* is off, it is in effect an open circuit and *C* is permitted to charge through *R*. The transistor *Q* is initially on, and then is switched off. Plot the output voltage at 10-μs intervals for 50 μs after *Q* is switched off. (*ans.* 0.396 V, 0.784 V, 1.165 V, 1.538 V, 1.9 V)

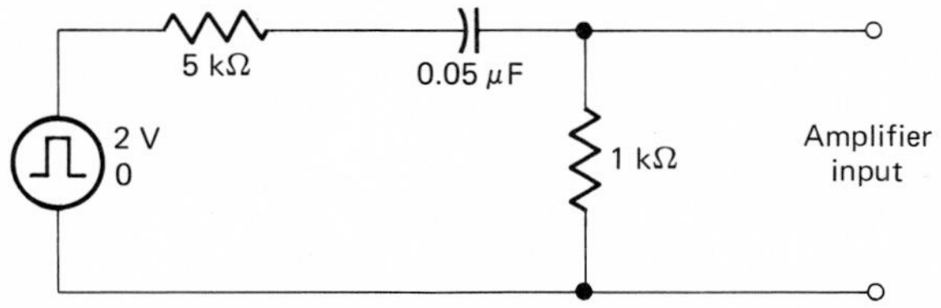

Fig. 10-3P Problems 10-8 and 10-9: circuit.

10-8 The circuit in Fig. 10-3P represents a capacitive coupling between a pulse generator and the input of an amplifier. With a 2-V 100-μs pulse applied, find the voltage at the amplifier input at
a. The beginning of the pulse.
b. The end of the pulse.

10-9 For the circuit in Fig. 10-3P and Prob. 10-8, to what must the value of *C* be changed so that the voltage at the input falls to only 90 percent of its initial value by the end of the 100-μs pulse? (*ans.* 0.158 μF)

Section 10-3

10-10 Using Fig. 10-1P with $E = 100$ V, $R = 5.6$ kΩ, and $C = 0.01$ μF, and Eq. (10-6), find the time required for i to fall to the following values after the switch is closed:
a. 16 mA *c.* 6 mA
b. 12 mA *d.* 3 mA

10-11 Using Fig. 10-1P with $E = 60$ V, $R = 120$ kΩ, and $C = 0.0022$ μF, and Eq. (10-7), find the time required for v_R to fall to the following values after the switch is closed:
a. 54 V (*ans.* 27.8 μs)
b. 6 V (*ans.* 608 μs)
c. 30 V (*ans.* 182.9 μs)
d. 22 V (*ans.* 264.8 μs)

10-12 Using Fig. 10-1P with $E = 25$ V, $R = 470$ Ω, and $C = 100$ pF, and Eq. (10-8), find the time required for v_C to reach the following values after the switch is closed:
a. 20 V *c.* 2.5 V
b. 9 V *d.* 15.8 V

10-13 The rise time of a waveform is the time required for the waveform to rise from 10 to 90 percent of its final amplitude. By finding the time to rise to 90 percent and subtracting the time required to rise to 10 percent, find a general expression for the rise time of an exponential waveform. Your answer should be in the form of a constant times a time constant. (*ans.* 2.2τ)

10-14 Find the time constant and the rise time of v_C for the circuit in Prob. 10-1.

10-15 Find the time constant and the rise time of v_C for the circuit in Prob. 10-2. (*ans.* 47 μs, 103.4 μs)

Section 10-5

10-16 Find the time required for the given capacitance to discharge through the resistance from the voltage V_0 to the voltage v_C in each of the following cases:
a. $V_0 = 45$ V, $v_C = 25$ V, $R = 4.7$ kΩ, $C = 0.001$ μF
b. $V_0 = 20$ V, $v_C = 5$ V, $R = 100$ kΩ, $C = 1000$ pF
c. $V_0 = 10$ V, $v_C = 0.5$ V, $R = 22$ kΩ, $C = 0.02$ μF
d. $V_0 = 100$ V, $v_C = 50$ V, $R = 10$ kΩ, $C = 0.2$ μF

10-17 Find the voltage across the capacitance when it has been discharging through the resistance from voltage V_0 for the given time t.
a. $V_0 = 45$ V, $R = 4.7$ kΩ, $C = 0.001$ μF, $t = 2$ μs (*ans.* 29.4 V)
b. $V_0 = 20$ V, $R = 100$ kΩ, $C = 1000$ pF, $t = 75$ μs (*ans.* 9.45 V)
c. $V_0 = 10$ V, $R = 22$ kΩ, $C = 0.02$ μF, $t = 1$ ms (*ans.* 1.03 V)
d. $V_0 = 100$ V, $R = 10$ kΩ, $C = 0.2$ μF, $t = 600$ μs (*ans.* 74.1 V)

Miscellaneous *RC* circuit problems

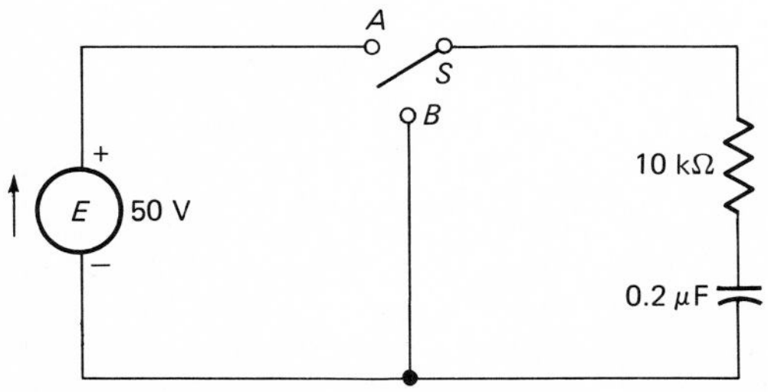

Fig. 10-4P Problem 10-18: circuit.

10-18 Using Fig. 10-4P and starting with the capacitor completely discharged, the switch is first placed to position A and the capacitor charges for time t_A. The switch is then moved to position B, and the capacitor discharges for time t_B. Find the capacitor voltage v_C at the end of time t_B when

a. $t_A = 1$ ms, $t_B = 1$ ms
b. $t_A = 5$ ms, $t_B = 5$ ms
c. $t_A = 4$ ms, $t_B = 2$ ms
d. $t_A = 2$ ms, $t_B = 4$ ms
e. $t_A = 10$ ms, $t_B = 10$ ms

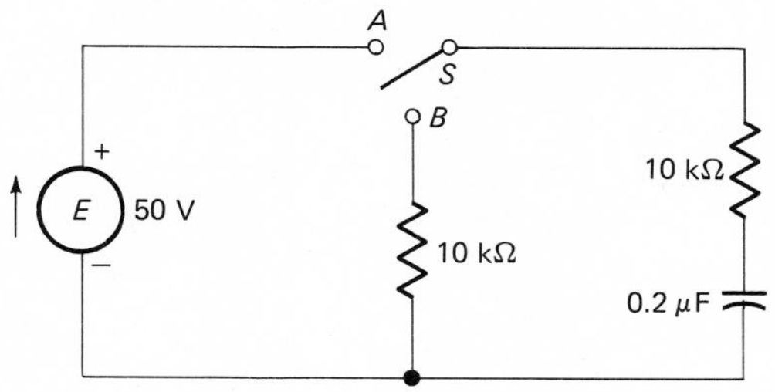

Fig. 10-5P Problem 10-19: circuit.

10-19 Repeat all the conditions in Prob. 10-18, using Fig. 10-5P.
(*ans. a.* 15.32 V; *b.* 13.15 V; *c.* 26.22 V; *d.* 11.63 V; *e.* 4.08 V)

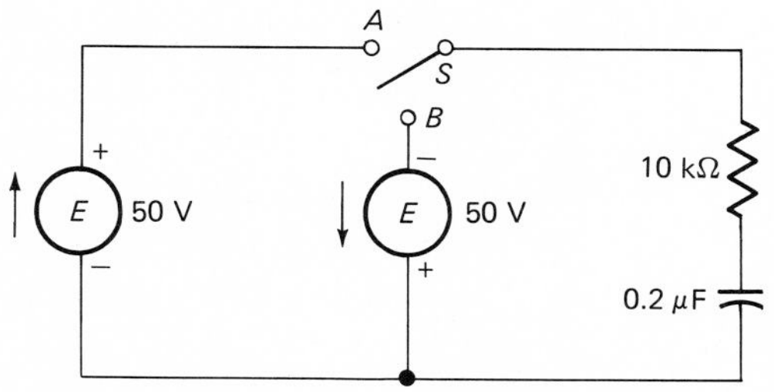

Fig. 10-6P Problem 10-20: circuit.

10-20 Repeat all the conditions in Prob. 10-18, using Fig. 10-6P.

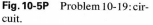

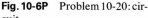

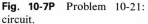

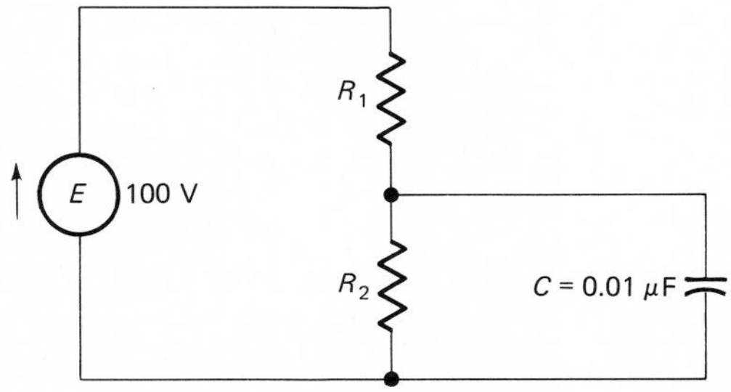

Fig. 10-7P Problem 10-21: circuit.

10-21 In the circuit in Fig. 10-7P find the time constant of the equivalent RC circuit, the rise time (10 to 90 percent), and the final v_C for each of the following cases:

a. $R_1 = 1000\ \Omega$, $R_2 = 1000\ \Omega$ (*ans.* 5 μs, 11 μs, 50 V)
b. $R_1 = 5\ \text{k}\Omega$, $R_2 = 500\ \Omega$ (*ans.* 4.545 μs, 10 μs, 9.09 V)
c. $R_1 = 500\ \Omega$, $R_2 = 5\ \text{k}\Omega$ (*ans.* 4.545 μs, 10 μs, 90.9 V)

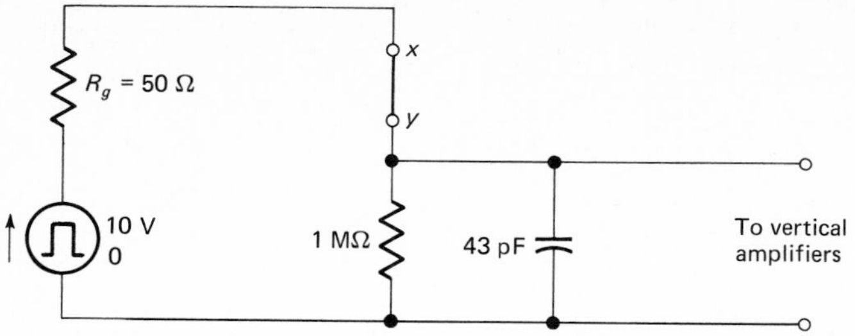

Fig. 10-8P Problems 10-22 and 10-23: circuit.

10-22 Figure 10-8P represents the input of an oscilloscope fed from a 50-Ω pulse generator E. Find the time constant of the equivalent RC circuit and the rise time.

10-23 In Prob. 10-22 if a resistance of 1 MΩ is placed between points x and y to reduce the signal to the oscilloscope input by 50 percent, find

a. The new time constant of the circuit. (*ans.* 21.5 μs)

b. The new rise time. (*ans.* 47.3 μs)

c. The final voltage to the input of the vertical amplifiers. (*ans.* 5 V)

Section 10-7

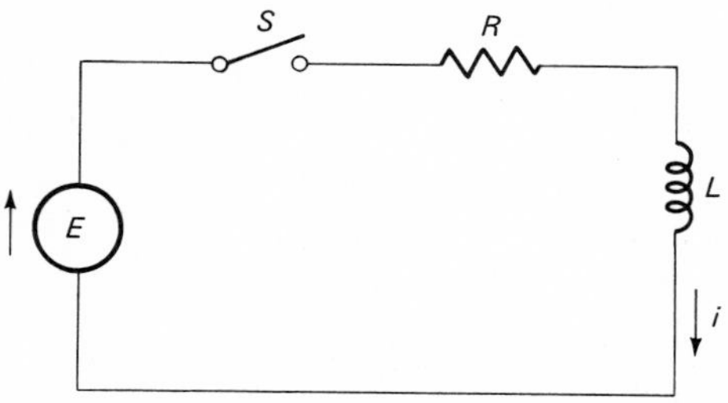

Fig. 10-9P Problems 10-24 to 10-26: circuit.

10-24 Using Fig. 10-9P with $E = 50$ V, $R = 100$ Ω, and $L = 100$ mH, and Eqs. (10-18) to (10-20), find i, v_R, and v_L at the following times after the switch is closed:

a. $t = 0+$ (a short time after time t_0)

b. $t = 1$ ms

c. $t = 2$ ms

d. $t = 500$ μs

10-25 Repeat Prob. 10-24, using the universal time-constant curves, at the following times after the switch is closed:

a. $t = 200$ μs (*ans.* 90 mA, 9 V, 41 V)

b. $t = 3$ ms (*ans.* 475 mA, 47.5 V, 2.5 V)

c. $t = 693$ μs (*ans.* 250 mA, 25 V, 25 V)

d. $t = 1500$ μs (*ans.* 390 mA, 39 V, 11 V)

10-26 Using Fig. 10-9P with $E = 10$ V, $R = 22$ Ω, and $L = 400$ μH, find the time required to reach the following voltages and currents after the switch is closed:

a. $v_R = 9$ V

b. $v_L = 9$ V

c. $i = 200$ mA

d. $i = 50$ mA

e. $v_L = 3$ V

f. $v_R = 5$ V

Sections 10-8, 10-9, and 10-10

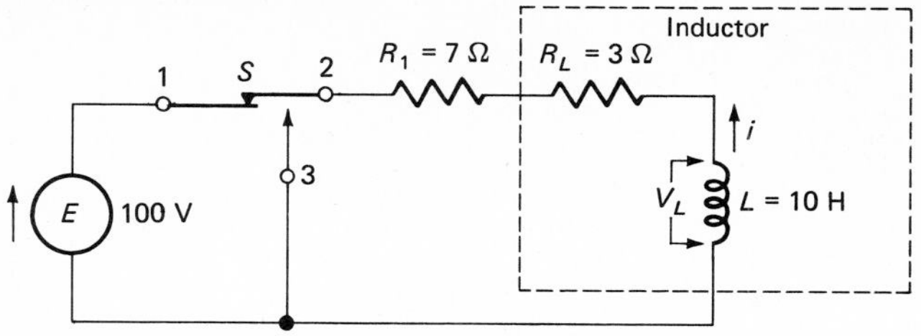

Fig. 10-10P Problem 10-27: circuit.

10-27 In the circuit in Fig. 10-10P switch S is a make-before-break switch. If switch contacts 1 and 2 are closed, allowing the inductor to charge for time t_A, then contacts 2 and 3 are closed, allowing discharge for time t_B. Find i and v_L at the end of time t_B when

a. $t_A = 0.5$ s, $t_B = 0.5$ s (*ans.* 2.39 A, 23.9 V)
b. $t_A = 1$ s, $t_B = 1$ s (*ans.* 2.33 A, 23.3 V)
c. $t_A = 3$ s, $t_B = 1$ s (*ans.* 3.5 A, 35 V)
d. $t_A = 1$ s, $t_B = 3$ s (*ans.* 0.315 A, 3.15 V)

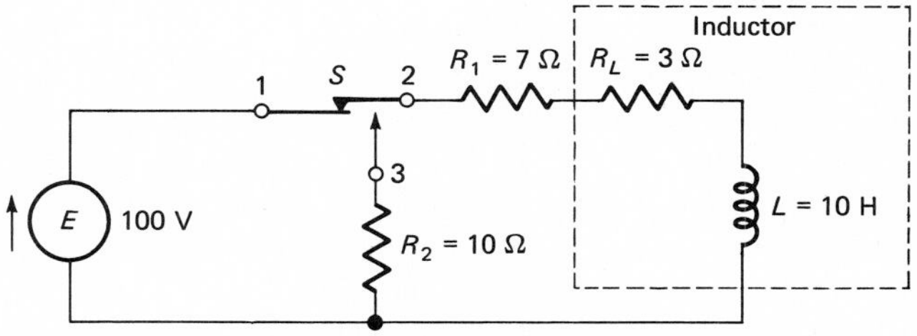

Fig. 10-11P Problem 10-28: circuit.

10-28 Repeat all the conditions in Prob. 10-27, using Fig. 10-11P.

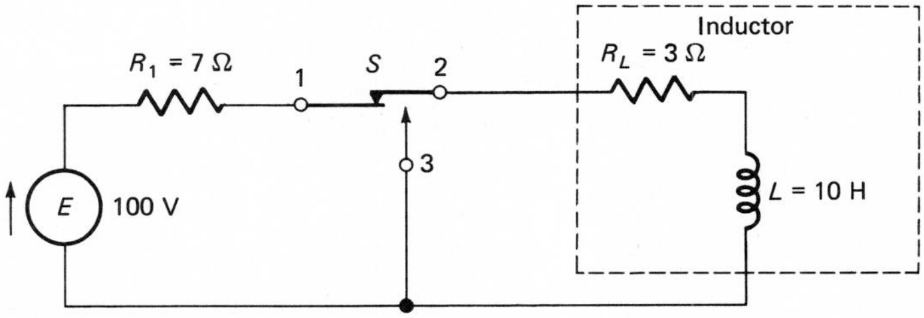

Fig. 10-12P Problem 10-29: circuit.

10-29 Repeat all the conditions in Prob. 10-27, using Fig. 10-12P.
(*ans. a.* 3.39 A, 10.16 V; *b.* 4.68 A, 14.05 V; *c.* 7.04 A, 21.11 V; *d.* 2.57 A, 7.71 V)

10-30 A large inductor is to be checked with a 6.6-V battery and an ammeter. Thirty seconds after the battery is connected, the current indicated by the ammeter is 1.2 A. If the final current is 1.65 A, find the inductance of the inductor.

10-31 If the inductor in Prob. 10-30 is discharged by placing a short circuit across the winding and then removing the battery, find the time required for the current to reach the accepted zero level. (*ans.* 115.5 s)

10-32 The battery across the inductor in Prob. 10-30 is disconnected while the inductor is fully charged, and the current through the ensuing arc falls to zero, at a linear rate, in 5 ms. Find the voltage across the open terminals of the inductor during that period.

FOUR
AC Circuits

INTRODUCTION

Steady dc circuits involve currents and voltages of constant magnitude, polarity, and direction. There is no current in a capacitance, and no voltage across an inductance; only resistance is of significance.

In dc transient circuits, in which a dc emf is suddenly applied to or removed from a network, capacitance, inductance, and resistance are equally significant. Here the circuit currents and voltages are not of constant magnitude. Some of the voltages and currents change polarity and direction.

Now we proceed to the investigation of ac circuits in which all voltages, including applied emfs, are of changing polarity and all currents are of changing direction.

While the principles applicable to circuits involving steady direct current and dc transients may relate to ac circuits, some entirely different viewpoints are encountered.

Phase, which is the time relationship between various voltages and currents, is important in ac circuits, and this aspect forms a common thread throughout this part of the text.

The sine waveform too is important, so it is given an early treatment in terms of its nature and its significance in ac circuits.

All the network theorems and laws used for dc circuit analysis apply to ac circuits. They must be interpreted in terms of the effects of varying voltages and currents in resistive, capacitive, and inductive networks. This interpretation involves calculation methods which are more complicated than those for their dc equivalents. It also introduces some new terms such as impedance, admittance, reactance, susceptance, frequency, and complex quantity.

11 Waveforms

Ac voltage, current, and power are time-varying quantities; their magnitudes change as time progresses. The investigation of the behavior of ac circuits is concerned with the relationships between the magnitudes of these quantities and time.

In previous chapters we mainly discussed circuits in which voltages and currents did not vary in magnitude and in which certain voltage-current relationships existed all the time. We did not need to concern ourselves with some of the concepts which arise in ac circuits as a result of voltage and current variation. Some introduction to such concepts was given in the treatment of dc transient circuits in which voltages and currents are also time-related. In dc transient circuits the magnitudes of current and voltage become steady if the emf is left connected to the circuit. In the ac circuits, however, current and voltage magnitudes continually change as long as the circuit continuity is maintained.

The treatment of ac circuits in this text starts with a comparison of steady direct current, varying direct current and alternating current on a time-related basis. After this comparison we view different types of ac waveforms commonly met in electrical and electronic practice and determine the special importance of the sine waveform. An introduction to some ac terminology lays the foundation for a study of the behavior of resistance, capacitance, and inductance in ac circuits.

"Ac" and "dc" are conventional symbols meaning "alternating" and "direct." These symbols may refer to current *or* voltage. When viewed in this way, *ac current* is understood to mean *alternating current,* and *ac voltage* is understood to mean *alternating voltage.* This symbolic use of ac and dc is common practice in the electrical and electronics industry.

11-1 DC AND AC CURRENT AND VOLTAGE

A dc voltage always has the same polarity, and direct current always has the same direction. Dc voltage and current do not necessarily have constant magnitude or value, steady direct current being a specific case.

On a time-related basis, direct current can be shown by waveforms such as those in Fig. 11-1.

Figure 11-1*a* shows a steady direct current. Its value is constant at 3 A throughout the 1-s period shown on the horizontal time axis.

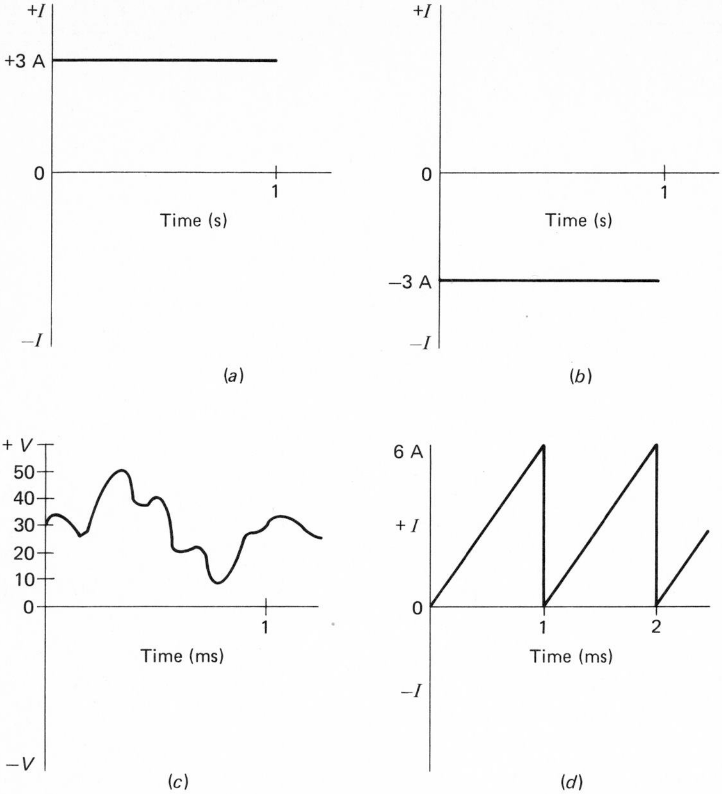

Fig. 11-1 Waveforms. (*a*) and (*b*) Steady direct currents of opposite direction; (*c*) randomly varying direct voltage; (*d*) sawtooth direct current.

Figure 11-1*b* also shows a steady 3-A direct current. Its value is shown as negative, while that in Fig. 11-1*a* is shown as positive. This usage implies that the currents represented by the two curves are of the same magnitude but have opposite directions. There is no fundamental meaning behind a positive current, it is an arbitrary choice. If in a particular circuit situation, it is decided to call clockwise currents positive, then counterclockwise currents will be negative.

In Fig. 11-1*c* a voltage always has the same polarity, because it is shown as positive for the whole 1-ms period. But its value starts at 30 V and then varies between 10 and 50 V in an apparently random manner. This too is a dc voltage.

The waveform in Fig. 11-1*d* depicts a dc current because, although its minimum value is zero, it never changes direction. It is shown as positive, rising in a linear manner from 0 to 6 A and then dropping back to 0 A. It repeats this movement at regular time intervals of 1 ms, thus constituting a periodic dc waveform.

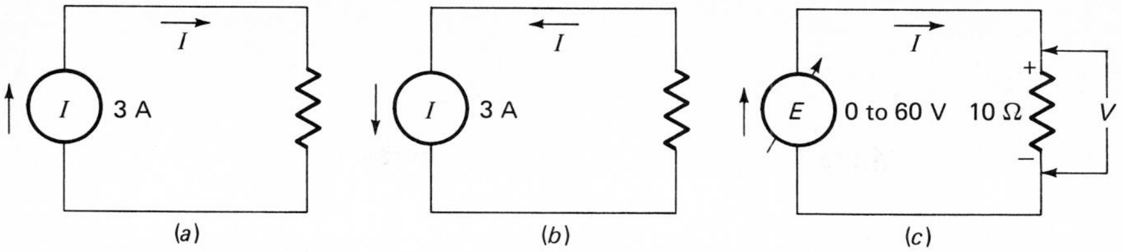

Fig. 11-2 Circuits for producing the dc waveforms in Fig. 11-1. (*a*) Steady 3-A clockwise current; (*b*) steady 3-A counterclockwise current; (*c*) varying direct voltage or current.

The circuits in Fig. 11-2 cover all the situations shown in Fig. 11-1.

In Fig. 11-2*a* a constant-current source produces a clockwise current of 3 A, as in Fig. 11-1*a*.

In Fig. 11-2*b* a constant-current source causes a 3-A current in a counterclockwise direction. Relative to Fig. 11-2*a* this is a negative current. If Fig. 11-1*a* represents a clockwise current, Fig. 11-1*b* represents a counterclockwise current.

The variable voltage source in Fig. 11-2*c* can result in either of the waveforms in Fig. 11-1*c* or *d*. The voltage applied to the resistance could vary in a random manner, as in Fig. 11-1*c*, or it could be controlled in a sawtooth manner to result in the current waveform in Fig. 11-1*d*.

Ac can be *periodic,* where alternations occur at regular time intervals, or *random,* where alternations do not occur at regular time intervals. Only periodic ac is discussed in this text, as it is of the greatest practical significance.

The polarity of a periodic ac voltage reverses at regular time intervals. The direction of a periodic ac current reverses at regular time intervals. Between alternations (reversals), the magnitude of an ac voltage or current can be constant or varying.

Figure 11-3 shows some typical ac waveforms.

Figure 11-3*a* shows a square-wave alternating current. Its magnitude stays constant at +3 A for 1 s, suddenly changes to −3 A, maintains this value for 1 s, and then changes back to +3 A. Thus this waveform represents a current which changes its direction at 1-s intervals but, between these reversals, has a constant magnitude of 3 A. The 1-s constant values of +3 A and −3 A, taken individually, are the same as the direct currents in Fig. 11-1*a* and *b*.

The sawtooth waveform in Fig. 11-3*b* represents a current which is the ac equivalent of the dc sawtooth in Fig. 11-1*d*. They both refer to currents which rise in a linear manner through a range of 6 A and then revert to the initial value. The ac sawtooth current starts at 3 A in one direction, goes to 0 A, and then continues to 3 A in the opposite direction.

In Fig. 11-3*c* is shown a sine-wave ac voltage. This waveform is of such special interest in electrical and electronic technology that almost all ac examples and problems in this text refer to sine waves. The sine wave is treated in detail in Secs. 11-4 to 11-6.

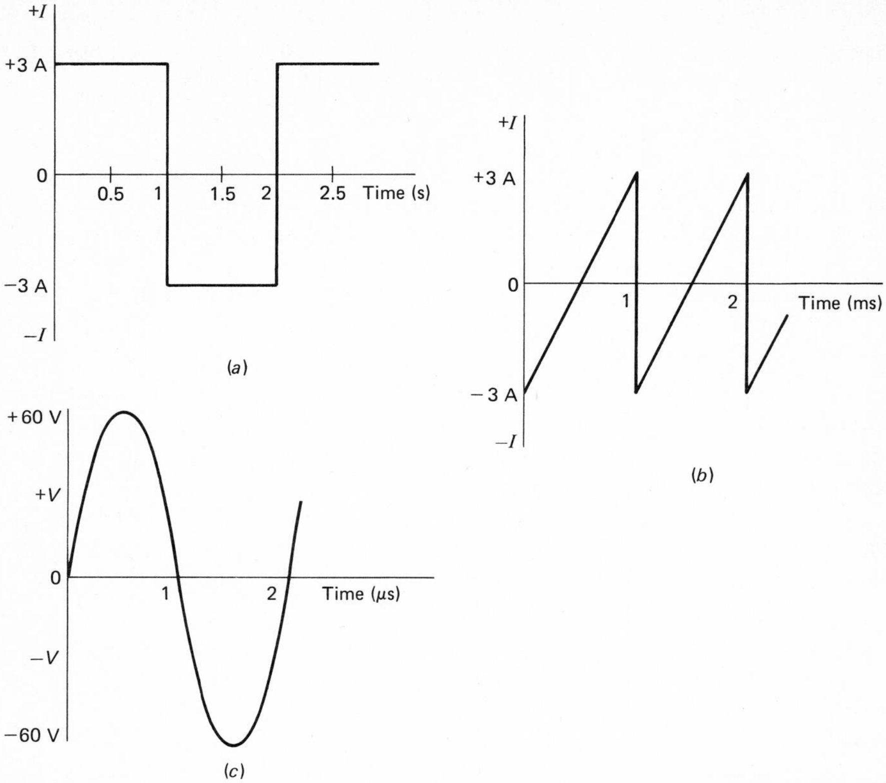

Fig. 11-3 Waveforms. (*a*) Square-wave alternating current; (*b*) sawtooth alternating current; (*c*) sine-wave alternating voltage.

The waveforms in Fig. 11-3 can be related to the respective circuits in Fig. 11-4.

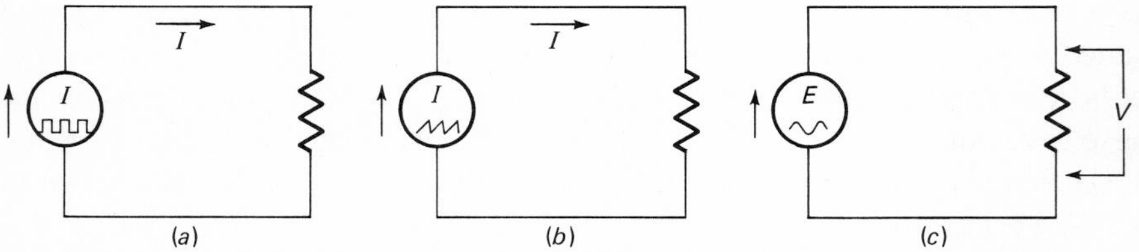

Fig. 11-4 Circuits for producing the ac waveforms in Fig. 11-3. (*a*) Square-wave current; (*b*) sawtooth current; (*c*) sine-wave voltage.

ENERGY SOURCE SYMBOLS

There appear to be no generally accepted standard symbols for energy sources. As for the dc sources in previous chapters, we adopt simple, clear symbols for ac sources. Current sources are shown in Fig. 11-4 as circles labeled I. Voltage sources are shown as circles labeled E. Both symbols show the type of waveform supplied by the source.

ARROWS ON AC CIRCUIT DIAGRAMS

Not only do ac voltage polarities and current directions change individually at regular time intervals but, if inductance or capacitance is involved, the polarities and directions of the various voltages and currents in an ac circuit change with respect to each other. Therefore arrows on ac circuit diagrams do not necessarily represent simple directional relationships as they do on dc circuit diagrams. In general, an arrow on an ac circuit diagram only indicates the *presence* of a current or a voltage, not its direction or its polarity.

Certain conditions in ac circuits are sometimes indicated by arrows on diagrams. We shall show later that it can be convenient to use arrows of opposite direction in each branch of a diagram for an ac parallel inductance-capacitance circuit. These arrows indicate the fact that, at a certain time instant, the currents in each branch are of opposite direction.

No values of voltage or current are given in the circuit diagrams in Fig. 11-4. This omission is deliberate, as the magnitude values for ac quantities have to be stated in special terms because of their varying nature.

11-2 WAVEFORM MAGNITUDE VALUES

Waveform shape influences the effects produced by a current or voltage in a given circuit. Measurements are also influenced by wave shape. To satisfy the demands of different functional and measuring needs, current and voltage magnitudes can be specified in terms of one or another of several values, each with its own area of application.

MAXIMUM OR PEAK VALUE

The *maximum* or *peak value* of an ac current or voltage is the maximum *magnitude* reached during a positive or negative excursion. If the waveform is symmetrical, as in Fig. 11-3, the positive and negative peak values are the same. The subscript "max" is used to denote a maximum or peak value.

Referring to Fig. 11-3*a* and *b*,

$$I_{\text{max}} = 3 \text{ A}$$

Referring to Fig. 11-3*c*,

$$V_{\text{max}} = 60 \text{ V}$$

Figure 11-5*a* shows a square-wave current which has the same overall 6-A swing as that in Fig. 11-3*a*.

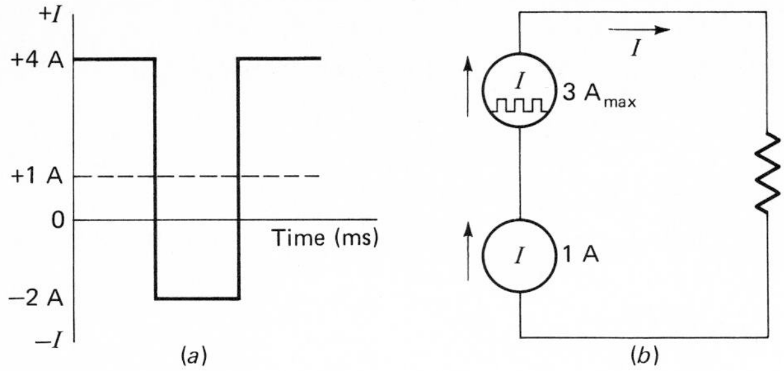

Fig. 11-5 (*a*) Asymmetrical square-wave current; (*b*) circuit for producing the asymmetrical square-wave current.

The waveform in Fig. 11-5*a* is asymmetrical, the negative and positive peaks being of different magnitude. In such cases it is necessary to name the peaks referred to:

Positive peak: $I_{max} = 4$ A

Negative peak: $I_{max} = 2$ A

This square-wave current can be considered to be made up of two components: a symmetrical ac square wave of 3 A maximum value superimposed on a 1-A steady direct current. From this viewpoint, the current swings 3 A positive and 3 A negative with respect to +1 A. The current is said to have a + 1-A dc level. The circuit in Fig. 11-5*b* shows this dc level as a 1-A dc source in series with a square-wave ac source of 3 A maximum value.

The minimum degree of insulation needed for the protection of wiring, components, equipment, and personnel is determined by the peak value of ac voltage applied to the circuit, since at least this value will be present at certain instants of time.

A peak voltage value is most commonly measured by displaying the waveform on a calibrated oscilloscope and then measuring the amount by which the waveform peak is displaced from the horizontal zero axis. The positive and negative peak values can be individually measured in this manner.

A peak current value can be measured by passing the current through a known resistance, thus developing a voltage of the same waveform as that of the current and of magnitude proportional to that of the current. Alternatively a current probe, a special transformer, can be used to produce the voltage. The current-derived voltage obtained by either method is applied to an oscilloscope for measurement and interpretation in terms of the known current-voltage relationship.

PEAK-TO-PEAK VALUE

The magnitude of an ac waveform taken from its most positive value to its most negative value is called the *peak-to-peak value*. The subscript "p-p" is used to identify peak-to-peak values.

In Fig. 11-3*a* and *b* the peak-to-peak value of current is

$$I_{\text{p-p}} = 6 \text{ A} \qquad \text{from } +3 \text{ to } -3 \text{ A}$$

In Fig. 11-3*c* the peak-to-peak value of the voltage is

$$V_{\text{p-p}} = 120 \text{ V} \qquad \text{from } +60 \text{ to } -60 \text{ V}$$

The peak-to-peak value of the waveform in Fig. 11-5*a* is the same as that in Fig. 11-3*a* and *b*:

$$I_{\text{p-p}} = 6 \text{ A}$$

In the case of Fig. 11-5*a*, the 6 A is from +4 to −2 A.

The concept of peak-to-peak value can be usefully applied to a varying dc waveform by finding the peak-to-peak value of its ac component. The ac component of a varying dc current or voltage is the actual variation in magnitude.

With reference to the random dc voltage in Fig. 11-1*c*, it can be seen that a 40-V variation takes place between +50 and +10 V. This waveform has a 40-$V_{\text{p-p}}$ ac component.

The sawtooth current in Fig. 11-1*d* is direct current and has the same peak-to-peak value as the ac currents in Figs. 11-3 and 11-5

$$I_{\text{p-p}} = 6 \text{ A}$$

Although it is only of academic interest, perhaps it should be mentioned that steady direct current, as in Fig. 11-1*a* and *b*, has a zero peak-to-peak value.

Peak-to-peak values are used only when the variation in magnitude is important. Consider the currents in Figs. 11-3*a* and 11-5. If, in a particular application, interest is only in their common 6-A excursion between their most positive and most negative values, this 6-A peak-to-peak value can be stated as an equality between the two currents. It must be understood that this value implies that the two currents are equivalent only for the purpose at hand.

Peak-to-peak values take no account of dc levels. Thus these values should not be stated when dc levels are important.

A peak-to-peak value of voltage or current is most readily measured by using an oscilloscope. The peak-to-peak value is determined by measuring the displacement between the most positive and most negative peaks of the display, and relating this displacement to the oscilloscope vertical range setting.

The ease with which peak-to-peak values associated with complex waveforms can be measured with an oscilloscope is the common justification for the extensive use of these values in the electronics industry.

INSTANTANEOUS VALUE

For any waveform an *instantaneous value* is a value of magnitude at a specified instant. The term "specified instant" means a selected point on the horizontal axis of the relevant waveform curve. Usually this axis represents time, but it may represent angle or distance. Figures 11-1, 11-3, and 11-5 all have time as the horizontal axis variable.

It is usual to imply that a specified instant occurs at the stated time after a reference time t_0. Time t_0 can have any of several meanings, such as:

1. The time a switch is closed.
2. The time observation begins.
3. The curve origin. This refers to a graphical presentation of the waveform, as in Figs. 11-1, 11-3, and 11-5. The origin is the point at which the horizontal and vertical axes intercept, the common zero point.
4. The time at which the waveform has zero magnitude. When this meaning is used, it is important to specify which zero is meant. For example, in Fig. 11-3*c*, at times $t = 0\ \mu s$ and $t = 2\ \mu s$, the voltage is zero swinging positive, while at time $t = 1\ \mu s$ the voltage is zero swinging negative.

Instantaneous values are given in lowercase symbols. Referring to Fig. 11-3*a*, and taking time t_0 as the curve origin, at time $t = 0.5\ \mu s$,

$$i = +3 \text{ A}$$

At time $t = 1.5\ \mu s$,

$$i = -3 \text{ A}$$

at time $t = 1\ \mu s$,

$$i = \pm 3 \text{ A}$$

The last example needs explanation. If the current changes from $+3$ to -3 A at a given time instant, then the magnitude must have both these values at that time. Although it is not physically possible for the current to change value in zero time, it can change rapidly enough for it to be considered so. For example, if the current actually takes 1 ns to change from $+3$ to -3 A, this time is virtually zero when compared with the 1 s between changes. If it is desired to specify an instantaneous value immediately prior to or after a rapid change, it can be shown as:
At the instant immediately prior to time $t = 1$ s,

$$i_{1\,s-} = +3 \text{ A}$$

At the instant immediately after time $t = 1$ s,

$$i_{1\,s+} = -3 \text{ A}$$

AVERAGE VALUE

The *average value* of an alternating current or voltage is the mean magnitude of the positive or negative part of its waveform. If the waveform is symmetrical, the positive and negative parts will be the same and their mean magnitudes will be equal. An asymmetrical waveform will have unequal positive and negative parts, and thus either the positive average value or the negative average value should be referred to.

It is important to distinguish between the average (practical) and the true average (theoretical) values.

The waveform in Fig. 11-6 is that of a symmetrical triangular voltage. During any 20-ms time interval the voltage magnitude varies in a linear manner over the range +10 to −10 V. In determining the average value over a 20-ms period, all positive instantaneous magnitudes are in effect canceled by equal and opposite negative instantaneous magnitudes, thus giving a true average value of 0 V. This zero result would be obtained for any symmetrical voltage waveform averaged over an interval between instants when the magnitudes have the same value and direction of change (swinging positive or swinging negative). Therefore the true average value has no practical use in specifying the relative magnitudes of symmetrical ac voltages and currents. An asymmetrical voltage or current has a true average value which is related to its positive and negative peak values.

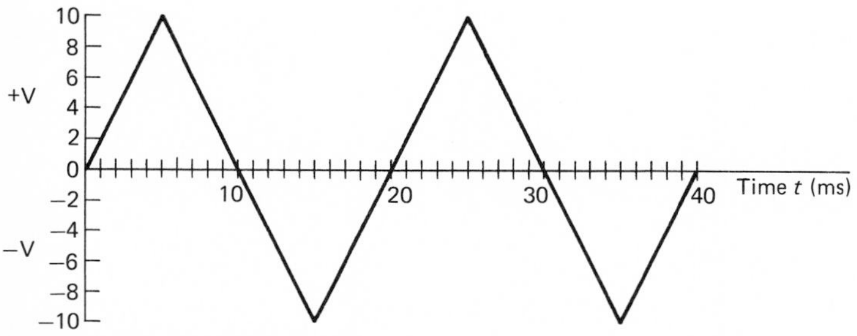

Fig. 11-6 Triangular-wave voltage.

Again with reference to the symmetrical voltage in Fig. 11-6, the mean magnitude over a time interval during which the voltage is either always positive or always negative is related to the peak value. Hence this interpretation of average value is of use in comparing the magnitudes of waveforms which have the same shape but differ in peak value. Over the 10-ms intervals $t = 0$ to $t = 10$ ms or $t = 20$ ms to $t = 30$ ms, the voltage is always positive and the average magnitude is +5 V. Over the 10-ms intervals $t = 10$ ms to $t = 20$ ms or $t = 30$ ms to $t = 40$ ms, the voltage is always negative and the average magnitude is −5 V. As 5 V is 50 percent of the 10 V peak value of the voltage, the positive or negative average indicates the waveform magnitude. This 5 V is the average value of the waveform as recognized in practical electrical work. The subscript "av" is used for average value.

Referring to Fig. 11-3*a*,

$$I_{av} = 3 \text{ A}$$

Referring to Fig. 11-3*b*,

$$I_{av} = 1.5 \text{ A}$$

Referring to Fig. 11-5,

$$\text{Positive } I_{av} = 4 \text{ A}$$

$$\text{Negative } I_{av} = 2 \text{ A}$$

The average value of a sine wave, as in Fig. 11-3*c*, is discussed in Sec. 11-6.

Almost all meters used for the measurement of current and voltage are based on a *moving coil* meter movement. This type of movement requires direct current to result in a meaningful indication. It indicates zero with ordinary symmetrical alternating current. The moving coil movement is fitted with a rectifier to adapt it for measuring alternating current and voltage. The rectifier causes the movement current to be of constant direction. When the movement is fitted with a rectifier, the deflection of its pointer is proportional to the average value of the ac current or voltage.

Although ac meters are usually calibrated in rms values, the fact that the moving coil type is the most popular makes measurement the most common application of ac average values.

ROOT MEAN SQUARE (RMS) VALUE

The rms value of an ac current or voltage is the value most widely used in practice. It is the value used to equate all waveforms, ac, steady dc, or varying dc. It relates waveforms of current or voltage on a power or energy basis. As it is concerned with the effect of energy transfer, such as occurs when heat is produced in resistance, rms value is also called *effective value*. Unless otherwise stated, ac currents or voltages are understood to be given in rms values.

The *rms value* of an ac current or voltage is that value which is equivalent, from a power or energy viewpoint, to steady direct current of the same magnitude.

If a steady direct current of 1 A will light a lamp, then an alternating current of 1 A rms will light the lamp to the same brilliance.

The concept of rms is arrived at as follows:

1. Power is expressed in terms of current or voltage for a given resistance value:

$$P = I^2R \qquad \text{or} \qquad P = \frac{V^2}{R}$$

Power is proportional to the *square* of either current or voltage.
2. In steady dc terms the power is constant, but where the current and voltage vary, the instantaneous power changes too. The steady equivalent of a varying power is represented by its average or *mean* value. When ac power is expressed in current or voltage terms, rms values are implied:

$$P_{av} = I_{rms}^2 R \qquad \text{or} \qquad \frac{V_{rms}^2}{R}$$

$$I_{rms}^2 = \frac{P_{av}}{R} \qquad \text{or} \qquad V_{rms}^2 = P_{av}R$$

3. To obtain the rms value of current or voltage, a square *root* operation must be performed:

$$I_{rms} = \sqrt{\frac{P_{av}}{R}} \qquad \text{or} \qquad V_{rms} = \sqrt{P_{av}R}$$

Items 1 to 3 refer to square, mean, and root. The term rms therefore aptly describes the overall operation of taking the square *root* of the *mean* value of the current or voltage *squared*.

In Fig. 11-3*a* the current is constant at +3 A for 1 s. The square of +3 A constant is +9 A^2, also constant. For the next 1 s, the current is constant at −3 A. The square of −3 A is +9 A^2 too. In this example, the square of the current is always +9 A^2, so the mean square of the current is 9 A^2. The square root of 9 A^2 is 3 A. Thus the rms value of a square wave is the same as its peak value, because the magnitude is constant. The fact that $(-3\text{ A})^2$ is the same as $(+3\text{ A})^2$ is not just mathematically correct, since it is logical that power is dissipated equally well by current in either direction (or voltage of either polarity).

Other waveforms do not share the simplicity of the square wave in the derivation of their rms values. Even the apparently simple sawtooth wave in Fig. 11-3*b* requires a calculus solution to obtain its rms value. The principle does not change, however; the rms value is the root of the mean of the squares of all the instantaneous values.

The rms values of sine-wave currents and voltages are dealt with in Sec. 11-6.

As ac voltage and current magnitudes are usually stated in rms terms, meters are usually required to indicate rms values. Most meters actually measure the average value of an ac waveform. Therefore the meter dial must be calibrated to register in rms values. The relationship between rms and average values is a function of the wave shape, thus limiting the calibration validity to only one particular wave shape. Most commercial ac meters are calibrated to indicate rms values of sine-wave currents and voltages, although meters which register the rms values of current or voltage waveforms of any shape are available.

Example 11-1 For the current waveform in Fig. 11-7, what is (*a*) the positive peak value, (*b*) the negative peak value, (*c*) the rms value, and (*d*) the instantaneous value 2.5 ns after time t_0.

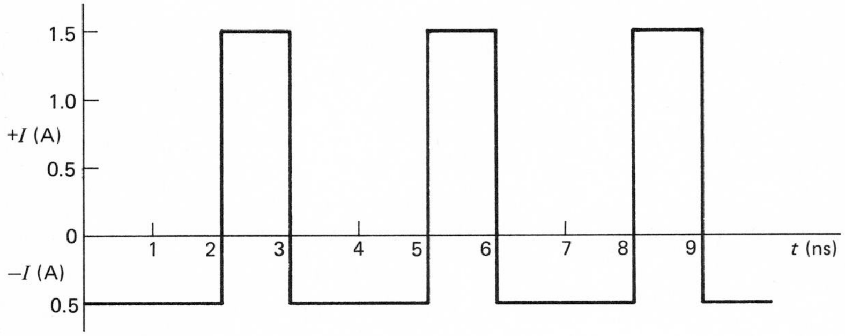

Fig. 11-7 Example 11-1: current waveform.

Solution Figure 11-7 shows a rectangular wave which is asymmetrical about the horizontal axis. Therefore the positive and negative peaks must be dealt with separately.

a. $$\text{Positive } I_{max} = \mathbf{1.5\ A}$$

b. $$\text{Negative } I_{max} = \mathbf{0.5\ A}$$

c. $$\text{Mean of positive values squared} = (1.5\ \text{A})^2 = 2.25\ \text{A}^2$$

$$\text{Mean of negative values squared} = (-0.5\ \text{A})^2 = 0.25\ \text{A}^2$$

In Fig. 11-8 is shown the graph of I^2 for a 3-ns period (a complete alternation). The mean magnitude of this I^2 curve will be that value obtained by taking the total area between the curve and the horizontal axis and spreading it out evenly over the whole 3-ns time occupied by the cycle.

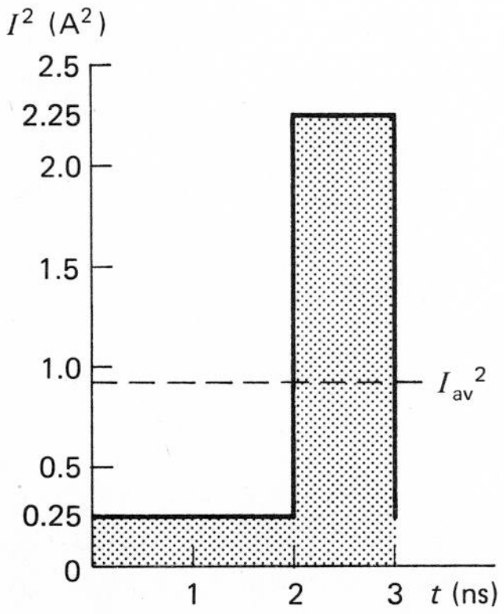

Fig. 11-8 Portion of the waveform in Fig. 11-7 with all instantaneous current magnitudes squared.

$$\text{Total area under curve} = 0.25\ \text{A}^2 \times 2\ \text{ns} + 2.25\ \text{A}^2 \times 1\ \text{ns}$$

$$\text{Total area} = 2.75\ \text{A}^2 \cdot \text{ns}$$

$$\text{Average magnitude} = \frac{\text{average area}}{\text{total time}}$$

$$I_{\text{av}}^2 = \frac{2.75\ \text{A}^2 \cdot \text{ns}}{3\ \text{ns}} = 0.9167\ \text{A}^2$$

$$I_{\text{rms}} = \sqrt{I_{\text{av}}^2} = \sqrt{0.9167\ \text{A}^2} = \mathbf{0.9574\ A}$$

d. The instantaneous value at time $t = 2.5$ ns is

$$i = \mathbf{+1.5\ A}$$

Figure 11-7 can be called a *pulse waveform* or *pulse train* because it consists of a sequence (or train) of relatively short-duration pulses of current in one direction with longer time intervals between the pulses. The pulse train in Fig. 11-7 consists of positive pulses, since the short-duration pulses are in a positive direction.

Pulse voltage waveforms and pulse current waveforms have similar characteristics.

Example 11-2 What is the average value of the trapezoidal waveform in Fig. 11-9?

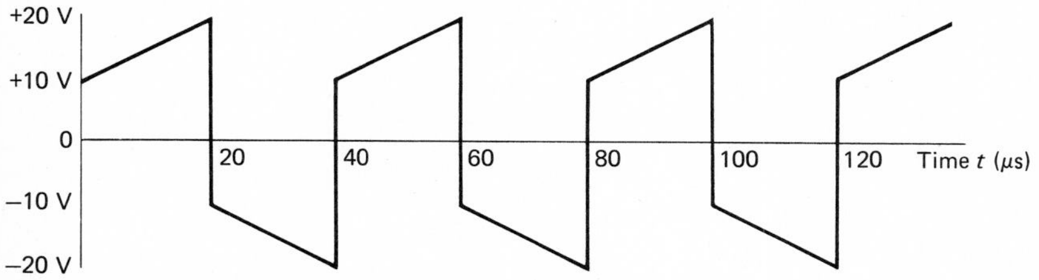

Fig. 11-9 Example 11-2: trapezoidal voltage waveform.

Solution Using the conventional meaning of the average value of a waveform, either the positive or the negative part of the symmetrical waveform in Fig. 11-9 can be selected to obtain the solution.

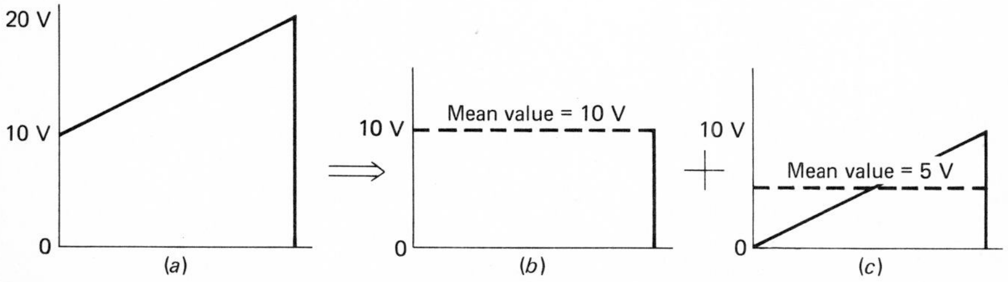

Fig. 11-10 Components of the waveform in Fig. 11-9.

The waveform in Fig. 11-9 may be regarded as being equivalent to square waves of 10 V magnitude with triangular waves of 10 V peak magnitude superimposed on them. Figure 11-10 shows how this concept can be applied to a positive section of the original waveform.

Figure 11-10*a* is represented as the sum of Fig. 11-10*b* and *c*. The mean value of the square-wave section in Fig. 11-10*b* is 10 V. The mean value of the triangular-wave section in Fig. 11-10*c* is 5 V. The average value of the waveform in Fig. 11-9 is the sum of these two mean values:

$$V_{av} = 10\text{ V} + 5\text{ V} = \mathbf{15\ V}$$

11-3 CYCLE, PERIOD, FREQUENCY, AND WAVELENGTH

CYCLE

The term "cycle" is used extensively in ac circuit work. A *cycle* is a complete alternation; it is the waveform of current or voltage between equivalent magnitudes. By "equivalent magnitude" is meant the same magnitude varying in the same direction.

Figure 11-11 shows a triangular waveform. Points $t = 0$, 8, and 16 ms are equivalent points, because here the waveform has zero magnitude and is changing in a positive direction. Points $t = 4$, 12, and 20 ms are also equivalent points; here the waveform has zero value and is changing in a negative direction. Points $t = 0$, 4, 8, 12, 16, and 20 ms are not all equivalent, even though they all represent zero values; in some the value is changing in a positive direction, and in others negative. Points $t = 2$, 10, and 18 ms are all positive peaks and are equivalent. The negative peaks, points $t = 6$ and 14 ms, are equivalent too. Points $t = 1$, 9, and 17 ms are equivalent, while points $t = 1$, 3, and 5 ms are not.

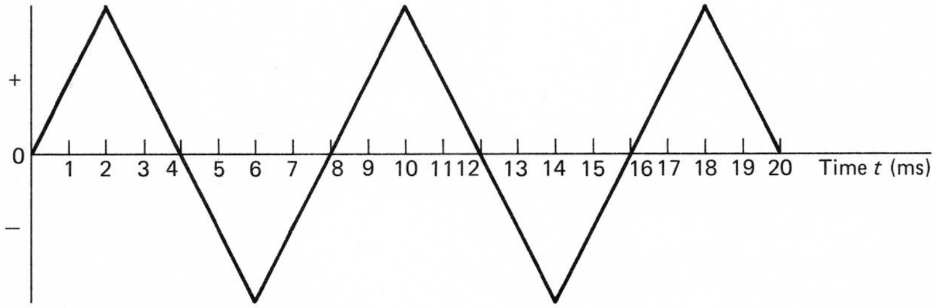

Fig. 11-11 Triangular waveform.

With reference to Fig. 11-11, typical cycles are the waveform existing between points

$$t = 0 \text{ and } 8\text{ ms}$$

$$t = 1 \text{ and } 9\text{ ms}$$

$$t = 2 \text{ and } 10\text{ ms}$$

$$t = 5 \text{ and } 13\text{ ms}$$

In Sec. 11-2 there were several references to part of the waveform. In each case this phrase meant a half-cycle. Although a half-cycle, as its name implies, is any half of a cycle, it is common to use the term in a limited sense to mean the waveform between adjacent zero values of magnitude. This usage results in the terms "positive half-cycle" and "negative half-cycle." Again with reference to Fig. 11-11, the positive half-cycles are the waveform between points

$$t = 0 \text{ and } 4 \text{ ms}$$

$$t = 8 \text{ and } 12 \text{ ms}$$

$$t = 16 \text{ and } 20 \text{ ms}$$

The negative half-cycles are the waveform between points

$$t = 4 \text{ and } 8 \text{ ms}$$

$$t = 12 \text{ and } 16 \text{ ms}$$

While recognizing that this limited meaning of half-cycle is common, it must be remembered that the waveform between, say, points $t = 1$ and 5 ms is also a half-cycle.

The term "cycle" is concerned with the waveform itself. It is not a measurement of amplitude or time; it has no symbol or unit.

PERIOD

The time occupied by one complete cycle is the *period* of the waveform. The symbol for period is T, and the unit for period is the second, symbol s.

In Fig. 11-11 the period is the time from point $t = 0$ ms to point $t = 8$ ms, or between any other two points representing one cycle. Thus the period T is 8 ms.

FREQUENCY

The number of complete cycles occurring each second is the *frequency* of the waveform. The symbol for frequency is f, and the unit for frequency is the hertz, symbol Hz.

Referring to Fig. 11-11, there are 2.5 complete cycles in 20 ms; there will be 125 cycles in 1 s. Therefore the frequency f of this waveform is 125 Hz.

There will be f cycles each second, so every cycle will occupy a time of $1/f$ second. This time is the period T.

$$T = \frac{1}{f} \quad \text{or} \quad f = \frac{1}{T} \tag{11-1}$$

where T = period, s
f = frequency, Hz

Equation (11-1) in its two forms, shows that period and frequency are reciprocal quantities.

WAVELENGTH

Sometimes the horizontal axis of a waveform is scaled in terms of physical distance. This is done when the waveform pertains to the voltage or current associated with propagated energy, such as sound or radio waves. There is no conflict in using time or distance as the horizontal axis quantity for an ac waveform. Distance is directly related to time and is used when it provides a convenient reference.

The distance occupied by one complete cycle of a propagated wave is a *wavelength*. The symbol for wavelength is λ (the lowercase Greek letter lambda). The unit of wavelength is the meter, symbol m.

Since one cycle occupies λ meters, f cycles will occupy $f\lambda$ meters. The number of cycles in 1 s is f. A propagated wave will travel at $f\lambda$ meters per second, so that $f\lambda$ is its velocity of propagation. This gives the equation

$$v = f\lambda \tag{11-2}$$

where f = frequency, Hz
λ = wavelength, m
v = velocity of propagation, m/s

In the case of electromagnetic radiation, such as radio waves, the velocity of propagation is the speed of light c which is approximately 3×10^8 m/s. This velocity gives

$$c = f\lambda \qquad \text{or} \qquad f\lambda = 3 \times 10^8 \text{ m/s} \tag{11-3}$$

From Eq. (11-2) it can be seen that frequency and wavelength are reciprocal:

$$f = \frac{v}{\lambda} \qquad \text{or} \qquad \lambda = \frac{v}{f}$$

This reciprocity is compatible with Eq. (11-1), since period and wavelength are both horizontal axis measures of one cycle.

Example 11-3 What is the frequency of the current in Example 11-1?
Solution The time occupied by one complete cycle is 3 ns. Therefore the period is

$$T = 3 \text{ ns}$$

The frequency is the reciprocal of the period

$$f = \frac{1}{T} \tag{11-1}$$

$$f = \frac{1}{3\text{ ns}} = \frac{1}{3 \times 10^{-9}\text{ s}} = \mathbf{333.3\ MHz}$$

Example 11-4 If in Example 11-3 the waveform represents a current pulse train propagated along a transmission line, what will be the scaling of the horizontal axis in Fig. 11-7? Assume that the pulse train is propagated along the transmission line at the speed of light.
Solution From Example 11-3,

$$f = 333.3\text{ MHz}$$

The wavelength will be

$$\lambda = \frac{c}{f} \tag{11-3}$$

$$\lambda = \frac{3 \times 10^8\text{ m/s}}{333.3\text{ MHz}} = 0.9\text{ m} \quad \text{or} \quad 90\text{ cm}$$

The physical length of one complete cycle will be 90 cm. Therefore the scaling of the horizontal axis will be **30 cm per division,** where one division in Fig. 11-7 is 1 ns.

The solution to Example 11-4 indicates that, on a propagation distance basis, Fig. 11-12 gives the waveform of the current pulse train.

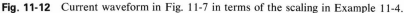

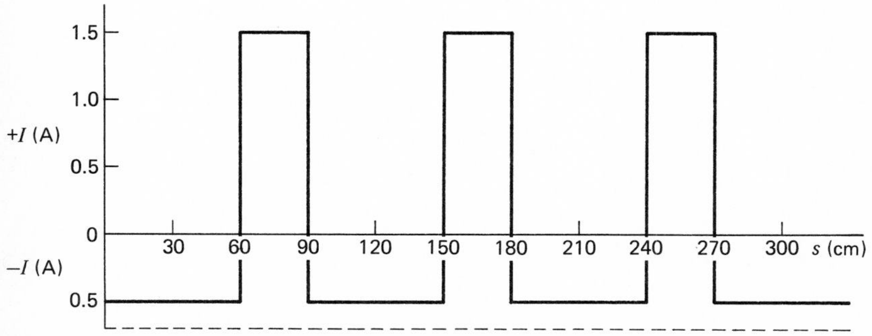

Fig. 11-12 Current waveform in Fig. 11-7 in terms of the scaling in Example 11-4.

Example 11-5 What is the frequency of an AM radio transmission if its wavelength is 350 m?
Solution

$$f = \frac{c}{\lambda} = \frac{3 \times 10^8\text{ m/s}}{350\text{ m}} = \mathbf{857.1\ kHz}$$

Example 11-6 The FM radio band covers the frequency range 88 to 108 MHz. What is the equivalent wavelength range?
Solution

$$\lambda = \frac{c}{f} = \frac{3 \times 10^8\text{ m/s}}{f\text{ Hz}} \quad \text{or} \quad \frac{300\text{ Mm/s}}{f\text{ MHz}}$$

At the lowest-frequency end of the FM band

$$\lambda = \frac{300 \text{ Mm/s}}{88 \text{ MHz}} = \mathbf{3.409 \text{ m}}$$

At the highest-frequency end of the band

$$\lambda = \frac{300 \text{ Mm/s}}{108 \text{ MHz}} = \mathbf{2.778 \text{ m}}$$

For the efficient reception of very high-frequency radio transmission, the length of an antenna is made approximately one-half the wavelength of the signal. Therefore the length of an antenna for the reception of FM-broadcast radio is typically of the order of 1.5 m (5 ft), one-half the average of 3.409 and 2.778 m, the results for Example 11-6.

Example 11-7 If the length of an organ pipe is approximately one-quarter wavelength, how long a pipe will be needed for a middle-C tone (frequency 256 Hz)? Take the velocity of sound in air as 336 m/s.
Solution

$$\lambda = \frac{v}{f} \tag{11-2}$$

$$\lambda = \frac{336 \text{ m/s}}{256 \text{ Hz}} = 1.313 \text{ m}$$

The organ pipe is to be approximately $\lambda/4$ in length:

$$\text{Pipe length} \simeq \frac{\lambda}{4} \simeq \frac{1.313 \text{ m}}{4} \simeq \mathbf{0.3282 \text{ m}} \qquad \text{(approximately 1 ft)}$$

Example 11-8 The waveform in Fig. 11-13 is displayed on an oscilloscope screen. The horizontal sweep speed is 2 ms per division. What is the waveform frequency?

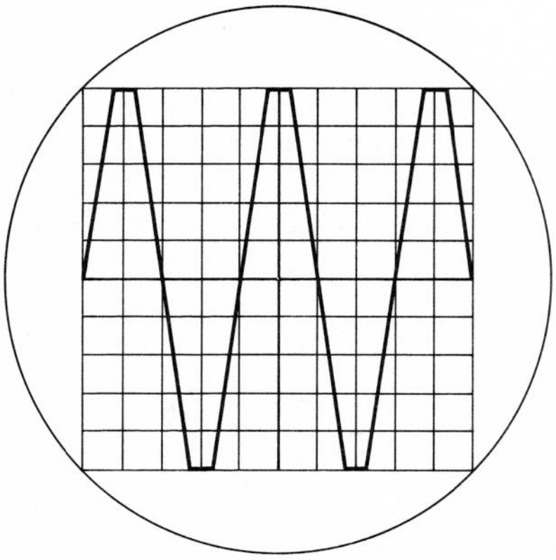

Fig. 11-13 Example 11-8: oscilloscope display.

Solution Figure 11-13 shows a display of 2.5 cycles of a trapezoidal waveform. These 2.5 cycles occupy 10 divisions of the horizontal scale. Ten divisions at 2 ms per division represents 20 ms. Therefore 2.5 cycles occupy a time interval of 20 ms. The period of the waveform T is the time occupied by one cycle:

$$T = \frac{20\text{ ms}}{2.5} = 8\text{ ms}$$

The frequency of the waveform f is the reciprocal of the period T:

$$f = \frac{1}{T} \tag{11-1}$$

$$f = \frac{1}{8\text{ ms}} = \mathbf{125\ Hz}$$

The solution to Example 11-8 could have been obtained more directly as follows:

$$f = \frac{\text{number of cycles displayed}}{\text{total time occupied}}$$

$$f = \frac{2.5\text{ cycles}}{20\text{ ms}} = 125\text{ Hz}$$

11-4 THE IMPORTANCE OF THE SINE WAVE

The *sine wave* is the waveform represented by the equation $y = k \sin x$. This rather uninspiring definition gives no clue to the reason for the dominance of the sine wave over all other waveforms in ac circuits. Before considering the details of sine-wave characteristics, we should obtain an idea of the significance of the sine wave in electrical technology.

The sine wave is important in ac circuits because:

1. It is the easiest waveform to generate.
2. It is the easiest waveform to distribute and use.
3. All other periodic waveforms can be synthesized from sine-wave components.

GENERATION OF SINE WAVES

Electrical energy used in industry, commerce, and the home is produced by electromagnetic generators. The functioning of these generators is based on Faraday's law of electromagnetic induction which was dealt with in Sec. 9-3.

Figure 11-14 shows a basic electromagnetic generator. It consists of a conducting loop of wire which is rotated within a magnetic field. The movement of the conducting loop sides across the field results in an emf being induced between the ends of the loop. Each loop end is connected to a metal slip ring. A conducting brush makes contact with each slip ring, and so connects the emf to an external network at all angular positions of the loop.

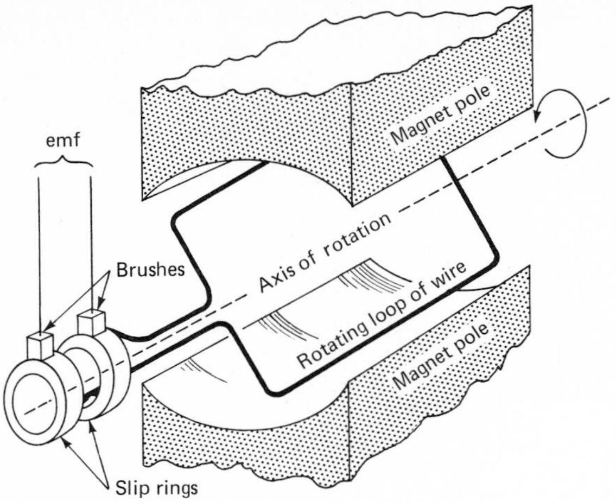

Fig. 11-14 Basic details of an ac generator or alternator.

In Fig. 11-15 is shown a sequence of events which occur while the loop is rotating. Although it is free to rotate either way, the loop is shown rotating in a counterclockwise direction. The small circles *A* and *B* represent sectional views of the loop sides.

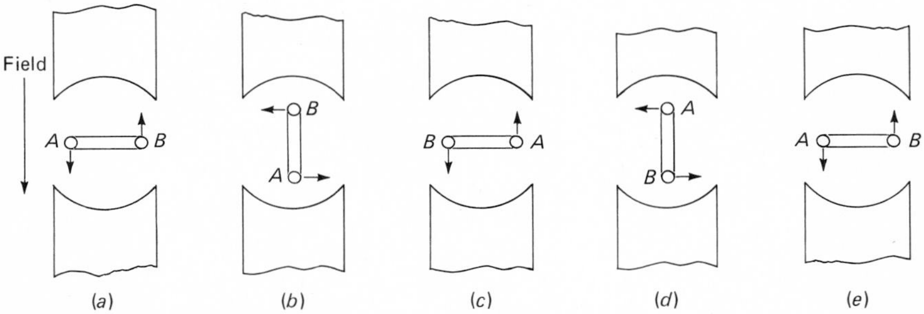

Fig. 11-15 Positions of the sides of a wire loop at 90° intervals.

Figure 11-15*a* At this instant the motion of the loop sides is parallel to the vertical direction of the magnetic field. There is no emf because there is no relative motion between the loop sides and the field; the conductor is not cutting the field.

Figure 11-15*b* Now the motion of each loop side is horizontal, perpendicular to the field direction. This is maximum relative motion which causes maximum emf to be generated.

Figure 11-15c Here the situation is the same as that in Fig. 11-15*a*, except that the sides of the loop, *A* and *B*, have changed places. The emf is again zero.

Figure 11-15*d* In this position the emf will have maximum magnitude, as in Fig. 11-15*b*. The loop side *A* is now moving to the left, whereas in Fig. 11-15*b* it was moving to the right. Similarly, side *B* has reversed its direction across the field. The effect of this reversal of direction of relative motion is a reversal of emf polarity.

Figure 11-15e The loop is now back in the same position as in Fig. 11-15*a*. The emf is zero again.

The complete rotation of the loop shown in Figs. 11-14 and 11-15 represents one complete cycle of alternation. A cycle of ac voltage has been generated. The voltage magnitudes have gone from zero to a maximum with one polarity, to zero, to a maximum with the opposite polarity, and then back to the original zero.

The diagram in Fig. 11-16 typifies the situation which exists when the generated voltage is neither zero nor maximum. The plane of the loop (a line through sides *A* and *B*) is at an angle θ to the horizontal. The instantaneous direction of a loop side is perpendicular to the radius of rotation and is at an angle of θ to the vertical.

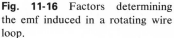

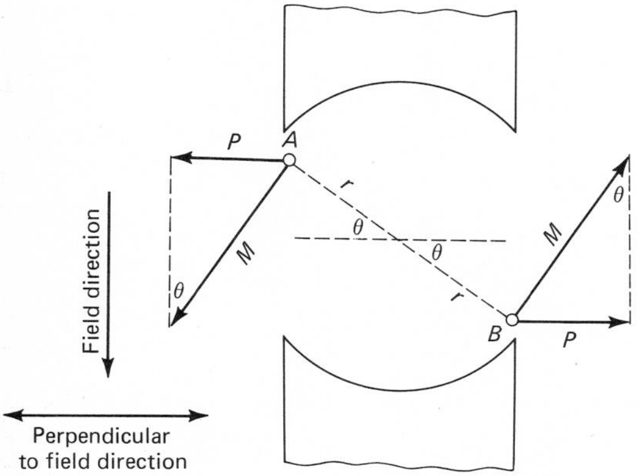

Fig. 11-16 Factors determining the emf induced in a rotating wire loop.

The generated emf is always proportional to the velocity of the component of conductor motion perpendicular to the field direction.

In Fig. 11-16 the actual directions of the instantaneous motion of each side of the loop (*A* and *B*) are shown by the arrows *M*. The directions of the components of motion perpendicular to the field direction are shown by the arrows *P*. The lengths of these arrows indicate the respective velocities.

The relationship between the actual and perpendicular velocities is

$$P \propto M \sin \theta$$

where θ is the angle between the instantaneous direction of a loop side and the direction of the field. θ can be specified in degrees or radians.

The generated emf e at any instant is proportional to the perpendicular velocity component P:

$$e \propto P \propto M \sin \theta$$

The actual velocity M of the conductors which form the loop sides A and B is the peripheral velocity of the loop and is related to the radius of rotation r by the equation

$$M = \omega r$$

where ω = angular velocity of rotation (the speed at which the loop turns), in radians per second.

Thus M is directly proportional to r, and the relationship between the generated emf and the radius of rotation is

$$e \propto r \sin \theta \qquad (11\text{-}4)$$

For our present purposes, the significant conclusion to be drawn from Eq. (11-4) is: The emf generated by an electromagnetic generator has a sine waveform.

Electronic generators of ac which incorporate inductance-capacitance resonant circuits are used extensively. They are called *oscillators* and typically are found as subsystems in radio transmitters and receivers and in measuring instruments. Resonant circuits naturally tend to develop sine-wave currents and voltages; so oscillators incorporating resonant circuits are sine-wave generators.

DISTRIBUTION AND USE OF SINE-WAVE AC

The presence of inductance and capacitance in significant quantities is inevitable in systems distributing and using ac energy. This is particularly true of inductance, which is an essential characteristic of distribution transformers, electric motors, and other electrical equipment.

The equations $e = L\, di/dt$ and $i = C\, dv/dt$, giving the voltage in an inductive circuit and the current in a capacitive circuit, show that, where L and C are present, the relationship between current and voltage is a function of rate of change in waveform.

The sine wave is the only waveform which has a rate-of-change curve with the same shape as itself (the waveform of the rate-of-change curve of a sine wave is sinusoidal).

The shape of a rate-of-change curve may seem to be a detail of purely theoretical interest, but actually it is of the utmost importance. It is so important that, even if sine waves were difficult to generate, they would have to be generated anyway. Why are they of such importance? Generating sine waves and supplying them to an electric network results in the same waveform for all currents and voltages throughout the network; there will be sine waves throughout.

Let us see what could happen if other current waveforms were used in an inductive circuit.

A square-wave alternating current, as in Fig. 11-17, in an inductance will only produce a voltage across the inductance during current reversals; there will be no voltage the remainder of the time because the current is constant.

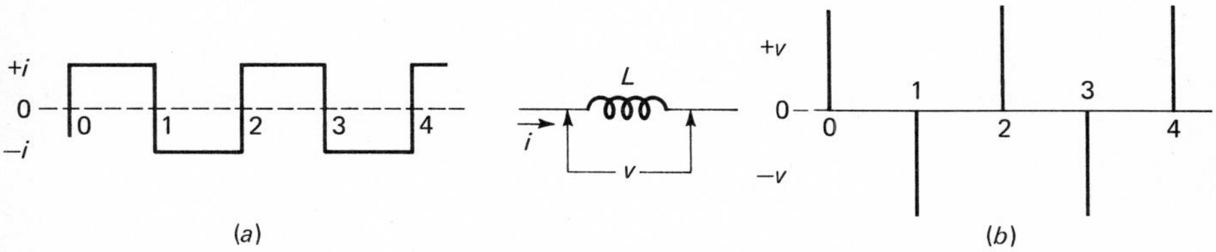

Fig. 11-17 Square-wave current in an inductance results in a pulse voltage waveform.

A triangular-wave current, as in Fig. 11-18, in an inductance will cause a constant magnitude of voltage to be developed across the inductance. The polarity of the voltage will reverse at each current peak.

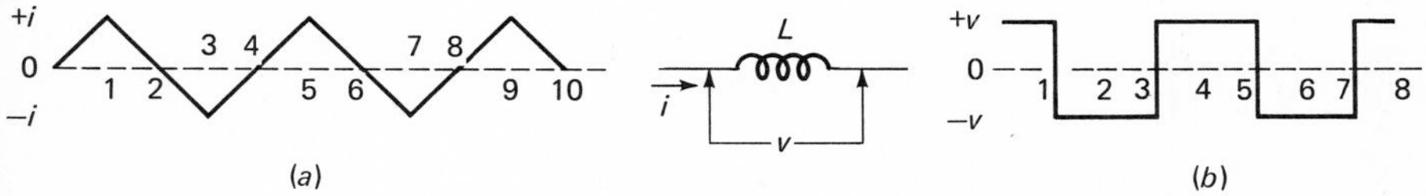

Fig. 11-18 Triangular-wave current in an inductance results in a square voltage waveform.

Figures 11-17 and 11-18 show just two possibilities. In a complex network involving many points of distribution and utilization of electrical energy, the waveforms will be difficult to predict and will render the system useless unless sine waves are used.

If an inductance carries the sine-wave current in Fig. 11-19*a*, the voltage across the inductance will have the waveform in Fig. 11-19*b*. Figure 11-19*b* is a cosine waveform, but sine and cosine waves are curves of the same shape displaced relative to each other on the horizontal time axis. Any linear network supplied with energy from a sine-wave ac source will contain only sinusoidal currents and voltages. These currents and voltages

Fig. 11-19 Sine-wave current in an inductance results in a sine-wave voltage.

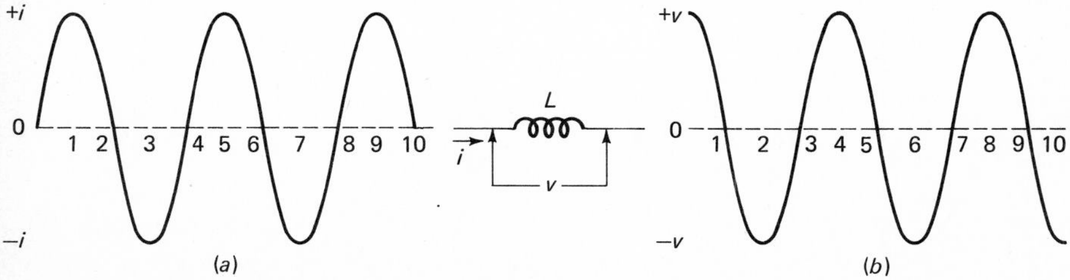

will all have the same frequency; their amplitudes and time relationships will depend on the resistance, inductance, and capacitance values within the network.

NONSINUSOIDAL PERIODIC WAVEFORMS

Any periodic wave can be reduced to a dc component and a number of sine-wave components of various frequencies, magnitudes, and time relationships. The mathematics of Fourier analysis is needed to fully prove that this is so, but the objectives of this text do not justify such a treatment. However, if you can be convinced that the principle is true, it will help you to understand the importance of sine waves. A demonstration that one non-sine wave can be synthesized from sine waves should assist you in accepting the fact that a general principle is involved.

A square wave, such as those in Figs. 11-17*a* and 11-18*b*, can be produced by adding together an infinite number of sine waves. The sine waves must start in phase; they must all have zero value and be swinging positive at the time instant that the square wave is to be zero swinging positive. If the frequency of the square wave is to be f Hz, then the frequencies of the sine-wave components must be odd multiples of f Hz; they must be f Hz, $3f$ Hz, $5f$ Hz, $7f$ Hz, etc. If the maximum value of the f-Hz sine wave is 1 V, the maximum values of the other sine-wave components must be $\frac{1}{3}$ V, $\frac{1}{5}$ V, $\frac{1}{7}$ V, etc.

The addition of sine waves can be shown graphically by adding the heights of the curves at each interval of time.

Figure 11-20 shows three waveforms. Two of these are sine waves and are shown as broken lines; the frequency of one is 1 kHz with a maximum value of 1 V, and the other has a frequency of 3 kHz and a maximum value of $\frac{1}{3}$ V. The solid-line curve is the result of adding the two sine waves. At each instant of time from 0 to 1 ms, the height of this solid curve is the sum of the heights of the two sine-wave curves. At time A, the sine waves have values of 0.5 and 0.33 V, so the addition curve has the

Fig. 11-20 A sine wave and its third harmonic added.

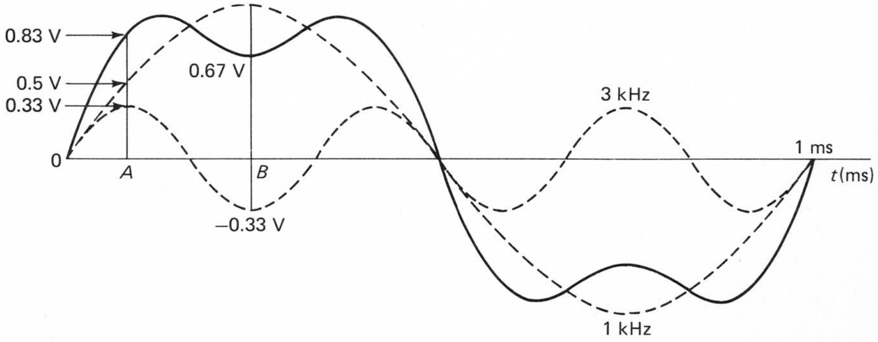

value of 0.83 V. At time *B*, the sine waves have values of +1 and −0.33 V, so the addition curve has the value 0.67 V.

The solid-line curve in Fig. 11-20 shows the beginning of the transformation from a 1-kHz sine wave to a 1-kHz square wave.

The solid-line curve in Fig. 11-20, representing the sum of the 1- and 3-kHz sine waves, is shown as a broken line in Fig. 11-21*a*. Also shown as a broken line is a 5-kHz sine wave of maximum value $\frac{1}{5}$ V. The solid-line curve is the sum of the two broken-line curves, and hence represents the sum of the 1-, 3-, and 5-kHz sine waves.

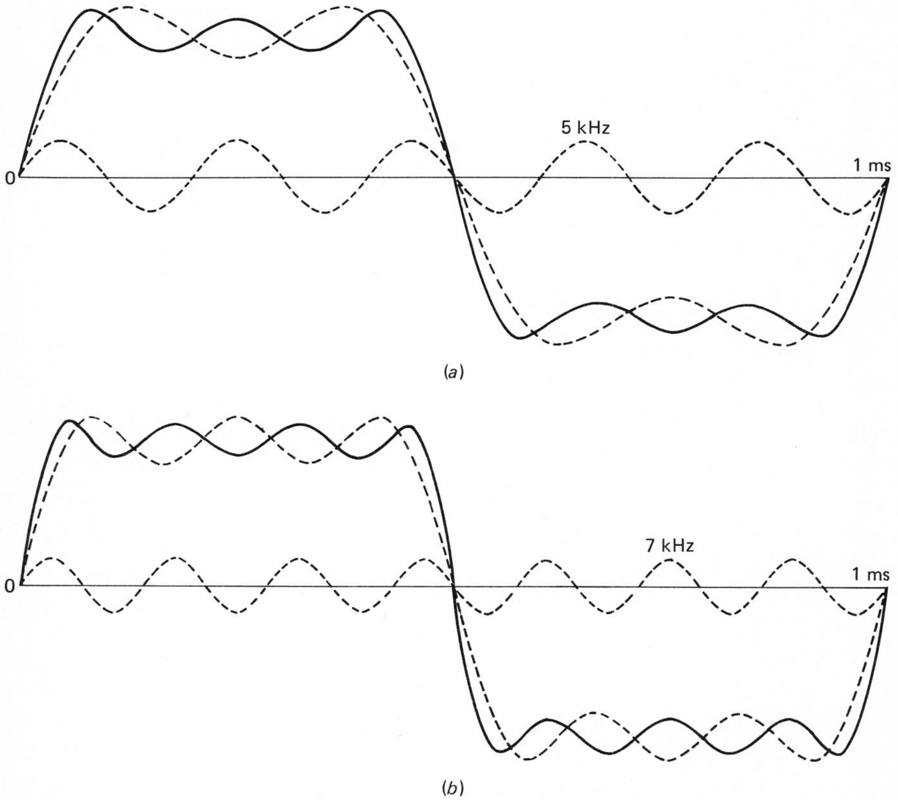

Fig. 11-21 (*a*) The addition waveform in Fig. 11-20 added to a fifth harmonic; (*b*) the addition waveform in Fig. 11-21*a* added to a seventh harmonic.

In Fig. 11-21*b*, the solid-line curve in Fig. 11-21*a* is shown as a broken line, as is a 7-kHz $\frac{1}{7}$-V maximum-value sine wave. The sum of these curves is the solid-line curve. This latter curve represents the addition of the 1-kHz (1-V), 3-kHz ($\frac{1}{3}$-V), 5-kHz ($\frac{1}{5}$-V), and 7-kHz ($\frac{1}{7}$-V) sine waves.

By comparing Figs. 11-20 and 11-21*a* and *b*, it can be seen that, as the higher frequencies are added, the total curve becomes closer to a square wave. The sides become steeper; the tops and bottoms become flatter, the ripples being more frequent but of smaller magnitude. It is not difficult to see that the addition of further sine waves of higher odd-multiple frequencies will result in even steeper sides and flatter tops and bottoms. The end result, if the number of sine waves added were to become infinite, would be the vertical sides and flat top and bottom of the perfect square wave.

The addition of sine waves of all multiple frequencies, odd and even, gives a sawtooth wave, if the amplitudes are again reciprocals of the frequency multiples (for example, the amplitude of the $4f$ sine wave is one-quarter that of the f sine wave). If sine waves which are even multiples of a given frequency are added, the result can be a sharp pulse of current or voltage.

The principle that all periodic waveforms can be regarded as sine waves added together is used in circuit analysis and design. For analysis, the sine-wave components may be considered to be produced by separate generators, the total circuit effect being determined by the superposition method. Transient waveforms such as square and sawtooth waves, containing very high-frequency sine-wave components, require circuit designs which take these frequencies into account.

11-5 THE SINE WAVE

The emf produced by a simple electromagnetic generator, at any time instant, is proportional to the sine of an angle. Basic trigonometric principles can be used to draw a sine waveform. If a circle is divided into a number of segments, beginning at the horizontal, the sines of each of the angles subtended to the horizontal are proportional to vertical projections from the circumference onto the horizontal.

In Fig. 11-22*a* a circle of radius E_{max} is divided into 12 equal segments, so that each segment represents an angle of 30° or $\pi/6$ rad. Figure 11-22*b* has a horizontal axis scaled so that each 30° angle is allocated an equal distance. This axis extends from 0 to 360° which is one complete revolution (2π rad), or one cycle of ac voltage.

For any angle to the horizontal in Fig. 11-22*a*, the vertical projection from the circumference to the horizontal gives a line of length representing

Fig. 11-22 Drawing a sine wave by projection from a circle.

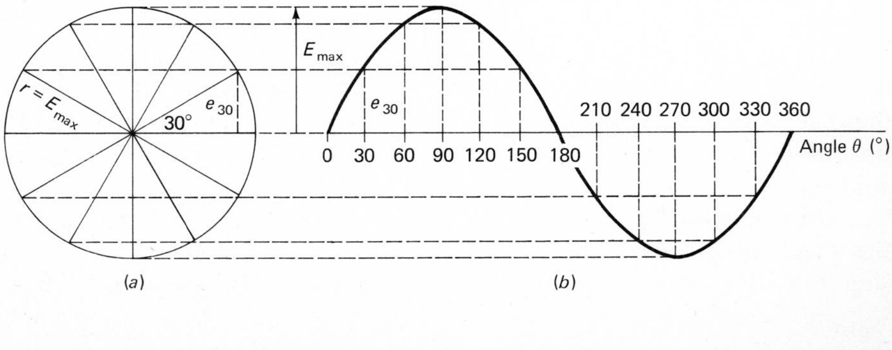

the instantaneous magnitude at that angle. Thus the line e_{30} represents the magnitude at 30°.

The waveform in Fig. 11-22*b* is drawn by erecting verticals at each 30° point and making their lengths equal to the vertical projections obtained from Fig. 11-22*a*. Since a vertical projection is equal to the circle radius multiplied by the sine of the relevant angle (for example, $e_{30} = E_{max} \sin 30°$), the resulting waveform is a sine wave representing all instantaneous voltage values for one ac cycle.

The maximum value of the ac voltage in Fig. 11-22 is E_{max} volts. The instantaneous value at any angle θ from the zero point will be

$$e = E_{max} \sin \theta \tag{11-5}$$

Equation (11-5) is the fundamental sine-wave ac relationship; it applies equally well to currents:

$$i = I_{max} \sin \theta \tag{11-5}$$

The shaft of an ac generator, or alternator, normally rotates at a constant rate. The circle in Fig. 11-22, together with its radius lines, should be regarded as rotating at a constant rate.

ANGULAR VELOCITY OR ANGULAR FREQUENCY

The rate of rotation is the *angular velocity,* the rate at which the angle changes. The symbol for angular velocity is ω (the lowercase Greek letter omega). The unit for angular velocity is the radian per second, symbol rad/s.

An ac frequency of f hertz can be related to Fig. 11-22 by saying that the radius rotates f revolutions per second. During each revolution the radius traverses 2π radians. Therefore the angular velocity or angular frequency can be expressed as

$$\omega = 2\pi f \tag{11-6}$$

where ω = angular velocity or angular frequency, rad/s
π = the numerical constant $\simeq 3.142$
f = frequency, Hz

Angular velocity is the angle traversed each second, so that

$$\omega = \frac{\theta}{t} \qquad \text{or} \qquad \theta = \omega t \qquad \text{or} \qquad t = \frac{\theta}{\omega} \tag{11-7}$$

where θ = angle, rad
ω = angular velocity, rad/s
t = time, s

Equation (11-7) gives the important link between angle and time. The sine wave in Fig. 11-22 can be redrawn with time as the horizontal axis by substituting $t = \theta/\omega$.

Figure 11-23 is the *same* as Fig. 11-22*b*. It is not merely the same shape. Appreciating this fact will help you to avoid confusion when faced with the various methods of expressing ac quantities used in electrical work. In the two figures the horizontal axis represents angle or time. From Eqs. (11-6) and (11-7) it can be seen that these two quantities are related by $2\pi f$ which is a numerical constant for any given periodic wave.

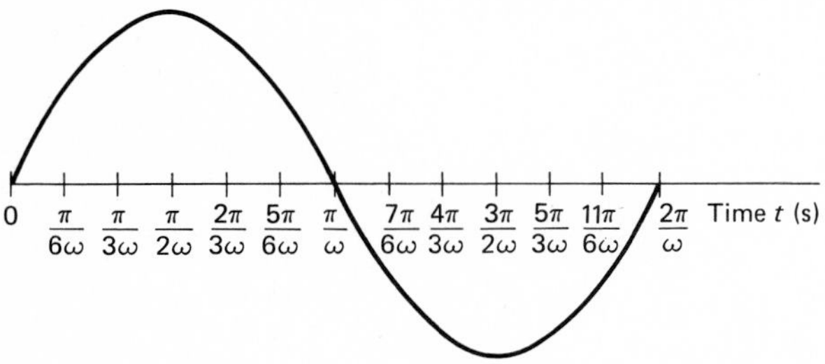

Fig. 11-23 Sine wave with respect to time.

As an example, consider an angle of π radians (180°) which is equal to one half-cycle of a sine wave:

$$\theta = \pi$$

but $$\theta = \omega t \qquad (11\text{-}7)$$

so $$\omega t = \pi$$

or $$t = \frac{\pi}{\omega}$$

To verify that π/ω really is time, substitute $\omega = 2\pi f$ from Eq. (11-6)

$$t = \frac{\pi}{\omega} = \frac{\pi}{2\pi f} = \frac{1}{2f}$$

The time occupied by one complete cycle is the period of the wave T

$$T = \frac{1}{f} \qquad (11\text{-}1)$$

We see that for an angle of π radians the time occupied is

$$t = \frac{1}{2f} = \frac{T}{2}$$

Therefore π/ω is a time equal to one half-period of the waveform.

Example 11-9 A wall outlet delivers 60-Hz, sine-wave ac at 170 V maximum value to domestic lamps and appliances. What is the instantaneous voltage for the following angles after angle θ_0: (a) $\pi/4$ rad, (*b*) $3\pi/4$ rad, (*c*) $5\pi/4$ rad, and (*d*) $7\pi/4$ rad.

Solution If it is not otherwise specified, it is normally assumed that angle θ_0 is the point on the horizontal axis at which the voltage magnitude is zero and is swinging in a positive direction. Figure 11-24 is drawn on this assumption.

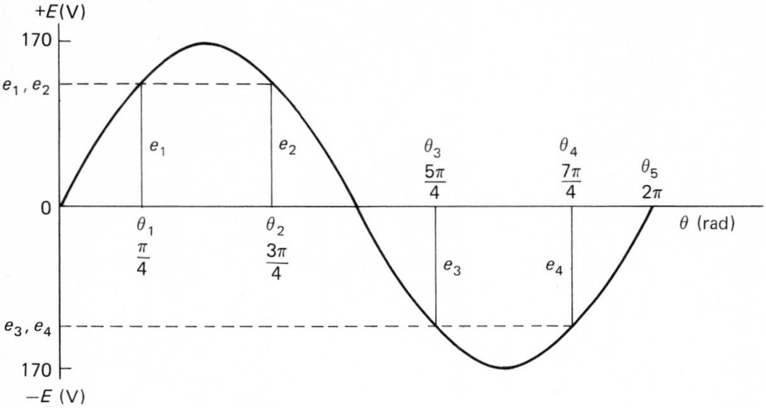

Fig. 11-24 Example 11-9: data.

The values of θ specified are all odd multiples of $\pi/4$ rad. This angle, $\pi/4$ rad, is 45°. By reference to a table of sines or a slide rule S scale, the value of sin 45° is found to be 0.7071.

a. $\theta_1 = \pi/4$ rad:

$$e_1 = E_{max} \sin \theta_1 \tag{11-5}$$

$$e_1 = 170 \text{ V} \times \sin 45° = 170 \text{ V} \times 0.7071 \simeq \mathbf{120\ V}$$

It is clear from Fig. 11-24 that the magnitudes of the voltages at all the specified angles are the same. At $\pi/4$ and $3\pi/4$ rad both voltages have the same positive polarity. At $5\pi/4$ and $7\pi/4$ rad both voltages have negative polarity. Therefore,

b. $\theta_2 = 3\pi/4$ rad:

$$e_2 \simeq \mathbf{120\ V}$$

c. $\theta_3 = 5\pi/4$ rad:

$$e_3 \simeq \mathbf{-120\ V}$$

d. $\theta_4 = 7\pi/4$ rad:

$$e_4 \simeq \mathbf{-120\ V}$$

It is important to note that for angles from 0 to 180° (0 to π rad) the sines are positive, while for angles between 180 and 360° (π to 2π rad) the sines are negative. If this is not clear to you, a review of basic trigonometry is necessary. Sines, cosines, tangents, and arc tangents will be met frequently in the discussions on ac circuits.

In the solutions to Example 11-9, each of the voltages has the magnitude $E_{max} \times 0.7071$, which is also the rms value of the voltage. This is because the sine of 45° and the relationship between rms and peak values for a sine wave are both $1/\sqrt{2}$.

Example 11-10 Draw the sine wave in Example 11-9 with time as the horizontal axis.
Solution The data required to scale the horizontal axis can be obtained by applying the equation

$$t = \frac{\theta}{\omega} \tag{11-7}$$

$$t = \frac{\theta}{2\pi f} = \frac{\theta}{2\pi 60\ \text{Hz}} \quad \text{or} \quad \theta \times \frac{1}{2\pi 60\ \text{Hz}}\ \text{s}$$

a. $\theta_1 = \pi/4$ rad:

$$t_1 = \frac{\pi}{4}\ \text{rad} \times \frac{1}{2\pi 60\ \text{Hz}} = \frac{1}{480}\ \text{s} \quad \text{or} \quad 2.083\ \text{ms}$$

The angles θ_2, θ_3, and θ_4 are, respectively, three, five, and seven times angle θ_1. Therefore the related times t_2, t_3, and t_4 will be three, five, and seven times t_1.
b. $\theta_2 = 3\pi/4$ rad:

$$t_2 = 3t_1 = \frac{3}{480}\ \text{s} \quad \text{or} \quad 6.25\ \text{ms}$$

c. $\theta_3 = 5\pi/4$ rad:

$$t_3 = 5t_1 = \frac{5}{480}\ \text{s} \quad \text{or} \quad 10.42\ \text{ms}$$

d. $\theta_4 = 7\pi/4$ rad:

$$t_4 = 7t_1 = \frac{7}{480}\ \text{s} \quad \text{or} \quad 14.58\ \text{ms}$$

e. The end of the cycle occurs at $\theta_5 = 2\pi$ rad $= 8\pi/4$ rad:

$$t_5 = 8t_1 = \frac{8}{480}\ \text{s} = \frac{1}{60}\ \text{s} \quad \text{or} \quad 16.67\ \text{ms}$$

The resulting sine-wave voltage, with time as the horizontal axis, is shown in Fig. 11-25.

Fig. 11-25 Example 11-10: result.

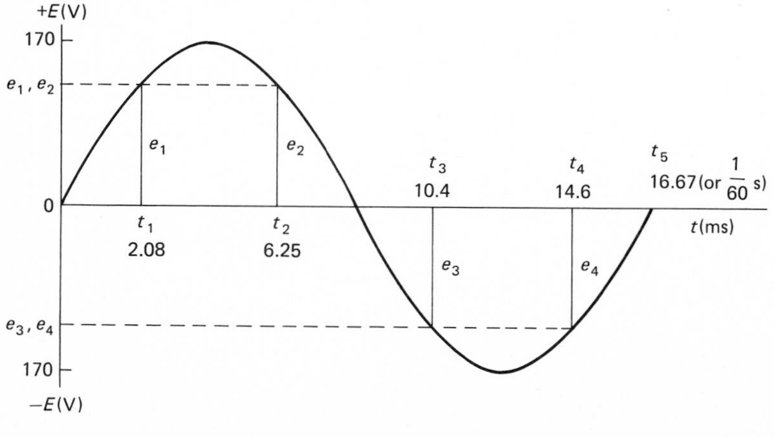

11-6 SINE-WAVE VALUES

The methods by which the magnitude of an ac wave can be specified apply to sine-wave voltages and currents. For the reasons given in Sec. 11-4, sine-wave alternating current is so common that the various means for specifying its magnitude deserve separate treatment.

In the following paragraphs describing sine-wave values, only voltage waves are shown; current waves follow the same principles.

MAXIMUM OR PEAK VALUE

The *maximum* or *peak value* of a sine wave is the greatest magnitude either side of zero.

A sine wave is symmetrical, its positive and negative half-cycles being the same. Thus its positive and negative peaks have the same magnitude. Figure 11-26 shows the maximum values of a sine-wave emf.

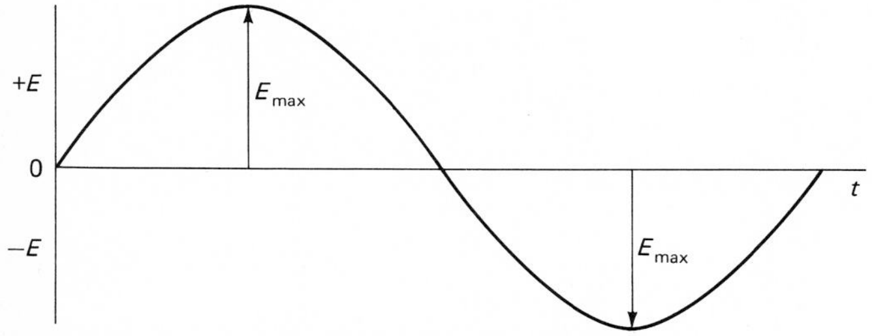

Fig. 11-26 Sine-wave maximum or peak value.

PEAK-TO-PEAK VALUE

The *peak-to-peak value* of a sine wave is the total waveform magnitude taken from the positive peak to the negative peak.

As you can see from Fig. 11-27, because of the symmetry of the sine wave,

$$E_{\text{p-p}} = 2E_{\text{max}} \tag{11-8}$$

Fig. 11-27 Sine-wave peak-to-peak value.

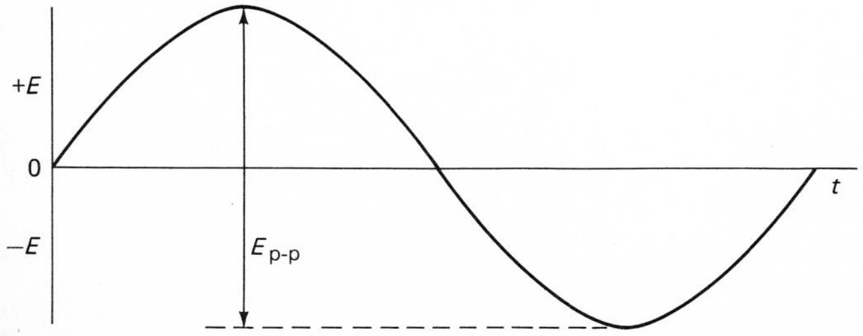

INSTANTANEOUS VALUE

The *instantaneous value* of a sine wave is the magnitude at a specified instant of time after a reference time. The reference time is usually called t_0, the time at which the magnitude is zero and is swinging in a positive direction.

Sines are expressed in terms of angles. The relationship between angle and time is given in Eq. (11-7) as

$$\theta = \omega t \tag{11-7}$$

Applying this relationship to Eq. (11-5) ($e = E_{\max} \sin \theta$) gives an expression for the instantaneous value of the emf shown in Fig. 11-28:

$$e = E_{\max} \sin \omega t \tag{11-9}$$

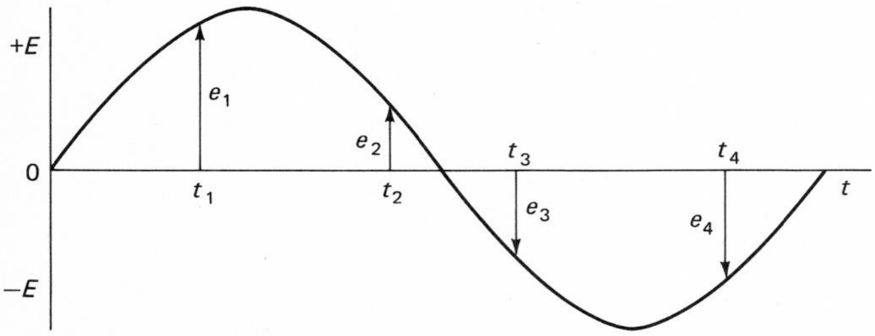

Fig. 11-28 Sine-wave instantaneous values.

AVERAGE VALUE

The true average or mean value of a sine wave is zero, the positive and negative half-cycles being identical in shape and magnitude. Thus, as a means of specifying the relative magnitudes of different sine waves, the true average value is of no practical use. By determining the mean magnitude of either the positive or negative half-cycle, a value is obtained which is useful for establishing the significance of the reading on an ac meter. An ac meter usually consists of a dc meter movement fitted with a rectifier to convert the alternating current being measured into dc half-cycles. The meter pointer indicates the average magnitude of these half-cycles.

All references to the average value of a sine wave in this text have the conventional meaning: the average magnitude of a positive or a negative half-cycle.

A half-cycle can be specified as beginning at any point in a cycle. To ensure that a full positive or negative half-cycle is chosen, it is usual to specify that it commences at a point of zero magnitude.

The *average value* of a sine wave is the average magnitude of a half-cycle which begins with zero magnitude.

In Fig. 11-29 the average value is related to the positive half-cycle, although it could as readily have been related to the negative half-cycle.

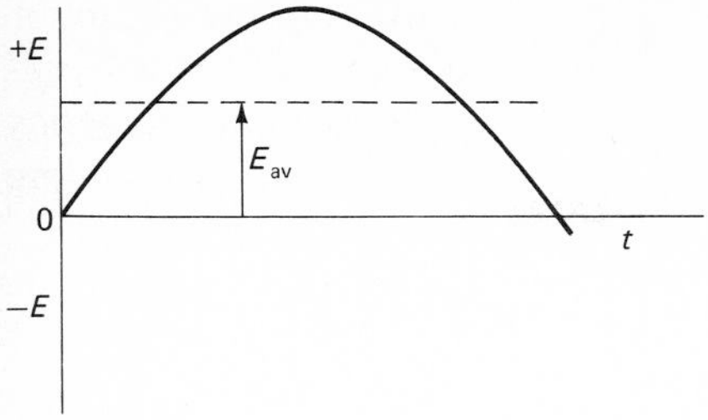

Fig. 11-29 Sine-wave average value.

Graphically the average value can be determined by dividing the half-cycle into many equal-width strips and then finding the average heights of the strips. This procedure is laborious and inaccurate. The preferred method is to use calculus. The calculus method is described in the appendix. Its result is

$$E_{av} = \frac{2E_{max}}{\pi} \quad \text{or} \quad 0.6366E_{max} \tag{11-10}$$

RMS VALUE

A sine wave of voltage or current in a network will result in a certain power level. The steady dc voltage or current which will result in the same power is the *rms* or *effective value* of the sine wave.

Figure 11-30 shows how the level of the rms value is related to a sine waveform of voltage.

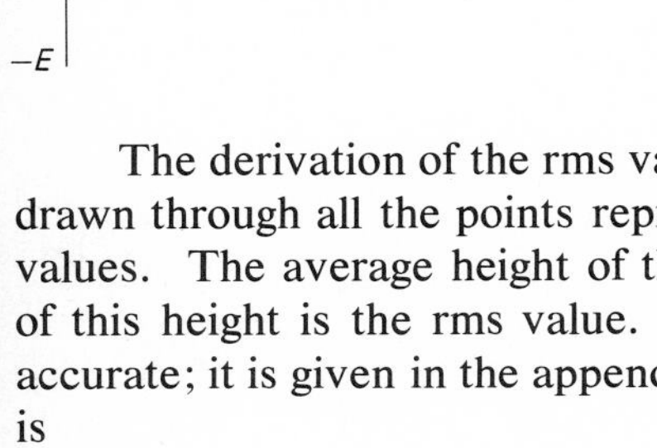

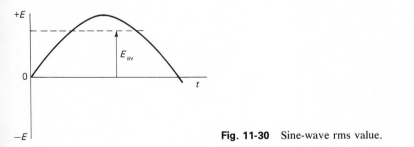

Fig. 11-30 Sine-wave rms value.

The derivation of the rms value can be shown graphically. A curve is drawn through all the points representing the squares of the instantaneous values. The average height of this curve is then found. The square root of this height is the rms value. The calculus method is faster and more accurate; it is given in the appendix. The rms value of a sine-wave voltage is

$$E = \frac{E_{max}}{\sqrt{2}} \quad \text{or} \quad 0.7071E_{max} \tag{11-11}$$

Note that the symbol E (no subscript) implies rms value.

Figure 11-31 shows one cycle of a sine-wave emf with maximum or peak, peak-to-peak, average, and rms values as defined above. The comparative magnitudes of these values can be clearly seen. Instantaneous values have been omitted because they can have any magnitude depending on the specified point in the cycle.

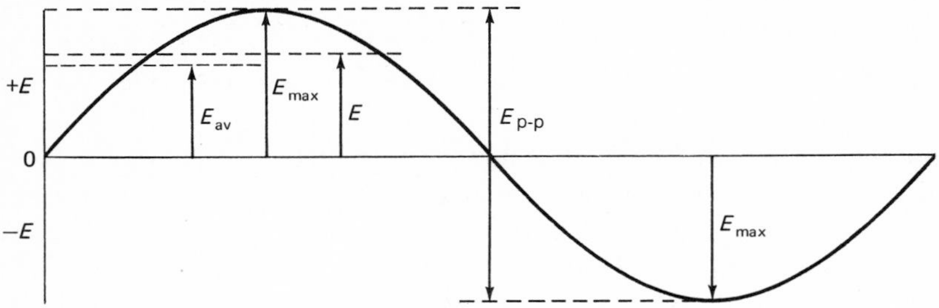

Fig. 11-31 Comparison of sine-wave magnitude values.

Example 11-11 For a domestic electricity supply rated at 120 V, 60 Hz, what is (*a*) the maximum value, (*b*) the average value, and (*c*) the instantaneous value 10 ms after time t_0? *Solution* Domestic electricity supplies are sine-wave ac for the reasons given in Sec. 11-4 (ease of generation, distribution, and use). A supply is rated in rms values, unless otherwise stated. Therefore the supply in this example can be more completely described as 120 V rms, 60 Hz, sine-wave ac. One cycle of this sine wave is shown in Fig. 11-32.

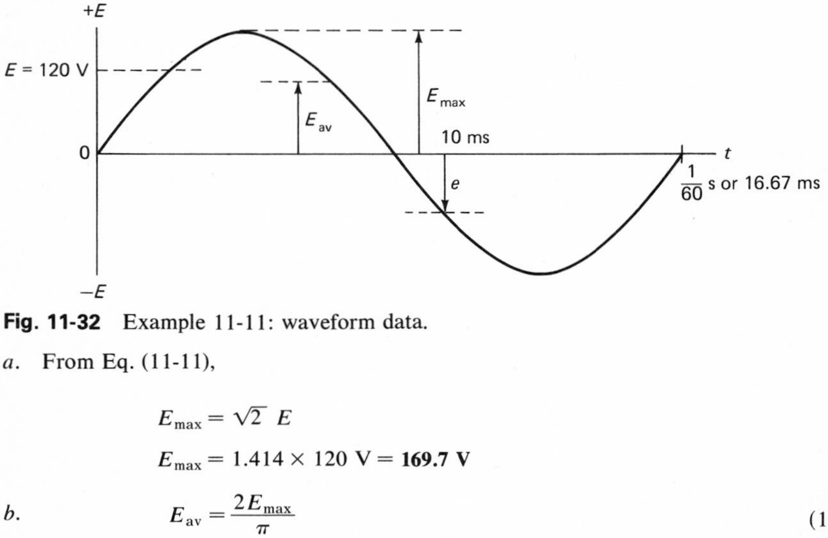

Fig. 11-32 Example 11-11: waveform data.

a. From Eq. (11-11),

$$E_{max} = \sqrt{2}\ E$$

$$E_{max} = 1.414 \times 120\ \text{V} = \mathbf{169.7\ V}$$

b.

$$E_{av} = \frac{2E_{max}}{\pi} \qquad (11\text{-}10)$$

$$E_{av} = 0.6366 \times 169.7\ \text{V} = \mathbf{108\ V}$$

c.

$$\omega = 2\pi f \qquad (11\text{-}6)$$

$$\omega = 2\pi \times 60\ \text{Hz} = 377\ \text{rad/s}$$

$$e = E_{max} \sin \omega t = 169.7\ \text{V} \times \sin 377\ \text{rad/s} \times 10\ \text{ms}$$

$$= 169.7\ \text{V} \times \sin 3.77\ \text{rad} \qquad \text{or} \qquad 169.7\ \text{V} \times \sin 216°$$

$$e = 169.7\ \text{V} \times (-0.5878) = \mathbf{-99.75\ V}$$

In North America the voltage of the domestic electricity supply has one of several different values, 110, 115, and 117 V being common. The value of 120 V, as given in Example 11-11, is the accepted standard.

It should be noted that the peak value of the ordinary household supply is about 170 V. The insulation of wiring and equipment must be able to safely withstand this peak voltage.

Example 11-12 The instantaneous value of a 5-kHz sine-wave current is 10 mA, 40° after angle θ_0. What are the (*a*) peak, (*b*) peak-to-peak, and (*c*) rms values of the current?
Solution See Fig. 11-33.

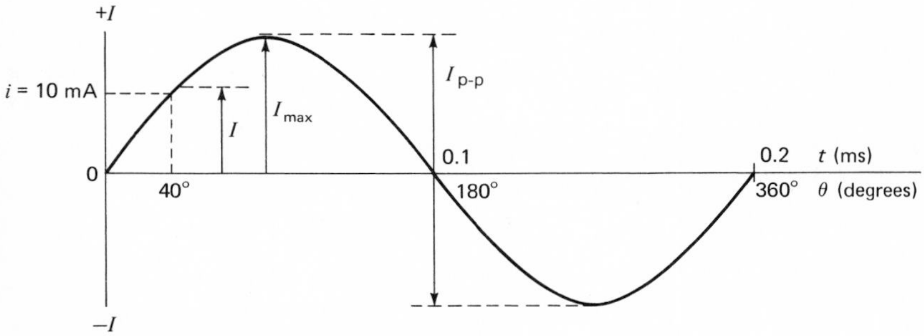

Fig. 11-33 Example 11-12: waveform data.

a. From Eq. (11-5),

$$I_{max} = \frac{i}{\sin\theta} = \frac{10\text{ mA}}{\sin 40°} = \frac{10\text{ mA}}{0.6428} = \mathbf{15.56\text{ mA}}$$

b.

$$I_{p\text{-}p} = 2\,I_{max} \qquad (11\text{-}8)$$

$$I_{p\text{-}p} = 2 \times 15.56\text{ mA} = \mathbf{31.12\text{ mA}}$$

c.

$$I = 0.7071\,I_{max} \qquad (11\text{-}11)$$

$$I = 0.7071 \times 15.56\text{ mA} = \mathbf{11\text{ mA}}$$

Example 11-13 A symmetrical ac sine-wave voltage when displayed on the screen of an oscilloscope appears as in Fig. 11-34. The vertical calibration of the oscilloscope is 5 mV per division. What is (*a*) the peak-to-peak voltage, (*b*) the peak voltage, and (*c*) the rms voltage?

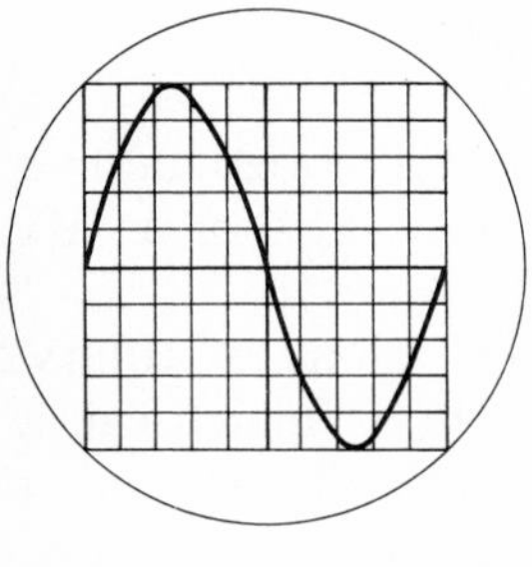

Fig. 11-34 Example 11-13: oscilloscope display.

Solution
a. The total height of the display is 10 divisions. The peak-to-peak voltage is the height multiplied by the vertical calibration.

$$V_{p\text{-}p} = 10\text{ divisions} \times 5\text{ mV per division} = \mathbf{50\text{ mV}}$$

b. From Eq. (11-8),

$$V_{\text{max}} = \frac{V_{\text{p-p}}}{2} = \frac{50 \text{ mV}}{2} = \mathbf{25\ mV}$$

c. From Eq. (11-11),

$$V = \frac{V_{\text{max}}}{\sqrt{2}} = \frac{25 \text{ mV}}{\sqrt{2}} = \mathbf{17.68\ mV}$$

The solutions given for Example 11-13 are only valid if it is known that the waveform displayed is a sine wave and symmetrical about the zero magnitude level. The display does not necessarily show whether the waveform is symmetrical or asymmetrical, or even whether it is alternating or direct current. To know, one must be aware of the vertical location of the zero voltage level on the display. Since we are told that Fig. 11-34 represents a symmetrical ac sine-wave voltage, we know that the zero voltage level is located at the central horizontal line.

When an oscilloscope is being used, it often happens that the nature of the waveform and the position of the zero voltage level are not known, resulting in situations similar to those depicted in Fig. 11-35.

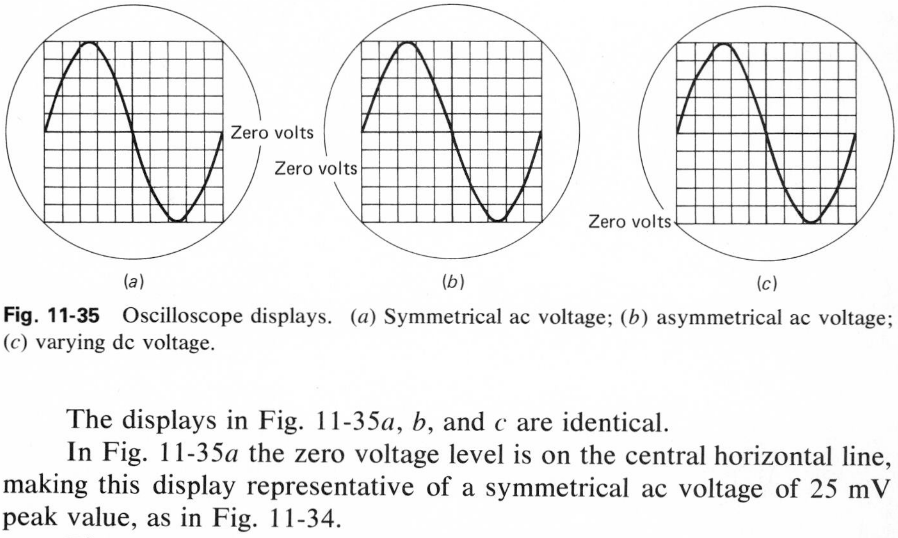

Fig. 11-35 Oscilloscope displays. (*a*) Symmetrical ac voltage; (*b*) asymmetrical ac voltage; (*c*) varying dc voltage.

The displays in Fig. 11-35*a*, *b*, and *c* are identical.

In Fig. 11-35*a* the zero voltage level is on the central horizontal line, making this display representative of a symmetrical ac voltage of 25 mV peak value, as in Fig. 11-34.

Figure 11-35*b* shows the zero voltage level two divisions, or 10 mV, below the center line. This display thus represents an asymmetrical ac voltage with a positive peak value of 35 mV and a negative peak value of 15 mV.

The display in Fig. 11-35*c* has its zero voltage level five divisions, or 25 mV, below the center line. The voltage represented is therefore varying direct current with a 50-mV peak-to-peak ac component and a dc level of +25 mV.

Example 11-14 The *form factor* of a waveform is the ratio of its rms value to its average value. What is the form factor of a sine-wave current?
Solution From Eq. (11-10),

$$I_{av} = \frac{2I_{max}}{\pi}$$

From Eq. (11-11),

$$I = \frac{I_{max}}{\sqrt{2}}$$

$$\text{Form factor} = \frac{I}{I_{av}} = \frac{I_{max}}{\sqrt{2}} \div \frac{2I_{max}}{\pi} = \frac{1}{\sqrt{2}} \times \frac{\pi}{2} = \mathbf{1.111}$$

11-7 SUMMARY OF EQUATIONS

$T = 1/f$ or $f = 1/T$	(11-1)
$v = f\lambda$	(11-2)
$c = f\lambda$ or $f\lambda = 3 \times 10^8$ m/s	(11-3)
$e \propto r \sin\theta$	(11-4)
$e = E_{max} \sin\theta$ or $i = I_{max} \sin\theta$	(11-5)
$\omega = 2\pi f$	(11-6)
$\omega = \theta/t$ or $\theta = \omega t$ or $t = \theta/\omega$	(11-7)
$E_{p\text{-}p} = 2E_{max}$	(11-8)
$e = E_{max} \sin\omega t$	(11-9)
$E_{av} = 2E_{max}/\pi$ or $0.6366\,E_{max}$	(11-10)
$E = E_{max}/\sqrt{2}$ or $0.7071E_{max}$	(11-11)

EXERCISE PROBLEMS

Section 11-2

11-1 A square-wave voltage has positive and negative maxima of 10 V, and the positive and the negative periods are each 1 ms. Find
a. The peak-to-peak voltage. (*ans.* 20 V)
b. The average voltage for the positive period. (*ans.* 10 V)
c. The average voltage for the negative period. (*ans.* 10 V)
d. The average voltage over one cycle. (*ans.* 10 V)
e. The rms voltage. (*ans.* 10 V)

11-2 A current square wave has a positive maximum of 5 A, a negative maximum of 10 A, a positive period of 1 ms, and a negative period of 2 ms. Find
a. The peak-to-peak current.
b. The average current for the positive period.
c. The average current for the negative period.
d. The average current over one cycle.
e. The rms current.

11-3 A voltage square wave has a maximum of +12 V for 1 ms and a minimum of +3 V for 3 ms. Find
a. The peak-to-peak voltage. (*ans.* 9 V)
b. The average voltage over one cycle. (*ans.* 5.25 V)
c. The rms voltage. (*ans.* 6.54 V)

Section 11-3

11-4 Find the frequency of the waveform when the period is

a. 1 ms *c.* 16.67 ms

b. 2 μs *d.* 50 ns

11-5 Find the period of the waveform when the frequency is

a. 10 kHz (*ans.* 100 μs)

b. 60 Hz (*ans.* 16.67 ms)

c. 30 MHz (*ans* 33.3 ns)

d. 256 Hz (*ans.* 3.91 ms)

11-6 The velocity of sound in air is given as 336 m/s. Find the wavelengths for the following sound waves:

a. 20 Hz *c.* 1000 Hz

b. 15 kHz *d.* 60 Hz

11-7 The velocity of electromagnetic radiation is given as 3×10^8 m/s. Find the wavelengths for the following radio frequencies:

a. 15 kHz (*ans.* 20 km)

b. 500 kHz (*ans.* 600 m)

c. 6 GHz (*ans.* 5 cm)

d. 1.2 MHz (*ans.* 250 m)

11-8 Find the frequency of electromagnetic radiations with the following wavelengths:

a. 2.5 m *c.* 80 m

b. 3 cm *d.* 2 km

11-9 The length of a dipole antenna for a given frequency is $\lambda/2$. Find the length of the antenna if the frequency is

a. 500 kHz (*ans.* 300 m)

b. 2170 kHz (*ans.* 69.12 m)

c. 88 MHz (*ans.* 1.7 m)

d. 450 MHz (*ans.* 33.3 cm)

11-10 Find the angular velocities for the following frequencies:

a. 60 Hz *c.* 1.5 MHz

b. 1000 Hz *d.* 2.8 GHz

11-11 For a 60-Hz sine wave (starting at zero and going positive) find the angle θ, in radians and in degrees, at the following times after t_0:

a. 1 ms (*ans.* 0.377 rad, 21.6°)

b. 5 ms (*ans.* 1.885 rad, 108°)

c. 8.33 ms (*ans.* π rad, 180°)

d. 13 ms (*ans.* 4.9 rad, 280.9°)

e. 0.55 s (*ans.* 0 rad, 0°)

Section 11-6

11-12 If the 60-Hz sine wave in Prob. 11-11 has a peak value of 170 V, find the instantaneous voltage at each of the times listed.

11-13 A sinusoidal current has a frequency of 1000 Hz and a maximum value of 5 A. Find the instantaneous current at the following times after t_0:

a. 1 ms (*ans.* 0 A)

b. 200 μs (*ans.* 4.76 A)

c. 2.25 ms (*ans.* 5 A)

d. 0.125 ms (*ans.* 3.536 A)

e. 35.625 ms (*ans.* −3.536 A)

f. 0.75 ms (*ans.* −5 A)

11-14 A 10-kHz sinusoidal voltage has a maximum value of 25 V. Find E_{av}.

11-15 A 10-kHz sinusoidal voltage has an average value of 25 V. Find $E_{p\text{-}p}$. (*ans.* 78.6 V)

11-16 A 10-kHz sinusoidal voltage has an instantaneous value of 25 V, 12.5 μs after t_0. Find E_{av}.

11-17 A 10-kHz sinusoidal voltage has an average value of 25 V. Find the instantaneous value of voltage at $t = 65$ μs. (*ans.* −31.8 V)

Miscellaneous sine-wave problems

11-18 Find the effective value of current when the peak value of an alternating current is 30 mA.

11-19 A signal generator is to be set to 5 V rms. If the generator is to be monitored by an oscilloscope, what peak-to-peak voltage should be read on the oscilloscope? (*ans.* 14.14 V)

11-20 If a 120-V 60-Hz line voltage is viewed on an oscilloscope, how many volts peak to peak will be indicated?

11-21 What is the average value of a 120-V 60-Hz line voltage? (*ans.* 108 V)

11-22 A 400-Hz generator has an output voltage of 200 V. Find the instantaneous voltage at the following times after t_0:

a. 1 ms *c.* 3 ms
b. 2 ms

11-23 $e = 212 \sin 377t$ describes a voltage.

a. What is the rms value? (*ans.* 150 V)
b. What is the frequency? (*ans.* 60 Hz)

11-24 The peak-to-peak voltage read across a 100-Ω resistor by an oscilloscope is 28.3 V. Find

a. The effective current in the resistor.
b. The power dissipated by the resistor.

11-25 What is the minimum voltage rating for a capacitor to be employed on a 240-V 60-Hz circuit? (*ans.* 340 V)

11-26 Write the equation for the instantaneous voltage when $E = 50$ V and $f = 1000$ Hz.

11-27 An ac ammeter in series with a 470-Ω resistor reads 35 mA. Find the peak voltage across the resistor. (*ans.* 23.26 V)

11-28 A current probe on an oscilloscope is used to read the current in a circuit consisting of a generator and a 220-Ω resistor. Find E when the peak-to-peak value of current is 18 mA.

11-29 An oscilloscope, using a current probe, indicates a peak-to-peak current of 30 mA and a wave period of 833 μs. If the circuit being measured contains a 3300-Ω resistor, find the instantaneous voltage across the resistor at 300 μs. (*ans.* 38.1 V)

11-30 An oscilloscope, using a current probe, indicates a peak-to-peak current of 20 mA in a circuit containing a 5600-Ω resistor. Find the power dissipated by the resistor.

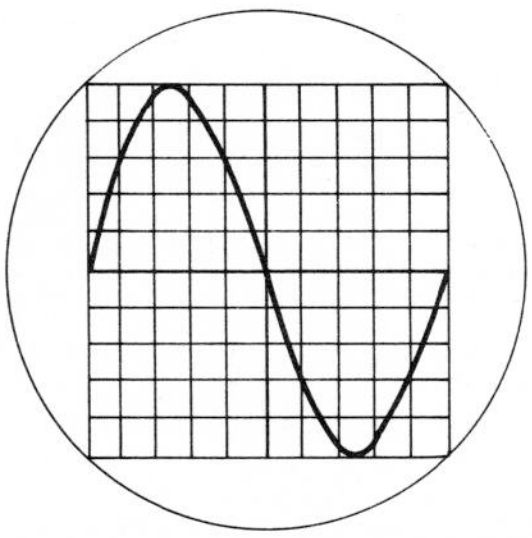

Fig. 11-1P Problem 11-31: oscilloscope display.

11-31 An oscilloscope is switched to a vertical range of 1 V per division and 5 μs per horizontal division when the display is as shown in Fig. 11-1P. What are the following waveform values?

a. Peak to peak. (*ans.* 10 V) *c.* Average. (*ans.* 3.183 V)
b. rms. (*ans.* 3.536 V) *d.* Frequency. (*ans.* 20 kHz)

12

Phases and Phasors

A time-varying quantity is one whose magnitude changes as time progresses. In any given circuit, the relationship between the magnitudes of a number of time-varying quantities may also change as time progresses, producing a need for special diagrams and calculation methods. The time-varying quantities in our discussion will be sine-wave voltages and currents.

12-1 PHASE

When two or more waveforms are present simultaneously in a circuit, the time relationship between them is called *phase*. When this relationship is expressed as an angle, it is called the *phase angle*.

Figure 12-1 shows two sine-wave voltages. They both have a peak magnitude of 100 V and a frequency of 250 Hz (the period is 4 ms). The equivalent magnitudes of each voltage have a time separation of 1 ms. The voltages have a 1-ms phase difference. As the 1-ms phase difference of v_1 and v_2 represents $\frac{1}{4}$ cycle, the phase angle between the two voltages is 90° (or $\pi/2$ rad).

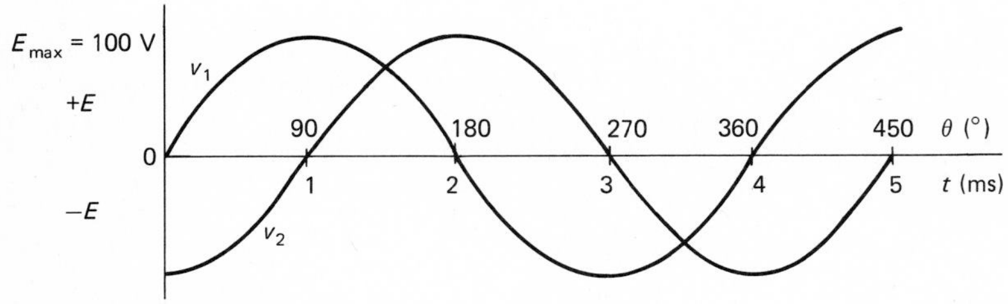

Fig. 12-1 Two out-of-phase sine-wave voltages.

The symbol for phase angle is ϕ or θ (lowercase Greek letter phi or theta). The more commonly used symbol is ϕ. The SI unit for phase angle is the radian (rad), although at present the degree (°) is more frequently used.

LAG AND LEAD

In a waveform diagram, such as Fig. 12-1, progression to the right implies successive time intervals or elapsed time. Time 1 (1 ms) occurs after time 0 (reference time); time 2 (2 ms) occurs after time 1.

Each point on the waveform of v_2 occurs after the equivalent point on the waveform of v_1. For instance, the positive peak of v_1 occurs at time 1, whereas the positive peak of v_2 occurs 1 ms later at time 2. It is said that voltage v_2 *lags* voltage v_1.

Each point on the waveform of v_1 occurs before the equivalent point on the waveform of v_2. Voltage v_1 *leads* voltage v_2.

A leading phase relationship is regarded as positive, and a lagging phase as negative. In Fig. 12-1 the phase angle of v_1 with respect to v_2 is $+\phi$ ($+90°$ or $+\pi/2$ rad). The phase angle of v_2 with respect to v_1 is $-\phi$ ($-90°$ or $-\pi/2$ rad).

It is important to realize that the phase between two ac quantities is only meaningful if they have the same frequency. If they have different frequencies, the time between equivalent points on the respective waveforms will continually change and a value cannot be specified.

IN-PHASE, ANTIPHASE, AND PHASE QUADRATURE

Three phase relationships encountered often in circuits are each given specific names. They are shown in Fig. 12-2.

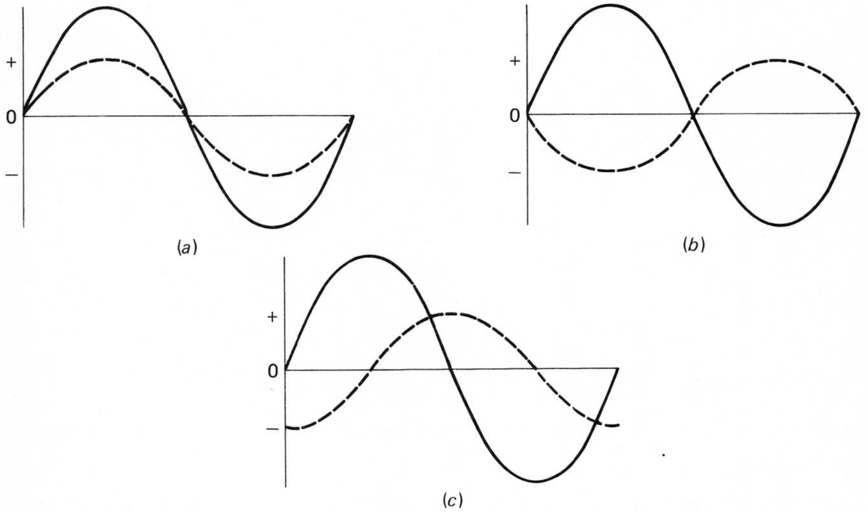

Fig. 12-2 Two sine waves. (*a*) In phase; (*b*) in antiphase; (*c*) in phase quadrature.

In phase (Fig. 12-2*a*) When two ac waveforms are exactly synchronized so that their zero values, swinging positive, occur at the same time, they are said to be *in phase*. The phase angle ϕ is 0° or 0 rad.

Antiphase (Fig. 12-2*b*) If two waveforms have their zero values occurring at the same time but one swings positive when the other swings negative, they are said to be in *antiphase*. The phase angle ϕ is 180° or π rad. Another common way of expressing antiphase is *180° out of phase*.

Phase quadrature (Fig. 12-2c) If two waveforms have their positive-going zero values displaced by one quarter-cycle, they are said to be in *phase quadrature*. The phase angle ϕ is 90° or $\pi/2$ rad.

The voltages in Fig. 12-1 are in phase quadrature, with v_1 leading v_2.

In waveform diagrams for in-phase and antiphase conditions (refer to Fig. 12-2*a* and *b*), there is no clear lag or lead. For all other phase angles between 0° and 180°, lag and lead show clearly. A phase angle of over 180° is confusing in waveform diagrams. For example, the diagram for 270° lead is the same as for 90° lag. It must be made clear that these statements refer only to waveform diagrams, and not to other ways of depicting phase relationships, such as phasor diagrams or trigonometric expressions.

CIRCUIT PHASE ANGLE

The angle between the total voltage and the total current, with voltage as the reference, is the *circuit phase angle*.

If the current lags the voltage, the circuit is said to have a lagging phase angle. Conversely, a leading phase angle occurs when the current leads the voltage.

12-2 PHASOR DIAGRAM

Although ac quantities can be drawn as waveforms, for circuit analysis this procedure is tedious. Furthermore, if several quantities are involved, it becomes difficult to distinguish between them. For these reasons it is convenient to use phasor diagrams to present ac relationships.

A phasor diagram consists of a number of straight lines, one for each quantity represented. Each line is drawn from a common origin. The length of each line is proportional to the magnitude of the quantity it represents. As the purpose of this scaling is to give a visual comparison of magnitudes, the values must be equivalent for each quantity. Rms values are usually adopted. The angles between lines represent the phases between the quantities.

The common current in a series circuit or the common voltage in a parallel circuit are typical reference quantities. The phasor lines for such reference quantities are usually placed in a horizontal position and point to the right.

It is a standard convention that counterclockwise rotation from one phasor line to another is regarded as positive motion and represents a leading phase. Conversely, clockwise rotation is negative motion representing a lagging phase.

Figure 12-3 shows some phasor conventions.

In Fig. 12-3*a* is shown a phasor for a current I. Its horizontal position with the arrowhead pointing to the right identifies it as being in the reference position.

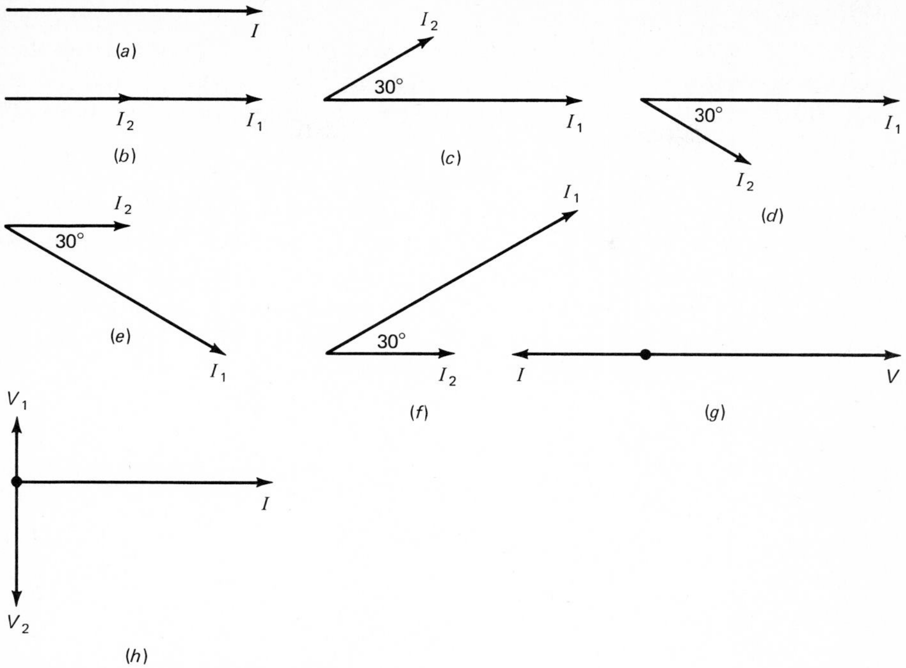

Fig. 12-3 Typical phasor diagrams.

In Fig. 12-3*b* two current phasors I_1 and I_2 are both in the reference position. They are in phase. The rms value of I_1 is twice that of I_2.

In Fig. 12-3*c* the current I_1 again has twice the magnitude of I_2, but now I_1 is a reference phasor with I_2 leading it by 30° or $\pi/6$ rad.

Figure 12-3*d* shows current I_1 with twice the magnitude of I_2. I_2 lags I_1 by 30°.

Figure 12-3*e* and *f* shows the same circuit conditions as Fig. 12-3*c* and *d*, respectively. I_2 is shown as the reference phasor in Fig. 12-3*e* and *f*. It may be said that Fig. 12-3*e* is Fig. 12-3*c* rotated 30° clockwise, and that Fig. 12-3*f* is Fig. 12-3*d* rotated 30° counterclockwise.

In Fig. 12-3*g* a voltage and a current are shown in antiphase.

Two voltages, each in phase quadrature with a current, are shown in Fig. 12-3*h*.

Ratios or products of voltage and current magnitudes (V/I, I/V, or VI) cannot be determined from phasor diagrams such as Fig. 12-3*g* and *h*, unless the magnitude scalings for voltage and current are specified.

12-3 POLAR AND RECTANGULAR PRESENTATIONS

A phasor diagram is drawn if it is desired to present a visual impression of circuit conditions. Whether or not the diagram is drawn, the phasor concept is used in circuit work by expressing the length and angular position of phasors in either polar or rectangular form.

PHASORS IN POLAR FORM

A phasor is represented in a diagram as a line of length proportional to the quantity magnitude, positioned so that its angle with respect to the reference position is the relative phase angle of the quantity. This type of presentation is known as *polar*.

In Fig. 12-4*a* is shown the basic polar coordinate system. The horizontal, or X, axis extends to the right (positive) and to the left (negative) of the origin 0. The vertical, or Y, axis extends up (positive) and down (negative) from the origin. The quadrants so formed are given the numbers 1 through 4 in a counterclockwise rotation from the reference position.

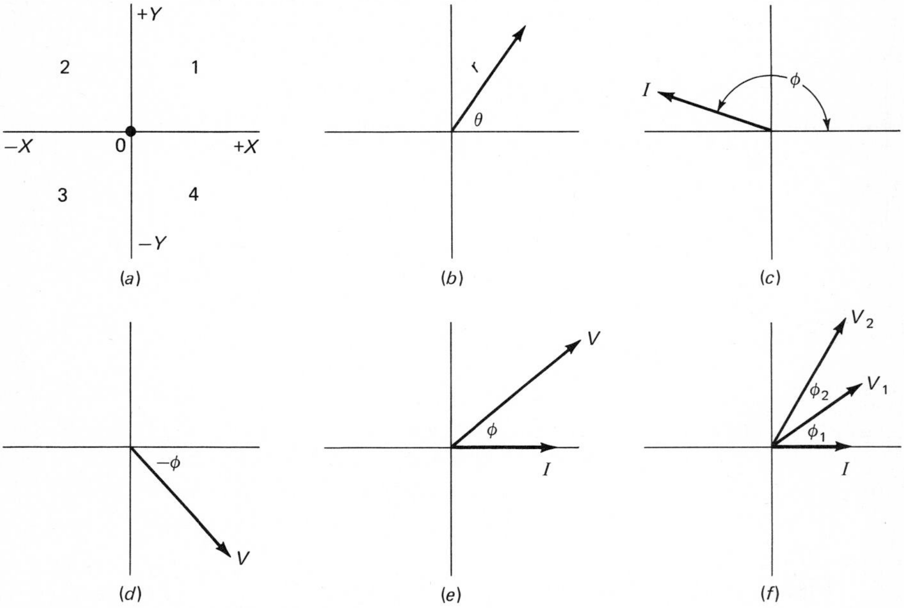

Fig. 12-4 Polar presentation of phasors.

Figure 12-4*b* shows a typical phasor. Its magnitude is represented by the length r, and its relative phase by the angle θ. The phasor is called $r\underline{/\theta}$ (magnitude r, angle θ).

Figure 12-4*c* shows a current phasor $I\underline{/\phi}$.

The voltage phasor in Fig. 12-4*d* is $V\underline{/-\phi}$.

The two phasors in Fig. 12-4*e* are $I\underline{/0}$ and $V\underline{/\phi}$.

The three phasors in Fig. 12-4*f* are $I\underline{/0}$, $V_1\underline{/\phi_1}$, and $V_2\underline{/\phi_1 + \phi_2}$.

It should be noted that all phasor magnitudes are measured from the common origin, and all phase angles from the reference position.

MAGNITUDE OR SCALAR VALUE

A phasor quantity is completely expressed in its polar form $r\underline{/\theta}$. In many applications of ac, only the current and voltage magnitudes are of interest.

For this reason, special notation is used to denote the magnitude of a phasor quantity while disregarding the phase angle. For the phasor quantity $r\angle\theta$, the magnitude alone can be expressed as $|r|$. Alternative terms for magnitude are "scalar value" and "modulus."

The practice of writing a quantity symbol between vertical lines to indicate a scalar value is not always rigidly adhered to. When, in a given situation, it is clear that magnitudes are referred to, the vertical lines can be omitted. This technique is adopted in the following chapters.

COMPLEX QUANTITIES—THE *j* OPERATOR

Phasors expressed in rectangular form use complex numbers to identify horizontal and vertical components. Complex quantities are not complicated. They are called "complex" because they have so-called real and imaginary parts. Before explaining how the rectangular form of phasor presentation is derived, an introduction to the *j* operator will be given. The *j*-operator concept is the basis of complex quantities and is the only difficult idea involved.

Signs like $+$, $-$, $\times$, $\div$, and $\sqrt{\ }$ are *operators,* because they convey instructions to carry out operations. The operator *j* gives the instruction, "rotate 90° or $\pi/2$ rad." Often it means, "change the direction from horizontal to vertical."

Unlike most operators, *j* has a numerical value, $\sqrt{-1}$, and can be operated on by other operators or even by *j* itself. To use *j* in ac circuit work it is necessary to get an idea of how to arrive at its numerical value.

In Fig. 12-5*a* the $+n$-unit phasor lies in the reference position. To move this phasor into the negative position to the left of the origin, it must be rotated through an angle of π rad (180°). It then becomes a $-n$-unit phasor. As $-n = (-1) \times (+n)$, it can be said that the original $+n$ phasor has been rotated through 180° or multiplied by -1.

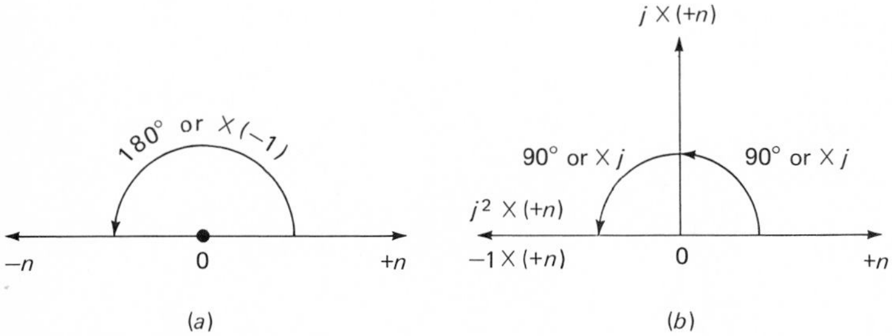

Fig. 12-5 The *j* operator.

In Fig. 12-5*b* the $+n$ phasor has been moved into the $-n$ position in two stages. First it was moved into the vertical position by a 90° rotation, and then rotated a further 90°. Each 90° rotation is defined as multiplying by *j*. To go from $+n$ to $-n$ two multiplications by *j* are needed, thus implying multiplication by j^2.

Combining the results of Fig. 12-5*a* and *b*, we see that rotation through 180° is the same as multiplying by −1 or multiplying by j^2. Therefore,

$$j^2 = -1 \qquad \text{or} \qquad j = \sqrt{-1} \tag{12-1}$$

Because it is not a rational quantity, j is called *imaginary*. However the use of this term is unfortunate because it gives the impression that j is not valid for use in calculating, which it certainly is since j^2 is a real number, −1.

Listing the effect of successive rotations through 90° ($\pi/2$ rad) leads to the rectangular coordinate system.

Reference position	$+n$
90° ($\pi/2$-rad) rotation	jn or $\sqrt{-1}n$
180° (π rad) rotation	j^2n or $-1n$ or $-n$
270° ($3\pi/2$-rad) rotation	j^3n or $-\sqrt{-1}n$ or $-jn$
360° (2π-rad) rotation	j^4n or $(-1)^2n$ or $+n$

PHASORS IN RECTANGULAR FORM

Figure 12-6*a* shows rectangular coordinates with j notation. In Fig. 12-6*b* to *e* phasors are shown in each of the four quadrants. They are

Figure 12-6*b*:	$a + jb$
Figure 12-6*c*:	$-a + jb$
Figure 12-6*d*:	$-a - jb$
Figure 12-6*e*:	$a - jb$

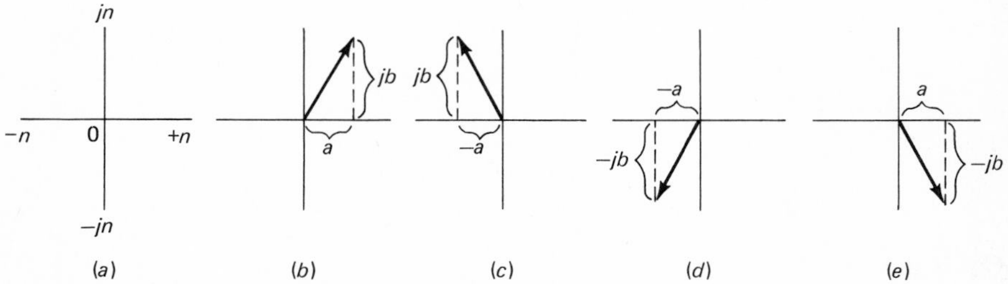

Fig. 12-6 Rectangular presentation of phasors.

CONVERSION OF PHASORS BETWEEN POLAR AND RECTANGULAR FORMS

A phasor expressed in polar form can be converted into its equivalent rectangular form by applying trigonometry.

The length r of the phasor is the hypotenuse of a right triangle of which the horizontal component a is the adjacent side and the vertical component jb is the opposite side. These symbols are shown in Fig. 12-7b.

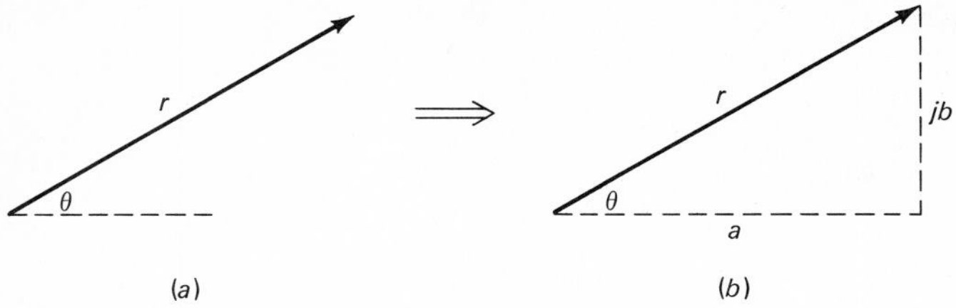

Fig. 12-7 Polar-to-rectangular conversion.

The horizontal component is sometimes called the real part of the phasor. Its magnitude is given by

$$|a| = r \cos \theta$$

The vertical component is sometimes called the imaginary part of the phasor. Its magnitude is given by

$$|b| = r \sin \theta$$

The rectangular expression for the phasor is therefore

$$a + jb = r \cos \theta + jr \sin \theta$$

Combining this with the polar expression, we obtain the equation for conversion from polar to rectangular form

$$r\underline{/\theta} = r \cos \theta + jr \sin \theta \qquad (12\text{-}2)$$

A phasor expressed in rectangular form can be converted into polar form by applying Pythagoras' theorem and trigonometry.

The magnitudes of the phasor and its rectangular components are r, a, and b, as shown in Fig. 12-8a.

$$r^2 = a^2 + b^2 \qquad \text{or} \qquad r = \sqrt{a^2 + b^2}$$

Fig. 12-8 Rectangular-to-polar conversion.

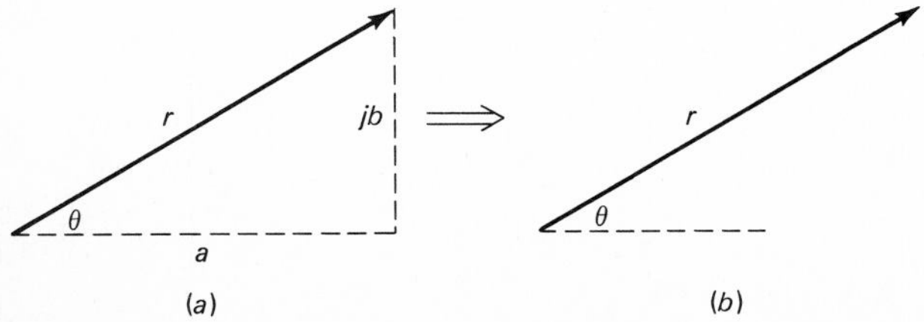

and $$\tan\theta = \frac{b}{a} \qquad \text{or} \qquad \theta = \arctan\frac{b}{a}$$

Therefore the equation for converting from rectangular to polar form is

$$a + jb = \sqrt{a^2 + b^2}\ \underline{/\arctan\frac{b}{a}} \tag{12-3}$$

Example 12-1 In a three-element series network the current I is 1 A. The voltages across the elements are $V_1 = 10$ V in phase with I; $V_2 = 20$ V in phase quadrature, leading I; and $V_3 = 25$ V in phase quadrature, lagging I. Draw to scale (*a*) a waveform diagram and (*b*) a phasor diagram.

Solution

a. See Fig. 12-9. The given current and voltage symbols I, V_1, V_2, and V_3 indicate rms values. The related maximum values are shown on the vertical axis in Fig. 12-9. The curves are labeled with instantaneous-value symbols. The current curve is shown as a broken line, and the voltage curves as solid lines.

b. See Fig. 12-10.

Fig. 12-9 Example 12-1: waveforms.

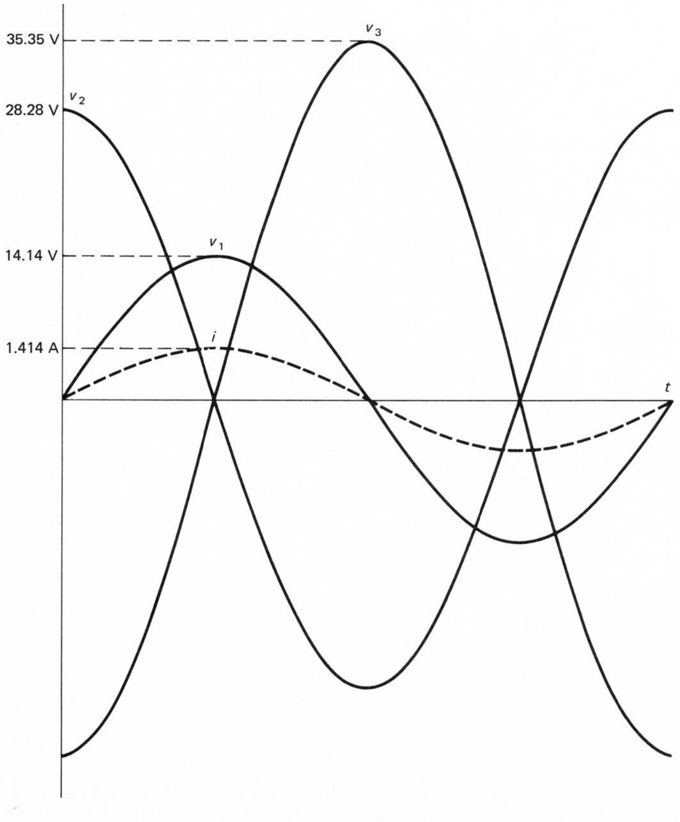

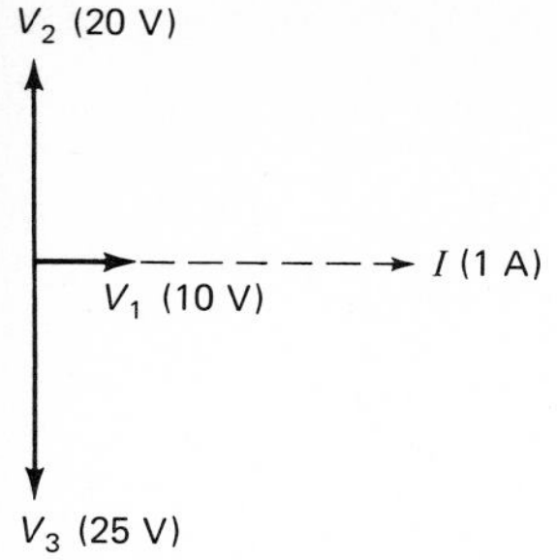

Fig. 12-10 Example 12-1: phasor diagram.

In waveform or phasor diagrams the relative scaling for voltage and current has no significance, as two different units are involved. Any convenient scaling can be chosen. In Fig. 12-9, the current magnitude is shown smaller than that of all the voltages. In Fig. 12-10, the current magnitude is shown greater. In both Figs. 12-9 and 12-10 the voltage magnitudes have been shown to scale, that is, in their correct proportions.

Example 12-2 Express the current and voltages in Example 12-1 in (*a*) polar form and (*b*) rectangular form.

Solution Refer to Fig. 12-10,

a.

$$I = \mathbf{1\ A\ \underline{/0^\circ}}$$

$$V_1 = \mathbf{10\ V\ \underline{/0^\circ}}$$

$$V_2 = \mathbf{20\ V\ \underline{/90^\circ}}$$

$$V_3 = \mathbf{25\ V\ \underline{/-90^\circ}}$$

b.

$$I = \mathbf{(1 + j0)\ A}$$

$$V_1 = \mathbf{(10 + j0)\ V}$$

$$V_2 = \mathbf{(0 + j20)\ V}$$

$$V_3 = \mathbf{(0 - j25)\ V}$$

Example 12-3 The voltage across a parallel network is 50 mV. The current in one branch is 100 mA and leads the voltage by 60°. The current in a second branch is 150 mA and lags the voltage by 45°. Draw a phasor diagram for this network.

Solution See Fig. 12-11.

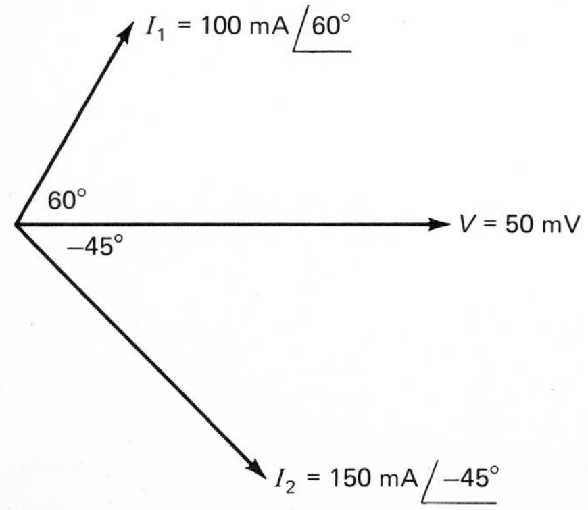

Fig. 12-11 Example 12-3: phasor diagram.

Example 12-4 Express the voltage and the currents in Example 12-3 in (*a*) polar form and (*b*) rectangular form.
Solution Refer to Fig. 12-11,

a.

$$V = \mathbf{50\ mV\ \angle 0^\circ}$$

$$I_1 = \mathbf{100\ mA\ \angle 60^\circ}$$

$$I_2 = \mathbf{150\ mA\ \angle -45^\circ}$$

b. Apply Eq. (12-2) ($r\angle\theta = r\cos\theta + j\sin\theta$):

$$V = 50\cos 0^\circ + j\sin 0^\circ = \mathbf{(50 + j0)\ mV}$$

$$I_1 = 100\cos 60^\circ + j100\sin 60^\circ = 100 \times 0.5 + j100 \times 0.866$$

$$I_1 = \mathbf{(50 + j86.6)\ mA}$$

$$I_2 = 150\cos(-45^\circ) + j150\sin(-45^\circ) = 150 \times 0.7071 + j150 \times (-0.7071)$$

$$I_2 = \mathbf{(106 - j106)\ mA}$$

Example 12-5 A network is connected to a 120-V 60-Hz commercial supply. Relative to this emf, there is a network pd of $(75 - j30)$ V and a network current of $(-2.5 + j3)$ A. Express the emf and the network pd and current in polar form.
Solution The supply voltage is, in a case like this, regarded as the reference. A phasor diagram helps to visualize the conditions.

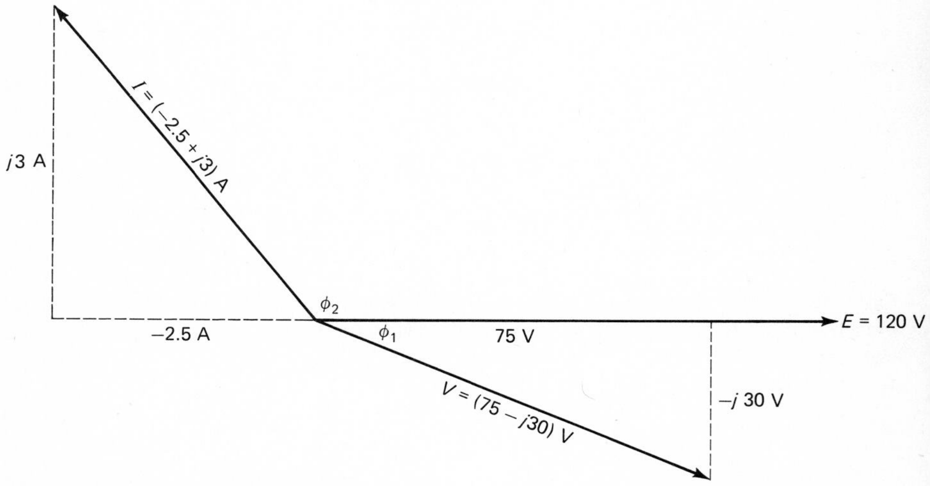

Fig. 12-12 Example 12-5: phasor diagram.

$$E = \mathbf{120 + j0\ V}$$

Apply Eq. (12-3)

$$V = \sqrt{75^2 + 30^2}\ \angle\arctan(-30)/75 = \mathbf{80.78\ V\ \angle -21.8^\circ} \quad \text{or} \quad \mathbf{\angle -0.3805\ rad}^*$$

$$I = \sqrt{2.5^2 + 3^2}\ \angle\arctan 3/(-2.5) = \mathbf{3.905\ A\ \angle 129.8^\circ} \quad \text{or} \quad \mathbf{\angle 2.265\ rad}$$

* Angles are given in both degrees and radians. For example, in this case, $\angle -21.8^\circ$ and $\angle -0.3805$ rad are the same angle.

It was previously mentioned that a knowledge of trigonometry is necessary for ac circuit work. This knowledge should include familiarity with the tangent forms for each quadrant.

A phasor diagram will give the location of a phasor's angle, as is evident from Fig. 12-12. Angle ϕ_1 [arctan (−30)/75] is in the fourth quadrant, and its value is thus between 270 and 360° (between $3\pi/2$ and 2π rad), or between −90 and 0°. Angle ϕ_2 [arctan 3/(−2.5)] is in the second quadrant, its value being between 90 and 180° (between $\pi/2$ and π rad).

It is likely that you will interconvert between tangents and angles by using tables or a slide rule, and that the angles will be in degrees. To convert degrees to radians, the SI angle units, divide by 57.3. For instance, in Example 12-5, the phase angle of the current is 129.8° which, divided by 57.3, is 2.265 rad.

If you are not accustomed to using the radian, it is recommended that you make an effort to develop some familiarity with this unit. At present the degree is used more often than the radian, and there is little evidence that the radian will become dominant in the near future. However, the radian is the SI unit by international agreement, and its use will probably increase.

12-4 PHASOR ARITHMETIC

Addition, subtraction, multiplication, division, and other arithmetic operations on phasor quantities are all encountered in ac circuit calculations. By the use of rectangular and polar phasor forms, any of these operations can be carried out with comparative ease. The following techniques should be memorized.

Addition and subtraction Use the rectangular phasor form. Using ordinary arithmetic methods, add or subtract the real parts of each phasor, and then add or subtract the imaginary (j) parts. Keep the sums or differences of each part separate.

The general operations are

Addition: $$(a + jb) + (c + jd) = (a + c) + j(b + d) \qquad (12\text{-}4)$$

Subtraction: $$(a + jb) - (c + jd) = (a - c) + j(b - d) \qquad (12\text{-}5)$$

a, b, c, and d are the magnitudes of the component parts of the phasors and may be either positive or negative.

Multiplication and division Use the polar phasor form. Multiply or divide the magnitudes, and add or subtract the phase angles.

The general operations are

Multiplication: $$r_1 \underline{/\theta_1} \times r_2 \underline{/\theta_2} = r_1 r_2 \underline{/\theta_1 + \theta_2} \qquad (12\text{-}6)$$

Division: $$r_1 \underline{/\theta_1} \div r_2 \underline{/\theta_2} = r_1/r_2 \underline{/\theta_1 - \theta_2} \qquad (12\text{-}7)$$

Powers and roots Raising to a power or extracting a root is rarely used in phasor work. However, when it is needed, the use of exponents can be regarded as an extension of multiplication.

$$(r \angle\theta)^n = r^n \angle n\theta \tag{12-8}$$

Squaring and taking a square root are good examples of the use of exponents.

$$(r \angle\theta)^2 = r^2 \angle 2\theta$$

$$\sqrt{r \angle\theta} = (r \angle\theta)^{1/2} = \sqrt{r} \angle\theta/2$$

Reciprocal of a phasor The reciprocal of phasor $r \angle\theta$ is $1/r$ ($\angle -\theta$). This can be determined as follows:

$$\frac{1}{r \angle\theta} = (r \angle\theta)^{-1} = r^{-1} \angle -1(\theta)$$

$$\frac{1}{r \angle\theta} = \frac{1}{r} \angle -\theta \tag{12-9}$$

Similarly,

$$\frac{1}{r \angle -\theta} = \frac{1}{r} \angle +\theta \tag{12-10}$$

Reciprocal of *j* The reciprocal of j is $-1/j$ (or the reciprocal of $-j$ is $1/j$). This can be shown to be true by applying the principle of the reciprocal of a phasor. It is more direct, though, to use the process of rationalization. This method involves the notion that multiplying a number by 1 will not change its value. Any number divided by itself equals 1; this includes j/j.

$$j = j \times \frac{j}{j} = \frac{j^2}{j} = \frac{-1}{j}$$

$$j = -\frac{1}{j} \tag{12-11}$$

or

$$-j = -j \times \frac{j}{j} = \frac{-j^2}{j} = \frac{-(-1)}{j}$$

$$-j = \frac{1}{j} \tag{12-12}$$

Example 12-6 There are two voltages in a series circuit, $V_1 = (10 + j15)$ V and $V_2 = (7 - j5)$ V. What is the total circuit voltage V_T?

Solution In phasor form, Kirchhoff's voltage law becomes,

$$V_T = V_1 + V_2 = (10 + j15) + (7 - j5) = (10 + 7) + j(15 - 5)$$

$$V_T = \mathbf{(17 + j10)\ V}$$

Example 12-7 The total current I_T in a two-branch series-parallel circuit is $(250 - j150)$ mA. One of the branch currents I_1 is $(-100 + j100)$ mA. What is the other branch current I_2?

Solution Apply Kirchhoff's current law,

$$I_T = I_1 + I_2 \qquad \text{or} \qquad I_2 = I_T - I_1$$

Use phasor subtraction,

$$\begin{aligned} I_2 &= (250 - j150) - (-100 + j100) \\ &= [250 - (-100)] + j[(-150) - 100] \\ &= (250 + 100) + j(-150 - 100) \\ I &= \mathbf{(350 - j250)\ mA} \end{aligned}$$

Example 12-8 The ratio of the voltage to the current in a two-branch parallel network is given by

$$\frac{V}{I} = \frac{(10 + j20)(5 + j3)}{(10 + j20) + (5 + j3)}$$

Express this ratio as a polar phasor.

Solution The numerator N involves multiplication, so each phasor should be converted to polar form.

$$10 + j20 = \sqrt{10^2 + 20^2}\ \underline{/\arctan \tfrac{20}{10}} = 22.36\ \underline{/63.43^\circ} \qquad \text{or} \qquad \underline{/1.107\text{ rad}}$$

$$5 + j3 = \sqrt{5^2 + 3^2}\ \underline{/\arctan \tfrac{3}{5}} = 5.831\ \underline{/30.96^\circ} \qquad \text{or} \qquad \underline{/0.5403\text{ rad}}$$

$$N = (10 + j20)(5 + j3) = 22.36\ \underline{/63.43^\circ} \times 5.831\ \underline{/30.96^\circ}$$

Using phasor multiplication,

$$N = 22.36 \times 5.831\ \underline{/63.43^\circ + 30.96^\circ}$$

$$N = 130.4\ \underline{/94.39^\circ} \qquad \text{or} \qquad \underline{/1.647\text{ rad}}$$

The denominator D involves addition, so the rectangular form is used as given:

$$D = (10 + j20) + (5 + j3) = 15 + j23$$

For division D should be converted to polar form:

$$D = \sqrt{15^2 + 23^2}\ \underline{/\arctan \tfrac{23}{15}} = 27.45\ \underline{/56.89^\circ} \qquad \text{or} \qquad \underline{/0.9928\text{ rad}}$$

$$\frac{V}{I} = \frac{N}{D} = \frac{130.4\ \underline{/94.39^\circ}}{27.45\ \underline{/56.89^\circ}} \qquad \text{or} \qquad \frac{\underline{/1.647\text{ rad}}}{\underline{/0.9928\text{ rad}}}$$

Using phasor division,

$$\frac{V}{I} = \frac{130.4}{27.45}\ \underline{/94.39^\circ - 56.89^\circ} \qquad \text{or} \qquad \underline{/1.647\text{ rad} - 0.9928\text{ rad}}$$

$$\frac{V}{I} = \mathbf{4.75\ \underline{/37.5^\circ}} \qquad \text{or} \qquad \mathbf{\underline{/0.6542\text{ rad}}}$$

12-5 IMPEDANCE AND ADMITTANCE

"Impedance" and "admittance" are general terms for the characteristics of an element or a network which determine current-voltage ratios. Although they can vary, impedance and admittance are normally regarded as parameters, as it is usual to be primarily interested in the current-voltage behavior of networks with impedances and admittances of constant value.

The words "impedance" and "admittance" are self-explanatory, referring to the ability to impede or restrict current and the ability to admit or permit current, respectively. Both impedance and admittance apply to all types of voltage and current, and so phasor values are relevant.

IMPEDANCE

The *impedance* of an element or a network is the voltage-current ratio. The symbol for impedance is Z, and the unit for impedance is the ohm, symbol Ω.

$$Z = \frac{V}{I} \qquad \text{or} \qquad V = ZI \tag{12-13}$$

ADMITTANCE

The *admittance* of an element or a network is the current-voltage ratio. The symbol for admittance is Y, and the unit for admittance is the siemens, symbol S.

$$Y = \frac{I}{V} \qquad \text{or} \qquad I = YV \tag{12-14}$$

Because they are general, the terms "impedance" Z and "admittance" Y can be used for any type of individual element or any combination of elements. However, it is usual to employ specific terms and symbols whenever it is convenient.

SPECIFIC IMPEDANCES AND ADMITTANCES

Specific terms are used to refer to the voltage-current behavior of resistance, which dissipates energy, and of capacitance and inductance, which store energy.

Resistance *Resistance* is the impedance of an element or a network in which there can be no energy storage. This view of resistance is the same as that adopted in the dc circuit sections of this text. The use of the word "impedance" does not conflict with, but adds a new dimension to, the concept of resistance. The symbol for resistance is R.

Reactance *Reactance* is the impedance of an element or a network in which all the energy can be stored in electric or magnetic fields. The general reactance symbol is X, the inductance reactance symbol is X_L, and the capacitive reactance symbol is X_C.

Conductance *Conductance* is the admittance of an element or a network in which there can be no storage of energy. The symbol for conductance is G.

Susceptance *Susceptance* is the admittance of an element or a network in which all the energy can be stored in electric or magnetic fields. The general susceptance symbol is B, the inductive susceptance symbol is B_L, and the capacitive susceptance symbol is B_C.

RECIPROCAL PARAMETERS

All impedances are voltage-current ratios, while all admittances are current-voltage ratios. Thus they are reciprocal parameters.

$$Z=\frac{1}{Y} \quad \text{or} \quad R=\frac{1}{G} \quad \text{or} \quad X_C=\frac{1}{B_C} \quad \text{or} \quad X_L=\frac{1}{B_L} \tag{12-15}$$

Example 12-9 A three-element series network carries a current of 2 A. The elements are (*a*) a resistance of 100 Ω, (*b*) a capacitive reactance of 50 Ω, and (*c*) an inductive reactance of 75 Ω. What is the magnitude of the voltage across each element?

Solution

a.

$$R = 100\ \Omega$$

$$V = IR = 2\ \text{A} \times 100\ \Omega = \mathbf{200\ V}$$

b.

$$X_C = 50\ \Omega$$

$$V = IX_C = 2\ \text{A} \times 50\ \Omega = \mathbf{100\ V}$$

c.

$$X_L = 75\ \Omega$$

$$V = IX_L = 2\ \text{A} \times 75\ \Omega = \mathbf{150\ V}$$

Example 12-10 A parallel network consisting of the same elements as those in Example 12-9 is supplied with 120-V 60-Hz ac. What is the magnitude of the current in each element?

Solution

a.

$$G = \frac{1}{R} = \frac{1}{100\ \Omega} = 0.01\ \text{S}$$

$$I = VG = 120\ \text{V} \times 0.01\ \text{S} = \mathbf{1.2\ A}$$

b.

$$B_C = \frac{1}{X_C} = \frac{1}{50\ \Omega} = 0.02\ \text{S}$$

$$I = VB_C = 120\ \text{V} \times 0.02\ \text{S} = \mathbf{2.4\ A}$$

c.

$$B_L = \frac{1}{X_L} = \frac{1}{75\ \Omega} = 0.0133\ \text{S}$$

$$I = VB_L = 120\ \text{V} \times 0.0133\ \text{S} = \mathbf{1.6\ A}$$

Example 12-10 could have been dealt with by using the $I = E/Z$ form of Ohm's law, thus avoiding separate computations for conductance and susceptance. In simple cases, using this technique is probably the sensible thing to do. Adopting the $I = VY$ form of Ohm's law is the start of a deliberate effort to get you thinking in terms of current being proportional to admittance. Such thinking develops an appreciation of parallel circuit characteristics and forms the basis of some techniques commonly used in advanced circuit work.

12-6 SUMMARY OF EQUATIONS

$$j^2 = -1 \quad \text{or} \quad j = \sqrt{-1} \tag{12-1}$$

$$r\,\underline{/\theta} = r\cos\theta + jr\sin\theta \tag{12-2}$$

$$a + jb = \sqrt{a^2 + b^2}\ \underline{/\arctan b/a} \tag{12-3}$$

$$(a + jb) + (c + jd) = (a + c) + j(b + d) \tag{12-4}$$

$$(a + jb) - (c + jd) = (a - c) + j(b - d) \tag{12-5}$$

$$r_1\,\underline{/\theta_1} \times r_2\,\underline{/\theta_2} = r_1 r_2\,\underline{/\theta_1 + \theta_2} \tag{12-6}$$

$$r_1\,\underline{/\theta_1} \div r_2\,\underline{/\theta_2} = r_1/r_2\,\underline{/\theta_1 - \theta_2} \tag{12-7}$$

$$(r\,\underline{/\theta})^n = r^n\,\underline{/n\theta} \tag{12-8}$$

$$1/(r\,\underline{/\theta}) = (1/r)\,\underline{/-\theta} \tag{12-9}$$

$$1/(r\,\underline{/-\theta}) = (1/r)\,\underline{/+\theta} \tag{12-10}$$

$$j = -1/j \tag{12-11}$$

$$-j = 1/j \tag{12-12}$$

$$Z = V/I \quad \text{or} \quad V = ZI \tag{12-13}$$

$$Y = I/V \quad \text{or} \quad I = YV \tag{12-14}$$

$$Z = 1/Y \quad \text{or} \quad R = 1/G \quad \text{or} \quad X_C = 1/B_C \quad \text{or} \quad X_L = 1/B_L \tag{12-15}$$

EXERCISE PROBLEMS

Section 12-4

12-1 Evaluate the following:

a. $(+j10)(+j20)$ (*ans.* -200)
b. $(+j15)(-j25)$ (*ans.* 375)
c. $(-j50)(-j30)$ (*ans.* -1.5×10^3)
d. $(-j15)(+j20)(+j30)$ (*ans.* $+j9 \times 10^3$)
e. $(-j10)^2 \times (+j15)^2$ (*ans.* -2.25×10^4)

12-2 Evaluate the following:

a. $(10 + j10) + (15 + j15)$
b. $(5 + j25) + (20 - j10)$
c. $(15 - j30) + (-10 - j20)$
d. $(10 + j50) + (-j50)$
e. $(30 + j30) - (20 - j10)$
f. $(15 + j20) + (15 - j20)$

12-3 Evaluate the following:

a. $(14\angle 35^\circ)(8\angle 15^\circ)$ (*ans.* $112\angle 50^\circ$)

b. $(25\angle 75^\circ)(6\angle -40^\circ)$ (*ans.* $150\angle 35^\circ$)

c. $(21\angle -24^\circ)(11\angle -12^\circ)$ (*ans.* $231\angle -36^\circ$)

d. $(45\angle 40^\circ)/(9\angle 12^\circ)$ (*ans.* $5\angle 28^\circ$)

e. $(28\angle 35^\circ)/(20\angle -28^\circ)$ (*ans.* $1.4\angle 63^\circ$)

f. $(60\angle 30^\circ)/(150\angle \pi/2 \text{ rad})$ (*ans.* $0.4\angle -60^\circ$)

12-4 Evaluate the following:

a. $14\angle 35^\circ + 8\angle 15^\circ$

b. $25\angle 75^\circ + 6\angle -40^\circ$

c. $22\angle -24^\circ + 11\angle -12^\circ$

d. $45\angle 40^\circ - 9\angle 12^\circ$

e. $28\angle 35^\circ - 20\angle -28^\circ$

f. $60\angle 30^\circ - 150\angle \pi/2 \text{ rad}$

12-5 Evaluate the following:

a. $(10+j10)(15+j15)$ (*ans.* $300\angle 90^\circ$)

b. $(5+j25)(20-j10)$ (*ans.* $570\angle 52.1^\circ$)

c. $(15-j30)(-10-j20)$ (*ans.* -750)

d. $(10+j50)(-j50)$ (*ans.* $2550\angle -11.3^\circ$)

e. $(30+j30)(20-j10)$ (*ans.* $948.7\angle 18.4^\circ$)

f. $(15+j20)(15-j20)$ (*ans.* 625)

12-6 Evaluate the following:

a. $(10+j10)/(15+j15)$

b. $(5+j25)/(20-j10)$

c. $(15-j30)/(-10-j20)$

d. $(10+j50)/(-j50)$

e. $(30+j30)/(20-j10)$

f. $(15+j20)/(15-j20)$

12-7 Convert the following quantities to rectangular form and draw each result as a phasor.

a. $120 \text{ V}\angle -6^\circ$ [*ans.* $(119.3-j12.54)$ V]

b. $30 \text{ mA}\angle 65^\circ$ [*ans.* $(12.68+j27.19)$ mA]

c. $240 \text{ V}\angle 48^\circ$ [*ans.* $(160.6+j178.4)$ V]

d. $25 \text{ V}\angle 85^\circ$ [*ans.* $(2.18+j24.9)$ V]

e. $100 \text{ A}\angle -32^\circ$ [*ans.* $(84.8-j53)$ A]

f. $50 \text{ V}\angle 42^\circ$ [*ans.* $(37.16+j33.46)$ V]

12-8 Convert the following quantities to polar form:

a. $(707+j707)$ V

b. $(190+j99)$ V

c. $(40-j13)$ A

d. $(20-j85)$ A

e. $(40-j360)$ V

f. $(300+j66)$ mA

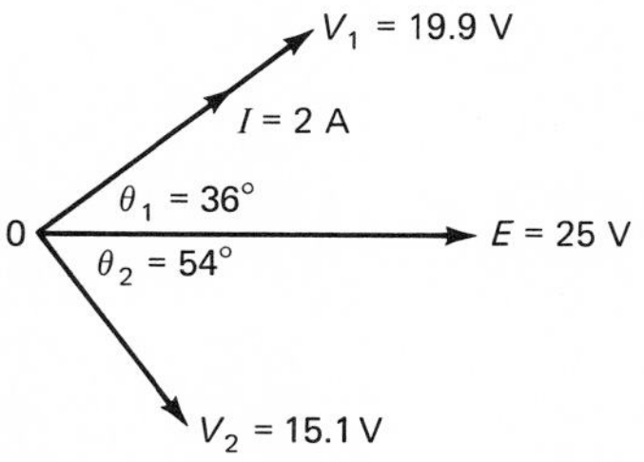

Fig. 12-1P Problem 12-9.

12-9 Express the current and voltages shown in Fig. 12-1P in

a. Polar form. (*ans.* $2 \text{ A}\angle 36^\circ$, $25 \text{ V}\angle 0^\circ$, $19.9 \text{ V}\angle 36^\circ$, $15.1 \text{ V}\angle -54^\circ$)

b. Rectangular form.

[*ans.* $(1.618+j1.176)$ A, $(25+j0)$ V, $(16.1+j11.7)$ V, $(8.876-j12.22)$ V]

12-10 Two devices A and B are connected in series across a 120-V 60-Hz energy source. If the phase angle of the current with respect to the emf is -51° and the voltage across device A is $95 \text{ V}\angle 50^\circ$, what is the voltage across device B?

12-11 Two inductors A and B are connected in series across a source of emf. $V_{L_A} = 75 \text{ V}\angle 80^\circ$ and $V_{L_B} = 65 \text{ V}\angle 61^\circ$. Find the magnitude of the emf and its phase angle with respect to the current. (*ans.* $138 \text{ V}\angle 71.2^\circ$)

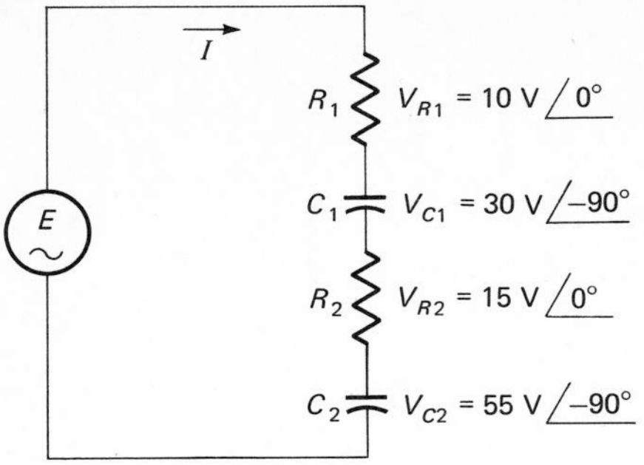

Fig. 12-2P Problem 12-12.

12-12 For Fig. 12-2P, express E with respect to I in
a. Polar form.
b. Rectangular form.

12-13 Two inductors A and B are connected in series across a 120-V 60-Hz generator. The angle of total voltage with respect to current is 80°. $V_{L_A} = 75\text{ V}\angle 80^\circ$. Find V_{L_B} in polar form. (*ans.* 45 V $\angle 80^\circ$)

12-14 Two motors are connected in parallel across a 120-V 60-Hz source. Find the total current with its phase angle referenced to voltage when the current drawn by the motors are 2.3 A $\angle -28^\circ$ and 3.3 A $\angle -38^\circ$.

12-15 A motor, drawing 1.7 A $\angle -42^\circ$, and two 100-W lights whose currents are in phase with the voltage, are connected across a 120-V 60-Hz source. Find the total current and its phase with respect to voltage. (*ans.* 3.143 A $\angle -21.2^\circ$)

12-16 A motor, drawing 1.7 A $\angle -42^\circ$, is connected in parallel with a capacitor across a 120-V 60-Hz source. Find the total current and the phase angle of this current with respect to voltage when the current through the capacitor is
a. 0.5 A $\angle 90^\circ$
b. 1.12 A $\angle 90^\circ$
c. 1.5 A $\angle 90^\circ$

Section 12-5

12-17 For Prob. 12-16, find the total admittance of the circuit when each of the capacitors is connected separately.
(*ans. a.* 0.01179 S $\angle -26.8^\circ$; *b.* 0.01053 S $\angle -0.8^\circ$; *c.* 0.01095 S $\angle 16^\circ$)

12-18 For Prob. 12-16, find the total impedance of the circuit when each of the capacitors is connected separately.

12-19 A series circuit with a current of 15 mA has a resistance of 1500 Ω $\angle 0^\circ$, a capacitive reactance of 2000 Ω $\angle -90^\circ$, and an inductive reactance of 3500 Ω $\angle 90^\circ$. Find the voltage across each element, the total voltage, and the phase angle of the total voltage with respect to the current. (*ans.* 22.5 V, 30 V, 52.5 V, 31.82 V, +45°)

12-20 A three-element series circuit has an emf of 120 V $\angle -12^\circ$. Element A is a 100-Ω $\angle 0^\circ$ resistance with a pd of 85 V $\angle 0^\circ$. Element B has a pd of 75 V $\angle -90^\circ$. Find the pd across the third element and its impedance.

12-21 A parallel circuit, fed from a 120-V 60-Hz source, consists of resistance of 60 Ω $\angle 0^\circ$, a motor with an impedance of $(200 + j120)$ Ω, and a capacitor with an impedance of 135 Ω$\angle -90^\circ$. Find the magnitude of the current in each element, the total current, and the phase angle of the total current with respect to the voltage.
(*ans.* 2 A, 0.5145 A $\angle -31^\circ$, 0.8889 A $\angle 90^\circ$, 2.519 A $\angle 14.3^\circ$)

12-22 The current drawn from a 240-V 60-Hz source by two motors connected in parallel is $(4.6 - j2.9)$ A. If one of the motors draws 1.9 A$\angle -36^\circ$, find the current drawn by the other motor.

13

AC Circuits with *R* or *L* or *C* Alone

In a resistive circuit, current and voltage exist together whether their magnitudes change or are constant. Energy is converted to some nonelectrical form and cannot be returned to the source.

In a capacitive circuit, current exists only when voltage changes. Energy can be stored and then returned to the source.

In an inductive circuit, voltage exists only when current changes. Energy can be stored and then returned to the source.

These basic differences in resistive, capacitive, and inductive circuits cause their reactions to sine-wave ac to be quite different. Change in voltage and current is inherent in sine-wave ac.

13-1 RESISTANCE

The relationship between voltage and current in any resistive circuit is given by Ohm's law ($V = IR$). In an ac resistive circuit this law applies for the values of voltage and current at any time instant (including peak values), for rms values, and for average values. The voltage and current magnitudes have a constant ratio (equal to the resistance) which is not a function of time.

In Fig. 13-1*a* a sine-wave ac source has an emf of E volts rms. This emf is applied across a resistance of R ohms (a conductance of G siemens). The circuit current is I amperes rms and the resistive pd is V volts rms.

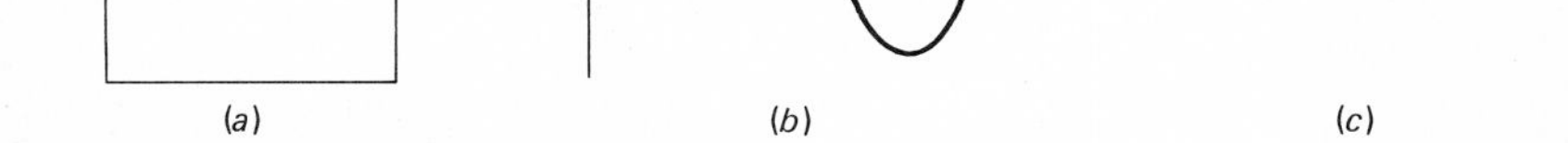

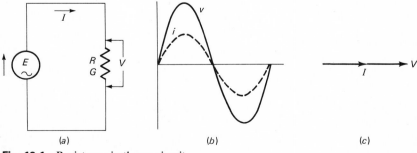

Fig. 13-1 Resistance in the ac circuit.

Ohm's law, expressed in rms values, for a resistive ac circuit is the same as for a dc circuit:

$$E = V = IR \qquad \text{or} \qquad I = EG = VG$$

At any instant of time, Ohm's law in a resistive ac circuit is:

$$e = v = iR \qquad \text{or} \qquad i = eG = vG$$

The waveforms of current and voltage at all time instants in one cycle are shown in Fig. 13-1*b*. A phasor diagram is given in Fig. 13-1*c*.

The characteristics of a resistive ac circuit are:

1. The current and the voltage are in phase ($\phi = 0$).
2. The impedance is the resistance ($Z = R$).
3. The admittance is the conductance ($Y = G = 1/R$).
4. The circuit dissipates average power ($P = VI = I^2R = V^2/R$).

13-2 CAPACITANCE

The relationship between voltage and current in a capacitive circuit at any time instant is

$$i = C\frac{dv}{dt}$$

For a sine waveform of voltage this becomes[1]

$$i = C\frac{dv_{\max} \sin \omega t}{dt}$$

The mathematical treatment of this equation is given in the appendix. Its result is

$$i = \omega C V_{\max} \sin\left(\omega t + \frac{\pi}{2}\right) \tag{13-1}$$

Analyzing Eq. (13-1), we see that

1. The current is of sine waveform.
2. The phase angle of the circuit is $\pi/2$ rad (90°). This conclusion is reached from $\sin(\omega t + \pi/2)$. The plus sign indicates a leading phase angle; the current leads the voltage.
3. The instantaneous value of the current is proportional to the maximum value.
4. The factor linking the magnitudes of the current and the voltage is ωC.

[1] We recall that ω is angular frequency in radians per second and that $\omega = 2\pi f$, where f is the frequency in hertz.

By replacing $\omega t + \pi/2$ with θ, Eq. (13-1) can be written:

$$i = \omega C V_{\max} \sin \theta$$

This can be rewritten in rms form:

$$I = \omega C V \tag{13-2}$$

Equation (13-2) is the Ohm's law of the capacitive circuit.

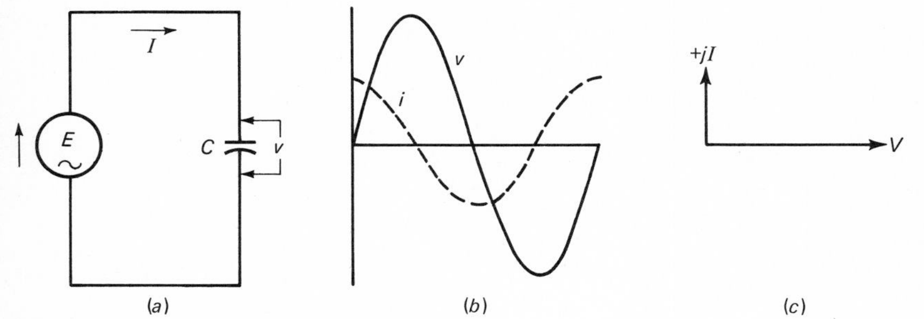

Fig. 13-2 Capacitance in the ac circuit.

Figure (13-2) shows a capacitive circuit with waveform and phasor diagrams. The phasor diagram in Fig. 13-2*c* can be shown with different symbols (Fig. 13-3).

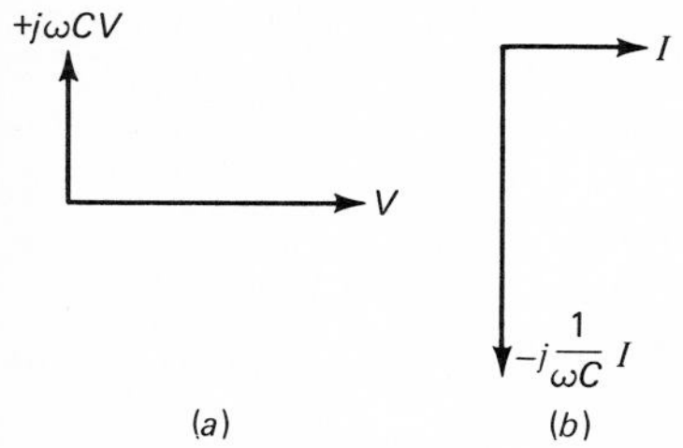

Fig. 13-3 Phasor diagram of a capacitive circuit. (*a*) In admittance form; (*b*) in impedance form.

Figure 13-3*a* is the same as Fig. 13-2*c*, except that the current phasor is labeled in terms of voltage, as in Eq. (13-2). The upward vertical position of the current phasor is indicated by $+j$.

In Fig. 13-3*b* the phasor diagram is shown with current as the reference phasor. It can be said that Fig. 13-3*b* is Fig. 13-3*a* rotated clockwise. Also, the voltage phasor in Fig. 13-3*b* is labeled in terms of current, from Eq. (13-2). The downward vertical position of the voltage phasor is indicated by $-j$.

From Fig. 13-3*a* it can be seen that

$$I = +j\omega C V \qquad \text{or} \qquad \frac{I}{V} = +j\omega C$$

In Sec. 12-5 it was stated that the ratio I/V is admittance, the ability of a circuit to permit current. In the capacitive circuit, the admittance is capacitive susceptance B_C.

$$B_C = +j\omega C \tag{13-3}$$

From Fig. 13-3*b* it can be seen that

$$V = -j\frac{1}{\omega C}I \qquad \text{or} \qquad \frac{V}{I} = -j\frac{1}{\omega C}$$

The ratio V/I is impedance, the ability of a circuit to restrict current. The impedance of a capacitive circuit is its capacitive reactance X_C.

$$X_C = -j\frac{1}{\omega C} \tag{13-4}$$

The sine-wave ac characteristics of a capacitive circuit can be summarized:

1. The current and the voltage are in phase quadrature, the current leading the voltage.
2. The circuit phase angle (the phase angle of the current with respect to the voltage) is $+90°$ or $+\pi/2$ rad.
3. The admittance is the capacitive susceptance ($Y = B_C = +j\omega C$ siemens).
4. The impedance is the capacitive reactance ($Z = X_C = -j1/\omega C$ ohms).
5. The circuit does not dissipate average power ($P = 0$).

13-3 INDUCTANCE

In a circuit which is purely inductive, the voltage and current at any instant of time are related by the equation

$$v = L\frac{di}{dt}$$

If the current is of sine waveform, this equation becomes

$$v = L\frac{dI_{\max}\sin\omega t}{dt}$$

The mathematical treatment of this equation is given in the appendix with the following result:

$$v = \omega L I_{\max}\sin\left(\omega t + \frac{\pi}{2}\right) \tag{13-5}$$

Equation (13-5) tells us that

1. The voltage is of sine waveform.
2. The phase angle of the circuit is 90° ($\pi/2$ rad). The plus sign in $\sin(\omega t + \pi/2)$ indicates that the voltage leads the current or that the current lags the voltage. Therefore the circuit has a lagging phase angle (−90°).
3. The instantaneous value of the voltage is proportional to the maximum value.
4. The factor linking the magnitudes of the voltage and the current is ωL.

Letting $\omega t + \pi/2 = \theta$, Eq. (13-5) can be written as

$$v = \omega L I_{\max} \sin \theta$$

which can be rewritten in rms form as

$$V = \omega L I \tag{13-6}$$

Equation (13-6) is the Ohm's law of the inductive circuit.

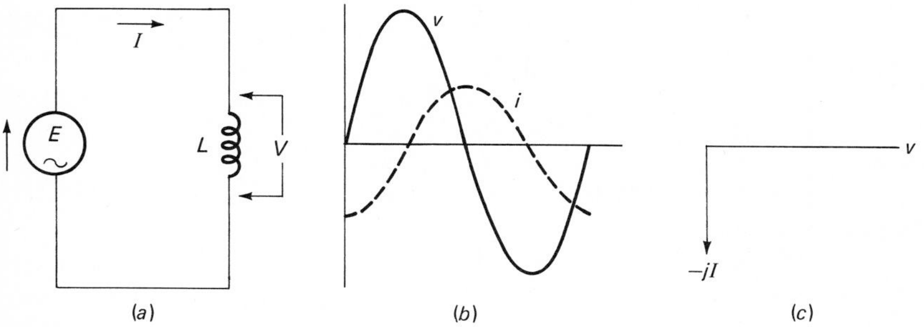

Fig. 13-4 Inductance in the ac circuit.

An inductive circuit with its waveform and phasor diagrams is shown in Fig. 13-4. The phasor diagram in Fig. 13-4*c* can be shown with different symbols (Fig. 13-5).

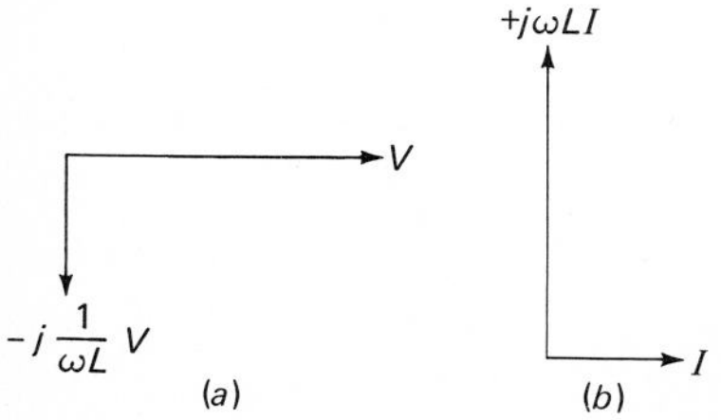

Fig. 13-5 Phasor diagram of an inductive circuit. (*a*) In admittance form; (*b*) in impedance form.

Figure 13-5*a* is the same as Fig. 13-4*c*, except that the current phasor is labeled in terms of voltage, from Eq. (13-6). The downward vertical position of the current phasor is indicated by $-j$.

In Fig. 13-5b the phasor diagram is shown with current as the reference phasor. The voltage phasor is labeled in terms of the current, as in Eq. (13-6). The upward vertical position of the voltage phasor is indicated by $+j$.

From Fig. 13-5a,

$$I = -j\frac{1}{\omega L}V \qquad \text{or} \qquad \frac{I}{V} = -j\frac{1}{\omega L}$$

The ratio I/V is admittance, the ability of a circuit to permit current. In the inductive circuit, the admittance is the inductive susceptance B_L:

$$B_L = -j\frac{1}{\omega L} \tag{13-7}$$

From Fig. 13-5b,

$$V = +j\omega LI \qquad \text{or} \qquad \frac{V}{I} = +j\omega L$$

The ratio V/I is impedance, the ability of a circuit to restrict current. The impedance of an inductive circuit is its inductive reactance X_L.

$$X_L = +j\omega L \tag{13-8}$$

The sine-wave characteristics of a circuit which is purely inductive can be summarized:

1. The current and the voltage are in phase quadrature, the current lagging the voltage.
2. The circuit phase angle (the phase angle of the current with respect to the voltage) is $-90°$ or $-\pi/2$ rad.
3. The admittance is the inductive susceptance ($Y = B_L = -j\,1/\omega L$ siemens).
4. The impedance is the inductive reactance ($Z = X_L = j\omega L$ ohms).
5. The circuit does not dissipate average power ($P = 0$).

Example 13-1 What is the resistance of a 2-kW electric stove hot plate when it is operating? The supply is 120-V 60-Hz ac.

Solution When ac voltages and currents are stated in rms values, a resistive element operating on ac is treated in exactly the same manner as if it were dc-operated:

$$P = \frac{V^2}{R}$$

$$R = \frac{V^2}{P} = \frac{(120\text{ V})^2}{2\text{ kW}} = \mathbf{7.2\ \Omega}$$

Example 13-2 The current in a 500-pF capacitor is 1 mA at 1 MHz. What is the magnitude of the voltage across it?
Solution The impedance of the capacitor is its capacitive reactance X_C. From Eq. (13-4),

$$X_C = \frac{1}{\omega C} = \frac{1}{2\pi fC} = \frac{1}{2\pi \times 1 \text{ MHz} \times 500 \text{ pF}} = 318.3 \ \Omega$$

Apply Ohm's law and omit j as it is not important when only magnitudes are of interest. From Eqs. (13-2) and (13-4),

$$V = IX_C = 1 \text{ mA} \times 318.3 \ \Omega = \mathbf{318.3 \ mV}$$

Example 13-3 What is the magnitude of the voltage across a pure inductance of 20 mH when its current is 80 mA at 1 kHz?
Solution The impedance of the inductance is its inductive reactance X_L. From Eq. (13-8),

$$X_L = \omega L = 2\pi fL = 2\pi \times 1 \text{ kHz} \times 20 \text{ mH} = 125.7 \ \Omega$$

Apply Ohm's law. From Eqs. (13-6) and (13-8),

$$V = IX_L = 80 \text{ mA} \times 125.7 \ \Omega = \mathbf{10.06 \ V}$$

The term "pure inductance" is used in Example 13-3 to indicate that, if a practical inductor is being considered, its resistance is negligible. This distinction is important because, unlike practical capacitors, real inductors often have so much resistance that it cannot be ignored. More will be said in later chapters about these resistive effects in inductors. Meanwhile it is assumed that, if the resistance of an inductor is not specifically mentioned, it is not significant to the circuit conditions.

Example 13-4 What is the magnitude of the current in a 50-μF capacitor when 50 V at 60 Hz is developed across it?
Solution The admittance of the capacitor is its capacitive susceptance B_C. From Eq. (13-3),

$$B_C = \omega C = 2\pi fC = 2\pi \times 60 \text{ Hz} \times 50 \ \mu\text{F} = 0.01885 \text{ S}$$

Apply the admittance form of Ohm's law. From Eqs. (13-2) and (13-3),

$$I = VB_C = 50 \text{ V} \times 0.01885 \text{ S} = \mathbf{0.9425 \ A}$$

Example 13-5 An electronic generator (an oscillator) applies 1 V at 5 MHz across a radio tuning coil. What is the inductance of the coil if its current magnitude is 0.5 mA?
Solution Ohm's law in admittance form is, from Eqs. (13-6) and (13-7),

$$I = VB_L$$

The admittance of the inductor is its inductive susceptance B_L:

$$B_L = \frac{I}{V} = \frac{0.5 \text{ mA}}{1 \text{ V}} = 0.5 \text{ mS}$$

From Eq. (13-7),

$$B_L = \frac{1}{\omega L}$$

$$L = \frac{1}{\omega B_L} = \frac{1}{2\pi f B_L} = \frac{1}{2\pi \times 5\text{ MHz} \times 0.5\text{ mS}}$$

$$L = \mathbf{0.06366\text{ mH}} \qquad \text{or} \qquad \mathbf{63.66\ \mu H}$$

Example 13-6 The inductance of a motor starting relay is 1 H. Assuming that its resistance is negligible, what is the phasor value of its current when the relay is supplied with 120-V 60-Hz ac?
Solution

$$X_L = j\omega L \tag{13-8}$$

$$X_L = j2\pi f L = j2\pi \times 60\text{ Hz} \times 1\text{ H} = j377\ \Omega$$

From Eqs. (13-6) and (13-8),

$$I = \frac{V}{X_L}$$

$$I = \frac{120\text{ V}}{j377\ \Omega} = \mathbf{-j0.3183\text{ A}}$$

Example 13-7 One of the components in the tuning circuit of a radio is a 35- to 300-pF variable capacitor. When the capacitor is set at 35 pF, the radio is tuned to the high end of the AM band, 1600 kHz. When the radio is tuned to the low end of the band, 550 kHz, the capacitor is in its 300-pF position. If the signal voltage across the capacitor is constant at 100 μV, what is the phasor value of its current at each end of the AM band?
Solution At the high-frequency end of the band, $f = 1600$ kHz and $C = 35$ pF. The phasor value of the capacitor's impedance is given by Eq. (13-4)

$$X_C = -j\frac{1}{\omega C} = -j\frac{1}{2\pi f C} = -j\frac{1}{2\pi \times 1600\text{ kHz} \times 35\text{ pF}} = -j2842\ \Omega$$

From Eqs. (13-2) and (13-4),

$$I = \frac{V}{X_C} = \frac{100\ \mu\text{V}}{-j2842\ \Omega} = \mathbf{j0.0352\ \mu A} \qquad \text{or} \qquad \mathbf{j35.2\text{ nA}}$$

At the low-frequency end of the band, $f = 550$ kHz and $C = 300$ pF:

$$X_C = -j\frac{1}{\omega C} = -j\frac{1}{2\pi \times 550\text{ kHz} \times 300\text{ pF}} = -j964.6\ \Omega$$

$$I = \frac{V}{X_C} = \frac{100\ \mu\text{V}}{-j964.6\ \Omega} = \mathbf{j0.1037\ \mu A} \qquad \text{or} \qquad \mathbf{j103.7\text{ nA}}$$

The solutions of Examples 13-6 and 13-7 used the reciprocal of *j* principle developed in Sec. 12-4. This technique of interchanging *j* and 1/*j* with a change in sign is used often in ac work.

In Example 13-7, the current at the low-frequency end of the band could have been obtained by applying proportionality to the high-frequency result. The proportionalities involved are:

Current is proportional to susceptance ($I \propto B$).

Capacitive susceptance is proportional to the product of the angular frequency and the capacitance ($B_C \propto \omega C$).

Angular frequency is proportional to frequency ($\omega \propto f$).

$$I \propto B_C \propto \omega C \propto fC$$

These proportionalities can be used to produce the following relationship between the low-frequency current and the high-frequency current (I_{550} and I_{1600}):

$$I_{550} = I_{1600} \times \frac{f_{550}C_{550}}{f_{1600}C_{1600}} = j35.2\text{ nA} \times \frac{550\text{ kHz} \times 300\text{ pF}}{1600\text{ kHz} \times 35\text{ pF}}$$

$$I_{550} = j103.7\text{ nA}$$

13-4 POWER

INSTANTANEOUS POWER

Whenever there is voltage *and* current, there is power. For sine-wave alternating current in any type of circuit, the power level will vary from instant to instant, sometimes being zero in magnitude.

Figure 13-6*a* shows that, for a resistive circuit, whenever there is voltage, there is also current. Power is zero when both i and v are zero.

In Fig. 13-6*b* and *c* it can be seen that, for inductive and capacitive circuits, there are instants when there is voltage with zero current and other instants when there is current with zero voltage. These are the instants of zero power.

Instantaneous power ($p = vi$) can be determined for resistive, inductive, or capacitive circuits for any point in the cycle. The maximum instantaneous power (peak power) is important in some electronics applications but, apart from this, instantaneous powers are rarely of interest. The average power over an extended period of time is of importance in most electric and electronic systems.

AVERAGE POWER

In ac circuit terminology "power" usually refers to average power, the true average taken over a complete cycle.

The average power in a resistive circuit carrying ac is the same as if the circuit carried steady dc of magnitude equal to the ac rms value. This

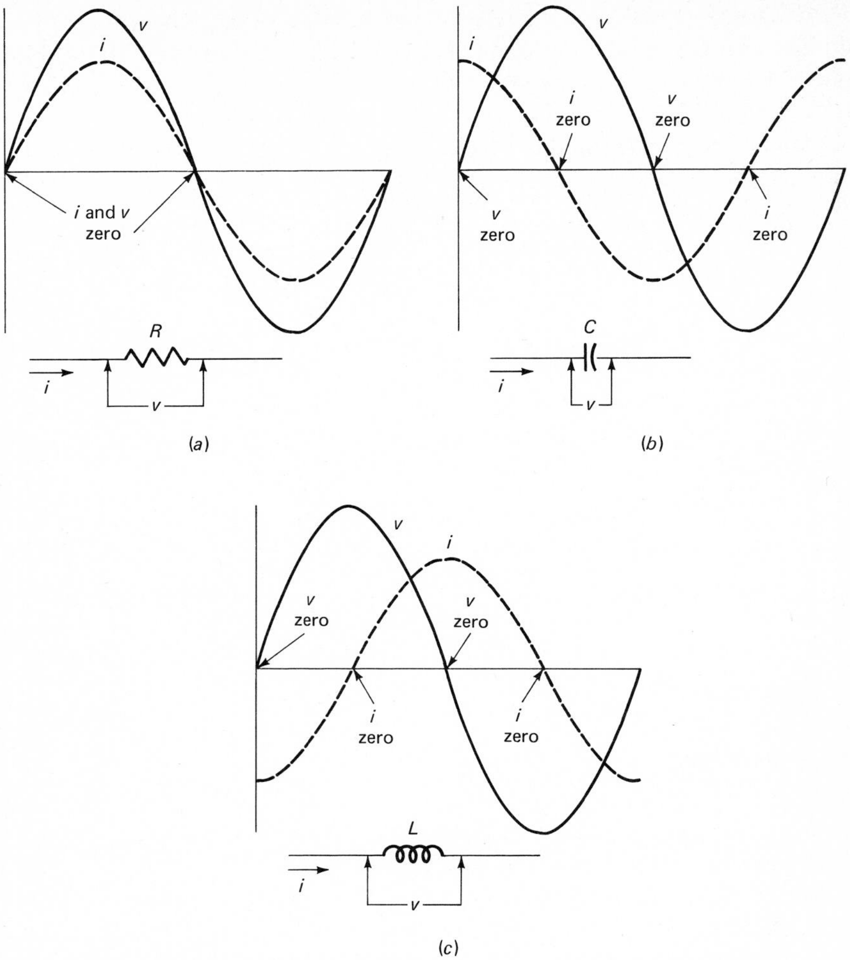

Fig. 13-6 Waveforms in ac circuits. (*a*) Resistive; (*b*) capacitive; (*c*) inductive.

is to be expected, since rms values are defined in terms of the power equivalence of steady direct current.

The power equations (2-19), (2-30), and (2-31) can be used for resistive ac circuits:

$$P = VI \qquad \text{or} \qquad I^2R \qquad \text{or} \qquad \frac{V^2}{R}$$

where P = power, W
V = emf or pd, rms V
I = current, rms A
R = resistance, Ω

The average power in a capacitive or inductive circuit is zero.

The power statements for resistance, capacitance, and inductance are derived by calculus in the appendix. At this point it is sufficient to refer to the basic logic of ac power.

As indicated in Fig. 13-7*a*, energy is supplied to a resistance all the time alternating current is present. The energy leaves the circuit, mainly as heat, and cannot be returned to the source.

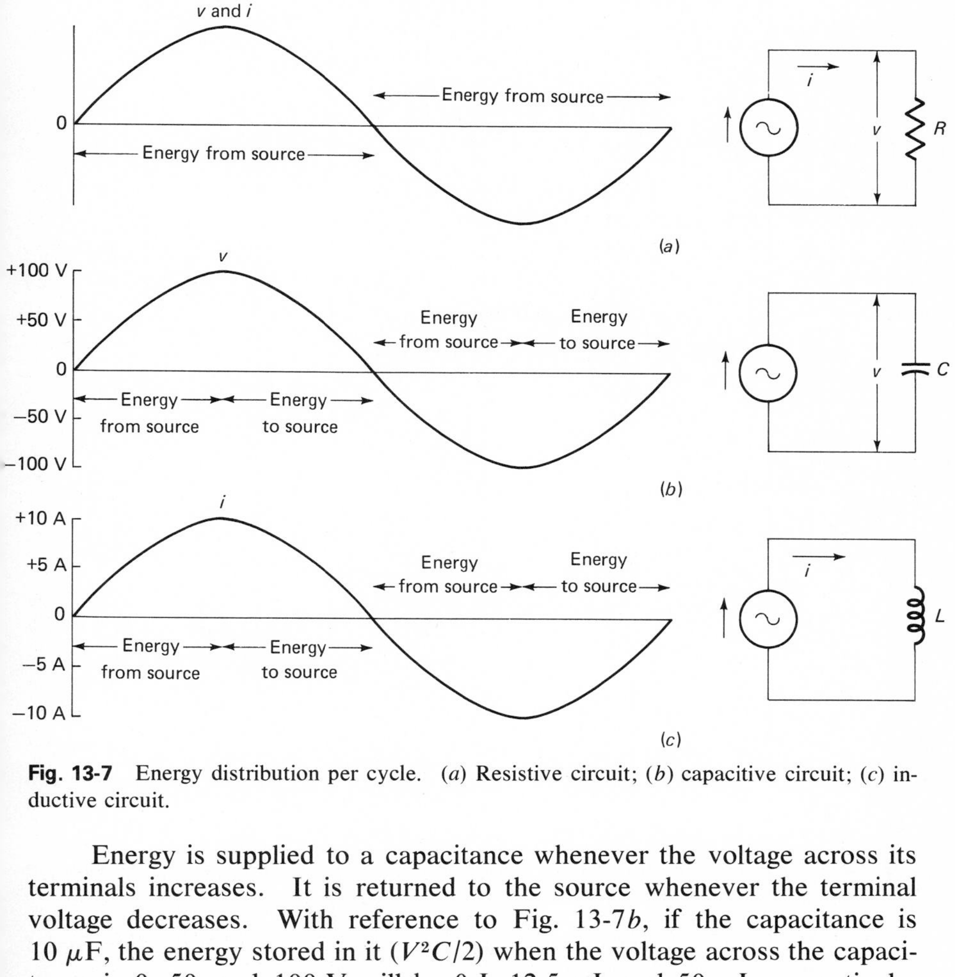

Fig. 13-7 Energy distribution per cycle. (*a*) Resistive circuit; (*b*) capacitive circuit; (*c*) inductive circuit.

Energy is supplied to a capacitance whenever the voltage across its terminals increases. It is returned to the source whenever the terminal voltage decreases. With reference to Fig. 13-7*b*, if the capacitance is 10 μF, the energy stored in it ($V^2C/2$) when the voltage across the capacitance is 0, 50, and 100 V will be 0 J, 12.5 mJ, and 50 mJ, respectively. Thus, during the first quarter-cycle, the energy will rise from 0 to 12.5 to 50 mJ, this energy coming *from* the source. During the second quarter-cycle the energy stored in the capacitance will drop from 50 to 12.5 to 0 mJ, this energy being returned *to* the source.

Energy is supplied to an inductance whenever the current through it increases. It is returned to the source whenever the current through the inductance decreases. The reasoning is essentially the same as for capacitance. With reference to Fig. 13-7*c*, during the first quarter-cycle when the current is rising from zero to 10 A, if the inductance is 1 mH, its energy *from* the source will rise from 0 to 50 mJ ($I^2L/2$). During the second quarter-cycle this 50 mJ of energy will be returned *to* the source while the current is dropping from 10 to 0 A.

When a capacitance or an inductance is connected to an ac source, no energy leaves the circuit; it is repeatedly transferred between the source and the element. No net energy is taken from the source, and hence there is no net or average power dissipation by the circuit.

13-5 FREQUENCY, IMPEDANCE, AND ADMITTANCE

The impedance and admittance of elements at different frequencies are often of importance in the operation of circuits. In some applications varying frequency characteristics are vital to the functioning of the equipment. An example is radio tuning which is carried out by controlling reactance. In other situations frequency effects can limit the usefulness of equipment. For instance, stray capacitive and inductive reactance can render equipment useless at high frequencies.

IMPEDANCE

Figure 13-8 shows the relationships between frequency and the magnitudes of resistance, inductive reactance, and capacitive reactance.

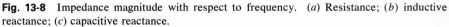

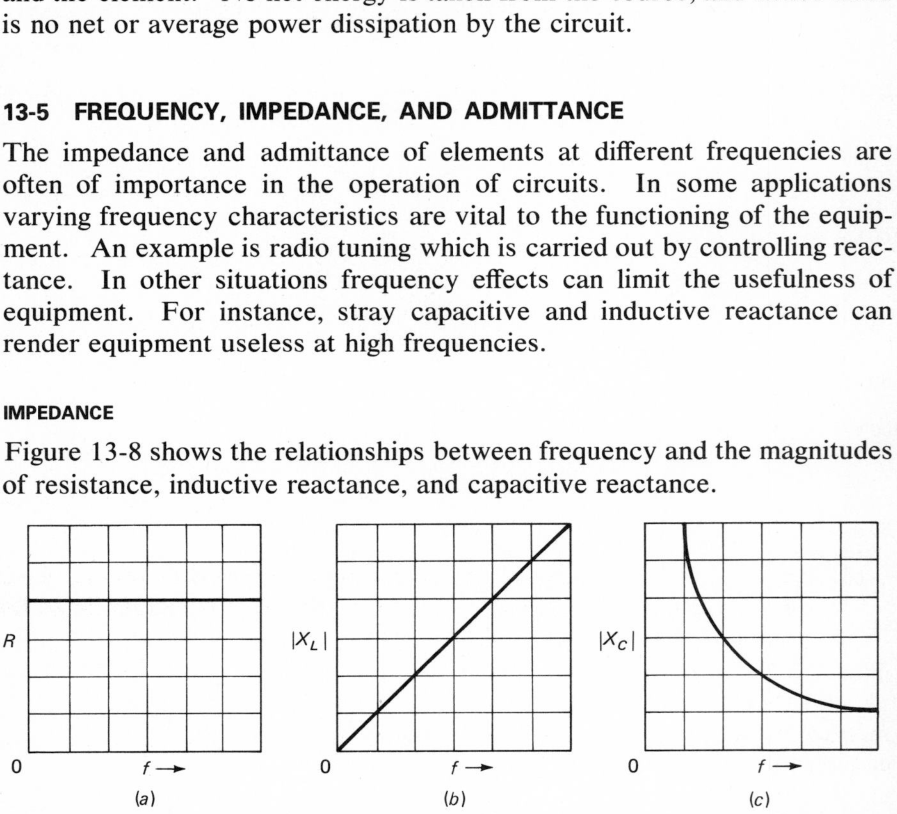

Fig. 13-8 Impedance magnitude with respect to frequency. (*a*) Resistance; (*b*) inductive reactance; (*c*) capacitive reactance.

Resistance is the same at all frequencies, as shown by the horizontal line in Fig. 13-8*a*.

The magnitude of inductive reactance ($|X_L| = 2\pi fL$) is directly proportional to frequency, as shown by the inclined line in Fig. 13-8*b*.

The magnitude of capacitive reactance ($|X_C| = 1/2\pi fC$) is inversely proportional to frequency. The exponential curve in Fig. 13-8*c* shows this relationship.

Figure 13-8*b* and *c* shows only the magnitudes of X_L and X_C. To depict the relative phases of inductive and capacitive circuits, it is usual to adopt a frequency-magnitude diagram which distinguishes between $+j$ and $-j$.

In Fig. 13-9 the frequency—inductive-reactance curve is shown above the horizontal axis to be related to $+j$. The frequency—capacitive-reactance curve is shown in the $-j$ location, below the horizontal axis.

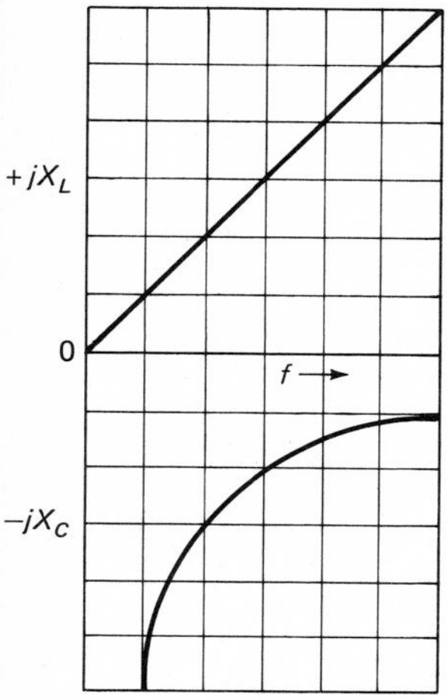

Fig. 13-9 Inductive and capacitive reactance curves showing the relative phases.

The magnitude of impedance at any frequency in diagrams such as Figs. 13-8 and 13-9 is indicated by the distance of the curve above or below the horizontal axis. There are no reactance differences between Figs. 13-8 and 13-9. Any change in magnitude away from zero is an increase. Any change in magnitude toward zero is a decrease. In Fig. 13-9 the magnitude of X_L is shown increasing with increasing frequency, as in Fig. 13-8*b* while the magnitude of X_C is shown decreasing, as in Fig. 13-8*c*.

ADMITTANCE

Admittance is the reciprocal of impedance, so the curves of admittance with respect to frequency are the reciprocals of Figs. 13-8 and 13-9.

Conductance is independent of frequency, as shown in Fig. 13-10*a*.

Fig. 13-10 Admittance magnitude with respect to frequency. (*a*) Conductance; (*b*) inductive susceptance; (*c*) capacitive susceptance.

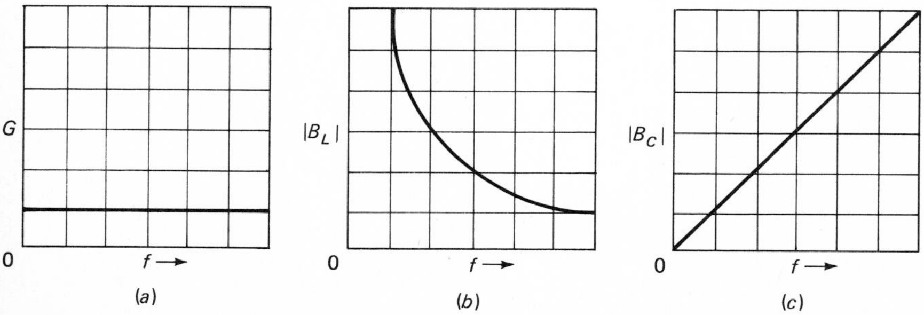

The magnitude of inductive susceptance ($|B_L| = 1/2\pi fL$) is inversely proportional to frequency, as indicated in Fig. 13-10*b*.

In Fig. 13-10*c* the magnitude of capacitive susceptance ($|B_C| = 2\pi fC$) is directly proportional to frequency.

Figure 13-11 shows the capacitive and inductive susceptance curves in their respective $+j$ and $-j$ positions.

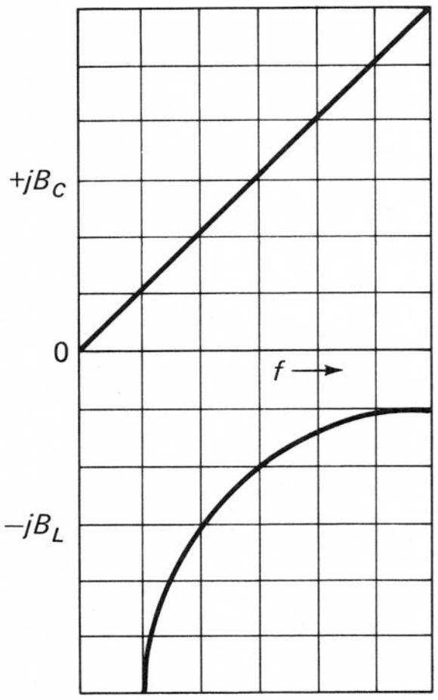

Fig. 13-11 Inductive and capacitive susceptance curves showing the relative phases.

It is evident that there are considerable differences in the behavior of resistive, inductive, and capacitive elements from a frequency point of view. These differences are among the most important of the resources at the disposal of the ac circuit designer.

13-6 SUMMARY OF EQUATIONS

$i = \omega CV_{max} \sin(\omega t + \pi/2)$	(13-1)	$v = \omega LI_{max} \sin(\omega t + \pi/2)$	(13-5)
$I = \omega CV$	(13-2)	$V = \omega LI$	(13-6)
$B_C = +j\omega C$	(13-3)	$B_L = -j1/\omega L$	(13-7)
$X_C = -j1/\omega C$	(13-4)	$X_L = +j\omega L$	(13-8)

EXERCISE PROBLEMS

Sections 13-2 and 13-3

13-1 Find the capacitive susceptance of the following capacitors at the frequencies stated.

a. 10 μF, 20 Hz (*ans.* 1.257 mS)
b. 10 μF, 60 Hz (*ans.* 3.77 mS)
c. 10 μF, 1000 Hz (*ans.* 62.8 mS)
d. 0.047 μF, 100 kHz (*ans.* 29.53 mS)
e. 33 pF, 100 kHz (*ans.* 20.73 μS)
f. 33 pF, 20 MHz (*ans.* 4.147 mS)
g. 0.0015 μF, 20 MHz (*ans.* 188.5 mS)
h. 220 pF, 20 MHz (*ans.* 27.65 mS)

13-2 Find the capacitive reactance of the capacitors in Prob. 13-1 at the frequencies stated.

13-3 Find the inductive susceptance of the following inductors at the frequencies stated.

a. 5 H, 20 Hz (*ans.* 1.592 mS)
b. 5 H, 100 Hz (*ans.* 318.3 μS)
c. 5 H, 1000 Hz (*ans.* 31.83 μS)
d. 200 mH, 50 kHz (*ans.* 15.92 μS)
e. 100 mH, 100 kHz (*ans.* 15.92 μS)
f. 5.6 μH, 10 MHz (*ans.* 2.842 mS)
g. 35 μH, 10 MH_2 (*ans.* 454.5 μS)
h. 160 μH, 10 MHz (*ans.* 99.5 μS)

13-4 Find the inductive reactance of the inductors in Prob. 13-3 at the frequencies stated.

13-5 Find the current in a circuit consisting of a 1000-Hz generator with an output of 50 V when each of the following capacitors is individually connected:

a. 5 μF (*ans.* 1.571 A)
b. 0.033 μF (*ans.* 10.37 mA)
c. 1000 pF (*ans.* 314.2 μA)
d. 0.0022 μF (*ans.* 691.2 μA)

13-6 Find the voltage across each of the following inductors when they are carrying a current of 30 mA at 10 kHz:

a. 0.2 H
b. 50 mH
c. 100 μH
d. 1 mH

13-7 An audio amplifier uses a 0.1-μF capacitor to couple two stages together. Find the reactance of the capacitor at the following audio frequencies:

a. 50 Hz (*ans.* 31.83 kΩ)
b. 1000 Hz (*ans.* 1.592 kΩ)
c. 15 kHz (*ans.* 106.1 Ω)

13-8 A 0.47-μF capacitor and a 0.27-μF capacitor are connected in parallel across a 120-V 400-Hz supply. Find

a. The admittance of each capacitor.
b. The total admittance.
c. The reactance of each capacitor.
d. The total reactance.
e. The total equivalent capacitance.
f. The total circuit current.

13-9 A 0.47-μF capacitor and a 0.27-μF capacitor are connected in series across a 120-V 400-Hz supply. Find

a. The total admittance. (*ans.* 431 μS)
b. The total reactance. (*ans.* 2.32 kΩ)
c. The total equivalent capacitance. (*ans.* 0.1715 μF)
d. The current in the circuit. (*ans.* 51.72 mA)
e. The pd across each capacitor. (*ans.* 43.78 V, 76.22 V)

13-10 A radio-frequency choke (inductor) permits the passage of low frequencies in preference to high frequencies. Find the reactance of a 300-μH radio-frequency choke at

a. 2 MHz
b. 400 Hz

13-11 Two relay coils, one 0.5 H and the other 1.0 H, are connected in parallel across a 120-V 800-Hz system. Find

a. The admittance of each coil. (*ans.* 398 μS, 199 μS)
b. The total admittance. (*ans.* 597 μS)
c. The reactance of each coil. (*ans.* 2.513 kΩ, 5.027 kΩ)
d. The total reactance. (*ans.* 1.676 kΩ)
e. The total equivalent inductance. (*ans.* 0.333 H)
f. The total current in the circuit. (*ans.* 71.61 mA)

13-12 A 20-mH inductor and a 100-mH inductor are connected in series across a 25-V 1000-Hz generator. Find

a. The total admittance.
b. The total reactance.
c. The total equivalent inductance.
d. The current in the circuit.
e. The pd across the 100-mH inductor.

Section 13-4

13-13 A 100-Ω resistance is fed from a 120-V 60-Hz source. Reference time t_0 is when the voltage is zero going positive.

a. Find the total power consumed by the circuit. (*ans.* 144 W)

b. Find the instantaneous power at the following times:

i. $t = 2.083$ ms (*ans.* 144 W)
ii. $t = 4.167$ ms (*ans.* 289 W)
iii. $t = 6.25$ ms (*ans.* 144 W)
iv. $t = 10.42$ ms (*ans.* 144 W)
v. $t = 12.5$ ms (*ans.* 289 W)
vi. $t = 14.58$ ms (*ans.* 144 W)

13-14 A 20-μF capacitor is fed from a 120-V 60-Hz source. Reference time t_0 is when the voltage is zero going positive.

a. Find the total power consumed by the circuit.

b. Find the instantaneous power at the following times:

i. $t = 2.083$ ms *iv.* $t = 10.42$ ms
ii. $t = 4.167$ ms *v.* $t = 12.5$ ms
iii. $t = 6.25$ ms *vi.* $t = 14.58$ ms

13-15 A circuit consists of a 5-μF capacitor connected across a 120-V 400-Hz generator. Find the energy stored in the electric field of the capacitor at the following times after t_0:

a. 0.2 ms (*ans.* 16.68 mJ)
b. 0.625 ms (*ans.* 72 mJ)
c. 0.937 ms (*ans.* 36 mJ)
d. 1.56 ms (*ans.* 35.5 mJ)
e. 1.875 ms (*ans.* 72 mJ)
f. 2.5 ms (*ans.* 0 J)

13-16 A circuit consists of a 5-H inductor connected across a 120-V 400-Hz generator. Find the energy stored in the magnetic field of the inductor at the times after t_0 listed for Prob. 13-15.

Section 13-5

13-17 A 0.001-μF capacitor and a 5-mH inductor are connected in series across a variable frequency source. Find the reactance of each at the following frequencies:

a. 0 Hz (*ans.* ∞ Ω, 0 Ω)
b. 1 kHz (*ans.* $-j159.2$ kΩ, $+j31.42$ Ω)
c. 20 kHz (*ans.* $-j7.958$ kΩ, $+j628.3$ Ω)
d. 100 kHz (*ans.* $-j1.592$ kΩ, $+j3.142$ kΩ)
e. 1 MHz (*ans.* $-j159.2$ Ω, $+j31.42$ kΩ)
f. 50 MHz (*ans.* $-j3.183$ Ω, $+j1.571$ MΩ)

13-18 When impedances are connected in series, the total impedance equals the phasor sum of the individual impedances. Find the total impedance for the circuit in Prob. 13-17 at each frequency.

13-19 Find the frequency at which $X_L = X_C$ in the circuit in Prob. 13-17. (*ans.* 71.18 kHz)

13-20 A 1500-pF capacitor and a 2-mH inductor are connected in parallel across a variable frequency source. Find the susceptance of each at the following frequencies:

a. 0 Hz *d.* 100 kHz
b. 1 kHz *e.* 1 MHz
c. 20 kHz *f.* 50 MHz

13-21 When admittances are connected in parallel, the total admittance equals the phasor sum of the individual admittances. Find the total admittance for the circuit in Prob. 13-20 at each frequency.
(*ans.* *a.* ∞ S; *b.* $-j97.57$ mS; *c.* $-j3.791$ mS; *d.* $+j146.7$ μS; *e.* $+j9.345$ mS; *f.* $+j471.2$ mS)

13-22 Find the frequency at which $B_L = B_C$ for the circuit in Prob. 13-20.

14

AC Circuits with *R* and *C* or *R* and *L*

Before detailing the behavior of ac circuits containing resistance and inductance or capacitance, a preview of some general characteristics of ac circuits will be helpful.

1. All the principles and laws of dc circuits, including Ohm's and Kirchhoff's laws and Thevenin's and Norton's theorems, apply to ac circuits.
2. Where a circuit contains both resistive and nonresistive components, there will be phase differences between voltages, between currents, and between voltages and currents. In these circumstances, the magnitudes of voltage or current occurring at a specified instant of time can be added arithmetically, but peak, peak-to-peak, average, and rms values cannot.
3. When Kirchhoff's laws are applied to ac circuits, voltages or currents are added, these usually being expressed in rms terms. The additions must be carried out using phasor methods.
4. Power is dissipated only in the resistive components of a circuit. Except in the special case of resonance, nonresistive components add to series voltage for a given current or to parallel current for a given voltage. Thus the presence of nonresistive components causes the product of voltage and current for the complete circuit to be greater than the amount of power dissipated by the circuit. It is therefore important to distinguish between the rate at which energy leaves the circuit from resistive components (called *active power*), the rate at which energy is alternately stored in and discharged from reactive components (*reactive power*), and the product of total voltage and total current (*apparent power*). The term "apparent power" refers to its similarity to dc circuit power, both being the product of total voltage and total current. But the real ac equivalent of dc power is active power, the actual power dissipated.
5. Often the production of active power is of particular significance in an ac circuit application; for example, in an electric motor the active power is responsible for the mechanical output power at the motor shaft. But apparent power can never be ignored. The apparent power determines the overall effect of the circuit on the supply source; the amount of current drawn for a given supply voltage.
6. The importance of both active power and apparent power has led to the adoption of a special term for the ratio between them. The *power factor* is the ratio of active power to apparent power; the power factor

describes the amount of power dissipated by the circuit in relation to the voltage and current supplied to the circuit.

7. Often electrical utilities base their rates on apparent power, since this determines the sizes of conductors, transformers, and other equipment. In such cases, the power factor tells the user how much useful power is available.

ADDITION OF PHASORS

Where an ac circuit contains inductance or capacitance, the currents and voltages do not rise and fall together; there are time lag or lead relationships between them, necessitating the use of phasor techniques in analysis.

As a common example arising in ac circuit work, consider the addition of two sine waveforms such as is involved in applying Kirchhoff's laws. Figure 14-1 shows two sine waves representing either two voltages or two currents.

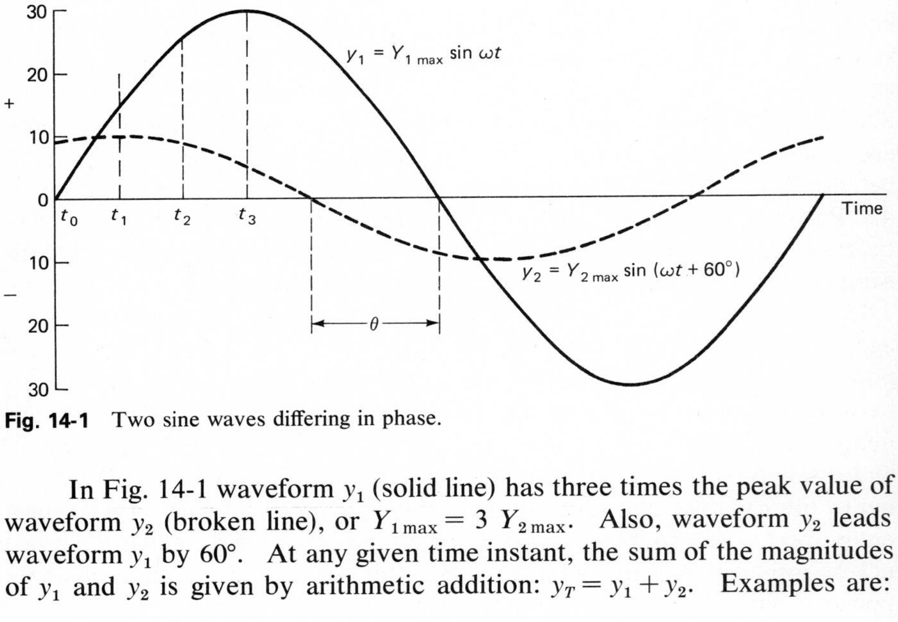

Fig. 14-1 Two sine waves differing in phase.

In Fig. 14-1 waveform y_1 (solid line) has three times the peak value of waveform y_2 (broken line), or $Y_{1\,max} = 3\ Y_{2\,max}$. Also, waveform y_2 leads waveform y_1 by 60°. At any given time instant, the sum of the magnitudes of y_1 and y_2 is given by arithmetic addition: $y_T = y_1 + y_2$. Examples are:

At time t_0, $y_1 = 0$, $y_2 = +8.66$, and $y_T = 0 + 8.66 = 8.66$.

At time t_1, $y_1 = +15$, $y_2 = +10$, and $y_T = 15 + 10 = 25$.

At time t_2, $y_1 = +26$, $y_2 = +8.66$, and $y_T = 26 + 8.66 = 34.66$.

At time t_3, $y_1 = +30$, $y_2 = +5$, and $y_T = 30 + 5 = 35$.

At time t_1, y_1 is approaching its maximum value, while y_2 is at its maximum value. At time t_3, y_1 is at its maximum value, but y_2 has passed its maximum value. At no time do the peaks of y_1 and y_2 occur simultaneously; thus it cannot be said that $Y_{T\max} = Y_{1\max} + Y_{2\max}$. Since the rms and average values of sine waves are fixed fractions of the peak values, it follows that it is not valid to add rms or average values arithmetically;[1] it cannot be said that $Y_T = Y_1 + Y_2$, or that $Y_{T\text{av}} = Y_{1\text{av}} + Y_{2\text{av}}$.

Arithmetic addition of each pair of instantaneous values y_1 and y_2 throughout the cycle in Fig. 14-1 results in the formation of a third sine wave y_T which represents the phasor sum of the two original sine waves. The three waveforms y_1, y_2, and y_T are shown in Fig. 14-2. You will see that the peak of y_T occurs slightly ahead of the peak of y_1. A phasor diagram helps visualize and determine the values of magnitude and phase angle of waveform y_T relative to those of y_1 and y_2. To keep the treatment general, the phase angle between Y_1 and Y_2 is initially given the symbol θ, and the phase angle between Y_T and Y_1 is given the symbol ϕ.

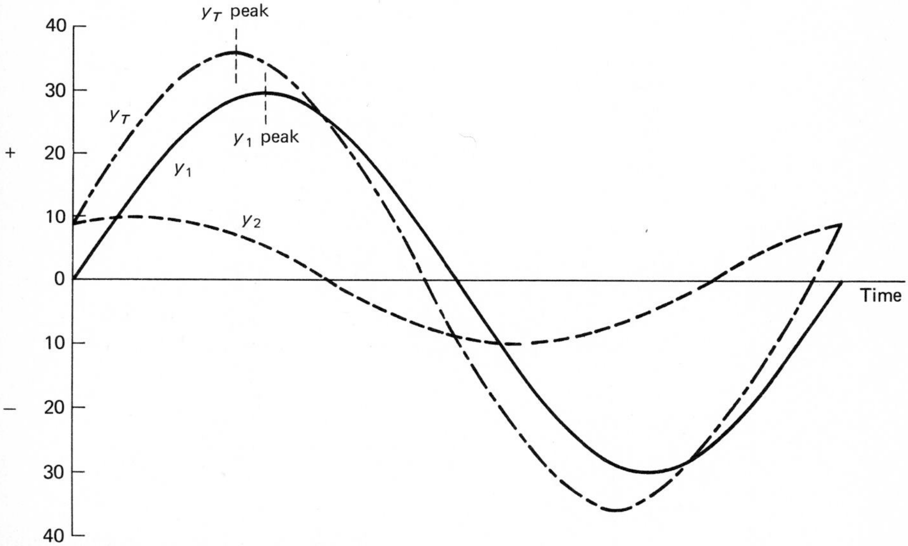

Fig. 14-2 Addition of two sine waves.

Figure 14-3*a* shows the phasors Y_1 and Y_2, with Y_1 as the reference phasor. The rectangular expressions for the phasors are

$$Y_1 = Y_1 + j0$$

$$Y_2 = Y_2 \cos\theta + jY_2 \sin\theta$$

[1] In the equation $Y_T = Y_1 + Y_2$ all three symbols are intended to represent rms values ($Y_{T\text{rms}} = Y_{1\text{rms}} + Y_{2\text{rms}}$). Often these same symbols are used as a simplified method of showing phasor values, the same equation meaning $Y_{T\max}\underline{/\phi_T} = Y_{1\max}\underline{/\phi_1} + Y_{2\max}\underline{/\phi_2}$. Where the context does not clearly identify values as phasor rather than rms, or where failure to identify phasor values would cause confusion, phasor terms must be shown in full phasor form.

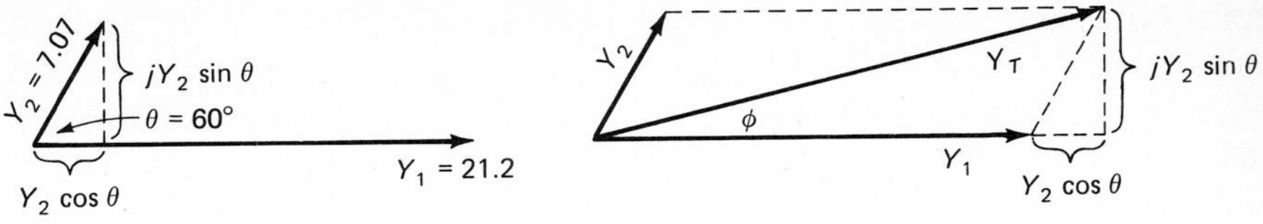

Fig. 14-3 (*a*) Sine waves in Fig. 14-2 as phasors; (*b*) adding the phasors.

The two phasors and their sum, the phasor Y_T, are shown in Fig. 14-3*b*. The projection of Y_T onto the horizontal reference axis is $Y_1 + Y_2 \cos \theta$. The vertical projection of Y_T is $jY_2 \sin \theta$. Thus the phasor Y_T is expressed in rectangular form as

$$Y_T = (Y_1 + Y_2 \cos \theta) + jY_2 \sin \theta$$

The magnitude of Y_T is

$$|Y_T| = \sqrt{(Y_1 + Y_2 \cos \theta)^2 + (Y_2 \sin \theta)^2}$$

$$|Y_T| = \sqrt{Y_1^2 + Y_2^2 + 2Y_1Y_2 \cos \theta}$$

The phase angle of Y_T with respect to Y_1 is

$$\phi = \arctan \frac{Y_2 \sin \theta}{Y_1 + Y_2 \cos \theta}$$

Therefore the polar form of the sum phasor appears as

$$Y_T = \sqrt{Y_1^2 + Y_2^2 + 2Y_1Y_2 \cos \theta} \;\Big/ \arctan \frac{Y_2 \sin \theta}{Y_1 + Y_2 \cos \theta}$$

The derivations of the rectangular and polar expressions for a phasor which is the sum of two other phasors are given to show how the basic principle of adding voltages or currents is applied in ac terms. This apparently simple task can involve rather extensive mathematics. In practice, matters are usually much simpler than this, since the original values are known and can be used throughout a calculation. As an example consider the situation in Figs. 14-1 to 14-3, where the rms values of Y_1 and Y_2 are 21.2 and 7.07, respectively, and the phase angle between them is 60°. The sum of these phasors is found in the following manner:

$$Y_1 = 21.2 \underline{/0°} = 21.2 + j0$$

$$Y_2 = 7.07 \underline{/60°} = 7.07(\cos 60° + j \sin 60°) = 3.54 + j6.12$$

$$Y_T = Y_1 + Y_2 = (21.2 + j0) + (3.54 + j6.12)$$

$$Y_T = 24.74 + j6.12$$

or, in polar form,

$$Y_T = \sqrt{24.74^2 + 6.12^2} \;\angle \arctan \frac{6.12}{24.74}$$

$$Y_T = 25.48 \angle 13.9^\circ$$

The value of Y_T shows that its magnitude is greater than the magnitude of either Y_1 or Y_2, but less than their arithmetic sum and its phase angle with respect to Y_1 is less than that of Y_2.

The characteristics of multielement ac circuits are essentially the same as those of steady dc, provided a phasor viewpoint is adopted. The application of phasor principles to ac circuits is introduced most simply by restricting the initial discussion to networks which contain only *R* and *C* or *R* and *L*. In adopting a practical approach to ac circuits, rms values of voltage and current are used unless otherwise stated.

14-1 SERIES *RC* CIRCUIT

The basic characteristics of any series ac circuit are:

1. The current in each element is the same.
2. The total circuit voltage is the phasor sum of the voltages of the individual elements (Kirchhoff's voltage law).

A two-element *RC* series circuit is shown in Fig. 14-4.

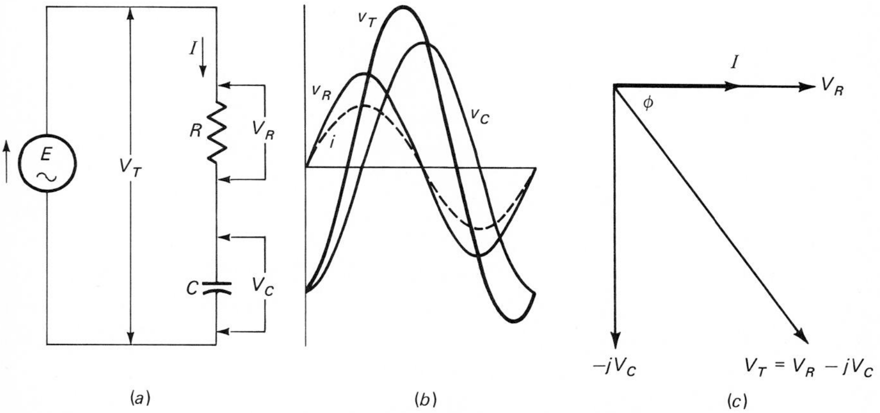

Fig. 14-4 Series *RC* circuit diagrams. (*a*) Schematic; (*b*) waveform; (*c*) phasor.

Figure 14-4*a* is a schematic diagram in which the current *I* is common to both elements. It is convenient to make this common current the reference phasor $I \angle 0^\circ$ or $I + j0$.

VOLTAGES

The current is shown as the reference phasor in the waveform diagram in Fig. 14-4*b* and in the phasor diagram in Fig. 14-4*c*.

The resistive voltage V_R is in phase with the current, so it has the phasor value $V_R \underline{/0^\circ}$ or $V_R + j0$. The capacitive voltage V_C lags the current by 90° or $\pi/2$ rad, making its phasor value $V_C \underline{/-90^\circ}$ or $0 - jV_C$.

The total circuit voltage V_T is the phasor sum of the resistive and capacitive voltages:

$$V_T = V_R + V_C = (V_R + j0) + (0 - jV_C)$$

$$V_T = V_R - jV_C \tag{14-1}$$

IMPEDANCES

By applying Ohm's law to each voltage term in Eq. (14-1), it can be rewritten:

$$IZ = IR - jIX_C$$

By dividing each term of this equation by the common current I, an impedance equation for the series *RC* circuit is obtained

$$Z = R - jX_C \tag{14-2}$$

POWERS

Each voltage term in Eq. (14-1) can be multiplied by the common current I:

$$V_T I = V_R I - jV_C I \tag{14-3}$$

In ac circuits containing reactance (X_C or X_L), the product of the total voltage and total current $V_T I$ is greater than the average power dissipated by the circuit.

$V_T I$ is called *apparent power.* From the viewpoint of the voltage-current demand on the supply source it appears that the circuit uses this much power. The symbol for apparent power is S, the initial letter of the word "sum," apparent power being the phasor sum of two other powers. The unit for apparent power is the volt-ampere (VA).

Average power dissipation takes place only in resistances, as explained in Sec. 13-4. In Eq. (14-3) the term $V_R I$ represents the power actively dissipated by the circuit and is thus called *active power.* The symbol for active power is P, the general power symbol. The unit for active power is the watt (W), the general power unit.

The quantity $V_C I$ in Eq. (14-3) is not a true average power because, as we have seen in Chap. 13, the average power dissipated in a capacitance is zero. However, since $V_C I$ is a voltage-current product associated with a reactance, it is called *reactive power*. The symbol for reactive power is Q. This is the initial letter of the word "quadrature," reactive power being the quadrature component of apparent power. The unit of reactive power is the volt-ampere reactive (var).

Inserting the symbols S, P, and Q in Eq. (14-3), we obtain

$$S = P - jQ$$

There is international agreement that power relationships in ac circuits of all types will be considered on a voltage-phasor-reference basis so that the resulting phase angle is the defined circuit phase angle: the angle of the total current with respect to the total voltage.

In Fig. 14-4*c* current is the reference phasor, and the phase angle is $-\phi$. If this phasor diagram is rotated counterclockwise through the angle ϕ, the phasor V_T will be in the reference position, I will be conventionally related to it, and the phase angle will be $+\phi$, the accepted phase angle of a capacitive circuit. This rotation is shown in Fig. 14-5. Figure 14-5*a* is in the same position as Fig. 14-4*c*, while Fig. 14-5*b* is Fig. 14-5*a* rotated counterclockwise through the angle ϕ.

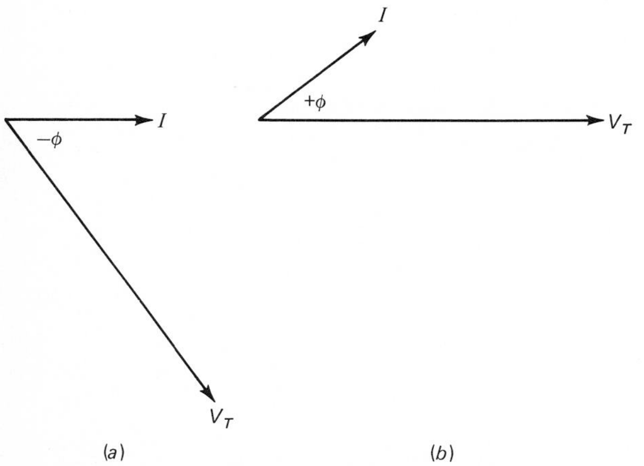

Fig. 14-5 (*a*) Current and voltage phasors in Fig. 14-4*c*; (*b*) phasors rotated $\phi°$ counterclockwise.

In Fig. 14-6*a* the phasor diagram in Fig. 14-5*b* is repeated with the addition of the horizontal and vertical components of the current.

The current I and its horizontal and vertical components, $I \cos \phi$ (resistive) and $jI \sin \phi$ (reactive), may each be multiplied by the total circuit voltage V_T to obtain the three powers in current terms. The results of this operation are:

Apparent power: $S = V_T I$ (14-4)

Active power: $P = V_T I \cos \phi$ or $S \cos \phi$ (14-5)

Reactive power: $jQ = jV_T I \sin \phi$ or $jS \sin \phi$ (14-6)

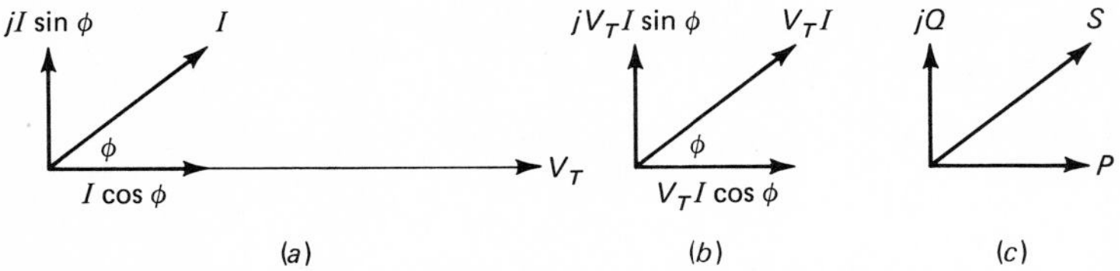

Fig. 14-6 Series *RC* circuit. (*a*) Phasor diagram with voltage reference; (*b*) current component phasors multiplied by the total voltage; (*c*) power phasors.

The powers given in Eqs. (14-4) to (14-6) are shown as phasors in Fig. 14-6*b* and *c*. In Fig. 14-6 the phase angle has a positive sign which is regarded as correct for a capacitive circuit. From this conventional point of view, the sign of the reactive power is positive, thus producing the usual form of the power equation for a capacitive circuit:

$$S = P + jQ \qquad (14\text{-}7)$$

In Eqs. (14-1) to (14-7),

V_T = total network pd, V
V_R = resistive pd, V
V_C = capacitive pd, V
ϕ = phase angle, deg or rad
Z = total circuit impedance, Ω
R = resistance, Ω
X_C = capacitive reactance, Ω
I = common circuit current, A
S = apparent power, VA
P = active power, W
Q = reactive power, var

VOLTAGE, IMPEDANCE, AND POWER TRIANGLES

When shown diagrammatically in the form of triangles, Eqs. (14-1), (14-2), and (14-7) become easier to analyze into polar and rectangular coordinates.

The relationships applicable to a series *RC* circuit can be summarized by reference to Eqs. (14-1), (14-2), and (14-7), and Fig. 14-7*a* to *c*.

Voltage triangle (Fig. 14-7*a*):

$$V_T = V_R - jV_C \qquad (14\text{-}1)$$

$$|V_T| = \sqrt{V_R^2 + V_C^2}$$

$$\phi = -\arctan \frac{V_C}{V_R}$$

$$V_R = V_T \cos \phi$$

$$-jV_C = -jV_T \sin \phi$$

Impedance triangle (Fig. 14-7*b*):

$$Z = R - jX_C \tag{14-2}$$

$$|Z| = \sqrt{R^2 + X_C{}^2}$$

$$\phi = -\arctan \frac{X_C}{R}$$

$$R = Z \cos \phi$$

$$-jX_C = -jZ \sin \phi$$

Power triangle (Fig. 14-7*c*):

$$S = P + jQ \tag{14-7}$$

$$|S| = \sqrt{P^2 + Q^2}$$

$$\phi = \arctan \frac{Q}{P}$$

$$P = S \cos \phi = V_T I \cos \phi$$

$$jQ = jS \sin \phi = jV_T I \sin \phi$$

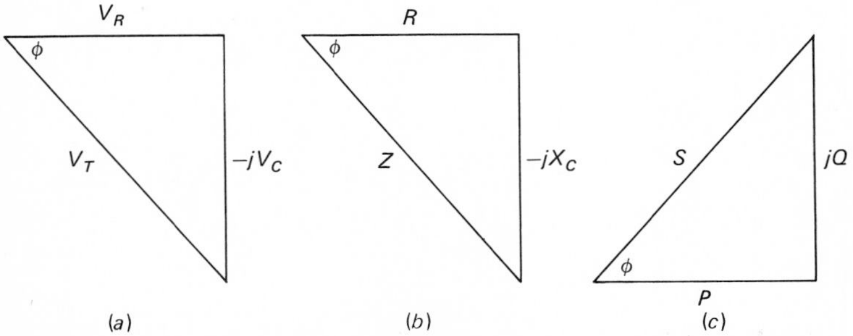

Fig. 14-7 Series *RC* circuit. (*a*) Voltage triangle; (*b*) impedance triangle; (*c*) power triangle.

POWER FACTOR

A useful figure of merit for electric components or systems is the *power factor*, the ratio of active power to apparent power.

For industrial electric systems it is usually desirable to obtain a required level of active power for minimum cost. This objective is achieved by designing the system so that its apparent power demand is as small as possible consistent with its useful active power. Another way of expressing this objective is to say that the aim is to make the apparent power equal to the active power, or to produce unity power factor (power factor = 1).

For some other applications, power factors less than unity may be the aim. For capacitors (and inductors) it is desirable that there be no power dissipation; there should be zero active power, or the apparent power should equal the reactive power. Thus these reactive devices should ideally possess zero power factor.

The power factor is the ratio P/S. Referring to Fig. 14-7*c*, it can be seen that this ratio is the cosine of the phase angle ϕ. The standard symbol for the power factor is $\cos\phi$. Since it is a ratio of powers, the power factor is a dimensionless quantity; it has no unit.

$$\cos\phi = \frac{P}{S} \tag{14-8}$$

From Fig. 14-7*a* and *b*, the power factor can be expressed in terms of voltage or impedance:

$$\cos\phi = \frac{V_R}{V_T} \quad \text{or} \quad \frac{R}{Z} \tag{14-9}$$

In Eqs. (14-8) and (14-9):

$\cos\phi$ = power factor
P = active power, W
S = apparent power, VA
V_R = resistive voltage, V
V_T = total circuit pd, V
R = resistance, Ω
Z = impedance, Ω

Example 14-1 A series circuit consists of a resistance of 1 kΩ and a capacitance of 0.022 μF connected to a 10-V 5-kHz source. What is (*a*) the voltage across the resistance, (*b*) the voltage across the capacitance, and (*c*) the phase angle of the circuit?

Solution See Fig. 14-8.

Fig. 14-8 Example 14-1: (*a*) circuit; (*b*) impedance or voltage triangle.

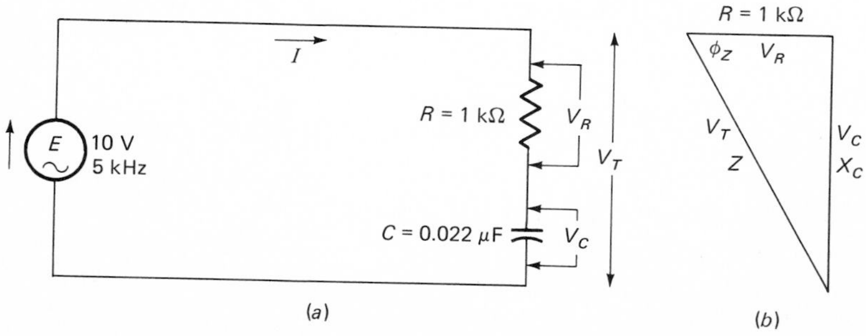

$$\omega = 2\pi f = 2\pi \times 5\ \text{kHz} = 31.42\ \text{krad/s}$$

$$X_C = \frac{1}{\omega C} = \frac{1}{31.42\ \text{krad/s} \times 0.022\ \mu\text{F}} = 1.447\ \text{k}\Omega$$

$$Z = R - jX_C \qquad (14\text{-}2)$$

$$Z = (1 - j1.447)\ \text{k}\Omega = \sqrt{1^2 + 1.447^2}\ \text{k}\Omega \ \angle \arctan\frac{-1.447\ \text{k}\Omega}{1\ \text{k}\Omega}$$

$$Z = 1.759\ \text{k}\Omega\ \angle -55.35^\circ \quad \text{or} \quad \angle -0.966\ \text{rad}$$

$$I = \frac{E}{Z} = \frac{10\ \text{V}\ \angle 0^\circ}{1.759\ \text{k}\Omega\ \angle -55.35^\circ}$$

$$I = 5.685\ \text{mA}\ \angle 55.35^\circ \quad \text{or} \quad \angle 0.966\ \text{rad}$$

a. $$V_R = IR = 5.685\ \text{mA}\ \angle 55.35^\circ \times 1\ \text{k}\Omega$$

$$V_R = \mathbf{5.685\ V\ \angle 55.35^\circ} \quad \text{or} \quad \mathbf{\angle 0.966\ rad}$$

b. $$V_C = I(-jX_C) = 5.685\ \text{mA}\ \angle 55.35^\circ \times 1.447\ \text{k}\Omega\ \angle -90^\circ$$

$$= 5.685\ \text{mA} \times 1.447\ \text{k}\Omega\ \angle 55.35^\circ - 90^\circ \quad \text{or} \quad \angle (0.966 - \pi/2)\ \text{rad}$$

$$V_C = \mathbf{8.226\ V\ \angle -34.65^\circ} \quad \text{or} \quad \mathbf{\angle -0.604\ rad}$$

c. From the phasor expression for the current I above,

$$\boldsymbol{\phi = 55.35^\circ = 0.966\ \text{rad}}$$

It should be noted from Example 14-1 that, if the emf or total voltage is made the reference phasor $E\ \angle 0^\circ$ or $V_T\ \angle 0^\circ$, the current phasor derived from it automatically gives the circuit phase angle. In the example the phase angle is a positive quantity, 55.35°, which is in agreement with the concept that a leading current phase angle exists in a capacitive circuit.

The phasor diagram for Example 14-1, with the total voltage as reference, is given in Fig. 14-9.

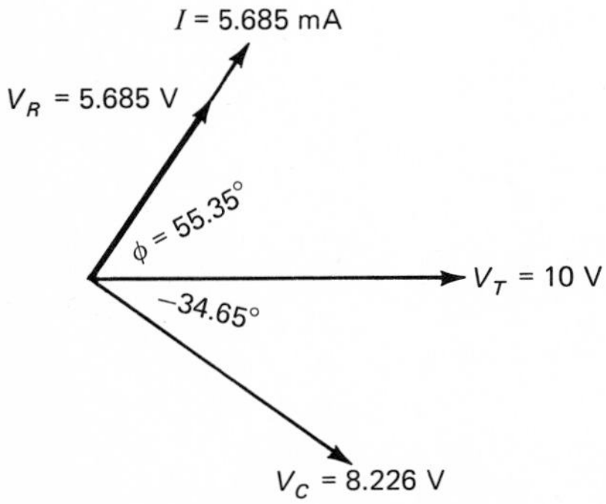

Fig. 14-9 Example 14-1: phasor diagram.

Example 14-2 A 50-pF capacitor has a power factor of 0.0001 at 10 MHz. What is its series equivalent resistance?

Solution

$$\omega = 2\pi f = 2\pi \times 10 \text{ MHz} = 62.83 \text{ Mrad/s}$$

$$X_C = \frac{1}{\omega C} = \frac{1}{62.83 \text{ M rad/s} \times 50 \text{ pF}} = 318.3 \ \Omega$$

$$\cos \phi = \frac{R_s}{Z} \tag{14-9}$$

$$\cos \phi = \frac{R_s}{\sqrt{R_s^2 + X_C^2}} = \frac{R_s}{\sqrt{R_s^2 + (318.3 \ \Omega)^2}}$$

$$\cos^2 \phi \ [R_s^2 + (318.3 \ \Omega)^2] = R_s^2$$

$$10^{-8}[R_s^2 + (318.3 \ \Omega)^2] = R_s^2$$

$$10^{-8} R_s^2 + 10^{-8}(318.3 \ \Omega)^2 = R_s^2$$

$$R_s \simeq \sqrt{(318.3 \ \Omega)^2 \times 10^{-8}} \simeq 318.3 \ \Omega \times 10^{-4} \simeq \mathbf{0.0318 \ \Omega}$$

From the results for Example 14-2, the capacitor appears to have a series equivalent circuit as shown in Fig. 14-10.

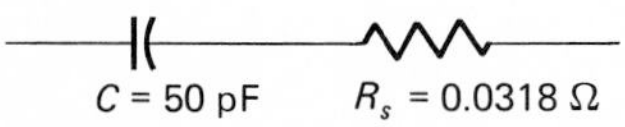

Fig. 14-10 Example 14-2: equivalent series circuit of the capacitor.

The term "equivalent" is used to qualify both the resistance and the circuit:

1. R_s is that value of resistance which, if connected in series with a theoretically perfect 50-pF capacitor, would result in the same power dissipation as occurs within this real imperfect capacitor. The components contributing to the power dissipation of the capacitor (connecting wire resistance, dielectric losses, radiation, etc.) may not all be series elements. However, as will be shown in Sec. 14-5, the series equivalent of any parallel element can be found, so that the grouping of all power-dissipating elements into a single equivalent series resistance is viable.
2. A parallel circuit for the capacitor can be derived by a method described in Sec. 14-5. Both the series and parallel circuits simulate the behavior of the capacitor, but neither can be said to be its exact replica. Each circuit is the electrical equivalent of the capacitor.

Example 14-3 A 500-W 120-V heater is to be operated from a 230-V 50-Hz supply. Assuming that the resistance of the heater remains constant with varying temperature, what value of capacitor could be used to drop the unwanted voltage?

Solution

For the heater alone on 120 V (Fig. 14-11*a*),

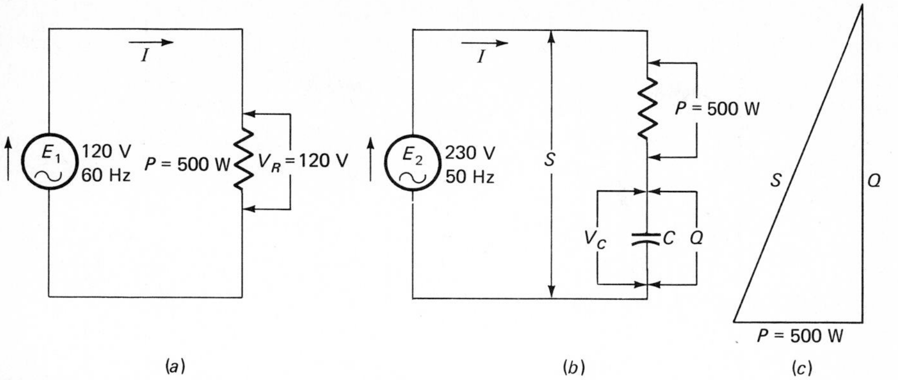

Fig. 14-11 Example 14-3: (*a*) circuit of the heater operating on 120 V; (*b*) circuit of the heater operating on 230 V; (*c*) phasor diagram of the heater operating on 230 V.

$$P = V_R I$$

$$I = \frac{P}{V_R} = \frac{500\ \text{W}}{120\ \text{V}} = 4.167\ \text{A}$$

For the heater and the capacitor on 230 V (Fig. 14-11*b* and *c*),

$$S = E_2 I = 230\ \text{V} \times 4.167\ \text{A} = 958.4\ \text{VA}$$

$$Q = \sqrt{S^2 - P^2} = \sqrt{(958.4\ \text{VA})^2 - (500\ \text{W})^2} = 817.6\ \text{var}$$

The reactive power Q is $V_C I$. By applying Ohm's law, $V_C = IX_C$, Q can be expressed as $I^2 X_C$:

$$Q = I^2 X_C$$

$$X_C = \frac{Q}{I^2} = \frac{817.6\ \text{var}}{(4.167\ \text{A})^2} = 47.09\ \Omega$$

$$X_C = \frac{1}{\omega C} = \frac{1}{2\pi f C}$$

$$C = \frac{1}{2\pi f X_C} = \frac{1}{2\pi \times 50\ \text{Hz} \times 47.09\ \Omega} = \mathbf{67.6\ \mu F}$$

Example 14-3 reveals some important ac circuit principles:

1. Capacitive reactance can be used as a series impedance to operate resistive equipment from an ac voltage higher than that for which it is designed. This result is achieved without dissipating extra power. In the example the heater is intended to operate on 120-V 60-Hz North American supply but is to be used on the 230-V 50-Hz supply common in Europe and other countries.
2. The frequency is only significant as far as the reactance is concerned. A change in frequency does not affect resistive equipment such as heaters.

3. The constancy of resistance with temperature referred to in the example is important. The resistance of a heater, or other resistive equipment, can rise considerably between the instant it is switched on cold and when it is operating hot. The reactance of the capacitor will not change appreciably with temperature. Any calculation, such as the example solution, is only valid for a given value of resistance, and hence for a given temperature.
4. If a reactive element is added in series with a resistive element to operate from a higher voltage, the active power can remain unchanged but the apparent power will increase (or the power factor decrease).

Example 14-4 A 100-V 400-Hz generator has an internal impedance equivalent to a resistance of 5 Ω. If it supplies a synchronous motor whose equivalent impedance is a 40-Ω capacitive reactance in series with a 10-Ω resistance, what is the magnitude of the motor's voltage? Refer to the diagram in Fig. 14-12.

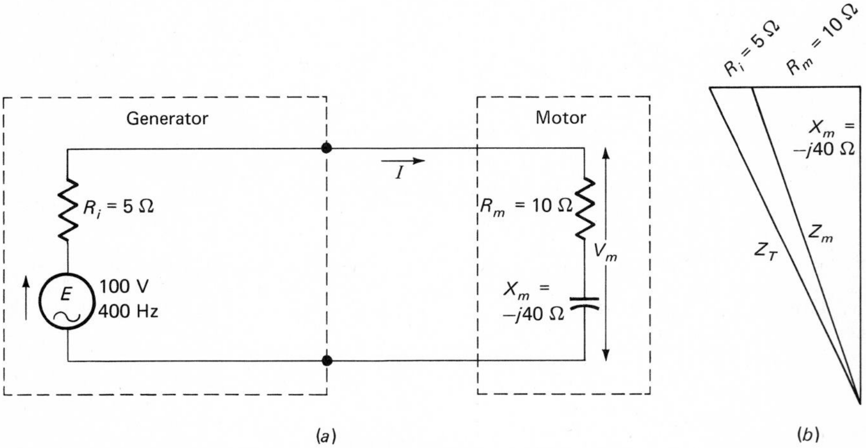

Fig. 14-12 Example 14-4: (*a*) circuit; (*b*) impedance diagram.

Solution The impedance of the motor is, from Eq. (14-2),

$$Z_m = R_m - jX_m = (10 - j40)\ \Omega \qquad \text{or} \qquad |Z_m| = 41.23\ \Omega$$

The impedance of the generator is its internal resistance:

$$Z_g = R_i = (5 + j0)\ \Omega$$

The total impedance of the circuit is the phasor sum of the motor and generator impedances:

$$Z_T = Z_g + Z_m$$

$$= (5 + j0)\ \Omega + (10 - j40)\ \Omega = (15 - j40)\ \Omega \qquad \text{or} \qquad |Z_T| = 42.72\ \Omega$$

$$|I| = \frac{|E|}{|Z_T|} = \frac{100\ \text{V}}{42.72\ \Omega} = 2.34\ \text{A}$$

$$|V_m| = |I||Z_m| = 2.34\ \text{A} \times 41.23\ \Omega = \mathbf{96.48\ V}$$

In Example 14-4 the load receives 96.48 percent of the available emf. If this were a dc circuit in which the generator resistance was 5 Ω and the load resistance was 10 Ω + 40 Ω = 50 Ω, the load would receive only 90.9 percent of the available emf. This difference is inherent in the difference between ac phasor methods and dc algebraic methods of calculation.

Example 14-5 The network in Fig. 14-13*a* is part of a transistor stereo amplifier circuit. If the ac voltage delivered to the *RC* circuit by transistor Q_1 is 5 V at any audio frequency, what voltage is passed on to transistor Q_2 at 20 Hz, 1 kHz, and 20 kHz? Assume that the input impedance to transistor Q_2 is infinite, so that there is no loading effect across resistor *R*.

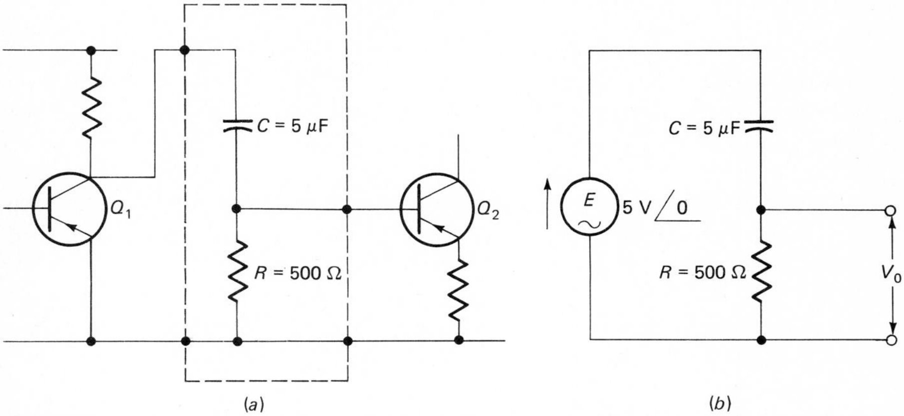

Fig. 14-13 Example 14-5: (*a*) given circuit; (*b*) equivalent circuit.

Solution Figure 14-13*b* shows the equivalent circuit of Fig. 14-13*a* from the viewpoint of the series *RC* circuit. Transistor Q_1 is represented by the 5-V emf generator. V_0 is the voltage supplied to transistor Q_2 if loading effects are ignored.

In this type of situation it is not necessary to calculate the current. The unloaded voltage divider principle applies to ac if a phasor approach is adopted.

$$V_0 = E\frac{R}{Z}$$

a. When $f = 20$ Hz,

$$\omega = 2\pi f = 2\pi \times 20 \text{ Hz} = 125.7 \text{ rad/s}$$

$$X_C = \frac{1}{\omega C} = \frac{1}{125.7 \text{ rad/s} \times 5\ \mu\text{F}} = 1.59 \text{ k}\Omega$$

$$Z = R - jX_C = (0.5 - j1.59) \text{ k}\Omega = 1.667 \text{ k}\Omega \underline{/-72.53^\circ} \quad \text{or} \quad \underline{/-1.265 \text{ rad}}$$

$$V_0 = E\frac{R}{Z} = 5 \text{ V} \underline{/0^\circ} \frac{500\ \Omega\underline{/0^\circ}}{1.667 \text{ k}\Omega \underline{/-72.53^\circ}} = \mathbf{1.5\ V\ \underline{/72.53^\circ}} \quad \text{or} \quad \mathbf{\underline{/1.265\ rad}}$$

b. When $f = 1$ kHz,

$$\omega = 2\pi f = 2\pi \times 1 \text{ kHz} = 6.283 \text{ krad/s}$$

$$X_C = \frac{1}{\omega C} = \frac{1}{6.283 \text{ krad/s} \times 5\ \mu\text{F}} = 31.83\ \Omega$$

$$Z = R - jX_C = (500 - j31.83)\ \Omega = 501\ \Omega \underline{/-3.65^\circ} \quad \text{or} \quad \underline{/-0.0637 \text{ rad}}$$

$$V_0 = E\frac{R}{Z} = 5\ \text{V}\ \underline{/0^\circ}\ \frac{500\ \Omega\underline{/0^\circ}}{501\ \Omega\underline{/-3.65^\circ}}$$

$$V_0 = \mathbf{4.99\ V}\ \underline{\mathbf{/3.65^\circ}} \quad \text{or} \quad \underline{\mathbf{/0.0637\ rad}}$$

c. When $f = 20$ kHz,

$$\omega = 2\pi \times 20\ \text{kHz} = 125.7\ \text{krad/s}$$

$$X_C = \frac{1}{\omega C} = \frac{1}{125.7\ \text{krad/s} \times 5\ \mu\text{F}} = 1.59\ \Omega$$

$$Z = R - jX_C = (500 - j1.59)\ \Omega \simeq 500\ \Omega\ \underline{/0^\circ}$$

$$V_0 = E\frac{R}{Z} \simeq 5\ \text{V}\ \underline{/0^\circ}\ \frac{500\ \Omega\ \underline{/0^\circ}}{500\ \Omega\ \underline{/0^\circ}} \simeq \mathbf{5\ V}\ \underline{\mathbf{/0^\circ}}$$

The results for Example 14-4 show that, for a series *RC* circuit as a coupling network to transmit ac between two other networks, voltage divider principles apply with the following effects:

1. When the frequency is low, such that the capacitive reactance is high compared with the resistance, the voltage across the resistance is low and it leads the applied voltage by an appreciable angle.
2. For high frequencies, where the capacitive reactance is low compared with the resistance, the voltage across the resistance is almost equal to the supply voltage and in phase with it.

When the value of V_0 and its phase angle relative to E are plotted over the complete audio-frequency spectrum from 20 Hz to 20 kHz, the results are as shown in Fig. 14-14.

Fig. 14-14 Example 14-5: frequency response; (*a*) magnitude; (*b*) phase.

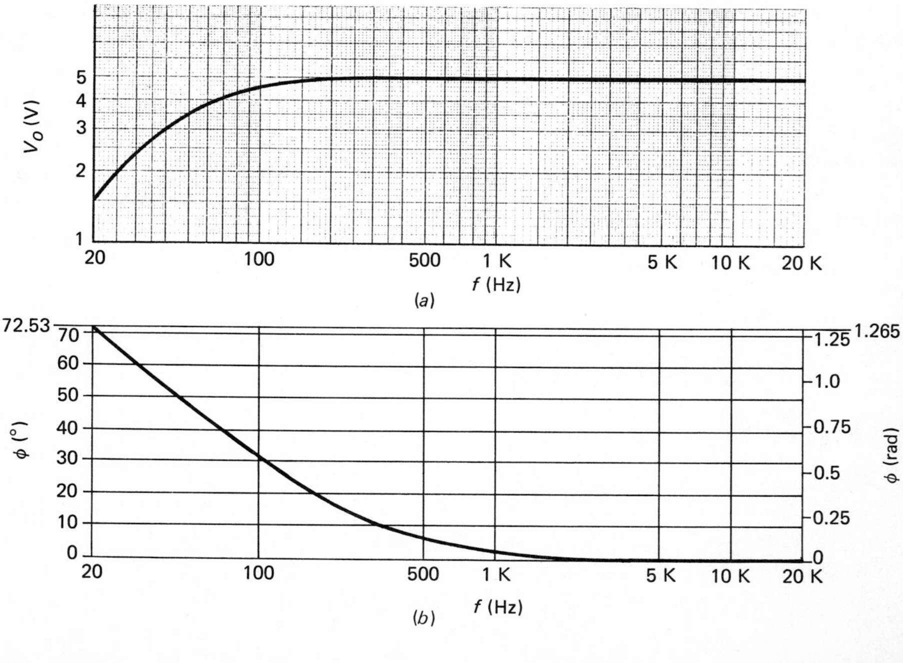

In Fig. 14-14 the horizontal frequency axis is logarithmic (equal distances represent equal frequency ratios). This method of display is commonly used so that all parts of the spectrum are shown with equal clarity. For instance, the frequency range 100 to 500 Hz occupies the same space as the 1 to 5 kHz range. In Fig. 14-14*a* the vertical voltage axis is also shown as logarithmic for the same basic reason.

In the case of audio-frequency equipment there is another reason for using logarithmic scaling for frequency and magnitude. The human ear tends to respond to changes in frequency (pitch) and sound level in a logarithmic manner. There is no equivalent relationship for phase angle magnitude; and it is usual to employ linear scaling, as shown on the vertical axis in Fig. 14-14*b*.

Example 14-6 An electric motor connected to a 120-V ac supply has a lagging power factor of 0.7. When a power factor improvement circuit is added (as discussed in Chap. 16), the resulting total load has an impedance of $(1 - j0.2)\ \Omega$. What is the (*a*) power factor, (*b*) apparent power, (*c*) active power, and (*d*) reactive power of the total load?

Solution Figure 14-15 shows (*a*) a block diagram of the motor together with the power factor improvement circuit, (*b*) the resulting equivalent circuit, (*c*) its impedance triangle, and (*d*) its power triangle.

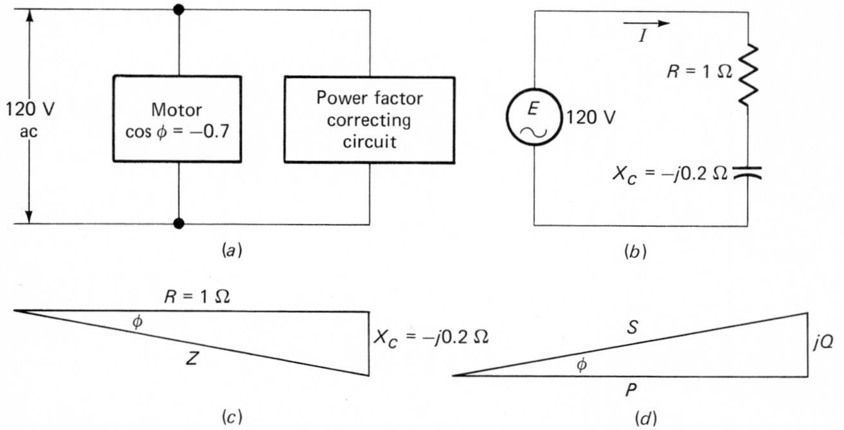

Fig. 14-15 Example 14-6: (*a*) block diagram; (*b*) equivalent circuit; (*c*) impedance triangle; (*d*) power triangle.

From Fig. 14-15*b* and *c*,

$$Z = R - jX_C = (1 - j0.2)\ \Omega = 1.02\ \Omega\ \underline{/-11.32^\circ} \quad \text{or} \quad \underline{/-0.197\ \text{rad}}$$

$$I = \frac{E}{Z} = \frac{120\ \text{V}\ \underline{/0^\circ}}{1.02\ \Omega\ \underline{/-11.32^\circ}} = 117.6\ \text{A}\ \underline{/11.32^\circ} \quad \text{or} \quad \underline{/0.197\ \text{rad}}$$

a. $$\cos\phi = \frac{R}{Z} \tag{14-9}$$

$$= \frac{1\ \Omega}{1.02\ \Omega} = \mathbf{0.9804}$$

Since the circuit is capacitive,

$$\cos \phi = \mathbf{0.9804\ leading}$$

b. $$S = I^2Z = (117.6\ \text{A})^2 \times 1.02\ \Omega = \mathbf{14\,110\ VA} \quad \text{or} \quad \mathbf{14.11\ kVA}$$

c. From Fig. 14-15*d* and Eq. (14-5),

$$P = S \cos \phi = 14\,110\ \text{VA} \times 0.9804 = \mathbf{13\,830\ W} \quad \text{or} \quad \mathbf{13.83\ kW}$$

d. From Eq. (14-6),

$$Q = \sqrt{S^2 - P^2} = \sqrt{14\,110^2 - 13\,830^2} = \mathbf{2780\ var} \quad \text{or} \quad \mathbf{2.78\ kvar}$$

In Example 14-6 the solutions have been given as scalar values, except for the power factor where it is indicated that it is leading. This use of scalar values is common when referring to ac circuit powers, because a knowledge of magnitudes and the phase angle in terms of the power factor is usually all that is required. Phasor solutions give the desired results but require more work:

a. Power factor, being the ratio of two similar quantities, is not itself a phasor quantity. Therefore the solution given in the example is complete.

b.

$$S = I^2Z = (117.6\ \text{A}\ \underline{/11.32^\circ})^2 \times 1.02\ \Omega\ \underline{/-11.32^\circ}$$

$$= (117.6\ \text{A})^2 \times 1.02\ \Omega\ \underline{/(2 \times 11.32^\circ) + (-11.32^\circ)}$$

$$S = 14.11\ \text{kVA}\ \underline{/11.32^\circ}$$

This expression for S gives the correct apparent power magnitude and its position in Fig. 14-15*d*.

c. In rectangular form,

$$S = 14.11 \cos 11.32^\circ + j14.11 \sin 11.32^\circ = (13.83 + j2.78)\ \text{kVA}$$

The rectangular expression for S gives the resistive and reactive components directly. Therefore,

$$P = 13.83\ \text{kW}$$

P is the real part of the expression for S, which is compatible with the reference position of P in Fig. 14-15*d*.

d. The reactive power can also be obtained directly from the rectangular expression for apparent power:

$$Q = +j2.78\ \text{kvar}$$

Q is the imaginary part of S, and so is correctly located in the $+j$ position in Fig. 14-15*d*.

Power factor improvement operates only on the reactive power. The active power, 13.83 kW, is the same before and after the addition of the power factor improvement circuit. The apparent power prior to improvement is 19.75 kVA ($S = P/\cos \phi = 13.83\ \text{kW}/0.7$). The power factor improvement circuit thus gives a 5.64-kVA (28.6 percent) reduction in apparent power.

14-2 SERIES *RL* CIRCUIT

The series *RL* circuit behaves in a manner which is the dual of the series *RC* circuit. Phasor quantities such as total voltage and impedance, which have real and imaginary parts, are conjugate[1] for each type of circuit. For example, the impedance of a series *RL* circuit is $R + jX$, while the impedance of a series *RC* circuit is $R - jX$. Thus the outstanding distinction between the two types of circuits is the sign of the imaginary part of the related rectangular expressions (or the sign of the phase angle).

The general characteristics of the series *RL* circuit are so similar to those of the series *RC* circuit that the treatment of the series *RL* circuit is essentially the same as that given in Sec. 14-1. Thus the following discussion summarizes the characteristics of the *RL* series circuit and notes the points of difference between it and the series *RC* circuit.

The common current I in the circuit in Fig. 14-16*a* is made the reference phasor $I \underline{/0°}$ or $I + j0$ in Fig. 14-16*c*.

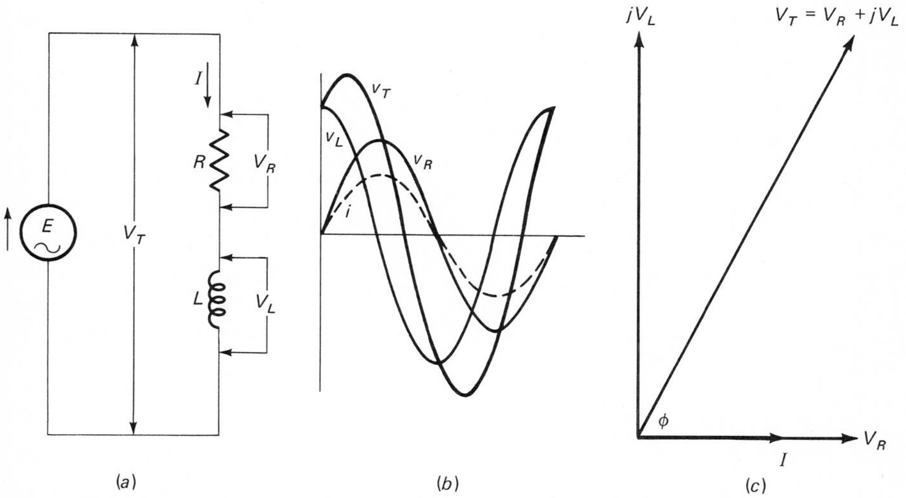

Fig. 14-16 Series *RL* circuit diagrams. (*a*) Schematic; (*b*) waveform; (*c*) phasor.

VOLTAGES

The resistive voltage V_R is in phase with the circuit current I and thus has the phasor value $V_R \underline{/0°}$ or $V_R + j0$.

The inductive voltage V_L leads the current by 90°, giving a phasor value of $V_L \underline{/90°}$ or $0 + jV_L$.

The total circuit pd V_T is equal to the phasor sum of the resistive and inductive voltages:

$$V_T = V_R + V_L = (V_R + j0) + (0 + jV_L)$$

$$V_T = V_R + jV_L \qquad (14\text{-}10)$$

[1] *Conjugate* quantities are equal except for the sign of the imaginary part. Thus $a + jb$ and $a - jb$ are conjugate, as are $5 + j4$ and $5 - j4$.

IMPEDANCES

The application of Ohm's law to each voltage in Eq. (14-10) gives

$$IZ = IR + jIX_L$$

Dividing each term of this equation by the common factor I results in an equation for the series RL circuit in terms of impedances:

$$Z = R + jX_L \tag{14-11}$$

POWERS

If each voltage in Eq. (14-10) is multiplied by the common current I, a power equation for the series RL circuit is obtained:

$$V_T I = V_R I + jV_L I \tag{14-12}$$

Equation (14-12) can be written in terms of apparent power, active power, and reactive power:

$$S = P + jQ$$

These power equations for the RL circuit have been derived from the phasor relationships in Fig. 14-16*c* which refers voltages to the common current. The conventional approach to ac powers is based on the circuit phase angle which refers current to voltage. Rotating Fig. 14-16*c* clockwise through an angle of ϕ degrees will place the total-voltage phasor in the required reference position. This rotation is shown in Fig. 14-17*a* and *b*.

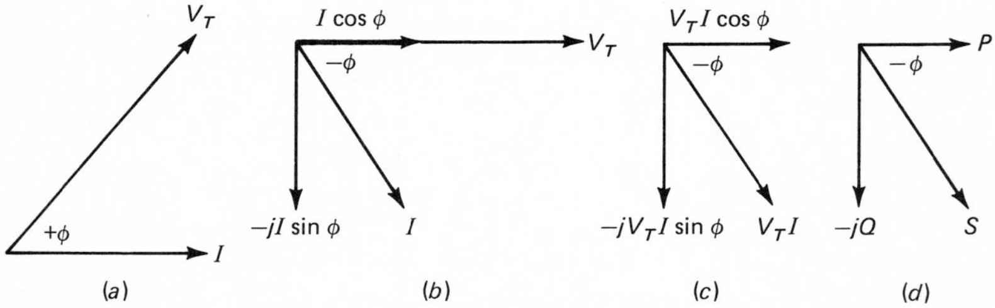

Fig. 14-17 Series RL circuit phasor diagrams. (*a*) and (*b*) Voltage and current; (*c*) and (*d*) power.

Figure 14-17*b* shows the current I lagging the total voltage V_T by the angle ϕ. This view gives the phase angle $-\phi$ which is usually associated with an inductive circuit. The rectangular components of the current, $I \cos \phi$ (resistive) and $-jI \sin \phi$ (reactive), are also given in Fig. 14-17*b*.

Multiplying each current phasor in Fig. 14-17*b* by the total voltage gives expressions for each of the circuit powers. The resulting power phasors are shown in Fig. 14-17*c* and *d*.

The individual power equations for the series RL circuit, as obtained from Fig. 14-17*c* and *d* are:

Apparent power: $$S = V_T I \tag{14-13}$$

Active power: $$P = V_T I \cos \phi = S \cos \phi \tag{14-14}$$

Reactive power: $$-jQ = -jV_T I \sin \phi = -jS \sin \phi \tag{14-15}$$

Figure 14-17*d* uses the standard ac power symbols and shows that, from the conventional viewpoint, the power equation for a series RL circuit is

$$S = P - jQ \tag{14-16}$$

In Eqs. (14-10) to (14-16),

V_T = total network pd, V
V_R = resistive pd, V
V_L = inductive pd, V
ϕ = phase angle, deg or rad
Z = total circuit impedance, Ω
R = resistance, Ω
X_L = inductive reactance, Ω
I = common circuit current, A
S = apparent power, VA
P = active power, W
Q = reactive power, var

VOLTAGE, IMPEDANCE, AND POWER TRIANGLES

Figure 14-18*a* to *c* shows Eqs. (14-10), (14-11), and (14-16) in triangle form. The series RL circuit relationships can conveniently be summarized from these triangles:

Fig. 14-18 Series RL circuit. (*a*) Voltage triangle; (*b*) impedance triangle; (*c*) power triangle.

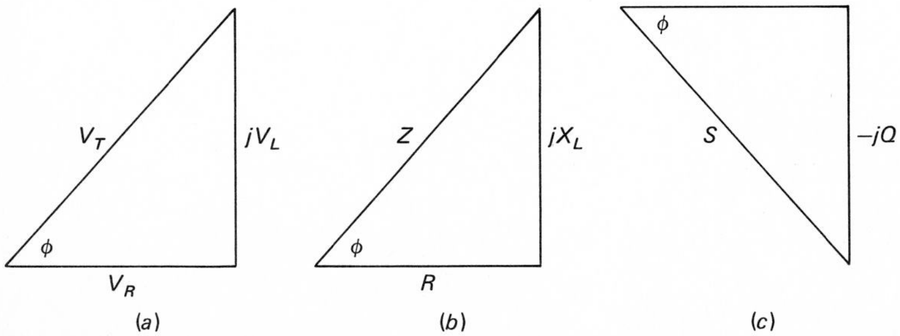

Voltage triangle (Fig. 14-18*a*):

$$V_T = V_R + jV_L \tag{14-10}$$
$$|V_T| = \sqrt{V_R^2 + V_L^2}$$
$$\phi = \arctan \frac{V_L}{V_R}$$
$$V_R = V_T \cos \phi$$
$$jV_L = jV_T \sin \phi$$

Impedance triangle (Fig. 14-18*b*):

$$Z = R + jX_L \tag{14-11}$$
$$|Z| = \sqrt{R^2 + X_L^2}$$
$$\phi = \arctan \frac{X_L}{R}$$
$$R = Z \cos \phi$$
$$jX_L = jZ \sin \phi$$

Power triangle (Fig. 14-18*c*):

$$S = P - jQ \tag{14-16}$$
$$|S| = \sqrt{P^2 + Q^2}$$
$$\phi = -\arctan \frac{Q}{P}$$
$$P = S \cos \phi = V_T I \cos \phi$$
$$-jQ = -jS \sin \phi = -jV_T I \sin \phi$$

POWER FACTOR AND Q FACTOR

The power factor $\cos \phi$ of any circuit is the ratio of active power to apparent power, P/S. Electric motors, transformers, and electromagnetic controls constitute the bulk of industrial electrical equipment. This equipment is usually inductive, and thus the total industrial load includes inductive reactive power, resulting in a lagging power factor. To obtain the best service rates from an electric supply utility, the reactive power should be zero, the active power and the apparent power being equal. This would give unity power factor. Industrial circuits usually deviate widely from this ideal situation, often justifying the power factor improvement techniques referred to in Chap. 16.

From the above, or from Fig. 14-18*c*,

$$\cos \phi = \frac{P}{S} \tag{14-17}$$

From Fig. 14-18*a* and *b*,

$$\cos \phi = \frac{V_R}{V_T} = \frac{R}{Z} \tag{14-18}$$

Inductors are used for a number of purposes in electrical and electronic equipment. Here the objective may be to obtain zero power factor, or reactive power without active power. This result cannot be achieved in practice, because inductors which do not dissipate power cannot be designed and manufactured. A frequently used figure of merit for an inductor is its Q factor. This term has been used for many years and originally meant quality. Today it is best to regard it as being related to the Q, or quadrature, component of power. An inductor with a perfect Q factor (Q = infinity) does not dissipate power; it has zero active power P. Its reactive power Q is the only power component and accounts for all the apparent power. Relating the quality factor to quadrature power in this manner should help to dispel some of the confusion caused by the fact that the symbol Q has two meanings.

Q factor is defined as the ratio of stored energy to dissipated energy.

$$Q^* = \frac{W_{\text{stored}}}{W_{\text{dissipated}}} = \frac{\text{reactive power} \times \text{time}}{\text{active power} \times \text{time}} = \frac{Qt}{Pt}$$

$$Q = \frac{\text{reactive power}}{\text{active power}} = \frac{Q}{P} = \frac{I^2 X_L}{I^2 R}$$

$$Q = \frac{X_L}{R} = \tan \phi \tag{14-19}$$

Substituting ωL for X_L in Eq. (14-19),

$$Q = \frac{\omega L}{R} \tag{14-20}$$

In Eqs. (14-17) to (14-20),

$\cos \phi$ = circuit power factor
P = active power, W
S = apparent power, VA

* Q on the left side of the following equations is Q factor. On the right side of the equations Q is reactive power.

V_R = resistive voltage, V
V_T = total circuit voltage, V
R = resistance, Ω
X_L = inductive reactance, Ω
Z = total circuit impedance, Ω
Q = circuit quality or quadrature factor
or Q = reactive power, var
W = energy, J
t = time, s
I = circuit current, A
$\tan \phi$ = tangent of the circuit phase angle
ω = angular velocity or angular frequency, rad/s
L = inductance, H

Example 14-7 A series circuit consists of a resistance of 1 kΩ and an inductance of 50 mH connected to a 10-V 5-kHz source. What is (*a*) the voltage across the resistance, (*b*) the voltage across the inductance, and (*c*) the phase angle of the circuit?
Solution See Fig. 14-19.

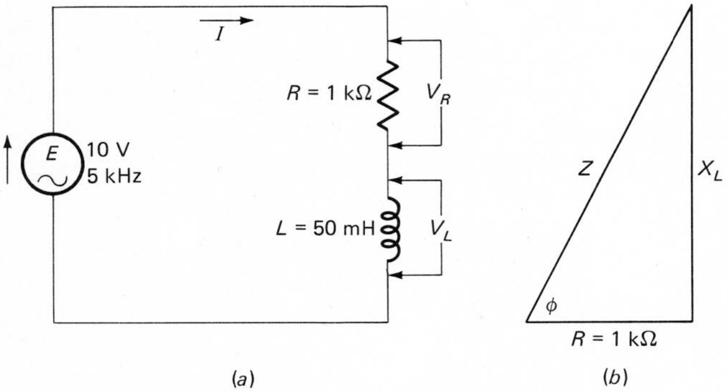

Fig. 14-19 Example 14-7: (*a*) circuit; (*b*) impedance triangle.

$$\omega = 2\pi f = 2\pi \times 5\text{ kHz} = 31.42\text{ krad/s}$$

$$X_L = \omega L = 31.42\text{ krad/s} \times 50\text{ mH} = 1.571\text{ k}\Omega$$

$$Z = R + jX_L \tag{14-11}$$

$$= (1 + j1.571)\text{ k}\Omega = \sqrt{1^2 + 1.571^2}\text{ k}\Omega \underline{/\arctan \tfrac{1.571}{1}}$$

$$Z = 1.862\text{ k}\Omega \underline{/57.5°} \quad \text{or} \quad \underline{/1.003\text{ rad}}$$

$$I = \frac{E}{Z} = \frac{10\text{ V}\underline{/0°}}{1.862\text{ k}\Omega\underline{/57.5°}}$$

$$I = 5.37\text{ mA}\underline{/-57.5°} \quad \text{or} \quad \underline{/-1.003\text{ rad}}$$

a. $$V_R = IR = 5.37\text{ mA}\underline{/-57.5°} \times 1\text{ k}\Omega$$

$$V_R = \mathbf{5.37\text{ V}\underline{/-57.5°}} \quad \text{or} \quad \mathbf{\underline{/-1.003\text{ rad}}}$$

b.

$$V_L = I(jX_L) = 5.37\ \text{mA}\ \underline{/-57.5^\circ} \times 1.57\ \text{k}\Omega\ \underline{/90^\circ}$$
$$= 5.37\ \text{mA} \times 1.571\ \text{k}\Omega\ \underline{/-57.5^\circ + 90^\circ}$$
$$V_L = \mathbf{8.43\ V}\ \underline{\mathbf{/32.5^\circ}} \quad \text{or} \quad \underline{\mathbf{/0.5678\ rad}}$$

c. From the phasor expression for the current I above,

$$\phi = \mathbf{-57.5^\circ = -1.003\ rad}$$

The phasor diagram for Example 14-7 with the total voltage as the reference phasor is shown in Fig. 14-20.

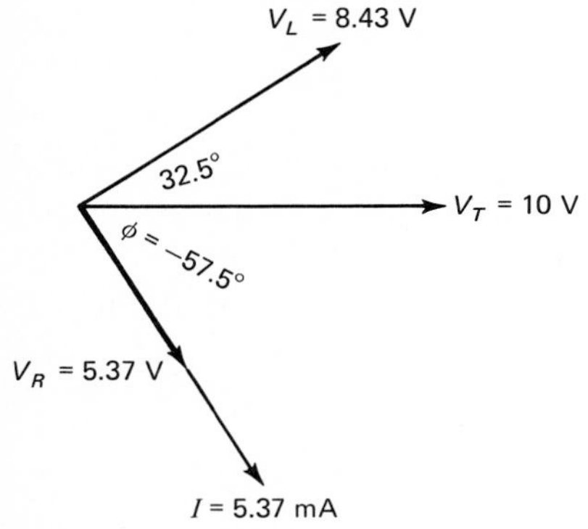

Fig. 14-20 Example 14-7: phasor diagram.

Example 14-8 A 17.5-μH radio tuning coil has a Q of 200 at 5 MHz. What is its (*a*) series equivalent resistance and (*b*) power factor?
Solution

$$\omega = 2\pi f = 2\pi \times 5\ \text{MHz} = 31.42\ \text{Mrad/s}$$
$$X_L = \omega L = 31.42\ \text{Mrad/s} \times 17.5\ \mu\text{H} = 549.9\ \Omega$$

a.

$$Q = \frac{X_L}{R} \qquad (14\text{-}19)$$
$$R = \frac{X_L}{Q} = \frac{549.9\ \Omega}{200} = \mathbf{2.749\ \Omega}$$

b.

$$Q = \tan\phi \qquad (14\text{-}19)$$
$$\phi = \arctan Q = \arctan 200 = 89.7^\circ = 1.565\ \text{rad}$$
$$\cos\phi = \mathbf{\cos 89.7^\circ = 0.0052}$$

In Example 14-8 the power factor is almost equal to the reciprocal of Q ($\frac{1}{200} = 0.005$). In an efficient inductor, the equivalent series resistance is so low compared with the reactance that the impedance is approximately the same as the reactance, giving this result.

For most purposes a practical inductor will have a Q value of at least 10. If an inductor is at least that good, the power factor will equal the reciprocal of the Q within 0.5 percent.

Unless otherwise specified, the resistance of an inductor is its total effective resistance. This resistance accounts for the many different forms in which power dissipation occurs in a practical inductor. They are:

1. An inductor is basically a coil of wire. The ordinary resistance of the wire is called the inductor's *ohmic resistance* or its *dc resistance.*
2. If the frequency is high, current tends to concentrate on the surface of a conductor rather than be evenly distributed throughout its cross section. This is called *skin effect* and acts to increase the resistance of an inductor's coil to a value higher than the ohmic value.
3. If the inductor has an iron core, currents will be developed within the core because it is a conductor in a varying magnetic field. These currents absorb energy from the inductor. This energy absorption is called *eddy current loss.*
4. The magnetization of an iron core does not exactly follow the variations in coil current. This lag between cause and effect is called *hysteresis* and constitutes a dissipation of power within an iron-cored inductor.
5. If the frequency is very high, the insulating materials used in the coil and the core can dissipate power in the form of *dielectric loss.*
6. At high frequencies energy may be emitted in the form of radio waves. This emission is called *radiation loss.*

The various forms of power dissipation which may be encountered in an inductor are listed to give reasons why there can be a large disparity between the resistance measured by an ohmmeter (dc) and the resistance determined by the phase angle (ac).

Example 14-9 Measurements made on an iron-cored inductor gave the following results:

	60 Hz	1 kHz
E	5 V	5 V
I	80 mA	5 mA
$\cos\phi$	0.27	0.34

What is the effective resistance and inductance of the inductor at 60 Hz and at 1 kHz?

Solution

a. At 60 Hz (Fig. 14-21*a*),

$$Z_1 = \frac{E}{I} = \frac{5\text{ V}}{80\text{ mA}} = 62.5\ \Omega$$

$$\cos\phi_1 = \frac{R_{L1}}{Z_1} \qquad (14\text{-}18)$$

$$\frac{R_{L1}}{62.5\ \Omega} = 0.27$$

$$R_{L1} = \mathbf{16.88\ \Omega}$$

$$\omega_1 = 2\pi f_1 = 2\pi \times 60 \text{ Hz} = 377 \text{ rad/s}$$

$$Z_1 = \sqrt{R_{L1}^2 + X_{L1}^2} = 62.5\ \Omega$$

$$R_{L1}^2 + (\omega_1 L_1)^2 = (62.5\ \Omega)^2$$

$$(16.88\ \Omega)^2 + (377 \text{ rad/s})^2\ L_1^2 = (62.5\ \Omega)^2$$

$$L_1 = \sqrt{\frac{(62.5\ \Omega)^2 - (16.88\ \Omega)^2}{(377 \text{ rad/s})^2}} = 0.1596 \text{ H} \simeq \mathbf{0.16\ H}$$

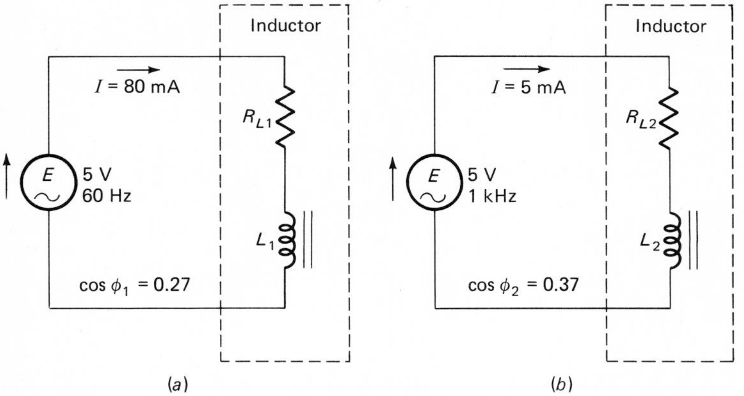

Fig. 14-21 Example 14-9: (*a*) circuit with 60-Hz values; (*b*) circuit with 1-kHz values.

b. At 1 kHz (Fig. 14-21*b*),

$$Z_2 = \frac{E}{I} = \frac{5 \text{ V}}{5 \text{ mA}} = 1 \text{ k}\Omega$$

$$\cos \phi_2 = \frac{R_{L2}}{Z_2}$$

$$\frac{R_{L2}}{1000\ \Omega} = 0.34$$

$$R_{L2} = \mathbf{340\ \Omega}$$

$$\omega_2 = 2\pi f_2 = 2\pi \times 1000 \text{ Hz} = 6283 \text{ rad/s}$$

$$Z_2 = \sqrt{R_{L2}^2 + X_{L2}^2} = 1000\ \Omega$$

$$R_{L2}^2 + (\omega_2 L_2)^2 = (1000\ \Omega)^2$$

$$(340\ \Omega)^2 + (6283 \text{ rad/s})^2\ L_2^2 = (1000\ \Omega)^2$$

$$L_2 = \sqrt{\frac{(1000\ \Omega)^2 - (340\ \Omega)^2}{(6283 \text{ rad/s})^2}} = 0.1497 \text{ H} \simeq \mathbf{0.15\ H}$$

The data given for Example 14-9 are typical of the values to be expected from an iron-cored inductor. They show the substantial increase in effective resistance as the frequency becomes higher.

Inductance is usually considered to be constant over a wide frequency spectrum. Some reduction in inductance can occur at higher frequencies as the result of a decrease in the incremental permeability of the core material and a change in effective wire dimensions caused by skin effect.

Example 14-10 A 500-W 120-V heater is to be operated from a 230-V 50-Hz supply. Assuming that the resistance of the heater remains constant with varying temperature, what value of inductance could be used to drop the unwanted voltage?

Solution This example is the same as Example 14-3, except for the use of an inductance instead of a capacitance. Therefore most of the data are the same. See Fig. 14-22.

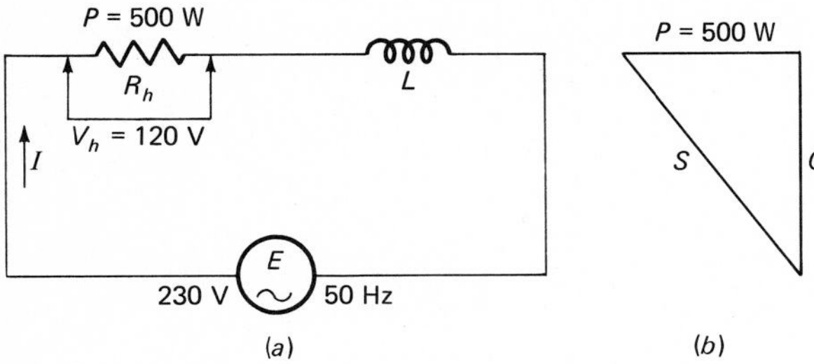

Fig. 14-22 Example 14-10: (*a*) circuit; (*b*) power triangle.

$$X_L = 47.09\ \Omega$$

$$X_L = \omega L = 2\pi f L$$

$$L = \frac{X_L}{2\pi f} = \frac{47.09\ \Omega}{2\pi \times 50\ \text{Hz}} = 0.1498\ \text{H} \simeq \mathbf{0.15\ H}$$

It might appear that the inductor in Example 14-9 could be used as the voltage-dropping element in Example 14-10. However, if the effective resistance of the inductor is taken into account, the power dissipated by the heater will be less than its rated 500 W. This power reduction is shown in the following calculation, which assumes that the inductor's effective resistance is approximately the same at 50 and at 60 Hz. (See Fig. 14-23.)

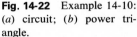

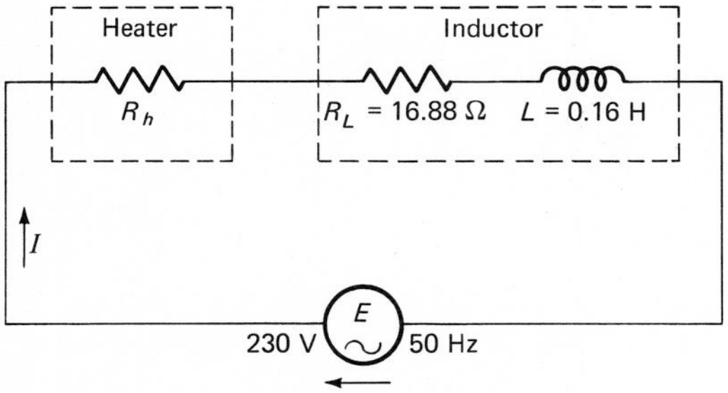

Fig. 14-23 Example 14-10: heater circuit including the inductor in Example 14-9.

Using the heater data in Example 14-10,

$$P = \frac{V_h^2}{R_h}$$

$$R_h = \frac{V_h^2}{P} = \frac{(120\ \text{V})^2}{500\ \text{W}} = 28.8\ \Omega$$

Using data from solution (*a*) for Example 14-9,

$$X_L = 2\pi f L = 2\pi \times 50 \text{ Hz} \times 0.16 \text{ H} = 50.26\ \Omega$$

$$Z = R_h + R_L + jX_L = (28.8 + 16.88)\ \Omega + j50.26\ \Omega$$

$$Z = 67.92\ \Omega$$

$$I = \frac{E}{Z} = \frac{230 \text{ V}}{67.92\ \Omega} = 3.386 \text{ A}$$

The actual power dissipated by the heater will be

$$P = I^2 R_h = (3.386 \text{ A})^2 \times 28.8\ \Omega = 330.2 \text{ W}$$

The voltage across the heater will be:

$$V_h = IR_h = 3.386 \text{ A} \times 28.8\ \Omega = 97.52 \text{ V}$$

Thus the inductor in Example 14-9 has far too much effective resistance to permit it to be used for the voltage-dropping job in Example 14-10.

Example 14-11 Figure 14-24*a* shows the basic circuit of an electronic light-dimming control. Increasing the value of the variable resistance *R* decreases the lagging phase of the

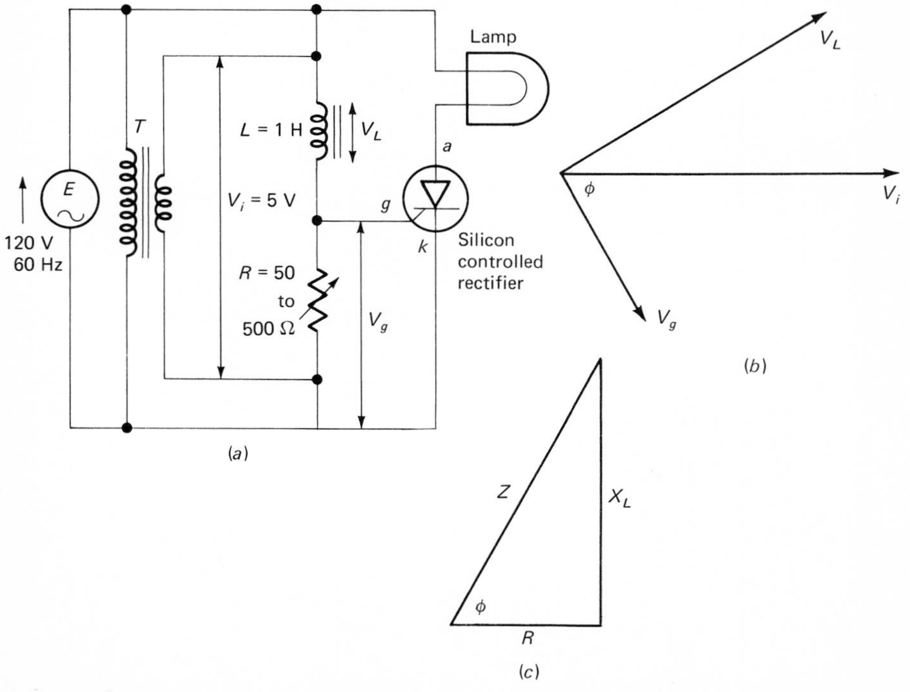

Fig. 14-24 Example 14-11: (*a*) circuit; (*b*) and (*c*) phasor diagram and impedance triangle for the silicon controlled rectifier gate circuit.

voltage V_g applied to the gate of the silicon controlled rectifier relative to the supply voltage E. The decreased phase angle causes the average lamp current to increase, thereby increasing the intensity of the light from the lamp. If the resistance range of R is 50 to 500 Ω, what is the phase angle range? Assume that the transformer T does not introduce phase shift.

Solution It is assumed that V_i, the 5-V ac voltage at 60 Hz applied to the RL series circuit, is in phase with the supply emf E. We are thus concerned with the phase angle ϕ of the voltage V_g (developed across the variable resistance R) with respect to V_i. A voltage phasor diagram of the RL circuit is given in Fig. 14-24*b*. The value of ϕ is obtained most readily from the impedance triangle (Fig. 14-24*c*):

$$X_L = \omega L = 2\pi f L = 2\pi \times 60 \text{ Hz} \times 1 \text{ H} = 377\ \Omega$$

When $R = 50\ \Omega$,

$$\phi = \arctan \frac{X_L}{R} = \arctan \frac{377\ \Omega}{50\ \Omega} = 82.45° = 1.438 \text{ rad}$$

When $R = 500\ \Omega$,

$$\phi = \arctan \frac{X_L}{R} = \arctan \frac{377\ \Omega}{500\ \Omega} = 37.01° = 0.6458 \text{ rad}$$

Thus the range of phase angle obtainable by varying R over its resistance range of 50 to 500 Ω is **−37.01° (−0.6458 rad) to −82.45° (−1.438 rad).**

CIRCLE DIAGRAMS

If the values of V_g and V_L are calculated for Example 14-11, an interesting method of displaying the total effect of varying the resistance can be obtained.

When $R = 50\ \Omega$,

$$Z = R + jX_L \tag{14-11}$$

$$= (50 + j377)\ \Omega = 380.3\ \Omega\ \underline{/82.45°}$$

$$V_g = V_i \frac{R}{Z} = \frac{5 \text{ V}\ \underline{/0°} \times 50\ \Omega\ \underline{/0°}}{380.3\ \Omega\ \underline{/82.45°}} = 0.6573 \text{ V}\ \underline{/-82.45°}$$

$$V_L = V_i \frac{X_L}{Z} = \frac{5 \text{ V}\ \underline{/0°} \times 377\ \Omega\ \underline{/90°}}{380.3\ \Omega\ \underline{/82.45°}} = 4.958 \text{ V}\ \underline{/7.55°}$$

When $R = 500\ \Omega$,

$$Z = R + jX_L = (500 + j377)\ \Omega = 626.2\ \Omega\ \underline{/37.01°}$$

$$V_g = V_i \frac{R}{Z} = \frac{5 \text{ V}\ \underline{/0°} \times 500\ \Omega\ \underline{/0°}}{626.2\ \Omega\ \underline{/37.01°}} = 3.992 \text{ V}\ \underline{/-37.01°}$$

$$V_L = V_i \frac{X_L}{Z} = \frac{5 \text{ V}\ \underline{/0°} \times 377\ \Omega\ \underline{/90°}}{626.2\ \Omega\ \underline{/37.01°}} = 3.011 \text{ V}\ \underline{/52.99°}$$

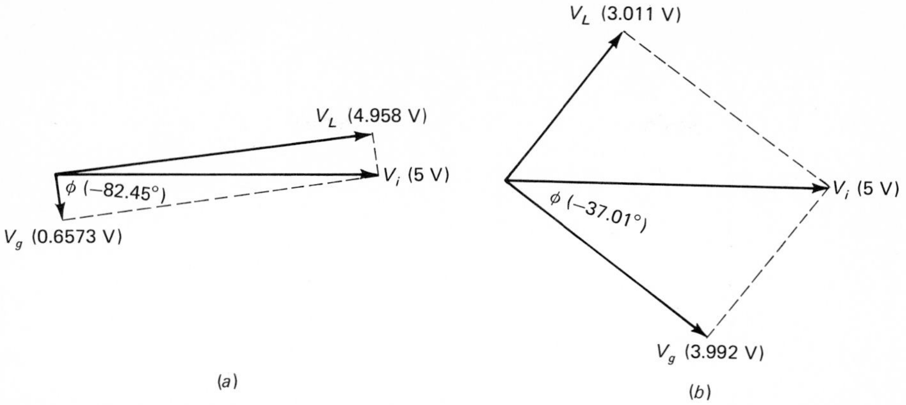

Fig. 14-25 Example 14-11: phasor diagrams for the gate circuit. (*a*) When $R = 50\ \Omega$; (*b*) when $R = 500\ \Omega$.

The phasor diagrams for $R = 50\ \Omega$ and $R = 500\ \Omega$, drawn to scale, are shown in Fig. 14-25*a* and *b*, respectively.

In Fig. 14-25 broken lines indicate how voltage triangles can be produced from the phasor diagrams. Such voltage triangles are given in Fig. 14-26 drawn to the same scale as Fig. 14-25.

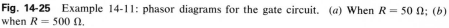

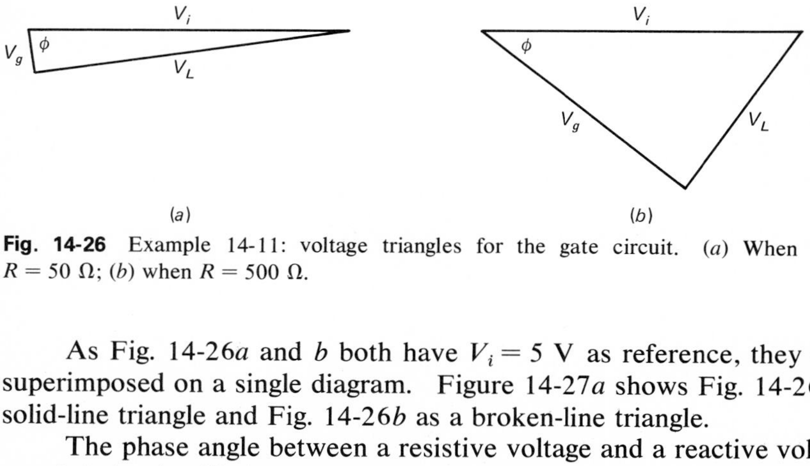

Fig. 14-26 Example 14-11: voltage triangles for the gate circuit. (*a*) When $R = 50\ \Omega$; (*b*) when $R = 500\ \Omega$.

As Fig. 14-26*a* and *b* both have $V_i = 5$ V as reference, they can be superimposed on a single diagram. Figure 14-27*a* shows Fig. 14-26*a* as a solid-line triangle and Fig. 14-26*b* as a broken-line triangle.

The phase angle between a resistive voltage and a reactive voltage in a series circuit will always be a right angle (90° or $\pi/2$ rad), so in Fig. 14-27 the apexes of the triangles are right angles. It is a well-known geometric principle that the locus of the apexes of all right triangles having a common hypotenuse is a circle. In Fig. 14-27*a* the points *A* and *B* are the apexes of the voltage triangles related to $R = 50\ \Omega$ and $R = 500\ \Omega$. The section of the semicircle between points *A* and *B* represents the positions of the apexes of all the voltage triangles related to all values of *R* between 50 and 500 Ω, its complete range of values.

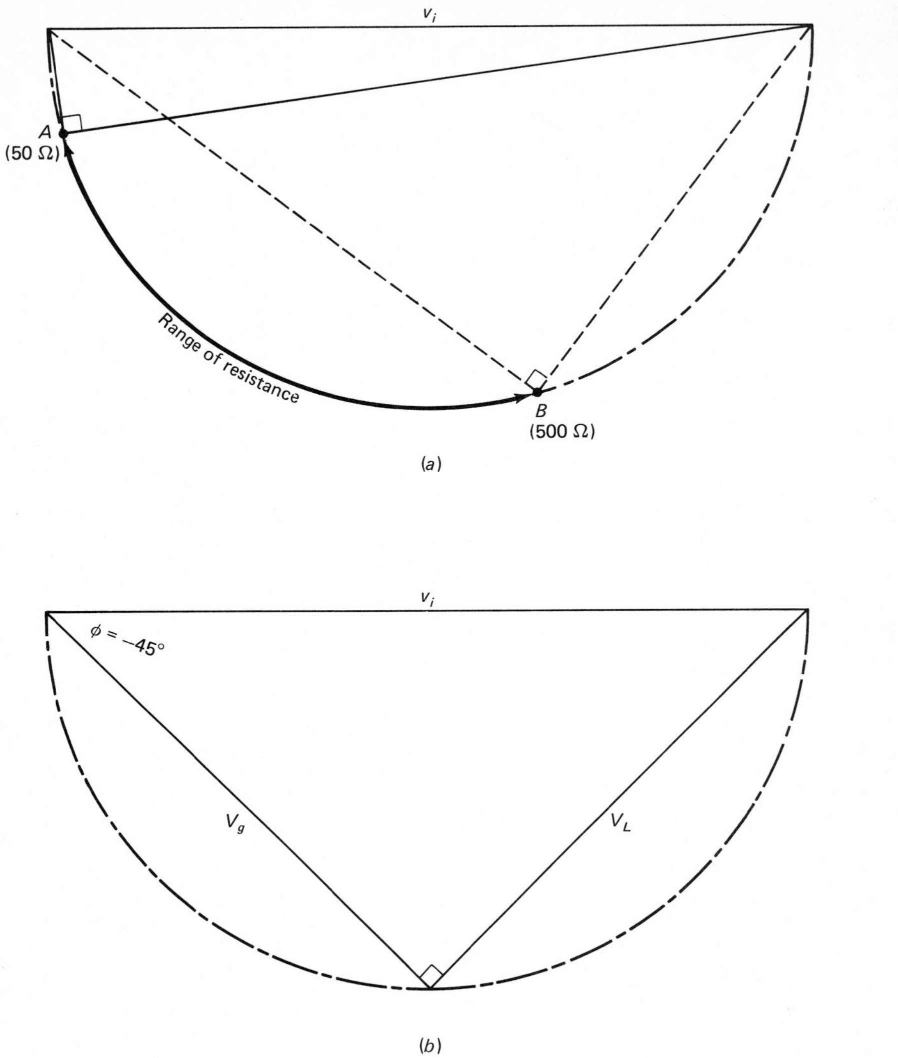

Fig. 14-27 Circle diagrams. (*a*) Relating to Example 14-11; (*b*) when $R = X$.

A value of R between 50 and 500 Ω is represented in Fig. 14-27*b*. Here $V_g = V_L$ or $R = X_L = 377$ Ω. As in any ac circuit where $R = X$, the phase angle is 45° ($\pi/4$ rad).

A diagram using a circle (or part of a circle) to depict an infinite number of different combinations of voltages or currents in a circuit when one parameter varies is called a *circle diagram*. In addition to showing the variation in phase angle and voltage values resulting from varying resistance value, as in Fig. 14-27, circle diagrams are used for such purposes as demonstrating the effect of varying frequency, inductance, or capacitance.

14-3 PARALLEL *RC* CIRCUIT

The basic characteristics of the parallel *RC* circuit are:

1. The voltage across each element is the same.
2. The total circuit current is the phasor sum of the currents in the individual elements (Kirchhoff's current law).

In the *RC* parallel circuit in Fig. 14-28*a*, V is the common circuit voltage; it is equal to the emf of the source E, and it also equals the pds across the resistance and the capacitance. It is convenient to make this common voltage the reference phasor $V\,\underline{/0^\circ}$ or $V+j0$.

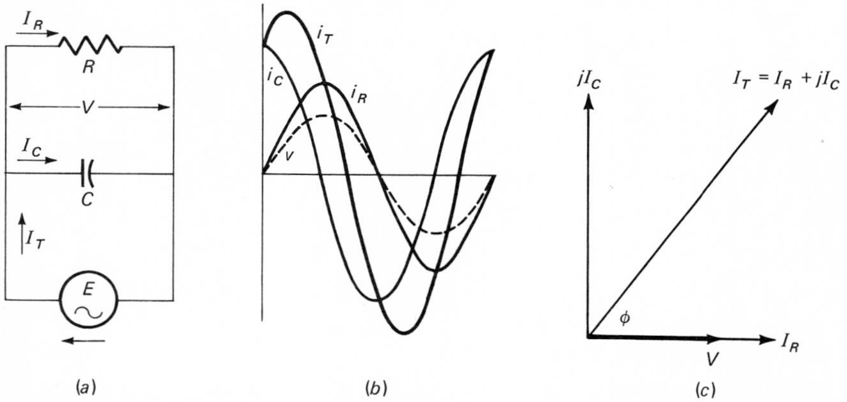

Fig. 14-28 Parallel *RC* circuit. (*a*) Schematic diagram; (*b*) waveform diagram; (*c*) phasor diagram.

CURRENTS

The resistive current I_R is in phase with the voltage. Its phasor value is $I_R\,\underline{/0^\circ}$ or $I_R + j0$.

The capacitive current I_C leads the voltage by 90° ($\pi/2$ rad). Its phasor value is $I_C\,\underline{/90^\circ}$ or $0 + jI_C$.

The total circuit current I_T is the phasor sum of the resistive and capacitive currents

$$I_T = I_R + I_C = (I_R + j0) + (0 + jI_C)$$

$$I_T = I_R + jI_C \qquad (14\text{-}21)$$

Figure 14-28*b* and *c* shows the voltage and the currents as waveforms and as phasors.

ADMITTANCES

By applying the admittance form of Ohm's law ($I = VY$) to each current in Eq. (14-21), it can be written:

$$VY = VG + jVB_C$$

If each term of this equation is now divided by the common factor V, an admittance equation for the RC parallel circuit is obtained:

$$Y = G + jB_C \qquad (14\text{-}22)$$

POWERS

A power equation for the parallel RC circuit results from multiplying each term of the current equation (14-21) by the common voltage V:

$$VI_T = VI_R + jVI_C \qquad (14\text{-}23)$$

In Eq. (14-23) the product of the voltage and the total current VI_T is the apparent power of the circuit S. The product of the voltage and the resistive current VI_R is the active power P dissipated by the circuit. The product of the voltage and the capacitive current VI_C is the reactive power Q. Substituting the symbols S, P, and Q into Eq. (14-23) gives

$$S = P + jQ \qquad (14\text{-}24)$$

Equations (14-21) to (14-24) are based on the use of voltage as the reference phasor in Fig. 14-28*c*, the currents in that figure being referred to the common voltage. Therefore the conventional view of phase angle, current referred to voltage, is implicit in this approach to parallel ac circuits. There is no need to rotate the phasor diagram, as was done for the series circuits, with the resultant change in sign of the reactive power component of apparent power.

Equations (14-21), (14-22), and (14-24) can be displayed as phasor diagrams, as shown in Fig. 14-29*a*, *b*, and *c*, respectively.

Fig. 14-29 Parallel RC circuit phasor diagrams. (*a*) Current; (*b*) admittance; (*c*) power.

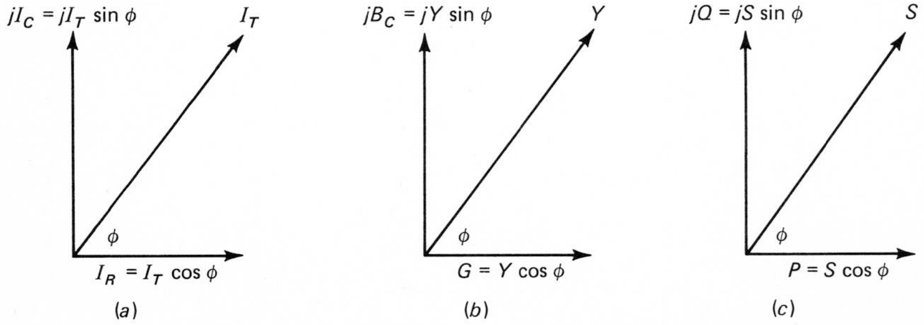

From Eqs. (14-23) and (14-24) and Fig. 14-29*c*, equations for each of the individual powers of a parallel *RC* circuit can be obtained:

Apparent power: $$S = VI_T \tag{14-25}$$

Active power: $$P = S \cos \phi = VI_T \cos \phi \tag{14-26}$$

Reactive power: $$jQ = jS \sin \phi = jVI_T \sin \phi \tag{14-27}$$

In Eqs. (14-21) to (14-27),

I_T = total circuit current, A
I_R = resistive current, A
I_C = capacitive current, A
ϕ = circuit phase angle, deg or rad
Y = total circuit admittance, S
G = conductance, S
B_C = capacitive susceptance, S
V = common circuit voltage, V
S = apparent power, VA
P = active power, W
Q = reactive power, var

It is interesting to note that the power relationships [Eqs. (14-7) and (14-24)] are exactly the same for the series *RC* circuit and the parallel *RC* circuit. This equality is a fortunate outcome of standardizing the requirement for drawing the power triangles: that the circuit phase angle (total current referred to total voltage) be shown in its correct orientation. In the case of a capacitive circuit a leading phase, or a positive phase angle, is shown regardless of the circuit configuration.

CURRENT, ADMITTANCE, AND POWER TRIANGLES

Figure 14-30 shows the current, admittance, and power triangles for the *RC* parallel circuit. These triangles can be used for visualizing the circuit relationships.

Fig. 14-30 Parallel *RC*. (*a*) Current triangle; (*b*) admittance triangle; (*c*) power triangle.

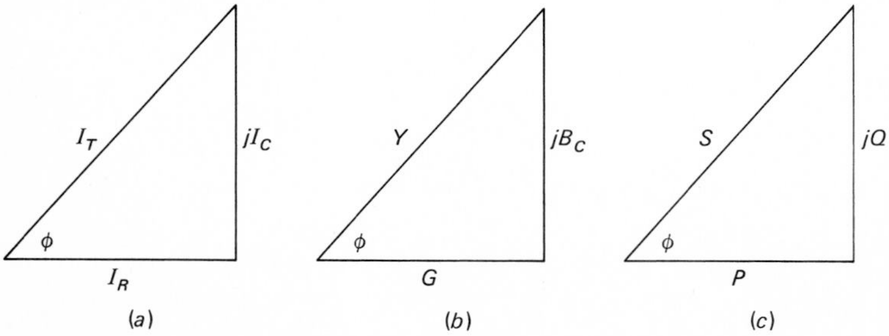

Current triangle (Fig. 14-30*a*):

$$I_T = I_R + jI_C \tag{14-21}$$

$$|I_T| = \sqrt{I_R{}^2 + I_C{}^2}$$

$$\phi = \arctan \frac{I_C}{I_R}$$

$$I_R = I_T \cos \phi$$

$$jI_C = jI_T \sin \phi$$

Admittance triangle (Fig. 14-30*b*):

$$Y = G + jB_C \tag{14-22}$$

$$|Y| = \sqrt{G^2 + B_C{}^2}$$

$$\phi = \arctan \frac{B_C}{G}$$

$$G = Y \cos \phi$$

$$jB_C = jY \sin \phi$$

Power triangle (Fig. 14-30*c*):

$$S = P + jQ \tag{14-24}$$

$$|S| = \sqrt{P^2 + Q^2}$$

$$\phi = \arctan \frac{Q}{P}$$

$$P = S \cos \phi = VI_T \cos \phi$$

$$jQ = jS \sin \phi = jVI_T \sin \phi$$

POWER FACTOR

The power factor of a circuit is cos ϕ, or the ratio of the active power to the apparent power. In the parallel *RC* circuit the power factor can be expressed in terms of the sides of the triangles in Fig. 14-30:

$$\cos \phi = \frac{P}{S} = \frac{I_R}{I_T} = \frac{G}{Y} \tag{14-28}$$

If Eq. (14-28) is compared with Eqs. (14-8) and (14-9), another aspect of the duality of parallel and series circuits is revealed.

From the power ratio viewpoint [Eqs. (14-8) and (14-28)], the power factors of series and parallel circuits are the same.

From Eqs. (14-9) and (14-28) it is seen that the power factor can be expressed as a voltage ratio for the series circuit, or as a current ratio for the parallel circuit.

An impedance ratio can be used to express the power factor of a series ac circuit [Eq. (14-9)], while an admittance ratio applies to a parallel circuit [Eq. (14-28)].

By substituting $1/R$ for G and $1/Z$ for Y in Eq. (14-28), the power factor for the parallel RC circuit can be written:

$$\cos\phi = \frac{Z}{R} \tag{14-29}$$

Equations (14-9) and (14-29) show that, from the impedance point of view, the power factor of the parallel circuit is the reciprocal of that of the series circuit.

In Eqs. (14-28) and (14-29):

$\cos\phi$ = circuit power factor
P = active power, W
S = apparent power, VA
I_R = resistive current, A
I_T = total circuit current, A
G = conductance, S
Y = total circuit admittance, S
Z = total circuit impedance, Ω
R = resistance, Ω

Example 14-12 A parallel network consisting of a 1-kΩ resistance and a 0.022-μF capacitance is connected to a 10-V 5-kHz source. What is (*a*) the current in the resistance, (*b*) the current in the capacitance, (*c*) the total circuit current, and (*d*) the circuit phase angle?
Solution See Fig. 14-31.

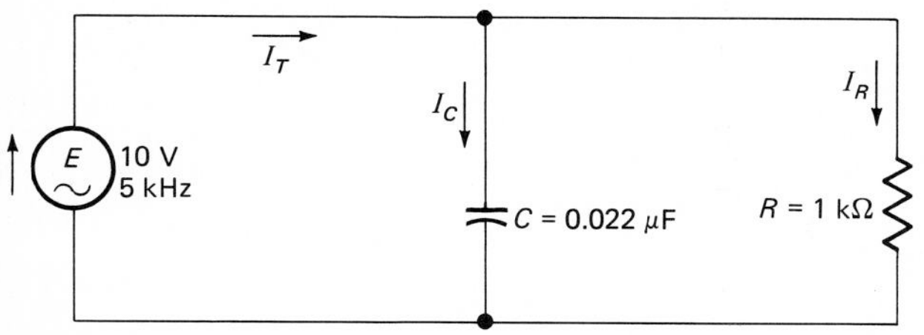

Fig. 14-31 Example 14-12: circuit.

$$\omega = 2\pi f = 2\pi \times 5\text{ kHz} = 31.42\text{ krad/s}$$

$$B_C = \omega C = 31.42\text{ krad/s} \times 0.022\ \mu\text{F} = 0.6912\text{ mS}$$

$$G = \frac{1}{R} = \frac{1}{1\text{ k}\Omega} = 1\text{ mS}$$

a. $$I_R = EG = 10\text{ V} \times 1\text{ mS} = \mathbf{10\ mA}$$

b. $I_C = EB_C = 10\text{ V}\angle 0° \times 0.6912\text{ mS}\angle 90°$

$\mathbf{I_C = 6.912\text{ mA}\angle 90°}$ or $\angle \pi/2\text{ rad}$

c. $I_T = I_R + jI_C$ (14-21)

$\mathbf{I_T = (10 + j6.912)\text{ mA}}$

or $\mathbf{I_T = 12.16\text{ mA}\angle 34.65°}$ or $\angle \mathbf{0.6047\text{ rad}}$

d. From the answer to part *c*,

$$\boldsymbol{\phi = 34.65° = 0.6047\text{ rad}}$$

In Example 14-12 the answer to part *a* could have been calculated as easily by using the resistance form of Ohm's law:

$$I_R = \frac{E}{R} = \frac{10\text{ V}}{1\text{ k}\Omega} = 10\text{ mA}$$

The conductance form was used for consistency in the adoption of an admittance view of parallel ac circuits.

You will have noticed that both the polar and the rectangular forms were given for the answer to part *c*. Either of these forms is acceptable, since each one specifies the circuit current as a complete phasor expression.

PARALLEL VIEW OF A PRACTICAL CAPACITOR

In Sec. 14-1 it was implied that a practical capacitor is shown as an equivalent series *RC* circuit. While this view is quite in order, it is just as valid to regard the capacitor as having an equivalent parallel circuit. From a logical approach, it is not easy to say for sure whether the various power losses within a capacitor should be represented as a series resistance component or as a resistance in parallel with a capacitance. We shall see later in this chapter that it is a fairly simple matter to convert a series equivalent circuit into parallel form or vice-versa. Thus there is no technical difficulty in handling these alternative views of the practical capacitor.

Example 14-13 A 50-pF capacitor has a power factor of 0.0001 at 10 MHz. What is its parallel equivalent resistance?
Solution

$$\omega = 2\pi f = 2\pi \times 10\text{ MHz} = 62.83\text{ Mrad/s}$$

$$B_C = \omega C = 62.83\text{ Mrad/s} \times 50\text{ pF} = 3.142\text{ mS}$$

$$\cos\phi = \frac{G_p}{Y} \qquad (14\text{-}28)$$

$$10^{-4} = \frac{G_p}{\sqrt{G_p^2 + B_C^2}}$$

$$10^{-8}(G_p^2 + B_C^2) = G_p^2$$

$$10^{-8}B_C^2 \simeq G_p^2$$

$$G_p \simeq 10^{-4}B_C \simeq 3.142 \text{ mS} \times 10^{-4} \simeq 0.3142\ \mu\text{S}$$

$$R_p = \frac{1}{G_p} \simeq \frac{1}{0.3142\ \mu\text{S}} \simeq \mathbf{3.183\ M\Omega}$$

The capacitor in Example 14-13 is the same as that in Example 14-2. Thus a 50-pF capacitor with a power factor of 0.0001 can be considered to have either a series equivalent resistance of 0.0318 Ω or a parallel equivalent resistance of 3.183 MΩ.

Figure 14-32 shows the series and parallel versions of this 50-pF, 0.0001–power factor capacitor.

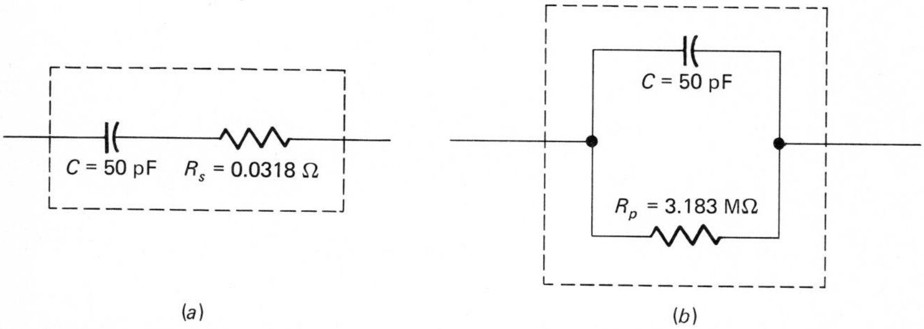

Fig. 14-32 Examples 14-2 and 14-13: capacitor series and parallel equivalent circuits.

As a check on the validity of the series and parallel views of equivalent resistance, consider the power dissipation within the capacitor if an emf of 1 V at 10 MHz is applied across it.

In the series circuit in Example 14-2 (Fig. 14-32*a*), an applied emf of 1 V will result in a current of 3.142 mA. The power dissipated [$I^2R = (3.142 \text{ mA})^2 \times 0.0318\ \Omega$] will be 0.314 μW.

In the parallel circuit in Example 14-13 (Fig. 14-32*b*), an applied emf of 1 V will give a power dissipation [$E^2/R = (1 \text{ V})^2/3.183\ \text{M}\Omega$] of 0.314 μW.

Thus, although the series and parallel equivalent resistances differ by a ratio of 10^8, the power dissipation is the same.

It is interesting to think of the effect of the series and parallel resistances as being the same. If the parallel resistance were the same as the series resistance (0.0318 Ω), the result would be a virtual short circuit. On the other hand, if the series resistance were the same as the parallel resistance (3.183 MΩ), the result would be a virtual open circuit.

Example 14-14 A constant-current source supplies 10 mA at 250 kHz to a network which consists of a 330-pF capacitance connected in parallel with a 3.3-kΩ resistance. What

is (*a*) the circuit voltage, (*b*) the current through the resistance, and (*c*) the current through the capacitance? Give the values with reference to the source current.
Solution See Fig. 14-33.

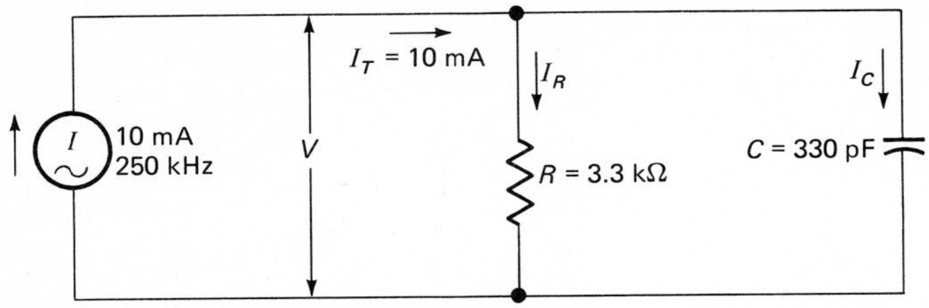

Fig. 14-33 Example 14-14: circuit.

$$G = \frac{1}{R} = \frac{1}{3.3\ \text{k}\Omega} = 0.303\ \text{mS} \quad \text{or} \quad 303\ \mu\text{S}$$

$$\omega = 2\pi f = 2\pi \times 250\ \text{kHz} = 1.57\ \text{Mrad/s}$$

$$B_C = \omega C = 1.57\ \text{Mrad/s} \times 330\ \text{pF} = 518.4\ \mu\text{S}$$

$$Y = G + jB_C \tag{14-22}$$

$$Y = (303 + j518.4)\ \mu\text{S} = 600.5\ \mu\text{S}\ \underline{/59.7^\circ} \quad \text{or} \quad \underline{/1.042\ \text{rad}}$$

a.
$$V = \frac{I_T}{Y} = \frac{10\ \text{mA}\ \underline{/0^\circ}}{600.5\ \mu\text{S}\ \underline{/59.7^\circ}} = \mathbf{16.65\ V}\ \underline{\mathbf{/-59.7^\circ}} \quad \text{or} \quad \underline{\mathbf{/-1.042\ rad}}$$

The current divider principle used for dc parallel circuits can be applied to ac circuits provided that phasor techniques are adhered to.

b.
$$I_R = I_T \frac{G}{Y} = 10\ \text{mA}\ \underline{/0^\circ}\ \frac{303\ \mu\text{S}\ \underline{/0^\circ}}{600.5\ \mu\text{S}\ \underline{/59.7^\circ}}$$

$$I_R = \mathbf{5.046\ mA}\ \underline{\mathbf{/-59.7^\circ}} \quad \text{or} \quad \underline{\mathbf{/-1.042\ rad}}$$

c.
$$I_C = I_T \frac{B_C}{Y} = 10\ \text{mA}\ \underline{/0^\circ}\ \frac{518.4\ \mu\text{S}\ \underline{/90^\circ}}{600.5\ \mu\text{S}\ \underline{/59.7^\circ}}$$

$$I_C = \mathbf{8.633\ mA}\ \underline{\mathbf{/30.3^\circ}} \quad \text{or} \quad \underline{\mathbf{/0.5288\ rad}}$$

The phasor diagram for Example 14-14, drawn to scale and with reference to the source current, is given in Fig. 14-34*a*.

Drawing the phasor diagram in the form which is conventional for a parallel circuit (with reference to the common voltage) results in Fig. 14-34*b*. Figure 14-34*b* is the same as Fig. 14-34*a* rotated counterclockwise through the circuit phase angle, 59.7°.

The common circuit voltage could have been calculated from either the resistive current or the capacitive current, once they had been obtained.

$$V = I_R R = 5.046\ \text{mA}\ \underline{/-59.7^\circ} \times 3.3\ \text{k}\Omega\ \underline{/0^\circ}$$

$$V = 16.65\ \text{V}\ \underline{/-59.7^\circ}$$

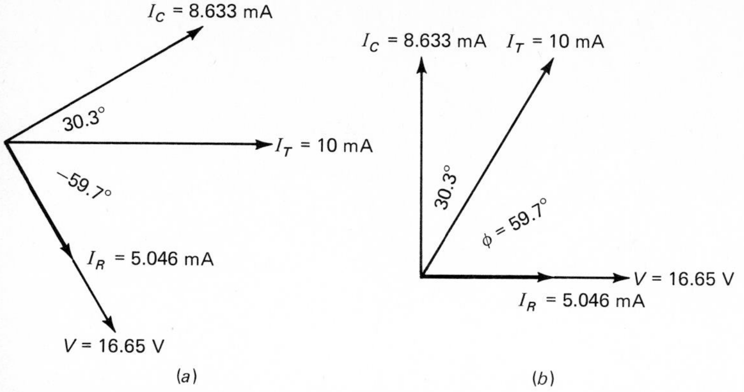

Fig. 14-34 Example 14-14: phasor diagrams. (*a*) Source current I_T as reference; (*b*) common voltage V as reference.

or

$$V = \frac{I_C}{B_C} = \frac{8.633 \text{ mA } \underline{/30.3^\circ}}{518.4\ \mu\text{S } \underline{/90^\circ}}$$

$$V = 16.65 \text{ V } \underline{/-59.7^\circ}$$

The individual element currents were found by applying the current divider principle. One test of the validity of this method is to add these currents, in phasor form, to ensure that their sum equals the source current (Kirchhoff's current law)

$$I_R = 5.046 \text{ mA } \underline{/-59.7^\circ}$$

$$= 5.046 \cos(-59.7^\circ) + j5.046 \sin(-59.7^\circ)$$

$$I_R = (2.546 - j4.355) \text{ mA}$$

$$I_C = 8.633 \text{ mA } \underline{/30.3^\circ}$$

$$= 8.633 \cos(30.3^\circ) + j8.633 \sin(30.3^\circ)$$

$$I_C = (7.454 + j4.355) \text{ mA}$$

$$I_R + I_C = (2.546 + 7.454) + j(4.355 - 4.355)$$

$$= (10 + j0) \text{ mA} \quad \text{or} \quad 10 \text{ mA } \underline{/0^\circ}$$

Therefore, $I_R + I_C = I_T$

Example 14-15 If the parallel RC network in Example 14-14 were connected to a signal generator with a source current of 10 mA at 250 kHz and an internal conductance of 500 μS, what would be the current through the 3.3-kΩ resistance?

Solution From Example 14-14,

$$G = 303\ \mu\text{S}$$

$$B_C = 518.4\ \mu\text{S } \underline{/90^\circ}$$

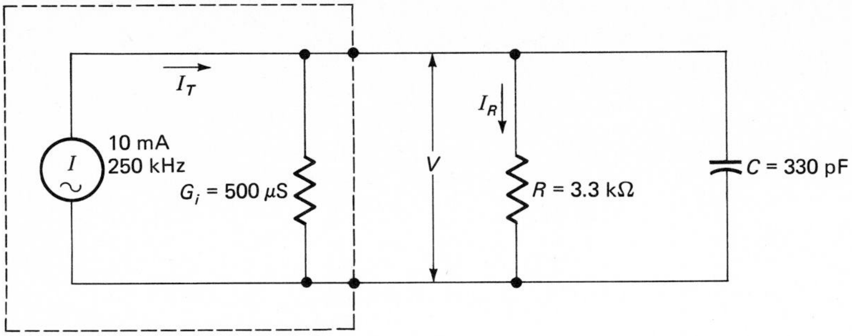

Fig. 14-35 Example 14-15: circuit.

The total admittance will be, from Eq. (14-22),

$$Y = (G_i + G) + jB_C = [(500 + 303) + j518.4]\ \mu\text{S}$$

$$Y = (803 + j518.4)\ \mu\text{S} = 955.8\ \mu\text{S}\ \underline{/32.85^\circ} \qquad \text{or} \qquad \underline{/0.5733\text{ rad}}$$

The circuit voltage and the current through R can be determined by the admittance form of Ohm's law:

$$V = \frac{I_T}{Y} = \frac{10\text{ mA}\ \underline{/0^\circ}}{955.8\ \mu\text{S}\ \underline{/32.85^\circ}} = 10.46\text{ V}\ \underline{/-32.85^\circ}$$

$$I_R = VG = 10.46\text{ V}\ \underline{/-32.85^\circ} \times 303\ \mu\text{S}\ \underline{/0^\circ}$$

$$I_R = \mathbf{3.17\ mA}\ \underline{\mathbf{/-32.85^\circ}} \qquad \text{or} \qquad \underline{\mathbf{/0.5733\ rad}}$$

Example 14-15 shows that the effect of the parallel internal conductance of a current source is to rob the load of current. This is analogous to the voltage robbing caused by the series internal resistance of a voltage source. The effect of each of these internal losses is to reduce the load voltage and current. A generator can be considered in relation to the voltage source viewpoint (Thevenin) or in relation to the current source viewpoint (Norton); the results are the same in either case.

Example 14-16 What is the impedance at 2 MHz of a parallel circuit consisting of a 820-pF capacitance and a 47-Ω resistance?

Solution See Fig. 14-36.

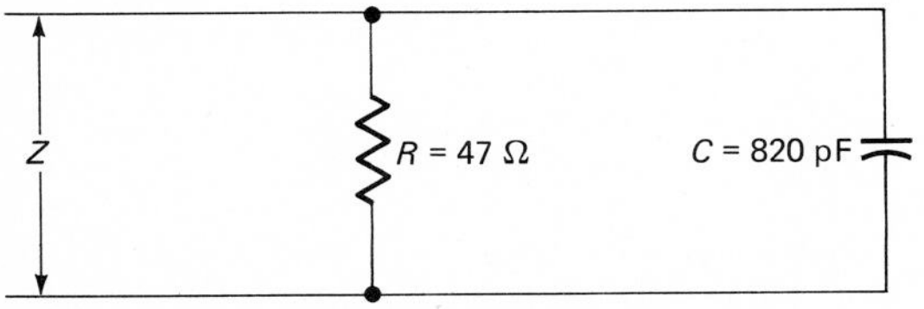

Fig. 14-36 Example 14-16: circuit.

The principle for calculating the total resistance of two parallel resistances [$R_T = R_1R_2/(R_1 + R_2)$] can be applied to impedances if a phasor approach is used.

$$|X_C| = \frac{1}{\omega C} = \frac{1}{2\pi fC} = \frac{1}{2\pi \times 2\text{ MHz} \times 820\text{ pF}} = 97.05\ \Omega$$

$$X_C = -j97.05\ \Omega \qquad \text{or} \qquad 97.05\ \Omega\ \underline{/-90^\circ}$$

$$Z = \frac{RX_C}{R + X_C} = \frac{47\ \Omega\ \underline{/0^\circ} \times 97.05\ \Omega\ \underline{/-90^\circ}}{(47 - j97.05)\ \Omega}$$

$$= \frac{(47 \times 97.05)\ \Omega^2\ \underline{/0^\circ + (-90^\circ)}}{107.8\ \Omega\ \underline{/-64.17^\circ}}$$

$$= \frac{4.561\ \text{k}\Omega^2\ \underline{/-90^\circ}}{107.8\ \Omega\ \underline{/-64.17^\circ}} = (4561 \div 107.8)\ \Omega\ \underline{/-90^\circ - (-64.17^\circ)}$$

$$Z = \mathbf{42.31\ \Omega\ \underline{/-25.83^\circ}} \quad \text{or} \quad \mathbf{\underline{/-0.4508\ rad}} \quad \text{or} \quad \mathbf{(38.08 - j18.43)\ \Omega}$$

In Example 14-16 the solution showed that

1. The total impedance of the two elements in parallel is numerically less than the impedance of either element. This is to be expected for a normal parallel circuit.
2. The nature of the total impedance is directly related to the nature of the elements. Thus, where the elements are resistive and capacitive, the total impedance is similarly resistive and capacitive ($R - jX$ or $\underline{/-\phi}$).

Example 14-17 Figure 14-37*a* shows the input circuit of a transistor amplifying stage in a domestic AM-FM radio. The voltage generator is a simulation of the ac source which drives the transistor. The input resistance R_{BE} of the transistor is constant at 1 kΩ. C is the stray capacitance of the input circuit. What is the voltage V_{BE} at the transistor input at frequencies of 1 MHz on the AM band and 100 MHz on the FM band?

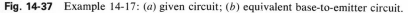

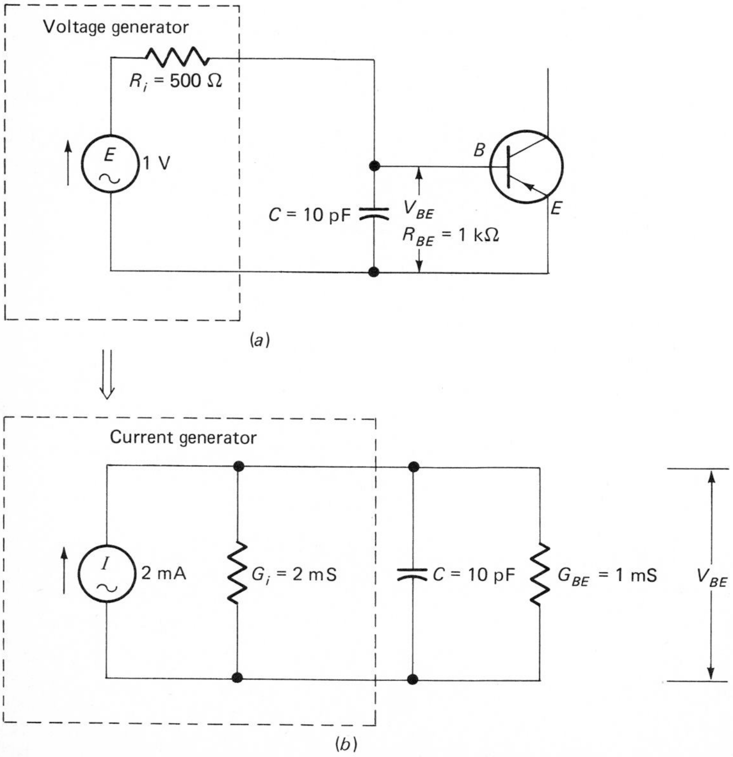

Fig. 14-37 Example 14-17: (*a*) given circuit; (*b*) equivalent base-to-emitter circuit.

Solution The circuit in Fig. 14-37*b* is the equivalent of that in Fig. 14-37*a*. The voltage generator is converted to a current generator using the Thevenin-to-Norton conversion principles developed for dc circuits.

The source current I is given by

$$I = \frac{E}{R_i} = \frac{1\ \text{V}}{500\ \Omega} = 2\ \text{mA}$$

The source conductance G_i is

$$G_i = \frac{1}{R_i} = \frac{1}{500\ \Omega} = 2\ \text{mS}$$

The transistor input conductance G_{BE} is

$$G_{BE} = \frac{1}{R_{BE}} = \frac{1}{1\ \text{k}\Omega} = 1\ \text{mS}$$

Now that a straightforward parallel circuit has been obtained, the transistor input voltage V_{BE} can be determined as the common circuit voltage.

When $f = 1$ MHz (AM band),

$$\omega = 2\pi f = 2\pi \times 1\ \text{MHz} = 6.283\ \text{Mrad/s}$$

$$B_C = \omega C = 6.283\ \text{Mrad/s} \times 10\ \text{pF} = 0.0628\ \text{mS}$$

The total admittance of the circuit is, from Eq. (14-22),

$$Y = (G_i + G_{BE}) + jB_C$$

$$Y = (3 + j0.0628)\ \text{mS} \simeq 3\ \text{mS}\ \underline{/0^\circ}$$

Applying Ohm's law gives the common circuit voltage which is the voltage developed at the transistor input.

$$V_{BE} = \frac{I}{Y} \simeq \frac{2\ \text{mA}\ \underline{/0^\circ}}{3\ \text{mS}\ \underline{/0^\circ}} \simeq \mathbf{0.6667\ V}\ \underline{/0^\circ}$$

When $f = 100$ MHz (FM band),

$$\omega = 2\pi f = 2\pi \times 100\ \text{MHz} = 628.3\ \text{Mrad/s}$$

$$B_C = \omega C = 628.3\ \text{Mrad/s} \times 10\ \text{pF} = 6.283\ \text{mS}$$

From Eq. (14-22),

$$Y = (G_i + G_{BE}) + jB_C = (3 + j6.283)\ \text{mS} \quad \text{or} \quad 6.962\ \text{mS}\ \underline{/64.48^\circ}$$

$$V_{BE} = \frac{I}{Y} = \frac{2\ \text{mA}\ \underline{/0^\circ}}{6.962\ \text{mS}\ \underline{/64.48^\circ}} = \mathbf{0.2873\ V}\ \underline{/\mathbf{-64.48^\circ}} \quad \text{or} \quad \underline{/\mathbf{-1.125\ rad}}$$

Example 14-17 demonstrates how a stray capacitance of 10 pF (which is a small value) can be significant at a high frequency but insignificant at a lower frequency.

If there were no stray capacitance, the transistor input voltage V_{BE} would be 0.6667 V and in phase with the source current (or voltage in Fig. 14-37*a*).

When the frequency is relatively low, as on the AM radio band, the effect of the 10-pF stray capacitance is negligible, and the transistor input voltage is approximately the same as if there were no capacitance.

At the high FM frequencies the susceptance of the 10-pF stray capacitance is great enough to affect the circuit admittance materially. The result is a reduced transistor input voltage which lags the source current (or voltage) by a substantial phase angle.

To summarize we may say that the effect of shunt capacitance is determined by frequency; high frequencies have a greater effect than low frequencies. "High" and "low" are terms which must be related to the context. One megahertz is low when compared with 100 MHz, but would be high if compared with 1 kHz.

14-4 PARALLEL *RL* CIRCUIT

Except for circuit phase conditions (lag instead of lead), the parallel *RL* circuit is treated in the same manner as the parallel *RC* circuit.

In Fig. 14-38, V is the common circuit voltage; it is the pd across R and L as well as the source emf. It is the convenient choice for the reference phasor, so we make its phasor value $V\,\underline{/0^\circ}$ or $V + j0$.

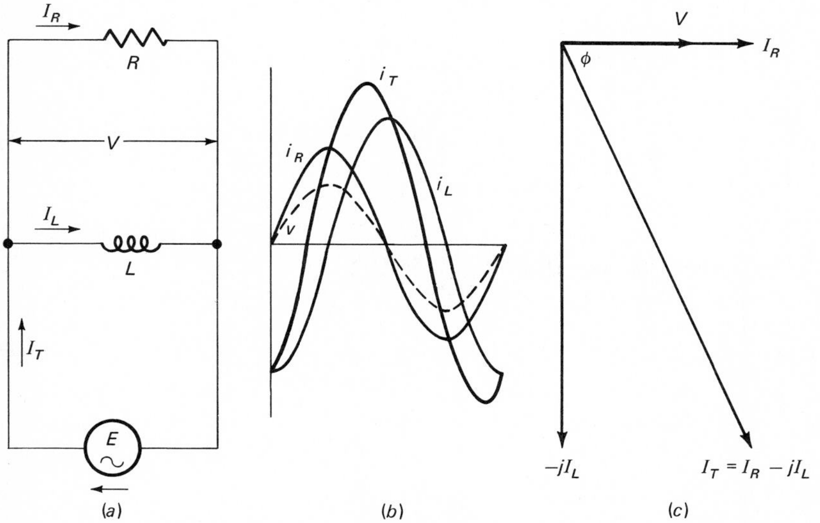

Fig. 14-38 Parallel *RL* circuit. (*a*) Schematic diagram; (*b*) waveform diagram; (*c*) phasor diagram.

CURRENTS

The resistive current I_R is in phase with the voltage. Its phasor value is $I_R\,\underline{/0^\circ}$ or $I_R + j0$.

The inductive current I_L lags the voltage by 90° or $\pi/2$ rad. Its phasor value is $I_L\,\underline{/-90^\circ}$ or $0 - jI_L$.

The phasor sum of the resistive and inductive currents is the total current I_T:

$$I_T = I_R + I_L = (I_R + j0) + (0 - jI_L)$$

$$I_T = I_R - jI_L \tag{14-30}$$

Figure 14-38*a* shows the *RL* parallel circuit, while Fig. 14-38*b* and *c* show its waveform and phasor diagrams.

ADMITTANCES

Applying the admittance form of Ohm's law ($I = VY$) to each term in Eq. (14-30) gives

$$VY = VG - jVB_L$$

Dividing each term of this equation by the common factor V results in an admittance equation for the parallel circuit:

$$Y = G - jB_L \tag{14-31}$$

POWERS

Multiplying each term of the current equation (14-30) by the common voltage V gives a power equation for the parallel *RL* circuit:

$$VI_T = VI_R - jVI_L \tag{14-32}$$

Equation (14-32) can be written in standard ac power symbol form:

$$S = P - jQ \tag{14-33}$$

Since the currents have been referred to the common voltage, Eq. (14-33) expresses the power relationship for the *RL* parallel circuit in conventional form.

Using Eqs. (14-30), (14-31), and (14-33), we can draw phasor diagrams to show the currents, admittances, and powers in the parallel *RL* circuit. These diagrams are shown in Fig. 14-39*a*, *b*, and *c*, respectively.

Fig. 14-39 Parallel *RL* circuit phasor diagrams. (*a*) Current; (*b*) admittance; (*c*) power.

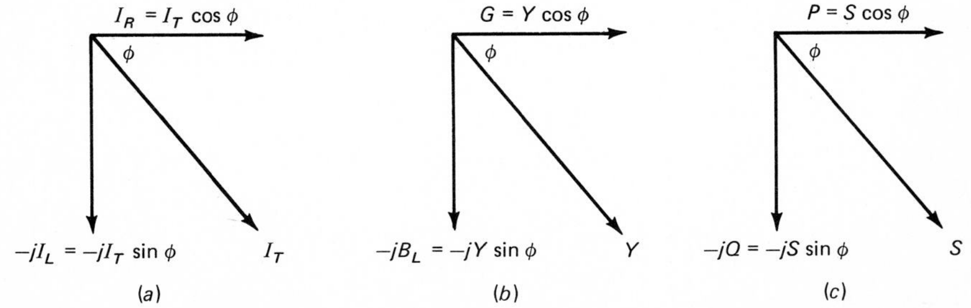

From Eq. (14-32) and Fig. 14-39*c*, expressions for each of the individual powers of a parallel *RL* circuit can be obtained:

Apparent power: $$S = VI_T \tag{14-34}$$

Active power: $$P = S \cos \phi = VI_T \cos \phi \tag{14-35}$$

Reactive power: $$-jQ = -jS \sin \phi = -jVI_T \sin \phi \tag{14-36}$$

In Eqs. (14-30) to (14-36):

I_T = total circuit current, A
I_R = resistive current, A
I_L = inductive current, A
ϕ = phase angle, deg or rad
Y = total circuit admittance, S
G = conductance, S
B_L = inductive susceptance, S
V = common circuit voltage, V
S = apparent power, VA
P = active power, W
Q = reactive power, var

As a result of the adoption of a standard convention for expressing ac power relationships (voltages as reference phasors), the power equations (14-16) and (14-33) for the series and parallel *RL* circuits are the same. In each case the lagging phase characteristic of inductive circuits is directly indicated.

CURRENT, ADMITTANCE, AND POWER TRIANGLES

The current, admittance, and power triangles for the *RL* parallel circuit are shown in Fig. 14-40. The circuit relationships applicable to these triangles are the following:

Fig. 14-40 Parallel *RL* circuit. (*a*) Current triangle; (*b*) admittance triangle; (*c*) power triangle.

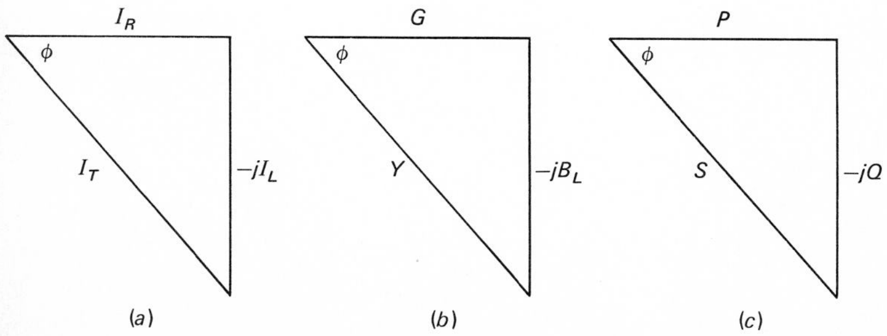

Current triangle (Fig. 14-40*a*):

$$I_T = I_R - jI_L \tag{14-30}$$

$$|I_T| = \sqrt{I_R^2 + I_L^2}$$

$$\phi = \arctan \frac{I_L}{I_R}$$

$$I_R = I_T \cos \phi$$

$$-jI_L = -jI_T \sin \phi$$

Admittance triangle (Fig. 14-40*b*):

$$Y = G - jB_L \tag{14-31}$$

$$|Y| = \sqrt{G^2 + B_L^2}$$

$$\phi = \arctan \frac{B_L}{G}$$

$$G = Y \cos \phi$$

$$-jB_L = -jY \sin \phi$$

Power triangle (Fig. 14-40*c*):

$$S = P - jQ \tag{14-33}$$

$$|S| = \sqrt{P^2 + Q^2}$$

$$\phi = \arctan \frac{Q}{P}$$

$$P = S \cos \phi = VI_T \cos \phi$$

$$-jQ = -jS \sin \phi = -jVI_T \sin \phi$$

POWER FACTOR

The power factor of a parallel *RL* circuit is expressed in essentially the same terms as for the parallel *RC* circuit:

$$\cos \phi = \frac{P}{S} = \frac{I_R}{I_T} = \frac{G}{Y} = \frac{Z}{R} \tag{14-37}$$

Q FACTOR

The quality factor Q of an inductor can be obtained from its parallel equivalent circuit using the same definition as for the series equivalent. The Q factor is the ratio of stored energy to dissipated energy:

$$Q^* = \frac{W_{\text{stored}}}{W_{\text{dissipated}}} = \frac{\text{reactive power} \times \text{time}}{\text{active power} \times \text{time}} = \frac{Qt}{Pt}$$

$$Q = \frac{\text{reactive power}}{\text{active power}} = \frac{Q}{P}$$

By referring to Fig. 14-40*b* and *c*, which are similar triangles, it can be seen that B_L/G is the same as Q/P. Therefore,

$$Q = \frac{B_L}{G} = \tan\phi \qquad (14\text{-}38)$$

Substituting $1/\omega L$ for B_L in Eq. (14-38),

$$Q = \frac{1}{\omega LG} \quad \text{or} \quad \frac{R}{\omega L} \quad \text{or} \quad \frac{R}{X_L} \qquad (14\text{-}39)$$

In Eqs. (14-37) to (14-39):

P = active power, W
S = apparent power, VA
I_R = resistive current, A
I_T = total circuit current, A
G = conductance, S
B_L = inductive susceptance, S
Y = total circuit admittance, S
Z = total circuit impedance, Ω
R = resistance, Ω
X_L = inductive reactance, Ω
Q = the quality or quadrature factor
or Q = reactive power, var
W = energy, J
t = time, s
$\tan\phi$ = tangent of the phase angle
ω = angular velocity or angular frequency, rad/s
L = inductance, H

A comparison of Eqs. (14-19), (14-20), (14-38), and (14-39) will show that:

1. In either series or parallel *RL* circuits the *Q* factor is equal to the tangent of the phase angle.
2. In terms of resistance and reactance, the *Q* factors of the series and parallel circuits are reciprocal.

* On the left side of these equations *Q* is quality factor, and on the right side *Q* is reactive power.

Example 14-18 A parallel network consisting of a 1-kΩ resistance and a 50-mH inductance is connected to a 10-V 5-kHz source. What is (*a*) the current in the resistance, (*b*) the current in the inductance, (*c*) the total circuit current, and (*d*) the circuit phase angle?
Solution See Fig. 14-41.

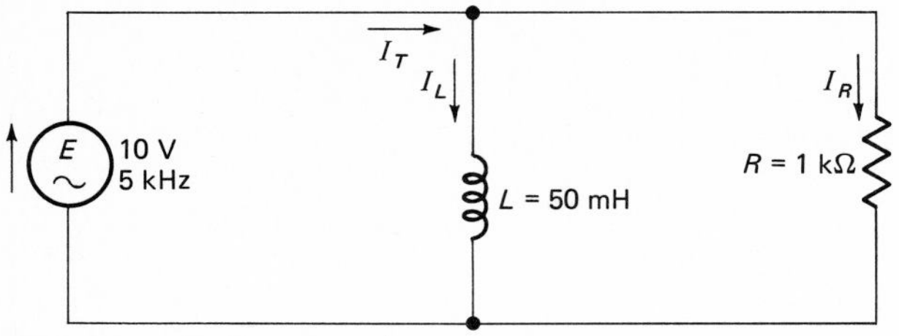

Fig. 14-41 Example 14-18: circuit.

$$\omega = 2\pi f = 2\pi \times 5\text{ kHz} = 31.42\text{ krad/s}$$

$$B_L = \frac{1}{\omega L} = \frac{1}{31.42\text{ krad/s} \times 50\text{ mH}} = 0.6365\text{ mS}$$

$$G = \frac{1}{R} = \frac{1}{1\text{ k}\Omega} = 1\text{ mS}$$

a. $I_R = EG = 10\text{ V} \times 1\text{ mS} = \mathbf{10\text{ mA}}$

b. $I_L = EB_L = 10\text{ V}\,\underline{/0^\circ} \times 0.6365\text{ mS}\,\underline{/-90^\circ}$

$I_L = \mathbf{6.365\text{ mA}\,\underline{/-90^\circ}}$

c. $I_T = I_R - jI_L$ (14-30)

$I_T = \mathbf{(10 - j6.365)\text{ mA}}$

$I_T = \mathbf{11.85\text{ mA}\,\underline{/-32.41^\circ}}$ or $\mathbf{\underline{/-0.5656\text{ rad}}}$

d. From the answers to part *c*,

$$\phi = \mathbf{-32.41^\circ = -0.5656\text{ rad}}$$

Example 14-19 A constant-current source supplies 10 mA at 250 kHz to a network consisting of a 2-mH inductance connected in parallel with a 3.3-kΩ resistance. What is (*a*) the circuit voltage, (*b*) the current through the resistance, and (*c*) the current through the inductance. Give the values with respect to the source current.
Solution See Fig. 14-42.

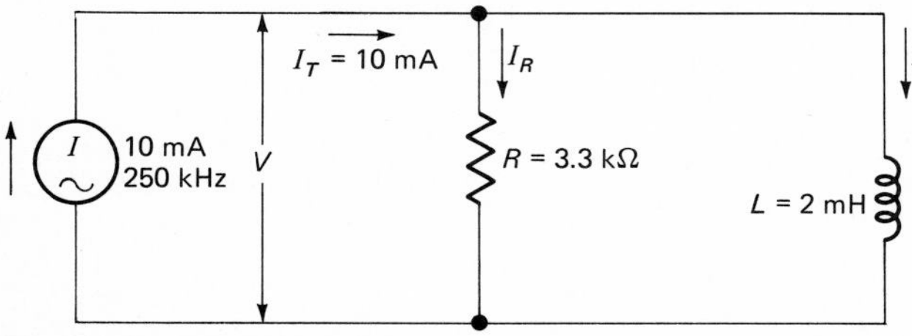

Fig. 14-42 Example 14-19: circuit.

$$G = \frac{1}{R} = \frac{1}{3.3\text{ k}\Omega} = 303\ \mu\text{S}$$

$$\omega = 2\pi f = 2\pi \times 250\text{ kHz} = 1.57\text{ Mrad/s}$$

$$B_L = \frac{1}{\omega C} = \frac{1}{1.57\ \text{Mrad/s} \times 2\ \text{mH}} = 318\ \mu\text{S}$$

The total admittance of the circuit is given by

$$Y = G - jB_L \tag{14-31}$$

$$Y = (303 - j318)\ \mu\text{S} = 439.2\ \mu\text{S}\ \underline{/-46.4^\circ} \quad \text{or} \quad \underline{/-0.8098\ \text{rad}}$$

a.

$$V = \frac{I_T}{Y} = \frac{10\ \text{mA}\ /0^\circ}{439.2\ \mu\text{S}\ /-46.4^\circ}$$

$$V = \mathbf{22.76\ V}\ \underline{/\mathbf{46.4^\circ}} \quad \text{or} \quad \underline{/\mathbf{0.8090\ rad}}$$

Using the current divider principle:

b.

$$I_R = I_T \frac{G}{Y} = 10\ \text{mA}\ /0^\circ \frac{303\ \mu\text{S}\ /0^\circ}{439.2\ \mu\text{S}\ /-46.4^\circ}$$

$$I_R = \mathbf{6.899\ mA}\ \underline{/\mathbf{46.4^\circ}} \quad \text{or} \quad \underline{/\mathbf{0.8098\ rad}}$$

c.

$$I_L = I_T \frac{B_L}{Y} = 10\ \text{mA}\ /0^\circ \frac{318\ \mu\text{S}\ /-90^\circ}{439.2\ \mu\text{S}\ /-46.4^\circ}$$

$$I_L = \mathbf{7.24\ mA}\ \underline{/\mathbf{-43.6^\circ}} \quad \text{or} \quad \underline{/\mathbf{-0.761\ rad}}$$

The phasor diagram for Example 14-19, drawn to scale and with reference to the source current, is given in Fig. 14-43*a*.

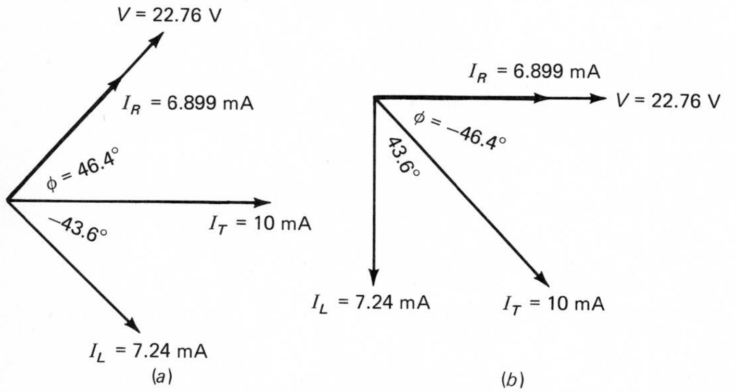

Fig. 14-43 Example 14-19: phasor diagrams. (*a*) Source current I_T as reference; (*b*) common voltage V as reference.

Figure 14-43*b* shows the phasor diagram with the common voltage as the reference phasor. This is the conventional way of representing parallel circuit phasors. Figure 14-43*b* can be considered to be Fig. 14-43*a* rotated clockwise through the circuit phase angle 46.4°.

The current divider principle was used in the solutions for Example 14-19 as a demonstration. The currents could have been found by Ohm's law; perhaps more easily.

Example 14-20 What is the impedance at 2 MHz of a parallel circuit consisting of a 47-Ω resistance and a 10-μH inductance?

Solution See Fig. 14-44.

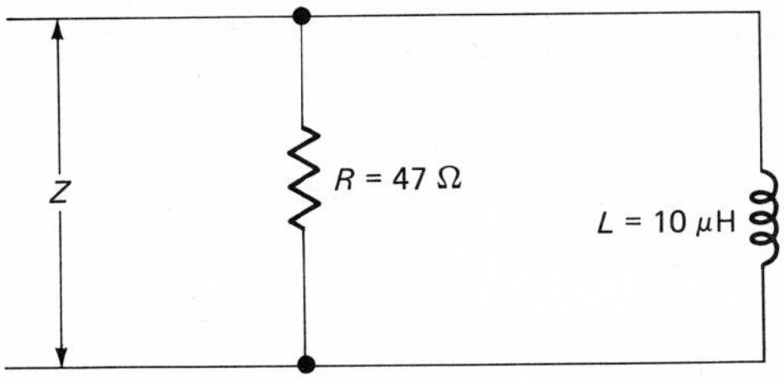

Fig. 14-44 Example 14-20: circuit.

$$|X_L| = \omega L = 2\pi f L = 2\pi \times 2\text{ MHz} \times 10\ \mu\text{H} = 125.7\ \Omega$$

$$X_L = j125.7\ \Omega \qquad \text{or} \qquad 125.7\ \Omega \angle 90^\circ$$

Apply the general relationship for the total impedance of two parallel impedances:

$$Z = \frac{Z_1 Z_2}{Z_1 + Z_2} = \frac{RX_L}{R + X_L} = \frac{47\ \Omega \angle 0^\circ \times 125.7\ \Omega \angle 90^\circ}{(47 + j125.7)\ \Omega}$$

$$= \frac{(47 \times 125.7)\ \Omega \angle 0^\circ + 90^\circ}{134.2\ \Omega \angle 69.5^\circ}$$

$$= \frac{5.908\text{ k}\Omega \angle 90^\circ}{134.2\ \Omega \angle 69.5^\circ} = (5908 \div 134.2)\ \Omega \angle 90^\circ - 69.5^\circ$$

$$Z = \mathbf{44.02\ \Omega \angle 20.5^\circ} \qquad \text{or} \qquad \mathbf{\angle 0.3578\ rad}$$

Example 14-21 Using data from the solutions for Examples 14-18 to 14-20, find the power factors of the circuits in Figs. 14-41, 14-42, and 14-44.

From Example 14-18,

$$P = EI_R = 10\text{ V} \times 10\text{ mA} = 100\text{ mW}$$

$$Q = EI_L = 10\text{ V} \times 6.365\text{ mA} = 63.65\text{ mvar}$$

$$S = \sqrt{P^2 + Q^2} = \sqrt{(100\text{ mV})^2 + (63.65\text{ mvar})^2} = 118.5\text{ mVA}$$

$$\cos\phi = \frac{P}{S} = \frac{100\text{ mW}}{118.5\text{ mVA}} = \mathbf{0.8439}$$

From Example 14-19,

$$\cos\phi = \frac{I_R}{I_T} = \frac{6.899\text{ mA}}{10\text{ mA}} = \mathbf{0.6899}$$

From Example 14-20,

$$\cos\phi = \frac{Z}{R} = \frac{44.02\ \Omega}{47\ \Omega} = \mathbf{0.9366}$$

14-5 SERIES ↔ PARALLEL CONVERSION

In network analysis it is sometimes convenient to determine the parallel equivalent of a series ac circuit, or vice-versa. For example, the series-parallel resonant network in Fig. 14-45*a* can be changed, as far as characteristics are concerned, to the totally parallel network in Fig. 14-45*b* by converting the series *RL* network L_sR_s into its parallel equivalent L_pR_p.

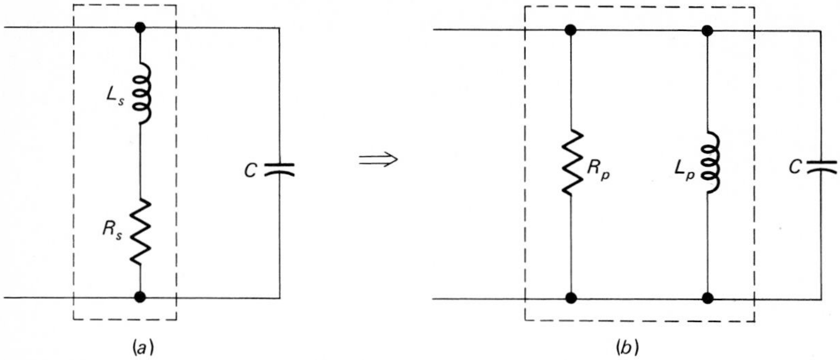

Fig. 14-45 (*a*) Series-parallel circuit; (*b*) parallel equivalent.

The circuit in Fig. 14-45*b* can be analyzed by direct parallel circuit methods. The resulting calculations and phasor diagrams are more simple than those of the original series-parallel circuit. This simplification can be helpful in understanding circuit behavior.

To perform the conversions it is necessary only to equate impedances or admittances. This procedure involves rationalization, which will be dealt with first.

RATIONALIZING COMPLEX QUANTITIES

Rationalizing is used to separate the real and imaginary parts of an expression which has a complex denominator. Unless this separation is made, the expression lacks clarity. For example, $Z = [1/(0.03 + j0.04)]\ \Omega$ is not a clear way of expressing an impedance. By rationalizing, this expression can be converted to $Z = (12 - j16)\ \Omega$, which is in the standard rectangular form for impedance. A circuit diagram and an impedance triangle can be drawn, as in Fig. 14-46.

Fig. 14-46 The impedance $Z = [1/(0.03 + j0.04)]\ \Omega$. (*a*) As an equivalent network; (*b*) as an impedance triangle.

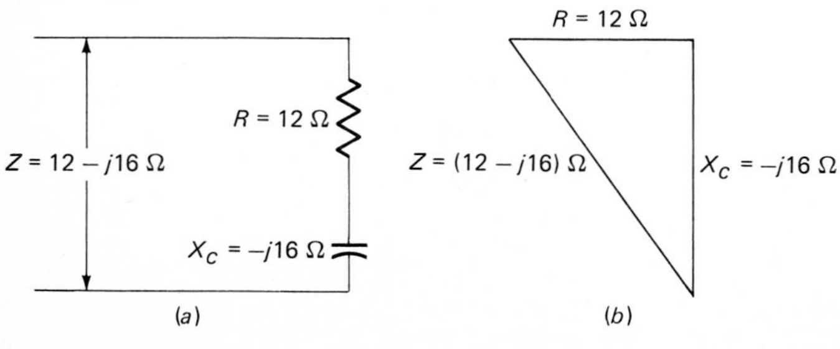

The technique of rationalization is based on the principle that a quantity is unchanged in value if it is multiplied by 1. The 1 can be any convenient quantity divided by itself. Another way of saying "multiplied by 1" is to say "numerator and denominator multiplied by the same quantity."

In complex number arithmetic the quantity chosen for division by itself is the conjugate of the denominator in the original expression. When a complex number is multiplied by its conjugate, the imaginary parts are eliminated, leaving just a real number. For example, consider the expression $1/(R+jX)$. The denominator is $R+jX$, so the expression will be multiplied by $(R-jX)/(R-jX)$,

$$\frac{1}{R+jX}=\frac{1}{R+jX}\times\frac{R-jX}{R-jX}=\frac{R-jX}{(R+jX)(R-jX)}$$

Concentrating first on the final denominator as the product of two binomials, each being rectangular-form complex numbers,

$$(R+jX)(R-jX)=R^2-jRX+jRX-j^2X^2$$

$$\text{Denominator}=R^2-j^2X^2$$

Since $j^2=-1$, it can be seen that $-j^2=+1$, so that

$$\text{Denominator}=R^2+X^2$$

We started with a denominator $R+jX$ which contained both real and imaginary parts. We now have a denominator which has only real parts. The latter denominator is an ordinary real number by which each part of the numerator can be divided. Therefore,

$$\frac{1}{R+jX}=\frac{R-jX}{R^2+X^2}=\frac{R}{R^2+X^2}-j\frac{X}{R^2+X^2}$$

The overall result of this rationalization is to convert a fraction, $1/(R+jX)$, which has a complex denominator to a rectangular phasor form consisting of a real part, $R/(R^2+X^2)$, and an imaginary part, $-jX/(R^2+X^2)$.

You should check to see that, if the original denominator had been $R-jX$, the final denominator would also have been R^2+X^2 but the imaginary part of the final expression would have been $+jX/(R^2+X^2)$.

Summarizing the results of rationalizing a fraction with a complex denominator,

$$\frac{1}{R+jX}=\frac{R}{R^2+X^2}-j\frac{X}{R^2+X^2}$$

$$\frac{1}{R-jX}=\frac{R}{R^2+X^2}+j\frac{X}{R^2+X^2}$$

The numerical example given above and displayed in Fig. 14-46 is thus evaluated:

$$Z = \frac{1}{0.03 + j0.04} \times \frac{0.03 - j0.04}{0.03 - j0.04}$$

$$= \frac{0.03}{0.03^2 + 0.04^2} - j\frac{0.04}{0.03^2 + 0.04^2}$$

$$Z = (12 - j16)\ \Omega$$

The application of rationalization to series ↔ parallel conversion can now proceed.

Series-to-parallel conversion, *RC* In Fig. 14-47*a* is shown a resistance R_s in series with a capacitive reactance $-jX_{Cs}$. The circuit impedance Z_s is given by

$$Z_s = R_s - jX_{Cs}$$

Its admittance Y_s will be

$$Y_s = \frac{1}{Z_s} = \frac{1}{R_s - jX_{Cs}}$$

Rationalizing, we obtain a rectangular expression for the admittance of the series circuit:

$$Y_s = \frac{R_s}{R_s^2 + X_{Cs}^2} + j\frac{X_{Cs}}{R_s^2 + X_{Cs}^2}$$

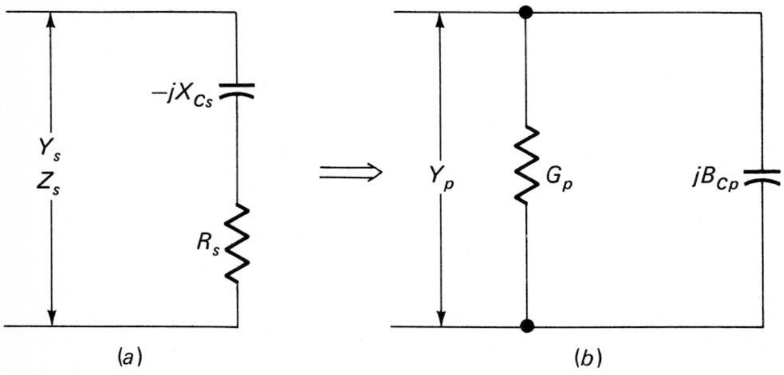

Fig. 14-47 (*a*) Series *RC* network; (*b*) its parallel equivalent.

The admittance of the parallel circuit in Fig. 14-47*b* is

$$Y_p = G_p + jB_{Cp}$$

For the parallel circuit and the series circuit to be equivalent, their admittances must be equal. Therefore,

$$Y_p = Y_s$$

$$G_p + jB_{Cp} = \frac{R_s}{R_s^2 + X_{Cs}^2} + j\frac{X_{Cs}}{R_s^2 + X_{Cs}^2}$$

This equality can only be valid if the real parts are equal *and* if the imaginary parts are equal. Thus,

$$G_p = \frac{R_s}{R_s^2 + X_{Cs}^2} \tag{14-40}$$

$$jB_{Cp} = j\frac{X_{Cs}}{R_s^2 + X_{Cs}^2} \tag{14-41}$$

Equations (14-40) and (14-41) give the necessary conditions for the conversion of the series circuit in Fig. 14-47*a* to its parallel equivalent, Fig. 14-47*b*.

Parallel-to-series conversion, *RC* With reference to Fig. 14-48:

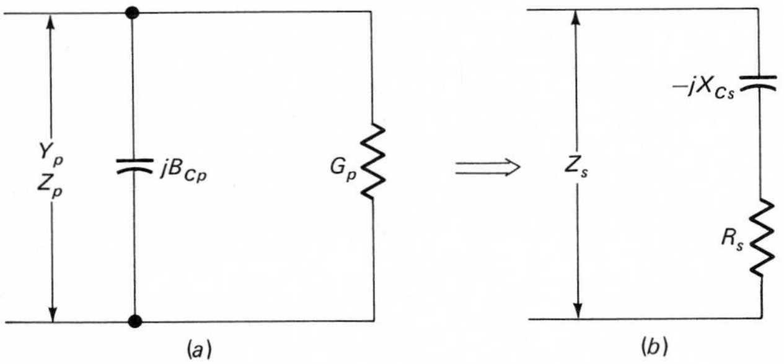

Fig. 14-48 (*a*) Parallel *RC* network; (*b*) its series equivalent.

$$Y_p = G_p + jB_{Cp}$$

$$Z_p = \frac{1}{Y_p} = \frac{1}{G_p + jB_{Cp}}$$

Rationalizing,

$$Z_p = \frac{G_p}{G_p^2 + B_{Cp}^2} - j\frac{B_{Cp}}{G_p^2 + B_{Cp}^2}$$

To obtain expressions for series elements, impedances are equated:

$$Z_s = Z_p$$

$$R_s - jX_{Cs} = \frac{G_p}{G_p^2 + B_{Cp}^2} - j\frac{B_{Cp}}{G_p^2 + B_{Cp}^2}$$

The values of the series elements are found by equating the real parts and the imaginary parts of the impedances:

$$R_s = \frac{G_p}{G_p^2 + B_{Cp}^2} \tag{14-42}$$

$$-jX_{Cs} = -j\,\frac{B_{Cp}}{G_p^2 + B_{Cp}^2} \tag{14-43}$$

Series-to-parallel conversion, *RL* With reference to Fig. 14-49,

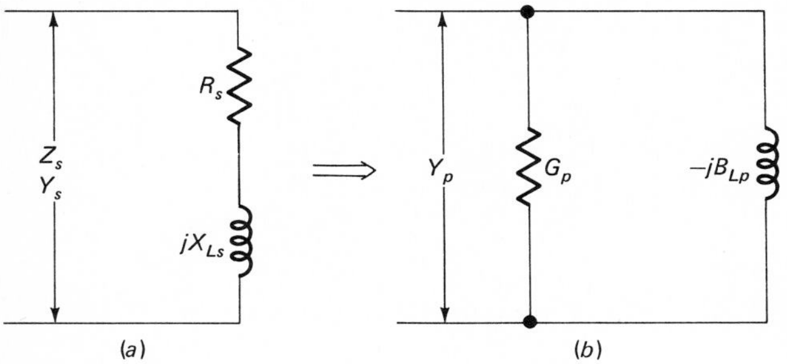

Fig. 14-49 (*a*) Series *RL* network; (*b*) its parallel equivalent.

$$Z_s = R_s + jX_{Ls}$$

$$Y_s = \frac{1}{Z_s} = \frac{1}{R_s + jX_{Ls}} = \frac{R_s}{R_s^2 + X_{Ls}^2} - j\,\frac{X_{Ls}}{R_s^2 + X_{Ls}^2}$$

Equating admittances,

$$Y_p = Y_s$$

$$G_p - jB_{Lp} = \frac{R_s}{R_s^2 + X_{Ls}^2} - j\,\frac{X_{Ls}}{R_s^2 + X_{Ls}^2}$$

Therefore, for Fig. 14-49*a* and *b* to be equivalent, the parallel elements must have the magnitudes

$$G_p = \frac{R_s}{R_s^2 + X_{Ls}^2} \tag{14-44}$$

$$-jB_{Lp} = -j\,\frac{X_{Ls}}{R_s^2 + X_{Ls}^2} \tag{14-45}$$

Parallel-to-series conversion, *RL* With reference to Fig. 14-50,

$$Y_p = G_p - jB_{Lp}$$

$$Z_p = \frac{1}{G_p - jB_{Lp}} = \frac{G_p}{G_p^2 + B_{Lp}^2} + j\,\frac{B_{Lp}}{G_p^2 + B_{Lp}^2}$$

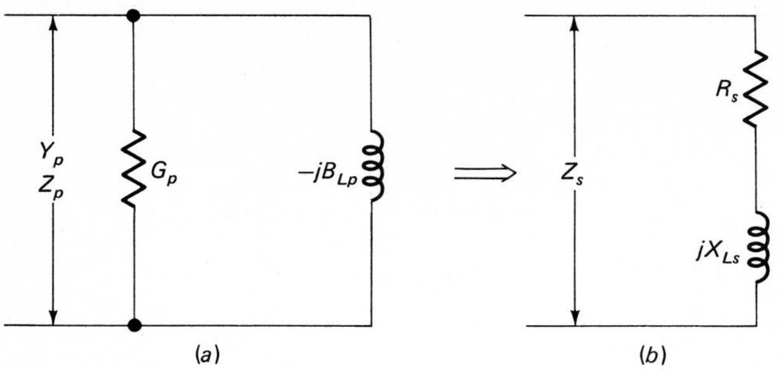

Fig. 14-50 (*a*) Parallel *RL* network; (*b*) its series equivalent.

Equating impedances,

$$Z_s = Z_p$$

$$R_s + jX_{Ls} = \frac{G_p}{G_p^2 + B_{Lp}^2} + j\frac{B_{Lp}}{G_p^2 + B_{Lp}^2}$$

The magnitudes of the elements in the equivalent series circuit will be

$$R_s = \frac{G_p}{G_p^2 + B_{Lp}^2} \qquad (14\text{-}46)$$

$$jX_{Ls} = +j\frac{B_{Lp}}{G_p^2 + B_{Lp}^2} \qquad (14\text{-}47)$$

Equations (14-40) to (14-47), for the values of impedance and admittance obtained as a result of series ↔ parallel conversion, can be summarized:

1. The nature of the elements does not change. For example a capacitive element in the original circuit results in a capacitive element in the equivalent circuit.
2. If the conversion is series to parallel, the resulting elements are admittances. For a parallel-to-series conversion the resulting elements are impedances.
3. The denominator for any derived element is the sum of the squares of the magnitudes of the original circuit elements.
4. The numerator for any derived element is the magnitude of the related element in the original circuit. For instance, the magnitude of an original resistance becomes the numerator of a conductance.
5. The sign of an imaginary quantity follows the general conventions (examples are $-jX_C$ and $+jB_C$).

Example 14-22 A series circuit consists of a resistance of 1 kΩ and a capacitor of 0.01 μF. What are the values of resistance and capacitance in the equivalent parallel circuit at a frequency of 16 kHz?

Solution See Fig. 14-51.

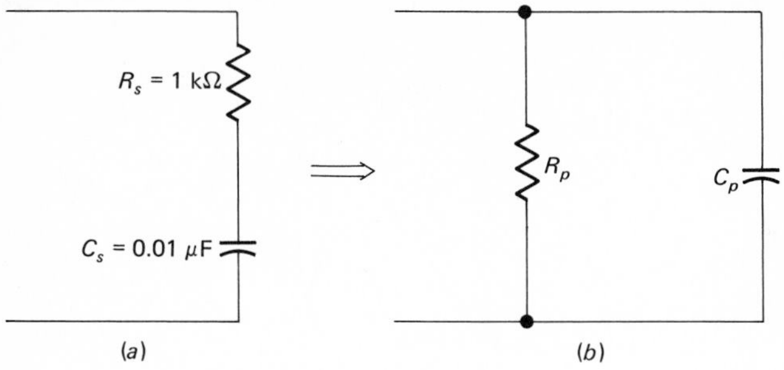

Fig. 14-51 Example 14-22: (*a*) given network; (*b*) its parallel equivalent.

$$\omega = 2\pi f = 2\pi \times 16 \text{ kHz} \simeq 100 \text{ krad/s}$$

$$X_{Cs} = \frac{1}{\omega C_s} = \frac{1}{100 \text{ krad/s} \times 0.01\ \mu\text{F}} = 1 \text{ k}\Omega$$

$$R_s^2 + X_{Cs}^2 = (1 \text{ k}\Omega)^2 + (1 \text{ k}\Omega)^2 = 2 \text{ M}\Omega^2$$

$$G_p = \frac{R_s}{R_s^2 + X_{Cs}^2} \tag{14-40}$$

$$G_p = \frac{1 \text{ k}\Omega}{2 \text{ M}\Omega^2} = 0.5 \text{ mS}$$

$$R_p = \frac{1}{G_p} = \frac{1}{0.5 \text{ mS}} = \mathbf{2\ k\Omega}$$

$$jB_{Cp} = j\frac{X_{Cs}}{R_s^2 + X_{Cs}^2} \tag{14-41}$$

$$B_{Cp} = \frac{1 \text{ k}\Omega}{2 \text{ M}\Omega^2} = 0.5 \text{ mS}$$

$$B_{Cp} = \omega C$$

$$C = \frac{B_{Cp}}{\omega} = \frac{0.5 \text{ mS}}{100 \text{ krad/s}} = \mathbf{0.005\ \mu F}$$

The numbers in Example 14-22 were kept simple to aid in relating the original series circuit to its parallel equivalent.

It is important to be aware that the actual capacitance value changes in conversion. Only by this awareness can you hope to understand why the value of the capacitance of a capacitor is different for the series and parallel viewpoints. Instruments for the measurement of the characteristics of capacitors differ, some using the series view and others the parallel. This difference between the series capacitance and the parallel capacitance of a practical capacitor is brought out in Example 14-23.

The series and parallel inductances of a practical inductor are also different.

You should note that the approximately equal sign (≃), is used only once in the solution to Example 14-22. Naturally, everything which follows is approximate if it depends on the value of ω. The only point in using the approximately equal sign is to indicate that the value given is close enough for our purposes. Therefore there is no need to maintain repetition of the sign.

Also note that the symbol distinction between scalar and phasor values was not made in Example 14-22.

It is not usual to maintain a rigid discipline in using the scalar value symbol unless clarity demands it. An example is $\sqrt{R^2 + X^2}$, which really ought to be written $\sqrt{|R|^2 + |X|^2}$. You will probably agree that the scalar symbol does nothing to clarify such a situation. Whether or not the scalar symbol should be used depends on whether or not the circumstances warrant it.

Example 14-23 An electrolytic capacitor has a power factor of 0.05 and a capacitance of 25 μF when measured on a 1-kHz bridge (an instrument for measuring resistance, capacitance, and inductance). The bridge uses the parallel approach to capacitor measurement. What is the capacitor's (*a*) parallel resistance, (*b*) series resistance, and (*c*) series capacitance?

Solution See Fig. 14-52.

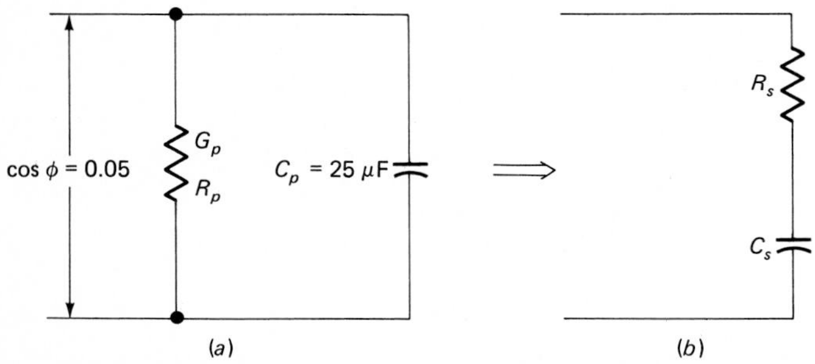

Fig. 14-52 Example 14-23: (*a*) parallel equivalent network of the capacitor; (*b*) its series equivalent.

$$\omega = 2\pi f = 2\pi \times 1\ \text{kHz} = 6.283\ \text{krad/s}$$

$$B_{Cp} = \omega C = 6.283\ \text{krad/s} \times 25\ \mu\text{F} = 157.1\ \text{mS}$$

From Eq. (14-28),

$$\cos\phi = \frac{G_p}{Y_p} = \frac{G_p}{\sqrt{G_p^2 + B_{Cp}^2}}$$

$$(G_p \cos\phi)^2 + (B_{Cp}\cos\phi)^2 = G_p^2$$

$$G_p^2(1 - \cos^2\phi) = (B_{Cp}\cos\phi)^2$$

$$G_p^2 = \frac{(B_{Cp}\cos\phi)^2}{1 - \cos^2\phi} = \frac{(157.1\ \text{mS} \times 0.05)^2}{1 - 0.05^2}$$

$$G_p \simeq 157.1\ \text{mS} \times 0.05 \simeq 7.855\ \text{mS}$$

a. $$R_p = \frac{1}{G_p} \simeq \frac{1}{7.855\ \text{mS}} \simeq \mathbf{127.3\ \Omega}$$

b.

$$G_p^2 + B_{Cp}^2 = (7.855 \text{ mS})^2 + (157.1 \text{ mS})^2 \simeq 24.74 \text{ mS}^2$$

$$R_s = \frac{G_p}{G_p^2 + B_{Cp}^2} \qquad (14\text{-}42)$$

$$R_s = \frac{7.855 \text{ mS}}{24.74 \text{ mS}^2} = \mathbf{0.3175\ \Omega}$$

c.

$$-jX_{Cs} = -j\frac{B_{Cp}}{G_p^2 + B_{Cp}^2} \qquad (14\text{-}42)$$

$$-jX_{Cs} = -j\frac{157.1 \text{ mS}}{24.74 \text{ mS}^2} = -j6.35\ \Omega$$

$$X_{Cs} = \frac{1}{\omega C_s}$$

$$C_s = \frac{1}{\omega X_{Cs}} = \frac{1}{6.283 \text{ krad/s} \times 6.35\ \Omega} = \mathbf{25.06\ \mu F}$$

The power factor of the capacitor in Example 14-23 is too high for many purposes. Such a high power factor could only be tolerated in an electrolytic capacitor used in high-power applications. This high value was given to show that it is possible to obtain a noticeably different capacitance value from the series and parallel viewpoints of the same capacitor. The difference between series and parallel capacitance is rarely an important matter, since the power factor of a capacitor is usually much lower than 0.05, resulting in the series and parallel equivalent capacitance values being virtually the same.

Example 14-24 The series-parallel network in Fig. 14-53*a* has the values $L_s = 100\ \mu H$, $R_s = 50\ \Omega$, and $C = 400$ pF. What are the element values in the equivalent parallel circuit in Fig. 14-53*b* at a frequency of 0.8 MHz?

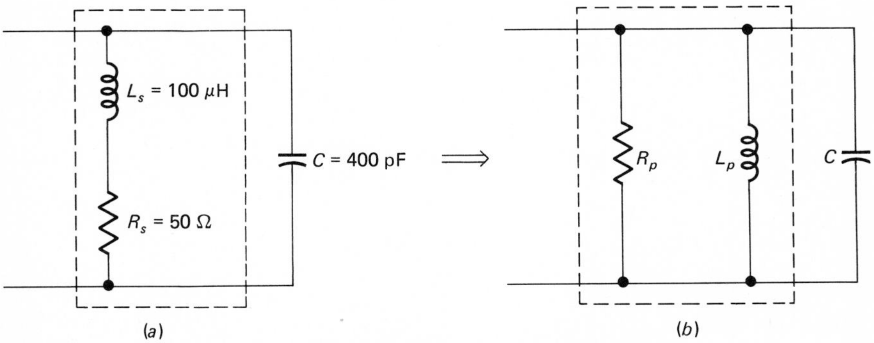

Fig. 14-53 Example 14-24: (*a*) given network; (*b*) parallel equivalent.

Solution Convert the series circuit L_sR_s to the parallel equivalent L_pR_p:

$$\omega = 2\pi f = 2\pi \times 0.8 \text{ MHz} = 5.027 \text{ Mrad/s}$$

$$X_{Ls} = \omega L = 5.027 \text{ Mrad/s} \times 100\ \mu H = 502.7\ \Omega$$

$$R_s^2 + X_{Ls}^2 = 255.2\ \text{k}\Omega^2$$

$$G_p = \frac{R_s}{R_s^2 + X_{Ls}^2} \qquad (14\text{-}44)$$

$$G_p = \frac{50\ \Omega}{255.2\ \text{k}\Omega^2} = 0.196\ \text{mS}$$

$$R_p = \frac{1}{G_p} = \frac{1}{0.196\ \text{mS}} = \mathbf{5.102\ k\Omega}$$

$$B_{Lp} = \frac{X_{Ls}}{R_s^2 + X_{Ls}^2} \qquad (14\text{-}45)$$

$$B_{Lp} = \frac{502.7\ \Omega}{255.2\ \text{k}\Omega^2} = 1.97\ \text{mS}$$

$$B_{Lp} = \frac{1}{\omega L_p}$$

$$L_p = \frac{1}{\omega B_{Lp}} = \frac{1}{5.027\ \text{Mrad/s} \times 1.97\ \text{mS}}$$

$$L_p = \mathbf{101\ \mu H}$$

The capacitance was not involved in the conversion process, so in the parallel circuit it is 400 pF, exactly the same as in the series circuit.

Figure 14-54 shows the values of the circuit before and after the conversion from series-parallel to parallel. The parallel equivalent inductance differs from the original series value by only 1 percent, because the Q of the RL circuit is fairly high (a little over 10).

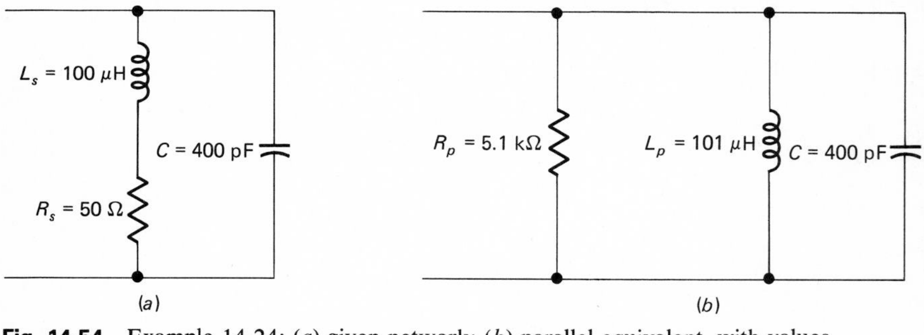

Fig. 14-54 Example 14-24: (*a*) given network; (*b*) parallel equivalent, with values.

14-6 SERIES-PARALLEL CIRCUITS (*RC* AND *RL*)

When the techniques for handling series and parallel ac circuits have been mastered, series-parallel circuits can be dealt with by adopting an approach similar to that used for dc circuits. This means analyzing the series-parallel circuit as series and parallel subcircuits.

As it is pointless to attempt to develop general rules which can be applied to the infinite variety of series-parallel networks, some numerical examples are given showing how to tackle typical problems. The techniques demonstrated are applied to practical situations in later chapters.

Example 14-25 What is the impedance of the network in Fig. 14-55?

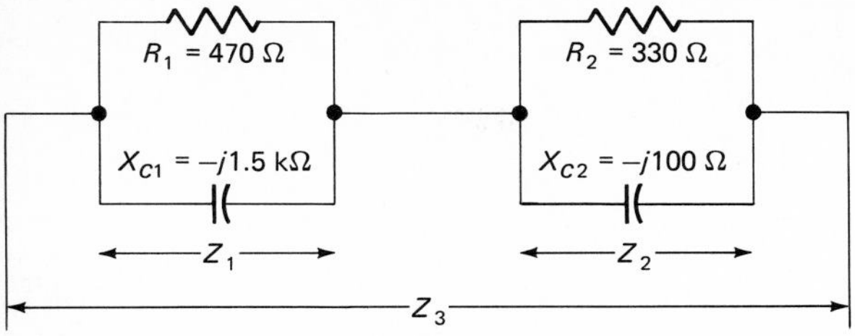

Fig. 14-55 Example 14-25: network.

Solution Z_1 is the impedance of the parallel combination of R_1 and X_{C1}. Applying the principle for the determination of the total impedance of two parallel impedances $[Z = Z_aZ_b/(Z_a + Z_b)]$,

$$Z_1 = \frac{R_1X_{C1}}{R_1 + X_{C1}} = \frac{470\ \Omega(-j1.5\ \text{k}\Omega)}{(470 - j1.5)\ \text{k}\Omega} = \frac{705\ \text{k}\Omega^2\ \underline{/-90°}}{1.572\ \text{k}\Omega\ \underline{/-72.6°}} = 448\ \Omega\ \underline{/-17.4°}$$

Z_2 is the impedance of the parallel combination of R_2 and X_{C2}

$$Z_2 = \frac{R_2X_{C2}}{R_2 + X_{C2}} = \frac{330\ \Omega(-j100\ \Omega)}{(330 - j100)\ \Omega} = \frac{33\ \text{k}\Omega^2\ \underline{/-90°}}{344.8\ \Omega\ \underline{/-16.88°}} = 95.71\ \Omega\ \underline{/-73.12°}$$

The complete network consists of the group impedances Z_1 and Z_2 connected in series. Therefore the total impedance Z_T is

$$Z_T = Z_1 + Z_2 = 448\ \Omega\ \underline{/-17.4°} + 95.71\ \Omega\ \underline{/-73.12°}$$

$$= (427.5 - j134)\ \Omega + (27.79 - j91.58)\ \Omega$$

$$Z_T = \mathbf{(455.3 - j225.6)\ \Omega = 508.1\ \Omega\underline{/-26.36°}} \quad \text{or} \quad \mathbf{\underline{/-0.46\ rad}}$$

Example 14-26 What is the admittance of the network in Fig. 14-56?

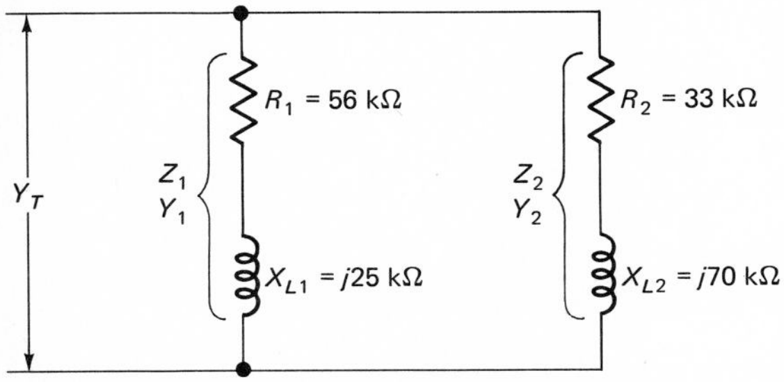

Fig. 14-56 Example 14-26: network.

Solution Z_1 and Y_1 are the impedance and admittance of the series combination of R_1 and X_{L1}.

$$Z_1 = R_1 + jX_{L1} = (56 + j25)\ \text{k}\Omega$$

$$Y_1 = \frac{1}{Z_1} = \frac{1}{(56 + j25)\ \text{k}\Omega} = \frac{56\ \text{k}\Omega}{(56\ \text{k}\Omega)^2 + (25\ \text{k}\Omega)^2} - j\frac{25\ \text{k}\Omega}{(56\ \text{k}\Omega)^2 + (25\ \text{k}\Omega)^2}$$

$$Y_1 = (14.89 - j6.647)\ \mu\text{S}$$

Z_2 and Y_2 are the impedance and admittance of the series combination of R_2 and X_{L2}.

$$Z_2 = R_2 + jX_{L2} = (33 + j70)\ \text{k}\Omega$$

$$Y_2 = \frac{1}{Z_2} = \frac{1}{(33 + j70)\ \text{k}\Omega} = \frac{33\ \text{k}\Omega}{(33\ \text{k}\Omega)^2 + (70\ \text{k}\Omega)^2} - j\frac{70\ \text{k}\Omega}{(33\ \text{k}\Omega)^2 + (70\ \text{k}\Omega)^2}$$

$$Y_2 = (5.51 - j11.69)\ \mu\text{S}$$

The complete network consists of the admittances Y_1 and Y_2 connected in parallel. The total admittance Y_T will be

$$Y_T = Y_1 + Y_2 = (14.89 - j6.647)\ \mu\text{S} + (5.51 - j11.69)\ \mu\text{S}$$

$$Y_T = \mathbf{(20.4 - j18.34)\ \mu S = 27.43\ \mu S\ \underline{/-41.97^\circ}} \quad \text{or} \quad \mathbf{\underline{/-0.7324\ rad}}$$

Example 14-27 What is the load voltage V_L in the circuit in Fig. 14-57?

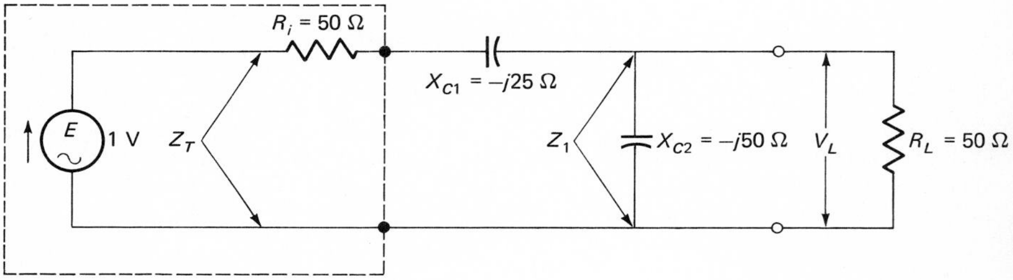

Fig. 14-57 Example 14-27: circuit.

Solution Z_1 is the impedance of the parallel network R_LX_{C2}.

$$Z_1 = \frac{R_LX_{C2}}{R_L + X_{C2}} = \frac{50\ \Omega(-j50\ \Omega)}{(50 - j50)\ \Omega} = \frac{2500\ \Omega\ \underline{/-90^\circ}}{70.71\ \Omega\ \underline{/-45^\circ}}$$

$$Z_1 = 35.36\ \Omega\ \underline{/-45^\circ} \quad \text{or} \quad (25 - j25)\ \Omega$$

Z_T is the total circuit impedance; it is Z_1 in series with R_i and X_{C1}.

$$Z_T = Z_1 + R_i + X_{C1} = (25 - j25)\ \Omega + (50 + j0)\ \Omega + (0 - j25)\ \Omega$$

$$Z_T = (75 - j50)\ \Omega \quad \text{or} \quad 90.14\ \Omega\ \underline{/-33.69^\circ}$$

The voltage divider principle can be applied to this example:

$$V_L = E\frac{Z_1}{Z_T} = 1\ \text{V}\ \underline{/0^\circ}\ \frac{35.36\ \Omega\ \underline{/-45^\circ}}{90.14\ \Omega\ \underline{/-33.69^\circ}}$$

$$V_L = \mathbf{0.3923\ V\ \underline{/-11.31^\circ}} \quad \text{or} \quad \mathbf{\underline{/-0.1974\ rad}}$$

Example 14-28 What is the value of the load current I_L in the circuit in Fig. 14-58? Assume that no mutual inductance is present.

Solution The impedance of R and X_{L1} in series is Z_1.

$$Z_1 = R + jX_L = (100 + j50)\ \Omega$$

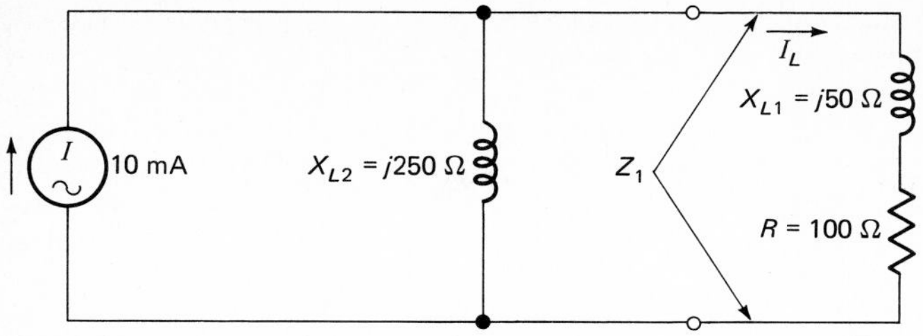

Fig. 14-58 Example 14-28: circuit.

The current divider principle can be applied to find the value of the current I_L:

$$I_L = I\frac{X_{L2}}{X_{L2}+Z_1} = 10\text{ mA}\ \underline{/0^\circ}\ \frac{250\ \Omega\ \underline{/90^\circ}}{j250\ \Omega + (100+j50)\ \Omega}$$

$$= \frac{10\text{ mA}\ \underline{/0^\circ} \times 250\ \Omega\ \underline{/90^\circ}}{(100+j300)\ \Omega}$$

$$I_L = \frac{10\text{ mA}\ \underline{/0^\circ} \times 250\ \Omega\ \underline{/90^\circ}}{316.2\ \Omega\ \underline{/71.57^\circ}} = \underline{\mathbf{7.906\ mA}\ /18.43^\circ} \qquad \text{or} \qquad \underline{\mathbf{/0.3216\ rad}}$$

Example 14-29 Figure 14-59*a* is a schematic diagram of a two-stage transistor amplifier. An equivalent circuit of the network coupling transistors Q_1 and Q_2 is given in Fig. 14-59*b*. R_L is the effective load coupled by capacitor C_c and includes the effects of R_2 and R_3 together with the input resistance of Q_2. C_s is the stray capacitance of the input circuit of Q_2 including its wiring. What is the bandwidth of the network coupling transistors Q_1 and Q_2? The *bandwidth* is defined as the difference in frequency between the points at which the magnitude of the voltage V_B fed to transistor Q_2 is $1/\sqrt{2}$ of its maximum value.

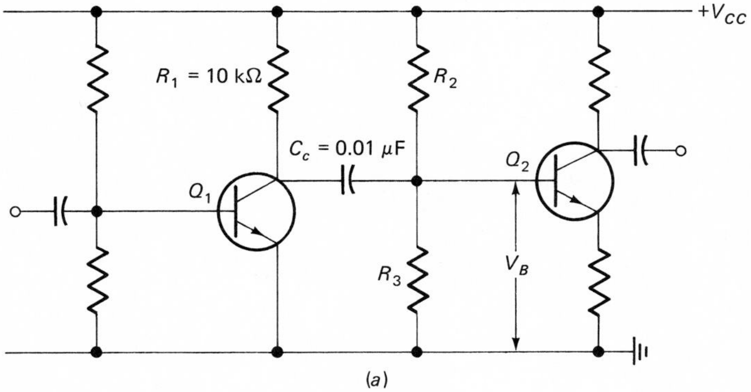

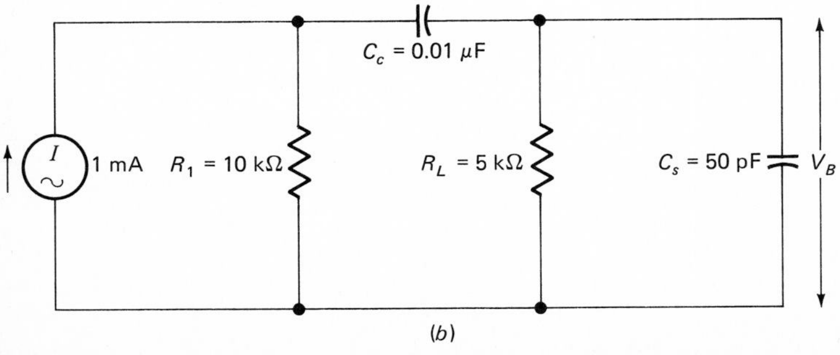

Fig. 14-59 Example 14-29: (*a*) given circuit; (*b*) equivalent of the base-to-emitter circuit of Q_2.

Solution The current generator in Fig. 14-59*b* represents transistor Q_1 in its effect on driving transistor Q_2. It is assumed that this source produces a constant ac current of 1 mA. One milliampere is chosen as a convenient current level since only the variation in voltage V_B, with respect to frequency, is of concern in this example.

MID-FREQUENCY The maximum possible value of voltage V_B would occur if the effects of both capacitances C_c and C_s could be neglected. This condition does occur at a so-called mid-frequency when the series reactance of C_1 and the shunt susceptance of C_s are both small compared with the values of the other circuit elements. Figure 14-60*a* shows this situation.

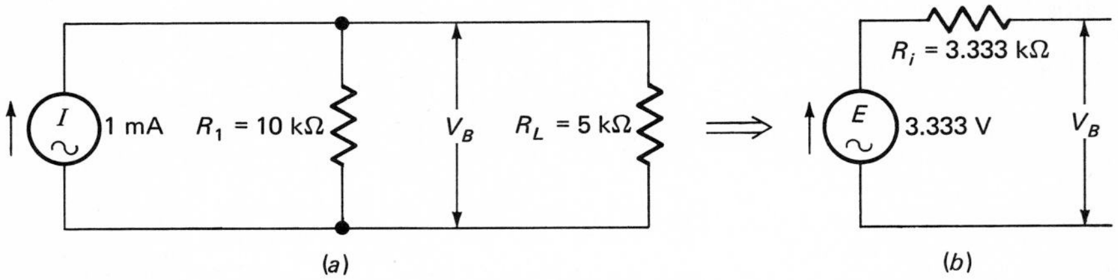

Fig. 14-60 Example 14-29: (*a*) mid-frequency equivalent of Fig. 14-59*b*; (*b*) Thevenin equivalent.

The complete circuit in Fig. 14-60*a* can be regarded as a current generator with a source current of 1 mA and an internal conductance consisting of R_1 and R_L in parallel. This current generator can be converted to the voltage generator in Fig. 14-60*b*. Its internal resistance R_i is the parallel combination of R_1 and R_L:

$$R_i = \frac{R_1 R_L}{R_1 + R_L} = \frac{10\ \text{k}\Omega \times 5\ \text{k}\Omega}{10\ \text{k}\Omega + 5\ \text{k}\Omega} = 3.333\ \text{k}\Omega$$

The voltage V_B will equal the source emf of the voltage generator:

$$V_B = E = IR_i = 1\ \text{mA} \times 3.333\ \text{k}\Omega = 3.333\ \text{V}$$

This mid-frequency value of V_B (3.333 V) becomes our reference voltage level. The task now is to determine two frequencies, one below the mid-frequency and one above it, at which the value of V_B is 3.333 V $\div \sqrt{2} =$ 2.357 V.

LOW FREQUENCY At low frequencies the series reactance of C_c is significant, but the shunt susceptance of C_s is negligible. The effective circuit is that in Fig. 14-61*a*.

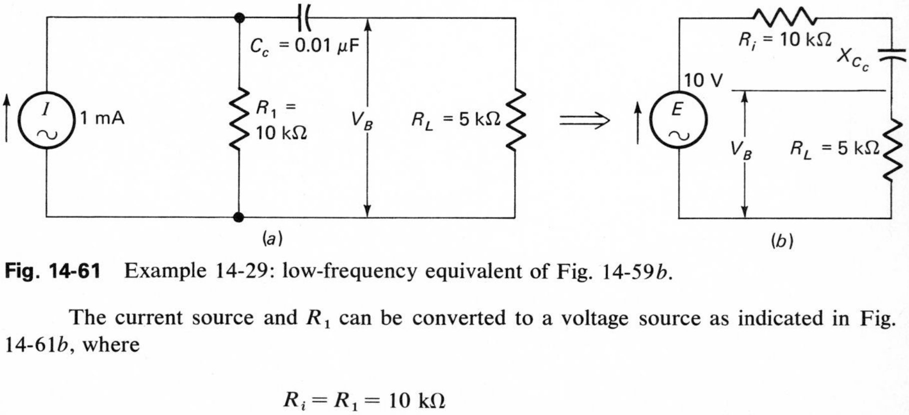

Fig. 14-61 Example 14-29: low-frequency equivalent of Fig. 14-59*b*.

The current source and R_1 can be converted to a voltage source as indicated in Fig. 14-61*b*, where

$$R_i = R_1 = 10\ \text{k}\Omega$$

and

$$E = IR_1 = 1\ \text{mA} \times 10\ \text{k}\Omega = 10\ \text{V}$$

V_B can be determined by the voltage divider principle:

$$V_B = E\frac{R_L}{Z_T} = E\frac{R_L}{R_L + R_i - jX_{Cc}} = 10\text{ V}\frac{5\text{ k}\Omega}{5\text{ k}\Omega + 10\text{ k}\Omega - jX_{Cc}}$$

The low frequency of interest occurs when $V_B = 2.357$ V.

$$2.357\text{ V} = 10\text{ V}\frac{5\text{ k}\Omega}{15\text{ k}\Omega - jX_{Cc}}$$

$$15\text{ k}\Omega - jX_{Cc} = 21.21\text{ k}\Omega$$

$$X_{Cc} = \sqrt{(21.21\text{ k}\Omega)^2 - (15\text{ k}\Omega)^2}$$

$$X_{Cc} = 15\text{ k}\Omega$$

$$\frac{1}{\omega C_c} = 15\text{ k}\Omega$$

$$\omega = \frac{1}{0.01\ \mu\text{F} \times 15\text{ k}\Omega} = 6.667\text{ krad/s}$$

$$2\pi f = 6.667\text{ krad/s}$$

$$f = 1.061\text{ kHz}$$

HIGH FREQUENCY At high frequencies the shunt susceptance of C_s is significant, but the series reactance of C_c is negligible. The equivalent circuit is as in Fig. 14-62.

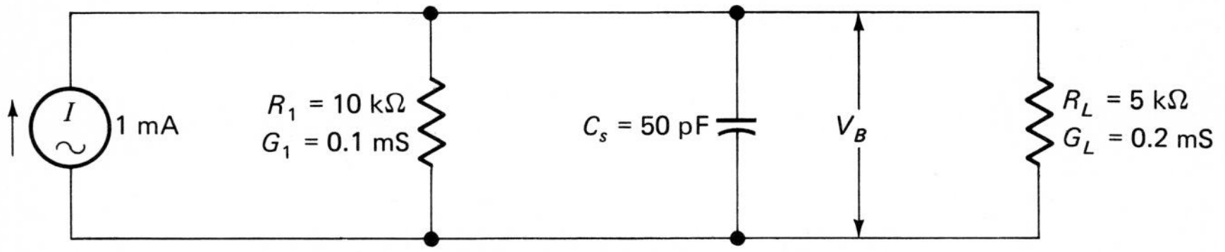

Fig. 14-62 Example 14-29: high-frequency equivalent of Fig. 14-59*b*.

The total load on the current source is

$$Y = G_1 + jB_{Cs} + G_L = 0.1\text{ mS} + jB_{Cs} + 0.2\text{ mS}$$

$$Y = 0.3\text{ mS} + jB_{Cs}$$

The voltage V_B will be given by Ohm's law in admittance form:

$$V_B = \frac{I}{Y} = \frac{1\text{ mA}}{0.3\text{ mS} + jB_{Cs}}$$

The high frequency we wish to determine occurs when $V_B = 2.357$ V.

$$2.357\text{ V} = \frac{1\text{ mA}}{0.3\text{ mS} + jB_{Cs}}$$

$$0.3\text{ mS} + jB_{Cs} = 0.4242\text{ mS}$$

$$B_{Cs} = \sqrt{(0.4242\text{ mS})^2 - (0.3\text{ mS})^2}$$

$$B_{Cs} = 0.3\text{ mS}$$

$$\omega C = 0.3\text{ mS}$$

$$\omega = \frac{0.3 \text{ mS}}{50 \text{ pF}} = 6 \text{ Mrad/s}$$

$$2\pi f = 6 \text{ Mrad/s}$$

$$f = \frac{6 \text{ Mrad/s}}{2\pi} = 0.955 \text{ MHz} = 955 \text{ kHz}$$

Therefore the bandwidth of the coupling network is the difference between the high frequency of 955 kHz and the low frequency of 1.061 kHz.

$$B = 955 \text{ kHz} - 1.061 \text{ kHz} = \mathbf{953.9 \text{ kHz}}$$

In Example 14-29 it can be seen that the lower bandwidth frequency occurs when X_{Cc} equals the net series resistance, and that the upper bandwidth frequency occurs when B_{Cs} equals the net shunt conductance. This is not a coincidence occurring in this particular example. The definition of bandwidth can be considered to be based on the 45° right triangles in Fig. 14-63.

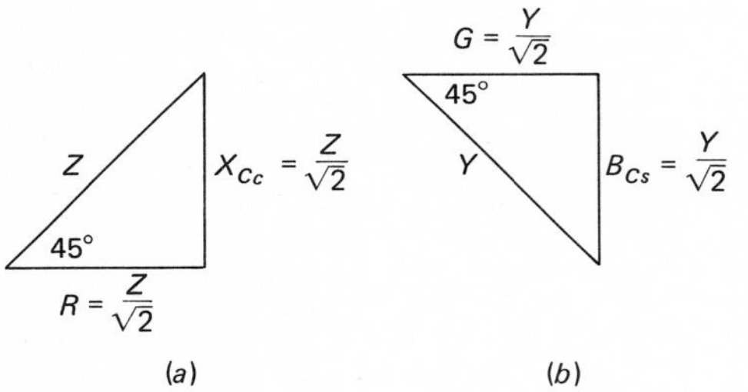

Fig. 14-63 Example 14-29: (*a*) lower bandwidth frequency impedance triangle; (*b*) upper bandwidth frequency admittance triangle.

In Fig. 14-63*a*,

$$R = X_{Cc} = \frac{Z}{\sqrt{2}}$$

In Fig. 14-63*b*,

$$G = B_{Cs} = \frac{Y}{\sqrt{2}}$$

Since the $1/\sqrt{2}$ points define bandwidth, the bandwidth frequencies of an *RC* coupling network can be described as occurring when

1. The series capacitive reactance equals the Thevenin equivalent resistance of the driving source or preceding network.
2. The shunt capacitive susceptance equals the Norton equivalent conductance of the driving source or preceding network.
3. The phase angle between the driving source and the coupled voltage is 45° or $\pi/4$ rad.

The terms "high" and "low" in the solution for Example 14-29 are relative terms. If one frequency is many times greater than another, "high" and "low" distinguishes between them. However, whether a frequency of 955 kHz is high depends on the circumstances.

Example 14-30 An outline diagram of a Wien bridge oscillator is shown in Fig. 14-64*a*. If the gains of amplifiers *A*1 and *A*2 are adequate, the circuit will oscillate; that is, it will act as an electronic ac generator. The frequency of oscillation will be determined by the frequency at which the phase angle of V_{AB}, the input to amplifier *A*1, is zero with respect to the voltage V_o. This frequency is controlled by the values of R_1, R_2, C_1, and C_2. What is the frequency of oscillation in terms of these elements?

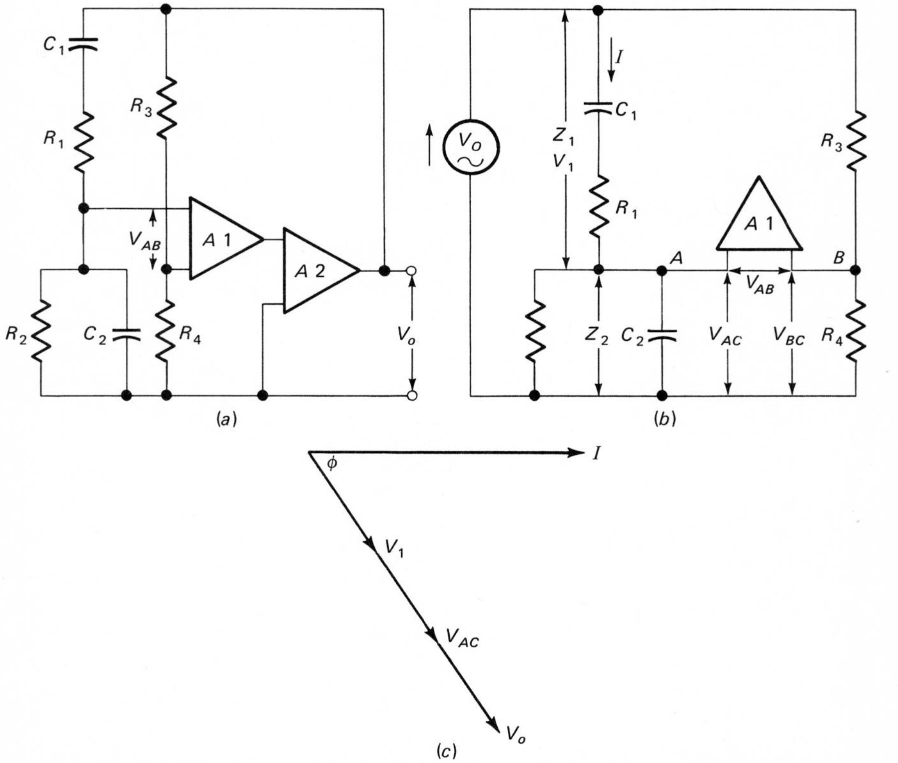

Fig. 14-64 Example 14-30: (*a*) given circuit; (*b*) part of the circuit rearranged with V_o as the input voltage; (*c*) phasor diagram.

Solution The phase-shift circuit controlling the input to amplifier *A*1 is given in Fig. 14-64*b*.

It is a general axiom that, if a number of networks, each with the same phase angle, are connected in series, the voltages across each network and across the complete series circuit are all in phase with each other. Figure 14-64*c* shows a phasor diagram in which the phase angles of the series network R_1C_1 and the parallel network R_2C_2 are each equal to $\phi°$. The current *I* common to both networks is given as the reference phasor. The voltage V_1 across the series network R_1C_1 lags the current by $\phi°$. The voltage V_{AC} across the parallel network R_2C_2 also lags the current by $\phi°$. The voltage V_o is the phasor sum of the in-phase voltages V_1 and V_{AC} and is in phase with each of them. Thus the voltage V_{AC} will be in phase with the voltage V_o if the phase angles of the series and parallel *RC* networks are equal. It will be obvious that, since the phase angles of R_3 and R_4 are both always zero, V_{BC} is in phase with V_o.

The input voltage V_{AB} of amplifier $A1$ is the phasor difference between V_{AC} and V_{BC}. Where V_{AB} and V_{BC} are each in phase with V_o, and hence in phase with each other, V_{AB} is in phase with V_o. The frequency of oscillation is the frequency at which these conditions exist, that is, the frequency at which the series network R_1C_1 has the same phase angle as the parallel network R_2C_2.

The phase angle of the series network R_1C_1 is given by

$$\phi_1 = \arctan \frac{X_{C1}}{R_1}$$

The phase angle of the parallel network R_2C_2 is given by

$$\phi_2 = \arctan \frac{B_{C2}}{G_2} = \arctan \frac{R_2}{X_{C2}}$$

The frequency of oscillation is the frequency at which ϕ_1 and ϕ_2 are equal, or when

$$\arctan \frac{X_{C1}}{R_1} = \arctan \frac{R_2}{X_{C2}}$$

or

$$\frac{X_{C1}}{R_1} = \frac{R_2}{X_{C2}}$$

$$\frac{1}{\omega C_1 \omega C_2} = R_1 R_2$$

$$\omega = \frac{1}{\sqrt{R_1 R_2 C_1 C_2}}$$

$$\boldsymbol{f = \frac{1}{2\pi \sqrt{R_1 R_2 C_1 C_2}}}$$

Example 14-31 If in the circuit in Fig. 14-64 capacitors C_1 and C_2 are both 0.1 μF, and resistors R_1 and R_2 are both variable with a range of 1 to 10 kΩ and are ganged so that both have the same resistance value at all times, what is the oscillation frequency range?
Solution If $C_1 = C_2 = C$ and $R_1 = R_2 = R$, the frequency of oscillation will be

$$f = \frac{1}{2\pi \sqrt{R^2 C^2}} = \frac{1}{2\pi RC}$$

When $R = 1$ kΩ,

$$f = \frac{1}{2\pi \times 1 \text{ k}\Omega \times 0.1\ \mu\text{F}} = 1.591 \text{ kHz}$$

When $R = 10$ kΩ,

$$f = \frac{1}{2\pi \times 10 \text{ k}\Omega \times 0.1\ \mu\text{F}} = 159.1 \text{ Hz}$$

The frequency of the oscillation is from **159.1 Hz** to **1.591 kHz.**

From the results for Example 14-31 it is clearly possible to make a Wien bridge oscillator of a chosen frequency range. In the example the capacitance value of 0.1 μF was selected for ease of calculation. Had the value been 0.1591 μF, the frequency range would have been 100 Hz to 1 kHz.

When capacitors of 0.01591, 0.1591, and 1.591 μF are connected into the circuit by selector switches, an oscillator with a frequency range of 10 Hz to 10 kHz will result. Such oscillators are commercially manufactured as signal generators for measuring and testing purposes. These generators are good examples of the application of ac circuit principles.

14-7 SUMMARY OF EQUATIONS

Series *RC* circuit

$V_T = V_R - jV_C$ (14-1)

$Z = R - jX_C$ (14-2)

$V_T I = V_R I - jV_C I$ (14-3)

$S = V_T I$ (14-4)

$P = V_T I \cos \phi$ or $S \cos \phi$ (14-5)

$jQ = jV_T I \sin \phi$ or $jS \sin \phi$ (14-6)

$S = P + jQ$ (14-7)

$\cos \phi = P/S$ (14-8)

$\cos \phi = V_R/V_T$ or R/Z (14-9)

Series *RL* circuit

$V_T = V_R + jV_L$ (14-10)

$Z = R + jX_L$ (14-11)

$V_T I = V_R I + jV_L I$ (14-12)

$S = V_T I$ (14-13)

$P = V_T I \cos \phi = S \cos \phi$ (14-14)

$-jQ = -jV_T I \sin \phi = -jS \sin \phi$ (14-15)

$S = P - jQ$ (14-16)

$\cos \phi = P/S$ (14-17)

$\cos \phi = V_R/V_T = R/Z$ (14-18)

$Q = X_L/R = \tan \phi$ (14-19)

$Q = \omega L/R$ (14-20)

Parallel *RC* circuit

$I_T = I_R + jI_C$ (14-21)

$Y = G + jB_C$ (14-22)

$VI_T = VI_R + jVI_C$ (14-23)

$S = P + jQ$ (14-24)

$S = VI_T$ (14-25)

$P = S \cos \phi = VI_T \cos \phi$ (14-26)

$jQ = jS \sin \phi = jVI_T \sin \phi$ (14-27)

$\cos \phi = P/S = I_R/I_T = G/Y$ (14-28)

$\cos \phi = Z/R$ (14-29)

Parallel *RL* circuit

$I_T = I_R - jI_L$ (14-30)

$Y = G - jB_L$ (14-31)

$VI_T = VI_R - jVI_L$ (14-32)

$S = P - jQ$ (14-33)

$S = VI_T$ (14-34)

$P = S \cos \phi = VI_T \cos \phi$ (14-35)

$-jQ = -jS \sin \phi = -jVI_T \sin \phi$ (14-36)

$\cos \phi = P/S = I_R/I_T = G/Y = Z/R$ (14-37)

$Q = B_L/G = \tan \phi$ (14-38)

$Q = 1/\omega LG$ or $R/\omega L$ or R/X_L (14-39)

Series $\leftrightarrow$ parallel conversion

$G_p = R_s/(R_s^2 + X_{Cs}^2)$ (14-40)

$jB_{Cp} = jX_{Cs}/(R_s^2 + X_{Cs}^2)$ (14-41)

$R_s = G_p/(G_p^2 + B_{Cp}^2)$ (14-42)

$-jX_{Cs} = -jB_{Cp}/(G_p^2 + B_{Cp}^2)$ (14-43)

$G_p = R_s/(R_s^2 + X_{Ls}^2)$ (14-44)

$-jB_{Lp} = -jX_{Ls}/(R_s^2 + X_{Ls}^2)$ (14-45)

$R_s = G_p/(G_p^2 + B_{Lp}^2)$ (14-46)

$jX_{Ls} = jB_{Lp}/(G_p^2 + B_{Lp}^2)$ (14-47)

The set of equations for Chap. 14 is the most formidable, mainly because of the differences in sign of the vertical phasor components of capacitive and inductive circuits. An inspection of the list reveals groups of equations which are either the same or similar. Recognition of these similarities helps to dispel the apprehension natural to an initial encounter with all these equations.

In *any* circuit

1. The apparent power is the product of total voltage and total current, $S = VI$ [Eqs. (14-4), (14-13), (14-25), and (14-34)].
2. The active power is the product of the apparent power and the power factor, $P = S \cos \phi$ [Eqs. (14-5), (14-8), (14-14), (14-17), (14-26), (14-28), (14-35), and (14-37)].
3. The magnitude of the reactive power is the product of the apparent power and the sine of the phase angle, $|Q| = S \sin \phi$ [Eqs. (14-6), (14-15), (14-27), and (14-36)].
4. The apparent power is the phasor sum of the active and reactive powers, $S = P \pm jQ$ [Eqs. (14-7), (14-16), (14-24), and (14-33)].

In a *series* circuit

5. The total voltage is the phasor sum of the resistive and reactive voltages, $V_T = V_R \pm jV_X$ [Eqs. (14-1) and (14-10)].
6. The total impedance is the phasor sum of the resistance and the reactance, $Z = R \pm jX$ [Eqs. (14-2) and (14-11)].
7. The power factor is the ratio of resistive voltage to total voltage or the ratio of resistance to impedance, $\cos \phi = V_R/V_T = R/Z$ [Eqs. (14-9) and (14-18)].

In a *parallel* circuit

8. The total current is the phasor sum of the resistive and reactive currents, $I_T = I_R \pm jX$ [Eqs. (14-21) and (14-30)].
9. The total admittance is the phasor sum of the conductance and the susceptance, $Y = G \pm jB$ [Eqs. (14-22) and (14-31)].
10. The power factor is the ratio of resistive current to total current, or the ratio of conductance to admittance, $\cos \phi = I_R/I_T = G/Y$ [Eqs. (14-28) and (14-37)].

In a *capacitive* circuit, series or parallel,

11. The power equation is $S = P + jQ$ [Eqs. (14-7) and (14-24)].

In an *inductive* circuit, series or parallel,

12. The power equation is $S = P - jQ$ [Eqs. (14-16) and (14-33)].
13. The quality factor is the tangent of the phase angle, $Q = \tan \phi$ [Eqs. (14-19) and (14-38)].

EXERCISE PROBLEMS

Section 14-1

14-1 Calculate V_R, V_C, Z, I, and ϕ for series circuits having the values

a. $V_T = 25$ V, $R = 1.2$ kΩ, $C = 0.01$ μF, $f = 10$ kHz
(*ans.* 15.05 V, 19.97 V, 1.993 kΩ, 12.54 mA, 52.99°)

b. $V_T = 120$ V, $R = 250$ Ω, $C = 2$ μF, $f = 400$ Hz
(*ans.* 93.9 V, 74.72 V, 319.5 Ω, 0.376 A, 38.51°)

c. $V_T = 6$ V, $R = 100$ Ω, $C = 10$ μF, $f = 60$ Hz
(*ans.* 2.117 V, 5.615 V, 283.5 Ω, 21.17 mA, 69.35°)

d. $V_T = 50$ mV, $R = 2.2$ kΩ, $C = 100$ pF, $f = 1.2$ MHz
(*ans.* 42.82 mV, 25.81 mV, 2.569 kΩ, 19.46 μA, 31.08°)
e. $V_T = 50$ V, $R = 10$ kΩ, $C = 0.0018$ μF, $f = 20$ kHz
(*ans.* 45.73 V, 20.22 V, 10.93 kΩ, 4.57 mA, 23.85°)

14-2 Find *S*, *Q*, and *P* for each of the sets of values given in Prob. 14-1.

14-3 Use each of the three methods shown in the text to determine the power factor for each of the sets of values given in Prob. 14-1.
(*ans. a.* 0.602; *b.* 0.7825; *c.* 0.3527; *d.* 0.8564; *e.* 0.9146)

14-4 A 20-μF capacitor with a power factor of 0.001 is connected across a 120-V 60-Hz supply. Find
a. The equivalent series resistance.
b. The current in the capacitor.
c. The apparent power.
d. The reactive power.
e. The active power.

14-5 A 1-μF capacitor is connected across the output of a signal generator. If the generator has an internal resistance of 600 Ω and is set for an open-circuit voltage of 10 V, find the circuit current, the voltage across the capacitor, and the phase angle of V_C with respect to *I* at each of the following frequencies:
a. 50 Hz (*ans.* 3.087 mA, 9.826 V, −79.32°)
b. 500 Hz (*ans.* 14.72 mA, 4.685 V, −27.95°)
c. 2000 Hz (*ans.* 16.52 mA, 1.315 V, −7.55°)
d. 15 kHz (*ans.* 16.66 mA, 0.1768 V, −1.01°)

14-6 The voltage across the 680-Ω resistance of an *RC* circuit is 4.6 V. The current in the circuit leads the emf by 0.91 rad. If the frequency is 1000 Hz, find
a. The voltage across the capacitor.
b. The total pd of the circuit.
c. The value of the capacitor.

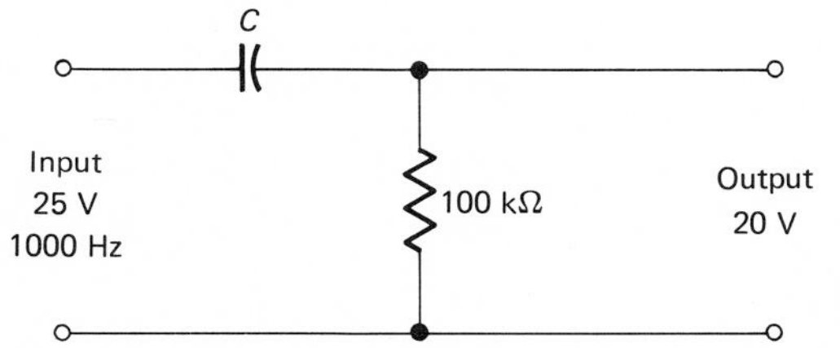

Fig. 14-1P Problems 14-7 to 14-9: circuit.

14-7 Figure 14-1P shows the circuit of a high-pass *RC* filter circuit. Find the value of *C* and the phase relationship of the output voltage to the input voltage.
(*ans.* 2122 pF, 36.87°)

14-8 For Prob. 14-7 find the output voltages and the phase angle of the output voltage with respect to the input voltage when the frequency is changed to
a. 100 Hz *b.* 10 kHz

14-9 For Prob. 14-7 what is the minimum input frequency if the output voltage is to be greater than 95 percent of the input voltage? (*ans.* 2.281 kHz)

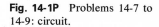

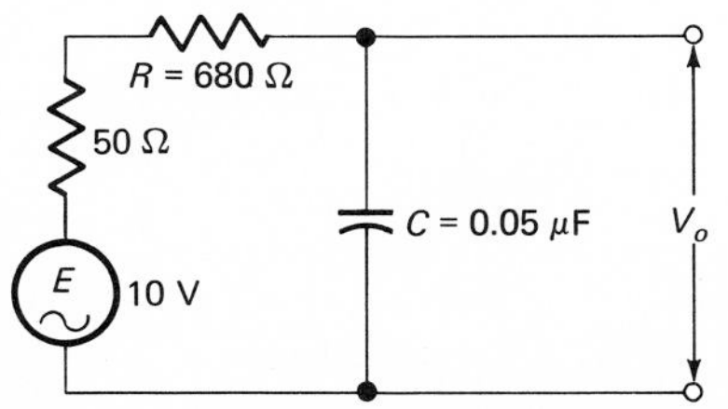

Fig. 14-2P Problems 14-10 and 14-11: circuit.

14-10 For the low-pass filter circuit shown in Fig. 14-2P, the cutoff frequency f_0 is reached when $R = X_C$,

a. Find f_0.
b. Find the magnitude of the output voltage V_o and its phase relative to the input voltage E at the frequency f_0.

14-11 For Prob. 14-10, find the output voltage and phase angle at
a. 500 Hz (*ans.* 9.934 V, 83.46°)
b. 20 kHz (*ans.* 2.13 V, 12.3°)

Section 14-2

14-12 Calculate V_R, V_L, Z, I, and ϕ for series circuits having the following values:
a. $V_T = 300$ V, $R = 1000\ \Omega$, $L = 50$ mH, $f = 800$ Hz
b. $V_T = 50$ V, $R = 20$ kΩ, $L = 0.1$ H, $f = 10$ kHz
c. $V_T = 120$ V, $R = 60\ \Omega$, $L = 0.16$ H, $f = 60$ Hz
d. $V_T = 50\ \mu$V, $R = 1500\ \Omega$, $L = 150\ \mu$H, $f = 1$ MHz
e. $V_T = 20$ mV, $R = 5$ kΩ, $L = 150\ \mu$H, $f = 20$ MHz

14-13 Find S, Q, and P for each of the sets of values given for Prob. 14-12.
(*ans. a.* 87.3 VA, 21.28 var, 84.68 W; *b.* 0.1192 VA, 0.0357 var, 0.1138 W; *c.* 169.2 VA, 120 var, 119.3 W; *d.* 1.4 pVA, 0.7448 pvar, 1.185 pW; *e.* 20.52 nVA, 19.83 nvar, 5.261 nW)

14-14 Use each of the three methods shown in the text to determine the power factor for each of the sets of values given for Prob. 14-12.

14-15 A 0.5-H inductor with a Q factor of 12 is connected across a 120-V 60-Hz supply. Find
a. The equivalent series resistance. (*ans.* 15.71 Ω)
b. The current in the inductor. (*ans.* 0.6344 A)
c. The apparent power. (*ans.* 76.13 VA)
d. The reactive power. (*ans.* 75.86 var)
e. The active power. (*ans.* 6.323 W)

14-16 A 120-V 60-Hz electric motor draws 2.5 A. The voltage leads the current by 60°. Find the impedance of the motor in
a. Rectangular form. *b.* Polar form.

14-17 For Prob. 14-16 find
a. The apparent power. (*ans.* 300 VA)
b. The reactive power. (*ans.* 259.8 var)
c. The active power. (*ans.* 150 W)

14-18 For Prob. 14-16 with time referenced to 0 V going positive, find the instantaneous current and instantaneous power at times
a. $t = 3.7$ ms
b. $t = 6.95$ ms
c. $t = 9.5$ ms

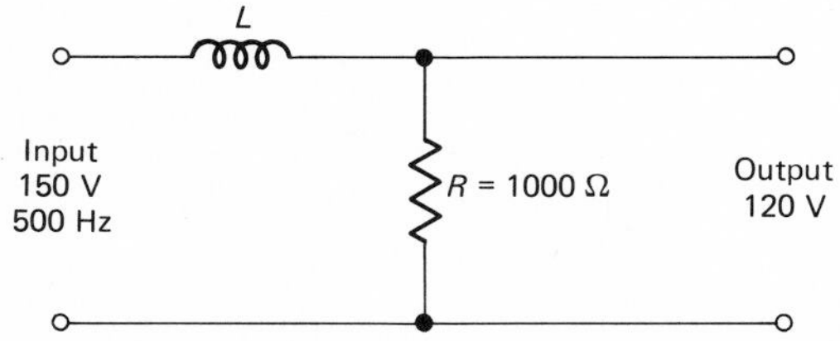

Fig. 14-3P Problems 14-19 and 14-20: circuit.

14-19 Figure 14-3P shows the circuit of a low-pass RL filter. Find
a. The value of L. (*ans.* 0.239 H)
b. The phase relationship of output voltage to input voltage. (*ans.* 36.9° lag)

14-20 For the circuit in Fig. 14-3P, find the output voltage and phase relationship of output to input at
a. 50 Hz *b.* 5 kHz

14-21 A 240-V 60-Hz electric motor has an impedance of $(55 + j35)\ \Omega$.
a. Express the impedance in polar form. (*ans.* $65.19\ \Omega\ \underline{/32.47^\circ}$)
b. Find the current drawn by the motor. (*ans.* 3.682 A)
c. Find the energy consumed by the motor in 1 h.
[*ans.* 745.4 Wh (2.684×10^6 J)]

14-22 An electromagnetic relay, when connected across a 120-V 60-Hz supply, consumes 207 J of energy per minute. If the current is 48 mA, find the series equivalent circuit.

14-23 An inductor with an inductance of 0.16 H has a dc resistance of 4 Ω. When connected across a 120-V 60-Hz source, the phase angle is 82.5°. Find
a. The ac resistance of the inductor at 60 Hz. (*ans.* 7.94 Ω)
b. The current in the inductor. (*ans.* 1.972 A)

Section 14-3

14-24 A resistor *R* and a capacitor *C* are connected in parallel across a source of emf. Find I_R, I_C, I_T, Y, and ϕ when
a. $E = 25$ V, $R = 1.2$ kΩ, $C = 0.01\ \mu$F, $f = 10$ kHz
b. $E = 120$ V, $R = 250$ Ω, $C = 2\ \mu$F, $f = 400$ Hz
c. $E = 6$ V, $R = 100$ Ω, $C = 10\ \mu$F, $f = 60$ Hz
d. $E = 50$ mV, $R = 2.2$ kΩ, $C = 100$ pF, $f = 1.2$ MHz
e. $E = 50$ V, $R = 10$ kΩ, $C = 0.0018\ \mu$F, $f = 20$ kHz

14-25 Find *S*, *Q*, and *P* for each of the sets of values given in Prob. 14-24.
(*ans. a.* 0.6515 VA, 0.3925 var, 0.52 W; *b.* 92.48 VA, 72.36 var, 57.6 W; *c.* 0.3847 VA, 0.1356 var, 0.36 W; *d.* 2.2 μVA, 1.886 μvar, 1.135 μW; *e.* 0.618 VA, 0.565 var, 0.25 W)

14-26 A 20-μF capacitor with a power factor of 0.001 is connected across a 120-V 60-Hz supply. Find
a. The equivalent parallel resistance.
b. The total current in the circuit.
c. The apparent power.
d. The reactive power.
e. The active power.
(Compare your answers to those for Prob. 14-4.)

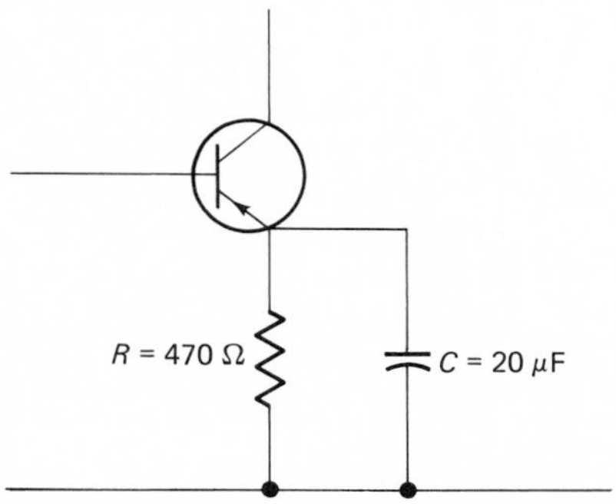

Fig. 14-4P Problems 14-27 and 14-28: circuit.

14-27 Figure 14-4P shows a parallel *RC* circuit connected to the emitter circuit of a transistor audio amplifier. Find the admittance, in polar form, of the circuit at the following frequencies.
a. 10 kHz (*ans.* $1.257\ \text{S}\ \underline{/89.9^\circ}$)
b. 1 kHz (*ans.* $0.1257\ \text{S}\ \underline{/89.03^\circ}$)
c. 50 Hz (*ans.* $6.632\ \text{mS}\ \underline{/71.28^\circ}$)
d. 20 Hz (*ans.* $3.293\ \text{mS}\ \underline{/49.74^\circ}$)

14-28 Find the impedance, in polar and rectangular form, of the *RC* circuit in Prob. 14-27 at the frequencies stated.

14-29 Calculate I_R, I_L, I_T, Y, and ϕ for parallel circuits having the following values:

a. $V = 300$ V, $R = 1000\ \Omega$, $L = 50$ mH, $f = 800$ Hz
(*ans.* 0.3 A, 1.194 A, 1.231 A, 4.103 mS, 75.9°)

b. $V = 50$ V, $R = 20$ kΩ, $L = 0.1$ H, $f = 10$ kHz
(*ans.* 2.5 mA, 7.958 mA, 8.345 mA, 166.9 μS, 72.56°)

c. $V = 120$ V, $R = 60\ \Omega$, $L = 0.16$ H, $f = 60$ Hz
(*ans.* 2 A, 1.989 A, 2.82 A, 23.51 mS, 44.86°)

d. $V = 50\ \mu$V, $R = 1500\ \Omega$, $L = 150\ \mu$H, $f = 1$ MHz
(*ans.* 33.33 nA, 53.05 nA, 62.66 nA, 1.253 mS, 57.85°)

e. $V = 20$ mV, $R = 5$ kΩ, $L = 150\ \mu$H, $f = 20$ MHz
(*ans.* 4 μA, 1.061 μA, 4.138 μA, 206.9 μS, 14.86°)

14-30 Find S, Q, and P for each of the sets of values given in Prob. 14-29.

Section 14-4

14-31 A 240-V 60-Hz electric motor has an admittance of $(0.01 - j0.006)$ S.

a. Express the admittance in polar form. (*ans.* 0.01166 S $\angle 30.96°$)

b. Find the current drawn by the motor. (*ans.* 2.799 A)

c. Find the power drawn by the motor. (*ans.* 576 W)

14-32 A 240-V 60-Hz electric motor has an impedance of $(55 + j35)\ \Omega$. Find

a. The parallel equivalent circuit.

b. The current drawn by the equivalent parallel resistance.

c. The current drawn by the equivalent parallel inductance.

d. The total current drawn by the equivalent parallel circuit.

14-33 For Prob. 14-32 find

a. The value of capacitance that would have the same reactance as the equivalent parallel inductance. (*ans.* 21.8 μF)

b. The total circuit current, and its phase angle, if the capacitor is connected across the equivalent parallel circuit. (*ans.* 3.106 A)

Section 14-5

14-34 Convert the series circuits in Prob. 14-1 to their parallel equivalents, expressing the equivalent circuits in terms of resistance and capacitance.

14-35 Convert the parallel circuits in Prob. 14-24 to their series equivalents, expressing the equivalent circuits in terms of resistance and capacitance.
(*ans. a.* 765.2 Ω, 0.0276 μF; *b.* 96.5 Ω, 3.25 μF; *c.* 87.56 Ω, 80.4 μF; *d.* 586.2 Ω, 136 pF; *e.* 1635 Ω, 0.00215 μF)

14-36 Convert the series circuits in Prob. 14-12 to their parallel equivalents, expressing the equivalent circuits in terms of resistance and inductance.

14-37 Convert the parallel circuits in Prob. 14-29 to their series equivalents, expressing the equivalent circuits in terms of resistance and inductance.
(*ans. a.* 59.4 Ω, 47 mH; *b.* 1797 Ω, 91 mH; *c.* 30.16 Ω, 79.58 mH; *d.* 424.6 Ω, 107.5 μH; *e.* 4.67 kΩ, 9.86 μH)

Miscellaneous *RC* and *RL* circuit problems

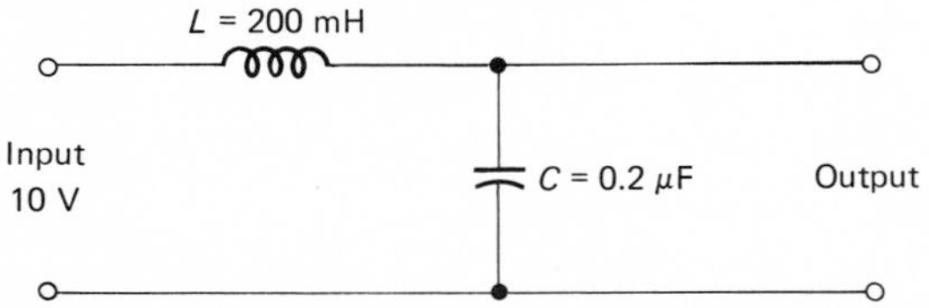

Fig. 14-5P Problem 14-38: circuit.

14-38 Figure 14-5P is the circuit of a low-pass *LC* filter. Find the output voltage at the following frequencies:

a. 500 Hz
b. 2000 Hz
c. 5 kHz
d. 20 kHz

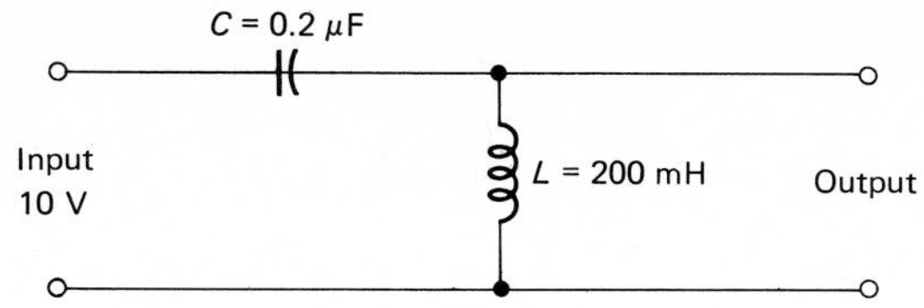

Fig. 14-6P Problem 14-39: circuit.

14-39 Figure 14-6P is the circuit of a high-pass *LC* filter. Find the output voltage at the following frequencies:

a. 500 Hz (*ans.* 6.52 V)
b. 2000 Hz (*ans.* 11.88 V)
c. 5 kHz (*ans.* 10.26 V)
d. 20 kHz (*ans.* 10.02 V)

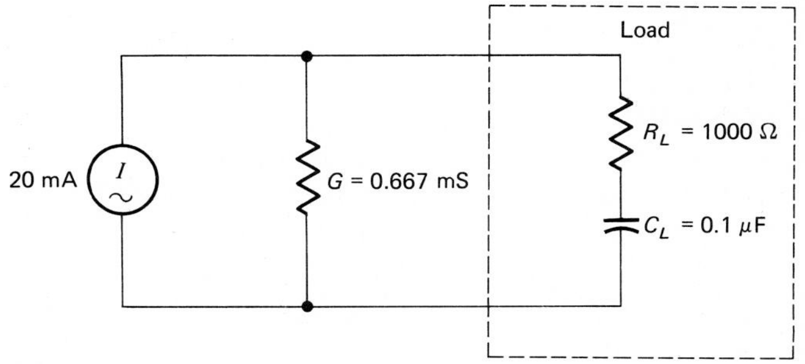

Fig. 14-7P Problems 14-40 and 14-41: circuit.

14-40 Figure 14-7P shows the equivalent circuit of a transistor amplifier feeding a load consisting of a 1000-Ω resistance in series with a 0.1-μF capacitor. Find the voltage across the load and the phase angle of this voltage with respect to the total current at

a. 50 Hz
b. 1000 Hz
c. 5 kHz

14-41 Repeat Prob. 14-40 when three similar loads are connected in parallel across the output of the amplifier. (*ans. a.* 29.56 V, 8°; *b.* 9.849 V, 41.72°; *c.* 5.71 V, 14.34°)

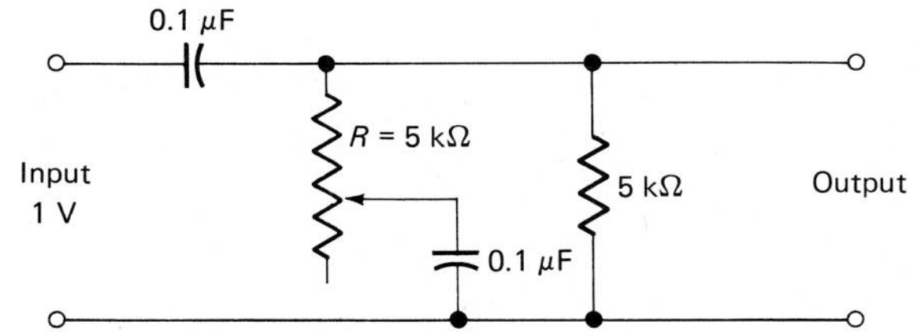

Fig. 14-8P Problem 14-42: circuit.

14-42 Figure 14-8P shows the circuit of a tone control for an audio amplifier. When *R* is set at zero, find the output voltage of the tone control circuit at the following frequencies:

a. 100 Hz
b. 1000 Hz
c. 10 kHz

14-43 Repeat Prob. 14-42 when *R* is set at 5000 Ω.

(*ans. a.* 0.2525 V; *b.* 0.8002 V; *c.* 0.997 V)

15

AC Circuits with *R*, *C*, and *L* (Nonresonant)

The principles and techniques developed for ac circuits containing two types of elements (resistance and capacitance or resistance and inductance) can be applied to circuits which contain all three types of elements, resistance, capacitance, *and* inductance, often called *RLC* circuits. The only new point arising in dealing with the phasor relationships in *RLC* circuits is that, in addition to two phasors in quadrature (resistive and reactive), there is another reactive phasor. The two reactive phasors are in antiphase with respect to each other, so they partially or completely cancel in effect.

Partial cancellation of a circuit's phasors occurs when the inductive and capacitive components are unequal. This inequality leaves a net inductive or net capacitive condition. Together with the resistive (or conductive) component, the net inductive or net capacitive component in effect reduces the *RLC* circuit to an *RL* or an *RC* circuit (series or parallel). As these two-element circuits were fully covered in Chap. 14, we need only consider the cancellation of phasors to deal with *RLC* circuits.

Complete cancellation of the reactive components of a circuit's phasors is called *resonance*, and raises some rather special issues which justify a separate chapter. This chapter therefore is confined to the general techniques of handling *RLC* circuits, deliberately avoiding situations in which complete cancellation of reactive phasors occurs.

15-1 *RLC* SERIES CIRCUIT

A series circuit containing resistance, capacitance, and inductance is shown in Fig. 15-1*a*.

Figure 15-1*b* and *c* shows partial phasor diagrams for the series *RLC* circuit in Fig. 15-1*a*. In each phasor diagram the common current I is shown as the reference phasor $I\,\underline{/0^\circ}$ or $I + j0$. The resistive pd V_R is in phase with the current and has the phasor value $V_R\,\underline{/0^\circ}$ or $V_R + j0$. The inductive pd leads the current by 90° and thus has the phasor value $V_L\,\underline{/90^\circ}$ or $0 + jV_L$. The capacitive voltage V_C lags the current by 90°, and has the phasor value $V_C\,\underline{/-90^\circ}$ or $0 - jV_C$.

As both V_L and V_C are quadrature components of the total voltage V_T, but are of opposite j sign, the magnitude of their phasor sum is the arithmetic difference between them. This phasor sum of the individual reactive voltages may be called the net reactive voltage V_X:

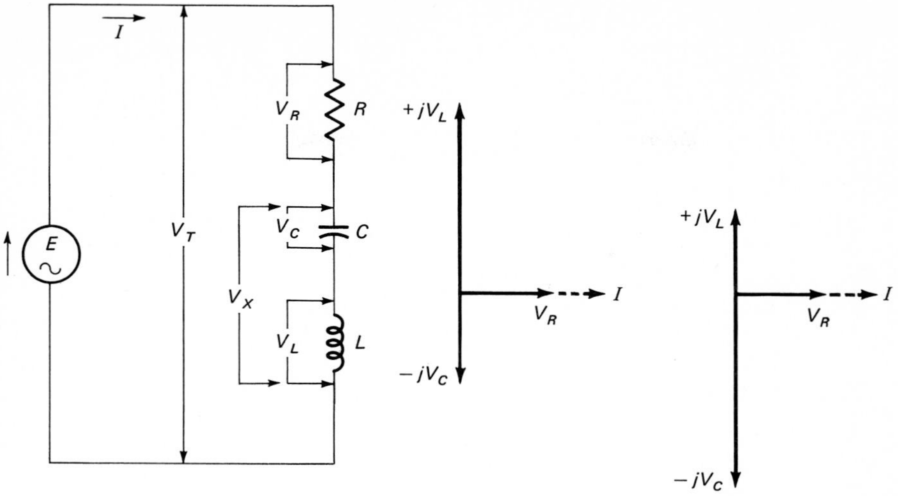

Fig. 15-1 *RLC* series circuit. (*a*) Schematic diagram; (*b*) partial phasor diagram when $V_L > V_C$; (*c*) partial phasor diagram when $V_L < V_C$.

$$V_X = (+jV_L) + (-jV_C)$$

$$V_X = jV_L - jV_C \qquad \text{or} \qquad j(V_L - V_C) \tag{15-1}$$

If we concern ourselves only with situations in which inductive and capacitive components are unequal, there are just two alternatives for the series *RLC* circuit; V_L is either greater than or less than V_C. We will therefore consider each of these alternatives.

When $V_L > V_C$ The partial phasor diagram in Fig. 15-1*b* shows the phasor line for $+jV_L$ longer than that for $-jV_C$, indicating that the magnitude of V_L is greater than that of V_C. This partial phasor diagram leads to the full phasor diagram in Fig. 15-2*a*.

In Fig. 15-2*a* the net reactive voltage V_X is shown as a $+j$ quantity since, with V_L greater than V_C, $V_L - V_C$ is a positive quantity.

The overall characteristics of the series *RLC* circuit can be determined by considering only the relationships between the current, the resistive voltage, the net reactive voltage, and the total circuit voltage. In Fig. 15-2*b* these four variables are shown in the conventional power position with the total voltage as the reference phasor.

From Fig. 15-2 it can be seen that, unless the characteristics of individual reactive elements are of interest, an *RLC* series circuit in which V_L is greater than V_C behaves as an *RL* series circuit.

All the characteristics of the *RL* series circuit referred to in Sec. 14-2 apply with V_X substituted for V_L, X_T for X_L, etc.

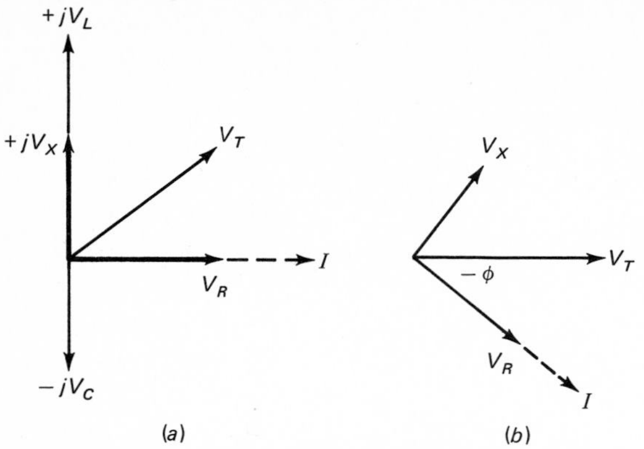

Fig. 15-2 *RLC* series circuit when $V_L > V_C$. (*a*) Complete phasor diagram with I as reference; (*b*) phasor diagram of the main variables with V_T as reference.

When $V_L < V_C$ In Fig. 15-1*c* the phasor line for $+jV_L$ is shorter than that for $-jV_C$, indicating that the magnitude of V_L is less than that of V_C. A full phasor diagram for this situation is given in Fig. 15-3*a*.

When V_L is less than V_C, $V_L - V_C$ is a negative quantity; Fig. 15-3*a* illustrates this, V_X being shown as a $-j$ quantity. The common circuit current, the resistive voltage, the net reactive voltage, and the total circuit voltage are shown in Fig. 15-3*b* with the total voltage as the reference phasor, the conventional power viewpoint.

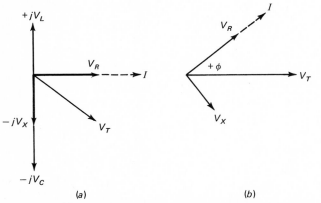

Fig. 15-3 *RLC* series when $V_L < V_C$. (*a*) Complete phasor diagram with I as reference; (*b*) phasor diagram of the main variables with V_T as reference.

When V_L is less than V_C, the net reactive voltage is capacitive and the circuit has a leading phase angle. These characteristics are typical of an *RC* circuit.

All the characteristics of the *RC* series circuit referred to in Sec. 14-1 apply to the overall behavior of an *RLC* series circuit in which V_L is less than V_C. Substitute V_X for V_C, X_T for X_C, etc.

Example 15-1 What is the value of the current in the circuit in Fig. 15-4*a*?

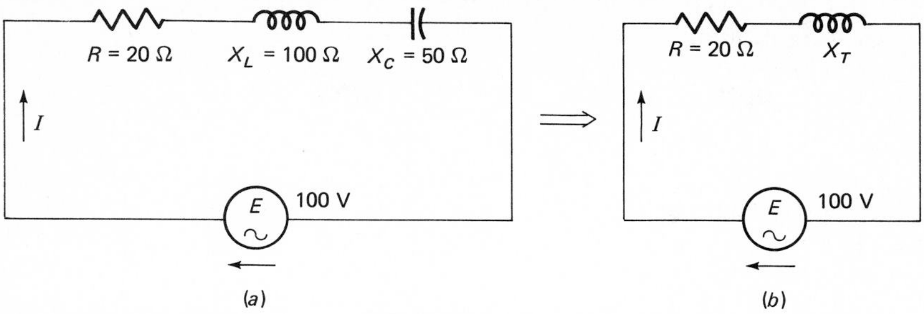

Fig. 15-4 Example 15-1: (*a*) circuit; (*b*) circuit with X_L and X_C combined as X_T.

Solution As X_L is greater than X_C, for any value of current I, V_L will be greater than V_C. Therefore the circuit in Fig. 15-4*a* is net inductive as shown in the equivalent circuit in Fig. 15-4*b*.

$$X_T = j(X_L - X_C) \qquad \text{extension of Eq. (15-1)}$$

$$X_T = j(100\ \Omega - 50\ \Omega) = j50\ \Omega$$

$$Z = R + jX_T = (20 + j50)\ \Omega = 53.85\ \Omega\ \underline{/68.2^\circ}$$

Taking the applied emf E as the reference phasor,

$$I = \frac{E}{Z} = \frac{100\ \text{V}\ \underline{/0^\circ}}{53.85\ \Omega\ \underline{/68.2^\circ}} = \mathbf{1.857\ A}\ \underline{\mathbf{/-68.2^\circ}} \qquad \text{or} \qquad \underline{\mathbf{/-1.19\ rad}}$$

The result for Example 15-1 gives the expected nature of the current in an *RL* circuit; the current lags the emf by less than 90°. The effect of the capacitive reactance is offset by the greater inductive reactance.

Phasor diagrams for the circuit in Example 15-1 are given in Fig. 15-5. Figure 15-5*a* has the common current as the reference phasor, as is

Fig. 15-5 Example 15-1: phasor diagrams. (*a*) With I as reference; (*b*) with E as reference.

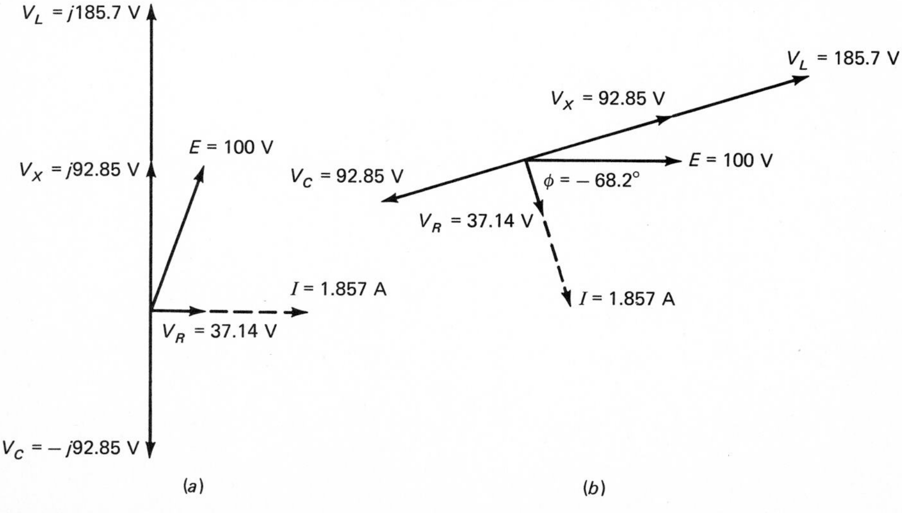

usual in deriving series circuit relationships, while Fig. 15-5*b* has the emf (or total circuit voltage) as the reference phasor, the conventional view for power considerations in any circuit.

Had the values of X_L and X_C in Example 15-1 been reversed ($X_L = 50\ \Omega$ and $X_C = 100\ \Omega$), the resulting current would have been numerically the same but its phase angle would have been positive. A positive current phase angle (leading current) indicates a net capacitive circuit.

Example 15-2 In the circuit in Fig. 15-6*a*, what is (*a*) each individual element pd, (*b*) the net reactive pd, and (*c*) the applied emf?

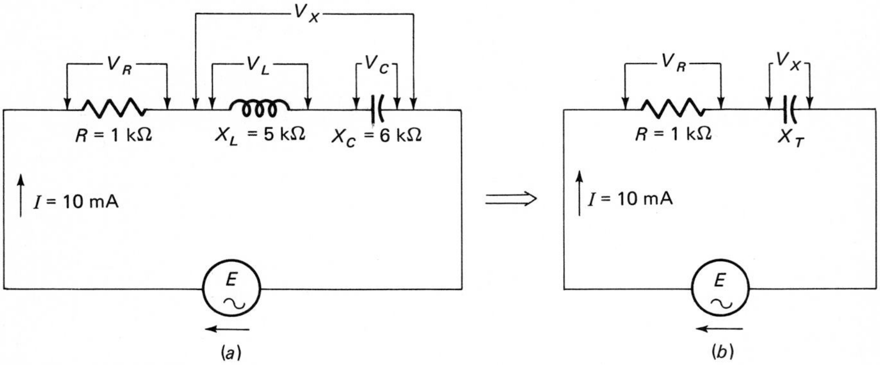

Fig. 15-6 Example 15-2: (*a*) circuit; (*b*) circuit with X_L and X_C combined as X_T.

Solution As X_L is less than X_C, V_L is less than V_C. The circuit in Fig. 15-6*a* is thus net capacitive. Figure 15-6*b* shows the net reactance X_T to be capacitive.

a. Applying Ohm's law to each element with the common current *I* as the reference phasor,

$$V_R = IR = 10\ \text{mA}\ \underline{/0^\circ} \times 1\ \text{k}\Omega\ \underline{/0^\circ} = \mathbf{10\ V\ \underline{/0^\circ}}$$

$$V_C = IX_C = 10\ \text{mA}\ \underline{/0^\circ} \times 6\ \text{k}\Omega\ \underline{/-90^\circ} = \mathbf{60\ V\ \underline{/-90^\circ}} = \mathbf{-}\boldsymbol{j}\mathbf{60\ V}$$

$$V_L = IX_L = 10\ \text{mA}\ \underline{/0^\circ} \times 5\ \text{k}\Omega\ \underline{/+90^\circ} = \mathbf{50\ V\ \underline{/+90^\circ}} = \mathbf{+}\boldsymbol{j}\mathbf{50\ V}$$

b.

$$V_X = jV_L - jV_C \qquad (15\text{-}1)$$

$$V_X = j50\ \text{V} - j60\ \text{V} = \mathbf{-}\boldsymbol{j}\mathbf{10\ V} \quad \text{or} \quad \mathbf{10\ V\ \underline{/-90^\circ}}$$

c.

$$E = V_R + jV_X = 10\ \text{V}\ \underline{/0^\circ} + 10\ \text{V}\ \underline{/-90^\circ}$$

$$E = \mathbf{(10 - }\boldsymbol{j}\mathbf{10)\ V = 14.14\ V\ \underline{/-45^\circ}} \quad \text{or} \quad \mathbf{\underline{/-0.7853\ rad}}$$

The element pds in Example 15-2*a* have the correct phase relationships with respect to their common current. The resistive pd V_R is in phase with the current, the capacitive pd V_C lags the current by 90°, and the inductive pd V_L leads the current by 90°.

The capacitive reactance of this series circuit is greater than the inductive reactance; the net reactive voltage V_X is capacitive, lagging the circuit current by 90°. This conclusion is shown by the result for Example 15-2*b*.

The complete circuit in Example 15-2 behaves as a resistance-capacitance network, the emf lagging the current by less than 90° (or the current leading the emf by less than 90°). The result for part *c* substantiates this conclusion.

Phasor diagrams for the circuit in Example 15-2 are given in Fig. 15-7 with both current and total voltage references.

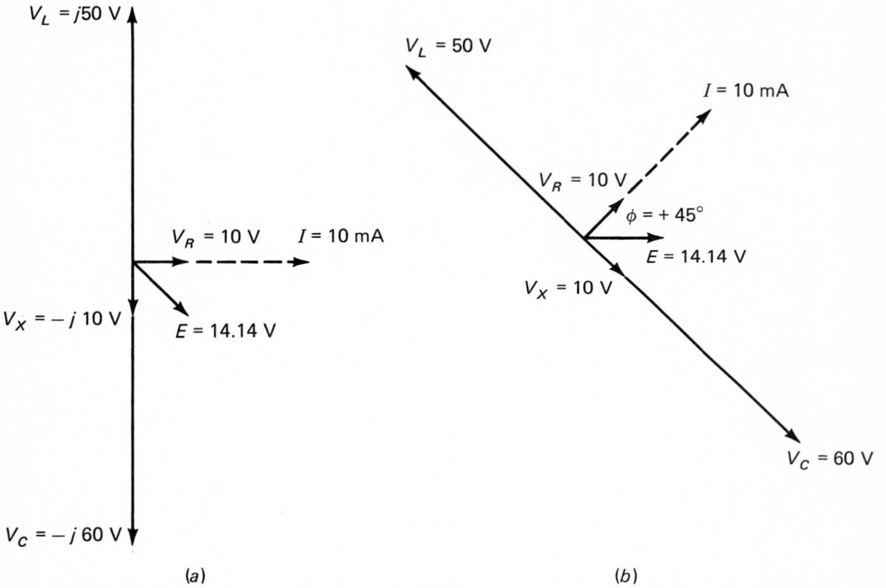

Fig. 15-7 Example 15-2: phasor diagrams. (*a*) With *I* as reference; (*b*) with *E* as reference.

INSTANTANEOUS REACTIVE POLARITIES

In Example 15-1, $V_L = 185.7$ V, which is greater than the 100-V applied emf. In Example 15-2, $V_L = 50$ V and $V_C = 60$ V, both of which are greater than the 14.14-V applied emf. It is a unique characteristic of *RLC* series circuits that individual element pds can be greater in magnitude than the emf applied to the circuit (or than the total pd across the circuit). This characteristic is due to the tendency for series inductive and capacitive voltages to cancel.

The waveforms of the reactive voltages in Example 15-2 are shown in relation to each other and to the circuit current in Fig. 15-8.

The waveform of the inductive voltage v_L leads the current i by 90°. The waveform of the capacitive voltage v_C lags the current by 90°. Therefore v_L and v_C have a relative phase of 180°; they are in antiphase; their polarities are always opposite.

Figure 15-9*a* shows the polarities of the reactive pds at time *A* in Fig. 15-8. The inductive pd v_L is at its positive maximum value, its polarity being indicated by + ⌒⌒⌒ −. The capacitive pd v_C is at its negative max-

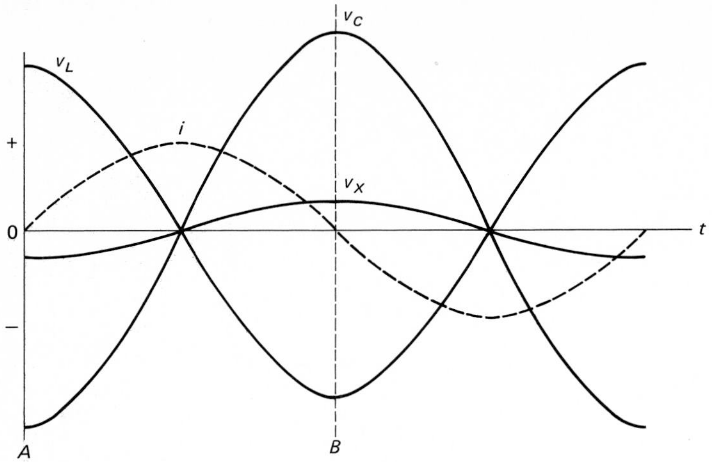

Fig. 15-8 Waveforms of current and reactive voltages in the series ac circuit of Example 15-2.

imum value, its polarity being shown as —⊣⊢+. Since the instantaneous polarities of v_L and v_C are opposite, the net reactive voltage v_X is the difference between them. The polarity of v_X is the same as that of v_C, the greater of the two individual reactive pds; its relative magnitude is similarly a negative maximum.

One half-cycle later, at instant of time B, the voltage values are the same as at time A but both of the polarities are reversed, as shown in Fig. 15-9b.

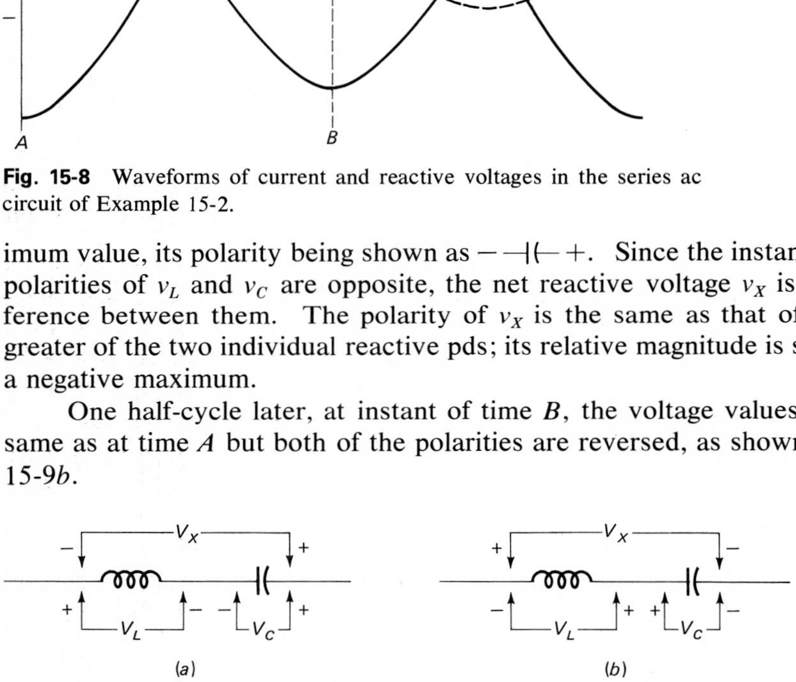

Fig. 15-9 Instantaneous reactive voltage polarities in a series ac circuit.

At instants of time other than A or B the inductive and capacitive voltages will still have opposite polarities. The magnitude of the net reactive pd will be the difference between the capacitive and inductive voltages, and its polarity will be that of the capacitive pd. The only exceptions are the time instants, two per cycle, when all three reactive voltages v_C, v_L, and v_X are zero.

Example 15-3 Figure 15-10a shows two transistors coupled by an LC series circuit. Transistor Q_1 amplifies the video-frequency signal from the input and passes this amplified signal on to transistor Q_2 in the form of a voltage V_{BE}. An equivalent circuit is given in Fig. 15-10b, where Q_1 is depicted as a voltage generator of 15 V emf and 200 Ω internal resistance. R_{BE} is the total effective resistance between the base B of transistor Q_2 and ground. What is the value of the voltage V_{BE} at a frequency of 2.5 MHz?

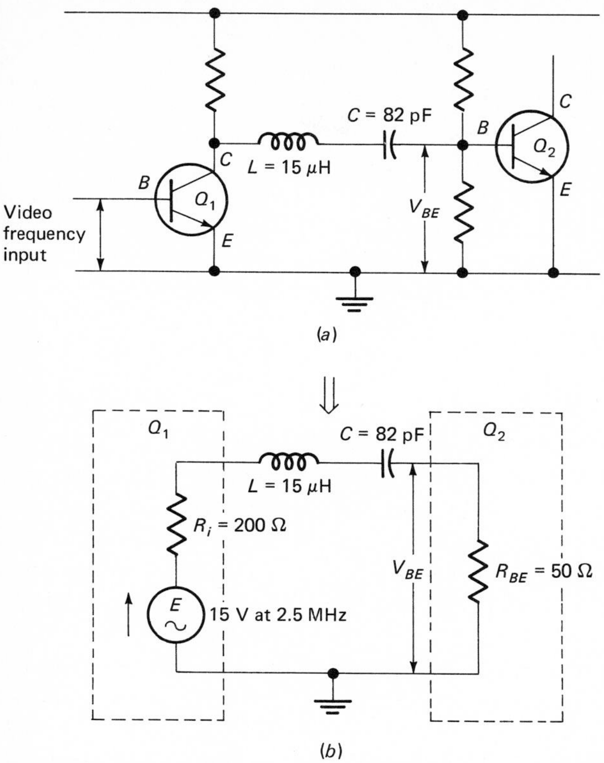

Fig. 15-10 Example 15-3: (*a*) given circuit; (*b*) equivalent of the circuit connecting Q_1 to Q_2.

Solution

$$\omega = 2\pi f = 2\pi \times 2.5 \times 10^6 \text{ Hz} = 1.571 \times 10^7 \text{ rad/s}$$

$$X_L = \omega L = 1.571 \times 10^7 \text{ rad/s} \times 1.5 \times 10^{-5} \text{ H} = 235.6 \ \Omega$$

$$X_C = \frac{1}{\omega C} = \frac{1}{1.571 \times 10^7 \text{ rad/s} \times 8.2 \times 10^{-11} \text{ F}} = 776.3 \ \Omega$$

The equivalent circuit can be redrawn as in Fig. 15-11.

The voltage divider principle can be applied to any ac series circuit if phasor techniques are used.

First calculating the impedance of the circuit Z,

$$Z = (R_i + R_{BE}) + j(X_L - X_C) = [(200 + 50) + j(235.6 - 776.3)] \ \Omega$$

$$Z = (250 - j540.7) \ \Omega = 595.7 \ \Omega \ \underline{/-65.19^\circ}$$

Now applying the voltage divider principle,

$$V_{BE} = E \frac{R_{BE}}{Z} = 15 \text{ V } \underline{/0^\circ} \times \frac{50 \ \Omega \ \underline{/0^\circ}}{595.7 \ \Omega \ \underline{/-65.19^\circ}}$$

$$V_{BE} = \mathbf{1.259 \text{ V }} \underline{\mathbf{/-65.19^\circ}} \qquad \text{or} \qquad \underline{\mathbf{/-1.138 \text{ rad}}}$$

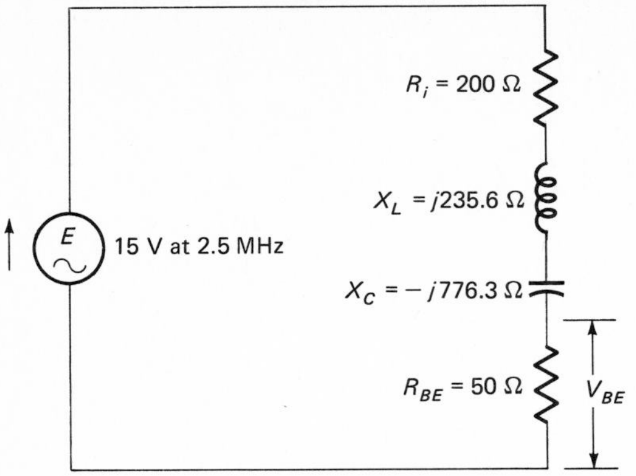

Fig. 15-11 Example 15-3: Fig. 15-10*b* redrawn.

In Example 15-3 the capacitive reactance exceeds the inductive reactance. The total circuit is essentially a series resistance-capacitance network. Therefore the circuit current and the resistive voltages lead the applied emf. Thus the voltage V_{BE} developed across the resistance R_{BE} has a leading phase with respect to the emf. The magnitude of V_{BE} is appreciably less than the emf, because R_{BE} (50 Ω) is low compared to R_i (200 Ω) and the net reactance (540.7 Ω).

At a frequency of 4.5 MHz, X_C and X_L in Fig. 15-11 would be equal, resulting in the complete cancellation of reactive effects. At this frequency the circuit would behave as a resistive voltage divider, so that V_{BE} would be the fraction $R_{BE}/(R_{BE} + R_i) = 0.2$ of the emf. This 3 V is the maximum possible value of V_{BE}. Circuits such as that in Fig. 15-10 are used in television sets to separate sound (audio) and picture (video) signals. The 4.5-MHz sound intercarrier signal is directed to the audio section of the set. Picture signals, mainly below 2.5 MHz, are attenuated (reduced in level) and so discouraged from entering the audio section.

Example 15-4 An electric motor designed to operate on a 60-Hz supply has equivalent series resistance and inductance of 2 Ω and 1 H, respectively. If the motor's power factor is 0.8 lagging, what is its equivalent series capacitance?

Solution Figure 15-12 is an equivalent series circuit of the motor connected to a 60-Hz supply.

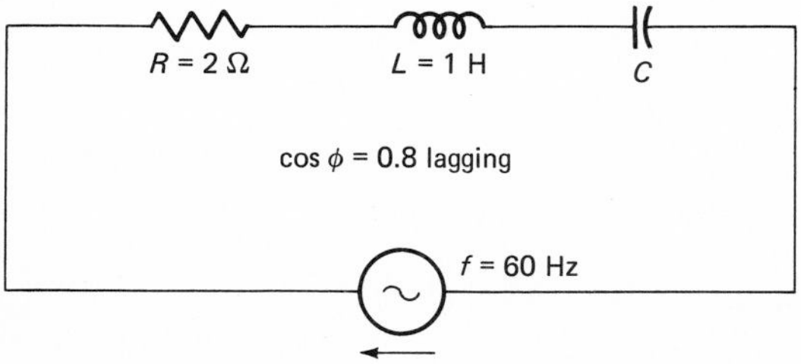

Fig. 15-12 Example 15-4: equivalent circuit.

$$\omega = 2\pi f = 2\pi \times 60 \text{ Hz} = 377 \text{ rad/s}$$

$$X_L = \omega L = 377 \text{ rad/s} \times 1 \text{ H} = 377 \ \Omega$$

Calculating the impedance magnitude,

$$\cos \phi = \frac{R}{Z} \qquad \text{or} \qquad Z = \frac{R}{\cos \phi}$$

The power factor $\cos \phi$ is 0.8, so

$$Z = \frac{2\ \Omega}{0.8} = 2.5\ \Omega$$

When the impedance magnitude is known, the net reactance can be found:

$$Z = \sqrt{R^2 + X_T^2}$$

$$Z^2 = R^2 + X_T^2$$

or

$$X_T^2 = Z^2 - R^2 = (2.5\ \Omega)^2 - (2\ \Omega)^2 = 2.25\ \Omega^2$$

$$X_T = \sqrt{2.25\ \Omega^2} = 1.5\ \Omega$$

From the net reactance and the inductive reactance, the capacitive reactance can be determined:

$$X_T = X_L - X_C \qquad \text{extension of Eq. (15-1)}$$

or

$$X_C = X_L - X_T = 377\ \Omega - 1.5\ \Omega = 375.5\ \Omega$$

The value of the series capacitance of the motor can be obtained from the capacitive reactance:

$$X_C = \frac{1}{\omega C}$$

or

$$C = \frac{1}{\omega X_C} = \frac{1}{377\ \text{rad/s} \times 375.5\ \Omega} = \mathbf{7.06\ \mu F}$$

For Example 15-4 an impedance diagram drawn in phasor diagram form may be helpful in visualizing the circuit conditions. Such an impedance diagram is shown in Fig. 15-13.

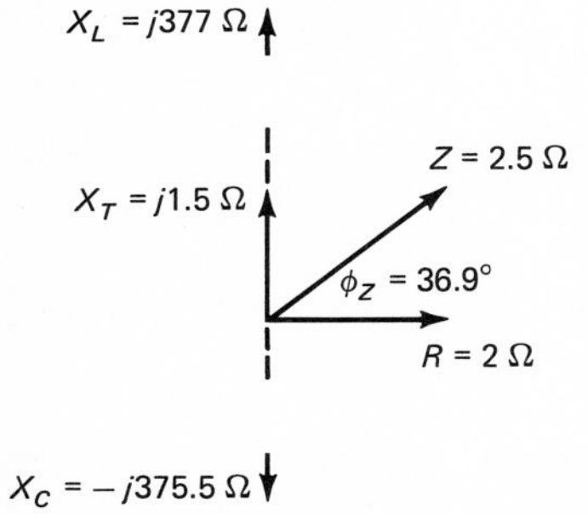

Fig. 15-13 Example 15-4: impedance diagram.

Figure 15-13 shows the phase angle of the impedance ϕ_Z as $+36.9°$. As impedances are proportional to voltages, the figure is the same in configuration as a voltage phasor diagram. A positive, or leading, voltage phase angle represents a lagging current phase angle. Thus the circuit in Example 15-4 has a lagging power factor.

Example 15-5 In the circuit in Example 15-4, if the supply voltage is 240 V, what is (*a*) the apparent power, (*b*) the active power, and (*c*) the reactive power?

Solution Each component of power can be determined by first calculating the circuit current.

$$I = \frac{E}{Z} = \frac{240\ \text{V}}{2.5\ \Omega} = 96\ \text{A}$$

a. $$S = I^2Z = (96\ \text{A})^2 \times 2.5\ \Omega = \mathbf{23.04\ kVA}$$

b. $$P = I^2R = (96\ \text{A})^2 \times 2\ \Omega = \mathbf{18.43\ kW}$$

c. $$Q = I^2X_T = (96\ \text{A})^2 \times 1.5\ \Omega = \mathbf{13.82\ kvar}$$

A power triangle for Example 15-5 is given in Fig. 15-14.

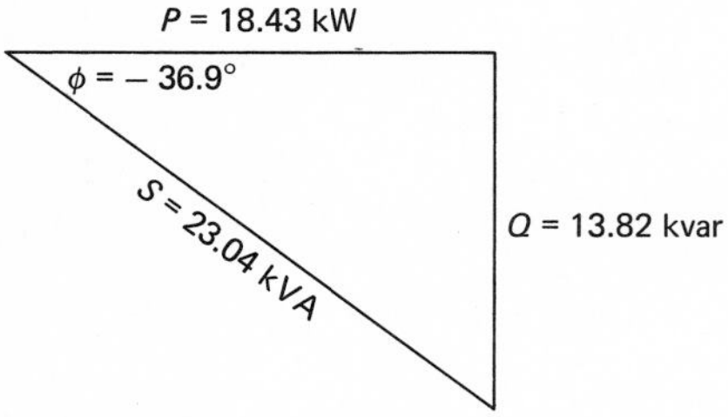

Fig. 15-14 Example 15-5: power triangle.

15-2 *RLC* PARALLEL CIRCUIT

Figure 15-15*a* is a schematic diagram of an *RLC* parallel circuit.

Partial phasor diagrams for the parallel *RLC* circuit in Fig. 15-15*a* are given in Fig. 15-15*b* and *c*. In each of these phasor diagrams the common voltage V is made the reference phasor $V\,\underline{/0°}$ or $V + j0$. The resistive current I_R is in phase with the voltage and has the phasor value $I_R\,\underline{/0°}$ or

Fig. 15-15 *RLC* parallel circuit. (*a*) Schematic diagram; (*b*) partial phasor diagram when $I_C > I_L$; (*c*) partial phasor diagram when $I_C < I_L$.

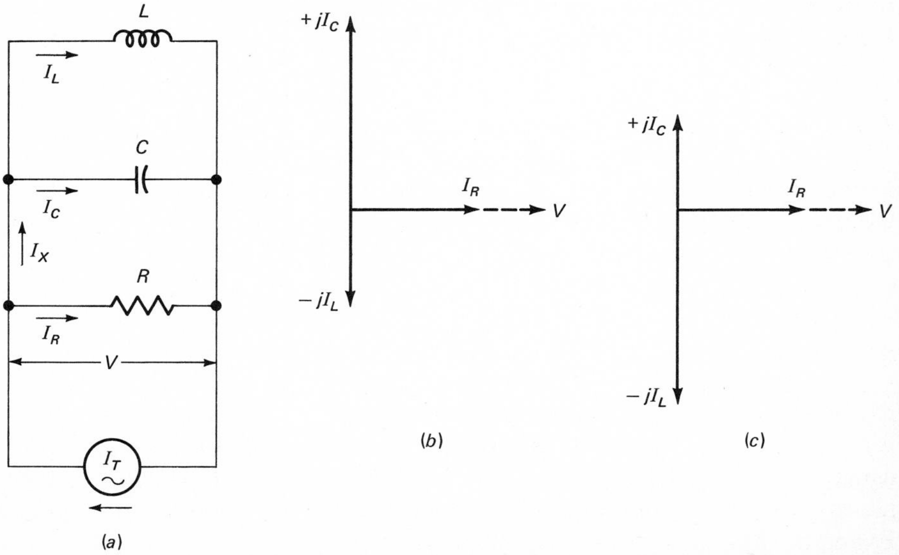

$I_R + j0$. The capacitive current I_C leads the voltage by 90° and has the phasor value $I_C \angle 90°$ or $0 + jI_C$. The inductive current I_L lags the voltage by 90° and has the phasor value $I_L \angle -90°$ or $0 - jI_L$.

The reactive currents I_C and I_L are of opposite j sign, so I_X, the net reactive current, is numerically equal to their difference:

$$I_X = jI_C - jI_L \qquad \text{or} \qquad j(I_C - I_L) \tag{15-2}$$

In this chapter we are concerned with situations in which reactive components are unequal. We will therefore consider only the *RLC* parallel circuit conditions in which I_C is greater than I_L and in which I_C is less than I_L.

When $I_C > I_L$ Figure 15-15*b* has the phasor line for $+jI_C$ longer than the line for $-jI_L$, illustrating the case in which I_C is greater than I_L. A full phasor diagram for a parallel *RLC* circuit in which I_C is greater than I_L is given in Fig. 15-16*a*.

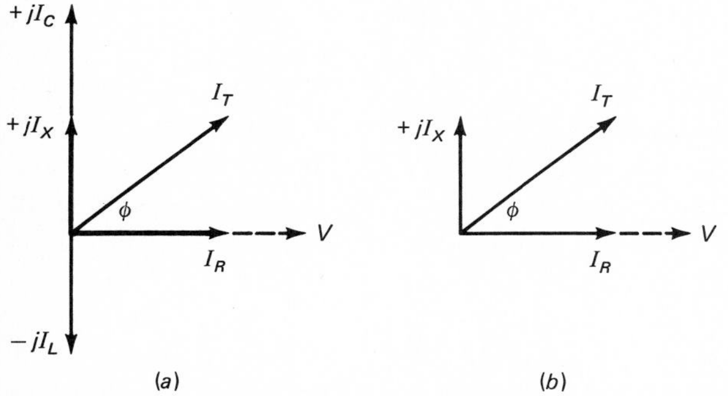

Fig. 15-16 *RLC* parallel circuit phasor diagrams when $I_C > I_L$. (*a*) Complete; (*b*) main variables.

When I_C is greater than I_L, $I_C - I_L$ is a positive quantity, so the net reactive current I_X is a $+j$ quantity, as shown in Fig. 15-16*a*.

Figure 15-16*b* is essentially the same as Fig. 15-16*a* but concentrates on the variables which determine the overall characteristics of the *RLC* parallel circuit: the common voltage V, the resistive current I_R, the net reactive current I_X, and the total circuit current I_T. It can be seen that Fig. 15-16*b* is the phasor diagram of a parallel *RC* circuit. Unless individual element characteristics are of interest, the behavior of a parallel *RLC* circuit in which $I_C > I_L$ can be regarded as the same as that of an *RC* parallel circuit. Refer to Sec. 14-3 for a treatment of the parallel *RC* circuit. When this approach is used, I_X is substituted for I_C, B_T for B_C, etc.

When $I_C < I_L$ The phasor line for $+jI_C$ in Fig. 15-15*c* is shorter than the line for $-jI_L$. This phasor diagram applies when I_C is less than I_L. Figure 15-17*a* is a full phasor diagram for the situation $I_C < I_L$.

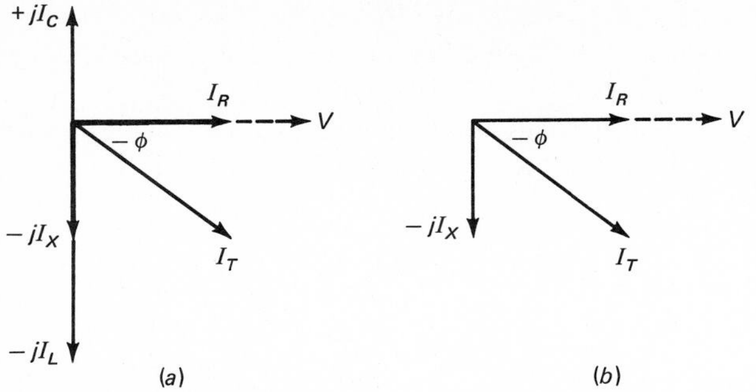

Fig. 15-17 *RLC* parallel circuit phasor diagrams when $I_C < I_L$. (*a*) Complete; (*b*) main variables.

If I_C is less than I_L, $I_C - I_L$ is a negative quantity. Figure 15-17*a* shows the net reactive current I_X as a $-j$ quantity.

The main variables controlling the behavior of the total *RLC* parallel circuit are the common voltage *V*, the resistive current I_R, the net reactive current I_X, and the total circuit current I_T. These variables for the $I_C < I_L$ condition are shown in the phasor diagram in Fig. 15-17*b*. From this phasor diagram it can be seen that a parallel *RLC* circuit with $I_C < I_L$ may be treated in total as a parallel *RL* circuit. Therefore, if individual element characteristics are not of direct concern, the principles developed in Sec. 14-4 can be applied with the substitution of I_X for I_L, B_T for B_L, etc.

It should be observed that in Figs. 15-16 and 15-17 no phasor diagram rotation was involved for the *a* and *b* diagrams, as was needed in Figs. 15-2 and 15-3. This was because the parallel circuit approach is oriented toward a voltage reference so that the conventional view for ac power considerations (the circuit phase angle is the angle of the total-current phasor with the total-voltage phasor as reference) is automatic when dealing with parallel circuits.

Example 15-6 What is the value of the voltage *V* in the circuit in Fig. 15-18?

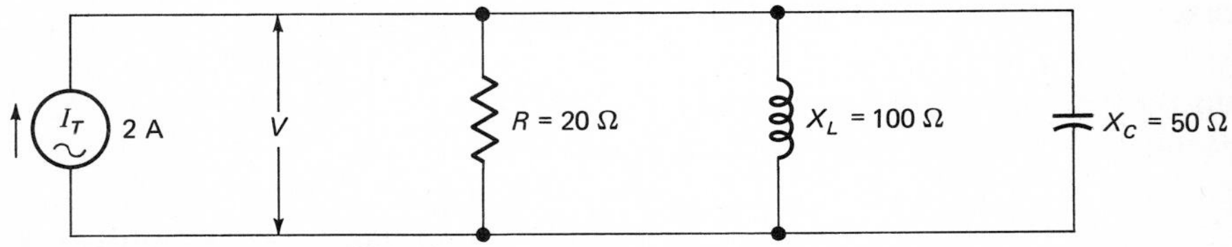

Fig. 15-18 Example 15-6: circuit.

Solution Convert the impedances of the individual elements into admittances:

$$G = \frac{1}{R} = \frac{1}{20\ \Omega} = 0.05\ \text{S}$$

$$B_L = \frac{1}{jX_L} = -j\frac{1}{100\ \Omega} = -j0.01\ \text{S}$$

$$B_C = \frac{1}{-jX_C} = j\frac{1}{50\ \Omega} = j0.02\ \text{S}$$

$$B_T = j(B_C - B_L) \qquad \text{extension of Eq. (15-2)}$$

$$B_T = j(0.02\ \text{S} - 0.01\ \text{S}) = j0.01\ \text{S}$$

The total circuit admittance is obtained by the phasor addition of the conductance and the net susceptance:

$$Y = G + jB_T = 0.05\ \text{S} + j0.01\ \text{S}$$

$$Y = (0.05 + j0.01)\ \text{S} \qquad \text{or} \qquad 0.051\ \text{S}\ \underline{/11.31^\circ}$$

The common circuit voltage V (with respect to the total current I_T) can now be obtained by applying the admittance form of Ohm's law ($I = VY$):

$$V = \frac{I_T}{Y} = \frac{2\ \text{A}\ \underline{/0^\circ}}{0.051\ \text{S}\ \underline{/11.31^\circ}} = \mathbf{39.22\ V}\ \underline{\mathbf{/-11.31^\circ}} \qquad \text{or} \qquad \underline{\mathbf{/-0.1974\ rad}}$$

The element values selected for Example 15-6 are the same as those for the series circuit in Example 15-1. Since the inductive reactance was greater than the capacitive reactance, the series circuit was predominantly inductive. The series circuit current lagged the applied emf. With the voltage of the parallel circuit common, and the capacitive susceptance greater than the inductive susceptance, the capacitive current in Example 15-6 is greater than the inductive current. The parallel circuit is thus predominantly capacitive. The circuit voltage V lags the source current I_T, or the current leads the voltage.

A phasor diagram for the circuit in Example 15-6 is given in Fig. 15-19.

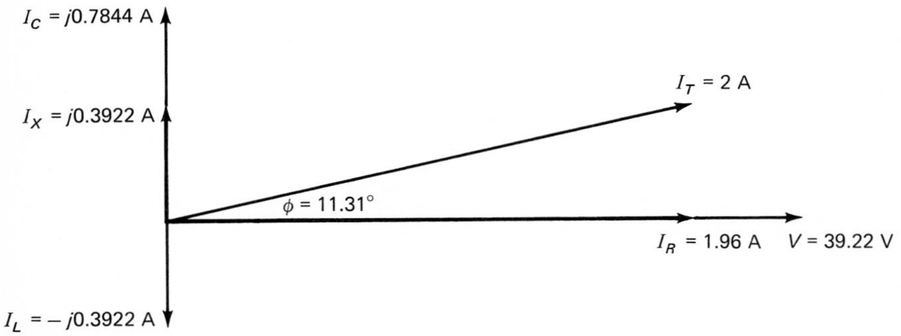

Fig. 15-19 Example 15-6: phasor diagram.

Example 15-7 In the circuit in Fig. 15-20, what is (*a*) each individual element current, (*b*) the net reactive current, and (*c*) the total circuit current?

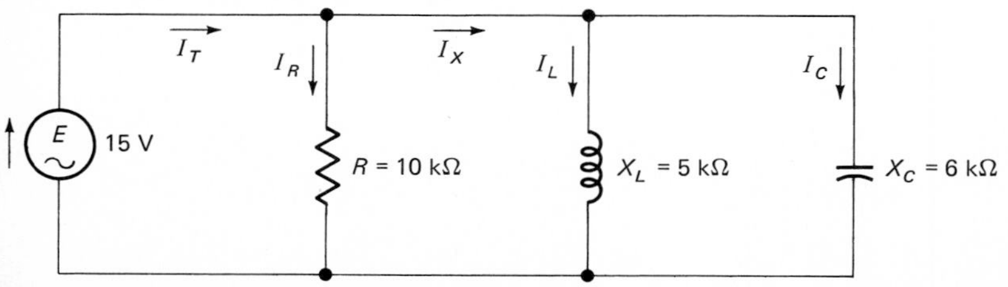

Fig. 15-20 Example 15-7: circuit.

Solution Calculating the admittances of the individual elements,

$$G = \frac{1}{R} = \frac{1}{10\ \text{k}\Omega} = 0.1\ \text{mS}$$

$$B_L = \frac{1}{X_L} = \frac{1}{j5\ \text{k}\Omega} = -j0.2\ \text{mS}$$

$$B_C = \frac{1}{X_C} = \frac{1}{-j6\ \text{k}\Omega} = j0.1667\ \text{mS}$$

a. Applying Ohm's law,

$$I_R = EG = 15\ \text{V} \times 0.1\ \text{mS} = \mathbf{1.5\ mA}$$

$$I_L = EB_L = 15\ \text{V} \times (-j0.2\ \text{mS}) = \mathbf{-j3\ mA}$$

$$I_C = EB_C = 15\ \text{V} \times j0.1667\ \text{mS} = \mathbf{j2.5\ mA}$$

b. The net reactive current can be obtained by the phasor addition of the inductive and capacitive currents:

$$I_X = j(I_C - I_L) \tag{15-2}$$

$$= j(2.5\ \text{mA} - 3\ \text{mA})$$

$$I_X = \mathbf{-j0.5\ mA = 0.5\ mA\ \underline{/-90^\circ}} \quad \text{or} \quad \mathbf{\underline{/-\pi/2\ rad}}$$

c. The total circuit current is the phasor sum of the resistive current and the net reactive current:

$$I_T = I_R + jI_X$$

$$I_T = \mathbf{(1.5\ mA - j0.5)\ mA = 1.58\ mA\ \underline{/-18.43^\circ}} \quad \text{or} \quad \mathbf{\underline{/-0.3216\ rad}}$$

In the parallel *RLC* circuit in Example 15-7 the inductive reactance is less than the capacitive reactance, so the circuit is net inductive. The inductive current is greater than the capacitive current, and the total circuit current lags the applied emf.

A phasor diagram for the conditions in Example 15-7 is given in Fig. 15-21.

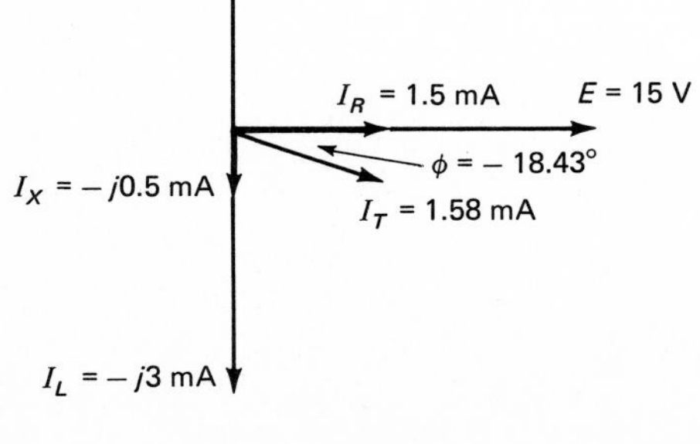

Fig. 15-21 Example 15-7: phasor diagram.

INSTANTANEOUS CURRENT DIRECTIONS

The total current in Example 15-7 (1.58 mA) is less than either its inductive component (3 mA) or its capacitive component (2.5 mA). It frequently happens that the values involved in an *RLC* parallel circuit result in branch currents which exceed the total circuit current in magnitude. This apparently illogical situation is due to the fact that, at any time instant, the inductive and capacitive currents have opposite directions. The waveform diagrams in Fig. 15-22 show the relationships between the reactive currents and the applied emf in the circuit in Example 15-7.

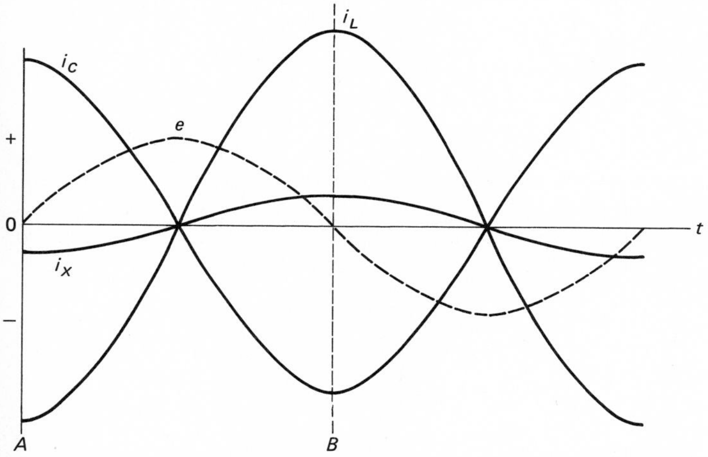

Fig. 15-22 Waveforms of the currents and the voltage in the parallel ac circuit of Example 15-7.

The waveform of the inductive current i_L lags that of the emf e by 90°. The waveform of the capacitive current i_C leads that of the emf by 90°. Thus i_L and i_C are in antiphase (180° phase relationship). At any instant in time, one of the reactive currents is positive while the other is negative. These positive and negative signs indicate opposite current directions.

At time *A* in Fig. 15-22 the capacitive current i_C is at its positive peak value, while the inductive current i_L is at its negative peak value, their directions being opposite. The magnitude of i_L being greater than that of i_C results in the net reactive current i_X having a direction the same as that of the inductive current. The conditions existing at instant *A* are shown in the circuit in Fig. 15-23*a*. Figure 15-23*b* shows the situation one half-cycle later, at instant *B*. At this time the current magnitudes are the same as in Fig. 15-23*a*, but their directions are reversed. At all instants (except when the currents are zero) the inductive and capacitive currents will be opposite in direction, the net reactive current direction being that of the inductive current.

It must be understood that the predominantly inductive conditions referred to above are related to specific circuit values such as those in Example 15-7.

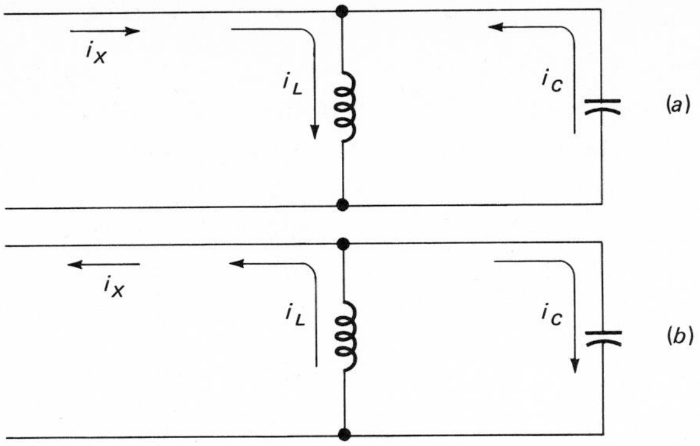

Fig. 15-23 Instantaneous current directions in a parallel *LC* circuit.

Example 15-8 An industrial electric load has a power factor of 0.5 lagging and draws 5 kW from a 208-V 60-Hz supply. What value of parallel capacitor would be needed to raise the circuit power factor to 0.95 lagging?

Solution An equivalent parallel circuit of the load is shown in Fig. 15-24 where G and B_L are, respectively, load conductance and inductive susceptance. The active power is dissipated in this circuit entirely within the resistive component G.

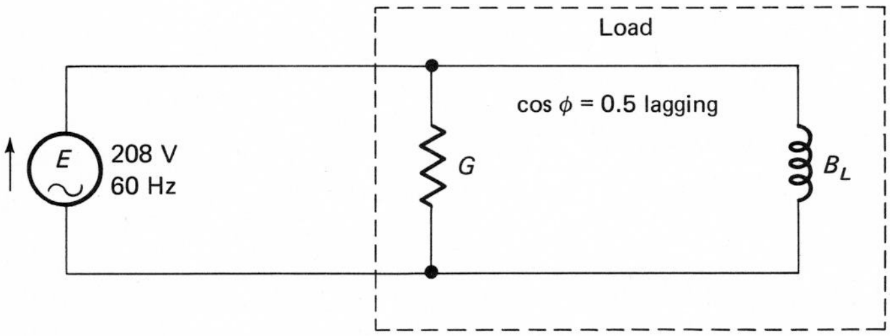

Fig. 15-24 Example 15-8: circuit.

$$P = \frac{E^2}{R} \quad \text{or} \quad E^2G$$

$$G = \frac{P}{E^2} = \frac{5\ \text{kW}}{(208\ \text{V})^2} = 0.1156\ \text{S}$$

The power factor of a parallel circuit is the ratio of its conductance to its admittance:

$$\cos\phi = \frac{G}{Y}$$

$$Y = \frac{G}{\cos\phi} = \frac{0.1156\ \text{S}}{0.5} = 0.2312\ \text{S}$$

Figure 15-25 is an admittance triangle for the load.

From the admittance triangle,

$$B_L = \sqrt{Y^2 - G^2} = \sqrt{(0.2312\ \text{S})^2 - (0.1156\ \text{S})^2} = 0.2002\ \text{S}$$

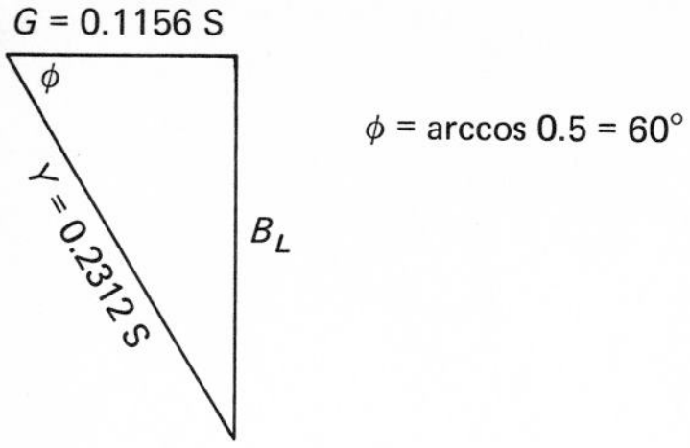

Fig. 15-25 Example 15-8: admittance triangle.

Thus the load has the characteristics of a parallel circuit consisting of a 0.1156-S conductance (8.651-Ω resistance) and a 0.2002-S inductive susceptance (4.995-Ω inductive reactance or 0.0132-H inductance).

When the capacitance is added in parallel, the circuit will appear as in Fig. 15-26.

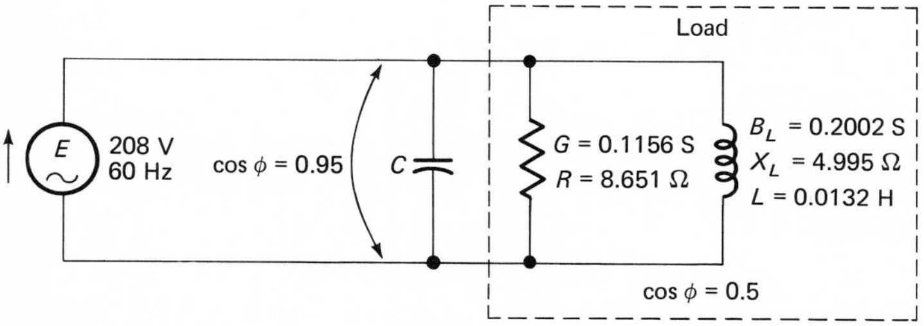

Fig. 15-26 Example 15-8: circuit with added capacitor.

For the power factor cos ϕ to equal 0.95 lagging, the admittance triangle must appear as in Fig. 15-27.

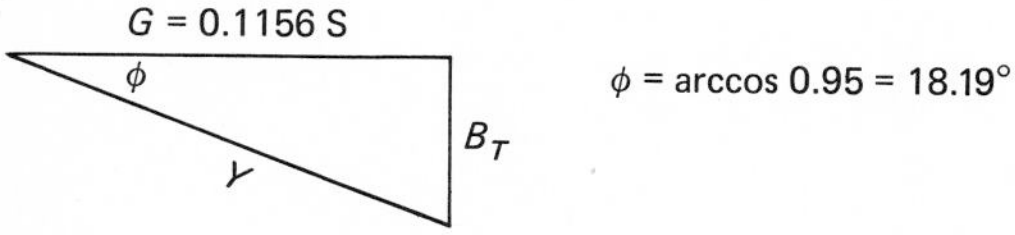

Fig. 15-27 Example 15-8: admittance triangle.

$$\cos\phi = \frac{G}{Y}$$

$$Y = \frac{G}{\cos\phi} = \frac{0.1156 \text{ S}}{0.95} = 0.1217 \text{ S}$$

The total susceptance B_T can now be found

$$B_T = \sqrt{Y^2 - G^2} = \sqrt{(0.1217 \text{ S})^2 - (0.1156 \text{ S})^2} = 0.038 \text{ S}$$

A complete admittance diagram, in phasor form, is shown in Fig. 15-28, and from it the value of the capacitive susceptance can be determined. The lagging power factor means that the net susceptance B_T is a $-j$ quantity (inductive). Therefore the capacitive susceptance, which is a $+j$ quantity, must be lower in value than the inductive susceptance of the load.

$$B_C = B_L - B_T = 0.2002 \text{ S} - 0.038 \text{ S} = 0.1622 \text{ S}$$

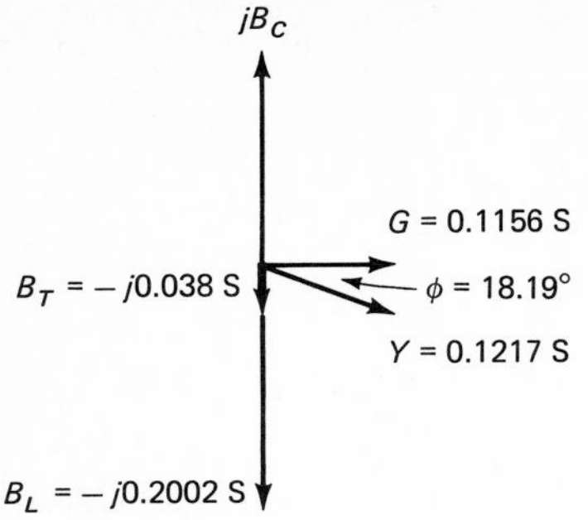

Fig. 15-28 Example 15-8: admittance diagram.

The value of the capacitance C can now be calculated:

$$\omega = 2\pi f = 2\pi \times 60 \text{ Hz} = 377 \text{ rad/s}$$

$$B_C = \omega C$$

$$C = \frac{B_C}{\omega} = \frac{0.1622 \text{ S}}{377 \text{ rad/s}} = \mathbf{430.2\ \mu F}$$

The final circuit for Example 15-8 is given in Fig. 15-29.

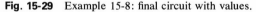

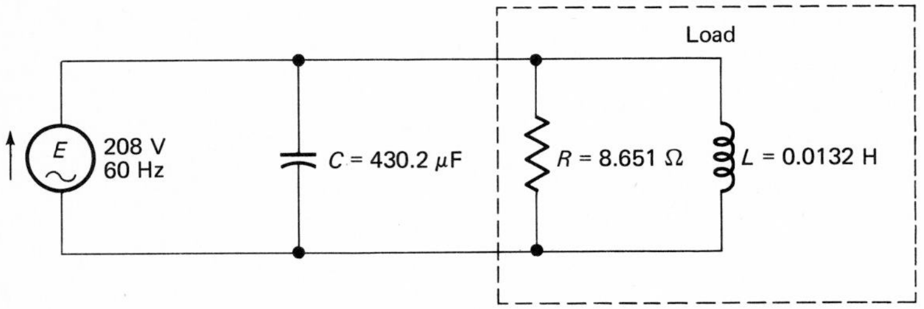

Fig. 15-29 Example 15-8: final circuit with values.

The power dissipated by the total circuit in Fig. 15-29 is 5 kW, the same as that of the original load (in Fig. 15-24). The addition of the parallel capacitor has not affected the active power but has reduced the apparent power from 10 to 5.263 kVA. In some industrial situations, electric supply costs are directly related to apparent power. If this were so in the case of Example 15-8, a saving of 52.6 percent of the cost of supplying the load power could result from connection of the 430-μF capacitor. This supply cost reduction could justify the installation of the capacitor.

The connection of a reactive element in parallel with a load to raise the circuit power factor toward unity ($\cos \phi = 1$) is called *power factor improvement.* It is an important subject which is dealt with again in Chap. 16.

Example 15-9 A tape recorder head has an inductance of 100 mH, a self-capacitance of 250 pF, and a parallel internal resistance of 10 kΩ. The recorder amplifier delivers 1 mA to the head at all audible frequencies. What magnitude of current is developed in the coil's inductance at 20 Hz, 1 kHz, and 20 kHz?

Solution (See Fig. 15-30)

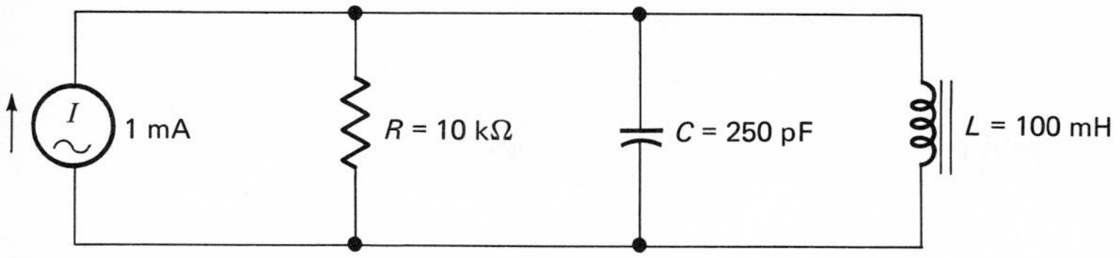

Fig. 15-30 Example 15-9: circuit.

a. When $f_1 = 20$ Hz,

$$\omega = 2\pi f_1 = 2\pi \times 20\text{ Hz} = 125.7\text{ rad/s}$$

$$X_L = \omega L = 125.7\text{ rad/s} \times 100\text{ mH} = 12.57\ \Omega$$

$$X_C = \frac{1}{\omega C} = \frac{1}{125.7\text{ rad/s} \times 2.5 \times 10^{-10}\text{ F}} = 31.82\text{ M}\Omega$$

At 20 Hz, X_L is so low compared with either R or X_C that the inductance can be considered to be the only significant element connected to the 10-mA source. Therefore,

$$I_L \simeq \mathbf{1\ mA}$$

b. When $f_2 = 1$ kHz, the ratio of the frequencies ($f_2/f_1 = 50$) can be used to compute the reactances at 1 kHz:

$$X_L\text{ at 1 kHz} = X_L\text{ at 20 Hz} \times 50 = 12.57\ \Omega \times 50 = 628.5\ \Omega$$

$$X_C\text{ at 1 kHz} = X_C\text{ at 20 Hz} \div 50 = 31.82\text{ M}\Omega \div 50 = 636\text{ k}\Omega$$

Again at 1 kHz, X_L is much lower than either R or X_C, so

$$I_L \simeq \mathbf{1\ mA}$$

c. When $f_3 = 20$ kHz, applying the ratio $f_3/f_1 = 1000$:

$$X_L\text{ at 20 kHz} = X_L\text{ at 20 Hz} \times 1000 = 12.57\ \Omega \times 1000 = 12.57\text{ k}\Omega$$

$$X_C\text{ at 20 kHz} = X_C\text{ at 20 Hz} \div 1000 = 31.82\text{ M}\Omega \div 1000 = 31.82\text{ k}\Omega$$

Thus at 20 kHz, X_L, X_C, and R are all of the same order of magnitude, and a detailed calculation of I_L is necessary.

Figure 15-31 shows the values of the element impedances. Converting these impedances to admittances,

Fig. 15-31 Example 15-9: circuit with reactance values at 20 kHz.

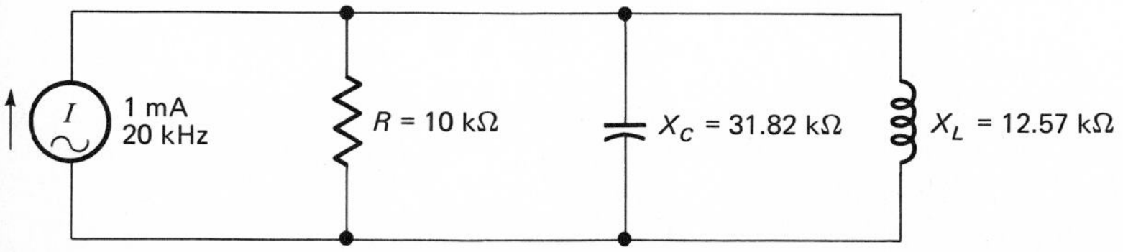

$$G = \frac{1}{R} = \frac{1}{10\ \text{k}\Omega} = 0.1\ \text{mS}$$

$$B_C = \frac{1}{X_C} = \frac{1}{31.82\ \text{k}\Omega} = 0.0314\ \text{mS}$$

$$B_L = \frac{1}{X_L} = \frac{1}{12.57\ \text{k}\Omega} = 0.0796\ \text{mS}$$

$$B_T = j(B_C - B_L) \qquad \text{extension of Eq. (15-2)}$$

$$B_T = j(0.0314\ \text{mS} - 0.0796\ \text{mS}) = -j0.0482\ \text{mS}$$

The total admittance Y is the phasor sum of the conductance and the net susceptance:

$$Y = G + jB_T = (0.1 - j0.0482)\ \text{mS} = 0.111\ \text{mS}\ \underline{/-25.73^\circ}$$

The magnitudes of the common voltage V and the current I_L can both be determined by applying Ohm's law:

$$V = \frac{I}{Y} = \frac{1\ \text{mA}}{0.111\ \text{mS}} = 9.009\ \text{V}$$

$$I_L = VB_L = 9.009\ \text{V} \times 0.0796\ \text{mS}$$

$$I_L = \mathbf{0.7171\ mA}$$

A tape recorder head consists of a coil of wire wound on an iron core. When the instrument is recording, the magnetic tape passes over the head while an alternating current in the coil produces an alternating magnetic field. This field magnetizes the tape by induction, resulting in a magnetic reproduction of the sound being recorded. The results for Example 15-9 show that, while the coil's self-capacitance and resistance play an insignificant role at lower frequencies, at the top of the audio spectrum, 20 kHz, these parameters of capacitance and resistance can noticeably reduce the inductive current. This current reduction will cause the high-frequency performance of the recording head to be lower than its performance at lower frequencies.

15-3 *RLC* SERIES-PARALLEL CIRCUIT

Series-parallel *RLC* circuits are dealt with by

1. Applying the technique of considering each subcircuit, series or parallel, and then combining the results
2. Adopting series-to-parallel (or parallel-to-series) conversion to produce a circuit which is simple series or parallel in configuration

Some examples will demonstrate how these alternative techniques may be used.

Example 15-10 What is the total current in the circuit in Fig. 15-32?

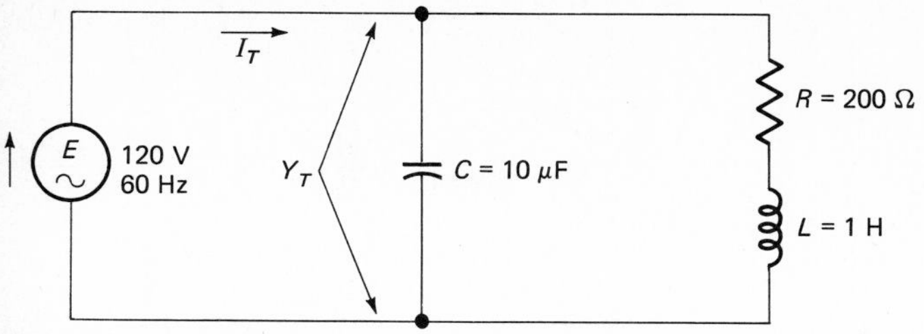

Fig. 15-32 Example 15-10: circuit.

Solution

$$\omega = 2\pi f = 2\pi \times 60 \text{ Hz} = 377 \text{ rad/s}$$

$$X_C = \frac{1}{\omega C} = \frac{1}{377 \text{ rad/s} \times 10\ \mu\text{F}} = 265.3\ \Omega$$

$$B_C = \frac{1}{X_C} = \frac{1}{265.3\ \Omega} = 3.77 \text{ mS}$$

$$X_L = \omega L = 377 \text{ rad/s} \times 1 \text{ H} = 377\ \Omega$$

Finding the impedance of the RL series branch:

$$Z_{RL} = R + jX_L = (200 + j377)\ \Omega = 426.8\ \Omega\ \underline{/62.05^\circ}$$

Converting the RL branch impedance to its equivalent admittance:

$$Y_{RL} = \frac{1}{Z_{RL}} = \frac{1}{426.8\ \Omega\ \underline{/62.05^\circ}} = 2.343 \text{ mS}\ \underline{/-62.05^\circ}$$

$$Y_{RL} = (1.098 - j2.07) \text{ mS}$$

The total admittance Y_T of the series-parallel network can be obtained by the phasor addition of the admittances of the capacitive and inductive branches:

$$Y_T = Y_C + Y_{RL} = j3.77 \text{ mS} + (1.098 - j2.07) \text{ mS}$$

$$= [1.098 + j(3.77 - 2.07)] \text{ mS}$$

$$Y_T = (1.098 + j1.7) \text{ mS} \qquad \text{or} \qquad 2.024 \text{ mS}\ \underline{/57.1^\circ}$$

The application of Ohm's law will give the total circuit current I_T:

$$I_T = EY_T = 120 \text{ V}\ \underline{/0^\circ} \times 2.024 \text{ mS}\ \underline{/57.1^\circ}$$

$$I_T = \mathbf{242.9 \text{ mA}\ \underline{/57.1^\circ}} \qquad \text{or} \qquad \mathbf{\underline{/0.9965 \text{ rad}}}$$

Example 15-11 Verify the result for Example 15-10 by series-to-parallel conversion. *Solution* Figure 15-33*a* shows the series RL branch in Fig. 15-32 with the elements identified as series by the symbols R_s and X_{Ls}. In Fig. 15-33*b* the equivalent parallel circuit elements have the symbols G_p and B_{Lp}. Applying the series-to-parallel relationships developed in Chap. 14,

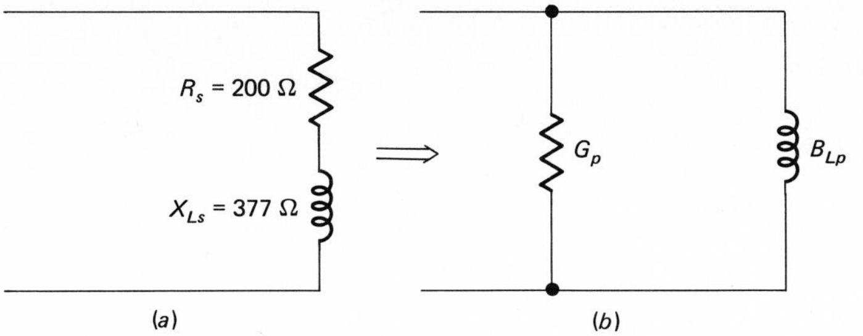

Fig. 15-33 Example 15-11: (*a*) the *RL* series circuit in Fig. 15-32; (*b*) its parallel equivalent.

$$G_p = \frac{R_s}{R_s^2 + X_{Ls}^2} \tag{14-44}$$

$$G_p = \frac{200\ \Omega}{(200\ \Omega)^2 + (377\ \Omega)^2} = 1.098\ \text{mS}$$

$$B_{Lp} = -j\frac{X_{Ls}}{R_s^2 + X_{Ls}^2} \tag{14-45}$$

$$B_{Lp} = -j\frac{377\ \Omega}{(200\ \Omega)^2 + (377\ \Omega)^2} = -j2.07\ \text{mS}$$

The capacitive susceptance, calculated in the solution to Example 15-10, is $B_C = 3.77$ mS.

Figure 15-34 is the parallel equivalent of the series-parallel circuit in Fig. 15-32.

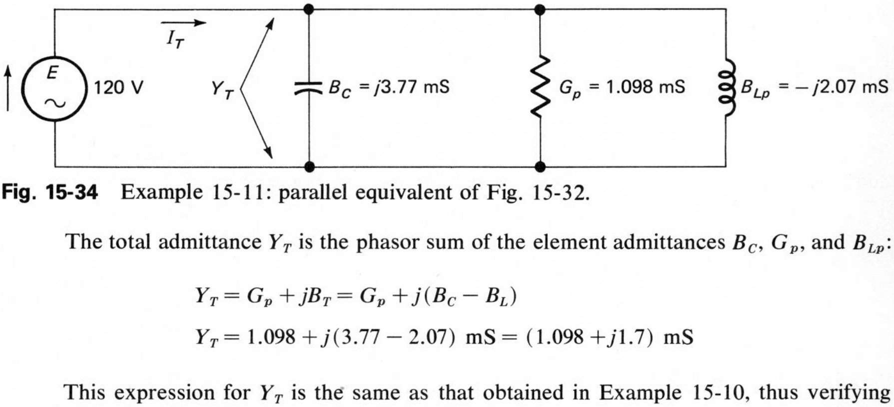

Fig. 15-34 Example 15-11: parallel equivalent of Fig. 15-32.

The total admittance Y_T is the phasor sum of the element admittances B_C, G_p, and B_{Lp}:

$$Y_T = G_p + jB_T = G_p + j(B_C - B_L)$$

$$Y_T = 1.098 + j(3.77 - 2.07)\ \text{mS} = (1.098 + j1.7)\ \text{mS}$$

This expression for Y_T is the same as that obtained in Example 15-10, thus verifying the result for that example.

Example 15-12 What is the active power in the circuit in Fig. 15-35?
Solution

$$\omega = 2\pi f = 2\pi \times 1.5\ \text{MHz} = 9.425\ \text{Mrad/s}$$

$$X_L = \omega L = 9.425\ \text{Mrad/s} \times 1\ \text{mH} = 9.425\ \text{k}\Omega$$

$$X_C = \frac{1}{\omega C} = \frac{1}{9.425\ \text{Mrad/s} \times 15\ \text{pF}} = 7.073\ \text{k}\Omega$$

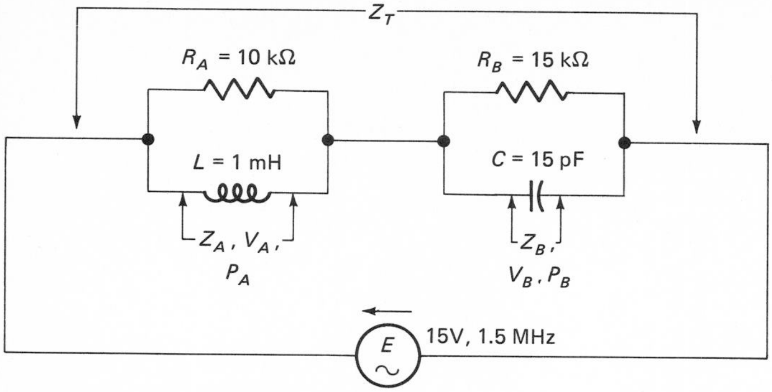

Fig. 15-35 Example 15-12: circuit.

Calculate the impedance Z_A of the RL parallel group by applying the method for two parallel elements $[Z_T = Z_1Z_2/(Z_1 + Z_2)]$:

$$Z_A = \frac{Z_{R_A} Z_L}{Z_{R_A} + Z_L} = \frac{R_A \times jX_L}{R_A + jX_L} = \frac{jR_A X_L}{R_A + jX_L}$$

$$= \frac{j(10\ \text{k}\Omega \times 9.425\ \text{k}\Omega)}{10\ \text{k}\Omega + j9.425\ \text{k}\Omega} = \frac{94.25\ (\text{k}\Omega)^2\ \underline{/90^\circ}}{13.74\ \text{k}\Omega\ \underline{/43.3^\circ}}$$

$$Z_A = 6.86\ \text{k}\Omega\ \underline{/46.7^\circ}$$

Calculate the impedance Z_B of the RC parallel group:

$$Z_B = \frac{Z_{R_B} Z_C}{Z_{R_B} + Z_C} = \frac{R_B \times (-jX_C)}{R_B - jX_C} = \frac{-jR_B X_C}{R_B - jX_C}$$

$$= \frac{-j(15\ \text{k}\Omega \times 7.073\ \text{k}\Omega)}{15\ \text{k}\Omega - j7.073\ \text{k}\Omega} = \frac{106.1\ (\text{k}\Omega)^2\ \underline{/-90^\circ}}{16.58\ \text{k}\Omega\ \underline{/-25.25^\circ}}$$

$$Z_B = 6.399\ \text{k}\Omega\ \underline{/-64.75^\circ}$$

The group impedances Z_A and Z_B are in series. Therefore the total impedance Z_T is the phasor sum of Z_A and Z_B:

$$Z_T = Z_A + Z_B = 6.86\ \text{k}\Omega\ \underline{/46.7^\circ} + 6.399\ \text{k}\Omega\ \underline{/-64.75^\circ}$$

$$= (4.705 + j4.993)\ \text{k}\Omega + (2.73 - j5.788)\ \text{k}\Omega$$

$$= [(4.705 + 2.73) + j(4.993 - 5.788)]\ \text{k}\Omega$$

$$Z_T = (7.435 - j0.795)\ \text{k}\Omega \quad \text{or} \quad 7.477\ \text{k}\Omega\underline{/-6.1^\circ}$$

The voltages across each of the two parallel groups can be found by applying the voltage divider principle:

$$V_A = E\frac{Z_A}{Z_T} = 15\ \text{V}\ \underline{/0^\circ}\ \frac{6.86\ \text{k}\Omega\ \underline{/46.7^\circ}}{7.477\ \text{k}\Omega\ \underline{/-6.1^\circ}} = 13.76\ \text{V}\ \underline{/52.8^\circ}$$

$$V_B = E\frac{Z_B}{Z_T} = 15\ \text{V}\ \underline{/0^\circ}\ \frac{6.399\ \text{k}\Omega\ \underline{/-64.75^\circ}}{7.477\ \text{k}\Omega\ \underline{/-6.1^\circ}} = 12.84\ \text{V}\ \underline{/-58.65^\circ}$$

The active power in the circuit is developed only in resistances R_A and R_B:

$$P_A = \frac{V_A^2}{R_A} = \frac{(13.76\ \text{V})^2}{10\ \text{k}\Omega} = 18.93\ \text{mW}$$

$$P_B = \frac{V_B^2}{R_B} = \frac{(12.84\ \text{V})^2}{15\ \text{k}\Omega} = 10.99\ \text{mW}$$

The total power in any circuit is the arithmetic sum of the individual powers:

$$P_T = P_A + P_B = 18.93\ \text{mW} + 10.99\ \text{mW} = \mathbf{29.92\ mW}$$

In Example 15-12 the active power for each parallel subnetwork was calculated and the resulting powers added to arrive at the total circuit power, as this technique is perhaps the easiest to visualize. Actually, the amount of calculation can be reduced appreciably by realizing what is implied by the rectangular expression for the total circuit impedance Z_T.

Figure 15-36 is a series circuit with impedance $Z_T = 7.435 - j0.795\ \text{k}\Omega$, the same as the impedance of the series-parallel circuit in Example 15-12. The power dissipated by the 7.435-kΩ resistance in Fig. 15-36 is the same as the total active power in Example 15-12:

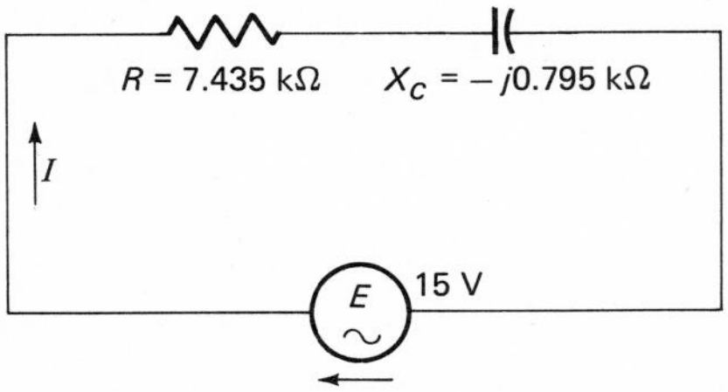

Fig. 15-36 Series circuit with the same overall characteristics as Fig. 15-35.

$$I = \frac{E}{Z_T} = \frac{15\ \text{V}}{7.477\ \text{k}\Omega} = 2.006\ \text{mA}$$

$$P_T = I^2R = (2.006\ \text{mA})^2 \times 7.435\ \text{k}\Omega = \mathbf{29.92\ mW}$$

Example 15-13 By applying parallel-to-series conversions, verify the result for Example 15-12.

Solution Figure 15-37*a* gives the parallel *RL* group in Fig. 15-35 with the elements identified as parallel by the subscript *p*. The series equivalent is given in Fig. 15-37*b* using the subscript *s*.

Fig. 15-37 Example 15-13: (*a*) parallel *RL* network in Fig. 15-35; (*b*) its series equivalent.

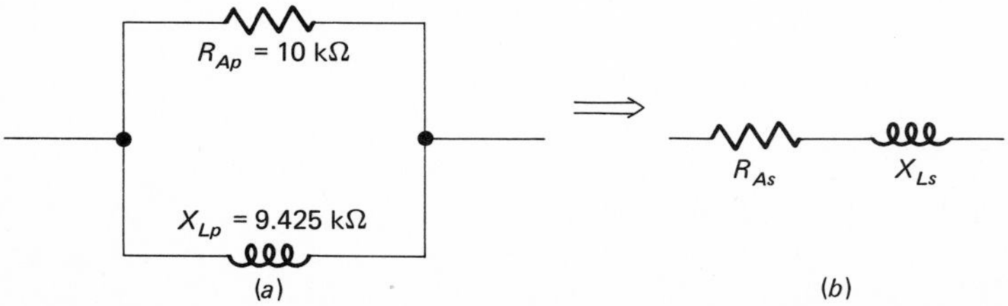

Applying the parallel-to-series conversion expressions in Chap. 14,

$$G_{Ap} = \frac{1}{R_{Ap}} = \frac{1}{10\ \text{k}\Omega} = 0.1\ \text{mS}$$

$$B_{Lp} = \frac{1}{X_{Lp}} = \frac{1}{9.425\ \text{k}\Omega} = 0.1061\ \text{mS}$$

From Eq. (14-46),

$$R_{As} = \frac{G_{Ap}}{G_{Ap}^2 + B_{Lp}^2} = \frac{0.1\ \text{mS}}{(0.1\ \text{mS})^2 + (0.1061\ \text{mS})^2} = 4.704\ \text{k}\Omega$$

From Eq. (14-47),

$$X_{Ls} = j\frac{B_{Lp}}{G_{Ap}^2 + B_{Lp}^2} = j\frac{0.1061\ \text{mS}}{(0.1\ \text{mS})^2 + (0.1061\ \text{mS})^2} = j4.991\ \text{k}\Omega$$

The parallel RC group in Fig. 15-35 is shown in Fig. 15-38*a* with its elements identified as parallel. The equivalent series circuit, with elements identified as series, is given in Fig. 15-38*b*.

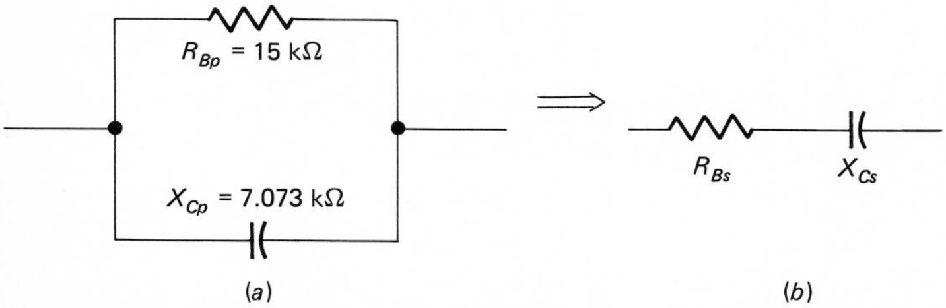

Fig. 15-38 Example 15-13: (*a*) parallel RC network in Fig. 15-35; (*b*) its series equivalent.

$$G_{Bp} = \frac{1}{R_{Bp}} = \frac{1}{15\ \text{k}\Omega} = 0.0667\ \text{mS}$$

$$B_{Cp} = \frac{1}{X_{Cp}} = \frac{1}{7.073\ \text{k}\Omega} = 0.1414\ \text{mS}$$

From Eq. (14-42),

$$R_{Bs} = \frac{G_{Bp}}{G_{Bp}^2 + B_{Cp}^2} = \frac{0.0667\ \text{mS}}{(0.0667\ \text{mS})^2 + (0.1414\ \text{mS})^2} = 2.729\ \text{k}\Omega$$

From Eq. (14-43),

$$X_{Cs} = -j\frac{B_{Cp}}{G_{Bp}^2 + B_{Cp}^2} = -j\frac{0.1414\ \text{mS}}{(0.0667\ \text{mS})^2 + (0.1414\ \text{mS})^2} = -j5.785\ \text{k}\Omega$$

The series-parallel circuit in Fig. 15-35 can now be redrawn as a series circuit using the calculated values of R_{As}, R_{Bs}, X_{Ls}, and X_{Cs}. Figure 15-39 shows this series circuit.

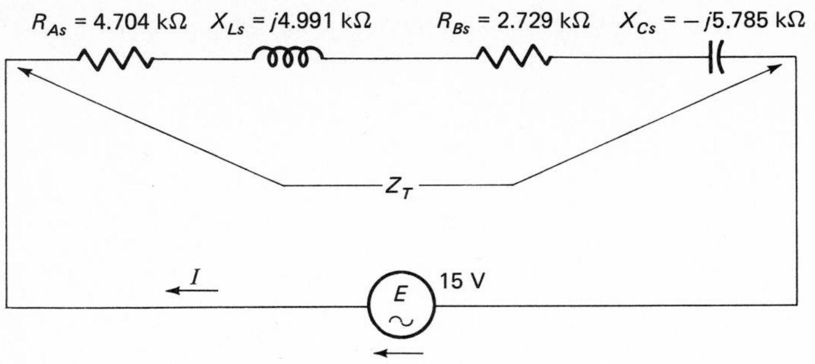

Fig. 15-39 Example 15-13: series equivalent of Fig. 15-35.

The total impedance of the equivalent series circuit in Fig. 15-39 is

$$Z_T = (R_{As} + jX_{Ls}) + (R_{Bs} - jX_{Cs})$$

$$= (4.704 + j4.991)\ \text{k}\Omega + (2.729 - j5.785)\ \text{k}\Omega$$

$$Z_T = (7.433 - j0.794)\ \text{k}\Omega \qquad \text{or} \qquad 7.476\ \text{k}\Omega\underline{/-6.1^\circ}$$

The circuit current I can be obtained by using Ohm's law:

$$I = \frac{E}{Z_T} = \frac{15\ \text{V}\ \underline{/0^\circ}}{7.476\ \text{k}\Omega\ \underline{/-6.1^\circ}} = 2.006\ \text{mA}\ \underline{/6.1^\circ}$$

The active power in the circuit in Fig. 15-39 is developed entirely in the series equivalent resistances R_{As} and R_{Bs}:

$$P_T = I^2(R_{As} + R_{Bs}) = (2.006\ \text{mA})^2 \times (4.704 + 2.729)\ \text{k}\Omega = \mathbf{29.91\ mW}$$

The solution for Example 15-13 verifies the result for Example 15-12 within the normal limits of four significant figures.

Example 15-14 A noise filter circuit in an audio amplifier is shown in Fig. 15-40. When the switch S is open, high audio frequencies are reduced in level relative to when the switch is closed. What is the magnitude of the ratio V_o/V_i at a frequency of 14 kHz when (*a*) S is open and (*b*) S is closed?

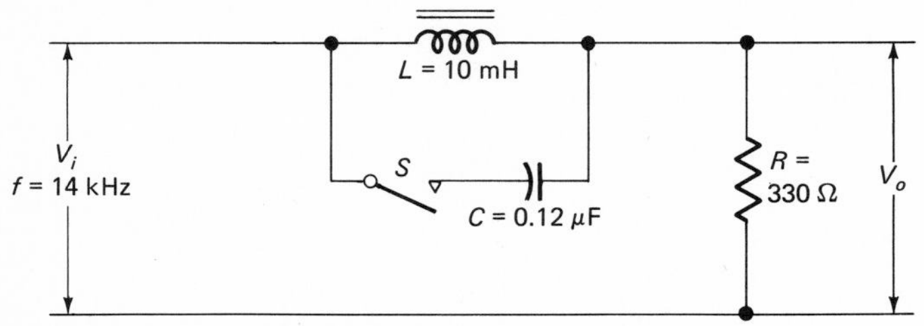

Fig. 15-40 Example 15-14: circuit.

Solution

a. When S is open the equivalent circuit of Fig. 15-40 is that in Fig. 15-41.

$$\omega = 2\pi f = 2\pi \times 14\ \text{kHz} = 87.96\ \text{krad/s}$$

$$X_L = \omega L = 87.96\ \text{krad/s} \times 10\ \text{mH} = 879.6\ \Omega$$

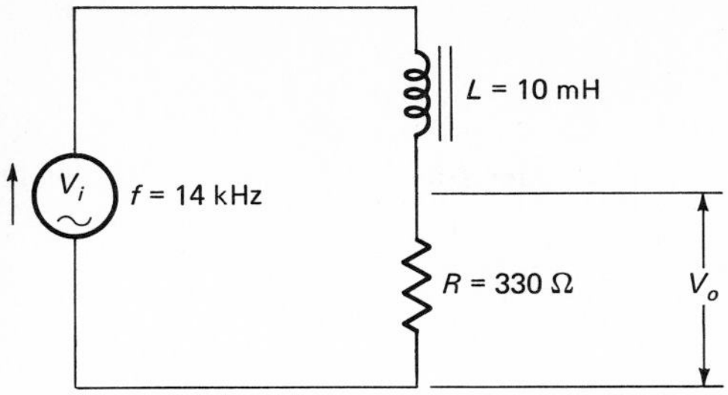

Fig. 15-41 Example 15-14: equivalent of Fig. 15-40 when the switch is open.

The impedance of the series *RL* network is

$$Z = R + jX_L = (330 + j879.6)\ \Omega \qquad \text{or} \qquad 939.5\ \Omega\ \underline{/69.42^\circ}$$

The magnitude of the voltage V_o can be obtained by applying the voltage divider principle:

$$V_o = V_i \frac{R}{Z} = V_i \frac{330\ \Omega}{939.5\ \Omega} = 0.3513\ V_i$$

Therefore when the switch is open, the ratio V_o/V_i is

$$\frac{V_o}{V_i} = \mathbf{0.3513}$$

b. When the switch is closed the equivalent circuit of Fig. 15-40 is that in Fig. 15-42.

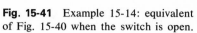

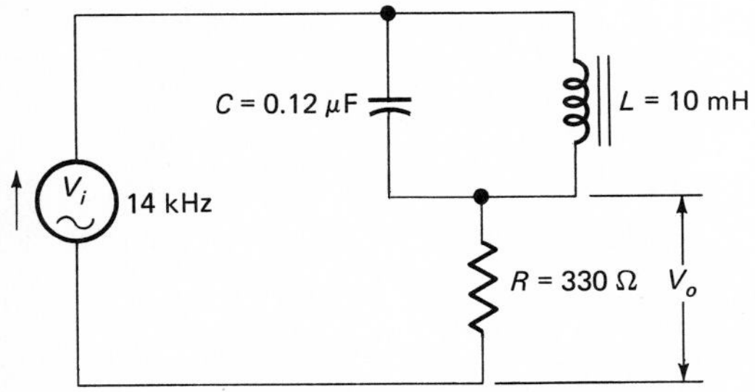

Fig. 15-42 Example 15-14: equivalent of Fig. 15-40 when the switch is closed.

$$\omega = 87.96\ \text{krad/s}$$

$$X_L = 879.6\ \Omega$$

$$X_C = \frac{1}{\omega C} = \frac{1}{87.96\ \text{krad/s} \times 0.12\ \mu\text{F}} = 94.74\ \Omega$$

The impedance of the *LC* parallel circuit is

$$Z_{LC} = \frac{Z_L Z_C}{Z_L + Z_C} = \frac{jX_L(-jX_C)}{j(X_L - X_C)} = \frac{-j^2 X_L X_C}{j(X_L - X_C)} = -j\frac{X_L X_C}{X_L - X_C}$$

$$Z_{LC} = -j\frac{879.6\ \Omega \times 94.74\ \Omega}{879.6\ \Omega - 94.74\ \Omega} = -j106.2\ \Omega$$

The total impedance of the series-parallel network is

$$Z_T = Z_R + Z_{LC} = (330 - j106.2)\ \Omega \qquad \text{or} \qquad 346.7\ \Omega \angle{-17.85°}$$

The magnitude of the voltage V_o obtained by applying the voltage divider principle is

$$V_o = V_i \frac{R}{Z_T} = V_i \frac{330\ \Omega}{346.7\ \Omega} = 0.9518\ V_i$$

Thus, when the switch is closed, the ratio V_o/V_i is

$$\frac{V_o}{V_i} = \mathbf{0.9518}$$

The results for Example 15-14 show that, with the circuit in Fig. 15-41, substantially less of the voltage V_i is transmitted through the circuit when the switch is open than when it is closed. Much of the irritating background noise in a poor-quality recording or radio transmission occurs at audio frequencies higher than about 12 kHz. A simple ac circuit, such as in Fig. 15-41, can be used to reduce the level of this type of noise. The switch would only be opened when the background noise is objectionable. At other times the higher audio frequencies would be transmitted for good sound fidelity.

15-4 SUMMARY OF EQUATIONS

$V_X = jV_L - jV_C$ or $j(V_L - V_C)$ (15-1)

$I_X = jI_C - jI_L$ or $j(I_C - I_L)$ (15-2)

EXERCISE PROBLEMS

Section 15-1

15-1 Using Fig. 15-1 in the text, find the magnitudes of I, V_R, V_C, V_L, and the circuit phase angle, and show that the phasor sum of V_R, V_C, and V_L equals the emf when

a. $E = 100$ V, $R = 100\ \Omega$, $X_C = 250\ \Omega$, $X_L = 1000\ \Omega$
(*ans.* 132.2 mA, 13.22 V, 33.04 V, 132.2 V, 82.4° lag)

b. $E = 12$ V, $R = 1000\ \Omega$, $X_C = 1000\ \Omega$, $X_L = 3000\ \Omega$
(*ans.* 5.37 mA, 5.37 V, 5.37 V, 16.1 V, 63.43° lag)

c. $E = 200\ \mu$V, $R = 20\ \Omega$, $X_C = 300\ \Omega$, $X_L = 280\ \Omega$
(*ans.* 7.071 μA, 141.4 μV, 2.121 mV, 1.98 mV, 45° lead)

d. $E = 200\ \mu$V, $R = 2\ \Omega$, $X_C = 300\ \Omega$, $X_L = 280\ \Omega$
(*ans.* 9.95 μA, 19.9 μV, 2.985 mV, 2.786 mV, 84.29° lead)

15-2 Repeat Prob. 15-1 when $E = 1$ V, $R = 10\ \Omega$, $C = 500$ pF, $L = 200\ \mu$H, and

a. $f = 400$ kHz *c.* $f = 500$ kHz
b. $f = 450$ kHz *d.* $f = 550$ kHz

15-3 Using Fig. 15-1 in the text, find the magnitudes of I, V_C, and the phase angle, and show that the phasor sum of V_C, V_L, and V_R equals the emf when $E = 4$ mV, $R = 8\ \Omega$, $L = 250\ \mu$H, $f = 1$ MHz, and C is

a. 60 pF (*ans.* 3.693 μA, 9.8 mV, 89.58° lead)
b. 90 pF (*ans.* 20.22 μA, 35.87 mV, 87.67° lead)
c. 110 pF (*ans.* 32.19 μA, 46.58 mV, 86.31° lag)
d. 200 pF (*ans.* 5.16 μA, 4.106 mV, 89.41° lag)

15-4 Using Fig. 15-1 in the text, find the magnitude of I and the phase angle of the circuit when $E = 120$ V, $R = 40\ \Omega$, $L = 0.15$ H, $f = 60$ Hz, and C is

a. Short-circuited *c.* 50 μF
b. 200 μF *d.* 20 μF

15-5 Find the apparent power, active power, and reactive power for each of the circuits in Prob. 15-4.
(*ans. a.* 208 VA, 169.8 var, 120.1 W; *b.* 244.3 VA, 179.4 var, 165.8 W; *c.* 358.6 VA, 31.23 var, 357.3 W; *d.* 167.5 VA, 148.3 var, 77.96 W)

Section 15-2

15-6 Using Fig. 15-15 of the text, find the magnitudes of I_T, I_R, I_L, I_C, and the phase angle, and show that the phasor sum of I_R, I_L, and I_C equals I_T when

a. $E = 120$ V, $R = 30\ \Omega$, $X_L = 60\ \Omega$, $X_C = 120\ \Omega$
b. $E = 120$ V, $R = 60\ \Omega$, $X_L = 120\ \Omega$, $X_C = 60\ \Omega$
c. $E = 1$ V, $R = 27$ kΩ, $X_L = 5$ kΩ, $X_C = 2$ kΩ
d. $E = 10$ V, $R = 100$ kΩ, $X_L = 1$ kΩ, $X_C = 2$ kΩ

15-7 Repeat Prob. 15-6 when $E = 1$ V, $R = 40$ kΩ, $C = 500$ pF, $L = 200$ μH, and

a. $f = 400$ kHz (*ans.* 733.1 μA, 25 μA, 1.99 mA, 1.257 mA, 88.05° lag)
b. $f = 450$ kHz (*ans.* 355.6 μA, 25 μA, 1.768 mA, 1.414 mA, 85.97° lag)
c. $f = 500$ kHz (*ans.* 32.52 μA, 25 μA, 1.592 mA, 1.571 mA, 39.76° lag)
d. $f = 550$ kHz (*ans.* 282.3 μA, 25 μA, 1.447 mA, 1.728 mA, 84.92° lead)

15-8 Using Fig. 15-15 in the text, find the magnitudes of I_T, I_L, and the phase angle, and show that the phasor sum of I_R, I_L, and I_C equals I_T when $V = 10$ mV, $R = 250$ kΩ, $L = 250$ μH, $f = 1$ MHz, and C is

a. 60 pF *c.* 110 pF
b. 90 pF *d.* 200 pF

15-9 Using Fig. 15-15 in the text, find the magnitude of I_T and the phase angle of the circuit when $V = 120$ V, $R = 40\ \Omega$, $L = 0.15$ H, $f = 60$ Hz, and C is

a. Short-circuited (*ans.* 3.675 A, 35.27° lag)
b. 200 μF (*ans.* 7.55 A, 66.58° lead)
c. 50 μF (*ans.* 3.003 A, 2.67° lead)
d. 20 μF (*ans.* 3.237 A, 22.08° lag)

15-10 Find the apparent power, active power, and reactive power for each of the circuits in Prob. 15-9.

Section 15-3

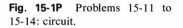

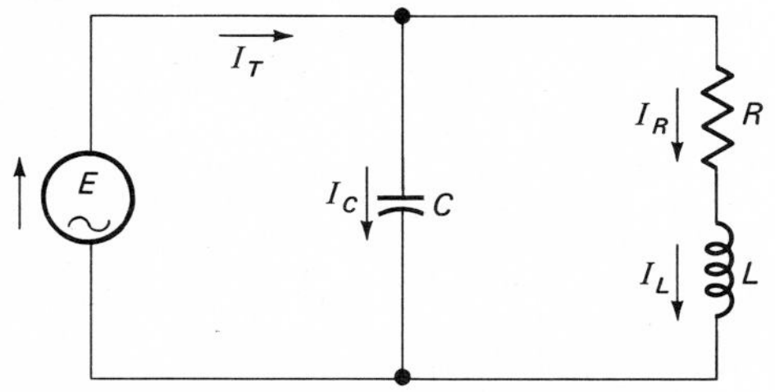

Fig. 15-1P Problems 15-11 to 15-14: circuit.

15-11 Find the magnitude of I_T, I_C, I_R, I_L, and the phase angle for the circuit in Fig. 15-1P when

a. $E = 120$ V, $R = 30\ \Omega$, $X_L = 60\ \Omega$, $X_C = 120\ \Omega$
(*ans.* 1 A, 1 A, 1.789 A, 1.789 A, 36.87° lag)
b. $E = 120$ V, $R = 60\ \Omega$, $X_L = 60\ \Omega$, $X_C = 120\ \Omega$
(*ans.* 1 A, 1 A, 1.414 A, 1.414 A, 0°)
c. $E = 1$ V, $R = 10\ \Omega$, $X_L = 200\ \Omega$, $X_C = 200\ \Omega$
(*ans.* 249.7 μA, 5 mA, 4.994 mA, 4.994 mA, 2.86° lead)
d. $E = 1$ V, $R = 1\ \Omega$, $X_L = 200\ \Omega$, $X_C = 200\ \Omega$
(*ans.* 25 μA, 5 mA, 5 mA, 5 mA, 0.286° lead)

15-12 Find the magnitude of I_T and I_L in Fig. 15-1P when $E = 10$ mV, $R = 10\ \Omega$, $C = 400$ pF, $L = 350\ \mu$H, and

a. $f = 400$ kHz *c.* $f = 500$ kHz
b. $f = 450$ kHz *d.* $f = 550$ kHz

15-13 Find the magnitude of I_T and I_L in Fig. 15-1P when $E = 10$ V, $R = 10\ \Omega$, $L = 50\ \mu$H, $f = 2$ MHz, and

a. $C = 100$ pF (*ans.* 3.35 mA, 15.91 mA)
b. $C = 120$ pF (*ans.* 867 μA, 15.91 mA)
c. $C = 140$ pF (*ans.* 1.7 mA, 15.91 mA)
d. $C = 160$ pF (*ans.* 4.2 mA, 15.91 mA)

15-14 Find the magnitude of I_T, I_L, S, P, and the circuit phase angle in Fig. 15-1P when $E = 120$ V, $R = 24\ \Omega$, $L = 0.1$ H, $f = 60$ Hz, and C is

a. Not connected *c.* 40 μF
b. 20 μF *d.* 60 μF

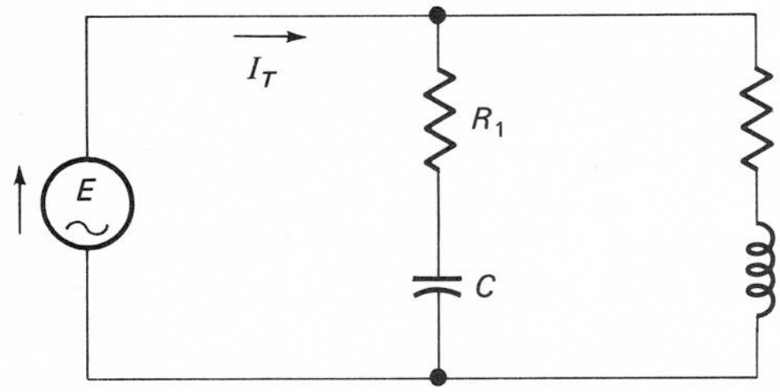

Fig. 15-2P Problem 15-15: circuit.

15-15 Find the magnitude of I_T, S, and P for the circuit shown in Fig. 15-2P when $E = 120$ V, $f = 60$ Hz, and

a. $R_1 = 5\ \Omega$, $C = 100\ \mu$F, $R_2 = 10\ \Omega$, $L = 0.2$ H (*ans.* 2.987 A, 358.5 VA, 123.7 W)
b. $R_1 = 5\ \Omega$, $C = 60\ \mu$F, $R_2 = 10\ \Omega$, $L = 0.2$ H (*ans.* 1.228 A, 147.26 VA, 61.34 W)
c. $R_1 = 5\ \Omega$, $C = 30\ \mu$F, $R_2 = 10\ \Omega$, $L = 0.2$ H (*ans.* 0.354 A, 42.48 VA, 34.13 W)

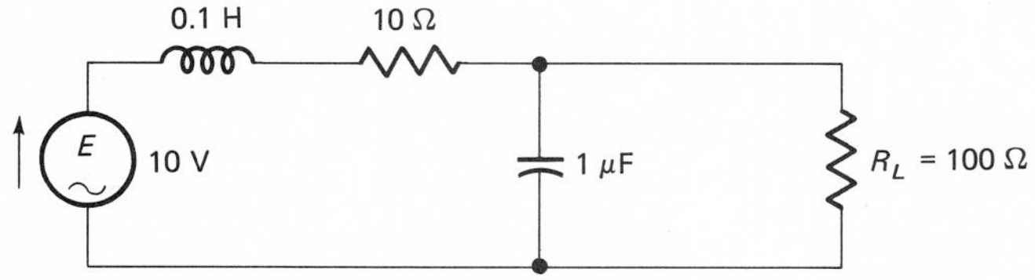

Fig. 15-3P Problem 15-16.

15-16 The *RLC* network in Fig. 15-3P is a low-pass filter. Find the voltage across R_L when the frequency of the generator is

a. 500 Hz
b. 1000 Hz
c. 1500 Hz

15-17 A 120-V 60-Hz electric motor has an effective resistance of 58 Ω in series with an effective inductance of 0.175 H.

a. Find the power factor, the line current, and the apparent power. (*ans.* 0.66 lag, 1.366 A, 163.9 VA)
b. Find the power factor, the line current, and the apparent power if a capacitor of 50 μF is connected across the motor. (*ans.* 0.587 lead, 1.53 A, 183.3 VA)

15-18 A 240-V 60-Hz electric motor has an output power of 1 hp, is 75 percent efficient, and has a power factor of 0.6 lagging. What value of capacitance connected across the motor will change the power factor to 0.95 lagging?

15-19 If the motor described in Prob. 15-18 has its power factor changed to 0.95 lagging by a series capacitor, find
a. The value of the capacitor. (*ans.* 126.7 μF)
b. The line current. (*ans.* 10.94 A)
c. The voltage across the capacitor. (*ans.* 229 V)
d. The voltage across the motor. (*ans.* 380 V)

15-20 The tank circuit of a radio-frequency amplifier contains a parallel *RLC* circuit. One branch of the circuit contains a 100-pF capacitor, the other branch contains a 150-μH inductance with an effective series resistance of 12 Ω. Find the impedance of the parallel circuit when *f* equals
a. 1280 kHz
b. 1305 kHz

15-21 If the emf across the circuit in Prob. 15-20 is 20 V at 1280 kHz, find
a. The line current. (*ans.* 520 μA)
b. The current in the inductor. (*ans.* 16.58 mA)

15-22 If a 120-kΩ resistor is shunted across the circuit in Prob. 15-20 when the voltage is 20 V at 1280 kHz, find
a. The line current.
b. The current in the inductor.

15-23 A 600-W load with a power factor of 0.65 lagging is connected in parallel with a 500-W resistive load across a 120-V 60-Hz line. Find
a. The total line current. (*ans.* 10.87 A)
b. The circuit phase angle. (*ans.* 32.53° lag)

15-24 If a 100-μF capacitor is connected across the circuit in Prob. 15-23, find
a. The new line current.
b. The new phase angle.

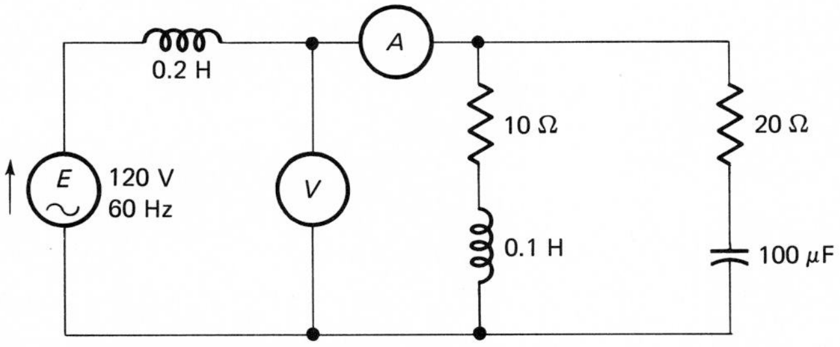

Fig. 15-4P Problems 15-25 and 15-26: circuit.

15-25 Find the current and voltage which would be indicated by the meters in Fig. 15-4P. (*ans.* 1.385 A, 56.07 V)

15-26 Find the power that would be indicated by a wattmeter if its current coil and potential coil replace the ammeter and voltmeter in Fig. 15-4P.

16

Resonant Circuits

The *RLC* circuits dealt with in Chap. 15 all had one characteristic in common: they had net reactance. Each of those circuits could be reduced to an equivalent *RL* or *RC* network as a result of partial cancellation of reactive effects. What if the circuit values were such that the inductive and capacitive effects completely canceled? The resulting equivalent circuit would be entirely resistive in character, and the power factor would be unity. This condition of unity power factor in a circuit containing *R*, *L*, and *C* is called *resonance,* the circuit which results in resonance being a *resonant circuit.*

The importance of resonance is not just academic. Many electric and electronic systems are dependent on resonant circuits for their operation. The most familiar example is radio. The radio-frequency spectrum is filled with countless transmissions. There are deliberate transmissions: domestic radio and television; marine, aircraft, and land mobile radio; radio for telephone systems; space communications; etc. In addition to these desirable transmissions there is a large amount of radio-frequency noise, transmissions which are not deliberate, serve no useful purpose, and interfere with desirable transmissions. Some of the sources of noise are lightning, electric machinery, radiation from space, medical equipment, and automobile ignition. Out of this miscellany of transmissions, desirable and undesirable, the radio receiver selects one transmission and rejects all other signals. Resonant circuits do the selecting and rejecting.

Like all electric circuit characteristics, resonance can cause problems in some circumstances. For instance, resonant effects can cause excessive voltages and currents, resulting in the breakdown of wiring and equipment.

Important resonant circuit concepts are introduced by concentrating on the cancellation of reactive effects in hypothetical, loss-free circuits. This treatment lays the foundation for dealing with practical resonant circuits.

16-1 RESONANT CIRCUITS WITHOUT RESISTANCE

Practical inductors and capacitors are imperfect; they dissipate power, and they have resistance. Therefore practical resonant circuits, which must contain inductors and capacitors, have resistance. Considering the effects of this resistance tends to obscure some of the important basic concepts involved in resonance. An initial view of resonant circuits without resistance can be helpful, even though such circuits are not available in practice.

SERIES *LC* CIRCUIT

Figure 16-1 shows a circuit in which an inductance of 100 μH and a capacitance of 250 pF are connected in series across a current source of variable frequency.

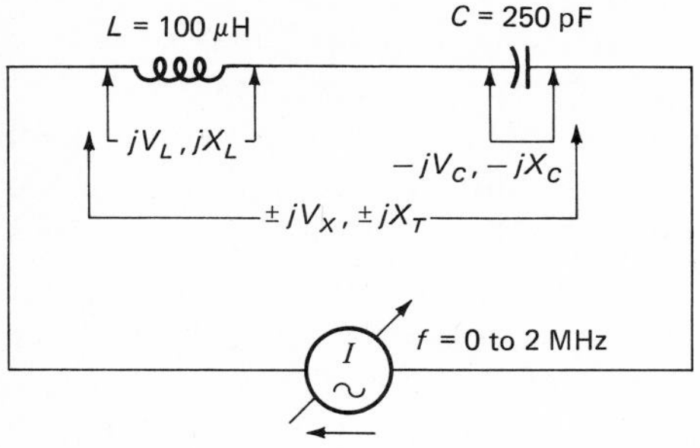

Fig. 16-1 Series *LC* circuit.

The reactance of the inductance X_L is directly proportional to the frequency, the capacitive reactance X_C being inversely proportional to the frequency. Plots of these reactances against frequency are given in Fig. 16-2*a*.

A curve for total reactance X_T can be obtained by adding the inductive and capacitive reactance curves in Fig. 16-2*a*. The algebraic sums of the values of X_L and X_C at each frequency are plotted. At frequency *A* (0.6 MHz), $-jX_C$ is greater than $+jX_L$, so the total reactance is a $-j$ quantity of magnitude $X_C - X_L$. At frequency *B* ($\simeq$ 1 MHz), $-jX_C$ equals

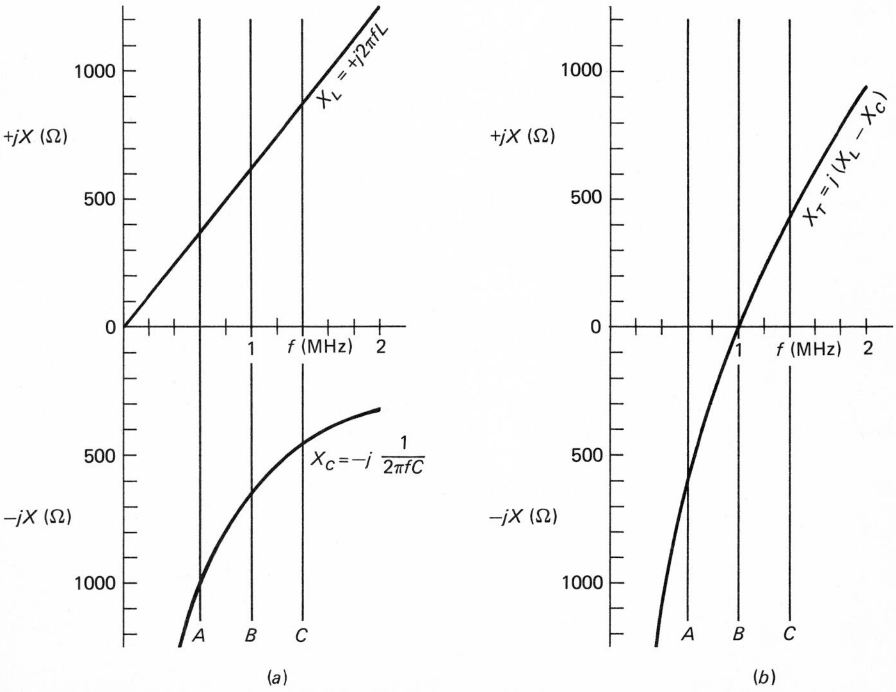

Fig. 16-2 Series *LC* circuit reactance-frequency curves. (*a*) Individual elements; (*b*) total circuit.

$+jX_L$, resulting in a total reactance of zero. At frequency C (1.4 MHz) $+jX_L$ is greater than $-jX_C$, giving a total reactance X_T which is a $+j$ quantity of magnitude $X_L - X_C$. These values of total reactance are shown in Fig. 16-2*b*, the curve for X_T for all frequencies from 0 to 2 MHz.

Figure 16-2 shows the magnitude and the phase (in terms of j sign) of the reactances X_L, X_C, and X_T. An appreciation of circuit behavior can be obtained by separating the magnitude and phase characteristics.

In Fig. 16-3 the magnitudes of the reactances are shown with respect to frequency, the broken lines depicting the individual reactances and the solid lines their total (or the total impedance Z). The values for Fig. 16-3 are taken directly from Fig. 16-2, observing the principle that the magnitude of a quantity shown graphically is represented by the distance between the curve and the zero line. This principle applies to both negative and positive quantities. It will be seen from Fig. 16-3 that the magnitude of the total reactance of a series resonant circuit decreases from infinity toward zero as the frequency increases from zero to the resonant frequency f_r. The total reactance then increases toward infinity after the resonant frequency is passed.

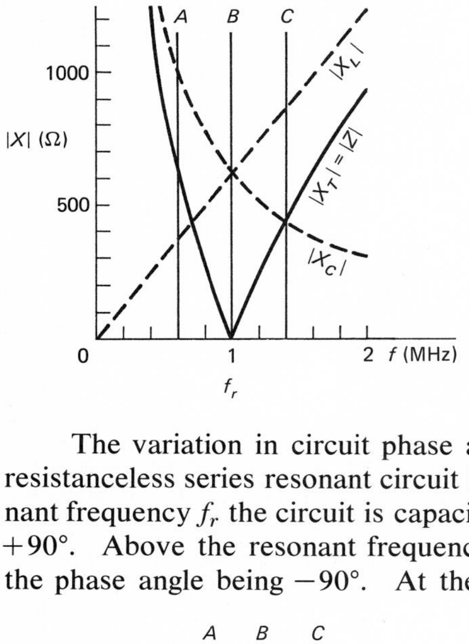

Fig. 16-3 Series LC circuit reactance magnitude-frequency curves.

The variation in circuit phase angle with respect to frequency for a resistanceless series resonant circuit is given in Fig. 16-4. Below the resonant frequency f_r the circuit is capacitive ($X_C > X_L$) and the phase angle is $+90°$. Above the resonant frequency the circuit is inductive ($X_C < X_L$), the phase angle being $-90°$. At the resonant frequency the circuit is in

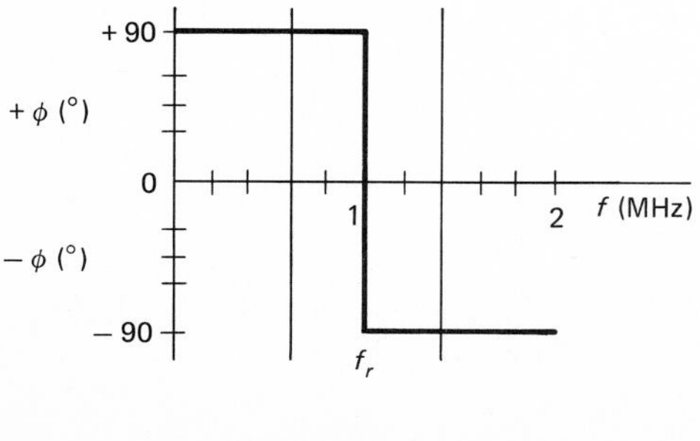

Fig. 16-4 Phase angle-frequency curve for a series LC circuit (no resistance).

transition between being capacitive and being inductive; the phase angle is in transition between +90° and −90°.

In Fig. 16-3 the reactance magnitudes apply to all series resonant circuits, with and without resistance. But Fig. 16-4 for phase angles applies only to series resonant circuits without resistance or energy losses.

In Fig. 16-5, phasor diagrams are shown for the circuit in Fig. 16-1 in relation to the frequencies *A*, *B*, and *C* in Figs. 16-2 to 16-4. As the source supplies a constant current to the circuit, each pd is directly proportional to the related reactance X_L, X_C, or X_T.

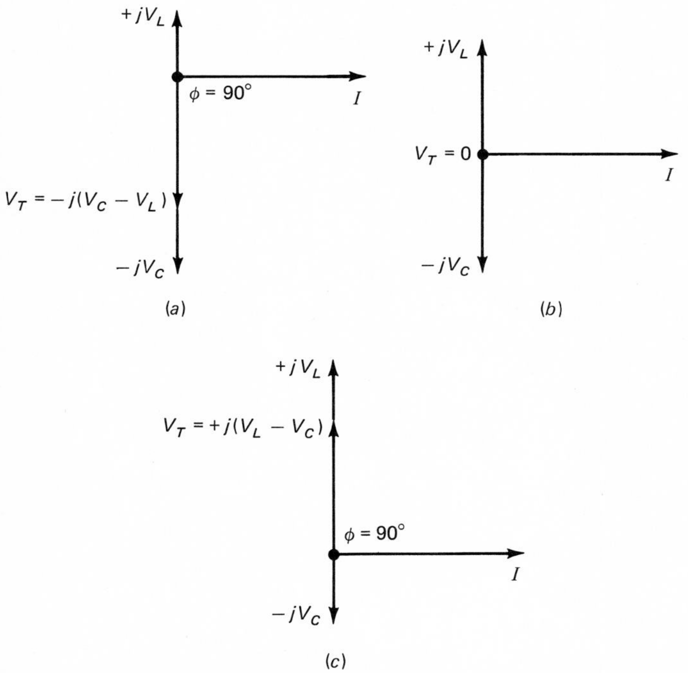

Fig. 16-5 Series *LC* circuit phasor diagrams. (*a*) Below resonance; (*b*) at resonance; (*c*) above resonance.

At frequency *A* (0.6 MHz), below resonance, $-jV_C$ is greater than $+jV_L$, so the total circuit voltage V_T is equal to $-j(V_C - V_L)$ and lags the current by 90°. These conditions are shown in Fig. 16-5*a*.

At frequency *B* (1 MHz), the resonant frequency f_r, V_L and V_C have the same magnitude. The magnitude of V_T is the difference between V_L and V_C and is thus zero at this frequency. Since there is no voltage, there is no circuit phase angle. Figure 16-5*b* shows this situation.

At frequency *C* (1.4 MHz), above resonance, $+jV_L$ is greater than $-jV_C$; V_T is equal to $+j(V_L - V_C)$ and leads the current by 90°, as shown in Fig. 16-5*c*.

The reason for choosing a constant-current source for the series resonant circuit in Fig. 16-1, rather than a constant-voltage source (a more usual choice for explaining series circuit action), can be seen in the descriptions of the phasor diagrams of Fig. 16-5. At resonance the total circuit voltage is zero, while at all other frequencies it has a finite value. This voltage variation renders a constant-voltage source meaningless for this convenient voltage approach to the behavior of the series resonant circuit.

The circuit conditions existing in a series *LC* circuit at the resonant frequency f_r can be summarized:

1. The magnitudes of the inductive reactance X_L and the capacitive reactance X_C are equal.
2. The total reactance X_T is zero.
3. There is no resistance, so the total impedance Z equals X_T and is therefore zero.
4. The magnitudes of the inductive voltage V_L and the capacitive voltage V_C are equal.
5. The total circuit voltage V_T is zero.
6. The ratio of either reactive voltage to the total voltage (V_L/V_T or V_C/V_T) is infinite. In other words, either of the reactive voltages is infinitely greater than the total circuit voltage.

At frequencies below the resonant frequency the following conditions exist:

1. The circuit is predominantly capacitive.
2. The phase angle is $+90°$.
3. The power factor is zero.

Conditions existing at frequencies above the resonant frequency are:

1. The circuit is predominantly inductive.
2. The phase angle is $-90°$.
3. The power factor is zero.

PARALLEL *LC* CIRCUIT

In Fig. 16-6 an inductance of 100 μH and a capacitance of 250 pF are shown connected in parallel across a voltage source of variable frequency.

Fig. 16-6 Parallel *LC* circuit.

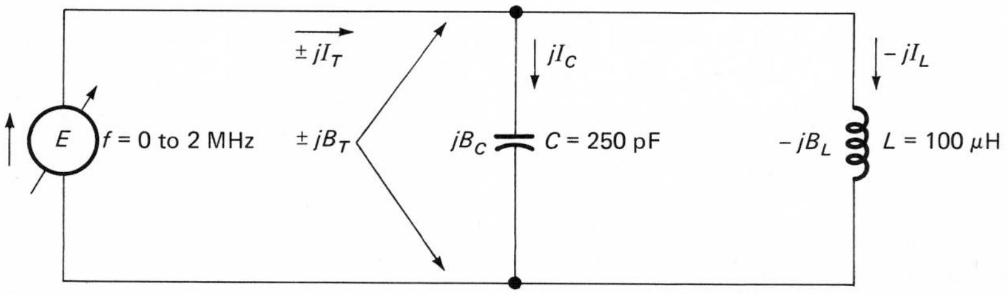

The capacitive susceptance B_C is directly proportional to the frequency, and the inductive susceptance B_L is inversely proportional to the frequency. Curves showing how these susceptances vary with respect to frequency are given in Fig. 16-7*a*.

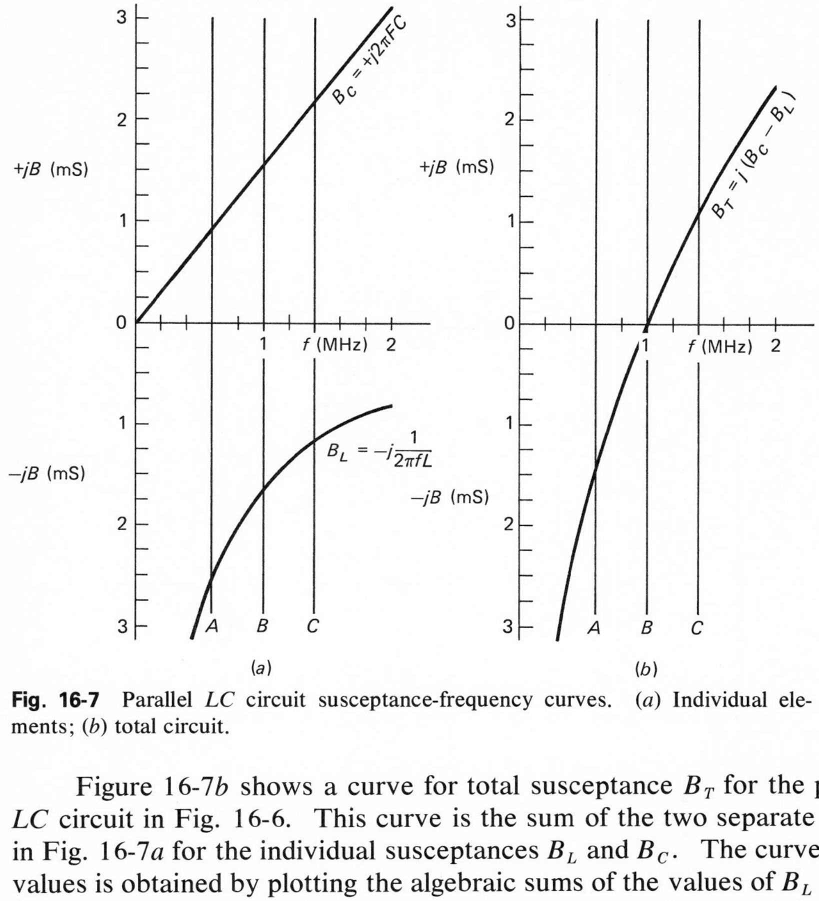

Fig. 16-7 Parallel *LC* circuit susceptance-frequency curves. (*a*) Individual elements; (*b*) total circuit.

Figure 16-7*b* shows a curve for total susceptance B_T for the parallel *LC* circuit in Fig. 16-6. This curve is the sum of the two separate curves in Fig. 16-7*a* for the individual susceptances B_L and B_C. The curve for B_T values is obtained by plotting the algebraic sums of the values of B_L and B_C for each frequency. At frequency *A* (0.6 MHz), $-jB_L$ is greater than $+jB_C$ and B_T is a $-j$ quantity of magnitude $B_L - B_C$. At frequency *B* ($\simeq$ 1 MHz), $-jB_L$ is equal to $+jB_C$ and B_T is zero. At frequency *C* (1.4 MHz), $+jB_C$ is greater than $-jB_L$ and B_T is a $+j$ quantity of magnitude $B_C - B_L$.

In Fig. 16-7 the variation in the magnitudes of the susceptances B_L, B_C, and B_T with frequency is shown, together with their phase angles (in terms of *j* signs).

The magnitudes of the susceptances and their variation with respect to frequency are given in Fig. 16-8. The broken lines represent the individual susceptances B_L and B_C, while the solid line represents the total susceptance B_T (or total admittance *Y*). The magnitudes are the same as in Fig. 16-7, the *j* positions being ignored. As the frequency increases from

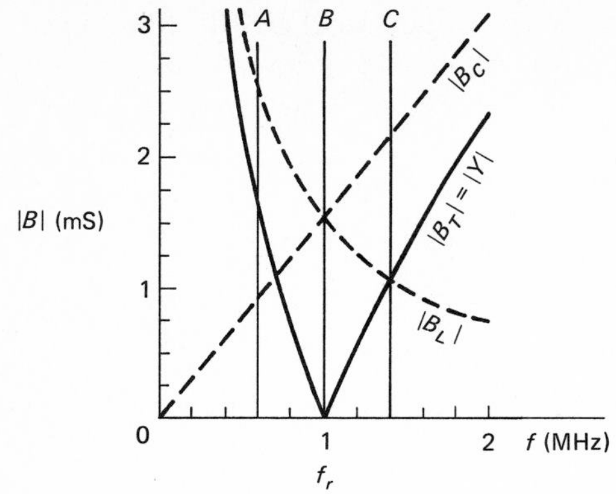

Fig. 16-8 Parallel *LC* circuit susceptance magnitude-frequency curves.

zero toward the resonant frequency f_r, the magnitude of the total susceptance decreases from infinity toward zero. As the frequency then increases beyond resonance, the susceptance increases toward infinity.

If you prefer to think in terms of impedance for all circuits, the statement for the variation in the total impedance magnitude of a parallel *LC* circuit is: As the frequency increases from zero toward the resonant frequency f_r, the magnitude of the total impedance increases from zero toward infinity. As the frequency then increases beyond resonance, the impedance decreases toward zero.

A curve for the total impedance magnitude of the *LC* parallel circuit in Fig. 16-6 is given in Fig. 16-9. This curve is the reciprocal of the total admittance curve in Fig. 16-8.

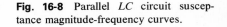

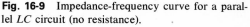

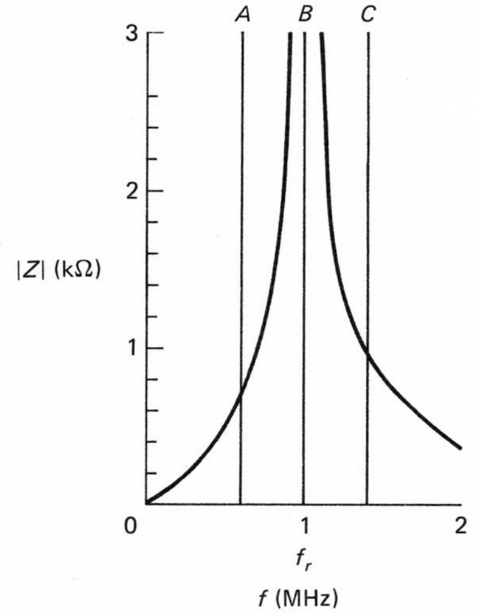

Fig. 16-9 Impedance-frequency curve for a parallel *LC* circuit (no resistance).

The manner in which the phase angle of a parallel *LC* circuit varies with respect to frequency is shown in Fig. 16-10. Below resonance the circuit is inductive ($B_L > B_C$), and the phase angle is $-90°$. Above resonance the circuit is capacitive ($B_L < B_C$), and the phase angle is $+90°$. At

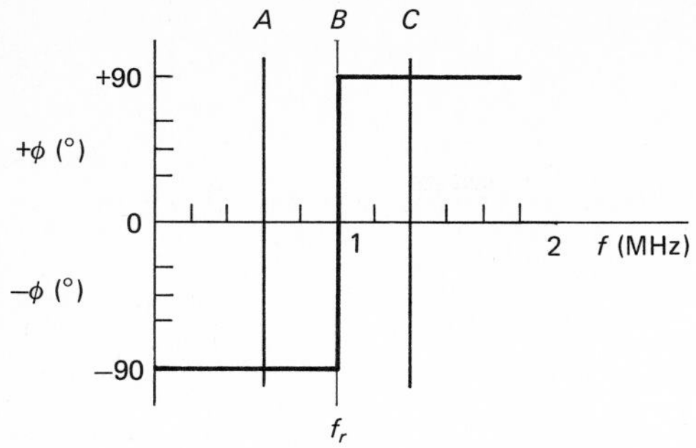

Fig. 16-10 Phase angle-frequency curve for a parallel *LC* circuit (no resistance).

the resonant frequency f_r the circuit is in transition between the net inductive and net capacitive states, the phase angle being in transition between −90 and +90°.

In Fig. 16-8, the magnitudes of the susceptances apply to all parallel resonant circuits, with and without resistance. Figure 16-9 (for impedance) and Fig. 16-10 (for phase angle) apply only to parallel resonant circuits without resistance or energy losses.

In Fig. 16-11 phasor diagrams are shown for the circuit in Fig. 16-6 in relation to frequencies *A*, *B*, and *C* in Figs. 16-7 to 16-10. The parallel *LC* network is fed with a constant voltage from the source. Each current is directly proportional to the associated susceptance.

At frequency *A* (0.6 MHz) $-jI_L$ is greater than $+jI_C$. The total current I_T is equal to $-j(I_L - I_C)$ and lags the emf *E* by 90°. These relationships are shown in Fig. 16-11*a*.

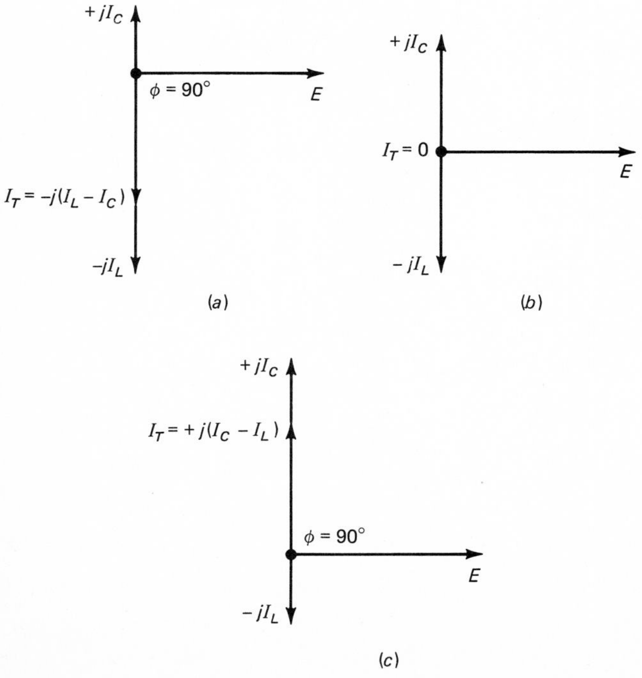

Fig. 16-11 Parallel *LC* circuit phasor diagrams. (*a*) Below resonance; (*b*) at resonance; (*c*) above resonance.

At frequency B (1 MHz), the resonant frequency f_r, $-jI_L$ equals $+jI_C$, the total current I_T is zero, and there is no circuit phase angle. The phasors for frequency B are shown in Fig. 16-11*b*.

At frequency C (1.4 MHz) $+jI_C$ is greater than $-jI_L$. The total current I_T equals $+j(I_C - I_L)$ and leads the emf E by 90°, as shown in Fig. 16-11*c*.

The circuit conditions present in a parallel *LC* circuit at the resonant frequency f_r can be summarized:

1. The magnitudes of the inductive susceptance B_L and the capacitive susceptance B_C are equal.
2. The total effective susceptance B_T is zero.
3. There is no conductance, so the total circuit admittance Y equals B_T and is therefore zero.
4. The circuit impedance Z is infinite.
5. The magnitudes of the inductive current I_L and the capacitive current I_C are equal; their directions are opposite.
6. The total circuit current I_T is zero.
7. The ratio of either reactive current to the total current (I_L/I_T or I_C/I_T) is infinite, or either of the reactive currents is infinitely greater than the total circuit current.

Below resonance the following conditions exist in the parallel *LC* circuit:

1. The circuit is predominantly inductive.
2. The phase angle is −90°.
3. The power factor is zero.

Above resonance the following conditions exist in the parallel *LC* circuit:

1. The circuit is predominantly capacitive.
2. The phase angle is +90°.
3. The power factor is zero.

It is well to compare the characteristics of series and parallel *LC* circuits:

1. Compare Figs. 16-2 and 16-7 and note that the actual curves are identical. The differences between the series reactance and parallel susceptance curves lie only in the scaling and units of the vertical axis and the labeling of the curves.
2. A comparison of Figs. 16-3 and 16-8 reveals identical curves. The scaling and units of the vertical axis and the curve labels are the only dif-

ferences between the series reactance and parallel susceptance scalar value curves. As the frequency increases from zero, the values of $|X_T|$ for the series circuit and $|B_T|$ for the parallel circuit decrease from infinity toward zero at the resonant frequency, and then increase again from zero toward infinity.

3. Figures 16-4 and 16-10 show that the series and parallel LC circuits both change their phase angles at resonance. The series circuit goes from $+90°$ (capacitive) to $-90°$ (inductive). The parallel circuit goes from $-90°$ (inductive) to $+90°$ (capacitive).
4. In Figs. 16-5 and 16-11 the usual duality of series and parallel ac circuits can be seen. The phasor diagrams are similar if the basic rule is followed; for series circuits, voltages are shown with the current as the reference phasor; for parallel circuits, currents are shown with the voltage as the reference phasor.
5. Below resonance the series circuit is capacitive; the parallel circuit is inductive.
6. Above resonance the series circuit is inductive; the parallel circuit is capacitive.

To emphasize the duality of series and parallel circuit behavior, parallel resonance is sometimes called *antiresonance*. Where this convention is used, series resonance is merely referred to as resonance.

When, as in a practical situation, a resonant circuit contains resistance, its characteristics will be similar to those of the simple LC circuit but with detail modifications. By bearing in mind the characteristics of theoretically resistanceless LC resonant circuits and noting how the presence of resistance modifies them, the behavior of practical RLC resonant circuits can be visualized.

16-2 SERIES *RLC* RESONANT CIRCUIT

A series RLC circuit is shown in Fig. 16-12. Its impedance Z is the phasor sum of the element impedances:

$$Z = R + jX_T = R + j(X_L - X_C)$$

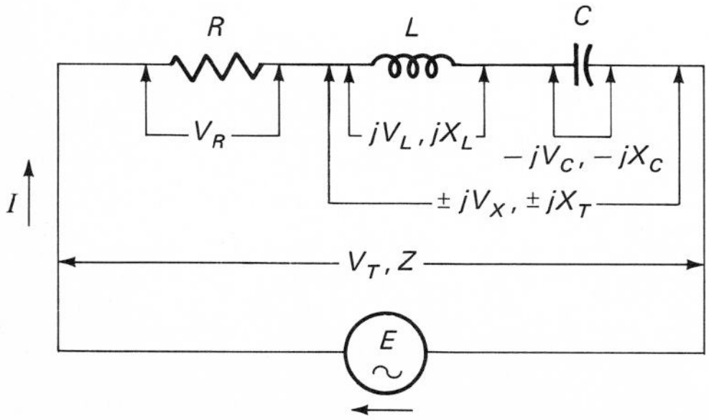

Fig. 16-12 Series RLC circuit.

The characteristics of the series resonant circuit are determined by this impedance equation. Initially three conditions will be considered.

At the resonant frequency f_r By definition, at the resonant frequency the power factor is unity. This implies that the circuit is purely resistive; the impedance consists only of the circuit resistance, the net reactance is zero, and the phase angle is zero.

$$Z = R$$

$$X_T = j(X_L - X_C) = 0 \qquad \text{or} \qquad X_L = X_C$$

Expanding X_L and X_C we obtain

$$\omega_r L = \frac{1}{\omega_r C}$$

or

$$\omega_r^2 = \frac{1}{LC} \tag{16-1}$$

Expanding ω_r gives

$$(2\pi f_r)^2 = \frac{1}{LC}$$

or

$$f_r = \frac{1}{2\pi\sqrt{LC}} \tag{16-2}$$

In Eqs. (16-1) and (16-2),

ω_r = resonant angular frequency, rad/s
L = inductance, H
C = capacitance, F
f_r = resonant frequency, Hz

Equation (16-2) is one of the most important electrical relationships. Not only does it accurately give the resonant frequency of a practical series circuit, but also an approximate value of the resonant frequency of a practical parallel circuit.

Figure 16-13*a* shows a phasor diagram of a series *RLC* circuit at resonance. The inductive and capacitive voltages jV_L and $-jV_C$ are equal in magnitude but are in antiphase (opposite j signs). Thus the net reactive voltage V_X is zero. This cancellation of reactive voltages leaves the resistive pd as the only active circuit voltage. The resistive voltage V_R equals the total circuit voltage V_T.

In Fig. 16-13*b* an impedance diagram is shown. The reactances jX_L and $-jX_C$ cancel, leaving only the resistance R to constitute the total circuit

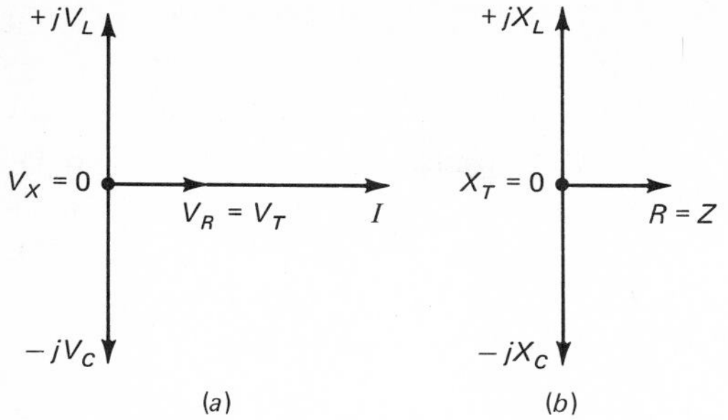

Fig. 16-13 Series *RLC* circuit at resonance. (*a*) Phasor diagram; (*b*) impedance diagram.

impedance Z. It is not possible to draw an impedance triangle for the resonant condition, because there is no net reactance.

VOLTAGE MAGNIFICATION FACTOR

It will be seen from Fig. 16-13 that, for a given value of resistance, no matter how large the magnitude of either reactance, the total circuit impedance Z will equal the resistance R, and the total circuit voltage V_T will equal the resistive voltage V_R. This observation leads to the special significance of the quality factor Q in a series resonant circuit. It has been mentioned previously that Q is defined as the ratio of stored energy to dissipated energy and that, for a series circuit, Q equals the ratio of reactance to resistance. In a resonant circuit the energy is stored alternately in the inductance and the capacitance, so at resonance Q becomes the ratio of either reactance to the resistance:

$$Q_r = \frac{X_{Lr}}{R} = \frac{X_{Cr}}{R}$$

or

$$X_{Lr} = X_{Cr} = Q_r R = Q_r Z \qquad (16\text{-}3)$$

Equation (16-3) can be written in voltage form:

$$Q_r = \frac{V_{Lr}}{V_T} = \frac{V_{Cr}}{V_T}$$

or

$$V_{Lr} = V_{Cr} = Q_r V_T \qquad (16\text{-}4)$$

In Eqs. (16-3) and (16-4),

Q_r = the quality factor at resonance
X_{Lr} = inductive reactance at resonance, Ω
X_{Cr} = capacitive reactance at resonance, Ω
R = equivalent resistance of circuit losses, Ω
Z = total circuit impedance, Ω
V_{Lr} = inductive voltage at resonance, V
V_{Cr} = capacitive voltage at resonance, V
V_T = total circuit voltage, V (at resonance the total voltage V_T is the same as the resistive voltage V_R)

For practical circuits designed to exhibit resonant conditions, Q_r will usually be greater than unity. Therefore in a series resonant circuit it is common for the voltage across either reactance to exceed the total circuit voltage. This statement, which is given in quantitative terms in Eq. (16-4), leads to another definition of Q_r:

In a series *RLC* circuit at resonance the quality factor Q_r is called the voltage *magnification factor*. The voltage across the entire circuit is magnified so that it appears across each reactance Q_r times as large.

At any frequency other than the resonant frequency If the circuit is not operating at its resonant frequency, its power factor will be less than unity. The phase angle and the net reactance will both be greater than zero.

$$Z > R$$

$$X_T = j(X_L - X_C) > 0 \qquad \text{or} \qquad X_L \neq X_C$$

In Fig. 16-14*a* an impedance diagram is shown for the situation in which the frequency is below resonance. Here the capacitive reactance X_C is greater than the inductive reactance X_L, resulting in a net reactance X_T equal to $-j(X_C - X_L)$.

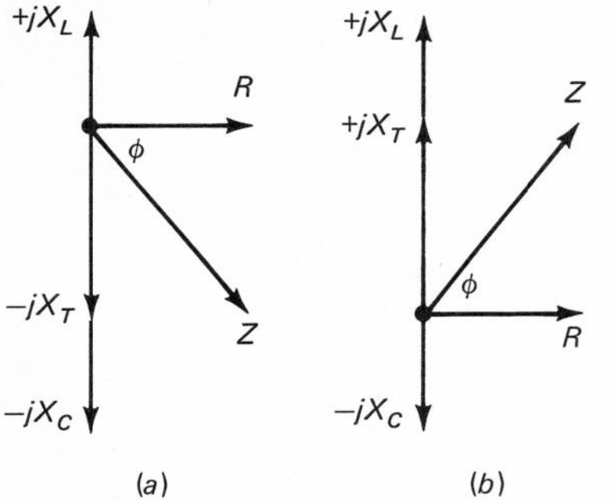

Fig. 16-14 Series *RLC* circuit impedance diagrams. (*a*) Below resonance; (*b*) above resonance.

When the frequency is above resonance, X_L is greater than X_C and X_T equals $+j(X_L - X_C)$. This condition is shown in Fig. 16-14*b*.

For Fig. 16-14*a* and *b* the circuit impedance *Z* is the phasor sum of the net reactance X_T and the resistance *R*. The phase angle ϕ is equal to arctan X_T/R.

It follows that the voltage relationships will be

$$V_L \neq V_C$$

$$V_X = (V_L - V_C) > 0$$

$$V_T = V_R + j(V_L - V_C) > V_R$$

Figure 16-15 shows typical phasor diagrams for a series resonant circuit operated at nonresonant frequencies. In Fig. 16-15*a* the frequency is

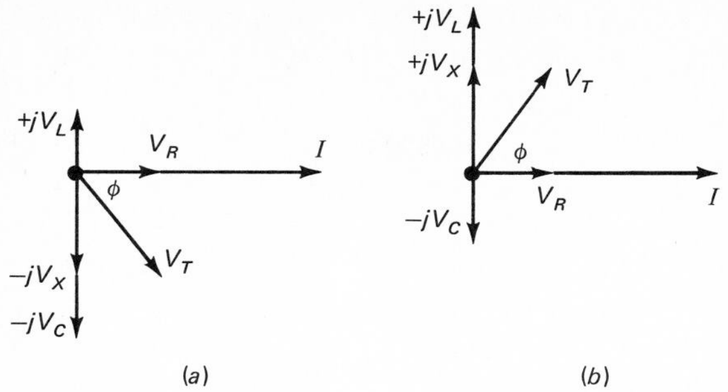

Fig. 16-15 Series *RLC* circuit phasor diagrams. (*a*) Below resonance; (*b*) above resonance.

below resonance, so $V_C > V_L$, resulting in a net reactive voltage V_X equal to $-j(V_C - V_L)$. In Fig. 16-15*b* the frequency is above resonance ($V_L > V_C$), and the net reactive voltage V_X is $+j(V_L - V_C)$. In both cases the total circuit voltage V_T is the phasor sum of the net reactive voltage V_X and the resistive voltage V_R. The phase angle ϕ is equal to arctan V_X/V_R.

As there is resistance in the *RLC* resonant circuit, impedance and voltage triangles can be drawn, as in Fig. 16-16.

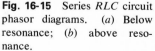

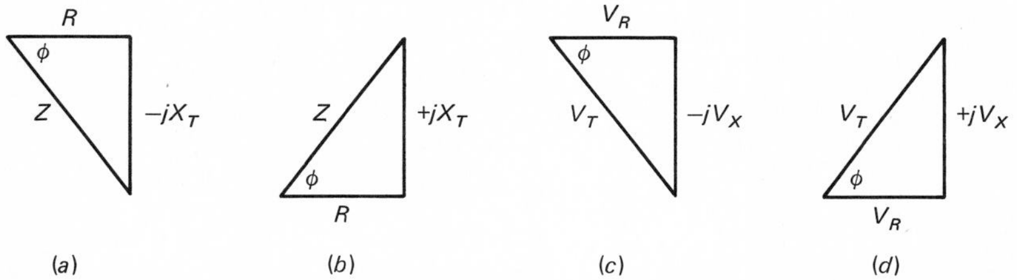

Fig. 16-16 Series *RLC* circuit. (*a*) Impedance triangle for below resonance; (*b*) impedance triangle for above resonance; (*c*) voltage triangle for below resonance; (*d*) voltage triangle for above resonance.

Figure 16-16*a* and *b* shows impedance triangles for below- and above-resonance frequencies. These triangles are derived from Fig. 16-14. In a similar manner, the voltage triangles in Fig. 16-16*c* and *d* refer to below- and above-resonance frequencies and are obtained from Fig. 16-15.

At the frequencies where $X_T = R$ "Bandwidth" is a term which is dealt with later in this chapter when the characteristics of resonant circuits over a range of frequencies are discussed. The term is introduced now, since its definition is related to two specific frequencies, one below and one above resonance.

Bandwidth in a series resonant circuit can be defined as the frequency difference between the two conditions at which the net reactance equals the resistance. If this definition is used, the specific frequencies are obtained from the equation

$$X_T = R$$

At the bandwidth frequency below resonance f_1, the capacitive reactance X_{C1} will be greater than the inductive reactance X_{L1}:

$$X_{T1} = X_{C1} - X_{L1} = R \tag{16-5}$$

At the bandwidth frequency above resonance f_2, the inductive reactance X_{L2} will be greater than the capacitive reactance X_{C2}:

$$X_{T2} = X_{L2} - X_{C2} = R \tag{16-6}$$

It follows that

$$X_{T1} = X_{T2} = R$$

Figure 16-17*a* shows the impedance triangle which refers to the below-resonance bandwidth frequency, while Fig. 16-17*b* refers to the above-resonance bandwidth frequency. From each of the triangles it can be seen that, for either bandwidth frequency,

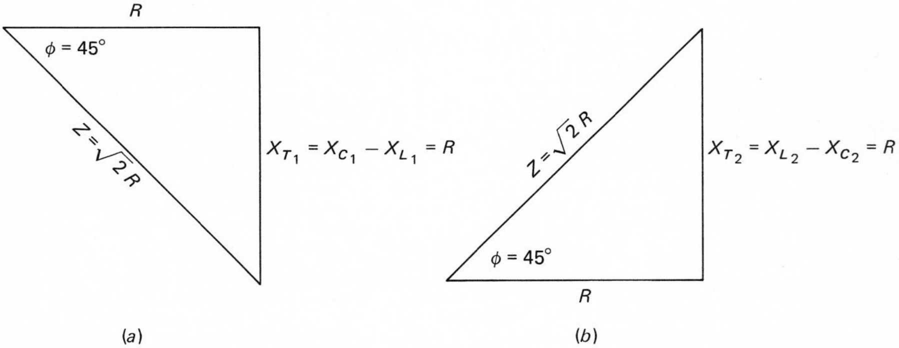

Fig. 16-17 Series *RLC* circuit, impedance triangles for the bandwidth frequencies. (*a*) Below resonance; (*b*) above resonance.

$$X_T = R$$

$$Z = \sqrt{R^2 + X_T^2} = \sqrt{2}R$$

$$\phi = \arctan \frac{X_T}{R} = \arctan 1 = 45° = \pi/4 \text{ rad}$$

IMPEDANCE AND PHASE ANGLE CURVES

We have seen that, for a series resonant circuit

1. At resonance the impedance equals the resistance, and the phase angle is zero.
2. Off resonance the impedance is greater than the resistance, and the phase angle is greater than zero.

3. At the bandwidth frequencies the impedance is $\sqrt{2}$ times the resistance, and the phase angle is 45° ($\pi/4$ rad).

Three more statements can be added:

4. At zero frequency the capacitive reactance is infinitely high and the inductive reactance is zero. Thus the circuit in effect consists of an infinitely high capacitive reactance in series with a resistance. The impedance is therefore infinite, and the phase angle is +90° ($+\pi/2$ rad).
5. If the frequency were to become infinitely high, the inductive reactance would become infinite and the capacitive reactance zero. The effective circuit would be that of an infinite inductive reactance in series with a resistance. The impedance would be infinite, and the phase angle would be −90° ($-\pi/2$ rad).
6. Between zero frequency and resonance, and between resonance and infinite frequency, the transition of impedance and phase angle values is gradual, the rate of change depending on the circuit values involved.

Figure 16-18 shows the effect of varying frequency on the impedance magnitude and the phase angle of a series resonant circuit which has the values $L = 100\ \mu H$, $C = 250$ pF, and $R = 100\ \Omega$.

Fig. 16-18 Series *RLC* circuit curves. (*a*) Impedance-frequency; (*b*) phase angle-frequency; (*c*) impedance-frequency—expanded close to resonance; (*d*) phase angle-frequency—expanded close to resonance.

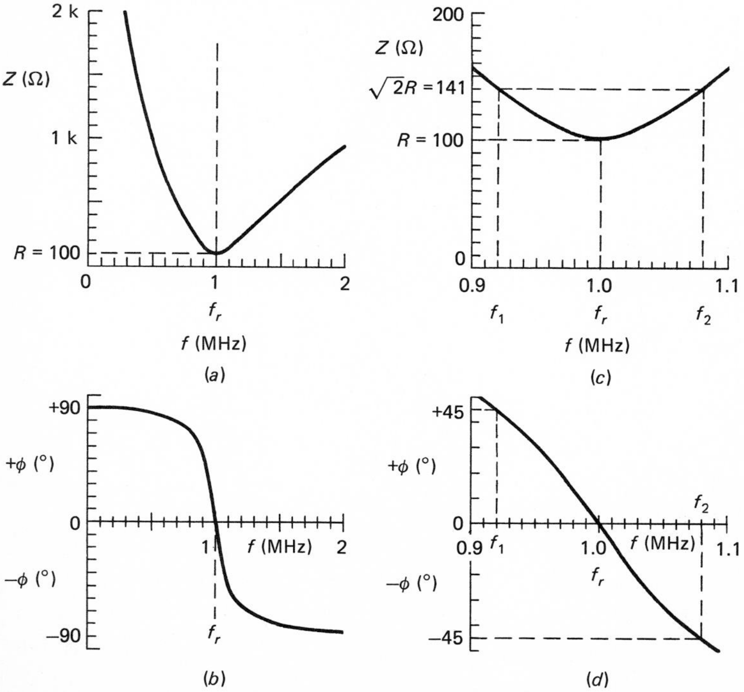

In Fig. 16-18*a* the impedance decreases from infinity to the value of the resistance ($R = 100\ \Omega$) at the resonant frequency f_r. Above f_r the impedance increases again toward infinity.

Figure 16-18*b* shows the circuit phase angle decreasing from $+90°$ (capacitive) toward zero (resistive) at the resonant frequency, and then increasing toward $-90°$ (inductive) above resonance (although the curve goes downward through zero, the change from zero toward $-90°$ is an increase in magnitude).

Figure 16-18*c* and *d* shows magnifications in horizontal and vertical scaling of the curves in Fig. 16-18*a* and *b* to show in greater detail the impedance and phase angle values close to the resonant frequency. In Fig. 16-18*c* it can be seen that the impedance is $\sqrt{2}R = 141\ \Omega$ at the bandwidth frequencies f_1 and f_2, while Fig. 16-18*d* shows the phase angle of 45° at these frequencies.

Figure 16-19 refers to Fig. 16-18 and shows how the current varies with respect to frequency in a series circuit with the values $L = 100\ \mu$H, $C = 250$ pF, and $R = 100\ \Omega$. If the voltage across the circuit is maintained constant at 100 mV while the frequency increases from zero, the circuit current ($I = V_T/Z$) will increase from zero to a maximum value of $I = V_T/R = 1$ mA at resonance. The current will then decrease toward zero at frequencies above resonance. The current at the bandwidth frequencies f_1 and f_2 will be $I = V_T/\sqrt{2}R = 0.7071$ mA, as shown in Fig. 16-19*b* which is a magnified view of the current variations at frequencies close to resonance.

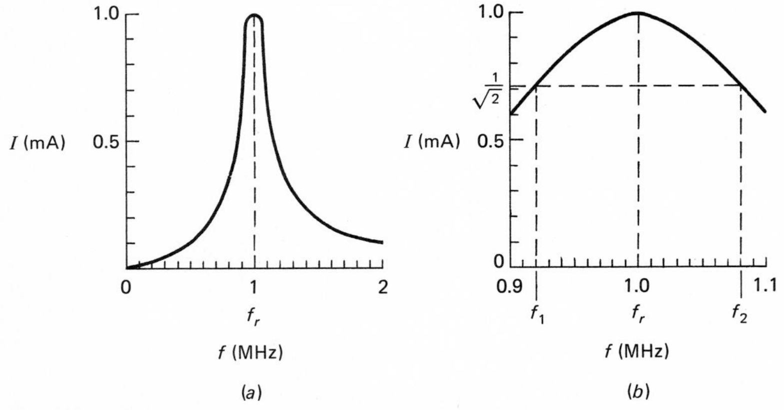

Fig. 16-19 Series *RLC* circuit current-frequency curves.

16-3 BANDWIDTH OF A SERIES *RLC* CIRCUIT

Bandwidth frequencies have been referred to, but what is the significance of bandwidth? By referring to Fig. 16-19, it can be seen that there is a distinct difference between the current at the resonant frequency and the current at any other frequency. The resonant circuit is thus able to discriminate between one frequency and all others. This capability is called *selectivity*. A series circuit is highly selective if it gives a high ratio of current at the resonant frequency to current at some other stated frequency.

The degree of selectivity required of a resonant circuit depends on the application. The selectivity of a resonant circuit in an oscillator should be high, as high selectivity is required to maintain the oscillator's frequency constant at a specific value. In a radio receiver a resonant circuit used to tune to a required station must be able to pass a band of frequencies on either side of the resonant frequency. Thus a tuning circuit in a radio could be expected to have lower selectivity than that of a resonant circuit in an oscillator.

Bandwidth is a measure of a circuit's selectivity. High selectivity is narrow bandwidth. Low selectivity is wide or broad bandwidth.

The bandwidth of a circuit is the frequency difference between points at which the power response is one-half that of the maximum.

In a series resonant circuit supplied with a constant voltage, maximum power is dissipated at the resonant frequency. From the definition of bandwidth, the relationship between the powers at the bandwidth frequencies and the power at resonance is

$$P_1 = P_2 = \frac{P_r}{2} \tag{16-7}$$

where P_1 and P_2 = power at the lower and upper bandwidth frequencies f_1 and f_2, W
P_r = power at the resonant frequency f_r, W

In any ac circuit power is dissipated in the resistance, and in a series circuit is given by $P = I^2R$. Therefore, Eq. (16-7) can be written in the form:

$$I_1^2R = I_2^2R = \frac{I_r^2R}{2}$$

$$I_1^2 = I_2^2 = \frac{I_r^2}{2}$$

or

$$I_1 = I_2 = \frac{I_r}{\sqrt{2}} = 0.7071 I_r \tag{16-8}$$

Equation (16-8) leads to another definition of bandwidth: The bandwidth of a series *RLC* circuit is the frequency difference between conditions where the magnitudes of the currents are $1/\sqrt{2}$, or 0.7071, of the magnitude at the resonant frequency.

In the series resonant circuit to which Fig. 16-19 refers, the current at resonance is 1 mA. At frequencies f_1 and f_2 the current is 0.7071 mA, so these frequencies fit the criteria for bandwidth frequencies. At frequencies f_1 and f_2

$$B = f_2 - f_1 \tag{16-9}$$

where B = bandwidth, Hz
f_1 = lower bandwidth frequency, Hz
f_2 = upper bandwidth frequency, Hz

At frequencies f_1 and f_2 the power dissipated by the circuit is one-half that dissipated at the resonant frequency f_r.

Curves for X_C and X_L, expanded to show the region close to resonance, are given in Fig. 16-20. Although these curves apply specifically to 250 pF and 100 μH, any other values of C and L will give curves of the same shape, only the scaling of the axes will be different. Thus the following discussion and conclusions may be considered general.

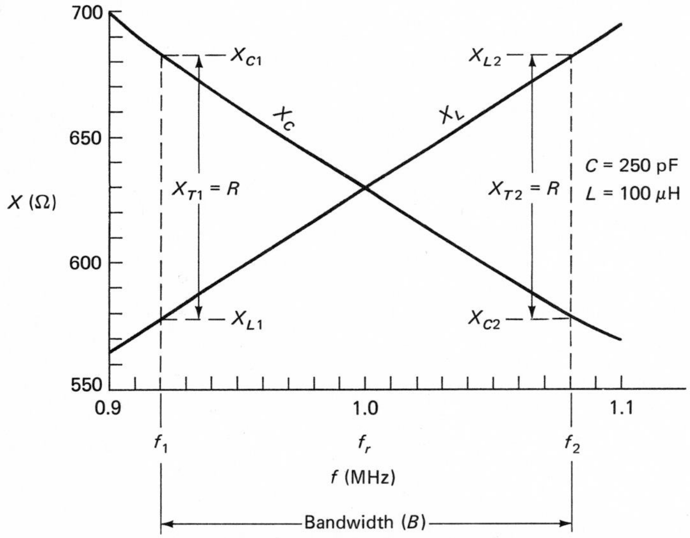

Fig. 16-20 Series resonant circuit reactance-frequency curves expanded close to resonance to show bandwidth conditions.

At the bandwidth frequencies the net reactance equals the resistance. Therefore, at frequency f_2,

$$X_{T2} = X_{L2} - X_{C2} = R \tag{16-6}$$

But from Fig. 16-20 it can be seen that

$$X_{C2} \simeq X_{L1}$$

Therefore,

$$X_{L2} - X_{L1} \simeq R$$
$$2\pi f_2 L - 2\pi f_1 L \simeq R$$
$$2\pi L(f_2 - f_1) \simeq R$$

Since $f_2 - f_1$ equals the bandwidth B,

$$B \simeq \frac{R}{2\pi L}$$

Dividing both sides of this equation by f_r gives

$$\frac{B}{f_r} \simeq \frac{R}{2\pi f_r L} \simeq \frac{R}{X_{Lr}}$$

X_{Lr}/R is the quality factor at resonance Q_r, so

$$\frac{B}{f_r} \simeq \frac{1}{Q_r}$$

or

$$B \simeq \frac{f_r}{Q_r} \tag{16-10}$$

where B = bandwidth frequency, Hz
f_r = resonant frequency, Hz
Q_r = the circuit quality factor

Equation (16-10) is used extensively for both series and parallel resonant circuits and is often memorized in statement form: The bandwidth of a resonant circuit is approximately equal to the ratio of the resonant frequency to the circuit Q at resonance.

In applying Eq. (16-10), the degree of approximation is dependent on the circuit values. If these values are such that Q_r is high, the degree of approximation is high. In practice it is common for Q_r to be at least 10, in which case Eq. (16-10) is accurate within 1 percent. For this reason the approximately equals sign is often omitted, treating Eq. (16-10) as a true equation rather than as an approximation. However, we will retain this sign to stress the fact that the use of Eq. (16-10) results in an approximation.

Example 16-1 A 500-pF capacitor and a 50-μH inductor (with an effective resistance of 20 Ω) are connected in series. What is the resulting (*a*) resonant frequency f_r and (*b*) bandwidth B?

Solution Figure 16-21 shows the circuit for this example.

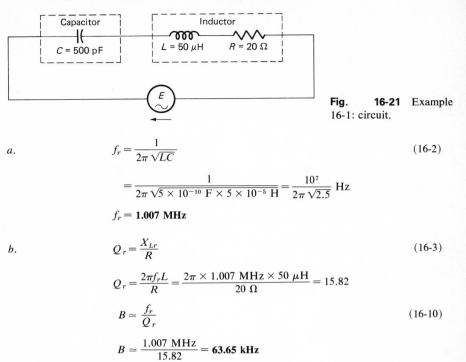

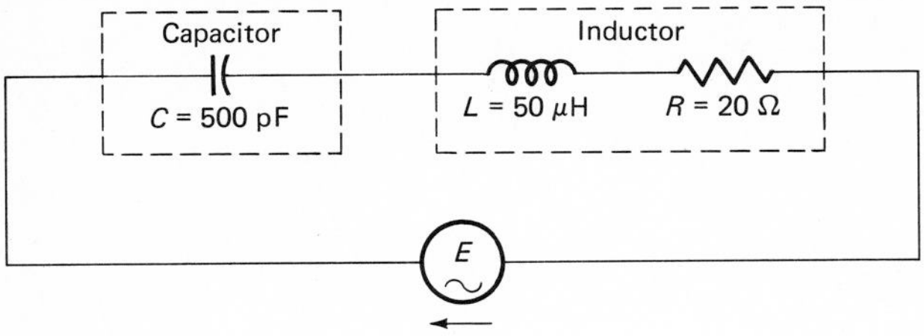

Fig. 16-21 Example 16-1: circuit.

a.

$$f_r = \frac{1}{2\pi \sqrt{LC}} \qquad (16\text{-}2)$$

$$= \frac{1}{2\pi \sqrt{5 \times 10^{-10}\ \text{F} \times 5 \times 10^{-5}\ \text{H}}} = \frac{10^7}{2\pi \sqrt{2.5}}\ \text{Hz}$$

$$f_r = \mathbf{1.007\ MHz}$$

b.

$$Q_r = \frac{X_{Lr}}{R} \qquad (16\text{-}3)$$

$$Q_r = \frac{2\pi f_r L}{R} = \frac{2\pi \times 1.007\ \text{MHz} \times 50\ \mu\text{H}}{20\ \Omega} = 15.82$$

$$B \simeq \frac{f_r}{Q_r} \qquad (16\text{-}10)$$

$$B \simeq \frac{1.007\ \text{MHz}}{15.82} = \mathbf{63.65\ kHz}$$

During the discussion on series resonance, reference was made to a circuit with the values $C = 250$ pF, $L = 100\ \mu$H, and $R = 100\ \Omega$, the implied resonant frequency being 1 MHz. Actually the resonant frequency for these values is 1.007 MHz, the same as that in Example 16-1. The two sets of element values result in the same resonant frequency because the product of L and C is the same in both cases:

$$250\ \text{pF} \times 100\ \mu\text{H} = 500\ \text{pF} \times 50\ \mu\text{H} = 2.5 \times 10^{-14}\ \text{F} \cdot \text{H}$$

In a series circuit, combinations of L and C which have the same product have the same resonant frequency.

In Example 16-1 the reactances at resonance are both 316.4 Ω, $Q_r = 15.82$, and $B = 63.65$ kHz. For the values $C = 250$ pF, $L = 100\ \mu$H, and $R = 20\ \Omega$, the resonant reactances are both 632.8 Ω, $Q_r = 31.64$, and $B = 31.83$ kHz. For any given value of resistance, the bandwidth of a series resonant circuit with an inductance of 50 μH will be wider than the bandwidth with 100 μH by a ratio of 100 μH/50 μH = 2. Figure 16-22 shows current response curves for the two sets of values with an assumed emf of 20 mV; the solid line represents the values $C = 500$ pF, $L = 50\ \mu$H, $R = 20\ \Omega$, and the broken line represents the values $C = 250$ pF, $L = 100\ \mu$H, $R = 20\ \Omega$.

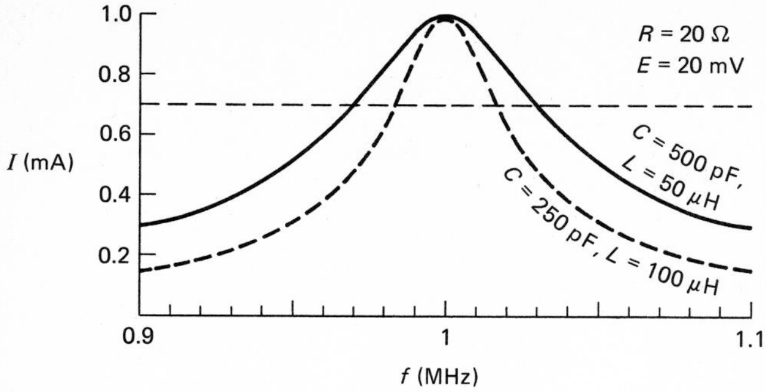

Fig. 16-22 Series *RLC* circuit current-frequency curves for two sets of values. Solid line for the values in Example 16-1; broken line for doubled inductance and halved capacitance.

It is a general principle for series circuits with the same resonant frequency that, for a given resistance value, low inductance will result in wide bandwidth.

If in Example 16-1 the resistance had been 40 Ω, the bandwidth would have been 127.3 kHz, twice as great as for 20 Ω. For series circuits with the same values of *L* and *C*, high series resistance results in wide bandwidth.

As the current at resonance with $R = 40\ \Omega$ will be one-half that with $R = 20\ \Omega$, the two current curves will be displaced vertically if drawn with same current axis. This is shown in Fig. 16-23*a*.

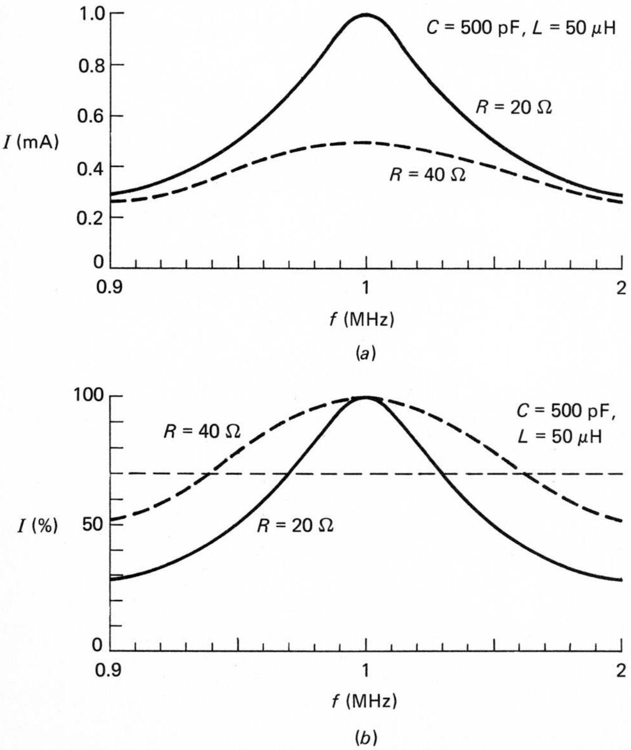

Fig. 16-23 Series *RLC* current-frequency curves for two resistance values. (*a*) Actual current values; (*b*) values normalized on a percentage basis.

A better comparison of the shapes of the curves in Fig. 16-23*a* is obtained by normalizing the resonant currents, that is, making them equivalent. In Fig. 16-23*b* this equivalence is achieved by using percentages of resonant current rather than actual current values. Note that Fig. 16-23*a* and *b* gives different presentations of the same conditions.

Example 16-2 What are the voltages at resonance across each element in the circuit in Example 16-1 if 100 mV is applied across the complete circuit as shown in Fig. 16-24?

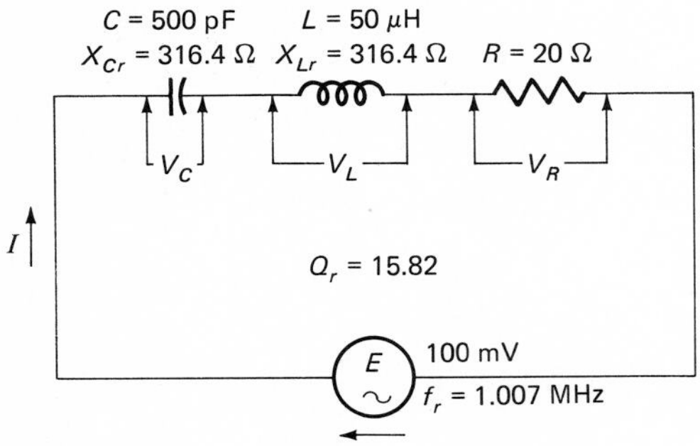

Fig. 16-24 Example 16-2: circuit.

Solution

a. At resonance the voltage across the resistance is the total voltage applied to the circuit.

$$V_R = E = \mathbf{100\ mV}$$

b. At resonance the voltages across the capacitance and across the inductance are Q_r times the applied voltage. From Eq. (16-4),

$$V_C = V_L = Q_r E = 15.82 \times 100\ \text{mV} = V_L = \mathbf{1.582\ V}$$

The result for Example 16-2*b* can be verified by Ohm's law:

$$I_r = \frac{E}{Z_r} = \frac{E}{R} = \frac{100\ \text{mV}}{20\ \Omega} = 5\ \text{mA}$$

$$V_C = I_r X_{Cr} = 5\ \text{mA} \times 316.4\ \Omega = 1.582\ \text{V}$$

$$V_L = I_r X_{Lr} = 5\ \text{mA} \times 316.4\ \Omega = 1.582\ \text{V}$$

Although maximum current occurs at resonance in a series circuit, it should be pointed out here that, strictly speaking, maximum voltage across inductance or capacitance does not occur at the resonant frequency. Maximum voltage across C occurs at a frequency slightly below resonance, and maximum inductive voltage slightly above resonance. However, the slight frequency differences are rarely of practical importance. If the circuit Q is high (over 10, say), the frequency differences are usually small enough to be ignored. If an application relies on maximum capacitive or inductive voltage for frequency discrimination, it is not likely that exact values of L or C will be calculated and selected. Probably either L or C will be made variable. A simple calculation of the resonant conditions will suffice, final adjustment of maximum element voltage being made during measurement.

The voltage across either reactance is 1.582 V, 15.82 times as great as the applied emf. Thus these particular circuit values result in a voltage magnification (sometimes referred to as voltage gain) of 15.82. The circuit Q (316.4 Ω/20 Ω) is also 15.82. Although there is a voltage gain, there is no power gain; the only power dissipation takes place in the resistive component of the circuit ($P_T = EI = V_R I$).

Example 16-3 Figure 16-25*a* shows an antenna feeding a signal to a radio receiver. The series resonant circuit, consisting of a variable capacitor and an inductor, is a wave trap installed for the purpose of rejecting an undesired signal. If the signal to be rejected originates from an aircraft navigation beacon and has a frequency of 400 kHz, to what value should the variable capacitor be set? Figure 16-25*b* is equivalent to the circuit in Fig. 16-25*a*.

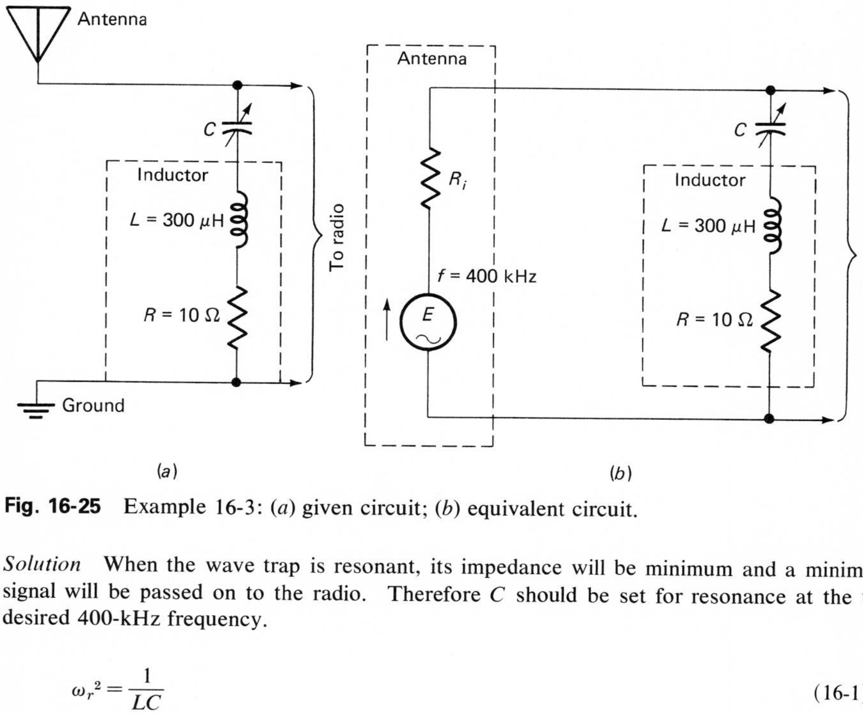

Fig. 16-25 Example 16-3: (*a*) given circuit; (*b*) equivalent circuit.

Solution When the wave trap is resonant, its impedance will be minimum and a minimum signal will be passed on to the radio. Therefore C should be set for resonance at the undesired 400-kHz frequency.

$$\omega_r^2 = \frac{1}{LC} \tag{16-1}$$

$$C = \frac{1}{\omega_r^2 L} = \frac{1}{(2\pi \times 4 \times 10^5\ \text{Hz})^2 \times 3 \times 10^{-4}\ \text{H}} = \frac{10^{-6}}{(2\pi \times 4)^2 \times 3}\ \text{F} = \mathbf{527.7\ pF}$$

In Example 16-3 the level of the signal fed to the radio at 400 kHz will be $ER/(R + R_i)$, the reactances not being relevant at resonance. Thus for efficient rejection, R should be low in value compared to R_i, the effective resistance of the antenna. This requirement implies that the wave trap inductor should have a high Q. A high Q also means narrow bandwidth, another way of viewing high ability to reject the undesired signal.

The Q at resonance in the case of the inductor in Fig. 16-25 is 75, a high value for a practical inductor operating at 400 kHz.

16-4 PARALLEL *RLC* RESONANT CIRCUIT

A circuit consisting of a resistor, a capacitor, and an inductor all in parallel is rarely met in practice. A practical parallel resonant circuit is usually made up of an inductor and a capacitor connected in parallel. An inductor will normally have losses which are significant and can be conveniently represented by a series resistance. The capacitor's losses will normally be negligible. Thus the practical parallel resonant circuit is usually depicted as a series-parallel network consisting of an inductance and a resistance in series and a capacitance connected across the RL series combination. Despite the lack of practicality of the true parallel RLC resonant circuit, the principles involved in its characteristics are important. Visualizing the duality of series and parallel resonant circuits is perhaps the best way to understand the behavior of all resonant circuits.

The development of series resonant circuit characteristics will be followed step by step, using parallel circuit terms. In this manner a direct comparison between series and parallel resonance can be obtained.

R, L, AND *C* ALL IN PARALLEL

In the parallel circuit in Fig. 16-26, the total admittance Y is the phasor sum of the element admittances:

$$Y = G + j(B_C - B_L)$$

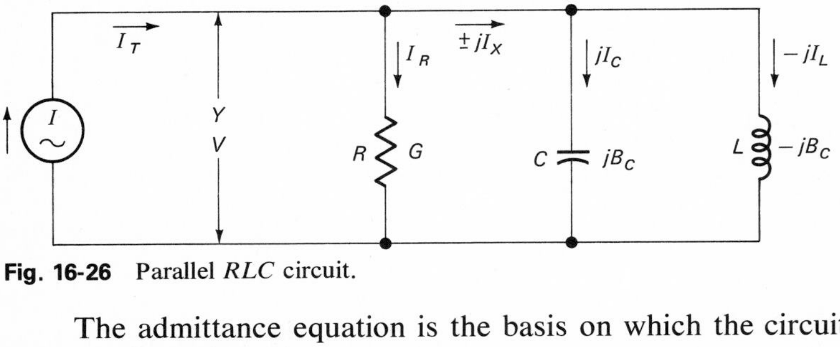

Fig. 16-26 Parallel RLC circuit.

The admittance equation is the basis on which the circuit characteristics are investigated.

At the resonant frequency f_r At the resonant frequency the power factor is unity, the circuit is purely resistive, the admittance consists only of the circuit conductance, the net susceptance is zero, and the phase angle is zero.

$$Y = G$$

$$B_T = j(B_C - B_L) = 0 \qquad \text{or} \qquad B_C = B_L$$

Substituting expressions for B_C and B_L gives

$$\omega_r C = \frac{1}{\omega_r L}$$

or

$$\omega_r^2 = \frac{1}{LC} \tag{16-1}$$

Expanding ω_r,

$$(2\pi f_r)^2 = \frac{1}{LC}$$

or

$$f_r = \frac{1}{2\pi\sqrt{LC}} \tag{16-2}$$

Thus it can be seen that the basic resonant conditions for both series and parallel circuits are the same. Although we have considered only the true parallel circuit, perhaps it should be added here that in many circumstances these conditions also apply to practical parallel resonant circuits to an adequate degree of accuracy.

A phasor diagram for the parallel *RLC* circuit at resonance is given in Fig. 16-27*a*. The capacitive current I_C is a $+j$ quantity with respect to the common voltage V. The inductive current I_L is a $-j$ quantity. Therefore, since their magnitudes are equal, the capacitive and inductive currents cancel, making the net reactive current I_X equal to zero. This cancellation of the reactive currents leaves the resistive current as the only component of the total circuit current. The resistive current I_R equals the total current I_T.

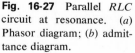

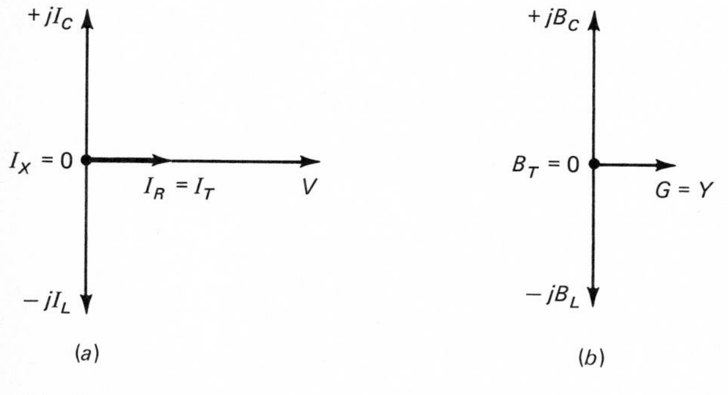

Fig. 16-27 Parallel *RLC* circuit at resonance. (*a*) Phasor diagram; (*b*) admittance diagram.

An admittance diagram for the circuit in Fig. 16-26 is given in Fig. 16-27*b*. Here are shown the equal magnitudes and opposite j signs of B_C and B_L. Cancellation of the susceptances results in the conductance of the resistive element being the total circuit admittance. Without net susceptance, an admittance triangle cannot be drawn.

CURRENT MAGNIFICATION FACTOR

At resonance the current in the *RLC* parallel circuit is determined solely by the resistance. Depending on the values of the susceptances, the equal capacitive and inductive currents can have any magnitude, their net effect being zero. Thus the capacitive and inductive currents can be many times greater than the total circuit current.

The quality factor Q of a parallel circuit is the ratio of either susceptance to the conductance

$$Q_r = \frac{B_{Cr}}{G} = \frac{B_{Lr}}{G}$$

or

$$B_{Cr} = B_{Lr} = Q_r G = Q_r Y \tag{16-11}$$

Writing Eq. (16-11) in current form,

$$Q_r = \frac{I_{Cr}}{I_R} = \frac{I_{Lr}}{I_R}$$

or

$$I_{Cr} = I_{Lr} = Q_r I_R = Q_r I_T \tag{16-12}$$

In Eqs. (16-11) and (16-12):

Q_r = the quality factor at resonance
B_{Cr} = capacitive susceptance at resonance, S
B_{Lr} = inductive susceptance at resonance, S
G = equivalent conductance of the circuit losses, S
Y = total circuit admittance, S
I_{Cr} = capacitive current at resonance, A
I_{Lr} = inductive current at resonance, A
I_R = resistive current, A
I_T = total circuit current, A (at resonance the total current I_T is the same as the resistive current I_R)

In practical parallel resonant circuits, Q_r is usually greater than unity. When this is so, the capacitive and inductive currents are each greater than the total circuit current.

In a parallel *RLC* circuit the quality factor is called the *current magnification factor.* The total current in the circuit is magnified so that it appears Q_r times as large in both capacitor and inductor.

At any frequency other than the resonant frequency When the circuit is supplied with current at a nonresonant frequency, the power factor is less than unity; the phase angle and the net susceptance are both greater than zero.

$$Y > G$$

$$B_T = j(B_C - B_L) > 0 \qquad \text{or} \qquad B_C \neq B_L$$

Figure 16-28 shows admittance diagrams for off-resonance conditions. In Fig. 16-28*a* the frequency is below resonance, so the inductive susceptance is greater than the capacitive susceptance, making the net susceptance $B_T = -j(B_L - B_C)$.

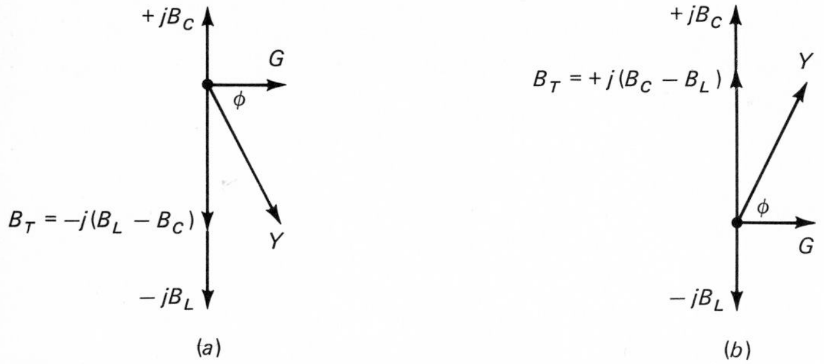

Fig. 16-28 Parallel *RLC* circuit admittance diagrams. (*a*) Below resonance; (*b*) above resonance.

At frequencies above the resonant frequency, B_C exceeds B_L and $B_T = +j(B_C - B_L)$, as shown in Fig. 16-28*b*.

In Fig. 16-28*a* and *b* the circuit admittance Y is the phasor sum of the net susceptance B_T and the conductance G. The phase angle ϕ is equal to arctan B_T/G.

The current relationships under off-resonance conditions will be

$$I_C \neq I_L$$

$$I_X = (I_C - I_L) > 0$$

$$I_T = I_R + j(I_C - I_L) > I_R$$

Phasor diagrams for an *RLC* parallel circuit operated at nonresonant frequencies are shown in Fig. 16-29. Figure 16-29*a* shows the conditions below resonance, where $I_L > I_C$ and the net reactive current $I_X =$

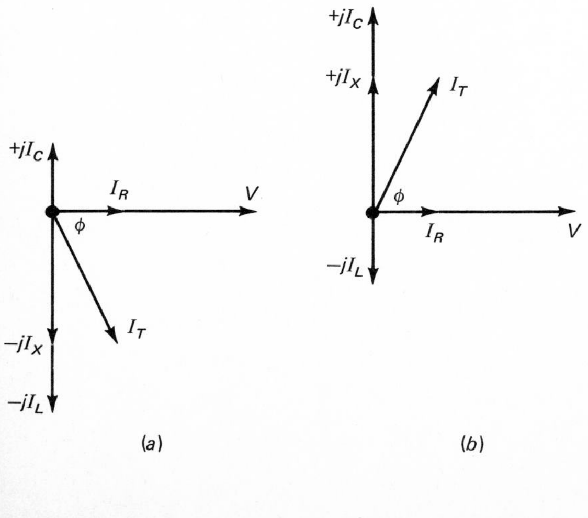

Fig. 16-29 Parallel *RLC* circuit phasor diagrams. (*a*) Below resonance; (*b*) above resonance.

$-j(I_L - I_C)$. In Fig. 16-29*b* the frequency is above resonance ($I_C > I_L$), and the net reactive current is $+j(I_C - I_L)$. In both situations the total circuit current I_T is the phasor sum of the resistive current I_R and the net reactive current I_X. The phase angle ϕ is arctan I_X/I_R.

Figure 16-30*a* and *b* shows admittance triangles for frequencies below and above resonance and is derived from Fig. 16-28. The current triangles in Fig. 16-30*c* and *d* also apply to frequencies below and above resonance and are derived from Fig. 16-29.

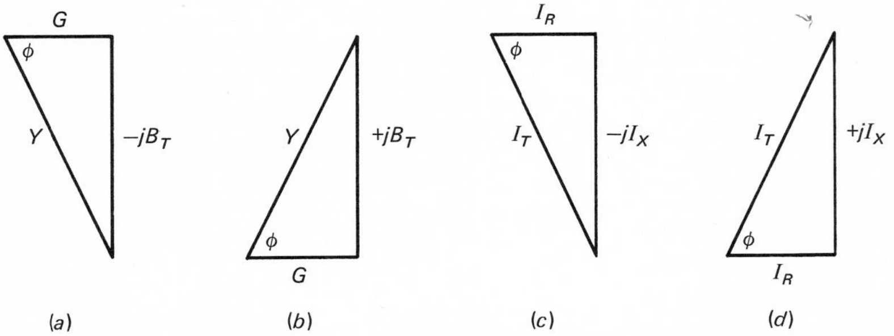

Fig. 16-30 Parallel *RLC* circuit. (*a*) Admittance triangle for below resonance; (*b*) admittance triangle for above resonance; (*c*) current triangle for below resonance; (*d*) current triangle for above resonance.

At frequencies where $B_T = G$ At two frequencies, one below resonance and one above resonance, the net susceptance equals the conductance. These frequencies are the bandwidth frequencies for the parallel *RLC* circuit.

Bandwidth in a circuit consisting of *R*, *L*, and *C* all in parallel can be defined as the frequency difference between the two conditions where the net susceptance equals the conductance. At these bandwidth frequencies,

$$B_T = G$$

At the bandwidth frequency below resonance f_1, the inductive susceptance B_{L1} will be greater than the capacitive susceptance B_{C1}:

$$B_{T1} = B_{L1} - B_{C1} = G \tag{16-13}$$

At the bandwidth frequency above resonance f_2, the capacitive susceptance B_{C2} will be greater than the inductive susceptance B_{L2}:

$$B_{T2} = B_{C2} - B_{L2} = G \tag{16-14}$$

From Eqs. (16-13) and (16-14) it is evident that

$$B_{T1} = B_{T2} = G$$

Figure 16-31a and b shows the admittance triangles for the bandwidth frequencies below and above resonance. For either bandwidth frequency

$$B_T = G$$

$$Y = \sqrt{G^2 + B_T^2} = \sqrt{2}G$$

and

$$\phi = \arctan \frac{B_T}{G} = \arctan 1 = 45° = \pi/4 \text{ rad}$$

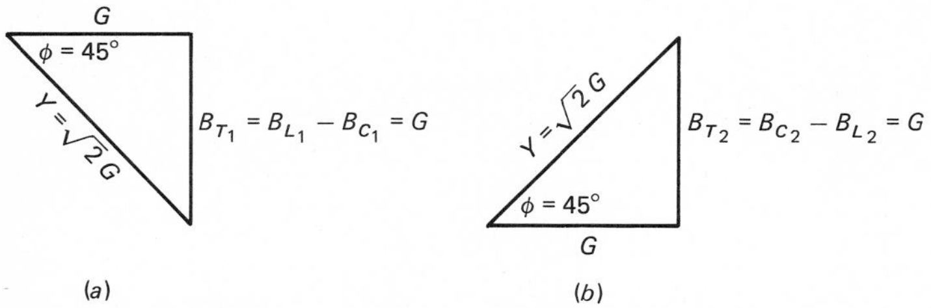

Fig. 16-31 Parallel RLC circuit. Admittance triangles for the bandwidth frequencies. (a) Below resonance; (b) above resonance.

ADMITTANCE AND PHASE ANGLE CURVES

From the foregoing discussion these characteristics of the parallel resonant circuit have emerged:

1. At resonance the admittance equals the conductance, and the phase angle is zero.
2. Off resonance the admittance is greater than the conductance, and the phase angle is greater than zero.
3. At the bandwidth frequencies the admittance is $\sqrt{2}$ times the conductance, and the phase angle is 45° or $\pi/4$ rad.

It can be added that

4. At zero frequency the inductive susceptance is infinitely high, and the capacitive susceptance is zero. Thus at zero frequency the circuit consists of an infinitely high inductive susceptance in parallel with a conductance. The admittance is infinite, and the phase angle is $-90°$ or $-\pi/2$ rad.
5. If the frequency could become infinitely high, the capacitive susceptance would become infinite and the inductive susceptance zero. In this case the circuit would be in effect an infinite capacitive susceptance in parallel with a conductance. The admittance would be infinite, and the phase angle would be $+90°$ or $\pi/2$ rad.
6. Between zero frequency and resonance, and between resonance and infinite frequency, the change in admittance and phase angle is gradual, the rate of change depending on the specific circuit values.

The curves in Fig. 16-32 show the effect of varying frequency on the admittance magnitude and the phase angle of a parallel resonant circuit which has the values $L = 100\ \mu H$, $C = 250$ pF, and $R = 5\ k\Omega$ ($G = 0.2$ mS).

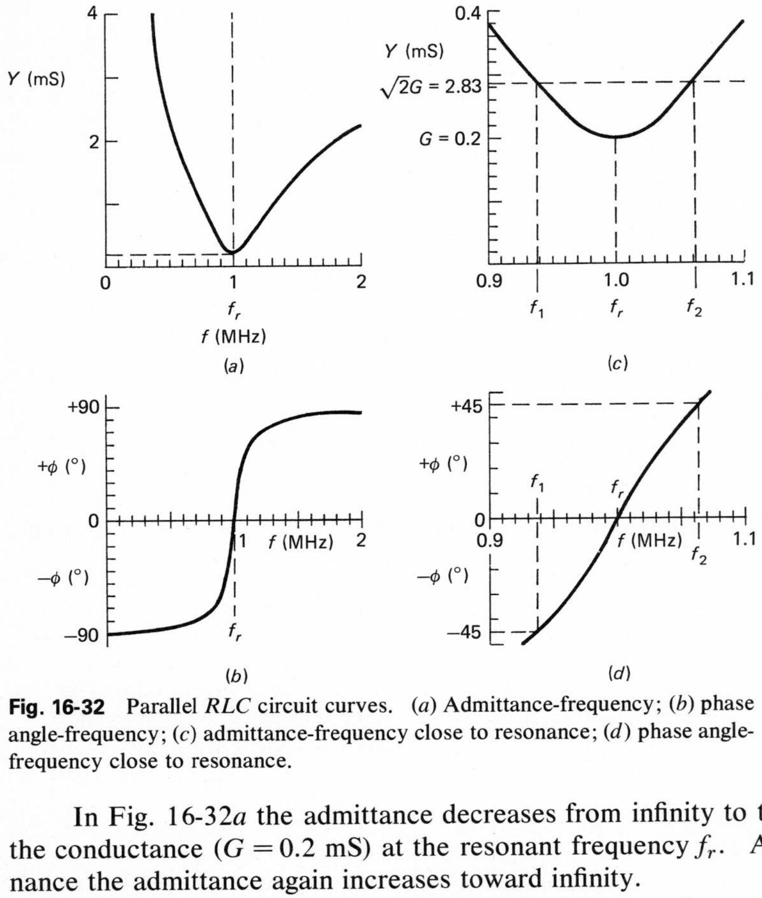

Fig. 16-32 Parallel *RLC* circuit curves. (*a*) Admittance-frequency; (*b*) phase angle-frequency; (*c*) admittance-frequency close to resonance; (*d*) phase angle-frequency close to resonance.

In Fig. 16-32*a* the admittance decreases from infinity to the value of the conductance ($G = 0.2$ mS) at the resonant frequency f_r. Above resonance the admittance again increases toward infinity.

In Fig. 16-32*b* the circuit phase angle decreases from $-90°$ (inductive) toward zero (resistive) at the resonant frequency, and then increases toward $+90°$ (capacitive) above resonance.

Figure 16-32*c* and *d* shows in detail the variations in admittance and phase angle close to the resonant frequency. Figure 16-32*c* shows the admittance magnitude at the bandwidth frequencies f_1 and f_2 to be $\sqrt{2}G = 2.83$ mS. Figure 16-32*d* shows a phase angle of 45° at f_1 and f_2.

It can be useful to refer to the impedance of parallel circuits, particularly when series and parallel network characteristics are being compared. Figure 16-33 shows how the impedance and voltage magnitudes of a parallel resonant circuit vary with changing frequency.

In Fig. 16-33*a* the impedance of the *RLC* parallel circuit rises from zero at zero frequency to a maximum value at resonance and then drops toward zero as the frequency increases beyond resonance. The impedance

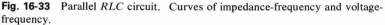

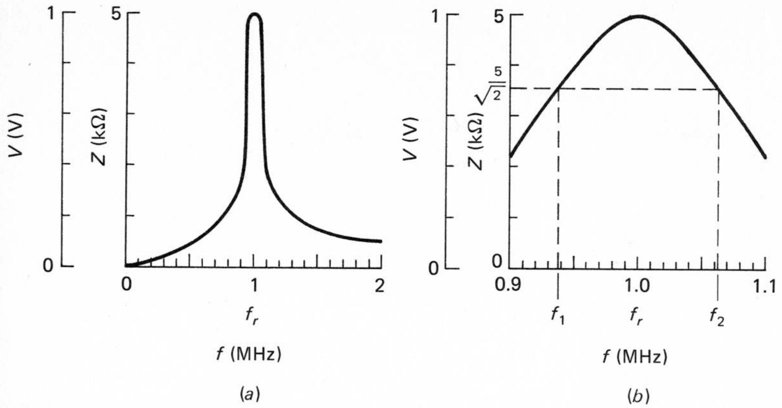

Fig. 16-33 Parallel *RLC* circuit. Curves of impedance-frequency and voltage-frequency.

curve is magnified near the region of resonance in Fig. 16-33*b*, showing the value of $Z_r/\sqrt{2} = 5\ \text{k}\Omega/\sqrt{2}$ at the bandwidth frequencies f_1 and f_2.

The pd across a network is directly proportional to its impedance. Therefore the curves in Fig. 16-33 show how the voltage across a parallel resonant circuit varies with frequency if the total current is maintained constant.

Subsidiary scales to the left of Fig. 16-33*a* and *b* give the voltage values with a constant current of 0.2 mA.

16-5 BANDWIDTH OF A PARALLEL RESONANT CIRCUIT

Bandwidth and related concepts, such as selectivity, have the same significance for all circuits. This is a convention which enables performance characteristics to be specified irrespective of circuit configuration. Thus the power definition of bandwidth for a parallel resonant circuit is the same as for a series circuit.

In a parallel resonant circuit supplied with constant current, maximum power is dissipated at the resonant frequency. From the definition of bandwidth,

$$P_1 = P_2 = \frac{P_r}{2} \tag{16-7}$$

In an ac circuit, power is only dissipated in resistive elements and in a parallel circuit is given by $P = V^2/R$ or $P = V^2G$. Thus Eq. (16-7) can be written as,

$$V_1^2G = V_2^2G = \frac{V_r^2G}{2}$$

$$V_1^2 = V_2^2 = \frac{V_r^2}{2}$$

or
$$V_1 = V_2 = \frac{V_r}{\sqrt{2}} = 0.7071 V_r \tag{16-15}$$

where V_1 = circuit voltage at bandwidth frequency f_1, V
V_2 = voltage at bandwidth frequency f_2, V
V_r = resonant frequency voltage, V

The bandwidth B is given by

$$B = f_2 - f_1 \tag{16-9}$$

Figure 16-34 shows, in expanded form, the curves for B_C and B_L close to the resonant frequency for the values $C = 250$ pF and $L = 100\ \mu$H. For any other values of C and L the appropriate curves will be the same except for the axis scaling.

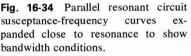

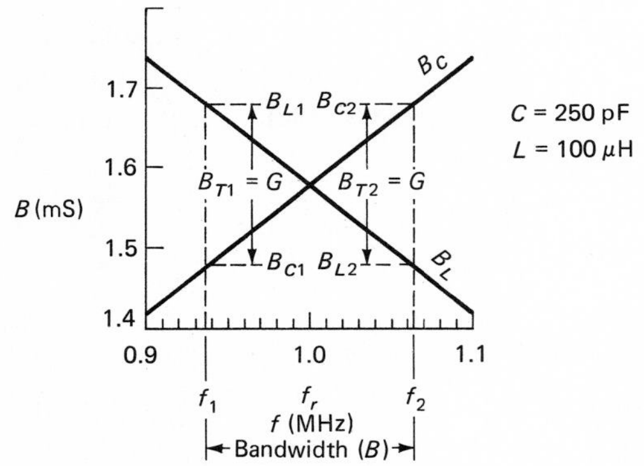

Fig. 16-34 Parallel resonant circuit susceptance-frequency curves expanded close to resonance to show bandwidth conditions.

At the bandwidth frequencies the net susceptance equals the conductance. Therefore, at frequency f_2,

$$B_{T2} = B_{C2} - B_{L2} = G \tag{16-14}$$

By applying to Eq. (16-14) and Fig. 16-34 the same logic as was applied to Eq. (16-6) and Fig. 16-20, the bandwidth for the parallel resonant circuit is found to be the same as for the series circuit:

$$B \simeq \frac{f_r}{Q_r} \tag{16-10}$$

Example 16-4 A capacitance of 500 pF, an inductance of 50 μH, and a resistance of 5 kΩ are connected in parallel. What is (*a*) the resonant frequency f_r and (*b*) the bandwidth B?

Solution The circuit applicable to this example is given in Fig. 16-35.

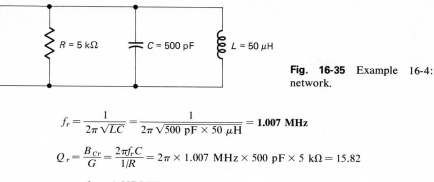

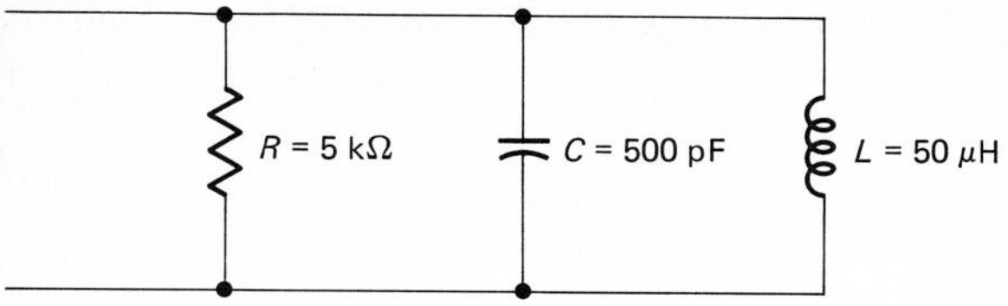

Fig. 16-35 Example 16-4: network.

a. $$f_r = \frac{1}{2\pi\sqrt{LC}} = \frac{1}{2\pi\sqrt{500\text{ pF} \times 50\ \mu\text{H}}} = \mathbf{1.007\ MHz}$$

b. $$Q_r = \frac{B_{Cr}}{G} = \frac{2\pi f_r C}{1/R} = 2\pi \times 1.007\text{ MHz} \times 500\text{ pF} \times 5\text{ k}\Omega = 15.82$$

$$B \simeq \frac{f_r}{Q} \simeq \frac{1.007\text{ MHz}}{15.82} \simeq \mathbf{63.65\ kHz}$$

The parallel circuit in Example 16-4 has the same resonant frequency, quality factor, and bandwidth as the series circuit in Example 16-1. The resonant frequency of each circuit is determined by the same relationship ($f_r = 1/2\pi\sqrt{LC}$) so, since the LC product is the same for each example, the resonant frequencies must likewise be the same.

The quality factor at resonance is the same for each example. Therefore,

$$\text{Series resonant } Q = \text{parallel resonant } Q$$

$$Q_{rs} = Q_{rp}$$

$$\frac{X_{rs}}{R_s} = \frac{B_{rp}}{G_p} \quad \text{or} \quad \frac{X_{rs}}{R_s} = \frac{R_p}{X_{rp}}$$

The two circuits have the same values of L and C, so X_r (the reactance at resonance of either L or C) is the same:

$$R_s R_p = X_{rs} X_{rp} = X_r^2$$

For series or parallel resonance ($X_{Lr} = X_{Cr}$):

$$X_r^2 = X_{Lr} X_{Cr} = \frac{\omega_r L}{\omega_r C}$$

therefore
$$R_s R_p = \frac{L}{C} \tag{16-16}$$

Substituting $R_s = 20\ \Omega$ from Example 16-1 into Eq. (16-16),

$$R_p = \frac{L}{C} \times \frac{1}{R_s} = \frac{5 \times 10^{-5}\text{ H}}{5 \times 10^{-10}\text{ F}} \times \frac{1}{20\ \Omega}$$

$$R_p = 5\text{ k}\Omega$$

This derivation of R_p explains why 20-Ω series resistance is equivalent to 5-kΩ parallel resistance if the capacitance and inductance values are 500 pF and 50 μH.

For any combination of L and C, Eq. (16-16) gives the relationship between the series and parallel circuits which will have the same resonant frequency and bandwidth.

In Example 16-4 the reactances are both 316.4 Ω, $Q_r = 15.82$, and $B = 63.65$ kHz. For the values $C = 250$ pF, $L = 100$ μH, and $R = 5$ kΩ, the resonant reactances are both 632.8 Ω, $Q_r = 7.91$, and $B = 127.3$ kHz. For any given value of resistance, the bandwidth of the parallel circuit with an inductance of 50 μH will be narrower than one with 100 μH by a ratio 100 μH/50 μH = 2. Figure 16-36 shows two voltage response curves (constant current); the solid-line represents the values $C = 500$ pF, $L = 50$ μH, $R = 5$ kΩ, and the broken-line curve the values $C = 250$ pF, $L = 100$ μH, $R = 5$ kΩ.

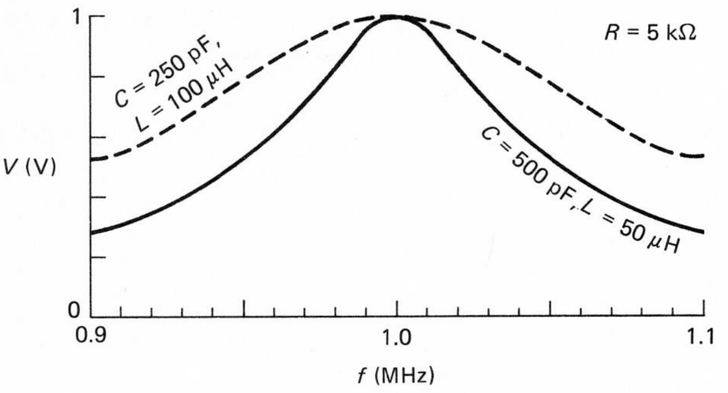

Fig. 16-36 Parallel *RLC* voltage-frequency curves for two sets of values. Solid line for the values in Example 16-4; broken line for doubled inductance and halved capacitance.

Figure 16-36 should be compared with Fig. 16-22. The solid-line curves represent parallel and series equivalent circuits. In each case the broken-line curves represent changes in capacitance and inductance by the same amount, the resistance remaining constant.

For parallel circuits with the same resonant frequency, low inductance will result in narrow bandwidth for a given resistance value.

For parallel resonant circuits with the same values of L and C, high parallel resistance results in narrow bandwidth.

Figure 16-37*a* and *b* shows voltage curves for the L and C values in Example 16-4 with the original 5-kΩ resistance and with the resistance increased to 10 kΩ. The curves in Fig. 16-37*a* show the actual voltages with 0.2-mA current supply. The curves are normalized on a percentage basis in Fig. 16-37*b*.

Compare Fig. 16-37 with Fig. 16-23. They both represent the same values of L and C. The solid-line curves apply to parallel and series circuits which are equivalent in terms of Q_r and bandwidth. However, it can be seen from the broken-line curves that the effect of resistance change in the parallel resonant circuit is opposite that in the series circuit.

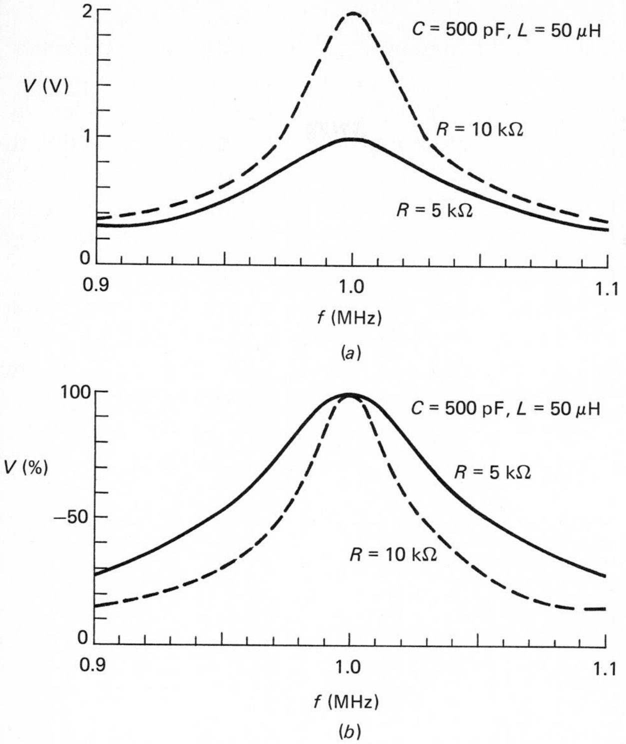

Fig. 16-37 Parallel *RLC* circuit voltage-frequency curves for two resistance values. (*a*) Actual voltage values; (*b*) values normalized on a percentage basis.

Example 16-5 What are the currents at resonance in each element in the circuit in Fig. 16-38 if 1 mA is supplied to the complete circuit?

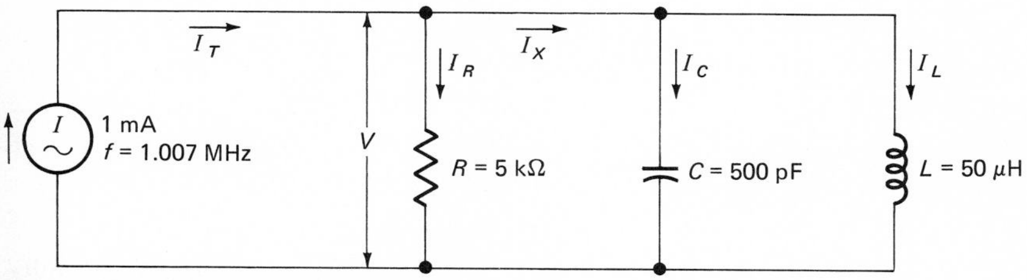

Fig. 16-38 Example 16-5: circuit.

Solution

a. At resonance the resistive current is the total current fed to the circuit:

$$I_R = I_T = \mathbf{1\ mA}$$

b. At resonance the currents in the capacitance and in the inductance are both Q_r times the total circuit current. The values of R, L, and C are the same as in Example 16-4, so the value of Q_r is the same.

$$I_C = I_L = Q_r I_T \qquad (16\text{-}12)$$

$$I_C = I_L = 15.82 \times 1\ \text{mA} = \mathbf{15.82\ mA}$$

CIRCULATING CURRENT

In Fig. 16-38 the arrows for I_C and I_L both point downward. But these two currents are in antiphase; their directions are opposite. If during one half-cycle the capacitive current is downward, the inductive current will be upward. During the next half-cycle the capacitive current will be upward and the inductive current downward. Thus a current arrow in Fig. 16-38 indicates the presence of current rather than its direction. In ac circuits, where many phase relationships can exist simultaneously, it is only feasible for arrows to indicate current presence.

Figure 16-39 shows the two alternate current directions in capacitance and inductance. In either situation $I_C = I_L = Q_r I_T$. Since $I_C = I_L$ and they are in the same loop direction (clockwise or counterclockwise), they may be said to form a single current which circulates within the loop.

$$I_{\text{circ}} = Q_r I_T \tag{16-17}$$

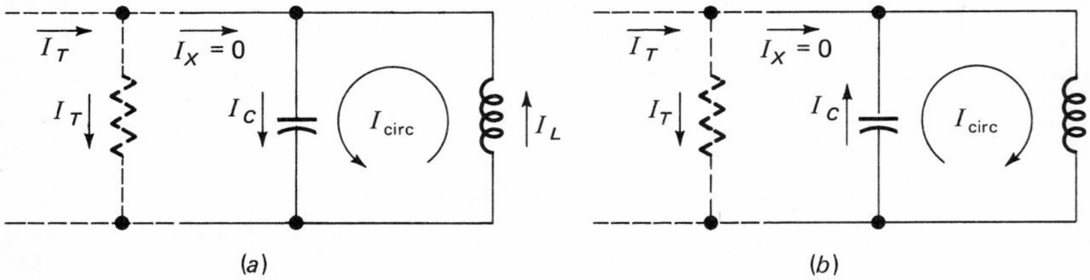

Fig. 16-39 Parallel *RLC* circuit. Instantaneous reactive current directions.

This concept of a circulating current, of value $Q_r I_T$, within the capacitance-inductance loop of a parallel resonant circuit brings a new aspect to the significance of the quality factor Q_r.

The magnitude of the circulating current in a parallel resonant circuit is Q_r times the magnitude of the total current fed to the circuit.

The current in either reactance is 15.82 mA, 15.82 times as great as the current fed to the circuit. These particular circuit values therefore result in a current magnification (sometimes referred to as *current gain*) of 15.82, the same as the value of Q_r. There is no power gain, since the only dissipation occurs in the resistive component of the circuit ($P_T = I_T^2 R = I_R^2 R$).

TANK CIRCUIT

Visualize the notion that the parallel resonant circuit contains energy which circulates back and forth within it via the circulating current. Energy is continually being interchanged between the inductance and the capacitance. The energy remains within the magnetic and electric fields. This concept leads to the view of a parallel resonant circuit as a container of electrical energy.

So evolves the term "tank circuit" used in radio work, a parallel resonant circuit being considered an energy storage tank. The current, and hence the energy, fed to the *RLC* parallel resonant circuit can be small, while the circulating energy is large. This notion suggests a tank of large capacity which is fed slowly, just fast enough to compensate for any loss of contents. The energy in L and C never needs replenishing because reactive elements do not dissipate power. Energy is supplied to the *RLC* circuit only to provide for the loss represented by the resistance.

Example 16-6 A circuit consists of a capacitance of 20 pF and a resistance of 10 kΩ connected in parallel. What inductance value is needed to make the circuit resonant at 16 MHz? What is the effect of connecting a 10-μH inductance in series with the first one?
Solution

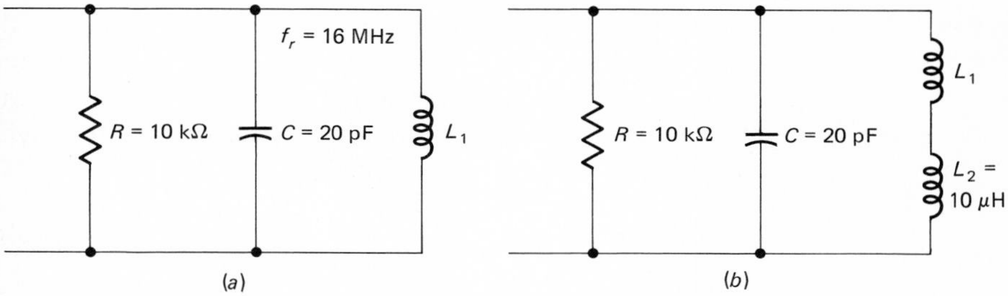

Fig. 16-40 Example 16-6: circuits. (*a*) Original; (*b*) with added inductance.

In Fig. 16-40*a*,

$$\omega_r = 2\pi f_r = 2\pi \times 16 \text{ MHz} = 100.5 \text{ Mrad/s}$$

Apply Eq. (16-1):

$$\omega_r^2 = \frac{1}{L_1 C}$$

$$L_1 = \frac{1}{\omega_r^2 C} = \frac{1}{(100.5 \text{ Mrad/s})^2 \times 20 \text{ pF}} = \mathbf{4.95\ \mu H}$$

In Fig. 16-40*b*, L_1 and L_2 are connected in series. Assuming no mutual inductance, the total inductance is:

$$L_T = L_1 + L_2 = 4.95\ \mu\text{H} + 10\ \mu\text{H} = 14.95\ \mu\text{H}$$

Apply Eq. (16-2):

$$f_r = \frac{1}{2\pi\sqrt{L_T C}}$$

$$f_r \propto \frac{1}{\sqrt{L_T}}$$

$$f_r = \sqrt{\frac{4.95\ \mu\text{H}}{14.95\ \mu\text{H}}} \times 16 \text{ MHz} = 9.207 \text{ MHz}$$

The addition of L_2 reduces the resonant frequency from **16 to 9.207 MHz.**

Figure 16-40*b* is a series-parallel circuit. However, it is not regarded as a series-parallel resonant circuit, since the total circuit inductance, capacitance, and resistance are connected in parallel.

The proportionality approach was used in the solution to Example 16-6. Not only does this technique give a simple solution, but an understanding of circuit behavior is aided by developing a sense of the significance of proportionality.

16-6 THE PRACTICAL PARALLEL RESONANT CIRCUIT

When a real inductor is connected in parallel with a real capacitor, the result is a practical parallel resonant circuit. The inductor and the capacitor will both have losses which add resistance to the inductance and capacitance. In many applications of parallel resonance it is desired to keep this resistance to a minimum, so as to achieve narrow bandwidth and/or high circulating current. An inductor's losses can be great enough to require consideration. By comparison with those of an inductor, the losses of a good capacitor can be ignored. Thus a practical parallel resonant circuit is regarded as a lossy inductor in parallel with a perfect capacitor.

The inductor can be represented by equivalent series or parallel *RL* networks. The series view is more commonly used, since many instruments for measuring the characteristics of inductors give the results in terms of series components.

A series-parallel version of the practical parallel resonant circuit is given in Fig. 16-41. To determine its resonant characteristics, the most direct method is to refer to its admittance:

$$Y_T = Y_C + Y_{RL}$$

$$= j\omega C + \frac{R_s - j\omega L}{R_s^2 + \omega^2 L^2} = j\omega C + \frac{R_s}{R_s^2 + \omega^2 L^2} - j\frac{\omega L}{R_s^2 + \omega^2 L^2}$$

$$Y_T = \frac{R_s}{R_s^2 + \omega^2 L^2} + j\left(\omega C - \frac{\omega L}{R_s^2 + \omega^2 L^2}\right) \tag{16-18}$$

Fig. 16-41 Practical parallel resonant circuit.

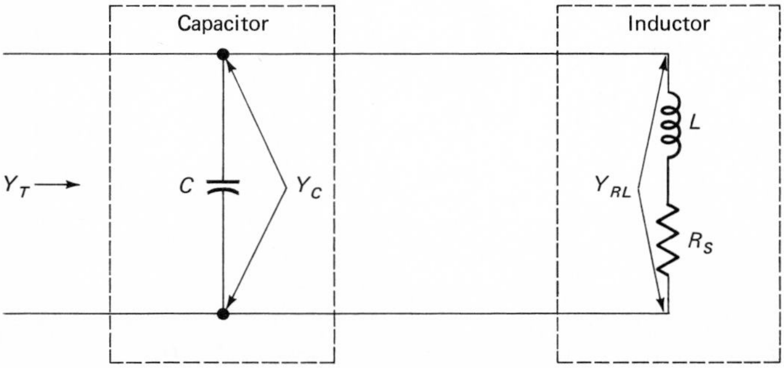

The circuit will be resonant at the frequency f_r, where the admittance is entirely resistive. This condition is met when the imaginary component of the admittance expression [Eq. (16-18)] becomes zero in magnitude:

$$\omega_r C - \frac{\omega_r L}{R_s^2 + \omega_r^2 L^2} = 0$$

$$\omega_r^2 = \frac{1}{LC} - \frac{R_s^2}{L^2} \tag{16-19}$$

or

$$\omega_r^2 = \frac{1}{LC}\left(1 - \frac{R_s^2 C}{L}\right)$$

When Eq. (16-19) is written in the frequency form

$$f_r = \frac{1}{2\pi\sqrt{LC}}\sqrt{1 - \frac{R_s^2 C}{L}} \tag{16-20}$$

it can be seen that, theoretically, the resonant frequency of a practical parallel resonant circuit differs from that of either a series circuit or a purely parallel circuit even if all three circuits have the same values of inductance and capacitance. The degree of difference is slight in practice, since $1 - R_s^2 C/L$ is usually close to unity ($R_s^2 C/L << 1$). For instance, consider the elements in the series circuit in Example 16-3. If the 527.7-pF capacitor were connected in parallel with the 300-μH 10-Ω inductor, the quantity $R_s^2 C/L$ would equal 1.757×10^{-4} and the resulting resonant frequency would be almost exactly equal to the 400 kHz of the series circuit.

The frequency of a radio transmitter must be maintained to a high order of accuracy to avoid interfering with other transmissions. For this reason and other precision frequency needs, frequencies are often quoted to seven or more significant figures. This order of accuracy is rarely needed in the calculation of resonant frequencies of *RLC* circuits. Resonant circuits cannot be manufactured to close frequency limits, final adjustment being achieved by the use of variable component parameters. Thus the term $\sqrt{1 - R_s^2 C/L}$ is often ignored, and Eqs. (16-1) and (16-2) are applied to practical parallel resonant circuits as well as to purely parallel and series *RLC* circuits. A commonly used rule is: If the Q of the inductor at the resonant frequency is 10 or more, the resistance can be considered to have an insignificant effect on the resonant frequency.

The admittance at resonance is the real part of Eq. (16-18), the imaginary part being zero:

$$Y_r = \frac{R_s}{R_s^2 + \omega_r^2 L^2}$$

From this equation we can obtain an expression for the impedance of the practical parallel resonant circuit at resonance:

$$Z_r = \frac{1}{Y_r} = \frac{R_s^2 + \omega_r^2 L^2}{R_s}$$

or

$$Z_r = R_s + \frac{\omega_r^2 L^2}{R_s} \tag{16-21}$$

Substituting for $X_{Lr} = \omega_r L$ in Eq. (16-21),

$$Z_r = R_s + \frac{X_{Lr}^2}{R_s}$$

As $X_{Lr}/R_s = Q_r$,

$$Z_r = R_s + Q_r X_{Lr} \tag{16-22}$$

where Z_r = total circuit impedance, Ω
R_s = series resistance of the inductor, Ω
Q_r = the quality factor of the inductor (also the quality factor of the complete circuit)
X_{Lr} = inductive reactance, Ω

All terms are to be taken at their resonant frequency values. This applies to the resistance R_s which represents the total losses of the inductor, these being influenced by frequency.

Equation (16-22) gives an expression for the circuit impedance which is more accurate than is usually necessary. The equation can be simplified by adopting reasonable practical assumptions:

$$X_{Lr} >> R_s$$

$$Q_r X_{Lr} >> R_s$$

Therefore,

$$Z_R \simeq Q_r X_{Lr}$$

or, since

$$X_{Lr} \simeq X_{Cr}$$

$$Z_r \simeq Q_r X_r \tag{16-23}$$

where X_r = reactance of either inductor or capacitor.

Substituting $X_{Cr}/R_s = 1/\omega_r C R_s$ for Q_r in Eq. (16-23),

$$Z_r \simeq \frac{\omega_r L}{\omega_r C R_s}$$

$$Z_r \simeq \frac{L}{C R_s} \tag{16-24}$$

The dynamic impedance Z_r of the practical parallel resonant circuit, as given in Eq. (16-24), is important in the design of circuits for maximum voltage ($V_r = IZ_r$) or maximum power transfer (Z_r is resistive and maximum power is transferred when Z_r equals the internal resistance of the source).

Practical parallel resonant circuits are used so extensively in commercial equipment that a summary of their characteristics is justified. You will see that a number of the items listed are the same as those for the purely parallel *RLC* circuit. Both true and practical parallel circuits behave in the same general manner; only some numerical details are different.

At resonance:

1. The power factor is unity, and the phase angle is zero.
2. The impedance is resistive and approximately of maximum value.
3. The impedance magnitude is given by $Z_r = R_s + \omega_r^2 L^2/R_s$, by $Z_r = R_s + Q_r X_r$, or approximately by $Z_r = L/CR_s$.
4. The resonant frequency is given by $f_r = 1/2\pi\sqrt{LC} \times \sqrt{1 - R_s^2 C/L}$, or approximately by $f_r = 1/2\pi\sqrt{LC}$.

Off resonance:

5. Below resonance the circuit is inductive.
6. Above resonance the circuit is capacitive.

At the bandwidth points:

7. The phase angle is 45°.
8. The impedance is $Z_r/\sqrt{2} \simeq L/CR_s\sqrt{2}$ or $0.7071L/CR_s$.
9. The bandwidth is given by $B \simeq f_r/Q_r$.

Example 16-7 The measured series characteristics of an inductor at 1 kHz are 200-mH inductance and 500-Ω resistance. A 0.1-μF capacitor is connected in parallel with the inductor. What is the resonant frequency of the circuit?

Solution (See Fig. 16-42)

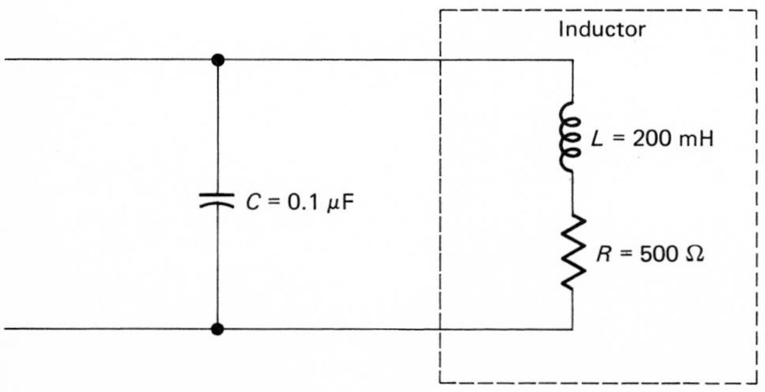

Fig. 16-42 Example 16-7: circuit.

The Q of the coil at 1 kHz is

$$Q = \frac{X_L}{R} = \frac{2\pi f L}{R} = \frac{2\pi \times 1\ \text{kHz} \times 200\ \text{mH}}{500\ \Omega} = 2.513$$

This Q value is well below 10, so Eq. (16-20) should be used to determine the resonant frequency:

$$f_r = \frac{1}{2\pi\sqrt{LC}}\sqrt{1 - \frac{R^2C}{L}} \tag{16-20}$$

$$= \frac{1}{2\pi\sqrt{200\ \text{mH} \times 0.1\ \mu\text{F}}}\sqrt{1 - \frac{(500\ \Omega)^2 \times 0.1\ \mu\text{F}}{200\ \text{mH}}}$$

$$= \frac{1}{2\pi\sqrt{2 \times 10^{-8}}}\sqrt{1 - 0.125} = 1125 \times \sqrt{0.875}$$

$$f_r = \mathbf{1053\ Hz}$$

As the resonant frequency in Example 16-7 is close to 1 kHz, the inductance and resistance values measured at 1 kHz were valid for use in the calculation.

If the general equation $f_r = 1/2\pi\sqrt{LC}$ had been used for this low-Q circuit, the resonant frequency would have been computed as 1.125 kHz, representing an error of 6.837 percent. With the resistance decreased to 50 Ω ($Q \simeq 25$), the error would be only 0.0626 percent.

Example 16-8 What voltage is developed across the circuit in Example 16-7 if a current of 1 A is supplied to it (*a*) at the resonant frequency and (*b*) at 60 Hz? Assume that the inductance and resistance values of the inductor are the same at 60 Hz as at 1 kHz.

Solution

a. At the resonant frequency, 1053 Hz, the Q at resonance will be

$$Q_r = \frac{\omega_r L}{R} = \frac{2\pi \times 1053\ \text{Hz} \times 200\ \text{mH}}{500} = 2.646$$

The impedance at resonance will be

$$Z_r = R + Q_r X_{Lr} \tag{16-22}$$

$$Z_r = 500\ \Omega + 2.646 \times 2\pi \times 1053\ \text{Hz} \times 200\ \text{mH} = 4.001\ \text{k}\Omega$$

The circuit voltage at resonance is the product of the total current and the resonant impedance:

$$V_r = I_T Z_r = 1\ \text{A} \times 4.001\ \text{k}\Omega = \mathbf{4.001\ kV}$$

b. At 60 Hz

$$\omega = 2\pi \times 60\ \text{Hz} = 377\ \text{rad/s}$$

$$X_L = \omega L = 377\ \text{rad/s} \times 200\ \text{mH} = 75.4\ \Omega$$

The impedance of the RL branch is

$$Z_{RL} = R + jX_L = (500 + j75.4)\ \Omega = 505.7\ \Omega\ \underline{/8.58^\circ}$$

$$X_C = -j\frac{1}{\omega C} = -j\frac{1}{377\ \text{rad/s} \times 0.1\ \mu\text{F}} = -j26.53\ \text{k}\Omega$$

The magnitude of X_C is very much greater than that of Z_{RL}. Therefore, since X_C is in parallel with Z_{RL}, its effect is negligible and the total circuit impedance at 60 Hz is approximately the same as the impedance of the RL branch:

$$V = I_T Z \simeq I_T Z_{RL} \simeq 1\ \text{A} \times 505.7\ \Omega\ \underline{/8.58^\circ}$$

$$V \simeq \mathbf{505.7\ V}\ \underline{\mathbf{/8.58^\circ}} \quad \text{or} \quad \underline{\mathbf{/0.1497\ rad}}$$

Rms values of voltage and current are assumed in Example 16-8 because no value types are specified. Thus the peak value of the circuit voltage at resonance is $4.001\sqrt{2} = 5.658$ kV. The 0.1-μF capacitor would have to be able to withstand this high voltage without sustaining damage.

Example 16-9 The Q of a 25-μH inductor is 75 at a frequency of 3 MHz. What is (*a*) the value of a parallel capacitor required to make the inductor resonant at 3 MHz, (*b*) the resulting bandwidth, and (*c*) the resonant impedance of the circuit?

Solution As the Q of the inductor (which is also the circuit Q at resonance) is well in excess of 10, the approximate equations can be used with reasonable accuracy. In circumstances such as these, it is not usual to adhere rigidly to the approximately equals sign.

a.

$$\omega_r = 2\pi f_r = 2\pi \times 3\ \text{MHz} = 18.85\ \text{Mrad/s}$$

$$\omega_r^2 = \frac{1}{LC} \tag{16-1}$$

$$C = \frac{1}{\omega_r^2 L} = \frac{1}{(18.85\ \text{Mrad/s})^2 \times 25\ \mu\text{H}} = \mathbf{112.6\ pF}$$

b.

$$B = \frac{f_r}{Q_r} \tag{16-10}$$

$$B = \frac{3\ \text{MHz}}{75} = \mathbf{40\ kHz}$$

c.

$$Z_r = Q_r X_r \tag{16-23}$$

$$Z_r = Q_r \omega_r L = 75 \times 18.85\ \text{Mrad/s} \times 25\ \mu\text{H} = \mathbf{35.34\ k\Omega}$$

As noted in the solution to Example 16-9, approximately equals signs are sometimes not used even though the results are only approximate. This is because resonant circuits, particularly high-Q ones, are subject to many variables which are difficult to predict accurately. Thus it is usual to adjust these circuits by measurement, after construction. In the circuit in Example 16-9, it is likely that the capacitor, calculated as 112.6 pF, will be made variable for final adjustment. Typically, this capacitance would be made up of a 100-pF fixed capacitor with a 25-pF variable capacitor connected in parallel with it. This arrangement would give a range of capacitance values from 100 to 125 pF, or about 12.5 pF on either side of the calculated value.

Example 16-10 The resonant network in Example 16-9 is part of an amplifier circuit. The transistor supplies 1 mA to the network at its resonant frequency of 3 MHz. What is the magnitude of the circulating current at resonance?
Solution (See Fig. 16-43)

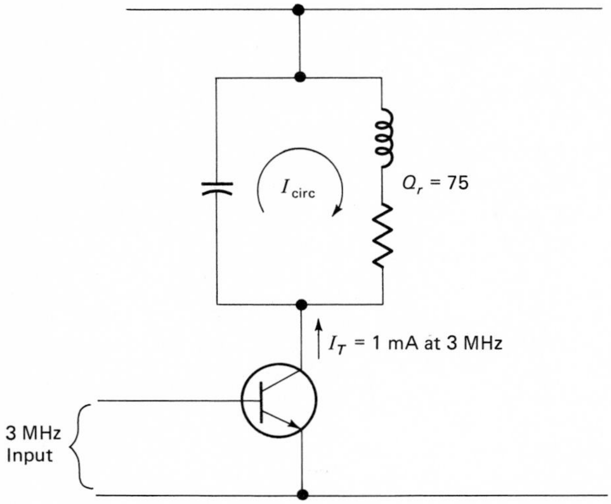

Fig. 16-43 Example 16-10: circuit.

$$I_{circ} = Q_r I_T \qquad (16\text{-}17)$$

$$I_{circ} = 75 \times 1 \text{ mA} = \mathbf{75\ mA}$$

Although the current supplied to the resonant circuit in Example 16-10 is only 1 mA, which is easily handled by a small transistor, the current in the inductor (75 mA) is sufficiently great to pose a design and construction problem if space is limited.

16-7 SOME APPLICATIONS OF RESONANCE

Resonance is a phenomenon which occurs, intentionally or otherwise, so often that only a small sampling of its many applications can be given. The following selection will serve to show the versatility of resonance and indicate how the principles we have dealt with can be applied to other resonant circuits.

DOUBLE-TUNING

Arranging for a circuit to be in resonance is often called tuning, the resonant circuit then being referred to as a tuned circuit. When a circuit resonates at two different frequencies, it is called a double-tuned circuit. It should be mentioned that the term "double-tuned" is also used for a transformer which has two of its windings resonant at the same frequency.

Example 16-11 The network in Fig. 16-44 is part of the circuit of a transistor radio which operates on either AM signals or on FM signals. On AM the input to the circuit is at a frequency of 450 kHz, while on FM it is at 10.7 MHz. How does this circuit discriminate between the two frequencies? Neglect the effects of any loading on the outputs.

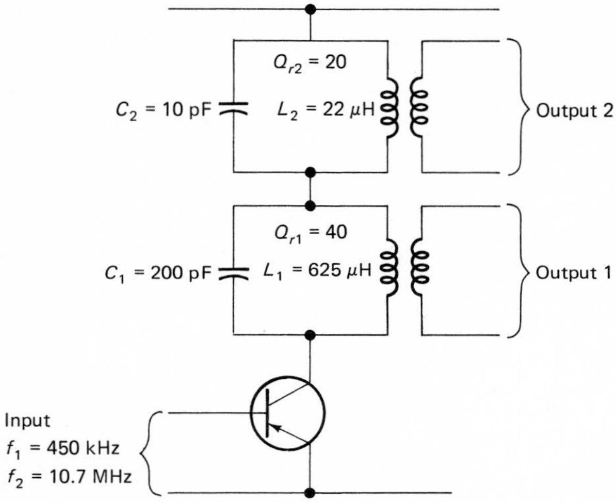

Fig. 16-44 Example 16-11: circuit.

Solution At frequency $f_1 = 450$ kHz ($\omega_1 = 2.827$ Mrad/s):
The parallel circuit L_1C_1 resonates at 450 kHz; its impedance is

$$Z_{r1} = Q_{r1}X_{r1} \tag{16-23}$$

$$Z_{r1} = Q_{r1}\omega_1 L_1 = 40 \times 2.827 \text{ Mrad/s} \times 625\ \mu\text{H} = 70.68 \text{ k}\Omega$$

$$X_{L2} = \omega_1 L_2 = 2.827 \text{ Mrad/s} \times 22\ \mu\text{H} = 62.19\ \Omega$$

$$X_{C2} = \frac{1}{\omega_1 C_2} = \frac{1}{2.827 \text{ Mrad/s} \times 10 \text{ pF}} = 35.37 \text{ k}\Omega$$

As X_{C2} is a very high impedance in parallel with X_{L2}, a low impedance, the impedance of circuit L_2C_2 at 450 kHz is approximately equal to X_{L2}:

$$Z_2 \simeq X_{L2} \simeq 62.19\ \Omega$$

L_2 and C_2 resonate at 10.7 MHz, so that the impedance of circuit L_2C_2 at this frequency is, from Eq. (16-23),

$$Z_{r2} = Q_{r2}X_{r2}$$

$$Z_{r2} = Q_{r2} \times \omega_2 L_2 = 20 \times 67.23 \text{ Mrad/s} \times 22\ \mu\text{H} = 29.58 \text{ k}\Omega$$

$$X_{L1} = \omega_2 L_1 = 67.23 \text{ Mrad/s} \times 625\ \mu\text{H} = 42.02 \text{ k}\Omega$$

$$X_{C1} = \frac{1}{\omega_2 C_1} = \frac{1}{67.23 \text{ Mrad/s} \times 200 \text{ pF}} = 74.37\ \Omega$$

X_{L1} is very much greater than X_{C1}. Thus the impedance of the circuit L_1C_1 at 10.7 MHz will be approximately equal to X_{C1}:

$$Z_1 \simeq X_{C1} \simeq 74.37\ \Omega$$

Using the impedance values calculated, the conditions at the two frequencies are summarized in Fig. 16-45.

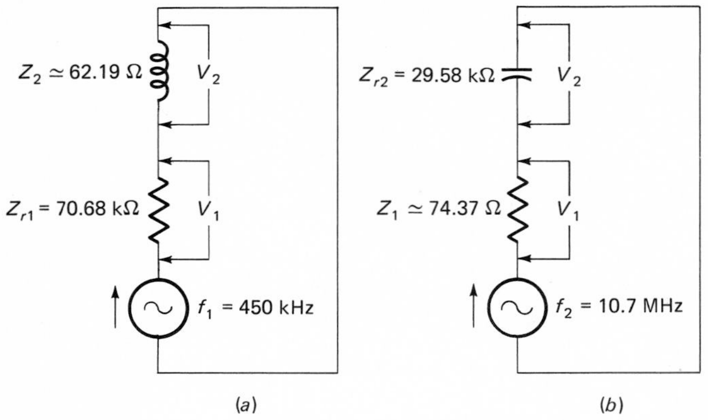

Fig. 16-45 Example 16-11: equivalent circuits. (*a*) At 450 kHz; (*b*) at 10.7 MHz.

Figure 16-45*a* shows the situation in which the input is $f_1 = 450$ kHz. At this frequency V_1 is approximately $70.68\ k\Omega/62.19\ \Omega = 1137$ times V_2. Therefore, in referring back to Fig. 16-44, at 450 kHz output 1 is higher than output 2.

When the input frequency is $f_2 = 10.7$ MHz, the conditions are as shown in Fig. 16-45*b*. At this frequency V_2 is approximately $29.58\ k\Omega/74.37\ \Omega = 397.8$ times V_1. Thus at 10.7 MHz, output 2 will be considerably greater than output 1.

Summarizing the action of the circuit in Fig. 16-44, we can say that on AM stations the signal is delivered at output 1, while on FM stations the signal leaves the circuit at output 2.

Example 16-12 Within a television receiver, the sound is carried on a 45.75-MHz signal f_s, the picture being carried on a 41.25-MHz signal f_p. The circuit in Fig. 16-46 passes the picture signal to the video circuits and prevents the sound signal from passing. How does the circuit achieve its purpose?

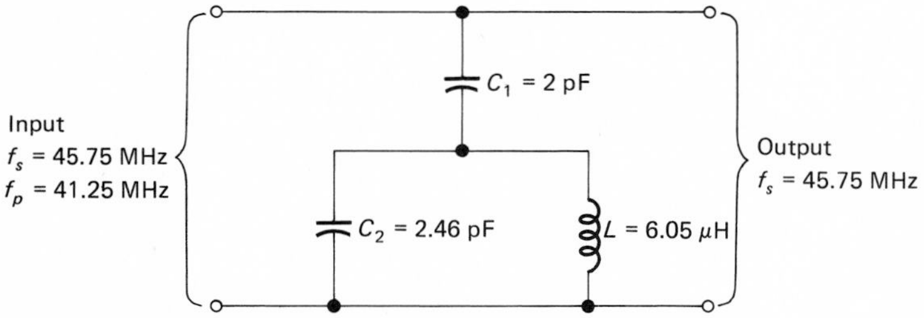

Fig. 16-46 Example 16-12: circuit.

Solution At frequency $f_s = 45.75$ MHz ($\omega_s = 287.5$ Mrad/s):

$$X_{C1s} = \frac{1}{\omega_s C_1} = \frac{1}{287.5\ \text{Mrad/s} \times 2\ \text{pF}} = 1.739\ k\Omega$$

$$X_{C2s} = X_{C1s}\frac{C_1}{C_2} = 1.739\ k\Omega \times \frac{2\ \text{pF}}{2.46\ \text{pF}} = 1.414\ k\Omega$$

$$X_{Ls} = \omega_s L = 287.5\ \text{Mrad/s} \times 6.05\ \mu\text{H} = 1.739\ k\Omega$$

At 45.75 MHz, C_1 and L are series resonant and act as a low-impedance shunt across the output terminals, preventing the development of substantial output voltage at this frequency.

At frequency $f_p = 41.25$ MHz:

$$X_{C1p} = X_{C1s}\frac{f_s}{f_p} = 1.739\text{ k}\Omega \times \frac{45.75\text{ MHz}}{41.25\text{ MHz}} = 1.929\text{ k}\Omega$$

$$X_{C2p} = X_{C2s}\frac{f_s}{f_p} = 1.414\text{ k}\Omega \times \frac{45.75\text{ MHz}}{41.25\text{ MHz}} = 1.568\text{ k}\Omega$$

$$X_{Lp} = X_{Ls}\frac{f_p}{f_s} = 1.739\text{ k}\Omega \times \frac{41.25\text{ MHz}}{45.75\text{ MHz}} = 1.568\text{ k}\Omega$$

At 41.25 MHz, C_2 and L are in parallel resonance, acting as a circuit of high shunt impedance which will not inhibit output at this frequency.

TRIMMING, TRACKING, AND PADDING

A radio usually receives transmissions over a wide range of frequencies. For example, the ordinary domestic radio receives signals broadcast on the 525 kHz to 1.6 MHz band. A tuned circuit used to select a required transmission has a variable capacitor to vary the resonant frequency. To allow for deviations in inductance and capacitance from their nominal values, which are the inevitable consequences of manufacturing tolerances, a trimming capacitor may be used. This trimmer is a preset variable capacitor which is connected in parallel with the main tuning capacitor. The capacitance value of the trimmer is set during the final adjustment of the tuning of the radio (called alignment).

Example 16-13 A resonant circuit in a radio is required to tune over the band 525 kHz to 1.6 MHz. The variable tuning capacitor has a range of 30 to 300 pF. What values of inductance and trimming capacitance are needed?

Solution (See Fig. 16-47)

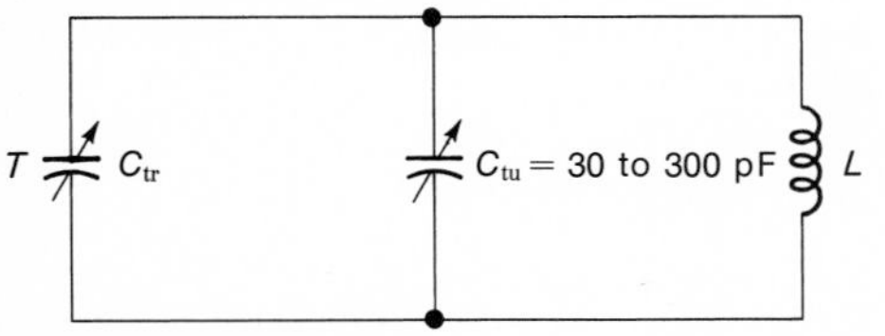

Fig. 16-47 Example 16-13: circuit.

Apply Eq. (16-1):

$$C = \frac{1}{\omega^2 L}$$

$$C \propto \frac{1}{f^2}$$

$$\frac{C_{\text{low}}}{C_{\text{high}}} = \left(\frac{f_{\text{high}}}{f_{\text{low}}}\right)^2 = \left(\frac{1.6\text{ MHz}}{525\text{ kHz}}\right)^2 = 9.288$$

where C_{low} and C_{high} are the total circuit capacitances at the low and high ends of the frequency band. The tuning capacitor C_{tu} will have maximum capacitance at the lowest frequency. At any point in the frequency band the total capacitance is the sum of the tuning and trimming capacitances.

$$C_{low} = C_{tu} + C_{tr} = (300 + C_{tr})\ \text{pF}$$

$$C_{high} = C_{tu} + C_{tr} = (30 + C_{tr})\ \text{pF}$$

$$\frac{(300 + C_{tr})\ \text{pF}}{(30 + C_{tr})\ \text{pF}} = 9.288$$

$$C_{tr} = \mathbf{2.582\ pF}$$

The inductance can be determined at any point on the frequency band. Taking the high end, where $f = 1.6$ MHz,

$$L = \frac{1}{\omega^2 C} \qquad (16\text{-}1)$$

$$L = \frac{1}{(2\pi \times 1.6\ \text{MHz})^2 \times (30 + 2.582)\ \text{pF}} = \mathbf{303.7\ \mu H}$$

There are two circuits in a radio which, for any transmission, have to be resonant to two different frequencies, the difference between these frequencies being a constant for any particular radio. The relationship between the two frequencies is

$$f_{osc} - f_{sig} = f_i$$

where f_{osc} = frequency of an oscillator, Hz
f_{sig} = frequency of the desired station, Hz
f_i = intermediate frequency of the radio, Hz

The frequencies in Examples 16-11 and 16-12 are intermediate frequencies.

The process of maintaining a constant frequency difference between the oscillator and signal frequency resonant circuits is called tracking. It can be accomplished by special tuning capacitor design or by padding. Padding is a circuit technique using a series padding capacitor, and is best illustrated by an example.

Example 16-14 The network in Fig. 16-47 is used for the signal frequency resonant circuit of a radio. The radio has an intermediate frequency of 450 kHz. If the capacitor which tunes the oscillator is the same type as that which tunes the signal frequency circuit, what values of inductance and padding capacitance are needed for the oscillator circuit?

Solution Figure 16-48 shows the oscillator resonant circuit, C_{tu} and C_p being the tuning and

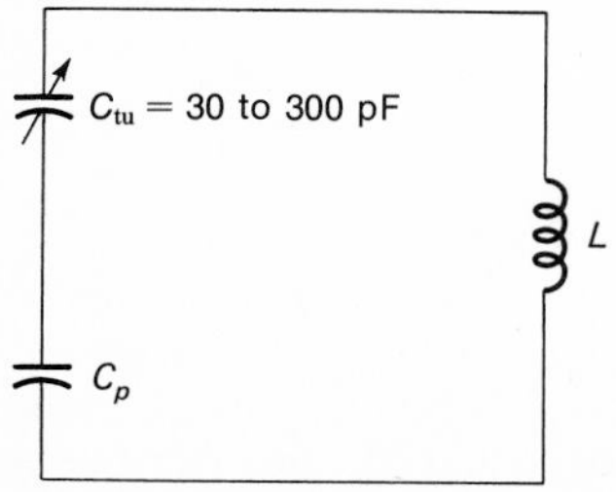

Fig. 16-48 Example 16-14: circuit.

padding capacitors. Values for C_p and L which will apply with reasonable accuracy over the entire frequency range of the radio can be obtained by considering each end of the band.

At the low-frequency end ($f_{sig} = 525$ kHz),

$$f_{osc} = f_{sig} + f_i = 525 \text{ kHz} + 450 \text{ kHz} = 975 \text{ kHz}$$

The total capacitance, C_{tu} and C_p in series, will be given by

$$C_{T,low} = \frac{C_{tu}C_p}{C_{tu} + C_p} = \frac{300C_p}{300 + C_p} \text{ pF}$$

At the high-frequency end ($f_{sig} = 1.6$ MHz),

$$f_{osc} = f_{sig} + f_i = 1.6 \text{ MHz} + 450 \text{ kHz} = 2.05 \text{ MHz}$$

The total capacitance will be

$$C_{T,high} = \frac{C_{tu}C_p}{C_{tu} + C_p} = \frac{30C_p}{30 + C_p} \text{ pF}$$

The ratio of the highest to the lowest oscillator frequency is

$$\frac{2.05 \text{ MHz}}{975 \text{ kHz}} = 2.103$$

The ratio of the greatest to the least total capacitance is the square of the frequency ratio:

$$\frac{C_{T,low}}{C_{T,high}} = 2.103^2 = 4.423$$

$$\frac{300C_p}{300 + C_p} \div \frac{30C_p}{30 + C_p} = 4.423$$

$$C_p = \mathbf{184.2 \text{ pF}}$$

The inductance can be found from the data for either end of the frequency range. Using the low-end data,

$$f_{osc} = 975 \text{ kHz}$$

$$\omega_{osc} = 2\pi \times 975 \text{ kHz} = 6.126 \text{ Mrad/s}$$

$$C_{T,low} = \frac{300 \times 184.2}{300 + 184.2} = 114.1 \text{ pF}$$

Apply Eq. (16-1):

$$L = \frac{1}{\omega_{osc}^2 C_{T,low}}$$

$$L = \frac{1}{(6.126 \text{ Mrad/s})^2 \times 114.1 \text{ pF}} = \mathbf{233.54\ \mu H}$$

POWER FACTOR IMPROVEMENT

A load on an electric supply system is usually inductive, its power factor being less than unity and lagging. This means that the apparent power is greater than the active power. For a given supply voltage the current drawn from the supply is greater than the current needed for the same useful power at unity power factor. On a domestic load the heavy load components, such as the kitchen stove and the clothes dryer, are resistive so that, although the total current can be about 100 A, the power factor is close to unity. However, an industrial load may consist mainly of electric motors, controls, fluorescent lights, and other equipment with a low power factor. Also the total current drawn by an industrial load may be several hundreds or even thousands of amperes. In such a situation the electric supply company may base its rates on the maximum current drawn by the load, since this determines the distribution transformer and conductor sizes. The result is that the cost of installing equipment which will raise the power factor toward unity may be more than offset by the savings in supply costs. Installation of equipment to increase the power factor is called *power factor improvement* or *power factor correction.* The additional equipment must have a leading power factor to oppose the lagging power factor of the load; it must be capacitive. Power factor improvement in no way affects the performance characteristics of the original load. It is still connected across the supply, so its apparent, active, and reactive powers are not changed. Only the effect on the electric supply is changed.

Power factor improvement is mentioned as an application of resonance, because it is concerned with opposing reactances, the ultimate aim being complete reactance cancellation, which is resonance.

The following examples demonstrate power factor improvement.

Example 16-15 An industrial load draws 500 A from a 440-V 60-Hz supply at a power factor of 0.6 lagging. What is (*a*) the reactive power of a capacitor which will raise the circuit power factor to 0.9 lagging, (*b*) the capacitor's current, and (*c*) its capacitance?

Solution (See Fig. 16-49)

Fig. 16-49 Example 16-15: circuit.

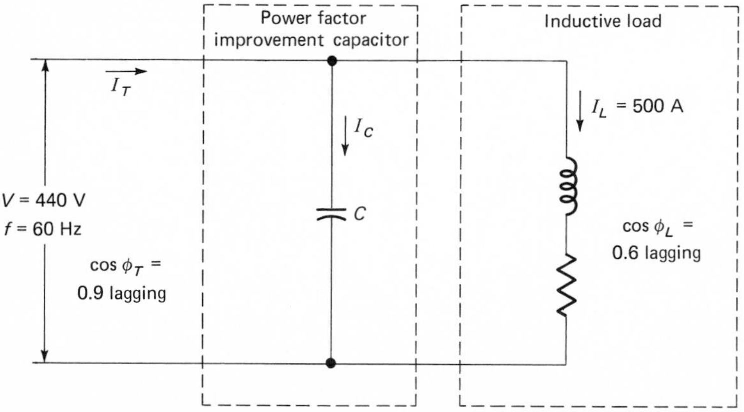

Consider the load alone.

The apparent power of the load is

$$S_L = VI_L = 440\ \text{V} \times 500\ \text{A} = 220\ \text{kVA}$$

The active power of the load is

$$P_L = S_L \cos \phi_L = 220\ \text{kVA} \times 0.6 = 132\ \text{kW}$$

The reactive power of the load is

$$Q_L = \sqrt{S_L^2 - P_L^2} = \sqrt{(220\ \text{kVA})^2 - (132\ \text{kW})^2} = 176\ \text{kvar}$$

Consider the load together with the power factor improvement capacitor.

The active power will be unchanged:

$$P_T = P_L = 132\ \text{kW}$$

The apparent power of the complete circuit will be

$$S_T = \frac{P_T}{\cos \phi_T} = \frac{132\ \text{kW}}{0.9} = 146.7\ \text{kVA}$$

The reactive power of the complete circuit will be

$$Q_T = \sqrt{S_T^2 - P_T^2} = \sqrt{(146.7\ \text{kVA})^2 - (132\ \text{kW})^2} = 63.9\ \text{kvar}$$

Figure 16-50 gives the power triangles relevant to this example. The power triangle for the load alone is shown in Fig. 16-50*a*, and that for the complete circuit in Fig. 16-50*b*. The power triangles in Fig. 16-50*a* and *b* are superimposed in Fig. 16-50*c*.

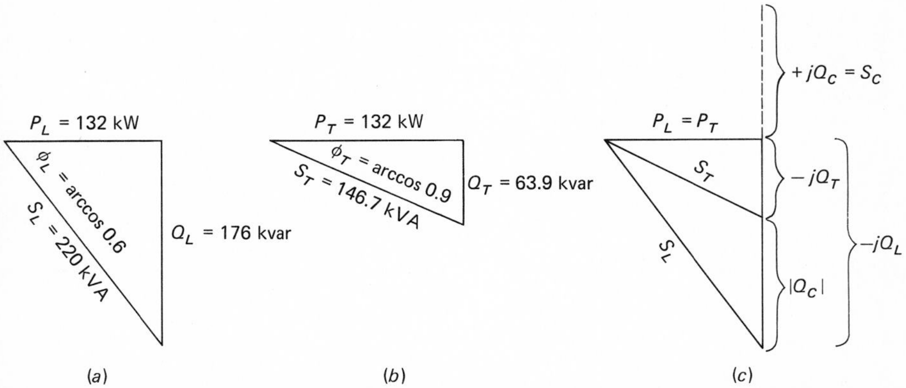

Fig. 16-50 Example 16-15: (*a*) power triangle for the load alone; (*b*) power triangle for the complete circuit; (*c*) power triangles superimposed.

a. Q_L and Q_T are the reactive powers associated with the lagging power factors of the load and the complete circuit. Q_C, shown by the broken line in Fig. 16-50*c*, is the reactive power of the power factor improvement capacitor.

The relationship between the reactive powers is

$$Q_C = Q_L - Q_T = 176\ \text{kvar} - 63.9\ \text{kvar} = \mathbf{112.1\ kvar}$$

b. The capacitor is connected across the 440-V supply and dissipates no power:

$$S_C = Q_C = 112.1 \text{ kVA}$$

$$S_C = VI_C$$

$$I_C = \frac{S_C}{V} = \frac{112.1 \text{ kVA}}{440 \text{ V}} = \mathbf{254.8\ A}$$

c. The reactance of the capacitor is the ratio of its voltage and its current:

$$X_C = \frac{V}{I_C} = \frac{440 \text{ V}}{254.8 \text{ A}} = 1.727\ \Omega$$

$$X_C = \frac{1}{\omega C}$$

$$C = \frac{1}{\omega X_C} = \frac{1}{2\pi \times 60 \text{ Hz} \times 1.727\ \Omega} = \mathbf{1536\ \mu F}$$

It can be seen from the results for Example 16-15 that capacitors used for power factor improvement in industrial situations are large power devices. A capacitor of 1536-μF capacitance with insulation to withstand a peak voltage of 623 V and capable of continuously passing a current of 254 A is a large, expensive device. But over a period of time the total cost saving produced by reducing the reactive power by 112.1 kvar can far exceed the cost of the capacitor.

Example 16-16 The standby power plant for a microwave radio relay station, capable of supplying 10 kVA at 220 V 60 Hz, was adequate for the station prior to the installation of additional equipment. With the new equipment installed the load is increased to 12 kVA at a power factor of 0.7 lagging. This is too great a load for the power plant. What is the minimum capacitor rating required to enable the power plant to handle the increased load?
Solution

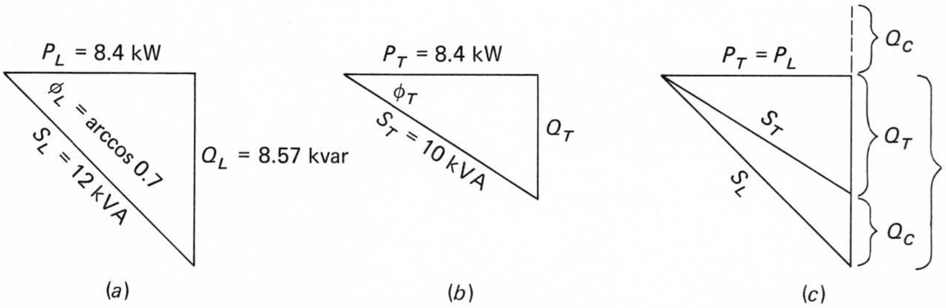

Fig. 16-51 Example 16-16: (*a*) power triangle for the load alone; (*b*) power triangle for the complete circuit; (*c*) power triangles superimposed.

For the relay station without the capacitor, refer to Fig. 16-51*a*

$$S_L = 12 \text{ kVA}$$

$$P_L = S_L \cos \phi_L = 12 \text{ kVA} \times 0.7 = 8.4 \text{ kW}$$

$$Q_L = \sqrt{S_L^2 - P_L^2} = \sqrt{(12 \text{ kVA})^2 - (8.4 \text{ kW})^2} = 8.57 \text{ kvar}$$

Figure 16-51*a* shows the power triangle for the relay station load.

Figure 16-51*b* gives the required power triangle with the total apparent power S_T equal to 10 kVA. The total active power P_T is the same as the load power P_L:

$$P_T = P_L = 8.4 \text{ kW}$$

The total permissible reactive power without overloading the power plant is

$$Q_T = \sqrt{S_T^2 - P_T^2} = \sqrt{(10 \text{ kVA})^2 - (8.4 \text{ kW})^2} = 5.43 \text{ kvar}$$

The capacitor must supply a correcting influence equivalent to the difference between the reactive power of the load Q_L and the total permissible reactive power Q_T:

$$Q_C = Q_L - Q_T = 8.57 \text{ kvar} - 5.43 \text{ kvar} = 3.14 \text{ kvar}$$

The current in the capacitor is

$$I_C = \frac{Q_C}{V} = \frac{3.14 \text{ kvar}}{220 \text{ V}} = 14.27 \text{ A}$$

$$X_C = \frac{V}{I_C} = \frac{220 \text{ V}}{14.27 \text{ A}} = 15.42\ \Omega$$

$$C = \frac{1}{\omega X_C} = \frac{1}{377 \text{ rad/s} \times 15.42\ \Omega} = 172\ \mu\text{F}$$

Thus the capacitor's minimum ratings are **172-μF** capacitance at a peak voltage of **311 V** and a reactive power of **3.14 kvar.**

16-8 SUMMARY OF EQUATIONS

Series *RLC* circuit

$$\omega_r^2 = 1/LC \qquad (16\text{-}1)$$

$$f_r = 1/2\pi\sqrt{LC} \qquad (16\text{-}2)$$

$$X_{Lr} = X_{Cr} = Q_r R = Q_r Z \qquad (16\text{-}3)$$

$$V_{Lr} = V_{Cr} = Q_r V_T \qquad (16\text{-}4)$$

$$X_{T1} = X_{C1} - X_{L1} = R \qquad (16\text{-}5)$$

$$X_{T2} = X_{L2} - X_{C2} = R \qquad (16\text{-}6)$$

$$P_1 = P_2 = P_r/2 \qquad (16\text{-}7)$$

$$I_1 = I_2 = I_r/\sqrt{2} = 0.7071 I_r \qquad (16\text{-}8)$$

$$B = f_2 - f_1 \qquad (16\text{-}9)$$

$$B \simeq f_r/Q_r \qquad (16\text{-}10)^*$$

Parallel *RLC* circuit

$$\omega_r^2 = 1/LC \qquad (16\text{-}1)$$

$$f_r = 1/2\pi\sqrt{LC} \qquad (16\text{-}2)$$

$$B_{Cr} = B_{Lr} = Q_r G = Q_r Y \qquad (16\text{-}11)$$

$$I_{Cr} = I_{Lr} = Q_r I_r = Q_r I_T \qquad (16\text{-}12)$$

$$B_{T_1} = B_{L_1} - B_{C_1} = G \qquad (16\text{-}13)$$

$$B_{T2} = B_{C2} - B_{L2} = G \qquad (16\text{-}14)$$

$$P_1 = P_2 = P_r/2 \qquad (16\text{-}7)$$

$$V_1 = V_2 = V_r/\sqrt{2} = 0.7071 V_r \qquad (16\text{-}15)$$

$$B = f_2 - f_1 \qquad (16\text{-}9)$$

$$B \simeq f_r/Q_r \qquad (16\text{-}10)^*$$

$$R_s R_p = L/C \qquad (16\text{-}16)$$

$$I_{\text{circ}} = Q_r I_T \qquad (16\text{-}17)$$

Practical parallel resonant circuit

$$Y_T = R_s/(R_s^2 + \omega^2 L^2) + j[\omega C - \omega L/(R_s^2 + \omega^2 L^2)] \qquad (16\text{-}18)$$

* When applying this equation, note that $Q_r = X_{Lr}/R = X_{Cr}/R$ for series circuits, and $Q_r = B_{Lr}/G = B_{Cr}/G = R/X_{Lr} = R/X_{Cr}$ for parallel circuits.

$$\omega_r^2 = 1/LC - R_s^2/L^2 = (1/LC)(1 - R_s^2C/L) \qquad (16\text{-}19)$$

$$f_r = [1/(2\pi\sqrt{LC})]\sqrt{1 - R_s^2C/L} \qquad (16\text{-}20)$$

$$Z_r = R_s + \omega_r^2L^2/R_s \qquad (16\text{-}21)$$

$$Z_r = R_s + Q_rX_{Lr} \qquad (16\text{-}22)$$

$$Z_r \simeq Q_rX_r \qquad (16\text{-}23)$$

$$Z_r \simeq L/CR_s \qquad (16\text{-}24)$$

EXERCISE PROBLEMS

Sections 16-2 and 16-3

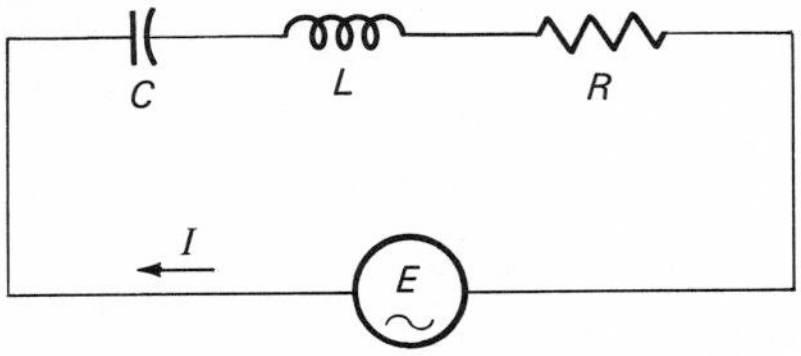

Fig. 16-1P Problems 16-1 to 16-7: circuit.

16-1 For the series *RLC* circuit shown in Fig. 16-1P, find the resonant frequency when
a. $E = 10\ \mu$V; $C = 150$ pF, $L = 250\ \mu$H, $R = 25\ \Omega$ (*ans.* 821.9 kHz)
b. $E = 1$ mV, $C = 350$ pF, $L = 500\ \mu$H, $R = 20\ \Omega$ (*ans.* 380.5 kHz)
c. $E = 1$ V, $C = 0.002\ \mu$F, $L = 1$ mH, $R = 2\ \Omega$ (*ans.* 112.5 kHz)

16-2 Using the circuits in Prob. 16-1, find the reactances of *L* and *C* at their resonant frequencies.

16-3 Find the *Q* of each of the circuits in Prob. 16-1. (*ans. a.* 51.64; *b.* 59.75; *c.* 353.6)

16-4 Find the voltage across the capacitors of each of the circuits in Prob. 16-1.

16-5 For each of the circuits in Prob. 16-1, find
a. The current *I*. (*ans. a.* 0.4 μA; *b.* 50 μA; *c.* 0.5 A)
b. IX_l. (*ans. a.* 516 μV; *b.* 59.75 mV; *c.* 353.6 V)

16-6 Find the bandwidth for each of the circuits in Prob. 16-1.

16-7 For each of the circuits in Prob. 16-1, find the reactances of *L* and *C* at the bandwidth frequencies and compare the difference of these two reactances to the circuit resistance. (*ans. a.* 1.304 kΩ, 1.279 kΩ, 24.9 Ω, 1.279 kΩ, 1.304 kΩ, 25.1 Ω; *b.* 1.205 kΩ, 1.185 kΩ, 19.9 Ω, 1.185 kΩ, 1.205 kΩ, 20.1 Ω; *c.* 708 Ω, 706 Ω, 2 Ω, 706 Ω, 708 Ω, 2 Ω)

16-8 The bandwidth of a series resonant circuit consisting of $C = 450$ pF, $L = 350\ \mu$H, and $R = 8\ \Omega$ is found to be too narrow. Find the amount of additional resistance required to increase the bandwidth to 10 kHz.

16-9 If the voltage across the circuit in Prob. 16-8 is 10 mV, find the voltage across the inductance
a. Before the additional resistance is added. (*ans.* 1.102 V)
b. After the additional resistance is added. (*ans.* 0.401 V)

16-10 It is desired to change the frequency of the circuit in Prob. 16-8 to 450 kHz.
a. Find a new value of *L* which will obtain this frequency with *C* remaining at 450 pF.
b. Find a new value of *C* which will obtain this frequency with *L* remaining at 350 μH.

Sections 16-4 to 16-6

16-11 For the parallel *RLC* circuit in Fig. 16-2P, find the resonant frequency using Eq. (16-20) when
a. $E = 10$ V, $C = 150$ pF, $L = 100\ \mu$H, $R = 20\ \Omega$ (*ans.* 1.299 MHz)
b. $E = 10$ V, $C = 150$ pF, $L = 100\ \mu$H, $R = 200\ \Omega$ (*ans.* 1.2599 MHz)
c. $E = 5$ V, $C = 0.022\ \mu$F, $L = 12$ mH, $R = 10\ \Omega$ (*ans.* 9.795 kHz)

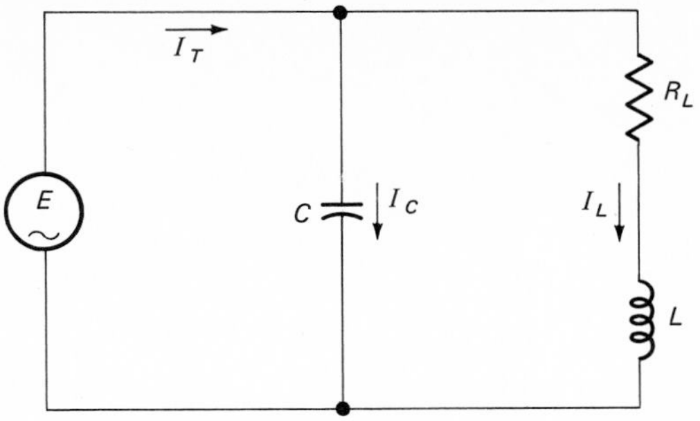

Fig. 16-2P Problems 16-11 to 16-16: circuit.

16-12 Which of the frequencies of the circuits in Prob. 16-11 would not be appreciably changed by solving with Eq. (16-2)?

16-13 Using the circuits in Prob. 16-11, find the reactances of L and C at the resonant frequencies.
(*ans. a.* $+j816.2\ \Omega$, $-j816.8\ \Omega$; *b.* $+j791.6\ \Omega$, $-j842.2\ \Omega$; *c.* $+j738.5\ \Omega$, $-j738.6\ \Omega$)

16-14 Find the Q and the total impedance of each of the circuits in Prob. 16-11.

16-15 Find I_T and I_L for each of the circuits in Prob. 16-11.
(*ans. a.* 300 μA, 12.24 mA; *b.* 3 mA, 12.24 mA; *c.* 91.7 μA, 6.77 mA)

16-16 Find the bandwidth for each of the circuits in Prob. 16-11.

16-17 A variable capacitor with a tuning range of 10 to 100 pF is to be used to tune over the band of frequencies 1.6 to 4 MHz. Find the value of trimming capacitor and the value of inductance required. (*ans.* 7.143 pF, 92.3 μH)

16-18 For Prob. 16-17, when the circuit is resonant at 4 MHz, if the power response to a frequency of 3.975 MHz is 50 percent of the response to a 4-MHz signal, what is the resistance of the inductor?

16-19 Find the bandwidth of a parallel resonant circuit consisting of a capacitor of 120 pF and an inductor of 180 μH with 18-Ω resistance. (*ans.* 15.92 kHz)

16-20 Find the bandwidth of the circuit in Prob. 16-19 if a resistance of 100 kΩ is shunted across the circuit.

16-21 What additional resistance placed in series with the inductor in Prob. 16-19 would give essentially the same results as Prob. 16-20? (*ans.* 15 Ω)

16-22 For Prob. 16-18, what shunt resistance would be required to increase the bandwidth to 70 kHz?

Section 16-7

16-23 A parallel resonant tank circuit in a radio transmitter contains an inductor of 350 μH and 15 Ω resistance and is turned to 500 kHz. When the antenna is coupled to the tank circuit, its loading effect increases the total circuit current and decreases the circuit Q by the same amount as connecting an additional 200-Ω resistance in series with the inductor. Find the resonant frequency of the circuit after the antenna is coupled. (*ans.* 490.35 kHz)

16-24 For Example 16-15 in the text, what value of capacitor would be required to correct the power factor to unity?

16-25 A 240-V 60-Hz motor with an output of 1 hp (746 W) has an efficiency of 85 percent and a power factor of 0.75 lagging. Find the value of the capacitor which will correct the power factor to unity. (*ans.* 35.6 μF)

16-26 If the power factor correction in Prob. 16-25 was made incorrectly by placing a capacitor in series so that unity power factor was obtained, find

a. The current through the motor.

b. The voltage across the motor.

17

AC Networks

All the techniques of dc network analysis apply to ac circuits. In employing these techniques, phasor methods must be used.

The principles of network analysis were developed in Chaps. 6 and 7. The complexity of phasor calculations makes it possible to lose sight of those principles when they are applied to ac circuits. Chapters 6 and 7 should be reviewed if there is any doubt.

The principles of network analysis having been established and phasor methods understood, the time is opportune to combine them in their application to ac networks.

Ac network analysis is demonstrated by circuit examples. Impedances and admittances are used as general element types. The resistance symbol -⁄\/\/- is used as a general symbol for elements or subnetworks, except when there is need for more specific symbols.

17-1 THE USE OF NETWORK ANALYSIS TECHNIQUES

Example 17-1 What is the magnitude of the voltage developed across the load Z_L in the circuit in Fig. 17-1? Use the current assumption technique (refer to Sec. 6-1).

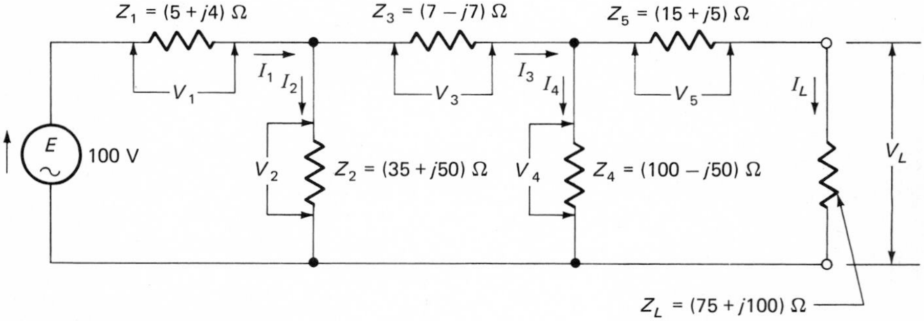

Fig. 17-1 Example 17-1: circuit.

Solution Assuming that the load current I_L is 1 A $\underline{/0^\circ}$:

$$V_L = I_L Z_L = 1\text{ A}\,\underline{/0^\circ} \times (75 + j100)\ \Omega = (75 + j100)\text{ V}$$

$$V_5 = I_L Z_5 = 1\text{ A}\,\underline{/0^\circ} \times (15 + j5)\ \Omega = (15 + j5)\text{ V}$$

$$V_4 = V_L + V_5 = (75 + j100)\text{ V} + (15 + j5)\text{ V} = (90 + j105)\text{ V}$$

$$I_4 = \frac{V_4}{Z_4} = \frac{(90 + j105)\text{ V}}{(100 - j50)\ \Omega} = \frac{138.3\text{ V}\,\underline{/49.4^\circ}}{111.8\ \Omega\,\underline{/-26.57^\circ}} = 1.237\text{ A}\,\underline{/75.97^\circ}$$

$$I_3 = I_4 + I_L = 1.237\ \text{A}\ \underline{/75.97^\circ} + 1\ \text{A}\ \underline{/0^\circ}$$

$$I_3 = (0.2999 + j1.2)\ \text{A} + (1 + j0)\ \text{A} = (1.299 + j1.2)\ \text{A}$$

$$V_3 = I_3 Z_3 = (1.299 + j1.2)\ \text{A} \times (7 - j7)\ \Omega$$

$$V_3 = 1.768\ \text{A}\ \underline{/42.73^\circ} \times 9.899\ \Omega\ \underline{/-45^\circ} = 17.5\ \text{V}\ \underline{/-2.27^\circ}$$

$$V_2 = V_3 + V_4 = 17.5\ \text{V}\ \underline{/-2.27^\circ} + (90 + j105)\ \text{V}$$

$$V_2 = (17.49 - j0.6932)\ \text{V} + (90 + j105)\ \text{V} = (107.5 + j104.3)\ \text{V}$$

$$I_2 = \frac{V_2}{Z_2} = \frac{(107.5 + j104.3)\ \text{V}}{(35 + j50)\ \Omega} = \frac{149.8\ \text{V}\ \underline{/44.13^\circ}}{61.03\ \Omega\ \underline{/55.01^\circ}} = 2.455\ \text{A}\ \underline{/-10.88^\circ}$$

$$I_1 = I_2 + I_3 = 2.455\ \text{A}\ \underline{/-10.88^\circ} + (1.299 + j1.2)\ \text{A}$$

$$I_1 = (2.411 - j0.4634)\ \text{A} + (1.299 + j1.2)\ \text{A} = (3.71 + j0.7366)\ \text{A}$$

$$V_1 = I_1 Z_1 = (3.71 + j0.7366)\ \text{A} \times (5 + j4)\ \Omega$$

$$V_1 = 3.782\ \text{A}\ \underline{/11.23^\circ} \times 6.403\ \Omega\ \underline{/38.66^\circ} = 24.22\ \text{V}\ \underline{/49.89^\circ}$$

$$E = V_1 + V_2 = 24.22\ \text{V}\ \underline{/48.89^\circ} + (107.5 + j104.3)\ \text{V}$$

$$E = (15.92 + j18.25)\ \text{V} + (107.5 + j104.3)\ \text{V} = (123.4 + j122.6)\ \text{V}$$

$$E = 173.9\ \text{V}\ \underline{/44.81^\circ}$$

Therefore for a load current of 1 A magnitude, the generator emf would have to be of magnitude 173.9 V. The actual emf magnitude is 100 V. Thus all the calculated voltage and current values are too great by a factor of 173.9/100 = 1.739.

Using correct values,

$$|I_L| = \frac{1\ \text{A}}{1.739} = 0.575\ \text{A}$$

$$Z_L = (75 + j100)\ \Omega$$

$$|Z_L| = 125\ \Omega$$

$$|V_L| = |I_L| \times |Z_L| = 0.575\ \text{A} \times 125\ \Omega = \mathbf{71.88\ V}$$

Example 17-2 What is the magnitude of the voltage V_L across the load in the circuit in Fig. 17-2 (same circuit as for Example 17-1). Use nodal analysis (refer to Sec. 6-2).

Fig. 17-2 Example 17-2: circuit.

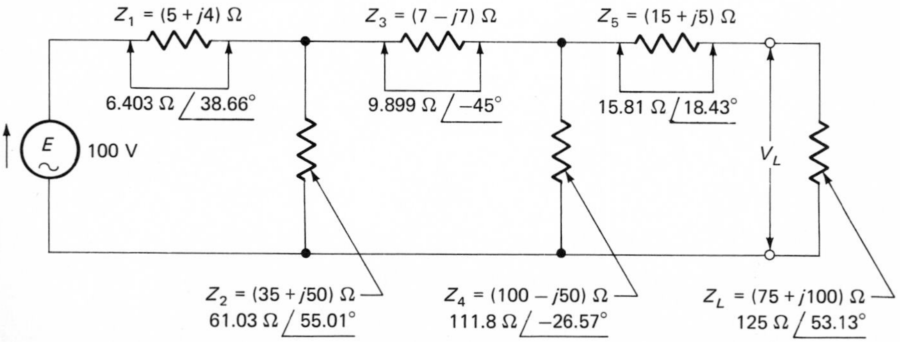

Solution Redrawing Fig. 17-2 with the voltage source converted to a current source and the impedance values converted to admittance values gives Fig. 17-3.

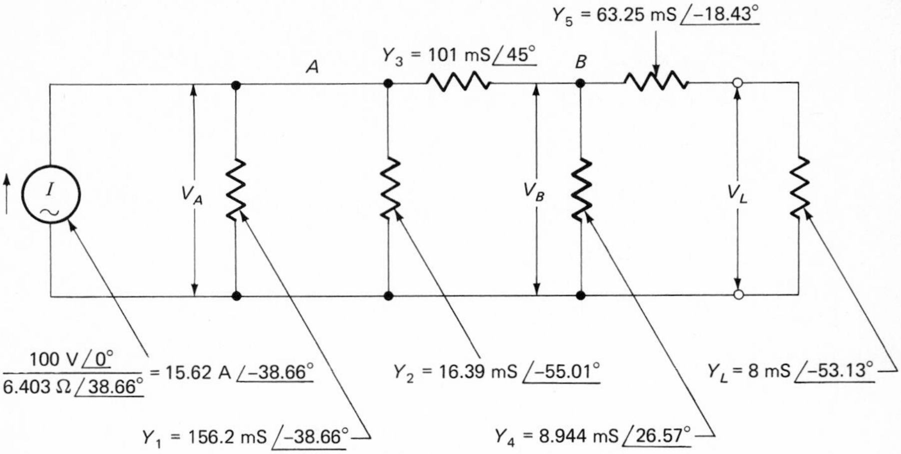

Fig. 17-3 Example 17-2: circuit with nodes and admittances identified.

In Fig. 17-3 two independent nodes A and B are shown. The reference node is selected to be the common connection of the two generators and the elements Y_1, Y_2, Y_4, and Y_L. The nodal voltages V_A and V_B are the pds between the independent nodes and the reference node.

AT NODE A

$$V_A(Y_1 + Y_2 + Y_3) - V_B Y_3 = I$$

$$V_A(156.2\,\underline{/-38.66^\circ} + 16.39\,\underline{/-55.01^\circ} + 101\,\underline{/45^\circ})\text{ mS} - V_B(101\text{ mS}\,\underline{/45^\circ}) = 15.62\text{ A}\,\underline{/-38.66^\circ}$$

$$V_A[(122 - j97.58) + (9.399 - j13.43) + (71.42 + j71.42)]\text{mS} - V_B(101\text{ mS}\,\underline{/45^\circ}) = 15.62\text{ A}\,\underline{/-38.66^\circ}$$

$$V_A(202.8 - j39.59)\text{mS} - V_B(101\text{ mS}\,\underline{/45^\circ}) = 15.62\text{ A}\,\underline{/-38.66^\circ}$$

$$V_A(206.6\text{ mS}\,\underline{/-11.05^\circ}) - V_B(101\text{ mS}\,\underline{/45^\circ}) = 15.62\text{ A}\,\underline{/-38.66^\circ} \quad (A)$$

AT NODE B The series combination of admittances Y_5 and Y_L is given by

$$Y_{5L} = \frac{1}{Z_5 + Z_L} = \frac{1}{(15 + j5)\,\Omega + (75 + j100)\,\Omega} = \frac{1}{(90 + j105)\,\Omega}$$

$$Y_{5L} = \frac{1}{138.3\ \Omega\,\underline{/49.4^\circ}} = 7.231\text{ mS}\,\underline{/-49.4^\circ}$$

$$V_B(Y_3 + Y_4 + Y_{5L}) - V_A Y_3 = 0$$

$$V_B(101\,\underline{/45^\circ} + 8.944\,\underline{/26.57^\circ} + 7.231\,\underline{/-49.4^\circ})\text{ mS} - V_A(101\text{ mS}\,\underline{/45^\circ}) = 0$$

$$V_B[(71.42 + j71.42) + (7.999 + j4) + (4.706 - j5.49)]\text{mS} - V_A(101\text{ mS}\,\underline{/45^\circ}) = 0$$

$$V_B(84.13 + j69.93)\text{ mS} - V_A(101\text{ mS}\,\underline{/45^\circ}) = 0$$

$$V_B(109.4\text{ mS}\,\underline{/39.73^\circ}) - V_A(101\text{ mS}\,\underline{/45^\circ}) = 0 \quad (B)$$

From Eq. (B),

$$V_A = \frac{V_B(109.4 \text{ mS} \angle 39.73°)}{101 \text{ mS} \angle 45°} = V_B(1.083 \angle -5.27°)$$

Substituting for V_A in Eq. (A),

$$V_B(1.083 \angle -5.27°) \times (206.6 \text{ mS} \angle -11.05°) - V_B(101 \text{ mS} \angle 45°) = 15.62 \text{ A} \angle -38.66°$$

$$V_B(223.7 \text{ mS} \angle -16.32° - 101 \text{ mS} \angle 45°) = 15.62 \text{ A} \angle -38.66°$$

$$V_B[(214.7 - j62.86) \text{ mS} - (71.42 + j71.42) \text{ mS}] = 15.62 \text{ A} \angle -38.66°$$

$$V_B(143.3 - j134.3) \text{ mS} = 15.62 \text{ A} \angle -38.66°$$

$$V_B(196.4 \text{ mS} \angle -43.14°) = 15.62 \text{ A} \angle -38.66°$$

$$V_B = \frac{15.62 \text{ A} \angle -38.66°}{196.4 \text{ mS} \angle -43.14°} = 79.53 \text{ V} \angle 4.48°$$

V_B is the voltage across Y_5 and Y_L in series. Using impedance values from Fig. 17-2,

$$V_L = V_B \frac{Z_L}{Z_5 + Z_L} = 79.53 \text{ V} \angle 4.48° \frac{125\ \Omega \angle 53.13°}{(90 + j105)\ \Omega}$$

$$= 79.53 \text{ V} \angle 4.48° \frac{125\ \Omega \angle 53.13°}{138.3\ \Omega \angle 49.4°} = 71.88 \text{ V} \angle 8.21°$$

$$|V_L| = \mathbf{71.88\ V}$$

Example 17-3 What is the value of the inductive current I_L in the circuit in Fig. 17-4? Use loop (mesh) analysis (refer to Sec. 6-3).

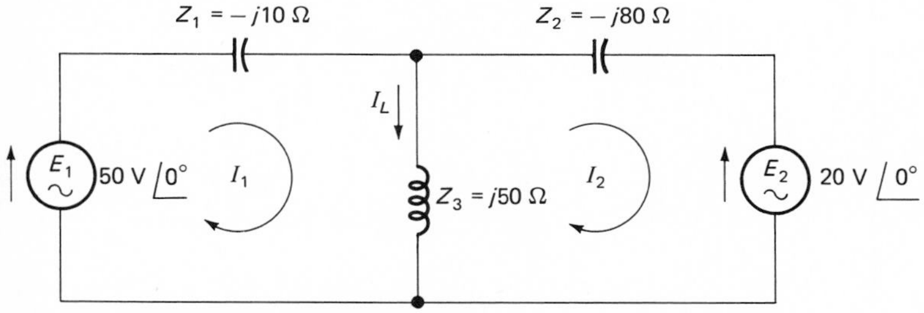

Fig. 17-4 Example 17-3: circuit.

Solution The loop equations are

$$(Z_1 + Z_3)I_1 - Z_3 I_2 = E_1 \qquad (A)$$

$$-Z_3 I_1 + (Z_2 + Z_3)I_2 = -E_2 \qquad (B)$$

Substituting values into Eqs. (A) and (B),

$$(-j10\ \Omega + j50\ \Omega)I_1 - j50\ \Omega\ I_2 = 50 \text{ V} \angle 0°$$

$$j40\ \Omega\ I_1 - j50\ \Omega\ I_2 = 50 \text{ V} \angle 0° \qquad (A)$$

$$-j50\ \Omega\ I_1 + (-j80\ \Omega + j50\ \Omega)I_2 = -20 \text{ V} \angle 0°$$

$$-j50\ \Omega I_1 - j30\ \Omega\ I_2 = -20 \text{ V} \angle 0° \qquad (B)$$

Put Eqs. (A) and (B) into matrix form for solution by determinants (see appendix):

$$\begin{bmatrix} 40\,\underline{/90^\circ} & 50\,\underline{/-90^\circ} \\ 50\,\underline{/-90^\circ} & 30\,\underline{/-90^\circ} \end{bmatrix} \begin{bmatrix} I_1 \\ I_2 \end{bmatrix} = \begin{bmatrix} 50\,\underline{/0^\circ} \\ -20\,\underline{/0^\circ} \end{bmatrix}$$

The solutions are

$$I_1 = 0.6757 \text{ A } \underline{/-90^\circ} \qquad \text{and} \qquad I_2 = 0.4595 \text{ A } \underline{/90^\circ}$$

The current I_L is the difference between I_1 and I_2:

$$I_L = I_1 - I_2 = -j0.6757 \text{ A} - j0.4595 \text{ A} = -j1.135 \text{ A}$$

Thus the current I_L with its phase relative to the phase of the two voltage sources is

$$I_L = \mathbf{1.135\ A\ \underline{/-90^\circ}} \qquad \text{or} \qquad \mathbf{\underline{/-\pi/2\ rad}}$$

In Example 17-3 the magnitude of the voltage across the inductor ($I_L X_L = 56.75$ V) is greater than the emf of either source. This is another reminder that voltages (or currents) in circuits containing *both* inductance and capacitance can be greater than the source values.

Example 17-4 What is the value of the current I_L in the circuit in Fig. 17-5 (a repeat of Fig. 17-4). Apply the superposition principle (refer to Sec. 6-4).

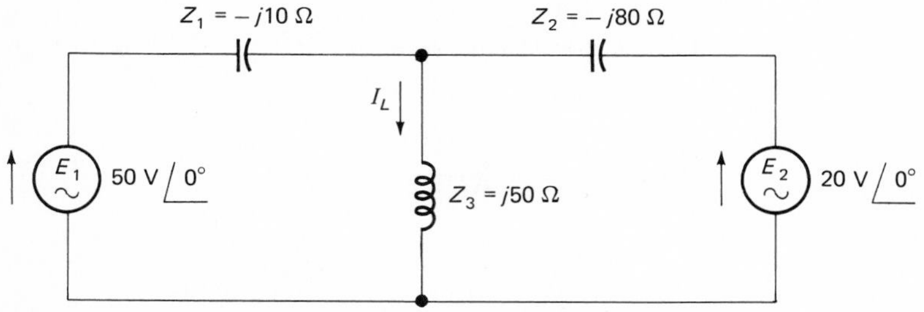

Fig. 17-5 Example 17-4: given circuit.

Solution In applying the superposition principle the circuit in Fig. 17-5 is reduced by the elimination of sources, so each voltage source is alternately the only source in the circuit. Figure 17-6a and b shows the resulting reduced circuits and the currents I_{La} and I_{Lb} which are components of the current I_L due to each of the two voltage sources.

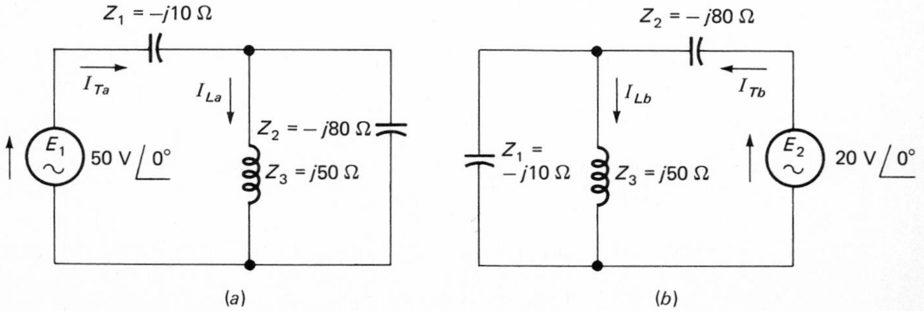

Fig. 17-6 Example 17-4: (a) circuit with E_2 removed; (b) circuit with E_1 removed.

From Fig. 17-6a,

$$Z_{Ta} = Z_1 + \frac{Z_2 Z_3}{Z_2 + Z_3} = -j10\ \Omega + \frac{(-j80\ \Omega)(j50\ \Omega)}{(-j80\ \Omega) + (j50\ \Omega)}$$

$$Z_{Ta} = (-j10 + j133.3)\ \Omega = j123.3\ \Omega = 123.3\ \Omega\ \underline{/90^\circ}$$

$$I_{Ta} = \frac{E_1}{Z_{Ta}} = \frac{50\ \text{V}\ \underline{/0^\circ}}{123.3\ \Omega\ \underline{/90^\circ}} = 0.4055\ \text{A}\ \underline{/-90^\circ}$$

$$I_{La} = I_{Ta}\frac{Z_2}{Z_2 + Z_3} = 0.4055\ \text{A}\ \underline{/-90^\circ}\ \frac{80\ \Omega\ \underline{/-90^\circ}}{30\ \Omega\ \underline{/-90^\circ}} = 1.081\ \text{A}\ \underline{/-90^\circ}$$

From Fig. 17-6b,

$$Z_{Tb} = Z_2 + \frac{Z_1 Z_3}{Z_1 + Z_3} = -j80\ \Omega + \frac{(-j10\ \Omega)(j50\ \Omega)}{(-j10 + j50)\ \Omega}$$

$$Z_{Tb} = (-j80 - j12.5)\ \Omega = -j92.5\ \Omega = 92.5\ \Omega\ \underline{/-90^\circ}$$

$$I_{Tb} = \frac{E_2}{Z_{Tb}} = \frac{20\ \text{V}\ \underline{/0^\circ}}{92.5\ \Omega\ \underline{/-90^\circ}} = 0.2162\ \text{A}\ \underline{/90^\circ}$$

$$I_{Lb} = I_{Tb}\frac{Z_1}{Z_1 + Z_3} = 0.2162\ \text{A}\ \underline{/90^\circ}\ \frac{10\ \Omega\ \underline{/-90^\circ}}{40\ \Omega\ \underline{/90^\circ}} = 0.0541\ \text{A}\ \underline{/-90^\circ}$$

As each voltage source tends to produce currents through Z_3 in the same direction, the total current I_L is the sum of the components I_{La} and I_{Lb}:

$$I_L = I_{La} + I_{Lb} = 1.081\ \text{A}\ \underline{/-90^\circ} + 0.0541\ \text{A}\ \underline{/-90^\circ} = -j(1.081 + 0.0541)\ \text{A}$$

$$I_L = \mathbf{1.135\ A}\ \underline{\mathbf{/-90^\circ}} \qquad \text{or} \qquad \underline{\mathbf{/-\pi/2\ rad}}$$

Example 17-5 What impedance values are necessary for the T network in Fig. 17-7b to be equivalent to the π network in Fig. 17-7a? Apply π-to-T conversion (refer to Sec. 7-1).

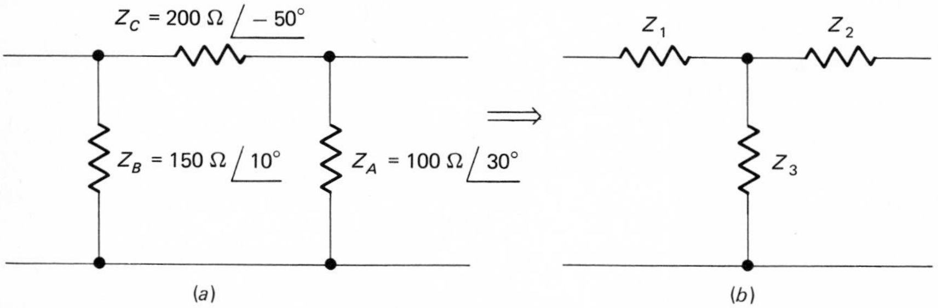

Fig. 17-7 Example 17-5: (a) given π network; (b) equivalent T network.

Solution

$$\Sigma Z_\nabla = Z_A + Z_B + Z_C = 100\ \Omega\ \underline{/30^\circ} + 150\ \Omega\ \underline{/10^\circ} + 200\ \Omega\ \underline{/-50^\circ}$$

$$= (86.6 + j50)\ \Omega + (147.7 + j26.04)\ \Omega + (128.6 - j153.2)\ \Omega$$

$$\Sigma Z_\nabla = (362.9 - j77.16)\ \Omega = 371.1\ \Omega\ \underline{/-12^\circ}$$

$$Z_1 = \frac{Z_B Z_C}{\Sigma Z_\nabla} = \frac{150\ \Omega\ \underline{/10^\circ} \times 200\ \Omega\ \underline{/-50^\circ}}{371.1\ \Omega\ \underline{/-12^\circ}} = \mathbf{80.84\ \Omega}\ \underline{\mathbf{/-28^\circ}} \qquad \text{or} \qquad \underline{\mathbf{/-0.4887\ rad}}$$

$$Z_2 = \frac{Z_C Z_A}{\Sigma Z_\nabla} = \frac{200\ \Omega\ \underline{/-50^\circ} \times 100\ \Omega\ \underline{/30^\circ}}{371.1\ \Omega\ \underline{/-12^\circ}} = \mathbf{53.89\ \Omega}\ \underline{\mathbf{/-8^\circ}} \qquad \text{or} \qquad \underline{\mathbf{/-0.1396\ rad}}$$

$$Z_3 = \frac{Z_A Z_B}{\Sigma Z_\nabla} = \frac{100\ \Omega\ \underline{/30^\circ} \times 150\ \Omega\ \underline{/10^\circ}}{371.1\ \Omega\ \underline{/-12^\circ}} = \mathbf{40.42\ \Omega}\ \underline{\mathbf{/52^\circ}} \qquad \text{or} \qquad \underline{\mathbf{/0.9075\ rad}}$$

Example 17-6 Obtain the original values of the π network in Example 17-5 from those calculated for the equivalent T network. Apply T-to-π conversion (refer to Sec. 7-1).

Solution Figure 17-8*a* and *b* is the same but with the values from Example 17-5 inserted.

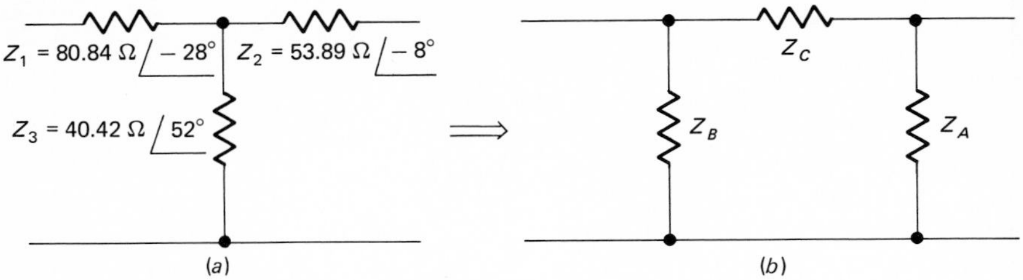

Fig. 17-8 Example 17-6: (*a*) given T network; (*b*) equivalent π network.

$$\Sigma Z_Y = Z_1Z_2 + Z_2Z_3 + Z_3Z_1$$

$$= 80.84\ \Omega\ \underline{/-28^\circ} \times 53.89\ \Omega\ \underline{/-8^\circ} + 53.89\ \Omega\ \underline{/-8^\circ} \times 40.42\ \Omega\ \underline{/52^\circ} + 40.42\ \Omega\ \underline{/52^\circ} \times 80.84\ \Omega\ \underline{/-28^\circ}$$

$$= 4356\ \Omega^2\ \underline{/-36^\circ} + 2178\ \Omega^2\ \underline{/44^\circ} + 3267\ \Omega^2\ \underline{/24^\circ}$$

$$= (3524 - j2560)\ \Omega^2 + (1567 + j1513)\ \Omega^2 + (2984 + j1329)\ \Omega^2$$

$$\Sigma Z_Y = (8075 + j282)\ \Omega^2 = 8080\ \Omega^2\ \underline{/2^\circ}$$

$$Z_A = \frac{\Sigma Z_Y}{Z_1} = \frac{8080\ \Omega^2\ \underline{/2^\circ}}{80.84\ \Omega\ \underline{/-28^\circ}} = \mathbf{99.95\ \Omega\ \underline{/30^\circ}} \quad \text{or} \quad \mathbf{\underline{/0.5236\ rad}}$$

$$Z_B = \frac{\Sigma Z_Y}{Z_2} = \frac{8080\ \Omega^2\ \underline{/2^\circ}}{53.89\ \Omega\ \underline{/-8^\circ}} = \mathbf{149.9\ \Omega\ \underline{/10^\circ}} \quad \text{or} \quad \mathbf{\underline{/0.1745\ rad}}$$

$$Z_C = \frac{\Sigma Z_Y}{Z_3} = \frac{8080\ \Omega^2\ \underline{/2^\circ}}{40.42\ \Omega\ \underline{/52^\circ}} = \mathbf{199.9\ \Omega\ \underline{/-50^\circ}} \quad \text{or} \quad \mathbf{\underline{/0.8726\ rad}}$$

The results for Example 17-6 all agree with the original values of Example 17-5 within 0.1 percent. That this accuracy is achieved after so many arithmetic operations is justification for trust in the principles of T $\leftrightarrow$ π conversions.

To bring into focus the results of Examples 17-5 and 17-6, Fig. 17-9 shows the equivalent π and T networks with individual element values for a frequency of 10 kHz ($\omega = 62.83$ krad/s).

Fig. 17-9 Networks of Examples 17-5 and 17-6 with individual element values for a frequency of 10 kHz.

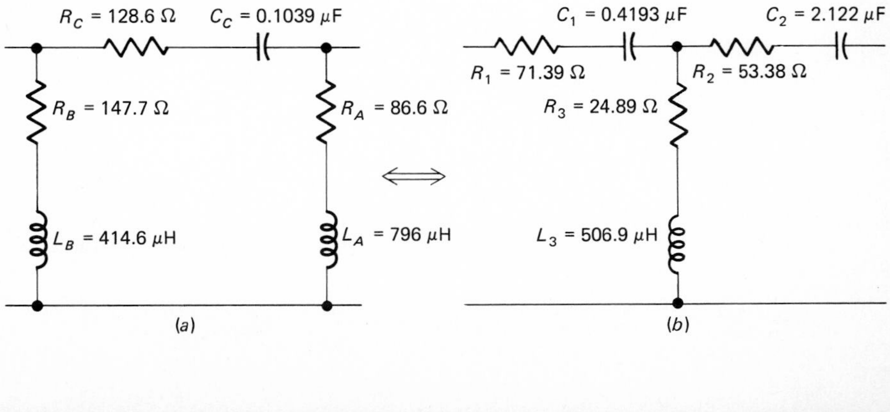

Example 17-7 The load resistance R_L in the circuit in Fig. 17-10 is variable over the range 500 Ω to 5 kΩ. What is the load voltage magnitude range? Employ the Thevenin principle (refer to Sec. 7-2).

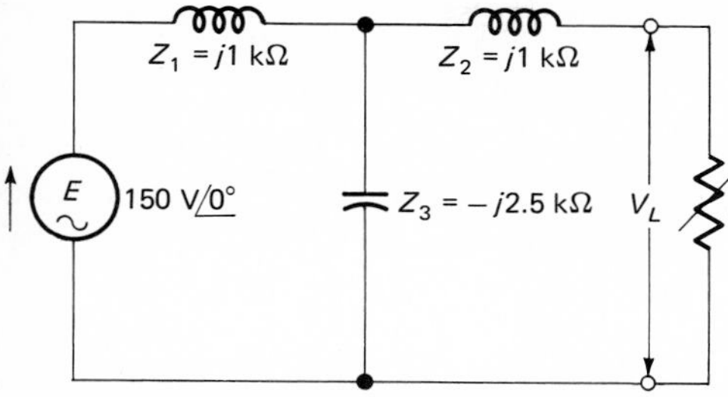

Fig. 17-10 Example 17-7: given circuit.

Solution The Thevenin equivalent of the circuit of Fig. 17-10 is given in Fig. 17-11.

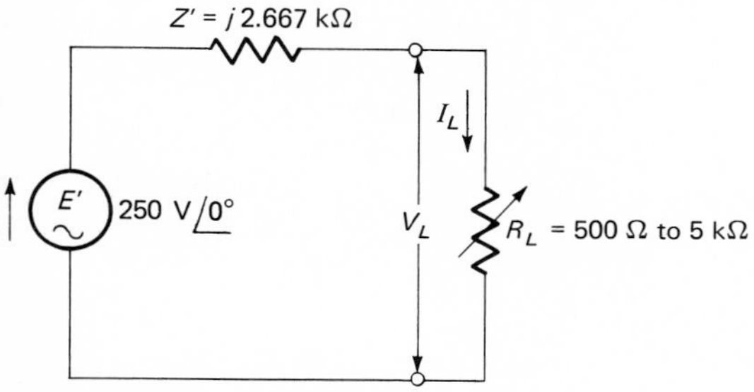

Fig. 17-11 Example 17-7: Thevenin equivalent circuit.

Obtain the Thevenin voltage E':

$$E' = E\frac{Z_3}{Z_1 + Z_3} = 150\text{ V}\,\underline{/0^\circ}\,\frac{2.5\text{ k}\Omega\,\underline{/-90^\circ}}{j1\text{ k}\Omega - j2.5\text{ k}\Omega} = 150\text{ V}\,\underline{/0^\circ}\,\frac{2.5\text{ k}\Omega\,\underline{/-90^\circ}}{1.5\text{ k}\Omega\,\underline{/-90^\circ}}$$

$$E' = 250\text{ V}\,\underline{/0^\circ}$$

Obtain the Thevenin impedance Z':

$$Z' = Z_2 + \frac{Z_1 Z_3}{Z_1 + Z_3} = j1\text{ k}\Omega + \frac{(j1\text{ k}\Omega)(-j2.5\text{ k}\Omega)}{j1\text{ k}\Omega - j2.5\text{ k}\Omega} = j1\text{ k}\Omega + \frac{2.5(\text{k}\Omega)^2}{-j1.5\text{ k}\Omega}$$

$$Z' = j1\text{ k}\Omega + j1.667\text{ k}\Omega = j2.667\text{ k}\Omega$$

The load current I_L is given by

$$I_L = \frac{E'}{R_L + Z'} = \frac{250\text{ V}\,\underline{/0^\circ}}{R_L + j2.667\text{ k}\Omega}$$

When $R_L = 500\ \Omega$ (0.5 kΩ),

$$R_L + j2.667\text{ k}\Omega = (0.5 + j2.667)\text{ k}\Omega = 2.713\text{ k}\Omega\,\underline{/79.38^\circ}$$

$$I_L = \frac{250\text{ V}\,\underline{/0^\circ}}{2.713\text{ k}\Omega\,\underline{/79.38^\circ}} = 92.15\text{ mA}\,\underline{/-79.38^\circ}$$

$$|V_L| = |I_L|\,R_L = 92.15\text{ mA} \times 0.5\text{ k}\Omega = \mathbf{46.07\ V}$$

When $R_L = 5$ kΩ,

$$R_L + j2.667\text{ k}\Omega = (5 + j2.667)\text{ k}\Omega = 5.667\text{ k}\Omega\,\underline{/28.08^\circ}$$

$$I_L = \frac{250 \text{ V} \angle 0^\circ}{5.667 \text{ k}\Omega \angle 28.08^\circ} = 44.12 \text{ mA} \angle -28.08^\circ$$

$$|V_L| = |I_L|\, R_L = 44.12 \text{ mA} \times 5 \text{ k}\Omega = \mathbf{220.6\ V}$$

The load voltage magnitude range is **46.07** to **220.6 V.**

Again in Example 17-7 we see the voltage magnification effect in an *LC* circuit. The magnitude of the Thevenin voltage is 250 V. When the load resistance is 5 kΩ, the load voltage magnitude is 220.6 V. Both of these voltages are higher than the 150-V applied emf.

Example 17-8 What load impedance would result in maximum load power in the circuit in Fig. 17-12*a*? Refer to Sec. 7-3 for the conditions for maximum power transfer in a circuit containing steady direct current.

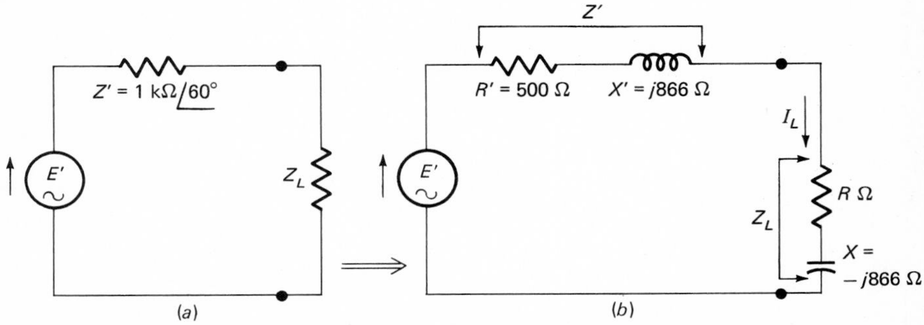

Fig. 17-12 Example 17-8: (*a*) given circuit; (*b*) circuit showing the components of the impedances.

Solution Figure 17-12*a* is the Thevenin equivalent of some unspecified circuit. Figure 17-12*b* is the same as Fig. 17-12*a*, but with the Thevenin impedance shown as a series equivalent *RL* network $[Z' = 1 \text{ k}\Omega \angle 60^\circ = (500 + j866)\ \Omega]$.

The circuit current I_L, and hence the load power, for any value of load resistance will be maximized if the circuit is arranged to be in series resonance. This can be achieved by including in the load impedance a capacitive reactance equal in magnitude to the inductive reactance component of the Thevenin impedance. Thus Fig. 17-12*b* shows the load impedance as $(R - j866)\ \Omega$.

After the circuit is made entirely resistive by resonance, the dc maximum power transfer conditions will apply: The load resistance should equal the Thevenin resistance. Therefore the load impedance for maximum power transfer will be:

$$Z_L = R - jX = \mathbf{(500 - j866)\ \Omega = 1\ k\Omega \angle -60^\circ} \quad \text{or} \quad \mathbf{\angle -1.047\ rad}$$

The conditions for maximum power transfer determined for Example 17-8 are general for ac circuits:

1. The load impedance should contain a reactance equal in magnitude and opposite in type to the reactive component of the Thevenin equivalent impedance of the remainder of the circuit.
2. The resistive components of the load and Thevenin impedances should be equal.

These conditions can be summarized by the statement: Maximum power is transferred from an ac network to a load if the load impedance is the conjugate of the Thevenin equivalent impedance of the network.

The above statement is written in equation form as:

$$\text{When } Z' = R' + jX',\ Z_L = R_L - jX_L \qquad |R'| = |R_L|$$

$$\text{When } Z' = R' - jX',\ Z_L = R_L + jX_L \qquad |X'| = |X_L|$$

where R_L and jX_L are the resistive and reactive components of the load impedance, and R' and X' are the resistive and reactive components of the Thevenin equivalent impedance of the network feeding the load.

Example 17-9 Find the range of load current values in the circuit in Fig. 17-13*a*. The load resistance is variable over the range 500 Ω to 5 kΩ (conductance range 2 to 0.2 mS). Apply the Norton principle (refer to Sec. 7-4).

Solution Figure 17-13*b* is the Norton equivalent of Fig. 17-13*a*.

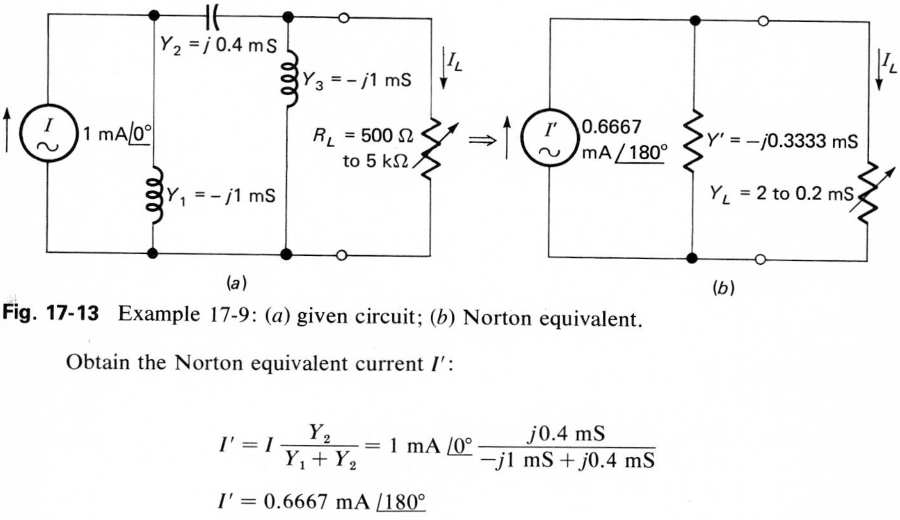

Fig. 17-13 Example 17-9: (*a*) given circuit; (*b*) Norton equivalent.

Obtain the Norton equivalent current I':

$$I' = I\frac{Y_2}{Y_1 + Y_2} = 1\text{ mA }\underline{/0^\circ}\,\frac{j0.4\text{ mS}}{-j1\text{ mS} + j0.4\text{ mS}}$$

$$I' = 0.6667\text{ mA }\underline{/180^\circ}$$

Obtain the Norton equivalent admittance Y':

$$Y' = Y_3 + \frac{Y_1Y_2}{Y_1 + Y_2} = -j1\text{ mS} + \frac{(-j1\text{ mS})(j0.4\text{ mS})}{-j1\text{ mS} + j0.4\text{ mS}}$$

$$Y' = -j1\text{ mS} + j0.6667\text{ mS} = -j0.3333\text{ mS}$$

$$I_L = I'\frac{Y_L}{Y_L + Y'} = \frac{0.6667\text{ mA }\underline{/180^\circ} \times Y_L}{(Y_L - j0.3333)\text{ mS}}$$

When $Y_L = 2$ mS ($R_L = 500\ \Omega$),

$$I_L = \frac{0.6667\text{ mA }\underline{/180^\circ} \times 2\text{ mS }\underline{/0^\circ}}{(2 - j0.3333)\text{ mS}} = \frac{0.6667\text{ mA }\underline{/180^\circ} \times 2\text{ mS }\underline{/0^\circ}}{2.028\text{ mS }\underline{/-9.46^\circ}}$$

$$I_L = 0.6575\text{ mA }\underline{/189.5^\circ} = 0.6575\text{ mA }\underline{/-170.5^\circ} \qquad \text{or} \qquad \underline{/-2.976\text{ rad}}$$

When $Y_L = 0.2$ mS ($R_L = 5$ kΩ),

$$I_L = \frac{0.6667 \text{ mA } \underline{/180^\circ} \times 0.2 \text{ mS } \underline{/0^\circ}}{(0.2 - j0.3333)\text{ mS}} = \frac{0.6667 \text{ mA } \underline{/180^\circ} \times 0.2 \text{ mS } \underline{/0^\circ}}{0.3887 \text{ mS } \underline{/-59.04^\circ}}$$

$$I_L = 0.343 \text{ mA } \underline{/239^\circ} = 0.343 \text{ mA } \underline{/-121^\circ} \quad \text{or} \quad \underline{/-2.112 \text{ rad}}$$

The range of load current values is **0.343 mA $\underline{/-121^\circ}$** to **0.6575 mA $\underline{/-170.5^\circ}$**, the phase angles being with respect to the source current.

Example 17-10 Obtain the Norton equivalent of the Thevenin circuit in Fig. 17-11 (refer to Sec. 7-5).
Solution Figure 17-14*a* is a repeat of Fig. 17-11. Figure 17-14*b* is the Norton equivalent of Fig. 17-14*a*.

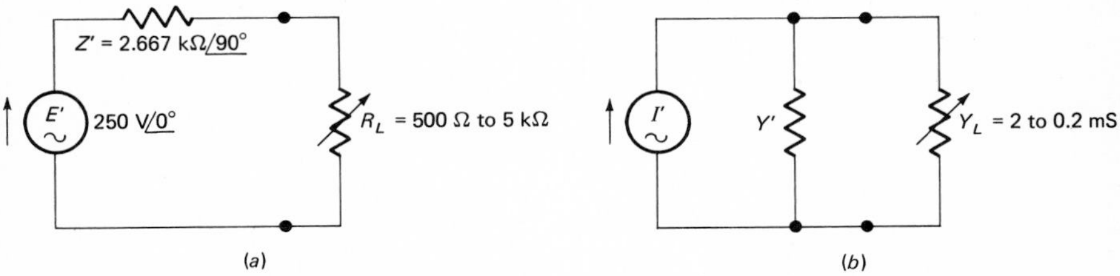

Fig. 17-14 Example 17-10: (*a*) given Thevenin circuit; (*b*) Norton equivalent.

$$I' = \frac{E'}{Z'} = \frac{250 \text{ V } \underline{/0^\circ}}{2.667 \text{ k}\Omega \underline{/90^\circ}} = \mathbf{93.74 \text{ mA } \underline{/-90^\circ}} \quad \text{or} \quad \mathbf{\underline{/-\pi/2 \text{ rad}}}$$

$$Y' = \frac{1}{Z'} = \frac{1}{2.667 \text{ k}\Omega \underline{/90^\circ}} = \mathbf{0.375 \text{ mS } \underline{/-90^\circ}} \quad \text{or} \quad \mathbf{\underline{/-\pi/2 \text{ rad}}}$$

Example 17-11 Obtain the Thevenin equivalent of the Norton circuit in Fig. 17-13*b* (refer to Sec. 7-5).
Solution Figure 17-15*a* is the same as Fig. 17-13*b*. Figure 17-15*b* is the Thevenin equivalent of Fig. 17-15*a*.

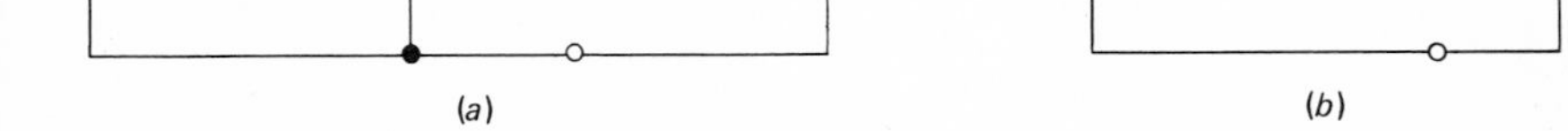

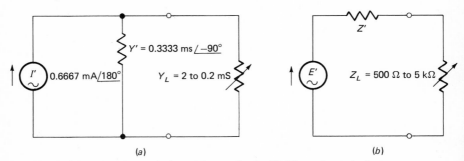

Fig. 17-15 Example 17-11: (*a*) given Norton circuit; (*b*) Thevenin equivalent.

$$E' = \frac{I'}{Y'} = \frac{0.6667 \text{ mA } \underline{/180^\circ}}{0.3333 \text{ mS } \underline{/-90^\circ}} = \mathbf{2 \text{ V } \underline{/270^\circ} = 2 \text{ V } \underline{/-90^\circ}} \quad \text{or} \quad \mathbf{\underline{/-\pi/2 \text{ rad}}}$$

$$Z' = \frac{1}{Y'} = \frac{1}{0.3333 \text{ mS } \underline{/-90^\circ}} = \mathbf{3 \text{ k}\Omega \underline{/90^\circ}} \quad \text{or} \quad \mathbf{\underline{/\pi/2 \text{ rad}}}$$

Example 17-12 What is the magnitude of the load voltage in the circuit in Fig. 17-16*a* (a repeat of the circuit in Fig. 17-1). Use the consecutive Thevenin approach (refer to Sec. 7-6).

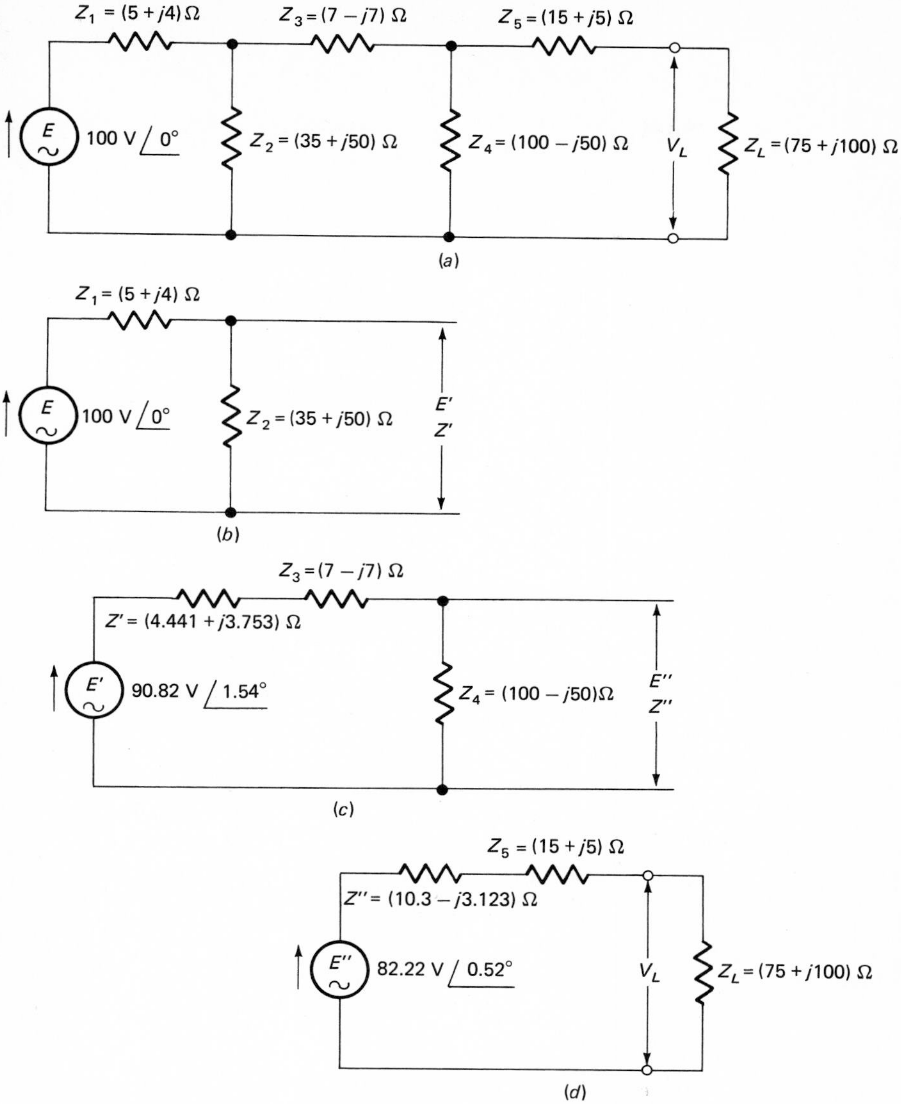

Fig. 17-16 Example 17-12: (*a*) given circuit; (*b*) to (*d*) steps in the solution.

Solution Refer to Fig. 17-16*b* to obtain the values of the first Thevenin emf E' and impedance Z':

$$E' = E\frac{Z_2}{Z_1 + Z_2} = 100\text{ V}\,\underline{/0^\circ}\,\frac{(35 + j50)\ \Omega}{(5 + j4)\ \Omega + (35 + j50)\ \Omega} = 100\text{ V}\,\underline{/0^\circ}\,\frac{(35 + j50)\ \Omega}{(40 + j54)\ \Omega}$$

$$E' = 100\text{ V}\,\underline{/0^\circ}\,\frac{61.03\ \Omega\,\underline{/55.01^\circ}}{67.2\ \Omega\,\underline{/53.47^\circ}} = 90.82\text{ V}\,\underline{/1.54^\circ}$$

$$Z' = \frac{Z_1 Z_2}{Z_1 + Z_2} = \frac{(5 + j4)\ \Omega(35 + j50)\ \Omega}{(5 + j4)\ \Omega + (35 + j50)\ \Omega} = \frac{6.403\ \Omega\,\underline{/38.66^\circ} \times 61.03\ \Omega\,\underline{/55.01^\circ}}{67.2\ \Omega\,\underline{/53.47^\circ}}$$

$$Z' = 5.815\ \Omega\,\underline{/40.2^\circ} = (4.441 + j3.753)\ \Omega$$

Refer to Fig. 17-16c to obtain the values of the second Thevenin emf E'' and impedance Z'':

$$E'' = E' \frac{Z_4}{Z' + Z_3 + Z_4}$$

$$= 90.82\ \text{V} \angle 1.54^\circ \frac{(100 - j50)\ \Omega}{(4.441 + j3.753)\ \Omega + (7 - j7)\ \Omega + (100 - j50)\ \Omega}$$

$$= 90.82\ \text{V} \angle 1.54^\circ \frac{111.8\ \Omega \angle -26.57^\circ}{(111.4 - j53.25)\ \Omega} = 90.82\ \text{V} \angle 1.54^\circ \frac{111.8\ \Omega \angle -26.57^\circ}{123.5\ \Omega \angle -25.55^\circ}$$

$$E'' = 82.22\ \text{V} \angle 0.52^\circ$$

$$Z'' = \frac{Z_4(Z' + Z_3)}{Z' + Z_3 + Z_4} = \frac{(100 - j50)\ \Omega[(4.441 + j3.753)\ \Omega + (7 - j7)\ \Omega]}{(4.441 + j3.753)\ \Omega + (7 - j7)\ \Omega + (100 - j50)\ \Omega}$$

$$= \frac{111.8\ \Omega \angle -26.57^\circ (11.44 - j3.247)\ \Omega}{(111.4 - j53.25)\ \Omega}$$

$$= \frac{111.8\ \Omega \angle -26.57^\circ \times 11.89\ \Omega \angle -15.85^\circ}{123.5\ \Omega \angle -25.55^\circ} = 10.76\ \Omega \angle -16.87^\circ$$

$$Z'' = (10.3 - j3.123)\ \Omega$$

The magnitude of the load voltage V_L can now be found from Fig. 17-16d:

$$V_L = E'' \frac{Z_L}{Z'' + Z_5 + Z_L} = 82.22\ \text{V} \angle 0.52^\circ \frac{(75 + j100)\ \Omega}{(10.3 - j3.123)\ \Omega + (15 + j5)\ \Omega + (75 + j100)\ \Omega}$$

$$V_L = 82.22\ \text{V} \angle 0.52^\circ \frac{125\ \Omega \angle 53.13^\circ}{143\ \Omega \angle 45.45^\circ} = 71.88\ \text{V} \angle 8.2^\circ$$

Therefore the magnitude of the load voltage V_L is **71.88 V.**

The same problem was posed in Examples 17-1, 17-2, and 17-12. By three completely different analysis techniques the same answer was obtained, thus validating the methods.

17-2 AC BRIDGE CIRCUITS

Measurement of some of the characteristics of a component or network under ac conditions may be carried out using an ac bridge. A number of ac bridge circuits appear in commercial measuring instruments. Each of these different circuits has its own field of usefulness, and all are ac adaptations of the dc Wheatstone bridge typified in Example 6-12.

The basic ac bridge circuit is shown in Fig. 17-17, page 657.

When the bridge arms Z_1 to Z_4 in the circuit in Fig. 17-17 have a definite relationship to each other, the bridge will be balanced. At balance there will be zero voltage across the detector. Thus the balance condition, a zero detector indication, can be used as a means for showing when a certain relationship exists between the arms. The impedance or admittance of three of the arms, at balance, can be used to determine the impedance or admittance of the remaining arm. A practical bridge will have three of its

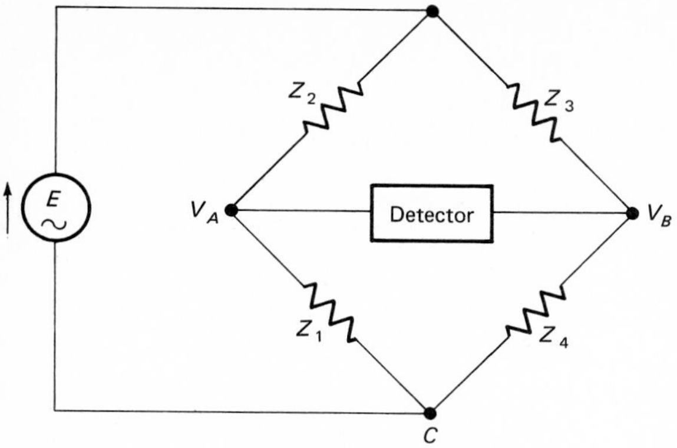

Fig. 17-17 Basic ac bridge circuit.

arms of known (fixed or variable) value, the fourth arm being the impedance or admittance to be measured and called the unknown.

Example 17-13 Derive an equation for determining the balance condition for the ac bridge circuit in Fig. 17-17.

Solution At balance there will be zero voltage across the detector. When this is so, the node voltages V_A and V_B, with respect to the reference node C, will be equal. There will be zero detector current, so that Z_1 and Z_2 will each carry the same current and thus be in series. Similarly, Z_3 and Z_4 will be in series. The voltage divider principle can be used.

$$V_A = V_B$$

$$E\frac{Z_1}{Z_1 + Z_2} = E\frac{Z_4}{Z_3 + Z_4}$$

$$Z_1Z_3 + Z_1Z_4 = Z_1Z_4 + Z_2Z_4$$

$$Z_1Z_3 = Z_2Z_4$$

The impedance relationship for ac bridge balance in the specific case of Fig. 17-17 can be couched in general terms applicable to any ac impedance bridge by the statement: A bridge is balanced when the products of the impedances of the opposite bridge arms are equal.

The statement for impedance bridge balance applies for any supply frequency, including zero frequency (steady dc).

Example 17-14 Express the resistance R_4 and inductance L in Fig. 17-18 in terms of R_1, R_2, R_3, and C when the bridge is balanced.

Solution From the conclusion for Example 17-13, the bridge circuit in Fig. 17-18 will be balanced if

$$Z_1Z_3 = Z_2Z_4$$

$$R_1R_3 = (R_2 - jX_C)(R_4 + jX_L)$$

$$R_1R_3 = (R_2R_4 + X_LX_C) + j(R_2X_L - R_4X_C)$$

This equation states that the impedance expression on the left (R_1R_3) equals the impedance expression on the right $[(R_2R_4 + X_LX_C) + j(R_2X_L - R_4X_C)]$. This equality will only be true if real (in-phase) components are equal and the imaginary (quadrature) components are also equal.

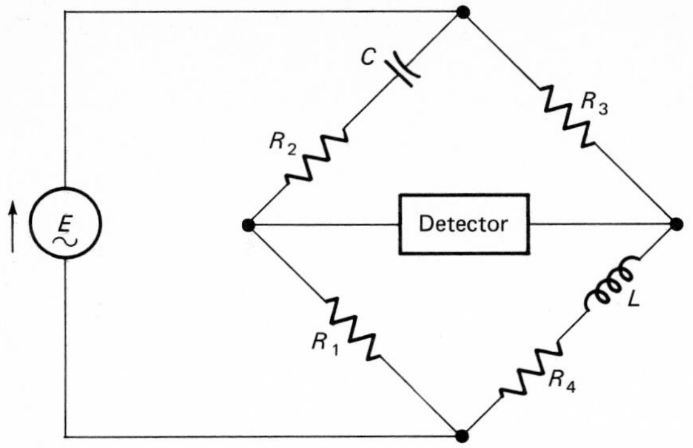

Fig. 17-18 Example 17-14: Hay bridge circuit.

Equating real components,

$$R_1R_3 = R_2R_4 + X_LX_C$$

$$R_4 = \frac{R_1R_3 - X_LX_C}{R_2} \quad \text{or} \quad X_L = \frac{R_1R_3 - R_2R_4}{X_C} \tag{A}$$

Equating imaginary components,

$$0 = j(R_2X_L - R_4X_C)$$

$$R_4X_C = R_2X_L$$

$$R_4 = \frac{R_2X_L}{X_C} \quad \text{or} \quad X_L = \frac{R_4X_C}{R_2} \tag{B}$$

Substituting for R_4 from Eq. (B) in Eq. (A),

$$\frac{R_2X_L}{X_C} = \frac{R_1R_3 - X_LX_C}{R_2}$$

$$R_2^2X_L = R_1R_3X_C - X_LX_C^2$$

$$X_L(R_2^2 + X_C^2) = R_1R_3X_C$$

$$X_L = R_1R_3\frac{X_C}{R_2^2 + X_C^2}$$

$$\omega L = R_1R_3\frac{1/\omega C}{R_2^2 + 1/(\omega C)^2} = R_1R_3\frac{\omega C}{(\omega CR_2)^2 + 1}$$

$$\mathbf{L = C\frac{R_1R_3}{(\omega CR_2)^2 + 1}}$$

Substituting for X_L from Eq. (B) in Eq. (A),

$$\frac{R_4X_C}{R_2} = \frac{R_1R_3 - R_2R_4}{X_C}$$

$$R_4X_C^2 = R_1R_2R_3 - R_2^2R_4$$

$$R_4(R_2^2 + X_c^2) = R_1R_2R_3$$

$$R_4 = R_1R_3\frac{R_2}{R_2^2 + X_C^2} = R_1R_3\frac{R_2}{R_2^2 + 1/(\omega C)^2} = R_1R_3\frac{(\omega C)^2R_2}{(\omega CR_2)^2 + 1}$$

$$\mathbf{R_4 = (\omega C)^2R_2\frac{R_1R_3}{(\omega CR_2)^2 + 1}}$$

The circuit in Fig. 17-18 is called a Hay bridge. It is one of several circuit configurations used in commercially produced instruments for measuring the characteristics of electric and electronic components. The Hay bridge is used for measuring the characteristics of inductors. Typically the values of R_1, R_2, R_3, and C in Fig. 17-18 are switched or varied to obtain a balance condition. Calibrated dials associated with these elements indicate the measured inductor values. The inductance value is usually indicated directly. As the Q value of an inductor gives a more graphic impression of its merit than does its resistance, it is usual for commercial instruments to read Q directly using a relationship derived as follows:

Using the symbols in Example 17-14 (Fig. 17-18),

$$Q = \frac{X_L}{R_4} = \omega C \frac{R_1 R_3}{(\omega C R_2)^2 + 1} \div (\omega C)^2 R_2 \frac{R_1 R_3}{(\omega C R_2)^2 + 1} = \frac{1}{\omega C R_2}$$

Example 17-15 What are the values of inductance and Q of the inductor in Fig. 17-19? The Hay bridge circuit is balanced with the values given for R_1, R_2, R_3, and C.

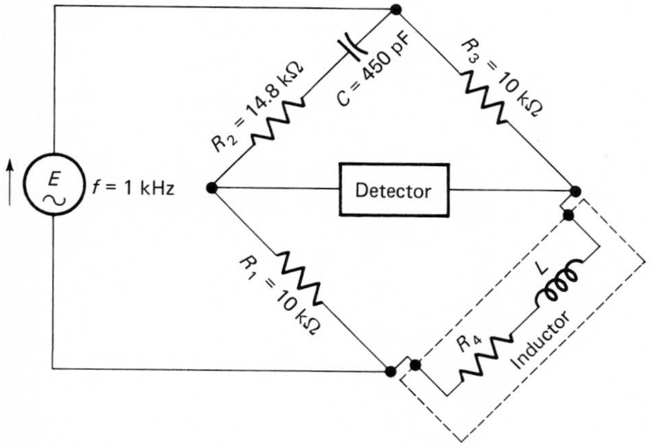

Fig. 17-19 Example 17-15: Hay bridge circuit.

Solution The term $\omega C R_2$ occurs in the equations for both L and Q.

$$\omega C R_2 = 2\pi \times 1 \text{ kHz} \times 450 \text{ pF} \times 14.8 \text{ k}\Omega = 4.185 \times 10^{-2}$$

$$L = C \frac{R_1 R_2}{(\omega C R_2)^2 + 1} = 450 \text{ pF} \frac{10 \text{ k}\Omega \times 10 \text{ k}\Omega}{(4.185 \times 10^{-2})^2 + 1} \simeq \mathbf{45 \text{ mH}}$$

$$Q = \frac{1}{\omega C R_2} = \frac{1}{4.185 \times 10^{-2}} = \mathbf{23.89}$$

17-3 POLYPHASE CIRCUITS

SINGLE-PHASE AND TWO-PHASE

Most of the ac circuits dealt with to this point in the text were single-phase. A *single-phase circuit* can be described as one in which there is either (1) one energy source, or (2) a number of sources all acting in phase with each other in supplying energy to the load system.

A single-phase circuit with one energy source is called *two-wire*, because there are just two conductors connecting the source to its load. A *three-wire*, single-phase circuit has in effect two energy sources connected in series and acting in phase across the load system.

Figure 17-20 illustrates the essential similarity of single-phase circuits.

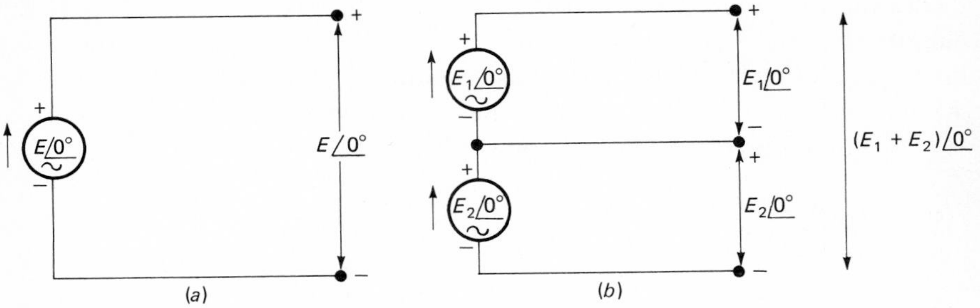

Fig. 17-20 (*a*) Single-phase, two-wire circuit; (*b*) single-phase, three-wire circuit.

In the two-wire circuit in Fig. 17-20*a* an instantaneous polarity is shown for the single generator, its emf being indicated as the reference phasor $E\,/0^\circ$.

The three-wire circuit in Fig. 17-20*b* also shows instantaneous polarities with each source producing in-phase emfs $E_1\,/0^\circ$ and $E_2\,/0^\circ$, both of reference phase. The two emfs, acting in series, result in a third emf $(E_1 + E_2)\,/0^\circ$, also of reference phase. This three-wire circuit forms the basis for domestic electric supplies, giving two 120-V sources and one of 240 V, all in phase with each other.

Compare the three-wire, single-phase circuit in Fig. 17-20*b* with the circuit in Examples 17-3 and 17-4. The latter circuit is redrawn in Fig. 17-21*a*.

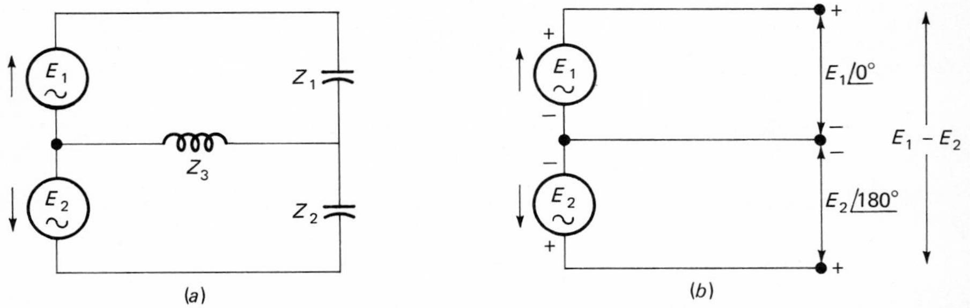

Fig. 17-21 (*a*) Circuit in Examples 17-3 and 17-4 redrawn; (*b*) the emfs of the circuit.

Figure 17-21*a* has two generators and three wires feeding the load system, as for a three-wire, single-phase circuit. But its two sources are connected in series opposition. The generators alone are shown in Fig. 17-21*b* for direct comparison with Fig. 17-20*b*. In assigning the reference phase to E_1 in Fig. 17-21*b*, the three circuit emfs are $E_1\,/0^\circ$, $E_2\,/180^\circ$, and

$(E_1 \underline{/0^\circ} + E_2 \underline{/180^\circ}) = E_1 - E_2$. If the magnitudes of E_1 and E_2 are equal, the third emf will be of zero magnitude.

The circuit in Fig. 17-21 is *two-phase*. A two-phase circuit with emfs in antiphase has little practical use, but a two-phase circuit with emfs in quadrature is found in some industrial control systems.

Single-phase and two-phase circuits are usually restricted to low-power applications such as electronic equipment and domestic installations.

For the generation, transmission, and distribution of electrical energy, and its industrial use, there are technical and economic advantages in using a three-phase electric system. As some of the details of these considerations involve equipment rather than circuits, they are not appropriate to this text. Three-phase circuits, however, raise some ac network issues which warrant our attention.

THREE-PHASE CONVENTIONS

In our treatment of three-phase circuits it will be assumed that two basic conventions of commercial electric systems apply:

1. The three phases originate within the circuit as three emfs of equal magnitude.
2. The three emfs have mutual phase differences of 120° or $2\pi/3$ rad.

The equal emf magnitudes imply a balanced three-phase energy source. The term $2\pi/3$ indicates that every cycle (2π rad) is effectively divided into three equal angle or time segments.

The three emfs required for operating three-phase (often symbolically abbreviated 3ϕ) have the waveform and phasor relationships given in Fig. 17-22.

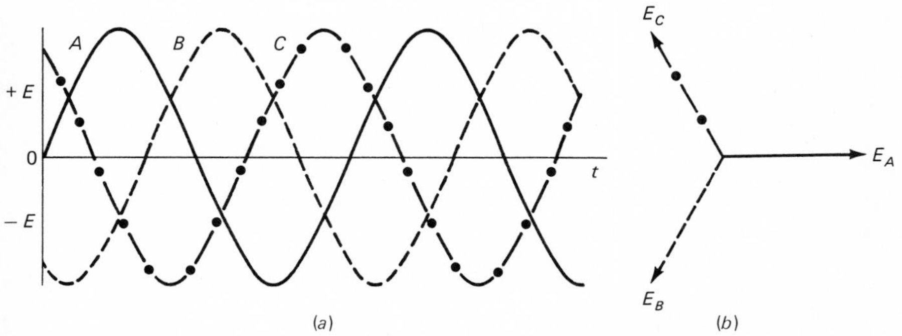

Fig. 17-22 Three-phase emfs. (*a*) Waveforms; (*b*) phasor diagram.

In Fig. 17-22*a* the waveforms of the emfs of phases *A*, *B*, and *C* are shown. The zero-going-positive points of the phase waveforms follow each other at intervals of time equivalent to 120° of angle. Thus the figure presents phase *B* lagging phase *A* by 120°, phase *C* lagging phase *B* by 120°, and phase *A* lagging phase *C* by 120°. The phasor diagram (Fig. 17-22*b*) shows the same mutual 120° lag between adjacent phases.

SOURCE CONNECTIONS IN 3ϕ CIRCUITS

The energy sources originate at an electric generating station. Typically, in a hydroelectric station energy is converted from the potential energy of elevated water to electrical energy by a turbine-generator combination. The actual output of the generator is unlikely to be fed to the electric system. Rather, it will be fed through a transformer. Furthermore, it is often desired to refer to circuits remote from the generating station, where again the effective energy source is likely to be a transformer. Despite the fact that it may be more realistic to show transformers as virtual energy sources, it simplifies the understanding of 3ϕ circuits if sources are represented by generator symbols.

In Fig. 17-23*a* the emfs of a three-phase supply are shown as originating from three generators connected in a *wye* or *star* configuration. The phase emfs E_A, E_B, and E_C are each connected between a *line* terminal *a*, *b*, or *c*, respectively, and a *neutral* terminal *n*. The line terminals connect to the 3ϕ load. The neutral terminal may or may not connect to the load; if it does, the 3ϕ source is called four-wire wye; if not, it is classed as three-wire wye. The neutral terminal is so called because (1) it represents an equipotential reference point for the three emfs, and (2) if the load is balanced (the only condition we shall deal with), zero current exists in a conductor connecting this terminal to the load.

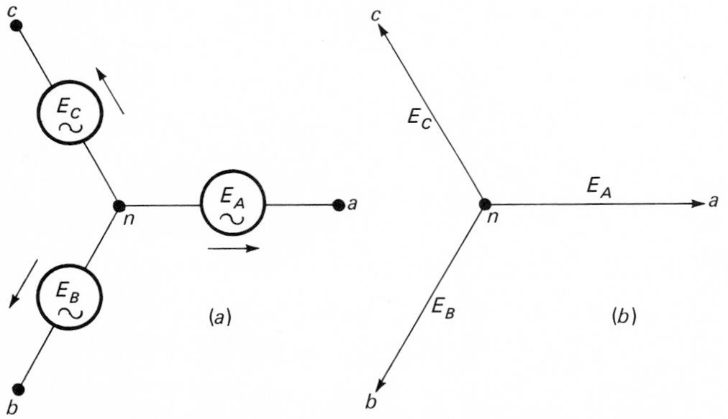

Fig. 17-23 Wye-connected three-phase supply source; (*a*) connections; (*b*) phasor diagram.

The emfs E_A, E_B, and E_C between each line terminal and the neutral are called phase voltages. Between each pair of line terminals exists a line voltage, E_{ab}, E_{bc}, or E_{ca}.

Taking the phase voltage E_A as having reference phase, and assigning the symbol E_p for the equal magnitudes of the phase voltages E_A, E_B, and E_C,

$$E_A = E_p \underline{/0^\circ} \qquad E_B = E_p \underline{/-120^\circ} \qquad E_C = E_p \underline{/-240^\circ}$$

The phasor lines in Fig. 17-23*b* are all directionally away from the neutral terminal. Thus, tracing from the end of one line to the end of another, the path goes through two opposing arrow directions, indicating that adjacent phase voltages are in phasor opposition. The line voltages, therefore, are the phasor differences between adjacent phase voltages.

$$E_{ab} = E_A - E_B = E_p \underline{/0^\circ} - E_p \underline{/-120^\circ}$$

$$= E_p - (-0.5 - j0.866)E_p = (1.5 + j0.866)E_p$$

$$E_{ab} = \sqrt{3}E_p \underline{/30^\circ}$$

$$E_{bc} = E_B - E_C = E_p \underline{/-120^\circ} - E_p \underline{/-240^\circ}$$

$$= (-0.5 - j0.866)E_p - (-0.5 + j0.866)E_p = -j1.732E_p$$

$$E_{bc} = \sqrt{3}E_p \underline{/-90^\circ}$$

$$E_{ca} = E_C - E_A = E_p \underline{/-240^\circ} - E_p \underline{/0^\circ}$$

$$= (-0.5 + j0.866)E_p - E_p = (-1.5 + j0.866)E_p$$

$$E_{ca} = \sqrt{3}E_p \underline{/-210^\circ}$$

All three line voltages are of magnitude equal to $\sqrt{3}$ times a phase voltage. Using the symbol E_l for the line voltage magnitude,

$$E_l = \sqrt{3}E_p$$

The three phase voltages E_A, E_B, and E_C are shown in the phasor diagram in Fig. 17-24*a* together with the three line voltages E_{ab}, E_{bc}, and E_{ca}. There is a mutual lagging phase angle of 120° between adjacent phase voltages and between adjacent line voltages, but a phase voltage lags by 30° the line voltage existing between that phase and the next adjacent phase; for example, E_A lags E_{ab} by 30°.

Figure 17-24*b* is the same as Fig. 17-24*a* except that it is labeled in terms of the common phase voltage E_p and the common line voltage E_l.

Fig. 17-24 Wye source phasor diagrams. (*a*) Terminal voltages; (*b*) line and phase voltages.

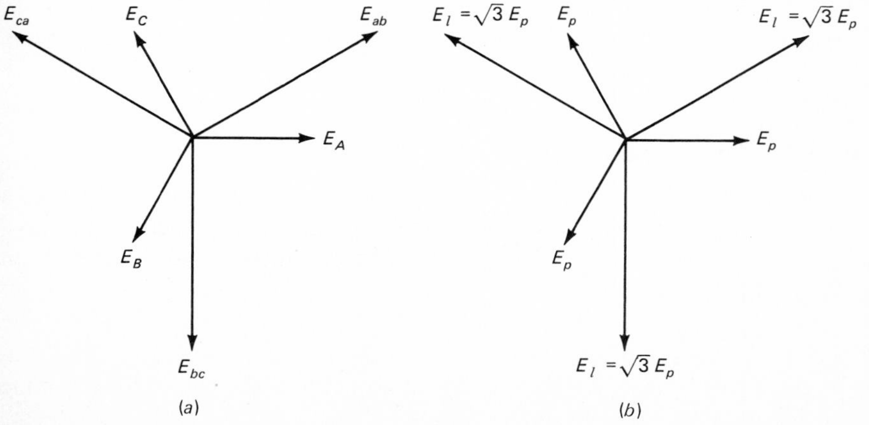

A *delta* three-phase source has the format of Fig. 17-25*a*. This type of source connection is less common than the wye. No neutral line is available, although it is possible to refer to a virtual or floating neutral. As the phasors in Fig. 17-25*b* and *c* have the same geometry as those of the line voltages of the wye connection in Fig. 17-24*a* and *b*, in the case of a balanced system the virtual neutral of a delta source will have the same relative potential as the true neutral of the wye source; the phase voltage from each line terminal to the virtual neutral will be $1/\sqrt{3}$ times that of the line voltage.

In a delta source the generator emfs are the line voltages. The potentials between each line terminal and the virtual neutral are the phase voltages. The relationship between the line and phase voltages is the same for the delta as for the wye connection:

$$E_l = \sqrt{3}E_p$$

The phasor diagram in Fig. 17-25*b* is shown in *funicular* form (each phasor line drawn from the end of the adjacent line), as this clearly relates to the formation of the delta circuit. To be consistent with all our other phasor diagrams, we will maintain the principle that phasor lines radiate from a common origin, as in Fig. 17-25*c*. This convention has the added advantages of presenting all 3ϕ diagrams as basically the same, and simplification where both currents and voltages are to be shown.

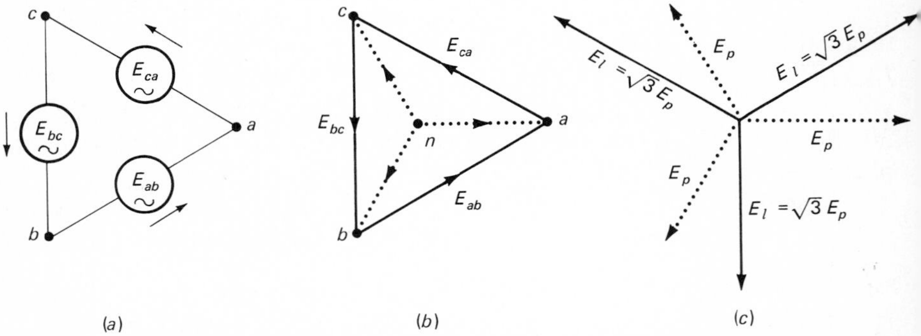

Fig. 17-25 Delta-connected three-phase supply source. (*a*) Connections; (*b*) terminal voltages; (*c*) line and phase voltages.

LOAD CONNECTIONS IN 3ϕ CIRCUITS

As for sources, loads may be connected in wye or delta configurations. In either case the load will consist of three elements, each of which can have any impedance value. This treatment will be confined to balanced loads in which the three elements are of the same impedance value.

A schematic diagram for a wye-connected load is given in Fig. 17-26*a*. The diagram is oriented in the same manner as the source diagrams in Figs. 17-23*a* and 17-25*a*; the symbols are also basically the same for direct comparison between a source and its load.

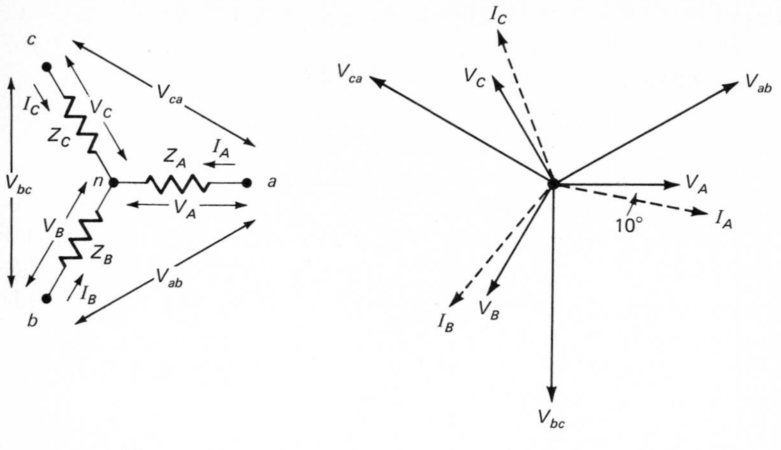

Fig. 17-26 Wye-connected three-phase load. (*a*) Schematic diagram; (*b*) phasor diagram for a 10° inductive load.

Figure 17-26*b* shows a phasor diagram for the wye load. The load is assumed to be fed from a balanced 3ϕ source, the line voltages V_{ab}, V_{bc}, and V_{ca} being of equal magnitude and the phase voltages V_A, V_B, and V_C of equal magnitude. The phasor lines for these voltages are the same as in Fig. 17-24*a* for the wye source. A common inductive phase angle of 10° was selected for the three impedances. The voltage across each impedance is a phase voltage, so the related current lags this voltage by 10°. Any other value of impedance phase angle, common or individual, will result in appropriately related current phase angles. The current in each element of a wye load is both phase current and line current.

The phasor diagrams in Fig. 17-27 refer to a wye load, the phasors being expressed in terms of phase and line voltages V_p and V_l, and phase

Fig. 17-27 Phasor diagrams of the line and phase voltages and currents for a wye load. (*a*) With a 10° inductive load; (*b*) with a 20° capacitive load.

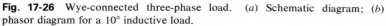

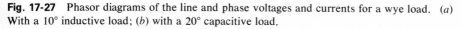

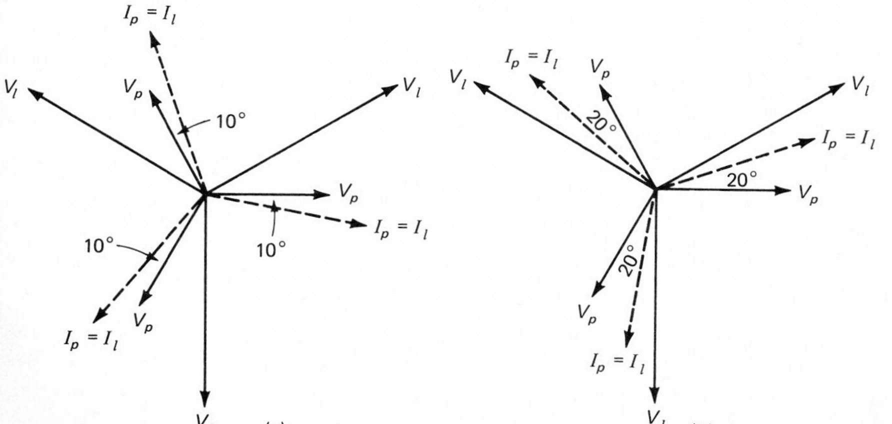

and line currents I_p and I_l. The common load impedances Z for Fig. 17-27*a* are inductive with a 10° phase angle, while those in Fig. 17-27*b* are capacitive with a 20° phase angle. In Fig. 17-27,

$$V_l = \sqrt{3}V_p \qquad \text{and} \qquad I_l = I_p = \frac{V_p}{Z}$$

Figure 17-28*a* gives a schematic diagram of a delta load. The voltages V_{ab}, V_{bc}, and V_{ca} are line voltages which, if the load is supplied from a balanced source, are equal in magnitude with a mutual 120° phase difference. Furthermore, if the load is balanced, each of the three impedances will be equal and their currents will be

$$I_a = \frac{V_{ab}}{Z_a} = \frac{V_l \angle 30^\circ}{Z}$$

$$I_b = \frac{V_{bc}}{Z_b} = \frac{V_l \angle -90^\circ}{Z}$$

$$I_c = \frac{V_{ca}}{Z_c} = \frac{V_l \angle -210^\circ}{Z}$$

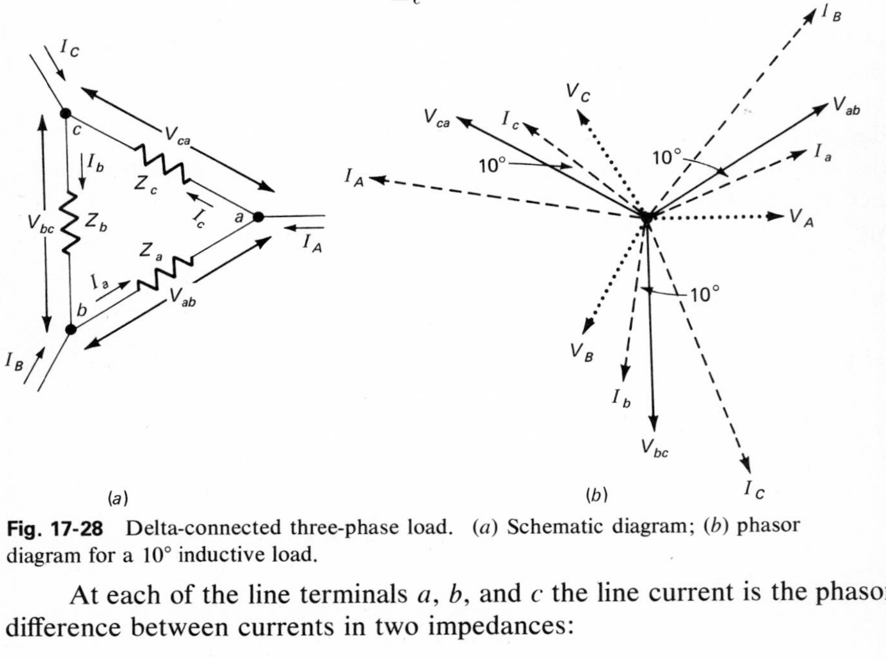

Fig. 17-28 Delta-connected three-phase load. (*a*) Schematic diagram; (*b*) phasor diagram for a 10° inductive load.

At each of the line terminals *a*, *b*, and *c* the line current is the phasor difference between currents in two impedances:

$$I_A = I_c - I_a = \frac{V_l \angle -210^\circ}{Z} - \frac{V_l \angle 30^\circ}{Z} = \sqrt{3}\,\frac{V_l \angle 180^\circ}{Z}$$

Similarly, $$I_B = \sqrt{3}\,\frac{V_l \angle 60^\circ}{Z} \qquad \text{and} \qquad I_C = \sqrt{3}\,\frac{V_l \angle -60^\circ}{Z}$$

To illustrate a typical case, in Fig. 17-28*b* the impedances are assumed to be inductive with a 10° phase angle, so their currents lag the line voltages by 10°:

$$I_a = \frac{V_l}{Z} \underline{/20^\circ} \qquad I_b = \frac{V_l}{Z} \underline{/-100^\circ} \qquad I_c = \frac{V_l}{Z} \underline{/-220^\circ}$$

$$I_A = \sqrt{3}\frac{V_l}{Z} \underline{/170^\circ} \qquad I_B = \sqrt{3}\frac{V_l}{Z} \underline{/50^\circ} \qquad I_C = \sqrt{3}\frac{V_l}{Z} \underline{/-70^\circ}$$

In Fig. 17-28*b* the phase voltages V_A, V_B, and V_C between the line terminals and the virtual neutral are shown as dotted lines.

Figure 17-29 gives the essentials of Fig. 17-28*b* with the line voltages shown as V_l, the line currents as I_l, and the currents in the individual impedances as I_p.

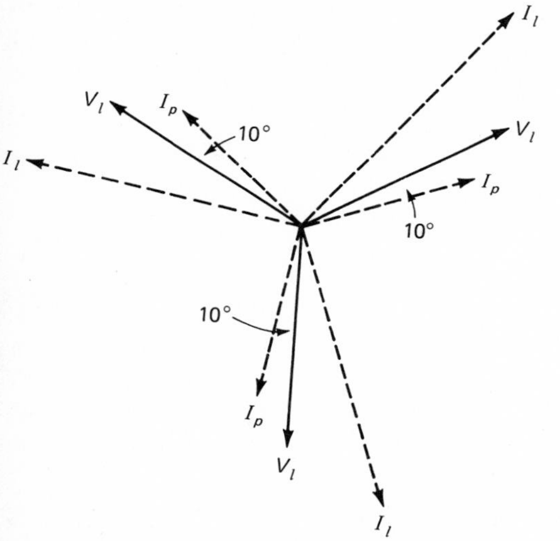

Fig. 17-29 Phasor diagram of the line and phase voltages and currents for a 10° inductive delta load.

SUMMARY OF THE MAIN 3ϕ DATA

The following summary of voltage and current magnitude data can be of help in understanding balanced 3ϕ networks.

1. Wye source: $E_l = \sqrt{3}E_p$

   $I_l = I_p$

2. Delta source: $E_l = \sqrt{3}E_p$

   $I_l = \sqrt{3}I_p$

3. Wye load: $V_l = \sqrt{3}V_p$

   $I_l = I_p = \dfrac{V_p}{Z}$

4. Delta load: $V_l = \sqrt{3}V_p$

   $I_l = \sqrt{3}I_p = \sqrt{3}\dfrac{V_l}{Z}$

In the above relationships the symbols represent phasor terms:

E_l = source line voltage, voltage between a pair of source terminals, V
E_p = source phase voltage, voltage between a line terminal and the neutral terminal or the virtual neutral, V
I_l = line current, current in a line conductor from a source or to a load, A
I_p = phase current, current in an element connected between a line terminal and neutral terminal or between two line terminals, A
V_l = load line voltage, V
V_p = load phase voltage, V

Example 17-16 A three-phase source consists of three 120-V transformer windings connected in wye formation. What are the phase and line voltages?
Solution Figure 17-30*a* is a schematic diagram of the wye 3ϕ source in which the transformer windings are represented as generators.

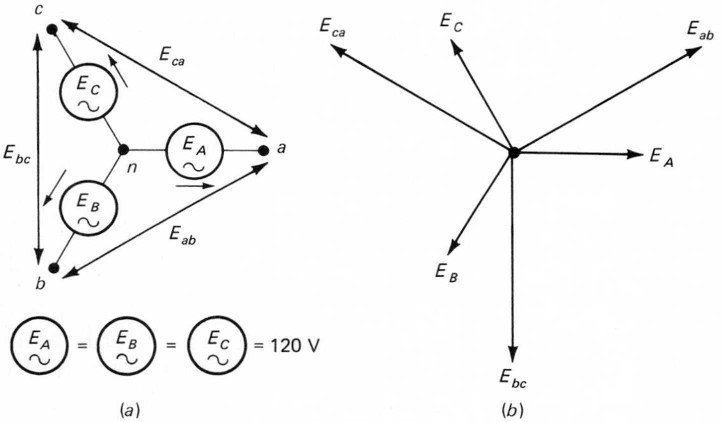

Fig. 17-30 Example 17-16: (*a*) source connections; (*b*) phasor diagram.

By definition, if E_A is made the reference phasor, the phase voltages are $E_A = \mathbf{120\ V\ \angle 0°}$, $E_B = \mathbf{120\ V\ \angle -120°}$, and $E_C = \mathbf{120\ V\ \angle -240°}$.

Each line voltage is the phasor difference between two phase voltages:

$$E_{ab} = E_A - E_B = 120\text{ V }\angle 0° - 120\text{ V }\angle -120° = \sqrt{3} \times 120\text{ V }\angle 30°$$

$$E_{ab} = \mathbf{207.8\ V\ \angle 30°} \qquad \text{or} \qquad \mathbf{\angle 0.5236\ rad}$$

$$E_{bc} = E_B - E_C = 120\text{ V }\angle -120° - 120\text{ V }\angle -240° = \sqrt{3} \times 120\text{ V }\angle -90°$$

$$E_{ab} = \mathbf{207.8\ V\ \angle -90°} \qquad \text{or} \qquad \mathbf{\angle -\pi/2\ rad}$$

$$E_{ca} = E_C - E_A = 120\text{ V }\angle -240° - 120\text{ V }\angle 0° = \sqrt{3} \times 120\text{ V }\angle -210°$$

$$E_{ca} = \mathbf{207.8\ V\ \angle -210°} \qquad \text{or} \qquad \mathbf{\angle -3.665\ rad}$$

The combination of phase and line voltages in Example 17-16 is commonly used in 3ϕ electric systems. In such systems it is usual to regard the line voltage magnitude as 208 V, rather than the true value of 207.8 V, resulting in the so-called 120/208-V system.

Example 17-17 Figure 17-31 shows a balanced delta load connected to a three-phase supply of 208-V line voltage. What are the magnitudes of the line currents if each load element has an apparent power of 1 kVA?

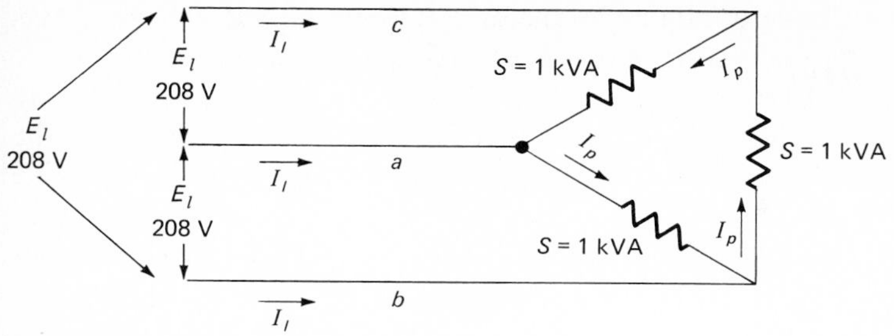

Fig. 17-31 Example 17-17: network.

Solution Each load element is directly connected across a 208-V line voltage. For each load element,

$$S = E_l I_p$$

$$I_p = \frac{S}{E_l} = \frac{1\ \text{kVA}}{208\ \text{V}} = 4.808\ \text{A}$$

The magnitude of each line current is the same in this balanced circuit:

$$I_l = \sqrt{3}\ I_p = \sqrt{3}\ \times 4.808\ \text{A} = \mathbf{8.327\ A}$$

For a single-phase 208-V circuit with the same 3-kVA total load, the line current would be 3 × 4.808 A = 14.42 A in each of two conductors. Thus if these two conductors were each of 1-Ω resistance, there would be individual line pds of 14.42 V and the source emf would have to be 236.8 V to deliver 208 V to the load. With the 3ϕ circuit in the example, there are three conductors, each carrying 8.327 A. A conductor resistance of 1 Ω means an individual line pd of 8.327 V. Two conductors connect each load element to the source, so the source emf need only be 224.7 V to develop 208 V at the load. The lower 3ϕ source voltage relative to the single-phase source voltage required to deliver the same voltage to the same total load implies a greater efficiency for the 3ϕ circuit. The greater efficiency can be viewed in several ways:

1. For a given conductor resistance, the load voltage regulation is better.
2. For a given load voltage regulation, line conductors can be of smaller gauge.
3. For a given conductor resistance and load, line losses are less.

From a circuit viewpoint, as distinct from an equipment viewpoint (for example, in electric motors), the advantages of three-phase operation are usually cited as greater efficiency and economy.

17-4 ELECTRONIC CIRCUITS

Dc network analysis can be applied to electronic circuits to determine the static or quiescent operating conditions. Typical cases embodying this approach are given in Examples 6-7, 6-13, 6-17, 7-10, and 7-18.

The dynamic operating conditions of electronic circuits can be determined by applying various ac network analysis methods.

This ability to treat separately the dc and ac characteristics of electronic circuits is a fortunate outcome of the superposition principle, the direct and alternating quantities being assumed to be due to separate energy sources.

The application of ac network analysis to electronic circuits is demonstrated by Example 17-18. Other examples are given in Chap. 18.

Example 17-18 Figure 17-32 shows the circuit of a transistor linear amplifier. Figure 17-33 gives an ac equivalent circuit of the amplifier. Figure 17-33 is shown as being in effect two separate circuits; Fig. 17-33*a* is the input circuit and Fig. 17-33*b* the output circuit. If the ac input voltage to the amplifier e_i is 10 mV rms, what is the output voltage v_o if the input frequency is (*a*) 100 Hz, (*b*) 1 kHz, (*c*) 10 kHz, and (*d*) any value, but the capacitor *C* is removed from the circuit?

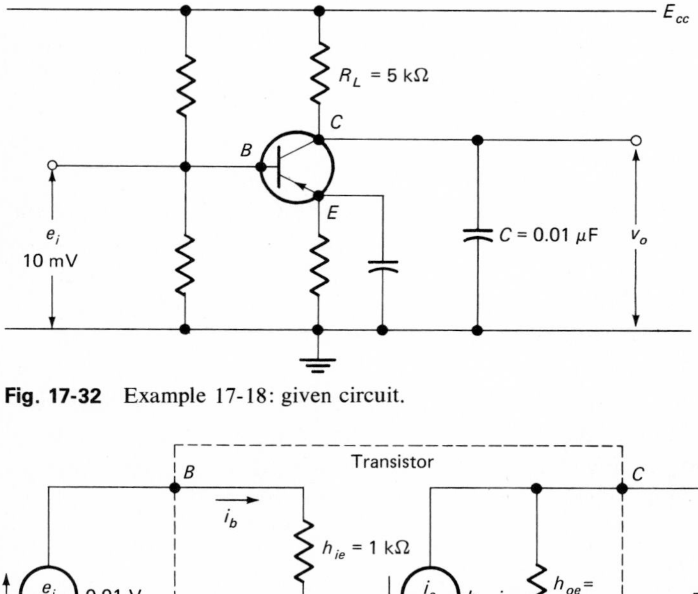

Fig. 17-32 Example 17-18: given circuit.

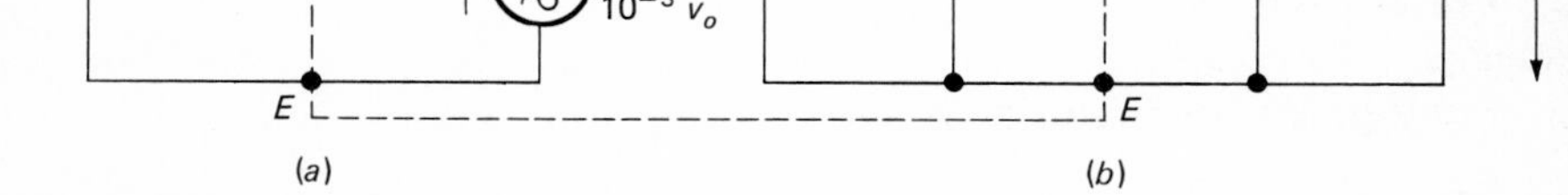

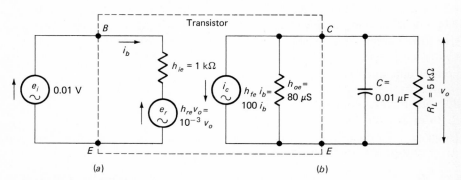

Fig. 17-33 Example 17-18: equivalent circuit.

The values of the transistor parameters are:

$$h_{ie} \text{ (input impedance)} = 1 \text{ k}\Omega$$

$$h_{re} \text{ (reverse voltage function)} = 10^{-3}$$

$$h_{fe} \text{ (forward current gain)} = 100$$

$$h_{oe} \text{ (output admittance)} = 80 \ \mu\text{S}$$

Solution For the input circuit (refer to Fig. 17-33*a*), $e_r = h_{re}v_o$ is a voltage fed back from the output circuit to the input circuit, and i_b is the input current.

The input loop equation is

$$h_{ie}i_b = e_i - e_r$$

$$h_{ie}i_b = e_i - h_{re}v_o$$

$$1\ \text{k}\Omega\, i_b = 0.01\ \text{V} - 10^{-3}v_o \qquad i_b \text{ in milliamperes; } v_o \text{ in volts} \qquad (A)$$

For the output circuit (refer to Fig. 17-33*b*), $i_c = h_{fe}i_b$ is a current fed forward from the input circuit to the output circuit.

With C (the collector) as an independent node and E (the emitter) as the reference node, the node voltage is v_o. The nodal equation is

$$(h_{oe} + G_L + jB_C)v_o = -i_c$$

$$\left(h_{oe} + \frac{1}{R_L} + jB_C\right)v_o = -h_{fe}i_b$$

$$(80\ \mu\text{S} + 200\ \mu\text{S} + jB_C)v_o = -100i_b$$

$$[(280 + jB_C)\ \mu\text{S}]v_o = -100i_b \qquad i_b \text{ in milliamperes; } v_o \text{ in volts} \qquad (B)$$

From Eq. (A),

$$i_b = \frac{0.01\ \text{V} - 10^{-3}v_o}{1\ \text{k}\Omega} = 0.01\ \text{mA} - 1\ \mu\text{S}\ v_o$$

Substituting for i_b in Eq. (B),

$$[(280 + jB_C)\ \mu\text{S}]v_o = -100(0.01\ \text{mA} - 1\ \mu\text{S}\ v_o) = -1\ \text{mA} + 100\ \mu\text{S}\ v_o$$

$$[(280 - 100 + jB_C)\ \mu\text{S}]v_o = -1\ \text{mA}$$

$$v_o = -\frac{1\ \text{mA}}{(180 + jB_C)\ \mu\text{S}}$$

a. When $f = 100$ Hz, $\omega = 628.3$ rad/s, and

$$B_C = \omega C = 628.3\ \text{rad/s} \times 0.01\ \mu\text{F} = 6.283\ \mu\text{S}$$

$$(180 + jB_C)\ \mu\text{S} = (180 + j6.283)\ \mu\text{S} \simeq 180\ \mu\text{S}\ \underline{/0^\circ}$$

$$v_o \simeq -\frac{1\ \text{mA}}{180\ \mu\text{S}\ \underline{/0^\circ}} \simeq \mathbf{-5.556\ V}\ \underline{\mathbf{/0^\circ}} \quad \text{or} \quad \underline{\mathbf{/0\ rad}}$$

b. When $f = 1$ kHz, $\omega = 6.283$ krad/s, and $B_C = 62.83\ \mu$S

$$(180 + jB_C)\ \mu\text{S} = (180 + j62.83)\ \mu\text{S} = 190.7\ \mu\text{S}\ \underline{/19.24^\circ}$$

$$v_o = -\frac{1\ \text{mA}}{190.7\ \mu\text{S}\ \underline{/19.24^\circ}} = \mathbf{-5.244\ V}\ \underline{\mathbf{/-19.24^\circ}} \quad \text{or} \quad \underline{\mathbf{/-0.3358\ rad}}$$

c. When $f = 10$ kHz, $\omega = 62.83$ krad/s, and $B_C = 628.3\ \mu$S

$$(180 + j628.3)\ \mu\text{S} = 653.6\ \mu\text{S}\ \underline{/74.02^\circ}$$

$$v_o = -\frac{1\ \text{mA}}{653.6\ \mu\text{S}\ \underline{/74.02^\circ}} = \mathbf{-1.53\ V}\ \underline{\mathbf{/-74.02^\circ}} \quad \text{or} \quad \underline{\mathbf{/-1.292\ rad}}$$

d. If C is removed from the circuit, $B_C = \omega C = 0$

$$(180 + jB_C)\ \mu\text{S} = 180\ \mu\text{S} \angle 0^\circ$$

$$v_o = -\frac{1\ \text{mA}}{180\ \mu\text{S} \angle 0^\circ} = \mathbf{-5.556\ V \angle 0^\circ} \quad \text{or} \quad \mathbf{\angle 0\ rad}$$

Example 17-18 shows the effect of shunt capacitance. At low frequencies (100 Hz in the example), the effect is negligible, so the output voltage with the capacitor in circuit is approximately the same as without the capacitor. As the frequency increases, the output voltage drops. At 1 kHz in the example the effect on the voltage is not great: $v_o = 94$ percent of the value without C. At 10 kHz v_o is only 28 percent of the value without C. When the frequency is low, the phase angle between the input and output voltages is small but becomes significant at higher frequencies.

In the example lowercase letters i, e, and v were used for the current and voltage symbols although they represented rms values. This use of lowercase symbols is common in electronic circuit work to distinguish between ac signal components (lowercase) and dc components (capital).

The values of the transistor parameters h_{ie}, h_{re}, h_{fe}, and h_{oe} were given in Example 17-18 with no attempt to explain how they might be obtained. Chapter 18 contains information on the definition and derivation of h parameters.

EXERCISE PROBLEMS

Section 17-1

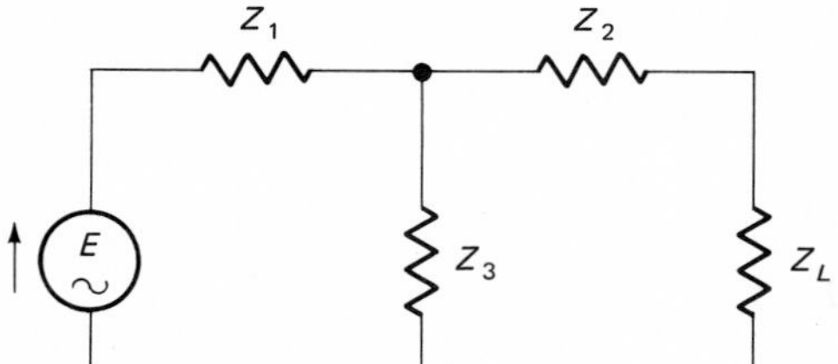

Fig. 17-1P Problems 17-1 to 17-4 and 17-14: circuit.

17-1 Find the power dissipated in Z_L in Fig. 17-1P, using current assumption, when

a. $E = 120$ V, $Z_1 = 15\ \Omega \angle{+25^\circ}$, $Z_2 = 10\ \Omega \angle{-30^\circ}$, $Z_3 = 20\ \Omega \angle{+40^\circ}$, $Z_L = 10\ \Omega \angle{+45^\circ}$ (*ans.* 58.58 W)

b. $E = 5$ V, $Z_1 = 350\ \Omega \angle{+50^\circ}$, $Z_2 = 400\ \Omega \angle{+50^\circ}$, $Z_3 = 450\ \Omega \angle{-60^\circ}$, $Z_L = 600\ \Omega \angle{+30^\circ}$ (*ans.* 7.36 mW)

17-2 Find the voltage across Z_L in Fig. 17-1P, using nodal analysis, when

a. $E = 50$ V, $Z_1 = 30\ \Omega \angle{+30^\circ}$, $Z_2 = 20\ \Omega \angle{-25^\circ}$, $Z_3 = 30\ \Omega \angle{+35^\circ}$, $Z_L = 40\ \Omega \angle{-40^\circ}$

b. $E = 30$ V, $Z_1 = 850\ \Omega \angle{-50^\circ}$, $Z_2 = 1000\ \Omega \angle{+40^\circ}$, $Z_3 = 650\ \Omega \angle{+30^\circ}$, $Z_L = 750\ \Omega \angle{-20^\circ}$

17-3 Find the current in Z_L in Fig. 17-1P, using loop analysis, when

a. $E = 1$ V, $Z_1 = 1500\ \Omega \angle{+45^\circ}$, $Z_2 = 1800\ \Omega \angle{-60^\circ}$, $Z_3 = 2000\ \Omega \angle{+55^\circ}$, $Z_L = 1200\ \Omega \angle{+60^\circ}$ (*ans.* 278 μA)

b. $E = 50$ V, $Z_1 = 10\ \Omega \angle{-75^\circ}$, $Z_2 = 6\ \Omega \angle{-65^\circ}$, $Z_3 = 5\ \Omega \angle{-45^\circ}$, $Z_L = 12\ \Omega \angle{-50^\circ}$ (*ans.* 0.81 A)

17-4 Find the current in Z_L in Fig. 17-1P, using Thevenin's theorem, when
  a. $E = 20$ V, $Z_1 = (8 - j8)\Omega$, $Z_2 = (11 + j15)\Omega$, $Z_3 = (15 + j12)\Omega$, $Z_L = (10 + j10)\Omega$
  b. $E = 100$ V, $Z_1 = (100 + j100)\Omega$, $Z_2 = (50 - j50)\Omega$, $Z_3 = (75 + j75)\Omega$, $Z_L = (25 - j25)\Omega$

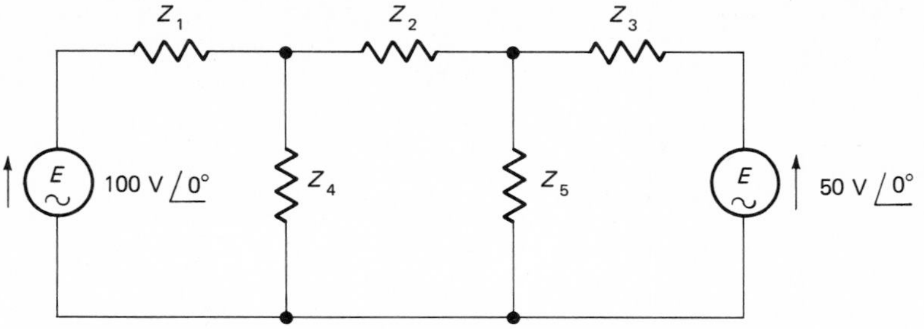

Fig. 17-2P Problems 17-5 to 17-8 and 17-15: circuit.

17-5 Find the voltage across Z_5 in Fig. 17-2P, using nodal analysis, when
  a. $Z_1 = 15\ \Omega\ \angle +18^\circ$, $Z_2 = 24\ \Omega\ \angle -24^\circ$, $Z_3 = 22\ \Omega\ \angle +30^\circ$, $Z_4 = 40\ \Omega\ \angle -30^\circ$, $Z_5 = 30\ \Omega\ \angle +40^\circ$ (*ans.* 43.5 V $\angle +7.6^\circ$)
  b. $Z_1 = 200\ \Omega\ \angle -15^\circ$, $Z_2 = 240\ \Omega\ \angle -30^\circ$, $Z_3 = 200\ \Omega\ \angle -15^\circ$, $Z_4 = 300\ \Omega\ \angle +20^\circ$, $Z_5 = 300\ \Omega\ \angle +20^\circ$ (*ans.* 39.5 V $\angle +16.8^\circ$)

17-6 Find the current in Z_5 in Fig. 17-2P, using loop analysis, when
  a. $Z_1 = (30 + j40)\ \Omega$, $Z_2 = (40 + j20)\ \Omega$, $Z_3 = (30 - j20)\ \Omega$, $Z_4 = (50 - j40)\ \Omega$, $Z_5 = (20 + j20)\ \Omega$
  b. $Z_1 = (90 - j150)\ \Omega$, $Z_2 = (200 + j100)\ \Omega$, $Z_3 = (60 + j50)\ \Omega$, $Z_4 = (60 - j100)\ \Omega$, $Z_5 = (90 + j75)\ \Omega$

17-7 Find the current in Z_5 in Fig. 17-2P, using the superposition theorem, when
  a. $Z_1 = 25\ \Omega\ \angle -30^\circ$, $Z_2 = 25\ \Omega\ \angle +45^\circ$, $Z_3 = 30\ \Omega\ \angle +30^\circ$, $Z_4 = 15\ \Omega\ \angle -30^\circ$, $Z_5 = 15\ \Omega\ \angle +60^\circ$ (*ans.* 1.592 A $\angle -130^\circ$)
  b. $Z_1 = 20\ \Omega\ \angle -45^\circ$, $Z_2 = 30\ \Omega\ \angle -45^\circ$, $Z_3 = 40\ \Omega\ \angle -45^\circ$, $Z_4 = 25\ \Omega\ \angle -45^\circ$, $Z_5 = 35\ \Omega\ \angle -45^\circ$ (*ans.* 0.954 A $\angle -45^\circ$)

17-8 Find the power dissipated by Z_5 in Fig. 17-2P, using Norton's theorem, when
  a. $Z_1 = (50 + j75)\ \Omega$, $Z_2 = (60 - j40)\ \Omega$, $Z_3 = (45 + j45)\ \Omega$, $Z_4 = (50 + j75)\ \Omega$, $Z_5 = (50 + j0)\ \Omega$
  b. $Z_1 = (50 + j75)\ \Omega$, $Z_2 = (60 - j40)\ \Omega$, $Z_3 = (45 + j45)\ \Omega$, $Z_4 = (50 + j75)\ \Omega$, $Z_5 = (50 + j100)\ \Omega$

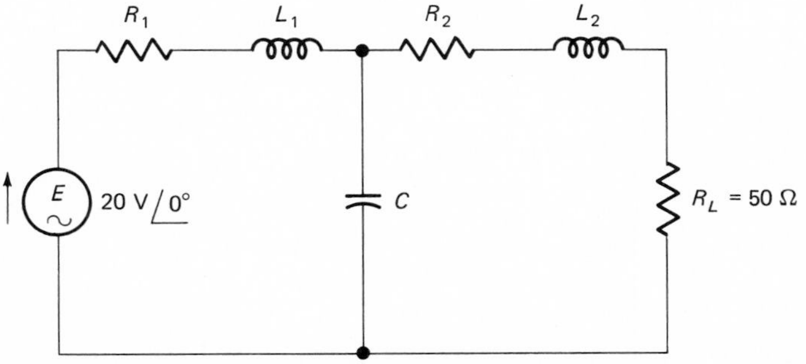

Fig. 17-3P Problems 17-9 and 17-10: circuit.

17-9 In the low-pass filter circuit in Fig. 17-3P, find the voltage across R_L when $R_1 = 0\ \Omega$, $R_2 = 0\ \Omega$, $L_1 = 100$ mH, $L_2 = 100$ mH, $C = 300\ \mu$F, and
  a. $f = 30$ Hz (*ans.* 55.84 V $\angle +100.7^\circ$)
  b. $f = 120$ Hz (*ans.* 0.719 V $\angle +125.3^\circ$)

17-10 In the low-pass filter circuit in Fig. 17-3P, find the voltage across R_L when $R_1 = 10\ \Omega$, $R_2 = 10\ \Omega$, $L_1 = 100$ mH, $L_2 = 100$ mH, $C = 300\ \mu$F, and
a. $f = 30$ Hz *b.* $f = 120$ Hz *c.* $f = 0$ Hz

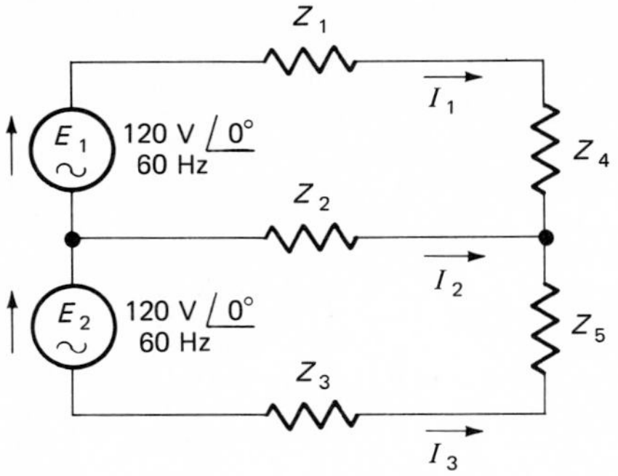

Fig. 17-4P Problems 17-11 and 17-12: circuit.

17-11 In the three-wire, single-phase circuit in Fig. 17-4P, find I_1, I_2, and I_3 when $Z_1 = Z_2 = Z_3 = 0\ \Omega$ and
a. $Z_4 = 8\ \Omega\ \underline{/+20^\circ}$, $Z_5 = 10\ \Omega\ \underline{/+35^\circ}$
(*ans.* 15 A $\underline{/-20^\circ}$, 4.61 A $\underline{/+22.3^\circ}$, 12 A $\underline{/-35^\circ}$)
b. $Z_4 = 12\ \Omega\ \underline{/-25^\circ}$, $Z_5 = 10\ \Omega\ \underline{/+30^\circ}$
(*ans.* 10 A $\underline{/+25^\circ}$, 10.31 A $\underline{/-82.6^\circ}$, 12 A $\underline{/-30^\circ}$)

17-12 In the three-wire, single-phase circuit in Fig. 17-4P, find I_1, I_2, and I_3 when $Z_1 = Z_2 = Z_3 = 2\ \Omega$ and
a. $Z_4 = 8\ \Omega\ \underline{/+20^\circ}$, $Z_5 = 10\ \Omega\ \underline{/+35^\circ}$
b. $Z_4 = 12\ \Omega\ \underline{/-25^\circ}$, $Z_5 = 10\ \Omega\ \underline{/+30^\circ}$

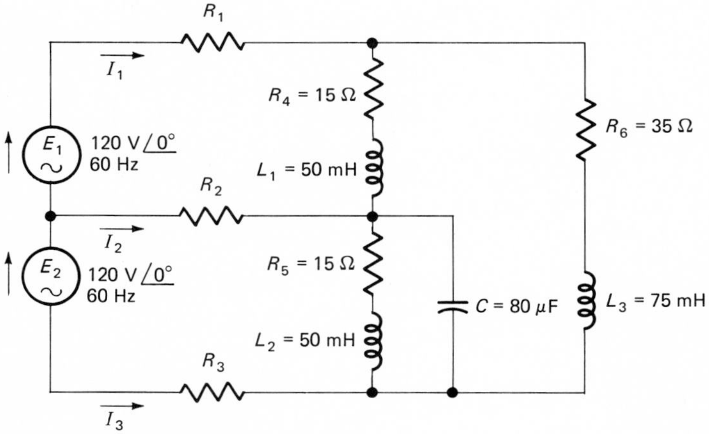

Fig. 17-5P Problem 17-13: circuit.

17-13 In the three-wire, single-phase circuit in Fig. 17-5P, find I_1, I_2, and I_3 when
a. $R_1 = R_2 = R_3 = 0\ \Omega$ (*ans.* 10.25 A $\underline{/-45^\circ}$, 3.618 A $\underline{/-90^\circ}$, 8.11 A $\underline{/-26.6^\circ}$)
b. $R_1 = R_2 = R_3 = 3\ \Omega$ (*ans.* 9.4 A $\underline{/-35.7^\circ}$, 3.11 A $\underline{/-67.2^\circ}$, 6.935 A $\underline{/-22.2^\circ}$)

17-14 For Fig. 17-1P, find the value of Z_L that will result in its power dissipation being maximum when
a. $Z_1 = 15\ \Omega\ \underline{/+25^\circ}$, $Z_2 = 10\ \Omega\ \underline{/-30^\circ}$, $Z_3 = 20\ \Omega\ \underline{/+40^\circ}$
b. $Z_1 = 350\ \Omega\ \underline{/+50^\circ}$, $Z_2 = 400\ \Omega\ \underline{/+50^\circ}$, $Z_3 = 450\ \Omega\ \underline{/-60^\circ}$

17-15 For Fig. 17-2P, find the value of Z_5 that will result in its power dissipation being maximum when $Z_1 = 15\ \Omega\ \underline{/+18^\circ}$, $Z_2 = 24\ \Omega\ \underline{/-24^\circ}$, $Z_3 = 22\ \Omega\ \underline{/+30^\circ}$, and $Z_4 = 40\ \Omega\ \underline{/-30^\circ}$.
(*ans.* 14.48 $\Omega\ \underline{/+13^\circ}$)

Section 17-2

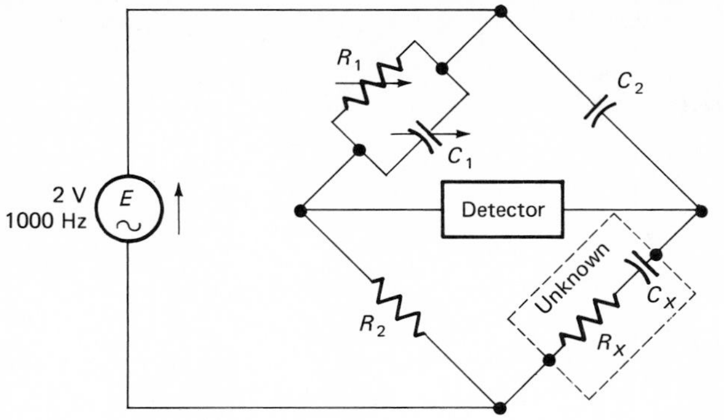

Fig. 17-6P Problem 17-16: circuit.

17-16 Derive the balance equations for the Schering bridge shown in Fig. 17-6P. For the unknown capacitor, find the values of R_x and C_x when $R_1 = 2300\ \Omega$, $R_2 = 1\ \text{k}\Omega$, $C_1 = 800$ pF, and $C_2 = 0.005\ \mu$F.

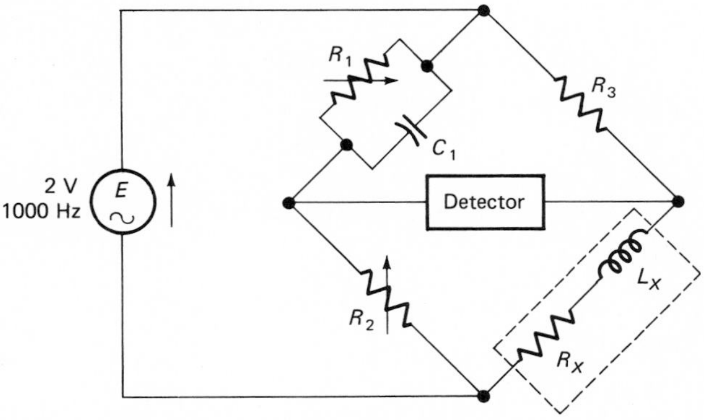

Fig. 17-7P Problem 17-17: circuit.

17-17 Derive the balance equations for the Maxwell bridge shown in Fig. 17-7P. Find the values of R_x, L_x, and Q when $R_1 = 8300\ \Omega$, $R_2 = 2420\ \Omega$, $R_3 = 100\ \Omega$, and $C_1 = 0.1\ \mu$F. (*ans.* $L_x = R_2R_3C_1$, $R_x = R_2R_3/R_1$, $Q = \omega C_1R_1$; $L_x = 24.2$ mH, $R_x = 29.15\ \Omega$, $Q = 5.215$)

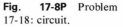

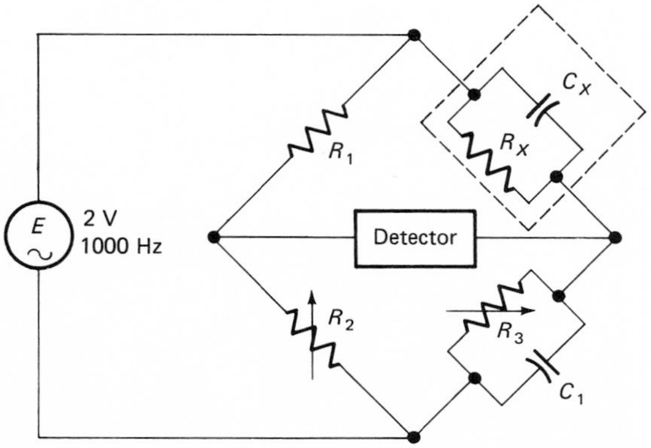

Fig. 17-8P Problem 17-18: circuit.

17-18 Derive the balance equations for the capacitance comparison bridge shown in Fig. 17-8P. Find the values of R_x and C_x when $R_1 = 10\ \text{k}\Omega$, $R_2 = 1800\ \Omega$, $R_3 = 5.5\ \text{k}\Omega$, and $C_1 = 0.1\ \mu$F.

18

Coupled Circuits

The preceding chapters introduced many circuits in which energy transfer takes place, for example, between a generator and a load. Energy is transferred between points in a circuit or between circuits when the points or circuits are coupled. Coupling can be described in terms of the response to a stimulus, as in the following definitions:

1. When a stimulus applied at one point in a circuit results in a response at another point in the circuit, the two points are said to be coupled and the circuit is a *coupled circuit*.
2. When a stimulus applied to one circuit results in a response in another circuit, the two circuits are said to be *coupled*.

The above two statements are extremely broad. To give more precise meaning, and hence value, to the term "coupled circuits," the two-port view is often adopted.

18-1 TWO-PORT CIRCUITS

A two-port circuit is shown in Fig. 18-1. Each of the two ports is a point of connection between the network and its external circuit. The generator is connected to terminals 1,1 of port 11. The load is connected to terminals 2,2 of port 22. Energy is transferred from the generator to the load through the network, or the generator is coupled to the load by the network.

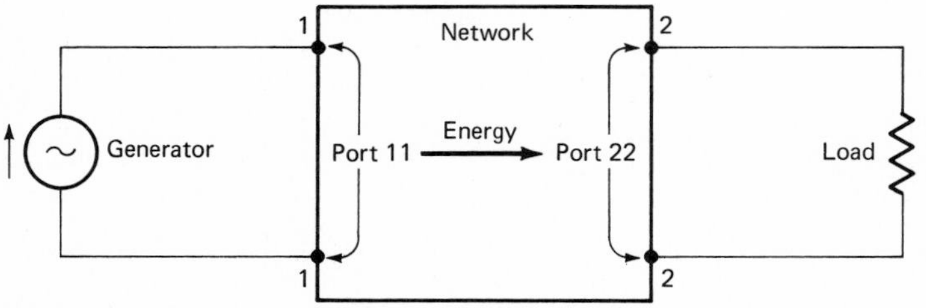

Fig. 18-1 Two-port circuit.

Coupling networks may be classified under two main headings, common impedance coupling and mutual inductance coupling.

COMMON IMPEDANCE COUPLING

Common impedance coupling, a generic term embracing common admittance coupling, occurs in two general forms.

1. When an element carries a current with components in two loops, or meshes, as in Fig. 18-2.

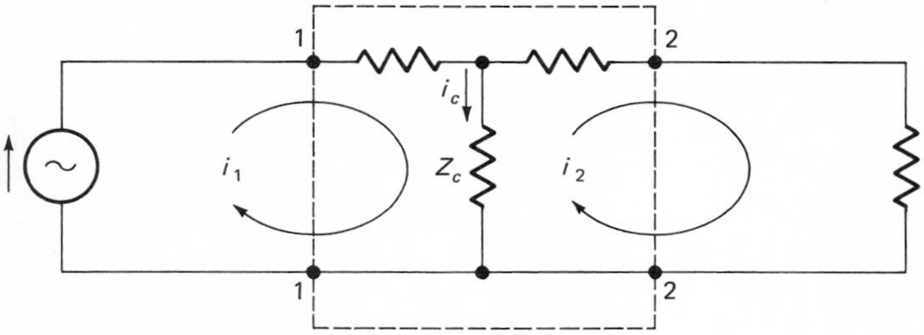

Fig. 18-2 Common impedance coupling.

In Fig. 18-2 the element Z_c has the loop currents i_1 and i_2 as components of its current i_c ($i_c = i_1 - i_2$). The value of Z_c influences the relationship between i_1 in the generator loop and i_2 in the load loop, hence influencing the amount of coupling between the generator and the load. Z_c is the coupling impedance.

2. When an element has a pd with two node voltages as its components.

The element Y_c in Fig. 18-3 has a pd v_c with components v_1 and v_2 ($v_c = c_1 - v_2$). The value of Y_c influences the relationship between v_1, the generator voltage, and v_2, the load voltage, hence influencing the coupling between the generator and the load. Y_c is the coupling admittance.

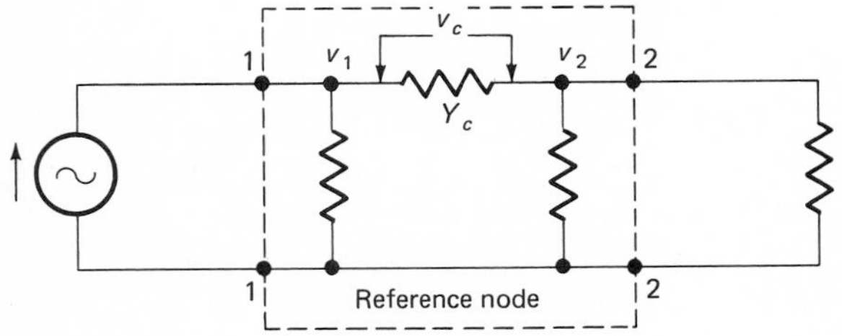

Fig. 18-3 Common admittance coupling.

It should be observed that lowercase letters i and v are used above. This use of lowercase symbols to identify ac or varying quantities is commonly adopted in practice and will be continued throughout this chapter. When specific values are intended, such as peak or rms, they will be identified.

Typical examples of common impedance coupling are given in Fig. 18-4. The coupling elements are

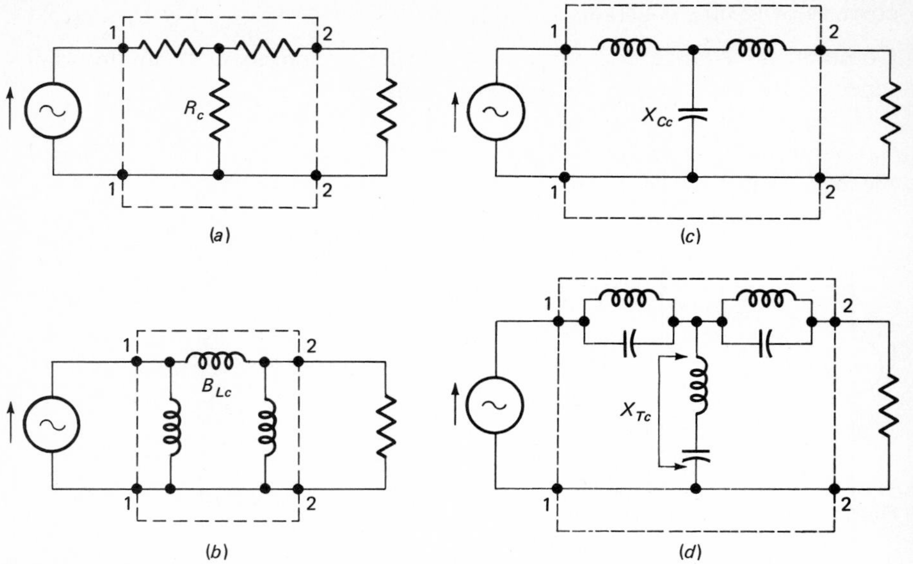

Fig. 18-4 Typical common impedance coupled circuits.

Figure 18-4*a*: resistance R_c.

Figure 18-4*b*: inductive susceptance B_{Lc}.

Figure 18-4*c*: capacitive reactance X_{Cc}.

Figure 18-4*d*: net reactance X_{Tc} of a series *LC circuit;* $Z_c = j(X_L - X_C)$; at resonance $Z_c = 0$.

In common impedance coupled circuits the coupling is conductive, there being a tangible connection between the coupled points.

MUTUAL INDUCTANCE COUPLING

Mutual inductance coupling occurs when energy is transferred as the result of a common magnetic field. In such a situation the coupling is inductive, and there is no tangible connection between the coupled points.

Figure 18-5 shows a *transformer* (a circuit device designed to provide mutual inductive coupling) as the conveyor of energy between ports 11 and

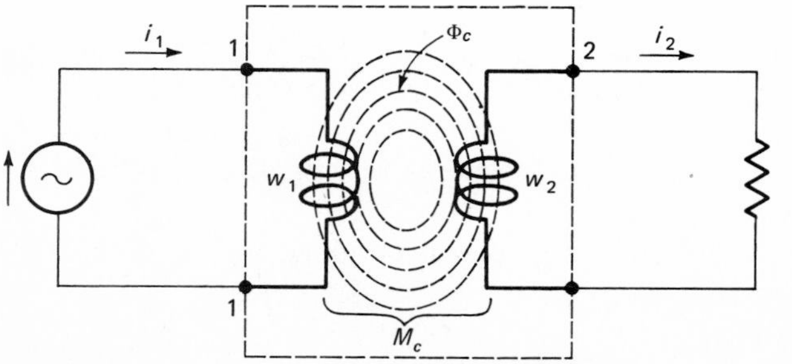

Fig. 18-5 Inductive coupling.

22. The alternating magnetic flux Φ_c is produced by the current i_1 in winding w_1. This flux links with winding w_2, producing the mutual inductance M_c. The flux Φ_c is thus common to the generator and load loops. There is no direct conductive connection between generator and load.

It should be mentioned that there is justification for classifying both mutual inductance coupling and series capacitance coupling as force-field coupling.

The alternating electric field within the capacitor in Fig. 18-6 is common to both the generator and the load. The coupling is by electric induction, the coupling medium being the common electric flux Ψ_c. There is no direct connection between the generator and the load. However, it is convenient to regard capacitance as an ac conductive parameter, and so to view Fig. 18-6 as a simple series circuit basically no different from a generator connected to a load through a resistance.

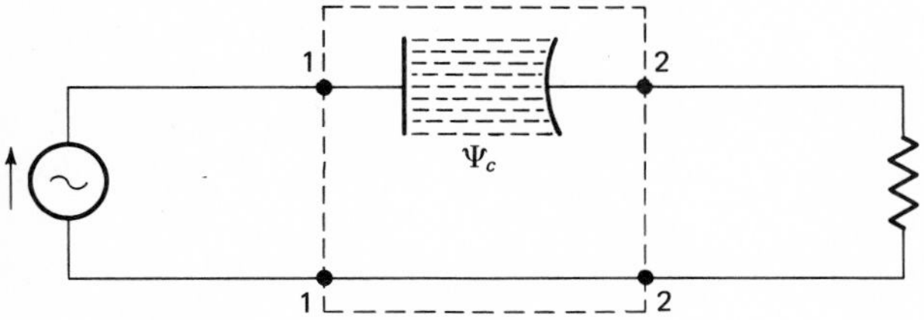

Fig. 18-6 Capacitive coupling.

MULTISTAGE COUPLING

There is no limit to the variety of circuit configurations obtainable by combining more than one stage or section of coupling, or more than one type of coupling. Two examples are given in Fig. 18-7.

Fig. 18-7 Multistage coupling. (*a*) Two stages; (*b*) three stages.

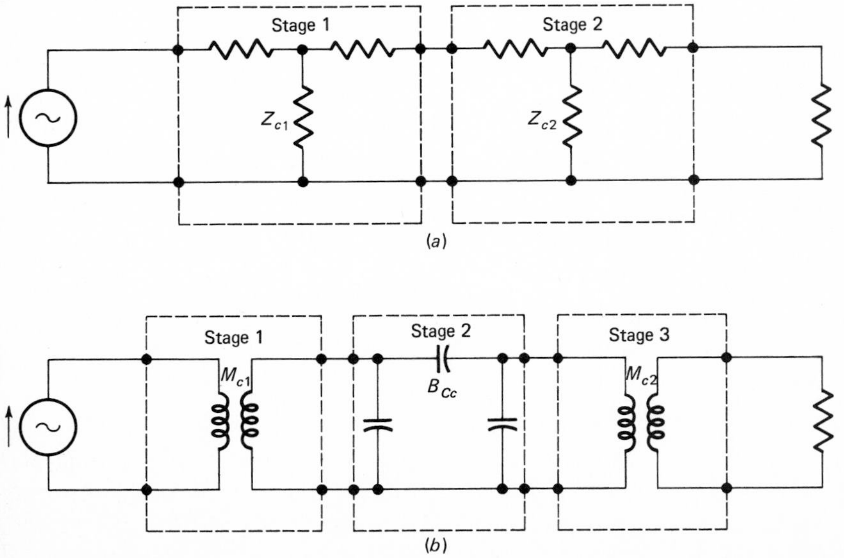

Two stages (two sections) of common impedance coupling are shown in Fig. 18-7*a*. Here the common impedance Z_{c1} of stage 1 couples the generator to stage 2. The common impedance Z_{c2} couples stage 1 to the load.

Two stages of mutual inductance coupling, together with one stage of mutual impedance coupling are shown in Fig. 18-7*b*. The generator is coupled to stage 2 by the mutual inductance coupling M_{c1} of stage 1. Stage 1 is coupled to stage 3 by the common susceptance B_{Cc} of stage 2. Stage 2 is coupled to the load by the mutual inductance M_{c2} of stage 3.

18-2 TWO-PORT NETWORK PARAMETERS

While the analysis of two-port networks may be carried out with the techniques dealt with in previous chapters, it can be convenient to adopt a system of parameters based on the two-port (four-terminal) concepts of Sec. 18-1. Two-port parameters are particularly useful when the measured characteristics of a network are more readily available than the details of its composition. A typical case is that of a transistor. To determine precisely the arrangement of resistances, capacitances, and inductances within a transistor is difficult, if not impossible. But to measure the characteristics of a transistor and express them in terms of a two-port equivalent network is not difficult.

GENERALIZED TWO-PORT CIRCUIT

Figure 18-8 shows a generalized two-port circuit with standard conventions. Two generators are shown. At a given instant each generator feeds a current i_1 or i_2 into the network via the upper terminal of one of its two ports. At the same instant there is a voltage v_{11} or v_{22} between each pair of port terminals, the voltage polarity being such that the upper terminal is positive with respect to the lower terminal.

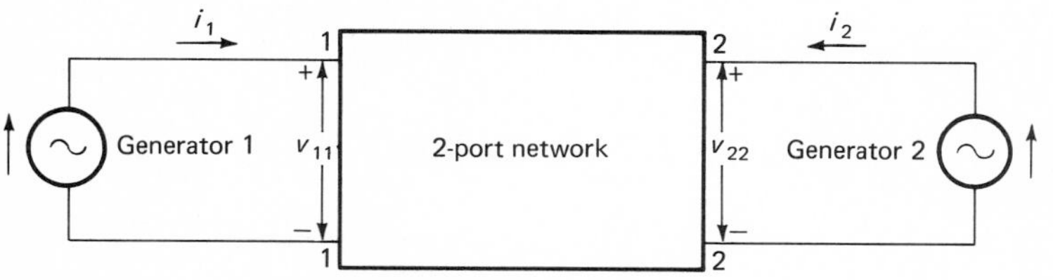

Fig. 18-8 Two-port circuit conventions.

In Fig. 18-8 lowercase symbols i and v are used to denote ac or varying currents and voltages.

INPUT-OUTPUT APPROACH

The conventions in Fig. 18-8 are deliberately arranged to avoid differentiating functionally between ports 11 and 22, which is desirable from a completely general viewpoint. But many applications of two-port principles relate to the transmission of energy through a network in one dominant direction, permitting the use of an input-output approach. This view involves simpler and more obvious symbols than the rigorous general approach, so we will adopt it in this introduction to two-port network parameters and their use in ac network analysis.

In the diagram in Fig. 18-9 energy is transmitted from the generator to the load through the two-port network. The energy enters the network at the input port and leaves at the output port. The input current i_i and the input voltage v_i are associated with the generator and the input port. The output current i_o and the output voltage v_o are associated with the output port and the load.

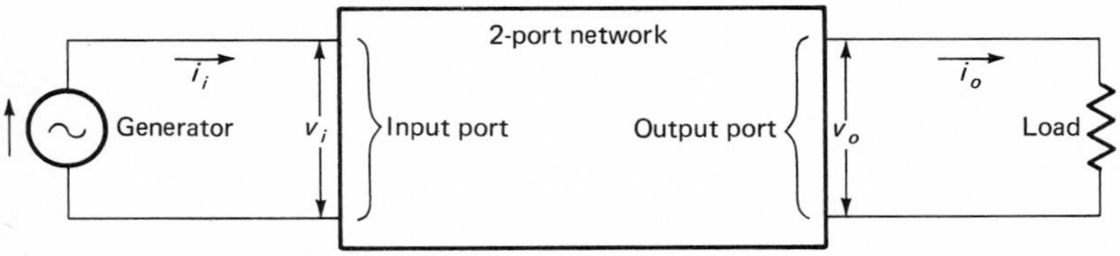

Fig. 18-9 Input-output form of a two-port circuit.

INPUT, OUTPUT, FORWARD, AND REVERSE PARAMETERS

The electrical characteristics of a two-port network can be described by four parameters shown pictorially in Fig. 18-10. These parameters are:

1. The input parameter, describing the characteristics determined at the input port.
2. The output parameter, describing the characteristics determined at the output port.
3. The forward parameter, describing the effect at the output port of a voltage or current applied at the input port. This effect is called *feed forward*.
4. The reverse parameter, describing the effect at the input port of a voltage or current applied at the output port. This effect is called *feedback*.

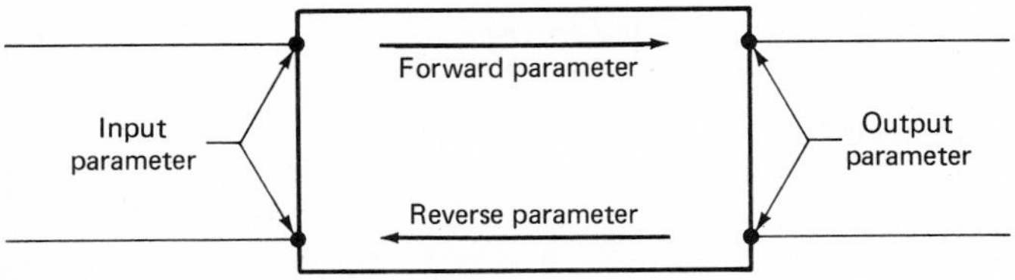

Fig. 18-10 Two-port network parameters.

PARAMETER SETS

Various ways of specifying the nature of network currents and voltages and the relationships between them are used for different circuit needs. This text will refer to impedance (z), admittance (y), and hybrid (h) parameters. These are three parameter sets, each consisting of four parameters (input, output, forward, and reverse). Each parameter is a specifically defined ratio of two quantities. The quantities we shall use are sine-wave voltages and currents, again adopting the convention of associating lowercase symbols v and i with ac voltages and currents of similar value, rms, peak-to-peak, etc.

It is important to realize that two-port parameters refer only to the characteristics of the two-port network, not to the characteristics of a complete circuit of which the network is part. The parameters are defined in terms of the reaction of the network to externally applied energy sources.

The two-port networks with which we shall concern ourselves will consist of basic resistances, capacitances, and inductances, and will thus be linear passive two-port networks.

18-3 IMPEDANCE (z) PARAMETERS

Consider the situation shown in Fig. 18-11. Generators simultaneously apply voltages v_i and v_o to the input and output ports of the network. Currents i_i and i_o enter the input and output ports at their upper terminals and are shown as loop currents. This is the basic setup for determining z parameters.

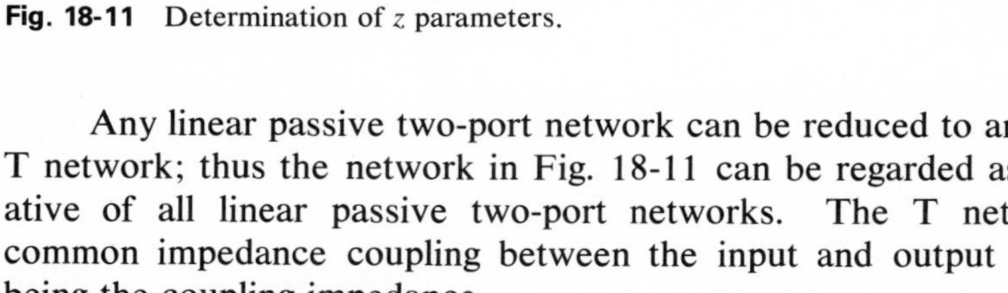

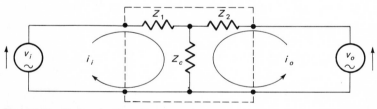

Fig. 18-11 Determination of z parameters.

Any linear passive two-port network can be reduced to an equivalent T network; thus the network in Fig. 18-11 can be regarded as representative of all linear passive two-port networks. The T network gives common impedance coupling between the input and output circuits, Z_c being the coupling impedance.

Applying loop analysis methods, the equations for the input and output circuits in Fig. 18-11 are

Input port

$$(Z_1 + Z_c)i_i + Z_c i_o = v_i \tag{18-1}$$

Output port

$$Z_c i_i + (Z_2 + Z_c)i_o = v_o \tag{18-2}$$

The impedance (z) parameters are described as follows:

1. The *input impedance parameter* z_i is the impedance of the network at the input port with the output port open circuit (disconnected from the external circuit).

$$z_i = Z_1 + Z_c \tag{18-3}$$

2. The *output impedance parameter* z_o is the impedance of the network at the output port with the input port open circuit.

$$z_o = Z_2 + Z_c \tag{18-4}$$

3. In Eq. (18-1), $Z_c i_o$ is a component of the input port voltage and represents a voltage fed back from the output to the input. Thus, in this context, Z_c can be regarded as an impedance controlling feedback. As such Z_c is called the *reverse impedance parameter* z_r.

$$z_r = Z_c \tag{18-5}$$

4. In Eq. (18-2), $Z_c i_i$ is a component of the output port voltage, representing a voltage fed forward from the input to the output. Therefore, in this context, Z_c can be regarded as an impedance controlling feed forward. From an input-to-output viewpoint, Z_c is called the *forward impedance parameter* z_f.

$$z_f = Z_c \tag{18-6}$$

It can be seen from Eqs. (18-1) and (18-2) that Z_c, the coupling impedance, is common to both input and output circuits. The effect of Z_c in controlling transmission in the forward and reverse directions (from input to output and from output to input) is indicated by the parameters z_f and z_r.

Equations (18-1) and (18-2) can be rewritten using the symbols z_i, z_o, z_f, and z_r:

$$z_i i_i + z_r i_o = v_i \tag{18-7}$$

$$z_f i_i + z_o i_o = v_o \tag{18-8}$$

The two-port impedance parameters are defined in open-circuit terms (for example, the input impedance parameter is the impedance at the input port with the output port open circuit). Applying these terms to Eqs. (18-7) and (18-8) results in expressions for each of the four parameters.

With the output port open circuit ($i_o = 0$), Eq. (18-7) becomes

$$z_i i_i = v_i \qquad \text{or} \qquad z_i = \left.\frac{v_i}{i_i}\right|_{i_o = 0} \tag{18-9}$$

and Eq. (18-8) becomes

$$z_f i_i = v_o \qquad \text{or} \qquad z_f = \frac{v_o}{i_i}\bigg|_{i_o = 0} \tag{18-10}$$

With the input port open circuit ($i_i = 0$), Eq. (18-7) becomes

$$z_r i_o = v_i \qquad \text{or} \qquad z_r = \frac{v_i}{i_o}\bigg|_{i_i = 0} \tag{18-11}$$

and Eq. (18-8) becomes

$$z_o i_o = v_o \qquad \text{or} \qquad z_o = \frac{v_o}{i_o}\bigg|_{i_i = 0} \tag{18-12}$$

Equations (18-9) to (18-12) correctly express the impedance parameters as ratios of voltage to current.

z-PARAMETER DEFINITIONS

The two-port, open-circuit, impedance parameters are defined as follows:

The *open-circuit input impedance parameter* z_i is the ratio of input voltage to input current with the output port open circuit.

The *open-circuit output impedance parameter* z_o is the ratio of output voltage to output current with the input port open circuit.

The *open-circuit forward impedance parameter* z_f is the ratio of output voltage to input current with the output port open circuit.

The *open circuit reverse impedance parameter* z_r is the ratio of input voltage to output current with the input port open circuit.

Each z parameter has the unit ohm.

Each of the open-circuit impedance parameters may be said to represent a local impedance (input or output) or transfer impedance (forward or reverse) which controls the relative magnitudes of a voltage and a current.

z-PARAMETER EQUIVALENT CIRCUIT

Equations (18-7) to (18-12) can be used to produce a general two-port network with the four z parameters, as shown in Fig. 18-12.

In the network in Fig. 18-12 two elements are associated with each port:

1. A local impedance (z_i or z_o) whose pd is a function of the local current (i_i or i_o)

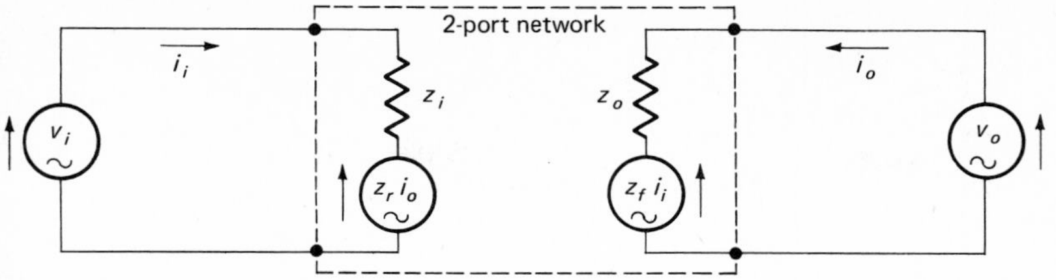

Fig. 18-12 z-parameter circuit.

2. A source of voltage equal to the product of a transfer impedance (z_r or z_f) and the other port current (i_o or i_i)

Example 18-1 What are the values of the input and output currents (i_i and i_o) in the circuit in Fig. 18-13*a*?

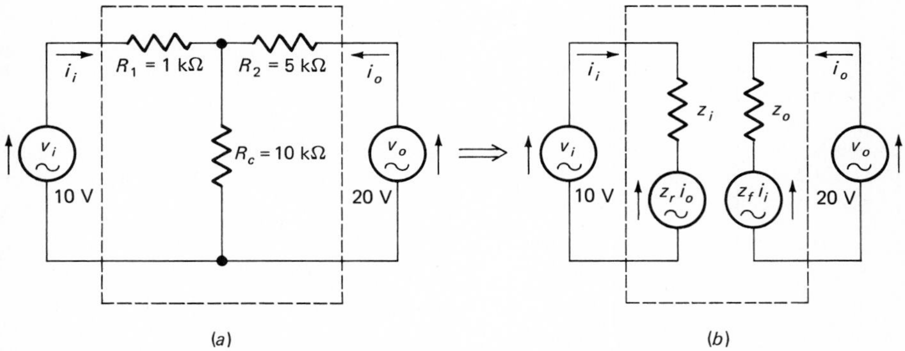

Fig. 18-13 Example 18-1: (*a*) given circuit; (*b*) z-parameter equivalent circuit.

Solution Figure 18-13*b* is a z-parameter equivalent of Fig. 18-13*a*.

The z parameters are, from Eq. (18-3),

$$z_i = R_1 + R_c = 11\ \text{k}\Omega$$

from Eq. (18-4),

$$z_o = R_2 + R_c = 15\ \text{k}\Omega$$

and from Eqs. (18-5) and (18-6),

$$z_f = z_r = 10\ \text{k}\Omega$$

The input port loop equation is

$$z_i i_i + z_r i_o = v_i \qquad (18\text{-}7)$$

$$11\ \text{k}\Omega\ i_i + 10\ \text{k}\Omega\ i_o = 10\ \text{V} \qquad (A)$$

The output port loop equation is

$$z_f i_i + z_o i_o = v_o \qquad (18\text{-}8)$$

$$10\ \text{k}\Omega\ i_i + 15\ \text{k}\Omega\ i_o = 20\ \text{V} \qquad (B)$$

From Eq. (A),

$$i_o = \frac{10\text{ V} - 11\text{ k}\Omega\ i_i}{10\text{ k}\Omega} = (1 - 1.1i_i)\text{ mA} \qquad (C)$$

Substituting i_o from Eq. (C) into Eq. (B),

$$10\text{ k}\Omega\ i_i + 15\text{ k}\Omega \times (1 - 1.1i_i)\text{ mA} = 20\text{ V}$$

$$10\text{ k}\Omega\ i_i + 15\text{ V} - 16.5\text{ k}\Omega\ i_i = 20\text{ V}$$

$$(10 - 16.5)\text{k}\Omega\ i_i = 5\text{ V}$$

$$i_i = -\frac{5\text{ V}}{6.5\text{ k}\Omega} = \mathbf{-0.7692\text{ mA}}$$

Substituting for i_i in Eq. (C),

$$i_o = (1 - 1.1i_i)\text{ mA} = (1 + 0.8462)\text{ mA} = \mathbf{1.846\text{ mA}}$$

Therefore $i_i = 0.7692$ mA with a direction opposite that shown in Fig. 18-13, and $i_o = 1.846$ mA with a direction as shown in Fig. 18-13.

Example 18-2 The T network in Fig. 18-14 acts as an attenuator so that the load voltage v_L is of smaller magnitude than the input voltage v_i. What is the ratio of the load voltage to the input voltage?

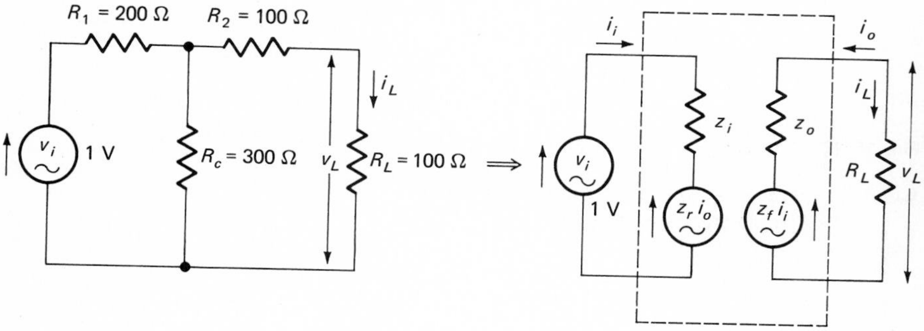

(a) (b)

Fig. 18-14 Example 18-2: (*a*) given circuit; (*b*) z-parameter equivalent circuit.

Solution Figure 18-14*b* is the z-parameter equivalent of Fig. 18-14*a*. In Fig. 18-14*b* the current i_o is shown entering the output port. The load current direction is actually opposite this convention ($i_L = -i_o$).

The z parameters are, from Eq. (18-3),

$$z_i = R_1 + R_c = 500\ \Omega$$

from Eq. (18-4),

$$z_o = R_2 + R_c = 400\ \Omega$$

and from Eqs. (18-5) and (18-6),

$$z_f = z_r = 300\ \Omega$$

The input port loop equation is

$$z_i i_i + z_r i_o = v_i \tag{18-7}$$

$$500\ \Omega\ i_i + 300\ \Omega\ i_o = 1\ \text{V} \qquad (A)$$

The output port loop equation is

$$z_f i_i + z_o i_o = v_o \tag{18-8}$$

The output port voltage is $v_o = R_L i_L = R_L(-i_o)$. Therefore the output port loop equation becomes

$$z_f i_i + z_o i_o = -R_L i_o$$

$$300\ \Omega\ i_i + 400\ \Omega\ i_o = -100\ \Omega\ i_o$$

$$300\ \Omega\ i_i + 500\ \Omega\ i_o = 0 \qquad (B)$$

The matrix form of Eqs. (A) and (B) is

$$\begin{bmatrix} 500 & 300 \\ 300 & 500 \end{bmatrix} \begin{bmatrix} i_i \\ i_o \end{bmatrix} = \begin{bmatrix} 1 \\ 0 \end{bmatrix} \qquad i_i \text{ and } i_o \text{ in amperes}$$

The solutions for i_i and i_o are

$$i_i = 3.125\ \text{mA} \qquad \text{and} \qquad i_o = -1.875\ \text{mA}$$

$$i_L = -i_o = 1.875\ \text{mA} \qquad \text{direction as shown in Fig. 18-14}$$

$$v_L = R_L i_L = 100\ \Omega \times 1.875\ \text{mA} = 0.1875\ \text{V}$$

Therefore the ratio $v_L/v_i =$ **0.1875.**

Example 18-3 A linear electronic network of unknown circuit design has the measured two-port z parameters $z_i = 500\ \Omega$, $z_o = 2\ \text{k}\Omega$, $z_f = 1\ \text{k}\Omega$, and $z_r = 5\ \text{k}\Omega$. What is the magnitude of the voltage developed across a 10-kΩ load resistor if the input voltage is 100 mV?
Solution Figure 18-15a shows the electronic network as an unspecified type of unit. The two-port equivalent network in Fig. 18-15b is based on the measured parameters.
The input loop equation is

$$z_i i_i + z_r i_o = v_i \tag{18-7}$$

$$0.5\ \text{k}\Omega\ i_i + 5\ \text{k}\Omega\ i_o = 100\ \text{mV} \qquad (A)$$

The output loop equation is, from Eq. (18-8),

$$z_f i_i + z_o i_o = -R_L i_o$$

$$1\ \text{k}\Omega\ i_i + 2\ \text{k}\Omega\ i_o = -10\ \text{k}\Omega\ i_o$$

$$1\ \text{k}\Omega\ i_i + 12\ \text{k}\Omega\ i_o = 0 \qquad (B)$$

In matrix form the loop equations are

$$\begin{bmatrix} 0.5 & 5 \\ 1 & 12 \end{bmatrix} \begin{bmatrix} i_i \\ i_o \end{bmatrix} = \begin{bmatrix} 100 \\ 0 \end{bmatrix} \qquad i_i \text{ and } i_o \text{ in microamperes}$$

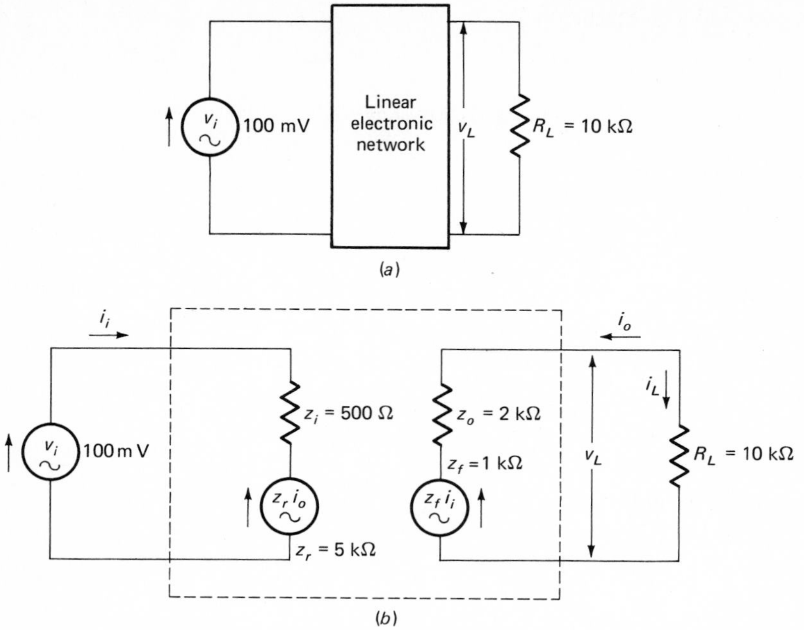

Fig. 18-15 Example 18-3: (*a*) block diagram; (*b*) *z*-parameter equivalent circuit.

The solution for i_o is

$$i_o = -100\ \mu\text{A}$$

The magnitude of i_L is the same as that of i_o, the directions being opposite.

$$i_L = 100\ \mu\text{A} \qquad \text{direction as shown in Fig. 18-15}$$

$$v_L = R_L i_L = 10\ \text{k}\Omega \times 100\ \mu\text{A} = \mathbf{1\ V}$$

18-4 ADMITTANCE (*y*) PARAMETERS

A set of four admittance (*y*) parameters can be developed in a manner similar to that adopted in Sec. 18-3 for the *z* parameters. A circuit consisting of a two-port π network fed by current sources applied to each port lends itself to a nodal equation derivation for the *y* parameters. As the method is essentially the same as for the *z* parameters, it will not be detailed here. A *y*-parameter equivalent of a generalized two-port circuit, expressed in input-output terms, is given in Fig. 18-16.

In Fig. 18-16 two current sources supply currents i_i and i_o to the input and the output ports. These currents each have two components within the network, one local ($y_i v_i$ and $y_o v_o$) and one transfer ($y_r v_o$ and $y_f v_i$). The input and output port nodal equations are

$$y_i v_i + y_r v_o = i_i \tag{18-13}$$

$$y_f v_i + y_o v_o = i_o \tag{18-14}$$

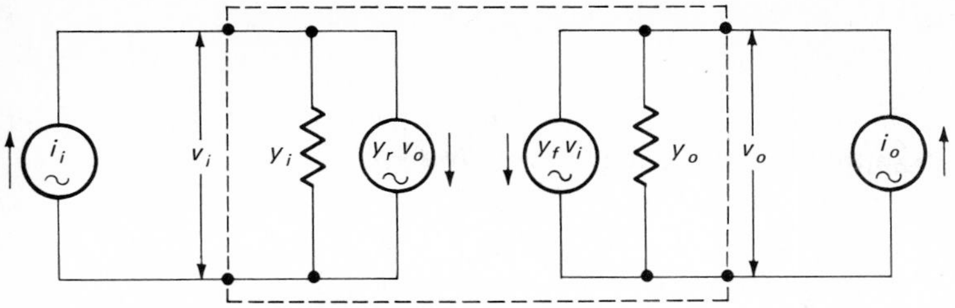

Fig. 18-16 y-parameter circuit.

y_i, y_o, y_f, and y_r are described as short-circuit admittance parameters. For the determination of a parameter it is assumed that one source is removed from the circuit and replaced by its internal admittance, resulting in a short circuit. The remaining source thus supplies current to a network terminated by a short-circuited port.

With the output port short circuit ($v_o = 0$), Eq. (18-13) becomes

$$y_i v_i = i_i \qquad \text{or} \qquad y_i = \left.\frac{i_i}{v_i}\right|_{v_o = 0} \tag{18-15}$$

and Eq. (18-14) becomes

$$y_f v_i = i_o \qquad \text{or} \qquad y_f = \left.\frac{i_o}{v_i}\right|_{v_o = 0} \tag{18-16}$$

With the input port short circuit ($v_i = 0$), Eq. (18-13) becomes

$$y_r v_o = i_i \qquad \text{or} \qquad y_r = \left.\frac{i_i}{v_o}\right|_{v_i = 0} \tag{18-17}$$

and Eq. (18-14) becomes

$$y_o v_o = i_o \qquad \text{or} \qquad y_o = \left.\frac{i_o}{v_o}\right|_{v_i = 0} \tag{18-18}$$

Equations (18-15) to (18-18) correctly express the admittance parameters as ratios of current to voltage.

y-PARAMETER DEFINITIONS

The two-port, short-circuit, admittance parameters are defined as follows:

> The *short-circuit input admittance parameter* y_i is the ratio of input current to input voltage with the output port short circuit.
>
> The *short-circuit output admittance parameter* y_o is the ratio of output current to output voltage with the input port short circuit.

The *short-circuit forward admittance parameter* y_f is the ratio of output current to input voltage with the output port short circuit.

The *short-circuit reverse admittance parameter* y_r is the ratio of input current to output voltage with the input port short circuit.

Each y parameter has the unit siemens.

It may be said that each of the short-circuit admittance parameters represents a local admittance (input or output) or transfer admittance (forward or reverse) which controls the relative magnitudes of a current and a voltage.

18-5 THE SENSE OF z AND y PARAMETERS

When applying two-port parameters it is important to be aware of the effect of the specific network on the sense (sign, polarity, or direction) of its parameters. Two examples will illustrate this.

Conductive transmission path Figure 18-17*a* and *b* shows a source (current or voltage) connected to the input port of a two-port conductive network.

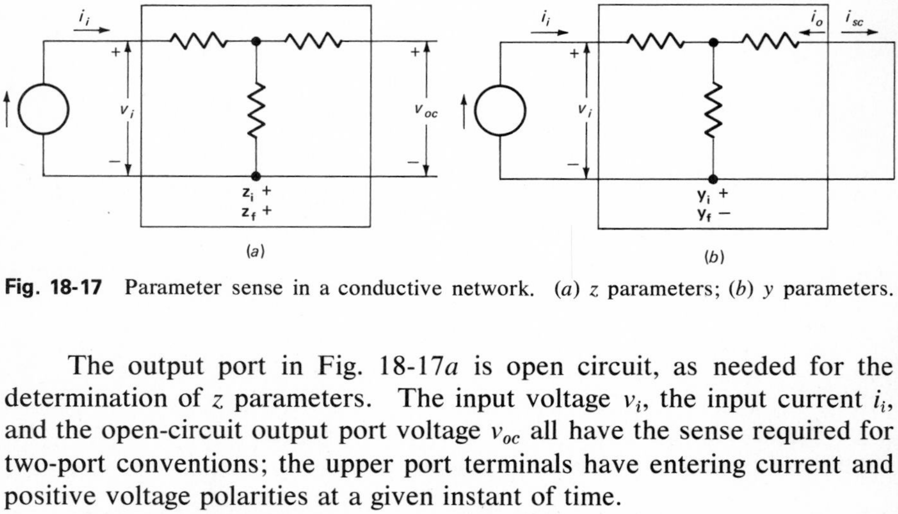

Fig. 18-17 Parameter sense in a conductive network. (*a*) z parameters; (*b*) y parameters.

The output port in Fig. 18-17*a* is open circuit, as needed for the determination of z parameters. The input voltage v_i, the input current i_i, and the open-circuit output port voltage v_{oc} all have the sense required for two-port conventions; the upper port terminals have entering current and positive voltage polarities at a given instant of time.

Figure 18-17*b* shows the output port short-circuited, as specified for determining y parameters. The input voltage v_i and the input current i_i have the sense required for the two-port conventions, but the short-circuit output port current i_{sc} *leaves* the upper terminal of the output port and is thus regarded as negative ($i_{sc} = -i_o$).

Applying a source at the output port and terminating the input port with an open circuit or a short circuit produce similar results.

The conclusions to be drawn for a conductive two-port network with input and output sources applied in the conventional sense are:

1. All z parameters are of positive sense.
2. Input and output y parameters are positive; transfer y parameters are negative.

Inductive transmission path In Fig. 18-18*a* to *d* a voltage or current source is connected to the input port of a two-port inductive network. The network is a transformer which couples the input port to the output port by means of a common magnetic field. The input current i_i and input voltage v_i both follow the two-port conventions in all four situations.

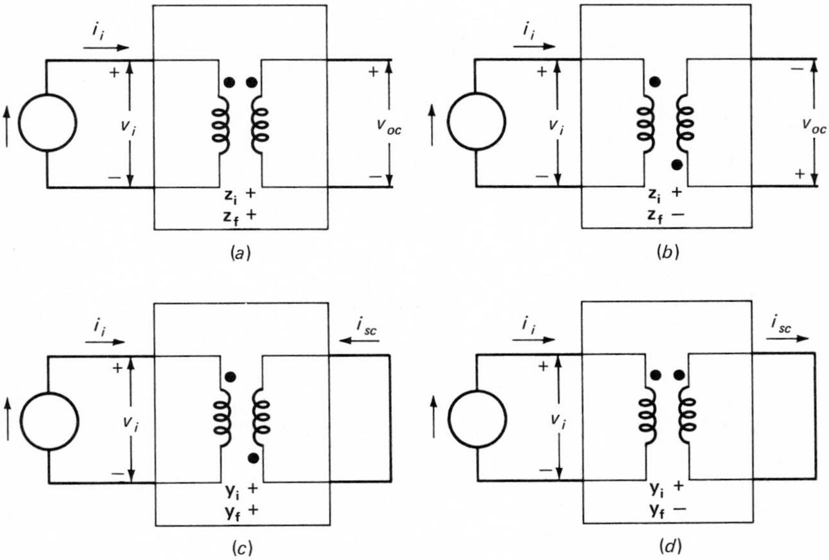

Fig. 18-18 Parameter sense in an inductive network. (*a*) and (*b*) z Parameters; (*c*) and (*d*) y parameters.

The dots associated with the transformer coils indicate the coil ends which normally have the same polarity at any instant of time. The physical construction of a transformer will determine the relative polarities of its port terminals; reversing the direction of winding of one coil will reverse the polarity of that coil's terminals.

Figure 18-18*a* and *b* shows the output port open circuit. The open-circuit output port voltage v_{oc} can be of conventional two-port polarity (Fig. 18-18*a*) or of opposite polarity (Fig. 18-18*b*).

In Fig. 18-18*c* and *d* the output port is short-circuited. The output port short-circuit current i_{sc} can have a conventional two-port direction (Fig. 18-18*c*) or the opposite direction (Fig. 18-18*d*).

A source applied to the output port would give similar results.

The conclusions to be drawn for a two-port inductive network are:

1. Input and output z and y parameters are of positive sense.
2. Transfer z and y parameters can be of either positive or negative sense, depending on the specific details of the inductive network.

Example 18-4 What are the values of the input and output voltages (v_i and v_o) in the circuit in Fig. 18-19*a*?

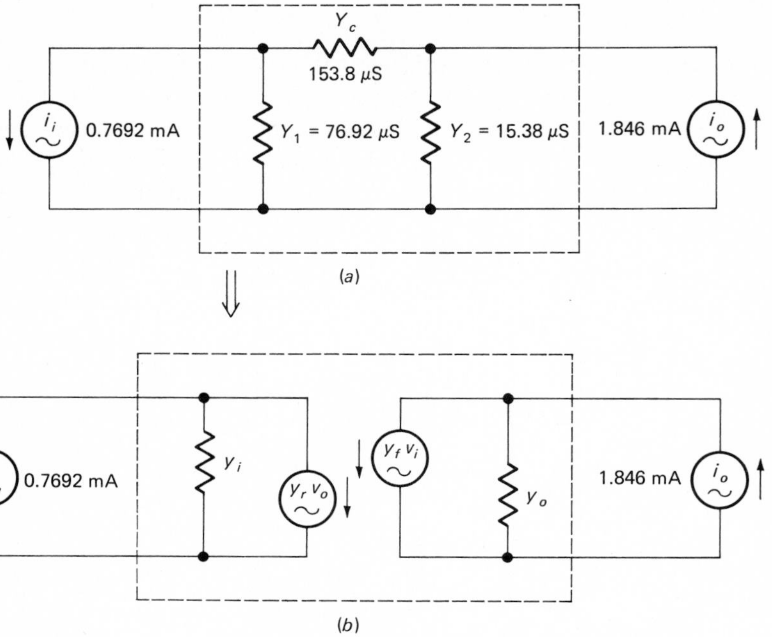

Fig. 18-19 Example 18-4: (*a*) given circuit; (*b*) *y*-parameter equivalent circuit.

Solution Figure 18-19*b* is the *y*-parameter equivalent of Fig. 18-19*a*. The sense of each *y* parameter can be determined by reference to Fig. 18-19*a*. With the output port short circuit, the input source and the network will appear as in Fig. 18-20*a*. From the viewpoint of two-port circuit conventions, in Fig. 18-20*a* the input current i_i has the wrong direction, the input voltage v_i has the wrong polarity, and the output current i_o has the correct direction.

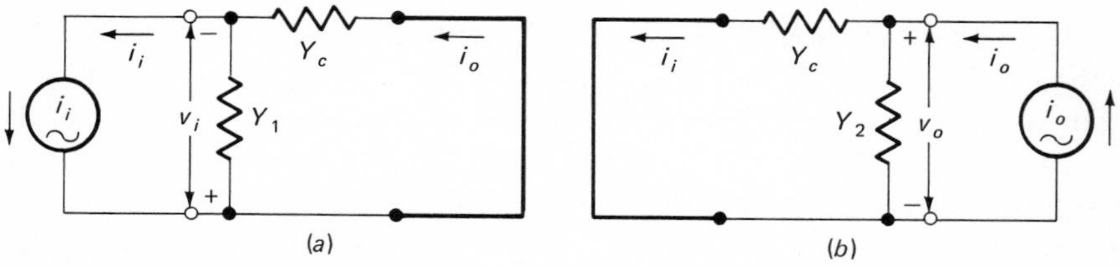

Fig. 18-20 Example 18-4: Fig. 18-19*a*. (*a*) With output short circuit; (*b*) with input short circuit.

$$y_i = \left.\frac{i_i}{v_i}\right|_{v_o = 0} \tag{18-15}$$

$$y_i = \frac{-}{-} = + \qquad y_i \text{ is positive}$$

$$y_f = \left.\frac{i_o}{v_i}\right|_{v_o = 0} \tag{18-16}$$

$$y_f = \frac{+}{-} = - \qquad y_f \text{ is negative}$$

With the input port short circuit, the output source and the network will be as shown in Fig. 18-20b. From the viewpoint of two-port circuit conventions, in Fig. 18-20b the output current i_o has the correct direction, the output voltage v_o has the correct polarity, and the input current i_i has the wrong direction. Thus,

$$y_r = \frac{i_i}{v_o}\bigg|_{v_i=0} \tag{18-17}$$

$$y_r = \frac{-}{+} = - \qquad y_r \text{ is negative}$$

$$y_o = \frac{i_o}{v_o}\bigg|_{v_i=0} \tag{18-18}$$

$$y_o = \frac{+}{+} = + \qquad y_o \text{ is positive}$$

The values of the y parameters are

$$y_i = Y_1 + Y_c = (76.92 + 153.8)\ \mu\text{S} = 230.7\ \mu\text{S}$$

$$y_f = -Y_c = -153.8\ \mu\text{S}$$

$$y_o = Y_2 + Y_c = (15.38 + 153.8)\ \mu\text{S} = 169.2\ \mu\text{S}$$

$$y_r = -Y_c = -153.8\ \mu\text{S}$$

With reference to Fig. 18-19b, the nodal equations for the input and output ports are

$$y_i v_i + y_r v_o = i_i \tag{18-13}$$

$$230.7\ \mu\text{S}\ v_i - 153.8\ \mu\text{S}\ v_o = -0.7692\ \text{mA} \tag{A}$$

$$y_f v_i + y_o v_o = i_o \tag{18-14}$$

$$-153.8\ \mu\text{S}\ v_i + 169.2\ \mu\text{S}\ v_o = 1.846\ \text{mA} \tag{B}$$

Equations (A) and (B) in matrix form are:

$$\begin{bmatrix} 230.7 & -153.8 \\ -153.8 & 169.2 \end{bmatrix} \begin{bmatrix} v_i \\ v_o \end{bmatrix} = \begin{bmatrix} -0.7692 \\ 1.846 \end{bmatrix} \qquad v_i \text{ and } v_o \text{ in kilovolts}$$

The determinant solutions for v_i and v_o are

$$v_i = \mathbf{10\ V}$$

$$v_o = \mathbf{20\ V}$$

The values for Example 18-4 were chosen to be the same as those for Example 18-1; the π network in Example 18-4 is the equivalent of the T network in Example 18-1.

The current sources in Example 18-4 produce the same input and output currents as were obtained in the solutions to Example 18-1. The solutions to Example 18-4 result in the same voltages (value and sign) as those of the voltage sources in Example 18-1.

Example 18-5 Figure 18-21a shows the circuit of a power supply filter. The input to the filter may be regarded as an 85-V steady dc with 60-V rms, 120-Hz ac superimposed on it. This combined input has an effective internal resistance of 100 Ω. The purpose of the filter is to reduce to a low value the magnitude of the ac component of the output voltage. What is the magnitude of this voltage?

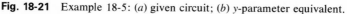

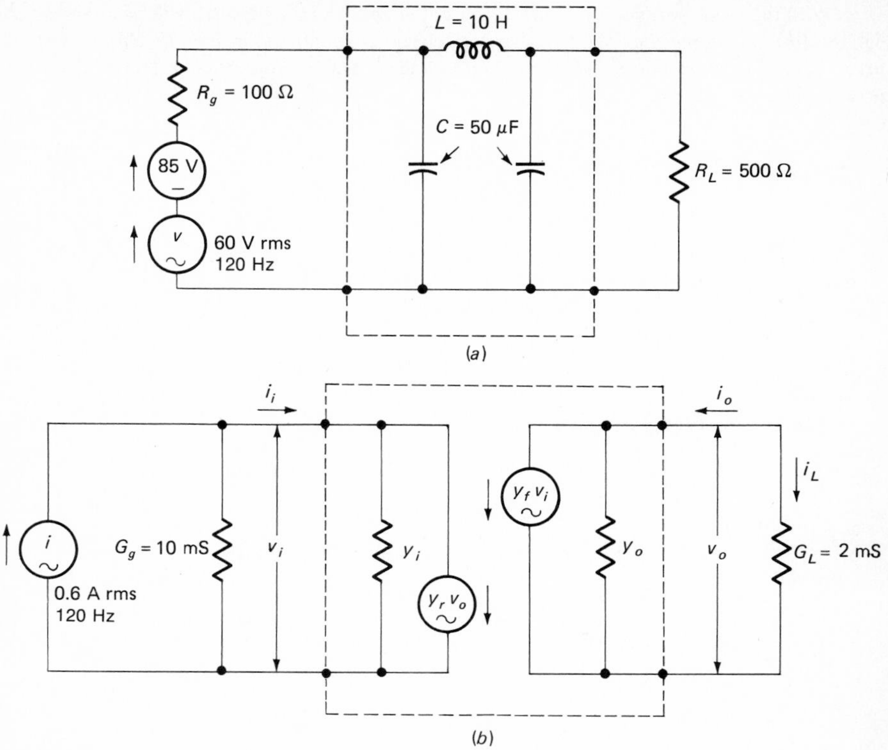

Fig. 18-21 Example 18-5: (*a*) given circuit; (*b*) *y*-parameter equivalent.

Solution Applying the superposition principle, we may consider the ac components of the circuit voltages and currents separately from the dc components. Figure 18-21*b* shows an ac equivalent circuit in admittance parameter terms. The ac voltage v has been converted to a source current $i = v/R_g$. The current source internal conductance is $G_g = 1/R_g$.

$$\omega = 2\pi f = 2\pi \times 120 \text{ Hz} = 754 \text{ rad/s}$$

$$B_L = -j\frac{1}{\omega L} = -j0.1326 \text{ mS}$$

$$B_C = j\omega C = j37.7 \text{ mS}$$

$$y_i = y_o = B_L + B_C = j(37.7 - 0.1326) \text{ mS} = j37.57 \text{ mS}$$

$y_r = y_f = +jB_L = +j0.1326$ mS   the sign of $-jB_L$ is changed because, in a conductive network, a transfer admittance parameter is the negative of the coupling admittance

The equation for the input port node is

$$y_i v_i + y_r v_o = i_i \tag{18-13}$$

where i_i = total current entering the input port upper terminal

$$i_i = i - G_g v_i$$

Substituting for i_i, the input nodal equation becomes

$$y_i v_i + y_r v_o = i - G_g v_i$$

$$j37.57 \text{ mS } v_i + j0.1326 \text{ mS } v_o = 0.6 \text{ A} - 10 \text{ mS } v_i$$

$$(10 + j37.57) \text{ mS } v_i + j0.1326 \text{ mS } v_o = 0.6 \text{ A}$$

The equation for the output port node is

$$y_f v_i + y_o v_o = i_o \tag{18-14}$$

where i_o = current entering the output port upper terminal

$$v_o = \frac{i_L}{G_L} \qquad \text{or} \qquad i_L = G_L v_o$$

But i_L *leaves* the output port upper terminal, so $i_o = -i_L = -G_L v_o$. The output nodal equation becomes

$$y_f v_i + y_o v_o = -G_L v_o$$

$$j0.1326 \text{ mS } v_i + j37.57 \text{ mS } v_o = -2 \text{ mS } v_o$$

$$j0.1326 \text{ mS } v_i + (2 + j37.57) \text{ mS } v_o = 0$$

The nodal equations in matrix form are

$$\begin{bmatrix} 38.88 \underline{/75.1^\circ} & 0.1326 \underline{/90^\circ} \\ 0.1326 \underline{/90^\circ} & 37.62 \underline{/87^\circ} \end{bmatrix} \begin{bmatrix} v_i \\ v_o \end{bmatrix} = \begin{bmatrix} 0.6 \\ 0 \end{bmatrix} \qquad v_i \text{ and } v_o \text{ in kilovolts}$$

The solution for v_o is

$$v_o = 54.42 \text{ mV } \underline{/-72.1^\circ} \qquad \text{or} \qquad \underline{/-1.258 \text{ rad}}$$

Therefore the magnitude of the ac component of the output voltage is **54.42 mV rms.**

18-6 HYBRID (*h*) PARAMETERS

In introducing Sec. 18-2 it was pointed out that two-port parameters are particularly useful when the characteristics of a network are readily determined only by measurement at its ports. The transistor was given as an important example. However, with the transistor it is often difficult to make reliable measurements of y and z parameters, because of its markedly different input and output characteristics. There are other networks which do not lend themselves to satisfactory z- and y-parameter treatment.

It may be roughly generalized that y and z parameters are suitable for networks in which the input and output characteristics are comparable. If input and output ports differ greatly in characteristics, h (hybrid) parameters may be more suitable.

h parameters are developed by combining the concepts involved in developing z and y parameters. The four h parameters are an impedance, an admittance, a current ratio, and a voltage ratio; hence the name hybrid.

Figure 18-22 shows the equivalent circuit of a two-port network drawn in conventional h-parameter form. As the circuits associated with the input and output ports in h-parameter terms are quite different, it is important to stress that the input-output approach has been adopted for convenience. For a particular network it might be desirable to reverse the format, making the input a parallel circuit and the output a series circuit. The generalized 1,1–2,2 approach, without specifying input and output, could be adopted to reduce confusion. But, since you will probably use h parameters mainly in transistor work and Fig. 18-22 applies to most transistors, the convenience of this configuration makes its choice reasonable. Although it should not be regarded as an invariable rule, the circuit in Fig. 18-22 may be considered to apply to networks which have low input impedance and high output impedance. Transistors usually fall into this category.

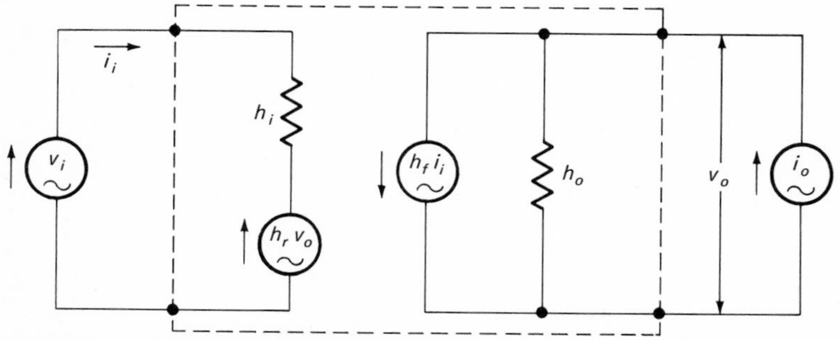

Fig. 18-22 h-Parameter circuit.

The input port elements within the network in Fig. 18-22 are:

1. An impedance h_i which, with the input current i_i, forms a pd of $h_i i_i$ volts.
2. A voltage source $h_r v_o$ formed by a voltage ratio h_r and the output voltage v_o. The voltage $h_r v_o$ is fed back from the output to the input.

The output port elements within the network are:

1. An admittance h_o which, with the output voltage v_o, forms a current of $h_o v_o$ amperes.
2. A current source $h_f i_i$ formed by a current ratio h_f and the input current i_i. The current $h_f i_i$ is fed forward from the input to the output.

The input loop equation and the output nodal equation together define the four h parameters

$$h_i i_i + h_r v_o = v_i \tag{18-19}$$

$$h_f i_i + h_o v_o = i_o \tag{18-20}$$

Expressions for the four h parameters are obtained by alternately considering the output port to be short circuit and the input port to be open circuit.

With the output port short circuit ($v_o = 0$), Eq. (18-19) becomes

$$h_i i_i = v_i \qquad \text{or} \qquad h_i = \left.\frac{v_i}{i_i}\right|_{v_o = 0} \tag{18-21}$$

and Eq. (18-20) becomes

$$h_f i_i = i_o \qquad \text{or} \qquad h_f = \left.\frac{i_o}{i_i}\right|_{v_o = 0} \tag{18-22}$$

With the input port open circuit ($i_i = 0$), Eq. (18-19) becomes

$$h_r v_o = v_i \qquad \text{or} \qquad h_r = \left.\frac{v_i}{v_o}\right|_{i_i = 0} \tag{18-23}$$

and Eq. (18-20) becomes

$$h_o v_o = i_o \qquad \text{or} \qquad h_o = \left.\frac{i_o}{v_o}\right|_{i_i = 0} \tag{18-24}$$

h-PARAMETER DEFINITIONS

The two-port h parameters are defined as follows:

> The *short-circuit input impedance parameter* h_i is the ratio of the input voltage to the input current with the output port short circuit.
>
> The unit for h_i is the ohm.
>
> The *open-circuit output admittance parameter* h_o is the ratio of the output current to the output voltage with the input port open circuit.
>
> The unit for h_o is the siemens.
>
> The *forward current gain* or *forward current parameter* h_f is the ratio of output current to input current with the output port short circuit.
>
> The *reverse voltage gain* or *reverse voltage parameter* h_r is the ratio of input voltage to output voltage with the input port open circuit.

18-7 THE SENSE OF *h* PARAMETERS

CONDUCTIVE NETWORK

Figure 18-23*a* shows a T network, typical of all conductive linear passive networks, with the output port short circuit. The short-circuit output current i_{sc} has a direction opposite that of the conventional two-port output current ($i_{sc} = -i_o$). The input current i_i and the input voltage v_i both have conventional sense. Figure 18-23*b* shows the T network with its input port open circuit. The open-circuit input port voltage v_{oc}, the output current i_o, and the output voltage v_o all have conventional sense. Thus an h parameter has positive sense unless it is related to the output port current in the short-circuit condition.

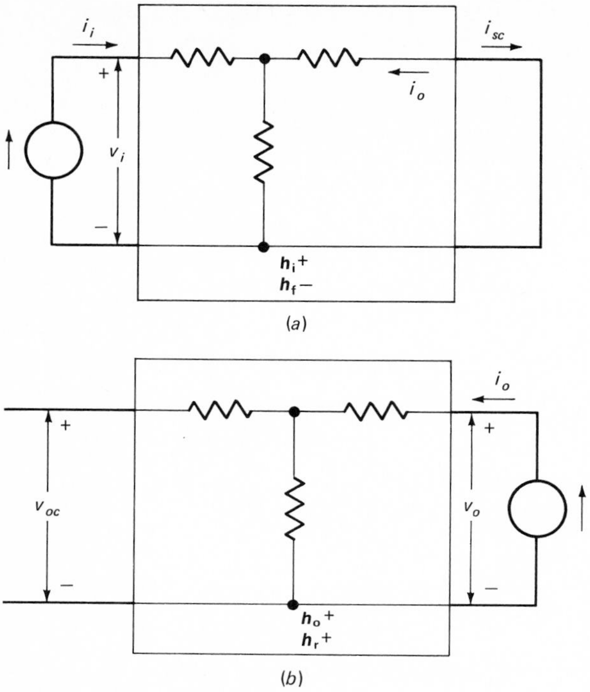

Fig. 18-23 *h*-Parameter sense in a conductive network. (*a*) Output port short circuit; (*b*) input port open circuit.

From Eqs. (18-21) to (18-24) it can be seen that only the forward current gain h_f involves the current i_o when the output port is short circuit ($v_o = 0$).

For a conductive network, h_f is negative, the other h parameters being positive.

INDUCTIVE NETWORK

Figure 18-24*a* to *d* shows a transformer as a typical inductive network.

In Fig. 18-24*a* and *b* the output port is shown short circuit. With the dot polarities in Fig. 18-24*a*, the input current i_i, the input voltage v_i, and the short-circuit output port current i_{sc} all have the required conventional senses for positive h parameters. With the dot polarities in Fig. 18-24*b*, the input current i_i and voltage v_i have conventional sense, but the short-circuit output port current i_{sc} has a direction opposite that required by two-port conventions ($i_{sc} = -i_o$).

In Fig. 18-24*c* and *d* the input port is shown open circuit. With the dot polarities in Fig. 18-24*c*, the output current i_o, the output voltage v_o, and the open-circuit input port voltage v_{oc} all have the two-port conventional sense. With the dot polarities in Fig. 18-24*d*, the output current i_o and output voltage v_o have conventional sense, but the open-circuit input port voltage v_{oc} has a polarity opposite the conventional two-port input voltage polarity ($v_{oc} = -v_i$).

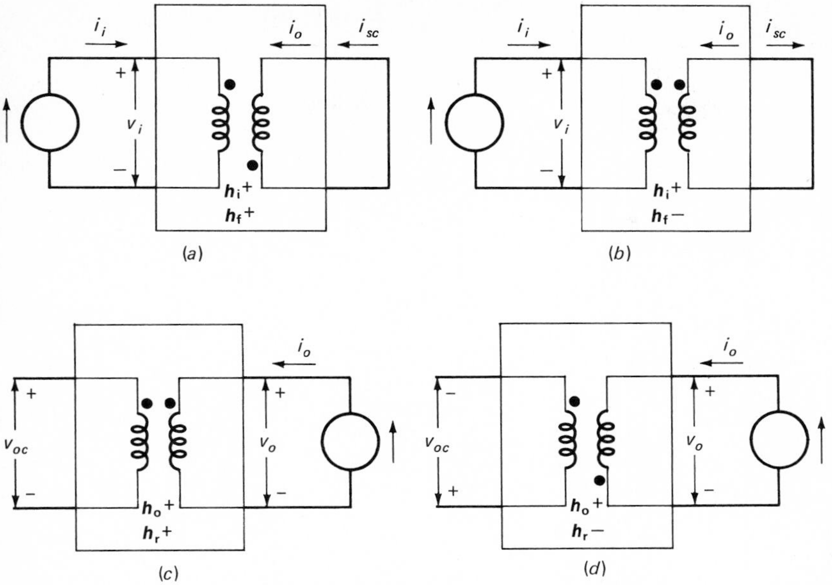

Fig. 18-24 *h*-Parameter sense in an inductive network. (*a*) Output port short circuit; (*b*) similar to (*a*) but with opposite dot polarity; (*c*) input port open circuit; (*d*) similar to (*c*) but with opposite dot polarity.

Summarizing the inductive network conditions from the two-port convention viewpoint, it may be said that

1. With the output short circuit ($v_o = 0$), v_i and i_i are both positive while i_o can be either positive or negative. Thus from Eqs. (18-21) and (18-22), h_i is always positive but h_f may be either positive or negative.
2. With the input port open circuit ($i_i = 0$), v_o and i_o are both positive while v_i can be either positive or negative. Applying this information to Eqs. (18-23) and (18-24), h_o is always positive but h_r may be either positive or negative.

For an inductive network the input and output *h* parameters are positive; the transfer *h* parameters may be positive or negative.

SUMMARY OF TWO-PORT PARAMETER SENSE CHARACTERISTICS

1. All input and output parameters are positive for both conductive and inductive networks.
2. Transfer *z* parameters are positive for conductive networks, and may be positive or negative for inductive networks.
3. Transfer *y* parameters are negative for conductive networks, and may be positive or negative for inductive networks.
4. For conductive networks h_r is positive and h_f is negative. For inductive networks both transfer *h* parameters may be either positive or negative.

Example 18-6 What are the values of the input current i_i and output voltage v_o in the circuit in Fig. 18-25*a*?

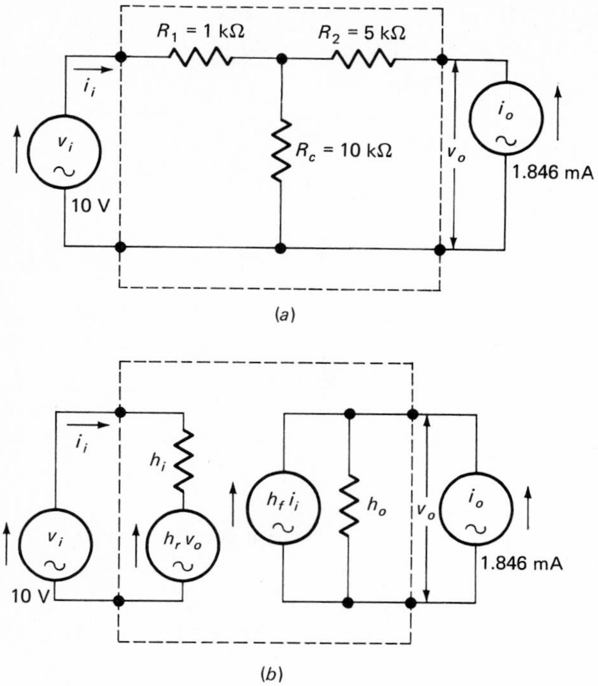

Fig. 18-25 Example 18-6: (*a*) given circuit; (*b*) *h*-parameter equivalent.

Solution Figure 18-25*b* is an *h*-parameter equivalent of Fig. 18-25*a*.

Figure 18-26*a* and *b* shows the short-circuit output and open-circuit input conditions.

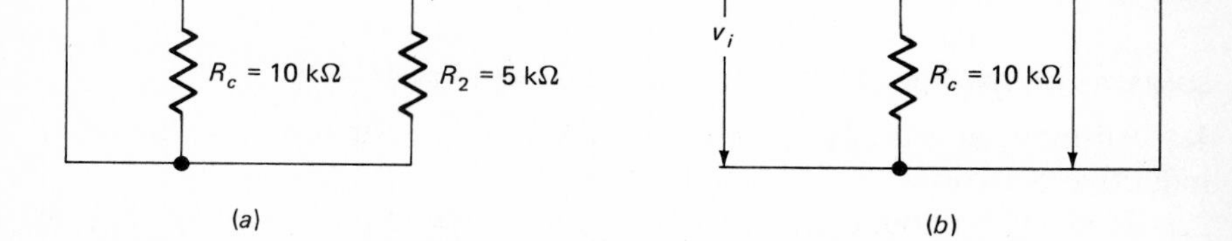

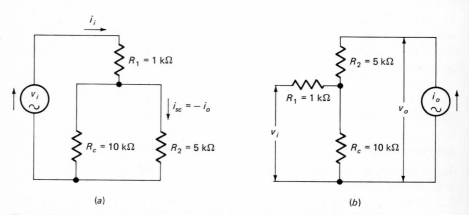

Fig. 18-26 Example 18-6: (*a*) output port short circuit; (*b*) input port open circuit.

With the output short circuit (Fig. 18-26*a*),

$$h_i = \frac{v_i}{i_i} = R_1 + \frac{R_2 \times R_c}{R_2 + R_c} = 1\ \text{k}\Omega + \frac{5\ \text{k}\Omega \times 10\ \text{k}\Omega}{5\ \text{k}\Omega + 10\ \text{k}\Omega} = 4.333\ \text{k}\Omega$$

Applying the current divider principle,

$$h_f = \frac{-i_o}{i_i} = -\frac{R_c}{R_2 + R_c} = -\frac{10\ \text{k}\Omega}{5\ \text{k}\Omega + 10\ \text{k}\Omega} = -0.6667$$

With the input open circuit (Fig. 18-26*b*),

$$h_o = \frac{i_o}{v_o} = \frac{1}{Z_o}$$

where Z_o = impedance seen by the source i_o

$$h_o = \frac{1}{R_2 + R_c} = \frac{1}{15\ \text{k}\Omega} = 0.06667\ \text{mS}$$

Applying the voltage divider principle,

$$h_r = \frac{v_i}{v_o} = \frac{R_c}{R_2 + R_c} = \frac{10\ \text{k}\Omega}{5\ \text{k}\Omega + 10\ \text{k}\Omega} = 0.6667$$

The input loop and output nodal equations are

$$h_i i_i + h_r v_o = v_i \tag{18-19}$$

$$4.333\ \text{k}\Omega\ i_i + 0.6667\ v_o = 10\ \text{V}$$

$$h_f i_i + h_o v_o = i_o \tag{18-20}$$

$$-0.6667 i_i + 0.06667\ \text{mS}\ v_o = 1.846\ \text{mA}$$

The matrix form of the equations is

$$\begin{bmatrix} 4.333 & 0.6667 \\ -0.6667 & 0.06667 \end{bmatrix} \begin{bmatrix} i_i \\ v_o \end{bmatrix} = \begin{bmatrix} 10 \\ 1.846 \end{bmatrix} \qquad i_i \text{ in milliamperes; } v_o \text{ in volts}$$

The solutions for i_i and v_o are

$$i_i = \mathbf{-0.7692\ mA}$$

$$v_o = \mathbf{20\ V}$$

If you will refer back to Example 18-1, you will see that the T network is the same as that in Example 18-6. The input voltage source is also the same ($v_i = 10$ V). The result ($i_o = 1.846$ mA) obtained for Example 18-1 was used for the value of the output current source in Example 18-6. The results for Example 18-6 complete the input and output port values as for Example 18-1.

The six diagrams in Fig. 18-27 compare Examples 18-1, 18-4, and 18-6. The π network in Example 18-4 (Fig. 18-27*c*) is the equivalent of the T network in Examples 18-1 and 18-6 (Fig. 18-27*a* and *e*). The *z*-, *y*-, and *h*-parameter configurations used in Examples 18-1, 18-4, and 18-6 are given beside the original circuit schematics as Fig. 18-27*b*, *d*, and *f*. All six diagrams are equivalent to each other.

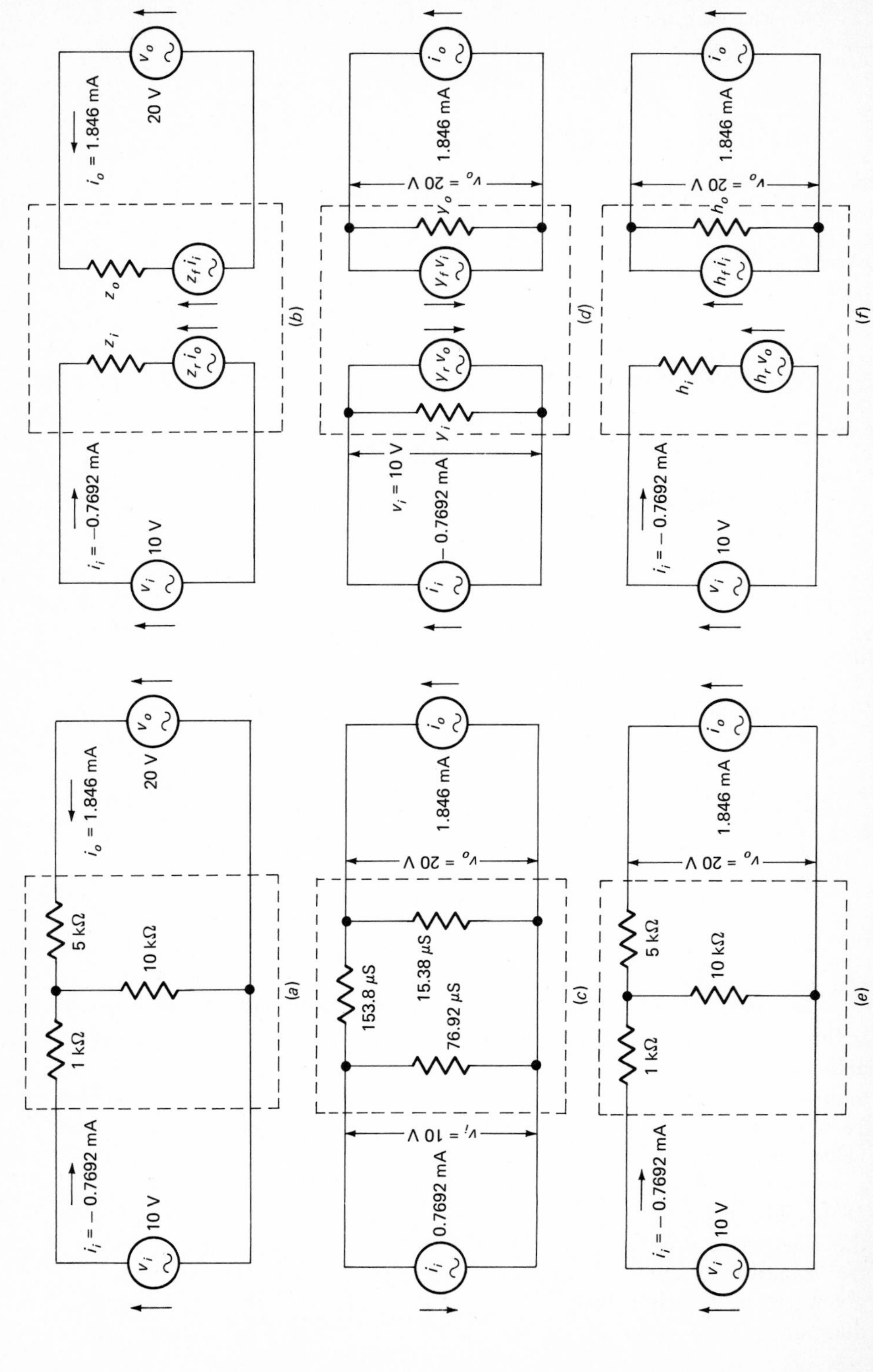

Fig. 18.27 Comparison of circuits (a) and (b) Example 18.1; (c) and (d) Example 18.4; (e) and (f) Example 18.6

Example 18-7 What are the h parameters for the π network in Fig. 18-28a at a frequency of 1 MHz?

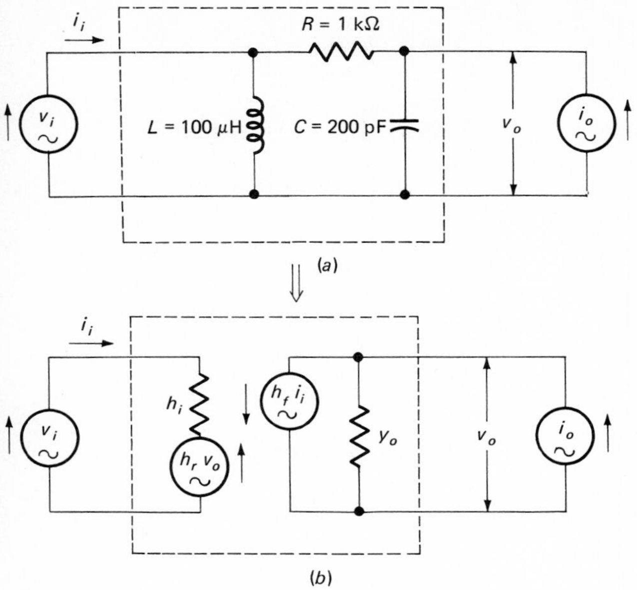

Fig. 18-28 Example 18-7: (a) given circuit; (b) h-parameter equivalent.

Solution Figure 18-28b is the h-parameter equivalent of Fig. 18-28a.

$$f = 1 \text{ MHz} \qquad \omega = 6.283 \text{ Mrad/s}$$

$$X_L = \omega L = 6.283 \text{ Mrad/s} \times 100\ \mu\text{H} = 628.3\ \Omega$$

$$B_C = \omega C = 6.283 \text{ Mrad/s} \times 200 \text{ pF} = 1.257 \text{ mS}$$

With the output short circuit, Fig. 18-28a becomes as shown in Fig. 18-29a. With the input open circuit, Fig. 18-28a becomes as shown in Fig. 18-29b.

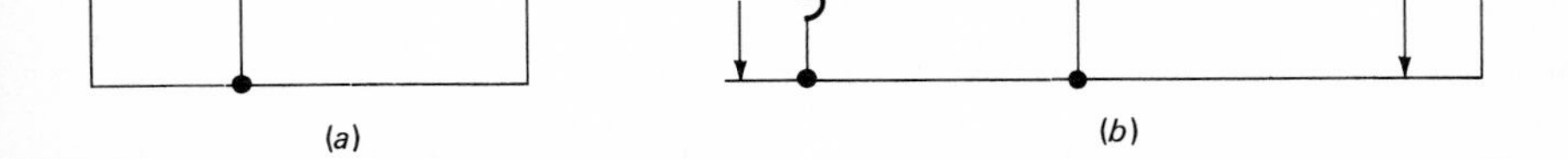

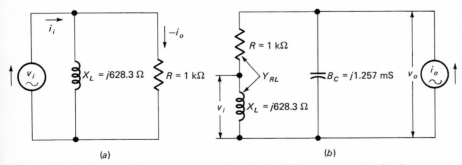

Fig. 18-29 Example 18-7: (a) output port short circuit; (b) input port open circuit.

For the output short circuit (Fig. 18-29a), h_i is the impedance seen by the voltage source v_i:

$$h_i = \frac{R \times jX_L}{R + jX_L} = \frac{1\text{ k}\Omega \times j628.3\ \Omega}{1\text{ k}\Omega + j628.3\ \Omega} = \frac{628.3\text{ k}\Omega^2 \underline{/90^\circ}}{1.181\text{ k}\Omega\ \underline{/32.14^\circ}}$$

$$h_i = \mathbf{532\ \Omega\ \underline{/57.86^\circ}} \qquad \text{or} \qquad \mathbf{\underline{/1.01 \text{ rad}}}$$

By current division,

$$h_f = \frac{-i_o}{i_i} = -\frac{jX_L}{R + jX_L} = -\frac{j628.3\ \Omega}{1\ \text{k}\Omega + j628.3\ \Omega}$$

$$h_f = -\frac{628.3\ \Omega\ \underline{/90^\circ}}{1.181\ \text{k}\Omega\ \underline{/32.14^\circ}}$$

$$= \mathbf{-0.532\ \underline{/57.86^\circ}} \quad \text{or} \quad \mathbf{0.532\ \underline{/-122.14^\circ}} \quad \text{or} \quad \mathbf{\underline{/-2.132\ rad}}$$

For the input open circuit (Fig. 18-29*b*), h_o is the admittance seen by current source i_o:

$$Y_{RL} = \frac{1}{Z_{RL}} = \frac{1}{R + jX_L} = \frac{1}{1\ \text{k}\Omega + j628.3\ \Omega} = \frac{1}{1.181\ \text{k}\Omega\ \underline{/32.14^\circ}}$$

$$Y_{RL} = 0.8467\ \text{mS}\ \underline{/-32.14^\circ} = (0.7169 - j0.4505)\ \text{mS}$$

$$h_o = B_C + Y_{RL} = j1.257\ \text{mS} + (0.7169 - j0.4505)\ \text{mS}$$

$$h_o = (0.7169 + j0.8065)\ \text{mS} = \mathbf{1.079\ mS\ \underline{/48.36^\circ}} \quad \text{or} \quad \mathbf{\underline{/0.844\ rad}}$$

By voltage division,

$$h_r = \frac{v_i}{v_o} = \frac{jX_L}{R + jX_L} = -h_f = \mathbf{0.532\ \underline{/57.86^\circ}} \quad \text{or} \quad \mathbf{\underline{/1.01\ rad}}$$

In the solutions for Example 18-7 the value of h_r is given as $-h_f$, $[h_r = jX_L/(R + jX_L)$ and $h_f = -jX_L/(R + jX_L)]$. This equality is not a coincidence. As a check back to each of the z, y, and h parameters indicates, it is a general principle that, for a bilateral two-port network (a network which operates equally well for transmission in either direction), the transfer parameters have the same magnitude. For conductive bilateral networks the y and z transfer parameters have the same sense ($y_f = y_r$ and $z_f = z_r$), but the h transfer parameters have opposing sense ($h_f = -h_r$).

Example 18-8 The circuit in Fig. 18-30*a* contains a network of unknown design. The h parameters of the network are measured and found to be $h_i = 500\ \Omega$, $h_o = 2$ mS, $h_f = 50$, and $h_r = -0.001$. (*a*) What type of network do these parameters suggest? (*b*) What is the voltage gain v_o/v_i of the complete circuit?

Fig. 18-30 Example 18-8: circuits. (*a*) With the network as a block; (*b*) h-parameter equivalent.

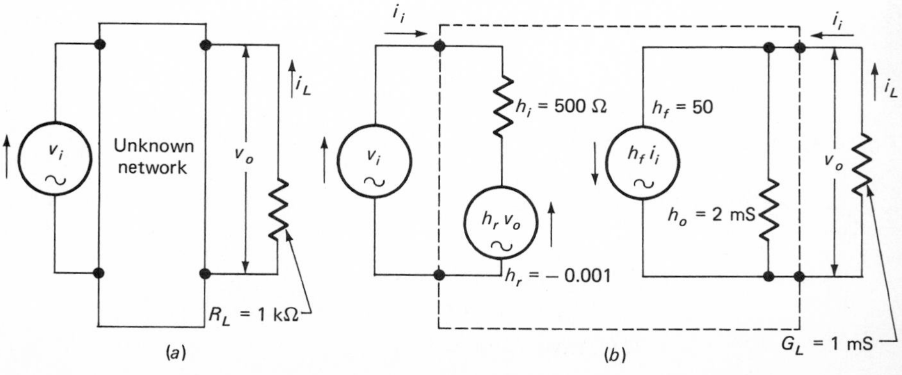

Solution

a. Figure 18-30*a* is a block diagram in which the unknown network is shown as an unspecified coupling between the generator and the load.

The given values of h_i, h_o, and h_f are all positive, while h_r is negative. Referring back to the descriptions of the sense of h parameters leads to the conclusion that this network is not conductive. A transformer with the dot polarities in Fig. 18-31 would meet the given sense requirements.

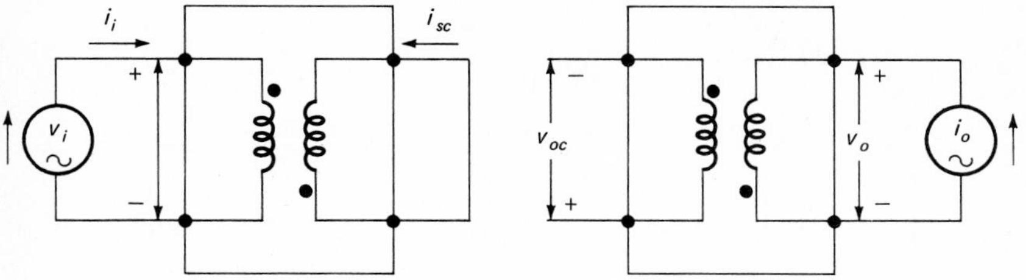

Fig. 18-31 Example 18-8, if network were a transformer. (*a*) Output port short circuit; (*b*) input port open circuit.

In Fig. 18-31*a* the conditions are set up for determining h_i and h_f. The input voltage v_i, the input current i_i, and the output port short-circuit current i_{sc} are all of conventional sense, so that h_i and h_f are both positive.

Figure 18-31*b* shows the conditions for determining h_o and h_r. The output port current i_o and the output voltage v_o are of conventional sense, but the input port open-circuit voltage v_{oc} is of opposite polarity to that of the two-port convention. Thus h_o is positive and h_r is negative.

The difference between the magnitudes of h_f and h_r implies that the unknown network has amplifying capability and thus may not be linear and bilateral.

b. If it is assumed that the network is virtually linear, the h-parameter loop and nodal equations can be applied, using the equivalent circuit in Fig. 18-30*b*.

AT THE INPUT LOOP The loop equation is

$$h_i i_i + h_r v_o = v_i \tag{18-19}$$

$$500\ \Omega\ i_i - 0.001\ v_o = v_i \tag{A}$$

AT THE OUTPUT NODE In Fig. 18-30*a* the load current direction is given as toward the upper output port terminal, the same as the conventional two-port i_o direction. Therefore $i_L = i_o$.

The load voltage is:

$$v_o = \frac{i_L}{G_L} = \frac{i_o}{G_L}$$

$$i_o = G_L v_o \tag{B}$$

The nodal equation is

$$h_f i_i + h_o v_o = i_o \tag{18-20}$$

Substituting for i_o from Eq. (*B*),

$$h_f i_i + h_o v_o = G_L v_o$$

$$50\ i_i + 2\ \text{mS}\ v_o = 1\ \text{mS}\ v_o$$

$$50\ i_i + 0.001\ \text{S}\ v_o = 0 \tag{C}$$

The matrix form of Eqs. (A) and (C) is

$$\begin{bmatrix} 500 & -0.001 \\ 50 & 0.001 \end{bmatrix} \begin{bmatrix} i_i \\ v_o \end{bmatrix} = \begin{bmatrix} v_i \\ 0 \end{bmatrix} \qquad i_i \text{ in amperes; } v_i \text{ and } v_o \text{ in volts}$$

The solution for v_o is

$$v_o = -90.91\ v_i$$

Therefore the voltage gain of the circuit is

$$\frac{v_o}{v_i} = \mathbf{-90.91}$$

The output voltage magnitude in Example 18-8 is 90.91 times as great as that of the input voltage, confirming the amplifying capability of the network. The negative sign of the voltage gain indicates that the output voltage is in antiphase (180°) with respect to the input voltage.

18-8 THE TRANSFORMER AS A TWO-PORT NETWORK

A transformer is a device which utilizes mutual inductance to couple circuits together. This text is not primarily concerned with the transformer as a device, but rather as a network which embodies some principles not met in the consideration of conductive two-port networks. The following introduction to the transformer as an electric network supplements descriptions of the transformer as a device, which can be found in other texts. The main transformer characteristics are as follows.

Voltage changing A transformer can be used to couple a source to a load when the voltage required by the load is different from that of the source. If the load voltage is required to be greater than the source voltage, the transformer steps up the voltage. If the load voltage must be less than the source voltage, the transformer steps down the voltage.

Current changing When a transformer steps up voltage, it steps down current in the same proportion (if voltage is stepped down, current is stepped up in proportion).

Impedance matching A transformer can produce the conditions needed for maximum power transfer by coupling a source to a load of different impedance. A suitable transformer will present to the source an impedance equal to its internal impedance, while appearing to the load as an energy source of internal impedance equal to the load impedance.

Isolation A transformer couples electrical energy from a source to a load by means of a magnetic field. Thus a transformer can be used to connect two circuits without direct contact between them. This isolation of a circuit by a transformer can be used for the protection of equipment or personnel or to prevent contact between two circuits operating at different dc levels.

THE PERFECT TRANSFORMER

For a transformer to be perfect it would have:

1. Perfect windings, wound with resistanceless wire
2. A perfect core on which the wire is wound, this core dissipating no energy
3. Perfect linkage, all the magnetic field produced due to the current in the primary coil (the input winding to which external energy is applied) linking with the secondary coil (the output winding to which the load is connected)

None of the criteria for a perfect transformer can be achieved in practice. However, the results obtained from considering such a hypothetical device can be used with acceptable accuracy for a number of practical applications.

Figure 18-32 shows a perfect iron-cored transformer as a two-port network. It should be noted that the standard symbol for a transformer implies perfection, this being consistent with the principle that component symbols represent perfect devices. For example, the symbol -/\/\/- represents a resistor which is entirely resistive with no stray capacitance or inductance. Significant imperfections are shown as separate equivalent elements.

The alternating current i_p in the primary winding w_p produces an alternating magnetic field. In this perfect transformer all the alternating magnetic flux Φ_c links with each turn of the primary winding and with each turn of the secondary winding w_s.

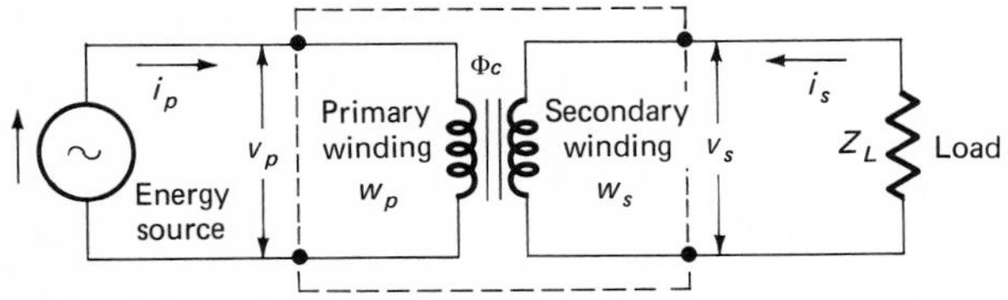

Fig. 18-32 A perfect transformer as a two-port network.

The alternating magnetic flux Φ_c is the mutual flux and is the coupling element in this two-port network.

Applying the definition of flux linkage Λ as the product of flux Φ and the number of turns N, the magnitude of the primary flux linkage is given by[1]

$$\Lambda_p = N_p \Phi_c \tag{18-25}$$

The magnitude of the secondary flux linkage is

$$\Lambda_s = N_s \Phi_c \tag{18-26}$$

[1] If necessary, refer to Chap. 9 for a review of the principles, terminology, and symbols of inductive circuits.

The voltage induced across the ends of a conductor is given by the rate of change in flux linkage ($v = d\Lambda/dt$).

The primary and secondary voltages v_p and v_s are

$$v_p = \frac{d\Lambda_p}{dt} \qquad v_s = \frac{d\Lambda_s}{dt}$$

$$= \frac{dN_p\Phi_c}{dt} \qquad = \frac{dN_s\Phi_c}{dt}$$

$$v_p = N_p\frac{d\Phi_c}{dt} \qquad v_s = N_s\frac{d\Phi_c}{dt}$$

The ratio of the secondary to primary voltages can be obtained from the expressions for v_s and v_p:

$$\frac{v_s}{v_p} = N_s\frac{d\Phi_c}{dt} \div N_p\frac{d\Phi_c}{dt}$$

$$\frac{v_s}{v_p} = \frac{N_s}{N_p} \tag{18-27}$$

Inductance is proportional to the square of the number of turns, or $N \propto \sqrt{L}$, so Eq. (18-27) can be written in inductance terms as

$$\frac{v_s}{v_p} = \sqrt{\frac{L_s}{L_p}} \tag{18-28}$$

The secondary-to-primary voltage ratio v_s/v_p of a perfect transformer is equal to the secondary-to-primary turns ratio N_s/N_p or the square root of the secondary-to-primary inductance ratio $\sqrt{L_s/L_p}$.

The voltage induced across an inductance is given by the product of the inductance value and the rate of change in current ($v = L\, di/dt$). Thus the primary voltage can be expressed in inductance terms as

$$v_p = L_p\frac{di_p}{dt} \tag{18-29}$$

The voltage induced across a coil by mutual induction is given by the product of the mutual inductance value and the rate of change in originating current ($v = M di/dt$). The secondary voltage of a transformer is produced by mutual induction with the primary current as its source:

$$v_s = M\frac{di_p}{dt} \tag{18-30}$$

Combining Eqs. (18-29) and (18-30) gives an expression for the voltage ratio in terms of self- and mutual inductance:

$$\frac{v_s}{v_p} = M\frac{di_p}{dt} \div L_p\frac{di_p}{dt}$$

$$\frac{v_s}{v_p} = \frac{M}{L_p} \tag{18-31}$$

An expression for mutual inductance in terms of the primary and secondary self-inductances can be obtained from Eqs. (18-28) and (18-31):

$$\frac{v_s}{v_p} = \sqrt{\frac{L_s}{L_p}} = \frac{M}{L_p} \qquad \text{or} \qquad M = L_p\sqrt{\frac{L_s}{L_p}}$$

$$M = \sqrt{L_pL_s} \tag{18-32}$$

The mutual inductance of a perfect transformer is equal to the geometric mean of the self-inductances of the primary and secondary windings.

A perfect transformer transfers power without loss. The power supplied by the source to the primary (input port) circuit equals that supplied by the secondary (output port) circuit to the load.

Equating the primary and secondary apparent powers for the circuit in Fig. 18-32 leads to important conclusions:

$$S_p = S_s$$

$$v_pi_p = v_si_s$$

$$\frac{i_s}{i_p} = \frac{v_p}{v_s} \tag{18-33}$$

Applying Eq. (18-27),

$$\frac{i_s}{i_p} = \frac{N_p}{N_s} \tag{18-34}$$

The secondary-to-primary current ratio of a perfect transformer is equal to the reciprocal of the secondary-to-primary voltage ratio or the reciprocal of the secondary-to-primary turns ratio.

Putting apparent powers in impedance terms and letting Z_{in} represent the input impedance (the impedance seen by the input source),

$$S_p = S_s$$

$$i_p^2Z_{\text{in}} = i_s^2Z_L$$

$$\frac{Z_L}{Z_{\text{in}}} = \left(\frac{i_p}{i_s}\right)^2$$

Substituting for i_p/i_s from Eq. (18-34),

$$\frac{Z_L}{Z_{\text{in}}} = \left(\frac{N_s}{N_p}\right)^2 \tag{18-35}$$

For a perfect transformer, the ratio of the load impedance to the impedance seen by the input source is equal to the square of the turns ratio.

Example 18-9 A perfect transformer has a primary inductance of 10 H and a secondary inductance of 70 H. If 120 V ac is applied to the primary and a 100-Ω resistive load is connected to the secondary, what is (*a*) the load voltage, (*b*) the load current, (*c*) the load power, and (*d*) the primary current?

Solution Figure 18-33 shows the conditions referred to in this example.

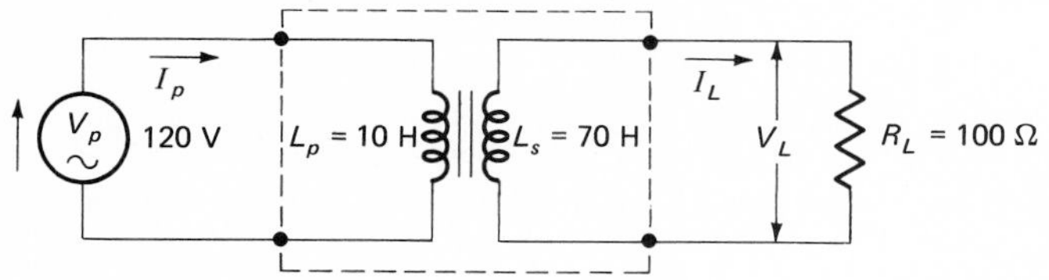

Fig. 18-33 Example 18-9: circuit.

a. From Eq. (18-28),

$$\frac{V_L}{V_p} = \sqrt{\frac{L_s}{L_p}}$$

$$V_L = V_p\sqrt{\frac{L_s}{L_p}} = 120 \text{ V}\sqrt{\frac{70 \text{ H}}{10 \text{ H}}} = \mathbf{317.5 \text{ V}}$$

b.
$$I_L = \frac{V_L}{R_L} = \frac{317.5 \text{ V}}{100 \ \Omega} = \mathbf{3.175 \text{ A}}$$

c.
$$P_L = I_L^2 R_L = (3.175 \text{ A})^2 \times 100 \ \Omega = \mathbf{1.008 \text{ kW}}$$

d. From Eq. (18-33),

$$\frac{I_p}{I_s} = \frac{V_s}{V_p}$$

$$I_p = I_s\frac{V_s}{V_p} = 3.175 \text{ A}\,\frac{317.5 \text{ V}}{120 \text{ V}} = \mathbf{8.4 \text{ A}}$$

Example 18-10 A television receiver designed for the North American 120-V 60-Hz power supply is to be used in Europe where the supply is 240 V 50 Hz. If the television set load is 100 VA, what turns ratio is required of a perfect transformer to perform the necessary conversion? What would be the relative sizes of wire needed for the primary and secondary windings?

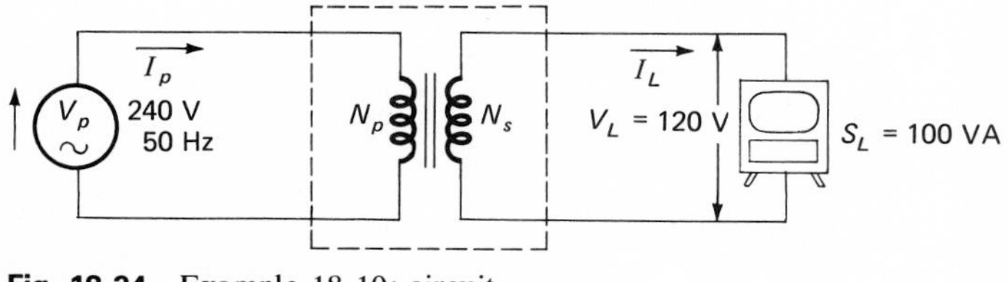

Fig. 18-34 Example 18-10: circuit.

Solution The most significant difference between iron-cored transformers for use on different supply frequencies is that for lower frequencies a more massive core is needed. In this example we are concerned with two frequencies, 60 and 50 Hz, which do not differ greatly. Furthermore, the apparent power involved, 100 VA, is low. In such cases a transformer designed for either 60 or 50 Hz would be suitable, so only voltages and currents need to be considered.

From Eq. (18-27),

$$\frac{N_s}{N_p} = \frac{V_L}{V_p}$$

$$\frac{N_s}{N_p} = \frac{120\ \text{V}}{240\ \text{V}} = \mathbf{0.5}$$

The secondary winding must have one-half as many turns as the primary winding. From Eq. (18-34),

$$\frac{I_L}{I_p} = \frac{N_p}{N_s}$$

$$I_L = I_p \frac{N_p}{N_s} = 2I_p$$

For the same current density (current per unit cross-sectional area), the wire used for the secondary winding must have twice the cross-sectional area of that used for the primary winding.

In Examples 18-9 and 18-10 capital letter symbols V and I were used for voltages and currents, since power supply values are usually stated as rms.

Example 18-11 An audio-frequency transformer is required to match a 4-Ω impedance loudspeaker to the 1-kΩ output impedance of a transistor amplifier. Assuming that both impedances are resistive and, at the frequencies involved, the transformer can be considered perfect, what turns ratio is needed?

Solution Figure 18-35*a* shows in simplified form the transistor amplifier connected to the loudspeaker via a transformer. An equivalent circuit, Fig. 18-35*b*, shows the transistor as a generator with an internal impedance $R_g = 1\ \text{k}\Omega$ and the loudspeaker as a resistive load $Z_L = 4\ \Omega$.

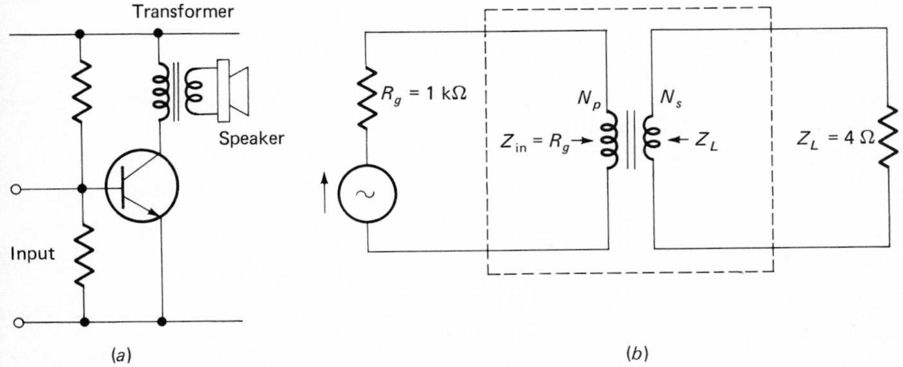

Fig. 18-35 Example 18-11: (*a*) given circuit; (*b*) two-port equivalent circuit of the transformer operation.

For matching to take place, the input impedance of the transformer Z_{in} must equal the generator internal impedance R_g. If this requirement is to be met, Eq. (18-35) must be satisfied:

$$\left(\frac{N_s}{N_p}\right)^2 = \frac{Z_L}{Z_{\text{in}}} \qquad (18\text{-}35)$$

$$\frac{N_p}{N_s} = \sqrt{\frac{Z_{\text{in}}}{Z_L}} = \sqrt{\frac{1\text{ k}\Omega}{4\ \Omega}} = 15.81$$

The primary winding should have **15.81** times as many turns as the secondary winding.

Example 18-12 If the primary winding of the transformer in Example 18-11 has 40 H inductance, what is the mutual inductance between the windings?

Solution The self-inductance of a coil is directly proportional to the square of the number of its turns. Therefore,

$$\frac{L_s}{L_p} = \left(\frac{N_s}{N_p}\right)^2$$

$$L_s = \left(\frac{N_s}{N_p}\right)^2 L_p = \left(\frac{1}{15.81}\right)^2 40\text{ H} = 0.16\text{ H}$$

$$M = \sqrt{L_p L_s} \qquad (18\text{-}32)$$

$$M = \sqrt{0.16\text{ H} \times 40\text{ H}} = \mathbf{2.53\ H}$$

The solution for Example 18-12 could have been obtained by using Eq. (18-27):

$$\frac{v_s}{v_p} = \frac{M}{L_p} \qquad (18\text{-}31)$$

$$M = L_p \frac{v_s}{v_p}$$

but

$$\frac{v_s}{v_p} = \frac{N_s}{N_p} \qquad (18\text{-}27)$$

so that

$$M = L_p \frac{N_s}{N_p} = 40\text{ H}\,\frac{1}{15.81} = 2.53\text{ H}$$

PRACTICAL TRANSFORMER

The four statements describing a perfect transformer refer to a device which is not realizable in practice. However, perfect transformer concepts are useful for many practical purposes. The results for Examples 18-9 to 18-12, for example, are adequate for dealing with the situations referred to. Furthermore, perfect transformer principles can be used to form a basis for the study of practical transformers since:

1. A high-quality power transformer approaches perfection to such a degree that it is usual to adopt these statements for calculations except when rigorous accuracy is required for design purposes.
2. A high-quality audio-frequency transformer approaches perfection over a wide frequency spectrum (typically 100 Hz to 5 kHz). At lower and higher frequencies, losses must be taken into account.
3. For use at radio frequencies (very high frequencies), transformers are

deliberately operated under nonperfect conditions. When this is done, transformer behavior can only be investigated by employing equivalent circuit techniques.

There are many factors which may be taken into account when analyzing the circuit behavior of a practical transformer. These include:

1. *Winding self-inductance.* Each winding is an inductor whose self-inductance results in an induced voltage when the winding is exposed to a magnetic field of varying strength.
2. *Winding resistance.* There is the obvious dc resistance, as measured by an ohmmeter, but there also is an added resistance factor caused by the skin effect. This effect is the tendency for alternating current to concentrate on the surface of a conductor, thus reducing its effective cross section.
3. *Leakage inductance.* It is not possible to get perfect flux linkage; the magnetic flux produced by a coil current does not link uniformly with each of that coil's turns, nor with each of the turns of a neighboring winding. This imperfect linkage is equivalent to inductance which fails to contribute usefully to transformer action.
4. *Eddy currents.* An iron transformer core is conducting material in which the magnetic field is concentrated. Currents within the core, due to induced emf, constitute energy loss.
5. *Hysteresis.* The crystal structure of an iron core changes with variation in magnetic flux. There is a form of inertia which inhibits the crystal structure in following the current changes. This inertia results in energy loss.
6. *Stray capacitance.* Between the turns of a given transformer winding, or between windings, there is capacitance resulting from the separation of conducting wire by insulating material.
7. *Radiation.* Some energy is radiated away from the transformer's magnetic and electric fields.

With the exception of dc resistance, each of the listed factors is dependent on frequency, often only being significant at high frequencies. Apart from capacitance and leakage inductance, all can be represented by equivalent resistance.

EQUIVALENT CIRCUIT OF A TRANSFORMER

Depending on the application and frequency of interest, the equivalent transformer circuit may take one of several forms. For our introductory purposes we will restrict our investigation to a two-port equivalent circuit which is valid for a linear, bilateral transformer, one with the same number of turns in its primary and secondary windings.

The transformer in Fig. 18-36*a* has coupling action, shown by the windings w_p and w_s (assumed to be inductanceless), coupled by the mutual

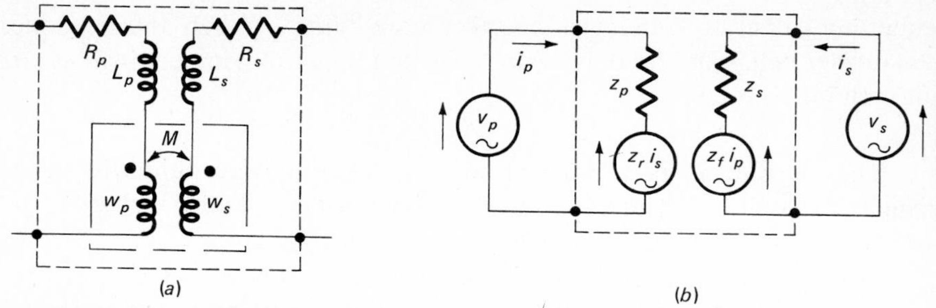

Fig. 18-36 Transformer equivalent circuits. (*a*) Basic; (*b*) *z* parameter.

inductance M. The inductances of the primary and secondary windings are shown by the elements L_p and L_s, and their losses by resistances R_p and R_s. An equivalent circuit in conventional two-port form is given in Fig. 18-36*b*. Sources v_p and v_s feed currents i_p and i_s into the upper terminals of the input and output ports. The mutual inductance coupling action is represented by reverse and forward transfer impedance parameters z_r and z_f. The total primary and secondary impedances are given by the local impedance parameters z_p and z_s:

$$z_r = z_f = jX_m \tag{18-36}$$

$$z_p = R_p + jX_p \tag{18-37}$$

$$z_s = R_s + jX_S \tag{18-38}$$

where X_m = mutual reactance (ωM), Ω
X_p = primary reactance (ωL_p), Ω
X_s = secondary reactance (ωL_s), Ω
R_p = resistance equivalent to primary losses, Ω
R_s = resistance equivalent to secondary losses, Ω

The mesh or loop equations are

$$z_p i_p + z_r i_s = v_p \tag{18-39}$$

$$z_f i_p + z_s i_s = v_s \tag{18-40}$$

TRANSFORMER OPERATING INTO A LOAD

Figure 18-37 shows the transformer as an equivalent two-port network with load Z_L connected to its output port. The load current i_L is of a direction opposite that of the conventional output port current $i_s (i_L = -i_s)$. The loop equations are obtained by applying Eqs. (18-39) and (18-40):

$$z_p i_p + z_r i_s = v_p \tag{18-39}$$

$$z_f i_p + z_s i_s = Z_L i_L = -Z_L i_s$$

or

$$z_f i_p + (z_s + Z_L) i_s = 0 \tag{18-41}$$

In matrix form Eqs. (18-39) and (18-41) become

$$\begin{bmatrix} z_p & z_r \\ z_f & (z_s + Z_L) \end{bmatrix} \begin{bmatrix} i_p \\ i_s \end{bmatrix} = \begin{bmatrix} v_p \\ 0 \end{bmatrix}$$

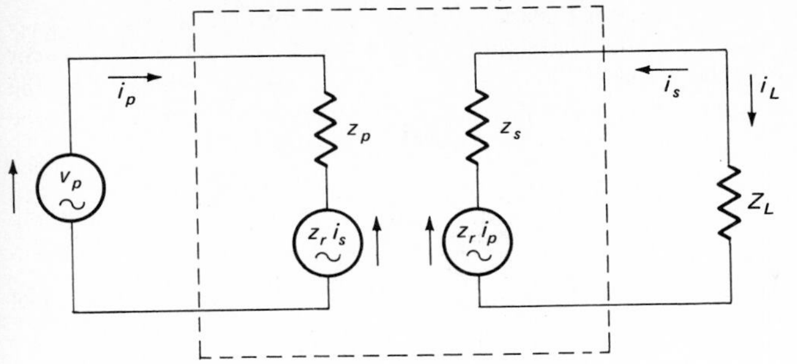

Fig. 18-37 Transformer operating into a load: z-parameter circuit.

The solutions for i_p and i_s are

$$i_p = \frac{z_s + Z_L}{z_p(z_s + Z_L) - z_f z_r} v_p \tag{18-42}$$

$$i_s = \frac{-z_f}{z_p(z_s + Z_L) - z_f z_r} v_p \tag{18-43}$$

Substituting for z_r, z_f, z_p, and z_s from Eqs. (18-36) to (18-38),

$$i_p = \frac{R_s + jX_s + Z_L}{(R_p + jX_p)(R_s + jX_s + Z_L) + X_m^2} v_p \tag{18-44}$$

$$i_s = \frac{-jX_m}{(R_p + jX_p)(R_s + jX_s + Z_L) + X_m^2} v_p \tag{18-45}$$

Example 18-13 An air-cored[1] transformer is used to couple radio-frequency energy in a radio receiver. The primary and secondary windings are identical, each having an inductance of 100 μH and a total effective resistance of 25 Ω. If 1 mV is supplied to the transformer at a frequency of 450 kHz, what voltage will it deliver to the 500-Ω resistive load presented by the input circuit of a transistor amplifier? Twenty percent of the primary flux links with the secondary winding.

Solution Figure 18-38 shows the conditions represented in the example.

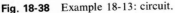

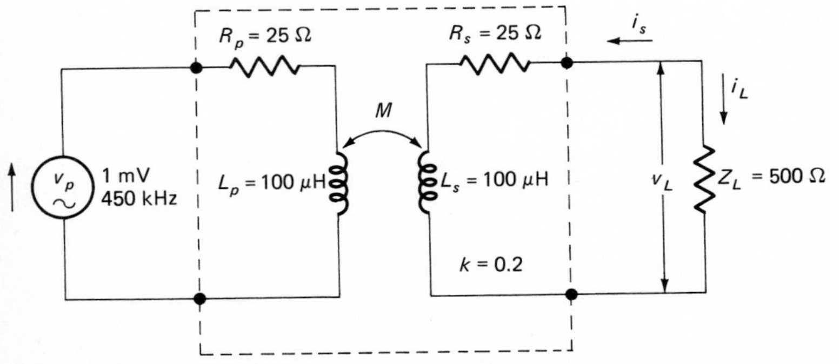

Fig. 18-38 Example 18-13: circuit.

[1] An air-cored transformer has coils wound on an insulator, such as a plastic tube, rather than on an iron core. This type of winding results in a low coefficient of coupling.

Equation (18-32) gave the mutual inductance of a perfect transformer as $M = \sqrt{L_p L_s}$, implying that this relationship gives the maximum possible value of mutual inductance for given primary and secondary inductances. The degree to which the flux generated by the primary current of a transformer links with the secondary winding is referred to as the *coupling coefficient k*. The coupling coefficient is defined by the equation

$$k = \frac{M}{\sqrt{L_p L_s}} \quad \text{or} \quad M = k\sqrt{L_p L_s} \tag{18-46}$$

In this example 20 percent of the primary flux links with the secondary winding. The coefficient of coupling is 0.2. Substituting this value into Eq. (18-46),

$$M = k\sqrt{L_p L_s} = 0.2\sqrt{100\ \mu\text{H} \times 100\ \mu\text{H}} = 20\ \mu\text{H}$$

$$\omega = 2\pi f = 2\pi \times 450\ \text{kHz} = 2.827\ \text{Mrad/s}$$

$$X_p = X_s = \omega L_p = \omega L_s = 2.827\ \text{Mrad/s} \times 100\ \mu\text{H} = 282.7\ \Omega$$

$$X_m = \omega M = 2.827\ \text{Mrad/s} \times 20\ \mu\text{H} = 56.55\ \Omega$$

Substituting the transformer resistance and reactance values into Eq. (18-45),

$$i_s = \frac{-jX_m}{(R_p + jX_p)(R_s + jX_s + Z_L) + X_m{}^2}\, v_p \tag{18-45}$$

$$= \frac{-j56.55\ \Omega}{(25 + j282.7)\ \Omega \times (25 + j282.7 + 500)\ \Omega + (56.55\ \Omega)^2} \times 1\ \text{mV}\ \underline{/0^\circ}$$

$$= \frac{56.55\ \Omega\ \underline{/-90^\circ} \times 1\ \text{mV}\ \underline{/0^\circ}}{283.8\ \Omega\ \underline{/84.95^\circ} \times 596.3\ \Omega\ \underline{/28.3^\circ} + 3198\ \Omega^2}$$

$$= \frac{56.55\ \Omega\ \underline{/-90^\circ} \times 1\ \text{mV}\ \underline{/0^\circ}}{169.2\ \text{k}\Omega^2\ \underline{/113.25^\circ} + 3.198\ \text{k}\Omega^2}$$

$$= \frac{56.55\ \Omega\ \underline{/-90^\circ} \times 1\ \text{mV}\ \underline{/0^\circ}}{(-66.79 + j155.5)\ \text{k}\Omega^2 + 3.198\ \text{k}\Omega^2}$$

$$= \frac{56.55\ \Omega\ \underline{/-90^\circ} \times 1\ \text{mV}\ \underline{/0^\circ}}{(-63.59 + j155.5)\ \text{k}\Omega^2} = \frac{56.55\ \Omega\ \underline{/-90^\circ} \times 1\ \text{mV}\ \underline{/0^\circ}}{168\ \text{k}\Omega^2\ \underline{/112.24^\circ}}$$

$$i_s = 0.3366\ \mu\text{A}\ \underline{/-202.24^\circ} = 0.3366\ \mu\text{A}\ \underline{/157.76^\circ}$$

The load current i_L has its direction out of the upper output port terminal, so its value is equal to $-i_s$:

$$i_L = -i_s = -0.3366\ \mu\text{A}\ \underline{/157.76^\circ} = 0.3366\ \mu\text{A}\ \underline{/-22.24^\circ}$$

The load voltage will be given by

$$v_L = i_L Z_L = 0.3366\ \mu\text{A}\ \underline{/-22.24^\circ} \times 500\ \Omega\ \underline{/0^\circ}$$

$$v_L = \mathbf{168.3\ \mu V\ \underline{/-22.24^\circ}} \quad \text{or} \quad \mathbf{\underline{/-0.3882\ rad}}$$

If the transformer in Example 18-13 had been perfect, R_p and R_s would both have been zero, the mutual inductance would have been the geometric mean of the primary and secondary inductances ($M = \sqrt{L_p L_s} = 100\ \mu\text{H}$), and the mutual reactance would have equaled each of the winding reactances ($X_m = \omega M = 282.7\ \Omega$). In such a situation the secondary current would have been

$$i_s = \frac{-j282.7\ \Omega}{j282.7\ \Omega \times (j282.7 + 500)\ \Omega + (282.7\ \Omega)^2} \times 1\ \text{mV}\ \underline{/0^\circ}$$

$$= \frac{282.7\ \Omega\ \underline{/-90^\circ} \times 1\ \text{mV}\ \underline{/0^\circ}}{282.7\ \Omega\ \underline{/90^\circ} \times 574.4\ \Omega\ \underline{/29.48^\circ} + (282.7\ \Omega)^2}$$

$$= \frac{282.7\ \Omega\ \underline{/-90^\circ} \times 1\ \text{mV}\ \underline{/0^\circ}}{162.4\ \text{k}\Omega^2\ \underline{/119.48^\circ} + 79.92\ \text{k}\Omega^2}$$

$$= \frac{282.7\ \Omega\ \underline{/-90^\circ} \times 1\ \text{mV}\ \underline{/0^\circ}}{(-79.92 + j141.4)\ \text{k}\Omega^2 + 79.92\ \text{k}\Omega^2}$$

$$= \frac{282.7\ \Omega\ \underline{/-90^\circ} \times 1\ \text{mV}\ \underline{/0^\circ}}{141.4\ \text{k}\Omega^2\ \underline{/90^\circ}}$$

$$i_s = 2\ \mu\text{A}\ \underline{/-180^\circ}$$

Therefore, $\quad i_L = -i_s = 2\ \mu\text{A}\ \underline{/0^\circ}$

and $\quad v_L = i_L Z_L = 2\ \mu\text{A}\ \underline{/0^\circ} \times 500\ \Omega = 1\ \text{mV}\ \underline{/0^\circ}$

Thus a perfect transformer with equal primary and secondary inductances transfers energy without loss, the load voltage being equal to the source voltage. This is clearly more effective energy transfer than that taking place with the transformer in Example 18-13, because of the losses and the low coefficient of coupling of that transformer. Despite this lack of effective energy transfer, air-core transformers are frequently used for radio-frequency coupling. The primary and/or secondary windings are often operated in a resonant condition, producing effects more significant in radio than efficient energy transfer.

EQUIVALENT T CIRCUIT OF A TRANSFORMER

Equations (18-39) and (18-40) are essentially the same as Eqs. (18-7) and (18-8) for the general two-port T network. Therefore the transformer equivalent circuit of Fig. 18-36 can be drawn as in Fig. 18-39.

Figure 18-39*a* shows a general two-port T network where $z_i = Z_1 + Z_c$ and $z_o = Z_2 + Z_c$. Figure 18-39*b* is a T equivalent circuit of a transformer. The coupling impedance in the transformer is the mutual reactance X_m, so Z_c in Fig. 18-39*a* becomes X_m in Fig. 18-39*b*.

In Fig. 18-39*a*,

$$z_i = Z_1 + Z_c \quad \text{or} \quad Z_1 = z_i - Z_c$$

$$z_o = Z_2 + Z_c \quad \text{or} \quad Z_2 = z_o - Z_c$$

Therefore, for Fig. 18-39*b*, using the expressions for z_p and z_s from Eqs. (18-37) and (18-38),

Z_1 becomes: $\quad z_p - Z_c = (R_p + jX_p) - jX_m = R_p + j(X_p - X_m)$

Z_2 becomes: $\quad z_s - Z_c = (R_s + jX_s) - jX_m = R_s + j(X_s - X_m)$

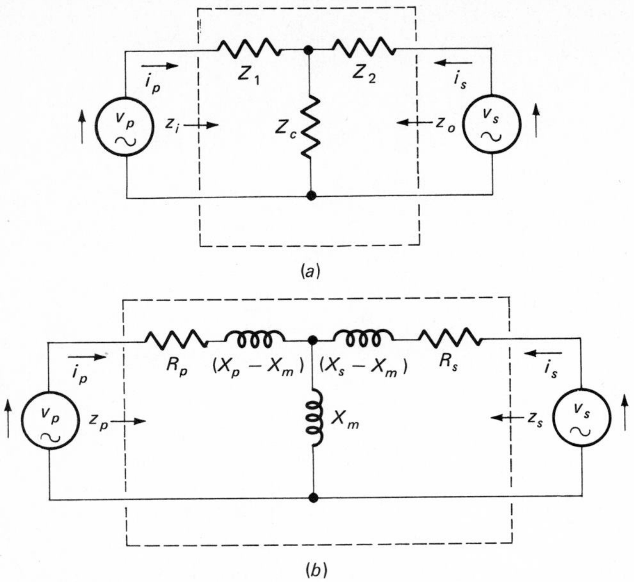

Fig. 18-39 Circuit containing (*a*) a generalized T network; (*b*) a T-network equivalent of a transformer.

To demonstrate the use of a T equivalent network of a transformer in circuit analysis, the problem in Example 18-13 will be repeated in Example 18-14.

Example 18-14 Using the data in Example 18-13, determine the value of the load voltage by applying an equivalent T network for the transformer.

Solution Figure 18-40*a* is Fig. 18-38 with reactance values inserted. A T equivalent of Fig. 18-40*a* is given in Fig. 18-40*b*. A Thevenin approach will give the value of the load voltage v_L.

Fig. 18-40 Example 18-14: (*a*) transformer circuit; (*b*) T-network equivalent.

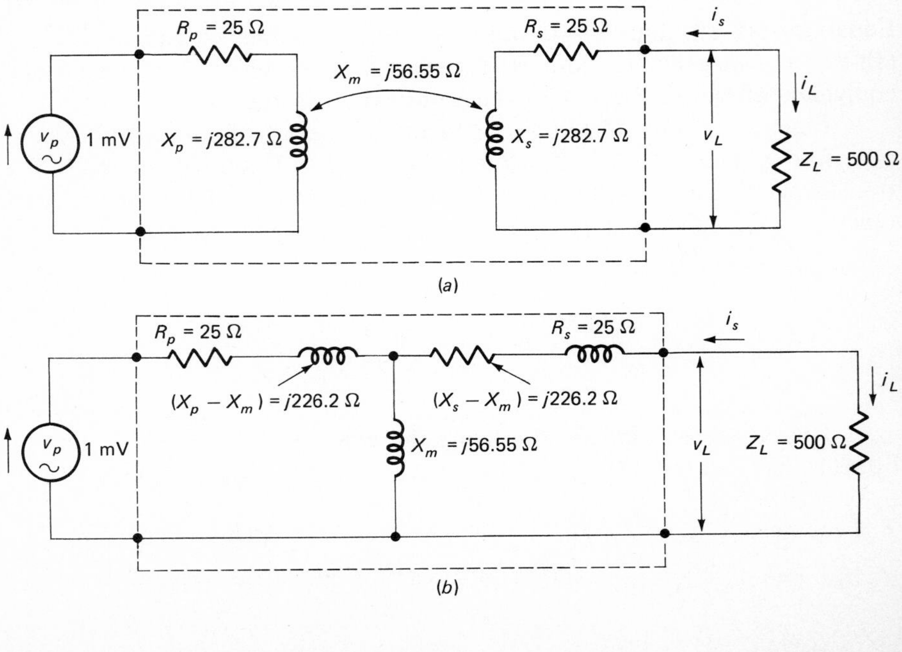

Figure 18-41 is a Thevenin equivalent of Fig. 18-40.

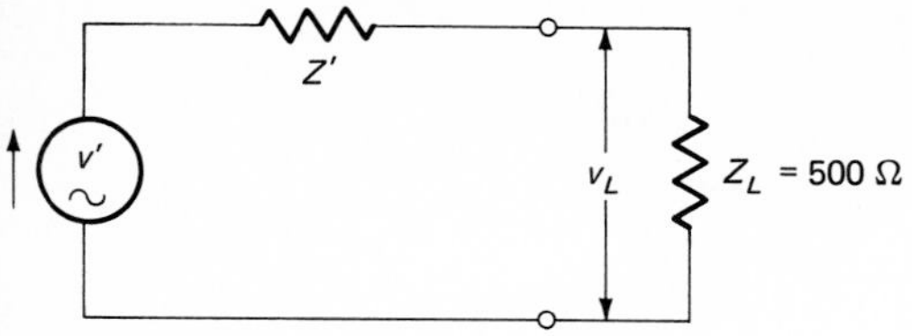

Fig. 18-41 Example 18-14: Thevenin equivalent.

Determine the value of the Thevenin voltage v':

$$v' = v_p \frac{X_m}{R_p + (X_p - X_m) + X_m} = 1 \text{ mV } \underline{/0^\circ} \frac{j56.55\ \Omega}{25\ \Omega + j226.2\ \Omega + j56.55\ \Omega}$$

$$v' = \frac{1 \text{ mV } \underline{/0^\circ} \times 56.55\ \Omega \underline{/90^\circ}}{283.8\ \Omega \underline{/84.95^\circ}} = 0.1993 \text{ mV } \underline{/5.05^\circ}$$

Determine the value of the Thevenin impedance Z':

$$Z' = [R_s + (X_s - X_m)] + \frac{X_m[R_p + (X_p - X_m)]}{X_m + [R_p + (X_p - X_m)]}$$

$$= (25 + j226.2)\ \Omega + \frac{j56.55\ \Omega \times (25 + j226.2)\ \Omega}{j56.55\ \Omega + (25 + j226.2)\ \Omega}$$

$$= (25 + j226.2)\ \Omega + \frac{56.55\ \Omega \underline{/90^\circ} \times 227.6\ \Omega \underline{/83.69^\circ}}{(25 + j282.7)\ \Omega}$$

$$= (25 + j226.2)\ \Omega + \frac{56.55\ \Omega \underline{/90^\circ} \times 227.6\ \Omega \underline{/83.69^\circ}}{283.8\ \Omega \underline{/84.95^\circ}}$$

$$= (25 + j226.2)\ \Omega + 45.35\ \Omega \underline{/88.74^\circ}$$

$$= (25 + j226.2)\ \Omega + (0.9972 + j45.33)\ \Omega$$

$$Z' = (26 + j271.53)\ \Omega$$

Apply the voltage divider principle:

$$v_L = v' \frac{Z_L}{Z_L + Z'} = 0.1993 \text{ mV } \underline{/5.05^\circ} \times \frac{500\ \Omega \underline{/0^\circ}}{500\ \Omega + (26 + j271.3)\ \Omega}$$

$$v_L = \frac{0.1993 \text{ mV } \underline{/5.05^\circ} \times 500\ \Omega \underline{/0^\circ}}{592\ \Omega \underline{/27.3^\circ}} = \mathbf{168.3\ \mu V\ \underline{/-22.25^\circ}} \quad \text{or} \quad \mathbf{\underline{/-0.3883 \text{ rad}}}$$

Example 18-15 For the circuits in Examples 18-13 and 18-14, what is (*a*) the primary current and (*b*) the input impedance seen by the source?

Solution

$$i_p = \frac{R_s + jX_s + Z_L}{(R_p + jX_p)(R_s + jX_s + Z_L) + X_m^2} v_p \tag{18-44}$$

The denominator of this equation is the same as that of Eq. (18-45). Therefore for this example we can use the calculated denominator in Example 18-13.

a.
$$i_p = \frac{(25 + j282.7 + 500)\ \Omega}{168 \text{ k}\Omega^2 \underline{/112.24^\circ}} \times 1 \text{ mV } \underline{/0^\circ}$$

$$i_p = \frac{596.3\ \Omega \underline{/28.3^\circ}}{168 \text{ k}\Omega^2 \underline{/112.24^\circ}} \times 1 \text{ mV } \underline{/0^\circ} = \mathbf{3.549\ \mu A\ \underline{/-83.94^\circ}} \quad \text{or} \quad \mathbf{\underline{/-1.465 \text{ rad}}}$$

b. The input impedance seen by the source is the ratio of the input voltage to the input current:

$$Z_{\text{in}} = \frac{v_p}{i_p} = \frac{1 \text{ mV} \underline{/0^\circ}}{3.549\ \mu\text{A} \underline{/-83.94^\circ}} = \mathbf{281.8\ \Omega\ \underline{/83.94^\circ}} \qquad \text{or} \qquad \mathbf{\underline{/1.465\ rad}}$$

The input current and impedance values which are the solutions to Example 18-15 indicate that the loosely coupled circuit in Examples 18-13 and 18-14, viewed from the source, appears to be almost the same as if the load were open circuit (if the load were open-circuited, Z_{in} would be 283.8 Ω $\underline{/84.95^\circ}$ and i_p would be 3.524 μA $\underline{/-84.95^\circ}$). The primary current is approximately 10 times as great as the secondary current. These conclusions illustrate the fact that loose coupling tends to isolate the load from the source, making the degree of energy transfer low.

18-9 THE TRANSISTOR AS A TWO-PORT NETWORK

A transistor is a semiconductor device used as an amplifier (magnifier) of direct and alternating voltages, currents, and powers. The transistor is a nonlinear device and in some applications must be considered as such. There are many applications, however, in which the transistor's action is quasi-linear, so near to linear that it can be so regarded.

As an example of the practical use of *h* parameters, the transistor will be viewed here as a linear amplifier of ac voltages.

The internal functioning of a transistor is the concern of electronics or physics textbooks. However, the following treatment will show how transistor circuits can be regarded as electric circuits if the transistor parameters are obtained by measurement or by manufacturer's specification, both of which are usual in industry.

To obtain the *h* parameters of a transistor by measurement, a setup such as that depicted in Fig. 18-42 may be used.

In Fig. 18-42 an NPN transistor is shown connected to a dc supply V_{cc} which, via resistors R_1 and R_2, provides the correct dc values of voltage and current to set the transistor operating conditions. If the characteristics of a PNP transistor are to be measured, the polarity of V_{cc} is reversed.

The ammeters and voltmeters are connected to the transistor by capacitors and thus measure ac values only.

The values of resistors R_1 and R_2 are selected to be high enough to avoid significant interference with the ac quantities being measured.

The meters measure the ac values of base-to-emitter voltage v_{be}, base current i_b, collector-to-emitter voltage v_{ce}, and collector current i_c.

Reference to Sec. 18-6 shows that the hybrid parameters h_i and h_f are defined under output short-circuit conditions. These conditions may be simulated for a transistor by connecting a large-value capacitor between the collector and the emitter (C_2 in Fig. 18-42*a*). The low reactance of C_2,

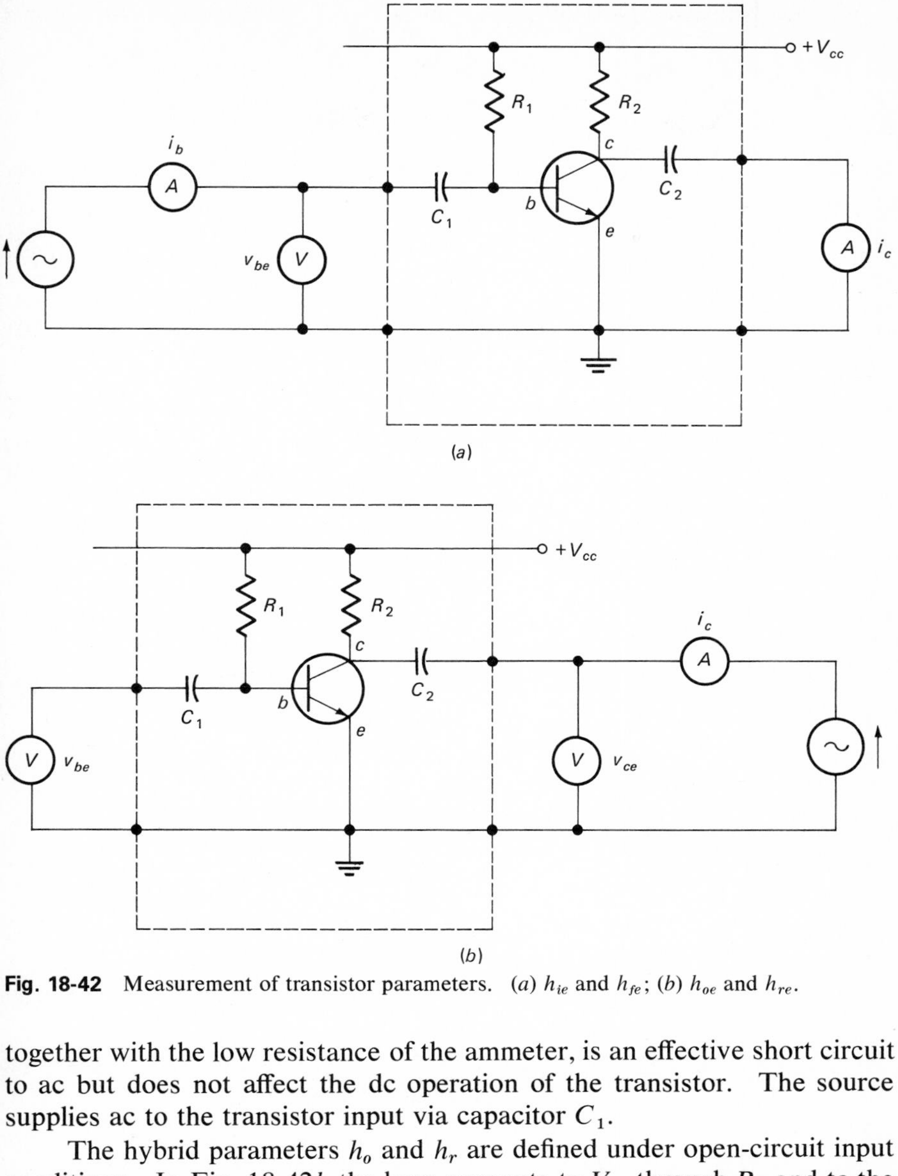

Fig. 18-42 Measurement of transistor parameters. (*a*) h_{ie} and h_{fe}; (*b*) h_{oe} and h_{re}.

together with the low resistance of the ammeter, is an effective short circuit to ac but does not affect the dc operation of the transistor. The source supplies ac to the transistor input via capacitor C_1.

The hybrid parameters h_o and h_r are defined under open-circuit input conditions. In Fig. 18-42*b* the base connects to V_{cc} through R_1 and to the emitter through the voltmeter. As both R_1 and the voltmeter are high resistances, the input is virtually open circuit. The source supplies ac to the output port through capacitor C_2.

In Fig. 18-42 the transistor is arranged for measurement of its common emitter characteristics, as is evidenced by the emitter *e* being connected to the common ground line. In these circumstances, the hybrid parameters are given an extra subscript, *e*. Manufacturers often give transistor hybrid characteristics in this form in their specifications.

With transistor symbols substituted, Eqs. (18-21) to (18-24) become

$$h_{ie} = \left.\frac{v_{be}}{i_b}\right|_{v_{ce}=0} \tag{18-47}$$

$$h_{fe} = \left.\frac{i_c}{i_b}\right|_{v_{ce}=0} \tag{18-48}$$

$$h_{re} = \left.\frac{v_{be}}{v_{ce}}\right|_{i_b=0} \tag{18-49}$$

$$h_{oe} = \left.\frac{i_c}{v_{ce}}\right|_{i_b=0} \tag{18-50}$$

The transistor is neither an ordinary conductive network, such as a T network, nor an inductive network, like a transformer. In Fig. 18-42 the input and output currents and voltages all have positive sense, so all four hybrid parameters have positive sense.

TRANSISTOR EQUIVALENT CIRCUIT

A hybrid-parameter equivalent circuit of a transistor is shown in Fig. 18-43.

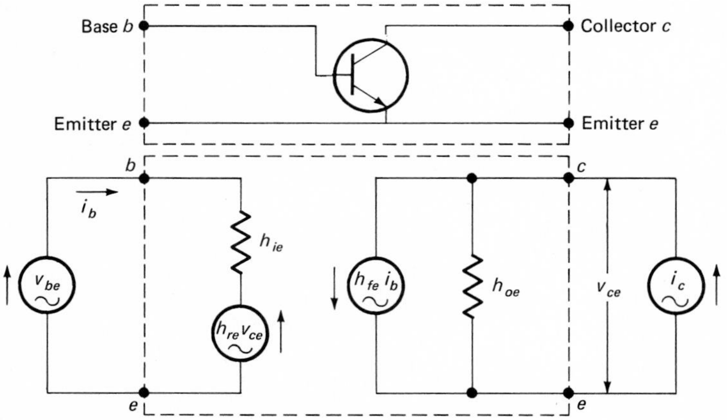

Fig. 18-43 Transistor h-parameter equivalent circuit.

The input loop equation and the output nodal equation are adapted from the h-parameter equations (18-19) and (18-20):

$$h_{ie}i_b + h_{re}v_{ce} = v_{be} \tag{18-51}$$

$$h_{fe}i_b + h_{oe}v_{ce} = i_c \tag{18-52}$$

Example 18-16 A transistor has these parameters: $h_{ie} = 1.5\ \text{k}\Omega$, $h_{oe} = 40\ \mu\text{S}$, $h_{re} = 0.0005$, and $h_{fe} = 100$. What is the voltage gain v_{ce}/v_{be} of the amplifier in Fig. 18-44a in which this transistor operates into a 4.7-kΩ load? Assume that the circuit elements R_1, R_2, R_3, C_1, C_2, and C_3 have negligible effect on the transistor's ac characteristics.

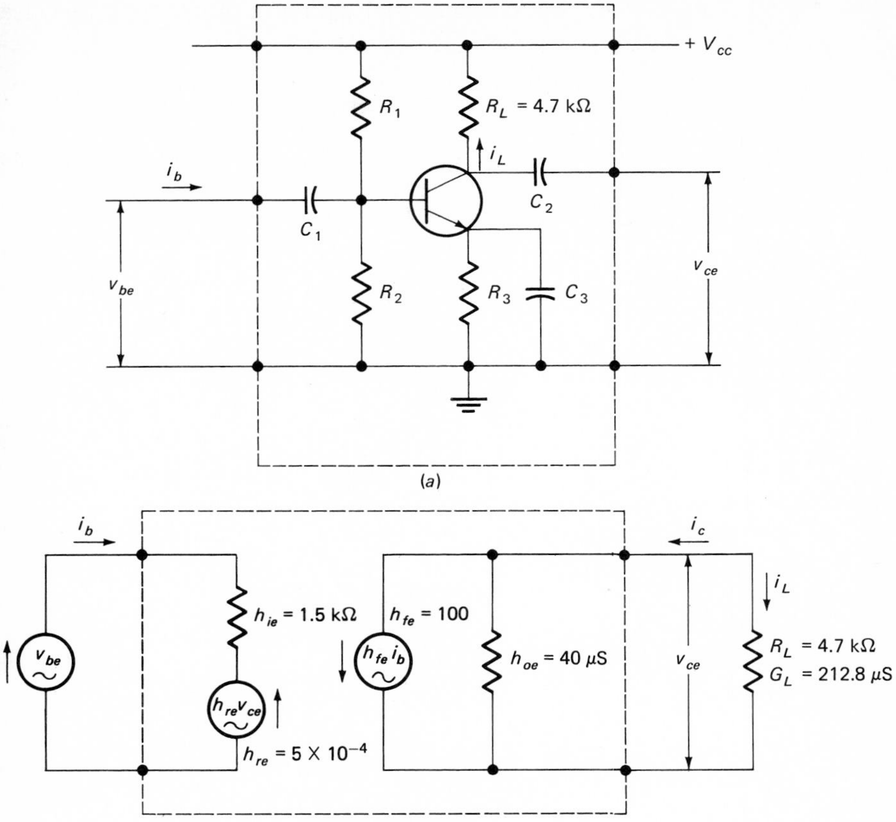

Fig. 18-44 Example 18-16: (*a*) given circuit; (*b*) *h*-parameter equivalent.

Solution Figure 18-44*b* shows an *h*-parameter equivalent to the circuit in Fig. 18-44*a*. The input loop and output nodal equations are

$$h_{ie}i_b + h_{re}v_{ce} = v_{be} \tag{18-51}$$

$$1.5\ k\Omega\ i_b + 0.0005v_{ce} = v_{be}$$

From Eq. (18-52),

$$h_{fe}i_b + h_{oe}v_{ce} = i_c = -i_L = -G_Lv_{ce}$$

$$100i_b + (40 + 212.8) \times 10^{-6}\ S\ v_{ce} = 0$$

The equations in matrix form are

$$\begin{bmatrix} 1.5 \times 10^3 & 5 \times 10^{-4} \\ 10^2 & 2.528 \times 10^{-4} \end{bmatrix} \begin{bmatrix} i_b \\ v_{ce} \end{bmatrix} = \begin{bmatrix} v_{be} \\ 0 \end{bmatrix} \qquad i_b \text{ in amperes; } v_{ce} \text{ and } v_{be} \text{ in volts}$$

The solution for v_{ce} is

$$v_{ce} = -303.8\ v_{be}$$

or

$$\frac{v_{ce}}{v_{be}} = \mathbf{-303.8}$$

The negative result for Example 18-15 indicates that there is a phase reversal between input and output. This phase reversal is a factual characteristic of a common emitter transistor amplifier.

Example 18-17 What is the input impedance of the transistor amplifier in Example 18-16?
Solution The input impedance of a network is the ratio of input voltage to input current. In the case of the common emitter transistor amplifier, this is $Z_{\text{in}} = v_{be}/i_b$.

The solution for i_b from the matrix in Example 18-16 is

$$i_b = (7.679 \times 10^{-4}\ \text{S})\, v_{be}$$

Therefore,
$$Z_{\text{in}} = \frac{v_{be}}{i_b} = \frac{v_{be}}{(7.679 \times 10^{-4}\ \text{S})\, v_{be}} = \mathbf{1.302\ k\Omega}$$

Example 18-18 A small power transistor, such as might be used in a car radio, has the following parameters: $h_{ie} = 250\ \Omega$, $h_{oe} = 4.5$ mS, $h_{fe} = 80$, and $h_{re} = 1.5 \times 10^{-3}$. What is the power gain of an amplifier which uses this transistor and operates into a 100-Ω resistive load?
Solution A typical common emitter power amplifier circuit is shown in Fig. 18-45*a*. In this circuit the output transformer connects the transistor to the loudspeaker load. If the loudspeaker is regarded as a resistive impedance of 3 Ω, the transformer (if assumed to be perfect) will have a primary-to-secondary turns ratio of 5.77, resulting in a collector load of 100 Ω. The hybrid parameter equivalent circuit is given in Fig. 18-45*b*.

Fig. 18-45 Example 18-18: circuits. (*a*) Typical common emitter power amplifier; (*b*) *h*-parameter equivalent.

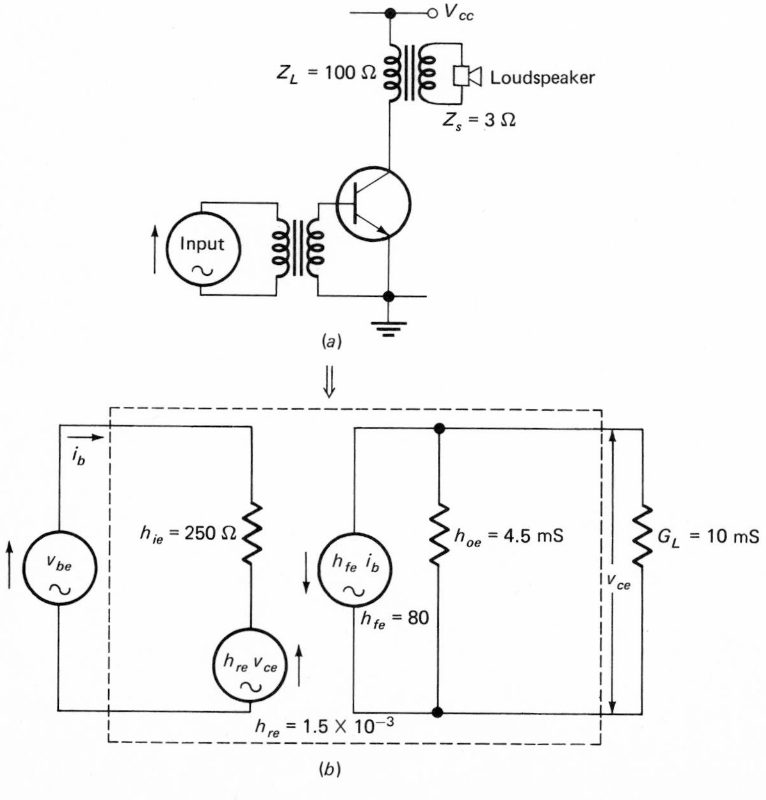

The input mesh and output nodal equations are

$$h_{ie}i_b + h_{re}v_{ce} = v_{be} \qquad (18\text{-}51)$$

$$250\ \Omega\ i_b + 1.5 \times 10^{-3}\ v_{ce} = v_{be}$$

From Eq. (18-52),

$$h_{fe}i_b + (h_{oe} + G_L)v_{ce} = 0$$

$$80\ i_b + (4.5 + 10) \times 10^{-3}\ \text{S}\ v_{ce} = 0$$

The matrix form for these equations is

$$\begin{bmatrix} 250 & 1.5 \times 10^{-3} \\ 80 & 1.45 \times 10^{-2} \end{bmatrix} \begin{bmatrix} i_b \\ v_{ce} \end{bmatrix} \begin{bmatrix} v_{be} \\ 0 \end{bmatrix} \qquad i_b \text{ in amperes; } v_{ce} \text{ and } v_{be} \text{ in volts}$$

The solutions for i_b and v_{ce} are

$$i_b = 4.137 \times 10^{-3}\ \text{S}\ v_{be}$$

$$v_{ce} = -22.82v_{be}$$

The input and output powers are computed:

$$P_{\text{in}} = i_b v_{be} = 4.137 \times 10^{-3}\ \text{S}(v_{be})^2\ \text{W}$$

$$P_{\text{out}} = (v_{ce})^2 G_L = (-22.82v_{be})^2 \times 10\ \text{mS} = 5.207\ \text{S}(v_{be})^2\ \text{W}$$

The power gain A_P is given by:

$$A_P = \frac{P_{\text{out}}}{P_{\text{in}}} = \frac{5.207\ \text{S}(v_{be})^2\ \text{W}}{4.137 \times 10^{-3}\ \text{S}(v_{be})^2\ \text{W}}$$

$$A_P = \mathbf{1259}$$

Some feeling for the quantities involved in the amplifier in Example 18-18 can be obtained. It is typical for the peak-to-peak voltage across the load impedance to be of the order of 12 V. For a sine wave, this would be 4.243 V rms. If it is assumed that this is indeed the load voltage, then the output power is

$$P_{\text{out}} = (V_L)^2 G_L = (4.243\ \text{V})^2 \times 10\ \text{mS} = 180\ \text{mW}$$

For a power gain of 1259 the input power would thus need to be 0.143 mW. The input impedance v_{be}/i_b is 241.7 Ω, so the input voltage would be 0.186 V rms or 0.526 V peak to peak, a suitable value for a silicon transistor.

18-10 SUMMARY OF EQUATIONS

z parameters

$$(Z_1 + Z_c)i_i + Z_c i_o = v_i \qquad (18\text{-}1) \qquad z_i = Z_1 + Z_c \qquad (18\text{-}3)$$

$$Z_c i_i + (Z_2 + Z_c)i_o = v_o \qquad (18\text{-}2) \qquad z_o = Z_2 + Z_c \qquad (18\text{-}4)$$

$z_r = Z_c$ (18-5)

$z_f = Z_c$ (18-6)

$z_i i_i + z_r i_o = v_i$ (18-7)

$z_f i_i + z_o i_o = v_o$ (18-8)

$z_i = v_i/i_i \; |i_o = 0$ (18-9)

$z_f = v_o/i_i \; |i_o = 0$ (18-10)

$z_r = v_i/i_o \; |i_i = 0$ (18-11)

$z_o = v_o/i_o \; |i_i = 0$ (18-12)

y parameters

$y_i v_i + y_r v_o = i_i$ (18-13)

$y_f v_i + y_o v_o = i_o$ (18-14)

$y_i = i_i/v_i \; |v_o = 0$ (18-15)

$y_f = i_o/v_i \; |v_o = 0$ (18-16)

$y_r = i_i/v_o \; |v_i = 0$ (18-17)

$y_o = i_o/v_o \; |v_i = 0$ (18-18)

h parameters

$h_i i_i + h_r v_o = v_i$ (18-19)

$h_f i_i + h_o v_o = i_o$ (18-20)

$h_i = v_i/i_i \; |v_o = 0$ (18-21)

$h_f = i_o/i_i \; |v_o = 0$ (18-22)

$h_r = v_i/v_o \; |i_i = 0$ (18-23)

$h_o = i_o/v_o \; |i_i = 0$ (18-24)

Transformer

$\Lambda_p = N_p \Phi_c$ (18-25)

$\Lambda_s = N_s \Phi_c$ (18-26)

$v_s/v_p = N_s/N_p$ (18-27)

$v_s/v_p = \sqrt{L_s/L_p}$ (18-28)

Transformer (continued)

$v_p = L_p \, di_p/dt$ (18-29)

$v_s = M \, di_p/dt$ (18-30)

$v_s/v_p = M/L_p$ (18-31)

$M = \sqrt{L_p L_s}$ (18-32)

$i_s/i_p = v_p/v_s$ (18-33)

$i_s/i_p = N_p/N_s$ (18-34)

$Z_L/Z_{in} = (N_s/N_p)^2$ (18-35)

$z_r = z_f = jX_m$ (18-36)

$z_p = R_p + jX_p$ (18-37)

$z_s = R_s + jX_s$ (18-38)

$z_p i_p + z_r i_s = v_p$ (18-39)

$z_f i_p + z_s i_s = v_s$ (18-40)

$z_f i_p + (z_s + Z_L) i_s = 0$ (18-41)

$i_p = \{(z_s + Z_L)/[z_p(z_s + Z_L) - z_f z_r]\} v_p$ (18-42)

$i_s = \{-z_f/[z_p(z_s + Z_L) - z_f z_r]\} v_p$ (18-43)

$i_p = \{(R_s + jX_s + Z_L)/[(R_p + jX_p)(R_s + jX_s + Z_L) + X_m^2]\} v_p$ (18-44)

$i_s = \{-jX_m/[(R_p + jX_p)(R_s + jX_s + Z_L) + X_m^2]\} v_p$ (18-45)

$k = M/\sqrt{L_p L_s}$ or $M = k\sqrt{L_p L_s}$ (18-46)

Transistor

$h_{ie} = v_{be}/i_b \; |v_{ce} = 0$ (18-47)

$h_{fe} = i_c/i_b \; |v_{ce} = 0$ (18-48)

$h_{re} = v_{be}/v_{ce} \; |i_b = 0$ (18-49)

$h_{oe} = i_c/v_{ce} \; |i_b = 0$ (18-50)

$h_{ie} i_b + h_{re} v_{ce} = v_{be}$ (18-51)

$h_{fe} i_b + h_{oe} v_{ce} = i_c$ (18-52)

EXERCISE PROBLEMS

Sections 18-3 and 18-5

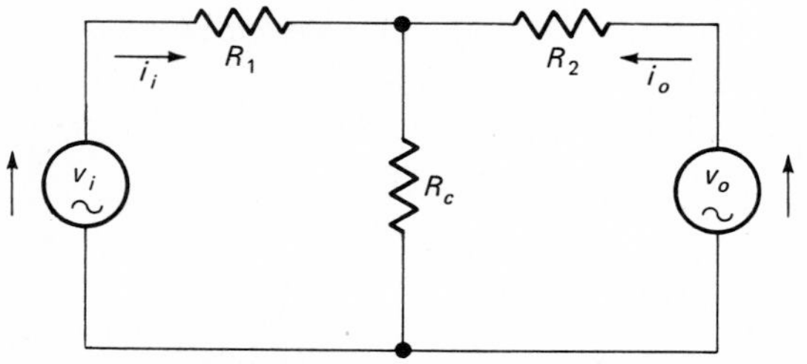

Fig. 18-1P Problem 18-1: circuit.

18-1 Using the input-output approach, find i_i and i_o in Fig. 18-1P when

a. $R_1 = 1.2$ kΩ, $R_2 = 1.5$ kΩ, $R_c = 2.2$ kΩ, $v_i = 10$ V, $v_o = 10$ V
(*ans.* 1.938 mA, 1.55 mA)

b. $R_1 = 3.3$ kΩ, $R_2 = 6.8$ kΩ, $R_c = 4.7$ kΩ, $v_i = 20$ V, $v_o = 5$ V
(*ans.* 2.954 mA, −0.7727 mA)

c. $R_1 = 220$ Ω, $R_2 = 470$ Ω, $R_c = 1$ kΩ, $v_i = 10$ V, $v_o = 20$ V
(*ans.* −6.68 mA, 18.15 mA)

d. $R_1 = 10$ kΩ, $R_2 = 15$ kΩ, $R_c = 18$ kΩ, $v_i = 10$ V, $v_o = -10$ V
(*ans.* 850.1 μA, −766.7 μA)

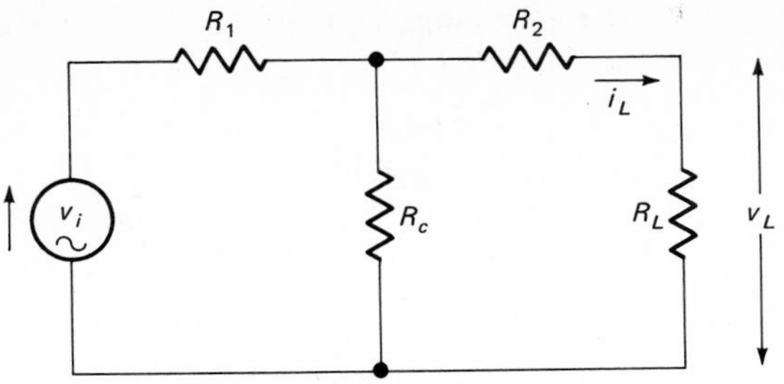

Fig. 18-2P Problem 18-2: circuit.

18-2 Using the input-output approach, find i_L and v_L in Fig. 18-2P when

a. $v_i = 10$ V, $R_1 = 1.2$ kΩ, $R_2 = 1.5$ kΩ, $R_c = 2.2$ kΩ, $R_L = 1$ kΩ
b. $v_i = 50$ mV, $R_1 = 960$ Ω, $R_2 = 360$ Ω, $R_c = 320$ Ω, $R_L = 600$ Ω
c. $v_i = 5$ V, $R_1 = 5$ kΩ, $R_2 = 1$ kΩ, $R_c = 2$ kΩ, $R_L = 2$ kΩ

18-3 Find v_L across a 1-kΩ load attached to the output of an electronic network which has an input of 5 V and z parameters of

a. $z_i = 8$ kΩ, $z_o = 8$ kΩ, $z_f = 5$ kΩ, $z_r = 6$ kΩ (*ans.* 0.595 V)
b. $z_i = 1$ kΩ, $z_o = 8$ kΩ, $z_f = 5$ kΩ, $z_r = 6$ kΩ (*ans.* −1.19 V)
c. $z_i = 2$ kΩ, $z_o = 5$ kΩ, $z_f = 25$ kΩ, $z_r = 2$ kΩ (*ans.* −3.29 V)

Sections 18-4 and 18-5

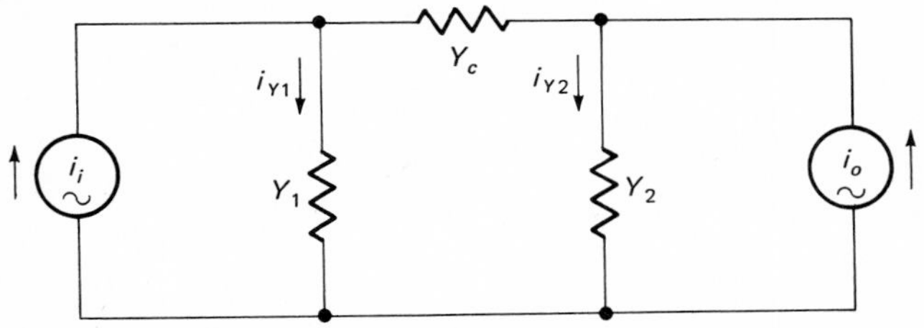

Fig. 18-3P Problem 18-4: circuit.

18-4 Using the input-output approach and y parameters, find i_{Y1} and i_{Y2} in Fig. 18-3P when

a. $Y_1 = 1$ mS, $Y_2 = 2$ mS, $Y_c = 3$ mS, $i_i = 10$ mA, $i_o = 10$ mA
b. $Y_1 = 1$ mS, $Y_2 = 2$ mS, $Y_c = 3$ mS, $i_i = 10$ mA, $i_o = -10$ mA
c. $Y_1 = 15$ mS, $Y_2 = 5$ mS, $Y_c = 1$ mS, $i_i = -5$ mA, $i_o = -15$ mA

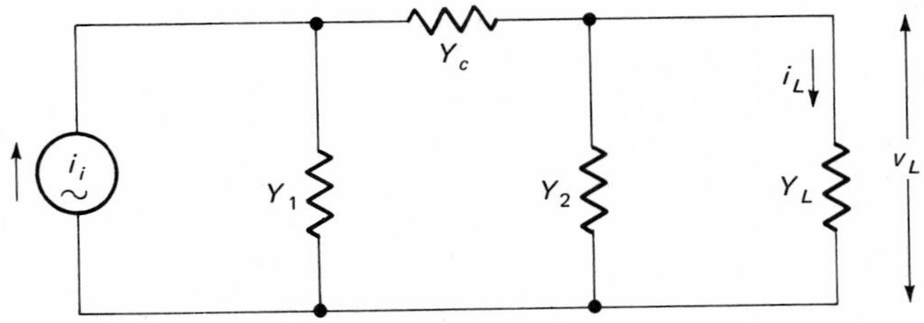

Fig. 18-4P Problem 18-5: circuit.

18-5 Using the input-output approach and y parameters, find i_L and v_L in Fig. 18-4P when

a. $i_i = 50$ mA, $Y_1 = 1$ mS, $Y_2 = 5$ mS, $Y_c = 4$ mS, $Y_L = 10$ mS
(*ans.* 25.32 mA, 2.532 V)
b. $i_i = 25$ mA, $Y_1 = 2.5$ mS, $Y_2 = 2.5$ mS, $Y_c = 4$ mS, $Y_L = 1.667$ mS
(*ans.* −4.495 mA, 2.697 V)
c. $i_i = 100$ μA, $Y_1 = 50$ μS, $Y_2 = 250$ μS, $Y_c = 250$ μS, $Y_L = 50$ μS
(*ans.* −12.2 μA, 0.2439 V)

18-6 Find v_L across a 1-mS load attached to the output of an electronic network which has an input current of 20 mA and y parameters:

a. $y_i = 250\ \mu S$, $y_o = 500\ \mu S$, $y_f = 150\ \mu S$, $y_r = 200\ \mu S$

b. $y_i = 200\ \mu S$, $y_o = 500\ \mu S$, $y_f = 150\ \mu S$, $y_r = 250\ \mu S$

c. $y_i = 2$ mS, $y_o = 2$ mS, $y_f = 5$ mS, $y_r = 3$ mS

18-7 In the circuit for Example 18-5 in the text, a defective rectifier caused the following conditions to occur. The dc component fell to 42.5 V, the ac component fell to 51.42 V, and the ripple frequency became 60 Hz. Find the magnitude of the output ripple ("ripple" refers to the ac component). (*ans.* 345 mV $\angle 124.4°$)

Sections 18-6 and 18-7

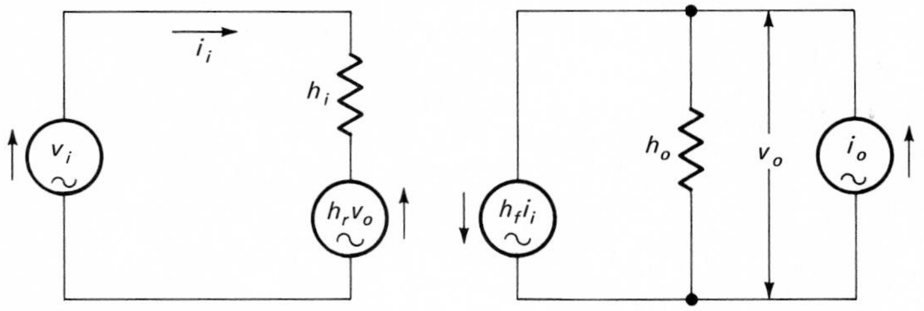

Fig. 18-5P Problem 18-8: circuit.

18-8 Find i_i and v_o for the circuit in Fig. 18-5P when $v_i = 10$ V, $i_o = 5$ mA, and

a. $h_i = 1\ k\Omega$, $h_o = 1$ mS, $h_f = 10$, $h_r = 0.2$

b. $h_i = 1\ k\Omega$, $h_o = 1$ mS, $h_f = 10$, $h_r = -0.2$

c. $h_i = 1\ k\Omega$, $h_o = 1$ mS, $h_f = -10$, $h_r = 0.2$

d. $h_i = 1\ k\Omega$, $h_o = 1$ mS, $h_f = -10$, $h_r = -0.2$

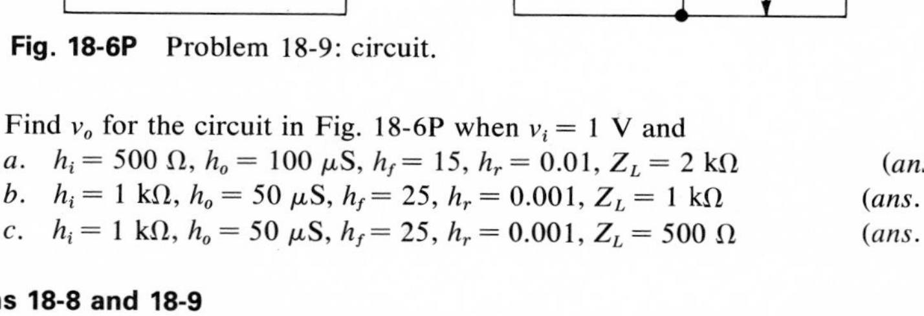

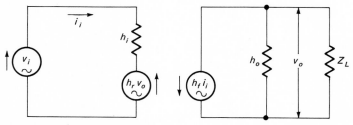

Fig. 18-6P Problem 18-9: circuit.

18-9 Find v_o for the circuit in Fig. 18-6P when $v_i = 1$ V and

a. $h_i = 500\ \Omega$, $h_o = 100\ \mu S$, $h_f = 15$, $h_r = 0.01$, $Z_L = 2\ k\Omega$ (*ans.* -100 V)

b. $h_i = 1\ k\Omega$, $h_o = 50\ \mu S$, $h_f = 25$, $h_r = 0.001$, $Z_L = 1\ k\Omega$ (*ans.* -24.39 V)

c. $h_i = 1\ k\Omega$, $h_o = 50\ \mu S$, $h_f = 25$, $h_r = 0.001$, $Z_L = 500\ \Omega$ (*ans.* -12.35 V)

Sections 18-8 and 18-9

18-10 A common emitter amplifier, using a 2N1305 transistor, has an equivalent circuit as shown in Fig. 18-7P. The characteristics of the transistor under the given operating conditions are $h_{ie} = 800\ \Omega$, $h_{oe} = 230\ \mu S$, $h_{fe} = 95$, and $h_{re} = 13 \times 10^{-4}$. If $v_i = 0.5$ V and $R_s = 1.5\ k\Omega$, find v_o when Z_L is

a. 560 Ω

b. 1000 Ω

c. 1500 Ω

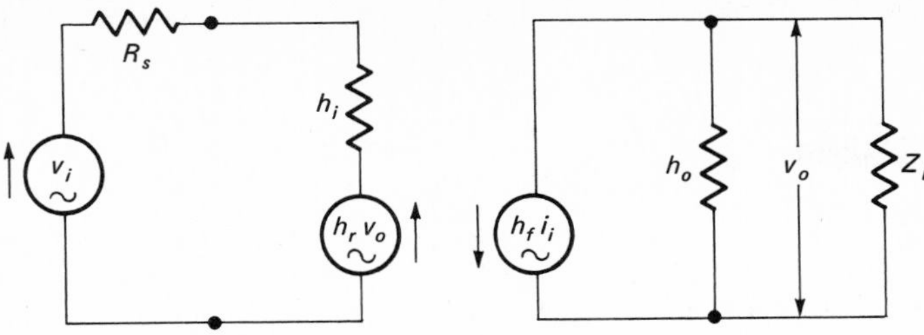

Fig. 18-7P Problems 18-10 to 18-13: circuit.

18-11 A common base amplifier, using a 2N1305 transistor, has an equivalent circuit as shown in Fig. 18-7P. The characteristics of the transistor under the given operating conditions are $h_{ib} = 20\ \Omega$, $h_{ob} = 1.3 \times 10^{-6}$ S, $h_{fb} = -0.99$, and $h_{rb} = 7 \times 10^{-4}$. If $v_i = 0.5$ V and $R_s = 100\ \Omega$, find v_o when Z_L is

a. 560 Ω (*ans.* 2.3 V)
b. 1000 Ω (*ans.* 4.096 V)
c. 1500 Ω (*ans.* 6.123 V)

18-12 A common emitter power amplifier, with an equivalent circuit as shown in Fig. 18-7P, uses an MJE2801 power transistor with the following characteristics: $h_{ie} = 170\ \Omega$, $h_{oe} = 320\ \mu$S, $h_{fe} = 160$, and $h_{re} = 9 \times 10^{-4}$. If $v_i = 0.2$ V and $R_s = 220\ \Omega$, find the power output to the load when Z_L is:

a. 200 Ω
b. 300 Ω
c. 400 Ω

18-13 If a perfect transformer is used to couple the output of the power amplifier in Prob. 18-12 to an 8-Ω speaker, find the turns ratio required, the speaker current, and the transformer primary current for each of the three cases.
(*ans.* *a.* 5:1, 414.3 mA, 82.86 mA; *b.* 6.124:1, 510.1 mA, 83.3 mA; *c.* 7.07:1, 591.6 mA, 83.67 mA)

BIBLIOGRAPHY

Adams, "Technical Mathematics," McGraw-Hill Book Company, New York.

ASME, "Letter Symbols for Quantities Used in Electrical Science and Electrical Engineering," USAS Y10.5-1968, American Society of Mechanical Engineers, New York.

BSI, "The Use of SI Units," British Standards Institution, London.

Carter, "Introduction to Electrical Circuit Analysis," Holt, Rinehart and Winston, New York.

Cooke and Adams, "Basic Mathematics For Electronics," McGraw-Hill Book Company, New York.

DeFrance, "Electrical Fundamentals," Prentice-Hall, Inc., Englewood Cliffs, N. J.

Dyson, "Introductory Physics for Electrical Engineers," McGraw-Hill Book Company, New York.

Edminister, "Electric Circuits," Schaum's Outline Series, McGraw-Hill Book Company, New York.

Godfrey and Siddons, "Four-Figure Tables," Cambridge University Press, New York.

Hayt and Kemmerly, "Engineering Circuit Analysis," McGraw-Hill Book Company, New York.

IEEE, "Standard and American National Standard Graphical Symbols for Electrical and Electronic Diagrams," ANSI Y32.2 or IEEE 315, Institute of Electrical and Electronics Engineers, New York.

Johnson, "Electric Circuits," Holt, Rinehart and Winston, New York.

Korn, "Basic Tables in Electrical Engineering," McGraw-Hill Book Company, New York.

Leach, "Basic Electric Circuits," John Wiley and Sons, New York.

Lurch, "Electric Circuits," John Wiley and Sons, New York.

Richmond, "Calculus for Electronics," McGraw-Hill Book Company, New York.

Romanowitz, "Introduction to Electric Circuits," John Wiley and Sons, New York.

Tedeschi, "Problems Book for Electric Circuits," Prentice-Hall, Inc., Englewood Cliffs, N.J.

Vitrogan, "Elements of Electric and Magnetic Circuits," Holt, Rinehart and Winston, New York.

APPENDIX

LINEAR AND NONLINEAR RESISTANCE

If the voltage and current associated with a resistive device are directly proportional to each other, the resistance is said to be *linear*. Conversely, *nonlinear* resistance implies lack of proportionality between voltage and current. "Constant resistance" is a better term than "linear resistance," since it is the relationship between current and voltage which is linear. Perhaps the term "constant" is not often used because there is no nice opposite term ("nonconstant" is an awkward word; "inconstant" might give a false impression).

Figure A-1 shows the relationship between voltage and current in linear and nonlinear resistances.

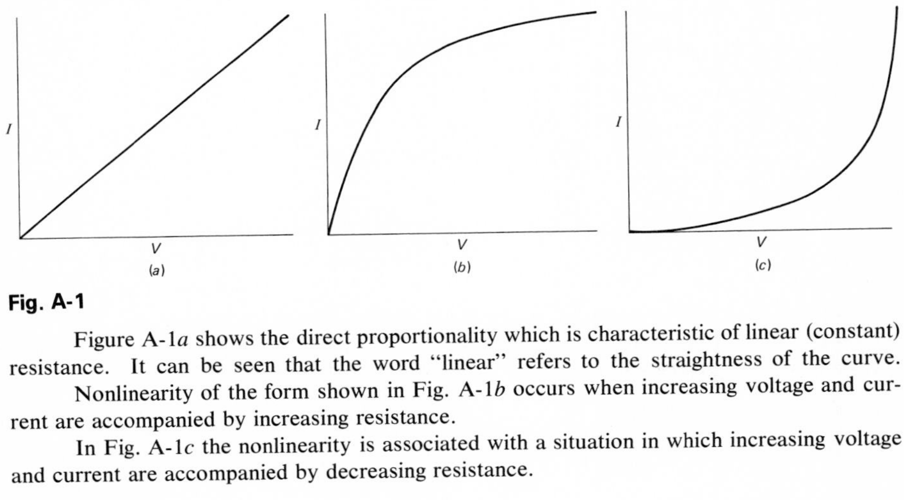

Fig. A-1

Figure A-1*a* shows the direct proportionality which is characteristic of linear (constant) resistance. It can be seen that the word "linear" refers to the straightness of the curve.

Nonlinearity of the form shown in Fig. A-1*b* occurs when increasing voltage and current are accompanied by increasing resistance.

In Fig. A-1*c* the nonlinearity is associated with a situation in which increasing voltage and current are accompanied by decreasing resistance.

RESISTORS

Ideally a conventional resistor would have constant resistance under all conditions. Such a resistor cannot be manufactured, because all materials change their resistivity (resistance per unit volume) with changes in temperature and all materials become hotter with increasing current. The temperature coefficient of resistance (TCR) of a resistor can be made close to zero by the suitable choice and combination of materials.

Ordinary conducting metals have a positive TCR, implying an increase in resistance with temperature rise. Carbon, a common material for making conventional resistors, has a *negative* TCR; its resistivity decreases with temperature rise.

SEMICONDUCTORS

Many of the materials used in the manufacture of semiconductor devices have a negative TCR. The most distinctive nonlinearity associated with these devices is not, however, due to the effects of temperature on the individual materials. It is, rather, due to the molecular effects produced when certain semiconductor materials are brought into intimate contact with each other. This very marked nonlinearity is responsible for much of the practical value of diodes, transistors, thermistors, varistors, and other semiconductor devices.

IMPEDANCE

Reactance can be linear or nonlinear. The reactance of an ordinary capacitor is linear (constant). The capacitance, and hence the capacitive reactance, of a semiconductor junction can be significantly changed by variation in applied voltage. The inductance and inductive reactance of an iron-cored inductor or transformer is dependent on coil current magnitude.

Nonlinearity of impedance is used deliberately in such devices as the varactor, a popular frequency multiplier for microwave radio.

DRIFT, DIFFUSION, AND RECOMBINATION

An electric current is a movement of charges in one general direction. Charges capable of movement can be obtained by one of a number of mechanisms, such as:

1. *Free electrons.* Some materials (conductors) have electrons which are so weakly attached to the atomic structure that little force is needed to make them move.
2. *Electric force.* An electric field applied across a material can cause atomic stresses which result in the liberation of charges.
3. *External bombardment.* The energy of light, heat, cosmic rays, x-rays, etc., directed at a material will, if the energy level is high enough, cause charges to be released from the atomic structure of the material.
4. *Internal bombardment.* Charges, liberated within a material by any method can, if their velocity is high enough, bombard other parts of the material and result in further charge release.

In ordinary circuit work, we are concerned with charge movement in conductors which takes place by the drift process involving free electrons and referred to in Sec. 2-6. In electronic circuits, which are nonlinear electric circuits, the production of movable charges frequently takes place by method 2, 3, or 4. Further, the movement of these charges may take place due to the diffusion or recombination processes. A simple comparison of drift, diffusion, and recombination can be obtained from a brief description of each.

1. *Drift.* Electrons drift from atom to atom in the general direction of the positive pole of an applied emf.
2. *Diffusion.* Charges of similar sign (negative or positive) repel each other and, if free to do so, will move apart. This action causes charges to diffuse within a material away from a region of high charge concentration.
3. *Recombination.* A force of attraction exists between charges of opposite sign (negative and positive) and, if free to do so, they will move toward each other. When a negative charge and an equal-value positive charge meet, they combine and cancel each other's charges; this results in zero total charge.

DEFINITIONS OF SI UNITS

The following are the official definitions of SI units as determined at various meetings of the General Conference of Weights and Measures (CGPM).

METER

The meter is the length equal to 1 650 763.73 wavelengths in vacuum of the radiation corresponding to the transition between the levels $2p_{10}$ and $5d_5$ of the krypton- 86 atom. (Eleventh CGPM, Resolution 6, 1960.)

KILOGRAM

The kilogram is the unit of mass; it is equal to the mass of the international prototype of the kilogram. (First CGPM, 1889; Third CGPM, 1901.)

SECOND

The second is the duration of 9 192 631 770 periods of the radiation corresponding to the transition between the two hyperfine levels of the ground state of the cesium-133 atom. (Thirteenth CGPM, Resolution 1, 1967.)

AMPERE

The ampere is the constant current which, if maintained in two straight parallel conductors of infinite length, of negligible circular cross section, and placed 1 meter apart in vacuum, would produce between these conductors a force equal to 2×10^{-7} newton per meter of length. (CIPM, Resolution 2, 1946; approved by the Ninth CGPM, 1948.)

KELVIN

The kelvin, unit of thermodynamic temperature, is the fraction 1/273.16 of the thermodynamic temperature of the triple point of water. (Thirteenth CGPM, Resolution 4, 1968.)

Note 1. The Thirteenth CGPM (Resolution 3, 1967) also decided that the unit kelvin and its symbol K should be used to express an interval or a difference of temperature.

Note 2. In addition to the thermodynamic temperature (symbol T) expressed in kelvins, use is made of Celsius temperature (symbol t) defined by the equation $t = T - T_0$, where $T_0 = 273.15$ K by definition. The Celsius temperature is in general expressed in degrees Celsius (symbol °C). The unit degree Celsius is thus equal to the unit kelvin, and an interval or a difference of Kelvin temperature may also be expressed in degrees Celsius.

DETERMINANTS

Unless an electronic computer or calculator is used, the solution of simultaneous equations is always laborious. This is particularly true if inconvenient numbers are involved, as is usually the case with equations obtained during electric circuit analysis. The use of determinant techniques minimizes the amount of work involved and renders the solution less liable to error than the more elementary substitution methods.

The following description of the use of determinants for solving simultaneous equations refers to the basic technique. It must be emphasized that, to gain the full advantages determinants provide, sophisticated methods beyond the scope of this introduction are needed. Refer to a mathematics textbook for a full treatment of determinant principles.

EQUATIONS WITH TWO UNKNOWNS

Two general simultaneous equations are

$$ax + by = c$$

$$dx + ey = f$$

The unknowns are x and y; a, b, d, and e are coefficients of the unknowns; c and f are constants. The procedure for solving these equations by determinants is, as follows.

1. Write the equations in *matrix* form:

$$\begin{bmatrix} a & b \\ d & e \end{bmatrix} \begin{bmatrix} x \\ y \end{bmatrix} = \begin{bmatrix} c \\ f \end{bmatrix}$$

The solutions will initially appear as fractions.

2. To obtain the common denominator D of these fractions,

a. Write the section of the matrix containing the coefficients as a determinant (between parallel lines):

$$\begin{vmatrix} a & b \\ d & e \end{vmatrix}$$

b. Cross-multiply the coefficients:

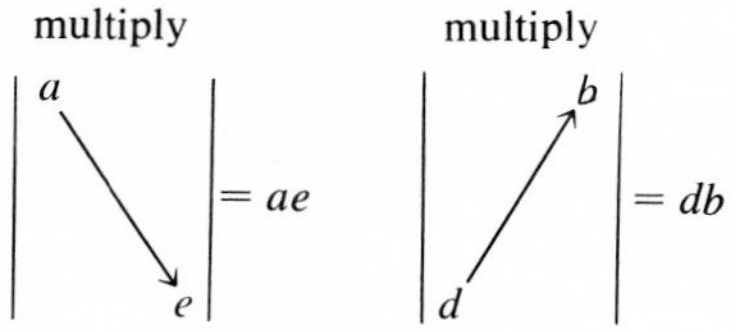

c. Give signs to the products:

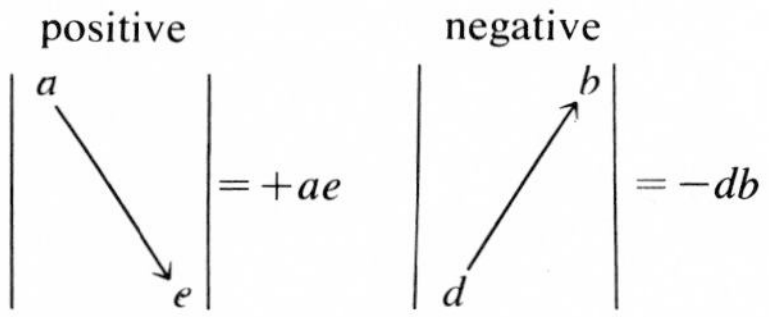

d. Write the demominator:

$$D = ae - db$$

3. To obtain the numerator N_x for the unknown x:

a. Write a determinant consisting of the coefficients of y in their original matrix locations and the coefficients of x replaced by the constants:

$$\begin{vmatrix} c & b \\ f & e \end{vmatrix}$$

b. Apply operations 2*b* to 2*d* to this determinant:

$$N_x = ce - fb$$

4. To obtain the numerator N_y for the unknown y, repeat operations similar to 3*a* and 3*b*:

$$N_y = \begin{vmatrix} a & c \\ d & f \end{vmatrix} = af - dc$$

The solutions for the pair of equations are

$$x = \frac{ce - fb}{ae - db} \quad \text{and} \quad y = \frac{af - dc}{ae - db}$$

Example 1

$$3x + 2y = 10$$

$$5x + 6y = 15$$

$$\begin{bmatrix} 3 & 2 \\ 5 & 6 \end{bmatrix} \begin{bmatrix} x \\ y \end{bmatrix} = \begin{bmatrix} 10 \\ 15 \end{bmatrix}$$

$$D = \begin{vmatrix} 3 & 2 \\ 5 & 6 \end{vmatrix} = 3 \times 6 - 5 \times 2 = 8$$

$$N_x = \begin{vmatrix} 10 & 2 \\ 15 & 6 \end{vmatrix} = 10 \times 6 - 15 \times 2 = 30$$

$$N_y = \begin{vmatrix} 3 & 10 \\ 5 & 15 \end{vmatrix} = 3 \times 15 - 5 \times 10 = -5$$

Solution

$$x = \tfrac{30}{8} = \mathbf{3.75} \qquad \text{and} \qquad y = -\tfrac{5}{8} = \mathbf{-0.625}$$

Example 2

$$3x - 2y = 10$$

$$-5x + 6y = 15$$

$$\begin{bmatrix} 3 & -2 \\ -5 & 6 \end{bmatrix} \begin{bmatrix} x \\ y \end{bmatrix} = \begin{bmatrix} 10 \\ 15 \end{bmatrix}$$

$$D = \begin{vmatrix} 3 & -2 \\ -5 & 6 \end{vmatrix} = 3 \times 6 - [(-5)(-2)] = 8$$

$$N_x = \begin{vmatrix} 10 & -2 \\ 15 & 6 \end{vmatrix} = 10 \times 6 - [15(-2)] = 90$$

$$N_y = \begin{vmatrix} 3 & 10 \\ -5 & 15 \end{vmatrix} = 3 \times 15 - [(-5) \times 10] = 95$$

Solution

$$x = \tfrac{90}{8} = \mathbf{11.25} \qquad \text{and} \qquad y = \tfrac{95}{8} = \mathbf{11.88}$$

EQUATIONS WITH THREE UNKNOWNS

The procedure is an extension of that used for two unknowns.

$$ax + by + cz = d$$

$$ex + fy + gz = h$$

$$jx + ky + lz = m$$

1. Write the equations in matrix form

$$\begin{bmatrix} a & b & c \\ e & f & g \\ j & k & l \end{bmatrix} \begin{bmatrix} x \\ y \\ z \end{bmatrix} = \begin{bmatrix} d \\ h \\ m \end{bmatrix}$$

2. To obtain the common denominator for the solutions:

a. Write the coefficient section of the matrix as a determinant:

$$\begin{vmatrix} a & b & c \\ e & f & g \\ j & k & l \end{vmatrix}$$

b. To simplify the cross-multiplication, repeat the first two columns to the right of the determinant:

$$\left|\begin{array}{ccc|cc} a & b & c & a & b \\ e & f & g & e & f \\ j & k & l & j & k \end{array}\right.$$

c. Cross-multiply the coefficients and give signs

$$\left|\begin{array}{ccc|cc} a & b & c & a & b \\ e & f & g & e & f \\ j & k & l & j & k \end{array}\right. = +(afl + bgj + cek)$$

$$\left|\begin{array}{ccc|cc} a & b & c & a & b \\ e & f & g & e & f \\ j & k & l & j & k \end{array}\right. = -(jfc + kga + leb)$$

d. Write the common denominator D

$$D = \begin{vmatrix} a & b & c \\ e & f & g \\ j & k & l \end{vmatrix} = (afl + bgj + cek) - (jfc + kga + leb)$$

3. Obtain the numerator N_x for x by applying methods 2*a* to 2*d* but with the coefficients of x replaced by the constants:

$$N_x = \begin{vmatrix} d & b & c \\ h & f & g \\ m & k & l \end{vmatrix} = (dfl + bgm + chk) - (mfc + kgd + lhb)$$

4. Repeat operation 3 for N_y:

$$N_y = \begin{vmatrix} a & d & c \\ e & h & g \\ j & m & l \end{vmatrix} = (ahl + dgj + cem) - (jhc + mga + led)$$

5. Repeat operation 3 for N_z:

$$N_z = \begin{vmatrix} a & b & d \\ e & f & h \\ j & k & m \end{vmatrix} = (afm + bhj + dek) - (jfd + kha + meb)$$

The solutions for the equations are

$$x = \frac{N_x}{D} = \frac{(dfl + bgm + chk) - (mfc + kgd + lhb)}{(afl + bgj + cek) - (jfc + kga + leb)}$$

$$y = \frac{N_y}{D} = \frac{(ahl + dgj + cem) - (jhc + mga + led)}{(afl + bgj + cek) - (jfc + kga + leb)}$$

$$z = \frac{N_z}{D} = \frac{(afm + bhj + dek) - (jfd + kha + meb)}{(afl + bgj + cek) - (jfc + kga + leb)}$$

Example 3

$$x + 3y + 4z = 14$$
$$x + 2y + z = 7$$
$$2x + y + 2z = 2$$

$$\begin{bmatrix} 1 & 3 & 4 \\ 1 & 2 & 1 \\ 2 & 1 & 2 \end{bmatrix} \begin{bmatrix} x \\ y \\ z \end{bmatrix} = \begin{bmatrix} 14 \\ 7 \\ 2 \end{bmatrix}$$

$$D = \begin{vmatrix} 1 & 3 & 4 \\ 1 & 2 & 1 \\ 2 & 1 & 2 \end{vmatrix} \begin{matrix} 1 & 3 \\ 1 & 2 \\ 2 & 1 \end{matrix} = (4 + 6 + 4) - (16 + 1 + 6) = -9$$

$$N_x = \begin{vmatrix} 14 & 3 & 4 \\ 7 & 2 & 1 \\ 2 & 1 & 2 \end{vmatrix} \begin{matrix} 14 & 3 \\ 7 & 2 \\ 2 & 1 \end{matrix} = (56 + 6 + 28) - (16 + 14 + 42) = 18$$

$$N_y = \begin{vmatrix} 1 & 14 & 4 \\ 1 & 7 & 1 \\ 2 & 2 & 2 \end{vmatrix} \begin{matrix} 1 & 14 \\ 1 & 7 \\ 2 & 2 \end{matrix} = (14 + 28 + 8) - (56 + 2 + 28) = -36$$

$$N_z = \begin{vmatrix} 1 & 3 & 14 \\ 1 & 2 & 7 \\ 2 & 1 & 2 \end{vmatrix} \begin{matrix} 1 & 3 \\ 1 & 2 \\ 2 & 1 \end{matrix} = (4 + 42 + 14) - (56 + 7 + 6) = -9$$

Solution

$$x = \frac{18}{-9} = \mathbf{-2} \qquad y = \frac{-36}{-9} = \mathbf{4} \qquad z = \frac{-9}{-9} = \mathbf{1}$$

Example 4 The following equations are from Example 6-10.

$$30\ \text{k}\Omega \times I_1 - 10\ \text{k}\Omega \times I_2 - 10\ \text{k}\Omega \times I_3 = 100\ \text{V}$$
$$-10\ \text{k}\Omega \times I_1 + 20\ \text{k}\Omega \times I_2 - 10\ \text{k}\Omega \times I_3 = -50\ \text{V}$$
$$-10\ \text{k}\Omega \times I_1 - 10\ \text{k}\Omega \times I_2 + 30\ \text{k}\Omega \times I_3 = 0$$

$$\begin{bmatrix} 30 & -10 & -10 \\ -10 & 20 & -10 \\ -10 & -10 & 30 \end{bmatrix} \begin{bmatrix} I_1 \\ I_2 \\ I_3 \end{bmatrix} = \begin{bmatrix} 100 \\ -50 \\ 0 \end{bmatrix} \qquad I_1,\ I_2,\ \text{and}\ I_3\ \text{in milliamperes}$$

$$D = \begin{vmatrix} 30 & -10 & -10 \\ -10 & 20 & -10 \\ -10 & -10 & 30 \end{vmatrix} = (18\,000 - 1000 - 1000) - (2000 + 3000 + 3000) = 8000$$

$$N_1 = \begin{vmatrix} 100 & -10 & -10 \\ -50 & 20 & -10 \\ 0 & -10 & 30 \end{vmatrix} = (60\,000 + 0 - 5000) - (0 + 10\,000 + 15000) = 30\,000$$

$$N_2 = \begin{vmatrix} 30 & 100 & -10 \\ -10 & -50 & -10 \\ -10 & 0 & 30 \end{vmatrix} = (-45\,000 + 10\,000 + 0) - (-5000 + 0 - 30\,000) = 0$$

$$N_3 = \begin{vmatrix} 30 & -10 & 100 \\ -10 & 20 & -50 \\ -10 & -10 & 0 \end{vmatrix} = (0 - 5000 + 10\,000) - (-20\,000 + 15\,000 + 0) = 10\,000$$

Solution

$$I_1 = \frac{30\,000}{8000} = \mathbf{3.75\ mA} \qquad I_2 = \frac{0}{8000} = \mathbf{0} \qquad I_3 = \frac{10\,000}{8000} = \mathbf{1.25\ mA}$$

Example 5 These equations are from Example 17-3.

$$j40\ \Omega\ I_1 - j50\ \Omega\ I_2 = 50\ \text{V}\ \underline{/0^\circ}$$

$$-j50\ \Omega\ I_1 - j30\ \Omega\ I_2 = -20\ \text{V}\ \underline{/0^\circ}$$

$$\begin{bmatrix} 40\ \underline{/90^\circ} & 50\ \underline{/-90^\circ} \\ 50\ \underline{/-90^\circ} & 30\ \underline{/-90^\circ} \end{bmatrix} \begin{bmatrix} I_1 \\ I_2 \end{bmatrix} = \begin{bmatrix} 50\ \underline{/0^\circ} \\ -20\ \underline{/0^\circ} \end{bmatrix}$$

$$D = \begin{vmatrix} 40\ \underline{/90^\circ} & 50\ \underline{/-90^\circ} \\ 50\ \underline{/-90^\circ} & 30\ \underline{/-90^\circ} \end{vmatrix} = 1200\ \underline{/0^\circ} - 2500\ \underline{/-180^\circ} = 3700\ \underline{/0^\circ} \quad \text{note that } \underline{/-180^\circ} = \underline{/+180^\circ}$$

$$N_1 = \begin{vmatrix} 50\ \underline{/0^\circ} & 50\ \underline{/-90^\circ} \\ -20\ \underline{/0^\circ} & 30\ \underline{/-90^\circ} \end{vmatrix} = 1500\ \underline{/-90^\circ} - (-1000\ \underline{/-90^\circ}) = 2500\ \underline{/-90^\circ}$$

$$N_2 = \begin{vmatrix} 40\ \underline{/90^\circ} & 50\ \underline{/0^\circ} \\ 50\ \underline{/-90^\circ} & -20\ \underline{/0^\circ} \end{vmatrix} = -800\ \underline{/90^\circ} - 2500\ \underline{/-90^\circ} = -j800 + j2500 = 1700\ \underline{/90^\circ}$$

Solution

$$I_1 = \frac{2500\ \underline{/-90^\circ}}{3700\ \underline{/0^\circ}} = \mathbf{0.6757\ A}\ \underline{\mathbf{/-90^\circ}} \quad \text{and} \quad I_2 = \frac{1700\ \underline{/90^\circ}}{3700\ \underline{/0^\circ}} = \mathbf{0.4595\ A}\ \underline{\mathbf{/90^\circ}}$$

SIGNIFICANT FIGURES

Consider the number 12.34. This number has four digits, ordered as to their significance. The digit corresponding to the highest power of 10 has the greatest significance. Thus the 1 has the greatest significance, as it corresponds to 10^1. Conversely the 4 has the least significance, as it relates to 10^{-2}.

For any number the number of digits from the highest significant nonzero digit to, and including, the least significant digit considered important is the number of significant digits or figures.

In our example there are four significant figures. Written as 0012.34, the number still has four significant figures; the two zeros to the left of the highest significant digit add nothing to our knowledge of the precision to which the number is known. But if written 12.3400, the number has six significant figures, the two zeros to the right of the least significant digit indicating that the number is known accurately to four decimal places (the number is neither 12.3401 nor 12.3399).

The number of significant figures can be reduced by rounding off. This means dropping the least significant digits until the desired number of digits remains. The new least significant digit may have to be changed as determined by the following rules.

If the highest significant digit dropped is

1. Less than 5, the new least significant digit is not changed.
2. Greater than 5, the new least significant digit is increased by 1.
3. 5, the new least significant digit is not changed if it is even.
4. 5, the new least significant digit is increased by 1 if it is odd.

These rules result in a rounding-off technique which, on the average of a large number of operations, gives the most consistent reliability.

Examples are:

1. 12 340 has five significant figures.
2. 12 340 written as 1.234×10^4 has four significant figures.
3. 0.001 234 has four significant figures.
4. 12.344 to four significant figures is 12.34.
5. 12.346 to four significant figures is 12.35.
6. 12.345 to four significant figures is 12.34.
7. 12.335 to four significant figures is 12.34.

CAPACITIVE ENERGY STORAGE[1]

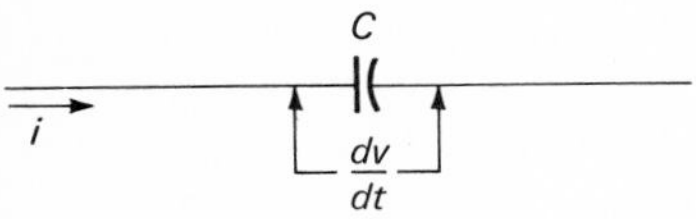

The power at any instant is given by

$$p = vi = vC\,\frac{dv}{dt}$$

The energy stored over a given time period is the integral of the power over that period:

$$W_C = \int_{t_0}^{t} p\,dt = C\int_{t_0}^{t} v\,\frac{dv}{dt}\,dt = C\int_{v(t_0)}^{v(t)} v\,dv$$

$$W_C = \frac{C}{2}\,\{[v(t)]^2 - [v(t_0)]^2\}$$

If the energy and the voltage are both zero at time t_0, and the voltage has the value V at time t,

$$W_C = \frac{CV^2}{2}$$

INDUCTIVE ENERGY STORAGE[2]

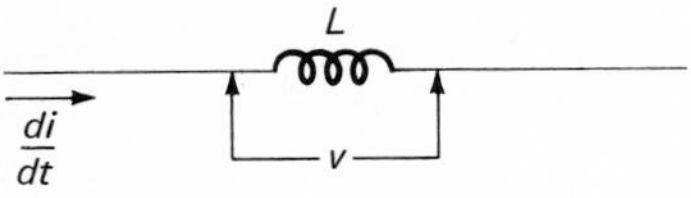

The power at any instant is given by

$$p = vi = L\,\frac{di}{dt}\,i$$

The energy stored over a given time period is the integral of the power over that period:

[1] From Hayt and Kemmerly, "Engineering Circuit Analysis," McGraw Hill Book Company, New York, with modifications.
[2] From Hayt and Kemmerly, "Engineering Circuit Analysis," McGraw Hill Book Company, New York, with modifications.

$$W_L = \int_{t_0}^{t} p\,dt = L \int_{t_0}^{t} i \frac{di}{dt}\,dt = L \int_{i(t_0)}^{i(t)} i\,di$$

$$W_L = \frac{L}{2} \{[i(t)]^2 - [i(t_0)]^2\}$$

If the energy and the current are both zero at time t_0, and the current has the value I at time t,

$$W_L = \frac{LI^2}{2}$$

GROWTH OF CURRENT IN AN *RC* CIRCUIT

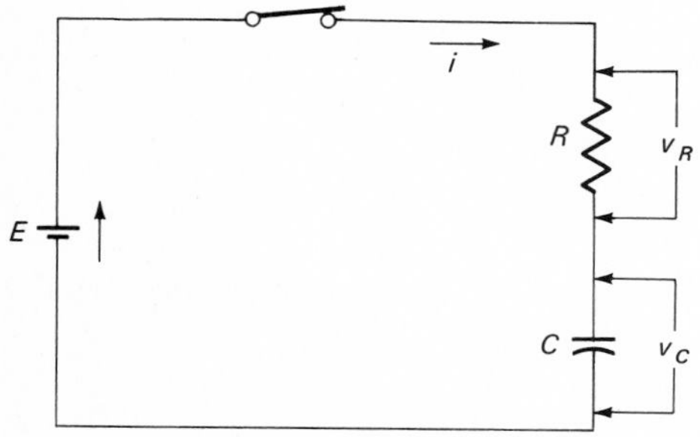

Assume the initial conditions before the switch is closed to be, at t_0, $v_C = 0$ and $q_C = 0$:

$$E = v_R + v_C = Ri + \frac{q_C}{C} = R\frac{dq_C}{dt} + \frac{q_C}{C}$$

$$\frac{dq_C}{q_C - CE} = \frac{dt}{CR}$$

Integrating,

$$\int_{q_c(t_0)}^{q_c(t)} \frac{1}{(q_C - CE)}\,dq_C = \frac{1}{CR}\int_{t_0}^{t} dt$$

$$\ln\,(q_C - CE) = -\frac{t}{CR} + k$$

Substituting the initial conditions, when $t = 0$ and $q_C = 0$,

$$k = \ln\,(-CE)$$

$$\ln \frac{q_C - CE}{-CE} = -\frac{t}{CR}$$

$$\frac{CE - q_C}{CE} = e^{-t/CR}$$

$$q_C = CE(1 - e^{-t/CR})$$

$$i = \frac{dq_C}{dt} = \frac{E}{R}e^{-t/CR}$$

DECAY OF CURRENT IN AN *RC* CIRCUIT

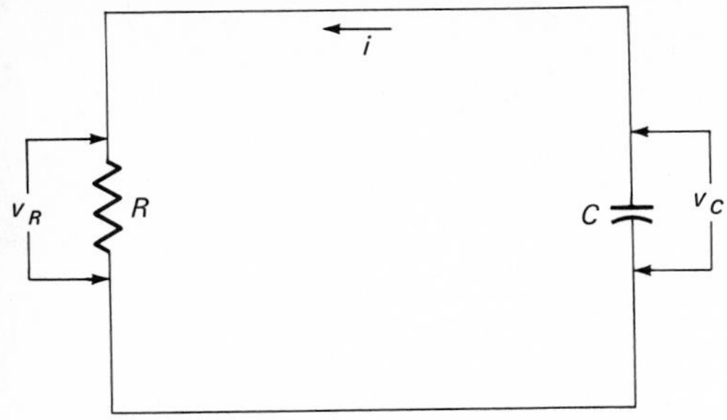

Assume the initial conditions to be, at t_0, $v_C = V_0$ and $q_C = V_0 C$:

$$v_R + v_C = 0$$

$$R\,\frac{dq_C}{dt} + \frac{q_C}{C} = 0$$

$$\frac{dq_C}{q_C} = -\frac{dt}{CR}$$

Integrating,

$$\int_{q_c(t_0)}^{q_c(t)} \frac{1}{q_C}\,dq_C = \frac{1}{CR}\int_{t_0}^{t} dt$$

$$\ln q_C = -\frac{t}{CR} + c$$

$$q_C = ke^{-t/CR}$$

Substituting the initial conditions, when $t = 0$ and $q_C = V_0 C$,

$$k = V_0 C$$

$$q_C = V_0 C e^{-t/CR}$$

$$i = \frac{dq_C}{dt} = -\frac{V_0}{R}\,e^{-tCR}$$

GROWTH OF CURRENT IN AN *RL* CIRCUIT

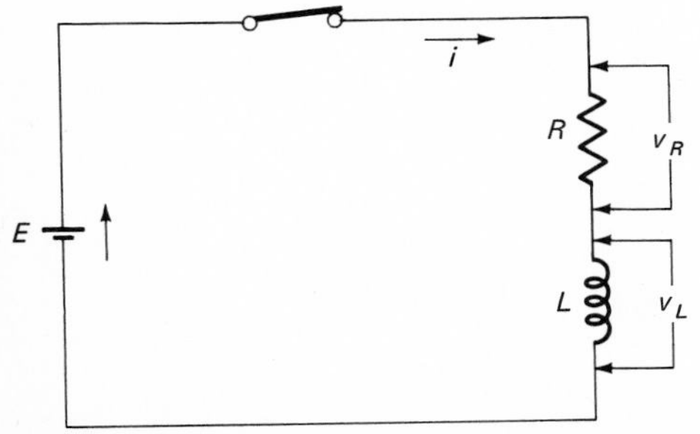

At the instant the switch is closed (time t_0), the current is zero. At time t, the current has the value i.

$$E = v_R + v_L = Ri + L\frac{di}{dt}$$

$$\frac{1}{(E - Ri)}\,di = \frac{1}{L}\,dt$$

Integrating,

$$\int_{i(t_0)}^{i(t)} \frac{1}{(E - Ri)}\,di = \frac{1}{L}\int_{t_0}^{t} dt$$

$$-\frac{1}{R}\ln(E - Ri) = \frac{t}{L} + c$$

$$\ln(E - Ri) = -\frac{Rt}{L} + k$$

Substituting for the initial conditions, at t_0, $i = 0$, $k = \ln E$.
Substituting for the constant k,

$$\ln(E - Ri) = -\frac{Rt}{L} + \ln E$$

$$\ln\left(1 - \frac{Ri}{E}\right) = -\frac{Rt}{L}$$

$$i = \frac{E}{R}(1 - e^{-Rt/L})$$

DECAY OF CURRENT IN AN *RL* CIRCUIT

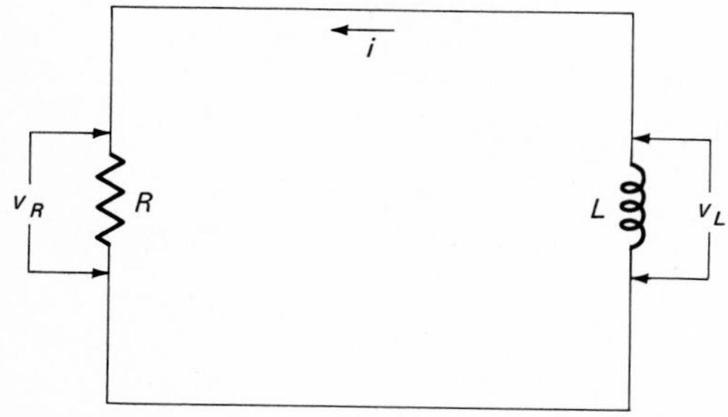

The resistance and inductance are connected at time t_0; the current has reached a value i_0. At time t, the current has the value i.

$$v_R + v_L = 0$$

$$Ri + L\frac{di}{dt} = 0$$

$$\frac{di}{i} = -\frac{R}{L}\,dt$$

Integrating,

$$\int_{i(t_0)}^{i(t)} \frac{di}{i} = -\frac{R}{L}\int_{t_0}^{t} dt$$

$$\ln i = -\frac{Rt}{L} + c$$

$$i = ke^{-Rt/L}$$

Substituting the initial conditions, at time t_0, $i = i_0$ and $k = i_0$,

$$i = i_0 e^{-Rt/L}$$

CAPACITIVE REACTANCE

In a capacitive circuit with sine-wave voltage,

$$i = C\frac{dv}{dt} = C\frac{V_{max}\sin\omega t}{dt} = \omega C V_{max}\cos\omega t$$

Converting to rms values,

$$I = \omega C V$$

$$X_C = \frac{V}{I} = \frac{1}{\omega C}$$

INDUCTIVE REACTANCE

In an inductive circuit with sine-wave current,

$$v = L\frac{di}{dt} = L\frac{I_{max}\sin\omega t}{dt} = \omega L I_{max}\cos\omega t$$

Converting to rms values,

$$V = \omega L I$$

$$X_L = \frac{V}{I} = \omega L$$

AVERAGE VALUE OF A SINE WAVE

Using current as the variable, the following refers to the average value in its ac context; the mean value taken over one half-cycle, commencing at zero magnitude.

$$I_{av} = \frac{\omega}{\pi}\int_{t=0}^{t=\pi/\omega} i\,dt = \frac{\omega}{\pi}\int_{t=0}^{t=\pi/\omega} I_{max}\sin\omega t\,dt = \frac{\omega I_{max}}{\pi}\left(\frac{\cos\omega t}{\omega}\right)_{t=0}^{t=\pi/\omega}$$

$$I_{av} = \frac{2I_{max}}{\pi} \simeq 0.636 I_{max}$$

RMS VALUE OF A SINE WAVE

Again using current as the variable,

$$I = \sqrt{\frac{\omega}{2\pi}\int_{t=0}^{t=2\pi/\omega} i^2\,dt} = \sqrt{\frac{\omega}{2\pi}\int_{t=0}^{t=2\pi/\omega} (I_{max}\sin\omega t)^2\,dt}$$

$$= \sqrt{\frac{\omega I_{max}^2}{4\pi}\int_{t=0}^{t=2\pi/\omega} (1-\cos 2\omega t)\,dt} = \sqrt{\frac{\omega I_{max}^2}{4\pi}\left[t - \frac{\sin 2\omega t}{2\omega}\right]_{t=0}^{t=2\pi/\omega}}$$

$$I = \sqrt{\frac{\omega I_{max}^2}{4\pi}\left(\frac{2\pi}{\omega} - \frac{\sin 2\omega \times 2\pi/\omega}{2\omega}\right)} = \sqrt{\frac{I_{max}^2}{2}}$$

$$I = \frac{I_{max}}{\sqrt{2}} \simeq \mathbf{0.7071}\boldsymbol{I}_{max}$$

POWER IN THE AC CIRCUIT

The power at any instant in an ac circuit is given by

$$p = vi = V_{max} \sin \omega t \; I_{max} \sin (\omega t + \phi)$$

where ϕ = phase angle between voltage and current.

Applying the trigonometry of compound angles,

$$p = \frac{V_{max} I_{max}}{2} [\cos \phi - \cos (2\omega t + \phi)]$$

The average power taken over one complete cycle between $t = 0$ and $t = 2\pi/\omega$ is given by

$$P = \frac{V_{max} I_{max}}{2} \frac{\omega}{2\pi} \int_{t=0}^{t=2\pi/\omega} [\cos \phi - \cos (2\omega t + \phi)] \, dt$$

Integrating, $$P = \frac{V_{max} I_{max}}{2} \cos \phi = \frac{V_{max}}{\sqrt{2}} \frac{I_{max}}{\sqrt{2}} \cos \phi$$

$$\boldsymbol{P = VI \cos \phi}$$

Thus the average power in any ac circuit is the product of the rms values of voltage and current and the power factor (cosine of the phase angle).

If the circuit is resistive ($\phi = 0°$ and $\cos \phi = 1$),

$$P = VI$$

If the circuit is inductive or capacitive ($\phi = 90°$ and $\cos \phi = 0$),

$$P = \mathbf{0}$$

Tables of trigonometric functions

	Natural sines										Add								
		.1	.2	.3	.4	.5	.6	.7	.8	.9	.01	.02	.03	.04	.05	.06	.07	.08	.09
0°	.0000	0017	0035	0052	0070	0087	0105	0122	0140	0157	2	4	5	7	9	11	12	14	16
1	.0175	0192	0209	0227	0244	0262	0279	0297	0314	0332	2	4	5	7	9	11	12	14	16
2	.0349	0366	0384	0401	0419	0436	0454	0471	0488	0506	2	4	5	7	9	11	12	14	16
3	.0523	0541	0558	0576	0593	0610	0628	0645	0663	0680	2	4	5	7	9	11	12	14	16
4	.0698	0715	0732	0750	0767	0785	0802	0819	0837	0854	2	3	5	7	9	10	12	14	16
5	.0872	0889	0906	0924	0941	0958	0976	0993	1011	1028	2	3	5	7	9	10	12	14	16
6	.1045	1063	1080	1097	1115	1132	1149	1167	1184	1201	2	3	5	7	9	10	12	14	16
7	.1219	1236	1253	1271	1288	1305	1323	1340	1357	1374	2	3	5	7	9	10	12	14	16
8	.1392	1409	1426	1444	1461	1478	1495	1513	1530	1547	2	3	5	7	9	10	12	14	15
9	.1564	1582	1599	1616	1633	1650	1668	1685	1702	1719	2	3	5	7	9	10	12	14	15
10	.1736	1754	1771	1788	1805	1822	1840	1857	1874	1891	2	3	5	7	9	10	12	14	15
11	.1908	1925	1942	1959	1977	1994	2011	2028	2045	2062	2	3	5	7	9	10	12	14	15
12	.2079	2096	2113	2130	2147	2164	2181	2198	2215	2233	2	3	5	7	9	10	12	14	15
13	.2250	2267	2284	2300	2317	2334	2351	2368	2385	2402	2	3	5	7	9	10	12	14	15
14	.2419	2436	2453	2470	2487	2504	2521	2538	2554	2571	2	3	5	7	8	10	12	14	15
15	.2588	2605	2622	2639	2656	2672	2689	2706	2723	2740	2	3	5	7	8	10	12	13	15
16	.2756	2773	2790	2807	2823	2840	2857	2874	2890	2907	2	3	5	7	8	10	12	13	15
17	.2924	2940	2957	2974	2990	3007	3024	3040	3057	3074	2	3	5	7	8	10	12	13	15
18	.3090	3107	3123	3140	3156	3173	3190	3206	3223	3239	2	3	5	7	8	10	12	13	15
19	.3256	3272	3289	3305	3322	3338	3355	3371	3387	3404	2	3	5	7	8	10	11	13	15
20	.3420	3437	3453	3469	3486	3502	3518	3535	3551	3567	2	3	5	7	8	10	11	13	15
21	.3584	3600	3616	3633	3649	3665	3681	3697	3714	3730	2	3	5	6	8	10	11	13	15
22	.3746	3762	3778	3795	3811	3827	3843	3859	3875	3891	2	3	5	6	8	10	11	13	14
23	.3907	3923	3939	3955	3971	3987	4003	4019	4035	4051	2	3	5	6	8	10	11	13	14
24	.4067	4083	4099	4115	4131	4147	4163	4179	4195	4210	2	3	5	6	8	10	11	13	14

Natural sines (Continued)

		.1	.2	.3	.4	.5	.6	.7	.8	.9	*Add* .01	.02	.03	.04	.05	.06	.07	.08	.09
25°	.4226	4242	4258	4274	4289	4305	4321	4337	4352	4368	2	3	5	6	8	9	11	13	14
26	.4384	4399	4415	4431	4446	4462	4478	4493	4509	4524	2	3	5	6	8	9	11	12	14
27	.4540	4555	4571	4586	4602	4617	4633	4648	4664	4679	2	3	5	6	8	9	11	12	14
28	.4695	4710	4726	4741	4756	4772	4787	4802	4818	4833	2	3	5	6	8	9	11	12	14
29	.4848	4863	4879	4894	4909	4924	4939	4955	4970	4985	2	3	5	6	8	9	11	12	14
30	.5000	5015	5030	5045	5060	5075	5090	5105	5120	5135	2	3	5	6	8	9	11	12	14
31	.5150	5165	5180	5195	5210	5225	5240	5255	5270	5284	1	3	4	6	7	9	10	12	13
32	.5299	5314	5329	5344	5358	5373	5388	5402	5417	5432	1	3	4	6	7	9	10	12	13
33	.5446	5461	5476	5490	5505	5519	5534	5548	5563	5577	1	3	4	6	7	9	10	12	13
34	.5592	5606	5621	5635	5650	5664	5678	5693	5707	5721	1	3	4	6	7	9	10	12	13
35	.5736	5750	5764	5779	5793	5807	5821	5835	5850	5864	1	3	4	6	7	9	10	11	13
36	.5878	5892	5906	5920	5934	5948	5962	5976	5990	6004	1	3	4	6	7	8	10	11	13
37	.6018	6032	6046	6060	6074	6088	6101	6115	6129	6143	1	3	4	6	7	8	10	11	12
38	.6157	6170	6184	6198	6211	6225	6239	6252	6266	6280	1	3	4	5	7	8	10	11	12
39	.6293	6307	6320	6334	6347	6361	6374	6388	6401	6414	1	3	4	5	7	8	9	11	12
40	.6428	6441	6455	6468	6481	6494	6508	6521	6534	6547	1	3	4	5	7	8	9	11	12
41	.6561	6574	6587	6600	6613	6626	6639	6652	6665	6678	1	3	4	5	7	8	9	10	12
42	.6691	6704	6717	6730	6743	6756	6769	6782	6794	6807	1	3	4	5	6	8	9	10	12
43	.6820	6833	6845	6858	6871	6884	6896	6909	6921	6934	1	3	4	5	6	8	9	10	11
44	.6947	6959	6972	6984	6997	7009	7022	7034	7046	7059	1	2	4	5	6	7	9	10	11
45	.7071	7083	7096	7108	7120	7133	7145	7157	7169	7181	1	2	4	5	6	7	9	10	11
46	.7193	7206	7218	7230	7242	7254	7266	7278	7290	7302	1	2	4	5	6	7	8	10	11
47	.7314	7325	7337	7349	7361	7373	7385	7396	7408	7420	1	2	4	5	6	7	8	9	11
48	.7431	7443	7455	7466	7478	7490	7501	7513	7524	7536	1	2	3	5	6	7	8	9	10
49	.7547	7559	7570	7581	7593	7604	7615	7627	7638	7649	1	2	3	5	6	7	8	9	10
50	.7660	7672	7683	7694	7705	7716	7727	7738	7749	7760	1	2	3	4	6	7	8	9	10

51	.7771	7782	7793	7804	7815	7826	7837	7848	7859	7869	1	2	3	4	5	7	8	9	10
52	.7880	7891	7902	7912	7923	7934	7944	7955	7965	7976	1	2	3	4	5	6	7	8	10
53	.7986	7997	8007	8018	8028	8039	8049	8059	8070	8080	1	2	3	4	5	6	7	8	9
54	.8090	8100	8111	8121	8131	8141	8151	8161	8171	8181	1	2	3	4	5	6	7	8	9
55	.8192	8202	8211	8221	8231	8241	8251	8261	8271	8281	1	2	3	4	5	6	7	8	9
56	.8290	8300	8310	8320	8329	8339	8348	8358	8368	8377	1	2	3	4	5	6	7	8	9
57	.8387	8396	8406	8415	8425	8434	8443	8453	8462	8471	1	2	3	4	5	6	7	7	8
58	.8480	8490	8499	8508	8517	8526	8536	8545	8554	8563	1	2	3	4	5	6	6	7	8
59	.8572	8581	8590	8599	8607	8616	8625	8634	8643	8652	1	2	3	4	4	5	6	7	8
60	.8660	8669	8678	8686	8695	8704	8712	8721	8729	8738	1	2	3	3	4	5	6	7	7
61	.8746	8755	8763	8771	8780	8788	8796	8805	8813	8821	1	2	2	3	4	5	6	7	7
62	.8829	8838	8846	8854	8862	8870	8878	8886	8894	8902	1	2	2	3	4	5	6	6	7
63	.8910	8918	8926	8934	8942	8949	8957	8965	8973	8980	1	2	2	3	4	5	5	6	7
64	.8988	8996	9003	9011	9018	9026	9033	9041	9048	9056	1	2	2	3	4	5	5	6	7
65	.9063	9070	9078	9085	9092	9100	9107	9114	9121	9128	1	1	2	3	4	4	5	6	6
66	.9135	9143	9150	9157	9164	9171	9178	9184	9191	9198	1	1	2	3	4	4	5	6	6
67	.9205	9212	9219	9225	9232	9239	9245	9252	9259	9265	1	1	2	3	3	4	5	5	6
68	.9272	9278	9285	9291	9298	9304	9311	9317	9323	9330	1	1	2	3	3	4	4	5	6
69	.9336	9342	9348	9354	9361	9367	9373	9379	9385	9391	1	1	2	2	3	4	4	5	5
70	.9397	9403	9409	9415	9421	9426	9432	9438	9444	9449	1	1	2	2	3	3	4	5	5
71	.9455	9461	9466	9472	9478	9483	9489	9494	9500	9505	1	1	2	2	3	3	4	4	5
72	.9511	9516	9521	9527	9532	9537	9542	9548	9553	9558	1	1	2	2	3	3	4	4	5
73	.9563	9568	9573	9578	9583	9588	9593	9598	9603	9608	1	1	2	2	3	3	4	4	5
74	.9613	9617	9622	9627	9632	9636	9641	9646	9650	9655	0	1	1	2	2	3	3	4	4
75	.9659	9664	9668	9673	9677	9681	9686	9690	9694	9699	0	1	1	2	2	3	3	4	4
76	.9703	9707	9711	9715	9720	9724	9728	9732	9736	9740	0	1	1	2	2	2	3	3	4
77	.9744	9748	9751	9755	9759	9763	9767	9770	9774	9778	0	1	1	1	2	2	3	3	3
78	.9781	9785	9789	9792	9796	9799	9803	9806	9810	9813	0	1	1	1	2	2	2	3	3
79	.9816	9820	9823	9826	9829	9833	9836	9839	9842	9845	0	1	1	1	2	2	2	3	3
80	.9848	9851	9854	9857	9860	9863	9866	9869	9871	9874	0	1	1	1	1	2	2	2	3

	Natural sines (Continued)										Add								
		.1	.2	.3	.4	.5	.6	.7	.8	.9	.01	.02	.03	.04	.05	.06	.07	.08	.09
81°	.9877	9880	9882	9885	9888	9890	9893	9895	9898	9900	0	0	1	1	1	1	2	2	2
82	.9903	9905	9907	9910	9912	9914	9917	9919	9921	9923	0	0	1	1	1	1	1	2	2
83	.9925	9928	9930	9932	9934	9936	9938	9940	9942	9943	0	0	0	1	1	1	1	1	2
84	.9945	9947	9949	9951	9952	9954	9956	9957	9959	9960	0	0	0	1	1	1	1	1	1
85	.9962	9963	9965	9966	9968	9969	9971	9972	9973	9974	0	0	0	0	1	1	1	1	1
86	.9976	9977	9978	9979	9980	9981	9982	9983	9984	9985	0	0	0	0	0	1	1	1	1
87	.9986	9987	9988	9989	9990	9990	9991	9992	9993	9993	0	0	0	0	0	0	0	0	0
88	.9994	9995	9995	9996	9996	9997	9997	9997	9998	9998	0	0	0	0	0	0	0	0	0
89	.9998	9999	9999	9999	9999	1.000	1.000	1.000	1.000	1.000	0	0	0	0	0	0	0	0	0

	Natural cosines										Subtract								
		.1	.2	.3	.4	.5	.6	.7	.8	.9	.01	.02	.03	.04	.05	.06	.07	.08	.09
0°	1.0000	1.000	1.000	1.000	1.000	1.000	9999	9999	9999	9999	0	0	0	0	0	0	0	0	0
1	.9998	9998	9998	9997	9997	9997	9996	9996	9995	9995	0	0	0	0	0	0	0	0	0
2	.9994	9993	9993	9992	9991	9990	9990	9989	9988	9987	0	0	0	0	0	0	0	0	0
3	.9986	9985	9984	9983	9982	9981	9980	9979	9978	9977	0	0	0	0	0	1	1	1	1
4	.9976	9974	9973	9972	9971	9969	9968	9966	9965	9963	0	0	0	0	1	1	1	1	1
5	.9962	9960	9959	9957	9956	9954	9952	9951	9949	9947	0	0	0	1	1	1	1	1	1
6	.9945	9943	9942	9940	9938	9936	9934	9932	9930	9928	0	0	0	1	1	1	1	1	2
7	.9925	9923	9921	9919	9917	9914	9912	9910	9907	9905	0	0	1	1	1	1	1	2	2
8	.9903	9900	9898	9895	9893	9890	9888	9885	9882	9880	0	0	1	1	1	1	2	2	2
9	.9877	9874	9871	9869	9866	9863	9860	9857	9854	9851	0	1	1	1	1	2	2	2	3
10	.9848	9845	9842	9839	9836	9833	9829	9826	9823	9820	0	1	1	1	2	2	2	3	3
11	.9816	9813	9810	9806	9803	9799	9796	9792	9789	9785	0	1	1	1	2	2	2	3	3
12	.9781	9778	9774	9770	9767	9763	9759	9755	9751	9748	0	1	1	1	2	2	3	3	3

15	.9659	9655	9650	9646	9641	9636	9632	9627	9622	9617	0	1	1	2	2	3	3	4	4
16	.9613	9608	9603	9598	9593	9588	9583	9578	9573	9568	1	1	2	2	3	3	4	4	5
17	.9563	9558	9553	9548	9542	9537	9532	9527	9521	9516	1	1	2	2	3	3	4	4	5
18	.9511	9505	9500	9494	9489	9483	9478	9472	9466	9461	1	1	2	2	3	3	4	4	5
19	.9455	9449	9444	9438	9432	9426	9421	9415	9409	9403	1	1	2	2	3	3	4	5	5
20	.9397	9391	9385	9379	9373	9367	9361	9354	9348	9342	1	1	2	2	3	4	4	5	5
21	.9336	9330	9323	9317	9311	9304	9298	9291	9285	9278	1	1	2	3	3	4	4	5	6
22	.9272	9265	9259	9252	9245	9239	9232	9225	9219	9212	1	1	2	3	3	4	5	5	6
23	.9205	9198	9191	9184	9178	9171	9164	9157	9150	9143	1	1	2	3	4	4	5	6	6
24	.9135	9128	9121	9114	9107	9100	9092	9085	9078	9070	1	1	2	3	4	4	5	6	6
25	.9063	9056	9048	9041	9033	9026	9018	9011	9003	8996	1	2	2	3	4	5	5	6	7
26	.8988	8980	8973	8965	8957	8949	8942	8934	8926	8918	1	2	2	3	4	5	5	6	7
27	.8910	8902	8894	8886	8878	8870	8862	8854	8846	8838	1	2	2	3	4	5	6	6	7
28	.8829	8821	8813	8805	8796	8788	8780	8771	8763	8755	1	2	2	3	4	5	6	7	7
29	.8746	8738	8729	8721	8712	8704	8695	8686	8678	8669	1	2	3	3	4	5	6	7	8
30	.8660	8652	8643	8634	8625	8616	8607	8599	8590	8581	1	2	3	4	4	5	6	7	8
31	.8572	8563	8554	8545	8536	8526	8517	8508	8499	8490	1	2	3	4	5	6	6	7	8
32	.8480	8471	8462	8453	8443	8434	8425	8415	8406	8396	1	2	3	4	5	6	7	7	8
33	.8387	8377	8368	8358	8348	8339	8329	8320	8310	8300	1	2	3	4	5	6	7	8	9
34	.8290	8281	8271	8261	8251	8241	8231	8221	8211	8202	1	2	3	4	5	6	7	8	9
35	.8192	8181	8171	8161	8151	8141	8131	8121	8111	8100	1	2	3	4	5	6	7	8	9
36	.8090	8080	8070	8059	8049	8039	8028	8018	8007	7997	1	2	3	4	5	6	7	8	9
37	.7986	7976	7965	7955	7944	7934	7923	7912	7902	7891	1	2	3	4	5	6	7	8	10
38	.7880	7869	7859	7848	7837	7826	7815	7804	7793	7782	1	2	3	4	5	7	8	9	10
39	.7771	7760	7749	7738	7727	7716	7705	7694	7683	7672	1	2	3	4	6	7	8	9	10
40	.7660	7649	7638	7627	7615	7604	7593	7581	7570	7559	1	2	3	5	6	7	8	9	10
41	.7547	7536	7524	7513	7501	7490	7478	7466	7455	7443	1	2	3	5	6	7	8	9	10
42	.7431	7420	7408	7396	7385	7373	7361	7349	7337	7325	1	2	4	5	6	7	8	9	11
43	.7314	7302	7290	7278	7266	7254	7242	7230	7218	7206	1	2	4	5	6	7	8	10	11

Natural cosines (Continued)

		.1	.2	.3	.4	.5	.6	.7	.8	.9	Subtract .01	.02	.03	.04	.05	.06	.07	.08	.09
44°	.7193	7181	7169	7157	7145	7133	7120	7108	7096	7083	1	2	4	5	6	7	9	10	11
45	.7071	7059	7046	7034	7022	7009	6997	6984	6972	6959	1	2	4	5	6	7	9	10	11
46	.6947	6934	6921	6909	6896	6884	6871	6858	6845	6833	1	3	4	5	6	7	8	10	11
47	.6820	6807	6794	6782	6769	6756	6743	6730	6717	6704	1	3	4	5	6	8	9	10	12
48	.6691	6678	6665	6652	6639	6626	6613	6600	6587	6574	1	3	4	5	7	8	9	10	12
49	.6561	6547	6534	6521	6508	6494	6481	6468	6455	6441	1	3	4	5	7	8	9	11	12
50	.6428	6414	6401	6388	6374	6361	6347	6334	6320	6307	1	3	4	5	7	8	9	11	12
51	.6293	6280	6266	6252	6239	6225	6211	6198	6184	6170	1	3	4	5	7	8	10	11	12
52	.6157	6143	6129	6115	6101	6088	6074	6060	6046	6032	1	3	4	6	7	8	10	11	12
53	.6018	6004	5990	5976	5962	5948	5934	5920	5906	5892	1	3	4	6	7	8	10	11	13
54	.5878	5864	5850	5835	5821	5807	5793	5779	5764	5750	1	3	4	6	7	9	10	11	13
55	.5736	5721	5707	5693	5678	5664	5650	5635	5621	5606	1	3	4	6	7	9	10	12	13
56	.5592	5577	5563	5548	5534	5519	5505	5490	5476	5461	1	3	4	6	7	9	10	12	13
57	.5446	5432	5417	5402	5388	5373	5358	5344	5329	5314	1	3	4	6	7	9	10	12	13
58	.5299	5284	5270	5255	5240	5225	5210	5195	5180	5165	1	3	4	6	7	9	10	12	13
59	.5150	5135	5120	5105	5090	5075	5060	5045	5030	5015	2	3	5	6	8	9	11	12	14
60	.5000	4985	4970	4955	4939	4924	4909	4894	4879	4863	2	3	5	6	8	9	11	12	14
61	.4848	4833	4818	4802	4787	4772	4756	4741	4726	4710	2	3	5	6	8	9	11	12	14
62	.4695	4679	4664	4648	4633	4617	4602	4586	4571	4555	2	3	5	6	8	9	11	12	14
63	.4540	4524	4509	4493	4478	4462	4446	4431	4415	4399	2	3	5	6	8	9	11	12	14
64	.4384	4368	4352	4337	4321	4305	4289	4274	4258	4242	2	3	5	6	8	9	11	13	14
65	.4226	4210	4195	4179	4163	4147	4131	4115	4099	4083	2	3	5	6	8	10	11	13	14
66	.4067	4051	4035	4019	4003	3987	3971	3955	3939	3923	2	3	5	6	8	10	11	13	14
67	.3907	3891	3875	3859	3843	3827	3811	3795	3778	3762	2	3	5	6	8	10	11	13	14
68	.3746	3730	3714	3697	3681	3665	3649	3633	3616	3600	2	3	5	6	8	10	11	13	15
69	.3584	3567	3551	3535	3518	3502	3486	3469	3453	3437	2	3	5	7	8	10	11	13	15

70	.3420	3404	3387	3371	3355	3338	3322	3305	3289	3272	2	3	5	7	8	10	11	13	15
71	.3256	3239	3223	3206	3190	3173	3156	3140	3123	3107	2	3	5	7	8	10	12	13	15
72	.3090	3074	3057	3040	3024	3007	2990	2974	2957	2940	2	3	5	7	8	10	12	13	15
73	.2924	2907	2890	2874	2857	2840	2823	2807	2790	2773	2	3	5	7	8	10	12	13	15
74	.2756	2740	2723	2706	2689	2672	2656	2639	2622	2605	2	3	5	7	8	10	12	13	15
75	.2588	2571	2554	2538	2521	2504	2487	2470	2453	2430	2	3	5	7	8	10	12	14	15
76	.2419	2402	2385	2368	2351	2334	2317	2300	2284	2267	2	3	5	7	9	10	12	14	15
77	.2250	2233	2215	2198	2182	2164	2147	2130	2113	2096	2	3	5	7	9	10	12	14	15
78	.2079	2062	2045	2028	2011	1994	1977	1959	1942	9125	2	3	5	7	9	10	12	14	15
79	.1908	1891	1874	1857	1840	1822	1805	1788	1771	1754	2	3	5	7	9	10	12	14	15
80	.1736	1719	1702	1685	1668	1650	1633	1616	1599	1582	2	3	5	7	9	10	12	14	15
81	.1564	1547	1530	1513	1495	1478	1461	1444	1426	1409	2	3	5	7	9	10	12	14	15
82	.1392	1374	1357	1340	1323	1305	1288	1271	1253	1236	2	3	5	7	9	10	12	14	16
83	.1219	1201	1184	1167	1149	1132	1115	1097	1080	1063	2	3	5	7	9	10	12	14	16
84	.1045	1028	1011	0993	0976	0958	0941	0924	0906	0889	2	3	5	7	9	10	12	14	16
85	.0872	0854	0837	0819	0802	0785	0767	0750	0732	0715	2	3	5	7	9	10	12	14	16
86	.0698	0680	0663	0645	0628	0610	0593	0576	0558	0541	2	4	5	7	9	11	12	14	16
87	.0523	0506	0488	0471	0454	0436	0419	0401	0384	0366	2	4	5	7	9	11	12	14	16
88	.0349	0332	0314	0297	0279	0262	0244	0227	0209	0192	2	4	5	7	9	11	12	14	16
89	.0175	0157	0140	0122	0105	0087	0070	0052	0035	0017	2	4	5	7	9	11	12	14	16

Natural tangents											Add								
		.1	.2	.3	.4	.5	.6	.7	.8	.9	.01	.02	.03	.04	.05	.06	.07	.08	.09
0°	0.0000	0017	0035	0052	0070	0087	0105	0122	0140	0157	2	4	5	7	9	11	12	14	16
1	0.0175	0192	0209	0227	0244	0262	0279	0297	0314	0332	2	4	5	7	9	11	12	14	16
2	0.0349	0367	0384	0402	0419	0437	0454	0472	0489	0507	2	4	5	7	9	11	12	14	16
3	0.0524	0542	0559	0577	0594	0612	0629	0647	0664	0682	2	4	5	7	9	11	12	14	16
4	0.0699	0717	0734	0752	0769	0787	0805	0822	0840	0857	2	4	5	7	9	11	12	14	16
5	0.0875	0892	0910	0928	0945	0963	0981	0998	1016	1033	2	4	5	7	9	11	12	14	16

	Natural tangents (Continued)										Add								
		.1	.2	.3	.4	.5	.6	.7	.8	.9	.01	.02	.03	.04	.05	.06	.07	.08	.09
6°	0.1051	1069	1086	1104	1122	1139	1157	1175	1192	1210	2	4	5	7	9	11	12	14	16
7	0.1228	1246	1263	1281	1299	1317	1334	1352	1370	1388	3	4	5	7	9	11	12	14	16
8	0.1405	1423	1441	1459	1477	1495	1512	1530	1548	1566	2	4	5	7	9	11	13	14	16
9	0.1584	1602	1620	1638	1655	1673	1691	1709	1727	1745	2	4	5	7	9	11	13	14	16
10	0.1763	1781	1799	1817	1835	1853	1871	1890	1908	1926	2	4	5	7	9	11	13	14	16
11	0.1944	1962	1980	1998	2016	2035	2053	2071	2089	2107	2	4	5	7	9	11	13	15	16
12	0.2126	2144	2162	2180	2199	2217	2235	2254	2272	2290	2	4	5	7	9	11	13	15	16
13	0.2309	2327	2345	2364	2382	2401	2419	2438	2456	2475	2	4	6	7	9	11	13	15	17
14	0.2493	2512	2530	2549	2568	2586	2605	2623	2642	2661	2	4	6	7	9	11	13	15	17
15	0.2679	2698	2717	2736	2754	2773	2792	2811	2830	2849	2	4	6	8	9	11	13	15	17
16	0.2867	2886	2905	2924	2943	2962	2981	3000	3019	3038	2	4	6	8	9	11	13	15	17
17	0.3057	3076	3096	3115	3134	3153	3172	3191	3211	3230	2	4	6	8	10	11	13	15	17
18	0.3249	3269	3288	3307	3327	3346	3365	3385	3404	3424	2	4	6	8	10	12	14	16	17
19	0.3443	3463	3482	3502	3522	3541	3561	3581	3600	3620	2	4	6	8	10	12	14	16	18
20	0.3640	3659	3679	3699	3719	3739	3759	3779	3799	3819	2	4	6	8	10	12	14	16	18
21	0.3839	3859	3879	3899	3919	3939	3959	3979	4000	4020	2	4	6	8	10	12	14	16	18
22	0.4040	4061	4081	4101	4122	4142	4163	4183	4204	4224	2	4	6	8	10	12	14	16	18
23	0.4245	4265	4286	4307	4327	4348	4369	4390	4411	4431	2	4	6	8	10	12	15	17	19
24	0.4452	4473	4494	4515	4536	4557	4578	4599	4621	4642	2	4	6	8	11	13	15	17	19
25	0.4663	4684	4706	4727	4748	4770	4791	4813	4834	4856	2	4	6	9	11	13	15	17	19
26	0.4877	4899	4921	4942	4964	4986	5008	5029	5051	5073	2	4	7	9	11	13	15	17	20
27	0.5095	5117	5139	5161	5184	5206	5228	5250	5272	5295	2	4	7	9	11	13	16	18	20
28	0.5317	5340	5362	5384	5407	5430	5452	5475	5498	5520	2	5	7	9	11	14	16	18	20
29	0.5543	5566	5589	5612	5635	5658	5681	5704	5727	5750	2	5	7	9	12	14	16	18	21
30	0.5774	5797	5820	5844	5867	5890	5914	5938	5961	5985	2	5	7	9	12	14	16	19	21
31	0.6009	6032	6056	6080	6104	6128	6152	6176	6200	6224	2	5	7	10	12	14	17	19	22

32	0.6249	6273	6297	6322	6346	6371	6395	6420	6445	6469	2	5	7	10	12	15	17	20	22
33	0.6494	6519	6544	6569	6594	6619	6644	6669	6694	6720	3	5	8	10	13	15	18	20	23
34	0.6745	6771	6796	6822	6847	6873	6899	6924	6950	6976	3	5	8	10	13	15	18	21	23
35	0.7002	7028	7054	7080	7107	7133	7159	7186	7212	7239	3	5	8	11	13	16	18	21	24
36	0.7265	7292	7319	7346	7373	7400	7427	7454	7481	7508	3	5	8	11	14	16	19	22	24
37	0.7536	7563	7590	7618	7646	7673	7701	7729	7757	7785	3	6	8	11	14	17	19	22	25
38	0.7813	7841	7869	7898	7926	7954	7983	8012	8040	8069	3	6	9	11	14	17	20	23	26
39	0.8098	8127	8156	8185	8214	8243	8273	8302	8332	8361	3	6	9	12	15	18	21	23	26
40	0.8391	8421	8451	8481	8511	8541	8571	8601	8632	8662	3	6	9	12	15	18	21	24	27
41	0.8693	8724	8754	8785	8816	8847	8878	8910	8941	8972	3	6	9	12	16	19	22	25	28
42	0.9004	9036	9067	9099	9131	9163	9195	9228	9260	9293	3	6	10	13	16	19	22	26	29
43	0.9325	9358	9391	9424	9457	9490	9523	9556	9590	9623	3	7	10	13	17	20	23	27	30
44	0.9657	9691	9725	9759	9793	9827	9861	9896	9930	9965	3	7	10	14	17	21	24	27	31
45	1.0000	0035	0070	0105	0141	0176	0212	0247	0283	0319	4	7	11	14	18	21	25	28	32
46	1.0355	0392	0428	0464	0501	0538	0575	0612	0649	0686	4	7	11	15	18	22	26	30	33
47	1.0724	0761	0799	0837	0875	0913	0951	0990	1028	1067	4	8	11	15	19	23	27	31	34
48	1.1106	1145	1184	1224	1263	1303	1343	1383	1423	1463	4	8	12	16	20	24	28	32	36
49	1.1504	1544	1585	1626	1667	1708	1750	1792	1833	1875	4	8	12	17	21	25	30	33	37
50	1.1918	1960	2002	2045	2088	2131	2174	2218	2261	2305	4	9	13	17	22	26	30	34	39
51	1.2349	2393	2437	2482	2527	2572	2617	2662	2708	2753	5	9	14	18	23	27	32	36	41
52	1.2799	2846	2892	2938	2985	3032	3079	3127	3175	3222	5	9	14	19	24	28	33	38	42
53	1.3270	3319	3367	3416	3465	3514	3564	3613	3663	3713	5	10	15	20	25	30	35	40	44
54	1.3764	3814	3865	3916	3968	4019	4071	4124	4176	4229	5	10	15	21	26	31	36	41	47
55	1.4281	4335	4388	4442	4496	4550	4605	4659	4715	4770	5	11	16	22	27	33	38	44	49
56	1.4826	4882	4938	4994	5051	5108	5166	5224	5282	5340	6	11	17	23	29	34	40	46	52
57	1.5399	5458	5517	5577	5637	5697	5757	5818	5880	5941	6	12	18	24	30	36	42	48	54
58	1.6003	6066	6128	6191	6255	6319	6383	6447	6512	6577	6	13	19	26	32	38	45	51	58
59	1.6643	6709	6775	6842	6909	6977	7045	7113	7182	7251	7	14	20	27	34	41	47	54	61
60	1.7321	7391	7461	7532	7603	7675	7747	7820	7893	7966	7	14	22	29	36	43	50	58	65
61	1.8040	8115	8190	8265	8341	8418	8495	8572	8650	8728	8	15	23	31	38	46	54	61	69

Natural tangents (Continued)

		.1	.2	.3	.4	.5	.6	.7	.8	.9	Add .01	.02	.03	.04	.05	.06	.07	.08	.09
62°	1.8807	8887	8967	9047	9128	9210	9292	9375	9458	9542	8	16	25	33	41	49	57	66	74
63	1.9626	9711	9797	9883	9970	0057	0145	0233	0323	0413	6	18	26	35	44	53	61	70	79
64	2.0503	0594	0686	0778	0872	0965	1060	1155	1251	1348	9	19	28	38	47	57	66	75	85
65	2.1445	1543	1642	1742	1842	1943	2045	2148	2251	2355	10	20	30	40	50	60	70	80	90
66	2.2460	2566	2673	2781	2889	2998	3109	3220	3332	3445	11	22	32	43	54	65	75	86	97
67	2.3559	3673	3789	3906	4023	4142	4262	4383	4504	4627	12	23	35	47	58	70	82	93	105
68	2.4751	4876	5002	5129	5257	5386	5517	5649	5782	5916	13	25	38	51	64	76	89	102	114
69	2.6051	6187	6325	6464	6605	6746	6889	7034	7179	7326	14	28	42	56	70	83	97	111	125
70	2.7475	7625	7776	7929	8083	8239	8397	8556	8716	8878	15	31	46	61	76	92	107	122	138
71	2.9042	9208	9375	9544	9714	9887	0061	0237	0415	0595	17	34	51	68	85	101	118	135	152
72	3.0777	0961	1146	1334	1524	1716	1910	2106	2305	2506	19	37	56	74	93	111	130	149	167
73	3.2709	2914	3122	3332	3544	3759	3977	4197	4420	4646	21	42	63	84	105	126	147	168	189
74	3.4874	5105	5339	5576	5816	6059	6305	6554	6806	7062	24	47	71	95	119	142	166	190	213
75	3.7321	7583	7848	8118	8391	8667	8947	9232	9520	9812	27	54	81	108	135	162	188	215	242
76	4.0108	0408	0713	1022	1335	1653	1976	2303	2635	2972	31	62	93	124	155	185	216	247	278
77	4.3315	3662	4015	4373	4737	5107	5483	5864	6252	6646	36	72	108	143	179	215	251	287	323
78	4.7046	7453	7867	8288	8716	9152	9594	0045	0504	0970	42	84	126	168	211	253	295	337	379
79	5.1446	1929	2422	2924	3435	3955	4486	5026	5578	6140									
80	5.671	5.730	5.789	5.850	5.912	5.976	6.041	6.107	6.174	6.243									
81	6.314	6.386	6.460	6.535	6.612	6.691	6.772	6.855	6.940	7.026									
82	7.115	7.207	7.300	7.396	7.495	7.596	7.700	7.806	7.916	8.028									
83	8.144	8.264	8.386	8.513	8.643	8.777	8.915	9.058	9.205	9.357									
84	9.51	9.68	9.84	10.02	10.20	10.39	10.58	10.78	10.99	11.20					Unreliable in this area				
85	11.43	11.66	11.91	12.16	12.43	12.71	13.00	13.30	13.62	13.95									
86	14.30	14.67	15.06	15.46	15.89	16.35	16.83	17.34	17.89	18.46									
87	19.08	19.74	20.45	21.20	22.02	22.90	23.86	24.90	26.03	27.27									
88	28.64	30.14	31.82	33.69	35.80	38.19	40.92	44.07	47.74	52.08									
89	57.29	63.66	71.62	81.85	95.49	114.6	143.2	191.0	286.5	573.0									

Index

Q

R